W0105982

Nondestructive Characterization of Materials VI

Nondestructive Characterization of Materials VI

Edited by

Robert E. Green, Jr.
The Johns Hopkins University
Baltimore, Maryland

Krzysztof J. Kozaczek
Oak Ridge National Laboratory
Oak Ridge, Tennessee

and

Clayton O. Ruud
The Pennsylvania State University
University Park, Pennsylvania

SPRINGER SCIENCE+BUSINESS MEDIA, LLC

Library of Congress Cataloging-in-Publication Data

On file

Proceedings of the Sixth International Conference on Nondestructive Characterization of Materials, held June 7–11, 1993, in Oahu, Hawaii

ISBN 978-1-4613-6100-8 ISBN 978-1-4615-2574-5 (eBook)
DOI 10.1007/978-1-4615-2574-5

© 1994 Springer Science+Business Media New York
Originally published by Plenum Press, New York in 1994

All rights reserved

No part of this book may be reproduced, stored in a retrieval system, or transmitted in any form or by any means, electronic, mechanical, photocopying, microfilming, recording, or otherwise, without written permission from the Publisher

The papers published in these proceedings represent some of the latest developments in nondestructive characterization of materials and were presented at the **Sixth International Symposium on Nondestructive Characterization of Materials** held June 7-11, 1993, at the Turtle Bay Hilton Hotel on the north shore of Oahu, Hawaii.

SYMPOSIUM CO-CHAIRMEN

Robert E. Green, Jr.
Center for NDE
The Johns Hopkins University
3400 N. Charles Street
Baltimore, MD 21218

Murli H. Manghnani
Hawaii Institute of Geophysics
University of Hawaii at Manoa
2525 Correa Road
Honolulu, Hawaii 96822

Teruo Kishi
The University of Tokyo
4-6-1 Komaba, Merguro-Ku
Tokyo 153, Japan

Clayton O. Ruud
The Pennsylvania State University
159 Materials Research Lab.
University Park, PA 26802

ORGANIZING COMMITTEE

Alfred Broz, Federal Aviation Administration, U.S.A.
Jean P. Bussiere, National Research Council, Canada
Joseph S. Heyman, NASA/Langley Research Center, U.S.A.
Katsuhiro Kawashima, Nippon Steel Corporation, Japan
Michael Kroning, IzfP, Fraunhofer-Institut, Germany
Soung-Nan Liu, Electric Power Research Institute, U.S.A.
Charles H. McGogney, Federal Highway Administration, U.S.A.
Yapa Rajapakse, Office of Naval Research, U.S.A.
Tetsuya Saito, National Research Institute for Metals, Japan
Donald O. Thompson, Iowa State University, U.S.A.
James W. Wagner, Johns Hopkins University, U.S.A.
F. Alan Wedgwood, NNDTC, Harwell Laboratory, England
H. Thomas Yolken, NIST, U.S.A.

FINANCIAL SUPPORTERS

Army Research Office
American Society for Nondestructive Testing
Federal Aviation Administration
Federal Highway Administration
National Aeronautics and Space Administration
National Institute of Standards and Technology
National Science Foundation
Office of Naval Research

PREFACE

Traditionally the vast majority of materials characterization techniques have been destructive, e.g., chemical compositional analysis, metallographic determination of microstructure, tensile test measurement of mechanical properties, etc. Also, traditionally, nondestructive techniques have been used almost exclusively for the detection of macroscopic defects, mostly cracks, in structures and devices which have already been constructed and have already been in service for an extended period of time. Following these conventional nondestructive tests, it has been common practice to use somewhat arbitrary accept-reject criteria to decide whether or not the structure or device should be removed from service. The present unfavorable status of a large segment of industry, coupled with the desire to keep structures in service well past their original design life, dramatically show that our traditional approaches must be drastically modified if we are to be able to meet future needs.

The role of nondestructive characterization of materials is changing and will continue to change dramatically. It has become increasingly evident that it is both practical and cost effective to expand the role of nondestructive evaluation to include all aspects of materials' production and application and to introduce it much earlier in the manufacturing cycle. In fact, the recovery of a large portion of industry from severe economic problems is dependent, in part, on the successful implementation of this expanded role. Currently, efforts are directed at developing and perfecting techniques which are capable of monitoring and controlling the materials production process; materials stability during transport, storage, and fabrication; and the amount and rate of degradation during the materials in-service life. To be more precise, the role of nondestructive testing has expanded far beyond its historical mission of detecting macroscopic defects in structures and devices which had already been constructed and most often had been in service for an extended period of time. Today, and ever increasingly in the future, using advanced sensors and modern measurement technology, along with signal/data processing techniques, information on the processing conditions and the properties and characteristics of the materials being processed can be continuously generated. Real-time process monitoring for more effective and efficient real-time process control and improved product quality and reliability will now become a practical reality.

The optimization of the processing and properties of polymers, ceramics and composites, the development of synthetically structured materials, the characterization of surfaces and interfaces, the measurement and character-characterization of amorphous metals and semiconductors, the growth of perfect electronic and optical crystals and thin films, and in all cases, the structures, devices and systems made from these materials demand the innovative application of modern nondestructive materials characterization techniques to monitor and control as many stages of the production process as possible. Simply put, intelligent manufacturing is impossible without integrating modern nondestructive evaluation into the production system.

Robert E. Green, Jr.
Center for Nondestructive Evaluation
The Johns Hopkins University

ACKNOWLEDGEMENTS

A special thanks is due Academician Leonid M. Lyamshev from the Acoustical Institute of the Russian Academy of Sciences, Moscow, for presentation of a special invited lecture on Radiation Acoustics and Nondestructive Evaluation. Although the complete manuscript of the presentation by Academician Lyanshev is not available for printing in the proceedings, a publication is available elsewhere which contain details of his subject matter, namely: L.M. Lyamshev, Radiation Acoustics, Sov. Phys. Usp. 162, 43-94 (1992) [in English].

The conference organizers are indebted to a number of individuals for their assistance in making the symposium a success. Thanks to: the authors for their excellent contributions and cooperation in providing manuscripts; to the session chairpersons for keeping the sessions on time and stimulating lively discussions; to the University of Hawaii for assistance with the poster sessions; to Moshe Rosen, Jim Spicer, Jim Wagner, and John Winter, all of the Center for Nondestructive Evaluation of The Johns Hopkins University, for manuscript reviews.

The symposium and the ensuing proceedings would have not been possible without the enthusiasm and extremely hard work of Debbie Harris, The Johns Hopkins University Center for Nondestructive Evaluation Center Coordinator, and her assistant, Debby Manley. All of us who enjoyed the symposium venue, the technical sessions, and the social events owe both of these women our most sincere vote of thanks.

CONTENTS

PROCESS CONTROL

ACOUSTIC TECHNIQUES I

MATERIALS CHARACTERIZATION I

ACOUSTIC TECHNIQUES II

CERAMICS

OPTICAL TECHNIQUES

RESIDUAL STRESS

ELECTRONIC MATERIALS AND COMPONENTS

AIRCRAFT/AEROSPACE

MATERIALS CHARACTERIZATION II

BIOMIMETIC and BIOTIC MATERIALS

POSTERS

MATERIALS CHARACTERIZATION III

MONITORING OF RESIN-TRANSFER MOLDING USING LASER-BASED ULTRASOUND

A.D.W. McKie, R.C. Addison, Jr., T.-L.T. Liao, and H.-S. Ryang

Rockwell International Science Center
1049 Camino Dos Rios
Thousand Oaks, CA 91360

INTRODUCTION

In both the aerospace and automotive industries, polymer-based composite materials are increasingly being used to replace metallic structures. To reduce the cost of fabricating complexly contoured polymer composite structures, manufacturers are actively developing low-cost alternatives to hand lay-up/autoclave curing and other labor and capital-intensive processes. One technique, that has proven to be an economical method for fabricating polymer composite structures containing complex contours, is resin-transfer molding (RTM). In the RTM process a dry fiber preform is placed between two faces of a mold. Resin is injected into the mold at low pressure and fills the mold cavity permeating the fiber reinforcement. After the part is fully cured it is removed from the mold. The key to efficient and cost-effective resin-transfer molding is to have an automated process in which the infusion of resin is controlled so that the flow front rapidly and evenly permeates the fiber preform. Several problems can occur during the transfer of the resin into the mold such as lack of penetration into all regions of the mold or incomplete infiltration of the fiber preform. Thus, *in situ* sensors are required to monitor the RTM process and ensure that pumping of resin is not stopped prematurely. Laser-based ultrasound (LBU) is well suited for this monitoring application since it can work with heated molds having nonplanar surfaces and can acquire ultrasonic C-scan images in times that are short compared to the total transfer time of the resin. These ultrasonic images can be used to enable rapid assessment of degree-of-completion of resin transfer, and indicate problems associated with the resin transfer or with defects in the mold. In this paper, results are presented which demonstrate the capability of laser-based ultrasound to monitor the resin flow front along the inner surface of the mold during the RTM process.

EXPERIMENTAL APPROACH

To verify correlation between the ultrasonic data acquired with the LBU system and the location of the resin flow front, a specially fabricated aluminum mold was used. A Pyrex

Nondestructive Characterization of Materials VI
Edited by R.E. Green, Jr. *et al.*, Plenum Press, New York, 1994

window on one side of the mold allowed the resin front to be visually monitored and recorded on video tape concurrently with ultrasonic imaging from the opposite side of the mold with the LBU system (Figure 1).

Resin was injected into the mold via a port located in the lower corner of the aluminum face, and the mold orientation was such that resin flow occurred in a vertical direction. An outlet port was located at the top of the aluminum face in the corner diagonally opposite the inlet port. The resin pot used could be evacuated to degas the resin, heated if desired to reduce the viscosity of the resin, and could be pressurized up to 100 psi to force the resin into the mold and through the fiber preform.

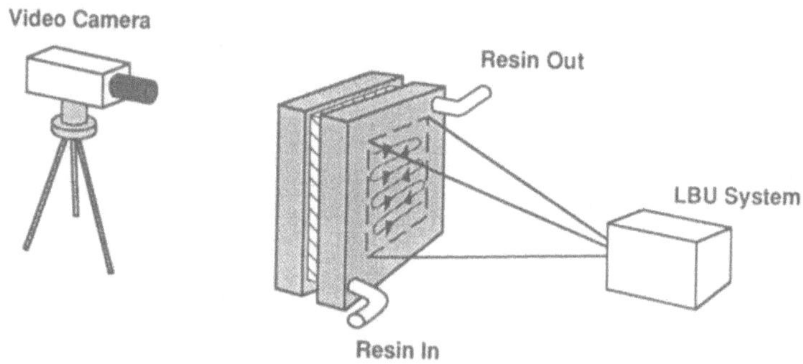

Figure 1. Experimental configuration for simultaneous monitoring of resin-transfer during the RTM process, using a video camera and the laser-based ultrasound system.

The essential features of the LBU system have been described previously[1,2]. To visualize the resin flow front during the RTM process, the generation (CO_2) and probe (argon-ion) laser beams were scanned over the aluminum mold face. The stand-off distance was ~ 1.5 m and the CO_2 laser pulse repetition rate was 40 Hz. With this configuration the scan time to acquire an ultrasonic C-scan image of a 205 x 335 mm area (42 x 70 pixels) was ~ 120 seconds. Signal averaging was not used although the aluminum surface was painted to provide a surface constraint which enhances the generation of longitudinal waves propagating in the forward direction in the metallic mold[3]. This is not viewed as a limitation of the usefulness of the technique since coatings are available that can be used at the temperatures required to transfer and cure the typical resins being considered for the RTM process.

The longitudinal pulse generated on the outside surface of the mold wall, by the LBU system, travels through 0.75 inch thick aluminum to the near surface of the mold inner cavity where it is reflected (echo A in Figure 2). If the resin flow has reached the inspection point the longitudinal pulse is partially transmitted into the resin, so that a reduction in pulse amplitude will be measured. If the resin has completely permeated the fiber preform, the longitudinal pulse will propagate through the composite material to the opposite side of the mold cavity where it will be reflected back towards the outside wall of the mold (echo B in Figure 2). Thus, two echoes can be monitored to determine the state of the resin-transfer. The amplitude of the echo from the near side of the mold cavity will sense whether the resin has flowed over the face of the mold, and the amplitude of the echo from the far side of the mold cavity will sense whether the fiber preform has been completely permeated by the resin. However, in the experiments reported here, only the pulse from the near wall of the mold cavity was detected (echo A, Figure 2).

EXPERIMENTAL RESULTS

The RTM process was recorded on video tape and simultaneously monitored with the LBU system for two different resin-fiber configurations. Measurements at room temperature were performed using a vinyl ester (Ashland Q6530) which was transferred through a preform comprised of five plies of high density randomly chopped glass fiber mats. A second experiment was performed in which the flow of a high-temperature epoxy resin (Shell Epon 862) was monitored as it infiltrated a woven graphite preform, at a temperature of 125°C. Since the mold was situated inside an oven for the high-temperature experiment, access was limited to one side and so the process was monitored with the LBU system only.

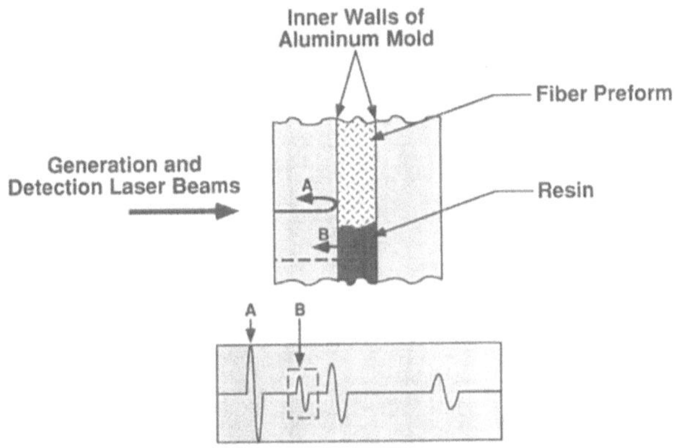

Figure 2. Key reflected ultrasonic pulses for monitoring the resin-transfer molding process.

For the vinyl ester configuration, the resin was injected into the mold under pressure, with a schedule which started at ~ 10 psi for the first 32 mins and was increased to 40 psi until the transfer of resin was completed at T = 44 mins. When the pumping of resin was stopped the resin flow front, seen through the Pyrex window, had reached the outlet port and apparently filled the mold. Previously, experiments were reported in which void formation occurred[4]. However, in this experiment the resin flowed uniformly up through the mold cavity forcing entrapped air out of the system so that voids were not present. The images made with the LBU system are based on changes in amplitude of the ultrasonic echo reflected from the near wall of the mold cavity (echo A in Figure 2). A series of ten ultrasonic C-scan images, acquired with the LBU system during the RTM process, clearly shows the resin front advancing uniformly through the mold from the inlet port on the lower right-hand side of the mold to the outlet port on the upper left (Figure 3). These images are in good agreement with the video images acquired of the resin flow up the opposite face of the preform (Figure 4), and suggests that the vertical resin flow is similar on both sides of the preform.

After completion of the resin injection, the charged mold was placed in an oven and cured at 60°C for 30 mins after which the manufactured part was removed from the mold. Even though both the video and LBU visualization of the resin flow front indicated completion of the process, subsequent transmission mode immersion C-scan inspection of the part after curing (Figure 5) shows that the panel contained a significant void region in the upper corner of the mold near the outlet port. Below this void region, wetting of the fiber reinforcement was nonuniform with large variations in transmissivity. It is evident that even though the vertical resin flow had uniformly covered both sides of the preform, resin flow in the transverse direction had not completely penetrated the preform resulting in incomplete fiber wetting during the resin transfer. This may have resulted from the loose fit of the low

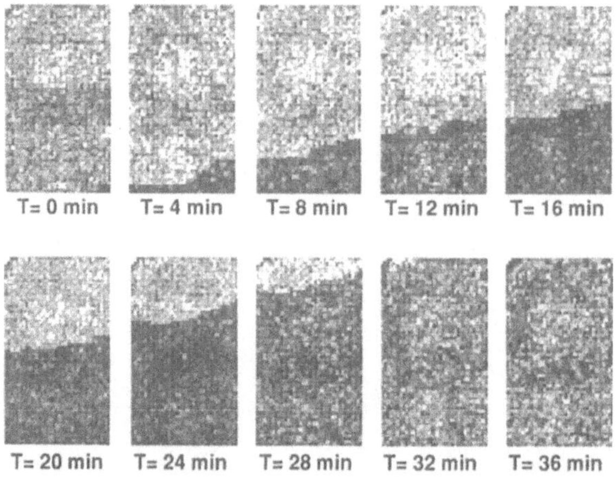

Figure 3. Series of ultrasonic C-scan images acquired using the laser-based ultrasound system to monitor the resin-transfer molding process.

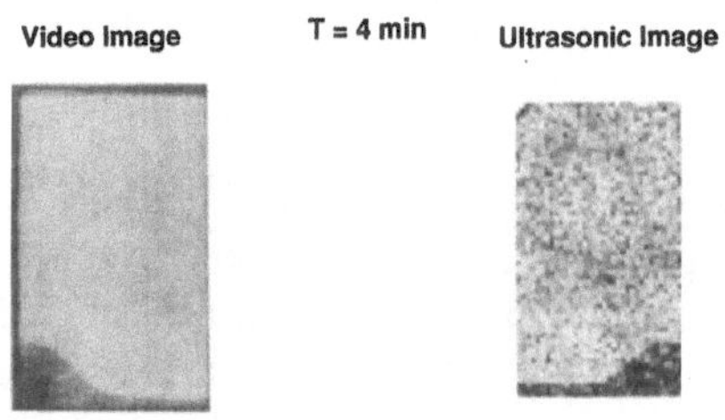

Figure 4. Comparison of the video and laser-based ultrasonic C-scan image acquired simultaneously at T = 4 min during the resin-transfer molding process.

fiber volume preform which could conceivably allow the resin to preferentially flow up the faces of the mold. In related experiments, high fiber volume braided graphite preforms which fit in the mold tightly seemed to wet satisfactorily.

For the high-temperature experiment, the mold was placed inside an oven and the LBU system was used to monitor the resin flow. The mold was preheated to 125°C to reduce the viscosity of the injected resin and facilitate easier infusion. A series of five ultrasonic C-scan

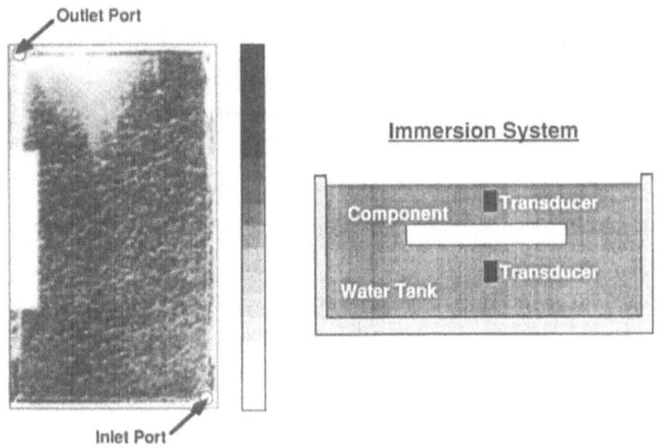

Figure 5. Transmission mode immersion C-scan image showing incomplete wetting of the fiber reinforcement after completion of resin-transfer molding.

images, acquired with the LBU system during the RTM process, again shows the resin front advancing through the mold (Figure 6). However, even with infusion pressures of 40 psi the resin flow was extremely slow and flow front "stall-out" occurred with the mold only partially filled after over 80 mins. The combination of tightly packed preform and deterioration in the chemical properties of the resin is believed to have caused the flow front to stop. Although this experiment was not successful with respect to RTM, it was a success in terms of *in situ* process monitoring at elevated temperature using the LBU system. As far as we are aware, these are the first reported laser-based ultrasonic C-scans performed at elevated temperatures.

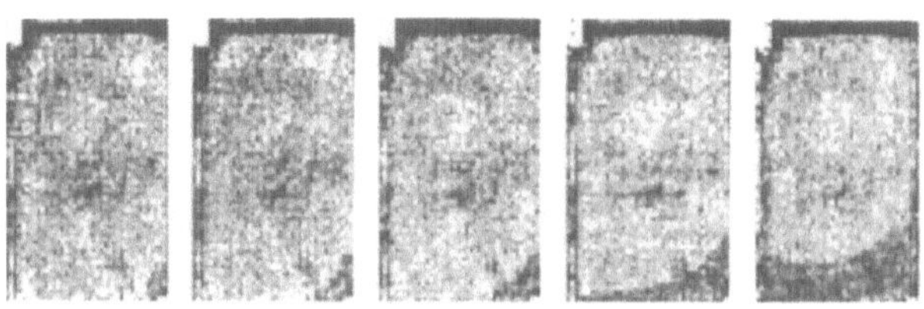

Figure 6. Series of ultrasonic C-scan images acquired using the laser-based ultrasound system to monitor the resin-transfer molding process at elevated temperature.

CONCLUSIONS

Experiments were performed in which the laser-based ultrasound system was used to noninvasively monitor the resin flow front during the resin-transfer molding process of several fiber composite panels. Video images of the resin flow front as it progressed vertically through the mold were in excellent agreement with ultrasonic C-scan images obtained by simultaneous monitoring using the LBU system. In addition, the noncontacting nature of LBU allowed the

RTM process to be monitored at elevated temperature. It was found that flow of resin through the thickness of the preform can significantly lag the vertical resin flow front occurring along the mold wall/preform interface, resulting in incomplete consolidation of the manufactured composite parts. This, however, illustrates the utility of using ultrasonic methods to monitor the RTM process

These preliminary experiments have demonstrated the applicability of the laser-based ultrasound system as an *in situ* process monitoring sensor. To realize the full potential of the LBU technique as a robust sensor for closed loop control of the RTM process, the signal-to-noise ratio of the LBU system must be improved to detect the ultrasonic echo that has traveled through the resin wetted fiber preform and been reflected back to the outer surface of the mold. Efforts are in progress to attain this improvement.

ACKNOWLEDGMENTS

This work has been supported by Rockwell Automotive Plastic Products Division and by Rockwell Independent Research and Development funds.

REFERENCES

1. A. D. W. McKie and R. C. Addison, Jr., Rapid inspection of composites using laser-based ultrasound, *in: "Review of Progress in Quantitative NDE,"* D.O. Thompson and D.E. Chimenti, eds., Plenum Press, New York (1993).

2. A. D. W. McKie and R. C. Addison, Jr., Inspection of components having complex geometries using laser-based ultrasound, *in: "Review of Progress in Quantitative NDE,"* D.O. Thompson and D.E. Chimenti, eds., Plenum Press, New York (1992).

3. C. B. Scruby and L. E. Drain. "Laser Ultrasonics – Techniques and Applications" Adam Hilger, New York (1990).

4. R. C. Addison, Jr., A. D. W. McKie T. -L. T. Liao and H. -S. Ryang, *In situ* process monitoring using laser-based ultrasound, in: *"IEEE 1992 Ultrasonic Symposium Proceedings,"* B. R. McAvoy, ed., IEEE, New York (1992).

EMBEDDED ACOUSTIC SENSORS FOR PROCESS CONTROL AND HEALTH MONITORING OF COMPOSITE MATERIALS

Michael J. Ehrlich, Christian V. O'Keefe,
B. Boro Djordjevic, and B.N. Ranganathan

Martin Marietta Laboratories
1450 South Rolling Road
Baltimore, Maryland 21227

INTRODUCTION

The past few years have seen a marked increase in the use of composite materials for structural applications, many of which put such high demands on the components that they operate just within their performance limits. To ensure that these composite components possess the high reliability and uniformity necessary for critical applications, it is desirable to carefully control the production environment. Although there are numerous facets to composite production, this paper focuses on in-situ monitoring of the cure process.

A number of techniques have been successfully demonstrated for composite cure monitoring, including ultrasonic, dielectric, and optical methods. Of these, only the ultrasonic techniques truly measure the mechanical properties of the composite, and thus may offer the most valuable information regarding the integrity of the part. Unfortunately, many ultrasonic techniques have proven difficult to implement under the harsh conditions of an industrial environment. In particular, contact ultrasonic transducers are poorly suited to operate at high temperatures, and rarely is it possible to mechanically couple the transducer to the part during processing. In an effort to provide large area, real-time cure monitoring capabilities, embedded acoustic "waveguides" have been investigated as a means to effectively couple ultrasound into and out of composite parts in the processing environment[Harrold & Sanjana 1986, Winfree & Parker 1985, Winfree & Sun 1989]. The original technique, developed by R. Harrold [Harrold 1986], used a metal or polymer filament, typically ranging from 0.005 to 0.020 inches in diameter, to guide ultrasonic energy through a composite part in which the filament was embedded. Historically, the amplitude of the received ultrasonic signal was monitored as a function of cure time, and this amplitude related to the cure state of the composite. During the course of the acoustic waveguide experiments reported here (hereafter referred to as an embedded acoustic sensor), it was determined that the reproducibility of the amplitude measurement technique was not high enough to merit its continued investigation. Rather, the cure monitoring work reported here has focused on time-of-flight measurements, which have produced far more reliable results and from which ultrasonic velocities and ultimately elastic moduli may be derived.

Nondestructive Characterization of Materials VI
Edited by R.E. Green, Jr. *et al.*, Plenum Press, New York, 1994

EXPERIMENT

To assess the applicability of using embedded acoustic sensors for cure monitoring, a number of experiments were performed aimed at correlating the sensor response with the degree of composite cure. For these experiments, the material used as the acoustic sensor was 0.020 inch Nichrome wire. The wire sensors were embedded in the mid plane of 16-ply graphite/epoxy panels during the prepreg lay-up process. "Acoustic horns" were used to couple ultrasonic tone bursts into the sensor and as a means to conveniently couple energy back out to the receiving transducer, as illustrated in Figure 1. The horns were designed to minimize ultrasonic back-reflection noise and maximize the useful energy transfer between the transducers and the wire. The prepreg panels with the embedded sensors were then cured using a hot-press. Full cure cycles (350°F) and partial cure cycles (250°F) were performed, so a comparison of the results could be made.

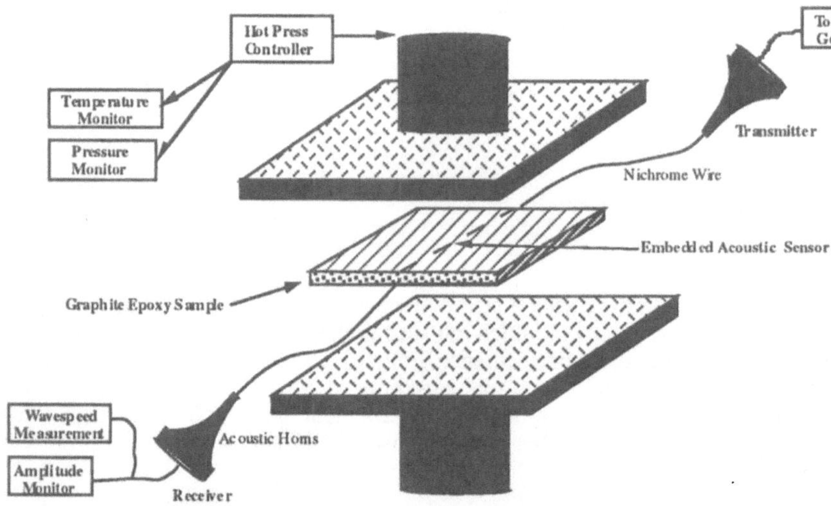

Figure 1. Schematic of the experimental setup used for composite cure monitoring. Specially designed "acoustic horns" were used to couple ultrasound into and out of the Nichrome wire sensor.

Throughout each experiment, the temperature, pressure, ultrasonic amplitude and ultrasonic time-of-flight were recorded. To perform the time-of-flight and amplitude measurements during the cure-cycle, ultrasonic tone bursts (nominally 50kHz) were transmitted into the embedded sensor and received at the other end. Every minute during the cure cycle, a received ultrasonic tone-burst was digitized and used to determine the signal amplitude and time-of-flight through the 12"x12" composite panels. At the same time, hot press temperature and pressure were monitored though a programming interface and recorded.

As stated previously, it was determined from these experiments that ultrasonic amplitude measurements did not exhibit the reproducibility required to perform as a cure-monitoring parameter. The ingress/egress locations where the wire entered and exited the composite panel were likely suspects for introducing variability into the system, as was the actual attachment of the wire to the acoustic horns. For these reasons, most attention was focused on time-of-flight measurements.

The time-of-flight for a 50kHz tone-burst propagating through the composite panel was determined by digitizing the received signal and comparing it to the original drive signal. Since the original prepreg lay up procedure influenced the time-of-flight (i.e. slight ply misorientation, varying hand pressure , etc.), it was necessary to normalize the data for a given experiment with respect to other experiments. A point ninety minutes into the cure cycle was chosen as the data normalization point, since all sample histories up to the ninety minute mark were identical. Figure 2 shows changes incurred in the time-of-flight during the full cure and partial cure cycles, as well as the cure cycle temperature and pressure profiles. For the 12 inch square by 0.1 inch thick samples, the time-of-flight excursions shown in Figure 2 correspond to velocities ranging from 3400m/s to 4000m/s. (Note: A *decrease* in time-of-flight corresponds to an *increase* in velocity).

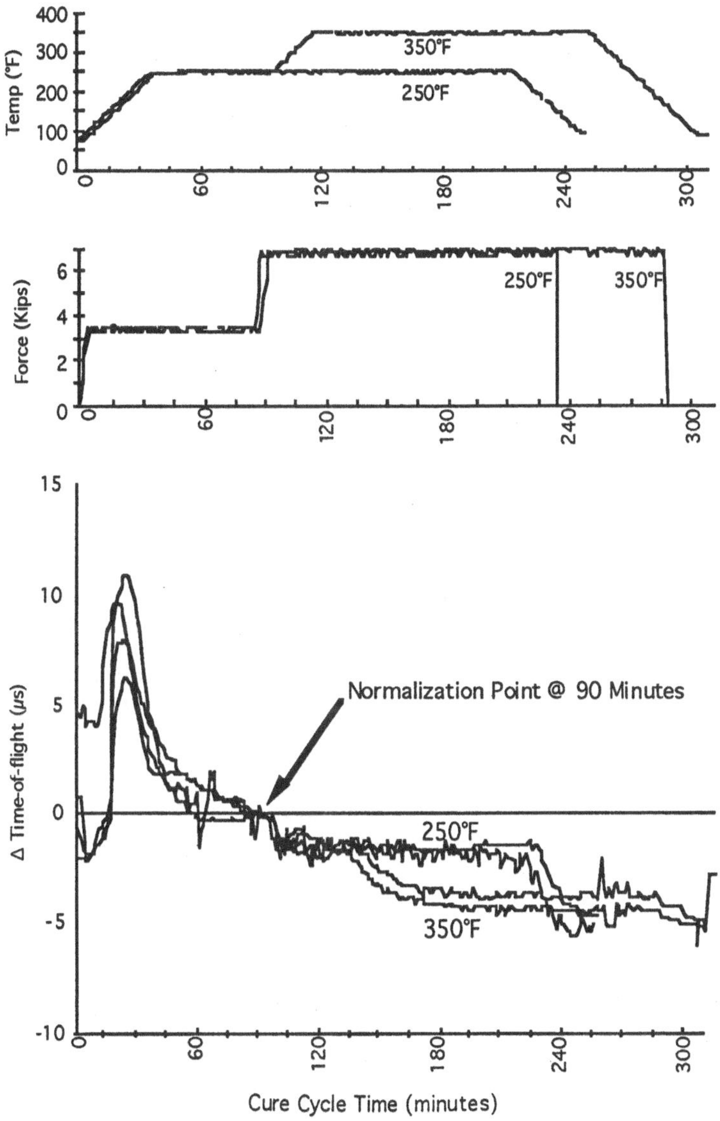

Figure 2. Results for the cure monitoring experiments are shown, detailing the change in ultrasonic signal time-of-flight as a function of cure cycle time. Also shown are the temperature and force profiles for both the 350°F full cure and the 250°F partial cure cycles.

The time-of-flight profiles follow the viscosity/stiffness of the composite panel during cure. Initially, the time-of-flight increases (velocity decreases) as the panel heats up and reaches its gelation point, the point of minimum viscosity. This occurs approximately 25 minutes into the cure-cycle. Following gelation, the time-of-flight decreases (velocity increases) as the epoxy begins to cure. Approximately ninety minutes into the cure cycle, the pressure is increased. An immediate decrease in time-of-flight (increase in velocity) is observed. Thus, at this point in the cure cycle, the sensor does respond to pressure. At the same time, the temperature for the full cure cycles is ramped to 350°F, while the partial cure cycle temperature remains at 250°F. As cure progresses, the difference in response between the 250°F cure and the 350°F cure panels becomes evident (see Figure 2). The 350°F cured panels exhibit higher ultrasonic velocities (shorter times-of-flight) indicative of their stiffer, more completely cured state relative to the 250° samples. In addition, there was no noticeable temperature effect on the fully cured panels once the process was complete, whereas the ultrasonic velocity in the under-cured panels did exhibit a temperature dependency during the cool-down procedure. It is unclear as to why the large temperature dependency exists, although some possible explanations include viscosity effects or relaxation phenomena.

WAVEGUIDE OR NOT ?

Some interesting observations were made regarding wave propagation in the embedded acoustic sensor during the cure cycle experiments described above. Most notable was the extent to which the received ultrasonic tone-burst signal was affected when an aluminum plate was positioned on the prepreg panel. The aluminum plate was coated with a release agent and placed on the composite panel so that, once cured, it would not adhere to the steel hot press. Interestingly, as the aluminum plate came in contact with the uncured prepreg composite panel, the received ultrasonic signal transmitted along the embedded sensor dropped in amplitude by as much as one order of magnitude. The amplitude reduction was attributed to ultrasonic energy coupling out of the composite panel and into the aluminum plate. Such a large deviation in amplitude however, would suggest that a large percentage of the ultrasonic energy was in fact not contained or guided by the embedded wire sensor, but rather was carried via plate modes in the host material itself. This would account for the tremendous amplitude reduction observed when the aluminum plate made physical contact with the composite specimen.

To test this theory, an experiment was performed where the Nichrome wire was terminated one inch into each side of the composite prepreg panel, so that only the composite itself could transmit the ultrasonic energy between transmitter and receiver. This is illustrated in Figure 3, and the results for a 350°F cure cycle using this configuration is shown in Figure 4. Results for a 350°F cure cycle using a continuous embedded wire are also shown for comparison. While small differences are noticeable between the continuous versus terminated wire experiments, it is clear that the overall response is quite similar. As such, the term "acoustic waveguide" often used for wires embedded in composite materials is a misnomer. The wire itself does not act as an ultrasonic waveguide, at least not over the frequency range from 50kHz to 200kHz. However, the wire does serve a useful purpose - it provides a convenient way to couple ultrasonic energy into and out of the composite panels.

FUTURE WORK

As described above, it would appear that the only function which the embedded wire serves is to couple energy into and out of the composite, but does not guide the energy within the composite. Since the embedded wire sensor is susceptible to ingress/egress variabilities and coupling effects at the acoustic horns, it would be prudent to eliminate the wire altogether, and replace it with a more robust means of coupling the energy into and out of the composite panels. It is anticipated that most future effort will be directed toward development of embedded acoustic sensors which do not suffer from the problems associated with the ingress/egress areas and which may also provide a controlled sensing area. Toward this end, a fiber-optic approach will be taken, using a laser to generate ultrasound within the sample, and a fiber-optic detector as the ultrasonic receiver. An all optical system would have the advantages of providing a localized sensing region, possible sensor multiplexing, interrogation of inaccessible areas, and would eliminate the signal variability arising from ingress/egress variabilities and mechanical coupling difficulties.

Lastly, sensors embedded during material manufacture for process control applications also offer a means by which the material health may be monitored throughout its useful lifetime. Changes in acoustic velocity may be indicate degradation of the material's elastic properties or a change in the material stress state. Similarly, a reduction in the transmitted acoustic amplitude may result from internal micro-cracking or delaminations. While these effects have yet to be investigated, it would seem a logical extension to in-situ process control monitoring, and would provide true "cradle-to-grave" material monitoring capabilities.

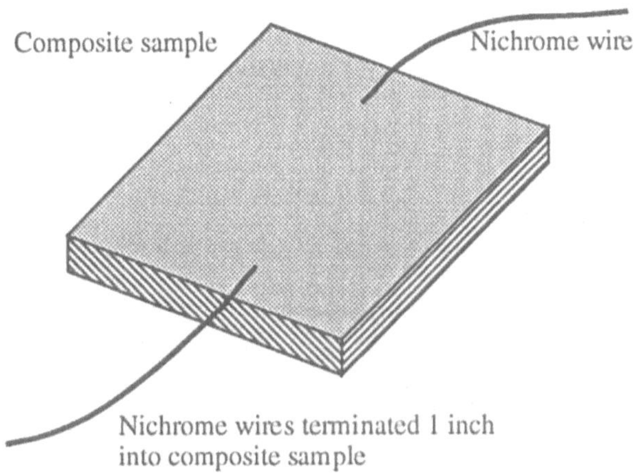

Figure 3. Schematic detailing the configuration used for the terminated "waveguide" experiment. Each piece of Nichrome wire was inserted only 1 inch into the test panels, and were not connected in any manner other than by the host Gr/Ep material.

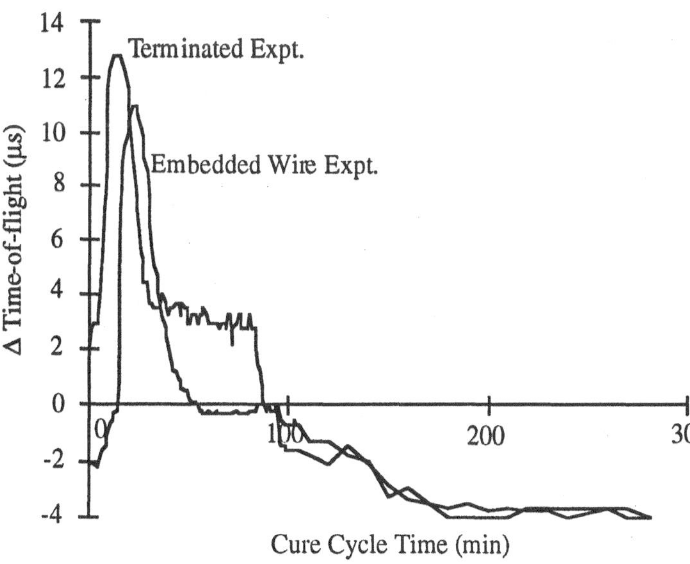

Figure 4. Results for the terminated "waveguide" experiment compared with those obtained from a continuous "waveguide". While there are slight discrepancies, the overall results are quite similar.

REFERENCES

R.T. Harrold. U.S. Patent No. 4,590,803, Acoustic Waveguide Monitoring, May 27, 1986.

R.T. Harrold and Z.N. Sanjana, Acoustic waveguide monitoring of the cure and structural integrity of composite materials, Polymer Eng. & Sci., v 26 no 5, 367-372 (1986).

W.P. Winfree and F.R. Parker, Ultrasonic characterization of changes in viscoelastic properties of epoxy during cure, Proc. QNDE, vol 4B, 1203-1208 (1985).

W.P. Winfree and K.J. Sun, Modelling guided acoustic waves for monitoring epoxy curing, Proc IEEE Ultrasonics Sym., 1227-1230 (1989).

SURFACE CONTROLLED MATERIALS EVALUATION

FOR Al-IMPLANTED Ni ALLOYS

T. Aizawa[1], J. Mitsuo[2] and J. Kihara[1]

[1] Department of Metallurgy, University of Tokyo
7-3-1 Hongo, Bunkyo-ku, Tokyo 113
[2] Tokyo Metropolitan Institute of Technology
Nishigaoka, Kita-ku, Tokyo

INTRODUCTION

In the design of structural materials, we often encounter various kinds of needs for improvement of surface material characteristics like hardness, oxidation toughness or wear resistance [1]. PVD/CVD ceramic coatings are the first approach to make surface control of mechanical properties [2]. These methods are widely used in fabrication of tool and die materials. Typical applications can be seen in TiN, TiC or Ti(CN) coated WC/Co and SKD materials. In parallel with this coating procedure, an ion implantation should become one of the most promising key technologies in the surface control engineering [1,3]. Active control of ion source, power and density of beam in use and its fluence enables us to change patterns of surface modification, to make different types of alloyment and to restructure the surface materials. In order to make full use of the above potential capabilities, however, it should be essentially indispensable to make mechanical and metallurgical evaluations on the synthesized or modified surface material. Such conventional methods as micro indentation and wear testing are difficult to obtain mechanical properties of surface materials with the thickness of submicrons.

Authors [4,5,6] have been developing the acoustic spectro microscopy (ASM) to make quantitative nondestructive evaluation (QNDE) on the elastic properties of ceramic coated materials with consideration of residual stresses inside coating. For PVD-TiN coatings with the thickness less than 10 μm, ASM has succeeded in elastic characterization of coatings and residual stress analysis. The present ASM provides a variety of QNDE methodologies to deal with elastic, elasto-plastic and toughness characterizations of actual materials system. Table 1 lists our developing family of ASM oriented QNDE method and new acoustic tools. In the present paper, our developed ASM is applied to QNDE of Al-implanted Ni-base alloys to investigate the effect of modified surface materials on the measured acoustic structure and to discuss the possibility of QNDE by ASM of surface-modified materials.

ACOUSTIC DIAGNOSIS FOR ION-IMPLANTED MATERIALS

Among various methodologies in the acoustic diagnosis, the present paper is concerning the dispersion analysis of the leaky surface wave velocity with use of the acoustic spectro microscopy (ASM).

Acoustic Spectro Microscopy

Acoustic spectro-microscopy (ASM) has been developed to make quantitative nondestructive evaluation (QNDE) on the mechanical properties of ceramic coated materials [3,4,5]. Several modes of acoustic diagnosis can be designed and constructed on ASM for each class of QNDE problems: 1) Rotational mapping for measurement of Young's moduli of the anisotropic materials and grain-controlled coating materials [6], 2) Two and three dimensional mapping for determination of elastic constant distribution in so small specified region as seen in microscopy, SEM or TEM with sufficient resolution [7], or 3) Insitu measurement of acoustic responses for deforming coated materials [8]. The original apparatus of ASM, which has been developed, is shown in Fig. 1 together with the fundamental measurement mode of ASM schematically illustrated in Fig. 2.

An ultrasonic pulse is injected from spherical lens into the target materials through coupler (water or Hg) with a controlled skew angle θ against the surface of materials. The whole reflected wave running back through the coupler can be received by the planar lens. Hence, even when the target materials could indicate high attenuation or dumping, the present approach should be free from inaccurate measurement or failure of measurement. In case that linear relation holds on the input/output acoustic signals, the planar lens can be utilized as a transmitter and spherical lens as a receiver.

Through the Fast Fourier Transform (FFT), spectra of both power and phase shift can be obtained from the reflected wave. The capacity of spatial resolution should be strongly dependent on the acoustic lens to be used. The point to be noted in the standard measurement by ASM is a sequential measurement mode by incremental increase or decrease $\Delta\theta$ of the incident angle θ. That is to say, the spherical - planar lens pair is controlled for the specified incident angle range $\theta_L \leq \theta \leq \theta_U$ to move by a constant $\Delta\theta$ in the circumferential direction with focussing point fixed on the surface of target materials. Fig. 3 depicts a typical configuration of received signals in one-shot measurement for PVD TiN coated WC/Co substrate. Through FFT, both the power P and the phase shift ϕ of the reflected waves can be obtained in the functional form of both the frequency and the incident angle: $P(f,\theta)$ and $\phi(f,\theta)$. Figs. 4 and 5 show these functional profiles of $P(f,\theta)$ and $\phi(f,\theta)$, respectively. When θ reaches to a critical angle θ_c where the Rayleigh wave is activated, P indicates a large dip and ϕ changes remarkably by 2π shift for a material with less dumping. Hence, we can distinguish this θ_c by these large dip or large phase shift. The leaky surface wave velocity V_{lsaw} is directly calculated from this θ_c by $V_{lsaw} = V_w / \sin(\theta_c)$, where V_w is the acoustic velocity of the coupler. As wellknown, V_w changes with temperature when pure water is employed as a coupler. Our developing system is all housed in the clean booth, and fluctuation of temperature is kept to be less than 0.1 K in measurement. Furthermore, pure lead plate with buffed surface was chosen as a reference material to yield homogeneous acoustic structure for comparison.

Estimate of Acoustic Structure for Al-implanted Ni Substrate

As has been investigated in Refs. (7) and (8), ceramic coated substrate material indicates characteristic dispersion of the leaky surface wave velocity V_{lsaw} with the frequency f. For the prescribed thickness of coating layer, this dispersion curve is uniquely determined by the elastic properties of coating and substrate materials and their interface. Furthermore, if the interface joining is perfect, or, both velocities and stresses are continuous across the interface, this dispersion curve is also predicted theoretically by the two dimensional elastodynamic analysis [(9)]. Hence, the acoustic structure of Al-coated or implanted Ni and Ni alloy can be estimated by this method.

Mechanical properties of pure Al and Ni is listed in Table 2 to be used for model materials in computation. Fig. 6 depicts the dispersion curves for Al-coated Ni substrate with different thickness. In this system of two cubic metals, its leaky Rayleigh wave velocity slightly increases with frequency, but its deviation from that for substrate when f = 120 MHz is only 3 to 4 m/s for the coating thickness of 200 nm. This slight dispersion is still within tolerance of the present ASM measurement with accuracy of 0.1 % for the leaky surface wave velocity.

As reported in Refs. (5) and (6), nondimensional dispersion curve should be inherent to the coating material where the leaky Rayleigh wave velocity is nondimensionalized by $(V_{lsaw} - V_{substrate}) / V_{substrate}$, and related with the nondimensional thickness d/λ or $d{\cdot}f$. Direct influences of both mechanical properties of substrate and coating thickness can be eliminated from this curve. Fig. 7 depicts the nondimensional dispersion curve of Al-coated Ni system. The obtained unique curve should reveal that single homogenous material is coated on the substrate with perfect joining.

The above calculation has assumed that no chemical reaction should take place and the

Table 1 Various methodologies of QNDE and acoustic tools based on ASM.

Item	Frame of Evaluation	Functions by Evaluation Tools
Nondestructive Quality Assurance	- Detection of flaws by imaging	- 2D acoustic microscopic imaging
	- Evaluation of interfacial defects between coating and substrate	- Dispersion analysis by ultrasonic spectroscopy
Elastic Characterization	- Evaluation of macroscopic and homogenized elstic properties	- Dispersion analysis by ultrasonic spectroscopy
	- Evaluation of anisotsopic elastic distribution	- Multi-Dispersion analysis on θ-φ plane
	- Description of grain growth in polycrystalline coating and substrate materials	- Two dimensional mapping method with dispersion analysis
	- Residual stress analysis of coating materials	- Micromechanics with dispersion analysis
Elastic-Plastic Characterization	- Evaluation of wrought zones and yielding/damage zones in materials	- Wide-Band Spectroscopy with acoustic structure analysis
	- Estimate of affected/alloyed zones on surface by ion implantation	- Acoustic structure analysis
Acoustic Emission	- Toughness, K_{Ic} and cracking mode evaluation - Analysis of micro cracking mode	- Ceramic acoustic fiber - Multi-sensor ultrasonic lens - Cracking mode analysis through wave-let analysis of received signals
On-Process Quality Evaluation	- Insitu monitoring of coating processing - On-line nondestructive materials evaluation	- Metallic acoustic fiber - Arrayed-sensor lens - Three dimensional mapping - QNDE at elevated temperture and high pressure

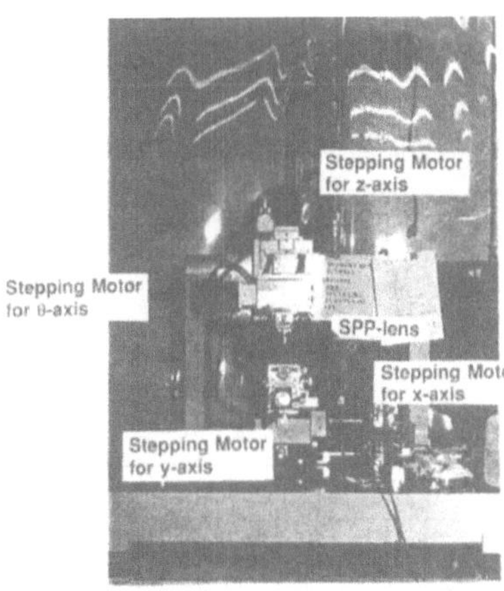

(Working in clean booth with class 1000)

Fig. 1 Acoustic spectro microscopy.

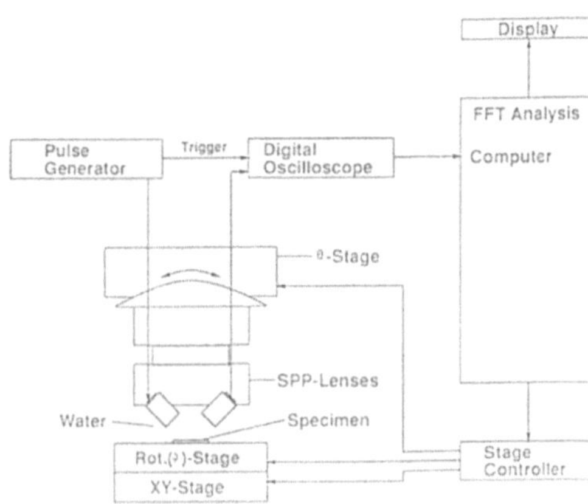

Fig. 2 Schematic view of ASM.

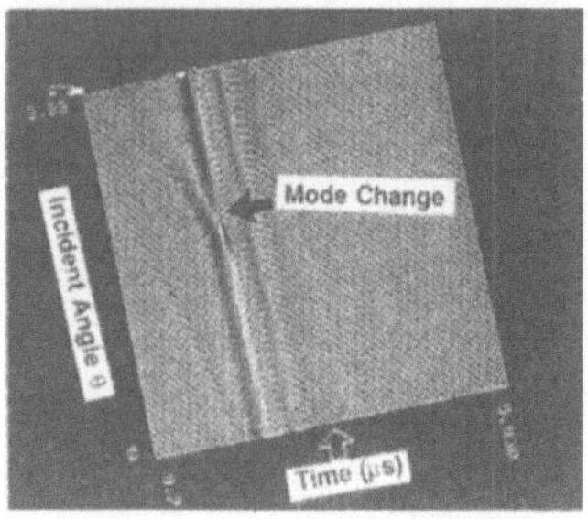

Fig. 3 Profile of received signals.

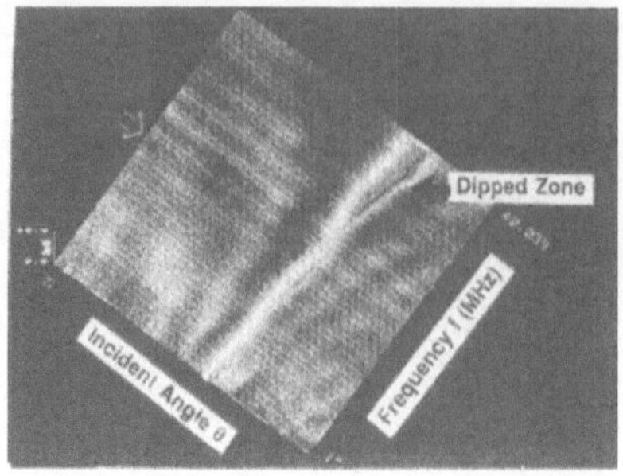

Fig. 4 Profile of power spectrum P(θ,f).

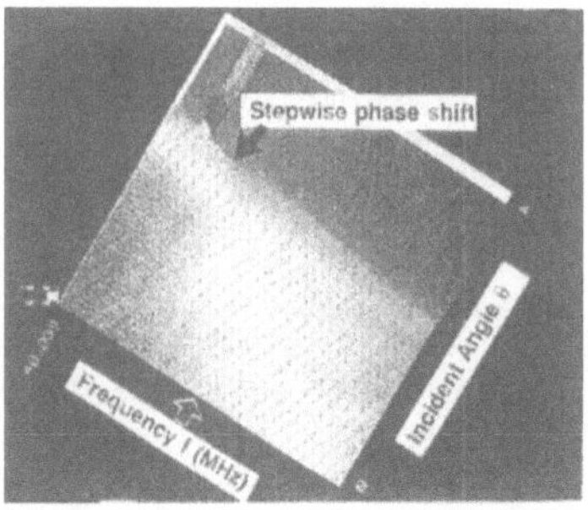

Fig. 5 Profile of phase shift ϕ(θ,f).

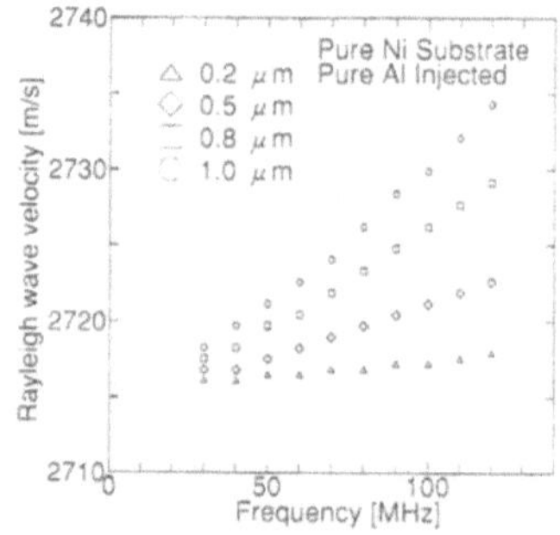

Fig. 6 Dispersion curve of Al-coated Ni.

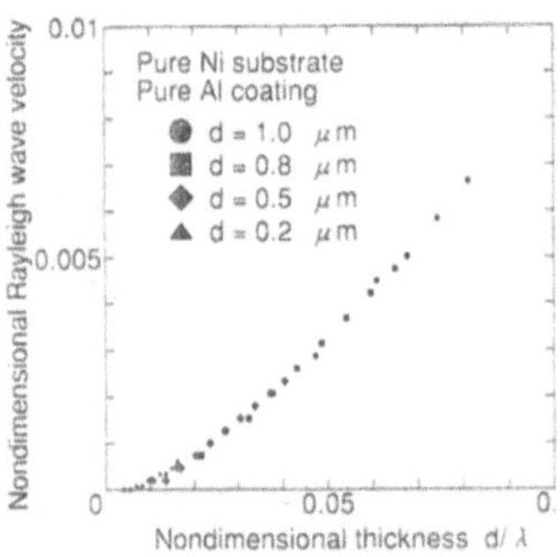

Fig. 7 Nondimensional dispersion curve.

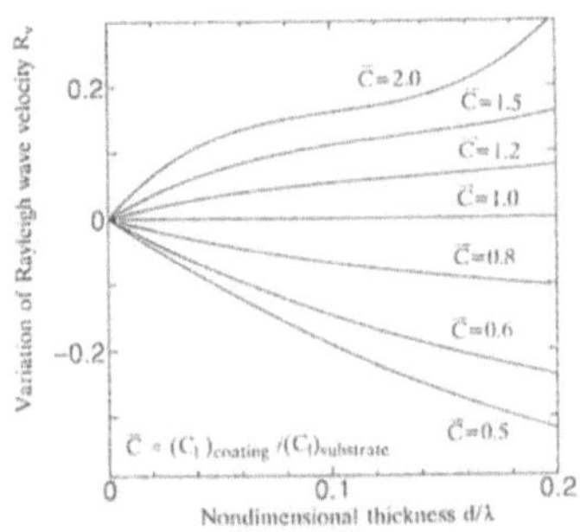

Fig. 8 Effect of surface stiffness on dispersion.

Table 2 Mechanical properties of pure Al and Ni for calculation.

Pure Al: $\rho = 2.70$ g/cm^3, $E = 70.6$ GPa, $\nu = 0.345$, $C_l = 6409$ m/s, $C_s = 3118$ m/s

Pure Ni(Substrate): $\rho = 8.90$ g/cm^3, $E = 199.5$ GPa, $\nu = 0.312$, $C_l = 5591$ m/s, $C_s = 2923$ m/s

Table 3 Chemical components of Ni-based alloy substrates.

	Ni	Mn	Cr	Al	Fe	Si	Cu
Inconel 600	Bal.	0.54	15.4	—	7.58	0.27	—
Alumel	Bal.	1.41	—	1.06	—	1.82	—
Chromel	Bal.	—	9.34	—	0.36	0.33	—
Constantan	42.9	0.89	—	—	—	—	56.0

surface structure be composed of pure Al layer and pure Ni substrate. In the actual implanted materials, alloyment takes place between implanted metal element and chemical components of substrate, so that the dispersion curve should change itself by this alloyment. Fig. 8 illustrates typical dispersion curves when mechanical properties of surface material change by alloyment. When the hardened material is synthesized at surface, convex type of curves must be observed. On the other hand, dispersion curve becomes of a concave shape for softened surface structure.

EXPERIMENTAL CONDITION

Four types of Ni-base alloys were employed as test specimens, and their bulk elastic moduli were obtained from the measured longitudinal and shear acoustic wave velocities by the sing-around method.

Testspecimen

The chemical components were listed in Table 3 for the employed four Ni-base alloys: Inconel-600, Alumel, Chromel and Constantan. Their dimensions of specimen are commonly set by 10 x 30 x 4 mm. As a preliminary treatment, the whole specimens are annealed in vacuum and ground by SiC-polishing paper with the mesh of 400. The surface 10 x 30 mm to be implanted is only buffed.

Al Implantation Condition

As wellknown, implantation capacity is controlled by both acceleration energy V_{ion} and radiation density ϕ_{rad}. In the present study, V_{ion} was fixed by V_{ion} = 50 KeV as the standard condition, and ϕ_{rad} was varied by 0.5 x 10^{17} and 1.0 x 10^{17} dpi/cm^2. Fig. 9 shows typical distributions of fundamental elements in the direction of depth for Inconel-600 by AES. Although the sputtering time depends on the target materials, Al-implanted penetration depth might be about 200 nm in total for V_{ion} = 50 KeV and ϕ_{rad} = 1.0 x 10^{17} dpi/cm^2.

On the very surface upto 10 nm, concentration of Al is varying by oxidation, but both Al and other constituent elements of substrate have indicated nearly constant distribution beyond 20 - 30 nm.

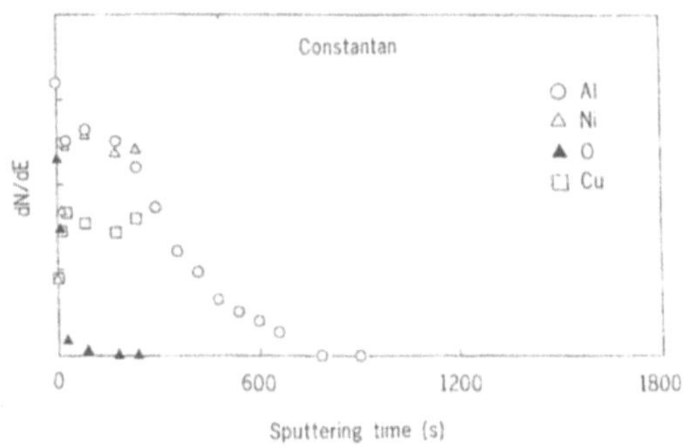

Fig. 9 AES depth profiles of Al implanted alloys.

Bulk Wave Analysis

For investigation of the elastic moduli of Ni-based alloy substrate materials, the sing-around method with 10 MHz was employed to measure both the longitudinal and the shear velocities, or, C_l and C_s. As listed in Table 4, Inconel-600, Alumel and Chromel indicated nearly the same velocities with each other, and their values are slightly shifted from those for the pure Ni substrate. From these measured velocities, both the Rayleigh wave velocities and the Young's moduli can be calculated by the following equations:

(1) $V_{lsaw} = \kappa \cdot C_s$ for $\kappa^6 - 8\kappa^4 + (24-16\alpha^2)\kappa^2 + (16\alpha^2-16) = 0$, $\alpha^2 = (1-2\nu)/(2-2\nu)$,

(2) $C_l = \{(K+(4/3)G)/\rho\}^{1/2}$, $C_s = \{G/\rho\}^{1/2}$, for $K = E/[3(1-2\nu)]$, and $G = E/[2(1+\nu)]$.

The estimated results are listed in Table 5.

Table 4 Measured longitudinal and shear velocities.

No	Substrate	C_l	C_s
A	Inconel-600	5860 m/s	3060 m/s
B	Alumel	6050 m/s	3160 m/s
C	Chromel	5860 m/s	3060 m/s
D	Constantan	5200 m/s	2712 m/s

Table 5 Estimated Rayleigh wave velocity and Young's moduli

No	Substrate	V_R	E
A	Inconel-600	2840 m/s	206 GPa
B	Alumel	2940 m/s	224 GPa
C	Chromel	2840 m/s	212 GPa
D	Constantan	2530 m/s	172 Gpa

ACOUSTIC DIAGNOSIS BY DISPERSION ANALYSIS

The acoustic spectro microscopy was applied to make dispersion analysis of Al-implanted Ni alloy substrates and to investigate whether the effective alloyment should take place.

Measurement Condition

The present measurement condition for ASM is listed in Table 6. The incident angle is controlled incrementally by $\Delta\theta = 0.1$ degree. For standardization in measurement, the location in the z-axis is fixed with its deviation less than 100 nm, and x-y stage is kept planar with skew angle less than 2.9×10^{-4}. Since these Ni-based alloys have remarkable acoustic dumping, location of critical angle θ_c was determined by both dip in power and large gradient in phase shift.

Table 6 The present measurement condition in ASM.

SPP lens System	Point Focussing
Frequency	$10 \leqq f \leqq 150$ MHz
Incident Angle	$20 \leqq \theta \leqq 40$ $\Delta\theta = 0.1°$
Reference Material	Lead

Experimental Results

Among four types of substrate materials, little dispersion was observed for Al-implanted Alumel and constantan. This is partially because 1) Ultrasonic signals are so weak in the present attenuation set-up that dip in power should not be noticeable, and 2) Large attenuation in materials reduces the amount of phase shift. The absence of dispersion assures that little alloyment must take place to make large influence on the mechanical properties of surface materials.

The measured data were depicted in Fig. 10 for unimplanted and Al-implanted Inconel-600 substrate. In case of unimplanted Inconel-600, the leaky surface wave velocity becomes nearly constant by $V_{lsaw} = 2860$ m/s, which is in good agreement with the estimated value in Table 5. In case of Al-implanted substrate with $\phi_{rad} = 1.0 \times 10^{17}$ dpi/cm^2, no dispersion was observed in the lower frequency range, but the Rayleigh wave velocity begins to increase with frequency and becomes constant by 2950 m/s, which is larger than V_{lsaw} for substrate by 90 m/s. Considering the correlation of sputtering time with thickness in AES, total sputtering time (upto the time when Al-level signal diminished into the background) 1000 s in this case might correspond to at most 250 to 300 nm. If

Al-implanted Inconel-600 could have the same structure as the Al-coated Ni substrate, amount of dispersion to be measured should be 3 to 4 m/s. Actually measured relatively large dispersion reveals that surface material should be hardened by alloyment to indicate large stiffness.

Dispersion curves were shown in Fig. 11 for Al-implanted Chromel. For unimplanted Chromel, nearly constant dispersion was observed, and the measured V_{lsaw} becomes 2850 m/s. When ϕ_{rad} = 0.5 x 10^{17} dpi/cm^2, no distinct dispersion was measured. However, ϕ_{rad} was increased to be 1.0 x 10^{17}, the dispersion of concave type was observed: V_{lsaw} monotonically decreases with frequency. When f = 120 MHz, V_{lsaw} becomes 2700 m/s, which is smaller than the bulk velocity by 150 m/s. This significant dispersion tells that 1) Implanted aluminum is not only coated on the substrate but alloyed with constituent elements of substrate material only on the surface upto 300 nm, and 2) Synthesized surface alloy indicates softened stiffness than the bulk Chromel.

Discussions

Owing to the ASM measurement and dispersion analysis, both Inconel-600 and Chromel are expected to have alloyed surface layer by Al-implantation among four Ni-base alloys. In general, surface modification has been reported to be effective to improve toughness against oxidation and to reduce wear rate. In the present study, oxidation testings were made for various surface modification procedures including N-implantation, Al-vaporization and their combined methods for comparison. The whole specimens were heated and held constant by T ≈ 970 K for 50 hours in O_2 atmosphere. The saturated H_2O(g) is mixed with O_2 at T ≈ 298 K and flown together into furnace with the flow rate of 2 l/min. The increase of weight ΔW for each surface treated material was measured and normalized by the increase of weight ΔW_0 for untreated material. The obtained normalized increases of mass are listed in Fig. 12 for four Ni-based alloys. In case of Alumel and Constantan, where no dispersion curves were observed, very little effect of surface modification on improvement of toughness against oxidation can be seen. This reveals that effective alloyment should never be synthesized at the vicinity of surface for these two substrate materials.

On the other hand, distinct difference can be seen between Al-deposited and Al-implanted materials. As before mentioned, the former has no possibility to change surface structure. Since alumina is easy to be synthesized and precipitate at the surface, addition of aluminum layer might lead to reduction of oxidation. In the latter case, essential modification should take place at the surface material so as to make further reduction of oxidation rate. In case of Inconel-600, significant influence of Al-implantation can never be seen. This is partially because 1) Thickness of effective alloyed layer is still too thin to have influence on the resistance against oxidation, and 2) The synthesized alloyment with stiff Young's moduli has little potential to reduce oxidation. On the other hand, oxidation toughness was improved than Al-implantation, since $\Delta W/\Delta W_0$ becomes lass than half of that for Al-deposited Chromel. This fact has precise correlation with the surface alloyed layer with softened stiffness, which was mechanically evaluated by ASM.

CONCLUSION

Our developed acoustic spectro microscopy was applied with success to nondestructive evaluation of Al-implanted Ni-base alloys. Acoustic structural analysis tells that little dispersion is expected for Al-coated Ni without surface modification. In case of Alumel and Constantan, although those materials indicate remarkable dumping, no dispersion curves were measured for the leaky Rayleigh wave velocity. This reveals that little surface modification by alloyment is expected for these materials. This prediction has been partially supported by the fact that oxidation toughness should be indifferent to Al-implantation for those two materials. Inconel-600 indicates increasing dispersion of the leaky surface wave velocity with frequency, so that alloyed material should have more stiff Young's modulus than Inconel-600 substrate owing to Al-implantation. The point to be noted is that relatively large decreasing dispersion of the leaky surface wave velocity with frequency for Al-implanted Chromel substrate. It was found that the oxidation toughness should be improved by this type of modification to reduce the elastic stiffness from substrate.

To make large step in progress of surface modification technology, nondestructive mechanical characterization studied in the present research should be indispensable. Authors are interested in Al and N implantation into Ni, Ti and Nb and O-implantation into Al to investigate the synthesized surface material structure from metallurgical aspect of analysis using AES, ESCA and X-diffraction and from mechanical aspect of acoustic analysis by ASM. The related paper will be reported in future.

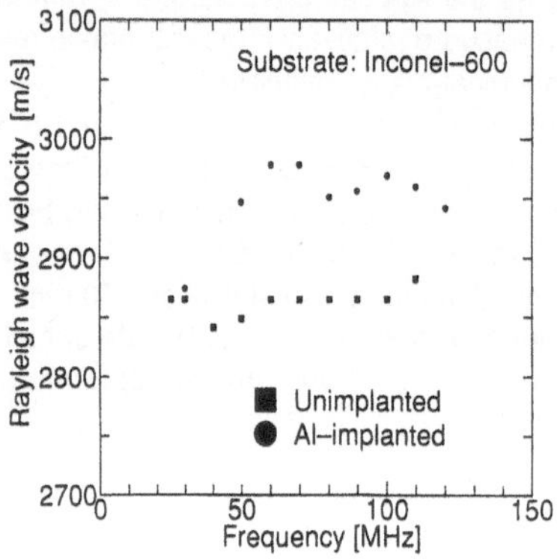

Fig. 10 Dispersion curves for Inconel-600.

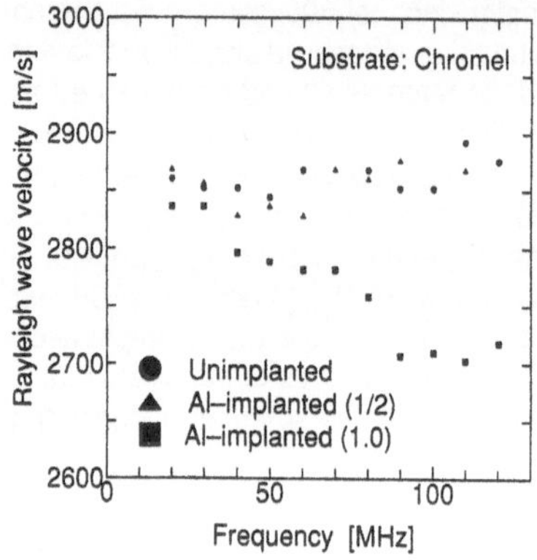

Fig. 11 Dispersion curves for Chromel.

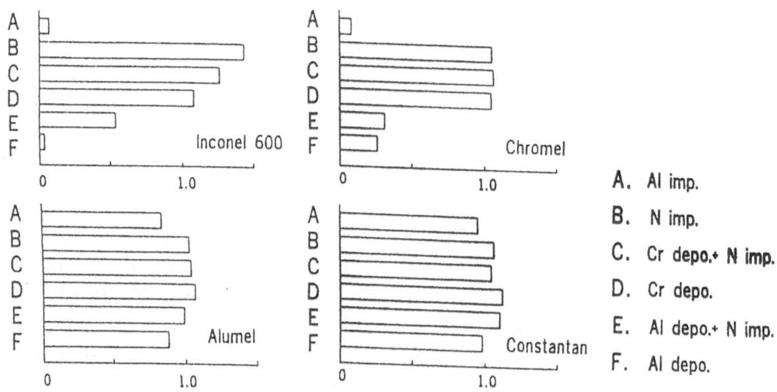

Fig. 12 Nondimensional loss of weight in the oxidation testings.

References

1. Bhushan, M.T.: Handbook of Tribology (1991) McGrow-Hill.
2. Takeda, H.: Ceramic Coating (1988) Nikkan-Kougyo.
3. Sudou, H.: Residual Stress and Distortion (1988) Uchida-Rokakuho.
4. Aizawa, T. and Kihara, J.: Ultrasonic Microscopic Evaluation of PVD/CVD Coated Hard Materials, Proc. Int. Conf. Residual Stress (1991) 346-351.
5. Aizawa, T. et al.: Residual Stress Evaluation of TiN Coated WC/Co Cermets in Hardness and Scratch Testings, Proc. Int. Conf. Residual Stress (1991) 692-697.
6. Aizawa, T. and Kihara, J.: Materials Evaluation of PVD/CVD Coated WC/Co Superhard Alloys by the Ultrasonic Microscopy, Nondestructive Testing Evaluation 9 (1992) 999-1011.
7. Aizawa, T. et al.: Nondestructive Evaluation of PVD/CVD TiN Coated WC/Co by Acoustic Spectro-Microscopy, AMD-140 (1992) 55-70.
8. Aizawa, T. and Kihara, J.: Nondestructive Measurement of Residual Stresses in PVD TiN Coatings (1993) (To be published in J. AmCer).
9. Aizawa, T. et al.: Prediction of Acoustic Structures for Multi-Layered Materials, Proc. COMMP'93 (1993, Sep. Tokyo).

DEVELOPMENT AND EVALUATION OF A WORKPIECE ANALYZER FOR INDUSTRIAL FURNACES

P. Kotidis,[1] J. Woodroffe,[1] J. Shah,[2] and T. Schultz[2]

[1]Textron Defense Systems
Everett, MA 02149

[2]Surface Combustion, Inc.
Maumee, OH 43537

INTRODUCTION

The traditional method of determining the temperature of heat-treated workpieces in industrial furnaces is to measure surface temperature with radiation pyrometers or local temperature with contact thermocouples. Contact thermocouples have limited use because they damage the workpiece in a continuous process or require drilling of a hole in the workpiece for batch operation. Radiation pyrometers are used for noncontact surface temperature measurements, but suffer from inherent inaccuracies because of interference by radiation from furnace hot walls and gases in the furnace atmosphere and varying emissivity of the workpiece during thermal processing. Dual-wavelength pyrometers are designed to provide independence from emissivity variations. However, they do not perform on non-gray-bodies, they have difficulty looking through non-gray windows, and they tend to measure background temperature when the background is hotter than the target. Multicolor pyrometers use algorithms to eliminate the variable emissivity problems with limited success and applicability. In addition, traditional nondestructive temperature measuring instruments lack the capability to measure temperature gradients in a workpiece, which are critical for uniform phase transformations. Moreover, the knowledge of the time at which this uniformity is achieved is important for process cycle time optimization. There is a need for an instrument which can measure surface temperature as well as average bulk temperature to provide knowledge of temperature uniformity in the workpiece. An instrument like that could provide substantial cost savings and near-term payback. The Surface Combustion, Inc. and Textron Defense Systems research team, under DoE support, has conceptualized such an instrument to be used as Workpiece Analyzer (WPA).[1,2] This paper describes results of an analytical study to determine the feasibility of this concept and preliminary experimental data of a bench scale demonstration.

PRINCIPLE OF OPERATION

The fundamental operating principle of the WPA is based on the generation, propagation, and subsequent detection of ultrasonic waves on the surface and through the body of the workpiece (Figure 1). A pulsed laser beam from the Impulse Laser (IL) is

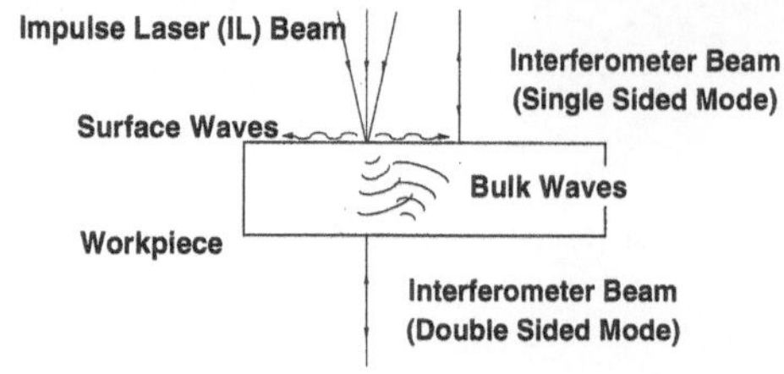

Figure 1. WPA operating principle.

directed to the workpiece to generate the waves. The arrival of the waves at an appropriately selected location (opposite or same side of the target) is monitored using a Polarizing Interferometer (PI). The speed of ultrasound is determined by measuring the time-of-flight of these waves. Since the speed of ultrasound is a function of the temperature of the solid, its bulk and surface temperatures can be measured by using calibration curves. Depending on the location of the PI with respect to the IL (same or opposite side), two WPA configurations can be realized: Single-Sided (SS) and Double-Sided (DS). Under the SS mode, both surface and reflected bulk waves can be simultaneously monitored, providing valuable information about temperature gradients in the workpiece. For thin materials, only surface wave measurements are required because the temperature gradients are negligible. In this case, a low energy impulse laser can be used leading to a much lower cost instrument. Moving workpieces (e.g., continuous strip lines) could also be monitored by the WPA due to its totally noncontact operating principle.

Metallurgical status of a workpiece during heat treatment could also be monitored by the WPA. For example, phase changes from ferrite to austenite in carbon steels have been detected. This is a result of the relationship between ultrasonic speed and crystalline structure of the material. The implication of this WPA feature is that excessive soaking times for workpieces can be eliminated, leading to major energy and cost savings.

BACKGROUND

Laser generation and detection of ultrasound ("Laser Ultrasonics") have been demonstrated and investigated for decades. The interaction of laser radiation with solid targets has been analyzed for years at TDS[3-6] and elsewhere.[7] The interaction regimes of interest to the WPA are: Thermoelastic (simple surface heating-no surface damage) and Plasma Generation (recoil effect due to confined plasma-possible surface damage). Laser pulses, or any other transient change in the structure of a solid, generate elastic waves (i.e., ultrasound). Four types can in general be generated: Longitudinal, Shear, Surface (Rayleigh), and Lamb.[8-10] All four types of these waves have been generated by many researchers[11-16] with various types of detection schemes, including contact piezoelectric transducers, Electromagnetic Acoustic Transducers (EMAT's), and interferometric or noninterferometric optical techniques. The dissipation of acoustic energy in polycrystalline metals, like steel, is important as far as signal strength is concerned and can in general be attributed to elastic hysteresis, thermoelastic relaxation, and scattering.[17] Strong dependence of the attenuation on temperature and magnetic properties of steel[18] has been observed, while the attenuation appears to be temperature independent for nonmagnetic steel.[17]

The list of references included in this paper is only a representative sample of the enormous body of literature in this field. A more complete list of references can be found in Ref. 1.

SEMI-EMPIRICAL MODEL FOR ULTRASOUND GENERATION AND PROPAGATION

Thermoelastic Regime

Consider a laser pulse of duration t_p with uniform and constant intensity impinging upon the surface of a metal target.

As the area of the laser spot is heated, the metal expands and some strain, ε ($\varepsilon = \phi \cdot \Delta T$), is generated where ϕ is the linear coefficient of expansion of the material and ΔT is the average temperature rise in the surface layer. Consequently, a displacement, $\Delta\ell_0$ ($\Delta\ell_0 = \varepsilon \cdot \delta$), is formed, where $\delta = (2Kt_p)^{1/2}$ is the penetration depth of heat and K the thermal diffusivity. The attenuation of the generated ultrasonic wave is due to spherical (cylindrical for Rayleigh waves) spreading of acoustic energy as well as elastic hysteresis, thermoelastic relaxation, and scattering. This attenuation is frequency dependent and scales with e^{ax}, where a is the attenuation coefficient and x is the distance traveled.[18]

Using farfield displacement functions[13] for both longitudinal and shear waves, the displacements at the detection point are:

$$\Delta\ell^L = \frac{\phi\,(1-r)\,\Phi \cdot t_p}{\rho C_v}\,\frac{d_s}{x}e^{a_L x}\frac{H_o}{\mu}\,, \quad \Delta\ell^S = \frac{\phi\,(1-r)\,\Phi \cdot t_p}{\rho C_v}\,\frac{d_s}{x}e^{a_s x}\frac{K_o}{\mu} \qquad (1)$$

where a_L, a_s are longitudinal and shear attenuation coefficients and $H_o = H(\theta=0)$, $K_o = K(\theta=0)$, where $H(\theta)$, $K(\theta)$ are directivity pattern equations. Φ is the laser flux (power per unit area), r is the surface reflection coefficient, d_s is the spot diameter, ρ is the mass density, C_v is the specific heat, and μ the Lame constant or modulus of rigidity. This solution corresponds to a system where the detection laser is located at the epicenter of the impulse point (i.e., $\theta=0$).

Figures 2-3 present comparisons between data[14,19] and model predictions for aluminum and steel targets. The agreement is quite good and ensures the validity of the model.

Plasma Regime

In this regime, the phenomenology of the laser-solid interaction is quite different from that of the thermoelastic regime and is dominated by the presence of plasma.[1,4] Extensive theoretical and experimental research has led to detailed modeling of the plasma characteristics and effects.[1,3-6] Combining the above information, the displacements a distance x away from the impulse point are:

$$\Delta\ell^L = \frac{p_D\sqrt{2Kt_p}}{E}\,\frac{d_s}{x}\,e^{a_L x}\,\frac{F_o}{\mu}\,, \quad \Delta\ell^S = \frac{p_D\sqrt{2Kt_p}}{E}\,\frac{d_s}{x}\,e^{a_S x}\,\frac{G_o}{\mu} \qquad (2)$$

where F_o, G_o account for the directivity pattern at $\theta = 0$ (epicenter) and p_D is the pressure delivered to the target surface by the plasma.

Figure 4 presents a comparison between model prediction and data.[20] The agreement is sufficient for the purpose of this simple model. High inaccuracies can be observed in the transition regime right before the plasma formation due to the inability of the plasma

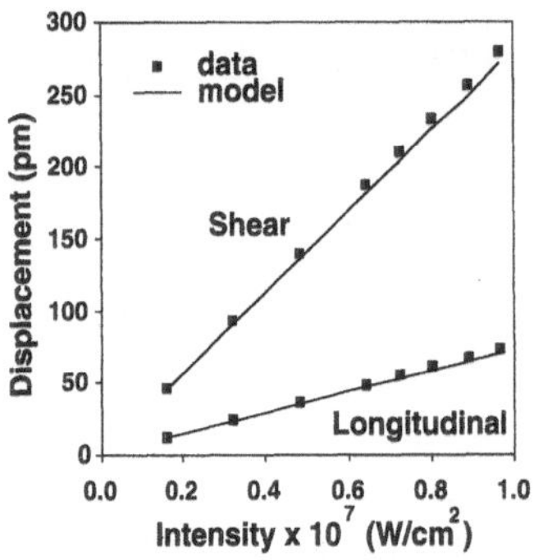

Figure 2. Comparison between data and model predictions for a 25 mm thick aluminum target.

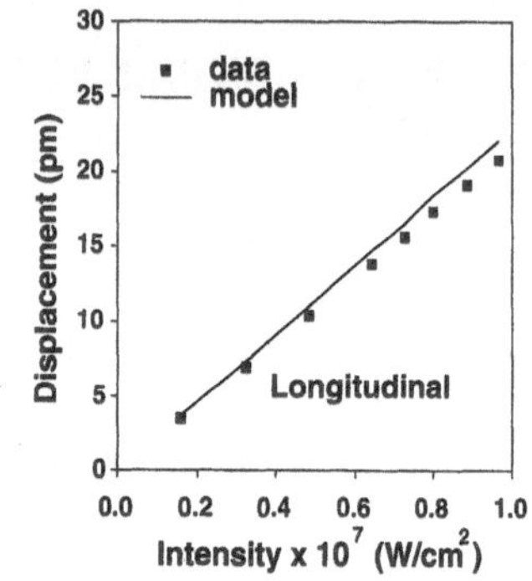

Figure 3. Comparison between data and model predictions for a 25 mm thick steel target.

model to predict thermoelastic effects. The model predicts a change in the slope of the displacement curve at about 26 x 10^7 W/cm^2, but it does not show any decrease in the amplitude of the displacements. These phenomena are a direct consequence of the interaction between two competing factors, i.e., the increase of the surface plasma pressure with laser flux[4] and the decrease of the delivered pressure to the workpiece as the plasma confinement time[4] increases. Another comparison between data and model prediction is also presented in Figure 5. The agreement is within the experimental error and is

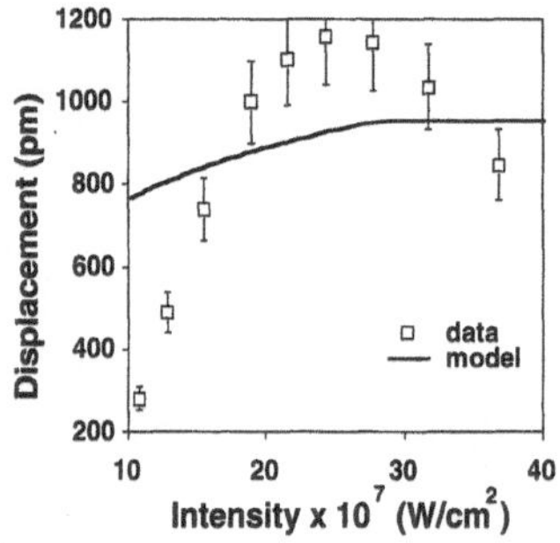

Figure 4. Comparison between data and model for a 25 mm thick aluminum target with constant impulse laser energy (33 mJ).

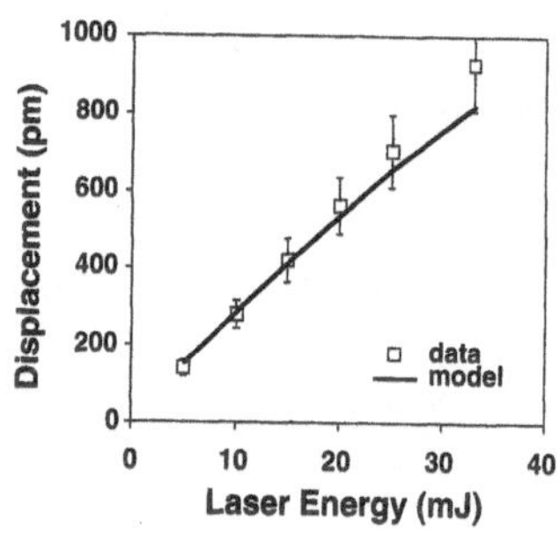

Figure 5. Comparison between data and model for a 25 mm thick aluminum target with constant impulse laser intensity (180 MW/cm^2).

considered satisfactory. Typical displacement achieved through a 25 mm thick aluminum is about 1100 pm or 11 x 10^{-10} m in the plasma regime, while for a similar steel specimen this value drops down to 55 pm or 55 x 10^{-12} m.

ANALYTICAL MODEL FOR THE POLARIZING INTERFEROMETER

The purpose of this model is to determine the minimum detectable displacement due to the arrival of ultrasonic waves. In describing the model of the PI, we will refer to Figure 6, where a schematic illustration is presented.

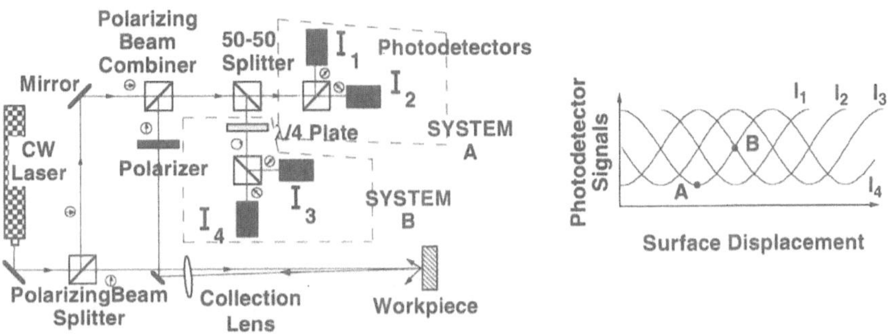

Figure 6. Polarizing Interferometer (PI) configuration and characteristic response curves.

Consider an ultrasonic displacement ξ of amplitude Δ and frequency ω_s as $\xi = \Delta \sin(\omega_s t)$. The corresponding light phase shift is $\Delta\phi = 2(2\pi\Delta/\lambda) \cdot \sin(\omega_s t)$ where λ is the light wavelength and the factor 2 is due to the double path for beam reflection. The signal beam from the workpiece and the reference beam are first linearly polarized, then combined and taken through a 50-50 splitter into systems A and B (Figure 6, dashed lines). After the 45° polarizing beam splitters and the $\lambda/4$ plate in system B, a set of signals 90° apart are generated from the four detectors, I_1, I_2, I_3, I_4. Signal differences (I_1-I_2, I_3-I_4) are then performed to generate two signals in quadrature, which are subsequently taken through high pass filters, followed by squaring and adding operations. A square root operation completes the signal processing. The final current, i, is given as:

$$i = \sqrt{\frac{I_R I_S}{2}} \left[\frac{4\pi\Delta}{\lambda} (RC)\ \omega_s\right] \cos(\omega_s t) \tag{3}$$

where RC is the time constant of the high-pass filter and I_R,I_S the light intensities in the reference and signal legs.

The minimum detectable displacement is determined by the level of photocurrent noise (shot noise) of the detector, its RMS value being, $\tilde{i}_N = \sqrt{2e\eta If}$ where f is the detector bandwidth, e the electronic charge, η a quantum efficiency, and I the average light on the photodiode. By setting the RMS signal-to-noise ratio $\tilde{i}/\tilde{i}_N$ equal to 1, the minimum detectable displacement Δ_{min} can be calculated, as:

$$\Delta_{\min} = \frac{\lambda \sqrt{ef(I_R+I_S)}}{\pi(RC)\ \omega_s\ \sqrt{2\eta I_R I_S}}\ \frac{1}{\sqrt{N}} \quad \text{for } N \text{ laser shots} \tag{4}$$

According to the model, minimum detectable displacements as small as 30 pm are theoretically feasible.

The signal processing procedure described above makes the interferometer insensitive to external vibrations (usually low frequency) and provides constant sensitivity, independent of the operating point. Since the ultimate goal of this effort is the development of a simple, rugged, low-cost detection scheme, a one-detector configuration has been assembled by eliminating three detectors, with some penalty to the signal-to-noise ratio. The insensitivity to low frequency external vibrations is again achieved through high-pass signal filtering, but the system relies upon multiple laser pulses. The presence of external vibrations (furnace environment) forces the PI to operate at random points, including those of minimum (point A) and maximum (point B) (Figure 6) sensitivity. Several algorithms have been developed to process the multiple signals (typically 10 to 50 laser pulses).

EXPERIMENTAL RESULTS

The objective of these tests was the demonstration of the WPA's ability to detect ultrasonic velocity changes due to temperature or metallurgical state (phase change) of a heated target.[2] A double-sided WPA configuration was used with a tube furnace rated for 1900°F. No paint or any type of coating was applied to the surface of the samples before or during the tests. A witness piece inside the furnace with an embedded thermocouple was used to monitor the sample temperature. Ultrasound was generated by an Excimer XeCl laser operating at 308 nm with a pulse duration of approximately 20 nsecs and typical energy per pulse of approximately 80 mJ. A CW, 10 mW He-Ne laser was used in the PI configuration.

Temperature tests were conducted with 1" thick 2" diameter samples of various types of carbon and stainless steels, aluminum, and brass. The thickness selection was driven by the need to formulate a data base representative of the most promising heat treating markets. A representative sample of the processed waveforms is shown in Figure 7 for stainless steel (Figure 7a) and brass (Figure 7b) samples. The highest temperature shown 1875°F) represents the limits of the furnace. The digitizer used throughout these tests had a 10 nsec resolution. Although the attenuation at high temperature levels increased, the signal-to-noise ratio was still adequate for the detection of the ultrasonic arrival.

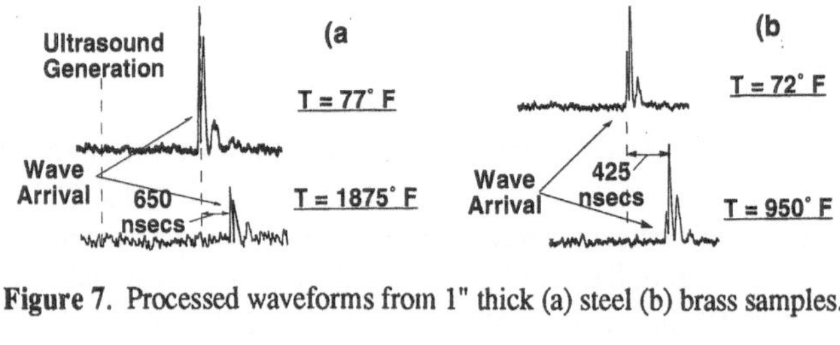

Figure 7. Processed waveforms from 1" thick (a) steel (b) brass samples.

Given the time-of-flight, the ultrasonic velocity can be measured if the sample thickness is known (assumed constant linear coefficient of expansion). Figure 8 presents a representative sample of the results for 1" thick 1010 carbon and 316 stainless steel samples. For the final WPA product, such curves will be generated during calibration with known samples, stored into the computer's memory and used to calculate the temperature of heat-treated workpieces. The discontinuity at 1250-1400°F represents the phase transformation of carbon steel from ferrite to austenite and should be accounted for in the data reduction scheme. However, this observation is a unique feature of the WPA concept

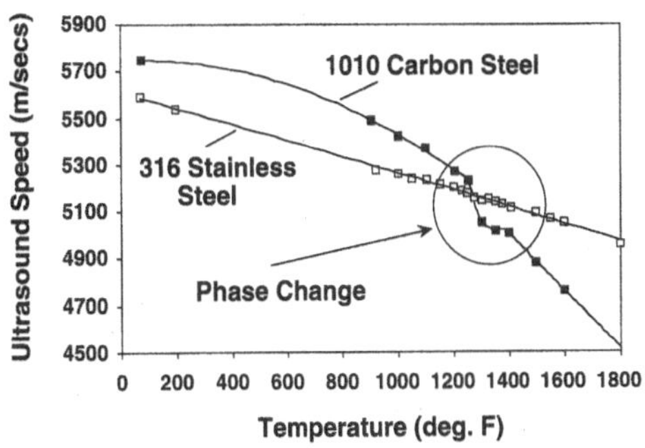

Figure 8. Phase transformation detection on 1010 carbon steel sample.

and can be used to detect the onset and completion of this phase transformation. Excessive soak times can thus be avoided during heat treating cycles. Data validation has been provided by the results of the stainless steel sample which does not go through such phase transformation and, hence, exhibits no curve discontinuity in that temperature range.

BENEFITS AND FUTURE APPLICATIONS

The WPA technology development has been driven by the market potential and the many benefits from the use of this instrument. The measurement is remote and noncontact with large standoff (at least 5 ft.) and the instrument can be placed outside the furnace, avoiding the harsh operating environment. Temperature and phase transformations can be measured on-line with moving workpieces, like those found in walking beam furnaces, continuous casters or continuous strip lines. The temperature readings are independent of the emissivity of the surface of the workpiece. Thus, the WPA can be effectively used for processes where emissivity variations during processing (like galvanneal lines) or low emissivity (like aluminum) hinder the temperature readings. Both bulk and surface temperatures can be monitored simultaneously, providing critical information on temperature gradients (especially with steel workpieces). A single instrument can be used to measure temperature at various points on a workpiece through scanning mirror arrangement for determination of temperature or phase transformation uniformity. Under the simple single detector configuration of the PI, the external low frequency vibrations encountered in normal industrial furnaces have no effect on the measurement. If only surface temperature readings are needed, as in thin strips or aluminum workpieces, then the WPA can be made into a very small and inexpensive unit, because the energy and cost of the impulse laser can be reduced. The TDS/SC team is currently developing a WPA prototype and pursuing its near-term commercialization.

ACKNOWLEDGMENTS

This work was supported by the U.S. Department of Energy, Office of Industrial Technologies, under the supervision of Mr. Robert N. Chappell (DoE Idaho Operating Office) and Mr. Joseph G. Keller (EG&G, Idaho). The authors wish to thank Mr. Paul Gozewski, Dr. Peter Rostler, Mr. Richard Gannon, and Dr. Daniel Klimek for their support.

REFERENCES

1. U.S. Department of Energy, Phase 1 Final Report, Development and evaluation of a Workpiece Temperature Analyzer for Industrial Furnaces (1990).
2. U.S. Department of Energy, Phase 1-A Final Report, Development and evaluation of a Workpiece Temperature Analyzer for Industrial Furnaces (1991).
3. J.P. Reilly, A. Ballantyne, and J.A. Woodroffe, Modeling of momentum transfer to a surface by laser-supported absorption waves, *AIAA J.* 17:1098 (1979).
4. J.A. Woodroffe, Pulsed-laser material interaction, in: "Gas Flow and Chemical Lasers," Michele Onorato, ed., Plenum Publishing Co., New York (1984).
5. J.A. Woodroffe, J.C. Hsia, and A. Ballantyne, Thermal and impulse coupling to an aluminum surface by a pulsed Krf laser, *Appl. Phys. Lett.* 36:14 (1980).
6. C. Duzy, J.A. Woodroffe, J.C. Hsia, and A. Ballantyne, Interaction of a pulsed XeF laser with an aluminum Surface, *Appl. Phys. Lett.* 37:542 (1980).
7. J.F. Ready, "Industrial Applications of Lasers," Academic Press, New York (1978).
8. H. Kolsky, "Stress Waves in Solids," Dover Publications, Inc., New York (1963).
9. A.E.H. Love, "A Treatise on the Mathematical Theory of Elasticity," Dover Publications, Inc., New York 4th ed. (1944).
10. I.A. Victorov, "Rayleigh and Lamb Waves," Plenum Press, New York (1967).
11. A.M. Aindow, R.J. Dewhurst, and S.B. Palmer, Laser-generation of directional surface acoustic wave pulses in metals, *Opt. Commun.* 42:116 (1982).
12. R.M. White, Generation of elastic waves by transient surface heating, *J. Appl. Phys.* 34:3559 (1963).
13. D.A. Hutchins, R.J. Dewhurst, and S.B. Palmer, Directivity patterns of laser-generated ultrasound in aluminum, *J. Acoust. Soc. Am.* 70:1362 (1981).
14. C.B. Scruby, R.J. Dewhurst, D.A. Hutchins, and S.B. Palmer, Quantitative studies of thermally generated elastic waves in laser-irradiated metals, *J. Appl. Phys.* 51:6210 (1980).
15. W. Kaule, Method and apparatus for receiving ultrasonic waves by optical means, U.S. Patent 4,388,832 (1983) (assigned to Krautkramer-Branson, Inc.).
16. J.P. Monchalin, Optical detection of ultrasound, *IEEE Trans. on Ultrasonics, Ferroelectrics and Frequency Control* 33:485 (1986).
17. G.S. Darbari, R.P. Singh, and G.S. Verma, Ultrasonic attenuation in carbon steel and stainless steel at elevated temperatures, *J. Appl. Phys.* 39:2238 (1968).
18. E.P. Papadakis, L.C. Lynnworth, K.A. Fowler, and E.H. Carnevale, Ultrasonic attenuation and velocity in hot specimens by the momentary contact method with pressure coupling, and some results on steel to 1200°C, *J. Acoust. Soc. Am.* 52:850 (1972).
19. A.M. Aindow, R.J. Dewhurst, D.A. Hutchins, and S.B. Palmer, Laser-generated ultrasonic pulses at free metal surfaces, *J. Acoust. Soc. Am.* 69:449 (1981).

SENSOR SYSTEM FOR INTELLIGENT PROCESSING OF HOT-ROLLED STEEL*

A.V. Clark, M.G. Lozev, B.J. Filla and L.J. Bond[1]

Materials Reliability Division
National Institute of Standards and Technology
Boulder, Colorado 80303

[1] Center for Acoustics, Mechanics and Materials
University of Colorado at Boulder
Boulder, Colorado 80309-0427

ABSTRACT

Microstructural engineering offers a means of reducing cycle time and improving product quality with reduced energy cost. Process modelling of hot-rolling of steel has advanced to the point where key metallurgical parameters (e.g. austenite grain size) can be predicted from computer models if other process parameters such as temperature are known. However, a need still remains to develop on-line sensors because sensors can rapidly acquire data for model validation, and could be combined with a process control strategy in feedforward or feedback loops.

We report a study to develop an on-line sensor for austenite grain size measurement at temperatures up to 1000°C. We are developing a system for measurement of attenuation in ultrasonic waves propagating through hot steel samples. Existing theories allow us to infer grain size for attenuation measurement. Our system consists of a high temperature transducer acoustically coupled to a high strength, low loss buffer rod. The rod is to be pressed against a hot steel specimen to achieve sufficiently good contact to transmit ultrasound into the specimen. To avoid plastic deformation of the specimen, we implement computer control of the load applied to the buffer rod.

Attenuation measurements were performed at room temperature on specimens of known grain size. Good agreement with optically measured grain size was obtained. The system is now being assembled for measurement at elevated temperatures.

*Contribution of US National Institute of Standards and Technology, not subject to copyright.

INTRODUCTION

To improve quality of steel plate there is a recognized need to improve the control of the various system and material parameters in the hot rolling process [Cook and Frock (1986)]. Improved control requires the availability of on-line measurements for parameters such as the mean austenite grain size. If the grain size can be measured early in the manufacturing process and these data combined with a process model [Cook and Frock (1986)], feedback systems can in principle be developed to give adjustments to process variables. This will result in reduced product variability, reduced reheating requirements, and a reduced need for post-processing heat treatment.

Ultrasonic techniques have been demonstrated in the laboratory to give a relationship between attenuation and grain size [Papadakis (1981, 1984)]. Excellent correlations have been shown in ferrous materials [Goebbels (1980)], and in-line monitoring measurements for grain size in steel mills have been performed near room temperature [Takafuji et al. (1985); Yada and Kawashima (1987)]. The measurements show a clear relationship between ultrasonic attenuation and an "ultrasonic mean grain diameter" which can be related to the optically measured mean grain size, if the sample microstructure is known. Several grain size instruments are already in use using these concepts [Takafuji et al. (1985), Yada and Kawashima (1987)].

This study considers the development of an ultrasonic attenuation measurement system. From this we propose to determine of the austenite grain size in hot-rolled steel at temperatures above 700°C. Here we envision an ultimate application in which a sensor system would measure austenite grain size during, for example, finish rolling. Ultrasonics may therefore become a practical tool for grain size measurement in the 10-200μm range, which can then become a process control parameter [Bridenbaugh et al. (1987)].

THEORY

The theory of the scattering of a wave propagating in a polycrystalline medium has been presented in a number of publications [Papadakis (1981, 1984)]. For a plate with a gradation of grain sizes, the attenuation is dominated by the contributions of the largest grains [Papadakis (1981)]. Steel above the austenite temperature is a single phase equiaxed polycrystalline metal. Measurements suggest that scattering contributes the greater part of the attenuation in this temperature range [Papadakis et al.(1971)]. Therefore, only a limited frequency range may be required for our purposes. This is in contrast to room temperature measurements, where a large frequency range is often necessary to separate the effect of scattering from other contributions to attenuation (such as internal friction).

The solution for the attenuation coefficient for the case of cubic crystallites, in the Rayleigh scattering (long-wavelength) was provided by Merkulov (1956) in a simplified form as:

$$\alpha = Tf^4S \tag{1}$$

where f is the ultrasonic frequency, S is a parameter of the material which is dependent on density and elastic moduli, and T is a measure of the average grain size, with the dimensions of volume. Values of S can be determined theoretically from single-crystal moduli or for other morphologies from experimental data [Takafuji et al. (1985)]. The functional dependence of the attenuation upon and grain size has been given by Papadakis (1984), and it is a weighted function depending on the grain size distribution.

MEASUREMENT METHOD AND SYSTEM DESIGN

Our experimental system employs the pulse-echo technique. Waves are generated by a $LiNbO_3$ crystal and are propagated down a buffer rod. $LiNbO_3$ was chosen because of its high Curie temperature and because it has an acoustic impedance close to that of the buffer rod. A hydraulic actuator presses the rod against the sample, which is heated (at temperature up to 1000°C) inside a furnace. This simulates the temperature range in hot rolling.

Three signals are required because there are three unknowns: amplitude of wave generated by the transducer; reflection coefficient R from buffer rod/sample interface; attenuation coefficient. Various echoes can be recorded: the buffer rod/air echo (A'); buffer rod/sample interface echo (A); first and second back wall echoes in the sample (B,C echoes). Two pulse-echo methods for attenuation determination have been considered. These methods are conventionally identified as the A'AB and ABC methods [Mak (1991)], and each employs three signal amplitudes.

A critical parameter in the evaluation of an attenuation measurement scheme is an "attenuation variance-noise number" ($N\alpha$) defined as the product of the (normalized) variance of the attenuation measurement σ_α/α and the signal to noise ratio S: $(\sigma_\alpha/\alpha)S$). This parameter is calculated using relationships given by Mak (1991) and is used to compare the two measurement schemes. The resulting values are plotted against total attenuation T_α in the sample ($T_\alpha = 2\alpha L$, where α is the attenuation coefficient and L the sample thickness). The results are shown in Fig 1. The A'AB method has better precision for given signal-to-noise ratio.

Fig. 1 can also be used to define a suitable operating frequency. Experimental uncertainty is minimized when T_α is approximately unity; for given sample thickness, this gives a "best" value of α. In practice, the expected grain size, thickness, and temperature are all approximately known. This allows us to calculate α from Equation (1), with frequency as a parameter. The frequency is then chosen so that $2\alpha L$ is approximately 1. In this study an optimized operating frequency between 4 and 5 MHz was determined, for grain sizes of about 50 μm, and specimen thickness about 50 mm; these values are typical of industrial practice.

To provide the measurement capability needed to operate at temperatures up to 1000°C, two designs of transducers are being considered. Both are based on a novel cooled buffer-rod design. The cooling is required to protect both the piezoelectric element and the hydraulics. The transducers are set onto the cooled end of the buffer rod.

The first design is a variation on a commercial unit which has been used for acoustic emission monitoring at elevated temperature [Arakawa et al. (1992)]. This has a lithium niobate transducer brazed to the buffer rod. This design is mechanically robust; however, its electrode thickness is fixed by brazing requirements and hence has a limited bandwidth. The second transducer has a more flexible design. (However, its long-term reliability at elevated temperatures is not well-established yet.) It employs a "sandwich" consisting of: a pressure plate, backing, gold foil electrodes and a lithium niobate crystal set onto the buffer rod. The electrode thickness can be chosen to give a wider bandwidth.

The response of these two transducers has been modelled [Clark et al. (1993)], and the expected output voltages are shown in Fig. 2. The larger bandwidth of the second design is clearly evident. These calculations will be compared with experimental data.

The buffer rod must take a significant mechanical load [Lu et al. (1990)] because the acoustic coupling into the sample is achieved using pressure coupling. The buffer rod must have low internal losses due to attenuation. This requires a small grain size. Grain growth must also be minimized. Mechanical strength must not be degraded at elevated

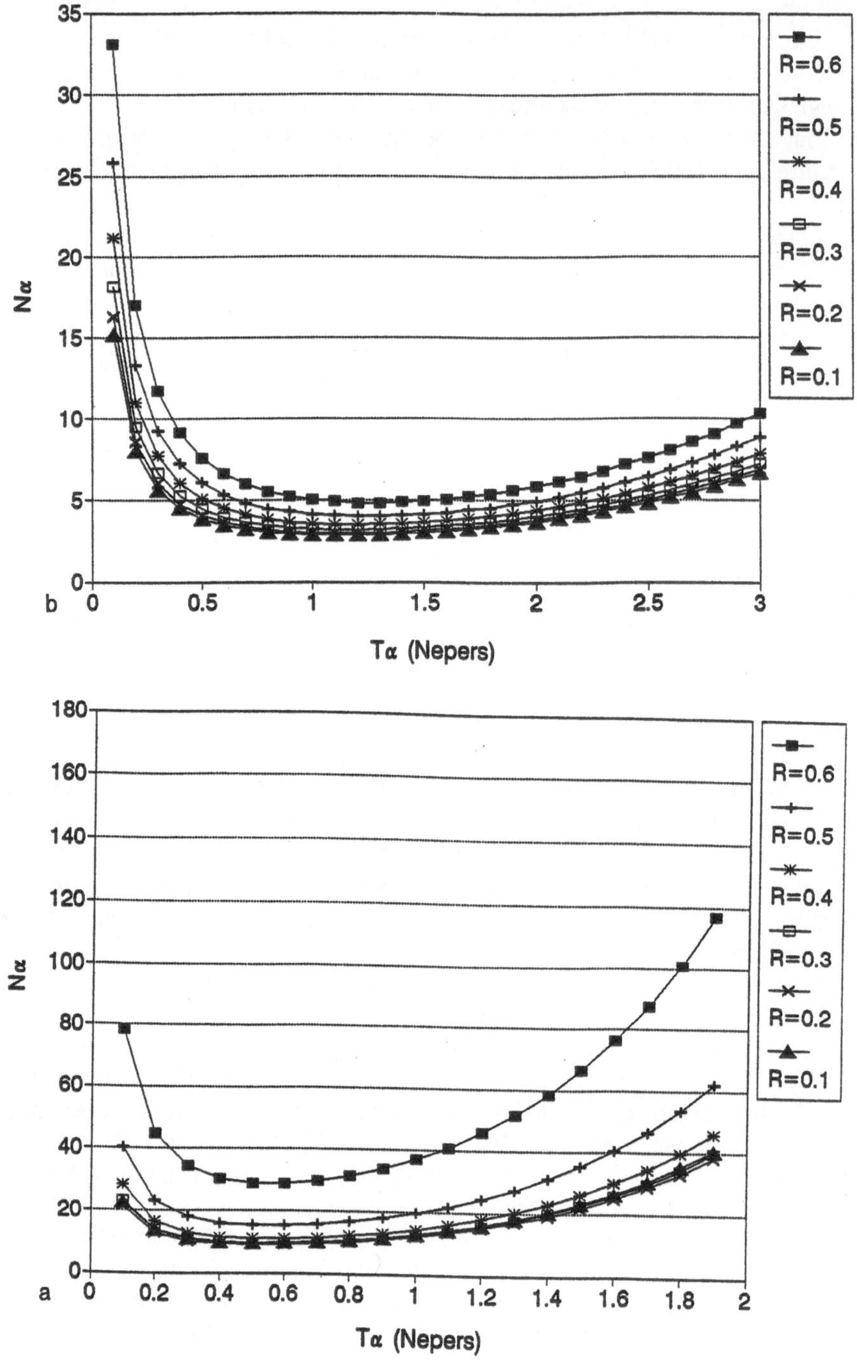

Figure 1. Attenuation variance-noise number, N_{α}, against total attenuation, T_{α}, (nepers). Calculations were done for case of 50 mm sample thickness with a 5 MHz, 25 mm diameter transducer is modelled for various interface reflection coefficients, R. Corrections for diffraction have been made.

(a) For A'AB method.

(b) For ABC method.

temperature, and oxidation must be minimized. Conventional stainless steels are not satisfactory, since they have large grain sizes for good mechanical properties at elevated temperatures (plastic deformation occurs by grain boundary motions at these temperatures, rather then by dislocation motion). To meet the various mechanical design requirements a nickel-based mechanical alloy was selected for the buffer rod material.

At elevated temperatures the stress-strain characteristics of the sample become highly dependent on the rate of strain. Typical data for microalloyed steel is shown as Fig. 3. Clearly, contacting the buffer rod to the sample for more than a few seconds at low strain rates will cause plastic deformation of the sample surface. (Momentary contact has been used by Papadakis et al., (1971) for attenuation measurements, with buffer rod

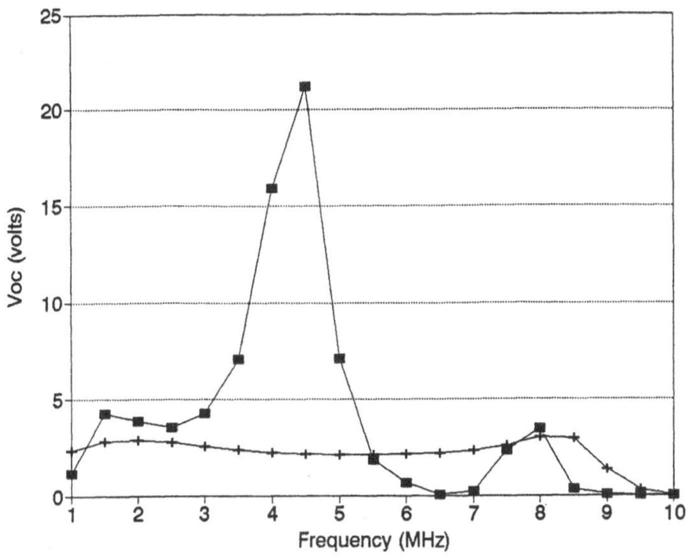

Figure 2. Theoretical transducer response characterized by output voltage (Voc) against frequency. The transducer is assumed driven by a 12 Ω pulser with toneburst of 10 V amplitude. A series inductance of 2.5 μH is used for tuning. The transducer response of brazed transducer indicated by squares; response of pressure plate transducer by crosses. Thickness of brazed Al electrodes is 200 μm. Pressure plate transducer has top and counter-electrodes of Au foil, thickness 100 μm.

completely removed from the furnace after attenuation measurement. The buffer rod is essentially at room temperature. However, this method does not appear feasible in a steel mill.) To avoid excessive plastic deformation computer control of the load applied to the buffer-rod-sample interface is being implemented.

The complete transducer/buffer rod/cooling jacket system, set in the furnace is shown as Fig 4. A methodology using a preheated buffer rod, computer controlled hydraulics to avoid sample deformation, and repeated measurements of the A'AB echoes is to be employed [Clark et al. (1993)].

Room temperature measurements have already been performed on the samples prepared for the high temperature study. The measured values of attenuation were in close agreement to values calculated from the optically measured grain sizes for the sample set [Lozev et al. (1993), Clark et al. (1993)].

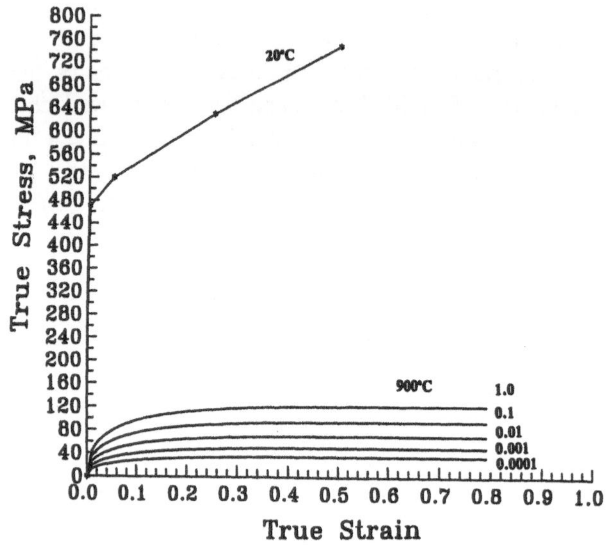

Figure 3. Stress-strain response for microalloyed steel at 20°C and 900°C with strain rate as parameter.

Figure 4. High temperature grain-size measurement system, showing open furnace with (top to bottom): cooling jacket, transducer/buffer rod assembly held above sample.

CONCLUSIONS

A prototype measurement system has now been designed, with the goal of high temperature measurements of attenuation in steel samples. From these attenuation measurements, the austenite grain size can be characterized.

The system consists of: a high temperature transducer acoustically coupled to a buffer rod; hydraulics (under computer control) to pressure-couple the buffer rod to the sample; a cooling jacket to protect the hydraulics from heat conducted through the buffer rod; a furnace to heat samples above the austenite temperature. The components have been chosen based on optimization studies. The system will be tested on hot steel specimens heated up to 1000°C. If these tests are successful, a sensor system with potential for on-line grain size estimation in a hot rolling mill would appear to be possible.

ACKNOWLEDGEMENTS

This work was supported in part by the Office of Intelligent Processing of Materials of the National Institute of Standards and Technology. We are grateful for many useful discussions with our colleagues Drs. Y. Cheng and C.M. Fortunko.

REFERENCES

Arakawa, T., Yoshikawa, K., Chiba, S., Muto, K., and Atsuta, Y. (1992), Applications of brazed-type ultrasonic probes for high and low temperature uses, Nondestructive Testing and Evaluation, Vol 6 pp. 263-272.

Bridenbaugh, P.R., Shabel, B.S., and Govada, A.K. (1987), Material characterization for process control for aluminum alloys and advanced materials, in "Nondestructive characterization of materials II," Ed. J.F. Bussiere et al., Plenum Press (New York), pp. 179-194.

Clark A.V., Lozev, M.G., Bond, L.J., Filla, B.J., and Reed, H. (1993), Austenite grain size characterization in hot rolled steel plate, Research in NDE (in preparation).

Cook, R.J., and Frock, B.G. (1986) Process control and materials characterization within the steel industry, in "Materials characterization for system performance and reliability" Ed. J.W. McCauley and V. Weiss, Plenum Press (New York) pp. 293-309.

Goebbles, K., (1980) Structure analysis by scattered ultrasonic radiation, in "Research Techniques in NDT" Vol. 4, ed. R.S. Sharpe, Academic Press (London) pp. 87 158.

Lozev, M.G., Bond, L.J., Clark, A.V., and Reed, H. (1993), Attenuation of ultrasonic waves in stainless steel at room temperature (in preparation).

Lu, G.L., Li, S.H., Chu, M.J. and Wang, B.X. (1990), The effect of pressure on ultrasonic testing sensitivity of hot steel blooms by pressure contact method, IEEE Trans. Ultrasonics, Ferroelectrics and Frequency Control 37 (6) pp. 587-589.

Mak, A.K. (1991), Comparison of various methods for the measurement of reflection coefficient and ultrasonic attenuation, British J. NDT 33 (9) pp. 441-449.

Merkulov, L.G. (1956), Sov. Phys-Tech. Phys. 1 (1) pp. 59-69.

Papadakis, E.P. (1981) Scattering in polycrystalline media, in Methods of Experimental Physics, Vol. 19, Ultrasonics, ed. P.D. Edmonds, Academic Press (New York) pp. 237-298.

Papadakis, E.P. (1984), Physical acoustics and microstructure of iron alloys, Int. Metals Reviews 29 (1) pp. 1-24.

Papadakis, E.P., Lynnworth, L.C., Fowler, K.A., and Carnevale, E.H. (1971), Ultrasonic attenuation and velocity in hot specimens by the momentary contact method with pressure coupling, and some results on steel to 1200°C, J. Acous. Soc. Amer. 52 (3) pp. 850-857.

Takafuji, H., Sekiguchi, T., Iuchi, T., and Matsuda, S. (1985), A new measurement method of the grain size of steel by ultrasonic attenuation technique, Proceedings, Ultrasonics International '85, Butterworth Scientific (Guildford) pp. 195-200.

Yada, H., and Kawashima, K. (1987), Important metallurgical parameters that must be determined to control the properties of steels during processing, in "Nondestructive characterization of materials II," ed. J.F. Bussiere et al., Plenum Press (New York) pp. 195-209.

ON-LINE ULTRASONIC CHARACTERIZATION OF POLYMER FLOWS

L. Piché, D. Lévesque, R. Gendron, and J. Tatibouët

Industrial Materials Institute, National Research Council Canada
75 de Mortagne, Boucherville, Qc, J4B 6Y4, Canada

INTRODUCTION

Although ultrasonic techniques have proven useful for investigating elasticity of solids and viscosity of fluids and gases[1], they have seldom been used for polymer studies. Notwithstanding, all reports[2,3,4] point out the distinctive behavior of ultrasound in polymers and suggest numerous prospects for fundamental studies, and industrial applications[5,6]. Thermomechanical properties of polymers are usually measured at low frequencies between 0.01 and 100 Hz, with deformations $\varepsilon \approx 10^{-4}$, while ultrasonic techniques involve frequencies in the MHz range and strains near $\varepsilon \approx 10^{-7}$. Whilst rheology measures global properties associated to long range diffusion of molecules, ultrasonic waves probe the mobility of short chain segments. In an attempt to relate the different measurements, we described experiments[7,8] using an apparatus[9] that measures the complex ultrasonic modulus with close control of the thermodynamic history. Although successful, the technique involves no macroscopic flow of molecules.

Processing, on the other hand, implies macroscopic flows; therefore there is a need for investigating the effect of large scale molecular movements on the short scale mobility. Here we describe an experiment for characterizing flowing polymer melts during processing. In continuum hydrodynamics, it is assumed that the component of the fluid velocity vanishes along a solid surface. This "no-slip" boundary condition may be argued intuitively since internal drag due to viscosity slows down the fluid near the interface. In the case of usual Newtonian fluids, this "no-slip" assumption is usually successful. Non-Newtonian fluids are different and for polymer melts it may be conjectured that near walls, the flow curves are modified due to alignment of molecular chains. In turn, this may lead to slippage at the walls, resulting in flow irregularities that are manifested by so-called melt fracture. This line of reasoning suggests that melt fracture is related to peculiarities of the polymer/wall interface region, which mostly agrees with experimental observations[10,11,12] from rheology. Recently, we described[13,14,15] an ultrasonic approach for investigating polymer/metal interfaces. We adapt this method to extrusion of polymer melts. We demonstrate that ultrasonic results correlate with materials properties obtained in rheology, and provide unique information on polymer/wall interactions that condition flow behavior. We conclude that the technique is a powerful tool for monitoring and controlling industrial processes.

Nondestructive Characterization of Materials VI
Edited by R.E. Green, Jr. *et al.*, Plenum Press, New York, 1994

METHODOLOGY

Ultrasonics and Fluid Mechanics

The Navier-Stokes equation is the basis for describing the movement of viscous fluids. For the one-dimensional case, the time evolution for the particle velocity, u is given through:

$$\rho\frac{\partial u}{\partial t}+\rho u\frac{\partial u}{\partial z}=-\frac{\partial p}{\partial z}+(\eta_B+\tfrac{4}{3}\eta)\frac{\partial^2 u}{\partial z^2} \tag{1}$$

where ρ is the density, p the pressure, η the shear viscosity and η_B the bulk viscosity. Solving Eq.(1) in an unbounded medium for small periodic displacements polarized along the z axis and having angular frequency ω, one recovers the usual wave equation, where the velocity of sound, v, and the attenuation, α, are given by:

$$v=\sqrt{\frac{\partial p}{\partial \rho}} \quad \text{and} \quad \alpha=\frac{1}{2\rho v^3}(\eta_B+\tfrac{4}{3}\eta)\,\omega^2=\frac{1}{2\rho v^3}\eta_L\,\omega^2 \tag{2}$$

However, it is often more useful to describe materials properties of the supporting media in terms of a complex longitudinal modulus, $L^* = L' + iL''$, such that:

$$L'=\rho v^2 \quad \text{and} \quad L''=2\,\alpha\,\rho v^3/\omega \tag{3}$$

For macroscopic displacements, such as laminar flow between two parallel planes, one usually assumes zero flow velocity at the walls. Solving Eq.(1) in the case of incompressible fluids, leads to a parabolic profile for particle velocity. Also, the shear stress is zero at mid-plane and reaches its maximum value at the walls:

$$\sigma_{12}=\frac{e}{2}\,\Delta p \quad \text{with} \quad \Delta p\equiv\frac{\partial p}{\partial x} \tag{4}$$

where e is the separation between plates and Δp the pressure drop per unit length along the flow channel. In application of Newton's law for fluids, σ_{12} may be related to the volumetric flow rate, Q through:

$$\sigma_{12}=-\eta\dot{\gamma} \quad \text{where} \quad \dot{\gamma}=\frac{\partial u}{\partial z}=6Q/we^2 \tag{5}$$

is the rate of shear, and w is the breadth of the channel. In principle therefore, ultrasonic measurement of viscosity of fluids can be correlated to properties of macroscopic flows.

Materials and Properties

We experimented on a commercial grade of polypropylene (PP) resin, that was provided by Himont Canada under the designation 6631FB. Specifications were given for the density, ρ = 890 kg/m^3, and the melt flow rate MFR = 2.0 g/10 min. We performed additional rheological measurements with capillary rheometry and dynamic mechanical spectrometry in order to characterize molecular weight, Mw = 415 000 g/mole, and determine the constitutive law for the viscosity. The results depended on temperature in accordance to an Arrhenius process. For illustrative purposes, at a temperature T= 220°C corresponding to our present experiments on extrusion, we found a Carreau dependence:

$$\eta = 9.9[1+(0.905\dot{\gamma})^2]^{-0.325} \tag{6}$$

with η in kPa.s. The fact that viscosity was shear rate dependent is in line with the trend for polymer fluids, which usually deviate from Newton's law.

Using an apparatus specially developed in our laboratory[9], we measured the specific volume, V_C, velocity of sound, v_C, and attenuation, α_C, of the confined material, in the absence of macroscopic flow. Operating at the ultrasonic frequency f= 5 MHz, the results in the range of pressure, p= 0 to 200 MPa and temperature T= 175 to 220°C, could be described through the following fitting functions:

$$V_C(p,T) = 1.115\ 10^{-3} - 7.269\ 10^{-7}\ p + 1.119\ 10^{-6}\ T - 5.029\ 10^{-9}\ pT \tag{7a}$$

$$v_C(p,T) = 1.383 + 3.781\ 10^{-3}\ p - 2.077 10^{-3}\ T + 1.288\ 10^{-5}\ pT \tag{7b}$$

$$\alpha_C(p,T) = 8.31 - 0.018\ T - 0.071\ p \tag{7c}$$

where V_C is expressed in m³/kg, v_C in km/s, α_C in dB/cm, p in MPa, and T in °C.

For T=220°C, and p=1.0 MPa, one finds $\eta_L \approx$ 0.05 Pa.s, while in the limit of low shear stress, Eq.(6) leads to $\eta \approx$ 9.9 kPa.s. The large difference between both values has its origin in the peculiar nature of the polymer liquid, where the long molecules form an entanglement network. The rheology measurement involves large displacements compared to those in ultrasonics so that, while η involves reptation of chains through entanglements, η_L mainly relates to small scale mobility of short molecular segments.

ULTRASONIC PROBE OF FLOWING POLYMERS

Technique

For the experiments on flowing melts, we used a single screw extruder (Flag 25D30V, with 6.35 cm diameter barrel, and length to diameter ratio 30) that we equipped with a slit die, Fig.1, enclosed in a thermostat sleeve for controlling the die temperature, T_D. The entrance channel of the die contained an immersion thermocouple for monitoring melt flow temperature, T_M. The slit section (length, l = 14.6 cm, width, w = 3.8 cm) had a thickness, e, that could be adjusted between 2.0 and 3.5 mm by means of calibrated spacers. Four pressure transducers(p), installed along the length of the slit served for rheological characterization of the flow by use of Eqs (4) and (5). Two identical ultrasonic probes(US), each consisting of a steel buffer rod with a transducer, were mounted in opposition, approximately at midstream and perpendicular to the slit, so the ultrasonic beam was normal to the flow channel. The free ends of pressure transducers and ultrasonic probes were flush with the die walls, in order not to disturb the flow. We used longitudinal ultrasonic broadband transducers, with a center frequency f = 5 MHz and 1 μs pulse width that we operated with a repetition rate ≈100 Hz. The signal from the receiving transducer was sent to a broadband amplifier connected to a fast digitizer. The numerized signals were then processed in a personal computer where attenuation and velocity were obtained by use of Fast Fourier Transforms and correlation techniques. The time duration for this routine was of the order of 1 s.

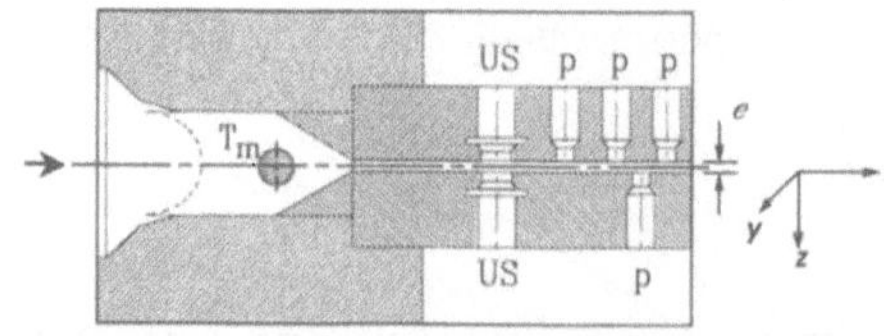

Figure 1. Schematic of slit die. Left hand part is convergence channel with immersion thermocouple for melt temperature, T_M. Right hand portion is instrumented slit having length 14.6 cm, breadth 3.8 cm, and adjustable thickness, *e* between 2.0 and 3.5 mm. Drawing shows position of pressure transducers, p, and ultrasonic probes, US.

Measurements and Results

In Figs 2a and 2b, we investigate the velocity, v_F, and attenuation, α_F, of sound in the polymer flowing through a slit of thickness e = 2.00 mm, and we plot the data as function of pressure, p(MPa), evaluated at the location of the ultrasonic probes. The squares, circles and triangles correspond to die temperatures T_D = 210, 220 and 230°C, respectively. The solid lines in Fig.2 are results of a model that will be described later on. For a given pressure, the velocity changed with temperature at a rate $[\partial \ln v_F / \partial T]_p \approx -2.15 \times 10^{-3}/°C$, close to $[\partial \ln v_C / \partial T]_p \approx -2.10 \times 10^{-3}/°C$ obtained from Eq.(7b) for the confined melt. On the other hand, for a given set of temperature and pressure, we observed that v_F always exceeded v_C. Also, at low pressure, the rate of change, $[\partial \ln v_F / \partial p]_T \approx 3.0 \times 10^{-8}$ Pa^{-1}, was more important than for the confined polymer, $[\partial \ln v_C / \partial p]_T \approx 0.69 \times 10^{-8}$ Pa^{-1}. Finally, instead of increasing steadily with p, the velocity would tend to level off near p≈ 3.0 MPa.

At low pressures, corresponding to small flow rates, the attenuation, α_F in Fig.2b, behaved similarly to that in the confined polymer: up to p ≈3.0 MPa, $\alpha_F \approx \alpha_C$ and $[\partial \ln \alpha_F / \partial T]_p \approx [\partial \ln \alpha_C / \partial T]_p \approx -4.4 \times 10^{-3}/°C$. However, as pressure increased due to higher flow rate, the behavior of α_F contrasted noticeably with that of α_C. Manifestly, temperature effects were more pronounced and $[\partial \ln \alpha_F / \partial T]_p >> [\partial \ln \alpha_C / \partial T]_p$. Moreover, the attenuation increased with pressure instead of decreasing slightly, as in the confined melt, from Eq.(7c).

The results in Fig.2 cannot be interpreted on the basis of Eqs(7). Obviously, the ultrasonic measurements contain information related not only to pressure and temperature, but also to the hydrodynamics of the macroscopic flow. This was confirmed in other experiments with different slits, which showed that v_F and α_F are dependent on the thickness, e. Although pressure, p, is a relevant parameter for describing flow behavior, one can choose to parametrize v_F and α_F in terms of any of the other variables in Eqs(4) or (5). However, the representation in Fig.(2) has the advantage of being more closely related to the ultrasonic data in Eq.(7) that are described in terms of p and T. On the other hand, given the geometry for the experiment, in Fig.2 one may use the approximation: $p \propto \sigma_{12}$.

BOUNDARY LAYERS

Our measurements on the confined melt, Eqs(7), indicate that attenuation is not highly dependent on pressure, p, or temperature, T. Then, one may suspect that the excess

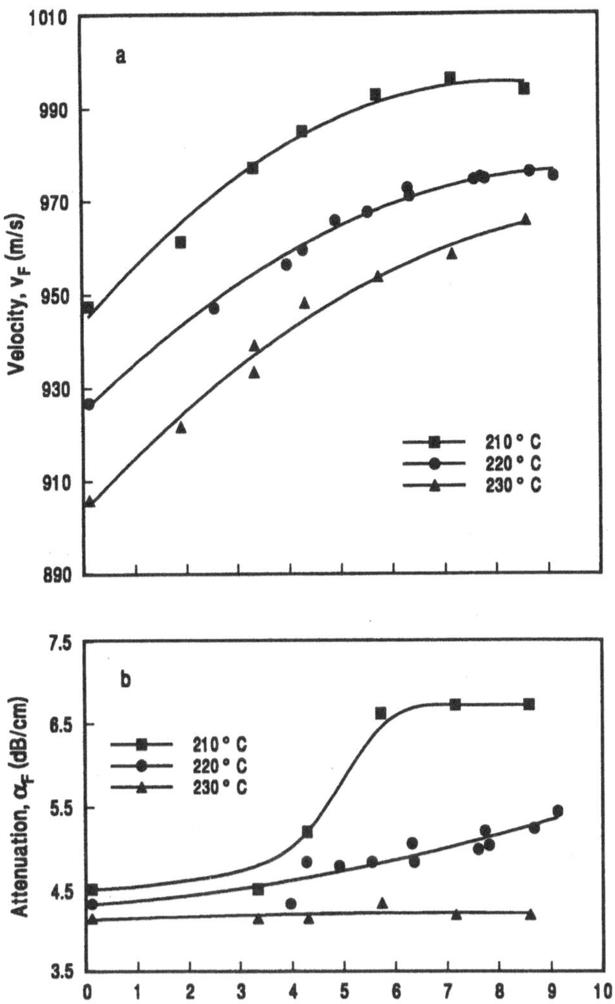

Figure 2. **(a)** Sound velocity, v_F(m/s) and **(b)** attenuation, α_F(dB/cm), for flow in slit with thickness e=2.00 mm. Temperature on the die was constant :T_D= 210 (squares), 220 (circles) and 230°C (triangles), and data were plotted versus pressure, p(MPa), related to flow rate.

attenuation in the flowing melt is related to effects due to shear stress, σ_{12}. These effects should be more important near the walls of the flow channel because of the larger shear stress. One is led to propose that the polymer morphology is modified by the flow process. Finally, it is well recognized that near interfaces, polymers exhibit distinctive features, whether or not they are undergoing flow. This brings up the idea of an interface layer where the material has properties different from those of the bulk polymer. This interfacial layer serves to transfer mechanical load between the channel walls, with which the polymer molecules interact, and the bulk polymer in the core, where the molecules arrange into an assembly of random coils.

Following along those lines, we make the hypothesis for two identical layers one adjacent to each wall. There results a multilayer structure for the the propagation of sound and, because of changes in acoustic impedance between constituents, calculation of ultrasonic velocity and attenuation develops into a complex problem. In a previous report[13], we presented a complete description for the ultrasonic response of multilayered absorbing media and discussed this relative to other approaches in the literature. The scheme is based on a recurrence relation for transferring stresses and displacements from one interface to another using the so-called Transfer Matrix Approach. The solution involves a great deal of analytical work, but the final operational result can only be obtained numerically.

We may now map the ultrasonic properties associated with the confined polymer onto the pressure and temperature profiles[16] for the different flow conditions. Both the thickness, d_i, and the (complex) modulus, L_i^*, of the interfacial material are unknown, and and constitute adjustable parameters. However, it may be shown[13] that, provided d_i is smaller than the acoustic wavelength, $\Lambda = v/f \approx 200\mu m$, the thickness and modulus are not independent parameters. Instead, the longitudinal specific stiffness, $S^* = L_i^*/d_i$, becomes the relevant variable. At this point however, we need to satisfy conditions of continuity, so that for vanishingly small flow rates the interfacial material and the confined melt have similar properties. For this, we set $L_i''/L_i' = L''/L' \approx 3x10^{-3}$, in which case $S^* \approx S' = L_i'/d_i$ is a real quantity.

We adjusted values of L_i^* and d_i in order to obtain the best fit to the data. The results for the velocity and attenuation are shown as solid lines in Fig.2. We found that this best fit corresponded to small values $d_i \ll \Lambda$, so that S' was the sole parameter of interest. Also, the best fit indicated that the relevent temperature in the problem was that of the melt T_M, which may exceed that for the die, T_D, by several degrees depending on flow rate[16]. The results at low flow rates show that the ultrasonic response is correlated to usual rheology., At high flow rates, the stiffness of the polymer in the boundary layer, S', gradually depreciates. There results a weak link between the core of the flow and the channel walls which acts to regulate flow behavior through feedback.

Conversely, one would like to use the ultrasonic measurements as a means of monitoring melt viscoelasticity during the process. However the results in Fig.2 for the velocity or the attenuation are not single valued in terms of pressure and temperature. For applications, one needs a unique parameter that may serve as a reference to describe or at least compare flow behavior. In this respect, our multilayer model suggests that flow behavior is actually regulated by the properties of the interface layer. In Fig.3 we plot the excess losses due to flow conditions, $A = 2e(\alpha_F - \alpha_C)$, versus the interface parameter, S'. The results, which pertain to numerous experiments over a wide range of flow conditions, demonstrate a unique power law correlation between A and S'. Such scaling behavior is characteristic of universality and constitutes a strong indication for the reality of an interfacial layer that controls the rheology. Also the power law correlation suggests an intervening critical behavior associated with the boundary layer, that may well be the origin of macroscopic flow instabilities, leading to so-called as melt fracture.

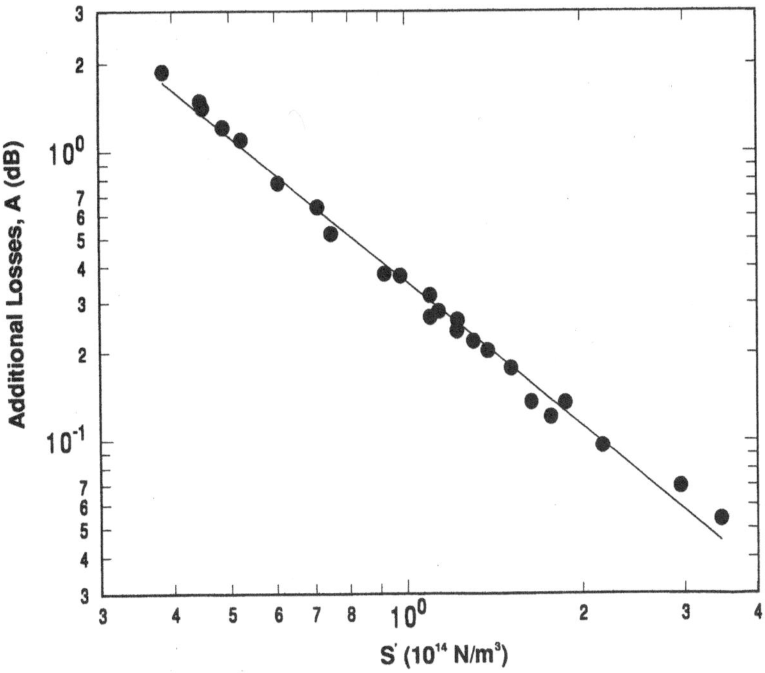

Figure 3. Universal master curve of additional acoustic losses due to flow conditions, A = $e(\alpha_F - \alpha_C)$, versus interface stiffness parameter, S'.

CONCLUSION

Through ultrasonic measurements as function of pressure and temperature we showed that velocity and attenuation of sound in a confined polymer melt correlated to the p,V,T equation of state and to the viscosity associated with small molecular moieties. We adapted the technique for measurements on melts during extrusion through a slit, under well controlled conditions. Compared to the confined melt, the ultrasonic behavior in the flowing polymer were quite anomalous. Making the assumption of a thin viscoelastic boundary layer with complex modulus, L_i^* and thickness d_i, we described the propagation in terms of the specific stiffness of the interface material $S^* \approx S' = L_i'/d_i$. Finally, our different results exhibit universal behavior which suggests that the interface layer governs flow behavior. A great more work is required and forthcoming reports will concern: i) interfacial mobility in relation to shear stress; ii) the influence of polymer architecture; iii) the non-linear flow regime where melt fracture occurs. Also, a theory is needed to describe the physical origin of the scaling behavior. The present work is open ended and leads to new avenues for novel developments in the field of hydrodynamics and applications. Indeed the technique provides a direct measurement of the "slip factor" that is needed for building representative models of polymer melt flows. On the other hand, the technique is non-invasive and may be easily implemented for on-line monitoring of polymer properties, during manufacturing of raw materials, compounding of blends and alloys, or processing for end-use purposes.

REFERENCES

1. R.T. Beyer, and S.V. Letcher, "Physical Ultrasonics", Academic Press, New York (1969).

2. B. Hartmann, Ultrasonic Measurements in "Methods of Experimental Physics", edited by R.A. Fava, Vol. 16-c, Chap. 12.1, pp. 59-90, Academic Press, New York (1980).
3. R.A. Pethrick, Acoustical Properties in "Comprehensive Polymer Science", edited by C. Booth, and C. Price, Vol. 2 : "Polymer Properties", Chap. 17, pp. 571-599, Pergamon, Oxford (1989).
4. L. Piché, *IEEE Utrasonics Symposium*, 599-608 (1989).
5. L. Piché, "Ultrasonic density of polymers", Can. Pat.: 1,268,536 (1990); US Pat.: 4,754,645 (1988), ASTM standard D.4883.
6. L. Piché, Ultrasonic characterization of polymers under simulated processing conditions, in "Nondestructive Characterization of Materials IV", Ruud C.O. et al. (Eds), pp.151-158, Plenum Press, New York (1991).
7. F. Massines, L. Piché, and C. Lacabanne, *Makromol. Chem., Macromol. Symp.*, 23: 121-137 (1989).
8. J. Tatibouët, and L. Piché, *Polymer*, 32: 3147-3152 (1991).
9. L. Piché, F. Massines, A. Hamel, and C. Néron, "Ultrasonic technique for characterizing polymers under simulated processing conditions", Can. Pat.: 1,264,195 (1990); US Pat.: 4,677,482 (1987).
10. S.G. Hatzikiriakos, and J.M. Dealy, *J. Rheol.*, 35: 497-523 (1991).
11. N. El Kissi, and J.M. Piau, *J. Non-Newtonian Fluid Mech.*, 37: 55-94 (1990).
12. A.V. Ramamurthy, *J. Rheol.*, 30: 337-357 (1986).
13. D. Lévesque, and L. Piché, *J. Acoust. Soc. Am.*, 92: 452-467 (1992).
14. D. Lévesque, A. Legros, A. Michel, and L. Piché, *J. Adh. Sci. Tech.*, (in press May 1993).
15. L. Piché, D. Lévesque, P. Deprez, A. Michel, and J. Tatibouët, *these Proceedings*.
16. **Note**: Due to shear heating, the true local temperature in the melt may be different from that which is measured on the die (T_D). As will be reported elsewhere, accurate determination of true melt temperature involves finite element analysis of the flow problem in order to solve the complete Navier-Stokes equation coupled with the equation for heat transport. Instead of a uniform temperature for the flow (T_D), the results demonstrate a temperature profile.

AN ULTRASONIC TESTING TECHNIQUE FOR MONITORING THE CURE AND MECHANICAL PROPERTIES OF POLYMERIC MATERIALS

E. C. Johnson, J. D. Pollchik and S. L. Zacharius

The Aerospace Corporation, M2/248
P. O. Box 92957
Los Angeles, CA 90009

INTRODUCTION

Large space booster solid rocket motors (SRMs) contain composite propellant materials based on either a hydroxyl terminated polybutadiene (HTPB) or the terpolymer of butadiene, acrylic acid, and acrylonitrile (PBAN) polymer matrix. Mechanical tests have revealed significant batch-to-batch variances and age related changes in the propellant moduli[1] probably due to continued crosslinking of the system. This paper documents a work in progress directed toward testing the feasibility of using ultrasonic measurements of the propellant mechanical properties to monitor the condition and the degree of cure of the propellant. If the ultrasonic shear and longitudinal velocities, c_S and c_L, and the density, ρ, of the propellant are known, the Young's modulus, shear modulus, bulk modulus, and Poisson's ratio, E, μ, K and σ, respectively, can be calculated via the familiar relations,

$$E = \frac{\rho c_L^2(1-2\sigma)(1+\sigma)}{1-\sigma},\ \mu = \rho c_s^2,\ K = \frac{E}{3(1-2\sigma)} \text{ and } \sigma = \frac{c_L^2 - 2c_s^2}{2c_L^2 - 2c_s^2}. \tag{1}$$

Considerable effort under the NASA Solid Propulsion Integrity Program (SPIP) has been directed along a similar vein.[2] The work presented here is unique in that a simple ultrasonic resonance technique which permits measurement of the acoustic velocities of thin adhesive material specimens is employed. The technique incorporates a slight modification of what was presented in earlier work[3] in that Fast Fourier Transforms (FFTs) are used to process the signals. The technique is characterized by a number of advantages. The same specimen and transducer pair are used to determine both the shear and longitudinal response. In addition, a fluid medium is employed to couple sound into the specimen, thereby eliminating many of the problems associated with the bonding of transducers. Preliminary SPIP results suggest that measurements of the propellant shear velocity are important, but difficult to perform, as the propellant material is very attenuative to shear waves. The hope is that use of thin specimens will serve to mitigate this difficulty.

TECHNIQUE

To understand the technique, consider first a thin medium having acoustic impedance, Z_2, sandwiched between two semi-infinite media of acoustic impedance, Z_1, as depicted in

Nondestructive Characterization of Materials VI
Edited by R.E. Green, Jr. *et al.*, Plenum Press, New York, 1994

Fig.1. Assuming a continuous plane wave stimulation in one of the semi-infinite media, the coefficient of acoustic power transfer, P_T, across the thin medium can be calculated;

$$P_T = \frac{4Z_2^2Z_1^2}{\left(Z_1^2+Z_2^2\right)^2 \sin^2\left(\frac{2\pi x_2 f}{c}\right) + 4Z^2Z_1^2 \cos^2\left(\frac{2\pi x_2 f}{c}\right)}, \tag{2}$$

where f is the frequency of the plane wave, c is the acoustic velocity (either shear or longitudinal) and x_2 is the thickness of the thin medium (see Ref. 3).

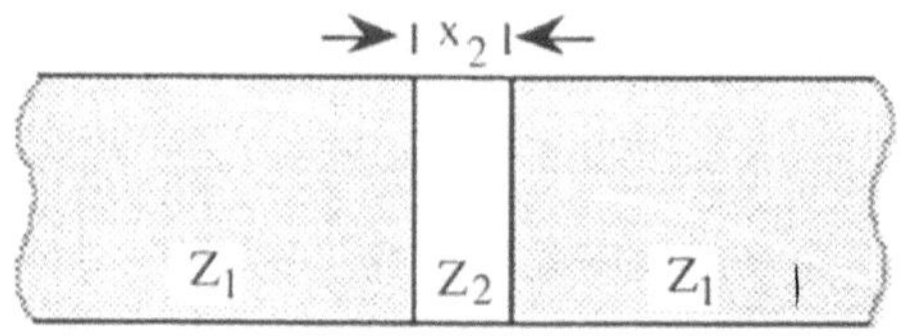

Figure 1. A thin medium of acoustic impedance, Z_2, and thickness, x_2, sandwiched between two semi-infinite media of acoustic impedance, Z_1.

In Fig. 2, P_T is plotted as a function of frequency for the case where $Z_1 = 17$ Rayls (aluminum) and $Z_2 = 1.48$ Rayls (water). It can be seen that the condition of maximum power transmission occurs when $f = f_R = nc/2x_2$, where n is an integer. Note that the thin medium was assumed to be lossless in the derivation of Eqn. 2. Taking attenuation into account, one would expect the amplitude of the local maxima to decrease with increasing n.

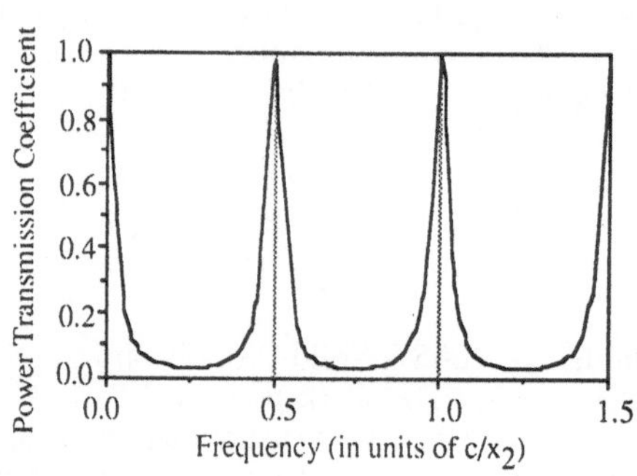

Figure 2. The coefficient of acoustic power transmission across the thin medium as a function of frequency (Eqn. 2) for the three layer system depicted in Fig. 1, where $Z_1 = 17$ Rayls (Al) and $Z_2 = 1.48$ Rayls (water).

The experimental apparatus is depicted in Fig. 3. Two aluminum blocks were cut from square rod stock, each having a 74° and 90° face with respect to one side of the block. A milled finish was determined to be adequate on the 74° faces. The 90° faces were lapped to insure that the surfaces were flat. A thin uniform layer of the specimen being tested was sandwiched between the 90° faces. The spacing between these faces and hence, the specimen thickness, was determined by two identical, stainless steel wire spacers. To hold this Al/specimen/Al sandwich together, a small rubber O-ring was stretched over a set of

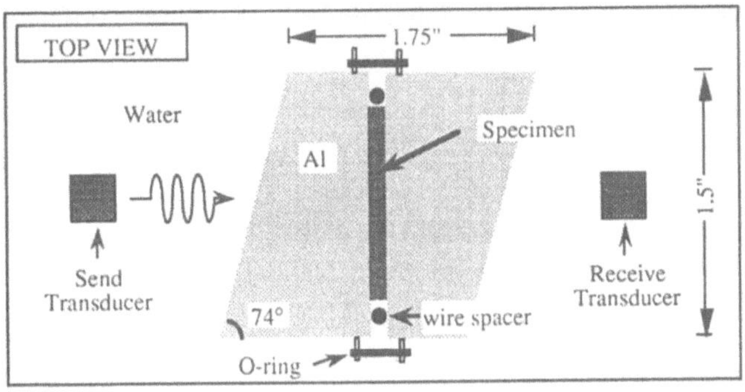

Figure 3. The experimental apparatus as described in the text. As indicated, this is a top view. The Al/specimen/Al sandwich extends 1.5 inches from top to bottom into the page. The transducers and O-ring are bisected by a plane parallel to that of the page and 0.75 inches into the page.

threaded pegs located on each of two opposite sides of the sandwich as shown. The depth of thread for these pegs was less than 0.125 inches. The resultant Al/specimen/Al sandwich was set upon a fixture (not included in Fig. 3 for clarity) and submerged in a water tank so as to be centered between a pair of plane wave, through-transmission, ultrasonic transducers. The fixture was designed to permit rotation of the specimen with respect to the transducers. The apparatus is designed to approximate the conditions leading to the derivation of Eqn. 2. The Al blocks correspond to the semi-infinite media, the specimen to the thin medium.

To perform measurements, the send transducer was stimulated with a spike pulse. The fixture was aligned so that this pulse would strike the first 74° face of the Al/specimen/Al sandwich at a particular angle of incidence, θ_i, to its normal, giving rise to both a longitudinal and a shear wave pulse within the aluminum. The longitudinal and shear wave emergence angles for a particular θ_i can be determined using Snell's law.

In the first phase of the experiment, θ_i was set at ~ 3.7° so that the longitudinal pulse transmitted through the first 74° face would strike the Al-specimen interface at normal incidence, and then travel through the remainder of the Al/specimen/Al sandwich to produce a pulse at the receive transducer. This pulse will be referred to as the longitudinal response. Because the longitudinal velocity in Al exceeds that of the shear, the longitudinal response was the first pulse detected after each drive pulse. This first pulse was followed by others due to the shear wave transmitted through the first 74° face and various internal reflections within the sandwich. Upon reception by the receive transducer, the longitudinal response was isolated and its FFT computed. To isolate the effect of the specimen, this FFT was divided by an FFT of the longitudinal response for a reference block comprised of solid Al and having the same dimensions as the Al/specimen/Al sandwich. The corrected data set was then normalized and stored for analysis. A computer was used to automate the system so that the longitudinal response could be measured repeatedly as a function of a specimen's cure time, t. Peaks in the longitudinal response FFT, which in accordance with Eqn. 2, occur when $f = f_{RL} = nc_L/2x_2$ could then used to determine c_L, the longitudinal acoustic velocity of the specimen for each measurement.

In the second phase of the experiment, θ_i was set to ~ 7.5° so that the direction of the transmitted shear wave was normal to the Al-specimen interface. The first received pulse resulting from this normal shear wave will be referred to as the shear response. Following each drive pulse, the shear response was preceded by not only the longitudinal response, but also other signals resulting from internal reflections involving the faster longitudinal pulse. To positively identify the shear response, one could increase θ_i beyond the critical angle for longitudinal wave production in the aluminum so that the shear response would be the first remaining pulse. One could then track this signal while decreasing θ_i to the appropriate value. The FFT of the shear response was then computed, normalized and stored in the same manner as that of the longitudinal response. Peaks in the shear response

FFT, occurring when $f = f_{RS} = nc_S/2x_2$, could then be used to determine the shear acoustic velocity of the specimen, c_S.

RESULTS

Three materials were tested: water; EPON 828, a Shell bisphenol-A/epichlorohydrin-based epoxy system, cured with diethylenetriamine (TETA); and an inert HTPB propellant mixture with sodium chloride substituted for the ammonium perchlorate found in live propellant.

Water was tested first to provide an end-to-end system check, as its acoustic velocity is well known. An Al/H_2O/Al sandwich with wire spacers of diameter $x_2 = 6$ mil was prepared. The longitudinal response for the 6 mil water specimen and its FFT are plotted in Figs. 4a and 4b, respectively. The longitudinal response and its FFT for the reference block under the same conditions are plotted in Figs. 4c and 4d. The FFT for the reference block (Fig. 4d) reflects the fact that 5 MHz transducers were used for the measurement. Comparison of Fig. 4a with Fig. 4c (and Fig. 4b with 4d) reveals that the 6 mil water specimen functioned like a bandpass filter for the input pulse. Dividing the reference block FFT (Fig. 4d) into that of the 6 mil water specimen (Fig. 4b) and normalizing the result yielded the plot depicted in Fig. 5. Note that as expected (compare with Fig. 2), multiple, equally-spaced peaks of diminishing amplitude are present. The results for frequencies outside the active range of the transducer, ~ 2 - 7.5 MHz (Fig. 4d), however, should be accepted with caution. From the position of the first non-zero maxima one can calculate; $c_L = 2f_{RL}x_2 = 0.056$ in/μs which compares favorably with the literature value for water of 0.058 in/μs. The slight difference was traced to the fact that the reference block was not cut from the same Al stock as the blocks used for the 6 mil water specimen.

Secondly, tests were performed on the system consisting of a 10:1 mixture of EPON 828 and diethylenetriamine (TETA) hardener. A typical plot of the longitudinal FFTfor this system is presented in Fig 6a. This plot was produced from data acquired following a cure time of t = 14.5 min. Three maxima are clearly evident. The third peak is distorted, but it occurs beyond the normal operating range of the 5.0 MHz transducers employed. The thickness of the longitudinal specimen was 10.25 mils. The position of the first maxima

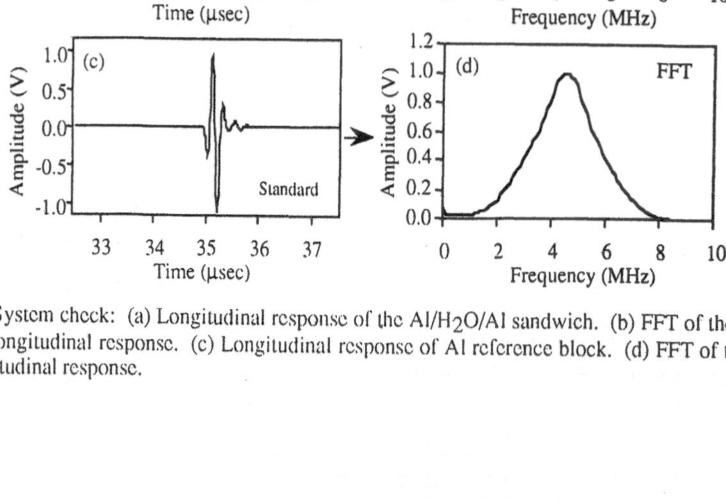

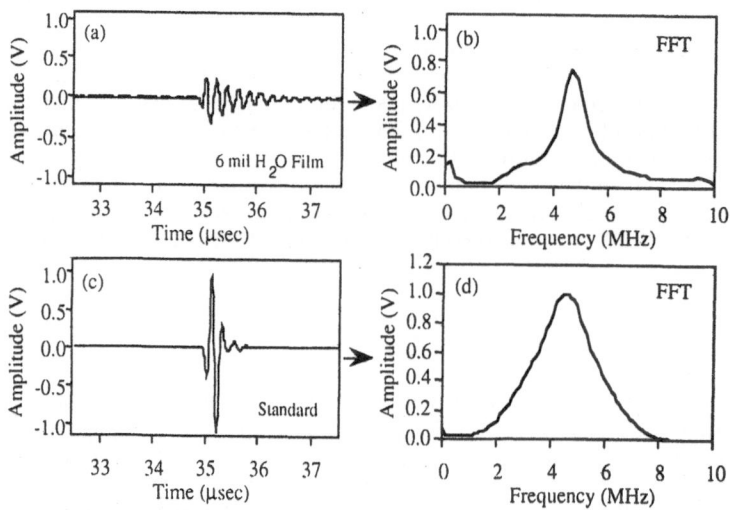

Figure 4. System check: (a) Longitudinal response of the Al/H_2O/Al sandwich. (b) FFT of the Al/H_2O/Al sandwich longitudinal response. (c) Longitudinal response of Al reference block. (d) FFT of the reference block longitudinal response.

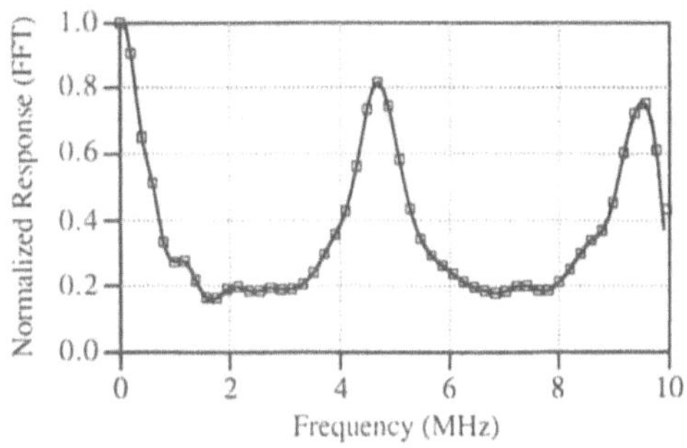

Figure 5. System check: Normalized results from division of the FFTs in Figure 4b and 4d. The first non-zero maxima occurs at 4.69 MHz.

implies that $c_L = 0.062$ in/μs. Measurements of the shear response for the EPON 828 system were more difficult. The FFTs obtained varied erratically during the early part of the cure. This is not surprising as liquids do not support shear waves, but at some point a transition must take place. The shear response FFT for t = 2 days is plotted in Fig. 6b. A pair of 2.25 MHz transducers were used to accumulate this data. Again, as expected, multiple peaks of diminishing amplitude are observed. The thickness of this specimen was 10.25 mils. The position of the first non-zero maxima implies that $c_S = 0.022$ in/μs. More work involving the shear response will be required before consistent trends can be identified.

The value of c_L for the EPON 828/TETA system changed significantly as the specimen cured, so that the peak positions shifted as depicted in Fig 7a, where plots for t = 6.3 min. and t = 96.3 min are presented. The position of the primary longitudinal peak as a function of cure time is plotted as a solid line in Fig. 7b. It can be observed that the peak position and hence, longitudinal velocity, increased by ~ 45 % during the first 6 hours of cure. The amplitude of the longitudinal FFT peaks also changed as a function of cure as indicated by the curve with open boxes in Fig. 7b. The amplitude change is not evident in Fig. 7a, because the data have been normalized. It can be seen that during the first 1.5 hrs of cure,the amplitude decreased, suggesting that the specimen became more attenuative to sound. The amplitude then rebounded, increasing to slightly above its original value before decreasing abruptly to a value which remained essentially constant for the last hour recorded.

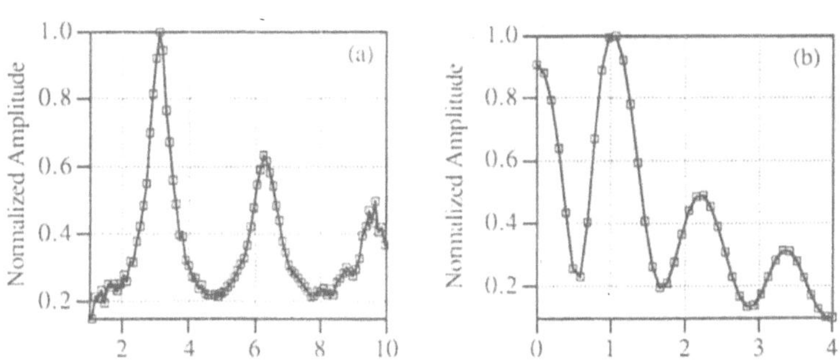

Figure 6. (a) Longitudinal response FFT of EPON 828/TETA mixture at t = 14.5 min. (b) Shear response FFT of EPON 828/TETA mixture at t = 2 days.

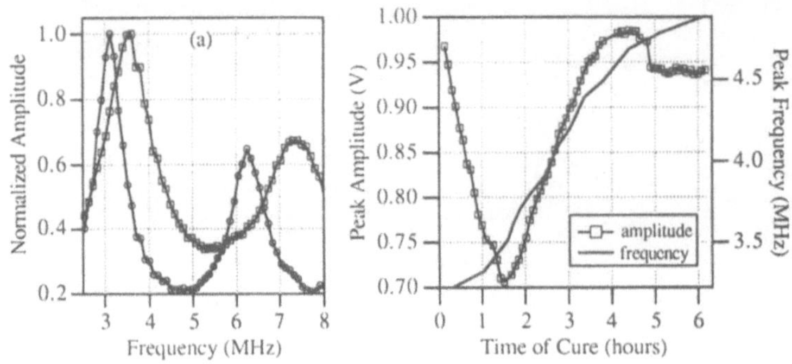

Figure 7. (a) Longitudinal response FFT at t = 6.3 min (circles) and t = 96.3 min (squares) and (b) the change in amplitude and resonant frequency as a function of t for an EPON 828/TETA specimen.

For comparative purposes, differential scanning calorimetry (DSC) was performed on a number of EPON 828 samples. By measuring the heat of reaction (ΔH) as a function of time beginning with the mix, one can estimate the degree of cure. The ΔH measured for a sample at t = 0 corresponds to 0% cure and when ΔH = 0 the system is considered 100% cured. Thus a plot of ΔH versus time provides a correlation between the degree of cure and time which can then be applied to the peak amplitude and frequency versus time data (Fig. 7b). The ΔH measurements were performed on a Mettler DSC 30 with a Mettler TC 10A TA processor. The DSC results for a sample taken from the same mix as that which led to Fig. 7 are plotted in Fig. 8. Each data point in Fig. 8 was measured on a separate sample and represents the residual heat of reaction remaining in the sample at time x. For the EPON 828/TETA system there is a significant amount of heat generated as the two oxirane groups of the EPON 828 react with the amine group of the TETA (an aliphatic polyamine) to produce a three-dimensional, crosslinked network. This reaction occurs at room temperature (RT) and requires several days (at RT) or one to two hours at 100 °C to reach complete cure.

In addition to DSC, a Rheometrics RDA II was employed to measure the epoxy system's viscosity, elastic modulus (G'), viscous modulus (G") and damping or loss tangent (tan δ = G"/G') as a function of time from initial mixing. The RDA measures these viscoelastic parameters in dynamic shear using a parallel plate fixture. The measurements were made at a constant temperature of 27° C. The RDA II results, which are presented in Fig. 9, give insight into the behavior exhibited by the peak amplitude curve of Fig. 7b. As expected,

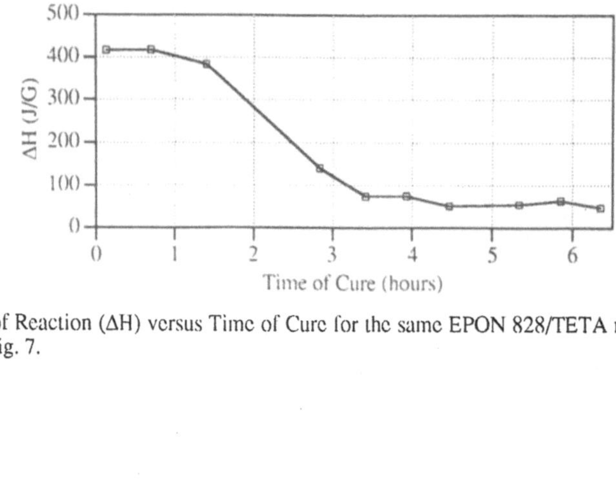

Figure 8. Heat of Reaction (ΔH) versus Time of Cure for the same EPON 828/TETA mixture as that used for the data of Fig. 7.

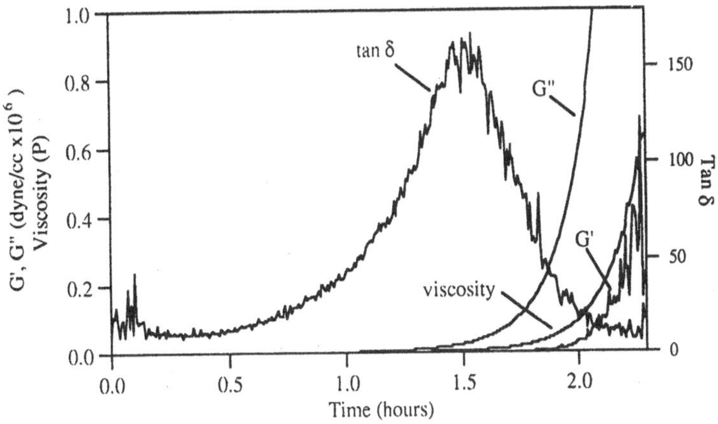

Figure 9. Viscosity, G', G" and Tan δ vs Time for the EPON 828/TETA mixture.

the viscosity increases with time. This is due to the constant increase in molecular weight of the epoxy as crosslinking occurs. Similarly, G' and G" (the real and imaginary parts of the dynamic shear modulus) increase with time. The G' begins it ascent at approximately the same time that the peak amplitude curve of Fig. 7b exhibits a minimum. The slight difference in time is probably related to fact that the ultrasonic measurements were performed at RT as opposed to 27° C. This same phenomenon is reflected as a maximum in the tan δ curve and is most likely associated with the onset of crosslinking for the epoxy. The gel point is traditionally defined as the point at which G' and G" cross.[4] The dynamic mechanical testing was terminated at ~ 2.2 hours as the viscosity of the sample exceeded the torque limits of the instrument.

Finally, measurements were performed on HTPB inert propellant specimens. The first difficulty encountered was that it was difficult to compress the material to form samples thinner than 10 mils. The longitudinal response FFT for a 10 mil specimen is plotted in Fig. 10a. Two peaks are clearly visible; the second riding on the tail of the first. The position of the primary peak implies that $c_L = 0.043$ in/μs, a value ~ 20% slower than that of water. This value remained essentially constant as a function of cure, perhaps a reflection of the high solid content of the inert propellant. The amplitude of the primary peak did, however, vary significantly with cure time as indicated by the plot depicted in Fig. 10b. The form of the plot is similar to that of the EPON 828/TETA mixture in that the amplitude decreases during the beginning stages of the cure and then rebounds to nearly its initial value. The results differ in that following the rebound, the signal drops monotonically, except for a curious upswing at t_c ~ 30 hrs, to a lower value than that achieved in the early stages of the cure. Preliminary attempts to perform a DSC of the inert propellant were unsuccessful. This is attributed to the high filler content. For the purposes of thermal analysis, the filler, which has a high specific heat , absorbs energy and reduces the exotherm temperature. Additionally, because the DSC sample size is small, on the order of 20 mg, the filler has the effect of reducing the amount of material involved in the curing reaction and thus reducing the exotherm. Measurements of the shear response of the inert propellant were less successful than those for the EPON 828/TETA mixture. Again, more work involving the shear response will be required before consistent trends can be identified.

DISCUSSION

The ultrasonic resonance technique described in this work exhibits potential as a means of monitoring the degree of cure as well as cure and age related mechanical property changes

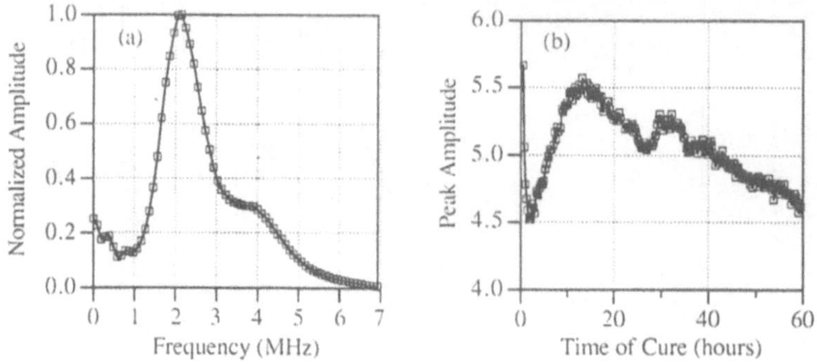

Figure 10. (a) Longitudinal response FFT and (b) the change in amplitude as a function of cure time for a HTPB inert propellant specimen.

in polymers. Preliminary results indicate that the changes in the peak frequency (ultrasonic velocity) data can be correlated with the degree of cure and that the changes in the peak amplitude can be related to the results of dynamic mechanical tests. The results presented in this report suggest that SRM propellants are particularly difficult materials to characterize, especially in the frequency range employed in these tests. Much of the difficulty probably stems from the high solid content of the propellant. It may be prudent to perform similar tests on unfilled HTPB specimens. In addition, future tests (particularly for the shear mode) should be performed at lower frequencies and, perhaps, different temperatures.

ACKNOWLEDGMENTS

The authors gratefully acknowledge the assistance of (1) R. C. Savedra with sample preparation and data acquisition, (2) R. M. Castaneda with the DSC measurements and (3) R. J. Zaldivar with the dynamic mechanical tests.

REFERENCES

1. P. H. Graham, "Analysis of the Variability of HTPB Propellant Mechanical Properties," Annual report for NASA Marshall Space Flight Center under Contract No. NAS8-37802, Report No. 443/1290/36, 12 Dec. 1990.
2. M. Rooney, C. L. Friant, C. V. O'Keefe, and W. M. Ferrell, "Determination of Modulus of HTPB Solid Rocket Propellant using Longitudinal and Shear Ultrasonic Waves," Annual report for NASA Marshall Space Flight Center under Contract No. NAS8-37802, Report No. 443/1291/37, pp. 4.3-55.
3. E. C. Johnson, J. D. Pollchik and J. N. Schurr, "An Ultrasonic Testing Technique for Measurement of the Poisson's Ration in Tin adhesive Layers," Review of Progress in Quantitative Nondestructive Evaluation (edited by D. O. Thompson and D. E. Chimenti, Plenum Press, New York, 1992), Vol 11B, 1291 - 1298 (1992).
4. C. M. Long, and P. J. Dynes, "Relationships Between Viscoelastic Properties and Gelation in Thermosetting Systems," J. App. Poly. Sci., Vol 27, 569 - 574 (1982).

ELASTICITY OF SINGLE-CRYSTAL Al_2O_3 (SAPHIKON) FIBER TO 1000°C BY BRILLOUIN SPECTROSCOPY

Murli H. Manghnani,[1] Vahid Askarpour,[1] and James A. DiCarlo[2]

[1]School of Ocean and Earth Science and Technology,
University of Hawaii, Honolulu, Hawaii 96822
[2]NASA Lewis Research Center, Cleveland, Ohio 44135

INTRODUCTION

Continuous fiber-reinforced metal and ceramic matrix composites have received wide applications in industry, engineering and medicine[1,2]. The mechanical and thermal performance of ceramic matrices in monolithic forms are enhanced by the addition of reinforcing fiber. The choice of the fiber material is very important in the deformation behavior of the composite particularly at high temperatures. In this study, we have characterized the elastic properties of single-crystal Al_2O_3 fibers (160 μm diameter) at elevated temperatures using the technique of Brillouin spectroscopy. Al_2O_3 fibers are employed to demonstrate the feasibility of such experiments on small diameter fibers where other methods to measure fiber elastic properties may not be conveniently available or larger specimen may not be readily accessible.

Single-crystal elastic properties of Al_2O_3 have been previously reported by both Brillouin spectroscopy[3] to 1800°C and ultrasonic interferometry[4] to 1500°C. However, no elasticity data for this material in fiber form and at high temperatures are available. Brillouin spectroscopy is perhaps, the only available technique at present for small sample elastic characterization at elevated temperatures.

THEORY

Brillouin scattering is the inelastic scattering of light by fluctuations in the optical dielectric constant of the scattering medium due to the propagation of thermally excited sound waves[5]. In solids, there are three acoustic bulk waves propagating in an arbitrary direction, one longitudinal (L) and two transverse (T_1 and T_2) acoustic modes.

When light interacts with the acoustic waves, the frequency of the incident light is Doppler shifted. The frequency shift $\Delta\nu$ in the incident light is related to the sound velocity, V , and q (the magnitude of the phonon wavevector **q** as defined by the scattering geometry) by $\Delta\nu = Vq$. In this study, platelet geometry (see Figure 1) was employed to study the elastic properties of Al_2O_3 fibers.

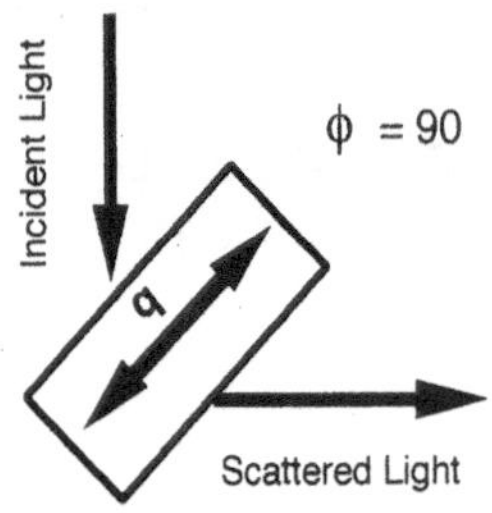

Figure 1. **q** is the phonon wavevector and represents the direction of acoustic wave propagation.

In platelet geometry, the acoustic velocity is given by:

$$V = \Delta\nu.\lambda_o/\sqrt{2} \qquad (1)$$

where, λ_o is the wavelength of the incident laser (514.5 nm in this study). The elastic constants are calculated from:

$$C_{ij} = \rho.V^2 \qquad (2)$$

where ρ is the density and C_{ij} are appropriate combinations of elastic constants. The elastic constants are summarized in Table 1.

Table 1. The elastic constants of hexagonal symmetry.

q	Longitudinal	Transverse
[100]	C_{11}	C_{44}
[001]	C_{33}	C_{44}
[011]	C_{13}[1]	C_{12}[2]

[1] $\{C_{11}+C_{33}+2C_{44}-[(C_{11}-C_{33})^2 + 4(C_{13}+C_{44})^2]^{1/2}\}/4$

[2] $(2C_{44}+C_{11}-C_{12})/4$

EXPERIMENTAL METHOD

Details of the experimental set up can be found elsewhere[6,7]. Briefly, an argon laser beam was incident on the sample in the platelet geometry. The scattered light was analyzed by a high-contrast high-resolution Brillouin spectrometer consisting of a 6-pass tandem Fabry-Pérot interferometer, designed by J. R. Sandercock, and photon counting units. The spectra were stored in a 1024 channels of a multichannel analyzer. Further analysis of the spectra were performed by a curve-fitting routine to calculate the frequency shifts. Due to the uncertainties in the frequency shift measurements (better than 0.5%) and the scattering angle (~0.5%), the estimated error in the calculated acoustic velocities were ~ 1%. The fibers were heated in a resistive heating furnace to temperatures of up to 1400 K (with an uncertainty of ±2 K).

The Al_2O_3 single-crystal (Saphikon) fibers of ~160 μm diameter were coated with a thin platinum coating and embedded in a Al_2O_3 ceramic matrix with the fibers aligned in one direction. Typically the orientation of the c-axis deviated from the fiber axis ±3°. Figure 2a shows a cross section of a fiber reinforced ceramic composite. Two samples for the present work were prepared. The first sample (A) was polished normal to the fiber length in the form of a 100 μm thick disc as shown in Figure 2b. The [100] axis was not known in this case and hence an estimate of C_{11} was obtained. The second sample (B) was polished with two faces parallel to the fiber length as shown in Figure 2c. This second sample allowed C_{33} and C_{44} to be determined.

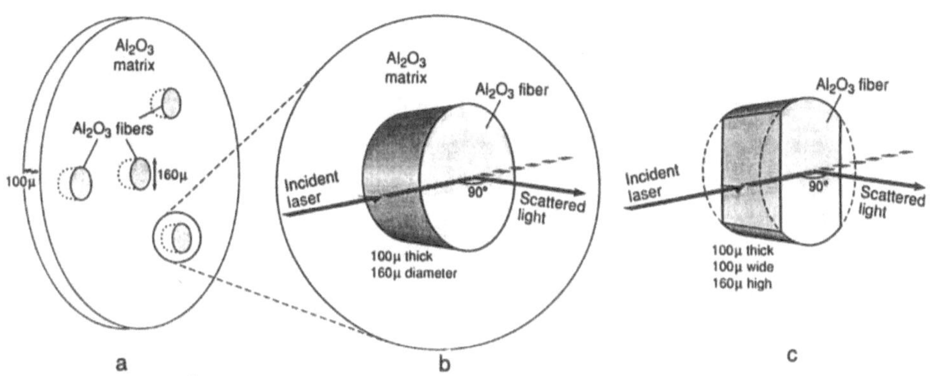

Figure 2. Al_2O_3 fibers in a ceramic matrix composite (2a). Sample (A) was polished normal to [001] axis (2b). Sample (B) was polished parallel to the fiber length (2c).

RESULTS

A typical Brillouin spectrum of the Al_2O_3 fiber is shown in Figure 3. The intense feature at the center of the spectrum is the elastic scattering from the fiber. A pair (upshifted and downshifted in frequency due to Doppler effect) of longitudinal (L) and two pairs of transverse components (T_1 and T_2) were observed. It is noted that the longitudinal peak has a doublet structure, indicating the presence of two crystals of slightly different orientations each contributing separately to the scattering process (i.e., the fiber is not completely single crystal).

Sample (A) was then mounted in the high-temperature furnace in platelet geometry (orientation of the a-axis is arbitrary). The Brillouin spectra of the fiber were recorded to 1000°C in intervals of 100°C. Brillouin spectra were also recorded for sample (B) to 900°C. From the measured velocities in [100] and [001], three elastic constants

C_{11}, C_{33} and C_{44} were determined (2 to 3% error). The temperature dependence of the density for Al_2O_3 needed for this calculation was obtained from Goto et al. (1989)[4].

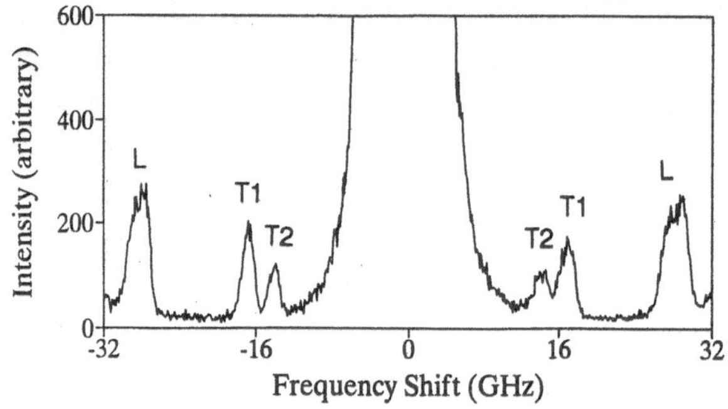

Figure 3. Brillouin spectrum of Al_2O_3 fiber. L and T's represent the longitudinal and transverse components of the spectrum, respectively.

The temperature dependences of the C_{11}, C_{33} and C_{44} are shown in Figure 4. The ultrasonic data[4] on large crystals are also shown by open squares. Since our fibers were not entirely single crystalline, we have plotted the elastic constants corresponding to all the observed peaks. It is noted that there is about 10% difference in the C_{11} values determined from the longitudinal doublet in Figure 2.

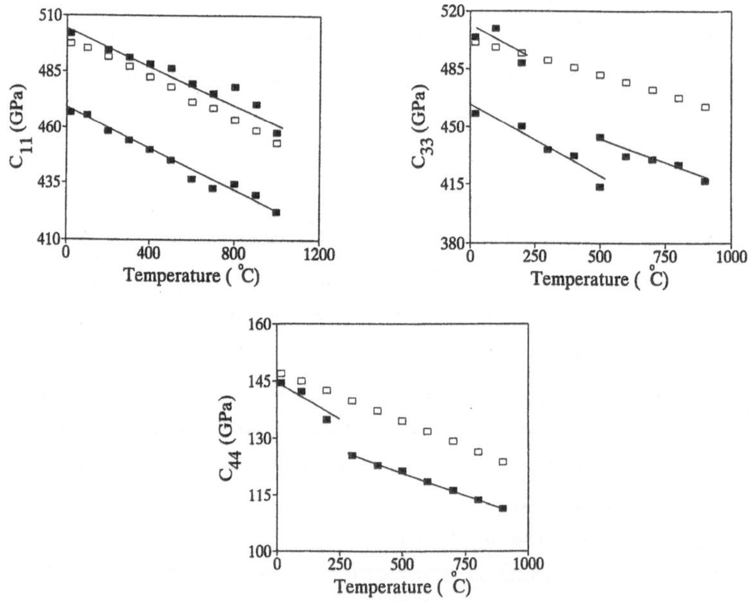

Figure 4. Temperature dependence of the three elastic moduli determined in this study.

DISCUSSION

It is noted from Figure 4 that there is good agreement (within 2-3% error) between our values of the elastic moduli and those determined by ultrasonic technique for appropriate fiber orientation. Crystal orientation and lack of perfect single-crystallinity along the fiber length are causing some discrepancies in the elastic moduli. For example, due to nonhomogeneity of the fiber, there is up to 10% difference in the elastic properties of the constituent microcrystals of the fibers in reinforced ceramic composites (see Figure 4) parallel and perpendicular to the fiber axis in the temperature range of this study. Such variations in the elastic properties are important factors in predicting their high-temperature performance.

SUMMARY

In this study, we have demonstrated the feasibility of carrying out Brillouin scattering measurements on Al_2O_3 single-crystal fibers (Saphikon) as a function of temperature to 1000°C. The results are in good agreement with previous resonance and Brillouin scattering measurements on large crystals.

The observed discrepancies as well as the variations in the C_{ij} values as a function of temperature are attributed to the departure of the fiber axis from true *c*-axis (±3°) and due to the fiber not being truly single-crystalline along the fiber length.

Brillouin scattering is the only available experimental method for investigating the elastic properties of fibers at high temperatures.

ACKNOWLEDGEMENT

The support for the facilities was provided by the W. M. Keck Foundation and the National Science Foundation (Grant 8720404). We are grateful to J. Balogh and O. Matthews for technical assistance and T. Bier for polishing the fiber samples. The composite samples were obtained from M. Jaskowiak at NASA Lewis Research Center. This is School of Ocean and Earth Science and Technology Contribution no. 3240.

REFERENCES

1. J. Doychak, Metals and intermetallic matrix composites for aerospace propulsion and power systems, *JOM* 44:46 (1992).
2. M.A. Karnitz, D.F. Craig, and S.L. Richlen, Continuous fiber ceramic composite program, *Ceram. Bull.* 70:430 (1991).
3. E.S. Zouboulis and M. Grimsditch, Refractive index and elastic properties of single-crystal corundum (α-Al_2O_3) up to 2100 K, *J. Appl. Phys.* 70:772 (1991).
4. T. Goto and O.L. Anderson, Elastic constants of corundum up to 1825 K, *J. Geophys. Res.* 94:7588 (1989).

5. G.B. Benedek and K. Fritsch, Brillouin scattering in cubic crystals, *Phys. Rev.* 149:647 (1966).
6. V. Askarpour, M.H. Manghnani, S. Fassbender, and A. Yoneda, Elasticity of single-crystal $MgAl_2O_4$ spinel up to 1273 K by Brillouin spectroscopy, *Phys. Chem. Miner.* 19:511 (1993).
7. J-A. Xu and M.H. Manghnani, Brillouin scattering studies of a sodium silicate glass in solid and melt conditions at temperatures of up to 1000°C, *Phys. Rev. B1* 45:640 (1992).

CHARACTERIZATION OF SHEET STEELS IN THE DEVELOPMENT OF ON-LINE SENSORS FOR QUALITY CONTROL MONITORING OF MECHANICAL PROPERTIES

L.J. Swartzendruber[1], Y. Rosenthal[2], and G.E. Hicho[1]

[1] National Institute of Standards and Technology
Gaithersburg, MD 20899
[2]Nuclear Research Center - Negev, P.O. Box 9001
Beer Sheva 84190, Israel

ABSTRACT

Magnetic and mechanical measurements of rolled sheet steel were performed as a part of a study on the feasibility of using magnetic measurements to supplement and improve the determination of sheet steel properties. Three low carbon steel sheets, having different mechanical properties, were selected and their microstructure, hardness, and grain size characterized. Each sheet was tested using tensile specimens cut from locations parallel, perpendicular, and 45° to the rolling direction in order to determine the extent of scatter in mechanical properties within each sheet. Additional specimens cut from these sheets adjacent to the tensile specimens were used for magnetic measurements. The results showed very low scatter (less than 10%) in the traditional mechanical properties as well as in the strain hardening parameters and in the plastic strain ratios within each sheet. Some correlations between the various mechanical parameters and between the magnetic and mechanical properties are shown.

INTRODUCTION

Both the mechanical and the magnetic properties of a material such as ferritic sheet steel depend sensitively on the microstructure. Thus magnetic properties may provide a probe for the nondestructive characterization of mechanical properties (see, for example, references 1-6). The magnetic properties which are sensitive to structure include initial permeability, maximum permeability, coercive force, remanance, and the Barkhausen effect. Some of these properties, such as coercive force and Barkhausen signal, may lend themselves to rapid on-line measurement as the steel is being processed.

In order to be useful, statistically valid correlations must be established between mechanical and magnetic properties that apply to specific types of steel. Establishing such correlations requires sets of well characterized samples that span a range of mechanical properties. Sets of sheet steel samples were obtained from a steel mill. Three sheets each from three different steels were obtained. The objective here is to determine the extent of mechanical property variation for samples within and between the sheets, and then to determine the same for the magnetic properties. Ultimately, it is desired that reproducible correlations between a mechanical property, or several mechanical properties, and selected magnetic properties be obtained. To achieve this goal, mechanical properties were determined for specimens that were parallel and perpendicular

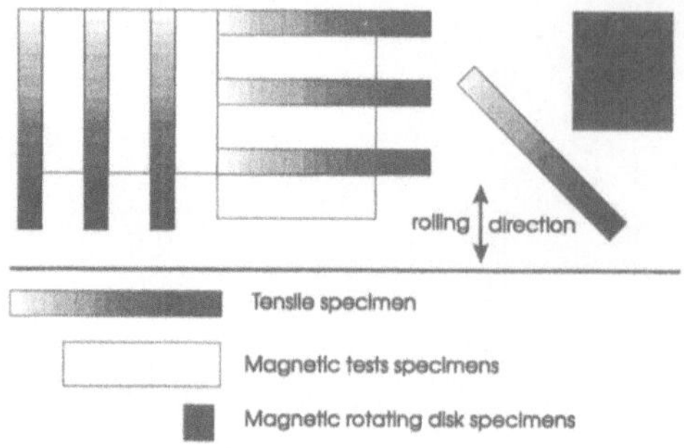

Figure 1. Schematic of relative locations of specimens taken. Three sets were taken from each sheet.

to the rolling direction, and 45° to this direction. Standard ASTM test methods were used to evaluate the mechanical properties of the three different steels. Additional specimens, adjacent to those used for the mechanical property measurements, were sectioned from the sheets and used for the magnetic property measurements. Chemical analysis, hardness, and microstructure evaluation were also performed on each sheet.

EXPERIMENTAL PROCEDURE

Materials

In order to be able to determine any mechanical parameters correlation with magnetic properties it was necessary to characterize the scatter in mechanical properties across a given steel sheet. Three different low-carbon steel sheets were used for this purpose: Cold rolled, interstitial free, Ti stabilized, coded as IF-18 (0.89 m x 0.91 m, 0.83 mm thick); cold rolled, interstitial free, coded as U-4 (0.84 m x 0.66 m, 1.12 mm thick); and cold rolled, electrogalvanized, coded as EI-2 (0.89 m x 0.91 m, 0.93 mm thick). These steels were chosen primarily because their yield strengths were uniquely different. The chemical composition of each steel is shown in Table 1.

TABLE 1. The chemical composition (in w/o) of the steel tested.

	C	Mn	P	S	Si	Al	N	Nb	Ti
IF18	0.004	0.17	-	0.007	-	0.048	0.0099	<0.002	0.72
U-4	0.0016	0.36	0.016	0.0088	0.007	0.077	0.0044	<0.002	0.002
E-I2	0.057	0.17	-	0.018	-	<0.004	0.0024	-	-

Mechanical Test Program

Twenty 50 mm gage standard ASTM flat tensile specimens were machined, eighteen magnetic testing specimens (50 mm x 150 mm), and three specimens for rotating disc magnetic measurements (150 mm x 150 mm) were cut from each sheet. In addition, specimens used for standard chemical analysis and microstructure evaluation were cut from each sheet. The location of the various specimens is shown in Fig. 1. It is emphasized that two tensile specimens were cut 45° to the rolling direction and the others were cut both parallel and transverse to the rolling direction.

Tensile testing was conducted using a universal mechanical testing machine. A fast personal computer and high-level data acquisition software were used to upgrade the existing mechanical testing system. The upgrading enables us to get a fully detailed set of mechanical parameters for any specimen immediately after each test. A loading rate of 2 mm min^{-1} was used for all the tensile tests .

The data acquisition procedure and analysis consists of four major steps: (1) The test itself during which instantaneous load and displacement (measured using an extensometer attached to the tensile specimen) data points are acquired at any desired rate and then saved in an ASCII stress-strain data file; (2) Determination of the mechanical elastic parameters (Young's modulus, proportional limit) using linear regression with a different graphics computer program having curve fitting capabilities; (3) Determination of the engineering elastic/plastic and the plastic deformation parameters (0.2% yield stress, ultimate tensile stress, uniform elongation, and elongation at fracture) and, the plotting of engineering stress-strain curve; and (4) Evaluation of the true stress-strain curve and determination of the strain hardening exponent and the strength coefficient as well as the plastic strain ratio.

Calibration of the testing machine's load cell and the extensometer used were conducted prior to the mechanical testing. The load was calibrated using a proving ring (traceable to NIST calibration apparatus) in accordance with section 6 of ASTM Method E74. The load average deviation was 0.2% in the range used for the mechanical testing.

The following equations were used for the evaluation of mechanical parameters:

Determination of Young's Modulus

Equation 1 below was used for linear regression fit to the first part of the stress-strain curve for the determination of the elastic modulus:

$$S = Ee \qquad (1)$$

where E is Young's Modulus, S is the engineering stress (equal to P/A_0 , where P is the applied load and A_0 is the initial cross sectional area), e is the engineering strain (equal to $\Delta l / l_0$, where Δl is the displacement increment, and l_0 is the initial gage length).

Determination of strain hardening parameters

The empirical Hollomon equation (Equation 2) was used for the evaluation of strain hardening parameters:

$$\sigma = K\epsilon^n \qquad (2)$$

where σ is the true stress (equal to S(1 + e), K is the strength coefficient, i.e. the Y intercept, ε the true strain (equal to ln(1 + e), and n is the strain hardening exponent. The value of n was calculated using linear regression fit for the $\ln \sigma$ vs. $\ln \varepsilon$ curve. This fit can be applied only to a certain range of the stress/strain curve starting from small plastic strains (that is beyond the yield point, and then any discontinuous yielding, where applicable) and ending at the ultimate tensile stress.

Determination of plastic strain ratio

The plastic strain ratio (according to ASTM Method E517) is considered a measure of sheet metal drawability and is obtained using Equation 3:

$$r = \frac{\ln(w_o l_o)}{\ln\left(\frac{l_f w_f}{l_o w_o}\right)} \qquad (3)$$

where r is the plastic strain ratio, w_0 is the initial width of the tensile specimen, w_f is the width of the specimen as measured just before the maximum load, l_0 is the initial gage length (also the extensometer gage length), and l_f is the gage length as measured before the maximum load is reached.

The average plastic strain ratio r_m is calculated as follows:

$$r_m = \frac{r_o + 2r_{45} + r_{90}}{4} \quad (4)$$

where r_0 is the plastic strain ratio parallel to the rolling direction, r_{45} is the plastic strain ratio 45^0 to the rolling direction, and r_{90} is the plastic strain ratio perpendicular to the rolling direction.

Determination of Magnetic Properties

The Barkhausen signal and hysteresis loops were obtained on representative samples using the apparatus illustrated in Fig. 2. The Barkhausen signal coil is a surface probe fabricated from a ferrite core. Its output is bandwidth limited to a range of 1 to 100 kHz. The pickup coil is a 300 turn encircling coil. A Hall probe is used to measure H, the tangential field at the surface of the sample. As the energizing yoke sweeps the sample through a hysteresis loop, the signal from the Hall probe, the pickup coil, and the signal coil are digitized and stored in computer memory. The output of the pickup coil is integrated to obtain the magnetic induction, B, as a function of the tangential field. The algorithms used to analyze the Barkhausen signal is discussed in detail in reference 6. Two types of curves will be shown here, the integrated Barkhausen intensity, and the Barkhausen signal emission rate, which is the derivative of the integrated signal after some smoothing has been performed.

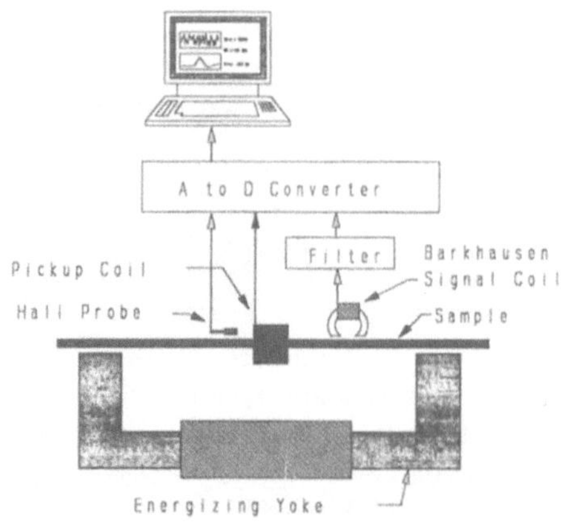

Figure 2. Schematic drawing of the magnetic property measurement system.

RESULTS AND DISCUSSION

Microstructure

The microstructures of three different sheet steels were examined, and Figures 3, 4, and 5 show the general microstructure over the full thickness, and at the quarter-thickness location. Figure 3, the cold rolled, interstitial free and titanium stabilized steel identified as IF-18, consists primarily of fine grained ferrite. The absence of any pearlite is noted, Figure 3a, but this is due to the rather low carbon content of 0.004 weight percent. Figure 4 shows the microstructure of the other cold rolled, interstitial free steel U-4, which was not titanium stabilized. The steel consists of uniform ferrite grains throughout the thickness, Figure 4a. The absence of any significant amount of pearlite is again noted. Interstitial free steels are melted in a way that

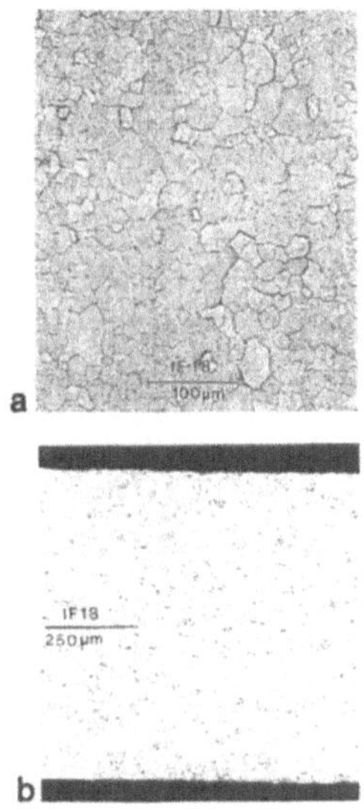

Figure 3. IF-18 microstructure: (a) 1/4 thickness, (b) entire thickness.

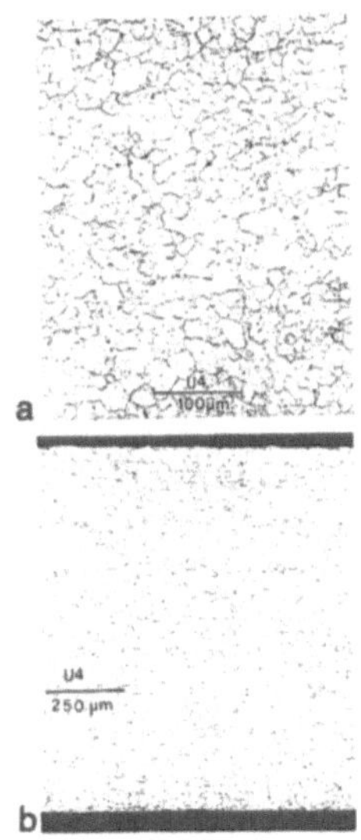

Figure 4. U-4 microstructure: (a) 1/4-thickness, (b) entire thickness.

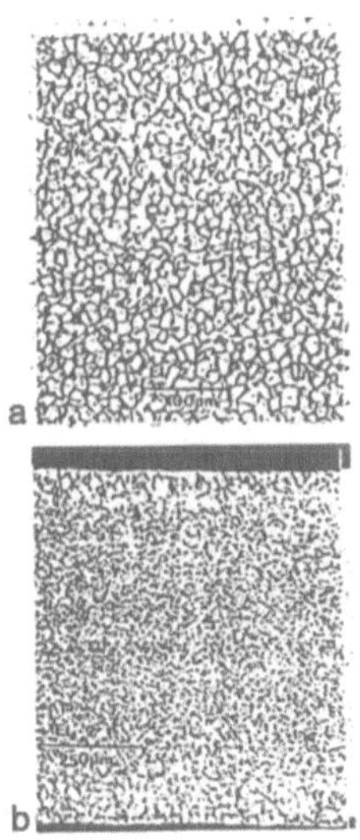

Figure 5. EI-2 microstructure : (a) 1/4 thickness (b) entire thickness.

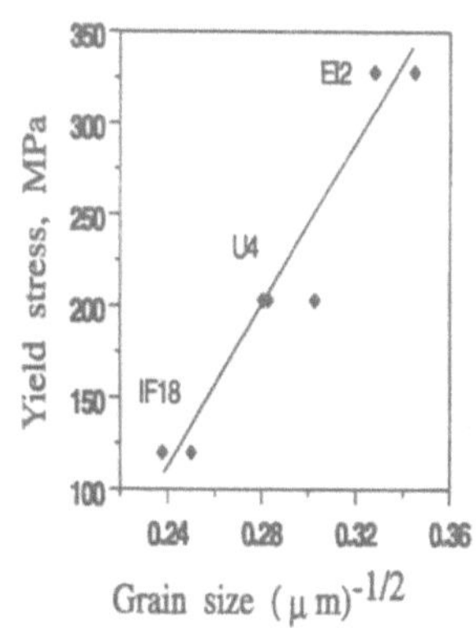

Figure 6. Yield stress vs. grain size relation as obtained for the sheets tested.

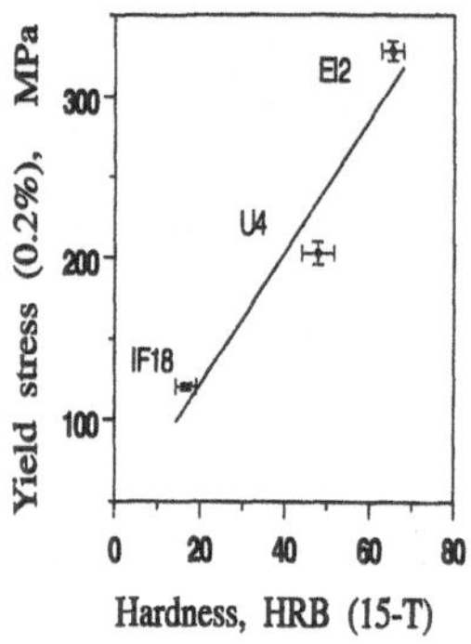

Figure 7. Linear fit of yield stress vs. hardness.

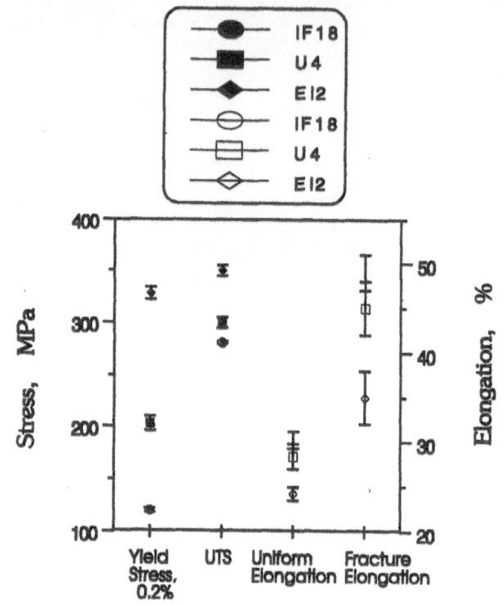

Figure 8. A summary of the differences in mechanical properties among the steels tested.

the levels of the interstitial free elements, carbon and nitrogen, are reduced to small amounts. The steel is normally deoxidized with aluminum and the residual aluminum in the melt ties up the nitrogen by forming nitrides. The residual carbon and nitrogen are removed by the addition of titanium, niobium, or zirconium to the melt. These elements, reportedly form stable carbides and nitrides that remove the interstitials from solid solution and therefore remove the yield point elongation, characteristic of interstitial carbon and nitrogen steels. The third steel examined in this study was a cold rolled, electrogalvanized steel that contained 0.057 weight percent carbon. Figure 5 shows the through thickness microstructure for this sheet that was coded EI-2. The figure shows that the microstructure near the surfaces was devoid of carbon and this was probably due to the galvanizing procedure on the sheet. The galvanizing temperature was such that it removed the carbon from the outside surfaces, essentially decarburizing these sheets. The microstructure of EI-2, Figure 5a, was also primarily ferrite with packets of pearlite dispersed throughout the ferrite and located along the grain boundaries.

The grain sizes of the steel sheets are shown in Table 2. A relation was obtained for the yield stress vs. grain size (Figure 6). The yield stress vs. hardness is plotted in Figure 7. Because of the sheets thickness hardness values were converted from the Rockwell 15 kg. 15-T scale to Rockwell scale B (HRB).

Mechanical Properties

The mechanical properties of the steel sheets are summarized in Table 3 and in Figures 8 (yield stress 0.2%, UTS, uniform elongation, and the total elongation at fracture) and 9 (strain hardening exponent and the strength coefficient). It can be seen that the steels differ in their properties, but the variations in the individual mechanical parameters within each sheet are very small. For example, standard deviations of 1.7%, 3.5%, and 1.8% were obtained for the yield stress values of IF-18, U-4, and EI-2 respectively. The standard deviations were 0.4%, 1.7%, and 1.4% as obtained for the ultimate tensile stress values of these steels. Considering the number of tensile specimens used, these standard deviations are very low suggesting that the sheets have uniform mechanical properties. Furthermore, the 20%-40% average differences in the yield stress and the UTS values (beyond the scatter within each sheet) among the sheets also suggest that they can be used in the search for any possible correlation with magnetic properties as the minimum desired sensitivity of the sensor developed was determined by its ability to reveal 10% changes in mechanical properties.

The plot of the yield stress vs. plastic strain ratio is shown in Figure 10. Because of the correlation between the magnetic characteristics of the steels and the yield stress, to be shown below, the plastic strain ratio, which is used as a measure of sheet metal drawability, can be related to the magnetic characteristics through its linear relation to the yield stress.

TABLE 2. Grain sizes in different locations within the sheets.
* Coarse surface grains, 100-150 μm in depth.
** Grains uniform through thickness.
*** Finer grains near surface, 50 μm in depth.

Sheet	Position	Grain Size, μm	ASTM #
EI-2	1/4 Thickness	8.4	10.5
	Parallel	14.1	9.0 *
	Perpendicular	9.3	10.2
U-4	1/4 Thickness	12.5	9.3
	Parallel	10.9	9.7
	Perpendicular	12.7	9.3 **
IF-18	1/4 Thickness	17.7	8.3
	Parallel	16.0	8.6
	Perpendicular	16.0	8.6
	Parallel	13.5	9.1 ***
	Parallel	12.8	9.3 ***

TABLE 3. Average mechanical properties in tension for the steel sheets tested.
* U.T.S. Uniform Strain
** Fracture Strain, Total in 50 mm

Property	EI-2	U-4	IF-18
Young's Modulus, GPa (standard deviation)	234 38	228 14	195 11
Yield Stress 0.2%,GPa (standard deviation)	331 4	203 7	120 2
U.T.S., MPa (standard deviation)	351 6	300 5	281 1
U.T.S.* , % (standard deviation)	23.9 0.8	28.3 1.4	30.2 0.9
Fracture Strain **, % (standard devaition)	36 3	45 3	49 2
n (standard devaiton)	0.18 0.003	0.21 0.007	0.26 0.003
K, MPa (standard deviation)	580 7	519 8	518 3
r (standard devaition)	0.61 0.05	0.86 0.31	1.06 0.11

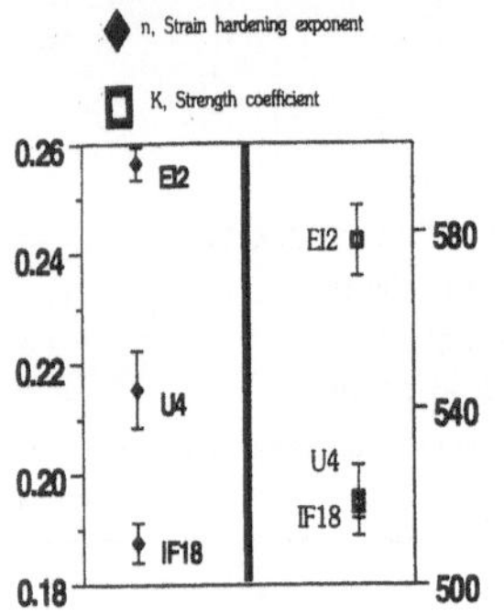

Figure 9. The average strain hardening results and their standard deviations for the steel tested.

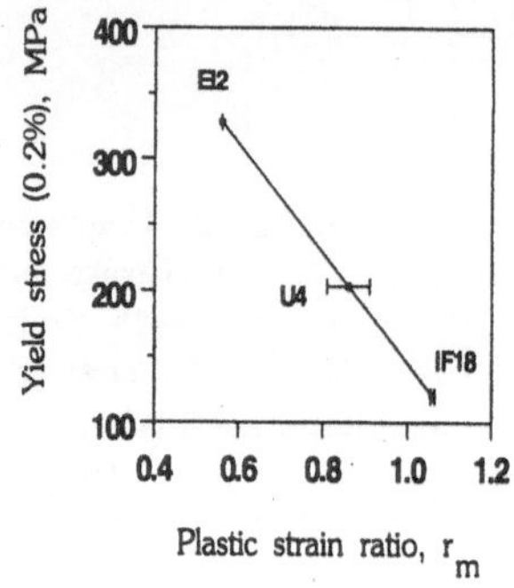

Figure 10. The yield stress vs. plastic strain ratio with a fitted linear regression line.

Yield stress related equations

Three linear equations were fitted to some of the parameters checked suggesting a correlation with the yield stress of the steels tested. The first correlation was found between the yield stress and the grain size, Figure 6, as follows:

$$S_y = 2157.7\ (d)^{-1/2} - 411 \qquad (5)$$

Where Sy is the yield stress in MPa, and d is the grain size in mm.

A second linear correlation found between the yield stress and the hardness of the steels (Figure 7):

$$S_y = 4.1(HRB) + 39.5 \qquad (6)$$

where HRB is the hardness on the Rockwell B scale.

The third linear relation is between the yield stress and the plastic strain ratio, r_m (Figure 10):

$$S_y = -416\ r_m + 561 \qquad (7)$$

All these three parameters (d, HRB, and r_m) might be useful, along with the actual values obtained for the traditional mechanical parameters, to evaluate quality control sensors of potential use in the rolling mill. The goal is to develop sensors which can rapidly evaluate when the process is out of control and enable rapid correction of the problem, thereby reducing scrap and enabling the production of a more uniform product.

Magnetic Measurements

The hysteresis loops obtained on the three different sheets, taken with the tangential field, H, parallel to the rolling direction, are shown in Figures 11, 12, and 13. Also plotted on these curves are the incremental premeabilities (the slope of the B vs H curve, dB/dH, in dimensionless units) and a small central loop taken at low applied field. The slope of the small loop gives a good estimation of the initial initial permeability. (An eddy current type measurement would be most affected by the initial permeability value). The field at which the value of B on the full hysteresis curve passes through zero is the coercive force (Hc) of the material. The coercive force and the width of the dB/dH curve will be affected by the presence of domain wall pinning centers, grain orientation, residual stresses, and the uniformity of these properties on a microscopic scale. The U-4 sample, which is from the material with the lowest carbon content, has a dB/dH curve with a very small width.

In the dB/dH curve for EI-2 (Figure 13) two nearly resolved peaks are evident. We attribute the peak occurring at a lower tangential field (i.e. the field closest to zero) to the surface of the sample, and the larger peak at a higher tangential field to the bulk of the material. This interpretation is consistent with the metallographic examination (see Figure 5) which shows a larger grain size and lower carbon content on the surface, resulting in a lower coercive force for the material at the surface.

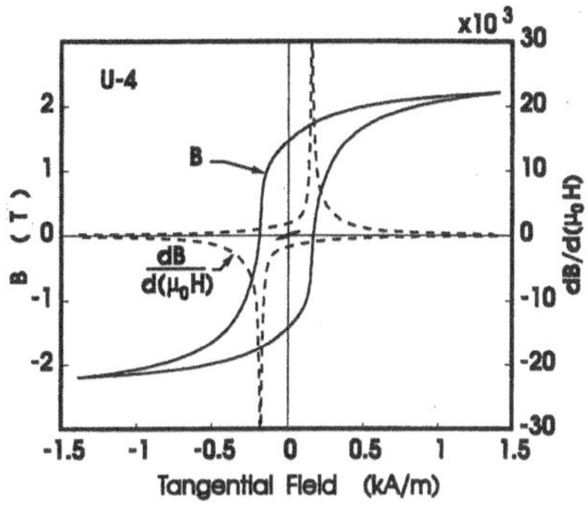

Figure 11. U-4 hysteresis curve, incremental permeability (peaked curve), and small minor loop.

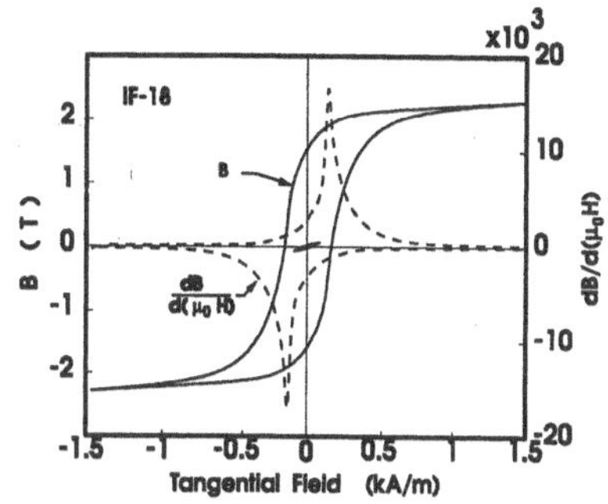

Figure 12. IF-18 hysteresis loop, incremental permeability (peaked curve), and small minor loop.

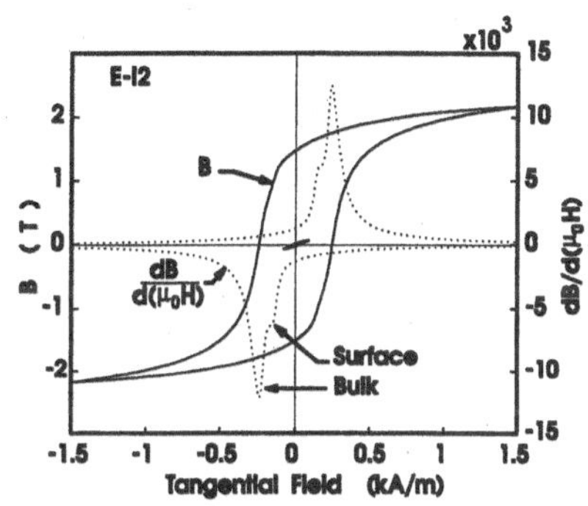

Figure 13. EI-2 hysteresis loop, incremental permeability (peaked curve), and small minor loop.

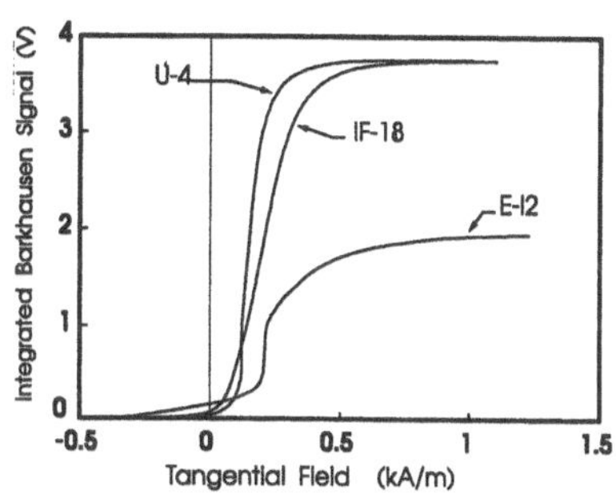

Figure 14. Integrated Barkhausen signals vs. applied tangential field for the three samples.

The integrated Barkhausen signal from the three samples is shown in Figure 14. Note that the signal from sample U-4 rises more sharply than from IF-18 or EI-2, consistent with the sharper incremental susceptibility seen for U-4 in the hysteresis loop curves. The total integrated intensity for EI-2 is much lower than for U-4 or IF-18. This is because the higher carbon - higher strength EI-2 contains a higher density of higher strength domain wall pinning centers. This configuration results in a large number of low intensity Barkhausen jumps which fall below the noise level of the signal coil and hence are not included in the total integrated signal.

The Barkhausen signal emission rates are plotted in Figure 15. The emission rate is the derivative of the filtered integrated Barkhausen signal. The filtering is performed by averaging over approximately 1 A/m to remove the fluctuations inherent in the Barkhausen jumps. For sample EI-2 the contributions from the surface and bulk contributions, which were clearly apparent in the incremental permeability curves of Figure 13, can be seen. In the Barkhausen signal, the contribution from the surface is enhanced because this measurement is more sensitive to the surface. Small additional peaks for U-4 and IF-18 are also evident and these also may be due to a surface layer which has different magnetic (and mechanical) properties that the bulk of the material. For U-4 and IF-18 these surface layers make a much smaller contribution and are not clearly evident in the micrographs.

The magnetic measurements made thus far show some interesting qualitative features. Quantitatively, the best correlation between magnetic and mechanical properties seen in these results is between the coercive force and the yield stress. This is displayed in Figure 16. In the same Figure it can be seen that any correlation between initial susceptibility and yield stress is poor. From Figure 14 it can be seen that the total Barkhausen signal clearly distinguishes the higher strength 0.06 w/o C steel EI-2 from the very low carbon U-4 and IF-18 samples. The total Barkhausen signal does not distinguish between the different strengths of the latter two steels. The higher strength of the U-4 sample is due to its finer and more uniform grain size. This finer, more uniform, structure is clearly reflected in both the Barkhausen signal and the hysteresis curves which give smaller widths and higher peaks for the U-4 sample.

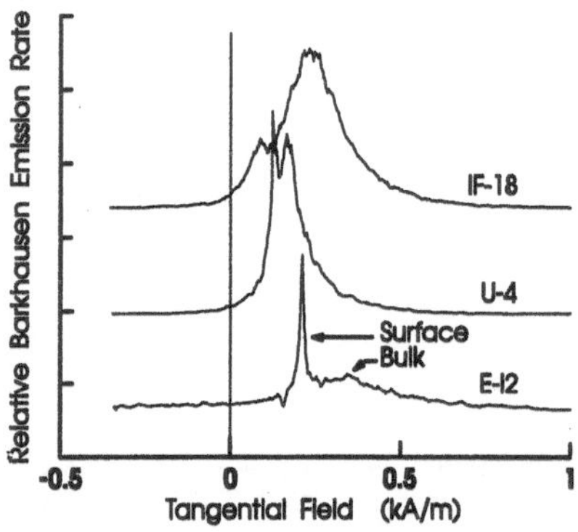

Figure 15. Barkhausen signal emission rates (relative units) vs tangential field. (Curves displaced for clarity).

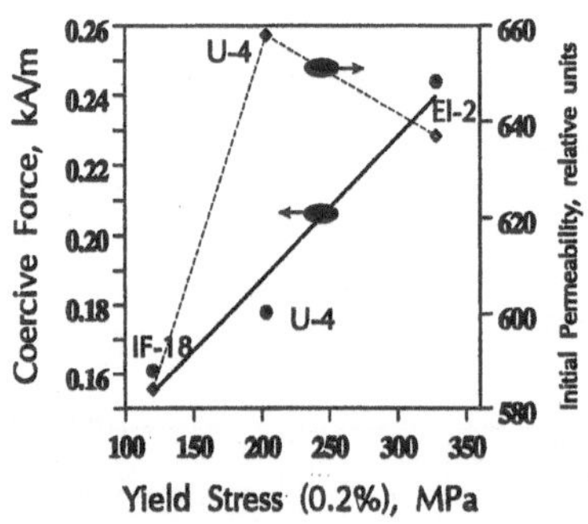

Figure 16. Coercive force and initial permeability vs. yield stress for the three samples.

CONCLUSIONS

(1) Three different low-carbon steel sheets were selected as the source materials for testing specimens in the evaluation of possible correlation between mechanical and magnetic properties.

(2) Mechanical properties and microstructures of the steels were characterized using a large number of specimens for each sheet. The amount of scatter in the parameters checked was found to be very small within each sheet.

(3) The steels differ from each other in their properties, much beyond the scatter within each steel, and as such can be used to obtain the magnetic response for different strength level materials.

(4) Best-fit methods were used in order to obtain the linear Correlations between the yield stress of the steels and other metallurgical parameters such as grain size, hardness, and the plastic strain ratio. Any correlation found between the yield stress and the magnetic properties can be used also as a measure of these other metallurgical parameters which have great importance during the production of steel sheets.

(5) The magnetic and mechanical properties are found to be clearly correlated but not is a simple way. The best magnetic indicator for this set of steels was found to be the coercive force. To obtain practical correlations it will be necessary to determine more than a single property.

ACKNOWLEDGMENTS

The authors would like to acknowledge Mr. Leonard C. Smith for metallographic assistance, Mr. Denzil Mathews for technical assistance, and Drs. R. Fields and R. DeWit for helpful discussions. We would also like to recognize the support of Dr. H. Thomas Yolken, Chief of the Intelligent Processing of Materials Division.

REFERENCES

[1] R. Ranjan, D.C. Jiles, O. Buck, and R.B. Thompson, *J. Appl. Phys.* **61**, 3199 (1987).
[2] K. Tiito, *Nondestr. Test. Eval.* **5**, No. 1, 27-37 (1989).
[3] G.L. Burkhardt and H. Kwun, Review of Progress in Quantitative NDE 8, D.O. Thompson and D.E. Chimenti, eds., Plenum Press, New York, (1989), p. 2043.
[4] D.C. Jiles, Review of Progress in Quantitative NDE 9, D.O. Thompson and D.E. Chimenti, eds., Plenum Press, New York, (1990), p. 1821.
[5] I. Altpeter and P. Holler, Review of Progress in Quantitative NDE **9**, 1837 (1990).
[6] L.J. Swartzendruber and G.E. Hicho, *Res. Nondestr. Eval.* **5**, 41 (1993)

QUANTITATIVE ULTRASONIC CHARACTERIZATION OF INTERFACIAL ADHESION IN METAL-POLYMER-METAL MULTILAYER COMPOSITES

L. Piché,[1] D. Lévesque,[1] P. Deprez,[2] A. Michel,[2] and J. Tatibouët[1]

[1]Industrial Materials Institute, Mortagne, Boucherville, Qc, J4B 6Y4, Canada
[2]Laboratoire Matériaux Organiques, CNRS, BP 24, Vernaison 69390, France

INTRODUCTION

Recent theoretical advances in adhesion science[1,2] have been accompanied by the development of new analytical tools[3] with sufficient resolution and sensitivity to probe structural details and forces near interfaces. These techniques, however, are not applicable to the characterization of adhesion in real structures. Indeed, for polymer adhesion[4,5], instead of only an interface plane, the adhesive and the adherend also interact through an interphase region where the homogeneous phases blend into one another. Interfacial adhesion therefore involves interplay of interface and interphase behavior. On the other hand, practical adhesion is defined by the energy required to separate the adherend and the adhesive and relates to complex nonreversible processes that concern the interface and the interphase, but also the bulk materials themselves. Obviously, nondestructive methods that could probe interfaces and interphases in situ, would constitute a great asset. Also, provided they discriminate between interfacial and bulk regions, these techniques could serve to correlate small scale properties near interfaces and large scale behavior in practical adhesion. In this context, ultrasonics stands out as most promising[6]. Although, ultrasonic techniques are well established for detection of delaminations and flaws[7], it is only recently that efforts were made at characterizing interfaces. Recently, the "Journal of Nondestructive Evaluation" devoted a complete issue at ultrasonic evaluation of microdefects near interfaces[8], and the "Journal of Adhesion Science and Technology" presented a series on probing mechanical properties of bonded joints[9]. Surprisingly however, few works addressed the question of materials physics, except for effects related to thickness or curing in epoxy joints[10-14]. Epoxies and thermosets constitute a major class of structural adhesives, but in this case, polymerization introduces an additional dimension not clearly related to adhesion per se. Likewise important for applications, thermoplastics also undergo structural changes but here the process occurs near interfaces and is intrinsic to the adhesion problem.

Recently[3,15], we presented an ultrasonic method for the quantitative and nondestructive characterization of interfacial adhesion that incorporates recent theories for polymers near surfaces. We demonstrated the method through experiments on metal/polymer/metal samples, where we modified interfacial adhesion by chemical action

Nondestructive Characterization of Materials VI
Edited by R.E. Green, Jr. *et al.*, Plenum Press, New York, 1994

on the polymer and on the substrates. We showed that the specific stiffness of the interphase, served as a materials constant for describing interfacial properties. Here we investigate adhesion in metal/polymer/metal multilayers with respect to materials properties. We discriminate between interfacial and bulk properties and identify interface and interphase behavior in relation to processing parameters and practical adhesion.

METHODOLOGY

The making of bonds

Adhesion is commonly taken in a broad sense to describe the sticking together of materials. Layered composites, where two materials are bonded through an intermediate layer of polymer adhesive, involve two interfaces prone to adhesive failure and three regions to cohesive failure. Here, we consider adhesion in reference to bonding at the interface between two materials in intimate contact: the adherend and the adhesive. We shall be concerned with the adhesion of thermoplastic polymers, namely polypropylene(PP). These are large molecules composed of $N \approx 1000$ monomer units having a length $a \approx 0.3$nm, that are covalently linked in a chain. Adhesives are applied as melts, where the molecules form interpenetrated Gaussian coils with average radius $R_g \approx a N^{1/2} \approx 10$nm. Intermolecular forces are associated with van der Waals interactions, but also with entanglements of the long chains. The polymer liquids may contour surfaces and, through capillary pressure, they may penetrate into pores, creating mechanical keying. Efficiency of mechanical interlocking depends on wetting, since unwetted areas cause stress concentration that initiate breakage. Wetting is related to interfacial tension which is determined by the nature of atomic/molecular binding. Metals feature ordered lattices, while polymers have disordered, entangled morphologies, resulting in complex interactions whereby the antagonistic structures accommodate. Molecules must diffuse or reptate (with characteristic times $\tau_D \propto N^2$ for short chains or $\tau_R \propto N^{3.4}$ for entangled chains)[16] so that segments become anchored on the substrate. At low grafting density, Σ, the anchored molecules form coils near the surface, with average thickness $d_i \approx R_g$. As Σ increases, the coils overlap and interchain stresses associated with attractive osmotic forces and repulsive entropic forces cause the molecules to stretch out, away from the surface. Finally, the hydrodynamic layer in the melt is frozen in during solidification, as suggested from results[17-19] with the surface force apparatus (SFA). This bonding layer serves to transfer mechanical load between the adherend and the adhesive per se.

Materials

Substrates made from sheets of stainless steel with thickness $d_s = 400\mu$m, were cleansed and etched in an acid solution. In some cases this was followed by anodizing in a sulfochromic solution. The rugosity was ≈200nm and, whilst we could not observe pores on the cleansed only substrates, we indentified mesopores with diameters between 4 and 10nm on the anodized surfaces. Also anodized surfaces were enriched with Cr and O and OH^-. Here, adhesion mostly relates to the chemical properties of the surfaces. Substrates made from aluminum sheets, $d_{Al} = 318\mu$m, were pretreated by etching in sodium dichromate acid and anodizing in phosphoric acid. The process removed original oxides and produced a new, cellular-like, oxide layer with a thickness ≈400nm and pore diameters ≈50nm. Molecular segments may penetrate the large pores and become embedded in the oxide layer, so that mechanical keying also contributes to adhesion.

Polypropylene (PP) is a non-polar and chemically non-reactive molecule with low wettability and very weak adhesion to metals. We investigate changes in adhesion consequent to grafting with 6.5% weight of glycidyl methacrylate (MAG), containing ester and epoxide functional groups. Films of PPMAG were prepared by molding at constant pressure, $p = 1.0$MPa, and temperature, $T = 190$°C, during 5min. The multilayer assemblies were pressed also at $p = 1.0$MPa and $T = 190$°C during 5min. The thickness of the polymer adhesive were measured carefully, but on the average $d_p \approx 80.0\mu$m.

Ultrasonic Techniques

First, we investigated[20-22] the ultrasonic properties of the PPMAG material with regards to viscoelasticity. Indeed, the ultrasonic strain modifies thermal equilibrium in the polymer, and irreversible internal processes occur on a time scale, τ, during which the system relaxes towards equilibrium, causing dissipation due to increased internal entropy. The small ultrasonic strains, probe small movements of segmental units in the polymer chains. Therefore τ is a Rouse relaxation time[23], quite different from that in usual rheology experiments, which is closer to τ_D, or τ_R. The steady state solution for the generalized complex dynamic modulus, $M^*(\omega) = M'(\omega) + iM''(\omega)$, is :

$$\frac{M'(\omega) - M_0}{M_\infty - M_0} = \frac{(\omega\tau)^2}{1 + (\omega\tau)^2} \qquad \frac{M''(\omega)}{M_\infty - M_0} = \frac{\omega\tau}{1 + (\omega\tau)^2} \tag{1}$$

where ω is the angular frequency, M_0 the relaxed modulus in the zero frequency limit and M_∞ the unrelaxed modulus at infinite frequency. In turn, M^* is related to the density, ρ, the sound velocity, v, and the attenuation, α, through :

$$M' = \rho v^2 \qquad M'' = 2\rho\alpha v^3 / \omega \tag{2}$$

Depending on polarization $M^*(\omega)$ represents the longitudinal, $L^*(\omega)$, or the shear, $G^*(\omega)$, modulus for longitudinal (P-wave) or transverse (S-wave) respectively. Measurements for density, sound velocity and attenuation were carried out at constant pressure, $p = 1.0$MPa in the temperature range from $T = 0$ to 220°C. We observed a strong relaxation feature with a maximum near 60°C corresponding to $\omega\tau \approx 1$. There was evidence of a broad distribution of relaxation times, but near $T = 23$°C, the results could be described through Eq. (1) using the nominal values from Table 1.

For the multilayers, we used a water immersion technique. The temperature was

Table 1. Ultrasonic data for bulk materials, in terms of complex modulus.

	Density, ρ kg/m³	Real Modulus Longitudinal L' GPa	Real Modulus Shear G' GPa	Loss Modulus Longitudinal L'' GPa	Loss Modulus Shear G'' GPa
Water	1000	2.20	—	≈ 0	—
Aluminum	2700	110.4	25.9	$2.2\ 10^{-9} f$	$0.8\ 10^{-9} f$
Steel	7930	263.6	75.42	$43.5\ 10^{-9} f$	$18.5\ 10^{-9} f$
PP relaxed	926	1.80	$0.5\ 10^{-3}$	τ =500 ns	
PP unrelaxed	"	7.53	2.00		

constant, T=23±0.5°C, and samples were protected from infiltration of humidity. We used broadband transducers with frequencies from f=5 to 50 MHz; effects due to diffraction were minimized and the beams were Gaussian, with 2.0mm half-widths. Transducers were energized with high voltage spikes and a broadband (100 MHz) amplifier served as a receiver. Signals were digitized in a computer and Fourier-transformed into the frequency domain. The resulting spectrum was deconvolved with a reference, which produced the true transfer function for the sample proper. We also used this technique to measure the P-wave and S-wave ultrasonic properties of the substrates, in Table 1.

MEASUREMENTS AND RESULTS

Specific Stiffness of the Boundary Layer

In Fig. 1, the heavy line illustrates the experimental frequency spectrum for the transmission coefficient at normal incidence, T_P (24dB between tick marks). Propagation of sound in multilayers is complex due to changes in acoustic impedance between layers. We accounted for boundary conditions and solved the equations[24] for the acoustic field in each layer. The solution[3,15] involves a great deal of analytical work, but the operational

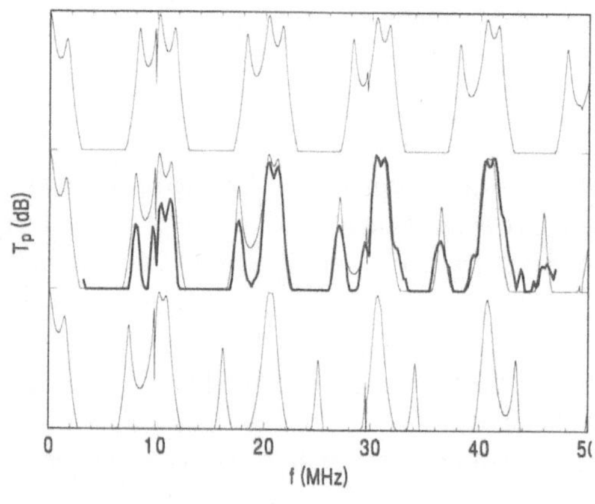

Figure 1. Transmission coefficient, T_P, (longitudinal waves) in Aluminum/PPMAG/Aluminum multilayer. Upper trace is a simulation relating to rigid bonding; center trace: light line is a simulation for an interface layer with S_P =11.2x10^{14} N/m^3 and heavy line is an experimental spectrum; bottom trace is a simulation for an interface layer with S_P =5.0x10^{14} N/m^3.

result is obtained numerically by using the Transfer Matrix Approach. For this, we built a computer code that is efficient, yet very robust. Based on data in Table 1, the top trace in Fig. 1, describes simulation results assuming that the layers are welded together. This approach cannot duplicate the experimental spectrum, and this is definitely not attributable to faulty values of bulk properties in Table 1. In line with theoretical ideas for polymers near interfaces, we add interfacial layers with unknown thickness, d_i, and modulus, M_i*. However, we find that, whenever d_i is smaller than the wavelength, the thickness and modulus are not independent, instead, the specific stiffness, $S* \equiv (M_i*/d_i)$, is the only relevant variable in the problem. To ensure continuity between the boundary layer and the bulk polymer, we set $M_i''/M_i' = M''/M' \approx 10^{-2}$, in which case, $S* \approx S \equiv (M_i'/d_i)$ is a real quantity. With only $S_P \equiv L_i'/d_i$ and $\tau_{i,P}$ as fitting parameters, the light line superimposed on the experimental spectrum in Fig. 1, corresponds to simulation results with $S_P = 11.2 \times 10^{14}$ N/m^3 and $\tau_{i,P} = \tau$. The bottom trace in Fig. 1 is a simulation with $S_P = 5.0 \times 10^{14}$ N/m^3 to illustrate weak bonding. In other experiments with shear waves, we found: i) $S_s = (G_i'/d_i) \approx (G'/L')S_P$, pointing out that adhesion is related to elongational forces, and not only to shear behavior, ii) $\tau_{i,S} \approx \tau_{i,P} \approx \tau$, suggesting that the details for small molecular movements of confined molecules remain essentially the same for longitudinal and shear waves, due to the random nature of the structure.

Interfacial Adhesion

Ultrasonic measurements provide average properties and therefore the specific stiffness parameter, $S* = M_i*/d_i$, is a mean field quantity that serves to match the different properties in the substrate and the adhesive. Near the interface, the modulus reflects the strength of chemical bonding of MAG moieties and the number density of grafted molecules, Σ, but also the monomer-wall interactions that may be attractive or repulsive. In Table 2, we show results using the same polymer but different substrates. For stainless steel substrates, changes in adhesion are governed by chemical properties of the interface which condition probabilities for grafting, i.e. Σ. In comparison the results for aluminum demonstrate the efficiency of mechanical interlocking in enhancing interfacial adhesion. When relating to interfaces, the natural length scale is $d_i \approx R_g \approx 10$nm, and we found that this also corresponds to the thickness of residue films after delamination. The results for L_i' calculated for $d_i \approx R_g \approx 10$nm in Table 2 represent very small moduli, but are within realistic bounds : i) L_i' remains larger than the van der Waals limit for adhesion forces, ii) L_i' compares well with SFA measurements[17-19] in confined molecular systems. Because all materials characteristics are constant, changes in adhesive behavior are essentially linked to the different surface properties of the substrates.

Table 2. Effect due to nature of substrate on interfacial adhesion parameter, S_P; and longitudinal modulus, L_i' calculated for the distance from the interface $d_i \approx R_g \approx 10$nm.

	Cleansed Steel	Anodized Steel	Aluminum
$S_P \equiv L_i'/d_i$, N/m^3	1.5×10^{14}	7.0×10^{14}	39.0×10^{14}
$L_i'(d_i \approx R_g)$, MPa	1.5	7.0	39.0

The Interphase Region

After attachment to the substrate, the arrangement of molecules near the surface slowly evolves until an equilibrium entanglement network is formed. This involves interdiffusion of molecules, with a time scale, τ_R, that is mostly governed by reptation of chains through entanglements. Compared to Rouse type relaxation, $\tau \approx 10^{-9}$s, the time scale for reptation is much longer, $\tau_R \approx 10^2$s, depending on temperature. While keeping the temperature constant at $T=190°C$, we prepared several multilayers using different processing times from 30.0 s to 45 min., above which the polymer would start degrading. In Fig. 2, we show the evolution of specific stiffness, S_P, as function of time, t, and the solid curve is a best fit to a classical Fickian diffusion equation. The comportment for $t>\tau_R$ refers to the cooperative movement of chains and describes the build up of the entanglement network in the interphase, a short distance away from the interface. In order to evaluate the thickness, δ of the interphase region, we assume continuity of stress, σ, at a short distance, d_i, from the wall. If x is the deformation, we may write $\sigma = L_i' x / d_i$, and correspondingly, for the bulk material $\sigma = L'\varepsilon$, with ε being the strain. Then $x = \delta\varepsilon$, where $\delta = L'/S_P$ is a linear interpolation length that describes the thickness of the interphase. In practice, we found $5.0 \times 10^{14} < S_P < 100 \times 10^{14} N/m^3$ for the limits of weak and strong adhesion; in turn, this serves to set bounds for the thickness: $15 > \delta > 0.7 \mu m$. Thus, for short processing times, the small values of S_P reflect the loose arrangement of molecules in the interphase material, hence the large density gradient near the substrate. At long times, the molecular density near the interface increases, and the interphase network more closely matches that in the bulk polymer.

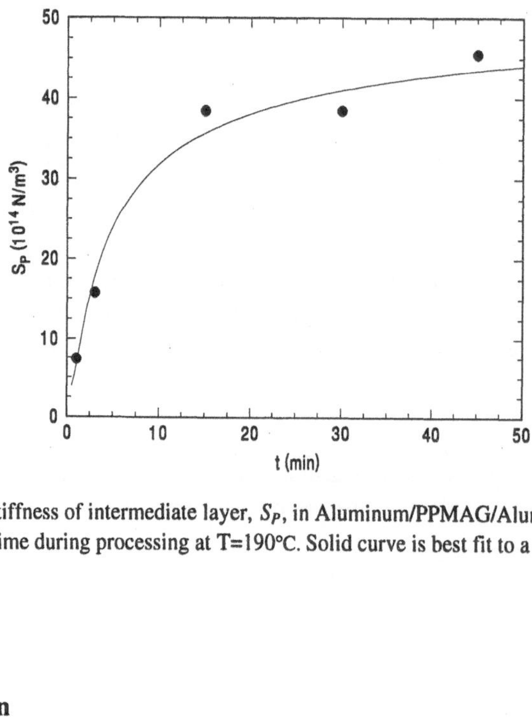

Figure 2. Specific stiffness of intermediate layer, S_P, in Aluminum/PPMAG/Aluminum multilayer as function of contact time during processing at T=190°C. Solid curve is best fit to a diffusion equation.

Practical Adhesion

The energy, for separating the adhesive from the substrate relates to complex phenomena, which can in principle take place without breaking of primary bonds. Intuitively, one expects values of the same order as typical surface energies, about 0.025 J/m^2, instead of about 10 J/m^2 for breaking chains. In practice, however one measures values that are much larger, by orders of magnitude. Also, adhesion energy is found to vary with the rate of separation. Therefore, adhesion is enhanced by energy dissipation within the adhesive, mostly because inelastic effects within the material help reduce stress intensity near crack tips. We performed "near-equilibrium" destructive peeling tests on the samples described above. The work, W, for separation per unit area of interface, was calculated from the peel force per unit width of the sample. In Fig. 3 we plot the results for the work of detachment, W, versus the specific stiffness, S_P. As seen, there exits a definite correlation between both quantities: from a best fit to the data, we found $W \propto S_P^2$, as illustrated by the dotted line in Fig. 3. Tentatively, we write the strain energy density near the interface as: $U \approx \sigma^2 / 2L_i'$. Then, as described above, the stress may be written: $\sigma \approx S_P x$, leading to $U \approx S_P^2(x^2/2L_i')$. The stored elastic energy may be used, for example, to pull the chain out from the interphase. Here, with the hypothesis that failure was mainly due to chain pullout, x is a criterion for detachment, such a chain length, and since the same material was used throughout, x is a constant for our samples. On the other hand, L_i' characterizes the region of $d_i \approx R_g \approx 10$nm near the substrate, and is independent of contact time. Also we may use the approximation that L_i' remained constant during delamination. Then $x^2/2L_i'$ is a constant for the experiment, so that the condition for chain pullout is simply $W \propto U \propto S_P^2$, in agreement with the results in Fig.3. Obviously, our approach is simplified, however, it satisfies intuitive reasoning and other more elaborate findings[25,26].

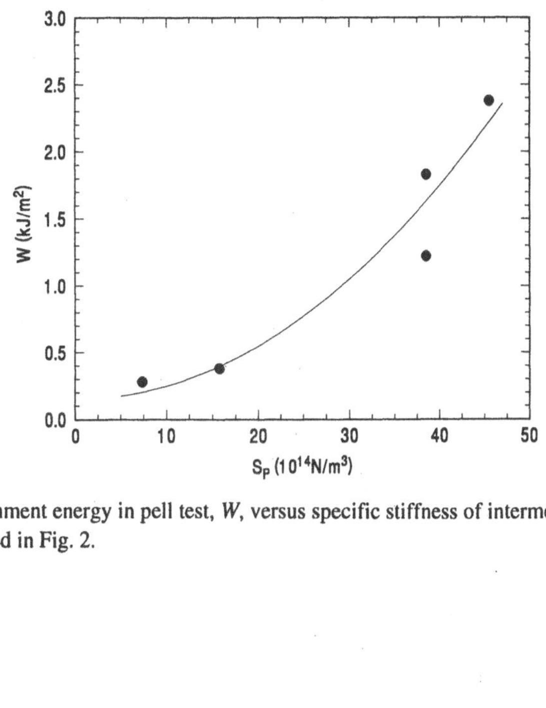

Figure 3. Detachment energy in pell test, W, versus specific stiffness of intermediate layer, S_P, for samples described in Fig. 2.

CONCLUSION

We presented a nondestructive technique for the quantitative characterization of interfacial adhesion. The method hinges on mapping data from ultrasonic measurements on a newly developed model for the propagation of sound in multilayered media. We included recent concepts for polymers near interfaces by allowing for a viscoelastic interface layer with a complex modulus, $M_i{}^*$, and a thickness d_i. We showed that the specific stiffness, $S^* = M_i{}^*/d_i$, was the true materials properties for describing interfacial behavior. Through experiments with different substrates we showed that S^* was related to surface forces with ranges of the order $d_i \approx R_g \approx 10$ nm. Measurements on multilayers made by varying the processing time, suggested the presence of an interphase region with a thickness $\delta \approx 0.5$ to 10 μm, where the molecular network accommodates the different properties of the substrate and the bulk polymer. Finally, we demonstrated that small scale behavior, described by $S*$, was correlated to practical adhesion, defined by the detachment energy, W, measured in peel tests. Here we mention that the problem for curing of epoxies[10-14] is different and more complicated, because most materials properties depend on the advancement of the crosslinking reaction. The present work is open ended and is expected to lead to novel applications for nondestructive evaluation. Already, the method has been successful for characterizing adhesive seals in the food packaging industry; in this case, $S*$ was directly correlated to processing parameters. However, a great deal of more fundamental work needs to be done, to fully understand the true meaning of $S*$, with respect to interfacial energy.

REFERENCES

1. P.G. de Gennes, *Macromolecules* 13:1069-1075 (1980).
2. J.M.H.M. Scheutjens, and G.J. Fleer, *J. Phys. Chem.* 84: 178-190 (1980).
3. D. Lévesque, A. Legros, A. Michel, and L. Piché, *J. Adhes. Sci. Tech*, in press.
4. J.D. Miller, and H. Ishida, Adhesive-adherend interface and interphase, in "Fundamentals of Adhesion", Chap. 10, pp. 291-324, L.-H. Lee (Ed.) Plenum Press, New York (1991).
5. K.L. Mittal, *Polym. Eng. Sci.* 17: 467-473 (1977).
6. G.M. Light, and H. Kwun, "Nondestructive Evaluation of Adhesive Bond Quality", Nondestructive Testing Analysis Center, Southwest Research Institute, San Antonio, TX. (1989).
7. R.B. Thompson, and D.O. Thompson, *J. Adhes. Sci. Tech.* 5: 583-599 (1991).
8. *J. Nondestr. Eval.* 11: 109-250 (1992).
9. *J. Adhes. Sci. Tech.* 5: 601-666 (1991).
10. P.B. Nagy, *J. Adhesion Sci. Technol.* 5: 619-630 (1991).
11. D. Jiao, and J.L. Rose, *J. Adhesion Sci. Technol.* 5: 631-647 (1991).
12. W. Wang, and S.J. Rokhlin, *J. Adhesion Sci. Technol.* 5: 647-666 (1991).
13. R.E. Challis, R.P. Cocker, A.K. Holmes, and T. Alper, *J. Appl. Polym. Sci.* 44: 65-81 (1992).
14. P. Fraisse, F. Schmit, and A. Zarembowitch, *J. Appl. Phys.* 78: 3264-3271 (1992).
15. D. Lévesque, and L. Piché, *J. Acoust. Soc. Am.* 92: 452-467 (1992).
16. P.G. de Gennes, "Scaling Concepts in Polymer Physics" Cornell University Press, Ithaca, NY, (1979).
17. A.M. Homola, H.V. Nguyen, and G. Hadziioannou, *J. Chem. Phys.* 94: 2346-2351 (1991).
18. J. Van Alsten, and S. Grannick, *Macromolecules* 23: 4856-4862 (1990).
19. J.C. Israelachvili, S.J. Kott, and L.J. Fetters, *J. Polym. Sci.: B: Polym.Phys.* 27: 489-502 (1989).
20. Massines, F., Piché, L. and Lacabanne, C., *Makromol. Chem. Macromol. Symp.* 23: 121-137 (1989).
21. Piché, L., Massines, F., Hamel, A. and Néron, C., "Ultrasonic technique for characterizing polymers under simulated processing conditions", Can. Pat. N^0 1,264,195 (1990); US Pat. N^0 4,677,482 (1987).
22. J. Tatibouët, and L. Piché, *Polymer* 32: 3147-3152 (1991).
23. J.D. Ferry, "Viscoelastic Properties of Polymers", John Wiley, New York (1980).
24. L.M. Brekhovskikh, "Waves in Layered Media", Academic Press, New York (1980).
25. P.G. de Gennes, *Can. J. Phys.* 68: 1049-1054 (1990).
26. A.N. Gent, *Int. J. Adh.Adhes.* April issue, 175-180 (1981).

SURFACE ROUGHNESS AND ULTRASONIC MATERIALS CHARACTERIZATION

Peter B. Nagy,[1] Gabor Blaho,[1] and James H. Rose[2]

[1]Department of Welding Engineering
The Ohio State University
Columbus, Ohio 43210, U. S. A.
[2]Center for NDE
Iowa State University
Ames, Iowa 50011, U. S. A.

INTRODUCTION

Ultrasonic materials characterization relies primarily on measurements of three quantities: (1) backscattered power from the microstructure, (2) frequency-dependent attenuation of the transmitted coherent wave, and (3) shifts in the sound velocity. Surface roughness tends to randomize the phase of the reflected and transmitted waves, which can significantly distort the ultrasonic measurements. For example, uncorrected surface roughness effects can cause substantial underestimation of the porosity in cast aluminum samples.[1] There are three basic effects of surface roughness which should be accounted for in evaluating the data measured by an unfocused transducer. First of all, the coherent reflection and transmission through the interface are attenuated by scattering at the rough surface. Second, the incoherent background scattering from inherent material inhomogeneities in the sample is relatively weakly affected by surface roughness. It does not decrease in proportion to the coherent signals and sometimes even increases slightly. Third, incoherent backscattering at the rough interface produces a slowly decaying tail in the reflection from the front surface, which adds to the incoherent background scattering, especially at small depths below the surface. It has been shown that, for the purposes of NDE, all three effects can be accurately described by the so-called phase-modulation technique. In this simple approximation, explicit analytical results can be derived for (i) the surface roughness induced attenuation of the coherent signals, (ii) the incoherent material noise, and (iii) the level of the additional incoherent surface noise. In this paper, we shall review the above mentioned three main effects of random surface roughness on ultrasonic materials characterization and present simple theoretical predictions and experimental data in order to quantify the resulting changes in the measured parameters.

COHERENT REFLECTION AND TRANSMISSION

Surface roughness tends to randomize the phase and to alter the magnitude of the transmitted and reflected waves. For roughnesses of relatively small curvature, i. e., where the r.m.s. height, h, of the roughness is small compared to its variation (correlation length) along the surface, the randomization of the phase primarily determines the effect on the transmission process.[2] The roughness induced change in the local wave amplitude across the interface is relatively unimportant. This insight leads us to examine our experimental results in terms of a method that considers only the phase variation, the so-called phase-screen approximation.[3] The experiments that will be reported in this section measure the loss (or attenuation) induced by the rough surface in the coherently transmitted and reflected waves. For the longitudinal transmitted wave, the loss is defined as the ratio of the transmission coefficients of the rough and a smooth, but otherwise identical, interface:

$$A^L(\omega,\theta_I) = ln\,\frac{T_o^L(\theta_I)}{T^L(\omega,\theta_I)}\,, \tag{1}$$

where ω is the angular frequency, θ_I denotes the angle of incidence measured with respect to the normal, and the superscript refers to the type of the transmitted wave (longitudinal) and the zero subscript indicates a smooth reference surface. The loss of the transmitted shear wave, A^T, and the reflected pressure wave in water, A^R, are defined in analogy with (1). In the phase-screen approximation, these losses can be written as[4]

$$A^{R,L,T} = h^2\,\omega^2\,C^{R,L,T}, \tag{2}$$

where

$$C^R = 2\,[cos(\theta_I)/c_w]^2, \tag{3}$$

$$C^L = \frac{1}{2}\,[cos(\theta_L)/c_L - cos(\theta_I)/c_w]^2, \tag{4}$$

and

$$C^T = \frac{1}{2}\,[cos(\theta_T)/c_T - cos(\theta_I)/c_w]^2, \tag{5}$$

where c_w, c_L, and c_T are the compressional wave velocity in the water and the longitudinal and shear velocities in the solid, respectively. θ_L and θ_T denote the longitudinal and shear refraction angles, which can be easily calculated from Snell's law.

Equations 1-5 provide simple explicit formulas for the scattering induced loss of the coherent reflection and transmission coefficients of a slightly rough interface in the phase-screen approximation. In order to verify the accuracy of these results, the same coefficients were numerically evaluated using the second-order approximation of Kuperman and Schmidt.[5] This approximation is expected to be good if the r.m.s. roughness is sufficiently small. The results of the second-order approximation can be cast in the same form as Eq.(2). However, the multiplicative functions $C^{R,L,T}$ now have to be changed to reflect the more complex nature of the calculations. We define the reduced losses (in the second-order approximation) by

$$A^{R,L,T} = h^2 \omega^2 K^{R,L,T}. \tag{6}$$

Explicit formulas cannot be given for the K's, rather they must be evaluated from certain integrals given in Ref. 5. Detailed calculations show that $K^{R,L,T}$ weakly depend on the ratio of the density of water to that of the solid and the ratio of the r.m.s. height to the surface-roughness correlation length (h/L).

Equations 2 and 6 show that the different losses depend on the r.m.s. height and the frequency as simple quadratics. This dependence can be removed naturally by normalizing the angular-dependent reflection and transmission losses to the reflection loss at normal incidence:

$$N^{R,L,T}(\omega,\theta_I) \equiv \frac{A^{R,L,T}(\omega,\theta_I)}{A^{R}(\omega,\theta_I = 0)}, \tag{7}$$

where

$$A^{R}(\omega,\theta_I = 0) = 2\,\frac{\omega^2 h^2}{c_w^2}. \tag{8}$$

In order to study the problem of surface roughness induced attenuation of the transmitted longitudinal and shear waves, we prepared a series of 5" x 5", 1/2"-thick aluminum plates roughened on one side by different surface preparation techniques. This large area facilitated more extensive spatial averaging, while the smaller thickness was necessary so that transmission measurements could be made at relatively high angles of incidence. Table I shows the list of the aluminum samples that we used to study the effects of surface roughness on the coherently transmitted components. Details of our experimental procedure are described in Ref. 4.

Figure 1 shows the surface roughness induced attenuation of the reflected ultrasonic wave at normal incidence as a function of frequency for the eight rough samples listed in Table I. In good agreement with our expectations, the measured attenuation seems to be proportional to the square of frequency. The best fitting f^2 curves (plotted with solid lines in Fig. 1) were used to estimate the r.m.s. surface roughness listed in Table I by using the attenuation coefficient of Eq. (8) at normal incidence. Figure 2 shows typical examples for the surface roughness induced attenuation of the transmitted longitudinal and shear waves at different angles of incidence as functions of frequency for sample #2. Again, the measured attenuation seems to be proportional to the square of frequency, but its value is significantly lower than the reflection loss at normal incidence, which is also shown for easier comparison. It is interesting that the transmission loss of the shear wave is somewhat lower than that of the longitudinal wave. This is due to the fact that the shear velocity is much closer to the compressional velocity in the fluid than the longitudinal velocity of aluminum. As a consequence, the scattering at the irregular liquid-solid interface is also smaller for the shear wave.

The main purpose of these experiments was to verify a single conclusion of great practical importance that follows from our theoretical results. Namely, the ratio of the surface roughness induced transmission loss at oblique incidence to the reflection loss at normal incidence is basically a single, universal function of the angle of incidence, which

Table I List of the aluminum samples used in this study.

Sample	Surface Preparation	Estimated Roughness (micron, r.m.s.)
#1	2 mm steel balls	45.6
#2	10 mm miller head	25.6
#3	>2 mm sand	15.2
#4	>2 mm quartz	12.8
#5	1.8 - 2 mm sand	11.4
#6	1 - 1.8 mm sand	9.9
#7	1 - 1.8 mm quartz	8.7
#8	1 mm steel balls	5.6
#9	smooth (reference)	0

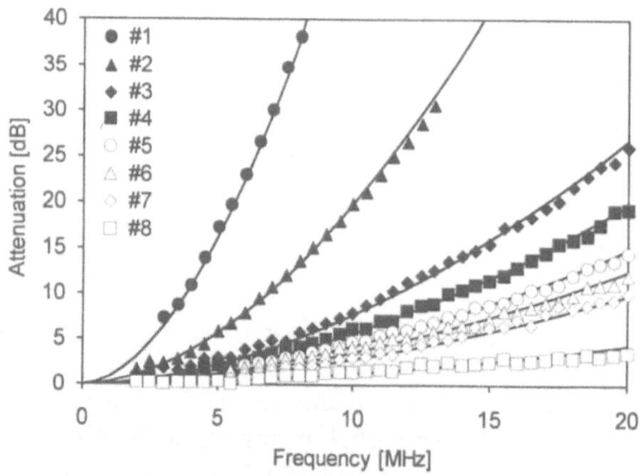

Figure 1 Surface roughness induced attenuation of the reflected wave at normal incidence (solid lines are best fitting f^2 curves).

is determined by the sound velocities in the fluid and the solid. The normalized transmission losses are independent of both frequency and r.m.s. roughness. They are also almost entirely independent of the densities of the fluid and the solid. In the two angular ranges of high practical importance in ultrasonic NDE, they are weakly sensitive to the auto-correlation length of the surface topography and to the angle of incidence.

INCOHERENT MATERIAL SCATTERING

Ultrasound backscatters from the microstructure of the sample. Acoustic backscatter can be used to characterize various material properties of structural solids but it is also the major source of noise in many ultrasonic inspections. In everyday practice of ultrasonic material characterization, we often take advantage of the fact that the average backscattered signal only weakly depends on the misalignment of the probe or on the shape and curvature of the sample. It has been shown experimentally that surface roughness attenuates the backscattered incoherent noise much less than the coherent transmission.[6] This lack of sensitivity can be readily explained by using the phase-screen approximation.[4] Interestingly, this simple theory also predicts that scattering at a rough surface can slightly

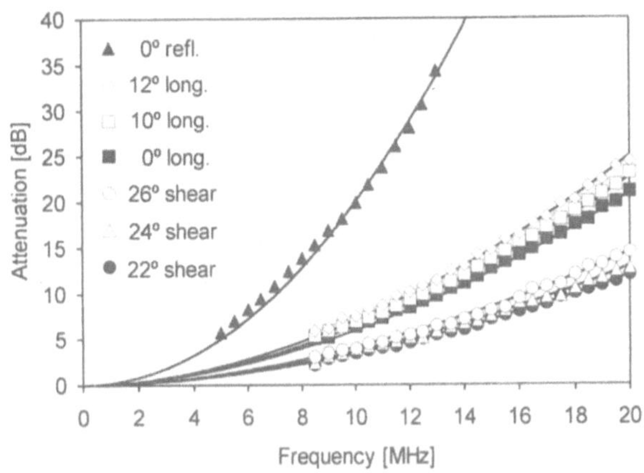

Figure 2 Surface roughness induced attenuation of the double-transmitted longitudinal and shear waves at different angles of incidence for Sample #2. (solid lines are best fitting f^2 curves).

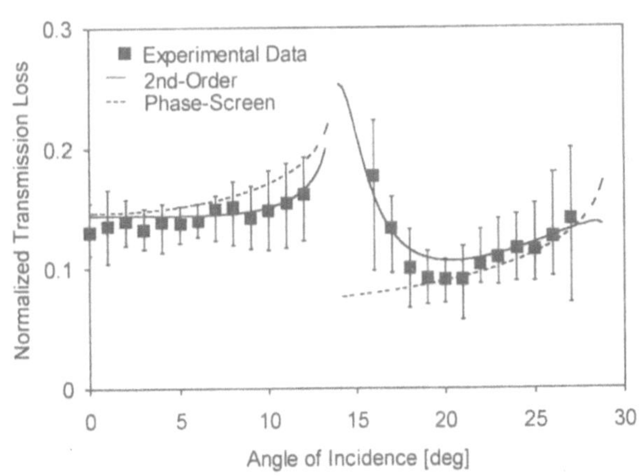

Figure 3 Normalized transmission loss versus angle of incidence averaged for all eight samples listed in Table 1 (in the 2nd-order approximation, $h/L = 100$ was assumed).

increase the incoherent material noise, which has been also experimentally observed.[7] However, as a general rule, surface roughness has a relatively weak effect on the incoherent backscattering from material inhomogeneities, which can be neglected in most typical applications.

Figure 4.a shows the frequency-dependent attenuation of the longitudinal wave backscattering due to the surface roughness at normal incidence in brass. Most of the apparent "attenuation" is due to the remanent random modulation. Although this ripple is greatly reduced by spatial averaging, the relatively small area available for such averaging limited our ability to further smooth the spectra. As can be seen, the attenuation is negligible throughout the whole frequency range of 4-22 MHz. In comparison, the coherently transmitted wave is attenuated by as much as 15 dB at the center of this range and by more than 40dB at the upper end of it. In order to study the surface roughness

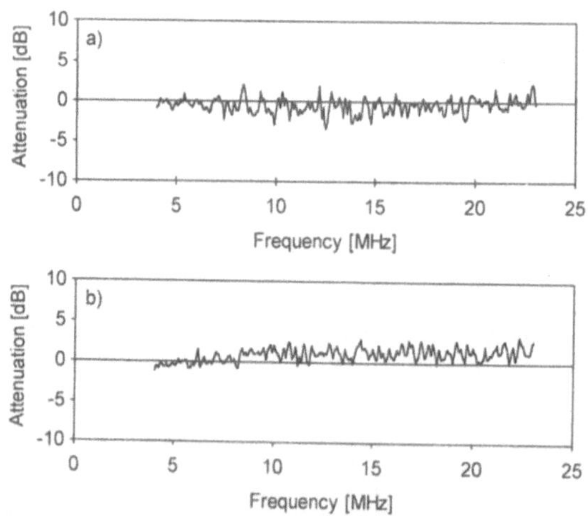

Figure 4 Surface roughness induced attenuation of grain noise in a brass sample of 40 μm roughness at normal (a) and 45° (b) angle of incidence.

induced attenuation of the incoherently transmitted shear component, a similar experiment was carried out above the first critical angle, too. Figure 4.b shows the frequency-dependent attenuation of shear wave backscattering from brass at 45° angle of incidence. Again, the surface roughness effect is negligible with respect to the much stronger attenuation of the coherently transmitted component. At low frequencies, the attenuation seems to be negative corresponding to an actual gain in the grain noise. This effect might be partially due to direct scattering at the surface, which produces an elongated "tail" in the front surface reflection. This "tail" can be observed at normal incidence, too, but there it is much easier to separate it from the grain scattering emanating from the interior of the specimen by proper time gating. This adverse effect of incoherent scattering from the surface on our ability to characterize the material at small depths below the rough interface will be discussed in more detail in the next chapter.

When the material noise is relatively weak, in a shallow region below the rough surface, the incoherent background noise is dominated by direct backscattering from the surface. This effect is demonstrated in Figure 5 showing the r.m.s. noise level in a low-carbon steel specimen as a function of propagation time. Both the grain size and the surface roughness were approximately 20 μm. The measurement was made by averaging the square of the received rf signal over a 1"-by-2" area in a 3 MHz wide frequency range centered around 10 MHz. From the smooth side of the specimen, the material noise slightly decreases between the front- and back-wall echoes, which is partly due to the spread of the acoustic beam and partly to scattering induced attenuation in the sample. From the rough side, the overall noise level is much higher. Close to the surface, the additional surface noise is significantly stronger than the inherent material noise. Due to its faster decay, surface noise becomes negligible with respect to the material noise at large depths.

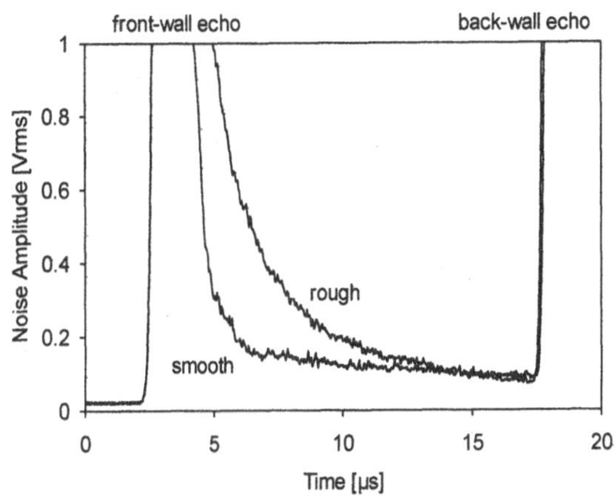

Figure 5 Example of increased near-surface noise in a low-carbon steel specimen.

The excess surface noise produced by direct ultrasonic backscattering from a rough liquid-solid interface was recently studied by both analytical and experimental means.[8] In weakly scattering materials and near the surface, this excess noise can easily overshadow the intrinsic material noise. At oblique incidence, the surface noise is primarily due to strong first-order scattering. At normal incidence, the first-order scattering is strong but very short and the excess noise observed at larger "depth" in the material is primarily due to higher-order scattering. The time-decay of the excess surface noise can be determined from purely geometrical considerations. For small roughness, the absolute level of the surface noise can be easily calculated as a function of frequency and angle of incidence from the r.m.s. surface roughness and the auto-correlation length assuming that the type of the correlation function is known.[8] The r.m.s. roughness is most easily assessed from the surface roughness induced loss of the coherent reflection at normal incidence. Finally, the correlation length and the most appropriate correlation distribution model can be estimated from the angular-dependence of the backscattered signal at a given frequency.

CONCLUSIONS

Three main effects of surface roughness on ultrasonic materials characterization by unfocused transducers were identified and discussed. First and most importantly, the coherent reflection and transmission through the interface are attenuated by scattering at the rough surface. The losses of all three coherent components are proportional to the squares of the r.m.s. roughness of the surface and the inspection frequency. Interestingly, the transmitted wave is usually much less attenuated than the reflected one. Experimental results from samples with r.m.s. roughness between 5 and 50 μm were found to be in good agreement with the predictions of the suggested phase-modulation approximation in the 2 to 20 MHz frequency range. Second, we found that the incoherent background scattering from inherent material inhomogeneities in the sample is relatively weakly affected by surface roughness. Third, incoherent backscattering at the rough surface adds a slowly decaying tail to the front surface reflection, which increases the incoherent background scattering, especially at small depths below the surface. These adverse effects should be taken into consideration in the optimization of ultrasonic characterization methods as well as during the quantitative evaluation of material properties of samples with rough surfaces.

ACKNOWLEDGMENT

This work was supported by the National Science Foundation under grant No. ECO-9008272 and the Center for NDE at Iowa State University. The authors greatly appreciate the valuable contributions of S. Meng, C. Pecorari, M. Bilgen, and L. Adler.

REFERENCES

1. P. B. Nagy, D. V. Rypien, and L. Adler, "Surface roughness effects in porosity assessment by ultrasonic attenuation spectroscopy," *in*: Review of Progress in Quantitative Nondestructive Evaluation, Eds. D. O. Thompson and D. E. Chimenti, Plenum, New York, 6B: 1435-1442 (1987).
2. J. A. Ogilvy, Theory of Wave Scattering from Random Rough Surfaces, Adam Hilger, Bristol (1991).
3. C. Eckhart, "The scattering of sound from the sea surface," J. Acoust. Soc. Am. 25: 556 (1953).
4. P. B. Nagy and J. H. Rose, "Surface roughness and the ultrasonic detection of subsurface scatterers" J. Appl. Phys. 73: 566 (1993).
5. W. A. Kuperman and H. Schmidt, "Self-consistent perturbation approach to rough surface scattering in stratified elastic media," J. Acoust. Soc. Am. 86: 1511 (1989).
6. P. B. Nagy and L. Adler, "Scattering induced attenuation of ultrasonic backscattering," *in*: Review of Progress in Quantitative Nondestructive Evaluation, Eds. D. O. Thompson and D. E. Chimenti, Plenum, New York, 7B: 1263- 1271 (1988).
7. M. Bilgen, J. H. Rose, and P. B. Nagy, "Ultrasonic inspection, material noise and surface roughness," *in*: Review of Progress in Quantitative Nondestructive Evaluation, Eds. D. O. Thompson and D. E. Chimenti, Plenum, New York, 12B: 1767- 1774 (1993).
8. P. B. Nagy, L. Adler, and J. H. Rose, "Effects of acoustic scattering at rough surfaces on the sensitivity of ultrasonic inspection," *in*: Review of Progress in Quantitative Nondestructive Evaluation, Eds. D. O. Thompson and D. E. Chimenti, Plenum, New York, 12B: 1775- 1782 (1993).

MICROSCOPIC DETERMINATION OF SURFACE WAVE VELOCITIES IN HEAT TREATED STEELS BY ULTRASONIC REFLECTIVITY MEASUREMENT

Ikuo Ihara,[1] Tatsuhiko Aizawa,[2] and Junji Kihara [2]

[1] Department of Mechanical Engineering
Nagaoka University of Technology
1603-1 Kamitomioka, Nagaoka, Niigata 940-21, Japan
[2] Department of Metallurgy
University of Tokyo
7-3-1 Hongo, Bunkyo-ku, Tokyo 113, Japan

INTRODUCTION

The measurement of mechanical properties on a microscopic scale is the key technology for the characterization of inhomogeneous materials, as well as small scale materials such as used for a micromechanism. Recently, many investigations on mechanical properties at a small area by surface waves have been carried out using an acoustic microscope and others. The surface wave provides useful information on the character near the surface of the materials. Generally, in investigations by the surface wave, it is very important to measure precisely the surface wave velocity because the velocity is closely related to the mechanical properties such as elastic properties and the state of stress. Since most wave-materials interaction is frequency dependent, an ultrasonic spectroscopy technique is particularly effective for such investigations. Authors have applied the spectroscopy technique based on the ultrasonic reflectivity measurement[1,2,3] for the microscopic characterization of mechanical properties on a surface layer.[4,5] However, basic research on the microscopic measurement by this technique is not yet sufficient. For practical applications of this technique, it is necessary to obtain the fundamental knowledge on the sensitivity and the reliability of the velocity measurement, and a clear physical interpretation for the measured value.

In the present paper, the resolution and the accuracy of the velocity measurement by the present technique are examined. Secondly, based on the investigations of the reflection coefficients for angles in the vicinity of the Rayleigh critical angle and the experimental results with various kinds of heat treated steels, the influence of absorption losses in materialson the measured velocity are considered. Finally, the paper closes with a discussion of problem associated with the microstructure of materials on the microscopic determination of the surface wave velocity of the polycrystalline materials.

Measurement Principle of Surface Wave Velocity

We suppose that the upper half space on a semi-infinite solid is filled with fluid, typically water, and an ultrasonic plane wave is injected on the solid with an incident angle θ. In general, the reflection coefficient for the liquid-solid interface can be given[6,7] as

$$R(k_x) = \frac{\begin{array}{l}[(2k_x^2 - k_t^2)^2 - 4k_x^2\{(2k_x^2 - k_l^2)(2k_x^2 - k_t^2)\}^{1/2}] \\ \quad - i(\rho_w/\rho_s)k_t^4\{(2k_x^2 - k_l^2)/(2k_w^2 - k_x^2)\}^{1/2}\end{array}}{\begin{array}{l}[(2k_x^2 - k_t^2)^2 - 4k_x^2\{(2k_x^2 - k_l^2)(2k_x^2 - k_t^2)\}^{1/2}] \\ \quad + i(\rho_w/\rho_s)k_t^4\{(2k_x^2 - k_l^2)/(2k_w^2 - k_x^2)\}^{1/2}\end{array}} , \tag{1}$$

where ρ_s and ρ_w are densities of the solid and the liquid, k_w is wavenumber in the liquid, k_l and k_t are wavenumbers for longitudinal and transverse waves in the solid, and k_x is tangential component of the wavevector in the fluid which matches the propagation vector of the surface wave. The effects due to absorption losses in materials can be incorporated into $R(k_x)$ by introducing the complex wavenumber k_l and k_t as defined below. Denoting d_l and d_t as attenuation per wavelength for the longitudinal and transverse waves in the solid, each wavenumber can be written as

$$\left.\begin{array}{l} k_w = \omega/v_w \\ k_l = (\omega/v_l)(1 + id_l/2\pi) \\ k_t = (\omega/v_t)(1 + id_t/2\pi) \\ k_x = (\omega/v_w)\sin\theta \end{array}\right\} , \tag{2}$$

where v_w is the sound velocity in the liquid, v_l and v_t the velocities for the longitudinal and transverse waves in the solid, and ω the angular frequency. As the incident angle increases, the reflection coefficient exhibits a phase change of nearly 2π and a minimum of a modulus in the vicinity of a critical angle. These phenomena are due to the generation of a leaky Rayleigh wave at the critical angle. For most liquid-solid interface, the angular position at which the phase change is π or the modulus is minimum is very closely equal to the so-called Rayleigh critical angle θ_c, when the absorption losses in materials are a small quantity.[7] With a known value for v_w, the Rayleigh wave velocity v_{sw} can then be determined by Snell's law:

$$v_{sw} = v_w/\sin\theta_c \ . \tag{3}$$

When the absorption losses in materials are significant, the angular position at which the phase change is π or the modulus is minimum is not always agree precisely with this Rayleigh critical angle.

Measurement of Reflection Coefficient at a Small Area

Figure 1 shows a block diagram of the equipment used for the measurement of the reflection coefficient.[3] The equipment has an ultrasonic sensor consisted of a transmitter with a planar lens and a receiver with a spherical lens. In this equipment, a broadband impulsive wave is incident on a specimen surface through water, and only a reflected wave component from the specimen surface in focus area of the spherical lens is selectively

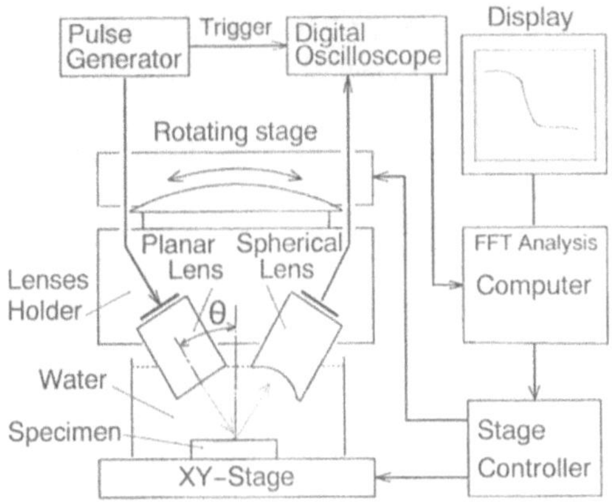

Figure 1. Block diagram of ultrasonic reflectivity measurement system.

received as an effective electrical signal. An output signal of the sensor is proportional to the reflection coefficient of the material even near the Rayleigh critical angle;[3] therefore, we can get the reflection coefficient at an arbitrary incident angle with high spatial resolution by the spherical lens. In the present measurement, the specimen surface held in the horizontal is precisely positioned at a focal point of the spherical lens; and tilting the sensor unit round the focal point, the reflected waves at the incident angle from 22° to 38° are measured at intervals of 0.1°. Applying fast Fourier transform (FFT) analysis to the received waveforms, then, phase and modulus of reflection coefficient at each incident angle are obtained in a frequency range from 40 to 120MHz.

MATERIALS

Cylindrical specimens(diameter 10mm, length 20mm) of structural carbon steel(JIS type S45C: 0.45% carbon content) were procured for this experiment. After homogenizing by heat treatments, nine types of specimens were prepared: quenched, tempered at various temperature and annealed. The specimen surface for the reflectivity measurement was buffed after grinding. Young's modulus and poisson's ratio of each specimen were calculated from the theory of elasticity for isotropic materials using longitudinal and transverse wave velocities measured by the pulse-echo method at 10MHz.

RESULTS AND DISCUSSION

Measurement Accuracy

In determining the Rayleigh wave velocity v_{SW} through the present technique, it is seen from Equation(3) that the measurement accuracy of the velocity v_{SW} is affected by the velocity v_W and the angle θ_c. The measurement error of the Rayleigh wave velocity δv_{SW} is, since v_{SW} is a function of v_W and θ_c, approximately given by

$$\delta v_{SW} = \left|\frac{\partial v_{SW}}{\partial v_W}\right|\delta v_W + \left|\frac{\partial v_{SW}}{\partial \theta_c}\right|\delta\theta_c = v_{SW}\left(\left|\frac{\delta v_W}{v_W}\right| + \left|\frac{\delta\theta_c}{\tan\theta_c}\right|\right). \tag{4}$$

Where δv_W and $\delta\theta_c$ are measurement errors of v_W and θ_c, respectively.

Experiments were made in order to estimate the $\delta\theta_c$; that is, the reflection coefficient measurements on the steel tempered at 973K were repeated 20 times under the same conditions(the temperature in water was maintained at 296.4±0.1K) and the Rayleigh critical angles were determined from the phase change of the reflection coefficient, where an incident angle at which the slope of the phase curve became steepest was defined as the Rayleigh critical angle. The standard deviation of the Rayleigh critical angle obtained from the above experiments had a tendency to increase with frequency. We regarded the standard deviation at 120MHz, 3.3664×10^{-4}, as the $\delta\theta_c$ in Equation(4).

In general, the sound velocity in water was functional in temperature. In a temperature range from 293K to 299K, the variation of the velocity in water with temperature is about $2.83ms^{-1}K^{-1}$. Therefore, fractional variation of the velocity is approximately given by $\delta v_w/v_w \approx 0.0019\delta T$, where δT is the variation of the temperature. Substituting $\delta v_w/v_w$ and $\delta\theta_c$ determined above into Equation(4), we obtain

$$\delta v_{sw} = v_{sw}\left(|0.0019\delta T| + \left|\frac{3.3664\times10^{-4}}{\tan\theta_c}\right|\right). \quad (5)$$

In measurements on steels, putting θ_c=30°, v_{sw}=2980ms^{-1}, and δT=0.1K as an error in the temperature measurement, then δv_{sw} is about 2.31ms^{-1} from Equation(5). Accordingly, in the results presented hereafter, It is considered that the measured velocities contain about ±2.31ms^{-1} of error.

Measurement Resolution

In measuring the velocity at a small area, it is important to understand quantitatively the spatial resolution of the measurement because the spatial resolution depends on the operating frequency and the degree of focusing ability of the spherical lens. To estimate the spatial resolution, an experiment was made using the quenched steel with a sharp edge; that is, the sensor was linearly scanned crossing the edge of the specimen at 5μm intervals keeping the incident angle at the Rayleigh critical angle(30.37°), and the reflected wave received at each scanning position was analyzed by the FFT analysis. Figure 2 shows the changes in the amplitude of the reflected wave with the displacement of the sensor. The rapid change in the amplitude curve means that the edge of the steel passed over the focusing area of the spherical lens. Since the slope of the curve in the rapid change region becomes steeper with increasing the frequency, it is clear that the spatial resolution becomes higher with the frequency. The sweep distances over which the amplitude increased from 15% to 85% were about 85μm, 38μm and 26μm for frequency at 40MHz, 80MHz and 120MHz, respectively. Thus, the frequency dependence of the spatial resolution was approximately estimated from the sweep distance.

Reflection Coefficients for Heat Treated Steels

Figure 3 shows the phase curves of the reflection coefficients measured for the quenched steel. The phase changes by 2π rad at a Rayleigh critical angle and no frequency dependence of the phase curve is observed; therefore, Rayleigh wave velocity at each frequency can be precisely determined from the phase curves. Results for tempered steels were almost the same as that obtained for the quenched. On the other hand, the phase curves for the annealed steel shown in Figure 4(a) are anomalous at frequency region higher than about 90MHz; that is, the phase does not fully change 2π rad but only wobbles at the

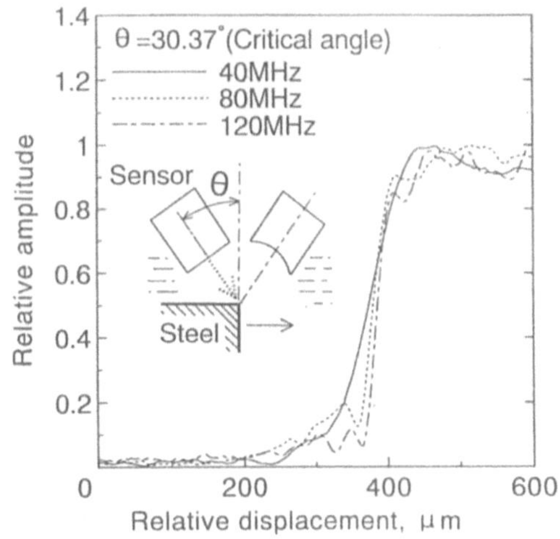

Figure 2. Changes of amplitude for the reflected wave when the sensor was scanned over the edge of steel.

Rayleigh critical angle. This is because the attenuation of the Rayleigh wave is significant at such high frequency owing to absorption losses in the solid.[6,7] In this case, we have difficulty in determining precisely the velocity from the phase curves at high frequency region. However, we can fortunately determine the velocity from the modulus curve of the reflection coefficient even at the high frequency region, because the modulus curves in the entire frequency region exhibit a sharp dip at the critical angle as shown in Figure 4(b).

Before examining the effect of the absorption loss, let us here estimate the attenuation coefficient of the annealed steel. In generally, the behavior of the reflection coefficient in the vicinity of the critical angle is determined by poles and zeros of $R(k_x)$ close to the real axis in a complex k_x plane;[7] and the value of the zero changes with the loss parameter d which is attenuation per wavelength. When the zero crosses the real axis of the complex plane, the loss parameter becomes the critical value d_c; and because the loss parameter d is a function of frequency, the critical value d_c corresponds to the critical frequency f_c which is the frequency of least reflection.[7] Then, the critical loss parameter for longitudinal wave d_{lc} is written as

$$d_{1c} = \lambda_{1c}\alpha_1 = v_1\alpha_1 / f_c . \tag{6}$$

Where λ_{lc} is wavelength corresponds to the critical frequency and α_l is the attenuation of the longitudinal wave. The frequency of least reflection f_c is about 86MHz from Figure 4(b); and the loss parameter d_{lc} was calculated to be 0.03328 on conditions that v_l=5900ms^{-1}, v_t=3200ms^{-1}, v_w=1495ms^{-1} and $\alpha_l=4\alpha_t$. Thus, α_l is calculated to be 485m^{-1} using Equation(6); and assuming that the attenuation coefficient is proportional to the forth power of the frequency, the attenuation coefficient can be estimated as 8.8664×10^{-30} s^{4}m^{-1}.

Figure 5 shows behaviors of the phase curve and the modulus curve calculated with the above estimated attenuation coefficient. These figures are similar in frequency dependence to those shown in Figure 4. This means that the estimated attenuation coefficient is adequate. The Rayleigh wave velocity calculated with an angular position of the minimum in Figure 5(b) agreed precisely with the theoretical value[8] calculated with elastic constants for isotropic material and density; therefore, it is considered that the Rayleigh wave velocity of the annealed steel determined from the modulus curve is independent of the absorption losses in the material.

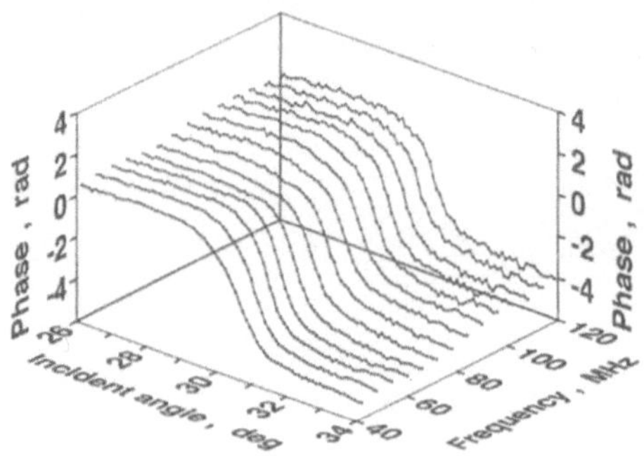

Figure 3. Measured phase curves of the reflection coefficients for the quenched steel.

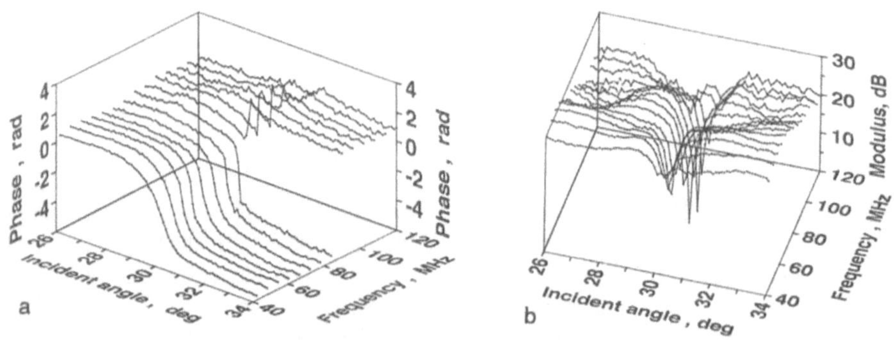

Figure 4. Measured reflection coefficients for the annealed steel: (a) phase curves; (b) modulus curves.

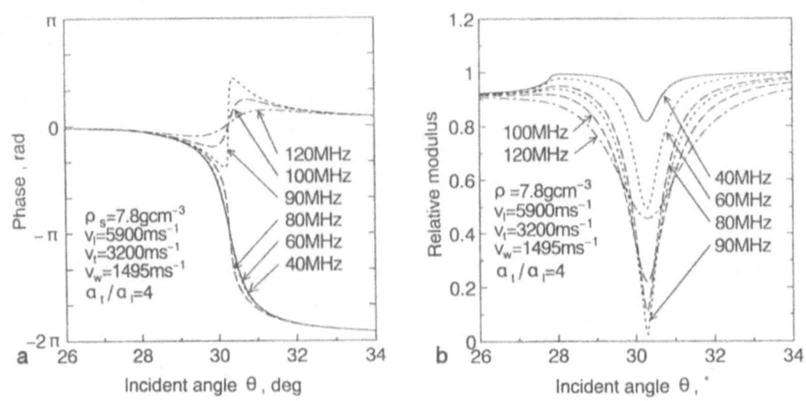

Figure 5. Frequency dependence on reflection coefficients calculated for the annealed steel: (a) phase curves; (b) modulus curves. Longitudinal wave attenuation coefficient of 8.86×10^{-30} s^4m^{-1} was used.

Rayleigh Wave Velocities for Heat Treated Steels

Figure 6 shows Rayleigh wave velocities measured at 120MHz for various kinds of heat treated steels. Theoretical values calculated by using Equations(1), (2) and (3) with

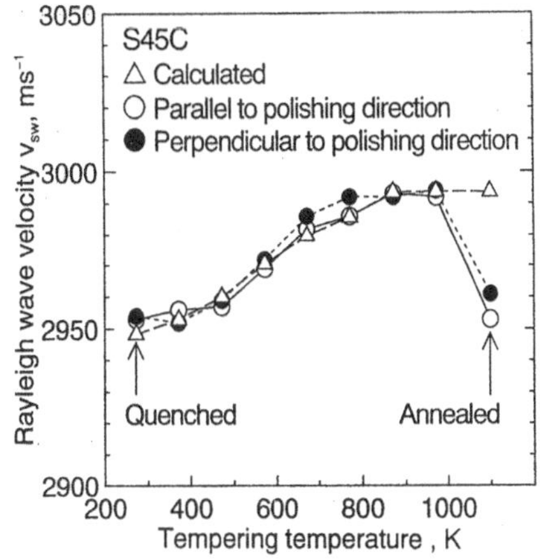

Figure 6. Measured at 120MHz and calculated results of Rayleigh wave velocities for heat treated steels.

macroscopic elastic constants for isotropic materials and densities also shown in Figure 6. For the quenched and the tempered steels, the measured values well agree with the theoretical values. This means that the velocities measured on these steels characterize exactly the macroscopic elastic properties of the materials. For the annealed, however, the measured and calculated value differ remarkably from each other. This suggests that the microstructure of materials should have great influence on the measured velocities. Figure 7 shows typical micrographs of heat treated steels. The quenched and the tempered steel have the fine structure which is composed of martensite or tempered structure with finely dispersed carbides; on the other hand, the annealed steel has the dual-phase structure of ferrite and pearlite. In the measurement at 120MHz for the annealed steel, where the spatial resolution or the size of the measurement area is comparable with the size of each phase, it is considered that effects of a difference in the velocity of each phase and/or an elastic anisotropy on the measured velocity should be remarkably enhanced. Figure 8 shows that changes of the measured velocity with the measuring position for the annealed steel and its frequency dependence. The velocity fluctuates with the measuring position and the fluctuation becomes larger with frequency. This is partially because the large fluctuation at high frequency was caused by inhomogeneity of the coarsely mixed structure as mentioned above. At frequency less than 60MHz, on the contrary, the velocities remain almost constant. This result indicates that the effect of the inhomogeneity of the dual-phase structure disappeared because the spatial resolution at such low frequency is larger than the size of each phase or the grain size.

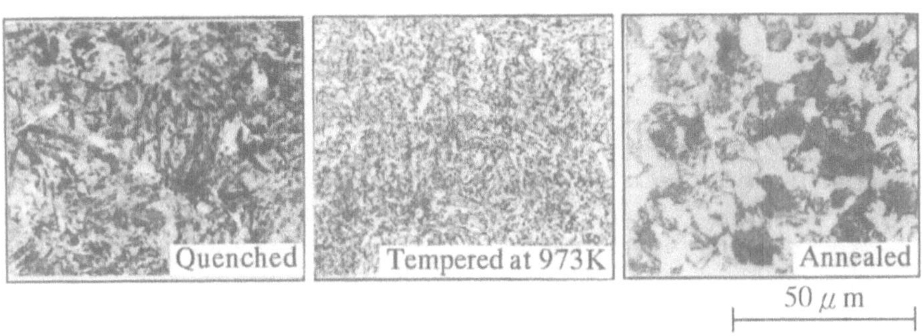

Figure 7. Typical microstructures of steel quenched, tempered at various temperatures and annealed.

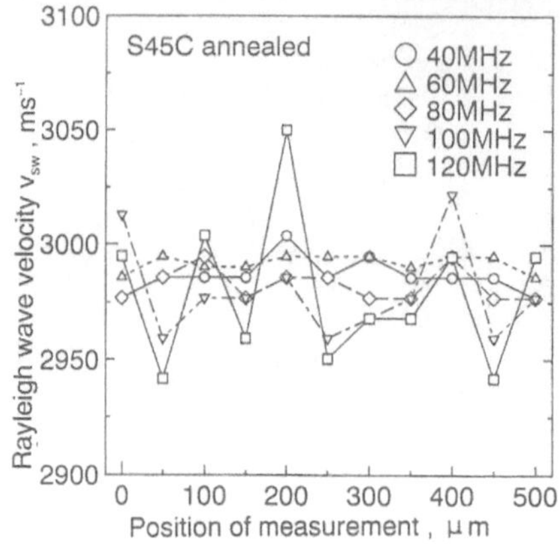

Figure 8. Variations in Rayleigh wave velocities of annealed steel with measuring position.

CONCLUSIONS

The Rayleigh wave velocity obtained by analyzing the ultrasonic reflectivity measured at a small area was found to be a useful parameter for the estimation of mechanical properties at the small area. For the quenched and the tempered steel specimens, the measured velocities characterized exactly macroscopic elastic properties of materials, and were independent of absorption losses of materials. Improving the resolution by using the higher frequency, it might be possible to estimate acoustic properties and mechanical properties of a fine grain by measuring the reflectivity.

Acknowledgements

The authors would like to thank Dr. Tsukahara, Mr. Nakaso and Mr. Ohira of Toppan Printing Co.,Ltd for valuable advice. This work was supported by a Grant-in Aid for Science Research from Iketani Science and Technology Foundation and the Ministry of Education, Science and Culture, Japan.

REFERENCES

1. F.R.Rollins,Jr., Ultrasonic Reflectivity at a Liquid-Solid Interface near the Angle of Incidence for Total Reflection, *Appl. Phys. Lett.*, 7-8, 212 (1965).
2. G.L.Fitzpatrick and B.P.Hildebrand, Near Surface Flaw Detection by Ultrasonic Critical Angle Imaging, *Journal of Nondestructive Evaluation*, 3-4, 201 (1982).
3. Y.Tsukahara, N.Nakaso and K.Ohira, Angular Spectral Approach to Reflection of Focused Beams with Oblique Incidence in Spherical-Planar-Pair Lenses, *IEEE Trans. Ultrason. Ferroelec. Freq. Contr.*, 38-5, 468 (1991).
4. T.Aizawa and J.Kihara, Ultrasonic Microscopic Evaluation of PVD/CVD Coated Hard Materials, in: "*Residual Stress III: Science and Technology* Vol.1," H.Fujiwara, ed., Elsevier Science Publisher, London (1992).
5. I.Ihara, A.Shimamoto, K.Tanaka, T.Aizawa and Y.Tsukahara, Microscopic Characterization of Elastic Property on Surface Damaged Layer of MnZn Ferrite by using an Ultrasonic Micro-Spectrometer, *Proc. 70th JSME Fall Annual Meeting* Vol.B, 522 (1992).
6. F.L.Becker and R.L.Richardson, Influence of Material Properties on Rayleigh Critical-Angle Reflectivity, *Journal of the Acoustical Society of America*, 51-5, 1609 (1972).
7. H.L.Bertoni and T.Tamir, Unified Theory of Rayleigh-Angle Phenomena for Acoustic Beams at Liquid-Solid Interfaces, *Appl. Phys.*, 2, 157 (1972).
8. I.A.Viktorov, "*Rayleigh and Lumb Waves, Physical Theory and Applications*," Plenum Press, New York, (1967).

ULTRASONIC BACKSCATTERING AS A FINGERPRINT TECHNIQUE FOR IDENTITY AND INTEGRITY VERIFICATION OF COMPONENTS

Herbert H. Willems[1] and Esther Wogatzki[2]

[1]Fraunhofer-Institute for Nondestructive Testing
University, Building 37
D-66123 Saarbrücken, Germany
[2]GNS Gesellschaft für Nuklear-Service
Lange Laube 7
D-30159 Hannover, Germany

INTRODUCTION

During the interim storage of spent fuel rods the verification of the identity as well as of the integrity of special shielding casks containing final disposal packages has to be guaranteed by applying appropriate safeguards measures over a period of up to 50 years until a geological repository goes into operation. The casks to be verified are made of cast iron (GGG 40) and consist of the actual final disposal package and a cylindrical shielding cask with a height of about 6 m and a diameter of about 1.5 m which is closed with a screwed-on lid. It is planned to link the lid to the cask by a short weld seam, so that the lid can only be removed by destroying the weld. The identification of the weld allows for statements on the integrity of the cask as well as on its identity. The latter is, in addition, supported by non-reproducible cask characteristics.

As the smooth surface of the cask without any openings prevents the application of conventional sealing methods like, for example, metal cap seals, other techniques have to be developed. In the case of storage cans (wall thickness: 1mm) for special nuclear material previously investigated[1], the use of guided ultrasonic waves was proposed using reflections signals obtained from a weld and its individual geometry as a fingerprint. Our approach is based on the material's microstructure as characterized by its grain and precipitation structure, which represents a non-reproducible fingerprint[2]. The microstructural characteristics can be converted into an ultrasonic signal using the backscattering technique. Backscattering signals provide an unambiguous representation of the microstructure considered and hence allow the recording of microstructural fingerprints in a reproducible way. In this contribution we present results on the influence of various parameters on the reproducibility of ultrasonic backscattering signals enabling us to define the conditions under which the backscattering technique can be used for the verification task.

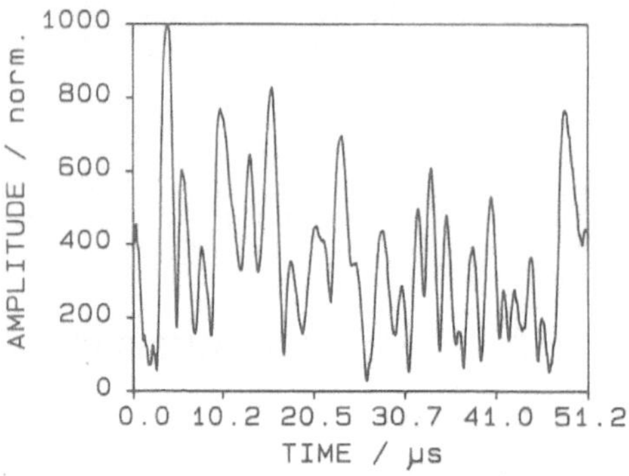

Figure 1. Ultrasonic backscattering signal from cast iron GGG 40 (2.5 MHz T-wave).

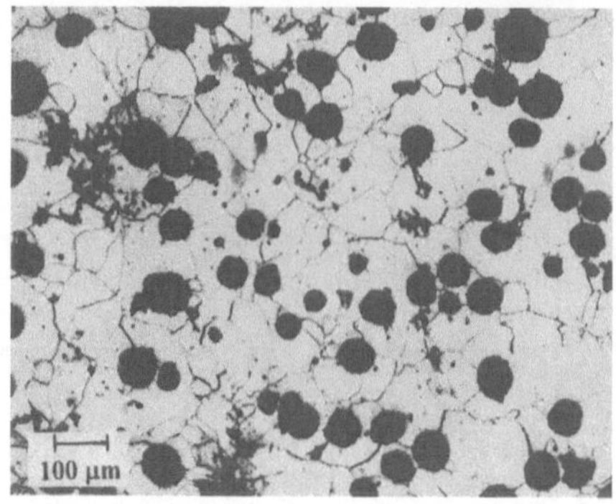

Figure 2. Metallographic image of the cask material GGG 40.

ULTRASONIC APPROACH

Background

In polycrystalline and/or multiphase materials ultrasonic scattering occurs at grain boundaries and phase boundaries due to a local mismatch in the acoustic impedance. Amplitude, phase and sound field characteristic of the scattered wavefield depend, apart from the microstructure, in particular on the frequency and the wave mode of the ultrasonic pulse. If the grain d is small compared to the ultrasonic wavelength λ, i.e., $d / \lambda << 1$, Rayleigh scattering is the dominant scattering mechanism. Then the scattering coefficient α_S is given by[3] $\alpha_S \sim d^3 \cdot f^4$ with f being the ultrasonic frequency. A rectified ultrasonic backscattering signal (Figure 1) measured in pulse-echo technique may be expressed in the form

$$A_S(x,z,t) = A_N(t) + A_S(x,z=0) \cdot \exp(-\alpha \cdot z) \cdot M(z). \quad (1)$$

$A_S(x,z,t)$ is the backscattered ultrasonic amplitude as a function of position x, of the sound path z and of time t. $A_N(t)$ is the contribution due to electronic noise which is negligible in case of sufficient signal amplitude ($A_S >> A_N$). Apart from the noise contribution the backscattering signal is stationary, i.e., independent of time. $A_S(x,z = 0)$ is the initial amplitude of the backscattering signal at position x = 0. The term $\exp(-\alpha \cdot z)$ accounts for the ultrasonic attenuation along the sound path z where the attenuation coefficient α comprises both the scattering losses and the absorption losses. The function M(z) in Eq. 3 describes the modulation pattern of the backscattering signal which is caused by the interference of all the backscattered waves (see Figure 1). This pattern represents an acoustic image of the icrostructure along the sound path and thus a microstructural fingerprint of the measuring position which may be used for the verification of its identity. Any signal recorded under the same conditions at the same position will always show the same interference pattern whereas signals recorded at different positions will look more or less different. Therefore, the problem of identification basically requires a method which allows the evaluation of the similarity (or non-similarity) of two backscattering signals. This can be achieved by correlation analysis.

Signal Evaluation

The similarity of two digitized signals i, j each with p sampling points can be numerically evaluated using their correlation coefficient CC. The correlation coefficient of two signals A_i, A_j is defined as[4]

$$CC_{ij} = \frac{p \sum_{k=1}^{p} A_{ik} A_{jk} - \sum_{k=1}^{p} A_{ik} \sum_{k=1}^{p} A_{jk}}{\sqrt{\left[p \sum_{k=1}^{p} A_{jk}^2 - \left(\sum_{k=1}^{p} A_{jk} \right)^2 \right] \left[p \sum_{k=1}^{p} A_{ik}^2 - \left(\sum_{k=1}^{p} A_{ik} \right)^2 \right]}} \quad (2)$$

Here, A_{ik} is the amplitude of signal i at sample point k. For similar signals the correlation coefficient is close to 1. For uncorrelated signals the correlation coefficient lies around zero.

Even otherwise uncorrelated backscattering signals show some similarity due to the exponentially decaying shape of the signals which is accounted for by the attenuation term $\exp(-\alpha \cdot z)$ in Eq. 2. This leads to an offset contribution in the correlation coefficient which can be compensated by multiplying the recorded signals with the correction term $\exp(+\alpha \cdot z)$. The attenuation coefficient α is obtained from a regression analysis.

EXPERIMENTAL PROCEDURE

Samples

In order to prove the suitability of the backscattering technique with regard to the verification task, it was necessary to determine the influence of the parameters which are relevant to the interference structure of the ultrasonic signal. Therefore, the dependence of the signal on the frequency, the pulse length, the transducer (including the probe wedge in the case of angular transmission) and the measuring position has been studied. The experiments were performed using longitudinal waves (L) in perpendicular transmission, shear (transversal) waves (T) in 45° and 60° transmission and surface waves also called Rayleigh waves (R). The nominal frequencies of the (commercially available) transducers used for the experiments were 2.5 MHz and 5 MHz.

For the experiments, a sample of the original cask material, with a size of 190 mm x 100 mm x 40 mm, was provided. The material is ductile cast iron GGG 40. A metallographic image of its microstructure is shown in Figure 2. The mean grain size of the ferrite phase is about 90 μm, the mean diameter of the graphite spheres is about 50 μm. The volume fraction of the graphite amounts to approximately 20 %. The graphite phase having a particle size comparable to the ferrite phase essentially contributes to the ultrasonic scattering because of the large difference in acoustic impedance between graphite and ferrite (impedance ratio 1:9). Additionally, experiments were performed using a sample of comparable size containing a weld seam.

Measuring Set-Up

The backscattering measurements were performed with an ultrasonic test device developed[5] at the IzfP. The system enables the generation of ultrasonic pulses with defined frequency and pulse-length. The frequency range is 2.5 - 30 MHz adjustable in steps of 0.1 MHz. The pulse-length can be set between 5 λ- 15 λ adjustable in steps of 1λ.

The signals received are rectified, amplified and finally digitized at a sampling rate of 20 MHz using an 8 bit A/D-converter. One signal consists of 1024 data points resulting in a ultrasonic time-of-flight 50.12 μs. This time-of-flight corresponds to an ultrasonic sound path of about 78 mm for shear waves and about 141 mm for longitudinal waves, respectively, in the material considered. In a subsequent averaging unit up to 1024 single signals can be added up for signal averaging in order to improve the signal-to-noise ratio. After recording the backscattering signals are transferred from the ultrasonic device to a personal computer in order to perform the signal evaluation. The total measuring time takes only a few seconds.

RESULTS

Frequency and Pulse-Length Variation

The excitation frequency of the ultrasonic pulse was varied in steps of 100 KHz in the range from (center frequency - 0.5 MHz) to (center frequency + 0.5 MHz). Figure 3a shows the signal decorrelation as a function of the frequency change Δf. For all test series a similar behaviour is obtained. Here, a frequency change of about 0.4 MHz leads to a decrease of the correlation coefficient from 1 to about 0.8.

The pulse-length of the excitation burst was varied in steps of one wavelength between 5λ and 15λ. The result of the pulse-length variation ΔP is shown in Figure 3b. ΔP refers to the initial pulse-length of 5 λ. A decrease of the correlation coefficient below 0.8 is found for ΔP > 3λ. By changing the pulse-length the volume of the microstructure contributing to the backscattered amplitude A (t) at the same time is changed as well. As a result the interference pattern changes, leading to the observed decrease in the correlation.

Position Variation

From Figure 4 follows that a change in position of 1 mm already causes a distinct decorrelation. This means that the interference pattern strongly depends on the measuring position. The decorrelation takes place the faster the higher the frequency because the wavelength and thus the spatial coherence length decreases with increasing frequency. A decrease of the correlation coefficient below 0.8 is obtained at a position change of 1/3 mm in the case of both 5 MHz longitudinal wave and 5 MHz shear wave whereas at a frequency of 2.5 MHz a change of about 0.5 mm is required.

Influence of Transducer and Wedges

In order to assess the impact of individual transducer characteristics, backscattering signals were recorded at the same position with up to 6 transducers of the same type and then compared to each other. Measurements were performed with 2.5 MHz respectively 5 MHz

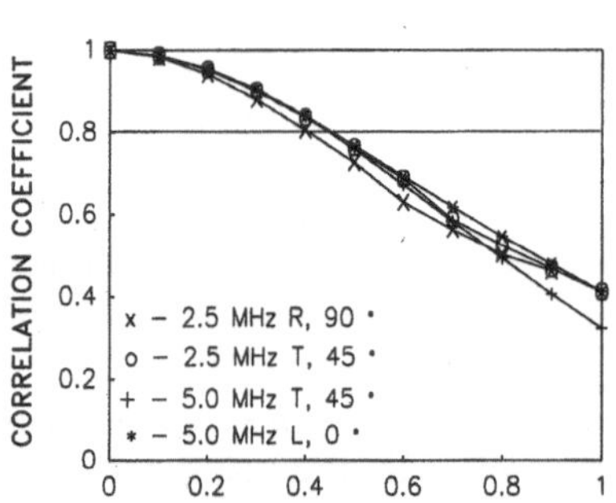

Figure 3a. Influence of frequency variation on the decorrelation of ultrasonic backscattering signals.

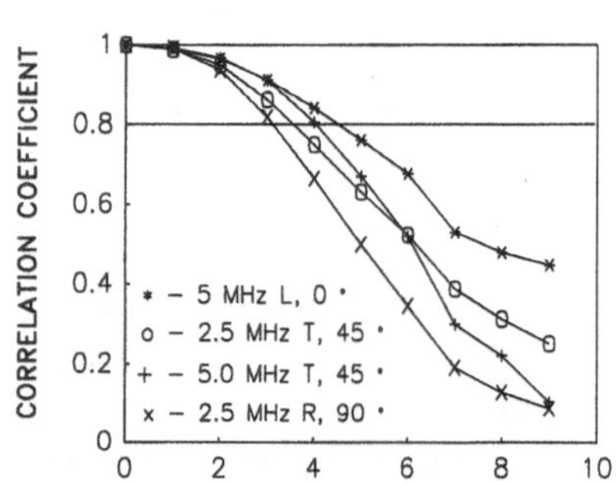

Figure 3b. Influence of pulse-length variation on the decorrelation of ultrasonic backscattering signals.

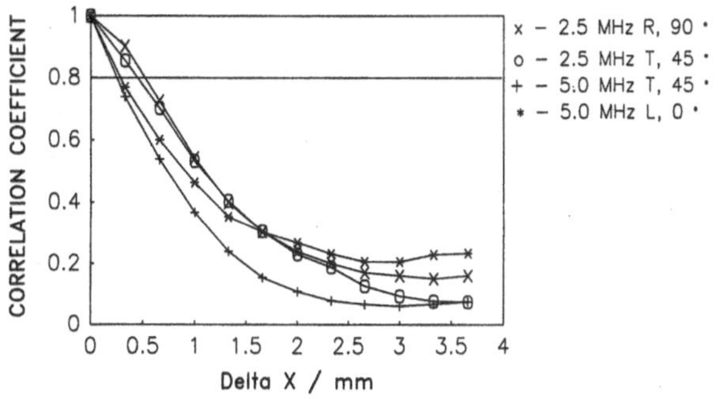

Figure 4. Decorrelation of ultrasonic backscattering signals by position variation.

T-waves using 45° transmission and with 5 MHz L-waves using perpendicular transmission. Only for the 2.5 MHz /45° T- wave transducers all correlation coefficients lie above 0.8. For the 5 MHz L-wave transducers, however, all values are below 0.8 indicating a large scatter of the individual transducer characteristics which may be attributed to variations in center frequency, radiation characteristic and sound field symmetry.

Additionally, the impact of the wedges to which the ultrasonic transducers are screwed on for angle transmission on the signal reproducibility was investigated for the case of 2.5 MHz / 45° - T-waves. Therefore, signals were recorded with the same transducer at the same position using five different wedges of the same type and all possible combinations of two signals were evaluated. The lowest correlation coefficient obtained was 0.926 (mean: 0.956) indicating that the wedges are almost identical in their ultrasonic properties.

Reproducibility and Individuality of the Backscattering Signal

In order to evaluate their reproducibility 20 signals were recorded at the same position. All possible signal combinations were evaluated resulting in 190 correlation coefficients for each test series. For every single measurement, the transducers were newly positioned. A precise positioning was ensured by using a stop bracket fixed to the sample. Mean, maximum, minimum and standard deviation of the correlation coefficient distribution of the individual testing modes are shown in Table 1. The reproducibility is excellent for the 2.5 MHz R- and T-wave as well as for the 5 MHz T-wave. All correlation coefficients are above 0.85. Only for 5 MHz L-waves in perpendicular transmission some data are below 0.8. This is attributed to the fact that in the case of perpendicular transmission the signals are much more influenced by small coupling variations than in the case of angular transmission.

The measurement and signal evaluation for the assessment of the signal individuality corresponds to the procedure described before with the difference that the backscattering signals were recorded at 20 different positions. The results are also given in Table 1. As expected, the mean values are now close to zero and the maximums are all below 0.51.

Tests of reproducibility and individuality were also performed on a sample containing a weld of 2 cm depth and 2 cm width. Ultrasound using 2.5 MHz T-waves was transmitted perpendicularly to the weld from the cask material across the weld (measuring geometry A) as well as in the weld direction along the weld (measuring geometry B). In order to obtain a sufficiently long sound path within the weld, 60° transmission was used. For both measuring geometries the results are essentially comparable to those obtained for the cask material as can be seen from Table 1.

Table 3. Statistical parameters obtained from the evaluation of the correlation coefficients for the tests with respect to reproducibility (R) and individuality (I).

Measuring Mode	Sample*	Mean		Standard Deviation		Minimum		Maximum	
		R	I	R	I	R	I	R	I
2.5 MHz R 90°	CM	0.980	0.033	0.013	0.13	0.935	-0.30	0.996	0.45
2.5 MHz T 45°	CM	0.991	0.046	0.005	0.12	0.967	-0.27	0.998	0.47
5.0 MHz T 45°	CM	0.954	0.093	0.026	0.09	0.856	-0.15	0.990	0.35
5.0 MHz L 0°	CM	0.885	0.115	0.079	0.15	0.614	-0.37	0.992	0.46
2.5 MHz T 60°	WS - A	0.969	0.090	0.010	0.15	0.938	-0.27	0.990	0.43
2.5 MHz T 60°	WS - B	0.954	0.083	0.022	0.14	0.892	-0.24	0.989	0.51

*CM - Cask Material, WS - Weld Seam (A, B: measuring geometries, see Text).

Statistical Reliability

In order to evaluate the reliability of the ultrasonic technique with respect to misinterpretations, it is necessary to know the number of independent characteristics contained in a backscattering signal. Obviously, the number of features is proportional to the total length of the signal (here 50 μs) on the one hand, and on the other it depends on the bandwidth of the electronic signal demodulation. The latter is 2.5 MHz corresponding to a time resolution of 0.4 μs. This means that, at most, 125 independent signal features can be expected within 50 μs. The real number is actually smaller due to the fact that the spatial coherence length of the ultrasonic pulse, that is the spatial region which contributes to the scattered signal at the same time and thus limits the time interval between independent scattering indications, is about half the pulse-length. At 2.5 MHz T-waves, the wavelength is 1.24 mm in the considered material. At a total pulse-length of 5 λ used for the measurements half the pulse-length comes to 3.1 mm corresponding to a time-of-flight of 1 μs yielding approximately 50 independent features.

To substantiate these considerations a computer programme was written enabling the simulation of backscattering signals. By means of a random number generator, 50 signals with a defined but variable number of independent characteristics were generated during each run providing 1225 signal combinations and the standard deviation σ of the distribution of the resulting correlation coefficients was determined. In Figure 5, σ is shown as a function of the number of independent characteristics which was varied between 5 and 100. If one compares the experimentally determined standard deviations ranging from 0.12 to 0.15 with the simulated data, one obtains in fact about 50 to 60 independent characteristics in agreement with the estimations given above.

Using the 2.5 MHz/45° - signals recorded at the cask material, the actual dependence of σ on the number of independent characteristics was checked for the measured signals. By taking only a part of the signals for the calculation of the correlation coefficients the number of independent characteristics was stepwise reduced assuming that the number is proportional to the signal length. As can be seen from Figure 5, the agreement with the result of the simulation is very good. The best fit is obtained assuming 53 independent characteristics for the full signal.

Based on these considerations it is possible to estimate the reliability of the ultrasonic technique with respect to the verification task in a more quantitative way. In all tests of individuality the experimentally determined standard deviation of the distribution of the correlation coefficients, which is found to be a normal distribution (Figure 6), lies below 0.15. Let us assume that a correlation coefficient ≥ 0.8 is required for correctly identifying a cask. Then, the gap between the required threshold value of 0.8 and the mean (≈ 0) becomes larger than 5 σ. Assuming a Gaussian distribution the probability P_I that another cask or an another location gives a signal similar to the identification signal of the cask under consideration comes to less than $5.7*10^{-7}$, i.e., the probability that a cask or a weld is wrongly identified or mistaken ('false acceptance') is negligible.

A similar consideration holds for the signal reproducibility. In this case, the correlation coefficients are, strictly speaking, not normally distributed since the distribution is asymmetric with regard to its mean. In the range below the mean, however, the actual distribution may be approximated by a Gaussian. Except for one test series, the gap between the mean and the threshold of 0.8 amounts to $> 6\sigma$ (see Table 1). This means that the probability P_{II} of not identifying a cask or a weld, i.e., 'false alarm', is less than $2.0*10^{-9}$. Consequently, the probability P_C of a correct identification is given by the complement to the sum of P_I and P_{II} yielding $P_C = 1 - (P_I + P_{II}) = 0.99999943$.

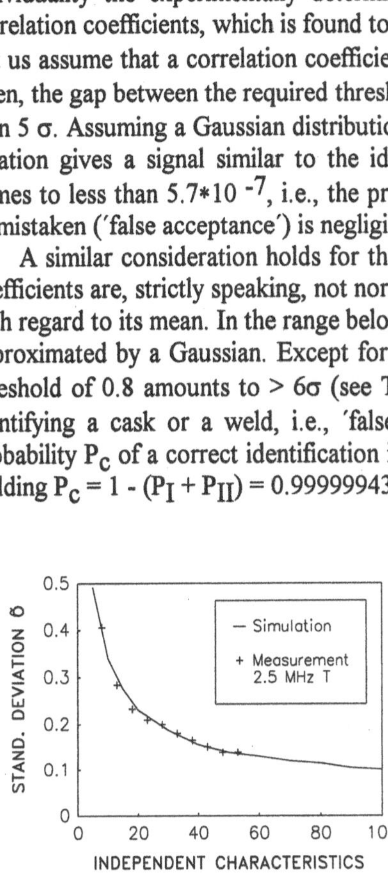

Figure 5. Standard deviation of the distribution of correlation coefficients as a function of independent signal characteristics.

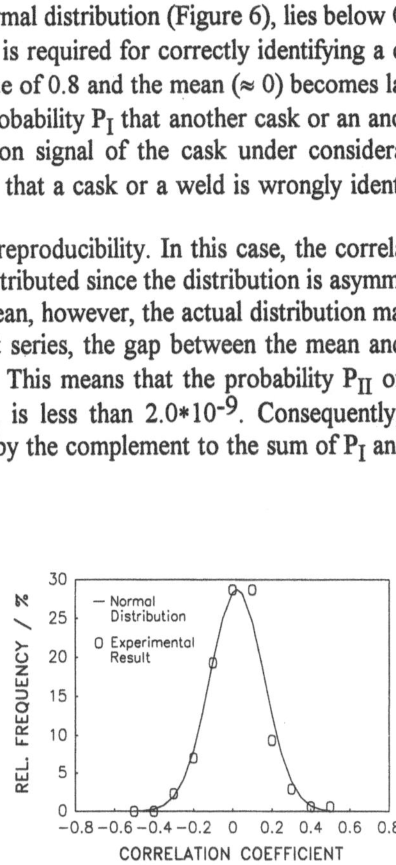

Figure 6. Distribution of correlation coefficients as obtained from individuality test.

DISCUSSION

The results presented allow for the assumption that the ultrasonic backscattering technique is suitable for the task considered. Proof of the cask integrity is achieved by the identification of the weld seam since a newly applied weld would give a different backscattering signal. Thus the proof of integrity would mean, if positive, a proof of identity at the same time, i.e., both requirements are met by just one measurement. A negative test result, however, implies the possibilities that either the weld seam has been tampered with or that the cask has been replaced. To avoid such ambiguity, it would seem useful to record two signals per cask, one from the cask and one from the weld. According to Figure 4 the positioning of the transducer must be as precise as 0.3 mm in order to reproduce the signal correctly. This can be attained easily by providing, for example, a stop bracket or a snug fit, thus forcing an exact positioning.

The results of frequency variation as well as pulse-length variation show the necessity of using well-defined electrical pulses for the ultrasonic excitation. The realization of suitable ultrasonic transmitters is state-of-the-art. Some development, however, would be needed in order to set up a portable, battery driven system. The effects of equipment aging, which may play a role in long-term applications, can be dealt with by using digital techniques for signal generation and signal processing

Though transducers of the same type have led to comparable signals in some of the tests one has to ensure in any case that the transducers to be used are interchangeable. Therefore, it is proposed that the correlation coefficient of signals recorded with different transducers at the same position should be at least > 0.9. This requirement can be achieved by applying appropriate selection criteria. The same argument holds for the angle wedges.

CONCLUSION

This work demonstrates that the ultrasonic backscattering technique is suitable for the identification and verification of components made of polycrystalline materials. Backscattering signals povide a non-reproducible fingerprint of the material's microstructure, which can be considered as absolutly proof against tampering. As a threshold value for the correct identification of a fingerprint signal, a correlation coefficient with the reference signal of at least 0.8 is recommended. In order to obtain reproducible signals a well defined excitation signal as well as precise positioning of the ultrasonic transducer is required. In the considered case of cast iron the use of 2.5 MHz shear waves in angular transmission turned out most appropriate. Apart from the high reliability, the ultrasonic technique provides easy operation and fast measuring times. It is planned now to develop a portable prototype system for inspection purposes.

ACKNOWLEDGEMENT

The work presented in this article was part of a feasibility study commissioned by GNS within the framework of a project sponsored by the German Ministry of Research and Technology under contract # KWA 7908/5.

REFERENCES

1. K.D. Bones, C.C. Holt, and R.P. Holt, Verification of nuclear material (SNM) containers by means of ultrasonic inspection and fingerprinting, Proc. of the 5th Annual Symp. on Safeguards and Nuclear Material Management, Versailles (1983).
2. P. Höller, German Patent DE 33 19 102 C2 (1987).
3. A.B. Bhatia, "Ultrasonic Absorption", Clarendon Press, Oxford (1967).
4. E. Walpole and R. H. Myers, "Probability and Statistics for Engineers and Scientists", The Macmillan Company, New York (1972).
5. Willems, R. Neumann, S. Hirsekorn, and K. Goebbels, Gefügeanalyse an Stählen mittels Ultraschall-Streuung, IzfP-Report # 820520-TW (1982).

AIR-COUPLED ULTRASONIC SYSTEM FOR DETECTING DELAMINATIONS AND CRACKS IN PAINTINGS ON WOODEN PANELS

A. Murray[1], E.S. Boltz[2], M.C. Renken[2], C.M. Fortunko[2], M.F. Mecklenburg[3], and R.E. Green, Jr.[1]

[1]Department of Materials Science and Engineering and the Center for Nondestructive Evaluation, The Johns Hopkins University, Baltimore, MD, 21218
[2]National Institute of Standards and Technology, Boulder, CO, 80303
[3]The Conservation Analytical Laboratory, The Smithsonian Institution, Washington, D.C., 20560

ABSTRACT

It has been established that the risk of damage to paintings on wood ("panel paintings") increases with the presence of cracks, delaminations, and their associated stress concentrations. Such flaws can originate and increase in size as a result of fluctuations in temperature and relative humidity, as well as shock and vibration. Many internal flaws cannot be detected either visually or by traditional testing techniques, and it is difficult, therefore, to assess the risk transportation poses to panel paintings.

Air-coupled ultrasound has been used to assess the condition of two panel paintings (*Parental Admonition* [a copy of the original] and *Women Gathering Yucca Plants*), in a non-contact, non-intrusive manner; this method provides information complementary to that given by radiography. It has been demonstrated that the ultrasonic system is clearly more suitable for detecting specific types of flaw, such as in-plane cracks and delaminations. The system enables measurements to be easily made of highly anisotropic and inhomogeneous materials such as wood.

The ultrasonic system used in this study has a superior signal-to noise ratio because it uses efficient transducers, low noise pre-amplifiers, and a phase-sensitive superheterodyne ultrasonic system that has analog signal averaging and filtering components. The signal can be exploited to yield both amplitude and phase information. The ultrasonic system also incorporates a mechanical scanner to produce easily interpreted two-dimensional images of large areas of paintings to give a clear indication of their condition. The results can be further enhanced by using image processing techniques.

INTRODUCTION AND BACKGROUND

Wood has often been used for a support for paintings. The structure of a typical panel painting may include the following layers: the wooden support (often oak or poplar); the ground or gesso layer (gesso is rabbit skin glue and calcium carbonate or calcium sulfate), which sometimes includes linen; the underdrawing; the paint layer; the varnish; and any retouchings. Later artists used both solid wood and wood products such as hardboard for supports, with surface layers that include gesso, gesso and linen, and paper. Throughout the centuries, some works of art and furniture, such as marquetry pieces, have a top surface of veneer (a thin layer of wood).

Catastrophic failure is much more likely in panels with severe stress concentrations caused by cracks and delaminations. For this reason it is imperative to detect, locate, and treat these flaws. By maintaining constant relative humidity, the lifetime of these artifacts can be prolonged; however, this high-maintenance environment cannot be achieved in all circumstances, for example during the transportation of works of art. If flaws can be detected, repairs can be made and predictive computer models (using finite element analysis) can anticipate the conditions that could lead to failure of the object. These conditions, including temperature and relative humidity changes and certain shock and vibration impacts, can then be avoided. It is possible to use computer models to predict the stresses in the individual components and in the entire painting, once any flaws have been detected.[1]

Detecting flaws in panels is not a trivial technical problem because wood is a highly anisotropic and inhomogeneous material and because conventional techniques cannot always locate cracks and delaminations. Moreover, working with art objects imposes restrictions that are not necessary when testing industrial materials. For example, contact ultrasonic methods should be avoided because they require pressure to transmit and receive the ultrasonic signals.

Air-Coupled Ultrasound

Much has been written about nondestructive testing methods used to examine paintings.[2] Flaws such as delaminations and cracks at certain angles have not been detected with traditional techniques such as radiography because the density change at these flaws is too small to be seen. The ultrasound technique is very useful in detecting delaminations and cracks oriented perpendicular to the sound beam because the signal is stopped by any delaminations and the amplitude is registered as zero. When a crack is at an angle, both ultrasonic and radiographic techniques can be used to advantage.

Air-coupled ultrasound uses air as a couplant, which does not saturate the material being examined, change its properties, or leave residues and therefore brings no danger of damage or contamination as other couplants do. This is true as long as the power or amplitude of the sound-wave is not too great.

The background and previous air-coupled ultrasound work has been discussed before.[3] The low signal-to-noise which is one of the major limitations of the air-coupled ultrasound technique, can now be overcome through the judicious selection of the critical components, including transducers, transmitting and receiving electronics, and the use of appropriate signal processing methods. Because of transducer limitations, operation is not possible in the pulse-echo mode, where only one transducer is used to send and receive the signal; this means that both sides of the painting or sample need to be accessible.

Materials with low acoustic impedances are more easily examined with air-coupled ultrasound because the acoustic impedance mismatch with air is low. Examples of materials that can be examined include: ligneous materials, foams, fibre-reinforced composites, rubber, paper, and non-metallic composites. Using non-contact ultrasound, the wood industry has made distance measurements and has located cracks and holes in wood, decay in lumber, and blows (delaminations) within hardboard, insulation board, and particle board.[4] Air-coupled ultrasound would be extremely advantageous in this field because of the coupling difficulties between the transducer and wood.[5]

The present work differs from past ultrasonic work on art objects, which used contact techniques with gels or pressure to examine stone, metal, and waterlogged wood, making measurements at specific points rather than over complete areas.[6] The technique used here allows entire paintings to be examined and comparisons to be very easily made between local and adjacent points, thus enabling contextual information to be obtained. This paper is the culmination of research in which air-coupled ultrasound has successfully examined samples used to mimic panel paintings. The samples were made with different flaws (delaminations and cracks), supports (oak, poplar, and hardboard), and surfaces (paint, gesso, gesso and linen, paper, and veneer); each sample varied one of these parameters at a time.

EXPERIMENTAL SET-UP

The main component of the air-coupled ultrasound system in this work was the phase-sensitive superheterodyne measurement system that generated and processed the ultrasonic signals.[7] It was linked to a commercial C-scan system which controlled the position of the two transducers on either side of the painting in the XY-plane. During the experiment, computer programs controlled the ultrasonic measurement and scanning systems. They also stored the data for the amplitude, phase, x and y positions, and the measurement system settings.

The system operated between 50 kHz and 5 MHz. The ultrasonic measurement system had modules that contained different components including: an IF oscillator and quadrature phase-sensitive detectors; a direct digital synthesizer; a high power gated RF amplifier; a broadband RF receiver; a mixer and IF amplifier; gated analog integrators; a coherent timer; and a 12-bit analog to digital converter (A/D converter).

The piezoelectric transducers used operated at 0.5 MHz and were spherically focused at around 5.08 cm (2.0 inches). During the experiments, the generating transducer was brought very close to the back surface of the sample, and was therefore focused within the wooden panel. The receiving transducer was focused on the sample-air interface of the front surface. This configuration ensured that the maximum energy entered the painting, to obtain the best resolution possible while maintaining the safety of the painting.

SAMPLES

The first painting examined came from the private collection of Dr. Hans Goedicke of Baltimore, Maryland. The painting, *Parental Admonition*, also known as *The Brothel Scene*, was a studiopiece (a copy from the same studio as the original). The original was painted by Gerard Terborch around 1654-55. The panel was oak and had a thickness of 3.6 mm (0.14 inches). Photographs of the front and back of the painting are shown in Figures 1a and 1b. The painting was of interest because of cracks in the wooden support, the craquelure or cracks in the paint layer, and because the painting had been "cradled". Cradling is an intrusive conservation treatment, used from the late 18th century until the early 20th century, where a secondary wooden support was added to constrain the panel dimensionally.

The second painting, *Women Gathering Yucca Plants*, is owned by the National Museum of American Art, the Smithsonian Institution (accession number 1969.64.9). It

a b

Figure 1. (a) Photograph of the painting *Parental Admonition*, (b) verso

Figure 2. Photograph of the painting *Women Gathering Yucca Plants*, National Museum of American Art, Smithsonian Institution, transfer from the U.S. Department of the Interior

was painted by an unidentified artist in the 20th century and is shown in Figure 2. It is a watercolour and ink painting on illustration board (laminated paper board with paper layers glued to its surface) mounted on hardboard. The painting was approximately 6 mm (0.24 inches) thick. The illustration board had delaminated from the hardboard in certain, hard-to determine areas.

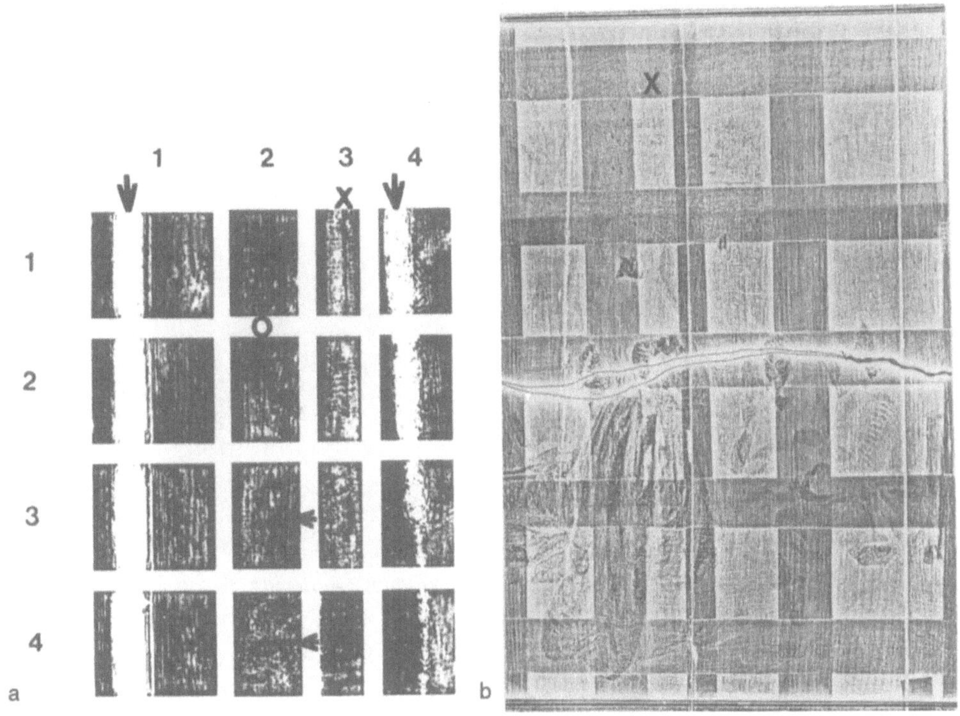

Figure 3. (a) Amplitude results from ultrasound scan of painting *Parental Admonition*, (b) xeroradiograph of the painting *Parental Admonition*: X is near closed crack which is hardly visible, and other features seen are the open cracks, the cradle, the hanging wire, and the image from the paint layer (x-ray tube was 130 cm or 51.2 inches above film, settings were between 40 and 60kV, 5 mA, and exposure time was one minute)

RESULTS

The results of air-coupled ultrasound investigations are shown using two-dimensional representations known as C-scans. Four ultrasonic parameters possible are: amplitude, phase, "processed" amplitude, and "processed" phase, where "processed" refers to the image processing technique of thresholding, where all the pixels above a certain grey level are shown as black and the pixels below the level are white. The first two types of amplitude scans will be shown in this paper. In the unprocessed amplitude images, the light areas indicate regions where the ultrasonic signal was easily able to penetrate the sample and the darker areas show where the signal was not been able to penetrate. Gradations of this can be seen in the unprocessed amplitude scan with the different grey levels. In the processed amplitude images, the light areas show regions where ultrasound has not been able to penetrate through.

Parental Admonition

Investigations were made of the cradled painting *Parental Admonition*. In this case, the cradle on the painting had not adhered well to the panel so that the signal-to-noise ratio was insufficient to allow an investigation of the entire painting. Thus ultrasonic scans could be performed only in the 16 areas between the battens of the cradle. A variety of artifacts can be seen in the ultrasonic C-scan of the amplitude results shown in Figure 3a.

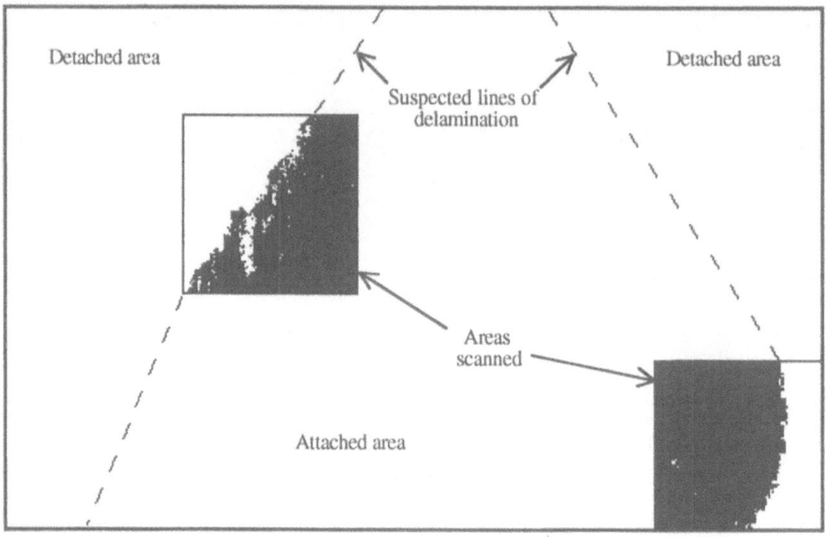

Figure 4. Outline of the front of the painting *Women Gathering Yucca Plants*, showing processed amplitude ultrasonic results from scanned areas (dotted lines show places at which it is suspected that the paperboard has become delaminated)

Two types of cracks exist: open cracks, where there is a visible opening to the other side, and closed cracks, where this is not the case. The open cracks appear as white lines (indicated by large arrows) and are larger on the ultrasonic image because of diffraction effects. Open cracks may mask other nearby artifacts, such as other cracks or paint craquelure. The closed cracks are dark lines (one is indicated by an "X").

Many of the vertical lines correspond to grain and closed cracks in the wood. Using the ultrasonic data only, it is difficult to distinguish the lines due to grain and those due to cracks. For example, in the two areas located in the second column and first and second rows, the lines could be indicative of either grain or cracks (adjacent to the circle).

The horizontal lines in the two areas in columns two and three and rows three and four are the craquelure running perpendicular to the grain. Some of the craquelure is identified by small arrows. Knots can also be seen in the first column, first row.

The closed crack marked by the X is detected with the ultrasound, but could not be detected visually or by xeroradiography (Figure 3b). In the positive xeroradiograph, the high density areas appear dark and those of low density appeared light. Visually it is very difficult to determine if the line indicates a crack running all the way through the panel, or merely an indentation in the wood.

Women Gathering Yucca Plants

Figure 4 shows the outline of the painting *Women Gathering Yucca Plants* (Figure 2), and the results from ultrasonic imaging of two areas within the painting. The processed amplitude scans are used as they show the delaminations most clearly.

The regions where the paperboard is detached from the hardboard are white. The dotted lines show the places at which it is suspected that the paperboard has become delaminated. The entire painting would need to be scanned to validate these lines. As expected, the xeroradiography (which is not shown) does not detect any delaminations.

IMPORTANCE AND FUTURE DIRECTIONS FOR ART CONSERVATION

Previous work has shown that air-coupled ultrasound can detect cracks and delaminations in panel paintings. The ultrasound can penetrate typical thicknesses of panel paintings, the thickest oak panel being 1.6 cm (0.63 inches), the thickest poplar panel being 2.0 cm (0.79 inches), and the thickest hardboard panel being 0.6 cm (0.24 inches). Thicker panels of all materials could probably be examined with additional signal processing such as time signal averaging.

The different ultrasonic results (amplitude, processed amplitude, phase, and processed phase) can be used to advantage to display certain types of flaws more clearly. Delaminations are most effectively shown with the processed amplitude scans, even when the paint layer has uneven thicknesses, while the amplitude scans show the degree of delamination. The amplitude scans better define cracks in radially cut panels, while the phase scans isolate them better in the tangentially cut panels. It is therefore useful to take both amplitude and phase measurements.[8]

Air-coupled ultrasound could also be used to examine paintings on canvas as the ultrasound easily penetrates paint, linen, and gesso layers. Flaws similar to those detected in panel paintings, such as delaminations between layers, could be found. Furniture pieces could also be investigated. It is conceivable that three-dimensional objects could be examined, although the results would be more complex and would need more interpretation.

As well as detecting delaminations and certain cracks, air-coupled ultrasound has other advantages over radiographic techniques. X-ray radiography and xeroradiography sometimes show the paint layer which may obscure other information found in a radiographic image. Also radiographic techniques cannot be used to examine a support with a lead white paint layer as x-rays aren't able to penetrate it, and therefore any information about the painting is masked. Image-processed ultrasonic amplitude scans, however, are not affected by changing paint layer thicknesses and ultrasound can penetrate the lead white paint layer.

The unique features of this system that make it of value in the conservation field should be emphasized. Ultrasonic components, such as the transducers, do not touch the material being examined. There is no evidence that the ultrasonic wave causes any damage. Moreover, a portable version of the system could be designed quite cheaply; for example, it could use less expensive transducers and a simpler transducer holder than were used in the experiments described here.

CONCLUSIONS AND FUTURE WORK

An air-coupled ultrasonic imaging system has been assembled for inspecting wooden panels. This system operates in the through-transmission mode, using two transducers, one on either side of the object; one transducer sends an ultrasonic signal and the other receives it. The signal has sufficient signal-to-noise and can be exploited to yield both phase and amplitude information, as well processed amplitude and phase versions using the image processing technique of thresholding where all the pixels above a certain grey level are black and the pixels below that level are white. Two-dimensional scans of samples are generated.

The ultrasound system reliably detected flaws in the panel paintings *Parental Admonition* and *Women Gathering Yucca Plants*. Cracks, delaminations, craquelure in the paint layer, wood grain, and knots found within the samples were all clearly imaged by the system.

The system maps flaws in wooden panel paintings that cannot be detected by other techniques such as radiography. The ultrasonic system was clearly shown to be more suitable for detecting certain flaws, such as in-plane cracks and delaminations.

Further work is needed in the area of image recognition in order to improve the reliability and ease of use of the air-coupled ultrasound system. In particular, it is important to develop methods that will distinguish between images of natural anomalies (grain) and flaws (cracks). Incorporation of information from visual examination and from other nondestructive techniques, such as radiography, will be useful here. As new transducers become available, the same transducer will be able both to send and to receive signals; this will remove the need to access both sides of the art work. Using more image processing may also be useful.

Air-coupled ultrasound is a potentially important, non-intrusive inspection method. The technique can yield information that will improve the long-term care of paintings on wood and, by implication, other works of art.

Acknowledgements

We would like to thank the Conservation Analytical Laboratory, the Smithsonian and the Center for Nondestructive Evaluation, The Johns Hopkins University for their financial support. We also grateful to Dr. Hans Goedicke and the National Museum of American Art, the Smithsonian Institution for lending us the paintings.

REFERENCES

1. Mecklenburg, M.F., and C.S. Tumosa, "Mechanical Behavior of Paintings Subjected to Changes in Temperature and Relative Humidity," *Art in Transit - Studies in the Transport of Paintings*, (Washington, D.C.: National Gallery of Art, 1991) 173-216.
2. Mairinger, F., and M. Schreiner, "Analysis of supports, grounds and pigments," *Art History and Laboratory Scientific Examination of Easel Painting, in the Journal of the European Study Group on Physical, Chemical and Mathematical Techniques Applied to Archaeology*, (Strasbourg, France: Council of Europe, Parliamentary Assembly, 1986) 13, 171-184.

   Wainwright, Ian N.M., "Examination of Paintings by Physical and Chemical Methods," *Shared Responsibility Proceedings, National Gallery of Canada, 26 to 28 October 1989* (Ottawa, Canada: National Gallery of Canada, 1990).

   Murray, A., R.E. Green, M.F. Mecklenburg, and C.M. Fortunko, "NDE Applied to the Conservation of Wooden Panel Paintings," edited by C.O. Ruud, J.F. Bussière, R.E. Green, Jr., *Nondestructive Characterization of Materials IV*, (New York: Plenum Press, 1991).
3. Murray, Alison, "Air-Coupled Ultrasound Used to Detect Flaws in Paintings on Wooden Panels," Ph.D. dissertation, The Johns Hopkins University, 1993.

   Fortunko, C.M., M.C. Renken, and A. Murray, "Examination of Objects Made of Wood Using Air-Coupled Ultrasound," *IEEE Ultrasonics Symposium 1990* (IEEE, 1991), 1099-1103.

   Fortunko, C.M., D.E. Chimenti, "Characterization of Pre-Impregnated Graphite Epoxy Lamina with Gas-Coupled Ultrasound," *IEEE Ultrasonics Symposium 1993* (IEEE, to be published).

   Fortunko, C.M., D.P. Dubé, J.D. McColskey, "Gas-Coupled Acoustic Microscopy in Pulse-Echo Mode," *IEEE Ultrasonics Symposium 1993* (IEEE, to be published).
4. Birks, A.S., "Particleboard Blow Detector," *Forest Products Journal* 22.6 (1972): 23-26.
5. Beall, F.C., "Acoustic Emission and Acousto-Ultrasonic Characteristics," *Wood Encyclopedia*, 1-2.
6. Murray, A., C.M. Fortunko, M.F. Mecklenburg, and R.E. Green, Jr., "Detection of Delaminations in Art Objects Using Air-Coupled Ultrasound," *Materials Issues in Art and Archaeology III*, Materials Research Society Volume 267, (Materials Research Society, 1992), 371-377.
7. Murray, Ph.D. dissertation, 1993.

   Murray, A., E.S. Boltz, C.M. Fortunko, M.F. Mecklenburg, R.E. Green, Jr., "Air-coupled Ultrasonic System for Nondestructive Evaluation of Wooden Panel Paintings," *3rd International Conference on Non-Destructive Testing, Microanalytical Methods and Environmental Evaluation for Study and Conservation of Works of Art, Viterbo, Italy, 4-8 October 1992.*
8. Murray, Ph.D. dissertation, 1993.

MODEL-BASED CALIBRATION OF ULTRASONIC SYSTEM RESPONSES FOR QUANTITATIVE MEASUREMENTS

Lester W. Schmerr, Jr.[1], Sung-Jin Song[2], and Huilian Zhang[3]

[1]Center for NDE and Dept. of Aerospace Engineering
and Engineering Mechanics
[2]Control and Instrumentation Dept.
Research Institute of Industrial Science and Technology
Pohang, 790-600 Korea
[3]Dept. of Electrical Engineering
Iowa State University
Ames, IA 50011

INTRODUCTION

In order to perform quantitative ultrasonic NDE measurements, it is necessary to make those measurements independent of the characteristics and settings of the pulser-receiver used and the choice of the transducer employed. The combined effects of those parts of the ultrasonic system on the measured signal can be described mathematically in terms of a single system "efficiency" factor.

Here, we will show that ultrasonic system efficiency factors can be experimentally determined with the use of recently developed ultrasonic models for a variety of calibration scatterers, including plane front and back surfaces, flat-bottom holes, and solid cylinders. Because of the generality of the models employed, these scatterers can be in either the near or farfield of a planar transducer. It is also demonstrated that the system efficiency factor determination can be made independent of the choice of the material in the sample being considered provided that suitable material attenuation corrections are made.

EFFICIENCY FACTOR DETERMINATION

First, consider a general ultrasonic immersion setup for making flaw or material measurements. In any of these applications, an input voltage pulse, $V_i(t)$, drives a piezoelectric transducer in a fluid which converts the electrical energy into mechanical motion. Here, we assume the motion generated is a spatially uniform velocity, $v_o(t)$, over the face of the transducer face (planar piston transducer model) whose frequency components, $v_0(\omega)$, are proportional to the frequency components of the input voltage, $V_i(\omega)$, i.e.

$$v_o(\omega) = \beta_i(\omega)V_i(\omega) \tag{1}$$

where ω is the frequency (in radians/sec) and $\beta_i(\omega)$ is an input proportionality factor. This motion generates a beam of ultrasound which interacts with the material or scatterer under examination and is partially scattered back to the same transducer, now acting as a receiver. If we also assume that the frequency components, $V_m(\omega)$, of the voltage pulse received are proportional to the spatially averaged velocity received at the transducer as a function of frequency, then we may write

$$V_m(\omega) = \beta_r(\omega) < v_m(\omega) > \exp\left(-\sum_{i=1}^{n} \alpha_i(\omega) d_i\right) \tag{2}$$

where $\beta_r(\omega)$ is a receiving proportionality factor and $< v_m(\omega) >$ represents the spatially averaged velocity received over the face of the transducer if the incident and scattered waves travel in ideal (non-attenuating) media. The exponential terms in eq.(2) account for the presence of material attenuation through the attenuation coefficients, $\alpha_i(\omega)$, of the ith media and the corresponding pathlengths, d_i. Combining eqs(1) and (2) we find

$$V_m(\omega) = \beta(\omega) < v_m(\omega) > \exp\left(-\sum_{i=1}^{n} \alpha_i(\omega) d_i\right) / v_o(\omega) \tag{3}$$

in terms of a single total "efficiency" factor $\beta(\omega) = \beta_r(\omega)\beta_i(\omega)V_i(\omega)$. For the moment, assume that we can, through a combination of models and measurements, obtain both $\beta(\omega)$ and the attenuation coefficients and pathlengths in eq.(3). Then the measurement of the voltage frequency components of a particular material measurement, $V_m(\omega)$, can be used to directly obtain, through eq.(3), the normalized average velocity, $< v_m > / v_o$. Thompson and Gray[1], for example, have shown how further modelling of $< v_m > / v_o$ and the concept of deconvolution can be used to directly determine the far-field response of a flaw. In contrast, Tang et al.[2] have shown how measurements and models can be used to obtain both $< v_m > / v_o$ and $\beta(\omega)$ so that measurements of $V_m(\omega)$ can be used to determine the unknown attenuation of the material under inspection. In both cases, $\beta(\omega)$ is obtained through the use of a separate reference experiment where the voltage frequency components, $V_r(\omega)$, are measured in a setup with the same transducer and system settings (so that $\beta_i, \beta_r,$ *and* V_i are the same as in the original measurement setup) and where the normalized average velocity and attenuation terms are known. Figs 1a-d, for example, show four different possible reference calibration setups. For any of these cases we have a relationship entirely similar to eq.(3), i.e.

$$V_r(\omega) = \beta(\omega) < v_r(\omega) > \exp(-\Sigma \alpha_i D_i) / v_o(\omega) \tag{4}$$

In principle, then, $\beta(\omega)$ can be obtained from this reference experiment through a simple division process. However, when V_r and $< v_r(\omega) > / v_0(\omega)$ are both small, noise contaminates this division process, making it unreliable. To overcome this difficulty, one approach has been to use the concept of a Wiener filter [3]. Using such a filter, we obtain

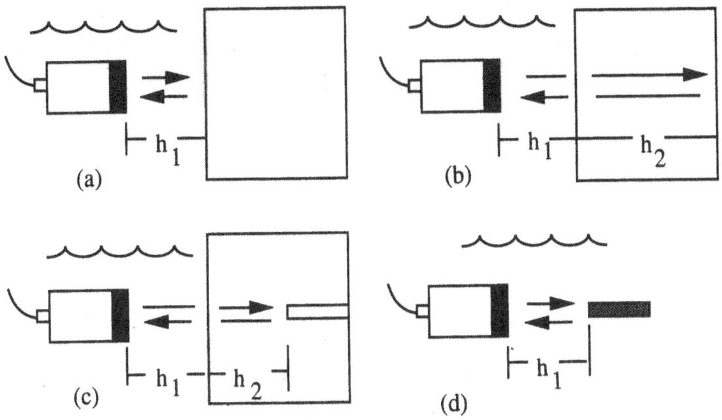

Fig. 1. Reference scattering geometries using (a) the reflection from the plane front surface of a specimen, (b) the reflection from the plane back surface of a specimen, (c) the backscatter response of an on-axis flat-bottom hole in a ASTM specimen, (d) the backscatter response of an on-axis solid cylinder in a fluid.

$$\beta(\omega)=\frac{V_r}{<v_r>/v_o \exp\left(-\sum\alpha_i D_i\right)}\left[\frac{\left|<v_r>/v_o \exp\left(-\sum\alpha_i D_i\right)\right|^2}{\left|<v_r>/v_o \exp\left(-\sum\alpha_i D_i\right)\right|^2+\varepsilon^2}\right] \qquad (5)$$

where ε is a small constant whose value is chosen so as to desensitize the division process. Since the use of such filters have been described elsewhere [3,4], we refer the reader to those sources for further details.

PLANE SURFACE REFERENCE MODELS

Front Surface Model

To turn eq.(5) into a useful expression for the determination of $\beta(\omega)$, it is necessary to have an explicit expression for the normalized average velocity received by the transducer, i.e. an explicit reference scattering model, and a knowledge of the material attenuation coefficients. One commonly used reference scattering configuration is that of a plane front surface (Fig. 1a) of a specimen oriented parallel to the face of the transducer. For the waves reflected once from the front surface and received by the transducer, we have[5]:

$$<v_r>/v_o=<v_{f.s.}>/v_o \qquad (6)$$

where

$$<v_{f.s.}>/v_0=R_{12}\exp(2ik_1h_1)C\left(k_1a^2/2h_1\right) \qquad (7a)$$

and

$$C(k_1a^2/2h_1)=\{1-\exp(ik_1a^2/h_1)[J_0(k_1a^2/2h_1)-iJ_1(k_1a^2/2h_1)]\} \quad (7b)$$

C is a diffraction coefficient and R_{12} is the ordinary plane wave reflection coefficient (at normal incidence) for the interface between the water and the specimen, k_1 is the wavenumber of the water, a is the radius of the transducer, and J_0, J_1, are Bessel functions. Although eq.(6) is strictly speaking only an approximate high frequency solution for the response from the interface between two fluid media, several studies[6,7] have validated its use for the fluid-solid interface. It should be noted that nothing in the derivation of eq.(6) restricts the location of the interface[5]. Thus, eq.(6) is applicable when the interface lies in either the near- or far-field of the transducer radiation field.

For this case the propagation distance in eq.(5) is $D_1 = 2h_1$ (Fig. 1a) and the attenuation term is that of water, i.e. $\alpha_1 = \alpha_w$. For water at room temperature we have[8]:

$$\alpha_w = 25.3\times10^{-15} f^2 \quad m^{-1} \quad (8)$$

where the frequency, f, is in Hz.

Back Surface Model

In a similar manner, one can use the plane back surface of a specimen as a reference. Here[9]

$$<V_r>/V_0 = <V_{b.s.}>/V_0 \quad (9)$$

where

$$<V_{b.s.}> V_0 = T_{12}R_{21}T_{21}\exp(2ik_1h_1+2ik_2h_2)C(k_1a^2/2h) \quad (10)$$

and $h = h_1 + c_2h_2/c_1$ (Fig. 1b). Here, c_1 and c_2 are the wavespeeds for the water and solid, respectively, and the R_{21} and T_{12}, T_{21} are the plane wave reflection and transmission coefficients. In this case one needs to know both the attenuation of the solid, α_s, as well as that of the water. In eq. (5), then, $\alpha_1 = \alpha_w$ and $D_1 = 2h_1, D_2 = 2h_2$. The water attenuation term can again be obtained from eq.(8). For the solid, we can use the ratio of measurements of the front and back surface echoes to eliminate the efficiency factor, β, and the common water attenuation coefficient to yield, from eqs. (4), (6), and (9), an expression that can be solved for the attenuation of the solid. Formally, we obtain

$$\exp(-2\alpha_s(\omega)h_2) = |B(\omega)|/|F(\omega)| \quad (11)$$

where

$$|F(\omega)| = |V_{b.s.}(\omega)R_{12}C(k_1a^2/2h_1)| \quad (12)$$

and

$$|B(\omega)| = \left|V_{f.s.}(\omega)T_{12}R_{21}T_{21}C\left(k_1a^2/2h\right)\right| \tag{13}$$

In eqs. (12),(13) $V_{b.s.}$ and $V_{f.s.}$ are the frequency components of the voltage pulses received from the back and front surface echoes, respectively. The division process of eq. (11), however, is also contaminated by noise at both high and low frequencies so that again using the Wiener filter approach, we replace eq.(11) by

$$\exp\left(-2\alpha_s(\omega)h_2\right) = \frac{|B(\omega)|}{|F(\omega)|}\left\{\frac{|F(\omega)|^2}{|F(\omega)|^2+\varepsilon^2}\right\} \tag{14}$$

which is now a well-behaved expression for the determination of α_s.

FLAT-BOTTOM HOLE MODEL

Equation (6) together with eq. (8), or eq. (9), together with eqs.(8) and (14), provide all the basic terms needed in eq.(5) to obtain the efficiency factor from either the plane front of back surface of a specimen. In a similar manner, other reference reflectors can, of course, be considered. A common calibration setup used in ultrasonic NDE involves a block containing a flat-bottom hole (Fig. 1c). In this case Song, Schmerr, and Sedov[6] have obtained an explicit expression for the received average velocity, which we write symbolically as:

$$< v_r > / v_0 = < v_{fbh} > / v_0 \tag{15}$$

Space does not permit us to give the rather lengthy expression for the normalized average velocity here so we refer the reader to Song, Schmerr, and Sedov[6] where the explicit form is given. The attenuation terms in this case are $\alpha_1 = \alpha_w$ and $\alpha_2 = \alpha_s$ and the propagation distances in eq.(5) are $D_1 = 2h_1, D_2 = 2h_2$. As in the case of the back surface response, the water and solid attenuation coefficients can be determined, if needed, from eqs.(8) and (14), respectively.

SOLID CYLINDER MODEL

The flat-bottom hole is a popular reference scatterer since it represents a highly specular reflector, like a crack, when viewed at normal incidence. However, when using the flat-bottom hole to obtain $\beta(\omega)$ through eq.(5), it is necessary, as just described, to measure the attenuation of the solid. This extra step can be eliminated if we use instead the flat end of a solid cylinder in an immersion setup (Fig. 1d). In this case, a high frequency Kirchhoff-like approximation for the scattering problem[10] gives the average received velocity symbolically as:

$$< v_r > / v_0 = < v_{cyl} > / v_0 \tag{16}$$

Again, for space reasons we do not write down the explicit form here but refer the reader to Sedov, Schmerr, and Song[10] where an explicit expression is given. The solid cylinder, like the flat-bottom hole, is a highly specular reflector but in this case the only attenuation coefficient needed is α_w, which can be obtained from eq. (8), and the propagation distance in eq.(5) is $D_1 = 2h_1$(Fig. 1d).

EXPERIMENTAL TESTS

The previous sections have discussed a variety of reference scatterers that can, in principle, be used in making system efficiency measurements. Here, we will demonstrate that consistent $\beta(\omega)$measurements are possible with any of these configurations by the use of the appropriate attenuation corrections as described previously.

Figure 2, for example, shows the results for the efficiency factor versus frequency for a 10 MHz, 1/4 inch diameter transducer, calculated from either the planar front or back surface of an aluminum specimen, using eqs. (6) and (9) and including the attenuation corrections for the water and aluminum through eq.(8) and eq.(14). As can be seen, there is excellent agreement between the two results, indicating that the efficiency factor can be calculated with either of these reference scattering configurations.

The system efficiency factor, β by definition is not dependent on the properties of the reflector being used. Thus, β should also be independent of the material properties of the reference scatterer being employed. Fig. 3 shows that this is indeed the case for a 10 MHz, 1/4 inch diameter plane transducer, using plane front surface reflections from aluminum, titanium, or steel specimens. Thus, β calculations, can, with careful measurements, also be transferred across materials.

Determination of β should also be independent of the geometry of the scatterer being used. Fig. 4 compares the efficiency factor calculations for a 10 MHz, 1/4 inch diameter plane transducer using four types of reference scattering geometries: 1) the front surface of a steel specimen, 2) the response from a flat-bottom hole in a steel block, 3) the response of a 4/64 inch diameter steel cylinder in water, and 4) the response of a 6/64 inch diameter steel cylinder in water. In each of these cases the appropriate attenuation corrections were included. As can be seen from Fig. 4, there is good consistency

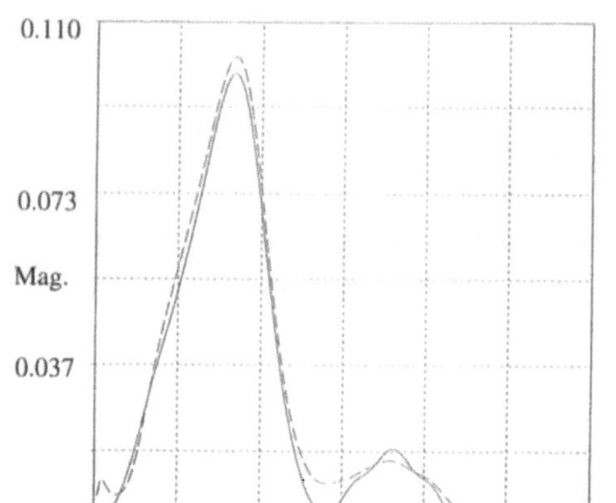

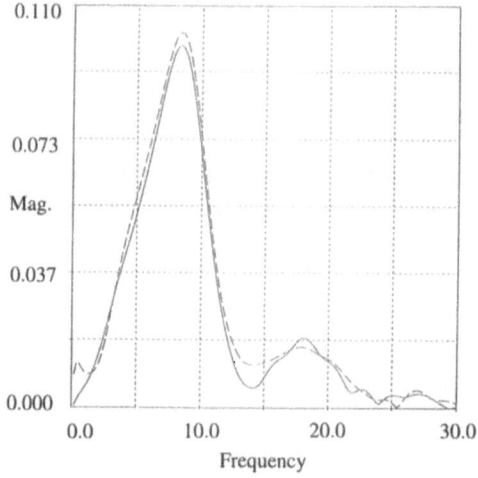

Fig. 2 Efficiency factor versus frequency for a 10 MHz, 1/4 inch diameter planar transducer calculated with (a) a planar front surface response ______, (b) a planar back surface response ------.

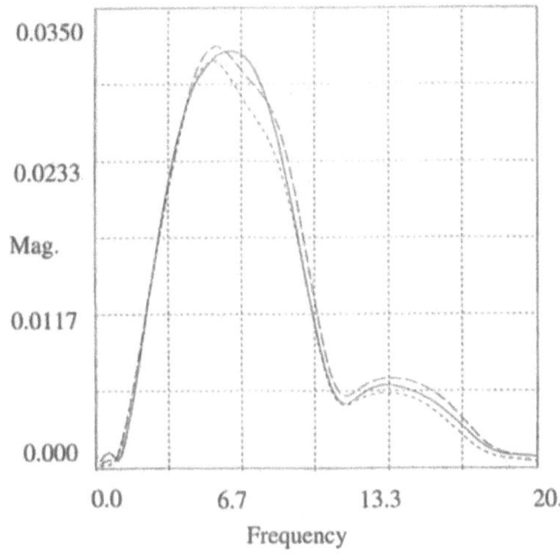

Fig. 3. Efficiency factor versus frequency for a 10 MHz, 1/4 inch diameter planar transducer using the front surface response of a specimen made of (a) aluminum ______, (b) titanium _ _ _ _ (c) steel ------.

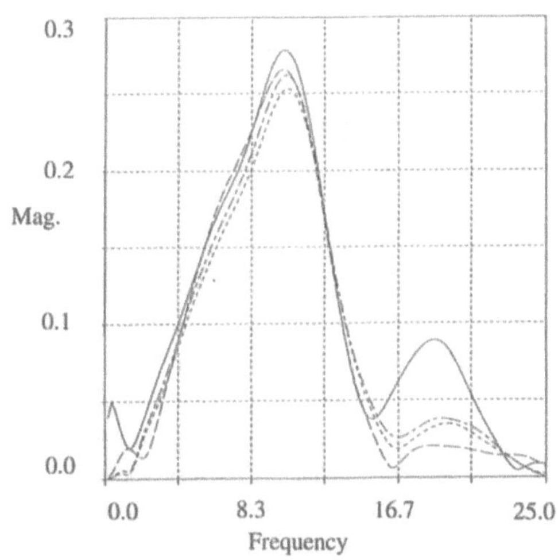

Fig. 4 . Efficiency factor versus frequency calculated for a 10 MHz, 1/4 inch diameter planar transducer using (a) the planar front surface of a steel specimen ______, (b) the backscatter response of a 5/64 inch diameter flat-bottom hole in an ASTM 4340-5-0038 specimen _ _ _ (c) the backscatter response of a 4/64 inch diameter steel cylinder in an immersion setup - - - (d) the backscatter response of a 6/64 inch diameter steel cylinder in an immersion setup _ - _ - _ .

(maximum differences are normally 10% or less across the major portion of the transducer spectrum) for all these configurations. Thus, the determination of β by these methods is transferable across different materials and geometric setups.

CONCLUSIONS

We have demonstrated a general procedure for determining the transducer system efficiency factor for an ultrasonic system. This determination requires both a model of the average velocity received by the transducer and measurements of the attenuation corrections for the materials under consideration. Although we have only shown these procedures for planar transducers, similar methods can also be used for focused probes[11]. We have also demonstrated that the results obtained from one material and geometrical setup can be transferred to a different material and/or geometry.

ACKNOWLEDGMENTS

This work was supported by the Center for NDE, Iowa State University.

REFERENCES

1. R.B.Thompson and T.A. Gray, "A model relating ultrasonic scattering measurements through liquid-solid interfaces to unbounded medium scattering amplitudes," J. Acoust. Soc.Am.74:1279 (1983).
2. X.M. Tang, M.N. Toksoz, and C. H. Cheng, "Elastic wave radiation and diffraction of a piston source," J. Acoust. Soc.Am. 87:1894 (1990).
3. D. Elsley, J. Richardson, and B. Addison,"Optimum measurement of broadband ultrasonic data," 1980 Ultrasonics Symposium Proceedings, IEEE, N.Y., 916 (1980).
4. Y.B. Murakami, B.T. Khuri-Yakub, G.S. Kino, J.M. Richardson, and A.G. Evans, " An application of Wiener filters to nondestructive evaluation,"Appl. Phys. Ltrs. 33:685 (1978).
5. P.H. Rogers and A.L. Van Buren,"An exact expression for the Lommel correction integral," J. Acoust. Soc.Am. 55:724 (1974).
6. S.J. Song, L.W. Schmerr, and A. Sedov,"DGS diagrams and frequency response curves for a flat-bottom hole: a model-based approach," Res. Nondestr. Eval. 3:201 (1991).
7. D.H. Green and H.F. Wang,"Shear wave diffraction loss for circular plane-polarized source and receiver," J. Acoust. Soc.Am. 90:2697 (1991).
8. J.M.M. Pinkerton, "A pulse method for the measurement of ultrasonic absorption in liquids: results for water," Nature 160:128 (1947).
9. A. Sedov, L.W. Schmerr, and S.J. Song,"Ultrasonic scattering by a flat-bottom hole in immersion testing: an analytical model," J. Acoust. Soc.Am. 92:478 (1992).
10. A. Sedov, L.W. Schmerr, and S.J. Song, "A bounded beam solution for the pulse-echo transducer response of an arbitrary on-axis scatterer in a fluid," Wave Motion (to appear), (1993).
11. L.W. Schmerr, T. Lerch, and A. Sedov,"A focused transducer/scatterer model for ultrasonic reference standards," Review of Progress in Quantitative Nondestructive Evaluation, D.O. Thompson and D.E. Chimenti, Eds., Plenum Press, N.Y., 12A, 925-931, 1993.

ULTRASONIC CHARACTERIZATION OF TEXTURE IN PURE AND ALLOYED ZIRCONIUM

A. Moreau[1], P.J. Kielczynski[2], J.F. Bussière[1], and J.H. Root[3]

[1] National Research Council of Canada
Industrial Materials Institute
75 de Mortagne Blvd.
Boucherville, Québec
Canada J4B 6Y4

[2] Same as above. Current address:
Research and Productivity Council
921 College Hill Road
Fredericton, New Brunswick
Canada E3B 6Z9

[3] AECL Research
Chalk River Laboratories
Chalk River, Ontario
Canada K0J 1J0

INTRODUCTION

Recently, we have developed a method to relate the angular dependence of ultrasonic velocities to the five expansion coefficients of the crystallographic orientation distribution function (CODF) of hexagonal materials with orthorhombic macroscopic symmetry.[1] The ultrasonic velocity measurements are performed with an acoustic microscope on one, two, or three of the principal planes of symmetry using one or more ultrasonic modes. In this paper, the theory of ultrasonic measurement of texture is reviewed, and we present data showing that it can be applied to two-phase alloys, when the primary phase makes up most of the material.

The paper begins with the expansion of the CODF in terms of generalized spherical harmonics and a discussion of the physical significance of the expansion coefficients W_{lmn}, in terms of the crystallite's c-axis orientations. Various methods used to obtain the W_{lmn} ultrasonically are summarized and discussed using a general theory. Results are presented for a sample of pure zirconium and six samples of Zr-2.5%Nb alloys containing 5 to 10 % beta (cubic) phase. The ultrasonic texture coefficients are compared to those obtained by neutron diffraction. A discussion of measurement errors, second phase effects, and averaging methods follows.

THE CRYSTALLITE ORIENTATION DISTRIBUTION FUNCTION

Texture is described by a crystallite orientation distribution function (CODF) that describes the probability of finding a single crystallite in a specific orientation. The crystallite orientation is described by the three Euler angles, ψ, θ, and ϕ. These angles relate the crystallite's internal coordinates to an external system of coordinates. This paper considers the particular case of hexagonal crystallites in an aggregate of orthorhombic symmetry. In this case, ψ and θ represent the "longitude" and the "latitude" of the crystallite's c axis with respect to the aggregate's axes, and ϕ represents the rotation of the crystallite around its own c axis.

The CODF, denoted here by $w(\psi,\theta,\phi)$, may be expanded in terms of generalized spherical harmonics,

$$w(\psi,\theta,\phi) = \sum_{l=0}^{\infty} \sum_{m=-l}^{l} \sum_{n=-l}^{l} W_{lmn} Z_{lmn}(\cos\theta)\, e^{-im\psi} e^{-in\phi} \quad , \tag{1}$$

where $Z_{lmn}(\cos\theta)$ is a generalization of the associated Legendre function, and W_{lmn} are the coefficients of the expansion.[2] The normalization condition for the CODF is

$$\int_0^{2\pi}\int_0^{2\pi}\int_{-1}^{1} w(\psi,\theta,\phi)\, d(\cos\theta)\, d\psi\, d\phi = 1 \quad . \tag{2}$$

It has been shown that the effective (second order) elastic constants of a polycrystalline aggregate depend only on expansion terms for which $l \leq 4$.[3] This very important result implies that fine details of the CODF will not be measurable with linear ultrasonics. Conversely, all linear acoustic properties and all linear mechanical properties of a polycrystalline aggregate only depend on a small number of texture coefficients.

Because of symmetry properties of orthorhombic aggregates of hexagonal crystallites,[2] the W_{lmn} are real, $W_{lm0} = W_{l\bar{m}0}$ and only the following texture coefficients with $l \leq 4$ are non zero: W_{000}, W_{200}, W_{220}, W_{400}, W_{420}, and W_{440}.[4] Moreover, Z_{lmn} always satisfies the relation $Z_{lmn} = (-1)^{m+n} Z_{l\bar{m}\bar{n}}$. Using these relations, we can write

$$w(\psi,\theta,\phi) \approx \sum_{l=0,2,4} \sum_{\substack{m=-l \\ m\ \text{even}}}^{l} W_{lm0} Z_{lm0}(\cos\theta) \cos m\psi \quad . \tag{3}$$

To better understand the meaning of the non-zero W_{lm0}, we plot $w_{lm0}(\theta,\psi) = Z_{lm0}(\cos\theta)\cos m\psi$ in spherical coordinates in Figure 1. In Figure 1, the vertical axis is the macroscopic z axis and the circle in the xy plane is labeled in degrees from the x axis. The new basis function, $w_{lm0}(\theta,\psi)$, is related to the probability of finding the c axis of a hexagonal crystallite pointing in the (θ,ψ) direction if all other texture coefficients are zero. Note that $w_{lm0}(\theta,\psi)$ can be positive or negative.

The w_{000} function is independent of θ and ψ, is equal to $\sqrt{2}/2$, and is trivially shown as a sphere in Figure 1. It represents an isotropic distribution. It is also the only term that integrates to a non zero value in the normalization integral (2). Therefore, W_{000} is always equal to $\sqrt{2}/8\pi^2$. The w_{200} function is axisymmetric and has two positive lobes in the $\pm z$ directions and a negative ring in the xy plane. Negative values of $w_{lm0}(\theta,\psi)$ are represented as positive radii in the $(\pi - \theta,\ \psi + \pi)$ direction. However, the $w_{lm0}(\theta,\psi)$ considered have point symmetry and Figure 1 can be thought of as $|w_{lm0}(\theta,\psi)|$. Although negative probabilities are unphysical, the individiual w_{lm0} are never observed individually,

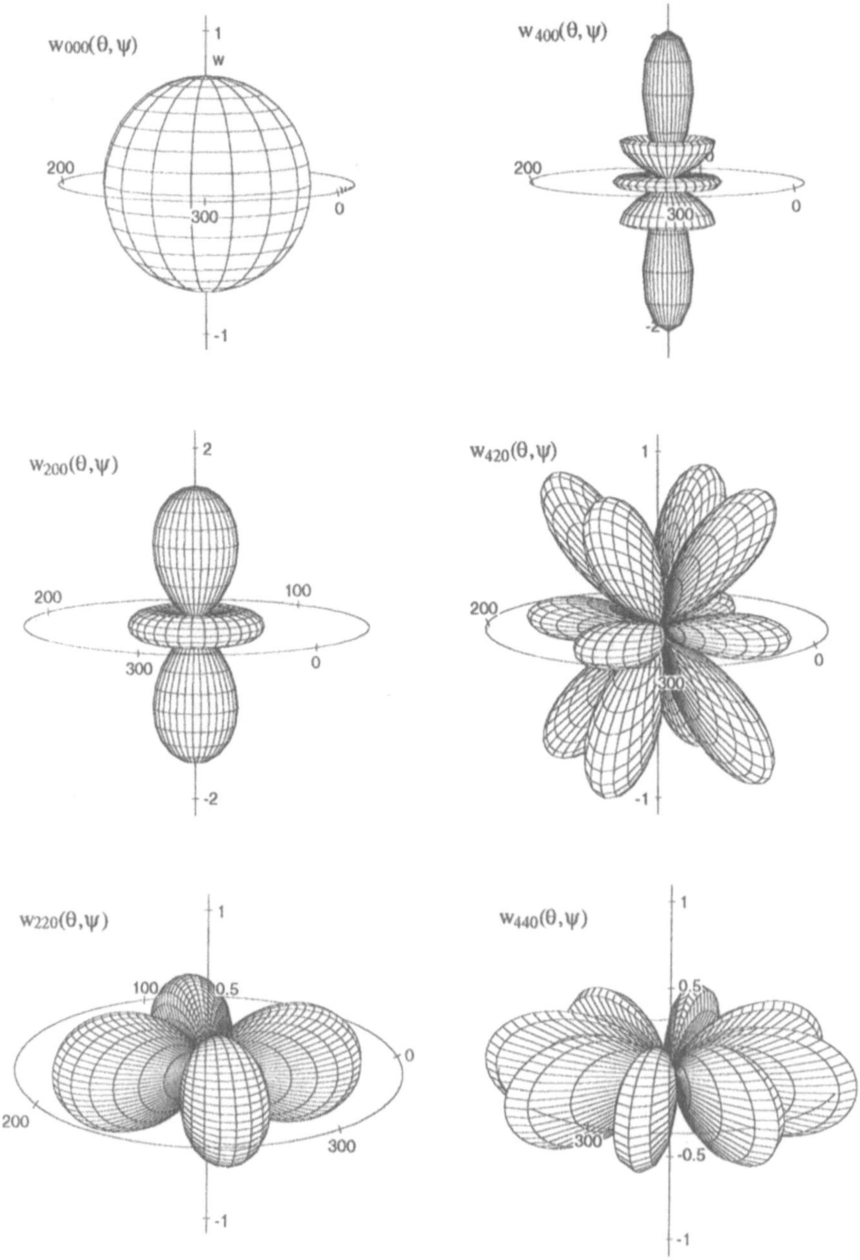

Figure 1. Three dimensional representation of the basis functions $w_{lm0}(\theta,\psi)$ for $l \leq 4$ and even values of m. The vertical axis is the z axis and a circle labeled in degrees from the x axis is drawn in the xy plane. The radii of the circles are: 1.0 for w_{000}, w_{420}, and w_{440}, 1.2 for w_{220}, and 2.0 for w_{200} and w_{400}. The sign of each lobe alternates, as described in the text.

but as part of the expansion (1). In particular, they must be added to the constant w_{000} term. Consequently, positive or negative values of W_{200} indicate that the c axes of the hexagonal crystallites are preferentially aligned along the z axis or in the xy plane, respectively. The w_{220} has two positive lobes along the x axis and two negative lobes along the y axis. Consequently, positive or negative values of W_{220} indicate that the c axes of the hexagonal crystallites are preferentially aligned along the x or y axes, respectively. The w_{400} function is axisymmetric and has two positive lobes along the z axis, one positive ring in the xy plane, and two negative "cones" around the z axis. The effect of w_{400} on texture can be quite complex, but one important effect is to sharpen the w_{200} distribution and to increase the probability of finding the crystallite's c axis aligned along the macroscopic z axis. Alternately, negative values of W_{400} may occur with positive values of W_{200} and a non zero value of W_{220}. This results in a twin peak about the z axis. This texture may be observed in Zr rolled plates when the z axis is defined as the direction transverse to the rolling direction. The function w_{420} is considerably more complicated and consists of 12 lobes of alternating sign, with negative lobes along the x axes. The function w_{440} consists of 8 lobes of alternating signs, all located in the xy plane, and with positive lobes along the x and y axes.

From the above representation of the w_{lm0} functions, it is clear that the expansion of the CODF in terms of the generalized spherical harmonics treats the z axis differently from the x and y axes. Although the basis functions form a complete set, the limited number of coefficients that can be obtained by acoustic measurements makes the choice of macroscopic axes important because a better choice can reduce the error sum-of-squares, i.e., the integral of the squares of the difference between the CODF and its truncated expansion. Conventional pole figures of the various basis functions and of the CODF can be obtained using a contour plot of a projection of the above representation. Given that only the probability distribution of the crystallite c axes can be obtained, the (0002) pole figure is the only relevant one. Pole figures of w_{400}, w_{420}, and w_{440} on a xy plane projection have been published elsewhere.[5]

ULTRASONIC MEASUREMENTS OF TEXTURE

Recently, we have shown how to relate the angular dependence of the acoustic velocity of various acoustic modes to the five texture coefficients W_{200}, W_{220}, W_{400}, W_{420}, and W_{440}.[1] The effective elastic constants of the aggregate, $\overline{C_{ij}}$, are expressed as[6]

$$\overline{C_{ij}} = C_{ij}^0 + \Delta C_{ij}, \qquad (4)$$

i.e., as a sum of an isotropic value,

$$C_{ij}^0 = C_{ij}^0(c_{sc}), \qquad (5)$$

and a texture dependent term,

$$\Delta C_{ij} = \Delta C_{ij}(c_{sc}, W_{lmn}), \qquad (6)$$

subject to the weak anisotropy condition

$$\Delta C_{ij} << C_{ij}^0. \qquad (7)$$

In (4) and (5), c_{sc} are the single crystal elastic constants. These expressions can be found using either Reuss, Voigt, or Hill averages.[7]

Experimentally, the angular dependence of the acoustic velocity in one to three principal planes of the orthorhombic aggregate is fitted to the approximate equation

$$v(\gamma) \approx v_0 + P + Q\cos 2\gamma + R\cos 4\gamma \,, \tag{8}$$

where v is the acoustic velocity of the aggregate, γ is the propagation angle in the principal plane, v_0 is the isotropic velocity, and P, Q, and R are the fitted parameters. In the weak anisotropy condition (7), equation (8) is a good approximation to the solution of the Christoffel's equation,[8] and the fitted parameters can be expressed in terms of the effective elastic constants. Therefore, we can write

$$\begin{aligned} P &= P(\overline{C}) = P(c_{sc}, W_{lmn}) \\ Q &= Q(\overline{C}) = Q(c_{sc}, W_{lmn}) \\ R &= R(\overline{C}) = R(c_{sc}, W_{lmn}) \,. \end{aligned} \tag{9}$$

These expressions have been obtained for bulk longitudinal and shear waves, Rayleigh surface waves, and Lamb symmetric plate waves.[1]

If enough acoustic modes are observed in one or several principal planes so that at least five different values of P, Q, and R are known, the W_{lmn} can be found by inverting (9).[1] (Note that if not all W_{lmn} are desired, fewer values of P, Q, or R may be required.) If more than five different values of P, Q, and R are known, equations (9) become overdetermined and a least squares fitting procedure can be used.[9] However, it has been pointed out that the W_{lmn} evaluated using the P values (absolute velocity measurements) are much more sensitive to errors in the single crystal elastic constants and density than those evaluated using only Q and R values (relative velocity measurements).[1] On the other hand, the R values are usually smaller and more difficult to measure than the Q values. Therefore, the choice of procedure to use for the inversion of (9) is very dependent on the amount and quality of the acoustic data.

THE ACOUSTIC MICROSCOPE

The measurements presented here were made with a 225 MHz line-focus acoustic microscope.[10] This method of measurement has the advantages that only small samples and relatively little machining are needed, and that the acoustic velocity of surface waves may be measured at any angle on the sample surface. However, the acoustic microscope requires careful polishing of the sample surfaces and scans of the elastic properties are made on a relatively small volume near the surface. Therefore, the technique is limited to samples with small grain sizes and it is sensitive to surface texture only. Moreover, the microscope measures leaky Rayleigh waves (LRW) and leaky surface-skimming compressional waves (LSSCW) instead of the pure Rayleigh waves and bulk longitudinal waves described in the above theory. The LRW are thought to behave essentially as pure Rayleigh waves.[11] We assume that the LSSCW behave as bulk longitudinal waves, but recent work indicates that this assumption may not always be valid.[12]

RESULTS

Pure Zr sample

To test the validity of the inversion of (9) and of our experimental technique, we have obtained the texture coefficient of a 99.8 % pure Zr rod (Goodfellow Corporation, Malvern,

PA, part #ZR007910) using both acoustic measurements and neutron diffraction. The neutron diffraction measurements were made at the NRU reactor at AECL Research, Chalk River Laboratories. The acoustic measurements were obtained using the LRW and the LSSCW propagating in the x = 0 and y = 0 planes (Acoustic 1) and in the x = 0 and z = 0 planes (Acoustic 2). One hundred velocities were measured in intervals of 2°. The single-crystal elastic constants used are from the literature[13] and the density was measured. The sample grain size was of the order of 20 μm but this large grain size did not seem to affect the quality of the data. The results are shown in Table 1. The numbers in parentheses are the error estimates on the least significant digits. For the neutron diffraction measurements, the errors were deduced from the consistency of five different pole figures. For the acoustic measurements, the errors are obtained from the statistical error of the fitted values of P, Q, and R. To this statistical error, one must add an unknown systematic error resulting from the choice of averaging method. In Table 1, Hill's averaging was used. The choice of averaging method will be discussed further below.

Table 1. Comparison of pure-zirconium texture coefficients obtained using neutron diffraction and acoustic measurements. The acoustic measurements were made using LRW and LSSCW in the x = 0 and y = 0 planes (Acoustic 1) and in the x = 0 and z = 0 planes (Acoustic 2). Hill's averaging is used.

	Neutron	Acoustic 1	Acoustic 2
W_{200}	-0.0156 (4)	-0.0147 (4)	-0.0141 (7)
W_{220}	0.0002 (1)	-0.0010 (5)	0.0017 (3)
W_{400}	0.0097 (2)	0.0136 (10)	0.0111 (7)
W_{420}	-0.0001 (1)	0.0006 (5)	-0.0011 (2)
W_{440}	0.0000 (1)	0.0007 (9)	0.0018 (5)

Zr-2.5%Nb alloys

Metals are usually alloyed to tailor various physical properties. Therefore, acoustic texture measurement techniques are of limited use unless they can be applied to alloys. Some complications arising in alloys are the presence of additional phases and possible changes in the single crystal elastic constants caused by the solute. We do not attempt to model each phase separately and obtain texture coefficients for each phase but restrict ourselves to alloys where the primary phase constitutes most of the material, so that the second phase can be ignored in first approximation.

To verify the method's applicability to alloys, we measured six samples of Zr alloyed with 2.5% Nb. The samples are made of 90 to 95% hexagonal α phase and 5 to 10% cubic β phase. The α phase contains up to 1% of Nb in solution and the β phase contains from 30 to 50% Nb. The β phase is known to have a much weaker texture than the α phase.[14] The grain sizes are of the order of 1 μm. The samples were annealed to remove possible residual stress effects. The acoustic texture was obtained using the same α-Zr single crystal elastic constants as for pure Zr, and the measured density of the alloy. The acoustic texture coefficients of these six samples and of the pure Zr sample are plotted in Figure 2 as a function of the α phase coefficients obtained by neutron diffraction. For the acoustic measurements, Hill's averaging and the two acoustic waves LSSCW and LRW propagating in the x = 0 and y = 0 planes are used.

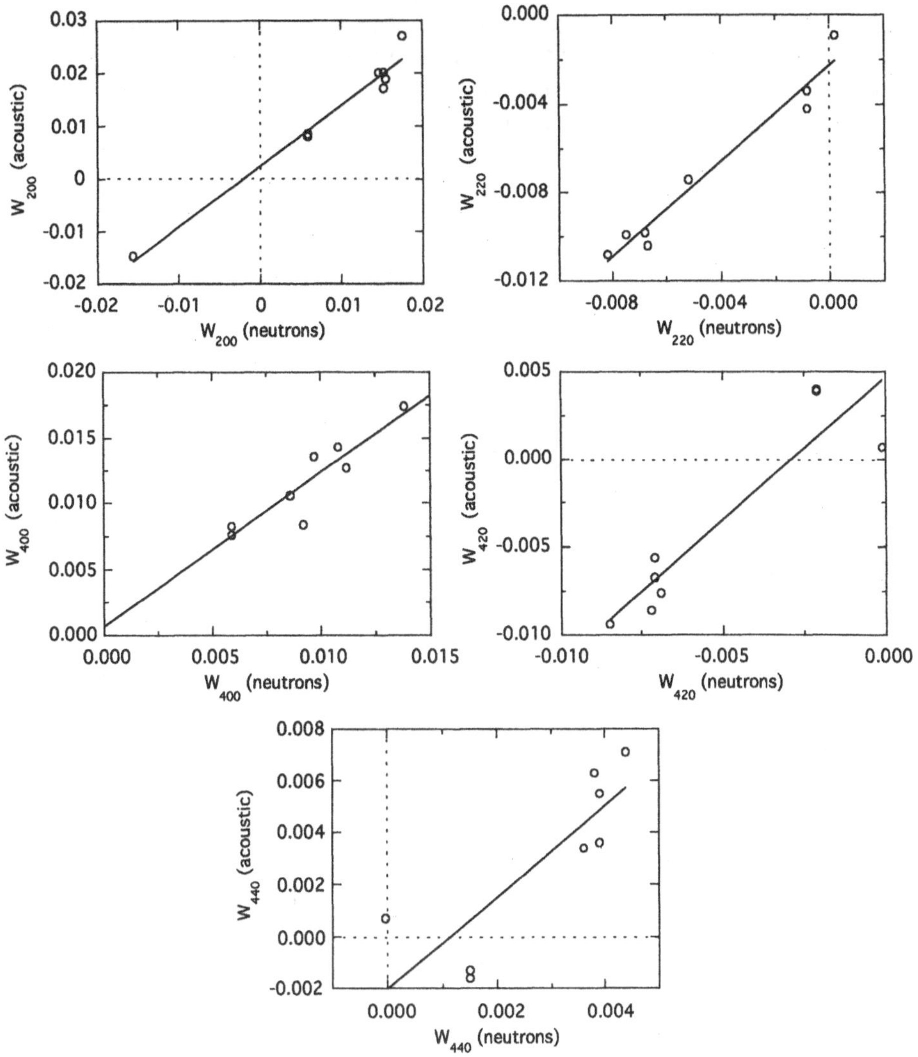

Figure 2. Texture coefficients as measured with ultrasound versus those obtained by neutron diffraction for one pure Zr and six Zr-2.5%Nb samples. The linear least squares fits shown include all data points.

In all cases (Fig. 2) the relation between the acoustic and neutron diffraction measurements is linear, the slope is nearly equal to one, and the y intercept is nearly zero. There occurred measurement difficulties for one sample: the measured acoustic velocities contained abnormally large errors, equation (8) was poorly satisfied, and some of the fitted values of P, Q, and R were unreliable. The measurements were repeated a second time without improvements and we do not know why this sample behaved differently. Consequently, the two points with neutron values of W_{420} near -0.002 and the two points with neutron values of W_{440} near 0.0015 are likely to be outliers. The linear relationship between neutron and acoustic measurements appears best for the W_{200} and W_{220} and, to a lesser extent, the W_{400} coefficients. Perhaps this should not be surprising because their corresponding basis functions are the simplest ones and have the largest amplitude (see Figure 1). Another reason for the better linearity of W_{200} and W_{220} might be that, because of symmetry, these coefficients must be equal to zero for orthorhombic aggregates of cubic materials[2] and consequently, the β phase is not expected to contribute.

These results show that the acoustic behavior of the samples is dominated by the texture of the primary α phase. This is not surprising because the effective elastic constants, in the Reuss or Voigt approximations, are volume averages of the individual phases and because the primary α phase represents 90 to 95 % of the material. On the other hand, the elastic constants and density of the β (cubic) crystallites can be substantially different from those of the α (hexagonal) crystallites[15] but because it is weakly textured, the β phase should affect mainly the isotropic component of the sample's effective elastic constants. Only relative velocity measurements were used in the texture measurement; therefore, the β phase is not expected to affect the results significantly. The results also show that the 1% Nb dissolved in the α phase does not affect the elastic properties of the crystallites sufficiently to prevent texture measurements.

Voigt, Reuss and Hill averaging

The influence of averaging method on the texture coefficients obtained using ultrasonic methods is best illustrated by Figure 3. The figure is similar to Figure 2 for the Hill average of W_{200} but, for each acoustic measurement, the Voigt and Reuss values are indicated. In all three cases, the linear relationship is recovered but the three slopes differ somewhat: the slope is equal to 0.97 for Voigt averaging, 1.16 for Hill averaging (Figure 2), and 1.43 for Reuss averaging. The three lines meet in a point at a W_{200} acoustic value near zero and a neutron diffraction value near -0.0025. Therefore, the choice of averaging method introduces a systematic relative error as large as ± 23% in the texture coefficient. The systematic absolute error is unexplained but may be due to measurement uncertainties or to the various approximations made: single phase material and LRW and LSSCW instead of pure Rayleigh and bulk longitudinal waves.

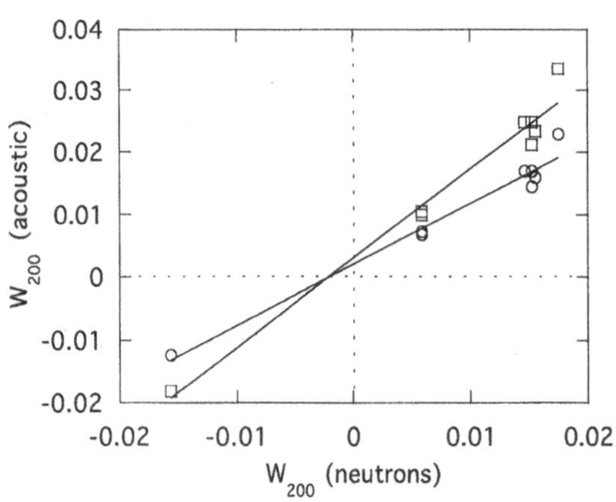

Figure 3. W_{200} as measured with ultrasound for Voigt (circles) and Reuss (squares) averaging methods versus W_{200} obtained by neutron diffraction for one pure Zr and six Zr-2.5%Nb samples.

CONCLUSION

We have used an acoustic microscope to measure the LRW and LSSCW velocities as a function of angle in two principal planes of one pure Zr and six Zr-2.5% Nb samples. The texture measurements of the pure Zr sample agree reasonably well with the neutron diffraction measurement. Discrepancies are related to the statistical error of the fitted P, Q, and R parameters, to the choice of averaging method, and to the use of two types of waves, LRW and LSSCW, that are only an approximation to the pure Rayleigh and bulk longitudinal acoustic modes. A linear relationship was found between the texture coefficients obtained by ultrasonics and those obtained by neutron diffraction for one pure Zr and six Zr-2.5%Nb samples. The presence of a 5 to 10% of a secondary phase and of up to 1% Nb in solution in the primary phase does not invalidate the measurement. This is an important result for the future use of acoustic methods to evaluate the texture of alloys. Our results also show systematic relative errors caused by the choice of averaging method.

ACKNOWLEDGEMENTS

This work was supported in part by the Fuel Channel NDE working party of the CANDU Owners Group.

REFERENCES

1 P. J. Kielczynski, A. Moreau, and J. F. Bussière, to be published in J. Acoust. Soc. Am.

2 Ryong-Joon Roe, J. Appl. Phys. **36**, 2024 (1965).

3 H.-J. Bunge, Kristall und Technik **3**, 431 (1968).

4 C. M. Sayers, Ultrasonics **24**, 289 (1986).

5 M. Hirao, K. Aoki, and H. Fukuoka, J. Acoust. Soc. Am. **81**, 1434 (1987).

6 C. M. Sayers, Ultrasonics **24**, 289 (1986).

7 Yan Li and R. Bruce Thompson, J. Appl. Phys. **67**, 1 (1990).

8 B. A. Auld, "Acoustic Fields and Waves in Solids," vol. 1, Krieger, Malabar, Florida, 1990.

9 P. J. Kielczynski, J. F. Bussière, J. H. Root, and D. Juul Jensen, Nondestr. Test. and Eval. **7**, 497 (1992).

10 Jun-Ichi Kushibiki and Noriyoshi Chubachi, IEEE Trans. Sonics Ultrason. **SU-32**, 189 (1985).

11 I. A. Viktorov, "Rayleigh and Lamb waves. Physical Theory and Applications," Plenum Press, New York, 1967.

12 Yusuke Tsukahara, Yongshen Liu, Christian Néron, C. K. Jen, and Jun-Ichi Kushibiki, unpublished.

13 E. S. Fisher, C. J. Renken, Phys. Rev. **135**, A482-A494 (1964).

14 J. H. Root and A. Moreau, unpublished.

15 E. Walker and M. Peter, J. Appl. Phys. **48**, 2820 (1977).

MICROTOMOGRAPHY USING CONVENTIONAL X-RAY SOURCES

Yasushi Yamauchi,[1] Naoki Kishimoto,[1] and Takashi Ikuta[2]

[1]National Research Institute for Metals
Sengen, Tsukuba 305, Japan
[2]Osaka Electro-Communication University
Hatsumachi, Neyagawa 572, Japan

INTRODUCTION

Microscopies such as optical and electronic types are mostly restricted to surface observation of objects. Information required for material research actually not only arises from a surface but is buried more in the bulk. Computerized x-ray tomography (CT) nondestructively provide us image of any cross section of the object, i. e., three dimensional image of the bulk.

Images of element distribution in objects can be obtained by absorption contrast using two monochromatic x-ray beams which straddle its absorption edge.[1] In order to implement this method monochromatic x-ray beams have been used. Monoenergetic sources are obtained mainly with the combination of SR and monochromator,[2-4] with radioisotopes[5] or with fluorescence from secondary targets.[6] The latter two sources are more accessible than SR but their intensity is mostly poor. Therefore we attempted to utilize polychromatic beams other than monochromatic beams of conventional x-ray sources. We have extended this critical absorption technique to polychromatic x-ray with filter modulation instead of crystal monochromators and have demonstrated its capabilities.

In the present paper the basic idea of our filter modulation technique for x-ray microtomography is described at first and the devices used in this study are briefly explained. Then the experimental results obtained through two different data subtraction processes are discussed.

FILTER MODULATION

Conventional x-ray sources such as sealed-off tubes generate a wide band of white radiation superimposed with characteristic lines of target elements. Filters have been used for crystal diffraction studies to select a suitable line from those characteristic lines. A single filter which has an absorption edge on the just higher energy side of K_α lines of a target suppresses a K_β line and a considerable part of white radiation. This filtered spectrum still contains continuum in the low energy region and in the far high energy. The remained continuum and the reduced K_β can be substantially eliminated by the Ross balanced filter technique, which is a combination of a β filter and an additional filter of proper thickness with an absorption edge on the just lower energy side of K_α lines. The difference between the two measurements obtained with each filter is ascribed to the narrow band between the absorption edges of filters, i. e., mainly to the K_α lines. We have applied these techniques to microtomography in the simpler condition.

Figure 1 shows schematic representation of continuum x-ray spectra from a conventional source. Characteristic lines are not essential and are omitted for simplicity. As illustrated in the figure, I_0 and I designate the intensities of the incident and transmitted x-ray fluxes, respectively. It is assumed that the object contains an element of an absorption edge at the energy denoted by s and that two filters of different absorption edges at the higher and the lower energies denoted by f_1 and f_2 are employed for the spectrum modulation of the incident beam.

Postreconstruction Subtraction

Spectra (a) are incident and transmitted ones without filtering. One absorption edge at s is illustrated in the transmitted spectrum of I. The contrast corresponding to this absorption should be involved in the CT image reconstructed from projections with these spectra. In the case of f_1-filter, (b) in the figure, the transmitted spectrum includes a part of absorption curve from the edge s but the transmitted spectrum of f_2-filter, (c) in the figure, does not contain any absorbed component due to the edge s. The reconstructed tomograph with filter f_1 should have the contrast due to the absorption edge s but the tomograph with f_2 should not. The contrast of reconstructed images are the results calculated from I/I_0, the ratio of transmitted to incident intensity, where I_0 and I are integrated intensities over the spectra. From the spectra (a), (b) and (c) it can be expected that the contrast due to the absorption edge s in reconstructed tomographs appears to be stronger in (b), moderate in (a) and weak in (c). By comparison of these three images, simply by subtraction between reconstructed images, the possibility for element-selective imaging arises. This may be termed "postreconstruction subtraction".

Prereconstruction Subtraction

The curves of (d), (e) and (f) in Figure 1 are narrow band spectra obtained by direct subtraction of (b) from (a), (c) from (a) and (c) from (b), respectively. This may be called "prereconstruction subtraction". Through this process the low energy envelopes are reduced in the spectra of incident beams. The transmitted spectra of (d) and (e) have to contain the absorption due to the edge s and the fraction of those components are nearly same. The most monoenergetic spectra are (f), which eliminate polychromatic artifact from CT images. This combination of subtractions provide highest contrast if the absorption edge s is equal to f_2 but least contrast will be given if the edge s is equal to f_1.

EXPERIMENTAI

The microtomography system used in the present study consists of an x-ray source, a filter changer, a sample stage, an image detector (a combination of a fluorescent screen and a cooled CCD still camera) and a computer, as shown in Figure 2. A personal computer (PC) was employed to control filtering the incident beam, rotation of the sample stage and to acquire image data of incident and transmitted x-ray. The PC is connected to a local area network and is able to access other powerful engineering work stations, which was used for the fast reconstruction of the cross section by filtered liner back projection using the Shepp-Logan

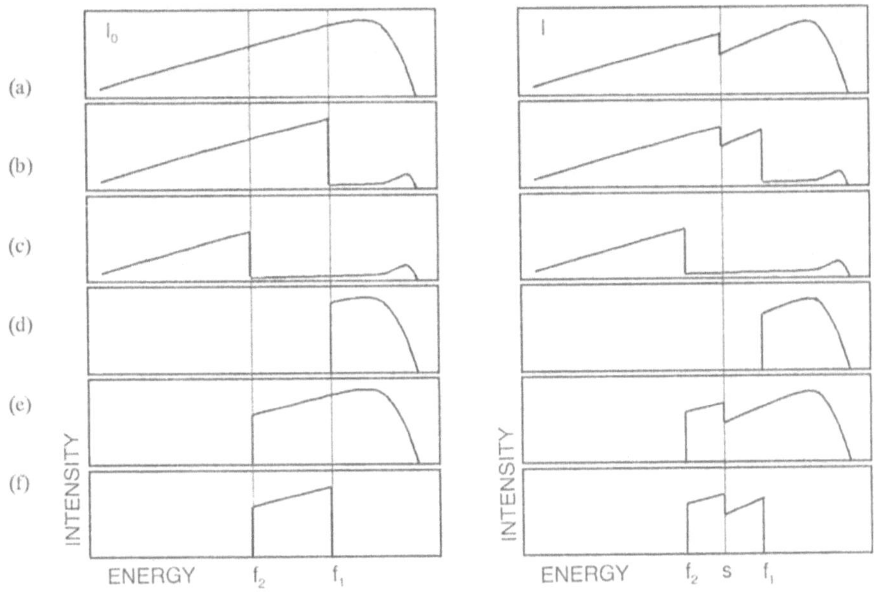

Figure 1. Schematic representation of spectral variation of incident (left) and transmitted (right) white x-ray modulated by filters. The object consists of an element of an absorption edge at *s*. Filters of different absorption edge at f_1 or f_2 is used in (b) and (c) to modulate incident beams but no filter is used in (a). Subtracted spectra of (a)-(b), (a)-(c) and (b)-(c) are also illustrated in (d), (e) and (f) respectively.

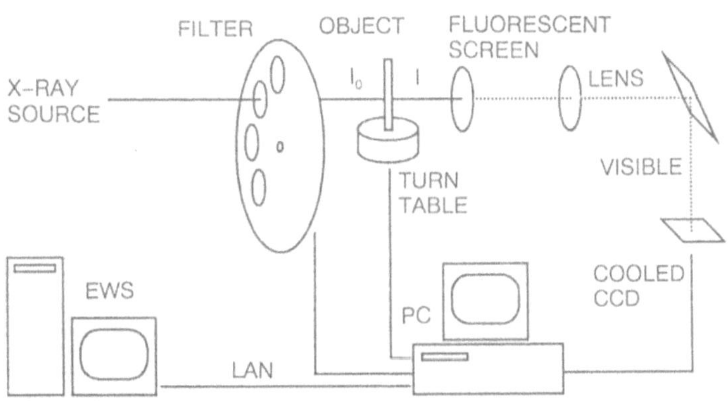

Figure 2. Block diagram of the microtomography system. The filter mechanism controlled by the PC exchanges the filters to modulate the spectrum of the incident x-ray beam.

Table 1. Transmission coefficients of filters.

		Transmission coefficient	
Material	Thickness(μm)	Cu K_α	Cu K_β
Copper	10	0.624	0.704
Nickel	10	0.668	0.089

These values are estimated from the attenuation coefficients listed in the reference (8).

filter. Most of the system has been described elsewhere.[7] Only the filter manipulation mechanism is added for this study.

A copper target tube of normal focus was used in the present study and was operated at a power of 20 kV x 40 mA. The equivalent source size was 1 mm or less. Number of pixels of the CCD is 576 x 384 but the data of neighboring two pixels in the vertical and the horizontal direction were joined to one and then they were again summed over 30 pixels in vertical direction along to the CT rotation axis. This data reduction provides a better signal to noise ratio with moderate spatial resolution.

RESULTS AND DISCUSSION

The test sample is made of two coaxial glass tubes of 0.7 mm in outmost diameter. The outer and inner tubes are respectively filled with $CuSO_4$ and $NiSO_4$ saturated solutions. Copper and nickel foils of 10 μm in thickness were used for beam modulation filter, whose transmission coefficients are given in Table 1. The combination of this sample and the filters comprises either of the special cases in which absorption edges s are equal to f_1 or f_2.

The reconstructed cross section of the test sample are shown in Figures 3. 3(a), 3(b) and 3(c) are measured without filter, with the cooper filter and with the nickel filter, respectively. At a glance there are no differences among these three photographs, but some difference is found on the CRT. In the original CRT images of 3(a) and 3(b), the nickel solution region appears slightly darker than the copper, while almost the same brightness are seen in 3(c). Information of element distribution are extracted from these delicate deviations by subtraction between the images.

Postreconstruction Subtraction

The image obtained by subtraction of Figure 3(b) from 3(a) is Figure 4(a), Figure 3(c) from 3(a) is Figure 4(b) and Figure 3(c) from 3(b) is Figure 4(c). The copper solution region is darker than the nickel region in Figure 4(a), while it is reversed in (b) and (c). These are in good agreement with the above consideration based on Figure 1(a), (b) and (c). Absorption edges f_1 and f_2 of Figure 1 correspond to copper and nickel, respectively, in this experiment. In the copper solution region the edge s is equal to f_1 and in the nickel solution region s is f_2.

Both of the copper and the nickel solution regions in Figure 3(a) exhibit characteristic absorption, and in 4(b) only the nickel region shows the absorption but no characteristic absorption occurs in 4(c). Therefore in Figure 4(a) [3(a)–3(b)] the copper solution region is more enhanced and becomes darker than the nickel region. This relation is found to be reversed in Figure 4(c).

Reconstructed absorption intensities in the copper an nickel solution in Figure 3(a) is not the same because nickel strongly absorbs copper K_β line. Since the tomograph is calculated from I/I_0 which includes the characteristic lines, it makes the nickel solution region slightly darker than the copper. This is confirmed in Figure 4(b) [3(a)–3(c)].

Prereconstruction Subtraction

To pick up properly a narrow band of spectra as shown in Figures 1(d), 1(e) and 1(f) intensities of the incident have to be adjusted over the envelope to be eliminated before the subtraction process. Intensities on the lower energy side of absorption edge of filters can be matched by considering transmission factors shown in Table 1.

Figure 5(a) is obtained by tomographic reconstruction after subtracting copper filter data multiplied by a factor of 1.43 from no filter data and is equivalent to a tomograph using x-ray of higher energy than the copper absorption edge. Also the other two images, Figures 5(b) and

(a)

(b)

(c)

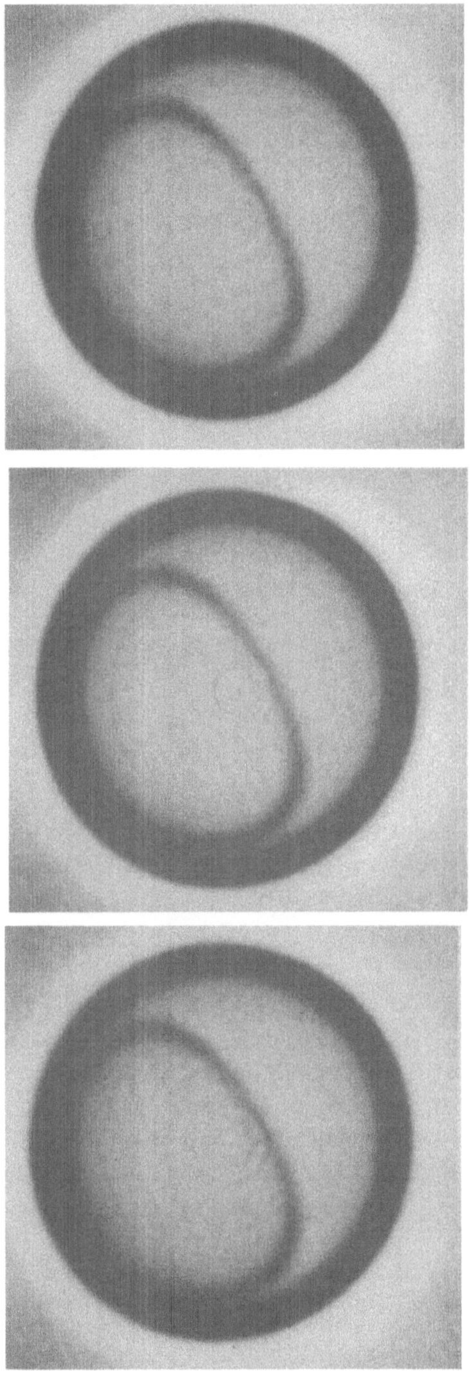

Figure 3. Reconstructed tomographs of the coaxial glass tubes of 0.7 mm in outmost diameter filled with $CuSO_4$ and $NiSO_4$ solutions. No filter is employed in (a). The copper and nickel filters are used in (b) and (c), respectively. Transmission coefficients of the filters are listed in Table 1.

(a)

(b)

(c)

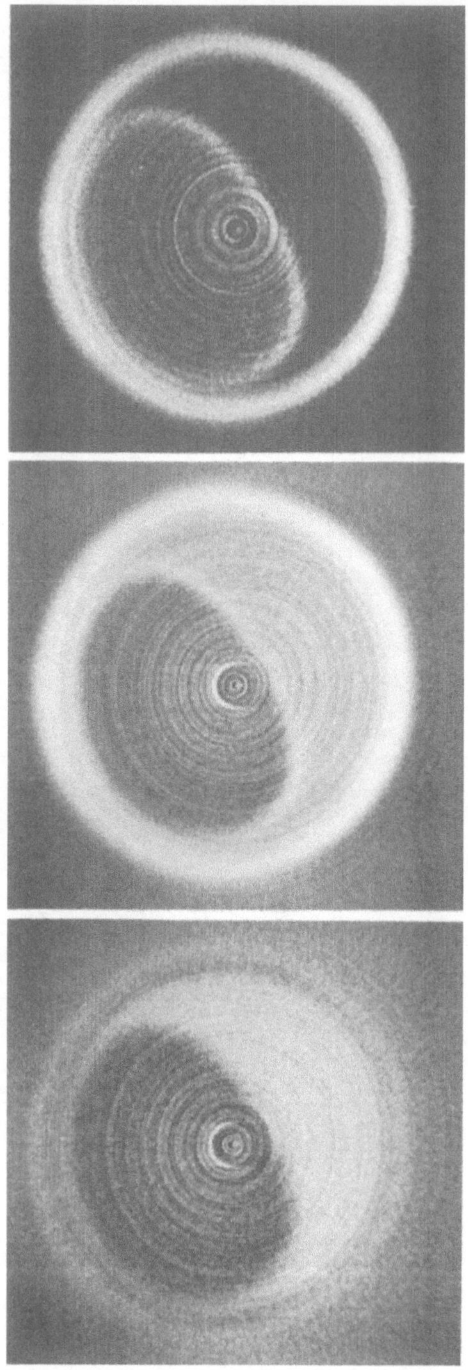

Figure 4. Images obtained by "postreconstruction subtraction" of the reconstructed tomographs in Figure 3. The images (a), (b) and (c) correspond to subtraction of Figure 3(b) from 3(a), of 3(c) form 3(a), and 3(c) form 3(b).

(a)

(b)

(c)

Figure 5. Reconstructed tomographs after "prereconstruction subtraction". The data of transmission images are subtracted each other after multiplied by correction factors which is estimated from the transmission coefficients in Table 1 before the reconstruction process. (a): (no filter) - 1.43x(copper filter), (b): (no filter) - 1.50x(nickel filter), (c): (copper filter) - 0.93x(nickel filter).

5(c), are reconstructed after subtraction of nickel filter by a factor of 1.50 from no filter and after subtraction of nickel filter by a factor of 0.93 from copper filter, i. e., correspond, respectively, to the higher region than nickel edge and the narrow band between copper and nickel edges.

Absorption by the copper and the nickel solution are nearly the same in the higher energy region than the copper edge. Thus no differences between both regions are found in Figure 5(a). The x-ray spectrum of higher energy than the nickel edge contains quite large K_β line of copper. This profile gives somewhat darker contrast to the nickel solution than to the copper in Figure 5(b). As for monoenergetic characteristics, high contrast between the nickel and the copper solution is seen in Figure 4(c), because x-ray having energy between copper and nickel edges is strongly absorbed by the nickel solution but not by the copper.

Figures 5(c) and 4(c) are different from each other even though they are obtained from the identical data set of the nickel filter and the copper filter. The "postreconstruction subtraction" extracts elemental information but it drops other contrast such as glass tubes in Figure 4(c). On the contrary, the "prereconstruction subtraction" provides quasimonochromatic CT which clearly exhibit elemental dependence but does not extract only elemental information. It still contains other contrasts such as glass tubes in Figure 5(c).

CONCLUSIONS

We have described an experimental approach to introduce filtering technique into x-ray microtomography using conventional sources for elemental evaluation of objects. Results are summarized as follows:

- Filter modulation of incident beam yields slight modification of contrast in tomographs.
- Postreconstruction subtraction between those tomographs extracts elemental information qualitatively.
- Prereconstruction subtraction with initial projection data provide quasimonochromatic features to polychromatic CT.

In this study incident x-ray contains not only continuum but also characteristic lines which seem to influence tomography both positively and negatively. Further investigation is necessary to generalize this technique for pure continuum spectra. The images processed by "prereconstruction subtraction" can be treated again by the "postreconstruction subtraction". Doubly processed images may be promising to effectively extract the elemental information.

REFERENCES

1. L. Grodzings, Critical Absorption Tomography of Small Samples, Nucl. Instr.and Meth. 206:547(1983).
2. B. P. Flannery, H. W. Deckman, W. G. Roberge, K. L. D'Amico, Three-Dimensional X-Ray Microtomography, Science 237:1439(1987).
3. J. H. Kinney, Q. C. Johnson, R. A. Sarayan, M. C. Nichols, U. Bonse, R. Nusshardt, and R. Pahl, Energy-Modulated X-Ray Microtomography, Rev. Sci. Instrum. 59:196(1988).
4. Y. Suzuki, K. Usami, K. Sakamoto, H. Kozaka, T. Hirano, H. Shinnno, and H. Kohno, X-Ray Computerized Tomography Using Monochromated Synchrotron Radiation, Jpn. J. Appl. Phys. 27:L461(1988).
5. J. Fryar, K. J. McCarthy and A. Fenelon, Differential X-Ray Absorptiometry Applied to Computerized X-Ray Tomography, Nucl. Instr. and Meth. A259:557(1987).
6. R. Cesareo, Principles and Applications of Differential Tomography, Nucl. Instr. and Meth. A270:572(1988).
7. Y. Yamauchi, T. Ikuta, and N. Kishimoto, Three-Dimensional High Resolution Tomography for Small Objects, Nondestr. Test. Eval., 7:309(1992).
8. J. A. Ibers and W. C. Hamilton, "International Tables for X-Ray Crystallography Vol. IV", Kynoch Press, Birmingham (1974).

X-RAY MEASUREMENT OF FATIGUE DAMAGE BY USING IMAGING PLATE

Satoru Yusa,[1] Kazuo Yoshida,[2] Yasuo Yoshioka[3]

[1] Research Institute, Ishikawajima-Harima Heavy Industries Co., Ltd.
3-1-15, Toyosu, Koto-ku, Tokyo 135 Japan
[2] Research and Development Group, Nuclear Power Division
Ishikawajima-Harima Heavy Industries Co., Ltd. 1
Shin-nakahara-cho, Isogo-ku, Yokohama 235 Japan
[3] Musashi Institute of Technology
1-28, Tamazutsumi, Setagaya-ku, Tokyo 158 Japan

INTRODUCTION

To improve reliability, safety and economy of structure or plant, needs of life assessment and residual life estimation are increasing. Therefore, the improvement of accuracy and expansion of the application range of damage measuring technique are requested.

Fatigue is one of the most important failure modes as well as corrosion, creep, etc. For the measurement of fatigue damage, various methods have been proposed, and most of them measure change of hardness(hardening or softening), initiation or growth of micro-crack, and change of substructure (dislocation density, subgrain size, total misorientation, etc.) The X-ray diffraction measures the third phenomenon and it is thought that fatigue damage measurement by using X-ray diffraction has the following three characteristics.

- Nondestructive
- Measured at surface
- Measured in small area

By the way, the imaging plate is a newly developed X-ray detector [1]. At first, it was used in the medical field instead of a radiographic film. As it was shown that imaging plate has an excellent characteristic and is useful for measurement of X-ray diffraction [2], the application of imaging plate to the field of material science has been tried recently [3,4].

Imaging plate combines following characteristics compared with old X-ray detectors.

- Two-dimensional detector
- High sensitivity
- Wide dynamic range
- Repetitious use
- Quantitative
- Quick development

Then, we measured the X-ray diffraction images of fatigue damaged material by using imaging plate, to study a new damage evaluation method for metallic material.

PRINCIPLE OF IMAGING PLATE

Imaging plate is a flexible plate on which fine particle of photostimulable phosphor crystal are spread uniformly (Figure 1). When the radiation such as X-ray or γ-ray is applied to imaging plate, photostimulable phosphor crystal absorb the energy of radiation and electrons are excited to semi-stable high energy level. After that, when the incitement such as heat or visible ray is applied to imaging plate, electron drop down to the ground

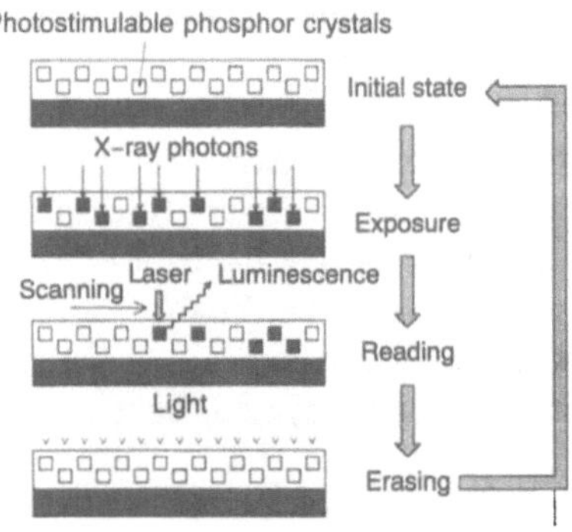

Figure 1. Principle of imaging plate.

level and discharge saved energy as fluorescence. As the number of excited electron is in proportion to the integral strength of applied radiation, the intensity of fluorescence is in proportion to it, too. That is, by measuring the position and intensity of the fluorescence brought by incitement, we can know the distribution of radiation applied to the imaging plate before. After reading X-ray image, imaging plate is erased by the visible ray for a new measurement.

CONSTRUCTION OF READOUT SYSTEM

As readout system is necessary to get the X-ray image from imaging plate, we constructed image reader (Figure 2) and used for the measurement of X-ray diffraction. We adopted the concentric circle type scanning for our image reader, because, we thought that Debye-Scherrer ring can be measured efficiently with this style.

Laser beam to incite the imaging plate passed through the optical filter, shutter, collimator, and was bent in the right angle with a beam splitter. And, laser beam was focused by objective lens to the small area with diameter of 100 μm.

We used the high-resolution type imaging plate with the thickness of 0.15 mm made by Fuji Film Co. The center hole of imaging plate was made to pass though the x-ray collimator through it, and this hole was used to fix the imaging plate on the turntable of image reader. The turntable was rotated by a DC servo motor at a constant speed of 60 rpm. The scanning point was moved 100 μm at a time in radial direction by a slide table that was set up under the turntable.

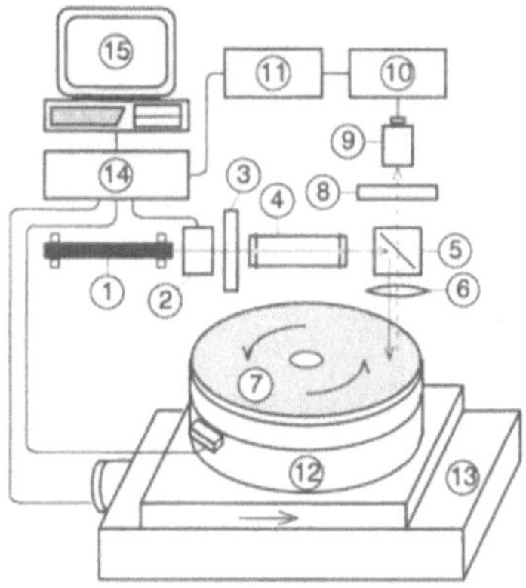

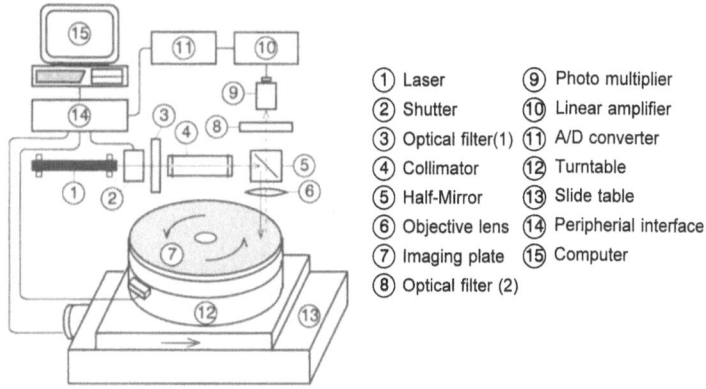

Figure 2. Schematic view of imaging plate readout system.

Fluorescence that the imaging plate emitted passed through the objective lens, beam splitter, optical filter, and was finally detected by the photomultiplier tube. The voltage, which the photomultiplier tube outputs, was amplified by a linear amplifier, converted into digital data by a A/D converter, and logged by a computer.

The movement of the slide table, opening and shutting of the shutter, and the fluorescence intensity logging were synchronized with the rotation of the turntable, and all of them were automatically controlled by a computer. The computer could log the intensity of fluorescence for 4096 point in a rotation of turn table, and the measured intensities of fluorescence were recorded in the form of the digital data of 12 bit.

To confirm the performance, we carried out the experiment using standard γ-ray source. We read out the imaging plate after applying the γ-ray to it for a proper time, and

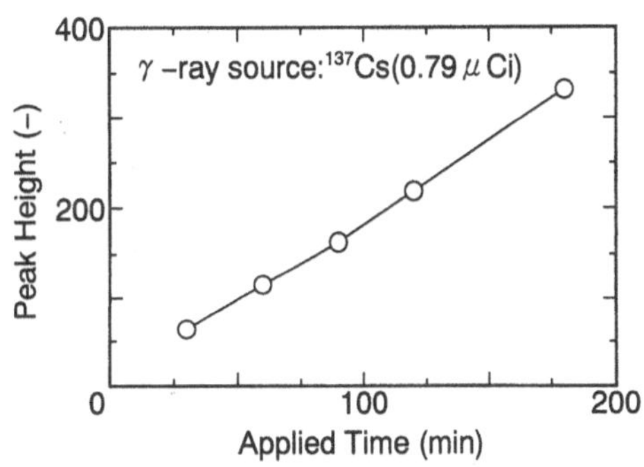

Figure 3. Change of peak height with exposure of γ-rays.

varied exposure time from 30 min to 180 min. As shown in Figure 3, the maximum peak value was in proportion to applying time, and we could confirm that the measured value by the image reader was in proportion to the integral dose of radiation.

EXPERIMENT

At first, we made samples which had already known amount of fatigue damage. Next we measured the X-ray diffraction image of them,and at last we calculated parameters from diffraction images to evaluate fatigue damages by them.

Making of Fatigue Damages Samples

To make the materials that have various fatigue damage, we carried out low-cycle fatigue tests. The specimen for low-cycle fatigue test is shown in Figure 4 and test .conditions are shown in Table 1.

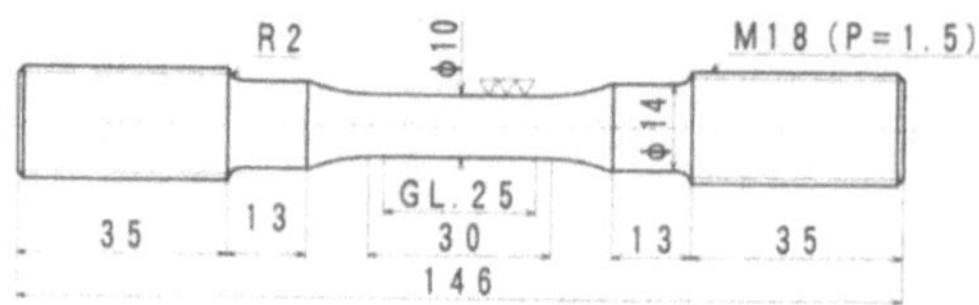

Figure 4. Specimen for low-cycle fatigue test (all dimensions in mm).

Table 1. Condition of low-cycle fatigue test.

Type of loading	Axial
Control mode	Strain
Waveform	Triangle
Strain rate $\dot{\varepsilon}$(%/sec)	0.4
Strain range $\Delta\varepsilon_t$(%)	1.2
Temperature	R.T.

At first, fatigue test was carried out to determine the fatigue life of the material. After that, other fatigue tests were interrupted with fewer cycle number than fatigue life to make materials that had a different amount of fatigue damage. The result of low-cycle fatigue test is shown in table 2.

Table 2. Results of low-cycle fatigue test.

Cumulated damage(%)	Cycle
100	3174 (failure)
75	2382
50	1588
25	794
10	315
1	31
0.1	3

Samples were cut out from the center of the specimen, polished by mechanical polishing, its surface layer was removed about 100 μm in thickness by electrolyte polishing.

Measurement of X-ray Diffraction

The X-ray diffraction patterns measured in the back reflection mode at the section of the sample. The conditions of diffraction measurement are shown in Table 3. The samples of diffraction image are shown in Figure 5.

Table 3. Conditions of X-ray diffraction.

Characteristic X-ray	Cu-Kα
Diffraction line	Fe-γ 420
Tube voltage	40 kV
Tube current	50 mA
Irradiated area	ϕl mm
Exposure time	1 ~ 20 min

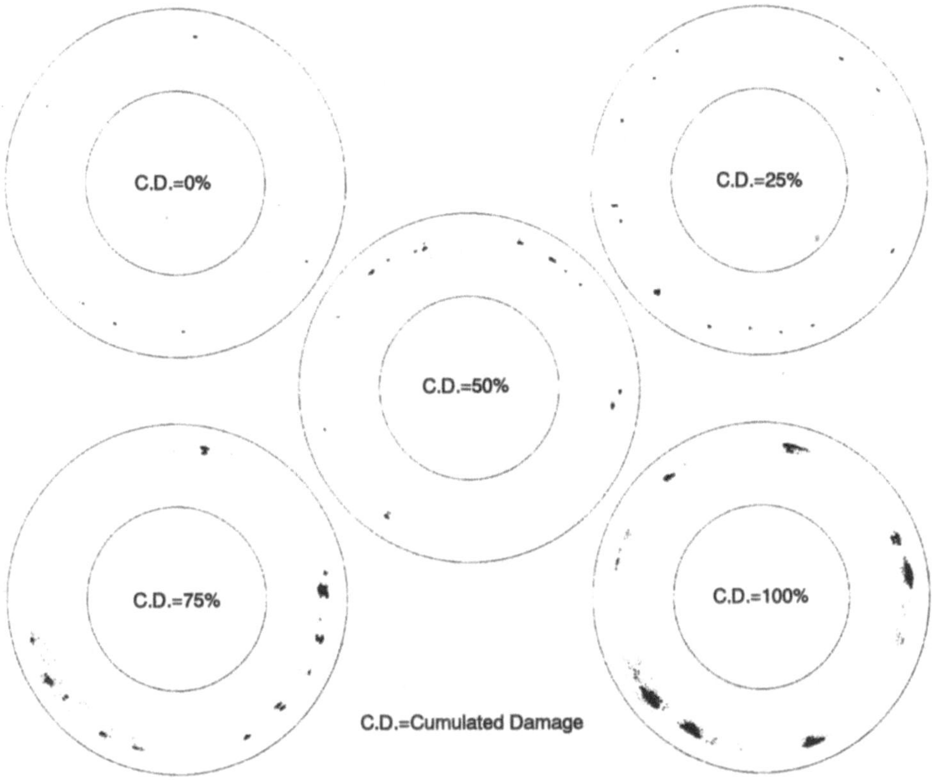

Figure 5. Change of Debye-Scherrenring with cumulated damage.

All of diffraction images were not continuous rings but formed with diffraction spots. Intensity of each diffraction spot seems to be decreasing, and the size of the spot looks increasing with the increase of damage. And we could observe the separation for some diffraction spots. The typical changes of diffraction spot with the increase of fatigue damage are shown in Figure 6. It was thought that such expansion of a diffraction spot was caused by the distortion of lattice a by large number of dislocations introduced into grain by fatigue, and, the separation was caused by the formation of subgrains. The diffraction image of a sample obtained from a failure specimen is significantly different from the image for other specimens, and it is thought that this big change of diffraction image is caused by the deformation of the sample that was added by the fracture of specimen.

Calculation of Parameter

In this study, we tried to get the parameters that can evaluate the change of diffraction spot, and examine the correlation of these parameters and the fatigue damage.

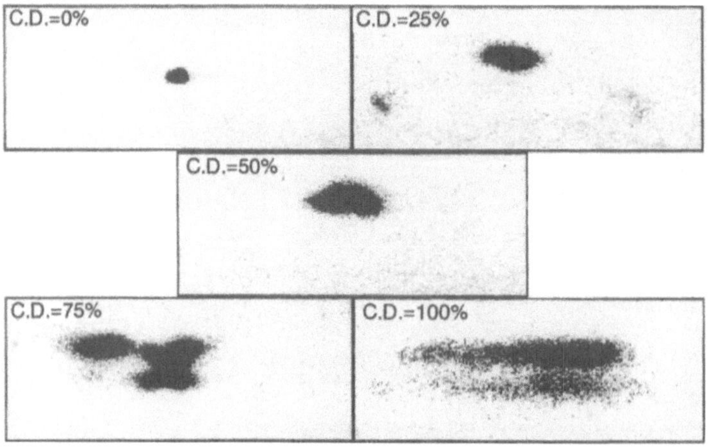

C.D.=Cumulated Damage

Figure 6. Typical change of diffraction spot.

As diffraction images in this study were not continuous, we could not apply the definition of half value breadth for the diffraction ring that is often used for the evaluation of fatigue damage. Then, we defined the following parameters (Figure 7) and calculated them for each diffraction stop in the Deby-Scherrer ring.

- Half value breadth for the tangential direction
- Half value breadth for the radial direction
- Half value aspect ratio
- Half (50%), 25%, 10% value area

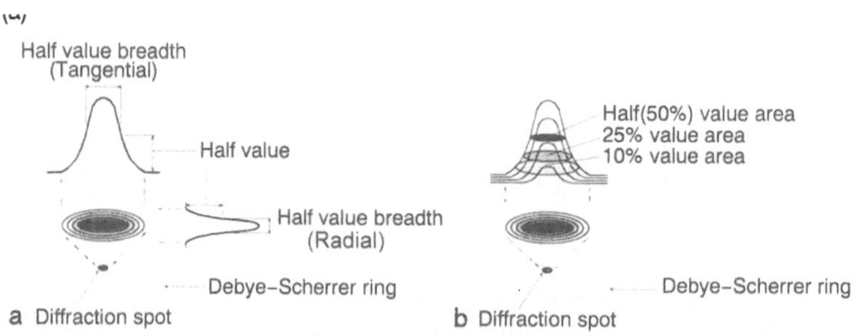

Figure 7. Definition of parameters. (a)Tangential and radial half value breadth. (b)Half(50%), 25%, 10% value area.

Half value breadth for the tangential direction and radial direction were the half value breadth of 2-dimensional peak that were calculated by projection of 3-dimensional intensity diagram of diffraction spot to each direction. Half value aspect ratio was a ratio of the tangential half value breadth and radial half value breadth. Half value area, 25% value area, and 10% value area were defined as area of the part where measured intensity exceeds 50%, 25%, and 10% of peak value respectively.

These values were measured in 100 of diffraction spots or more for each sample. And, the value of the spots that have peak intensity less than 500 was screened out. Next, the measured values were arranged in order, and, the value of 10% from maximum value and 10% from minimum value were screened out. The average of these values were calculated at the end.

RESULTS

The change of the averaged half value breadth for tangential and radial direction with damage is shown in Figure 8.

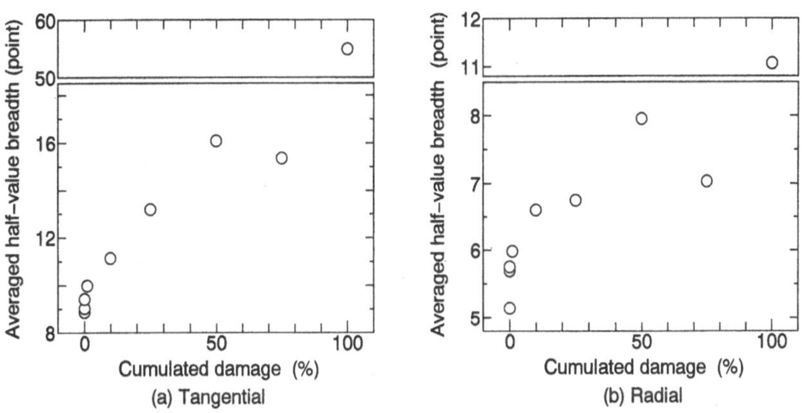

Figure 8. Change of averaged half value breadth with cumulated damage. (a) Tangential half value breadth. (b) Radial half value breadth.

Both half value breadth are increasing with the increase of fatigue damage, and the parameter of 100% damaged sample is larger than other samples. As mentioned in the observation of diffraction spot, it was thought that this difference was caused by the plastic deformation added by failure of sample.

In the comparison of the tangential half value breadth and the radial half value breadth, the tendency of value change is almost similar, but, radial half value breadth have a little larger scattering.

Although, the value for the sample that have 75% damage is less than that of 50% damage, we guessed that this is caused by some errors of low-cycle fatigue test condition such as total strain range etc.

The value change of tangential half value breadth from initial state is larger than that of radial half value breadth.

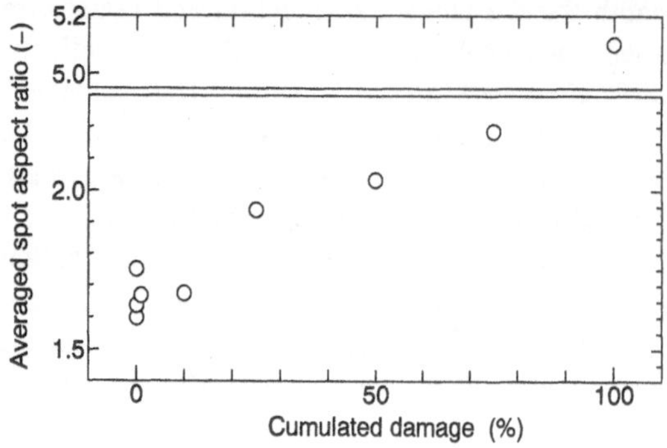

Figure 9. Change of averaged spot aspect ratio with cumulated damage.

The change of aspect ratio of the diffraction spot with the fatigue damage is shown in Figure 9, and the tendency of aspect ratio change looks similar to the change of half value breadths. But, as the statistic scatter of two half value breadth was multiplied, the scattering of the aspect ratio became large. The change of half value area, 25% value area, and 10% value area with fatigue damage are shown in Figure 10. The tendency of their changes are similar to the change of the tangential half value breadths, too. The change of the spot area is surely related to the product of the change of the tangential spot size and the change of the radial spot size. However, as the change of the tangential spot size is

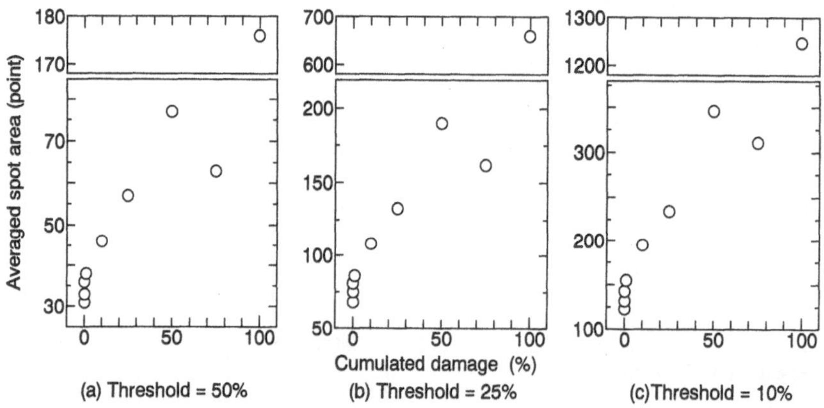

Figure 10. Change of averaged spot area with cumulated damage. (a) Threshold level = 50% of peak height. (b) Threshold level = 25% of peak height. (c) Threshold level = 10% of peak height.

larger than that of radial spot size, the change of tangential spot size affects the change of spot area remarkably.

The tendency of the area change is almost the same in any threshold level. In other words, the diffraction spot expand similarly with the increasing of fatigue damage. When the threshold level is low, as the area change from initial value becomes large, it seems easy to evaluate damage from it. However, when the threshold level is too low, there is the possibility that the background noise affects the value of the parameter, and it becomes difficult to separate the close diffraction spots. Therefore, we have to define the threshold level carefully to calculate these parameters.

As mentioned above, the tangential half value breath and the area of the diffraction spot are hopeful as parameter to evaluate the change of diffraction spot with fatigue damage. Especially, as area of diffraction spot is easy to get from diffraction image directly by data processing, we thought that it seems suitable for practical use.

CONCLUSIONS

We made an image reader to apply the imaging plate to fatigue damage evaluation by using X-ray diffraction. And, we measured the diffraction image of type 304 stainless steel that had various amount of fatigue damage. As a result, we observed that diffraction spots expanded with the increase of fatigue damage, and it was shown that fatigue damage can be evaluated by a parameter which indicate the change of spot size.

REFERENCES

1. M.Sonoda,M.Takano,J.Miyahara and H.Kato, Computed radiography utilizing scanning laser stimulated luminescence, *Radiography* 148:833(1983).

2. Y.Amamiya,N.Kamiya and J .Miyahara, Application of photostimulable phosphor film to X-ray diffraction studies ," *OYO BUTURI* " 55:957(1986) (In Japanese).

3. Y.Yoshioka,T.Shinkai and S.Ohya, The use of 2-D detector utilizing laser-stimulated luminescence for X-ray diffraction studies on mechanical behavior of materials, *Adv. X-Ray Anal.*33:339(1990).

4. Y.Yoshioka and S.Ohya, X-ray analysis of stress in a localized area by use of imaging plate, *Adv. X-Ray Anal* 35:537(1992).

NONDESTRUCTIVE CHARACTERIZATION OF METALS SUBJECTED TO HIGH-POWER ULTRASOUND

Kirsten A. Green and Robert E. Green, Jr.

Center for Nondestructive Evaluation
The Johns Hopkins University
Baltimore, Maryland 21218

X-ray diffraction topography and infrared imaging techniques were used in situ with aluminum and zinc specimens undergoing high-power insonation using a 20 kHz ultrasonic horn. X-ray topographic techniques allowed for a nonintrusive study of the mechanical alterations in the structure of the metal specimens, while the infrared system allowed for a study of the specimens' thermal properties. It should be noted that all testing was performed in real-time with x-ray and infrared images recorded simultaneously on video tape.

INTRODUCTION

In 1955, Blaha and Langenecker reported a softening effect in zinc crystals which were undergoing tensile deformation with superimposed ultrasonic vibrations [1]. This phenomenon has become known as the "Blaha effect" and has stimulated scientific interest. Subsequent investigators also noted work softening in specimens insonated below a certain sound intensity threshold [2-4]. However, above a certain sound intensity threshold a work hardening effect was observed [4,5]. In addition, application of high-power ultrasound to metal specimens during wire and strip drawing and bar and tube bending has been shown to reduce the internal (volume) and external (surface) frictional forces required to plastically form the specimens [6-8].

Several "theories" have been proposed to explain the above mentioned phenomena. Some investigators believe that work softening occurs due to additional energy supplied by the ultrasonic field which causes dislocations to be created and moved [1]. Others attribute the work softening to localized heating which takes place in regions around dislocations and other imperfections when ultrasonic waves are scattered [3,4]. The high intensity ultrasonic treatment (work hardening) has been likened to fatigue testing [9]. The internal and external friction reductions have been explained in terms of the superposition of alternating ultrasonic stress waves on an externally applied static stress [2,6-8]. For a more detailed account of past work conducted in this field one is referred to a review by Green [10].

Speculation surrounds what is in fact happening to the internal structure of a metal subjected to high-power insonation. It was the objective of this research to arrive at some conclusions regarding this matter. X-ray topographic and infrared imaging techniques were used in situ with aluminum and zinc specimens undergoing high-power insonation using a 20 kHz ultrasonic horn.

BACKGROUND

X-ray topography is the name given to several x-ray diffraction techniques which permit direct observation of lattice defects both on the surface (back-reflection) and in the bulk of single crystals (transmission). Topographic techniques are unique in that they yield information about the defect structure, down to the size of individual dislocations, throughout the volume of fairly thick crystals; these techniques have been reviewed in several publications [13,14]. Among the various possibilities, the Asymmetric Crystal Topography (ACT) technique [15] was used to study the effects of high-power ultrasound on aluminum and zinc single and polycrystalline specimens. In the ACT set-up used in this investigation, a slit collimated, white radiation x-ray source was incident upon a high quality asymmetrically cut silicon crystal. This crystal served as both a monochromator and beam expander, resulting in a x-ray beam of approximate dimensions of 1 1/2 inches high by 1/2 inch wide. Specimens of interest were placed in the path of this monochromated and expanded x-ray beam and diffraction information was detected using an image intensifier with a flourescent screen faceplate. Due to the relatively weak x-ray source available in the laboratory (copper tube operated at 50 kV, 32 mA), as well as to the thickness of the specimens examined (1/2 inch square or 1/2 inch diameter), the back-reflection mode was utilized in this research. Figure 1 shows a schematic diagram of the ACT experimental arrangement in the back-reflection mode.

In the ACT technique each individual topographic image is essentially a large Laue "spot" generated by diffraction from a particular set of "parallel" lattice planes covering a large area of the crystal. The x-ray beam incident on the specimen illuminates a large area unlike conventional Laue pin-hole techniques and, because of the special beam expanding monochromizing silicon crystal, this large incident beam experiences minimal divergence.

EXPERIMENTAL METHOD

Aluminum and zinc single and polycrystalline specimens of 1/2 inch square (or 1/2 inch diameter) were machined to various lengths depending on whether it was desired for the specimen to be of resonant (a half wavelength) or non-resonant length. Resonance was determined by accounting for the frequency of the ultrasound used (20 kHz) and the longitudinal sound velocity that was measured in each specimen by traditional pulse-echo techniques. Typical specimen lengths were about 4-5 inches; it should also be noted that polycrystalline specimens had a grain size of about 1/2 to 1 inch. The reason for using large grained polycrystalline materials is that a substantial amount of x-ray diffraction information can be gained (as is also true for single crystal specimens) while at the same time studying the influence of grain boundaries.

After specimens were cut to length, they were tapped at one end to facilitate coupling the specimen to the ultrasonic horn. In addition, a series of grinding and polishing clothes were used to polish one side of each specimen to a 5 um finish. Surface preparation was completed by chemically polishing the specimen in a Tegart acid solution. Early testing determined that surface finish plays a role in the quality of the x-ray diffraction topographs one obtains in the back-reflection mode.

Following surface preparation, specimens were photographed at low magnification for documentation purposes. In addition, topographs and Laue x-ray diffraction photographs of specimens were made and recorded on film prior to testing.

Specimens were screwed into a tapped catenoidal 20 kHz ultrasonic horn which was mounted in Newport rotation and translation stages allowing for two degrees of rotation and two translational motions. The third translational motion was controlled by a lab jack on which the

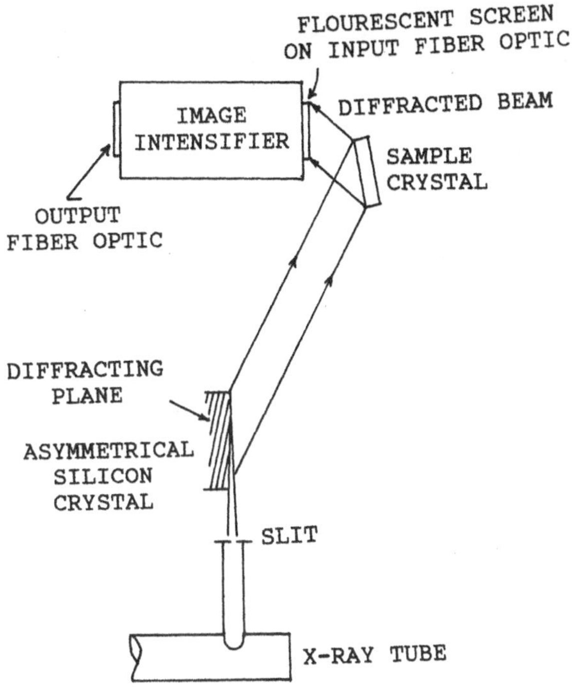

Figure 1. Schematic diagram of the asymmetriccrystal topography system in the back-reflection configuration.

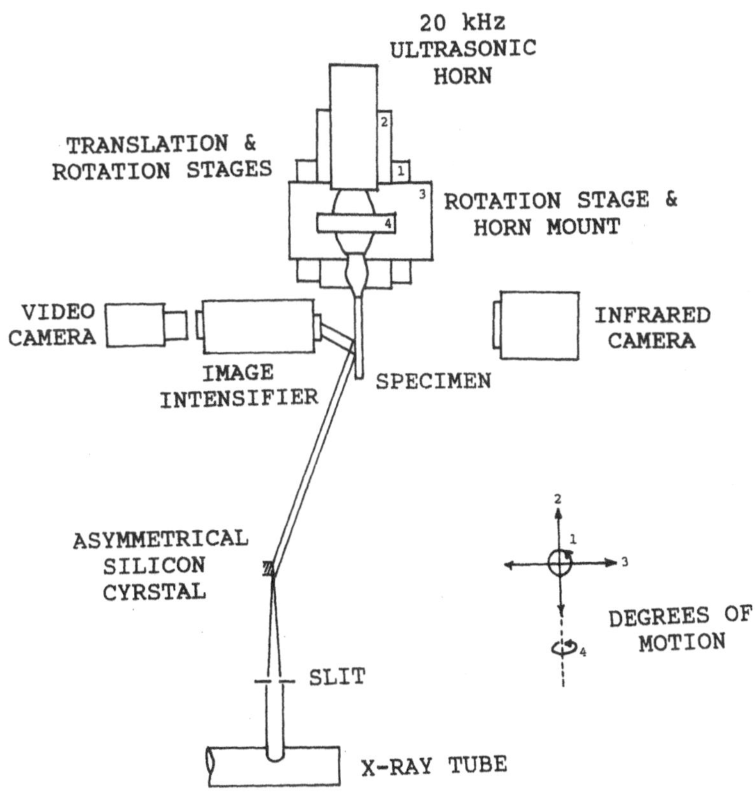

Figure 2. Schematic diagram of the experimental set-up used for analyzing the effects of high-power ultrasound in aluminum and zinc.

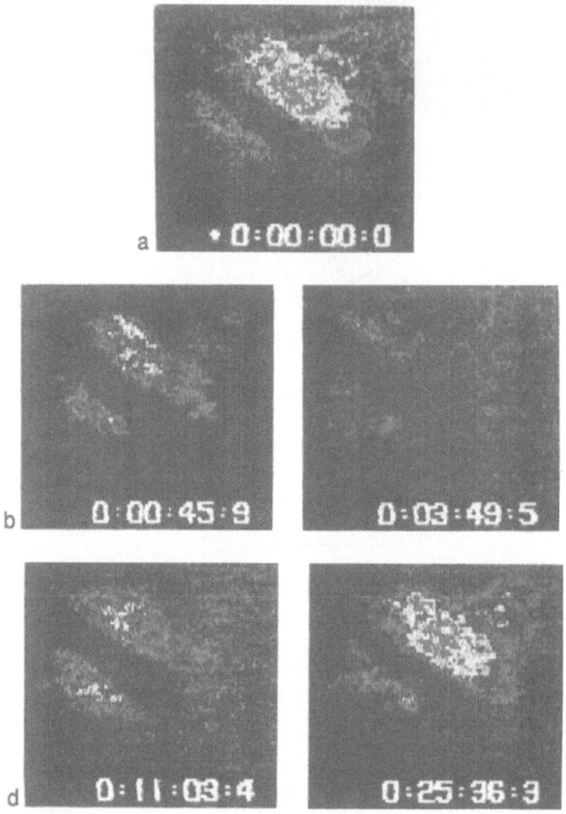

Figure 3. Sequence of frames taken from a videotape of a real-time x-ray run for an aluminum polycrystalline specimen of resonant length. (Only one grain is imaged.)

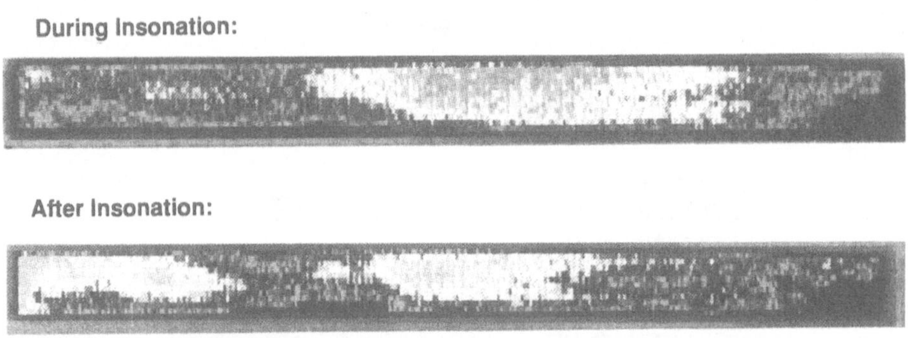

Figure 4. Infrared data for the same resonant length, polycrystalline aluminum specimen shown topographically in Figure 3.

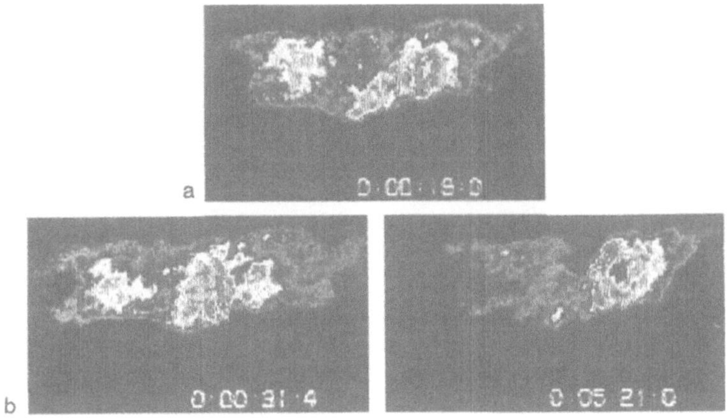

Figure 5. Sequence of frames taken from a video-tape of a real-time x-ray run for an aluminum polycrystalline specimen of non-resonant length. (Only one grain is imaged.)

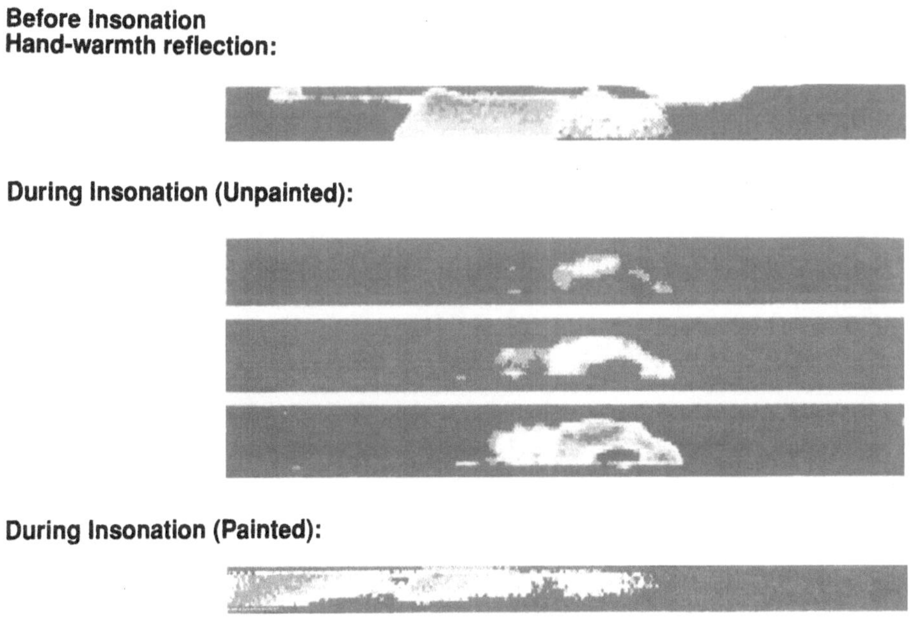

Figure 6. Infrared data for the same non-resonant length, polycrystalline aluminum specimen shown topographically in Figure 5.

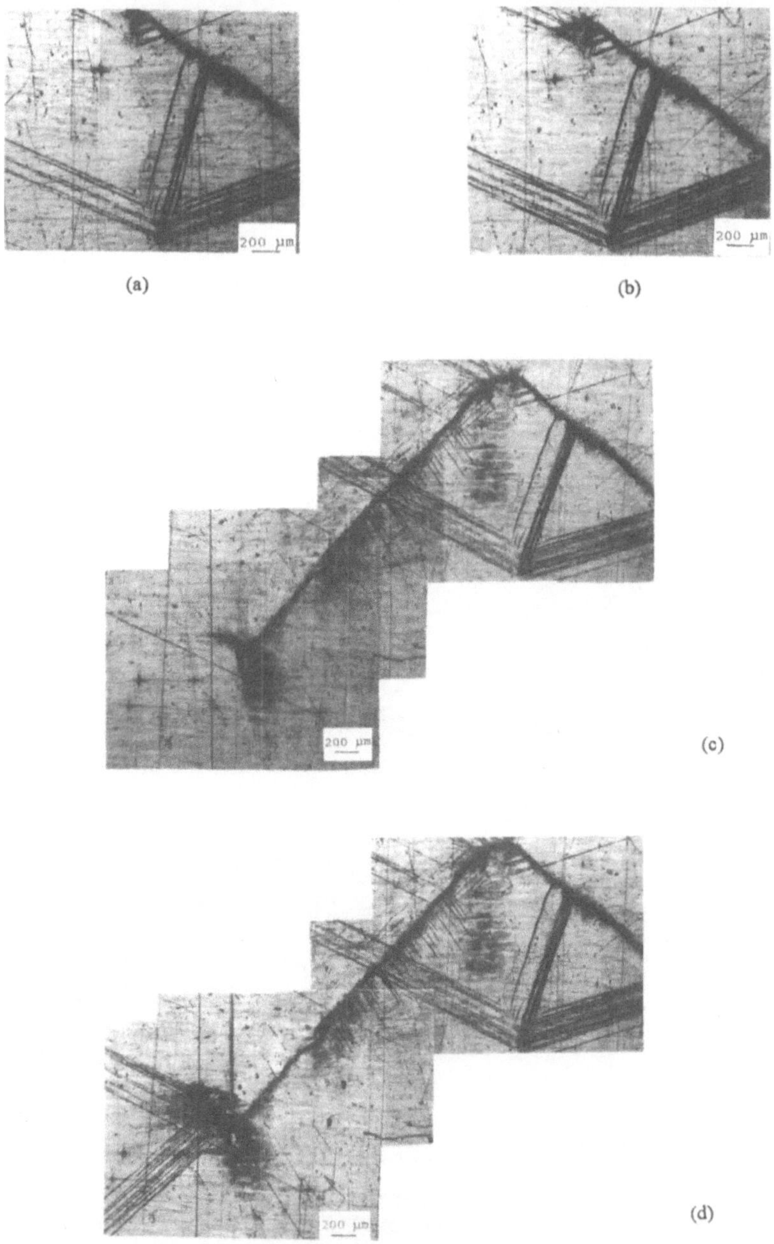

Figure 7. A series of metallographic montages of a crack propagating in a non-resonant length zinc bicrystal. Total duration of insonation was (a) 198 minutes, (b) 203 minutes, (c) 207 minutes, and (d) 249 minutes.

Newport stages were mounted. The horn-specimen configuration was placed in the asymmetric crystal topography (ACT) set-up such that the specimen was in the path of the expanded, monochromated x-ray beam. An image intensifier with flourescent screen face-plate was used to detect diffracted x-ray images and these images were recorded by a video camera/VCR which was focused on the output image of the image intensifier. In addition an infrared (IR) camera was placed opposite the image intensifier to detect thermal differences in the specimens undergoing insonation and this information was also recorded on a VCR. In order that the x-ray and IR data recorded on videotape could be directly compared upon analysis it was necessary to have a synchronized event occur which could be detected by both the IR camera and the video camera; therefore, a simple circuit consisting of a light-emitting diode and a light bulb were connected in series with a battery and a push button switch. The light bulb was placed on the front faceplate of the image intensifier in the field of view of the IR camera and the light-emmiting diode was placed on the rear faceplate of the image intensifier in the field of view of the video camera. Immediately preceeding insonation of a specimen the lights were triggered several times while the VCRs were running and then the horn was turned on. Specimens were insonated for periods of 1 1/2 minutes for the resonant length specimen to as much as 20 minutes for non-resonant length specimens. The resonant specimen was insonated for much shorter time periods due to the 250°C temperature obtained. A schematic diagram of the experimental set-up used for analyzing the effects of high-power ultrasound in aluminum and zinc is shown in Figure 2.

RESULTS AND DISCUSSION

Real-time x-ray topographs of aluminum specimens showed intensity shifts occurring in some areas the instant the ultrasonic horn was turned on. A 3-4°C temperature increase was also detected at the onset of insonation. It is speculated that dislocations are being moved/created by the applied ultrasonic energies as witnessed by the x-ray intensity shift. That is, the x-ray intensity variations in the specimens are likely due to strain fields associated with dislocations, and when these dislocations move so do the associated strain fields. In addition, the temperature increase may be attributed to the fact that it takes more energy to start dislocations moving than it does to keep them moving and this excess energy shows up in the form of thermal energy.

Figure 3 shows a sequence of frames taken from a videotape of a real-time x-ray run for an aluminum polycrystalline specimen of resonant length. Frame (a), at time zero, shows a picture of an x-ray diffraction topographic image of a single grain (dimensions 3/4 inch wide by 1/2 inch high) prior to testing. Frame (b) shows the same grain immediately following the onset of insonation; note the dramatic shift in intensity. Frame (c) shows that after three minutes of insonation the Bragg condition was no longer satisfied (the lattice planes in the crystal had expanded) due to the 250°C temperature reached during testing. Once the ultrasonic horn was turned off and the specimen was allowed to cool down to room temperature the topographic image reappeared, frames (d) and (e).

Figure 4 shows the IR data acquired during and after insonation for the same resonant length, polycrystalline aluminum specimen discussed in Figure 3. As stated above, the resonant length specimen obtained temperatures of 250°C within minutes of the onset of testing. During the insonation process the hot spot was seen to occur at the center of the specimen, as expected in the resonant length specimen, since this is the area of maximum stress (antinode). Upon cooling, it was observed that the specimen did not cool uniformly and may be dependent upon the orientation of the grains within the specimen (there were four grains in this particular specimen).

Figure 5 shows a sequence of frames taken from a videotape of a real-time x-ray run for an aluminum polycrystalline specimen of non-resonant length. Frame (a) shows a picture of an x-ray diffraction topographic image of a single grain (dimensions 1 1/2 inches wide by 1/2 inch high) prior to testing. Frame (b) shows the same grain immediately following the onset of insonation; note the dramatic shift in intensity. The dramatic shift in intensity at the onset of insonation was similar to that noted in the resonant length specimen, however, topographic images were not observed to disappear as in the case of resonant length specimens even for insonation times of five minutes as is seen in frame (c), as well as for longer time periods of testing (more than 20 minutes). This can be explained because non-resonant length specimens did not reach the temperatures that resonant length specimens achieved (approximately 100°C after 20 minutes of testing versus 250°C after 2 minutes of testing, respectively) and therefore the Bragg condition for diffraction was maintained.

Figure 6 shows the IR data acquired for the same non-resonant length, polycrystalline specimen discussed in Figure 5. The top picture in Figure 6 shows the affect that orientation plays in the emissivity of heat in aluminum specimens; the specimen shown was unpainted and warmed by heat radiated from a hand and clearly shows the four distinct grains in the specimen. The middle sequence of pictures in Figure 6 is the unpainted specimen as a function of heat produced while the specimen was being insonated. From this picture it appears that the grain which was the hottest is the second from the right. In fact, the grain which was the hottest is the one on the far left as shown in the last picture in Figure 6, the painted non-resonant length specimen. These observations indicate the effect of orinetation on emissivity and justifies the need for painting specimens black so that discontinuities within the specimen, such as grains of different orientation, are alleviated and results from one grain to another can be compared directly.

Figure 7 shows a series of metallographic montages of a crack propagating in a non-resonant length zinc specimen. The crack formed at a void within the crystal and propagated across one grain to the grain boundary where it was arrested. With further insonation the crack propagated across the second grain and continued to grow until it arrested again. The cause for the second arrest spot is unclear since the specimen was a bicrystal.

CONCLUSIONS

Aluminum specimens of resonant and non-resonant lengths and zinc specimens of non-resonant lengths were studied to determine the effects of high-power ultrasound on the internal structure of the crystals. X-ray topographic and infrared imaging techniques were used in situ with specimens undergoing insonation from a 20 kHz ultrasonic horn. It was determined at the onset of insonation in the aluminum specimens that topographic image contrast was seen to dramatically shift in areas, possibly due to a change in the strain fields associated with dislocation motion. In addition, a 3-4 °C temperature increase was also noted at the onset of testing in these aluminum specimens. The maximum temperature was seen to occur at the antinodal point in the aluminum resonant length specimen, while the position of maximum temperature in the aluminum non-resonant length specimens varied from specimen to specimen and appears to be dependent on crystal orientation.

Deformation in the form of slip bands has been noted in both resonant and non-resonant length specimens of aluminum and zinc. The most striking evidence of deformation occurred in the resonant length aluminum specimen which fractured at the antinodal point (point of maximum stress) due to insonation alone. In addition, stress concentrators aid in the deformation process as witnessed in the zinc specimen where a crack was generated at a void and in a cylindrical aluminum specimen where a possible notch in the threaded region of the specimen caused failure after 2 1/2 minutes of insonation.

REFERENCES

1. F. Blaha and B. Langenecker, "Dehnung von ZinkKristallen unter Ultraschalleinwirkung," Naturwissenschaften 42:556 (1955).
2. G.F. Nevill, Jr. and F.R. Brotzen, "The effect of vibrations on the static yield strength of low-carbon steel," *Proc. Am. Soc. Test. Mat.* 57:751 (1957).

3. B. Langenecker, W.H. Frandsen, C.W. Fountain, S.R. Colberg and J.A.M. Langenecker, "Effects of ultrasound on deformation characteristics of structural metals," US Naval Ordnance Test Station, China Lake, California, *NAVWEPS Rept.* 8482, NOTS TP 3447 (March 1964).
4. B. Langenecker, "Effects of ultrasound on deformation characteristics of metals," *IEEE Trans Sonics Ultrasonics* SU-13:1 (1966).
5. B. Lagenecker, "Work hardening of zinc crystals by high-amplitude ultrasonic waves," *Proc. Am. Soc. Test. Mat.* 62:602 (1962).
6. R. Pohlman and F. Lechfeldt, "Influence of ultrasonic vibration on metallic friction," *Ultrasonics* 4:178 (1966).
7. R. Pohlman, The reduction of friction and the forming of metals by generating high frequency bending stresses, *in*: "Ultrasonics Conference Papers," IPC Science and Technology Press Ltd., Guildford (1971).
8. A.G. Rozner, "Effect of ultrasonic vibration on coefficient of friction during strip drawing," *J. Appl. Phys.* 49:1368 (1971).
9. "Proceedings of the First International Symposium of High-Power Ultrasonics", Graz, Austria, September 1970, IPC Science and Technology Press Ltd., Guildford (1972).
10. R.E. Green, Jr. "Non-linear effects of high-power ultrasonics in crystalline solids," *Ultrasonics* 13:117 (1975).
11. R.B. Mignona and R.E. Green, Jr., "Multiparameter system for investigation of the effects of high-power ultrasound on metals," *Rev. Sci. Instr.* 50:1274 (1979).
12. R.B. Mignona, R.E. Green, Jr., J.C. Duke, Jr., E.G. Henneke II and K.L. Reifsnider, "Thermographic investigation of high-power ultrasonic heating in materials," *Ultrasonics*, 159 (July 1981).
13. R.W. Armstrong and C. Cm. Wu, X-ray diffraction microscopy *in* "Microstructural Analysis: Tools and Techniques," J.L. McCall and W.M. Mueller, eds., Plenum Press, New York (1973).
14. B.K. Tanner, "X-ray Diffraction Topography," Pergamon Press, New York (1976).
15. W.J. Boettinger, H.E. Burdette, M. Kuriyama and R.E. Green, Jr., "Asymmetric crystal topographic camera," *Rev. Sci. Instrum.* 47:906 (1976).

A STUDY OF SUBSTRUCTURES IN WELDED BETA TITANIUM ALLOY BY MICROBEAM X-RAY DIFFRACTION ANALYSIS

Y. Shirasuna, A. Nozue, T. Okubo[1], K. Kuribayashi[2],
S. Ishimoto, H. Sato[3] and Y. Yoshioka[4]

[1] Sophia University, 7-1 Kioi-cho, Chiyoda-ku, Tokyo 102, Japan
[2] The Institute of Space and Astronautical Science, Sagamihara, Kanagawa 229, Japan
[3] Nissan Motor Co., Ltd., Momoi, Suginami-ku, Tokyo 167, Japan
[4] Musashi Institute of Technology, Tamazutsumi, Setagaya-ku, Tokyo 158, Japan

INTRODUCTION

Ti-15V-3Cr-3Sn-3Al alloy has been used for welded structures such as an upper stage motor case of launch vehicle for space-study satellite [1] because of its good formability and high strength [2]. The fabrication process of the case is as follows: Two half spherical shells, which are formed by cold deep drawing from hot rolled plate, are welded to make a sphere. The welded shell is aged to have desired strength and fracture toughness. The fracture toughness of weld joint is found not to meet the design criterion when the welded case was heat-treated conventionally.

Figure 1 shows a schematic hardness distribution near the weld joint after heat treatment. With the conventional heat treatment process, the hardness of the weld metal is higher than that of base metal. The reason is that the age hardening rate is higher in weld metal than in the base metal. This excessive hardening resulted in low fracture toughness in the weld joint. Two kinds of welding and heat treatment processes have been developed to enhance the fracture toughness of the weld joint by the present authors [3],[4]. The idea behind the new methods is to make the hardness of the weld metal a little lower than of the base metal as shown by a solid line in Fig. 1.

Table 1 summarizes heat treatment, welding method and fracture toughness for conventional process, new process I and new process II. The design criterion [5] for the weld joint is also shown at the bottom of this table. The first process is to avoid excessive hardening in the weld metal by two-stage heat treatment [3]. The second process is to suppress the age hardening rate by enriching vanadium in the weld metal [4]. The fracture toughness of the weld joint obtained by the two new processes becomes higher compared with the conventional method, and satisfies the design criterion.

The present paper shows observations on substructures in vanadium enriched weld metal by microbeam X-ray analysis, and discusses the role of the substructures in the age hardening behavior with the results on conventionally welded joints so far reported [6].

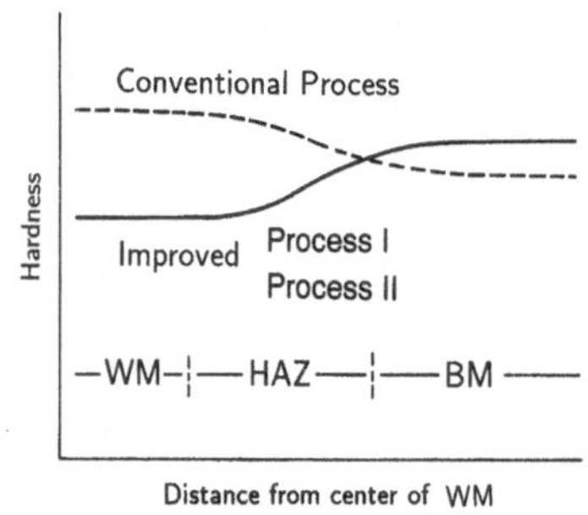

Figure 1. Schematic representation of hardness distribution of welded materials. Hardness distributions of dotted and solid lines are obtained by conventional and new processes. WM, HAZ and BM denote weld metal, heat-affected-zone and base metal respectively.

Table 1. Heat treatment and welding conditions of three processes and fracture toughness K_{IC} obtained.

	Solution treatment	Pre-aging	Welding Filler	Aging	K_{IC} $MPam^{1/2}$
Conventional		none	15V-3Cr-3Sn-3Al		25
Process I	1073K 1.8ks	748K 57ks	15V-3Cr-3Sn-3Al	748K 36ks	34
Process II		none	21V-3Cr-3Sn-3Al		38
Design criterion					30

MATERIALS AND METHOD

Materials, Welding and Heat treatment

Hot rolled plates of standard Ti-15V-3Cr-3Sn-3Al alloy were welded by multilayer gas arc method after solution treatment at 1073K. The weld joint was double U-groove shape and vertical to the rolling direction of the plate. A welding current was 100-120A, voltage 11-12V, and speed 2.0-2.5mm/s. The number of the layer was 20 for each side of the plate. The layer temperature was kept below 375K for each layer welding. For welding filler, four titanium alloys with different vanadium content of 15, 17, 21, and 25% were used to enrich vanadium in the weld metal. The amount of Cr, Sn, Al in filler materials was about 3% as in the base metal. The chemical analysis after welding showed that the vanadium content in the weld metal was almost the same as in filler material. After welding, the substructures in the weld metal were studied by an X-ray microbeam diffraction method and an optical microscope. The welds were aged at 738K for 1ks to 360ks and subjected to hardness measurement.

X-ray Microbeam Diffraction Analysis

The X-ray microbeam diffraction method [7] was used to study the substructures in the weld metal. Diffraction spots along Debye ring are obtained from polycrystalline materials when the beam is sufficiently fine as shown in Fig. 2. The shape of each spot reflects the structure of irradiated region of polycrystalline materials. A tangential length of a spot *St* comes from the orientation scattering in subgrains. A radial length *Sr* can be correlated to lattice strain in subgrains.

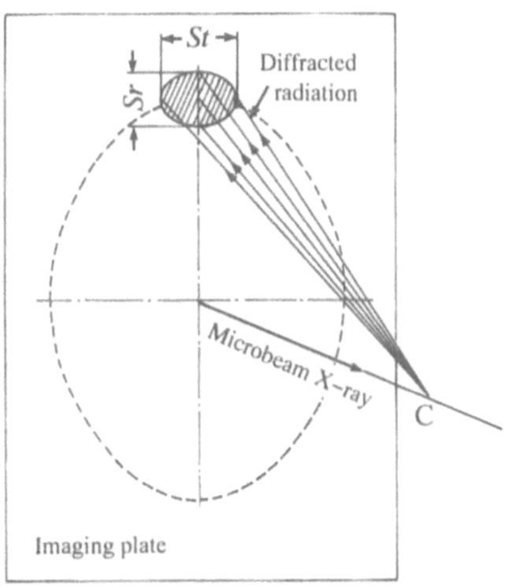

Figure 2. Diffraction spot in microbeam X-ray analysis. Specimen is located at C.

Copper K-α characteristic X-ray was used for the analysis. The diffraction plane is (321) of beta titanium. Imaging plate was used to detect diffraction spots instead of conventional X-ray film. By using the imaging plate, the shape of spots is observed with high resolution under shorter exposure time.

Figure 3 shows the system for the data retrieval from the imaging plate [8]. The photostimulated luminescence on the imaging plate activated by collimated Helium-Neon laser beam was detected with a photomultiplier tube. The signal is sent to a personal computer for data processing.

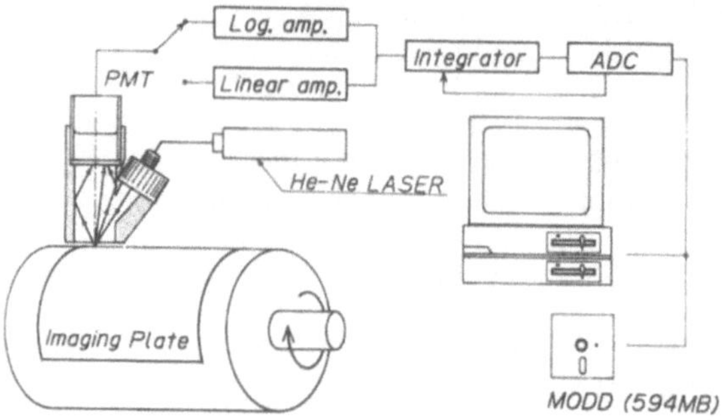

Figure 3. Schematic layout of imaging plate readout system.

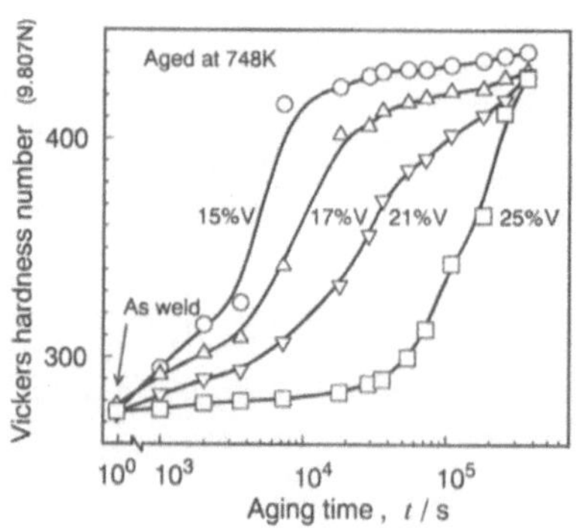

Figure 4. Hardness of weld metal as a function of aging time for different vanadium content.

RESULTS AND DISCUSSION

Aging Hardening Behavior

Figure 4 shows the hardness in weld metal as a function of aging time for four welds of different vanadium content. This clearly shows that the age hardening rate was suppressed by enriching vanadium in the weld metal. The filler of 21% vanadium was found suitable to have desired strength and the fracture toughness as shown in Table 1.

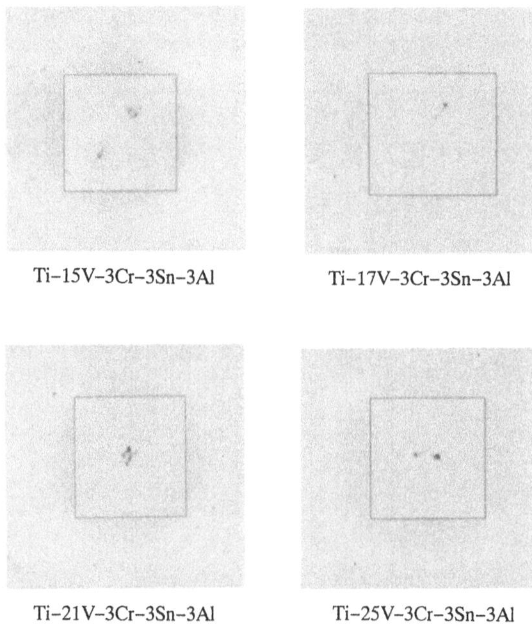

Figure 5. Back-reflection Debye patterns from weld metal of different vanadium content.

Microbeam X-ray Diffraction Patterns

Figure 5 shows X-ray diffraction patterns from the weld metal of different vanadium content. A typical diffraction spot was enlarged and shown in a square at the center of each figure. Diffuse diffraction spots are observed for the case of 15 %V where the vanadium content in weld metal is the same as in base metal. With increasing vanadium content in the weld metal, the diffraction spots become sharper.

Figure 6 schematically shows substructures in a grain formed during welding. Subgrains can be formed by reordering of dislocations which formed under thermal stresses during welding. The orientation of each subgrains differs slightly by introduction of a small angle boundary. Each diffraction spot comes from one grain in irradiated region. When this grain contains subgrains, the spot is diffused along tangential direction to the Debye ring as shown in Fig. 2. The range of orientation or total misorientation β is estimated from the tangential size of the spot. The accumulation of defects such as dislocations in subgrains causes the distortion of crystal lattice and the scatter in lattice parameter d. The amount of distortion or micro-lattice-strain is estimated from the radial size of the diffracted spot.

The total misorientation and micro-lattice-strain calculated from the size of the spots were plotted in Fig. 7 as a function of vanadium content. The open symbols in these figures are for weld metal and closed symbols are for the solution-treated but not welded material. Both parameters decreased with increasing vanadium content for the weld metal, while remained constant for solution-treated material.

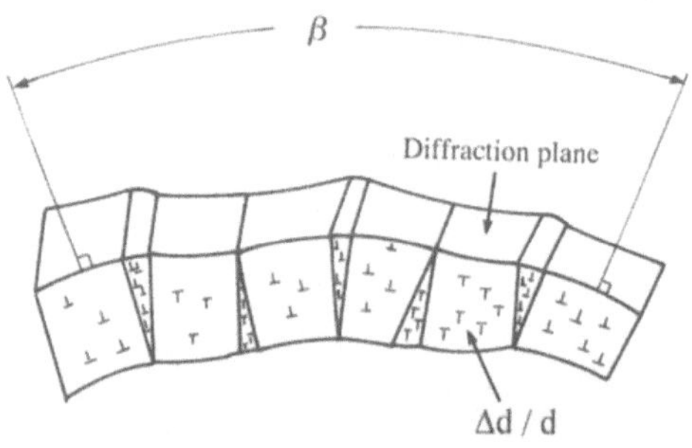

Figure 6. Schematic representation of substructures formed during welding. β and Δd/d denote total misorientation and micro-lattice-strain.

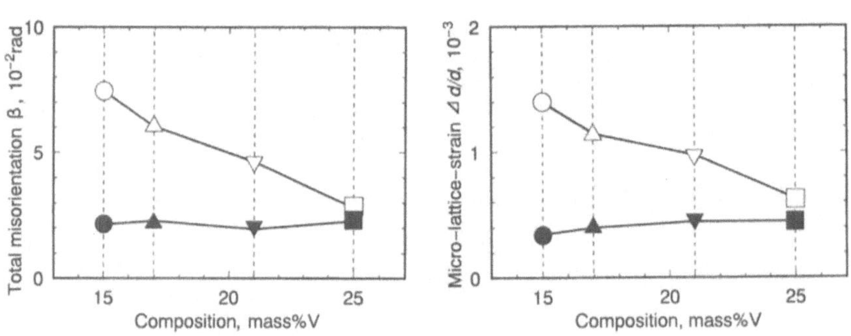

Figure 7. Total misorientation and micro-lattice-strain as a function of vanadium content. Open symbol is from weld metal and closed symbol from material without welding.

Optical Micrographs

Figure 8 shows optical micrographs of the weld metal with different vanadium content. Within the β-grains formed during welding, subgrains are observed. The size of subgrains becomes fine for 15%V and increases with vanadium content. For solution treated but not welded material, subgrains are not observed.

CONCLUSIONS

Table 2 compiles the results of this study so far reported [4]-[6]. The main findings are as follows:

(1) The age hardening rate of the weld metal becomes higher than that of the base metal when the vanadium content is 15%, which results in the low fracture toughness of the weld joint. The fracture toughness is increased by two-stage heat treatment which suppresses the excessive hardening in the weld joint.

(2) With increasing vanadium content in the weld metal, the age hardening rate decreases. Using the filler of 21%V, the desired fracture toughness can be achieved by one-stage heat treatment.

(3) Under the conditions where the enhancement of the age hardening occurs, the microbeam X-ray diffraction spots are diffused and the optical micrograph reveals fine subgrains.

(4) When the age hardening rate is low, the diffraction spots is sharp and subgrains are coarse or not observed.

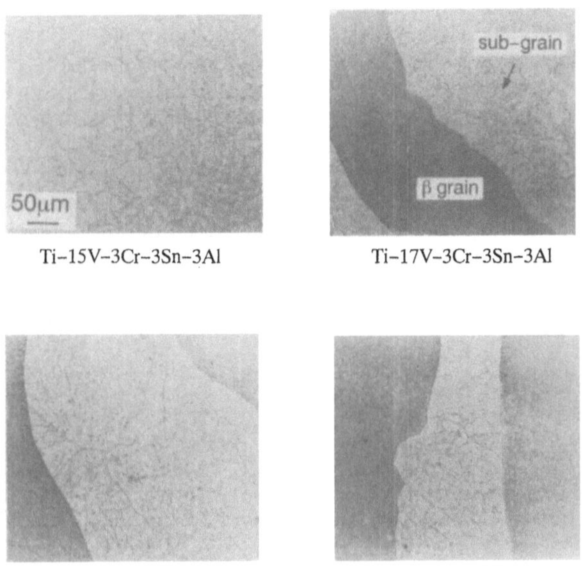

Ti-15V-3Cr-3Sn-3Al   Ti-17V-3Cr-3Sn-3Al

Ti-21V-3Cr-3Sn-3Al   Ti-25V-3Cr-3Sn-3Al

Figure 8. Microstructure of weld metal of different vanadium content.

The consistency of these findings suggests that the age hardening enhancement in the weld metal can be attributed to substructures formed under thermal cycles during welding. The enrichment of vanadium affects the age hardening rate through the suppression of the formation of such substructures. The detailed discussion on the relation between the substructures and the age hardening in the weld metal is left for further studies.

Table 2. Summary of relation between vanadium content, aging characteristics and substructure in weld metal.

	V content	Aging rate	X-ray diffraction pattern	Subgrain
Weld	15%	Rapid	Diffuse	Fine
	25%	Slow	Spotty	Coarse
Base Metal	15%	Slow	Spotty	Not Observed

REFERENCES

1. J. Onoda, N. Watanabe, K. Ichida, Y. Hashimoto and A. Nakata, Reports of The Institute of Space and Astronautical Science of Japan, Special Issue, No.29, 87(1991).
2. P. J. Bania, G. A. Lenning and J. A. Hall, in: "Beta Titanium Alloy in The 1980's", R. R. Boyer and H. W. Rosenberg, ed., TMS / AIME, New York, 209(1984).
3. Y. Shirasuna, A. Nozue, T. Okubo, K. Kuribayashi, R. Horiuchi, S. Ishimoto and H. Sato, TETSU-TO-HAGANE, The Iron and Steel Institute of Japan, 76, 614(1991).
4. Y. Shirasuna, A. Nozue, T. Okubo, K. Kuribayashi, S. Ishimoto, H. Sato, and Y. Yoshioka, Abstracts of the Japan Institute of Metals, 112, 370(1993).
5. Y. Shirasuna, A. Nozue, T. Okubo, K. Kuribayashi, R. Horiuchi, S. Ishimoto and H. Sato, Nondestructive Testing and Evaluation, 9, 1085(1992).
6. Y. Shirasuna, A. Nozue, T. Okubo, K. Kuribayashi, S. Ishimoto, H. Sato, and Y. Yoshioka, Journal of the Society of Materials Science, Japan, 42, 600(1993).
7. K. Hayashi and T. Konaga, in: "X-Ray Studies on Mechanical Behavior of Metals", Committee on Mechanical Behavior of Materials, ed, The Society of Materials Science, Japan, 351(1974).
8. Y. Yoshioka and S. Ohya, in: "Residual Stresses-III", H. Fujiwara, T. Abe and K. Tanaka, ed., Elsevier Applied Science, London, 2, 985(1992).

STUDY OF EFFECT OF METHANE CONCENTRATION IN ARGON PLASMA ON TANTALUM COMPOUND SPUTTERING DEPOSITION PROCESS

S.L. Lee[1] and C.S. Lee[2]

[1]U.S. Army Armament, Research, Development and Engineering Center
Benet Laboratories, Watervliet, NY 12189-4050
[2]Department of Mechanical Engineering, Rensselaer Polytechnic Institute
Troy, NY 12180

ABSTRACT

There are increasing demands for advanced materials which meet the erosion, corrosion resistant properties, possess desirable thermal and electrical properties, and which can endure the high pressure, high temperature and aggressive chemical environment of the bore in future projectile launchers. In this work, tantalum and tantalum compounds were sputter deposited in argon plasma containing methane of varying concentrations. X-ray diffraction analysis demonstrated that *bcc* alpha tantalum was formed at low methane concentrations, and *fcc* tantalum carbide at high methane concentrations. The steep transition occurred at 22 molar percent of methane and 78 percent argon plasma mixture. As percentage methane in argon increases, drastic changes in coating composition, crystalline structure, particle size and preferred orientations occur. Deposition rate, Knoop hardness and temperature coefficient of resistivity have also been found to be sensitive function of methane concentration. The successful deposition of refractory ceramic constituents, such as tantalum carbide, promises new tough ceramics for future refractory coating applications.

INTRODUCTION

A new generation of advanced refractory bore coating materials are being developed for future launch technology. Coating the bore surfaces with refractory metals and ceramics greatly improves the wear and erosion life of pressure vessel systems. However, in addition to the erosion and corrosion resistant requirements of coating deposition, the new materials must endure the aggressive high temperature, high pressure conditions, possess desirable thermal and electrical properties, and have low reactivity to the plasma environment. Active research programs are being conducted in refractory bore coatings by electrochemical deposition in aqueous solution and molten salt bath, and by physical vapor deposition through sputtering and evaporation processes. Objectives of the coating project are: (1) to develop refractory metals, alloys, and ceramics of superior physical and chemical properties, (2) to optimize deposition process parameters to obtain strong adhesion, low stress and uniform deposits, (3) to characterize coating processes by nondestructive X-ray diffraction, energy dispersive X-ray analysis, and electron microscopy techniques. Sputtering deposition of tantalum carbide was achieved in a triode sputtering chamber, where the plasma density was controlled by adjusting the electron emission from a hot tantalum filament. Advances in weapon sputtering coating technology have been reported[1,2]. As methane concentrations increase, Knoop hardness increases, and temperature coefficient of resistivity decreases[3].

Melting point temperatures of common refractory metals and alloys above 1000°C are given in Figure 1. Many of the metals and ceramic constituents of Cr, Ta, Mo, W, Re etc. have been considered for advanced coating materials for future projectile launchers. Tough ceramics, such as tantalum carbide and tantalum nitride, are among the most promising. Due to the low atomic weight of carbon in heavy tantalum matrix, energy dispersive X-ray microanalysis of specimens obtained at 0, 22, 33 and 50 percent methane concentrations in argon showed only tantalum, but no carbon peak. In this work, X-ray diffraction determined that *bcc* tantalum was deposited at methane concentrations below 20%, and *fcc* tantalum carbide was deposited at methane concentrations above 25%, and a mixture of tantalum and tantalum carbide at the transitional 22% methane concentration. For the 22% methane specimen, recent wavelength dispersive X-ray fluorescence spectrometer analysis further demonstrated both tantalum and carbon contents in the specimen. Grain size in polycrystalline materials has pronounced effects on strength and hardness, increasing strength and hardness accompanies a decrease in grain size. The introduction of methane (CH_4) gas in argon plasma in the sputtering chamber caused the coating deposits to transform from Ta to TaC. The Ta/TaC deposits consisted of very fine particle sizes. Coating composition, crystalline structure, particle size and preferred orientations were found to be very sensitive to methane concentrations in argon in the sputtering deposition process.

TANTALUM, TANTALUM CARBIDE AND TANTALUM HYDRIDE PHASES

Binary phase diagrams of Ta-C and Ta-H have been compiled by Hansen in 1958, Shank in 1969 and Massalski in 1986[4]. They disclose phase dependence on temperature and atomic percent of the constituents. Tantalum-carbon phase diagram discloses Ta, TaC and Ta_2C phases at temperatures above 1800°C, and tantalum-hydrogen phase diagram is not well defined. Data base search for tantalum/tantalum carbide/tantalum hydride phases in ICDD[5] results in- two phases of tantalum: alpha-Ta *(bcc)*, beta-Ta *(tetragonal)*; two phases of tantalum carbides: TaC *(fcc)*, Ta_2C *(hexagonal)*, and intermediate phases with various degrees of nonstoichiometry- zeta-$TaC_{0.47}$, metastable zeta-$TaC_{0.6}$, and xi-$C_{0.71}Ta$ *(rhombohedral-hex)*; several phases of tantalum hydrides: beta-TaH *(hexagonal)*, Ta_2H *(orthorhombic)*, and intermediate phases with various degrees of nonstoichiometry: $TaH_{0.8}$ *(orthorhombic)*, $TaH_{0.9}$ *(orthorhombic)*.

X-RAY SPECTROSCOPY CHEMICAL ANALYSIS

Our energy dispersive X-ray (EDX) microanalysis of sputtered specimens was not conclusive. Spectra obtained for sputtered deposits at 0, 22, 33 and 50 percent methane concentrations in argon plasma showed only tantalum peak, but no carbon peak. Atomic weight of carbon is 12.01, atomic weight of tantalum is 180.95, it is difficult to detect low Z materials in a heavy matrix. Our Tracor x-ray fluorescence (XRF) analyzer has capability of chemical analysis of low-Z elements down only to sodium. The coating deposit obtained at 22% methane concentration is of particular interest because it represents a state of co-existence of Ta and TaC. Wavelength dispersive spectrometer measurements were made of the 22% methane, 78% argon deposit during our recent visits to Rigaku USA and Philips Electronics. Strong carbon as well as tantalum peaks were observed.

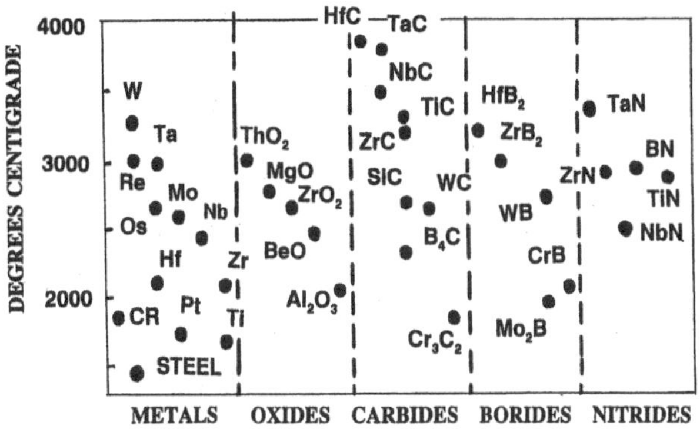

Figure 1. Melting point temperatures of refractory materials for bore coating applications.

X-RAY DIFFRACTION ANALYSIS

X-ray diffraction (XRD) analysis was first performed using a semi-quantitative Philips diffractometer, then with a recently procured Scintag diffractometer. The Philips diffractometer has a LiF crystal monochromator, NaI scintillation detector and strip chart recorder. Molybdenum radiation was used in the study. The four-axis Scintag diffractometer has a Peltier-cooled Si(Li) detector, multi-channel spectrum analyzer, and optimized divergent, receiving and scattering slits. Both molybdenum and copper radiations were used in the study.

In Figure 2, Philips diffractometer scans of sputtering deposits obtained at 0, 22 and 25 percent methane concentrations in argon are displayed along with diffraction scans of tantalum and tantalum carbide powders obtained from Semi-Elements Inc. A sieve with opening of 0.074 mm (0.0029 inch) was used to prepare tantalum and tantalum carbide powders. The tantalum powder pattern is attributed to alpha Ta, and tantalum carbide powder pattern to TaC. Comparison of pattern for deposit obtained at 0% methane in argon (thickness 23.24 microns on 0.05 mm Al foil substrate) with tantalum powder diffraction pattern discloses predominantly alpha-Ta in the sputtered sample. The pattern obtained at 22% methane in argon (thickness 22.73 microns on 3 mm Al_2O_3 substrate) is difficult to interpret because of the very broad and diffused diffraction peaks. The pattern indicates possible co-existence of tantalum and tantalum carbide. Comparison of pattern for deposit obtained at 25% methane in argon (thickness 4.8 microns on 3 mm Al_2O_3) with tantalum carbide powder diffraction pattern discloses predominately TaC phase.

In Figure 3, Philips diffraction scan for sputtered sample obtained at 50% methane, 50% argon (thickness 1.73 microns on 3 mm Al_2O_3 substrate) is displayed along with scans for tantalum carbide powder and aluminum oxide substrate. The pattern is characterized by very broad TaC peaks superimposed on sharp diffraction peaks from Al_2O_3 substrate through TaC layers. When methane concentration in argon plasma varies from 0 to 50%, deposition rate decreases from 2.5 to 0.3 microns/hr. Because of the slow deposition rate at high methane concentration, a long time is required to obtain thicker coatings.

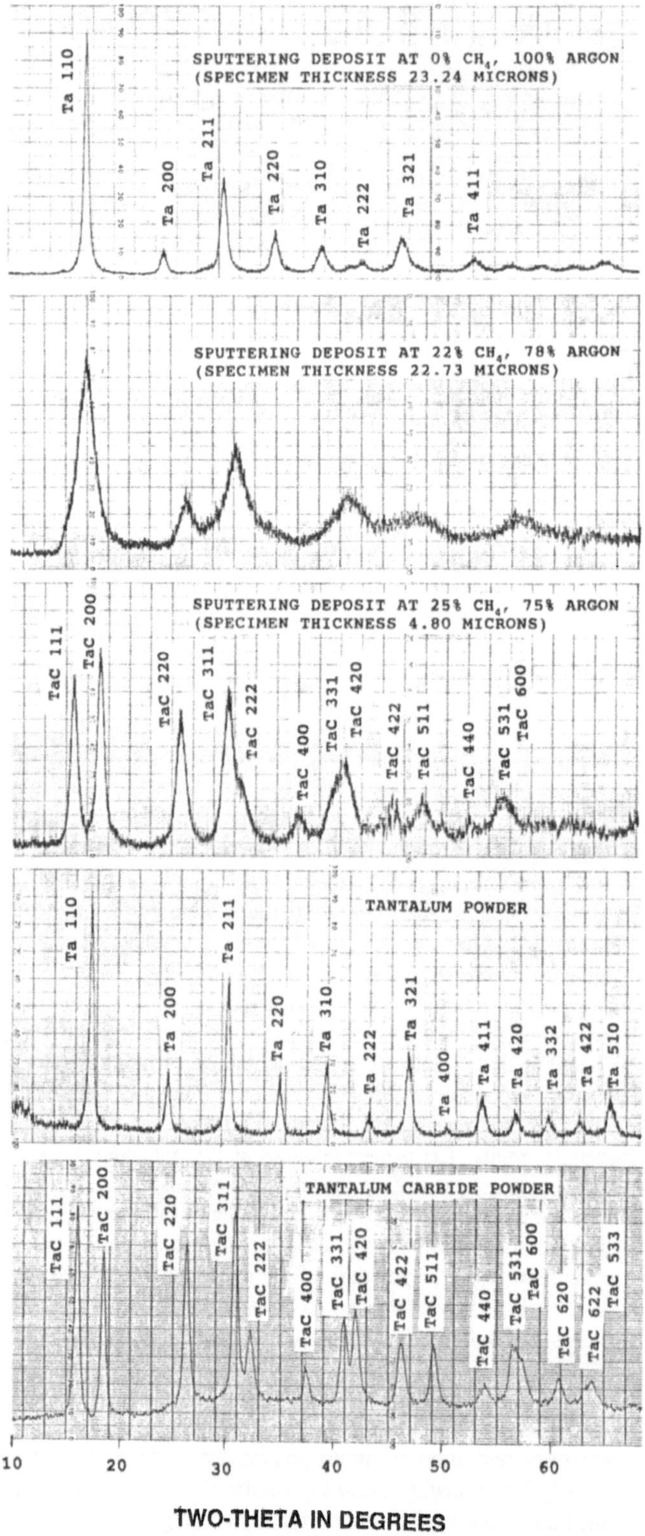

Figure 2. Diffraction scans of sputtering deposits obtained at 0, 22 and 25 percent methane concentrations in argon plasma, compared with scans of tantalum and tantalum carbide powder specimens using Philips diffractometer.

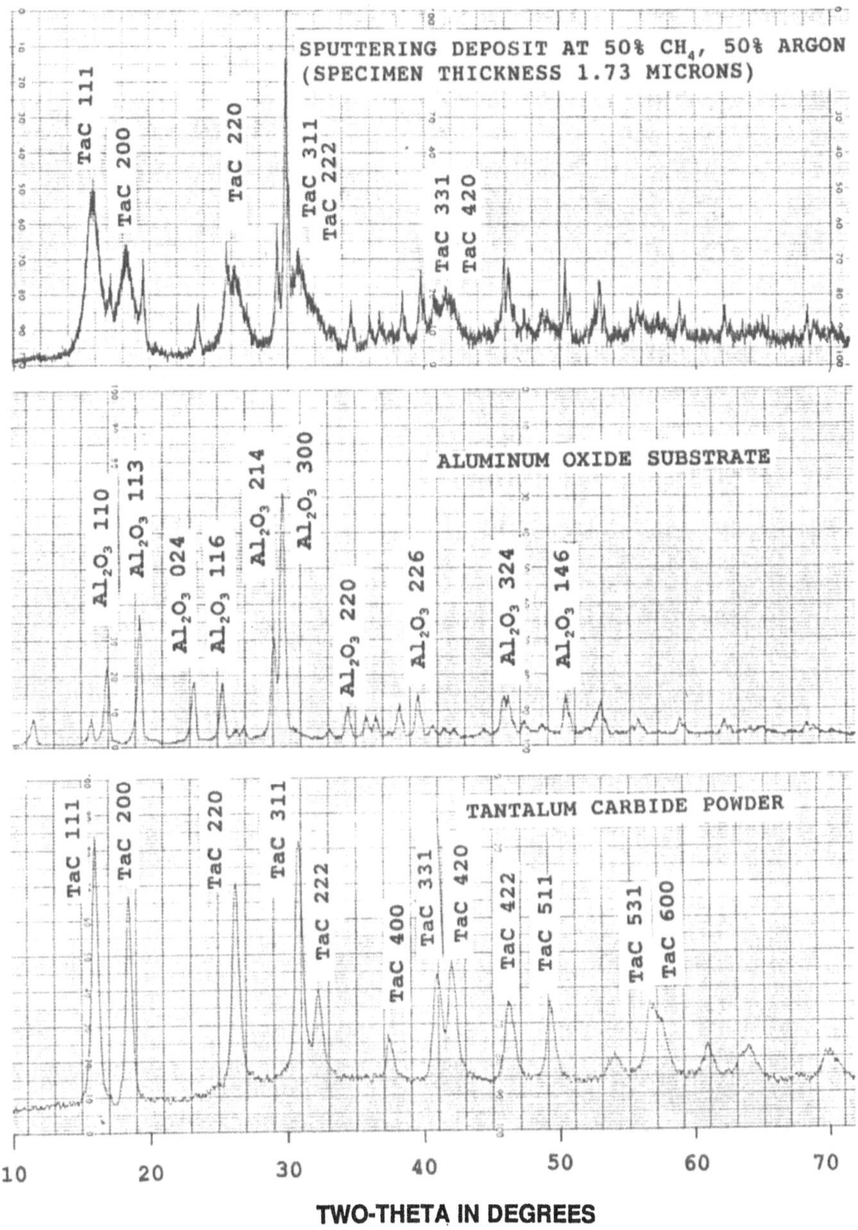

Figure 3. Diffraction scan of sputtering deposit obtained at 50% methane concentration compared to scans of aluminum oxide substrate and tantalum carbide powder specimen.

In Figure 4, Scintag diffraction patterns of tantalum target metal and sputtered specimen obtained at 0 % methane are displayed along with tantalum powder scan and simulated pattern from ICDD diffraction data base. The patterns were background subtracted and K-alpha2 corrected, the simulated pattern included only K-alpha1 peak. The results are summarized: (1) The randomly oriented Ta powder has diffraction peak intensities and peak locations in agreement with the simulated Ta pattern. (2) Tantalum target metal lines are from alpha Ta, with major diffraction peaks shifted 0.2° towards lower two-theta compared to Ta powder. (3) Full width at half maximum (fwhm's) of major peaks in Ta powder range from 0.03° to 0.07° for two-theta from 17° to 50°, Ta target metal peaks fwhm's range from 0.10° to 0.20° in similar two-theta range. (4) Strong texture with preferred <200>, <211> and <411> crystalline orientations, and diminished <110> orientation in the Ta target metal. (5) The 0% methane deposit Ta pattern exhibits a 0.5° shift towards higher two-theta, broadened lines which range from 0.3° to 0.6° in the same two-theta range. The 0% deposit was strongly textured with <211>, <222>, <321> and <332> preferred orientations.

In Figure 5, Scintag diffraction scan of sputtering deposit obtained at 25% methane concentration in argon is compared with diffraction scans of tantalum carbide powder and simulated tantalum diffraction pattern. The results are summarized: (1) The diffraction pattern of fine tantalum carbide powder is consistent with pattern for TaC. Diffraction peak intensities and peak locations indicate that TaC powder was randomly oriented with peak intensities and locations identical to TaC from ICDD data base. TaC powder lines have fwhm's of 0.03° to 0.06° in the range of interest. (2) The 25% methane deposit exhibits a 0.2° to 0.3° peak shift towards lower two-theta compared to TaC pattern. (3) The fwhm's of the broadened diffraction peaks range from 0.6° to 0.8° for two-theta range of 15° to 30°. (4) Enhanced <200> preferred orientation in the 25% methane coating sample.

In Figure 6, Scintag diffraction scan of sputtering deposit obtained at 22% methane concentration in argon is compared to simulated Ta and TaC diffraction data base. The very broad and diffused peaks are difficult to interpret. Co-existence of Ta and TaC is possible, as well as other carbide and hydride phases. The two strongest lines have fwhm's of 0.75° and 0.95°.

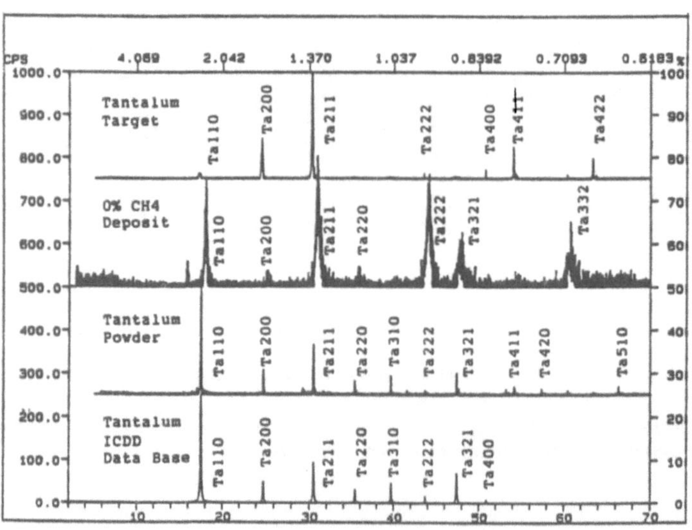

Figure 4. Scintag diffractometer scan of specimen obtained at 0% methane concentration as compared with scan of tantalum powder specimen and ICDD simulated diffraction pattern for tantalum.

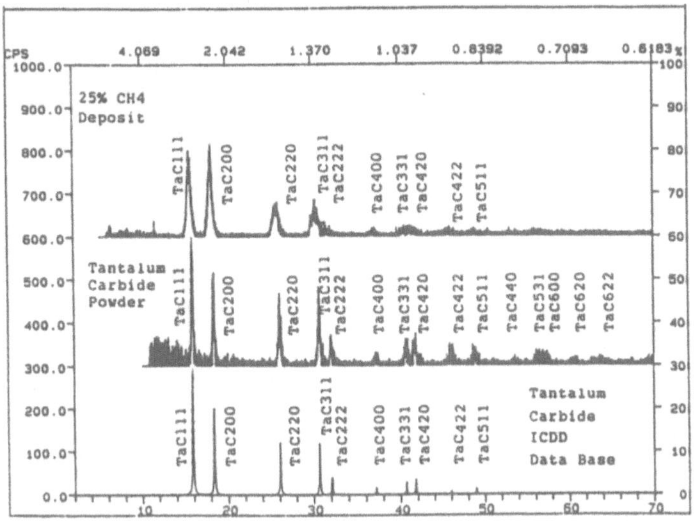

Figure 5. Scintag diffractometer scan of specimen obtained at 25% methane concentration as compared with scan of tantalum carbide powder specimen and ICDD simulated diffraction pattern for tantalum carbide.

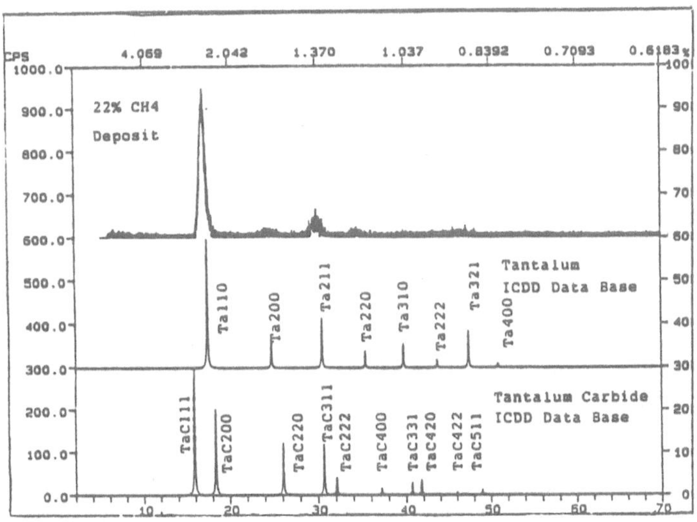

Figure 6. Scintag diffractometer scan of specimen obtained at 22% methane concentration as compared with simulated diffraction patterns for tantalum and tantalum carbide.

Table 1 summarizes major diffraction peaks in tantalum target metal, sputtering deposition specimens obtained at 0, 22 and 25 percent methane concentrations, and tantalum and tantalum carbide powders. Diffraction line shift reflect residual stresses in the specimens, diffraction peak broadening reflects a reduction in crystalline size, by fault of certain *hkl* planes and by microstrains in the coherently diffracting domains[6]. In addition, carbon and hydrogen dissolution interstitially in the Ta/TaC deposits can also cause shifts in the diffraction peaks. In the 0% deposition specimen, peak shifts to smaller d-spacings, which represents tensile stresses. Alpha Ta (***bcc***) has lattice parameter of 3.3058 angstroms, TaC (***fcc***) has lattice parameter 4.4547 angstroms. In the deposit obtained at 25% methane concentration, peak shifts to larger d-spacings, which may be due to the effect of carbon dissolved in Ta/TaC, or compressive residual stresses. In addition to internal residual stresses generated in the deposition process, thermal stresses should also be considered. Furthermore, the 0% deposit was made on Al foil substrate, while the 25% deposit was made on Al_2O_3 substrate, the shifts can be caused by the differences in thermal expansion coefficients of tantalum metal compared to the aluminum substrate, and tantalum carbide ceramic compared to the aluminum oxide substrate. For diffraction peak around 17° two-theta, fwhm for Ta target was 0.16°, for deposition samples at 0, 22% and 25% were 0.5°, 0.9° and 0.5°, while fwhm for Ta and TaC powder samples were of the order of 0.03°. These results suggest small particles sizes were generated during the deposition process, and in particular, very small particles were produced in the 22% transition range. According to Klug and Alexander[7], when diffraction lines are very broad and diffuse, crystalline size broadening is the major cause of peak broadening. At particle sizes much less than 100 angstroms, the back-reflection lines disappear, the low angle lines become very wide and diffuse. Scherrer and Warren-Auerbach[8,9] methods for particle size and microstrain broadening analysis have been implemented. Small microstrains and average particle sizes of approximately 85, 55 and 65 angstroms were observed for specimens obtained at 0, 22 and 25 percent methane concentrations.

CONCLUSIONS

Advances in x-ray techniques and instrumentation allow successful characterization of the effect of methane concentration on tantalum and tantalum compound sputtering deposition process. In the investigation of Ta/TaC sputtering deposition, tantalum was formed at low methane concentrations, at the threshold of 22% methane and 78% argon, there was an onset for tantalum carbide formation. As the methane concentration reached 25%, the coating deposit was predominately tantalum carbide. Ta/TaC formation was found to be a sensitive function of methane concentrations, with steep transition at 22% methane concentration.

Summarizing our current and previous investigations of Ta/TaC sputtering deposition, as methane concentration in argon increases from 0, 22, 25 to 50 percent, the following transitions occur in the coating deposits: (1) crystalline structure changes from *bcc* Ta to *fcc* TaC, (2) diffraction peaks change from broad, to very broad and diffused, to broad, (3) fine particle size of the order of 55-85 angstroms are observed, (4) strong deposition and thermal residual stresses, and strong preferred orientations, effect of solid solution are observed, (5) deposition rate decreases from 2.5 to 1 microns thickness per hour, (6) Knoop hardness increases from 820 to 1500, (7) temperature coefficient of resistivity decreases from 1.7 to -0.1 (1/Deg. C. x 10^{-3}), and the successful deposition and characterization of tantalum and tantalum carbide demonstrates a new tough ceramic for future refractory coating applications.

ACKNOWLEDGEMENTS

The authors would like to thank Joe Cox for the helpful discussions during the present and previous investigations of refractory coating project.

Table 1. Major x-ray diffraction peaks in the tantalum target, sputtering deposition specimens obtained at 0%, 22% and 25% methane concentrations, tantalum powder and tantalum carbide powder samples.

Sample Ident	Two-Theta (deg)	d-Space (angstrom)	I/I_o	FWHM (deg)	HKL
Ta target	24.5533	1.6679	26	0.19	Ta 200
	30.2467	1.3593	100	0.14	Ta 211
	53.9819	0.7814	25	0.10	Ta 411
	63.2812	0.6761	16	0.10	Ta 422
0 % methane	17.9943	2.2678	80	0.28	Ta 110
	30.9738	1.3282	100	0.47	Ta 211
	43.9933	0.9469	92	0.53	Ta 222
	47.7592	0.8761	26	0.60	Ta 321
	60.7076	0.7018	29	0.59	Ta 332
22% methane	17.0604	2.3909	100	0.75	
	29.9441	1.3728	9	0.95	
25% methane	15.6740	2.6006	99	0.57	TaC 111
	18.1642	2.2468	100	0.59	TaC 200
	25.7890	1.5893	38	0.80	TaC 220
	30.3966	1.3528	32	0.83	TaC 311
Ta powder	17.4625	2.3363	100	0.03	Ta 110
	24.7859	1.6525	16	0.06	Ta 200
	30.4813	1.3491	35	0.05	Ta 211
	35.3300	1.1687	10	0.03	Ta 220
	39.6631	1.0454	14	0.07	Ta 310
	47.3372	0.8834	15	0.06	Ta 321
TaC powder	15.8675	2.5694	100	0.03	TaC 111
	18.4718	2.2161	73	0.08	TaC 200
	26.0900	1.5712	61	0.03	TaC 220
	30.6388	1.3424	60	0.06	TaC 311

REFERENCES

1. Lee, S.L., Heffernan, W., and Walden,J.A., "Nondestructive characterization of sputtering deposition of tantalum and tantalum carbide by X-ray diffraction", American Society for Nondestructive Testing Paper Summaries of the 1992 Spring Conference Professional Program, Orlando, FL., pp. 185-187 (1992).
2. Walden, J.A., Heffernan, W. and Lee, S.L., "Advances in weapon coating technology by physical vapor deposition", US Army Technical Report in Basic and Applied Research, pp. 2-9 to 2-17 (1993).
3. Heffernan W. and Walden, J.A., "Sputtering deposition of tantalum and tantalum compound from argon plasmas containing methane", US Army Technical Report in Basic and Applied research, pp. 3-19 to 3-28 (1991).
4. Massalski, T. "Binary Alloy Phase Diagrams", vol. 1 & 2, American Society for Metals, p. 592, p. 1282, (1986); Francis Shunk, "Constitution of Binary Alloys", Second Supplement, McGraw-Hill Book Company, p. 158, p. 409 (1969); Phil M. Hansen, "Constitution of Binary Alloys", McGraw-Hill Book Company, p. 380-381, p. 796-797 (1958).
5. ICDD, International Center for Diffraction Data, Swarthmore, PA (1993).
6. Wagner, C.N.J., Boldrick, M.S., and Keller, L., "Microstructural characterization of thin polycrystalline films by x-ray diffraction", Advances in X-Ray Analysis, Plenum Press, pp. 129-142 (1988).
7. Klug, H.P. and Alexander, L.E., "X-Ray Diffraction Procedures for Polycrystalline and Amorphous Materials", John Wiley and Sons (1974).
8. Noyan, I.C. and Cohen, J.B., "Residual Stress Measurement and Interpretation", Springer-Verlag (1987).
9. Warren, B.E., "X-Ray Diffraction", London: Addison Wesley (1969).

QUANTITATIVE NONDESTRUCTIVE EVALUATION OF DENSITY

OF GREEN STATE COMPRESSED PRODUCTS

J. Muller[1], L. Ackermann[2], D. Babot[3], G. Peix[3], P. Zhu[3]

[1]PECHINEY CRV, BP 27, 38340 Voreppe (France)
[2]SINTERTECH, Route des Collines, 38800 Pont de Claix (France)
[3]INSA DE LYON, Bat.303, 20 av Albert EINSTEIN, 69621Villeurbanne Cédex (France)

INTRODUCTION

Industrial P/M products shapes are often complex and do not permit any density measurement by X-ray transmission. The local density at different locations (on several mm^3, just after powder compression) is an essential parameter of the process, which governs the future mechanical properties of the final product (after sintering). In order to set up compression presses, operators need rapid assessments after each tool modification that has influenced local density distribution within the compressed piece. Currently, a statistical approach is used on the basis of manual measurements on broken products. Drawbacks are evident : operator dependent results, insufficient accuracy on the density (±0.05), time consuming procedure (20 minutes).

FUNDAMENTAL APPROACH

There are several types of X-ray interaction process[1] , but only four are of concern in industrial radiology. Each of these interaction occurs with a probability depending on X-ray energy and atomic number Z of the interacting material (Figure 1, calculated for iron). In *Rayleigh scattering*, the incident X-ray photon is scattered with no change in energy, and no release of electrons. In *photoelectric interaction,* the X-ray photon is absorbed and its energy is used in removing an electron or electrons from the inner shells of the interacting atom. In *Compton scattering*, the incident X-ray photon is scattered non-elastically and a recoil electron is produced out of the interacting atom. With most of the X-ray energies used in industrial radiology, this is the dominant contribution to total attenuation. The result of this interaction process is a scattering photon of lower energy than the incident photon, travelling in a different direction, and a recoil electron. *Pair production* (electron-positron) occurs only if the primary photon has an energy greater than 1.02 MeV.

Nondestructive Characterization of Materials VI
Edited by R.E. Green, Jr. *et al.*, Plenum Press, New York, 1994

The proportion of the X-ray beam scattered in a small volume of the object tends to be relatively small, and the X-rays are scattered in all directions [2]. A portion of this radiation scattered is measured by detectors typically placed near the entry point of the X-ray beam. A finely collimated detector allows measurement of the number of photons N that are scattered inside the volume V, delimited by the intersection of the source beam with the focal area of the detector collimator (Figure 2). If the geometrical arrangement is kept constant, and if the same radiation (either mono- or poly-chromatic) is used, it can be shown that N is proportional to the electronic density. Because the Z/A ratio has roughly the same value for all elements except hydrogen, it can be assumed that N is proportional to the density of that part of the material included in the above-mentioned volume V. The value of N is given by equations 1 and 2, which contain a specific apparatus constant related to the main parameters of both the material and the installation.

local density — function of all interactions photon-material at the location x

$$N(x, \rho, \tau) = \dot{K} \cdot \rho(x) \cdot \tau \cdot F(x, \bar{\rho}, E) \qquad (1)$$

Measured scattered photons number — **Apparatus constant** — exposure time (fixed) — Photon energy

Collision differential cross section (constant) — Avogadro Number (constant) — Atomic number / atomic mass (constant)

$$\dot{K} = \left(\frac{d\sigma}{d\Omega} \cdot \Delta\Omega \cdot \Delta L \cdot N_0 \right) \left(N_a \cdot \frac{Z}{A} \right) \qquad (2)$$

Apparatus constant — *solid angle under which the detector is seen from the volume V* — *length of the primary beam path along the volume V* — *Incident photon flow*

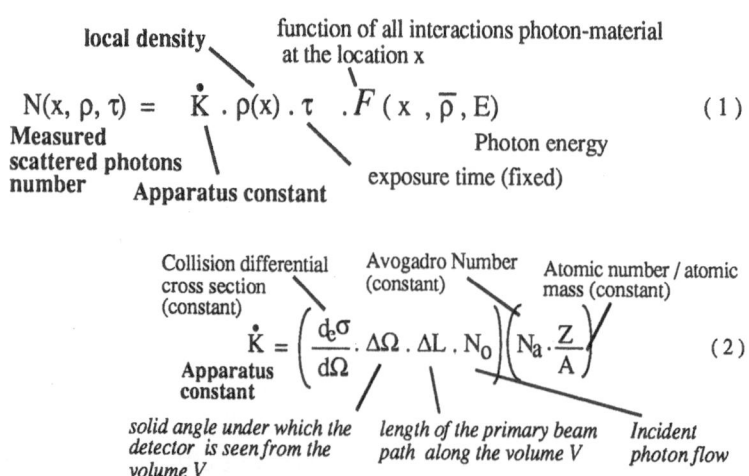

N also depends on the depth x (Figure 3), at which is located the measurement volume, because of the total attenuation. So, when a quantitative measurement is to be made, it is necessary to give a fixed value to x, generally the abscissa at which N is maximum. Such an assumption is satisfactory as long as the photoelectric effect can be neglected, as is the case for elements belonging to the four upper rows of the periodic chart of the elements.

SOME APPLICATIONS FROM IMAGING TO DENSITOMETRY

Backscatter Compton imaging [3] is a technique in which the object is irradiated by a narrow X-ray beam, and the Compton scattering of this radiation by the object is monitored[5,6]. The first applications were in the medical field, and were based on γ-ray radiation for bone and tissue density evaluation. [7,8,9,10] In general, radiography or computed tomography is preferred for internal inspection because of it is more practical to use[11]. The major advantage offered by backscatter imaging is that it can be implemented on one side of the object. Backscatter imaging provides a means of obtaining near-subsurface information, particularly if the object is either very massive, but does not permit X-ray transmission through it, or does not permit access to the opposite side[12]. Like computed tomography, backscatter imaging can also generate images of planar sections through the testpiece. Some authors[4,13,14] have described systems able to simplify scanning of the whole object by moving a X-ray beam ; Philips GbmH has recently proposed an industrial apparatus (ComScan) based on a 160 kV X-ray source, which scans over 50x100 mm areas with a 0.4

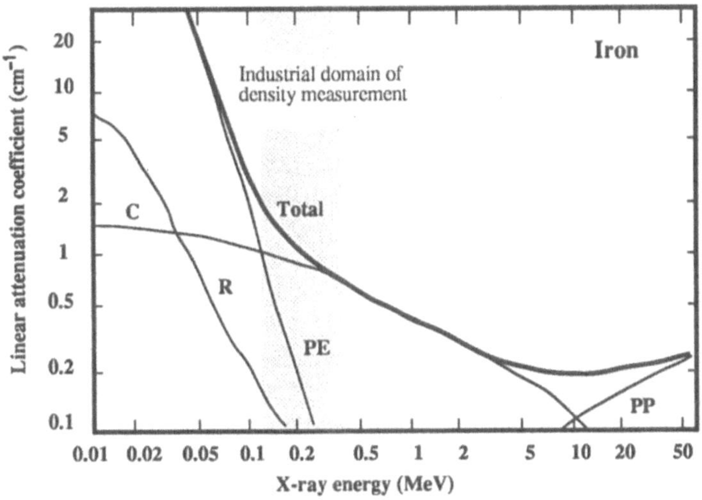

Figure 1. Linear interaction coefficients for iron, plotted against X-ray energy, showing different components of the total scatter: Compton scatter (C), Photoelectric scatter (PE), Pair production (PP), Rayleigh scatter (R). [1]

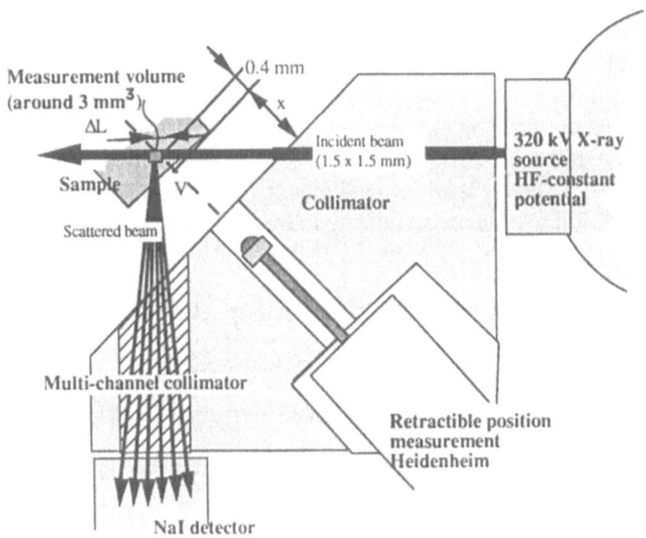

Figure 2. Principle of Compton radiation scattered measurement. A detector measures all the radiation scattered coming from a small volume (measurement volume V). The main difficulty is to provide enough radiation scattered in a short time and in usual industrial conditions (160 kV), for the required accuracy on the density. The technology of the multi-channel collimators governs the performance of the technique.

mm resolution. Other workers have constructed an apparatus suitable for *thickness measurements* of shells, coatings and intervals between two objects [2,15,16,17].

EXPERIMENTAL APPROACH FOR QUALIFICATION

Ten values of density between 6.10 and 7.30 have been targeted for evaluating quantitative measurements (three samples per value). Compressed test pieces (10x10x50mm) have been made in a classical P/M alloy : Distaloy AB + 0.3%C + 0.7% zinc stearate (Fe-1.5%wt Cu - 1.7%wt Ni - 0.5%wt Mo). The mean density of each sample is measured by the pycnometric method with a resolution of ± 0.05.

A high stabilized 320 kV X-ray generator (PANTAK HF) has been used (Figure 2), in order to ensure the highest stability of X-ray beam. The detector, a 2" diameter sodium iodide (thallium activated) crystal is connected to a selector and scaler through an amplifier. After the preliminary tests at the 320 kV level, we set the level at 160 kV in order to achieve a compromise between the highest X-ray photon flow and the limited cost of investment for a quality control apparatus. The P/M testpieces are quite small. They are placed on a three axis table, which is controlled by a computer, enabling explorations of samples to be performed automatically.

Since the measurement must be quick and precise, the detector must detect as many scattered photons as possible. The measurement volume has been targeted at a few mm^3. We have reached a compromise with a measurement configuration at a 90° Compton angle, and manufacturing of a multi-channel collimator (around 40 channels of ø 1.5 mm, all centered on the measurement volume). This particular design permits a high level of counting : 450000 counts in 40 seconds, for a density of 6.7, with standard radiographic conditions (160 kV, 19 mA).

When the sample is moved near the location of the measurement volume, the scattering process follows the law described on figure 3 : the number of photons scattered increases quickly when the measurement volume penetrates into the sample, so reaches its value given by equation (1) and decreases because of the interaction processes within the matter. The abscissa of the maximum radiation scattered is specific to the material tested.

At the beginning of the campaign, a number of test parameters must be fixed : voltage and intensity of the X-ray source (160 kV, 19 mA), minimum counting duration for the required accuracy on the scattered level (around 40 s for ±0.3%), optimum depth of the measurement volume according to the mean density to be measured (around 0.4±0.010 mm).

RESULTS : FROM DENSITY STANDARDS TO INDUSTRIAL PRODUCTS

Figure 4 shows the excellent fit of the linear regression between pycnometric densities d and radiation scattered measurements N, made on the ten standard samples (eqn.3) :

$$d = -14{,}835 + 4{,}5090.10^{-5} . N \quad (R^2 = 0{,}999) \qquad (3)$$

Of course, numerical values are extremely dependent on the configuration and test parameters. In day-to-day use, it is not necessary to verify so completely the calibration, but it must be made as often as possible (every half-hour) on one standard.

Advanced tests have been made on industrial samples, given by SINTERTECH : synchronizing rings for automotive industry. Those products are very representative of the industrial problem : how to measure as fast as possible the density on each tooth (21 teeth) with the best resolution ? We have asked SINTERTECH to modify partially the press in order to create an alteration in the density distribution of teeth. Two samples, one up, the oth-

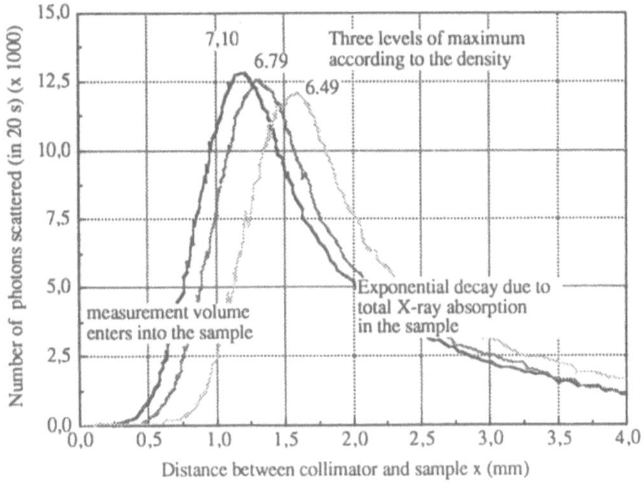

Figure 3 . Experimental curves of radiation scattered vs depth obtained by moving the samples in the x direction before the measurement volume. The level of maximums is reported for density calibration on standard samples.

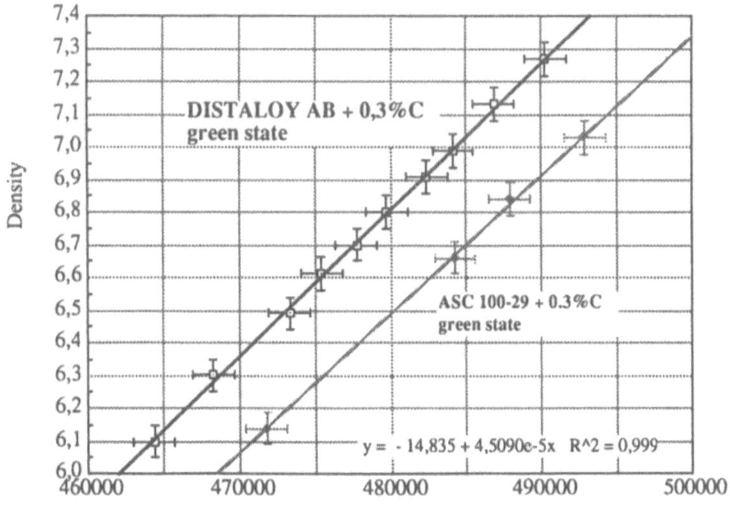

Figure 4 . pycnometric density vs photons scattered number for two P/M alloys.

er one down have been provided. Using an automatic rotating table, the two samples have been completely characterized within half an hour with a accuracy of ±0.070 for the density (± 0.3% on the scattered radiation). Figure 6 shows these results. It gives the operator the exact locations to be modified to go back to normal press settings.

That example shows that it is not necessary to have the relationship between density and the radiation scattered ; it is enough to calibrate the apparatus with a standard sample (the good ring), and to compare other samples on the basis of the scattered radiation.

DISCUSSION : AN ADVANCED TECHNIQUE FOR A MORE PRECISE AND QUICK MEASUREMENT

The measured dispersion of ±0.07 on the density is still too large compared with ±0.05 of the pycnometric method, even if the measurement is conderably quicker. First, we shall outline the main sources of that dispersion, before discussing how to reduce it.

Physical Origins of the Main Sources of Dispersion

Intrinsic Heterogeneity of Standard Samples : green state P/M products always have a relative heterogeneity. It is well known that a density gradient exists according to the direction of compression (this apparatus could characterize it later ...). That is why, each measurement on density standard has been conducted at the same location of the surface.

Multi-scattering Inside the Measurement Volume and Edge Effects : this phenomenon has already been reported[18] , up to 47% along the side of the samples when the measurement volume exits the sample. Industrial density evaluation also begins with verifying the integrity of the measurement (no edge effect).

Position Uncertainty and Maximum Counting Location Procedure : Figure 3 shows that it is important to determine very precisely the optimum depth at which the measurement volume is to be placed, corresponding to the maximum scattered radiation. This abscissa is found and fixed at the beginning of the campaign, according to an automatic procedure using a retractible position detector.

Intrinsic Statistical Process of the Photon Production : photon emission is a random phenomenon, well described by Poisson's law (large number of events with weak probability for each). In pratice, the number of photons scattered N is related to the number of emitted photons with the same law ; in fact, it can be considered that N is known with the dispersion of $\pm 2.\sqrt{N}$; in the case studied, N is around 450 000 counts ± 1341. The resolution of this measurement is also ± 0.3%.

X-ray Source Instability : each X-ray tube needs to warm up before use. In spite of the technology used (high frequency and constant potential generator), the incident beam always undergoes some evolution, when a constant flow is needed during a long time ; half an hour is enough to observe a drift of 0.5%, which cannot be due to natural dispersion. So, each campaign is punctuated by calibration on a density standard.

Industrial Tests Are in Progress for a Better Accuracy Than ± 0.05

All the above mentioned sources of dispersion are quite difficult to isolate. In order to evaluate the global accuracy of the technique, we have performed the same operation 150 times on the same sample (d around 6.7) following this procedure : measurement of photons scattered on the three standard samples, calculation of the linear regression law, measurement of the photons scattered number on the sample, calculation of the density. Figure 6 shows the results for the radiation scattered distribution and the calculated density distribution. Dispersion on results is given by :

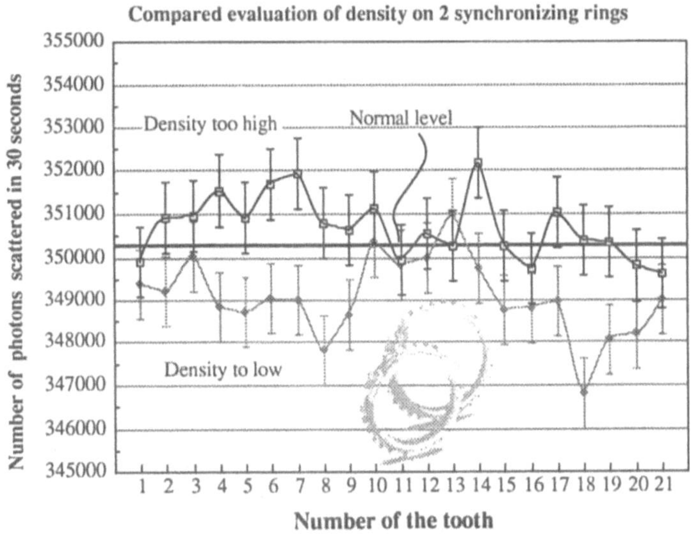

Figure 5 . Example of the sensitivity of the Compton measurement technique on industrial P/M product. (courtesy of SINTERTECH company) : synchronizing ring for the automotive industry. Each tooth of the ring (21 teeth) must satisfy a minimum density, in order to eliminate risk of breakage in normal use. Density characterization by Compton effect takes around 25 minutes, (21 measurements on each tooth) against 20 minutes (5 measurements on 5 teeth). This technique increases quality assurance for the customers.The press parameters have been slightly modified between the two rings (on two sectors). In less than half an hour, all teeth are characterized by the technique and the operator can directly see on which parameter he must act to set up correct production.

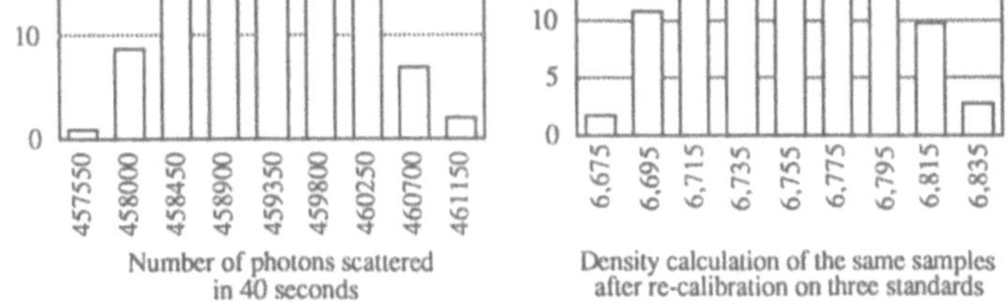

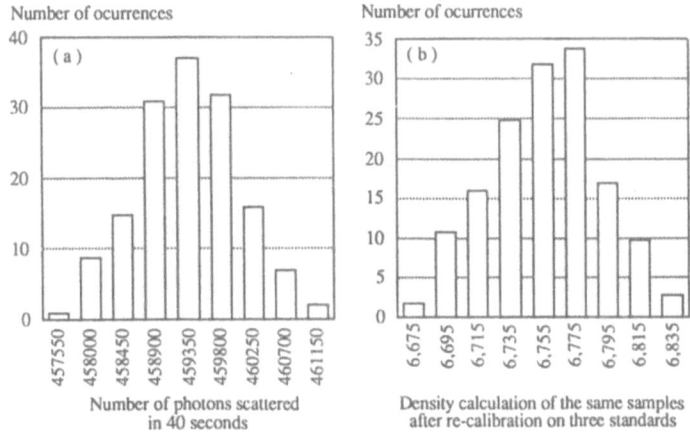

Figure 6 . Characterization of the distribution of radiation scattered measurement (a) and its effect on the evaluation of the density of a same sample (b) . (150 measurements have been made with 150 re-calculations of the linear regression with three standard samples).

$$N = 459353 \pm 1407 \text{ counts} \qquad (\pm 0.3\% \text{ like the Poisson statistic})$$

$$\text{and} \quad d = 6.757 \pm 0.068 \text{ in density .}$$

The main source of uncertainty is again the photon production governed by the Poisson's law. That means that better accuracy on the density is directly a function of : the flow of incident X-ray photons (already at the maximum level), the increase of measurement duration (from 40 to 60 seconds will give a accuracy of ± 0.055), the increase of the solid angle $\Delta\Omega$ (eqn.2), by changing the size of the detector (from 2" to 3"), and the technological control of multi-channel collimator design, which is the most critical piece in the apparatus. Most of our efforts are targeted on this topic.

In 1993, a MECASERTO prototype, developed from laboratory INSA apparatus, is being tested in France at the SINTERTECH plant.

CONCLUSIONS

This paper has demonstrated a new industrial development of the backscattered Compton effect in the powder metallurgy field. Evaluation of the local density of P/M products (a few mm^3) has been performed with an excellent resolution of ±0.070 in 40 seconds, with a laboratory apparatus. Industrial developpement is targeting a better resolution than ±0.050, based on multi-channel colllimator technology. This industrial apparatus is now tested in France as a prototype, and will be available in 1994 for industrial use.

REFERENCES

1. R.Halmshaw, "Non-DestructiveTesting", 2nd Edition (1991) pp 22-24
2. G.Berodias, G.Peix, *Materials Evaluation*, n°46, (August 1988), pp 1209-1213
3. "Radiographic inspection" in Nondestructive Evaluation and Quality Control, Metals Handbook 9th ed. Vol.17 (1989) p 309.
4. R.S.Holt, M.J.Cooper, *British Journal of NDT*, (March 1988) pp 75-80
5. C.Le Floch, P.Sarrazin, D.Babot, G.Peix, D.Duvauchelle, 17th Review of Progress in Quantitative NDE, SanDiego, California (USA), (15-20 Jul 1990), Ed. D. O. Thompson & D. E. Chimienti, Plenum
6. P.G..Lale, *Phys.Med.Biol.*, Vol.4 (1959)
7. S.S.Shulkla, A.Karellas, I.Leichter, J.D.Craven, M.A.Greenfield, *Medical Physics*, vol.12, n°4 pp 447-448 (Jul-Aug 1985).
8. K.H.Reiss, B.Steinle, *Siemens Forschungs- und Entwicklungberichte*, vol.2 n°1 pp 16-25.
9. H.Olkkonen, P.Karjalainen, *British Journal of Radiology*, N°48 (1975) 594-597.
10. J.T.Stalp, R.B.Mazess, *Medical Physics* vol.7 n°6, pp 723-726 (Nov-Dec 1980)
11. J.A.Stockes, K.R.Alvar, R.L.Corey, D.G.Costello, J.John, S.Kocimski, N.A.Lurie, D.D.Thayer, A.P.Trippe and J.C.Young, *Nuclear Instruments and Methods*, N°193 (1982) pp 261-267.
12. R.H.Rossi, K.D.Fridell, J.M.Nelson, *Mater. Eval.*, Vol.46(1988) pp 1462-1467.
13. B.C.Towe, A.M.Jacobs, *IEEE Transactions on Biomedical Engineering*, Vol.BME-28, (Sep1981) pp 646-654.
14. G.Harding, H.Strecker, R.Tischler, *Philips Technical Review*, Vol.41 NO 2, (1983/84) pp 46-59.
15. C.Bachmann, Report INI-MF-10652 (24 May 1985) Techniche Hochschule Aachen.
16. S.Teller, J.Meyer, F.K.Kristtofersen, J.M.Farley, 4th European Conference on Nondestructive Testing, London UK (1988) pp 2156-2163.
17. T.Dudzus, C.Segebade, *Materialprüfung*, Vol.18 (Sep 1976) pp 336-338.
18. J.J.Battista, L.W.Santon, M.J.Bronskill, *Physics in Medecine and Biology* Vol. 22, n°2 (1977) pp 229-244.

A NEURAL NETWORK FOR PREDICTING ULTIMATE STRENGTHS OF ALUMINUM-LITHIUM WELDS FROM ACOUSTIC EMISSION AMPLITUDE DATA

Eric v. K. Hill[1] and Gregory L. Knotts[2]

[1]Embry-Riddle Aeronautical University
Aerospace Engineering Department
600 S. Clyde Morris Boulevard
Daytona Beach, FL 32114

[2]Acoustic Emission Consultants
13490 Shelley Drive
Madison, AL 35758

ABSTRACT

Acoustic emission (AE) amplitude data have been shown to contain information concerning failure mechanisms and their correlation to ultimate strengths in both metallic and composite materials. As such, AE flaw growth activity was monitored in a set of eleven aluminum-lithium weld specimens from the onset of tensile loading to failure. The amplitude data from the beginning of loading up to 25% of the expected ultimate strength for five of the specimens were used along with the actual measured ultimate strengths to train a backpropagation neural network to predict ultimate strengths. Architecturally, the fully interconnected network consisted of an input layer for the AE amplitude data, two hidden layers for mapping, and an output layer for ultimate strength. The trained network was then applied to the prediction of ultimate strengths in the remaining six specimens where the worst case prediction error was found to be 4.3%.

INTRODUCTION

The 2195 aluminum-lithium alloy is being considered as a replacement material for the 2219 aluminum currently in use on the Space Shuttle External Tank (ET). Both materials exhibit good weldability and strength, but the aluminum-lithium is lighter in weight and stronger, thereby providing extra payload capacity; hence, the incentive for change. Since variable polarity plasma arc welding is the principal method of joining, and the welds are typically the weakest link, weld strength is paramount to ET structural integrity. A method is developed here for predicting ultimate weld strengths at proof loads as low as 25% of the expected ultimate using acoustic emission (AE) data.

Ultimate strengths have been predicted in both composites (Kalloo, 1988; Hill, 1992; and Walker, 1992) and in metals (Hill and Knotts, 1993) using the acoustic emission (AE) data taken during proof loading. These predictions included the use of AE amplitude, energy, and event rate data in combination with multivariate statistical analysis which in each case resulted in a linear prediction equation. Thus it may be concluded that the AE data contain information concerning failure mechanisms which can be correlated to ultimate strengths in materials. However, it should be noted that these correlations could *only* be obtained using subsets of the original data, e.g., high energy, high amplitude, or low amplitude AE data. No correlation could be obtained using all the energy or amplitude data for a given test. This is, in part, because all the AE data are not equally important to the prediction of ultimate strength.

In fiber reinforced plastic composites, matrix cracks are less damaging than fiber breaks, while delaminations can in some instances provide stress relief and lead to higher strengths (Hill, 1992). Plastic deformation in metals is less damaging than crack growth. Rubbing noises (friction emission) have little or no effect on the ultimate strength and heretofore have had to be removed from the AE data in order to accurately assess damage. In short, the various failure mechanisms contribute in varying degrees to the structural integrity, depending upon the material involved and the geometry of the part. Therefore, in an ultimate strength prediction equation each mechanism must be weighted differently (have a different coefficient). Mechanisms such as rubbing and plastic deformation, that contribute little or nothing to the ultimate strength, would have weighting functions (coefficients) approaching zero, while those that contribute significantly, such as fiber breaks in composites or intermetallic precipitate fractures in alloys, would have large coefficients. One technique for determining these coefficients or weighting functions is neural networks. In this research the input for the neural network was provided by the AE amplitude data.

THEORY

Acoustic Emission Amplitude Distributions

The acoustic emission amplitude parameter, A [dB], is a logrithmic representation of the peak signal voltage, V [V], of the AE waveform:

$$A = 20 \log(V/V_i) .$$

For most applications, $V_i = 1\ \mu V$ at the sensor output is chosen as the 0 dB reference because it is slightly above the noise level of the system electronics. Here the sensor contained a built-in 40 dB preamplifier; therefore, 0 dB was referenced to 100 μV at the preamplified sensor output.

Acoustic emission amplitude distributions have been shown to contain information that allows the identification of failure mechanisms in materials (Pollock, 1981). The typical (differential) amplitude distribution (histogram) can represent peak signal voltages of the AE waveforms from 100 μV (0 dB) to

10 V (100 dB). The various failure mechanisms are grouped together as characteristic humps or bands in the amplitude distribution, and while the amplitude bands for such mechanisms as plastic deformation and crack growth are widely separated, there are other mechanisms whose characteristic amplitude bands overlap. This overlap in the AE failure mechanism amplitude bands is increased by attenuation effects, especially dispersion. However, because the specimens used herein were small, the attenuation effects in the AE waveforms were expected to be minimal. It was therefore hoped that the amplitude distributions would have enough separation in the failure mechanism bands to allow accurate prediction of ultimate strengths in the aluminum-lithium weld specimens.

The amplitude distribution for specimen 01-5 is shown in Figure 1. Note that there are five distinct humps or failure mechanism bands and 307 total AE events. Since the logarithms of measurements can be expected to have normal distributions (Tennant-Smith, 1985), the various failure mechanism humps in the amplitude distribution should be normally distributed. This being the case, the first mechanism probably has amplitudes between 4-12 dB; the second mechanism ranges from approximately 8-20 dB; the third from 20-29 dB; the fourth from 34-35 dB; and the fifth (specimen failure) at 91 dB. It can be seen that there is some overlap between the first and second and the second and third mechanisms; moreover, inasmuch as the first hump does not appear to be normally distributed, there may well be more than one mechanism buried within it. This is where neural networks come into play.

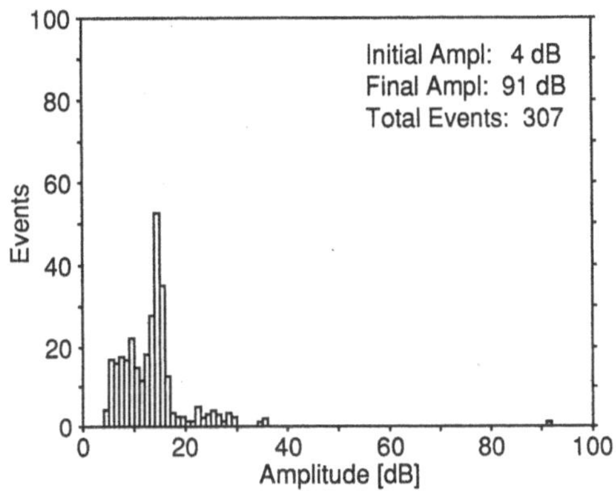

Figure 1. Differential amplitude distribution for aluminum-lithium weld specimen 01-5.

Neural Networks

One of the beauties of neural networks lies in the fact that they can take into account the effects of data overlap from the various failure mechanisms without explicitly identifying and/or isolating them from the rest of the data. Moreover, given the appropriate architecture and input, neural networks can learn all the pertinent data interactions and make accurate predictions using the entire data set, regardless of extraneous data (such as AE from rubbing noises) and without having to resort to multivariate statistical analysis (Kalloo, 1988; Hill, 1992; and Walker, 1992).

In this application the failure mechanism amplitudes were to be used as inputs with the output being the ultimate strength. If all the amplitude distributions were plotted, it could be seen that most of the failure mechanisms occur at amplitudes below 50 dB. Those AE events with higher amplitudes were, in this application, associated with ultimate specimen failure and were therefore not included. Only the amplitudes from 1 to 50 dB were input to the network.

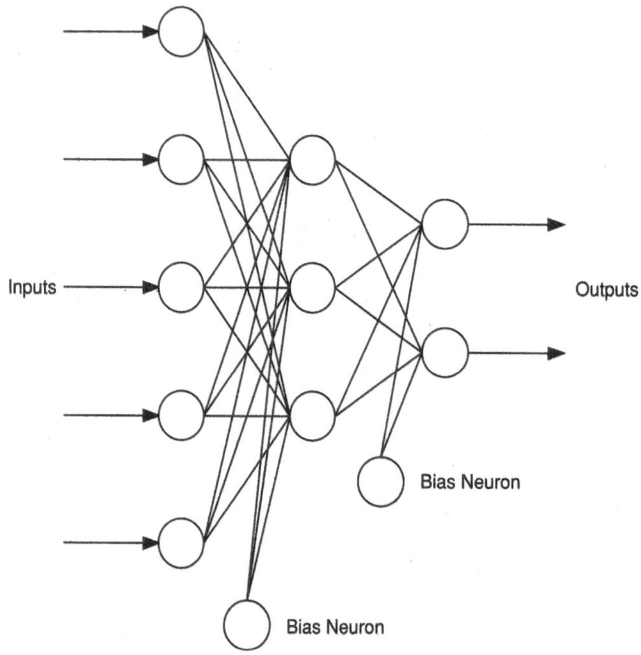

Figure 2. A typical three layer neural network with five inputs, a three neuron hidden layer, a two neuron output layer, and bias neurons.

A typical network is shown in Figure 2. Notice that a bias neuron is included in every layer except the input layer. The biases, like the interconnection weights, are learned during backpropagation. The bias neuron provides a constant threshold term in the weighted sum of the neurons in the succeeding layer, which translates their sigmoid activation functions to the left or right depending upon its sign. This bias term helps the network train faster by forcing the sigmoid function to operate at its midrange values where the error convergence is optimum.

Biases and weight adjustments in the various layers are calculated using the generalized delta rule (ANSim, 1988). This is a steepest descent method of computing the interconnection weights that minimizes the total squared output error over a set of training vectors (pairs). After each pass through the training set (cycle), the total normalized root mean square (RMS) error is calculated for the output. This is used to calculate the weight matrix changes, which are then backpropagated through each layer within the network.

Backpropagation

Backward error propagation or backpropagation is a method used to train multilayer networks. It compares the actual output of the network with the expected or target output, then backpropagates adjustments in the weights proportional to the calculated error. Network training is a two step procedure -- forward and backward -- propagating the inputs and their concomitant activations forward to the output layer, then propagating the error adjustments backward through the middle or hidden layers to upgrade the interconnection weights.

This means that inputting the amplitude distribution data for specimen 01-5 (Figure 1) should yield its ultimate strength of 51.5 ksi as the output. The known input (amplitude distribution) and target output (ultimate strength) constitute a training pair in this supervised learning scenario. Typically, several training pairs (specimens) and numerous cycles are needed to train the network to arrive at the appropriate weight components.

Once trained, a set of test data is used to verify the prediction accuracy of the neural network. The test data set, like the training set, consists of known inputs with known outputs. Thus, a prediction error can be calculated.

EXPERIMENTAL PROCEDURE

Eleven 5/16x1x12 inch tensile test specimens were cut from butt welded sheets. Both sheets were made of 2195 aluminum-lithium with a 2319 filler. In each case the weld was in the center of the specimen and transverse to the direction of pull.

A 10 kip/min linear load ramp was generated by a computer controlled Tinius Olsen tensile test machine. The load cell output ranged from 0.00 to 1.00 V (0.00 to 36.3 kips). A Physical Acoustics Corporation (PAC) 3000/3004 acoustic emission analyzer and R15I piezoelectric sensor (with integral preamplifier) were used to collect the AE data. The AE sensor was taped to the specimen next to the weld and acoustically coupled using Sonotrace 40 ultrasonic couplant. The PAC 3000/3004 threshold was set at 0.3 V with a total system gain of 81 dB. In order to develop the technique, all eleven specimens were taken to failure with the AE sensor attached. Otherwise, the sensor would have been removed after the application of the proof load.

DATA INPUT AND NETWORK ARCHITECTURE

Prior to training the network, the interconnection weights were initialized to small random numbers within the - 0.5 and + 0.5 range of the sigmoid activation function (ANSim, 1988), and the amplitude and ultimate strength data were normalized to fit into the same range. This was done to facilitate training.

As was mentioned previously, several training pairs (specimens) are needed to train the network. It is also necessary to have sufficient variety in the training inputs such that the network will be able to make valid correlations for unfamiliar cases -- a good distribution of possible inputs and outputs, both good and bad results, plus any border cases.

Specimens 01-5, 01-6, 01-7, 01-9, and 01-15 were chosen as a representative training set.

Some general rules apply when developing the network architecture for optimum performance. To improve accuracy on the training set, increase the size of the hidden layer(s). Alternately, to improve generalization capabilities and thus improve performance on new cases, reduce the size of the hidden layer(s). The optimal size for the hidden layer(s) is a balance between the objectives of accuracy and generalization for each application.

The network architecture was developed through trial and error using the above stated general rules. This resulted in an input layer of 50 neurons for the AE amplitude data, two hidden layers of 12 neurons each, and an output layer consisting of a single neuron for ultimate strength, all fully interconnected.

RESULTS

A summary of the neural network training is provided in Table 1. Altogether six networks were trained using the normalized amplitude data and the normalized ultimate strengths from the five specimen training set. In each case, the learning rate was set at $\eta = 0.4$ and the momentum at $\alpha = 0.9$. Two networks were trained using 100% of the AE amplitude data from load inception to specimen failure, one to a 3% maximum training error and the other to a 1% maximum training error. This procedure was then repeated for two networks using the AE data up to 50% of the expected failure load and for two using the data up to 25% of the expected failure load. (The value for the expected failure load was obtained by taking the average of the ultimate loads over the training set.)

Table 1. Summary of neural network training.

Network Name	Percentage Load Data	Training Cycles Req'd	Training Time Req'd*	Network Training
ASNT_01.NET	100%	1211 Cycles	1 Min., 43 Sec.	3%
ASNT_02.NET	100%	1625 Cycles	2 Min., 35 Sec.	1%
ASNT_03.NET	50%	1333 Cycles	1 Min., 55 Sec.	3%
ASNT_04.NET	50%	2077 Cycles	3 Min., 1 Sec.	1%
ASNT_05.NET	25%	2157 Cycles	3 Min., 18 Sec.	3%
ASNT_06.NET	25%	2807 Cycles	3 Min., 58 Sec.	1%

*486DX2-50 Computer

Because the ultimate strength data were normalized prior to input, the output prediction values from the neural network were also in a normalized format. As such, they had to be denormalized before any comparisons could be made. Table 2 presents a summary of the actual versus predicted ultimate strengths and the associated errors. These results were given by the three networks that were trained to a 1% maximum training error. The AE data from all eleven specimens were applied to these three networks. (This included the five training set specimens plus the six test set specimens.)

Table 2. Summary of the actual versus predicted errors for the three neural networks trained to a 1% maximum training error.

Specimen	Actual Ult Str [ksi]	100% Load Data		50% Load Data		25% Load Data	
		Predicted Ult Str [ksi]	% Error	Predicted Ult Str [ksi]	% Error	Predicted Ult Str [ksi]	% Error
01–5*	51.5	51.3	–0.39	51.3	–0.39	51.3	–0.39
01–6*	51.4	51.3	–0.20	51.4	0.00	51.4	0.00
01–7*	51.2	51.4	0.39	51.4	0.39	51.3	0.20
01–8	51.0	52.3	2.55	53.5	4.90	53.2	4.31
01–9*	50.8	50.7	–0.20	50.8	0.00	50.8	0.00
01–10	50.8	50.4	–0.79	50.4	–0.79	50.5	–0.59
01–11	50.6	50.5	–0.20	50.5	–0.20	50.5	–0.20
01–12	49.9	51.7	3.61	50.9	2.00	51.8	3.80
01–13	49.1	51.1	4.07	51.2	4.27	51.1	4.07
01–14	50.4	51.9	2.98	52.2	3.57	52.4	3.97
01–15*	49.5	49.7	0.40	49.6	0.20	49.6	0.20

*Training set

Note that the prediction errors on the five training set specimens were very small, as expected, while the prediction errors on the six test set specimens were an order of magnitude larger. Overall, the percent errors for the networks trained to a 1% maximum training error were only slightly less on average than those obtained from the networks trained to a 3% maximum training error. The 1% network had a worst case error of 4.3% for specimen 01-8.

Because the 25% load data contained the least amount of AE data, they were expected to yield the largest error, with the 50% data having the next largest error, and the 100% data having little or no error. This, however, was not found to be the case. Instead, the resulting errors were approximately the same for the 100%, 50%, and 25% load data, with the 50% data having the largest errors.

Acoustic emission due to gripping noise was experienced for the first few hundred pounds of load before each specimen became seated in the test machine grips. The amount of AE activity associated with grip slip varied from test to test. While it was anticipated that this might prevent accurate predictions, the neural networks were able to account for this effect and accurately predict ultimate strengths.

CONCLUSIONS

Ultimate strengths can be predicted in aluminum-lithium welds using acoustic emission amplitude data taken at loads up to 25% of the expected ultimate. This prediction is accomplished through the use of a backpropagation neural network with two hidden layers. The network automatically accounted for the AE activity associated with grip slip (through the interconnection weights) without having to remove this extraneous data *a priori* from the data set. It also adjusted for the overlap in the failure mechanism amplitudes. All of this was accomplished with a relatively small training set of only five specimens.

The fact that the prediction errors were essentially equal for the three AE data sets (taken to 25%, 50%, and 100% of ultimate load) suggested two things. First, the source of the error was the same in each case, that being the limited accuracy (± 1.7%) of the load data, and second, that the same basic ultimate strength information was inherent in all three data sets. This suggests the possibility of obtaining accurate ultimate strength predictions from the AE data at proof loads even lower than 25% of the expected ultimate.

Finally, the prediction errors being an order of magnitude larger than the training errors meant that the network had memorized the training set because it was trained to closely. It should have been trained to a maximum error of 3% to 5% rather than the 1% value that was used. Also, the architecture could probably have been optimized further. The previously mentioned optimization rules indicate that in order to increase the prediction or generalization capability of the network, there should be a reduction in the number of neurons in the hidden layers. Better predictions might also have been obtained by increasing the number of training sets, but probably the greatest need was to increase the accuracy of the load data (parametric input voltage).

Acknowledgements

This research was performed while on sabbatical leave at the NASA Marshall Space Flight Center under the auspices of the NASA/JOVE program, a jointly funded venture between Embry-Riddle Aeronautical University and NASA.

REFERENCES

"ANSim User Manual," 1988, Scientific Applications International Corporation, San Diego, CA, 4-13 to 4-17.

Hill, E.v.K., 1992, Predicting burst pressures in filament-wound composite pressure vessels by using acoustic emission data, *Materials Evaluation*. 50:1439.

Hill, E.v.K., and Knotts, G.L., 1993, Predicting ultimate strengths of aluminum-lithium (Al-Li) welds using acoustic emission, in: "ASNT 1993 Spring Conference," American Society for Nondestructive Testing, Columbus, OH.

Kalloo, F.R., 1988, "Predicting Burst Pressures in Filament Wound Composite Pressure Vessels Using Acoustic Emission Data," MS Thesis, Embry-Riddle Aeronautical University, Daytona Beach, FL.

Pollock, A.A., 1981, Acoustic emission amplitude distributions, *International Advances in Nondestructive Testing*. 7:215.

Tennant-Smith, J., 1985, "BASIC Statistics," Butterworth & Co. Ltd., London, UK. 106.

Walker, J.L., 1992, Ultimate strength prediction of ASTM D-3039 tensile specimens from acoustic emission amplitude data, Paper No. 92-0258, American Institute of Aeronautics and Astronautics, New York, NY.

ACOUSTIC EMISSION MONITORING OF THICK COMPOSITE LAMINATES UNDER COMPRESSIVE LOADS

Chris Byrne and Robert E. Green, Jr.

Center for Nondestructive Evaluation
The Johns Hopkins University
Baltimore, MD 21218

INTRODUCTION

Fiber reinforced polymer composites have become quite popular in recent years and have been used for a wide variety of applications ranging from sports equipment to advanced aircraft components. There has recently been considerable interest in the behavior of thick composite laminates under compressive loads[1]. This interest stems from the development of novel materials for applications in submersed vessels. A submersible vessel hull is one of the few structures in which the material would experience nearly pure compressive loads. For a structure which is designed to be brought to the depths normally encountered by submarines, a skin of material several inches thick is required. In order to utilize composites for this purpose research into the compressive strength and fatigue life of these materials is needed. In addition, in service inspection techniques must be developed to assess the integrity of such a structure.

The work described in this paper was part of a research effort undertaken which was aimed at determining the strength and fatigue life of thick graphite-epoxy composites subjected to compressive loads. Goals were to characterize the material mechanical response and to apply nondestructive techniques for the characterization of those materials. Towards those ends several NDE methods were applied. Contact and fluid coupled ultrasound, laser speckle decorrelation, infrared thermography penetrant enhanced radiography and optical microscopy were used to locate and identify specific damage modes in the composite specimens tested. Results from these techniques can be found in references 2, 3 and 4.

Mechanical characteristics of composite laminates are known to undergo changes as a result of cyclic tensile loading. Stiffness and strength have been shown to diminish as damage in the material accumulates[5,6]. In cross ply laminates, as was studied in this work, damage takes the form of matrix microcracks which extend parallel to the reinforcement fibers. Transverse matrix cracks (TMCs) occur in off-load axis plies at stresses well below the ultimate strength of the laminate. These TMCs accumulate rapidly during the first 20% of fatigue life and tend to reach a saturation spacing, referred to as the characteristic

Nondestructive Characterization of Materials VI
Edited by R.E. Green, Jr. *et al.*, Plenum Press, New York, 1994

damage state. Longitudinal matrix cracks (LMCs) occur in plies aligned with the load axis and often result from the different Poisson ratio of adjacent plies. Other damage modes consist of ply delaminations and fiber breaking. These damage modes have been shown to occur in a certain causative sequence as fatigue life progresses[5,6]. Along with this degradation there is a reduction in the measured stiffness. Stiffness drop in the early stages of tensile fatigue is associated with the accumulation of TMCs while an additional drop near the end of life is commonly associated with the failure of individual lamina or fiber bundles.

Acoustic emission (AE) monitoring of composite laminates has been utilized to detect both the presence of damage and the occurrence of damage progression[7]. It has been shown that AE occurs from the strain relief mechanisms resulting from the damage modes mentioned above. It has also been shown that the presence of damage in a composite can lead to rubbing internal surfaces which generate acoustic waves. Though specific AE fingerprints have yet to be recognized for the different modes of damage there is common agreement among researchers as to some of the characteristics of each of the damage mechanisms. High amplitude, short duration events are thought to originate from fiber breaks. Delamination progression, depending on the length of the crack front, generally gives longer duration events of lower amplitudes. Often long duration events are associated with all forms of matrix damage including crazing and cracking.

The load rate at which the composite is subjected as well as the load history of the specimen has been shown to influence the rate at which AE is generated. Awerbuck *et al.* present the results from the monitoring of graphite-epoxy composites during tensile fatigue[8,9,10]. They found that the initial stages of fatigue resulted in relatively intense AE at the higher loads in the fatigue cycle. These events were attributed to LMCs originating from stress concentration notches in the specimen. Later in the fatigue life a great many events were detected during the low load portions of the cycle. These were attributed to rubbing crack faces. A methodology for distinguishing between this internal friction and damage progression was implemented which took into account the applied load and the intensity of the acoustic event. Using this method to separate AE sources the researchers were able to track the matrix crack front using a location algorithm. It was also apparent that the presence of AE at low loads in the fatigue cycle indicated the presence of a specific form of matrix cracks in the laminate. This phenomena could clearly be used as a tool for the inspection of the structural integrity of a composite structure, while continuous monitoring would be useful in the detection of damage progression and possibly as a means to predict impending failure.

Experimental results from acoustic emission monitoring and mechanical characterization are presented in this paper. Mechanical tests were performed both in a quasi-static manner and by cyclic fatigue. Correlations were found between measured mechanical changes and detected acoustic activity. Damage accumulation early and late in specimen life was detected by acoustic emission monitoring. Composite failure was preceded by a high rate of high amplitude, high energy events at the high load portion of the fatigue cycle. Similar AE results were obtained from quasi-static loading.

EXPERIMENTAL

Composite panels of AS4/3501-6 prepreg were made with a $(0_2,90)_{32,S}$ stacking sequence giving a thickness of approximately one inch. Specimens 3 inches high by 1 inch wide were cut from these panels. Stress concentration notches of 1/8 in radius were cut in the through thickness direction at mid height of the specimens to localize fatigue induced damage. Specimens were epoxied into recessed steel endcaps to prohibit brooming and end splitting while allowing the load to be applied through the specimen ends. A 110 kip capacity MTS servo hydraulic load frame equipped with flat parallel platens and interfaced with a PC was used for mechanical testing and data acquisition. A one inch gauge length extensometer was mounted symmetrically about the stress concentration notch to measure axial strain. Specimens were then set in the load frame and a 200 lb compressive preload was applied. Fatigue cycles were load controlled using a 1 Hz haversine fatigue cycle. Quasi-static tests were performed by ramp loading to a predetermined load, holding for twenty seconds and then unloading to the preloaded state (considered no load). Additional load sequences were performed, after another hold period, to higher load values until the specimen failed. For example, load to 5 kip, unload to preload, load to 10 kip, unload to preload, then repeat until failure. Typical failure loads were 78 kip. A 10 kip/min load rate was used.

Acoustic emissions were monitored by a Physical Acoustic Corporation acoustic emission resonant transducer having a nominal center frequency of 125 kHz. Sensor preamplification was 40 dB with a signal band pass of 100 kHz to 300 kHz. A PAC 3000/3104 acoustic emission analyzer was used for data storage and manipulation and was set to provide a threshold of 58 dB (1 microvolt base) to omit background noise. Emissions were monitored continuously during fatigue and quasi-static testing A schematic of the test specimen and equipment is shown in Fig. 1.

RESULTS AND DISCUSSION

In previous work[4] damage accumulation was found to go through three stages during compressive fatigue cycling. The first stage consisted of the development of LMCs above and below the stress concentration notches along with some localized TMCs. Stage two, lasting for some 80% of fatigue life, consisted of the maturation of the LMCs and TMCs found in stage one. The final stage in damage accumulation of the thick composite lasted for a relatively short time and was characterized by the development and progression of an interlaminar crack which started at of the root of the stress concentration notches. This crack would propagate rapidly through the laminate breaking fibers and

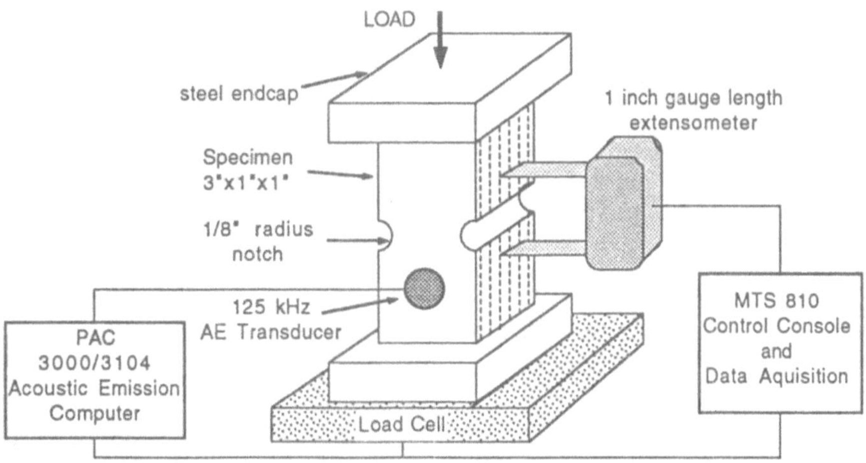

Figure 1. Specimen and equipment schematic.

matrix along the way. This crack often had delaminations associated with it and would lead to complete failure of the load bearing capabilities of the specimen. A schematic of the evolution of damage is shown in Fig. 2.

The results from monitoring specimen mechanical response indicated three stages of change occurred in the three parameters of secant stiffness, permanent deformation and hysteresis. Considering the type of damage found in the specimens the measured strain readings were not indicative of the response of the entire thickness of the specimen but were a superposition of the deformation of the central portion and the relative degree of detachment of the material above and below the stress concentration notches. Figure 3 is a plot of the measured mechanical response of a specimen fatigued to failure. Clearly a great deal of permanent deformation occurred with the first few cycles which is consistent with the development of LMCs above and below the notches. Concurrently, significant

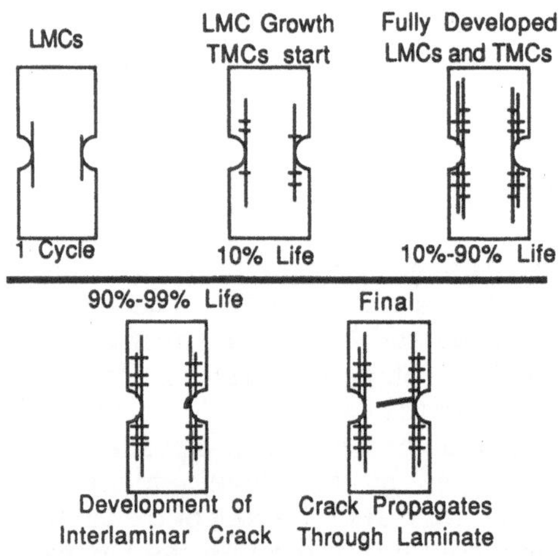

Figure 2. Compression-compression fatigue damage development (from ref. 4).

mechanical losses occurred as indicated by the development of hysteresis in the load cycle. Both of these parameters demonstrated dramatic changes early in the fatigue test after which a gradual change was recorded until just prior to failure when the mechanical loss increased at a renewed rate. Specimen stiffness was also observed to change in a three stage manner, but to a lesser degree.

The measured changes in specimen mechanical response suggested that different damage modes were active during each of the three stages of fatigue. It was shown[4] that during the first few cycles LMCs were created which would have the effect of creating a plane of strain discontinuity defined by the notch root and the load axis. Then prior to failure an interlaminar crack would develop, often in concert with a delamination, which would rapidly cause failure of the specimen. This type of damage progression could have been responsible for both the measured mechanical response and the acoustic emission results which are presented next.

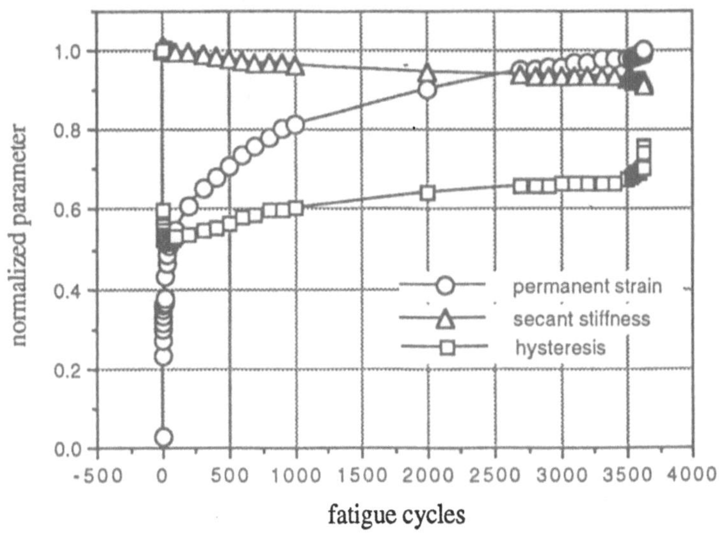

Figure 3. Mechanical characteristics during compressive fatigue.

Previously we have shown that when all detected acoustic events are considered no trends during fatigue are apparent[3]. However, filtering out events which occurred at the low load portion of the fatigue cycle indicated that the highest amplitude events occurred at the greatest loads. In addition, a pattern was detected in the rate of events occurring at the higher loads in the fatigue cycle. Both the beginning and end of fatigue life were marked by the highest rate of events. Little change in the rate of acoustic events at the higher loads was found during the majority of fatigue life. This three stage trend is also demonstrated when looking at just the higher amplitude events in the fatigue test.

Figure 4 shows the AE monitoring results from the specimen for which mechanical data has been presented. Both Hits and Energy of detected high amplitude events (greater than 64 dB) are plotted on a cumulative sum basis versus fatigue cycles. It can be seen that the first few cycles were marked by a high rate of high amplitude events which contained a considerable amount of energy. The majority of fatigue life was marked by little appreciable high amplitude or high energy acoustic activity. Just prior to failure an increase in the rate of high amplitude, high energy events occurred.

The results from the AE monitoring were consistent with the three stages of damage development and corresponding mechanical characteristics. The initial high energy events were associated with the rapid development of LMCs above and below the notches. This damage was also largely responsible for the measured permanent deformation and associated mechanical losses found during the first few fatigue cycles. Later on, less damage occurred with each cycle and the measured mechanical parameters showed little change. The monitoring of AE during this second stage gave few high amplitude events but many low amplitude events at the lower loads in the fatigue cycle. These low amplitude events often occurred at the same load in each cycle and repeated for many subsequent cycles indicating that the same source was responsible for them. Rubbing of crack faces associated with LMCs are proposed as the source of these events and therefore indicate the presence of damage and not the progression of damage. The final stage in the fatigue test was characterized by the development and propagation of an interlaminar crack in the notch region which caused a reduction in measured stiffness and increased mechanical hysteresis. This damage was responsible for the high rate of high amplitude and high energy acoustic events detected just prior to failure.

Several tests were performed with a ramp load, unload sequence to progressively higher loads while mechanical response and AE were monitored. These were done to identify whether the Kaiser effect would be found in the composite from compressive loading. Figure 5 is a graph of the measured permanent strain after unloading versus the stress level prior to unloading. Also shown is the number of detected AE during the loading portion of the load, unload sequence. It was found that little permanent deformation occurred from stresses less than 35 ksi and no acoustic events were detected. Above that stress level both the number of detected acoustic events and the amount of permanent deformation increased dramatically. The data showed that events were often detected during loading before surpassing the prior maximum load level. Thus,

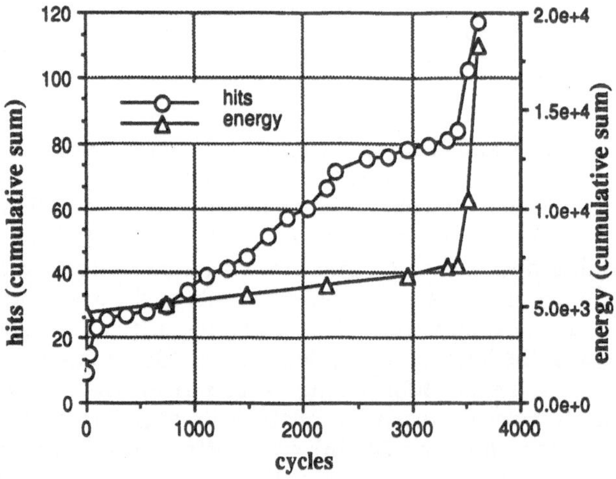

Figure 4. High amplitude (>64 dB) events during compressive fatigue.

the Kaiser effect was not observed to occur with any repeatability. Calculation of a Felicity ratio showed no apparent trend. When detected events were separated based upon their intensity (as with fatigue testing) different observations were made. The lower energy events occurred before surpassing the previous maximum load level. A higher rate of events were detected at higher loads and these tended to be of greater intensity. This behavior is attributed to the creation and presence of LMCs which generate AE by the rubbing of crack faces. The data in Fig. 5 also demonstrates that some mechanical recovery occurred when permanent deformation was high. Though not presented in the figure, AE was detected when the recovery was recorded. It was also found that the unloading portion of the test generated AE. These observations are all consistent with the concept of internal rubbing surfaces generating acoustic events at low loads from a cyclic loading condition.

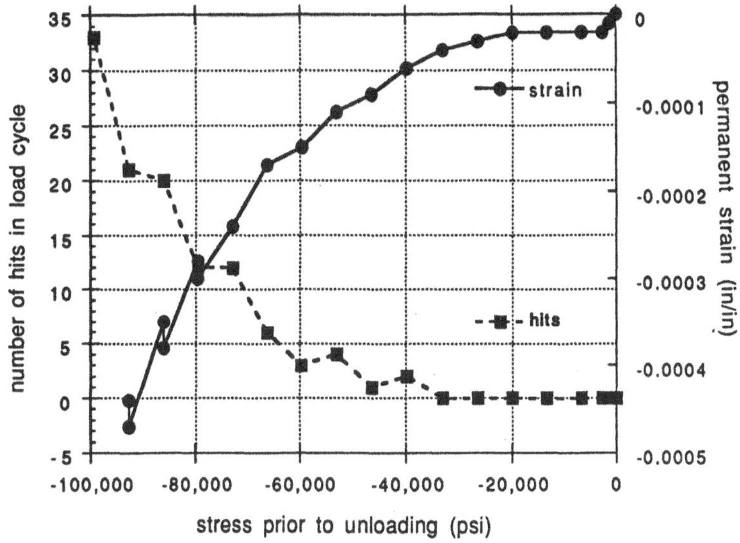

Figure 5. Acoustic emission and permanent deformation from a ramp load-unload sequence.

CONCLUSION

Acoustic emission monitoring of thick composites subjected to compressive loads was performed while specimen mechanical characteristics were recorded. It was found that the three stages in damage state were characterized by changes in both mechanical response and AE parameters. Stage one was identified by substantial permanent deformation and development of hysteresis losses. Specimen stiffness was also observed to drop early in the fatigue test. Damage in stage one was previously found to consist of a rapid development of localized LMCs and TMCs. Acoustic activity during this stage was typically of high amplitude and high energy. As fatigue continued into stage two the detected AE was of lower intensity with considerable activity during the low load portion of the load cycle. Since the load cycle did not pass through a null point it is suggested that the low amplitude events came from rubbing crack faces and not the load frame as has been seen in other test configurations. Indeed, separate monitoring of the machine noise verified this. Specimen mechanical response went through little change during this second stage of fatigue and no new damage mechanisms were found. The third stage was characterized by rising mechanical hysteresis, stiffness reduction, increasing permanent deformation, and an increasing rate of high amplitude and high energy AE. Damage during this stage consisted of development and propagation of an interlaminar crack originating at the stress concentration notch with delaminations emanating from it.

Acoustic emission monitoring of quasi-static tests indicated no trends in Felicity ratio. The Kaiser effect was not observed except when low intensity events were filtered out. The generation of AE while unloading and with no applied load suggested the mechanism of internal friction at crack faces as a source.

ACKNOWLEDGMENTS

This research was funded by the Defense Advanced Research Projects Agency, program manager Mr. James J. Kelly, through the Office of Naval Research, Mechanics Division, Dr. Yapa D.S. Rajapakse, DARPA agent. Additional support was provided by The Johns Hopkins Center for Nondestructive Evaluation. Recognition is given to Joseph W. Krynicki for his participation in the acoustic emission testing. The authors would also like to thank the many faculty and students at the Johns Hopkins University who have made numerous contributions to this research.

REFERENCES

1. G.M. Wood, Proceedings of the Fourth Annual Thick Composites in Compression Workshop, iii, 1990.
2. N.J. Gianaris and R.E. Green, Jr., Proceedings of the Symposium on Nondestructive Evaluation and Material Properties of Advanced Materials, TMS Annual Meeting, New Orleans (1991).
3. C. Byrne, J.W. Krynicki and R.E. Green, Jr., in Review of Progress in Quantitative Nondestructive Evaluation, Vol. 12B, edited by D.O. Thompson and D.E. Chimenti (Plenum Press, New York, 1992).
4. C. Byrne, Masters Essay, The Johns Hopkins University, 1993.
5. K.L. Reifsnider, Fatigue of Composite Materials, Edited by K.L. Reifsnider, (Elsevier Science, New York, 1991).
6. R. Talreja, Fatigue of Composite Materials, edited by R. Talreja, (Technomic Publishing, Pennsylvania, 1987).
7. M. Arrington in Non-Destructive Testing of Fiber Reinforced Plastics Composites, Vol. 1, edited by J. Summerscales, (Elsevier Science, New York, 1987).
8. J. Awerbuck and S. Ghaffari, Monitoring Progression of Matrix Splitting During Fatigue Loading Through Acoustic Emission in Notched Unidirectional Graphite/Epoxy Composite, J. Reinforced Plastics and Composites, Vol 7, pp. 245-264, 1988.
9. W. Eckles and J. Awerbuck, Monitoring Acoustic Emission in Cross-Ply Graphite/Epoxy Laminates During Fatigue Loading, in J. Reinforced Plastics and Composites, Vol 7, pp. 265-283, 1988.
10. S. Ghaffari and J. Awerbuck, On the Correlation Between Acoustic Emission and Progression of Matrix Splitting in a Unidirectional Graphite/Epoxy Composite, in Acoustic Emission: Current Practice and Future Directions, ASTM STP 1077, W. Sachse, J. Roget, and K. Yamaguchi, Eds., ASTM 1991.

ULTRASONIC EVALUATION OF COMPOSITE FATIGUE DAMAGE

Andrew J. Gavens and Robert E. Green, Jr.

Center for Nondestructive Evaluation
The Johns Hopkins University
Baltimore, Maryland 21209

INTRODUCTION

Damage in composite materials has been detected using ultrasonic nondestructive techniques.[1-5] These studies were all performed on thin composite samples subjected to either a quasi-static tensile force or a constant amplitude tensile fatigue load. Changes in basic ultrasonic parameters, attenuation or velocity, were used as an indicator of damage. Damage was detected once a sufficient amount was present. Attenuation was observed to be dependent on the frequency of the ultrasonic transducers that provides an indication of the dispersive nature of these materials.

The ability to detect damage ultrasonically becomes more difficult on thick composite materials. Being a highly attenuative material makes it hard to obtain multiple echoes for the accurate calculation of attenuation and velocity. The ultrasonic signal evaluates the material through the thickness of the specimen. Any damage must be sufficient to alter the ultrasonic parameters when compiled with the remaining undamaged material. For thick composites the amount of damage may need to be extensive to alter the ultrasound.

In the present study an investigation was initiated to evaluate the ability of standard contact ultrasonic techniques to determine fatigue damage in a thick composite sample subjected to compression-compression fatigue loading. Different pulse wave forms, transducer frequencies and ultrasonic methods were evaluated.

The stages of damage progression occurring in thick composites subjected to compressive-compressive fatigue loads were determined by Byrne.[6] The materials and specimen geometry were the same as those in the present study. Using double edge notched specimens it was observed that longitudinal matrix cracks occur in the first few load cycles, blunting the notches. Growth of localized regions of longitudinal and transverse matrix cracks then occurs which progresses away from the notch until close to failure. Just before failure, a transverse matrix crack initiates from the root of the notch. This crack then grows quickly and interlaminar cracking reduces the stability of the specimen leading to failure.

EXPERIMENTAL PROCEDURE

Double edge notched compression fatigue specimens were machined from autoclave fabricated plates of AS4/3501-6 prepreg to the dimensions shown in Figure 1. The composite plates had a stacking sequence of $[90_2/0]_s$ and were 196 plies thick (25 mm nominal thickness). The specimens were mounted in steel endcaps with a titanium filled epoxy to eliminate brooming of the specimen ends during testing. By using double edge notched specimens, damage was anticipated to initiate in the region of the notches allowing damage detection to be performed on a localized region of the specimen.

Compression-compression fatigue tests were performed on a MTS servo-hydraulic test machine. A sinusoidal load was applied to the specimens at a rate of 1 Hz. The specimens were loaded from -45 N to -8,990 N (~82 percent of the quasi-static failure strength) between two flat parallel platens. These loads were chosen based on earlier test results to obtain fatigue failures around 100,000 cycles.[6]

The fatigue tests were periodically stopped to inspect the specimen for damage. At each inspection the specimen was radiographed and ultrasonically evaluated. The notched region of the specimen was coated with zinc iodide penetrant before x-raying with a Hewlett Packard 43855A Faxitron. Radiographs were obtained with the x-ray beam directed normal and parallel to the lamina.

Both pulse-echo and through-transmission ultrasonic techniques were used to evaluate damage in the specimens. Two ultrasonic pulse systems were used. A Panametrics Pulser

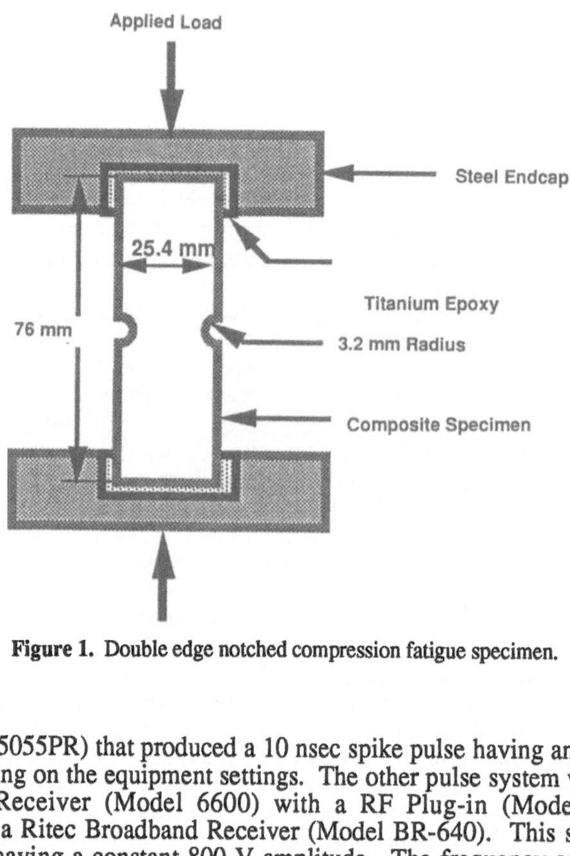

Figure 1. Double edge notched compression fatigue specimen.

Receiver (Model 5055PR) that produced a 10 nsec spike pulse having an amplitude of 200 to 380 V, depending on the equipment settings. The other pulse system was a Matec Pulse Modulator and Receiver (Model 6600) with a RF Plug-in (Model 950B) used in conjunction with a Ritec Broadband Receiver (Model BR-640). This system produces a tone burst pulse having a constant 800 V amplitude. The frequency and duration of the tone burst were adjusted to match the transducers and materials being used.

The equipment settings on both ultrasonic systems were adjusted to provide the largest amplitude received signal without an obvious distortion of the waveform. These settings were determined on each specimen prior to fatigue testing and were used for the duration of the fatigue test.

Ultran Laboratories Standard Miniature Contact Transducers (12.7 mm diameter) having nominal center frequencies of 0.5, 1.0, and 2.0 MHz were used. The 0.5 MHz transducers were only used with the Panametrics pulse system. Transducers designed with a narrow bandwidth (KC models) were used as the pulsing transducers to introduce more ultrasonic energy to the specimen. Wider bandwidth transducers (WC models) were used for receiving the through-transmission signals.

The transducers were mounted on the specimen between the notches with the ultrasound directed normal to the lamina. A fixture was designed to consistently position the transducers on the specimen each time it was evaluated. Sonotech Soundsafe Industrial Couplant was used to couple the transducers to the specimen. Elastic bands were placed around the fixture to keep the transducers in place.

The received ultrasonic signal was analyzed on a Data Precision digital oscilloscope (Model 6000A) with a 620 Plug-in unit. The signal was sampled at a rate of 100 MHz and averaged over 10 pulses to improve the signal to noise ratio. The averaged signal was rectified and the peak amplitude, RMS voltage, area under the curve, and the time of flight to the peak amplitude of each echo were recorded. The transducers were then recoupled to the specimen and the measurements were replicated a total of 5 times. The measurements were averaged and attenuation and velocities were calculated. A coefficient of variation of 5 to 10 percent was typical for these measurements.

RESULTS AND DISCUSSION

Longitudinal cracks at the root of the notch were observed radiographically after the first cycle. As the fatigue test progressed the number and length of longitudinal cracks increased. Away from the notch, the longitudinal cracks were often connected by fine transverse cracks. After approximately 75 percent of the fatigue life, transverse cracks developed near the root of the notch. Radiographs obtained within a few percent of failure showed that there was a significant increase in the number of longitudinal cracks and an increase in the length of the transverse cracks, as shown in Figure 2. The transverse cracks often extended to the surface of the specimen, although they initiated several plies deep. In the notch, a shear crack coupled several of the longitudinal cracks from which it is believed final failure resulted.

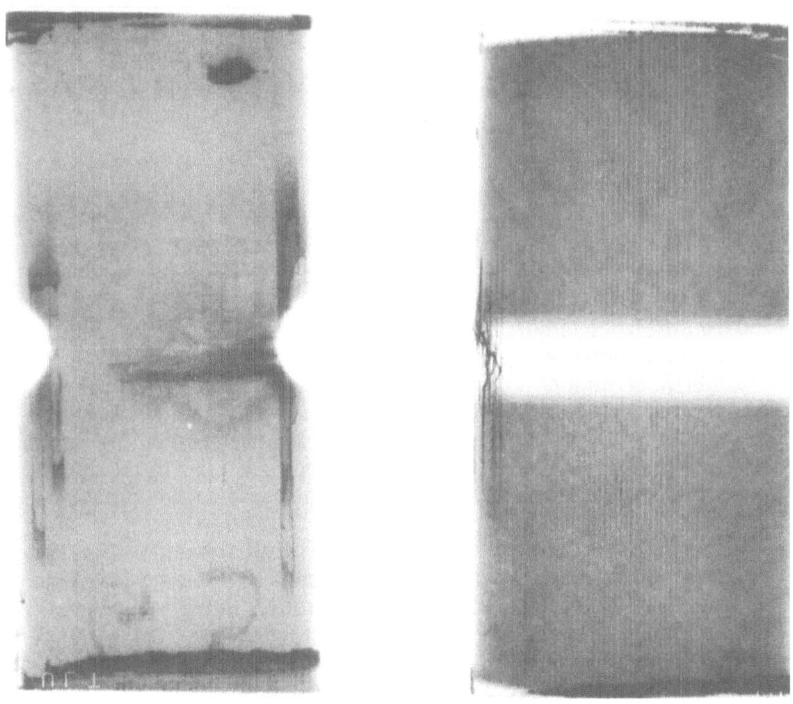

a) normal to the lamina   b) parallel to the lamina

Figure 2. Radiographs of a specimen at 98 percent of the fatigue life a) normal and b) parallel to the lamina.

Only two echoes were obtained using any of the ultrasonic methods and transducers, even prior to fatigue testing. The basic shape of the curve of attenuation versus fatigue cycles was the same regardless of whether attenuation was calculated using the peak amplitude, RMS voltage or the area under the curve. No significant change in attenuation was observed until just before failure, as shown in Figure 3. Typically there was a decrease in attenuation followed by an increase before failure. The decrease occurred when the transverse crack was first observed radiographically. Higher frequency transducers tended to yield higher attenuations which agrees with previous studies.[1, 5] Through-transmission produced greater attenuation than obtained using a pulse-echo method.

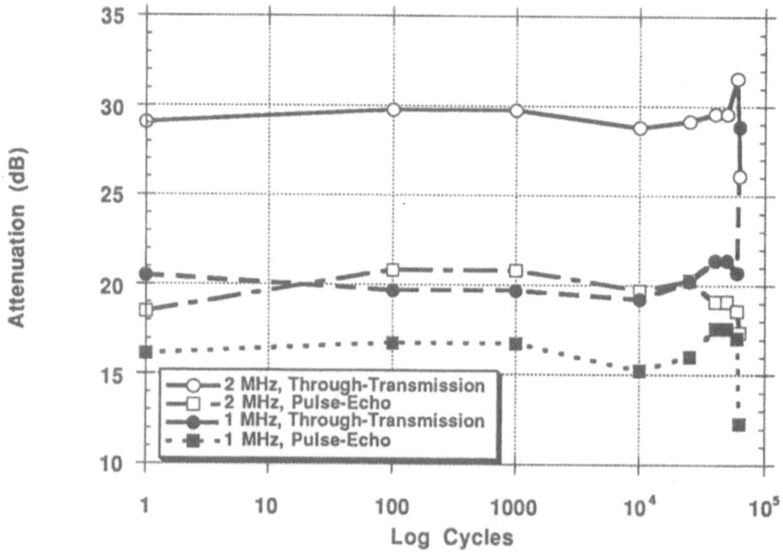

Figure 3. Attenuation measured using the Matec ultrasonic system.

The velocity was dependent on the method by which it was calculated. Velocities were calculated using three different values for the time of flight (tof), as shown in Figure 4. The first method used the tof from the excitation pulse to the maximum amplitude of the first echo (tof1), the second method used the tof to the second echo (tof2), and the third method used the difference in time between these two methods (tof1-2). Methods tof1 and tof2 were significantly more dependent on the ultrasonic method and transducer frequency than method tof1-2. Although the values were different, similar indications of fatigue damage were obtained with any of the velocity measurement methods, as shown in Figure 5. Method tof1-2 was still dependent on transducer frequency (Figure 6) but not on either through-transmission or pulse-echo ultrasonic methods. Similar to the attenuation there was no significant change in ultrasonic velocity until just before failure, as shown in Figures 5 and 6.

Some of the differences observed with the different frequency transducers, pulse systems, and ultrasonic methods could be due to differences in the spread of the ultrasonic beam. The transducers could have different beam profiles that may be dependent on the type of excitation pulse. This would result in different regions of the composite being evaluated.

In this study there was only 3.1 mm between the edge of the specimen and the transducer. Due to the thickness of the specimens the ultrasonic beam could have interacted with the edge of the specimen. To determine the effect of this on the ultrasonic parameters, a plate of the same material and thickness as the fatigue specimens was evaluated. The plate had a surface area significantly greater than the test specimens such that it could be considered as an infinite plate. The plate was ultrasonically tested in the same manner as the fatigue specimens. Figures 7 and 8 show the results obtained from these tests. Attenuation remains dependent on the frequency of the transducer. When the Matec system was used the attenuation decreased with an increase in the transducer frequency. The opposite occurred when the Panametrics pulser was used. Attenuation values for the plate were similar to the initial values measured on the fatigue specimens. The ultrasonic velocity of the plate decreases with decreasing transducer frequency for both the Matec and Panametrics pulsers. The velocities measured on the unfatigued specimens were slightly less than those measured on the plate. These results indicate that either the ultrasonic beam did not interact with the edge of the specimen or the interaction had minimal effect. The type of excitation pulse and ultrasonic method still had an affect on the measurements.

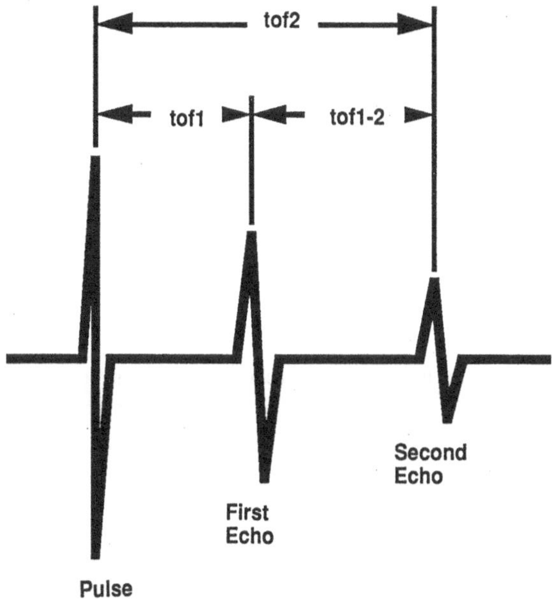

Figure 4. Different measurements of time of flight for calculating velocity.

Fatigue damage in the composite was not detected until just before failure for a variety of reasons. At every inspection interval, the transducers had to be recoupled to the specimen resulting in there being a change in the exact placement and coupling of the transducers to the specimen. This increases the variation in the data making it difficult to distinguish changes that are due to fatigue damage. In addition to this variation, the effect of any damage that is present is averaged with the undamaged material. This occurs both through the thickness of the specimen and across the area that the ultrasonic beam investigates. Until damage extends into the region of the ultrasonic beam and is sufficient to significantly alter the ultrasonic parameters there will not be any change in the detected attenuation or velocity. The attenuation measured in this study did begin to change when transverse cracks emanating from the root of the notch were observed radiographically to extend into the region under the transducers. Prior to this, although there was damage present in the specimen, there was not any damage in the direct path of the ultrasonic beam.

CONCLUSIONS

Ultrasonic measurements on a composite specimen subjected to compressive fatigue loading did not detect the presence of fatigue damage until very close to failure. Attenuation and velocity were both observed to be dependent on the frequency of the transducer, the type of excitation pulse and the ultrasonic method. Further work to reduce experimental error and utilize more advanced ultrasonic techniques should improve the damage detection capabilities.

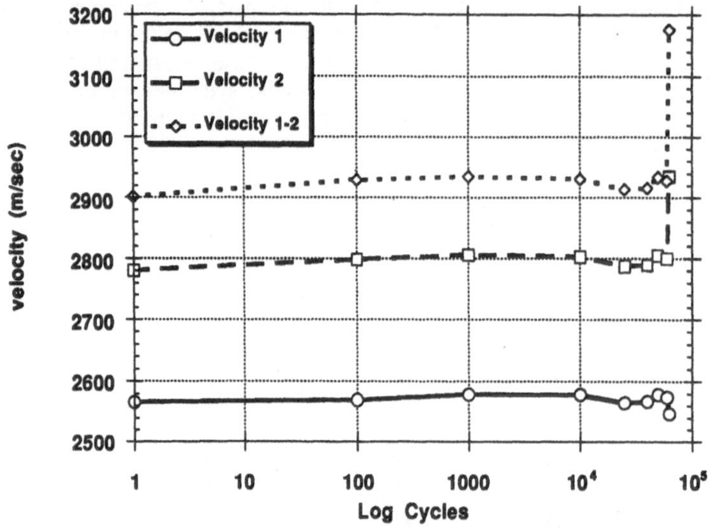

Figure 5. Ultrasonic velocity calculated using three different methods to determine the time of flight from results with the Panametrics ultrasonic system.

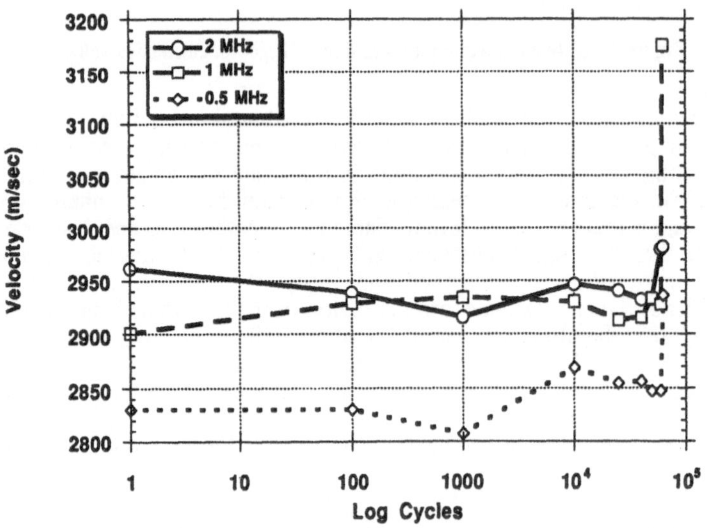

Figure 6. Ultrasonic velocity measured with different frequency transducers, tof1-2, and the Panametrics ultrasonic system.

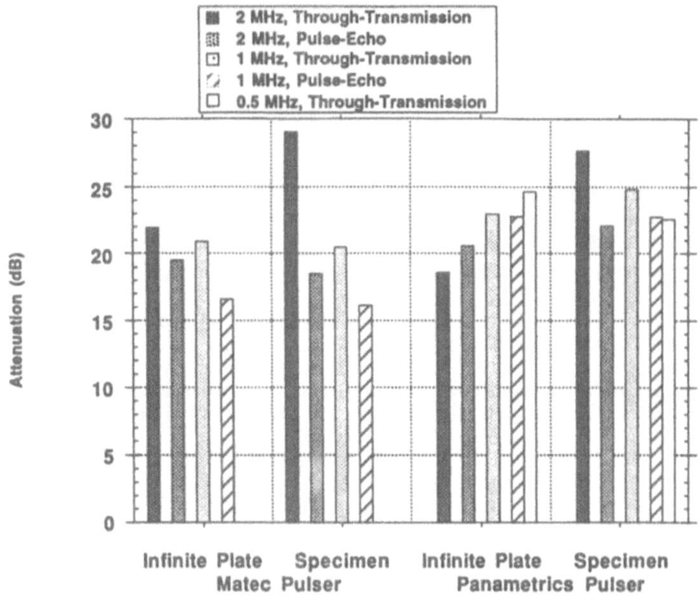

Figure 7. Effect of specimen size on attenuation.

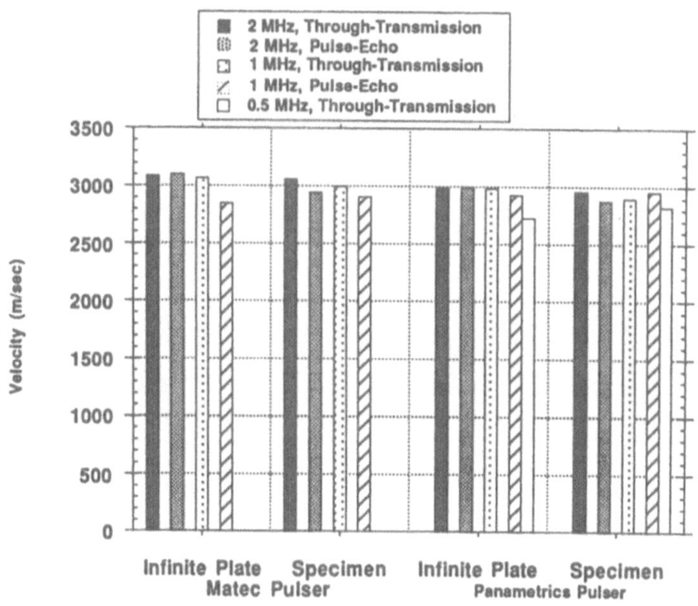

Figure 8. Effect of specimen size on ultrasonic velocity.

ACKNOWLEDGMENTS

This research was funded by the Defense Advanced Research Projects Agency, program manager Mr. James J. Kelly, through the Office of Naval Research, Mechanics Division, Dr. Yapa D.S. Rajapakse, DARPA agent. Additional support was provided by The Johns Hopkins Center for Nondestructive Evaluation.

REFERENCES

1. J. H. Cantrell Jr., W. P. Winfree, and J. S. Heyman, Profiles of fatigue damage in graphite/epoxy composites from ultrasonic transmission power spectra, *in*: "Recent Advances in Composites in the United States and Japan, ASTM STP 864," J. R. Vinson and M. Taya, Eds., American Society for Testing and Materials, Philadelphia, (1985).
2. V. Dayal, V. Iyer, and V. K. Kinra, Advances in Fracture Research, Proceedings of the 7th International Conference on Fracture (ICF7), 5, 3291 (1989).
3. D. T. Hayford, and E. G. Henneke II, A model for correlating damage and attenuation in composites, *in*: "Composite Materials: Testing and Design (Fifth Conference), ASTM STP 674," S. W. Tsai, Eds., American Society for Testing and Materials, Philadelphia, Pennsylvania, (1979).
4. B. Hosten, and S. Baste, Ultrasonics measurements of anisotropic damage in polymeric matrix/glass fibers composite laminates subjected to tensile stresses, to be published in Review of Progress in Quantitative Nondestructive Evaluation, 12, (1993).
5. J. H. Williams, and B. Doll, Ultrasonic attenuation as an indicator of fatigue life of graphite fiber epoxy composite, *Materials Evaluation*. 38:33, (1980).
6. C. Byrne, "Damage accumulation in thick graphite-epoxy composites during compressive fatigue," Master of Science, The Johns Hopkins University (1993).

A STUDY ON THE ACOUSTIC EMISSION CHARACTERISTICS OF THE CARBON FIBER REINFORCED PLASTICS

Y.K. Ji[1] and J.W. Ong[2]

[1]Korea Atomic Energy Research Institute, Daeduk Science Town
[2]Mechanical Eng. Department, Chungnam National Univ.
Daejon, 305, Republic of Korea

INTRODUCTION

Carbon fiber reinforced plastics(CFRP) have good characteristics; high specific strength and stiffness, corrosion–resistance, and excellent insulation. In addition, they can be designed and fabricated to get adequate strength and stiffness by changing the angle and sequence of their laminations. Therefore, the composite materials are increasingly used in industries of aerospace, automobile, ship, leisure and sports. Many investigations on their performance improvement and material evaluation have been studied, but the failure mechanism is difficult to clarify because the composite materials have complex failure process caused by their non–homogeneous and anisotropic characteristics.[1,2]

To evaluate their performance and analyze their failure mechanism, Acoustic Emission (AE), one of the non–destructive testing methods has been performed. AE is used for detection of initial defects, clarification of failure mechanism, estimation of final fracture stress, back data to integrity evaluation. AE parameters are count, event, amplitude distribution, frequency spectrum, etc. Primary sources for AE of the composite materials are plastic deformation of matrix, matrix cracking, debonding of matrix and fiber, fiber pull–out, delamination, fracture of fiber, and secondary sources are friction of fiber, disclosure of cracks, etc. Extensive studies on various NDT as well as the relation of AE parameters and AE sources are needed to synthetically evaluate and clarify failure characteristics and damage region of the composite materials, but they are restricted by experimental equipments, materials, etc.

In this report, we are trying to compare fracture behaviour and AE characteristics generated during tensile test using a laminate of carbon/epoxy prepreg and hence clarify availability of AE method and failure characteristics of the CFRP.

EXPERIMENTAL METHOD

Material of specimens used in this experiment is 8 layer–laminated carbon/epoxy prepreg according to the designated sequence and angle of lamination, which is fabricated

Nondestructive Characterization of Materials VI
Edited by R.E. Green, Jr. *et al.*, Plenum Press, New York, 1994

in dry oven to a fiber volumetric fraction of 0.6. In accordance with ASTM D 3039-76,[3] laminates are fabricated to total 14 kinds of specimen in shape and dimension shown in Figure 1 and free edges of the specimen are ground in order to prevent their damage region from occurring on cutting. To protect the specimen from tensile grip, epoxy tab in thickness of 3 mm is glued at both ends.

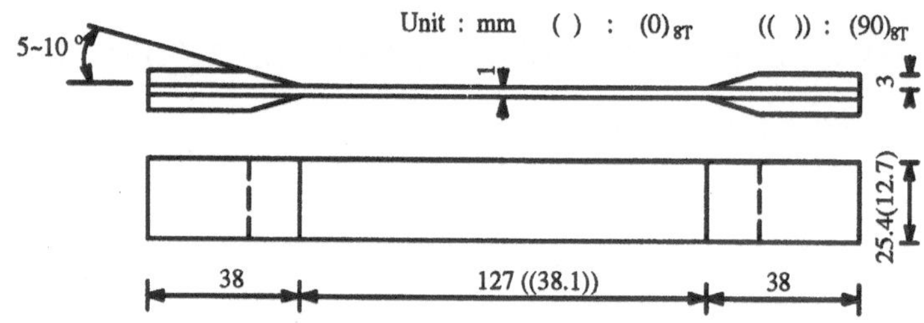

Figure 1. Configuration of test specimen

As shown in Figure 2, experimental equipments are composed of universal testing machine (Tensilon/UTM1-10000C) containing 5 ton-load cell and AE equipment (AET 5000) for detection of failure process in the test specimen. Also, dynamic strainmeter, X-Y recorder, PC/AT, and strain gauge are used. Cross head speed of UTM is 0.5 – 1.0 mm /min and variation of micro resistance detected by strain gauge is amplified in dynamic strainmeter and then load-displacement curve is plotted in X-Y recorder.

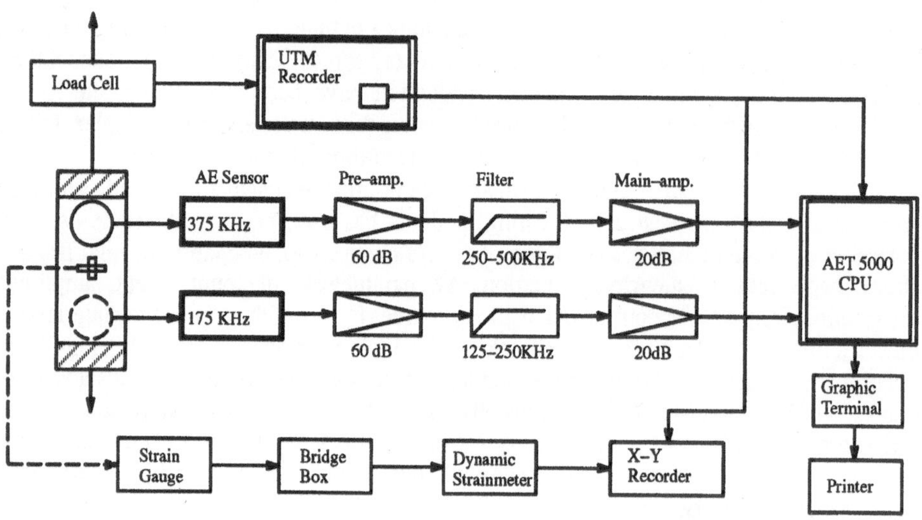

Figure 2. Block diagram of AE measuring system

Resonance type transducers, 175 kHz and 375 kHz transducers are provided as a AE sensor. For adhesion of transducers, specimen and transducer are greased and pressed using ring spring to collect generating AE signals sensitively. Signals detected from transducers are amplified to 60 dB through pre-amplifier and then to 20 dB through main amplifier to get required data in CPU of AET 5000. Since environmental noise levels are 50 μV for 375 kHz transducer and 80 μV for 175 kHz transducer, the threshold voltages are 1.0 V (fixed) and 1.5 V (fixed) respectively. Experiment for measurement of variation of Felicity ratio is performed with increase of load ratio step by step until fracture of specimen i. e., loading–unloading–reloading cycle. Felicity ratio is calculated on the basis of the generating load of AE signals.

EXPERIMENTAL RESULTS AND DISCUSSION

Activity of Ringdown Counts (RC) and Total Ringdown Counts (TRC)

Figures 3(a) and 3(b) are plots of $(0)_{8T}$ specimen using 375KHz and 175KHz transducer. Comparing to the two figures, AE signals are significantly detected from 175KHz transducer. Therefore we tried to analyze failure mechanism with AE signals of 175KHz transducer which is mainly used in the composite materials.

Relations between RC(TRC) and load–displacement of typical one–directional CFRP are shown in Figures 3(a)–(e). In the case of $(0)_{8T}$ specimen, AE signals are initiated around 10 percent of maximum load.[4] RC increases continuously from the initiation of AE signals and then decreases after it peaks about 60 percent of maximum load. As load increases, TRC increases almost linearly, but transition occurs near 60 percent of maximum load. It is considered that AE signals at 10 percent of maximum load are caused by matrix cracking or deformation and those at 50 percent of maximum load, initiating point of transition, are due to intra–delamination. Many researchers have reported[5–9] that CFRP failure processes have matrix cracking, delamination, fiber fracture as load increases and, especially, delamination takes place at the point of rapid increase of RC and TRC. Since the matrix cracking which could occur in the region of extremely low load is not propagated before fiber fracture in other region and remains stable isolated, it is known to have no effects on the final fracture.

However a few fibers may fail in the region of extremely low load due to statistical characteristics of fiber strength, these phenomena do not nearly affect tensile strength while they have influence on the final fracture.[5] Therefore, it is considered that, as load increases continuously, many matrix cracks are generated and developed along vertical fiber direction and final fracture results from gradual increase of fiber–matrix debonding, delamination, and fiber pull–out. $(90)_{8T}$ specimen fails just after the AE generation at the 75 percent region of maximum load. It represents aspect of brittle fracture compared to AE generation of the 20 to 30 percent region of maximum load, which could be caused by weakness of matrix itself. Behaviour of $(45)_{8T}$ specimen is similar to that of $(90)_{8T}$ specimen . Most of the AE signals are generated when final fracture occurs and are caused by matrix cracking.As angle of unidirectional specimen increases, RC, TRC and maximum load decrease.

Figures 3(f)–3(j) show relations between typical load–displacement and RC for $(\pm\theta)_{2S}$, where maximum load, RC, and TRC are reduced as angle of stacking sequence for $(\pm\theta)$ increases. As shown in figures 3(f)–3(j), $(\pm 15)_{2S}$, $(\pm 30)_{2S}$, $(\pm 45)_{2S}$ specimens failed not due to fiber fracture, but due to transverse crack and delamination. $(\pm 15)_{2S}$ specimen led to fail at 90 percent of maximum load accompanying many signals, after AE signals appear at around 30 percent of maximum load and increase more or less in spite of increase of comparatively high load. It is indicated that RC increases rapidly by delamination near 90 percent of maximum load after generation of transverse crack along fiber direction and matrix cracking and specimen fail suddenly without fiber damage. For $(\pm 45)_{2S}$ specimen, AE signals are generated at about 14 percent of maximum load and increase gradually up to 40 percent and exponentially after that. Matrix cracking is initiated at about 14 percent of maximum load and delamination is generated and superimposed during long time compared to other specimens, with propagated in the transverse direction. It is considered that $(\pm60)_{2S}$ specimen and $(\pm75)_{2S}$ specimen are observed to have matrix cracking under the very small load less than 100 kg, and failed by fiber pull–out after propagation of transverse crack along the fiber direction, not by delamination.

Figures 3(k)–3(o) represent phenomena of three–directional specimens.For $(90/0/\pm45)_S$, $(90/\pm45/0)_S$ specimens, outer stacking laminate of 90 degree is observed to fail as first step during tensile test. That is, TRC is very small due to that. After that, aspect of AE signal is similar to $(0)_{8T}$ specimen. Rapid increase in TRC curve is due to 0 degree laminate fiber fracture. Effect of ±45 degree laminate debonding is insignificant compared to fiber fracture. For $(\pm45/0/90)_S$ specimen, also we expect initial failure of 90 degree laminate and then complicated failure mechanism is affected. Especially, delamination occurs between 45 degree–0 degree and 0 degree–0 degree laminates caused by failure of 90 degree laminate and hence twisting appears considerably. Final fracture load is similar, but TRC is very different according to arrangement of stacking sequence in the three directional CFRP. Also, three steps can be classified by the slope of TRC curve. The first step with slow slope of TRC is considered causing generation and propagation of matrix cracking. Delamination is initiated with audible sound at the second step which shows the steep slope of TRC. Fiber pull–out and final specimen breakage take place at the last step.

Felicity Effect

Felicity effect is defined as the opposite concept of Kaiser effect.[10–12] During the cycle of loading–unloading–reloading, AE signals generated by Felicity effect are generated at reloading less than maximum load at the previous loading step while AE signals generated by Kaiser effect are initiated over that. Especially CFRP has the stringent characteristics of Felicity effect because the material has anisotropic property and an effect on inter–relation of complicated fiber fracture. Felicity ratio is introduced to explain Felicity effect as follows:

Felicity Ratio (R) = P_i / P_o

where: P_i = load initiating AE signals at reloading, and
P_o = maximum load at loading.

Figure 4 shows variations of representative R value for each direction as a function of load ratio. For $(0)_{8T}$ specimen, R remains 1.0 up to a load ratio of 0.15 as shown in Figure 4 (a). For load ratios of 0.15–0.75, R value decreases slowly while over a load ratio of 0.75 it decreases sharply. For $(45)_{8T}$, $(90)_{8T}$ specimens, R remains 1.0 up to a load ratio of 0.9 and then decreases sharply. Compared to AE activity, matrix cracking occurs as R get to a value of 1.0 and more failure occurs because R value decreases rapidly with increase of angle of stacking sequence.

Figure 4(b) shows the characteristics of $(\pm\theta)_{2S}$ specimen, which indicates that R value does not change and the detection of AE signals decreases with increase of angle.

For three directional CFRP of figure 4(c), R value remains 1.0 up to a load ratio of 0.3. They show different values depending on stacking sequences and, especially, R value decreases clearly with ±45 lamina arranged outside.

Matrix cracking occurs when R value is around 1.0, that is, Kaiser effect is dominant. Then, more failure occurs with decrease of R value which is changed due to the effect of complicated failure mechanism. As discussed above, relation between R value and failure mechanism is difficult to define clearly. It is considered that 1.0 to 0.85 of R is caused by

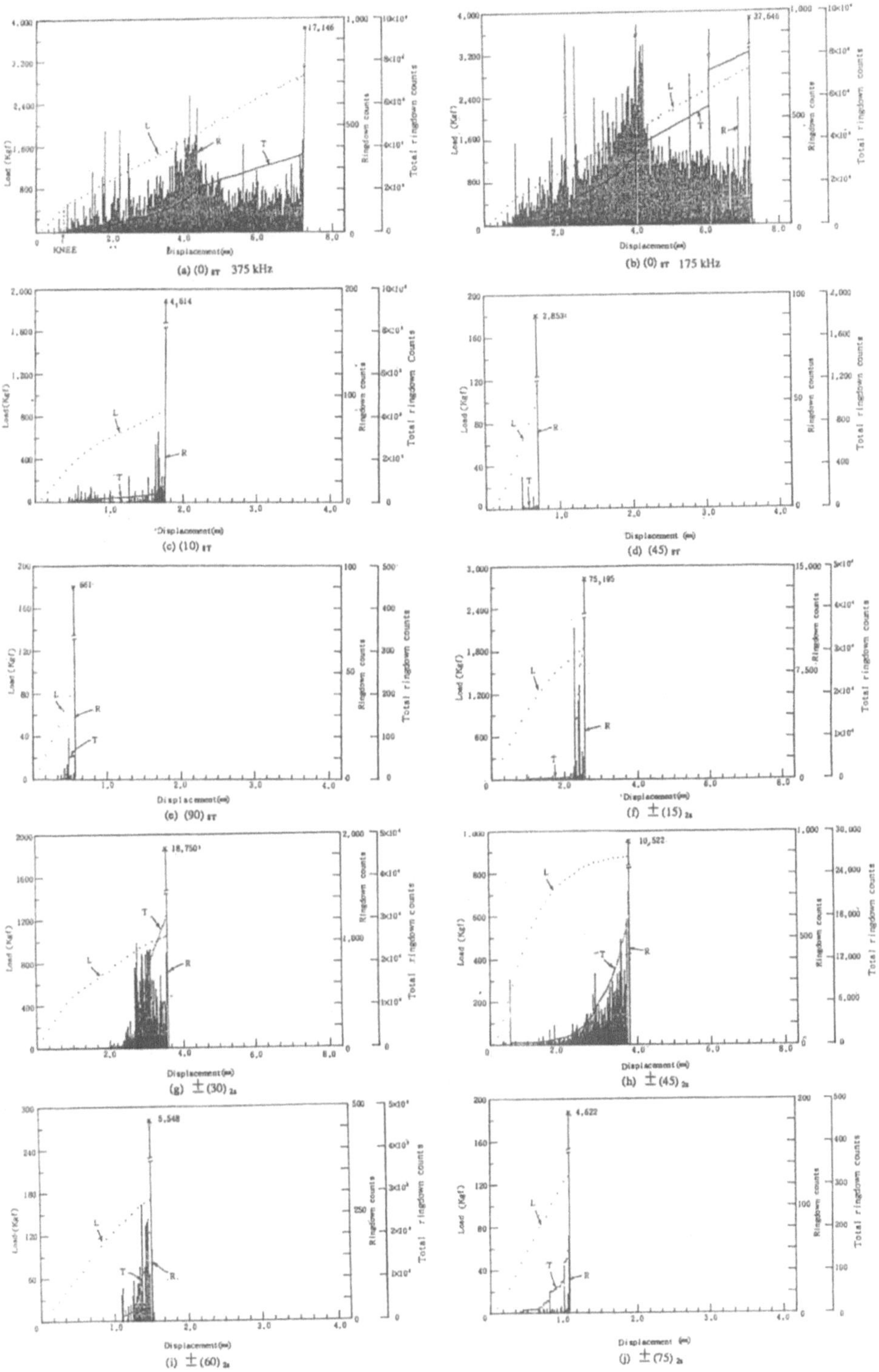

Figure 3. Plot of displacement vs. load , counts, total counts

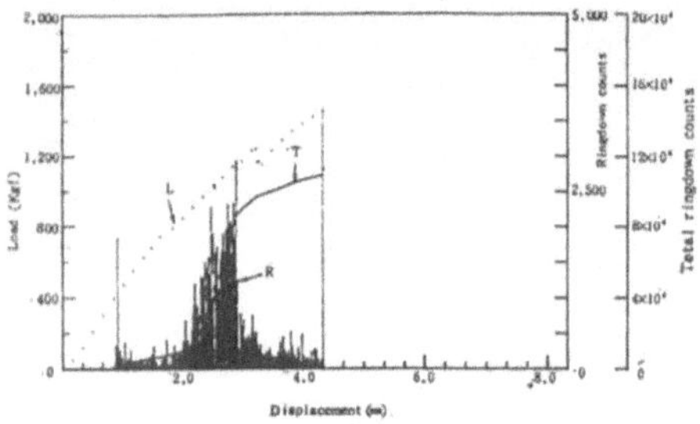

(k) (90/0/±45)s

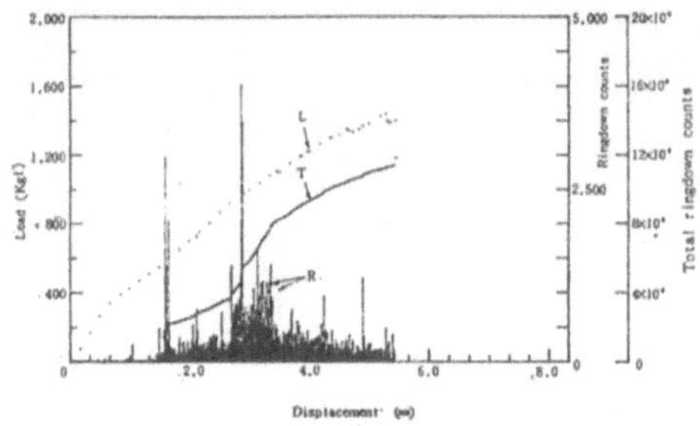

(l) (90/±45/0)s

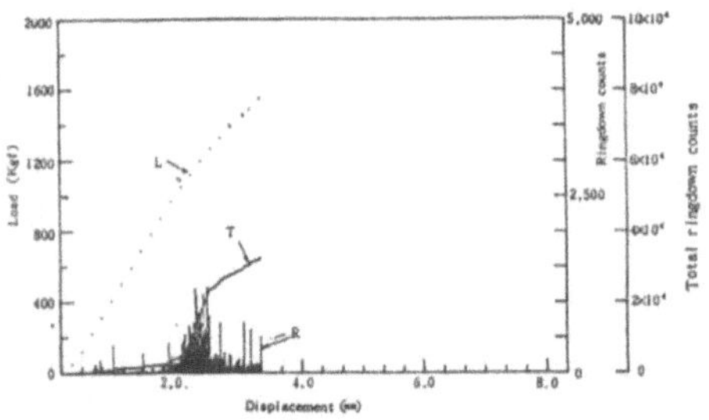

(m) (±45/0/90)s

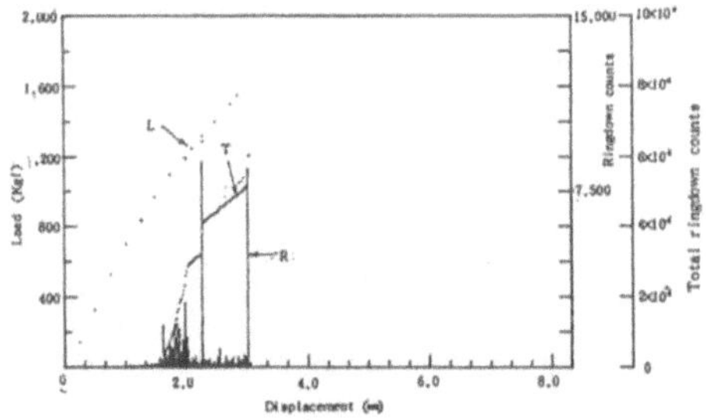

(n) (±45/90/0)s

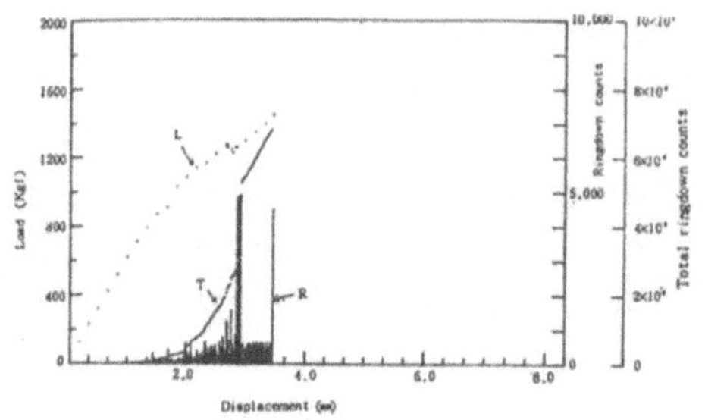

(o) (0/90/±45)s

Figure 3. (continued)

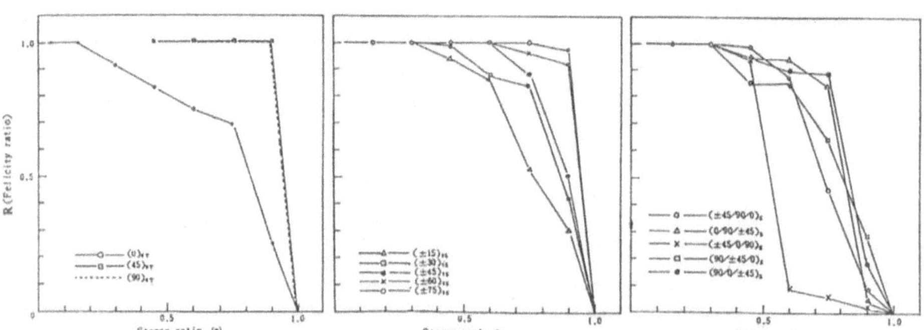

Figure 4. Plot of R vs. stress ratio : (a) one–direction (b) two–direction (c) three–direction

matrix cracking, 0.85 to 0.40 by generation and propagation of the delamination, and below 0.4 by fiber fracture, but failure mechanism is acted on complicated process.

Peak Amplitude Distribution

It is known that brittle materials, containing defects, of high strength, anisotropy and non–homogeneousness have high amplitude and amplitude distribution is related to failure mechanism. Figure 5 shows typical experimental results obtained from relation between peak amplitude distribution and events.

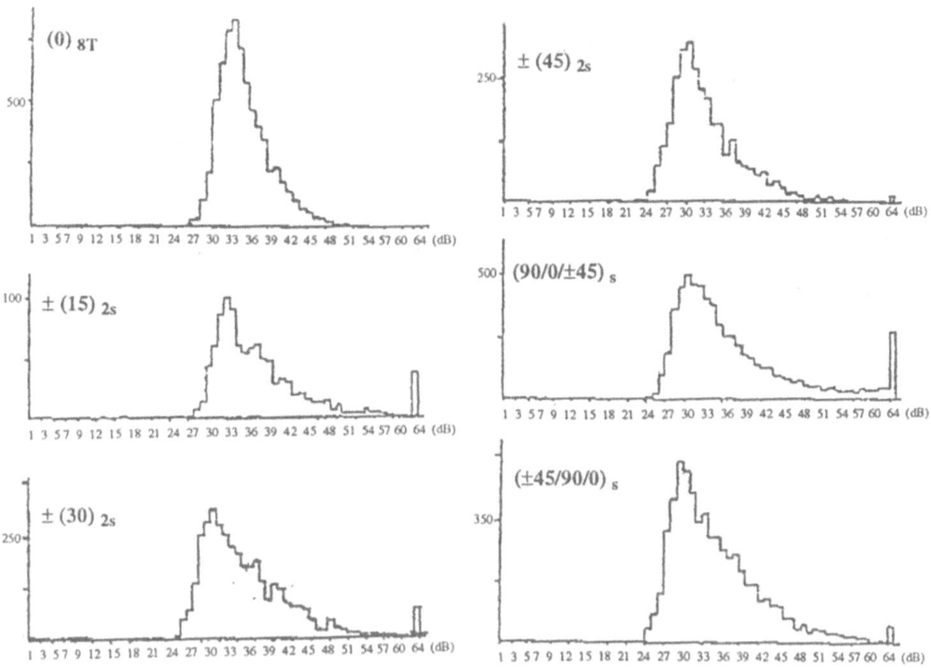

Figure 5. Plot of AE peak amplitude distribution and events

For $(0)_{8T}$ specimen, minimum amplitude appears at 26 dB and most of amplitude distribution is detected at range of 30 to 39 dB. AE signals, events peak at 34 dB and then tend to decrease exponentially. For $(90)_{8T}$ specimen, a few events, if any, are initiated at 25 dB since this specimen fails at once. Results of $(\pm 60)_{2S}$, $(\pm 75)_{2S}$ are similar to that of $(90)_{8T}$, they are not shown in Figure 5..

For $(\pm\theta)_{2S}$ specimens, $(\pm 15)_{2S}$ shows relatively high amplitude distribution and $(\pm 30)_{2S}$, $(\pm 45)_{2S}$ show similar tendency, that is, peak event occurs near 30 dB. According to total events, $(\pm 45)_{2S}$, $(\pm 30)_{2S}$, $(\pm 15)_{2S}$, $(\pm 60)_{2S}$, $(\pm 75)_{2S}$ are shown in turn.

Three directional specimens are inclined to give peak event at 30 dB regardless of stacking sequence; for 24 dB – 30 dB, rapid increase and for 30 dB – 60 dB, exponential decrease.

Also, previous researchers have reported AE events caused by delamination below peak amplitude of 30 dB, by matrix cracking 30 to 40 dB, and by fiber fracture 40 to 80 dB through compression testing and tensile testing of matrix and fiber only. However, they show disagreement with results in this study. With increase of load, events showing peak amplitude are increasingly accumulated in a wide range of amplitude distribution.

CONCLUSIONS

According to angle and sequence of stacking, relations between AE signals and failure mechanism of CFRP during tensile testing are as follows:

(1) For $(0)_{8T}$ specimen, matrix deformation and cracking are initiated around 10 percent of maximum load and delamination takes place at 50 percent of maximum load. AE signals caused by fiber fracture are dominant. For $(90)_{8T}$ specimen, AE signals caused mainly by matrix cracking are detected.

(2) For $(\pm\theta)_{2S}$ specimens, TRC decreases with increase of θ (angle) and debonding of fiber and matrix occurs by the effect of shear force parallel to fiber direction.

(3) For three directional specimens, different failure mechanisms appear depending on stacking sequence of the weakest 90 degree laminate. They have stringent differences of TRC associated with stacking sequence while having similar fracture strength, about 1600 kg_f.

(4) Failure mechanisms are considered having a relation with Felicity ratio (R); matrix cracking for 1.0–0.85, generation and propagation of delamination for 0.80–0.40, and fiber fracture below 0.40.

REFERENCES

1. Robert M.Jones, "Mechanics of Composite Materials",McGraw–Hill Book company, (1975).
2. Stephen W.Tasi, "Composite Design",3rd Ed., Think Composites,Dayton,Ohio,(1987).
3. ASTM D 3039–76, " Standard Test Method for Tensile Properties of Fiber–Resin Composites", Annual Book of ASTM Standards,(1982).
4. E.G. Henneke, C.T. Harakovich, G.L.Jones and M.P. Renieri, "Acoustic mission from Composite–reinforced Metals", VPI–E–73–27, NASA Report,(1975).
5. A. Rotem, "The Discrimination of Microfracture mode of Fibrous Composite Material by Acoustic Emission Technique", Fiber Science and Technology, Vol. 10, pp. 101–121,(1977).
6. R.L. Mehan and J.V. Mullin, " Analysis of Composites Failure Mechanisms using Acoustic Emission", Journal of composite Material 5, pp. 256–269,(1971).
7. Pedro Feres Filbo, "Results Evaluation of Acoustic Emission Tests Applied in Industrial Equipment", Journal of Composite Materials, pp. 489–491,(1985).
8. M. Shiwa and T. Kishi, "Acoustic Emission during Load–Holding and Unload–Reload in Fiberglass–Epoxy Composites", The second International Conference on Acoustic Emission, pp. 195–198,(1985).
9. Yuan Zhenming, " Acoustic Emission Characterization of internal Damage for GFRP", The First International Conference on Acoustic Emission, pp. 458–463,(1985).
10. Jin Zhou Geng, Song Guo Gui and Zhu Cheng, "Acoustic Emission Characteristics of Carbon Epoxy Laminates Beams during Bending Failure", Progress in Acoustic Emission II, pp. 472–479,(1984)
11. H.C.Kim, A.P.Ripper Neto and R.W.B. Stephens, "Some Observations on Acoustic Emission during Continuous Tensile Cycling of a Carbon Fiber/Epoxy Composite", Nature Physical Science, 237, pp78–80,(1972).
12. D.E.W.Stone and P.F.Dingwall, "The Kaiser Effect in Stress Wave Emission Testing of Carbon Fiber Composites", Nature Physical Science, 241, pp63–69,(1973).

TOOL WEAR MONITORING AT TURNING AND DRILLING

Eckhard Waschkies, Christoph Sklarczyk, and Eckhardt Schneider

Fraunhofer-Institute for Nondestructive Testing (IzfP)
University, Building 37
D-66123 Saarbrücken, Germany

ABSTRACT

To improve economy and as a contribution in quality assurance in the domain of machining metallic materials a method for automatic tool wear monitoring at turning and drilling has been developed based on the analysis of the acoustic emission (AE) generated on the tool at machining procedure (machining noise). Different wear types (tool flank face and tool chipping) result in changes of the different characteristic values of the continuous part of AE. In case of a uniform abrasion of the insert, e.g. flank face or crater wear, an increased mean signal level is observed, whereas for microbreakage at tool edge, an increase of the crest factor with nearly constant mean signal level is found. The burst-like signals from collision between chip and tool and from chip breakage have to be eliminated from analysis to avoid the distorsion of the signal parameters of continuous AE. This method should be well suited especially for monitoring of finishing processes (with small depth of cut).

1. INTRODUCTION

There is an industrial need on a method for automatic untended monitoring of tool wear during machining of metallic materials especially in the domain of finishing with small depth of cut. The method described below shall meet this requirement. It is based on the analysis of the acoustic emission (AE) generated during cutting process and consisting of continuous AE and transient (burst) AE. This monitoring method has been developed in the frame of a research project supported by the European Community in cooperation between the Fraunhofer-Institute for Nondestructive Testing (IzfP), Robert Bosch GmbH, Germany, and Danfoss A/S, Denmark [1].

2. THEORETICAL CONSIDERATIONS

2.1 Definition of the measuring quantities

The characterization of continuous AE can be performed by the following signal parameters:

Average Signal Level $ASL = 1/T \cdot {}_0\int^T |A(t)|\, dt$ (1)

Root Mean Square $RMS = \{1/T \cdot {}_0\int^T A^2(t)\, dt\}^{1/2}$ (2)

Crest factor $C = A_{max}/ASL$ (3)

with A(t): deflection of AE-signal as function of time, A_{max}: maximum amplitude of AE-signal within integration time T.

2.2. Sources of the AE during machining

Fig. 1 shows schematically the cutting procedure on a tool edge, especially for turning. The possible sources of the AE are [2]:

continuous AE (machining noise):

- friction contact between the tool flank face and the work piece resulting in flank wear
- friction contact between the tool rake face and the chip resulting in wear, e.g. crater wear
- plastic deformation during cutting process in the work piece
- plastic deformation in the chip

transient (burst) AE:

- collisions between chip and tool or tool holder
- crack formation in the chip (chip breakage)
- tool edge chipping

According to some researchers plastic deformation and friction have comparable importance with regard to the generation of the continuous AE [2]. However, Uehara has found that under turning conditions the sound amplitude of the signals from workpiece is reduced decisively during sound transfer from workpiece to tool possibly by reflection of the sound at the interface [3]. That means that the friction between workpiece and tool has to be regarded as the most important source of continuous AE. In the present investigation corresponding measurements have yielded the same results: the signal amplitude loss at the interface was about 35 dB (factor of 50 - 60). These findings are supported by more indirect experiments by Heiple et al. [4]. Therefore the authors of this paper support the view that the friction between workpiece and tool is the essential source of the continuous AE.

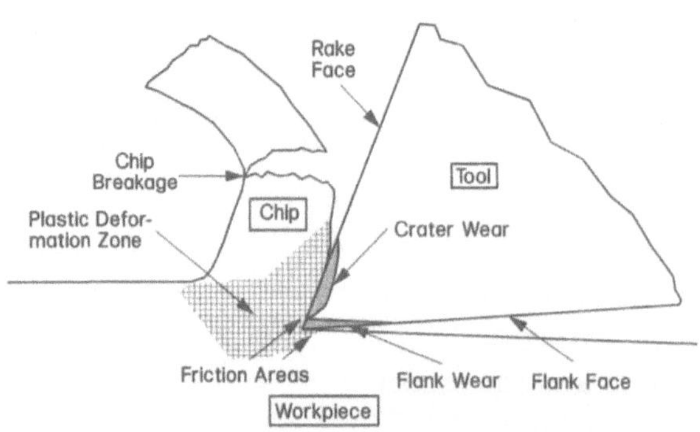

Fig. 1. Acoustic emission sources at cutting process

2.3. Continuous AE and tool wear

Starting from the experimentally well established view that the continuous AE is generated essentially by the friction between work piece and tool (see 2.2), the dependence of the signal parameters of continuous AE on abrasive tool wear can be described by a simple model. In microscopic scale the surface contact between work piece and tool can be regarded as a big number of point contacts [1]. During relative movement of the contact areas local stresses σ_i arise at the contact points which spontaneously release elastic waves and hence AE bursts. The signal amplitudes A_i are approximalety proportional to σ_i. Since the rate of the AE bursts is very high, these bursts interfere and their superposition results in a continuous or quasi-continuous AE signal with undistinguishable individual bursts. The ASL-level and the crest-factor of this continuous AE can be calculated from the machining and tool wear parameters [1]:

$$ASL = n_0^{1/2} \cdot a_p \cdot VB \cdot v_c \tag{4}$$

$$C = 1/(n_0^{3/2} \cdot a_p \cdot VB \cdot v_c) \tag{5}$$

where a_p is the depth of cut, v_c the velocity of cut, VB the wear land and n_0 the area density of contact points. A similar relationship like (4) holds for RMS, too.

The density of the contact points n_0 can be locally different and depends on the structure (roughness) of the surface [5]. In the equations (4) and (5) only the flank face area ($F = a_p \cdot VB$) is considered, since predominantly flank wear was found at the experiments reported below. In case of additional wear types (crater wear, notch on the rake face produced by the chip) the equations have to be modified by additional terms.

From the equations given above in principle two cases can be distinguished:

Case 1: n_0 remains nearly constant with increasing wear. This is true for uniform flank wear if only VB increases, but the structure of the wear contact area is unchanged. According to equation (4) an increase of the ASL-value proportional to VB is to be expected.

Case 2: n_0 changes to lower value, e.g. by chipping of the cutting edge. According to equation (4) and (5) ASL decreases and C increases.

3. EXPERIMENTAL CONDITIONS

The experiments have been performed with computer based AE equipment designed, developed and built by IzfP. The signals have been digitized with 10 MHz/12 bit, processed with regard to the signal parameters like peak amplitude or duration and stored on mass storage for evaluation. Furthermore the ASL- and RMS-values have also been determined. Commercial broadband piezoelectric ultrasound transducers have been used as sensors. The frequency domain was chosen between 0.1 and 1 MHz to avoid the influence of structure born noise which mainly occurs in the lower frequency range. The sensor with a diameter of about 5 mm was integrated in the tool holder directly underneath the insert.

One part of the experiments has been performed by using a cooling lubricant. Here no sensor cooling was necessary. Another part was performed without cooling lubricant. For these experiments a special cooling device has been constructed, and the sensor was cooled by flowing water. In both cases sensor temperature was held constant during the whole test.

The experiments described below have been carried out during NC-turning of a big variety of work materials (roller-bearing-steel 100 Cr 6, case-hardened-steel 16 MnCr 5, machining steel 9 SMn 28, cast iron (Meehanite)), inserts (coated carbide metal tools with and without chip breaking flute like CNMM, ceramics CBN) and cutting parameters (v_c = 80 to 600 m/min., a_p = 0.1 to 2.5 mm, feed f = 0.04 to 0.4 mm/rev.).

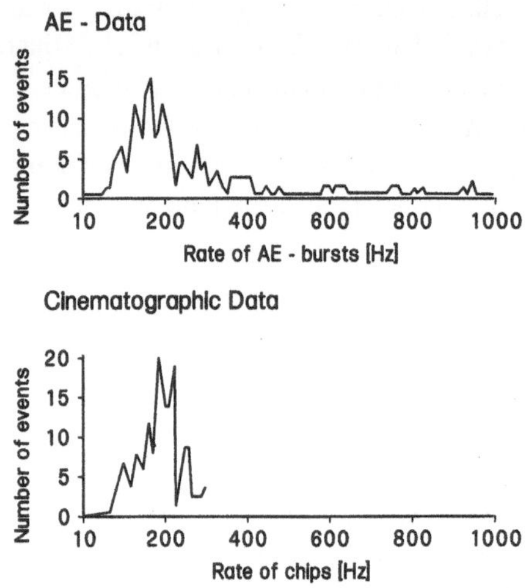

Fig. 2. Comparison of AE and cinematographic data for the occurence of chips

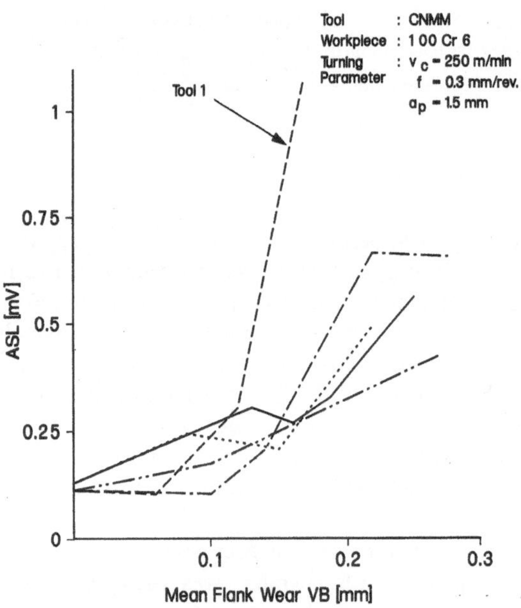

Fig. 3. Mean noise level versus mean flank wear for different tools

4. EXPERIMENTAL RESULTS

4.1. Chip collisions

From section 2.3 it can be expected that the friction between work piece and tool generates a continuous or quasi-continuous AE which can be characterized by ASL- or RMS-value. However, the experimental results show that sometimes burst-signals with high peak amplitudes interfere with the continuous AE. To clarify the origin of these bursts, simultaneously both AE has been recorded and the machining process has been filmed with high speed. The comparison of the AE and cinematographic data - by analysing chip behaviour slide by slide - has shown that the bursts were generated by collisions of the chips with the tool or tool holder and by chip breakage. This is shown in Fig. 2: both maxima of the rates of AE-bursts and chips, determined from the reciprocal values of time intervals between two AE-bursts resp. chips, are found at about 200 Hz indicating the coincidence of both processes.

It has been found that the AE related to the chips - especially the frequency of occurence of the chips - provides some information on the chip behaviour but not on tool wear. However these AE bursts distort the signal parameter ASL and C described above. Therefore for a reliable tool wear monitoring these bursts have to be filtered out from the continuous AE before further analysis is performed. Ref. [7] describes this procedure in detail.

4.2. Continuous AE and cutting parameters

In agreement to the theoretical relations in equations (4) and (5) the experiments have given following dependences of the continuous AE on cutting parameters [1]:

- ASL is independent of feed.
- ASL is proportional to v_c.
- ASL is proportional to a_p only, if wear occurs uniformly over the whole depth of cut and if it is not concentrated on certain areas of the flank face.

4.3. Dependence of continuous AE on wear

Fig. 3 shows an experimental example for the dependence of ASL on wear, characterized here by the width of flank wear VB. For predominant flank wear (= all curves except for tool 1) the level of continuous AE is approximately proportional to VB. The results displayed in Fig. 3 show that a flank wear of VB = 0.2 mm, which is often regarded as no more acceptable, is detectable by an increase of ASL by a factor of about 2-3. This result is in accordance with the theory given above (case 1 in 2.3.).

Additionally to flank wear, tool 1 exhibited a deep notch generated by chip edge sliding along the rake face (Fig. 3). The stronger increase of ASL compared to the other curves can be explained by the coincidence of both types of wear which resulted in a bigger contact area between workpiece and tool. The insert has become unusable at VB = 0.17 mm.

Similar results as shown in Fig. 3 have been obtained by a variety of other tools and workpieces. Fig. 3 shows that continuous AE gives an integral information on tool wear condition and not only on one single wear type like flank wear.

If microscopic fracture of tool (tool edge chipping) occurs, the proportion of signals with high amplitudes in continuous AE increases. This results in a distinct increase of C. In Fig. 4 (bottom) C, which is normalized on the starting value (new tool) is shown as a function of the number of turning parts machined with one tool. At about 100 parts an abrupt increase of C can be seen which coincides with the rise of chipping of the cutting edge found by microscopic examination. Fig. 4 (top) shows that the mean level of continuous AE (noise) is

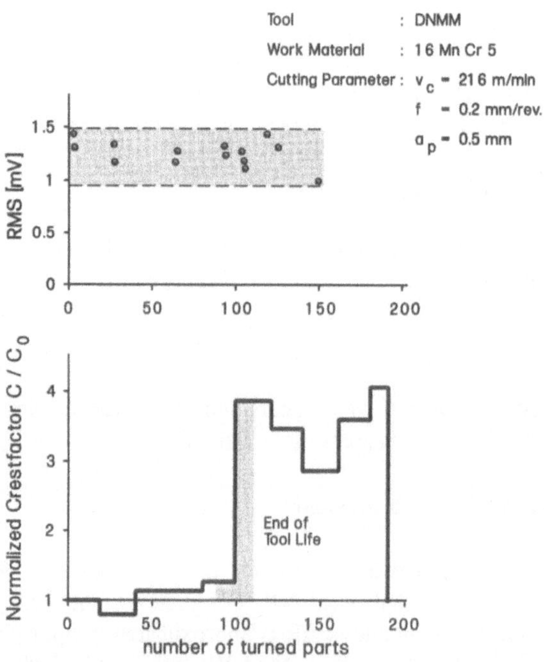

Fig. 4: AE-noise parameters versus number of turned parts
top: mean noise level
bottom: normalized crest factor

approximately constant or decreases slightly as wear increases. This result is in accordance with case 2 in 2.3. From these facts it is evident that the tool wear monitoring solely on the basis of increase of ASL is not sufficient, since it does not detect all types of wear.

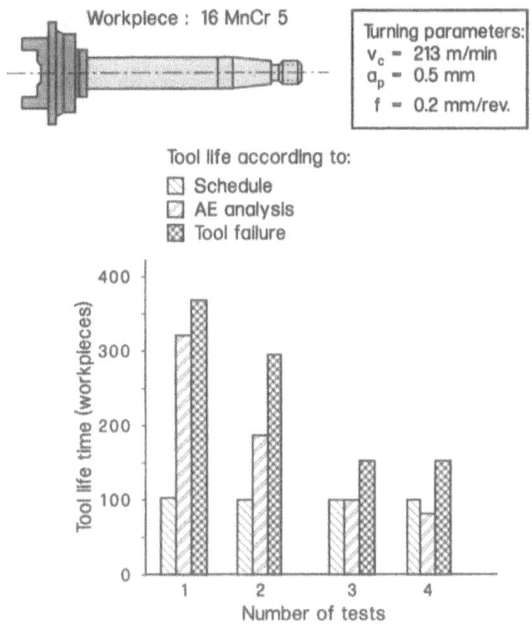

Fig. 5. Production parts with and without tool wear monitoring

4.4 An example for tool wear monitoring under industrial conditions

To verify the conception of AE-monitoring of tool wear given above, a big number of workpieces (driving shafts of a fuel pump) has been turned under industrial conditions using cooling lubricant. An increase of C by a factor of 2 has been regarded as clear indication for beginning tool wear. Without tool wear monitoring the tool would have been changed by schedule after 60 workpieces to prevent premature failure. However, in this test the tools have been used far beyond this limit. For tool no. 1 (Fig. 5) a clear increase of C has been observed after about 320 workpieces. This tool failed after ≥ 370 workpieces. That means that with the help of AE-monitoring this tool could be exploited much more better (> 500 %) than without it. This holds partly for the other tools in test (no. 2 - 4 in Fig. 5), too.

4.5 Tool wear monitoring at drilling

During drilling experiments a clear increase of ASL has been found as wear of the drilling tool increased (Fig. 6). The drill exhibited a uniform flank wear. In this test the continuous AE has been captured by a sensor fixed on the workpiece.

5. CONCLUSIONS WITH REGARD TO PROCESS MONITORING

The theoretical considerations concerning the change of the continuous AE due to tool wear are in good accordance with the experimental results. For a reliable determination of tool wear both ASL and C have to be measured and evaluated from the continuous AE. Bursts from chip collisions and breakage have to be filtered out if they exhibit high amplitudes (see 4.). The relations between ASL and C and the characteristic values of wear, e.g. VB, depend on the details of the machining process and have to be determined by a teach-in process taking place before or at the beginning of tool monitoring. As already mentioned in section 3 the sensor should be placed as near as possible to the insert in the tool holder. This may restrict the application area of the method. However, the detection sensitivity and the reliability of tool wear detection is increased distinctively compared to the monitoring systems presently available where the sensor is placed on the machine housing or spindle (e.g. at cutting force measurement). This is particularly true for the machining process with small cutting forces which cannot be monitored up to now in satisfactory way [6].

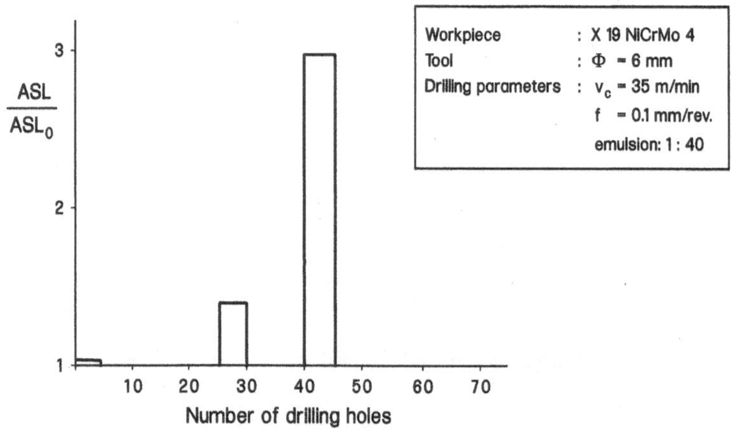

Fig. 6: ASL versus number of drilling holes at drilling

6. REFERENCES

[1] E. Waschkies, K. Hepp, J. Fisker, H. Pitsch: "Acoustic emission for process monitoring during turning and drilling", *Final report of the BRITE-project* 1300-4-85, 1991

[2] D.A. Dornfeld: "Tool wear sensing via acoustic emission analysis", *Proc. 8th NSF Grantee's Conf. on Production, Research and Technology*, Stanford Univ., 1981, A1-A8

[3] K. Uehara: "Identification of chip formation mechanism through acoustic emission measurement", *Annals of the CIRP*, Vol. 33 (1), 71-74

[4] C.R. Heiple, S.H. Carpenter, D.L. Armentrout: "Changes with material properties of acoustic emission produced during single point machining", *Progress in Acoustic Emission V*, The Japanese Society for NDI, 1990, 44-50

[5] I. Kragelski, M.N. Dobycin, V.S. Kombalov: "Grundlagen der Berechnung von Reibung und Verschleiß", Carl Hanser Verlag, München, Wien, 1983

[6] K.-D. Vöhringer: "Zerspanprozeß-Überwachung in der Großserie sinnvoll einsetzen", *Werkstatt und Betrieb*, Vol. 123 (1990), 763-766

[7] E. Waschkies, K. Hepp: "Process and device for monitoring the chip-removing treatment of a workpiece", *United States Patent 5,159,836*, Nov. 3, 1992

NONDESTRUCTIVE EVALUATION OF MICROMECHANISM OF DEFORMATION PROCESS DURING FATIGUE TESTING ON POLYMER BY ETFuM (ELASTIC-WAVE TRANSFER FUNCTION METHOD)

Y. Higo[1], H. Kawabe[2], Y. Natsume[2] and S. Nunomura[1]

[1]Precision and Intelligence Laboratory,
Tokyo Institute of Technology
4259 Nagatsuta Midori-ku Yokohama 227, Japan
[2]NIPPONDENSO Co.,Ltd.
1-1 Showa-cho Kariya 448, Japan

INTRODUCTION

According to comparison between fatigue phenomena on metals and on polymer, intrusion or extrusion were generated in slip bands at early stage of fatigue life on metals[1]. On the other hand, it is considered that fracture of molecular chains is equivalent to these phenomena on polymer. But micromechanism of deformation process during fatigue is not clear[2]. One of the reasons is that the micromechanism of nucleation and growth of microdefects is not clear. If polymer has perfect elastic property, fatigue does not generate during cyclic deformation. Therefore, the accumulation of microdefects is needed for fatigue. But micromechanism of stress concentration equivalent to conversion on metals is not clear.

Most investigations concerning the micromechanism of deformation process are studied under no-load. It is hard to find the researches concerning "dynamic evaluation of micromechanism of deformation process during fatigue". As for micromechanism of a deformation process which ends up fracture, it is considered that elastic heterogeneous area will be nucleated like as density, acoustic impedance and elastic modulus. Therefore, if elastic heterogeneous area can be evaluated dynamically and quantitatively on real time, we can investigated the dynamic microanalysis during fatigue process on polymer.

We have been investigated that new nondestructive evaluation method of ETFuM (Elastic–wave Transfer Function Method) in order to make clear the micromechanism of deformation process on polymer quantitatively[3,4]. This method has advantage of monitoring the specimen nondestructively, furthermore evaluating it under operating condition without removing the structural components. The change of the transfer function is caused by the existence of crazes and microcracks in the materials[5,6]. Furthermore, quantitative analysis of the damaged area on the tensile or fatigue specimen by scanning acoustic microscope was performed[7,8].

In this study, ETFuM is applied to make clear the micromechanism of a deformation process during fatigue testing dynamically in amorphous polymer (ABS; acrylonitrile butadiene styrene) and crystalline polymer (POM; acetal homopolymer), and discussed.

EXPERIMENTAL PROCEDURE

MATERIALS

Rods of commercially available ABS (ET–70 of Mitsui Touatsu Chemicals Inc.) and POM (M–25 of Polyplastics Co. Ltd.) were used in this study. Both materials are thermoplasticity polymer. But ABS is amorphous and POM is crystalline polymer. Crystallinity of POM was 63% by measuring the density at amorphous or crystalline domain. Mechanical properties of the materials in this study used are given in Table 1. Specimens were machined from extruded rods of ABS and POM which have 40mm diameter. Figure1 shows the dimension of test specimen used. The residual stress of the machined specimens was removed by annealing them at the condition of 80° in ABS and 140° in POM for 3hours, and then cooled in the furnace[9,10]. Specimens are prepared by buffing to obtain smooth surface enough for SAM observation. After that, Specimens are exposed in thermo–hygrostat at 25° and relative humidity of 5, 42, 97% for over 2000 hours and then used on fatigue test[7].

FATIGUE TEST

Fatigue tests were performed by closed loop servohydraulic fatigue test machine (Shimazu Co.,Ltd.) at 1Hz in load control. The R ratio (minimum load/maximum load) was maintained at R = –1 for all of the test. Environmental conditions are shown in Table 2, which are the same as exposure conditions of specimens.

Table 1. Mechanical properties of polymer tested

		ABS	POM
Specific gravity		1.05	1.41
Tensile strength	MPa	45.0	62.0
Modulus of elasticity	GPa	2.0	2.8
Deflection temperature	°C	93	158

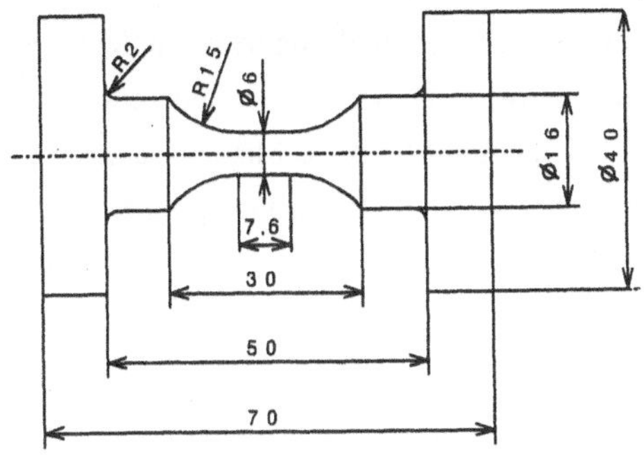

Figure 1. Shapes and dimensions of test specimen

Table 2. Fatigue testing conditions

	ABS		POM	
No.	Temperature(°C)	Humidity(%)	Temperature(°C)	Humidity(%)
1	25	5	25	5
2	25	42	25	42
3	25	97	25	97

EXPLANATION OF ETFuM

The respective functions in time domain are defined as follow[5].

$$g(t)= \int_{-\infty}^{+\infty} h(t-\tau)f(t)d\tau =h(t)\otimes f(t) \quad (1)$$

Then, they are transformed to frequency domain by performing a Fourier transform operation on the integral equation.

$$G(\omega)=H(\omega)\bullet F(\omega) \quad (2)$$

Where, $G(\omega)$, $H(\omega)$ and $F(\omega)$ are the respective Fourier transform of g(t), h(t) and f(t). The input signal path is expected and defined as follow.

$$G_A(\omega)=M\bullet\beta_2\bullet h_A\bullet\beta_1\bullet S\bullet F(\omega) \quad (3)$$

Where, M and S are the received and transmitted transducer sensitivity, and β_1 and β_2 is transfer function of coupling condition between transducer and specimen. The transfer function of elastic signal through the specimen is expressed as follows.

$$H_A(\omega)=G_A(\omega)/F(\omega)=S\bullet\beta_1\bullet h_A\bullet\beta_2\bullet M \quad (4)$$

The transfer function, H_A is found by multiplying the transfer functions of the system components in frequency domain as indicated in this equation. We only want to obtain information on the polymer specimen h_A. So we need to separate that information from the characteris–

tics of the transducers, and the coupling condition between the transducers and specimen expressed as β_1 or β_2. We will use the same transducers shown here as S and M. In addition, we have developed a method of coupling the transducer to the specimen so as to obtain good acoustic reproducibility. The transfer function of H_{AN} including other specimen is defined in the same way as H_{A0} including standard specimen. In this experiment, we used the transfer function on virgin specimen as H_{A0}. Then all components of the transfer function that reflect the characteristics of the transducers S and M, and coupling conditions β_1 and β_2 are canceled out, leaving only the information on the specimen as follows.

$$\begin{aligned} H_{AN}/H_{A0} &= (G_{AN}/F_N)/(G_{A0}/F_0) \\ &= (S_N/S_0)\bullet(\beta_{1N}/\beta_{10})\bullet(h_{AN}/h_{A0})\bullet(\beta_{2N}/\beta_{20})\bullet(M_N/M_0) \\ &= (h_{AN}/h_{A0}) \end{aligned} \quad (5)$$

Also, the signal source F provides good reproducibility. The ratio of the output signal G_{AN} to G_{A0} directly explains the difference corresponding to dynamic property of specimen.

$$G(\omega)=G_{AN}/G_{A0}=h_{AN}/h_{A0} \quad (6)$$

Therefore, the effects of specimen shape, transducer sensitivity and so on were eliminated on equation (6). Then, the evaluation function Δh_{Af12} was defined as follows.

$$\Delta h_{Af12}= \int_{f2}^{f1} |G(\omega)|\, d\omega \quad (7)$$

Here, Δh_{Af12} is integrated value in a frequency range from f_1 to f_2 on equation (7).

METHOD OF MEASUREMENT

The block diagram of ETFuM measurement system is shown in Fig.2. Two piezoelectric transducers were used as transmitter and receiver, respectively. They put in direct contact with the specimen with couplant grease. Signal F is generated by a function generator, input to the transmitting transducer S, passes through the polymer specimen and is picked up by the receiving transducer M and amplifier during fatigue test. It is then fed into a FFT analyzer along with the input signal. The signal are compared and only information on the specimen is extracted by previous equation (5) or (6). Also, cyclic stress-strain curve was measured in same specimens. During fatigue test, load was measured by load cell and displacement was measured by clip gauge.

A flat type PZT transducers (0.01-5.0MHz) were employed and the contact agent between the transducers and the specimen was W-400 (Nippon Steel Co.,Ltd.). It provides good reproducibility of the mounting condition, for both the amplitude and phase components[11]. The measured data obtained with these transducers were sufficiently reliable for comparison without performing any special sensitivity calibration. In this study, f_1 and f_2 were selected from 0.01 to 5.0 MHz on equation (7).

In order to measure the transfer function at anytime during fatigue test, special measuring system was constructed using the trigger generator. In this study, the transfer function was measured at four points in 1 fatigue cycleas shown in Fig.3; (1) maximum tensile stress, (2) applied stress=0 (under no-load from tension to compression), (3) maximum compression stress, (4) applied stress=0 (under no-load from compression to tension). In this system, measurement time of the transfer function for each point was 640μs. The change of applied stress during measurement is biggest at points of (2) and (4). But the applied stress changes within 0.4% of stress amplitude. Therefore, it is considered that the change of applied stress is negligible small during measurement.

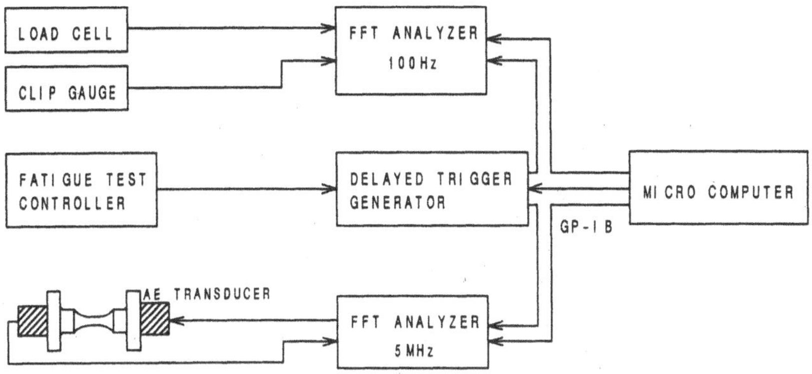

Figure 2. Block diagram of ETFuM measuring sysytem

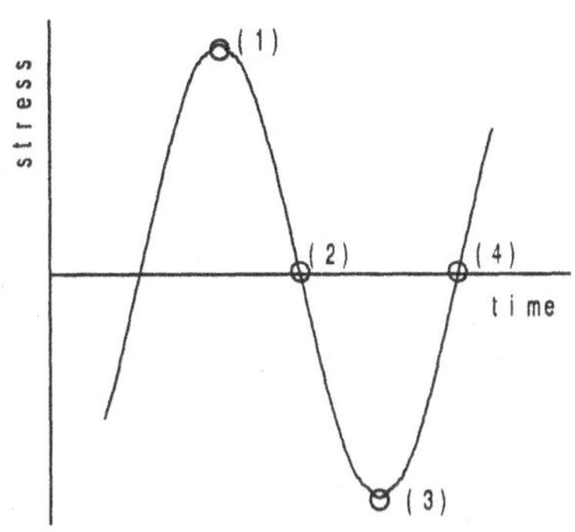

Figure 3. Measurement points of the transfer function

EXPERIMENTAL RESULTS AND DISCUSSION

In order to evaluate the transfer function quantitatively, difference of attenuation Δh_{Af12} was analyzed from equation (7). Figures4(a)-(d) show the relation between the number of cycles and Δh_{Af12} of 0.5-1.0MHz in frequency domain in ABS. Because we analyzed the transfer function from 0.5 to 5MHz at every 0.5MHz from equation (7), and the change of the transfer function was clearest in 0.5-1.0MHz. In Figs.6(a)(b), Δh_{Af12} decreased at the end of fatigue life.

But it began to change from n=1×10^4, which was approximately half of the fatigue life n=2×10^4. The sense of the results were the same in both ABS and POM.

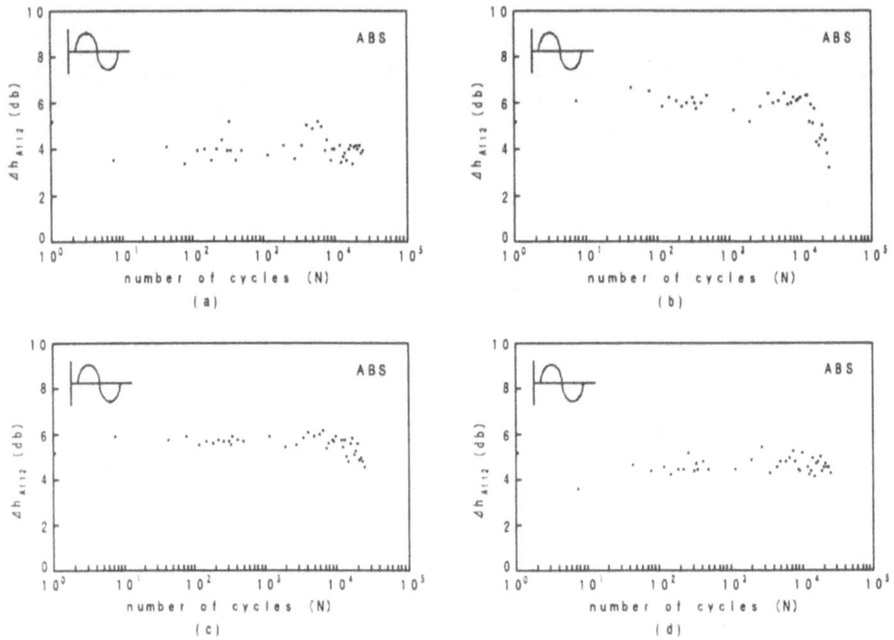

Figure 4. The relation between the number of cycles and Δh_{Af12} of 0.5−1.0MHz in frequency domain in ABS

During fatigue test, cyclic stress−strain curve was measured at the same time on same specimen measured the transfer function. Relation between the number of cycles and strain amplitude, elastic modulus analyzed from hysteresis loop are shown in Fig.5. Strain amplitude began to increase after n=1×10^4 and elastic modulus began to decrease in ABS. These trend are correspond to the results of the transfer function in Fig.4(b). In POM, the sense of the results was same.

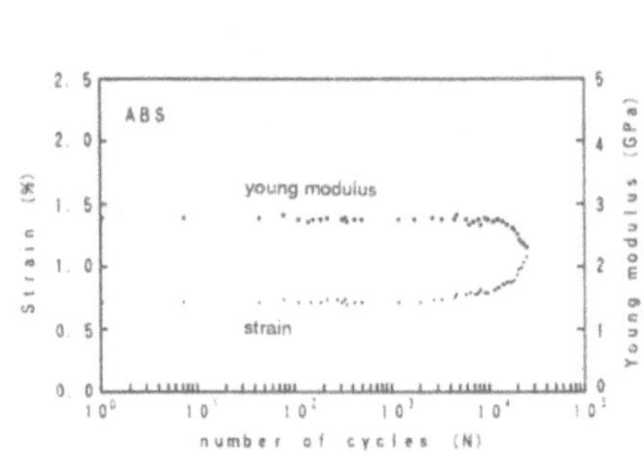

Figure 5. The variation of strain amplitude and elastic modulus during fatigue testing in ABS

We discuss the relation between the transfer function and the micromechanism of defor− mation process during fatigue on polymer. We had reported the micromechanism of a deforma−

tion process before crazing as follows[12]. Molecular chains locally were oriented in a very small area with increasing stress. it is considered that a fibril structure is constructed in the local oriented area, which are thought to be embryo crazing. The embryo crazing was produced from the early stage of plastic deformation, and developed with increasing plastic strain. And in the precursor stage of degradation process in tensile test, the variation of the transfer function indicated the dependence of molecular chain and initiation and growth of craze[5,6]. From scanning acoustic microscopy, it was found that the damaged area of the specimen was widely dispersed in tensile testing. On the other hand, in fatigue testing, the damaged area was concentrated near the fracture surface[7,8]. One of the reasons was explained by the nucleation and coalescence of crazes and microcracks. Crazes will be nucleated easily because of temperature rise in a very small area during fatigue testing.

Micromechanism of a deformation process under monotonic applied stress is discussed. At first, texture and stress under each measuring point in Fig.3 were considered as follows. (1) Maximum tensile stress; Crazes or microcracks open due to tensile stress, and texture becomes heterogeneous and anisotropic. (2) Applied stress=0 (under no-load from tension to compression); Crazes and microcracks are transition from open to close. Though stress is not applied, texture becomes anisotropic a little. (3) Maximum compression stress; Crazes and microcracks close. Texture becomes homogeneous due to compression stress. (4) Applied stress=0 (under no-load from tension to compression); Crazes or microcracks are transition from close to open, and texture becomes homogeneous because compression stress is released.

Finally, the transfer function was influenced by applied stress, loading hysteresis and micromechanism of deformation process during fatigue testing. In spite of stress was not applied at points of (2) and (4), micromechanism of deformation process was different due to stress hysteresis. It is considered that the fatigue on polymer depends on the accumulation of microdefects nucleated from the beginning of fatigue life. The transfer function is correspond to the micromechanism of deformation process on polymer dynamically.

CONCLUSIONS

New nondestructive evaluation method of ETFuM (Elastic-wave Transfer Function Method) has been applied to make clear the micromechanism of deformation process during fatigue testing dynamically in ABS and POM.

(1) The transfer function measured under no-load was very sensitive for the micromechanism of deformation process during fatigue testing in both ABS and POM. Because the oriented fibril was nucleated in crazes, and the influence of crazes or microcracks was negligible small, because the transfer function was measured under no-load.

(2) In spite of the transfer function was measured under no-load, the results of two measuring points (under no-load from tension to compression, and from compression to tension) were different. It suggested that the micromechanism of deformation process depended on loading hysteresis.

(3) ETFuM was good correspond to the micromechanism like as local orientation of molecular, nucleation of crazes and microcracks during fatigue process. It has been observed that ETFuM may be appropriate in evaluating the micromechanism of cyclic deformation on polymer dynamically and nondestructively.

REFERENCES

1. T.Yokobori, "Materials strength (in Japanese)", Gihodo press:153(1958) Tokyo
2. I.Narisawa, "Materials strength in plastics (in Japanese)", Ohm press:301(1982) Tokyo
3. Y.Higo, S.Nunomura, "Nondestructive evaluation of adhesive strength by ETFuM", Proceedings of 2nd symposium of nondestructive evaluation in new material and products:131(1988)
4. Y.Hushimi, A.Wada, "Time domain measurement of dielectric dispersion as a response to pseudorandom noise", Review of Scientific Instruments, vol.47,No.2:213(1976)
5. H.Kawabe, Y.Higo, "Nondestructive evaluation of craze and microcrack on polymer by ETFuM", Journal of Materials Science, in press
6. H.Kawabe, Y.Higo, "Nondestructive evaluation of precursory stage of fracture on polymer by ETFuM", Nondestructive Testing Evaluation, vol.8–9:745, Gordon and Breach S.A.(1992)
7. H.Kawabe, Y.Higo, "The influence of environmental conditions on the fatigue strength of plastics", Japan Society of Mechanical Engineers International Journal, in press
8. H.Kawabe, Y.Higo, "Micromechanism of deformation process in polymer during fatigue test (in Japanese)", Japan Society of Mechanical Engineers, vol.58,No.554:1759(1992)
9. "Technical paper of Santac (in Japanese)", Mitsui Touatsu Chemicals Inc.(1986)
10. "Design data of Duracon (in Japanese)", Polyplastics Co.Ltd.(1987)
11. Y.Higo and M.Ono, "On a real time calibration of AE transfer function", Progress in A.E., vol.3, Japan Society of N.D.I.:685(1986)
12. H.Kawabe and Y.Higo, "Micromechanism of a deformation process before crazing in a polymer during tensile testing", Journal of Materials Science, vol.27:5547(1992)

APPLICATION OF ULTRASONIC IMAGING TO WRAP ANALYSIS THROUGH FLOW PATTERN OBSERVATION OF INJECTION MOLDED FRTP PRODUCT

Toshihiko Abe[1], Isao Hanada[2], Takashi Kuriyama[3] and Ikuo Narisawa[3]

[1]Government Indst. Res. Inst., Tohoku, Nigatake 4-2-1, Sendai
[2]Hitachi Kasei Mold, Higtashitaga 1-1-1, Hitachi
[3]Yamagata University, Jounan 4-3-16, Yonezawa, Japan

INTRODUCTION

Fiber reinforced thermoplastics, FRTP, are widely used in many fields of industry because the strength of thermoplastics can be very much increased by fiber addition. Although most of the FRTP products are produced by the injection mold method, there are several problems which relate to the shape or size accuracy such as warp, fin and missrun. All of these problems relate to the flow path of the molten FRTP in a metal mold and also relate to the fiber density and orientation in the product. For this reason many inspection methods, destructive and nondestructive, were applied to observe the flow path and the fiber distribution. Conventional nondestructive methods, including X-ray and optical images proved to be insufficient for this purpose. Scanning acoustic microscopy using high frequency ultrasound is not applicable to internal observation of the polymer composites. However, relatively low frequency, 5 to 25MHz, ultrasound can reach to positions deeper than 1mm from the surface of polymers and polymer composites.

At the present time we are trying to observe the internal structure of many kinds of materials by ultrasonic imaging with this frequency range and found that the flow path of injection molded neat thermoplastics and the fiber distribution in FRTP or CFRP can be clearly shown. We therefore applied this ultrasonic imaging technique to warp analysis to produce high quality FRTP products.

INSTRUMENTATION

Ultrasonic images were obtained using an immersion type point focusing pulse wave probe possessing frequencies of either 5, 10 or 25MHz. After a specimen was immersed in water, the probe was scanned mechanically keeping at a constant distance from the surface so as to maintain focus at the measuring depth. All the images shown in this paper were measured using longitudinal waves, without image processing by sharpening or edge enhancement.

Fig. 1 shows the positional relation between a reflected ultrasonic pulse and the electrical gate which selects a received signal. In this figure, T is the trigger delay time to eliminate noise signals arriving before the main signal. A and D are surface and back echoes which were reflected at the specimen surface and the back surface. B and C are gate position and width, that is, measuring depth and thickness, respectively.

During measurements, the first pulse which appeared after T was regarded as the surface echo, i.e. A to be the surface. Therefore the reflected signal from a constant depth could be measured even if the specimen surface curved a little. A signal selected by the gate was converted to digitized to a 256 bit level, stored in memory, and used for intensity modulation of the image at the measuring point. The ultrasonic internal image, maximum 750 x 500 bits, was formed by repeating this process while the probe was scanned mechanically to X-Y direction up to 140mm.

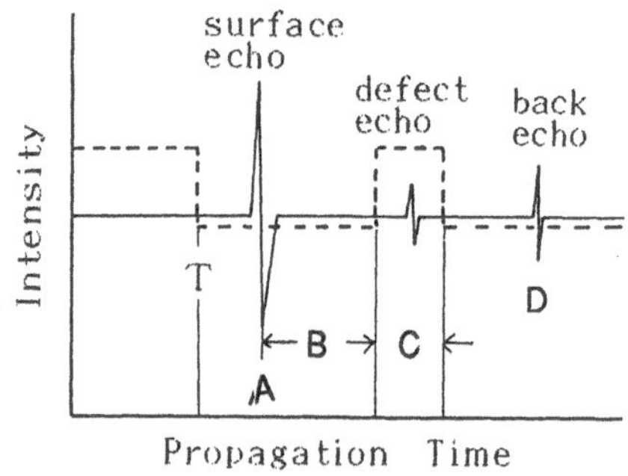

Fig. 1 Relation between echoes and gate position

SPECIMENS

To investigate the depth resolution of the ultrasonic image, laminate type carbon fiber reinforced plastics, CFRP, were prepared. The specimen was 32 ply with $(+45/0/-45/90)_{4s}$ of 4.5 mm in thickness. For the purpose of flow pattern observation, the following two types of neat thermoplastics were injection molded into a bar or a tensile test piece of thickness from 2 to 4 mm. (1) Crystalline type: polyethylene, PE, polypropylene, PP, liquid crystal polymer, LCP. (2) Amorphous type: polymethyl methacrylate, PMMA, polyvinyl chloride, PVC, polycarbonate, PC.

Injection molded FRTP product made by polybutyrene terephtalate, PBT, with 30% glass fiber, and also no fiber, were prepared for warp analysis.

ULTRASONIC IMAGING OF CFRP

Figs. 2a and 2b show the fiber orientation and delamination area in a buckling tested 32 ply CFRP (laminate composite) measured at 25MHz. Fig. 2a shows 1-2 ply in which a delaminated area appeared as a white part due to a strongly reflected echo at the delaminated interface. The fiber orientation (+45°) was recognized in a non-delaminated area.

Fig. 2b is a delamination image in 7-8 ply (-45°). Delaminated area expanded at inner ply. Gray zone (indicated by white arrow) shows water penetration into the delaminated area. Individual images separated plies were obtained up to the 12th ply using 25MHz, and up to more than the 20th ply using 10MHz. These figures show that the depth resolution of the image is of the order of the measured wavelength.

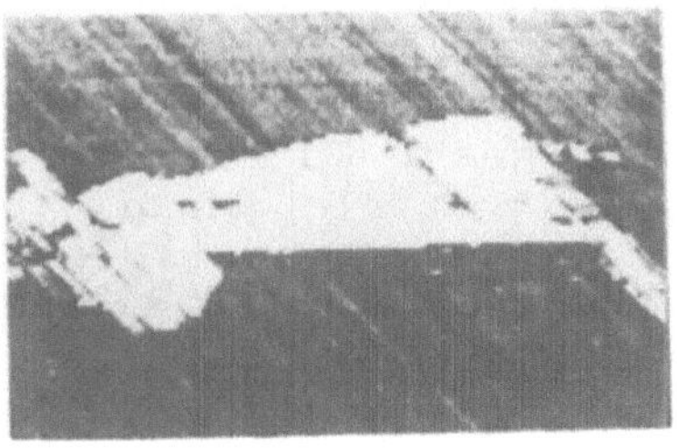

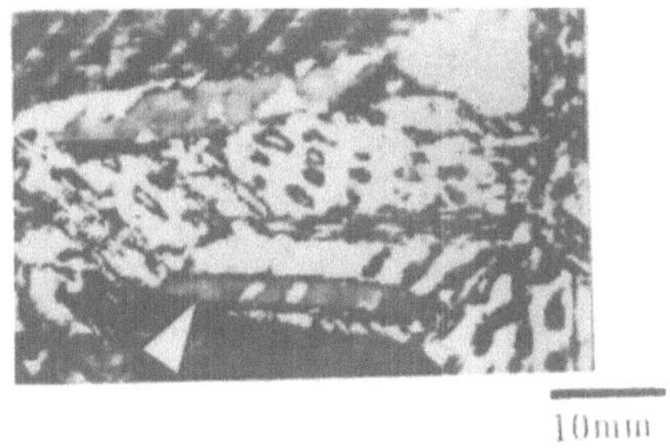

Fig. 2a Ultrasonic image buckled CFRP (1-2 ply)

Fig. 2b The same area (7-8 of ply) an arrow indicates water penetrates

ULTRASONIC IMAGING OF INJECTION MOLDED NEAT THERMOPLASTICS

Figs. 3a to 3d are ultrasonic flow pattern images at different depths in a polyethylene tensile test piece, 175x20x3mm, which was injected from left to right. All images were taken from the same area using 10MHz at a depth

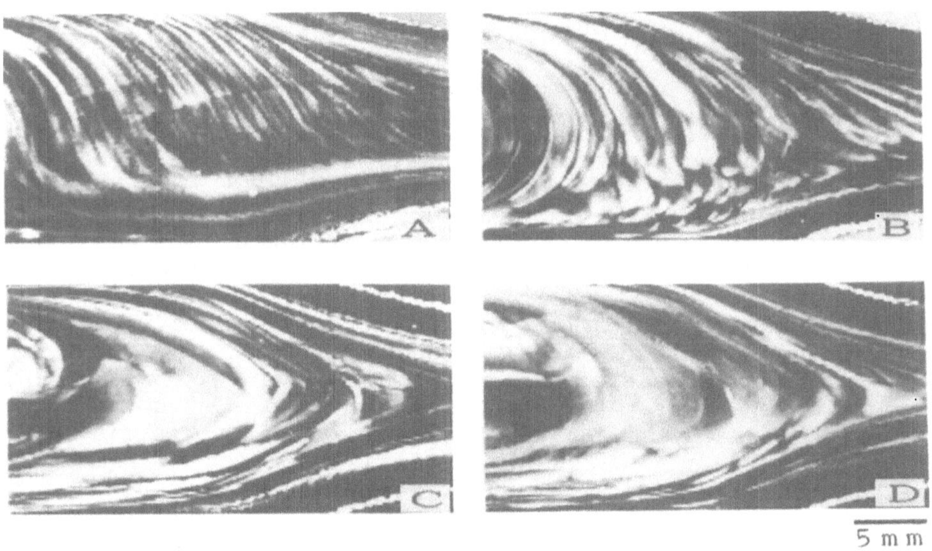

Fig. 3. Ultrasonic flow pattern imaged at 10MHz in an injection molded PE specimen. A~D are the same area at a depth of A: 1.0~1.2mm, B: 1.3~1.4mm, C: 1.4~1.5mm, D: 1.5~1.7mm.

of 1.1 to 1.2mm (A), 1.3 to 1.4mm (B), 1.4 to 1.5 mm (C), and 1.5 to 1.7mm (D), respectively. These images change continuously with measuring depth. This result suggests an inhomogeneous flow along the depth direction as well as the horizontal direction.

Flow pattern images just like Figs. 3 were observed in all other injection molded crystalline type specimens by using 5 to 25MHz. Amorphous type specimens, however, did not show the flow pattern image at all. It is clear from this fact that the flow pattern image reflects an inhomogeneous internal structure which was formed during solidification. In general, the crystal size is too small to be resolved using 5 MHz ultrasound corresponding to a wavelength of 0.5mm in many thermoplastics. A similar situation exists in the carbon fiber images of Figs. 2. The diameter of the fiber, 8μm, is much smaller than the wave length of 25MHz ultrasound, i.e., 120μm. Thus the fiber orientation image does not represent the fiber itself but reflects the fiber orientation.

In the case of an optical image, an expanded pattern will appear when periodic patterns overlapped one another. This is well known as a moire pattern. As about twenty carbon fibers are piled up in the same orientation in one ply of CFRP, an ultrasonic moire pattern was thought to be formed in the ultrasonic image. Thus the fiber orientation images in Figs. 2 were considered as an ultrasonic moire pattern image, or an ultrasonic interference image. A similar image formation mechanism was predicted in the flow pattern image because the image changed as the measuring depth changed. This result implies that multilayer structure exists in injection molded crystalline thermoplastics. This idea corresponds well with the reflected echo profile in Fig. 4 measured at 10MHz in the same specimen as Figs. 3. In Fig. 4, S and B are reflected echoes from the front surface and the back surface. C is the characteristic echo observed in injection molded crystalline type polymers. Even though the incident beam was a pulse, the reflected echo, C, from the middle depth became a continuous wave which was formed by multiple reflections at the central core of the part. The flow pattern images in Figs. 3 were obtained by the wave from C, whose profile altered widely as the measuring point was shifted. This change gave rise to a strong contrast of the flow pattern image. As the reflected intensity was dependent on the acoustic impedance, sound velocity x density, of the reflecting interface, it was thought that the core part of the specimen consisted of many thin layers of different acoustic impedance. The predicted multilayer structure was not recognized by an optical microscope on a crosssection of the cut specimen. However, a polarized optical image of a microtomed thin foil, Fig. 5, clearly shows the multilayer structure at the core part of the PE specimen. By comparing Fig 3 and 4, it can be seen that the disordered part of the layer was a location where the flow pattern image widely changed.

Injection molded transparent, i.e., amorphous, polymers such as PMMA, PVC, PC did not show either the ultrasonic flow pattern image or the multilayer structure. Even in a

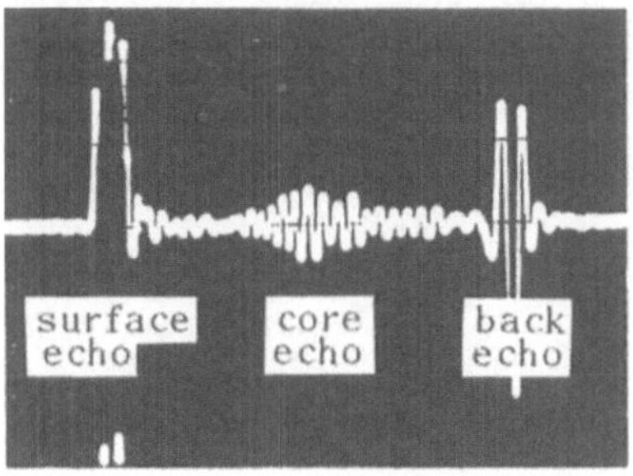

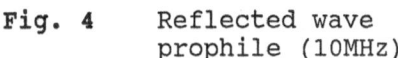

Fig. 4 Reflected wave prophile (10MHz)

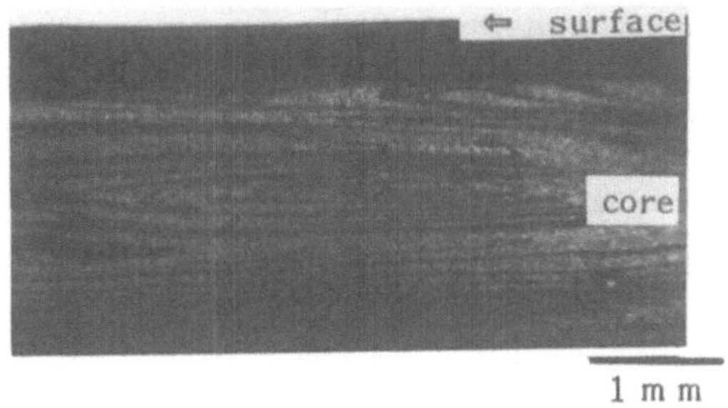

Fig. 5 Optical image of microtomed PE specimen cut from the crossection

crystalline type polymer, the degree of crystallization widely changes as the cooling rate changed, that is, rapidly cooled parts becomes amorphous and slowly cooled parts become well crystallized. The acoustic impedance, density x sound velocity, in the amorphous state of the same polymer is different from the crystallized state by several percent. Because the continuous wave reflected from the core is a result of multiple reflections, it is thought that the core consists of layers with different degrees of crystallization. Thus the effect of temperature, thickness, injection speed, which affect the degree of crystallization, can be determined from the flow pattern image.

WARP ANALYSIS OF FRTP

FRTP products with short glass fiber are mainly produced by the injection molded method. There are several problems in this method relating to the shape of the product, such as warp, fin, missrun etc., in thin products. All of these problems are responsible for the flow path, cooling rate and fiber distribution. Complicated relations exist between these factors because solidification tends to occur at a thin region where fibers are easily stopped. Solidification is accompanied by volume reduction which is affected by the degree of crystallization, or cooling rate, and a distributed fiber interrupts the volume reduction. Inhomogeneous polymer structure and fiber segregation frequently generate a warp of injection molded FRTP products. It is quite important to know the cause of the warp to produce a high quality product. For this purpose, the ultrasonic imaging technique was applied to the injection molded PBT with 30% glass fiber product of dimensions 47x57x1.2mm and the maximum warp was more than 0.3mm. This product was injection molded through a pin gate at the center of the surface plane. It was first attempted to observe the difference in the flow path in the warped and unwarped specimens. A specimen containing fibers, however, did now show the flow path because an incident pulse was reflected much stronger by the fiber than by the core structure. Consequently specimens not containing fibers were made from the same material, PBT, with the same injection condition and the flow pattern image was measured at 10MHz. Figs. 6 are the flow pattern images both of warped and unwarped specimens.

Fig. 6 Ultrasonic images of PBT specimens without fibers measure at 10MHz
A: warped, B: unwarped

A concentric circular flow pattern is recognized in both Fig. 6A and B. However, at the arrow mark, no clear pattern is seen in the warped specimen, A, whereas the flow pattern is seen in unwarped specimen, B, as indicated by the white arrow. It is considered from these difference that the degree of crystallization at the arrow mark on the warped specimen is lower than that of the unwarped specimen, which therefore did not show a clear image. As the degree of crystallization increases with decrease in cooling rate, and also increase in thickness, the effect of cooling rate was investigated.

At first, the thickness and sound velocity of the warped and unwarped specimens were measured at four corners in each of three specimens. The measured thickness and sound velocity is compared with those from a specimen without fibers in Table 1.
fibers in Table 1.

Table 1. Thickness and sound velocity (10MHz) of warped and unwarped specimens made by PBT with 30% glass fiber as compared with a specimen containing no fibers. (n=12 for each specimen)

specimen	thickness, mm	sound velocity, km/s
warped	1.199 (σ =0.017)	2.43 (σ =0.020)
no warped	1.224 (σ =0.011)	2.45 (σ =0.015)
no fiber	1.191 (σ =0.007)	2.31 (σ =0.012)

As would be expected from the flow pattern images in Figs. 6, the warped specimen was 0.025mm thicker on the average than the unwarped one. From the fact that no specimen without fibers was thinner than specimens containing fibers, solidification shrinkage of PBT was thought to be interrupted by the fiber. Reduction in the PBT volume fraction of specimens containing fibers also affected the increase in thickness. The sound velocity of PBT was increased more than

0.1km/s by adding 30% glass fibers. Standard deviation, sigma, of both the thickness and sound velocity was larger in specimens containing fibers, especially the warped ones, than in specimens without fibers. This result indicates that the homogeneity of fiber distribution was lower in warped specimens. This sound velocity increased as the thickness increased. This figure shows that the fiber density is high in the thick part, or the fiber orientation changed where the thickness changed because sound velocity is higher in the fiber direction than in the perpendicular direction even if the fiber content is the same.

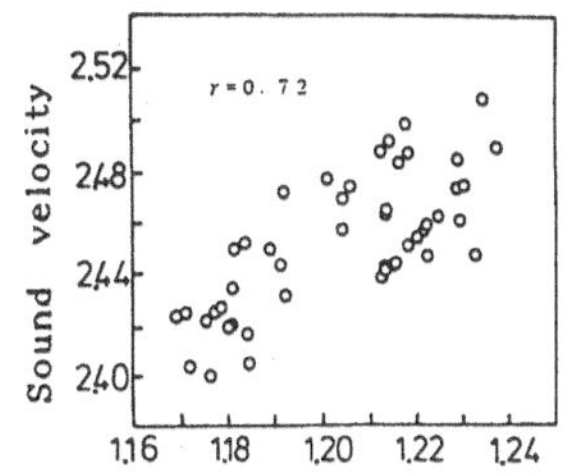

Fig.7 Sound velocity vs thickness of 30% fiber containing specimen.

Subsequently, the fiber distribution in warped and unwarped specimens was investigated using ultrasonic imaging. The image measured by the near surface echo did not show any clear difference between warped and unwarped specimens. Figs. 8 are ultrasonic images made by the characteristic echo from the middle part of the thickness where glass fibers were concentrated in a thin layer. Fig. 8A is the image of the warped specimen and Fig. 8B is that of the unwarped one. Both images were obtained at 10MHz. The right corner of the image from the warped specimen, Fig. 8A, is brighter than the other corners. On the other hand, the brightness of the image at all four corners are nearly equal in the image from the unwarped specimen, Fig. 8B. A lattice shaped pattern in both images corresponds to a strengthening rib structure on the back surface.

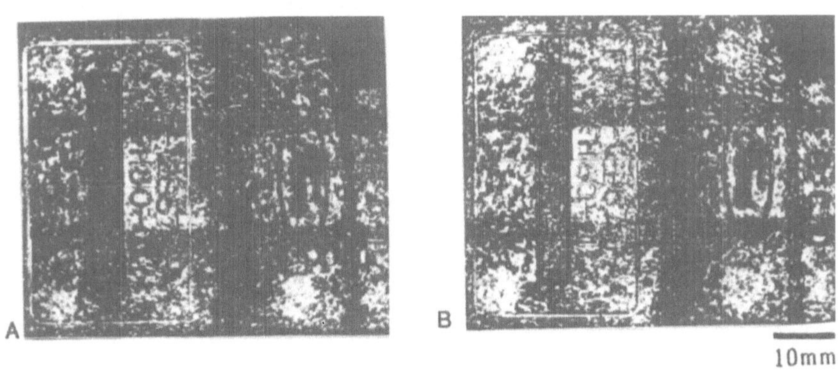

Fig. 8 Ultrasonic image of PBT containing 30% glass fiber. A: warped specimen B: unwarped specimen

As the bright areas in Figs. 8 are considered to indicate a fiber concentration, the difference in density was measured by Archimedes' method. The average densities are; warped specimen: 1.545g/cm^3 (sigma = 0.0008), unwarped specimen: 1.539g/cm^3 (sigma = 0.0020), no fiber: 1.311g/cm^3 (sigma = 0.0005). Densities were clearly separated into two parts, that is above 1.54g/cm^3 for warped specimens, and below 1.54g/cm^3 for unwarped specimens. This difference in density means that the fiber packing condition affects warp generation. To summarize the warp analysis by means of flow pattern images, fiber images, sound velocity and density, the causes of the warp was considered to be related to the cooling rate, or thickness, and with in injection pressure. Consequently cavity depth of the metal mold was precisely modified to the same depth, the cooling method of the mold was modified, and the injection pressure was decreased to avoid an over charge of the fiber. As a result of these modification, the maximum warp was decreased up to half of the initial amount (Fig. 9).

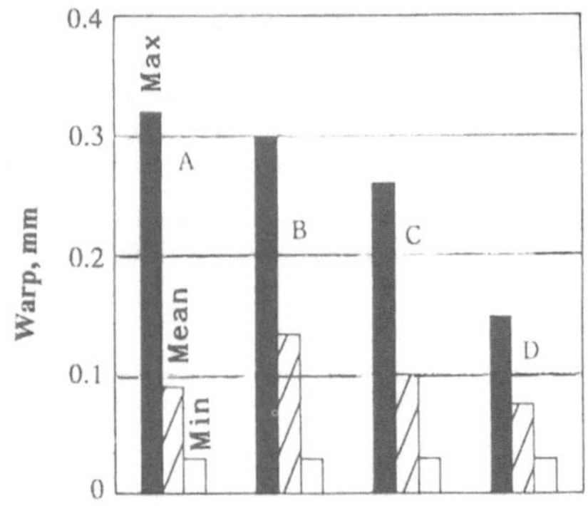

Fig. 9 Modification of Warp
A: initial warp B: thick modified C: modified cooling process D:decreased injection pressure

Flow pattern images of injection molded thermoplastics, and fiber orientation images of FRTP can be observed by using relatively low frequency, 5 to 25MHz, ultrasound. Depth resolution of the image was an order of the wavelength used. The flow patterns image changed with measuring depth due to the multilayer structure in the core of the part. This structure was observed by an optical microscope in the microtomed specimen. Both images of the flow pattern and fiber were applied to the analysis to prevent a warp of an injection molded PBT with 30% glass fiber content. From the flow pattern image of the specimen without fibers, the difference in cooling rate and thickness, between the warped and unwarped specimen was indicated. Differences in fiber distribution in the warped and unwarped specimen containing 30% fiber was also shown. Based on this information, the thickness of the product, the cooling method, and the injection pressure were modified. As a result of these modifications, the maximum amount of warp decreased up to half of the initial amount.

ACKNOWLEDGMENTS

The authors thank Mr. K. Tobukuro of Toray Research Institute for preparing the CFRP specimens.

CONVERSION OF NDE INSPECTION SIGNALS INTO AURAL SIGNALS FOR ENHANCED MATERIAL CHARACTERIZATION

Glenn M. Light, Amos E. Holt,
Kent D. Polk, and William T. Clayton

Southwest Research Institute
6220 Culebra Road
San Antonio, TX, U.S.A.

INTRODUCTION

Nondestructive evaluation (NDE) technologies are often used to characterize materials for quality. Usually they rely on visual interpretation of electronic signals generated from conventional NDE instrumentation [e.g., ultrasonic testing (UT) A-scan trace on an oscilloscope or a Lissajous figure on an eddy current testing (ET) instrument]. In some inspections, such as ultrasonic wall thickness measurement, the signals can be easily interpreted. In many other inspections, however, the signals observed on the oscilloscope have low signal-to-noise and are difficult to visually interpret.

One method to improve the accuracy of data evaluation would be to utilize other human senses such as the ears to simultaneously interpret the data. In the past, cursory attempts have been made to employ this concept. NDE instruments have taken electronic signals and generated alarms if the time signal appeared in a certain time window and had an amplitude above some threshold. This type of technology, however, was only an "off/on" process and did not make use of any information inherently contained in the NDE signal except amplitude (UT/ET) or phase (ET). Inspectors believed that if the material characterization information contained in the NDE signal could be converted into audio (aural) signals while keeping the content of the NDE signal intact, then the ear as well as the eye could be used to enhance interpretation of the data.

Although this aural technology has been applied to UT, ET, and acoustic emission,[1] only the concept, techniques, and application results of auralization of UT are presented.

AURALIZATION AND AURAL CONVERSION OF NDE SIGNALS

Since 1985, Southwest Research Institute (SwRI) has been conducting internally funded research on developing the capability to convert NDE signals from conventional NDE instrumentation into aural signals with the intent of improving the quality and accuracy of NDE inspections and material characterization. This paper discusses work done to convert visual inspection signals into aural signals for improved material characterization. Auditory studies show that principles used in listening perception are analogous to visual perception principles. These, in turn, are parallel to classic Gestalt perception[2-6] concepts used in psychology. The Gestalt concepts (1) interpret the perception processes by the relationships among all human senses and (2) allow each of the human senses its own unique way of interacting with the surrounding environment. Applying the Gestalt concepts to the aural perception process has yielded some important results. These include the following.

(1) All attributes of sound such as pitch, timbre, loudness, and frequency are not independent; that is, each aspect of sound production influences another.
(2) Sounds are automatically separated into discrete sound sources or auditory streams.
(3) Individual sound attributes are not normally perceived separately.
(4) Processes used to perceive speech, music, noise, and everyday environmental sound appear to be equivalent.
(5) Listener training and a priori knowledge of aural sources play an important role in successful identification of aural information.
(6) Human auditory system is very sensitive to changes in the sensory input and can detect and discriminate these changes from background noise. Evidence suggests that the auditory mechanisms that allow discrimination and detection of these changes are fundamental to auditory operations and even excel with certain types of noise stimuli.

These and other concepts were used to determine how to best use sound to represent NDE signal information from different inspection methods.

Conversion of UT Signals

Developing algorithms to convert high-frequency electronic signals from UT instruments into meaningful aural signals is the most difficult challenge. The task requires first an appropriate data acquisition and analysis system (DAAS). The prototype DAAS designed for this work uses a conventional UT instrument appropriately interfaced to the computer decision system, as illustrated in Figure 1.

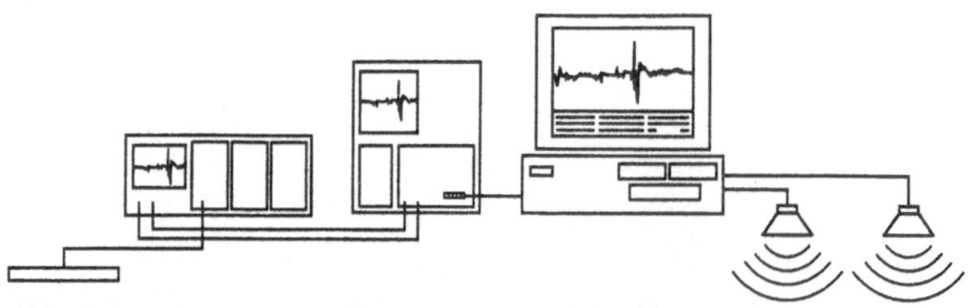

Figure 1. Aural NDE system configuration consisting of conventional ultrasonic equipment, signal digitizer transducer, and Amiga 2500 computer

The computer system is used to (1) digitize and analyze signals that could be developed into an aural/visual perception system and (2) improve the reliability of operator decision processes.

The selected computer was a realtime, icon-driven, modular, DAAS decision tool (software/hardware package). Its modular construction allows graphical selection and control of individual processing modules including gating, filtering, scaling, data source, and audio playback, depending on the specific design of the NDE process under study. Current programming modules allow selection of up to 16 channels of waveform data with the capability of changing the number of channels and some channel content dynamically. The inspector also can use one or two audio output channels for either mono or stereo (binaural) presentation of the aural information.

Auralization of Ultrasonic Signals

One basic form of aural conversion involves time dilation. This process increases the time interval between the digitized samples of UT information and a new waveform with frequencies in the audible range. For example, if a waveform is digitized at a 10-MHz sampling rate, then it is sampled every 0.1 microsecond (see Figure 2). By taking each of these sample points and increasing the time between them from 0.1 microsecond to 0.1 millisecond, the waveform appears to have a frequency 1000 times lower than the actual waveform before time dilation. Using the user-interface display, the operator can hear this time-dilated signal information through the two audio channels in mono or stereo. When presented in two separate channels, the operator can take advantage of binaural hearing capabilities.

To develop the aural UT system, a trained UT operator is asked to position a transducer on a test sample and locate a region with a signal of interest. The waveform (A-scan, amplitude-time) is acquired and digitized. It can be displayed three different ways for operator interpretation: on oscilloscope and computer screens and through two audio output channels. A typical UT computer display is shown in Figure 3. To form an initial set of data to begin algorithm development, signals from other areas, both good and defective, are then acquired and digitized

The modular elements of the system software are key to the developing technology because they allow the operator to produce a variety of waveform processing techniques. Using realtime interactive computer feedback, the operator can experiment with different features to enhance aural perception of the aural signal information. The acquired UT signal information is played back audibly in several different analysis forms so the operator can begin to discern differences aurally in the data obtained from good and defective regions.

A tone presentation option is also available, which translates a signal's time-of-arrival information into a representative tone. One use of this variable is to represent early signal arrivals with higher frequency tones. Another use is its application to gated portions of the waveform; that is, specific notes can be assigned to specific arrival times. The operator also can mentally relate specific notes to physical parameters of the part under inspection.

DISCUSSION OF RESULTS

During recent work at SwRI, the aural UT technique was used to characterize metals for corrosion, composite panels for debond and impact damage, and stainless steel piping for integranular stress corrosion cracking (IGSCC). The preliminary results obtained for each experiment are discussed in this section.

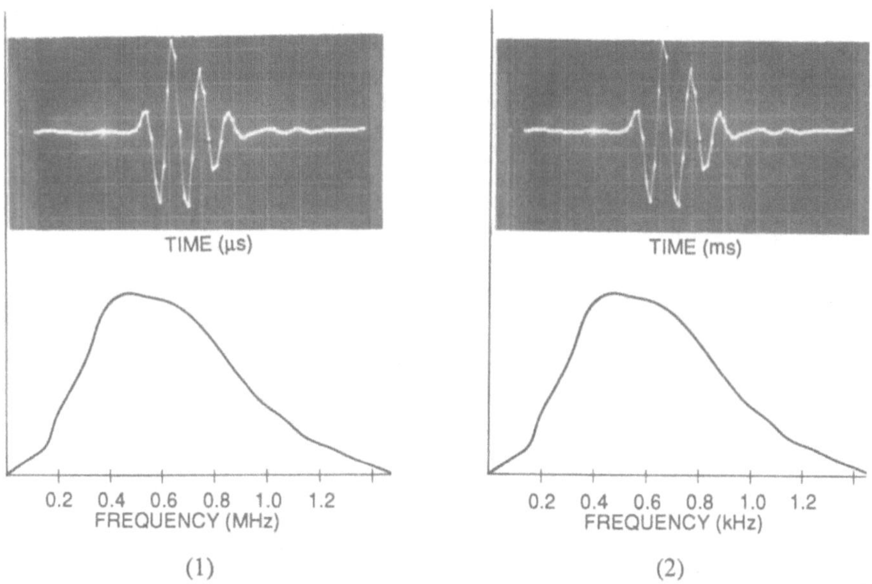

Figure 2. Illustration of a digitized signal being downscaled in frequency. Waveform (1) has points taken every 0.1 microsecond with a corresponding frequency of approximately 0.5 MHz. Waveform (2) is identical to (1) except the time between points has been time dilated or artificially stretched in the computer to 0.1 millisecond per point. The latter time corresponds to a frequency of approximately 0.5 kHz or 500 Hz, which is in the audible range.

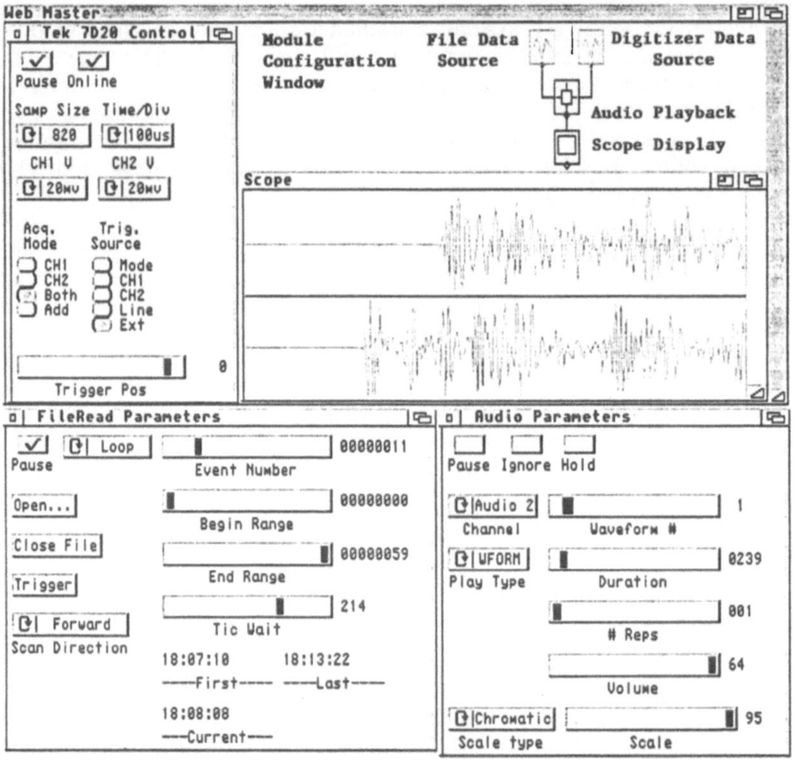

Figure 3. Typical user interface for the aural NDE system

Hidden Corrosion

Detection of hidden corrosion in aircraft structures is very difficult using conventional techniques. Figure 4 shows an ultrasonic A-scan of two regions from an aluminum wingskin mockup, with and without corrosion. The UT signals are very similar. By using multiple reflections from the internal surface of the material and converting that signal into aural signals by time dilation, the operator can easily hear the difference between good and corroded material. The UT waveform used for the time dilation is shown in Figure 5.

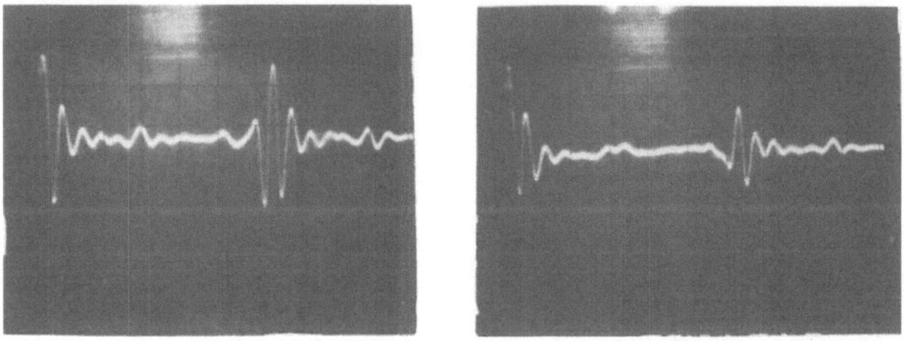

(1) (2)

Figure 4. Ultrasonic A-scans obtained from a wingskin mockup which has (1) no corrosion and (2) corrosion

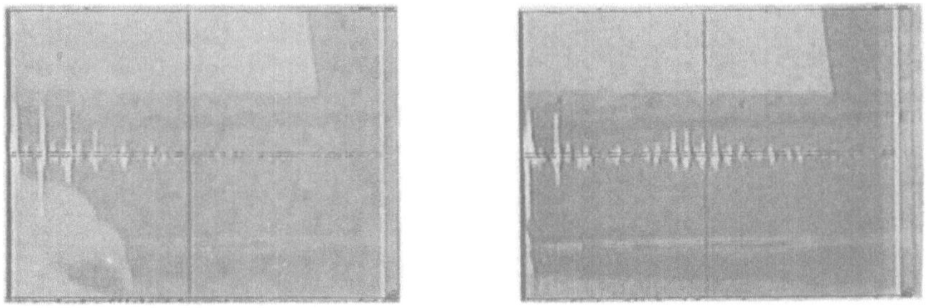

(1) (2)

Figure 5. Ultrasonic data that were time dilated into aural data and used to aurally determine (1) no corrosion and (2) corrosion

When both the A-scan and aural UT data are used simultaneously, the presence of the hidden corrosion is easily detected.

Characterization of Composite Material

In the aerospace industry, aircraft aluminum structure is often replaced by composite material because of the latter's stiffness and light weight. The composite material, adhesively attached to the aluminum parts, must be well bonded for aircraft safety. The quality of the bond, therefore, has to be verified throughout service life. Composite material also can lose much of its strength if its layers delaminate or the matrix cracks. These two problems often are caused by an impact of the aircraft with objects such as runway debris, hail, or birds.

To inspect composites, data from conventional UT using pulse-echo contact techniques can be difficult to interpret. Examples of 0-degree, longitudinal, pulse-echo signals from a composite panel containing good and defective regions are shown in Figure 6 for debond and Figure 7 for impact damage. When the UT signal was time dilated to audible signals using the UT waveforms shown in Figures 8 and 9, the difference between good and defective material was easily heard. Again, by using both the UT and aural data, the quality of the inspection was increased.

Detection and Discrimination of IGSCC

Detection and discrimination of intergranular stress corrosion cracks (IGSCC) in stainless steel recirculation piping in boiling water reactors (BWR) is of major importance to the nuclear power-plant industry. IGSCC is particularly difficult to discriminate correctly from root geometry because IGSCC develops in the heat-affected zone (HAZ) of the weld at the root region. Examples of root geometry and IGSCC signals obtained using a 1.5-MHz, 45-degree, shear-wave transducer are shown in Figure 10. The Electric Power Research Institute (EPRI) NDE Center has been working with ultrasonic inspectors that have had many years' experience in training and testing their abilities to detect and discriminate IGSCC successfully. The first-time failure rate on this test by these highly trained inspectors is as high as 60 percent.[7] This attests to how difficult evaluating stainless steel material for IGSCC is.

One goal of the SwRI aural NDE development is to improve detection and discrimination of IGSCC. Only preliminary work has been conducted so far, but results seem promising. By properly setting up the auralization system, both mono and stereo data can be obtained. Two channels of the four-channel aural system have been used to present the data to the inspector. One channel has been separated into each ear. To test this concept for IGSCC, a special transducer fixture was developed that contained three transducers. One provided for 1.5-MHz, 45-degree, shear-wave pulse-echo data to be obtained normal to the weld, and two for 1.5-MHz, 45-degree, shear-wave received-only data to be obtained at angles between approximately 20 and 45 degrees of normal to the weld. Figure 11 shows data obtained using this fixture on a pipe sample with actual IGSCC. These data were time dilated and separated into right and left channels on the audio output presented to the inspector. The discrimination of IGSCC was discernible.

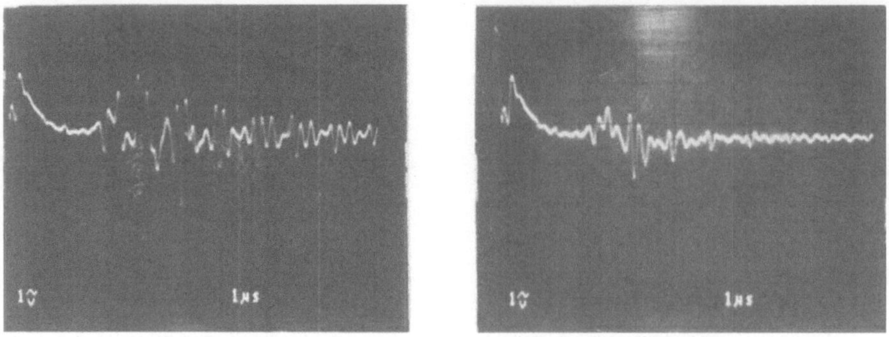

(1) (2)

Figure 6. Ultrasonic A-scan data obtained from a composite-to-aluminum test sample containing (1) no defects and (2) debonds using conventional 0-degree longitudinal ultrasonic inspection techniques

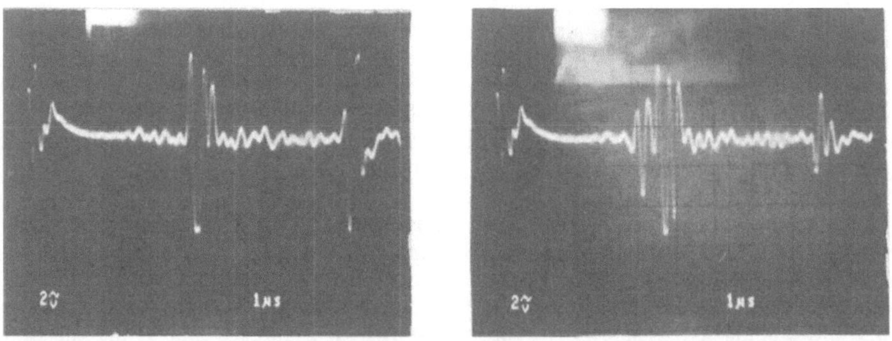

(1) (2)

Figure 7. Ultrasonic A-scan data obtained from a composite test sample containing (1) no defects and (2) impact damage using conventional 0-degree longitudinal ultrasonic inspection techniques

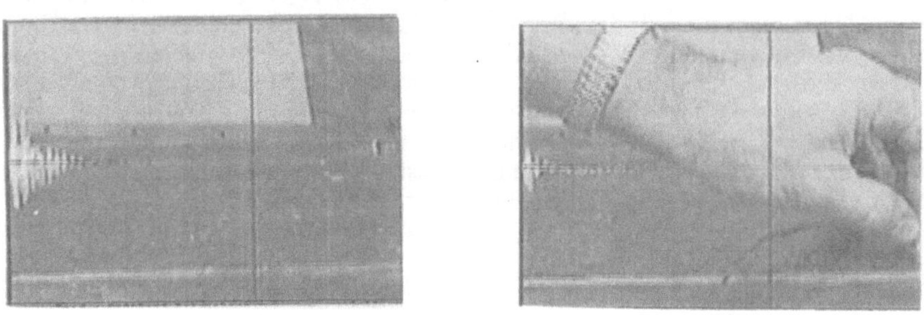

(1) (2)

Figure 8. Ultrasonic data obtained from a composite-to-aluminum test sample containing (1) no defects and (2) impact damage, which were time dilated into aural data so that the differences could be audibly detected

CONCLUSIONS AND FUTURE WORK

An inspection method has been developed by SwRI to provide simultaneously to the eye and ears of an inspector material-characterization information obtained through

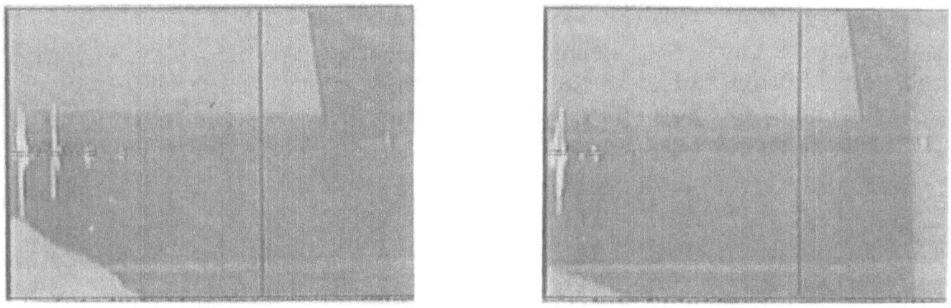

(1) (2)

Figure 9. Ultrasonic data obtained from a composite test sample containing (1) no defects and (2) impact damage, which were time dilated into aural data so that the differences could be audibly detected

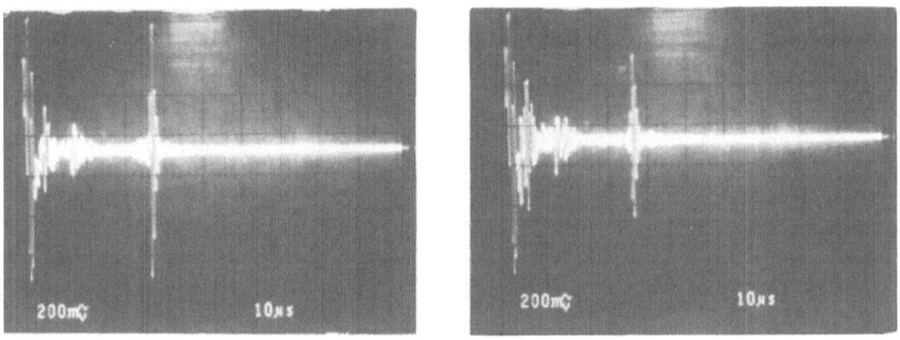

(1) (2)

Figure 10. Ultrasonic A-scan data from (1) root geometry and (2) IGSCC obtained using a 1.5-MHz, 45-degree, shear-wave transducer

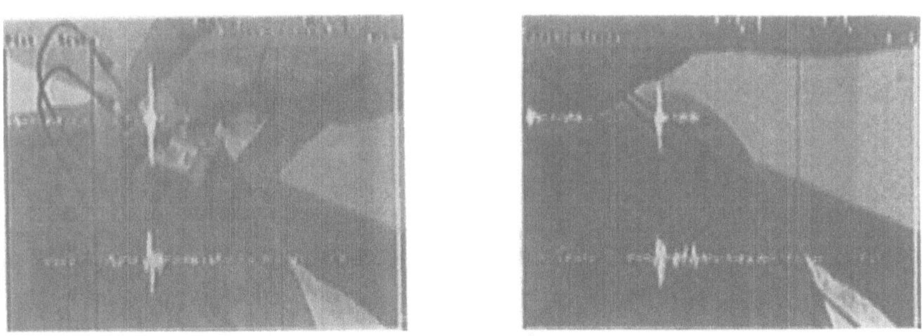

(1) (2)

Figure 11. Ultrasonic data obtained using the SwRI-designed fixture on a pipe sample with (1) geometry and (2) actual IGSCC

UT techniques. This method converts the actual ultrasonic signal into a aural signal so that it can be presented to the ear of the inspector in conjunction with the visual signal. The method, called Aural NDE, has been successfully demonstrated for ultrasonics, eddy current, and acoustic emission.[1] This paper illustrated uses of the Aural NDE method for ultrasonic detection and discrimination of hidden corrosion in aerospace material, delaminations and impact damage in composites, and IGSCC in stainless steel piping.

Aural NDE should play a major role in significantly improving the quality of NDE inspections performed in a wide variety of industries.

SwRI plans to transfer this technology to the power and aerospace industries for ultrasonics, eddy current, and acoustic emission applications. Successful transfer will involve development of appropriate data acquisition procedures and aural conversion techniques specific to each class of inspection.

REFERENCES

1. K. D. Polk, G. M. Light, R. H. Peterson, and J. L. Jackson, "Investigation of Audible Recognition Benefits for Ultrasonic Echo Detection and Characterization," Final Report, 17-1400-609 and 17-3131-606 (extension), Southwest Research Institute, San Antonio, Texas (1990).
2. "The Comparative Psychology of Audition: Perceiving Complex Sounds," R. J. Dooling and S. H. Hulse, eds., L. Erlbaum Associates, Hillsdale, New Jersey (1989).
3. S. Handel, "Listening: An Introduction to the Perception of Auditory Events," MIT Press, Cambridge, Massachusetts (1989).
4. R. M. Warren, "Auditory Perception: A New Synthesis," Pergamon Press, New York (1982).
5. B. C. J. Moore, "Introduction to the Psychology of Hearing," University Park Press, Baltimore, Maryland (1977).
6. R. M. Warren, "Auditory Perception: A New Synthesis," Pergamon Press, New York (1982).
7. H. Stephens, Training activities at the NDE Center, *in*: "Nondestructive Evaluation Research Progress in 1992: Proceedings: Thirteenth Annual EPRI NDE Information Meeting," NP-7560-S, Electric Power Research Institute, Palo Alto, California (1992).

DEVELOPMENT OF REAL TIME MONITORING OF DAMAGE BY ACOUSTO-ULTRASONIC NDE TECHNIQUE TO STUDY FAILURE MECHANISMS IN SiC/CAS CERAMIC COMPOSITES

Anil Tiwari and Edmund G. Henneke II

Engineering Science and Mechanics Department,
Virginia Polytechnic Institute & State University, Blacksburg, VA

INTRODUCTION

Future materials used in fuel efficient engines and certain space power applications must be able to withstand higher operating temperatures than required for present materials. Furthermore, future materials should offer the distinct advantages of higher strength to weight ratio, higher stiffness to weight ratio and better damage tolerance. Although ceramic materials possess these attractive features, they have inherent shortcomings limiting their usefulness in many applications. Poor material reliability make these materials unacceptable for use, even though these materials have excellent mechanical integrity and chemical stability at high temperatures. There is a need to develop NDE techniques that can assess and quantify the damage present in ceramic composites and relate a quantifiable parameter directly to strength or long time serviceability of the component. The present challenge in the field of NDE is to assess initial integrity and also damage accumulation and residual strength during service.

The present research effort is directed towards developing a near real time acousto-ultrasonic (AU) nondestructive evaluation (NDE) tool to study the failure mechanisms of ceramic composites. Recent work has verified the capability of the AU technique to assess the damage state in SiC/CAS ceramic composites. The development of near real time continuous monitoring of AU parameters under quasi-static ramp loading in tension has been accomplished to provide continuous monitoring of damage initiation and progression of unidirectional and cross-ply SiC/CAS ceramic composite specimens [1]. This paper will present our results to-date of the application of the AU technique to study failure mechanisms of SiC/CAS ceramic composite under fatigue loading.

Acoustic emission and ultrasonics have been widely used to evaluate material performance. Vary and co-workers [2-4] developed a technique, viz., the acousto-ultrasonic (AU) method, as a hybrid of acoustic emission and ultrasonics. Many studies [1-10] reveal that there is a correlation between stress wave propagation characteristics determined by AU and various mechanical properties of the material. The AU technique introduces an ultrasonic signal into a specimen and detects it using AE methods. A conceptual configuration of an AU set-up is shown in Fig. 1. One of the major advantages of this technique is that it gives an integrated effect of overall flaw/damage present in the structure between the transmitting and receiving transducers.

Early researchers analyzed the AU signal in the time domain. An alternative approach for analyzing the received AU signal has been developed in our laboratory [5-10]. This approach converts the received digitized time-voltage signal into an amplitude-frequency spectrum by means of a Hartley's transform algorithm [11]. Various statistical moments of the frequency spectrum are then calculated and defined as different stress wave factors (SWF). The zeroth moment of the frequency spectrum of the received AU signal is the area under the spectral density distribution [12]. Since this is indicative of the total energy content of the received signal, the SWF(M_0) value is a measure of stress wave energy transmission. The SWF(M_0) values rate the efficiency with which dynamic strain energy transfer takes place in the material between the two AU transducers. Studies have shown that the values of these moments are affected by damage/flaws present in the structure in a particular manner, such that they can be correlated to changes in mechanical properties.

Researchers [1-10] have shown acousto-ultrasonic (AU) parameters to be more sensitive than stiffness measurements as a means of tracking damage. Although the physical understanding of the effect of damage on wave propagation and on the AU signal is not yet completely understood, researchers have suggested that the local stress state or displacement field is modified by the damage present. By recording information on stress wave propagation characteristics one can hope to obtain information about the value of stress concentration or impending failure modes. Kiernan et. al. [8] gave a general physical explanation for AU results in composite plates based on elasticity solutions, Lamb wave theory and through-the-thickness transverse resonance (TTR). Velocity measurements and frequency content of AU signals were compared to Lamb wave theory predictions. Higher order Lamb waves were identified as a dominant mode of propagation. Tiwari, et al. [9] have used the moments method for assuring adhesive bond quality.

REAL TIME AU

Recent work has verified the capability of the AU technique to assess the damage state in SiC/CAS ceramic composite [1]. Continuous monitoring of damage initiation and progression under quasi-static ramp loading in tension to failure of unidirectional SiC/CAS ceramic composite specimens has been accomplished by monitoring real time AU parameters [10].

The received AU signals are stored every second during cyclic loading. The AU signals are averaged over ten load cycles and hence represent the damage state over those cycles. The AU parameter from the averaged signal are normalized with

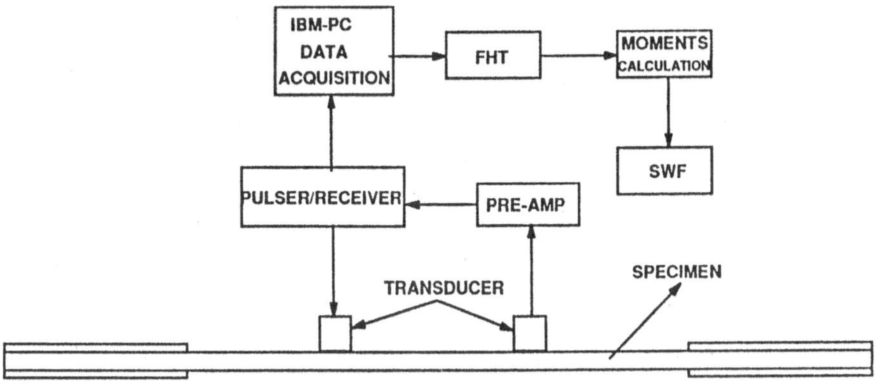

Figure 1. Conceptual configuration of AU set-up.

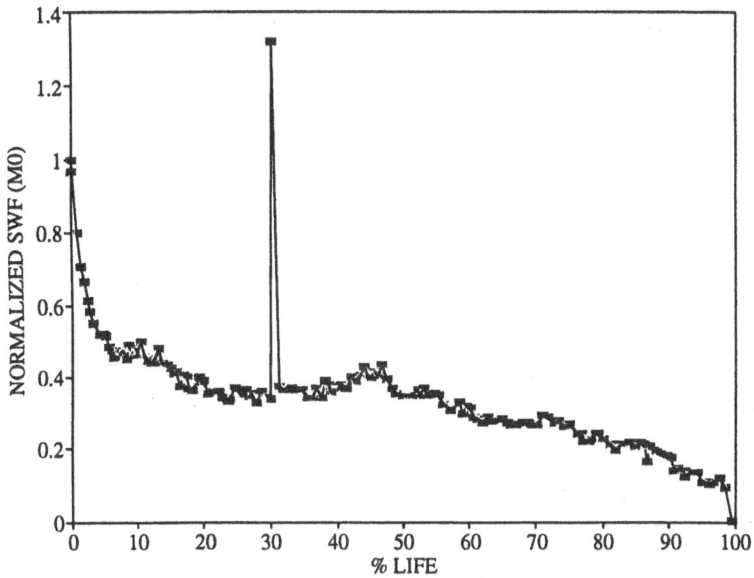

Figure 2. Graph of normalized SWF (M0) vs % life of sample C18. (Fatigue at 90 % UTS and R = 0.1)

the initial AU parameter obtained from the first load cycle. The normalized AU parameter represents the damage level at each cycle. Real time AU can provide better insight into failure mechanisms in materials subjected to cyclic loading. The work reported in this paper includes AU results obtained from cross-ply SiC/CAS specimens.

EXPERIMENTAL SET-UP

A block diagram of the experimental set up for AU is shown in Fig 1. Ultrasonic pulses were introduced into the mechanically loaded specimen through a broadband transducer, Panametrics model V133RM (2.25 Mhz/6.35 mm). The couplant used was Sonotrace-30. A Panametrics model 5052-AU pulser/receiver was used to generate and receive the ultrasonic signal. The receiving transducer was the same type as the transmitter.

A transducer holder was designed and fabricated to maintain a constant distance of 38.1 mm (1.5") between the centers of the AU transducers. The transducers were mounted on the opposite sides of the specimen from the extensometer. A constant pressure was maintained at the center of the transducer holder throughout the test with rubber bands. The received signal was amplified with gains set for each individual sample. After a group of signals was saved, the signals were input to a Fortran program developed at Virginia Tech [5-10] to calculate the frequency spectrum and its various moments. The AU data were normalized with respect to the initial value and plotted against stress level. The zeroth moment was found to be the most sensitive parameter of the ones studied for detection of matrix cracks and is referred to here as AU stress wave factor (SWF).

Cross-ply ceramic composites (SiC/CAS) were made by Corning, Incorporated, with approximately 40% fiber content and less than 1% porosity. Dimensions of a typical specimen were 152.4 mm x 12.7 mm x 2.92 mm. End tabs of glass-epoxy (12.7 mm x 2.92 mm) were glued to each specimen using FM-300 adhesive and cured at 300^0 F. End tabs were attached to prevent crushing damage when held in the grips of the testing machine. Samples C11, C12 and C18 were cross-ply $[0/90]_{4S}$. Samples C12 and C18 were subjected to cyclic loading (R = 0.1 ; f = 10 Hz) @ 90% of ultimate strength and sample C11 was subjected to cyclic loading (R = 0.1 ; f = 10 Hz) @ 60% of ultimate strength. Damage accumulation was continuously monitored by the real time AU system. Testing of samples C11 and C12 were interrupted at different stages of the fatigue test and reloaded in quasi static tension to 100 lbs to measure the static stiffness at that stage. Edge replicas were also taken along with static stiffness measurements to make a qualitative comparison of the damage and thereby complement and verify AU results.

RESULTS AND DISCUSSION

Figure 2 shows the graph of normalized SWF (M0) vs % life of sample C18. This specimen was subjected to cyclic load (f = 10 Hz ; σ_{max} = 90% ultimate strength and R = 0.1). The specimen failed in the gage section after 4053 cycles and a significant amount of fiber pullout was observed. The SWF (M0) values decreased by 60% in the first 8% of its fatigue life. The majority of the matrix cracking had

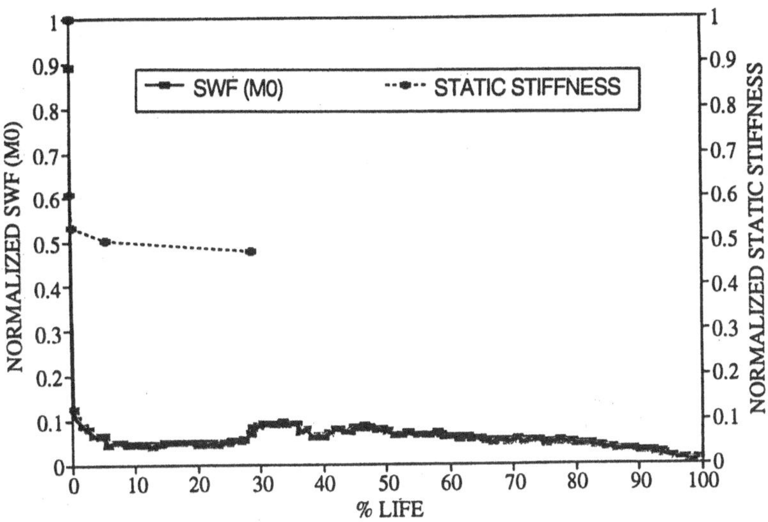

Figure 3. Graph of normalized SWF (M0) and static stiffness vs % life of sample C12. (Fatigue at 90 % UTS and R = 0.1)

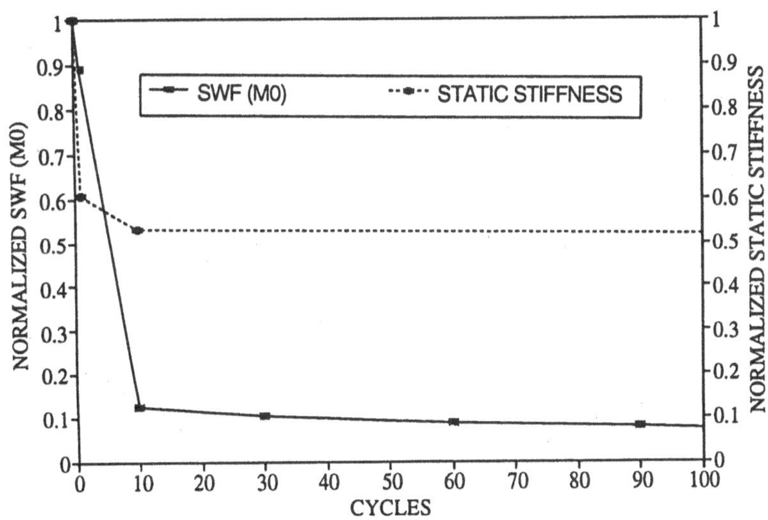

Figure 4. Graph of normalized SWF (M0) and static stiffness vs cycles of sample C12. (Fatigue at 90 % UTS and R = 0.1)

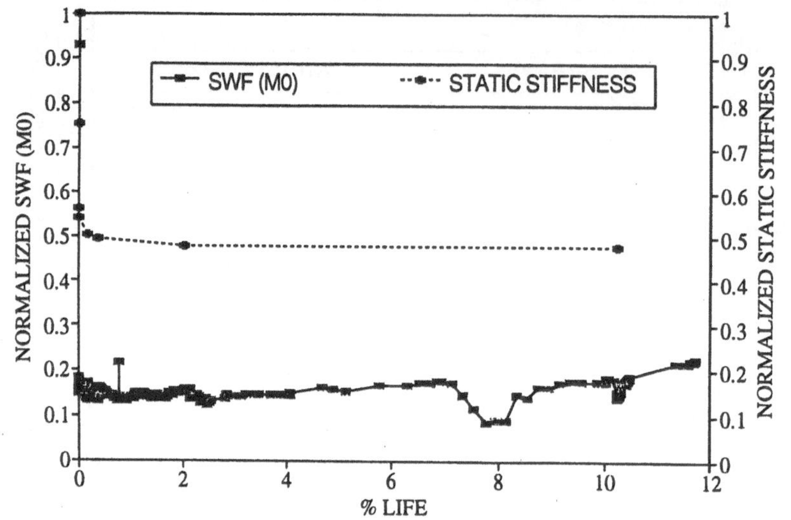

Figure 5. Graph of normalized SWF (M0) and static stiffness vs % life of sample C11. (Fatigue at 60 % UTS and R = 0.1)

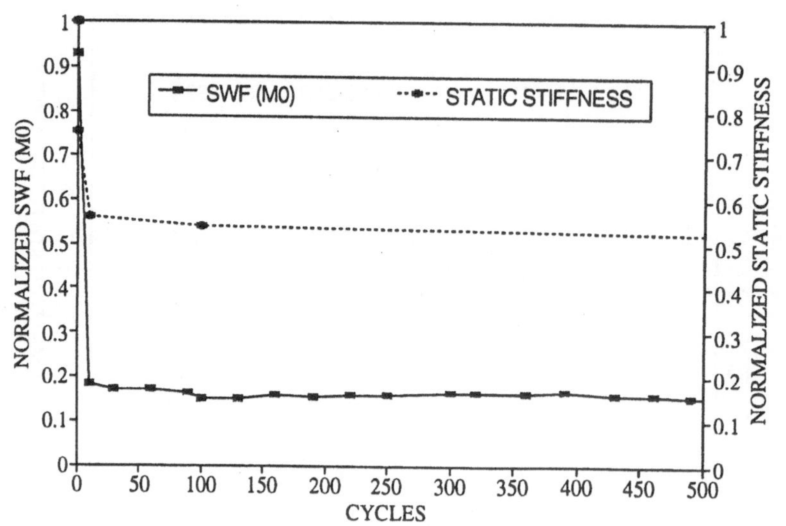

Figure 6. Graph of normalized SWF (M0) and static stiffness vs cycles of sample C11. (Fatigue at 60 % UTS and R = 0.1)

already taken place by this time. A very slow decline in SWF values was observed till just prior to failure. The peak seen at 30% of the fatigue life can be attributed to fiber breakage. The receiving AU transducer acts as an AE transducer receiving additional high energy signals from the fiber breakage.

Figure 3 shows the graph of normalized SWF (M0) and normalized static stiffness vs % of life of sample C12. The SWF values drop by 95% at 8% of the fatigue life. The sharp drop is caused by matrix cracking occurring perpendicular to the load axis as observed in sample C18 and other brittle matrix composites. The SWF values show a gradual decline in their values until failure at 3482 cycles. Significant fiber pullout was also observed at the failed section. The normalized static stiffness follows the same trend as the AU data. Figure 4 shows an enlarged view till 100 cycles to better illustrate this point.

Sample C11 was subjected to cyclic loading at 60 % of its ultimate strength. Figure 5 shows the graph of normalized SWF (M0) and normalized static stiffness vs % of life of sample C11. The static stiffness shows a trend similar to AU data. An enlarged view showing the variation of SWF (M0) and static stiffness till 500 cycles is plotted in Fig. 6. The majority of the damage occurred in the first few cycles.

The failure characteristics of cross-ply SiC/CAS for the loadings described here consist of three major regions. The first region shows a sharp decline in the static stiffness and real time AU parameter caused by matrix cracks. The load amplitude for 60 % ultimate strength is well above the onset of matrix cracking for these laminates and hence matrix cracking occurs at the very first cycle for all these samples. The mid region is a plateau with no observable damage. The damage mechanisms in this region are dominated by the failure mechanisms of the fibers and their statistical behavior.

CONCLUSIONS AND FUTURE WORK

The real time AU parameter is sensitive to damage progression during fatigue. SWF (M0) can be extrapolated to predict the residual stiffness of the specimen. The predicted stiffness or SWF (M0) itself can be used in existing models to predict life and residual strength of the composites.

This research effort has been directed towards developing a near real time acousto-ultrasonic NDE technique to monitor damage progression and thereby study the failure mechanisms in ceramic composites subjected to fatigue loading. The real time AU method is extremely sensitive in monitoring damage for fatigue loading. Edge replicas and static stiffness taken at regular intervals complement and verify the AU results. The real time AU parameter SWF (M0) can be used in critical element modelling philosophy to study failure mechanisms of ceramic composites. Future work includes developing laser based generation and detection of AU stress waves for monitoring damage development at elevated temperatures.

ACKNOWLEDGEMENTS

This work has been supported by the Structural Integrity Branch of NASA

Lewis Research Center, Cleveland. The authors acknowledge the support provided by Alex Vary and his co-workers at NASA Lewis.

REFERENCES

1. A. Tiwari, E. G. Henneke II, A. Vary and A. Chulya, " In-situ Acousto-ultrasonic Technique to Monitor Damage in Ceramic Composites," Fifth Annual HITEMP conference proceedings, Cleveland, Ohio, Oct. 27-28, 1992.

2. A. Vary, "Acousto-ultrasonic characterization of fiber reinforced composites", Materials Evaluation Vol. 40, pp. 650-662, 1982.

3. A. Vary, and K.J.Bowles,"An ultrasonic-acoustic technique for non-destructive evaluation of fiber composite quality,"Polymer Engineering and Science, Vol. 19, pp. 373-376, 1979.

4. A. Vary, and R.F.Lark,"Correlation of fiber composite tensile strength with the ultrasonic stress wave factor," Journal of Testing and Evaluation, Vol. 7, pp. 185-191, 1979.

5. E.G.Henneke II, J.C.Duke Jr., W.W.Stinchcomb, A.Gowda, and A.Lemascon,"A study of stress wave factor technique for the characterization of composite materials," NASA Contractor Report-3670, Feb.1983.

6. Grosskopf P.P. and Duke J.C. Jr., " Mechanical Behavior of a ceramic matrix composite", NASA CR 187072, Feb. 1991.

7. J.C.Duke Jr., E.G.Henneke II, M.T.Kiernan, and P.P.Grosskopf, "A study of stress wave factor technique for evaluation of composite materials," NASA Contractor Report 4195, Jan. 1989.

8. M.T.Kiernan, "A physical model for the acousto-ultrasonic method," Doctoral Dissertation, College of Engineering, Virginia Polytechnic Institute and State University, Blacksburg, VA, U.S.A., Aug. 1989.

9. A.Tiwari, E.G.Henneke II and J.C.Duke Jr., "Acousto-ultrasonic (AU) Technique for Assuring Adhesive Bond Quality," Journal of Adhesion, Vol. 34, 1991, pp. 1-15.

10. A. Tiwari and E. G. Henneke II, " Real Time Acousto-ultrasonic NDE Technique to Monitor Damage in SiC/CAS Ceramic Composites under dynamic loads," presented at second symposium on cyclic deformation fracture and NDE of advanced materials, Nov. 16-17, 1992, Miami, Florida, ASTM STP (in press).

11. R.N.Bracewell, <u>The Fourier Transform and its Applications</u>, second rev.ed.(McGraw-Hill, New York, 1986).

12. R.Talreja, "On fatigue life under stationary gaussian random loads, Engineering Fracture Mechanics, Vol. 5, pp. 993-1007, 1973.80.

QUANTITATIVE NONDESTRUCTIVE EVALUATION OF CERAMIC FIBERS

Renee M. Kent

University of Dayton Research Institute
Dayton, OH 45469

INTRODUCTION

Brittle ceramic materials typically require nonintrusive testing and evaluation. Though ceramics exhibit desirable properties such as high specific strength, high elastic modulus, and structural integrity at elevated temperatures which make them viable candidates for high temperature applications, they have a characteristically low fracture toughness. Therefore, current trends in materials research and development have emphasized the advancement of innovative methods for engineering the microstructure of these materials, such as developing ceramic fiber reinforced ceramic composites, to provide enhanced strength and fracture toughness. Concurrent development of improved nonintrusive measurement methodologies for the quantitative characterization of the ceramic fibers and composites has been essential to accurately and reliably predict the performance of these materials under their operating conditions.

Typical ceramic fibers have diameters on the order of tens of microns and exhibit tensile strains to 1 percent. Therefore, effective nonintrusive techniques for the investigation of small diameter ceramic fibers require high resolution, accuracy, and precision.

The scanning laser acoustic microscope (SLAM) can be effectively used to reliably measure the uniaxial tensile strain and longitudinal elastic modulus of small diameter (to 15 μm) ceramic fibers [1]. Tensile strain measurement using SLAM is based on the analysis of the acousto-optical interaction at the specimen interface [2,3].

THEORY

The operation of the SLAM system is described in detail in the literature [3,4]. Fundamentally, a piezoelectric transducer is used to generate a 100 MHz longitudinal wave in the fiber test specimen. The ultrasonic wave propagating through the specimen is altered as a function of interaction with elastic inhomogeneities in the specimen and a surface undulation, characteristic of the internal structure of the material, is created. This surface

Nondestructive Characterization of Materials VI
Edited by R.E. Green, Jr. *et al.*, Plenum Press, New York, 1994

displacement is detected by means of an optical wave generated by a HeNe laser (wavelength 632.8 nm; spot size 22 μm) scanned over the surface of the test specimen. The optical wave is modulated by the ultrasonic wave undulation. Demodulation and analysis is performed internally with a knife edge detection system [5]. The detected acousto-optical response is added to an electronically generated phase reference signal to produce interference fringes characteristic of the structure of the material test specimen.

A uniaxial tensile stress applied to the fiber test specimen results in a deformation of the modulation of the optical wave by the ultrasonic phase front [6]. The modulation deformation is a function of the linear superposition of the strain induced in the test specimens due to the elastic stress from the ultrasonic wave and the strains induced by the applied stress. The magnitude of the modulation is mathematically described by a deformation of the indicatrix of the material as [6,7]

$$\Delta B_j = P_{jk} \epsilon_k \tag{1}$$

where ΔB_j represents the change in the relative dielectric permeability tensor, P_{jk} is a strain coefficient determined by the crystal class symmetry of the material [8], and ε_k is the strain. This formulation assumes that the strains produced in the materials are elastic and that there is a negligible rotation component to the strain.. The resulting response is a sinusoidal modulation of the detected optical wave which effects a shift in the observed interference fringes.

In order to discriminate between strain due to the ultrasonics and the strain induced by the applied stress, the relationship between the internal strain in the system and the magnitude of the fringe shift is found by determining the difference in the acousto-optic path length between the two stress states of interest. As strain accumulates, the path difference changes due to macroscopic changes in the length of the stressed region and due to alteration of the index modulation described by equation (1). The tensile strain due to the total path difference can then be computed using the relation [9]

$$\epsilon = \frac{Km\Lambda}{L} \tag{2}$$

where K is a calibration constant which can be computed from material parameters [9,10], m is the normalized interference fringe shift, Λ is the effective wavelength of the ultrasonic wave propagation, and L is the effective gauge length of the sample.

EXPERIMENTAL PROCEDURE

Selected ceramic fibers (Nicalon SiC, Carborundum SiC, Saphikon Al_2O_3) were investigated for their in situ tensile strain and elastic modulus using the scanning laser acoustic microscope (SLAM). Each fiber was epoxied to a one mil thick tab assembly such that the test specimen gauge length was one inch, as shown in Figure 1. The tab assembly was designed such that, after placing the specimen in the SLAM system, the support tabs on either side of the fiber could be cut so that the applied load was effectively transferred to the fiber test specimen. The entire tab / fiber test specimen assembly was placed in the SLAM system (operating at 100 MHz, 10 degree insonification) as shown in Figure 2. Three arbitrarily selected static loads (less than the mean tensile strength of the fiber) were applied.

At each stress level, the normalized fringe shift was measured and used to calculate the tensile strain induced in the fiber. The resulting data were analyzed using a complete randomized block experimental design of fifteen individual Nicalon SiC fibers. The elastic modulus calculated from the stress-strain data at each point formed a three by fifteen experimental matrix. In this design, the fiber to fiber variation was normalized and the residual variance was used to estimate the measurement uncertainty. For each fiber, the strain data were plotted on stress-strain curves and the elastic modulus of the individual fibers was calculated by the method of linear regression.

The experiment was repeated using smaller statistical sample sets of Carborundum SiC (28 μm diameter) and Saphikon sapphire (140 μm) fibers.

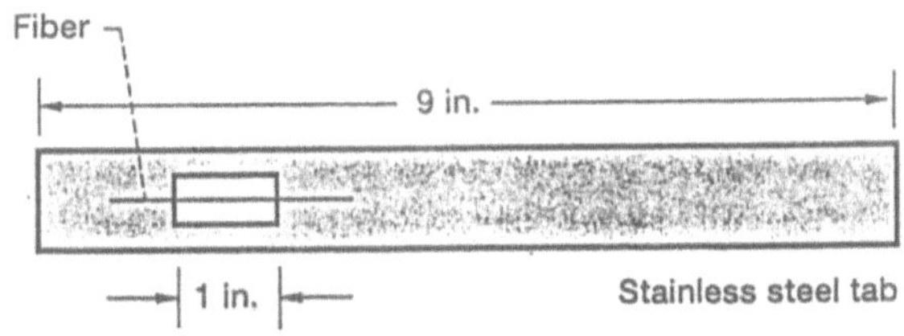

Figure 1 Fiber test specimen tab assembly.

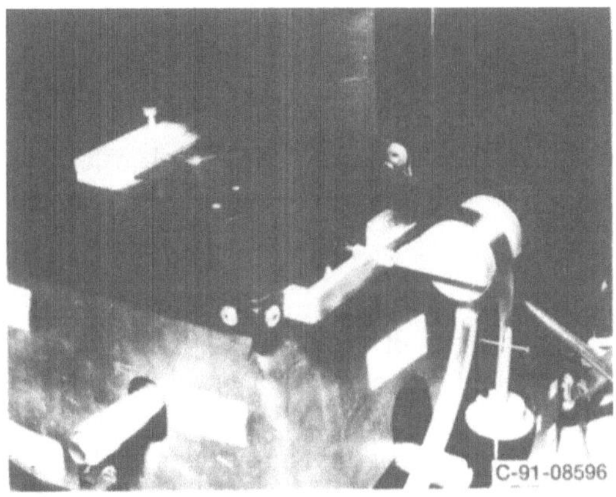

Figure 2 Experimental configuration for in situ strain measurement.

RESULTS AND DISCUSSION

The tensile strains and corresponding elastic modulus of fibers having diameters to 15 μm were determined in this work. An optical micrograph showing the typical interference fringe resulting from the signal response of the Nicalon fiber is shown in Figure 3.

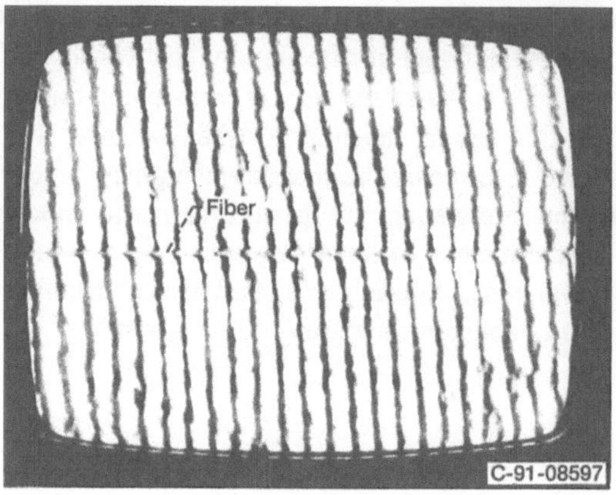

Figure 3 Interference fringes from Nicalon SiC fiber.

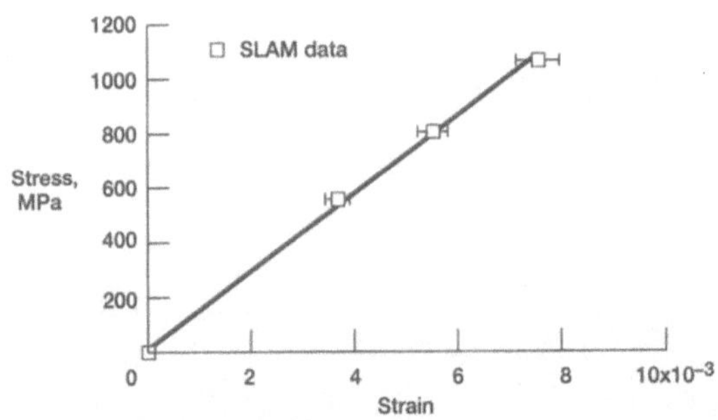

Figure 4 Stress-strain behavior for 15 μm Nicalon SiC fiber.

Figure 4 shows a typical plot of the empirically determined stress-strain behavior for an individual Nicalon fiber. The error bars shown were propagated from the measurement uncertainties and the errors due to assumptions about the available constants. The propagated measurement error was less than fiber percent of the absolute strain measurement. The elastic modulus measurements obtained from the empirical stress-strain data for the fifteen individual 15 μm diameter Nicalon fibers measured using the SLAM technique are tabulated in Table 1. For the fifteen fibers studied, the mean elastic modulus was 185.3 ± 2.6 GPa.

The standard deviation of the elastic modulus measurements was 22 GPa, approximately 12 percent about the mean. This standard deviation includes effects of the real fiber to fiber variations and random variations in the measurements. The complete randomized block analysis of the raw data was used to assess the statistical significance of any real differences between the fibers. This analysis indicated that there was a statistically significant variation in the measured elastic modulus from fiber to fiber. This variance was primarily attributed to subtle differences in the local microstructure due to nonstoichiometry

Table 1 Elastic modulus measurements for Nicalon SiC fibers

Fiber	**Modulus**
1	185.3
2	189.2
3	187.9
4	172.1
5	153.4
6	210.3
7	145.5
8	210.4
9	199.6
10	193.8
11	214.6
12	151.6
13	187.9
14	183.3
15	194.0
Mean ± Sigma	185.3 ± 22.6

inherent in Nicalon SiC fibers [11,12]. Variations in the measured modulus at the different load levels for a particular fiber was statistically insignificant and may be attributed to random measurement fluctuations. Using a scaled t-distribution as a reference distribution, the analysis indicated that the estimated standard error of the mean due solely to the measurement was 2.6 GPa at a 95 percent confidence level. This indicated that the measurement technique was reliable across a statistical sample set.

The stress-strain curves derived from the empirical data for the Carborundum SiC (28 μm diameter) and Saphikon Al_2O_3 (140 μm diameter) are shown in Figures 5 and 6, respectively. The linear regressional analysis of these curves indicated that the elastic modulus was 401.0 GPa for Carborundum SiC and 466.7 GPa for Saphikon Al_2O_3. In the case of Carborundum SiC, the measured elastic modulus was lower than the theoretical elastic modulus predicted for stoichiometric SiC (420.0 GPa). This difference was attributed to the presence of 3-4 percent internal porosity inherent to the particular fibers analyzed for this study. In the case of Saphikon sapphire fiber, the reported elastic modulus was measured for C-axis fibers, i.e. the longitudinal axis of the fibers were aligned with the c-axis of the hexagonally close-packed structure of sapphire. However, it should be noted

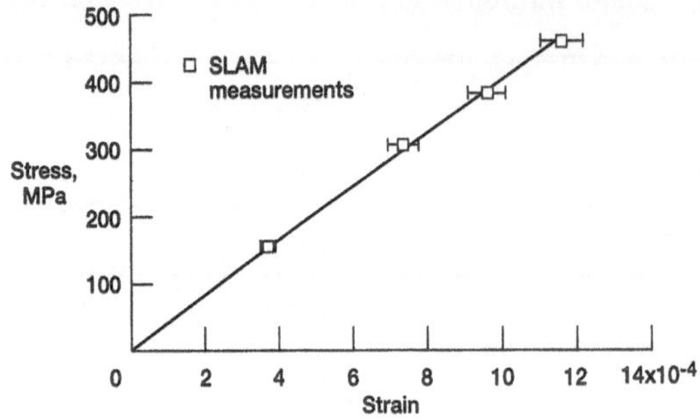

Figure 5 Stress-strain behavior for Saphikon Al_2O_3 fiber.

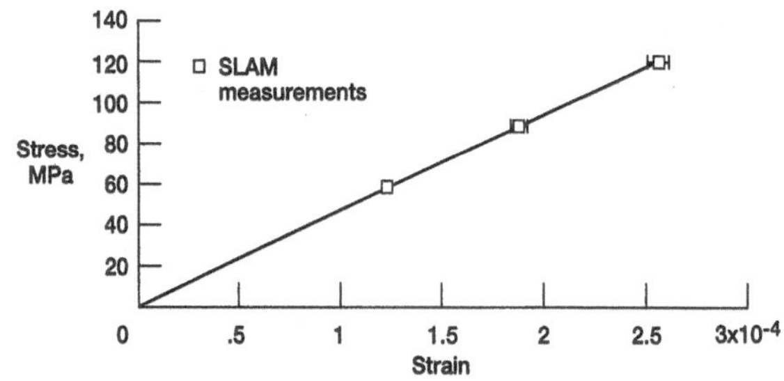

Figure 6 Stress-strain behavior for Carborundum SiC fiber.

that in the case when the sapphire fiber was misaligned by 20 degrees, the elastic modulus was empirically determined to be 420.0 GPa.

CONCLUSIONS

The scanning laser acoustic microscope provided a highly reliable and accurate methodology for measuring the in situ tensile strain (and elastic modulus) of ceramic fibers having diameters to 15 μm. The elastic modulus data generated was shown to be consistent and provide unique information regarding specific characteristics of the fiber material.

Extension of these analyses to elevated temperatures is a critical phase in the development of fibers. However, this presents one limitation for the application of SLAM to fiber analyses. The requirement for an acoustic couplant makes elevated temperature studies difficult using this technique. However development of laser generated ultrasonic nonintrusive methodologies for evaluating the material properties and behavior of small diameter fibers. Laser in /laser out ultrasonic capability is proving to be a potentially useful tool for the analysis of small diameter (15-140 μm) fibers at elevated temperatures. These experiments are described in further detail in the literature [13].

REFERENCES

1. R.M. Kent and A. Vary, "Tensile strain measurements of ceramic fibers using scanning laser acoustic microscopy", *NASA TM 105589* (1992).
2. A. Korpel, "Acousto-optics" in Appl. Solid State Sci., 3: Advances in Materials and Device Research, ed. R. Wolfe, New York: Academic Press (1972).
3. M. Born and E. Wolf, Principles of Optics 6th ed., New York: Pergamon Press, Inc. (1980).
4. E.P. Ippen, "Diffraction of Light by Surface Acoustic Waves", *Proc. IEEE*, 55 (1967).
5. R. Adler, A. Korpel, and P. Desmares, *IEEE Trans. Sonics Ultra.*, SU-15 (1968).
6. L.W. Kessler and D.E. Yuhas, "Principles and analytical capabilities of the scanning laser acoustic microscope", *Scanning Electron Microscopy*, 1 (1978).
7. S.F. Nye, Physical Properties of Crystals, London: Oxford Press (1954).
8. J.M. O'Hare, "Optical Radiation and Matter", internal publication, University of Dayton, Dayton, OH (1988).
9. R.M. Kent, "Internal Strain Analysis of Ceramics using Scanning Laser Acoustic Microscopy", *NASA CR 191118* (1993).
10. J.A. Powell, "Refractive Index and Birefringence of 2H Silicon Carbide", *NASA TN D-6635* (1972).
11. F. Hurwitz, NASA Lewis Research Center, Cleveland, OH (1991), personal communication.
12. J.A. DiCarlo, NASA Lewis Research Center, Cleveland, OH (1991), personal communication.
13. P. Stibich, D. Stubbs, and R.M. Kent, "Load-displacement and ultrasonic methods for determining the elastic modulus of SiC fibers", submitted to the *J. Exp. Mech.* (1993).

EVALUATION OF MICROFRACTURE DUE TO STRESS CORROSION OF PARTICULATE GLASS COMPOSITE BY ACOUSITC EMISSION

Kensuke Kageyama, Manabu Enoki, and Teruo Kishi

Research Center for Advanced Science and Technology
The University of Tokyo
4-6-1 Komaba, Meguro-ku, Tokyo 153, Japan

ABSTRACT

Acoustic emission method is useful for the characterization of the mechanical reliability of the materials controlled microfracture. We can detect the microcrack occurred in the brittle materials by using acoustic emission method. In our study, we tried to evaluate the change of microfracture due to stress corrosion of alumina particulate glass composite. Stress corrosion of glass causes a decrease in strength and fracture toughness. Alumina particulate glass composite is also subject to the influence of stress corrosion. The strengthening by the precipitation in addition to the particle has been tried, but the effect of precipitation is not clear. We measured AE signals of alumina particulate glass composite during the four point bending test in air, vacuum and water.

In the case of the non-precipitated material, we observed concentrated AE just before the break point in air. We observed dispersed AE in vacuum. However, we also observed dispersed AE in air in the case of the precipitated material and these microfracture processes were almost same in vacuum. It can be explained that the precipitation at the interface between alumina particle and glass matrix changed the unstable crack extension due to stress corrosion, which corresponds to concentrated AE, to the stable crack extension, which corresponds to dispersed AE. As a result, the precipitated material has also the higher strength and fracture toughness than the non-precipitated. Therefore we could clearly find the change of microfracture due to stress corrosion by acoustic emission method. Acoustic emission method is a good technique for the evaluation of the mechanical reliability against humidity in particulate and precipitated glass composite.

INTRODUCTION

There are few techniques for the evaluation of microfracture of brittle materials like ceramics and glass because microcracks are formed in the size of 1 to 100 μm and it is difficult to detect these microcracks.[1] Acoustic emission method is able to detect microcracking in a material and useful for monitoring the microfracture process and evaluation of the mechanical properties of brittle materials.[2] Quantitative evaluation of microfracture mechanism by AE method has been performed to know location, size and mode of microcracks by Enoki et al.[3] This method has been already applied to the evaluation of toughening mechanism of ceramics and ceramics matrix composite and the important data has been reported.[4] However, we have to consider the influence of humid environment on mechanical properties in the case of evaluating the microfracture mechanism of glass matrix composite because stress corrosion occurs at a crack tip in humid environment, the crack growth resistance of glass decreases and bending strength and fracture toughness also decrease.[5] The stress corrosion resistance is an important problem for the improvement of mechanical property in air, however, we found that the decrease of bending strength due to stress corrosion was improved by precipitation of anorthite in alumina particulate glass matrix composite.[6] Then we need to evaluate microscopic factors to control stress corrosion. The purpose of our study is to evaluate the microfracture due to stress corrosion by AE method and separate it from other microfractures. We carried out four point bending test in air and vacuum, and measured AE signals during the test. As a result, in non-precipitated samples, we observed only concentrated AE that corresponded to the unstable extension of microcrack due to stress corrosion in air. In the other hand, we observed dispersed AE as well as concentrated AE in precipitated samples. The reason is that microcrack extended into precipitated crystal (anorthite) and the bond at the crack tip was not cut by a water molecule. The extension of microcrack into anorthite corresponded to dispersed AE and we could evaluate the effect of the improvement due to precipitation by total events of dispersed AE.

EXPERIMENTS PROCEDURE

Materials

Table 1 shows the composition of the samples in this study. $PbO-Al_2O_3-SiO_2-B_2O_3$ glass was used as matrix glass for samples A1 and A2, and $CaO-Al_2O_3-SiO_2-B_2O_3$ glass as matrix glass for samples A3 and A4. These glasses were reduced by ball milling to powder of which the average diameter was 2 μm and were mixed with alumina powder (and anorthite in the case of A4) of which the average diameter was 1 μm. A slurry consisting of a mixture of glass, alumina and binder was cast by a docter-blade machine to make green sheets.[7] Green sheets were laminated, uniaxial pressed and sintered at 1173K in air. In samples A2 and A3, anorthite was precipitated at the interface between glass matrix and alumina particle by controlling alumina powder. Schematic morphology of anorthite in each sample is shown in Figure 1. Samples A1 had little precipitated anorthite, however, samples A2 and A3 had much precipitated anorthite at the interface between glass and alumina. There is also little precipitated anorthite in samples A4, however, anorthite particles was dispersed in glass matrix because anorthite powder were mixed at processing. To decide the proportions of precipitation of samples A1, A2 and A3, quantitative analysis by XRD was performed. We measured the (116) peak of α-alumina and the (114), (220), (040), (204), (004) and (200) peaks of anorthite, and then calculated the proportion of precipitation from the ratio of intensity of peaks on the basis of the peaks of A4 in which the proportion of anorthite was known. As anorthite increased, glass matrix decreased and a sintered body had the lower densification. The high porosity due to precipitation may cause to decrease bending strength, therefore, too much precipitation is not effective to improve mechanical properties.

Table 1. Composition of tested materials

	Anorthite (mass%)	Glass (mass%)	Alumina (mass%)	Density (Mg/m^3)
A1	2	44	54	3.18
A2	7	40	53	3.16
A3	13	37	50	2.91
A4	11	34	55	2.87

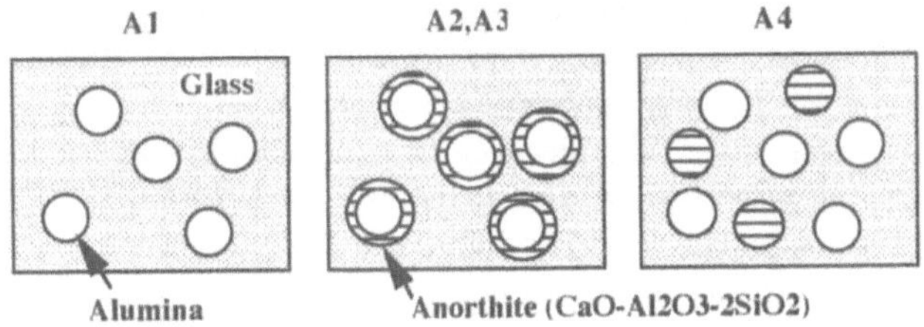

Figure 1. Schematic of the microstructure of alumina and anorthite.

Bending Test and AE Measurement

Samples (3x3x4mm) were cut and we carried out four point bending test in air and vacuum (10^{-5}Torr). Cross head speed was set to 0.01mm/min. Furthermore, we attached the two AE sensors on the both ends of specimens and measured AE signals during the test (Figure 2). The AE sensor with head amplifier has the gain of 54 dB (Fuji Ceramics, M31) and a resonance frequency of 300 kHz with about 2.5mm diameter. AE waveforms of two channels were recorded by the wave memory (NF, AE9620) with sampling rate of 50ns and 2kwords each channel. A threshold of 73dB and deadtime of 1ms were selected. Also conventional AE parameters such as event and amplitude with the load were analyzed by AE processor (NF, AE9600). Microcomputers (HP, model 216 and 310) were used to record the AE parameters and waveforms via GP-IB interface and performed one dimensional AE location.

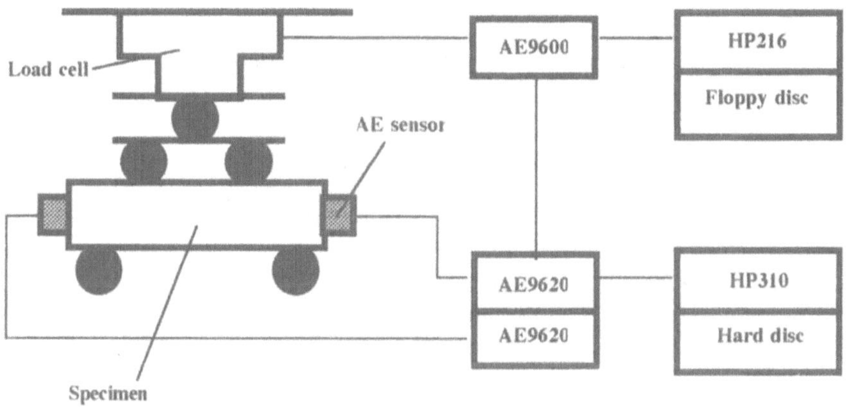

Figure 2. Schematic of the system for measuring AE.

RESULTS

Bending Strength

Figure 3 shows the relationship between average bending strength and proportion of anorthite crystal. In the case of samples A1, A2 and A3 in which anorthite was precipitated in the interface between glass and alumina, as proportion of anorthite increased, bending strength decreased in vacuum but increased in air. The decrease of bending strength in vacuum is attributive to porous sintered body due to the decrease of glass matrix. Furthermore the difference of bending strength between in vacuum and air corresponds to the decrease of bending strength due to stress corrosion, and that was improved as anorthite was precipitated.

The total effects of the precipitation to bending strength increased after all because the improvement of stress corrosion overcame the decrease of bending strength due to porous sintered body. As compared with Weibull constants in air, it was 6.9 in samples A1 and 17.5 in samples A2. This results shows the precipitation of anorthite improved the dispersion of bending strength as well as the average value. In the other hand, bending strength of samples A4 that had dispersed anorthite particles decreased in both environments as compared to samples A1. In samples A4, we recognized the only effect of the decrease of bending strength due to porous sintered body. The reason can be explained that microcrack could extend into anorthite precipitated at the interface between glass and alumina but could not extend into dispersed anorthite particle in glass matrix and deflected around anorthite particle. It is supposed that this behavior is attributive to interface morphology between anorthite and glass. Dispersed anorthite particles have distinct interface but precipitated anorthite have indistinct interface because there is amorphous anorthite near the interface. Therefore crack extension of microcrack at the interface between precipitated anorthite and glass is easier than dispersed anorthite particle and glass.

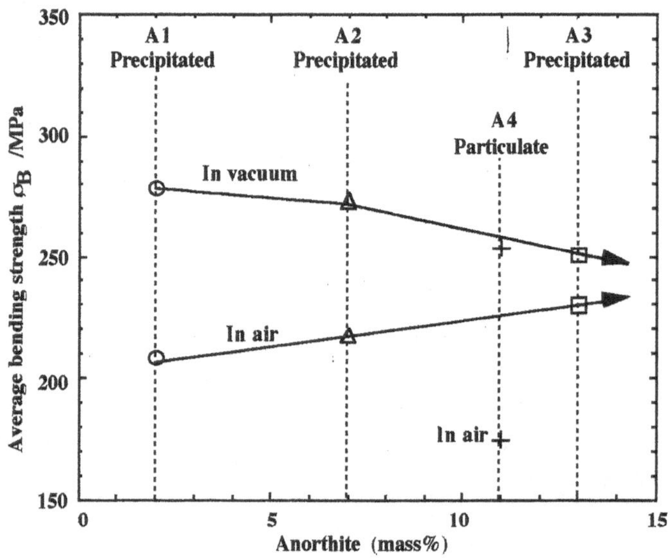

Figure 3. Relationship between bending strength and proportion of anorthite in air and vacuum. In samples A2 and A3, anorthite is precipitated at the interface between glass and alumina particles. In samples A4, anorthite particles are dispersed.

AE Behavior

Nucleation, extension and connection of microcracks are expected to be sources of AE of ceramic particulate glass matrix composite during bending test. However, we observed the preexisting microcracks that nucleated due to mismatch of thermal expansion of alumina and glass at processing (Figure 4). So we can expect AE due to not nucleation of microcracks but extension and connection. Furthermore the microcracks that we observed in glass matrix can't extend into alumina particle because alumina has fracture toughness (about 3MPa) much higher than glass (0.3-0.7MPa) and microcracks deflect around alumina particle. Therefore microcracks can extend into glass matrix and anorthite.

Figure 5 shows the results of AE location in vacuum and air. A horizontal axis shows the location of AE and zero is the center of specimen, and a vertical axis shows the stress at occurring AE. In the case of samples A1, we recognized the concentrated AE just before final fracture in air. This concentrated AE corresponds to the cumulative and unstable extension of microcrack due to stress corrosion because we could not recognized it in vacuum. On the other hand, in the case of samples A2 and A3, we recognized the dispersed AE in air. Samples A3 had

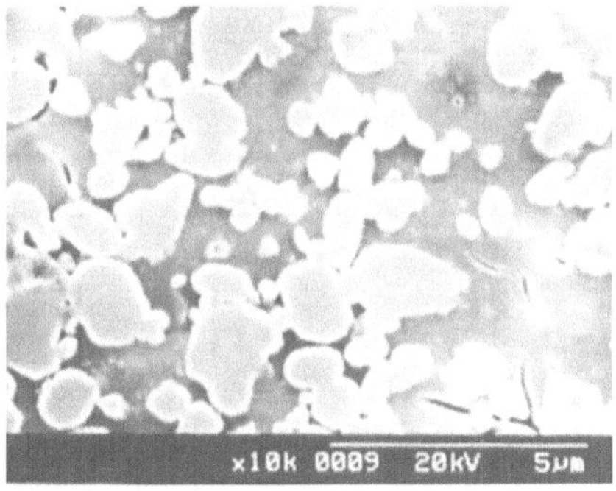

Figure 4. Observation of preexisting microcracks on the polished and hard etched surface of samples A1 by SEM.

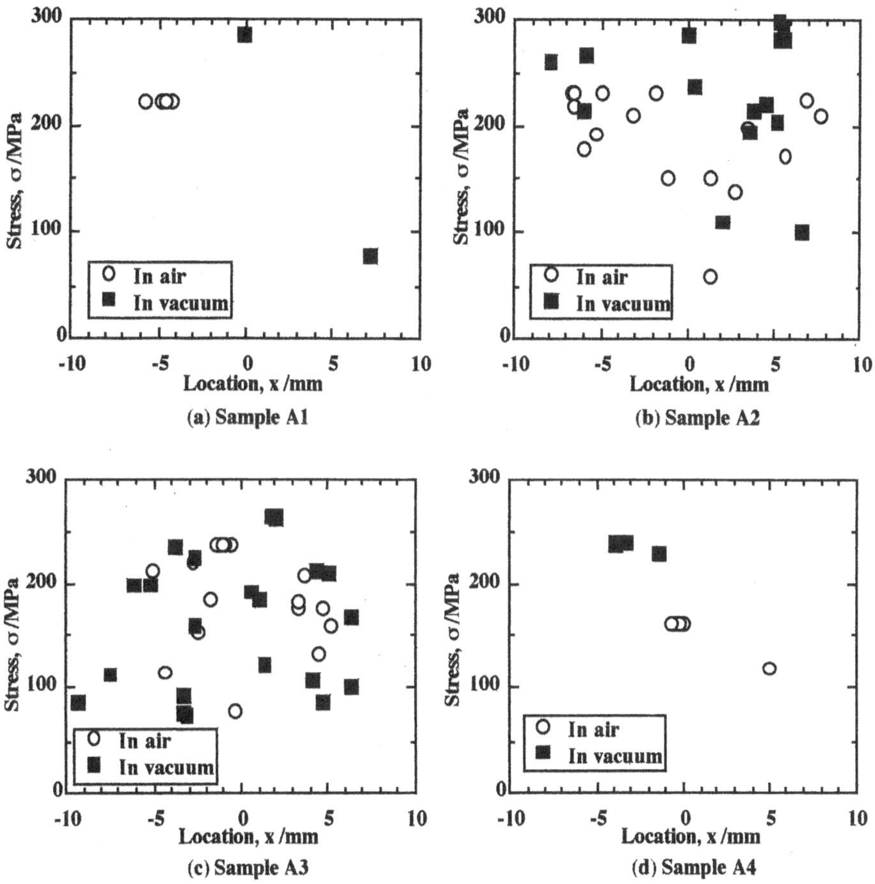

Figure 5. Result of the evaluation of the AE-location in air and vacuum during four point bending test.

more total events of dispersed AE than A3. Therefore, dispersed AE is related to anorthite precipitation and corresponds to the extension of microcrack into anorthite. If microcrack extends into anorthite, microcrack will be able to stop the unstable extension due to stress corrosion because stress corrosion resistance will increase. However even as samples A2 and A3, the concentrated AE due to stress corrosion was also observed. So the influence of stress corrosion on crack growth resistance was not perfectly removed.

In the case of samples A4, we recognized few dispersed AE in addition to concentrated AE. Therefore, it means that microcracks in glass can hardly extend into dispersed anorthite and anorthite particles can't contribute to improve stress corrosion resistance.

DISCCUSION

As mentioned above, microfracture process has changed specially in air. Anorthite was precipitated at the interface between alumina and glass, the cumulative and unstable extension of microcrack due to stress corrosion changed to the stable extension of microcrack into anorthite precipitated at the interface. Figure 6 shows the concept of this behavior. In the case of samples A1 and A4, even if a particle of alumina or anorthite is in front of microcrack which is extending due to stress corrosion, microcrack will deflect around particle and continue the unstable extension. In the case of samples A2 and A3, even if microcrack starts the unstable extension due to stress corrosion, microcrack tip will be trapped into anorthite precipitated at the interface between alumina and glass and stop the unstable extension due to stress corrosion.

As more microcracks extends into anorthite, stress corrosion resistance will be higher. Therefore the effect of anorthite precipitation on stress corrosion resistance can be evaluated by total events of dispersed AE because they correspond to number of microcracks which extend into anorthite. Figure 7 shows that as total events of dispersed AE increased, both of dispersion and average value of bending strength were improved.

Some specimens of samples A2 and A3 showed the concentrated AE in vacuum as well as air. However, the concentrated AE in vacuum had slower occurring rate than in air (Figure 8) and means the stable fracture. Dispersed AE in vacuum must be related to high porosity because we observed dispersed AE in vacuum in the case of precipitated samples which had more porous sintered body than non-precipitated. So we can separate AE due to high porosity from AE due to stress corrosion. If we can produce samples that has much the precipitated anorthite and finer sintered body, the samples will have higher mechanical reliability.

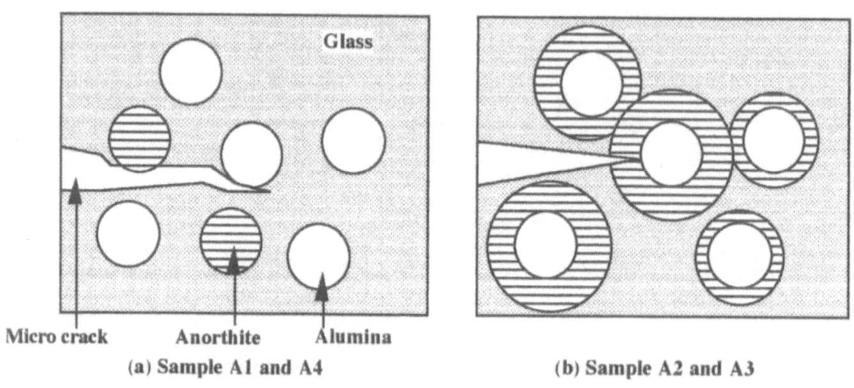

Figure 6. Schematic of the change of microfracture process by precipitation. Microcrack can be stopped only in the case that anorthite is precipitated at the interface between glass and alumina.

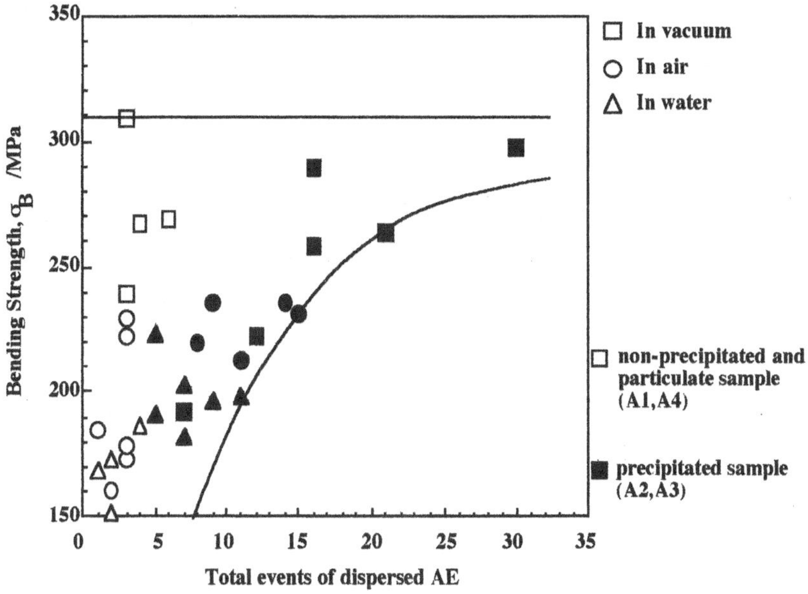

Figure 7. Relationship between total events of dispersed AE and bending strength, As total AE events increased, both of dispersion and average value of bending strength also increased.

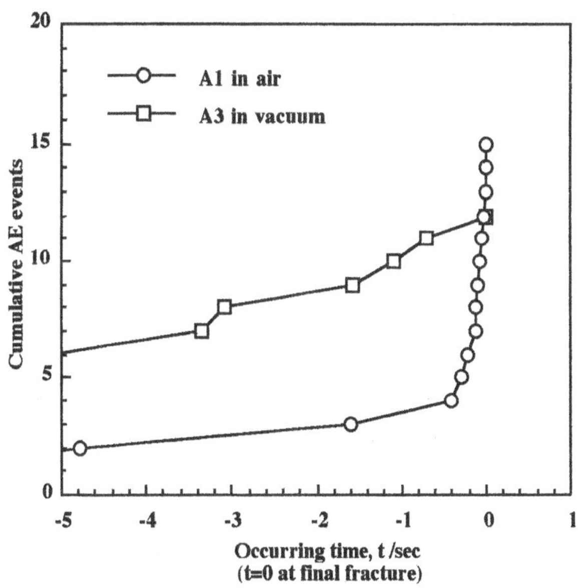

Figure 8. Difference of occurring rate of concentrated AE between the stable fracture due to high porosity and the unstable fracture due to stress corrosion.

CONCLUSION

We carried out four point bending test in vacuum and air, and evaluated the microfracture process due to stress corrosion of alumina particulate glass matrix composite by AE method. Then we could separate the measured AE into three microfracture processes:

1) the cumulative and unstable extension of microcracks due to stress corrosion,

2) the extension into precipitated anorthite,

3) the cumulative extension due to high porosity.

As a result, we confirmed that if anorthite was precipitated at the interface between alumina and glass, microcracks would extend into anorthite, and the cumulative and unstable extension of microcracks would be improved. Furthermore, the dispersion and average value of bending strength in air were improved and we could evaluate the effect of anorthite precipitation on stress corrosion resistance by total events of dispersed AE.

REFERENCES

1. R..W. Davidge and T.J. Green, The strength of two-phase ceramics-glass materials, *J. Mater. Sci.*, 3, 629 (1968).
2. M. Shiwa H. Inaba and T. Kishi, Development of the high sensitivity and low noise integrated AE sensor, *J. JSNDI*, **35**, (5), 374-382 (1990).
3. M. Enoki and T. Kishi, Theory and analysis of deformation moment tensor due to microcracking, *Int. J. Fracture*, **38**, 295-310 (1988).
4. T. Kishi, S. Wakayama and S. Kohara, Microfracture process during fracture toughness testing in Al_2O_3 ceramics evaluated by AE source characterization, *Fracture Mechanics of Ceramics vol.8*, Plenum Press, New York, 85-100 (1986).
5. B.C. Bunker and T.A. Michalske, Effect of surface corrosion on glass fracture, *ibid*, 391-413.
6. K. Kageyama, M Enoki, T Kishi, K Ikuina and M. Kimura, Controlling of microfracture by precipitation in alumina particulate glass composite, *J. Japan Institute of Metals*, **57**, (7), 756-760 (1993).
7. S. Nishigaki and J. Fukuta, Low-temperature cofireable, multilayered ceramics bearing pure-Ag conductors and their sintering behavior, *Advances in Ceramics vol. 26*, The American Ceramic Society, Westerville, 199-215 (1987).

AE STUDIES ON THE MICROFRACTURE PROCESS

IN STRUCTURAL CERAMICS

Shuichi Wakayama and Masanori Kawahara

Department of Mechanical Engineering, Faculty of Technology,
Tokyo Metropolitan University
1-1 Minami-Ohsawa, Hachioji-shi, Tokyo 192-03, Japan

INTRODUCTION

Silicon nitride is one of the most important ceramics for the structural materials. It has a microstructure consisting of acicular and equiaxed grains, which are formed by the $\alpha \rightarrow \beta$ phase transformation during fabrication. It has been reported that the fracture toughness of silicon nitride is enhanced by the contribution of the same microstructural features as whisker reinforced ceramic composites, where crack bridging mechanisms are effective[1,2]. Therefore, for reliability assessment of silicon nitride ceramics as structural materials, it is important to understand the microfracture process and the toughening mechanism.

On the other hand, many acoustic emission studies on ceramics[3,4], including authors'[5-8], have been carried out because of its simplicity in the detection of the nucleation time and location of microcracks in the material. For an example, the microfracture process during bending tests of alumina ceramics has been investigated by authors[7,8]. In particular, the critical stress for the maincrack formation, prior to the final failure, was evaluated and its feasibility as the advanced evaluation parameter for brittle materials was suggested.

In this study, the microfracture process during bending tests of silicon nitride was investigated. The locations of microcracks were determined using the difference of arrival times between 2 channel sensors and the crack growth behaviors were observed by a scanning electron microscope. The purpose of this work is to understand the influence of microstructural features on the fracture behavior.

EXPERIMENTS

AE Measurement

A schematic diagram of the AE measuring system used in the present study is shown in Figure 1. For the economical reason, two piezo-electric elements were used as AE sensors and attached directly at the both ends of the specimen. Therefore, special techniques for both the sensitivity calibration and the reduction of the electromagnetic noise are needed. In this

Nondestructive Characterization of Materials VI
Edited by R.E. Green, Jr. *et al.*, Plenum Press, New York, 1994

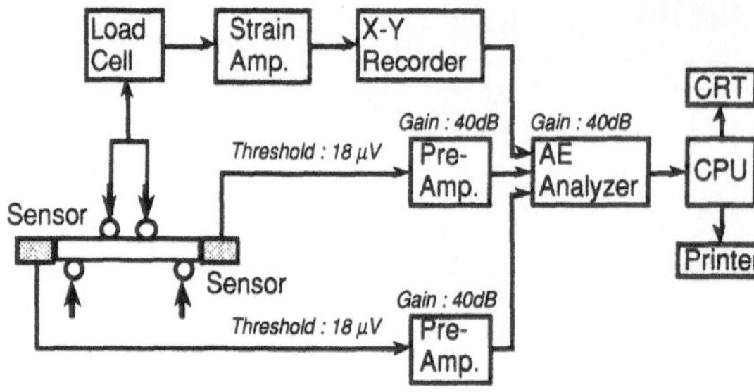

Figure 1. AE Measuring System

study, AE source locations were determined using the difference of arrival times between two channel sensors. Therefore, the sensitivities of two sensors, especially their equivalences, were calibrated carefully enough using pencil lead breakings as simulated sources before each testing.

Since the AE activity of ceramic materials is low, reduction in the noise level of the system are essential for AE measurement of such materials. In this study, noise-filter-transformers were used and the connections between direct sensors and pre-amplifiers were modified similar to differential type transducers. Consequently, the noise level at the input terminal of the pre-amplifier has been decreased to 14 μV. Threshold level was then selected at 18 μV. AE signals were measured using the AE system along with load signals, sent to a personal computer through an RS-232C interface and analyzed using a computer.

Materials

The materials used were silicon nitride ceramics with various microstructures. Figure 2 shows SEM fractographs of the materials. It can be seen that the materials have well-grown acicular β-phase Si_3N_4 among a matrix of equiaxed micrograins of the same phase. The sizes of acicular grains had been varied according to the sintering time and temperature. The materials can then be considered as *in-situ composite*, i.e. the acicular grains and micrograin matrix correspond to the whiskers and matrix in the whisker reinforced ceramics, respectively. In order to investigate the influence of microstructure on the microfracture

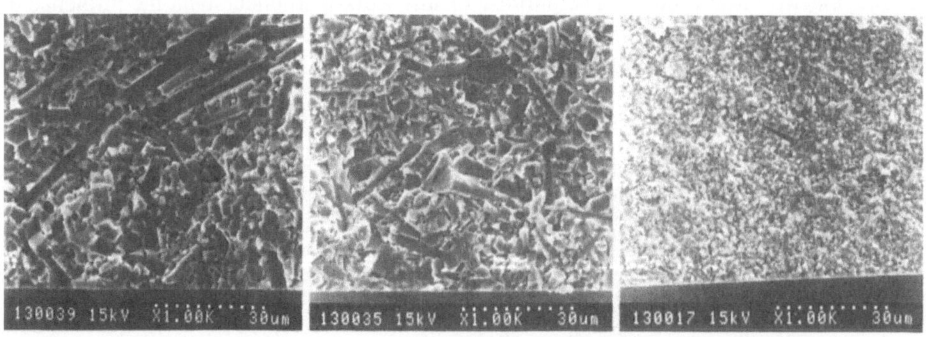

(a) H (b) M (c) L

Figure 2. SEM Fractographs of Materials

Table 1. Material Properties

Symbols	H	M	L
Grain Size (μm)	9	5	2.5
Fracture Toughness (MPa√m)	10	8	6
Young's Modulus (GPa)	295		
Density (g/cm³)	3,268		
Longitudinal Wave Velocity (m/s)	10,600		

process, three types of silicon nitride with different acicular grain size ("H", "M" and "L" materials) were prepared, as shown in Figure 2.

The material properties of the three materials are tabulated in Table 1. As shown in the table, depending on the size of the acicular grain, the fracture toughness, which was measured by SEPB method (JIS R1607), was able to be modified.

Bending Tests

Specimens were cut by diamond saw to dimensions of 3 mm × 4 mm × 40 mm, chamfered to 0.1 mm and polished with diamond powder of 1 μm. 4-point bending tests (upper span : 10 mm, lower span : 30 mm) were carried out, using an Instron-type tensile testing machine in air. Crosshead speed was carefully controlled so that strain rate was constant as 4×10^{-6} /s.

RESULTS

AE Behavior

In this study, electromagnetic noise was minimized as mentioned above and mechanical noise was removed by teflon sheets between apparatus and specimen. Furthermore, the AE source location was situated to reduce mechanical noise, especially for 4-point bending, and the events located out of the upper span were neglected.

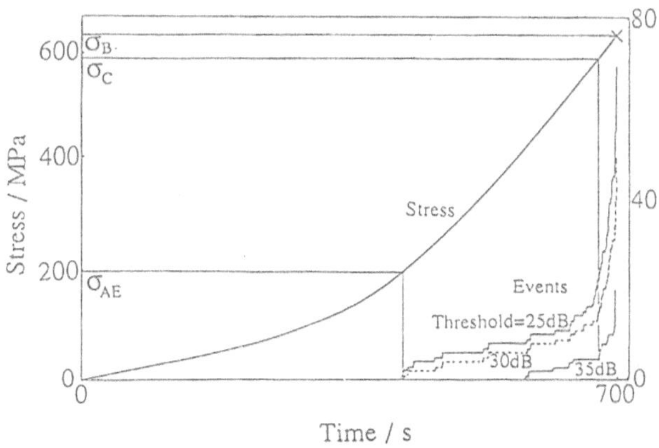

Figure 3. Typical Result of Bending Test and AE Generation Pattern

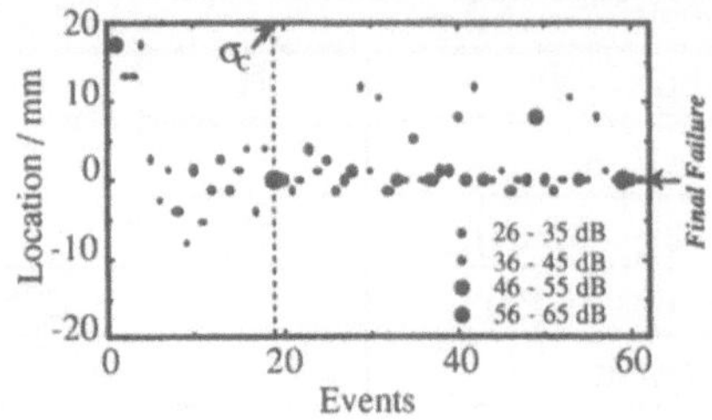

Figure 4. Results of AE Source Location

Figure 3 shows a typical AE generation pattern during 4 point bending tests of "H" material. In the figure, cumulative AE event counts are shown for threshold levels of 25 dB, 30 dB and 35 dB, respectively. It can be understood that the apparent bending stress, at which AE events increase rapidly, is independent of the threshold level. This stress is then determined from each specimen, as σ_C. The apparent stresses, σ_{AE}, where the first events were detected on the upper span, were also determined, as shown in Figure 3. On the other hand, AE events were detected only at the final failure in "L" material and the σ_{AE} and σ_C values have not been determined for this materials.

AE Source Location

In this study, the AE source locations were determined by obtaining the product of the longitudinal wave velocity of the material and the differences of the arrival times between the two channel transducers. Figure 4 shows the microcrack locations, which were nucleated during the 4-point bend test of the "H" material. The sizes of circles indicate the amplitude of each events. It can be seen from the figure that although microcracks distribute widely on the specimen surface before stress reached σ_C, the microcrack locations are concentrated within a narrow band after σ_C. It is important that the narrow band has close agreement with the origin of final fracture shown in the right side of the figure. Therefore, the figure shows the excellent ability of the AE technique to monitor the time and location of final failure.

Characteristic Stresses

Figure 5 shows the Weibull plots of the bending strength σ_B obtained from H, M and L materials. From the figure, it can be seen that the material with the highest fracture

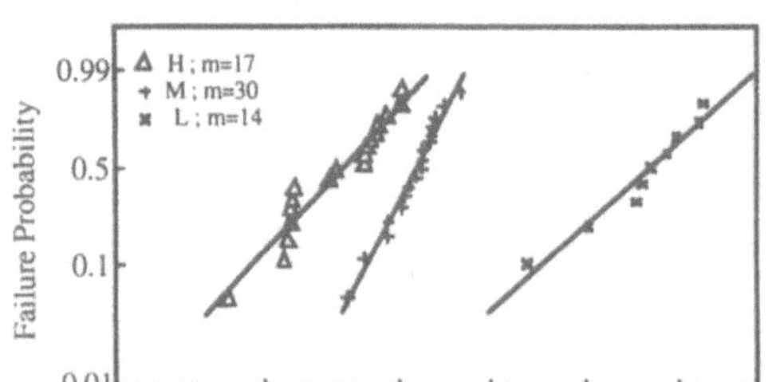

Figure 5. Weibull Plots of Bending Strength

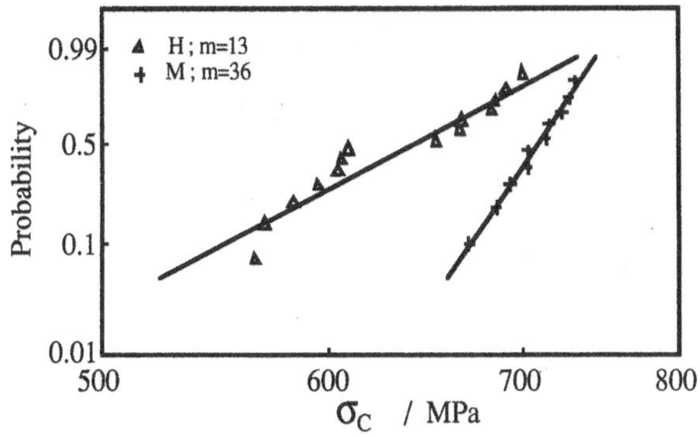

Figure 6. Weibull Plots of σ_C

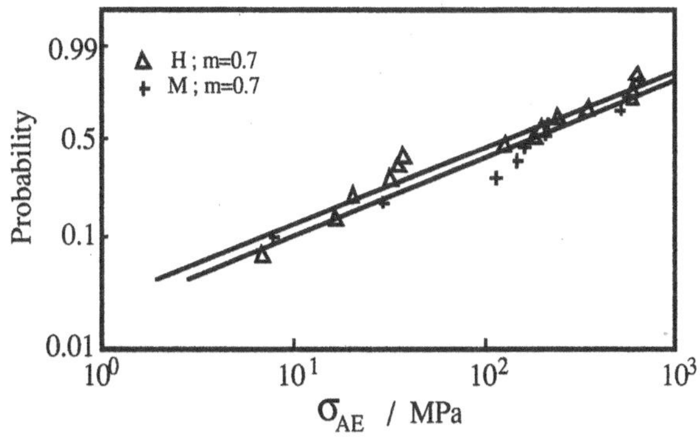

Figure 7. Weibull Plots of σ_{AE}

toughness has lowest fracture strength. It is known that the slopes of fitted lines, i.e. Weibull moduli, m, indicate the scattering range of the data. Weibull moduli of the M materials were relatively higher than the others.

The Weibull plots of σ_C obtained from H and M materials are shown in Figure 6. The σ_C values could not been obtained from L materials, since AE activities in those materials were quite low. It can be seen from the figure that the Weibull moduli obtained using σ_C were similar to those obtained using σ_B in the materials H and M. On the other hand, the Weibull plots using σ_{AE} are shown in Figure 7. It is interesting that the Weibull moduli using σ_{AE} are much lower than those using σ_B and σ_C.

DISCUSSION

Crack Growth Behavior

Since the fracture process during a bending test initiates on the surface under tensile stress, the tensile surface around the finally formed maincrack was observed by SEM after fracture. In the case of the bending tests, the specimens crashed into the bending apparatus and the fractured surface damaged. But the specimens used for the SEM observation were

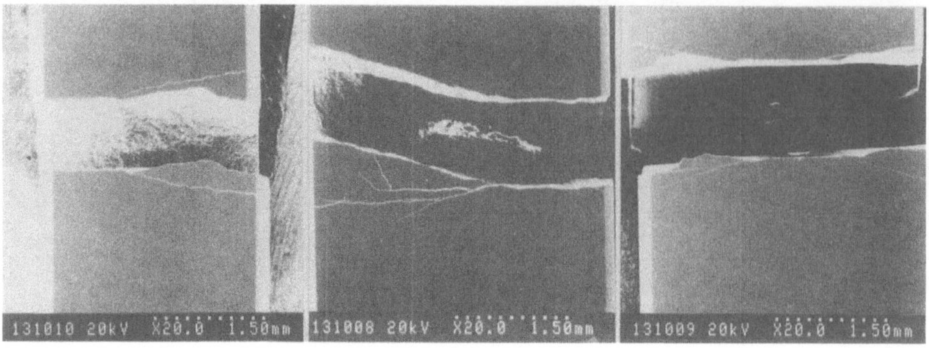

(a) H (b) M (c) L

Figure 8. Macroscopic Crack Growth Behavior during Bending Tests

fractured with the protection by soft cotton and fracture surfaces undamaged. Figure 8 show the results of the SEM observation on both sides for (a) H, (b) M and (c) L materials, respectively. It is important that the several branching cracks other than the maincrack, which had formed on the final fracture surface, are observed in all materials.

SEM observation under higher magnification shows the microscopic features of crack growth. Figure 9 shows the crack bridges observed in the H material. These bridges were observed at the edge of acicular grains (a) and several extrusions of acicular grains were also observe on the crack path (b). While crack bridges were observed in the H and M materials, they were not notable in the L material.

Microfracture Process in Silicon Nitride

In the present study, AE activities in L materials were quite low and AE signals were detected only at the final failure, therefore σ_{AE} and σ_C could not be obtained. Authors have applied AE source characterization to alumina ceramics elsewhere[5,6]. From those results, the detectability was 15 μm and that of the conventional AE method, using a narrow frequency band system, was estimated as ≈ 5 μm. Since the elastic properties, i.e. elastic modulus, density and/or wave velocity, of silicon nitride are equivalent to those of alumina, the detectability in the present study can be estimated as ≈ 5 μm. Considering that the acicular

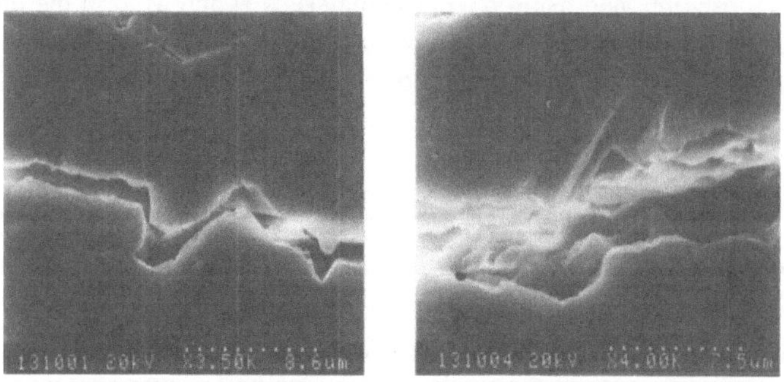

(a) Edge of acicular grain (b) Extrusion of acicular grain

Figure 9. Crack Bridgings Observed in H Materials

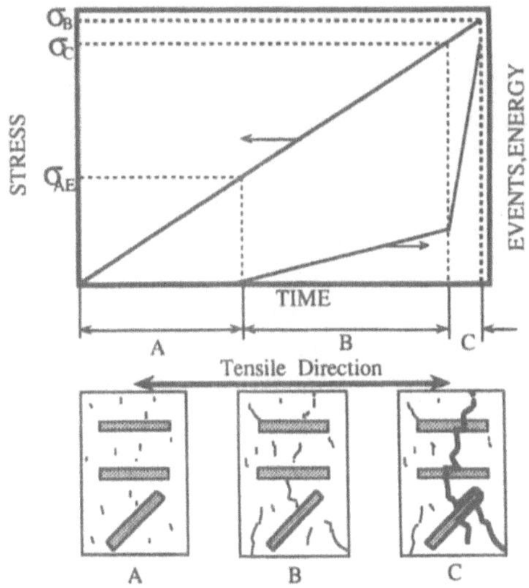

Figure 10. Microfracture Process during Bending Test of Silicon Nitride

grain sizes are 9, 5 and 2.5 μm in H, M and L materials, respectively, it can be concluded that σ_C corresponds to the cracking of the acicular grains.

On the other hand, assuming an elliptical surface crack with a large major to minor axis ratio, the criteria for the final failure is expressed as:[9]

$$K_I > K_C = \beta \bullet \sigma_B \sqrt{(\pi \bullet a_C)} \tag{1}$$

where K_I and K_C are the stress intensity factor and its critical value, respectively, and a_C is the critical crack length at a final failure. The factor β is for the correction of the surface effect and is ≈ 1.12. If K_C is replaced with the fracture toughness, the values of a_C are:

$$\begin{aligned} a_C(H) &= 54\ \mu m \\ a_C(M) &= 31\ \mu m \\ a_C(L) &= 9.9\ \mu m \end{aligned} \tag{2}$$

Therefore, it is consistent that AE signals were detected only at the final failure during the bending tests of L materials.

Consequently, the microfracture process during the bending tests of silicon nitride has been understood as shown schematically in Figure 10.

1. When the applied stress is less than σ_{AE}, microcracks, which are not large enough to emit the detectable AE signals, are nucleated among the equiaxed micrograin matrix.(Stage A)
2. When the applied stress reached σ_{AE}, larger intergranular microcracks, which can emit the detectable signals, were nucleated among the micrograins matrix.(Stage B)
3. Although the growth of those microcracks were impeded by acicular grains in Stage B, the maincrack, which was constituted by the coalescence of those microcracks and transgranular cracks of acicular grains, was formed and propagated at the stress beyond σ_C and caused the final failure.(Stage C)

CONCLUSIONS

In this study, AE measurement was carried out during the bend testing of silicon nitride materials of various microstructures. Three characteristic stresses were measured: the apparent stress, σ_{AE}, where the first AE event was detected, the apparent stress, σ_C, where AE generation was most active and the bending strength, σ_B. This enabled a sequence the events comprising the microfracture process to be determined. Firstly, intergranular microcracks, smaller than the acicular grains, form among equiaxed grains when the applied stress reaches σ_{AE}. Although these microcracks are impeded by the acicular grains, a maincrack forms at a stress beyond σ_C through the coalescence of these microcracks and transgranular cracks of the acicular grains and propagates to final failure.

ACKNOWLEDGEMENT

The materials used in this study were made by Advanced Technology Research Center of NKK Corporation. The authors would like to thank Mr. Kawashima and Dr. Nishio, NKK Corporation, for their kindness in offering the materials. Authors also appreciate financial support from the Iketani Science and Technology Foundation.

REFERENCES

1. T. Kawashima, H.Okamoto, H.Yamamoto and A.Kitamura, J. Ceram. Soc. Japan, 99:302(1991).
2. P. F.Becher, C.H.Hsueh, P.Angelini and T.N.Tiegs, J. Am. Ceram. Soc., 71:792(1989).
3. G.A.Gogotsi, A.V.Drozdov and A.N.Negovskii, Proc. Ultrason. Int., 83:67(1983).
4. A.Katagiri, T.Nishiyama, T.Fukuhara and Y.Nozue, Proc. National Conf. on AE, JSNDI, Tokyo, 110(1983).
5. T.Kishi, S.Wakayama, and S.Kohara, "Fracture Mechanics of Ceramics", Vol. 8, R.C.Bradt et al. ed., Plenum Press, New York, 85(1985).
6. S.Wakayama, T.Kishi and S.Kohara, "Progress in AE III", K.Yamaguchi et al. ed., JSNDI, Tokyo, 653(1986).
7. S.Wakayama and H.Nishimura, "Fracture Mechanics of Ceramics", Vol. 10, R.C.Bradt et al. ed., Plenum Press, New York, 59(1992).
8. S.Wakayama and H.Nishimura, "Nondestructive Characterization of Materials V", T.Kishi et al. ed., 717(1992).
9. H.Okamura, "Introduction to Linear Fracture Mechanics", Baifukan, Tokyo, 34(1976).

ELASTIC PROPERTIES OF THIN FILM SILICON NITRIDE BY BRILLOUIN SPECTROSCOPY

Vahid Askarpour,[1] Murli H. Manghnani,[1] Michael Mendik,[2] and Peter Wachter[2]

[1]School of Ocean and Earth Science and Technology,
University of Hawaii, Honolulu, Hawaii, USA 96822
[2]Laboratorium für Festkörperphysik, Eidgenössische Technische Hochschule Zürich, 8093 Zürich, Switzerland

INTRODUCTION

Elastic properties of thin films can be determined by ultrasonic techniques (using piezoelectric transducers) and Brillouin spectroscopy. Brillouin spectroscopy is a non-contact and purely optical technique in which the sample is not in physical contact with any probing device (Sandercock, 1982; Nizzoli and Sandercock, 1990).

In this study, Brillouin scattering is employed to determine the elastic properties of thin film Si_3N_4 on GaAs substrate. Si_3N_4/GaAs was chosen since ultrasonic values of the Si_3N_4 elastic constants (C_{11} and C_{44}) for the same specimen (Hickernell et al., 1990) are available for comparison.

THEORY

In an infinite elastic continuum, bulk phonons propagate in all directions and are chracterized by their particle displacements, i.e., longitudinal or transverse. At surface, longitudinal phonons and transverse phonons polarized normal to the surface are mixed and form corrugations at the surface, refered to as surface acoustic waves (SAW's). Since the SAW's are made up of bulk phonons, they are completely characterized by the elastic tensor C_{ij} and density ρ of the bulk material. In case of an isotropic material, the elastic tensor reduces to C_{11} and C_{44} due to symmetry considerations.

In case of a film on a substrate, if the V_{shear} of the film is higher than that of the substrate (stiffening case), the SAW velocity increases monotonically with qh. The dispersion of the SAW propagating on a thin film is often written as a function of qh, where q is the magnitude of the wavevector of the SAW and in this experiment is given by:

$$q = \frac{4\pi sin\theta}{\lambda_o} \tag{1}$$

where, θ is the angle between the normal to the film surface and the incident (same as scattered in this experiment) light direction and λ_o is the wavelength of the incident light. It is the characteristic of the dispersion curve to start for qh=0 at the substrate SAW velocity and increases until the substrate shear velocity is reached at a particular value of qh. No higher modes are expected in this range (Farnell and Adler, 1972). Si_3N_4 film on GaAs substrate represents a stiffening case since the V_{shear} of the Si_3N_4 (Slobodnik et al., 1973) is higher than that of GaAs (Hickernell et al., 1990).

EXPERIMENTAL TECHNIQUE

The details of the experimental set up can be found elsewhere (Askarpour et al., 1993). In brief, light from an argon ion laser (λ_o=5145 Å) was focused on the film with a lens (*f*=50 mm) and beam power of ~ 60 mW was employed. The incident laser light was polarized in the plane of incidence to minimize reflection losses. The scattered light was collected with the same lens (backscattering geometry) and analyzed using a high contrast and high resolution Brillouin spectrometer, which included a tandem six-pass Fabry-Pérot interferometer (Sandercock, 1982). The light was detected by a cooled photomultiplier tube and its output was stored in a multichannel analyzer for further analysis. Each spectrum was accumulated for 2-3 hours.

Isotropic Si_3N_4 film was grown by Hickernell et al. (1990) using low-temperature plasma-enhanced chemical-vapor-deposition (PECVD) technique on {100} plane of GaAs single-crystal wafer. Details of the sample preparation are described elsewhere (Hickernell et al., 1990). The thickness and density of the Si_3N_4 film were 4970 Å and 2500 kg/m^3, respectively (Hickernell et al., 1990).

RESULTS

A typical Brillouin spectrum of Si_3N_4 film is shown in Figure 1. The intense feature at the center of the spectrum is the elastically scattered light from the film surface. Five distinct spectral components were observed on both sides of the central peak. This is the first observation in a stiffening case where more than one SAW velocity is observed. We designate these peaks as R1, R2, R3, R4 and R5, with R1 being closest to the central peak.

Brillouin spectra of Si_3N_4 films were recorded for θ values ranging from 20° to 70°

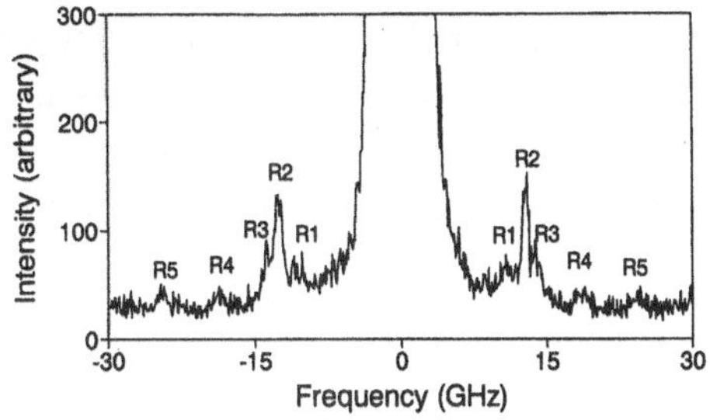

Figure 1. Brillouin spectrum of Si_3N_4 film (4970Å).

in steps of 5°. The velocities corresponding to each of the five peaks were determined from the following relation:

$$V_{SAW} = \frac{\Delta\nu.\lambda_o}{2sin\theta} \tag{2}$$

where, V_{SAW} is the SAW velocity and $\Delta\nu$ is the frequency shift of the spectral feature.

The measured surface wave velocities are plotted as function of qh for R1 and R2 modes only as shown in Figure 2. By keeping the density of the film constant, we fitted the data using the nonlinear least-squares fitting procedure of Levenberg and Marquardt (Press et al., 1986) to obtain C_{11} and C_{44}. We were able to fit R1 (Rayleigh) and R2 (first Sezawa) modes. We believe that the other three modes R3, R4, and R5 belong to longitudinal guided modes and do not follow the SAW mode treatment (Hillbrands et al., 1988).

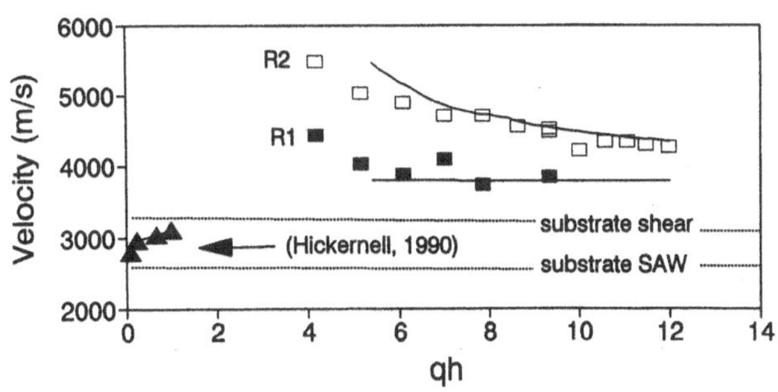

Figure 2. Dispersion relation for the SAW velocities as a function qh. The ultrasonic data by Hickernell et al. (1990) are also shown.

The best-fit elastic constants for Si_3N_4 thin film determined in this study are listed in Table 1. In addition, elastic constants from ultrasonic studies of Hickernell et al. (1990) on the same Si_3N_4 film specimen of 4970Å thickness (and of two other thickesses, 2500Å and 1020Å) and from bulk samples of isotropic Si_3N_4 (Fate, 1975) are included in Table 1.

Table 1. Comparison of the elastic constants of Si_3N_4 films by Brillouin scattering (this study) and ultrasonic techniques with those for the bulk ceramic specimens.

	This study	Hickernell et al. (1990)			Fate (1975)		
h (Å)	4970[1]	4970[1]	2500[2]	1020[2]			
ρ (kg/m^3)	2500	2500	2600	2800	2500[3]	2700[3]	3050[4]
C_{11} (GPa)	185±18	189	169	192	154	189	277
C_{44} (GPa)	40±2	56	77	92	61	72	101

[1] Si_3N_4 thin film (PECVD) employed in this study

[2] Si_3N_4 thin films (PECVD)

[3] Si_3N_4 ceramic (slip cast)

[4] Si_3N_4 ceramic (hot pressed)

DISCUSSION

In Figure 2, we also show the data obtained in the MHz frequency range by Hickernell et al. (1990) for the same Si_3N_4 film. At qh=0, the velocity corresponds to the SAW velocity on GaAs substrate. By increasing qh until ~1.2, the SAW velocity approaches the shear bulk velocity (V_{shear}) of the substrate. No higher order modes are expected (Farnell and Adler, 1972). However, from qh=4 to 12, the velocities were measureable by Brillouin scattering as shown in Figure 2. It is interesting that some higher order modes also appear in the present stiffening case which have not been expected (Farnell and Adler, 1972). From qh=4 to higher values of qh, the SAW velocity dispersion is similar to that of SAW waves propagating on free standing film (Farnell and Adler, 1972).

By comparing the elastic constants obtained from best-fit of the Brillouin scattering data with the C_{ij} obtained from ultrasonic SAW experiments (Hickernell et al., 1990), we find that the values from Brillouin scattreing are consistently lower. This can be attributed to the different sampling depth of the SAW's: MHz SAW samples the whole film and therefore provides an average value for the C_{ij} while Brillouin scattering samples a region about 2000 Å below the surface. It should be emphasized that we fitted C_{11} and C_{44} only and kept ρ of the film constant. The sensitivity of V_{SAW} is high; increasing ρ by 5% decreases V_{SAW} by more than 15%. Therefore, a change of the velocity of SAW's can be attributed to changes in both C_{ij} and ρ.

CONCLUSIONS

We have, for the first time, observed Rayleigh and higher order modes (first Sezawa and other three modes) in the stiffening case by Brillouin scattering. The assignment of Rayleigh and first Sezawa modes are consistent with the theoretical model for Si_3N_4 on GaAs substrate. We suggest that SAW's propagating on films (stiffening case) show a free standing film behavior for large qh while are well described by a layer model for small qh values. The differences between the MHz and GHz elastic constants are attributed to the different sampling depths of the ultrasonic and Brillouin scattering techniques.

ACKNOWLEDGEMENTS

The support for the facilities was provided by the W. M. Keck Foundation and the National Science Foundation (Grant 8720404). We thank T. S. Hickernell for providing the samples, J. Balogh and O. Matthews for technical assistance. This is School of Ocean and Earth Science and Technology Contribution no. 3241.

REFERENCES

Askarpour, V., Manghnani, M.H., Fassbender, S., and Yoneda, A., 1993, Elasticity of single-crystal $MgAl_2O_4$ spinel up to 1273 K by Brillouin spectroscopy, *Phys. Chem. Mineral.* 19:511.

Farnell, G.W., and Adler, E.L., 1972, Elastic wave propagation in thin layers, *in*: "Physical Acoustics," Vol. IX, W.P. Mason and R.N. Thurston, eds., Academic Press, New York.

Fate, W.A., 1975, High-temperature elastic moduli of polycrystalline silicon nitride, *J. Appl. Phys.* 46:2375.

Hickernell, T.S., Fliegel, F.M., and Hickernell, F.S., 1990, The elastic properties of thin-film silicon nitride, *1990 IEEE Ultrason. Symp.* 2:445.

Hillbrands, B., Lee, S., Stegeman, G.I., Cheng, H., Potts, J.E., and Nizzoli, F., 1988, Evidence for the existence of guided longitudinal acoustic phonons in ZnSe films on GaAs, *Phys. Rev. Lett.* 60:832.

Nizzoli, F., and Sandercock, J.R., 1990, Surface Brillouin scattering from phonons, *in*: "dynamical Properties of Solids," G.K. Horton, ed., Elsevier, New York.

Press, W.H., Flannery, B.P., Teukulsky, S.A., and Vetterling, W.T., 1986, "Numerical Recipes," Cambridge, New York.

Sandercock, J.R., 1982, Trends in Brillouin scattering: studies of opaque materials, supported films, and central modes, *in*: "Light Scattering in Solids," M. Cardona and G. Güntherodt, eds., Springer, Berlin.

Slobodnik, A.J., Conway, E.D., and Delmonico, R.T., 1973, "Microwave Acoustic Handbook," air Force Cambridge Research Laboratories.

Tsubouchi, K., and Mikoshiba, N., 1985, Zero-temperature-coefficient SAW devices on AlN epitaxial films, *IEEE Trans. Sonics Ultrason.* SU-32:634.

MEASUREMENT OF RESIDUAL STRESSES ON CERAMIC MATERIALS WITH HIGH SPATIAL RESOLUTION

K.J. Kozaczek[1], C.O. Ruud[2], and J.D. Fitting[2]

[1]Oak Ridge National Laboratory
Oak Ridge, TN 37831-6064
[2]The Pennsylvania State University
University Park, PA 16802

INTRODUCTION

Ceramic materials are often used in high performance semiconductor devices. Brazing is used as a common technique of joining the metal pins with the ceramic chip carriers. Brazing of thermally dissimilar materials results in high stress/strain gradients in the ceramic material which ultimately leads to the inferior performance of the device. Steep stress gradients and constantly decreasing size of the chips dictate the necessity of measuring the stresses with the spatial resolution of 100 μm or higher. The standard large area diffraction is not suitable for this purpose. The present research is concerned with the residual stress measurement in brazed kovar-alumina joints using x-ray micro-diffraction technique.

X-RAY DIFFRACTION STRESS MEASUREMENT

The x-ray diffraction technique actually measures the average elastic strains (interatomic spacing) by measuring the shift in the angle at which x-rays are diffracted (Bragg angle). Stress values are obtained from these elastic strains assuming that the material is elastically linear, homogeneous, and the effective diffraction elastic constants are known. Two techniques were used to measure stresses in the present research: "$\sin^2\psi$" and single exposure (SET) techniques.[1,2] Both techniques are valid for the plane stress state which is a reasonable assumption since the x-rays penetrate approximately 8 μm into the alumina. The $\sin^2\psi$ technique is recommended when high confidence is required on specimens likely to display pronounced crystallographic texture; SET assumes no texture present in the material, which reduces the number of necessary measurements.

TESTING PROCEDURE AND INSTRUMENTATION

X-Ray Micro-diffraction Technique

A back reflection diffractometer incorporating a position sensitive scintillation detector (PSSD)[3] was used for both SET and $\sin^2\psi$ techniques. A microfocus (focal spot 0.5 x 8 mm^2) x-ray tube with a copper target was used for measurements. CuKα radiation (wavelength λ = 1.541838 Å) provided for the analysis of the diffraction peaks in the range of Bragg's angle 2θ = 140° to 155° as shown in Table 1.

Table 1. Diffraction planes and Bragg angles for α-alumina (Cu radiation).

Diffraction plane (hk·l)	Bragg angle 2θ
(11·15)	142.7
(40·10)	145.5
(05·4)	149.6
(10·16)	150.6
(33·0)	152.8

The (10·16) CuKα reflection was used for stress measurements, since it assured the highest intensity of x-rays diffracted from alumina powder. A double-pinhole collimator and two position sensitive detectors were mounted on a x-y precision stage so that their position with respect to the x-ray target could be adjusted in order to provide for the maximum intensity of the incident x-ray beam. The sample was positioned using a rotary table and a translation stage; this arrangement allowed rotation of the sample with respect to the incident x-ray beam with the measurement point serving as a center of rotation and was suitable for $\sin^2\psi$ measurements.

Experimental Determination of Diffraction Elastic Constants

In order to convert the measured strains to stresses one needs to know the effective x-ray elastic constants. The bulk elastic constants mechanically measured or theoretically calculated may differ from the x-ray effective elastic constants.[1] Table 2 shows the average macroscopic elastic constants calculated from the alumina single crystal elastic constants published by Wachtman[4] using Voigt[5], Reuss[6], and Hashin-Shtrikman[7] averaging methods. For a fully dense material, the experimental data should fall within Hashin-Shtrikman upper (HS$^+$) and lower (HS$^-$) limits. However, for ceramic materials the elastic moduli depend strongly on the porosity and the shapes of pores[8-10], decreasing rapidly as percent porosity increases. The material used in this research had 8% - 10% porosity. Therefore, a four-point bending calibration method[1] was used to determine the effective diffraction elastic constants for alumina (10·16) planes. The value of $[E/(1+\upsilon)]_{(1016)}$ = 200 GPa was experimentally determined; mechanically measured bulk Young's modulus was E = 229 GPa.

Table 2. Macroscopic elastic moduli of fully dense α-alumina: E-Young Modulus, υ-Poisson ratio.

	Voigt	HS$^+$	HS$^-$	Reuss
E [GPa]	408.2	403.5	402.4	397.2
υ	0.229	0.232	0.233	0.236

Effect of X-Ray Beam Size on Accuracy of Measurements

It is generally accepted that at least 10,000 grains are required in the irradiated volume so that a sufficient number of grains contribute to the diffraction pattern so as to achieve a continuous cone of diffracted x-rays. If this condition is not satisfied, the Debye rings are composed of discrete, discontinuous spots, and this will hinder accurate measurements.[11] As the size of the incident x-ray beam decreases, the number of irradiated grains also decreases. In order to determine the effect of the incident beam size on the accuracy of measurements (and, therefore, to establish the minimum size of x-ray beam which can be used), the following procedure was applied. A double-pinhole collimator with replaceable apertures was built, which allowed control of the size of the x-ray beam in the range of 10 μm to 500 μm. The four-point bending stress on the surface of the alumina bar was measured simultaneously by the x-ray diffraction technique (SET) and calculated from the strain gauge response. The accuracy of x-ray diffraction measurements was assessed from the correlation between both techniques. It has to be recognized, however, that a strain gauge measures the strain averaged over its size (i.e. 10 mm^2 in this case), whereas the x-ray technique measures stress in the irradiated volume, whose cross section with the sample surface is much smaller than the size of the strain gauge.

The initial size of the x-ray beam was 470 μm (irradiated area 0.02 mm^2) and the detectors were placed 35 mm from the sample surface. Data collection time was 50 seconds. The good correlation between the XRD and strain gauge data is shown in Figure 1. The standard deviation of XRD measured stresses was ± 15 MPa. Subsequently, the beam size was reduced to approximately 170 μm. The data collection time was increased to 300 seconds. The results of the simultaneous four-point bending strain gauge and XRD measurements are shown in Figure 2. Two strain gauges and two independent strain indicators were used for measurements. The x-ray diffraction measurements were taken at various points in the region between the strain gauges at the same strain level. From Figure 3 it is seen that the XRD stresses differ depending on the location. This stems from the fact that different groups of grains can be subjected to different strains due to their orientation and grain interactions. The smaller the beam size, the smaller is the number of grains over which the strain is averaged. For the beam size of 170 μm and mean grain size of 6 μm, the estimated number of diffracting grains is approximately 20. In a polycrystalline aggregate the stresses are not uniformly distributed among the grains. The variance can be as high as ±50%, depending on the elastic anisotropy of single crystals.[12,13] However, the average strain measured by XRD correlates well with the strain gauge measurements (correlation factor $R^2 = 0.998$). The precision of measurements was approximately ± 15 MPa. The beam size of 170 μm was taken as the limit since it allowed one to average the measured strains over approximately 20 grains. If one assumed that the minimum number of grains could be as low as 10, then the initial beam size would be 120 μm. However, measuring stresses with a beam size less than 170 μm (for alumina mean grain size 6 μm) would not provide for measurement from a statistically significant number of grains. In general, the macroscopic state of stress (or strains) cannot be uniquely determined from measurements on a small number of grains.

Measurement of Residual Stresses in Alumina in the Vicinity of a Singular Pin

Initially, the residual stresses around the singular pin were mapped using a beam size of 470 μm. The stresses were measured in two perpendicular directions (Figure 3). The linear dependence of d-spacing on $\sin^2\psi$ indicated that there was no preferred crystallographic orientation (texture) and, therefore, the SET could be applied without sacrificing accuracy. Subsequently, the SET with the beam size of 170 μm was used to map the residual stresses around the singular Kovar pin brazed to an alumina substrate. The radial and tangential stresses were measured at locations showed in Figure 4: points #1,2 correspond to the location adjacent to the pin; points #3,4 are the midpoints between two brazing pads. The results of measurements are presented in Table 3.

Table 3. Residual stresses in alumina in the vicinity of the brazed kovar pin.

Point #	Radial stress (MPa)	Tangential stress (MPa)
1	-7.7 ± 2.4	296.1 ± 12.6
2	-34.0 ± 5.0	0.7 ± 7.7
3	-43.4 ± 20.3	154.0 ± 14.7
4	-4.2 ± 10.5	35.7 ± 25.2

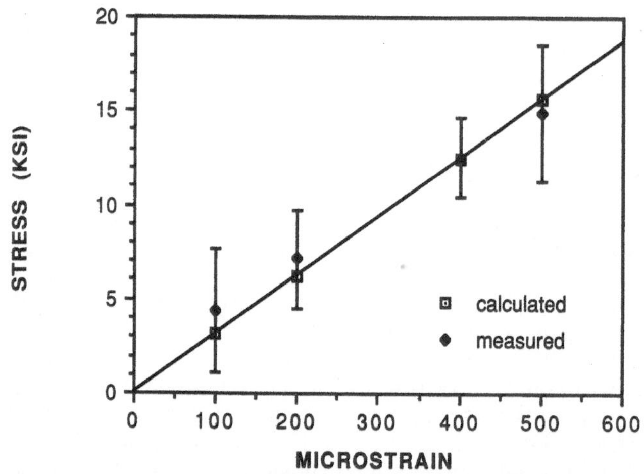

Figure 1. Correlation between the x-ray diffraction (SET) residual stress measurements and stress calculated from the strain gauge measurements; beam size 470 μm.

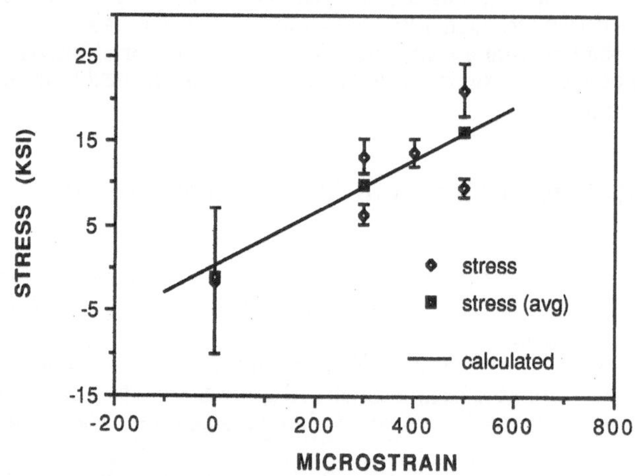

Figure 2. Correlation between the x-ray diffraction (SET) residual stress measurements and stresses calculated from strain gauge measurements; beam size 170 μm.

The radial stresses are slightly compressive in all cases: tangential stresses are tensile and much higher in one direction than in the other. The residual stress measured at the edge of the sample (far away from the pin) was 0.1 ± 1.1 ksi. The stress field around a singular pin is not symmetric due to the fact that the pins usually are not brazed at the center of the brazing pad, and this eccentricity varies from pin to pin. The measured stress field can also be affected by the orientation distribution and size of grains in the alumina in the vicinity of the pin. The diffraction patterns showed that for a given volume penetrated by x-rays, the number of grains (i.e. grain size) and their orientation vary, hindering, therefore, the determination of a unique stress, causing the particular strain state in a small group of grains.

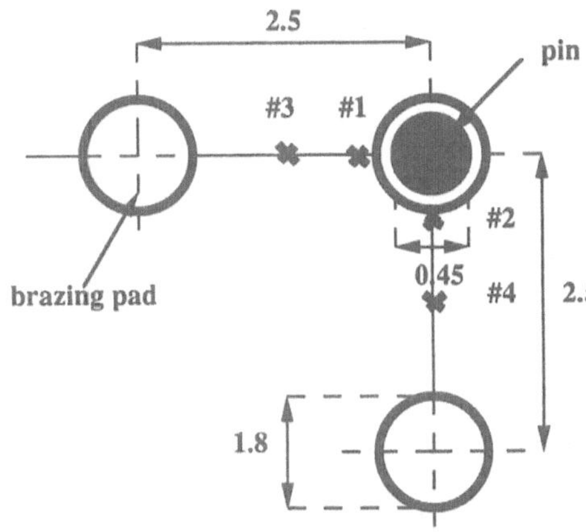

Figure 3. Locations of measurement points around a singular pin; dimensions in mm.

CONCLUSIONS

A fast x-ray diffraction technique has been developed for measuring the residual stresses with high spatial resolution in ceramic materials. This resolution is limited by the mean size of grains and the radiation type. The effective diffraction elastic constants were experimentally determined for alumina as $(E/1+\upsilon)_{(10\bar{1}6)} = 200$ GPa. The accuracy of XRD measurement of residual stresses with the spatial resolution of 170 μm and precision ± 15 MPa was verified experimentally by strain gauge measurements. The stress field around a singular Kovar pin brazed to alumina was asymmetric with high tangential stresses in the vicinity of the pin decreasing with the distance from the pin.

REFERENCES

1. SAE. "Residual Stress Measurement by X-Ray Diffraction, SAE J784a," SAE Inc., Warrendale, PA (1971).
2. I.C. Noyan and J.B. Cohen, "Residual Stress Measurement by Diffraction and Interpretation," Springer-Verlag, New York (1987).
3. C.O. Ruud, "Position Sensitive Detector Improves X-Ray Powder Diffraction," *Industrial Research and Development*: 84 (January 1983).
4. J.B. Wachtman, Jr., W.E. Teft, D.G. Lam, Jr., and R.P. Stinchfield, "Elastic Constants of Synthetic Single Crystal Corrundum at Room Temperature," *J. Research Natl. Bur. Standards* 64A (3):213 (1960).
5. W. Voigt, Lehrbuch der Kristallphysik, Taubner, Leipzig/Berlin (1928).
6. A. Reuss, *Z. Angew Math. Mech.* 9:49 (1929).
7. Z. Hashin and S. Shtrikman, "A Variational Approach to the Theory of the Elastic Behavior of Polycrystals," *J. Mech. Phys. Solids* 10:343 (1962).

8. F.P. Knudsen, "Effect of Porosity on Young's Modulus of Alumina," *J. Amer. Cer. Soc.* 45(2):94 (1962).
9. R.W. Rice, "Extension of the Exponential Porosity Dependence of Strength and Elastic Moduli," *J. Amer. Cer. Soc.* 59(11-12):536 (1976).
10. R.W. Rice, "Comparison of Stress Concentration Versus Minimum Solid Area Based Mechanical Property-Porosity Relations," *J. Mat. Science* 28(8):2187 (1993).
11. B.D. Cullity, Elements of X-Ray Diffraction, Addison-Wesley, Reading (1978).
12. L.D. Dikusar, E.F. Dudarev, and V.E. Panin, "Statistical Theory of the Microdeformations of Polycrystals. I.", *Izv. Vyssh. Zaved.*, Fizika 8:96 (August 1971).
13. K.J. Kozaczek, et al., "Microstructural Modeling of Grain Boundary Stresses in Alloy 600 ," submitted to *J. Mat.Sci.*

CHARACTERIZATION OF CERAMIC/CERAMIC MATRIX COMPOSITE MATERIALS FROM ELASTIC WAVE SCAN IMAGES

W. Sachse[2], M. Shiwa[1], T. Kishi[1] and M. O. Thompson[3]

[1] Research Center for Advanced Science and Technology
University of Tokyo, 4–6–1, Komaba, Meguro–ku, Tokyo 153 Japan
[2] Department of Theoretical and Applied Mechanics
[3] Department of Materials Science and Engineering
Cornell University, Ithaca, New York - 14853

INTRODUCTION

The need for materials capable of high–temperature operation in fuel–efficient engines and other applications points to ceramic materials. In a search for materials of high specific strengths and stiffnesses and better damage tolerance, ceramic fiber–reinforced, ceramic composite materials are being considered. But if these materials are to be used to fabricate structural components, measurement techniques which can be used to characterize their elastic and microstructural properties need to be developed. Here we focus on the elastic properties of these materials, specified in terms of their matrix of elastic moduli.

Because of the direct connection between elastic wavespeeds and material stiffnesses, ultrasonic techniques will likely play an essential role in determining these constants in a composite part or structure, either during fabrication, after assembly or during service. To be successful, any proposed procedure must overcome a number of material- and geometry-related measurement problems which have been summarized in a previous paper [1]. It is well-known that the geometric dispersion that accompanies the propagation of any transient pulse in a bounded structure, profoundly affects the ultrasonic signals. But, in addition, composite materials are generally highly attenuative of ultrasound and inherently inhomogeneous with a microstructure consisting of fibers, laminates and their stacking sequence, each possessing a characteristic dimension, which, when comparable to the wavelengths of the ultrasonic waves used to inspect them, results in multiple scattering, wave dispersion and dissipation that distort the ultrasonic signals. Furthermore, the fabrication of ceramic/ceramic matrix composite materials often results in the formation of microcracks. These, as well as the material heterogeneities inherent in these materials, result in a significant scattering of the elastic waves as they propagate through a specimen.

Nondestructive Characterization of Materials VI
Edited by R.E. Green, Jr. *et al.*, Plenum Press, New York, 1994

The elastic moduli of a material are often determined using pulsed ultrasonic measurements on samples cut from the material such that the equations relating elastic moduli and ultrasonic wavespeeds are easily inverted. Our goal here is the development of an ultrasonic technique for the *in situ* determination of the elastic stiffness properties of ceramic fiber–reinforced, ceramic composite specimens. We consider procedures for recovering the elastic constants of ceramic/ceramic matrix composite specimens that utilize ultrasonic signals propagating in arbitrary directions in an unprepared specimen, thus requiring no cutting or other preparation of the original specimen.

The basis of the measurements is the recently–developed ultrasonic *point–source/ point–receiver* measurement technique. In it, a pulsed Nd:YAG laser is used as a scanned ultrasonic point source while a small–aperture piezoelectric transducer is used as a detector. The collection of all the detected waveforms comprise a *scan image* which provides a view of the complete elastic wavefield in a test specimen. These images clearly reflect a material's anisotropy and its heterogeneities. Algorithms will be described which permit the processing of such scan images to remove noise and other extraneous information and to facilitate an unambiguous determination of the elastic properties of the material.

PS/PR ULTRASONIC SYSTEM

We have reported over the past several years the development and application of the ultrasonic *point-source/point-receiver (PS/PR)* measurement system which, because it can be accurately modelled, can be used to make quantitative ultrasonic measurements [2]–[4]. The basis of the technique is a small-aperture source and receiver in which knowledge of the wave propagation is used to interpret the detected ultrasonic signals. The requirements and operational characteristics of a number of sources and receivers which can be used in a *PS/PR* measurement system have previously been described [5]. For example, if a step–type source time–function is desired, the point source may be a normal force that can be obtained from the fracture of a capillary or a focused, pulsed laser beam of sufficient energy that it results in material ablation from the specimen surface. The result is a strong first arrival, longitudinal wave mode (or *P*–wave) signal in the material. A horizontal double force is realized when the pulsed laser beam is operating in the thermoelastic regime. In that case, a strong shear wave signal is generated in the specimen. As illustrated in Fig. 1, such a source can also form the basis of a repetitive and scanned ultrasonic measurement system.

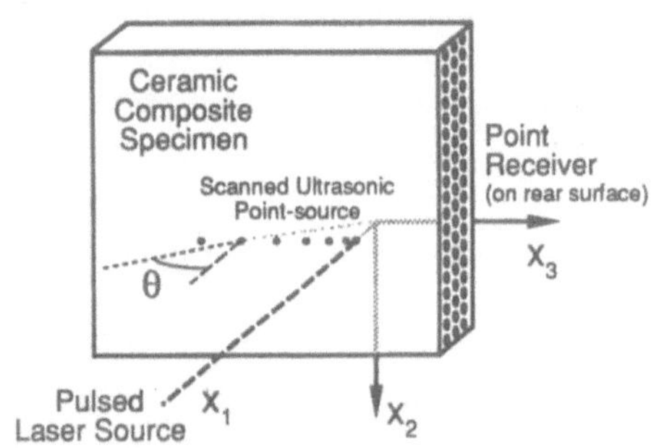

Figure 1. Measurement setup of a pulsed laser to make *PS/PR* measurements in a specimen.

The *PS/PR* technique has shown great promise in several materials inspection and characterization applications because it exhibits a number of distinct advantages. Sources generating signals possessing a broad frequency spectrum are easily realizable.

Except in special cases, a point source generates both longitudinal and shear waves in a specimen, therefore information about each of these wave modes can, in principle, be extracted from each waveform. Further, since *PS/PR* signals are simultaneously propagated in a wide range of directions in a specimen, one can use an array of sensors or scan either the source or the receiver over the specimen surface to determine the directional dependence of the speeds of propagation and amplitudes of various wave modes in a material. By viewing a large number of signals measured at adjacent source/receiver configurations, one obtains, a so-called *scan image* which represents the detailed spatial and temporal characteristics of the elastic wave field in a material that can be directly related to the material's anisotropy and macrostructure [6, 7]. We demonstrate in this paper that the first signal arrivals, corresponding to the *P*–wave, measured along a number of directions from a source in a composite specimen can be processed to recover a subset of the elastic constants of the ceramic/ceramic matrix composite material.

RECOVERY OF ELASTIC MODULI

Procedures for recovering the complete matrix of elastic constants of an anisotropic composite specimen from wavespeed measurements made along non-principal directions in it have been the focus of much recent research. Details of the wavespeed data inversion procedures have been published, hence we summarize here only the barest essentials. In an experiment, one measures the time taken for a particular signal to propagate from a source to a receiver point and these data can be used to calculate a 'speed' of propagation of the wave mode associated with this signal through the material.

The propagation of energy through a material from a point source occurs at the *group velocity* of the waves in the material. In an anisotropic material, the magnitude of this velocity is a function of propagation direction and its direction is generally not coincident with the wave normal. This complicates the wavespeed inversion procedure. There is no explicit equation which relates the group velocities to the elastic constants, therefore the inverse problem of recovering the elastic constants from the group velocity data measured along arbitrary directions in a sample needs to be approached indirectly. We recently described a two-step optimization algorithm for solving this problem [10]–[12].

Because the arrival of the first longitudinal signal is usually easily identified when the source/receiver separation is not too large, recent efforts have focused on ascertaining which of the elastic constants of an anisotropic material can be most accurately recovered when only longitudinal wavespeed data are input into the inversion scheme [13]. For a material of arbitrary elastic symmetry, the longitudinal wavespeed v_P depends most strongly on the following elastic constants and their combinations [13]

$$\rho v_P^2 = \mathrm{f}\,\{C_{11}, C_{22}, C_{33}, (C_{12}+2C_{66}), (C_{23}+2C_{44}), (C_{13}+2C_{55})\} \qquad (1)$$

In this equation, ρ refers to the density of the material.

In a transversely isotropic composite material, the longitudinal wavespeeds along various propagation directions, specified by angle θ to the axis of transverse isotropy, depend on the elastic constants, C_{11}, C_{33}, and the combination $(C_{13}+2C_{44})$. The form of the functional dependence is similar for phase and group velocity data [13]. The explicit form of the function appearing in Eq. (1) for transversely isotropic materials was shown by Musgrave to be given by [14]

$$v_P(\theta) = \sqrt{[C_{11}\sin^4\theta + C_{33}\cos^4\theta + 2(C_{13}+2C_{44})\sin^2\theta\cos^2\theta]/\rho} \qquad (2)$$

We will use this equation to process our P–wave arrival–time data.

MEASUREMENTS

Specimens

The specimens were SiC/fibers imbedded in a SiC matrix and were from two manufacturers. One material, denoted as SiC/SiC, was manufactured by DuPont using a chemical vapor infiltration (CVI) method. The finished material was a laminate of woven cloth, carbon–coated SiC fibers in a CVI SiC–matrix whose fiber volume fraction was 40%. The second material, denoted as Nicaloceram^R, was obtained from Nippon Carbon, Co., Ltd. This material was an 8–ply laminate of woven cloth consisting of carbon–coated SiC fibers (NL607) in a SiC matrix comprised of β–SiC and Polycarbosilane. The fiber volume fraction of this material was 30%.

During fabrication, Nicaloceram^R is subjected to molding, curing and burning cycles under pressure and it is not surprising that the resultant material is anisotropic and heterogeneous. The thickness of both plate–like specimens was approximately 3 mm.

Measurement System

The essential components of the ultrasonic measurement system were the laser, the specimen scanning system, the signal detector, amplifier and waveform recorder and a PC to control the entire system and to acquire waveform data.

The laser source was operating at 1.064 μm and generating pulses approximately 4 ns in duration. The incident energy density was adjusted using neutral density filters to be in the range from 1 to 50 mJ with a spot size of 100 μm. At each source location, signals from a fixed number of laser pulses (typically 2–10) were averaged to improve the signal to noise ratio.

The signals were detected with a piezoelectric transducer whose active element was 1.3 mm in diameter. Signals were amplified by 40 dB using a low–noise preamplifier whose bandwidth extended from 20 kHz to above 2 MHz. The signals were recorded using a digital oscilloscope, with the entire measurement system under control of a PC running OS/2.

MEASUREMENT RESULTS

Scan Image Data

It was found that short–duration laser pulses can be used to generate ultrasonic signals possessing good signal–to–noise ratio in ceramic/ceramic matrix composite materials. The scan images collected on the smaller specimens of SiC/SiC and Nicaloceram^R are compared in Figs. 2(a) and (b). As in the scan images of single crystal specimens and polymeric matrix composite materials [6, 7], a pattern of wave arrivals can be discerned in these images. However, here, the detected signal arrivals are not smooth functions of position. The SiC/SiC specimen had polished surfaces and it is unlikely that the differences in observed first arrivals are a consequence of different path lengths between

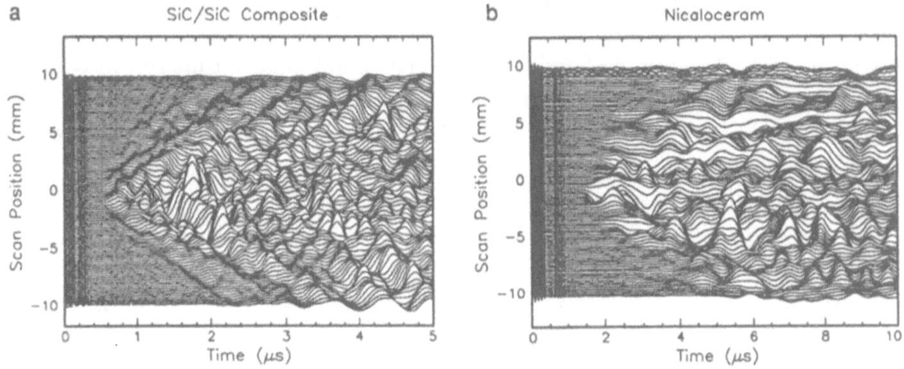

Figure 2. Scan images of ceramic/ceramic matrix composite specimens. (a) SiC/SiC; (b) Nicaloceram^{R} (For clarity, in each image only 101 of 201 lines are shown.)

source and receiver along the scan line. We thus conclude that the scatter observed in the wave arrivals is a consquence of the local inhomogeneities in these materials.

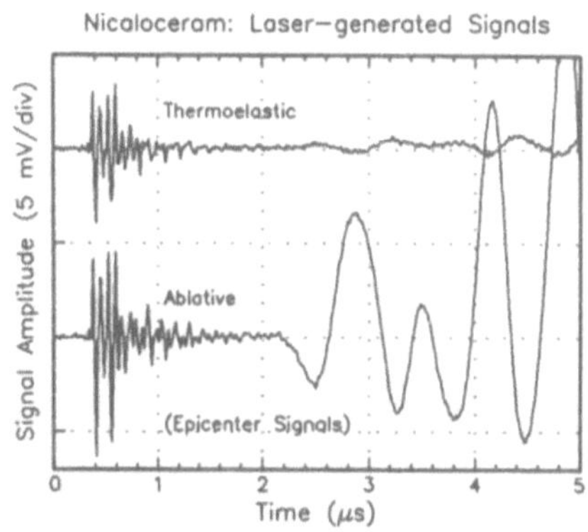

Figure 3. Nicaloceram^{R}: Epicentral signals; Thermoelastic and ablative laser sources.

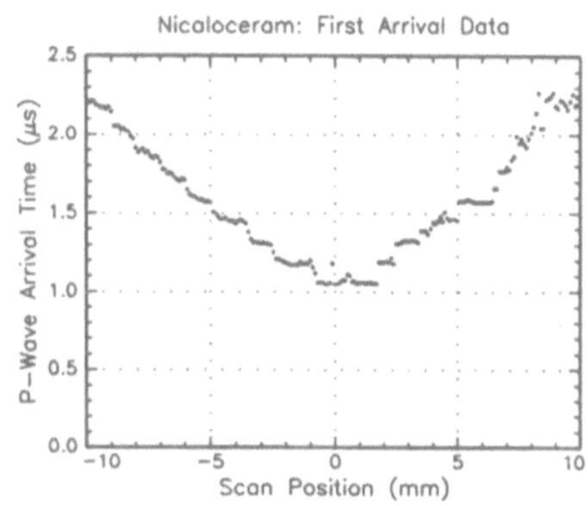

Figure 4. *P*–wave arrival–time data.

An enhancement of the first arrival is obtained when using the laser source in the ablative regime, although this results in visible surface damage. A comparison of data line #101 in the two scan images shown in Fig. 2, is shown in Fig. 3. The initial signal appearing in both records corresponds to the noise associated with the firing of the laser pulse and its initiation serves as a convenient initial time reference. It is similar in both traces. As expected, the first arrival signal generated by the ablative source, is significantly stronger that that generated by the thermoelastic source. However, the subsequent portions of the signals, which correspond to the elastic wave signals that have propagated through the specimen, appear to be of inverted polarity from each other. Not surprisingly, the signal–to–noise ratio of the ablative source signal is much better than the other.

Processing of First Arrival–Time Data

The determination of the time of arrival of the first signal in each time record comprising a scan image is not straightforward.The effects of the specimen boundaries, wave attenuation and scattering contribute to distorting the first arrival signal as a function position. This complicates any automtic arrival–time determination algorithm. After numerous unsuccessful attempts, the signal arrivals in the results reported here, were determined manually with a cursor on the digitized waveform and are estimated to be accurate within about 38 ns (± 5 sampling intervals). An example of the measured *P*–wave arrival–time as a function of source position is shown in Fig. 4. Reflecting the features visible in the scan images, the arrival–time of the first signal is not a smooth function of scan position, but rather, it changes in discrete steps whose spatial frequency appears to be similar to the warp and weft spacings of the composite.

Group Velocity Determination and Recovery of Elastic Moduli

In order to implement Eq. (2) to revover the partial set of elastic moduli of the material, we require the *P*–wave group velocity as a function of angle to the axis of transverse isotropy in the material. The group velocity corresponds to the ratio of the propagation path distance and the *P*–wave arrival–time. To verify the proper operation of the entire measurement system, measurements were made on a glass plate of known thickness and wavespeed. The plate had a 3000 Å thick Chromium coating sputtered onto its front surface to serve as a transducer film. It was found that when the thickness of a specimen is approximately equal to the transducer aperture, then the measured group velocities at certain source and receiver locations are highly sensitive to the effective area of the detecting transducer and precise knowledge of the source/receiver separation. The specific parameters are: (1) The scan zero position (relative to the center of the detecting transducer); (2) The effective aperture of the transducer; (3) The amount of post–trigger of the waveform recorder; (4) The precise thickness of the plate–like specimen. A non–linear fitting routine was used to extract from measured arrival–time data, such as that shown in Fig. 4, the optimal values of the aforementioned parameters (except for the specimen thickness). In one example, it was found that a scan zero position of only 161.6 μm resulted in a 4% difference in group velocity determined from the arrival–time data measured at the two ends of the scan, i. e. at -10 mm and $+10$ mm.

The measured *P*–wave arrival–time data for four scans on the Nicaloceram^{R} plate, with two scans in the two principal directions – denoted as 'longitudinal' and 'transverse' – are shown in Fig. 5(a). Within the scatter in the arrival–time data, the results are indistinguishable and when the corresponding group velocity values are evaluated, they exhibit a clear dependence on source/receiver angle, with the lowest values measured normal to the plate. (The glass plate was verified to have a uniform group velocity as a function of source/receiver angles between $-71°$ and $+71°$ to the plate normal.) The group velocities of two first arrival signals in the scan images were verified by a separate measurement using a glass capillary fracture source. The through–thickness longitudinal wavespeed was found to be 3.77 mm/μs while the shear wavespeed was 2.57 mm/μs. The longitudinal wavespeed in the plane of the specimen was measured to be 5.62 mm/μs. Based on all of the above observations, we conclude that these ceramic/ceramic matrix composite materials are adequately modelled as being elastically transversely isotropic, with their axis of transverse isotropy aligned with the normal to

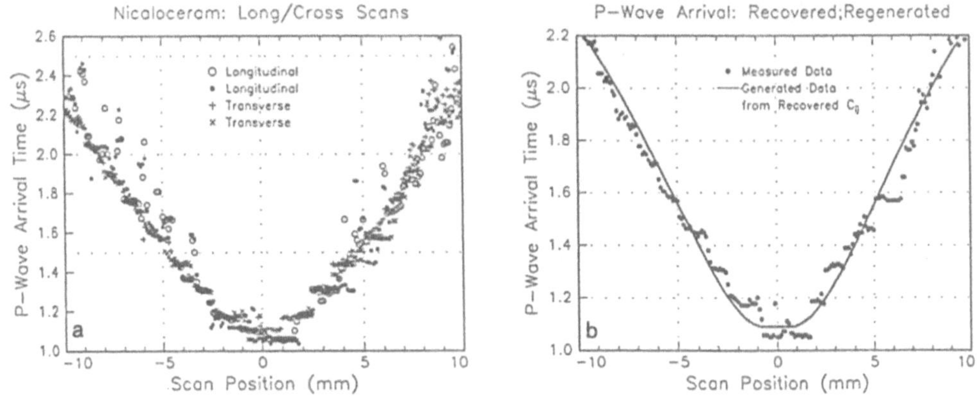

Figure 5. Nicaloceram[R] ceramic composite. (a) Measured *P*–wave arrival–time data; (b) Measured and recovered arrival–times from determined elastic moduli.

the specimen plane. This is consistent with the processing operations used to fabricate the Nicaloceram[R].

Performing the corrections outlined in the previous paragraph and fitting Eq. (2) to the resulting group velocity data, permits recovering the partial set of elastic moduli, C_{11}, C_{33} and $(C_{13}+2C_{44})$. The results from four scan images are listed in the accompanying Table. Fig. 5(b) shows the result obtained for one of the data sets of (a).

Table I. Elastic moduli of Nicaloceram[R]

Elastic Constant(s)	*Value* [GPa]
C_{11}	36.6 ± 3.3
C_{33}	22.4 ± 0.8
$C_{13} + 2C_{44}$	18.4 ± 3.8

As a check, the speeds of the longitudinal and shear waves through the thickness of the Nicaloceram[R] plate were evaluated to give C_{33} and C_{44}, respectively, while the longitudinal wavespeed along the axis of the plate was used to evaluate C_{11}. The other two elastic constants, i. e. C_{12} and C_{13}, were then adjusted to obtain an optimal fit between the measured and re–calculated group velocity data. The results were: $C_{11} : 40.0$; $C_{33} : 23.2$; $C_{44} : 8.7$; $C_{13} : 6.0$; $C_{12} : 3.0$ [GPa].

CONCLUSIONS

We have described in this paper the application of the ultrasonic *point–source/point–receiver* measurement technique to characterize the elastic and microstructural characteristics of composite/composite ceramic composite materials. In this approach, a pulsed Nd:YAG laser is used as a scanned ultrasonic point source while a small–aperture piezoelectric transducer is used as a detector. The collection of all the detected waveforms comprise a *scan image* which provides a view of the complete elastic wavefield in a test specimen. SiC/SiC and Nicaloceram[R] ceramic/ceramic composite specimens have been tested to date. The measured images clearly reflect these materials' anisotropy and heterogeneities. Algorithms were developed which permit the processing of the first arrival signal in the scan images to remove noise and other extraneous information and to enable an unambiguous determination of a partial set of elastic moduli of these materials.

ACKNOWLEDGEMENTS

It is a pleasure to acknowledge the many scan image arrival–time determinations carried out by Eric Haywiser. The capillary source wavespeed measurements were made by Dr. K. Y. Kim. This work was supported in part by RCAST, the University of Tokyo and the Nippon Steel Corporation. Additional support has been obtained from the Office of Naval Research, Physical Acoustics Program, and the Materials Science Center at Cornell University which is funded by the National Science Foundation.

REFERENCES

1. W. Sachse, "Towards a quantitative ultrasonic NDE of thick composites", in *Review of Progress in Quantitative Nondestructive Evaluation*, **10B**, D. O. Thompson and D. E. Chimenti, Eds., Plenum Press, New York (1991), pp. 1575–1582.
2. W. Sachse and K. Y. Kim, "Point-source/point-receiver materials testing", in *Review of Quantitative Nondestructive Evaluation*, **6A**, D. O. Thompson and D. E. Chimenti, Eds., Plenum Press, New York (1986), pp. 311-320. Also: 1987a, in *Ultrasonic Materials Characterization II*, J. Boussière, J. P. Monchalin, C. O. Ruud and R. E. Green, Eds., Plenum Press, New York (1986), pp. 707-715.
3. W. Sachse and K. Y. Kim, "Quantitative acoustic emission and failure mechanics of composite materials", *Ultrasonics*, **25**, 195-203 (1987).
4. W. Sachse, B. Castagnede, I. Grabec, K. Y. Kim and R. L. Weaver, "Recent developments in quantitative ultrasonic NDE of composites", *Ultrasonics*, **28**, 97-104 (1990).
5. W. Sachse, "Transducer considerations for point-source/ point-receiver materials measurements", in *Ultrasonics International '87 – Conference Proceedings*, Butterworths, Guildford, UK (1987), pp. 495-501.
6. A. G. Every, W. Sachse and M. O. Thompson, "Materials characterization from elastic wave anisotropy images", in *Non-destructive Characterization of Materials, IV*, C. O. Ruud and R. E. Green, Jr., Eds., Plenum Press, New York (1991), pp. 493–500.
7. A. G. Every and W. Sachse, "Imaging of laser-generated ultrasonic waves in silicon", *Physical Review B*, **44**(13), 6689–6699 (1991).
8. B. Castagnede, J. T. Jenkins, W. Sachse and J. Baste, "Optimal determination of the elastic constants of composite materials from ultrasonic wavespeed measurements", *J. Appl. Phys.*, **67**(6), 2753-2761 (1990).
9. B. Castagnede, K. Y. Kim, W. Sachse and M. O. Thompson, "Determination of the elastic constants of anisotropic materials using laser-generated ultrasonic signals", *J. Appl. Phys.*, **70**(1), 150-157 (1991).
10. A. G. Every and W. Sachse, "Determination of the elastic constants of anisotropic solids from acoustic wave group velocity measurements", *Physical Review B*, **42**(13), 8196-8205 (1990).
11. A. G. Every, W. Sachse, K. Y. Kim and L. Niu, "Determination of elastic constants of anisotropic solids from group velocity data", in *Review of Progress in Quantitative Nondestructive Evaluation*, **10B**, D. O. Thompson and D. E. Chimenti, Eds., Plenum Press, New York (1991), pp. 1663-1668.
12. L. Niu, *The Determination of the Elastic Constants of Composite Materials from Ultrasonic Group Velocity Data*, Ph. D. Dissertation, Cornell University, Ithaca, NY (May 1992). Manuscript in preparation.
13. A. G. Every and W. Sachse, "Sensitivity of inversion algorithms for recovering the elastic constants of anisotropic solids from longitudinal wavespeed data", *Ultrasonics*, **30**(1), 43–48 (1991).
14. M. J. P. Musgrave, *Crystal Acoustics*, Holden–Day, San Francisco (1970), p. 102.

NONDESTRUCTIVE CHARACTERIZATION OF ADHESIVE BOND STRENGTH IN LAMINATED SAFETY GLASS

Henrique Reis

Department of General Engineering
University of Illinois
117 Transportation Bldg.
104 S. Mathews
Urbana, Illinois 61801

INTRODUCTION

Laminated safety glass is widely used in windshields for automobiles and other moving vehicles; architectural applications, as in windows for skyscrapers; and bulletproof glass for military and security uses. Laminated safety glass consists of two or more glass sheets bonded with an interlayer of transparent, adherent plastic. The glass sheets in windshield laminates are made from either plate glass, float glass, chemically tempered glass or sheet glass and are approximately 0.1 inches (2.54 mm) thick.[1-2] The plastic interlayer is made of a thermoplastic material called polyvinyl butyral (PVB) with thicknesses of 0.015 inches (0.381 mm) or more commonly used.[3,4]

Regardless of the application, the adhesive bond strength between the flexible plastic interlayer(s) and the adjacent sheets of glass is crucial to laminated safety glass performance.[5] For architectural applications, adhesion must be strong to provide structural stability. In bulletproof glass, which is composed of several alternating layers of glass and plastic, the adhesion must be strong at the outer layers to prevent the release of large shards of glass while the inner layers must have lower adhesive bond strength to provide a mechanism to dissipate the kinetic energy of the projectile. Laminated safety glass for windshield applications is manufactured to improve the impact and penetration resistance against flying objects and to minimize the danger of flying glass after impact. Penetration is prevented mainly by absorbing the kinetic energy through stretching the plastic interlayer, delamination between the plastic interlayer and the two adjacent glass plates, and fracture of the adjacent glass plates. The plastic interlayer must delaminate partially to allow consumption of larger amounts of energy. Yet, to minimize the danger of flying glass after impact, delamination must not be excessive. Therefore, the best impact performance requires an optimal level of adhesion between the plastic interlayer and the two adjacent glass plates. A good review of the impact and penetration resistance mechanisms of laminated safety glass is given by Huntsberger.[6]

There exists several destructive test methods to evaluate the overall impact performance of laminated safety glass. These methods include the 90° peel test and the compressive shear test,[7] and the impact test method which is a commonly used method to rate overall laminate performance as described in safety standards test procedures.[8-10] However, the most widely used destructive test method for quality control monitoring of laminated safety glass is the destructive 0° Fahrenheit pummel test.[11] In this test, a 12 inch by 12 inch (30.5 cm by 30.5 cm) laminate is cut into four nominal 6 inch by 6 inch (15.3 cm by 15.3 cm) specimens. These are then conditioned at 0° Fahrenheit (-18° C) for at least one hour. The specimens are removed from the conditioning cabinet one at a time and pummeled with a one-pound hammer immediately after being removed. The specimen is held at about a 5 degree angle from a 1/2 inch (12.7 mm) steel plate so that only the edge of the unbroken glass contacts the plate. The laminate is then pounded progressively in 1/2 inch (12.7 mm) increments along the bottom 3/4 inch (19.1 mm) of the laminate. When the bottom edge has been completely pulverized, the next 3/4 inch is pulverized. This process is repeated until about 1/2 inch of the laminate remains unpummeled. All of the smooth glass in the pummeled section must be completely pulverized, and all loose glass removed. The pummeled samples are allowed to reach room temperature and the condensed moisture is allowed to evaporate before grading begins. The specimens are then compared to standards, which are pummeled specimens representing the range of possible adhesive bond strength. There are ten standards rated from 1 through 10, with 1 corresponding to very low and 10 to very high adhesive bond strength. The specimens with lower adhesive strength will have less pulverized glass remaining, and more of the plastic interlayer will be exposed. The amount of exposed interlayer can be determined by holding the specimens in fluorescent light and estimating how much light is reflected by the plastic. More exposed plastic means more light reflected which in turn means weaker adhesive bond strength. The specimens are then given a rating between 1 and 10 depending on which of the standards most nearly reflects the same amount of light. If the specimen falls between two of the standards, it is given a half pummel rating, i.e., if a specimen is rated between 4 and 5, it is given a pummel rating of 4.5. The ten pummel test standards are calibrated using the impact test method.[11]

The fact that the above-mentioned tests are intrinsically destructive precludes their use as viable on-line quality control techniques. In addition, these tests raise questions as to their practicality as adhesive bond strength testing procedures. For example, the peel test specimens are not fabricated in the same manner as typical safety glass laminates, and therefore they may not have the same characteristics as typical laminates. The problem with the compressive shear test is that the level of adhesion is not constant over the interfaces between the glass and plastic interlayer, i.e., pockets of relatively low and/or high adhesive bond strength are present in all laminates. As such, the local area of a specimen which has the lowest level of adhesion, i. e., the weak link, will fail first. Therefore, the test will only measure the load at which the area with the lowest adhesive bond strength will fail. A specimen which has a relatively high level of adhesion across the interfaces but has one pocket of lower adhesion will fail at a lower load, and therefore will be inaccurately assessed. Several drawbacks also exist with the pummel test procedure. The temperature at which the test is conducted does not represent the wide range of temperatures at which the safety glass is used. In addition, it allows for operator subjectivity. For instance, one operator's definition of hitting the glass with the hammer "very hard" might differ considerably from that of another. Furthermore, the pummel rating depends on how closely the operator thinks the specimen matches a particular standard. A different operator may give the same specimen a different rating. Finally, the drawback with impact testing is that it does not directly measure adhesive bond strength. Because of all of these drawbacks and because adhesion plays a key role in the performance of safety glass, the need for nondestructive testing methods for the evaluation/characterization of adhesive bond strength in laminated safety glass is apparent.

Analytical ultrasonics implies the measurement of material microstructure and associated factors that govern mechanical properties and dynamic response. It goes beyond flaw detection, flaw imaging and defect characterization and includes assessing the inherent properties of material environments in which the flaws reside. Acousto-ultrasonics is an analytical ultrasonic NDE technique which measures the relative efficiency of energy transmission in the specimen. An ultrasonic pulse is injected with a transmitting transducer mounted on the surface of the specimen. A larger amount of damage (i.e., flaws, changes in the microstructure, etc.) in the specimen produces a higher signal attenuation, resulting in lower stress-wave-factor (SWF) readings. Traditionally, the SWF has been evaluated as the number of oscillations higher than a chosen threshold in the ring down oscillations in the output signal from the receiving transducer. The stress-wave-factor does not yet have a standard definition. In this study, a stress-wave-factor is any stress wave parameter in any domain, such as the time and frequency domains, that help to characterize the acousto-ultrasonic signal. The SWF has already been correlated with the adhesive bond strength between rubber and steel,[12,13] and with the adhesive bond strength of connections in wood structures.[14] A good review of analytical ultrasonics in materials research and testing is given in references.[15-17] The purpose of this study is to investigate the applicability of the acousto-ultrasonic stress-wave-factor techniques to the nondestructive characterization of the adhesive bond strength of laminated safety glass.

EXPERIMENTAL PROCEDURE

To determine the feasibility of using the acousto-ultrasonic technique to nondestructively evaluate the adhesive bond strength in safety glass, laminated safety glass specimens with increasing levels of adhesive bond strength were manufactured. Each specimen consisted of a 12 inch by 12 inch by 0.030 inch (30.5 cm by 30.5 cm by 0.0762 cm) PVB layer laminated between two 12 inch by 12 inch by 0.090 inch (30.5 cm by 30.5 cm by 0.229 cm) sheets of float glass. A total of eighteen specimens were prepared. The adhesion was controlled by varying the moisture content of the interlayer and by rinse-washing the glass sheets with two different levels of water hardness.[18-25] The glass sheets in half of the specimens were rinsed with a standard demineralized water, while the sheets in the remaining specimens were washed with a harder water, i.e., water with added salts. A total of six different groups, with three specimens each, were manufactured. Each group corresponds to a different level of adhesive bond strength as shown in Table 1. The specimens are labeled by the moisture content (in % moisture) for each adhesion group. The adhesion groups which were washed with salts added to the final rinse water are also labeled with "salt". The specimens in each group were cut into 6 inch by 6 inch (15.3 cm by 15.3 cm) specimens, and their adhesive bond strength evaluated using the pummel test method. Table 1 lists the pummel test rating for each of the three-specimens group used in the acousto-ultrasonic measurements.

Table 1. Glass Laminate Specimens with Controlled Adhesion.

Adhesion Group	Number of Specimens	Rinse Type	Moisture Content (%H_2O)	Pummel Rating	Specimen Name
1	3	salt	0.70	2.0	0.70s-1 0.70s-2 0.70s-3
2	3	standard	0.60	3.0	0.60-1 0.60-2 0.60-3
3	3	salt	0.44	3.5	0.44s-1 0.44s-2 0.44s-3
4	3	standard	0.46	4.5	0.46-1 0.46-2 0.46-3
5	3	salt	0.20	7.5	0.20s-1 0.20s-2 0.20s-3
6	3	standard	0.20	9.0	0.20-1 0.20-2 0.20-3

Figure 1 represents the configuration of the acousto-ultrasonic data acquisition system used in this study. The system consisted of the portable Acoustic Emission Technology (AET) model 206 AU instrument, two transducers, an AST computer equipped with an STR*825 analog-to-digital board, and Digiscope software developed by Sonix, Inc. Both the transmitting and receiving transducers were AET Model FC-500 having an approximate sensitivity of -85 dB (relative to 1V/μbar) from 0.1 to 3 MHz. Both the transmitting and the receiving transducers were mounted on waveguides as shown in Figure 1. A thin layer of silicone rubber attached to the end of the waveguide was used as a dry couplant. The area of contact between the waveguide and the laminated glass specimens was a circle with the diameter of 6.4 mm (1/4 in.). The center-to-center spacing between the waveguides was 76.2 mm (3.0 in.). Waveform signals were recorded at one hundred random locations on each specimen to represent an average of the adhesion over the laminate. To determine if the specimens were isotropic or if any directional effects were introduced during the manufacturing processes, fifty of the waveforms were recorded with the transducers aligned parallel to a particular edge of the specimen, and the remaining fifty waveforms were recorded with the transducers aligned perpendicular to the same edge.

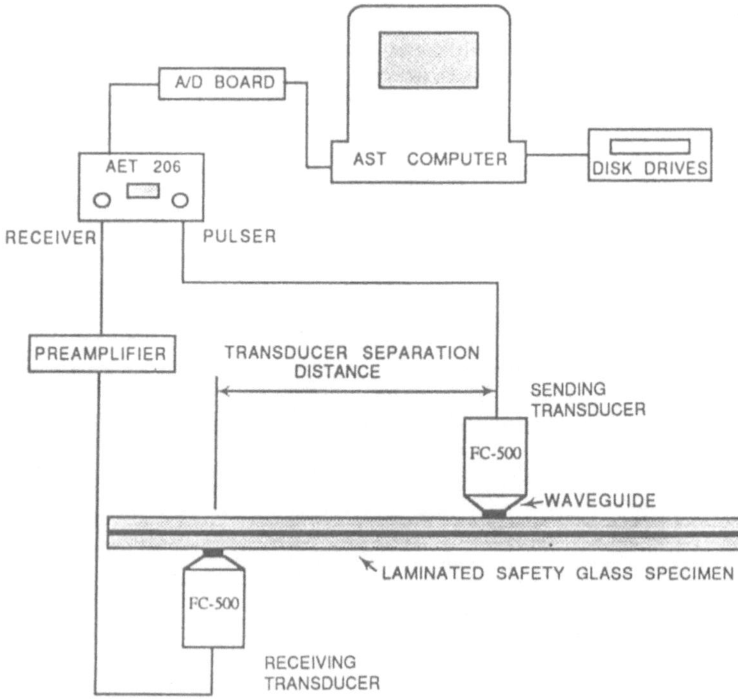

Figure 1. Configuration of the acousto-ultrasonic data acquisition system.

The transmitting transducer was excited by the Acoustic Emission Technology (AET) model 206 AU instrument which was set at a pulsing rate of 250 pulses/s, with an energy setting of 1 to provide -250 volt pulses to the transmitting transducer. The output signal of the receiving transducer was amplified 40 dB by a preamplifier (AET model 140B) with a plug-in filter having a passband between 30 kHz and 2 MHz. The signal was then further amplified 34 dB in the 206 AU instrument for a total amplification of 74 dB. A synchronous trigger, as well as the received signal,was also sent to an AST computer equipped with an STR*825 analog-to-digital converter board which digitized the signal. The Digiscope software package developed by Sonix, Inc. was used to display the signal on the AST monitor and to store the signal on a diskette. To reduce noise, Digiscope was set to average 16 waveforms into every signal that was recorded. A sampling rate of 0.781 MHz was used to save 512 points for each waveform. Once the acousto-ultrasonic signals were stored, they were quantified using the stress-wave-factor approach by the methods described in the waveform measurement and analysis program (WAVEMAP).[26]

EXPERIMENTAL RESULTS AND CONCLUDING REMARKS

Acousto-ultrasonic stress-wave-factor measurements have been conducted on laminated safety glass samples manufactured with controlled adhesive bond strength between the plastic interlayer and the two adjacent glass plates. Stress-wave-factor measurements were recorded and the results correlated with the results obtained using the destructive pummel test method.

The stress-wave-factor does not yet have a standard definition. In this study the stress-wave-factor is assumed to be any useful ultrasonic parameter in any domain such as the time and frequency domains. A variety of stress-wave-factors is discussed by Reis at al.[13,14,26] The discussion of experimental results given here will focus on the stress-wave-factors having the highest correlation with the pummel rating test data.

Figures 2, 3 and 4 show the average rectified area, signal energy, and maximum peak amplitude in the time domain, respectively, as a function of the pummel rating for the test specimens. The signal energy is defined as the square of the amplified transducer output voltage integrated over the time of the sweep. In the frequency domain, Figures 5 and 6 represent the average zeroth moment of the power spectral density (area under the curve) and the average maximum amplitude, respectively, also chosen as stress-wave-factors, as a function of the pummel rating. In Figures 2-6, for each of the eighteen data points, the average SWF represents the average of one hundred measurements (fifty measurements parallel to each side of the square plate). The average stress-wave-factor value was regressed on the average pummel rating and the results are provided on each figure. The correlation coefficient is reported as a qualitative measure of the linear relationship between the average stress-wave-factor and the average pummel rating. It was observed that higher values of the SWF measurements correspond to higher values of the pummel rating test data.

Typical standard deviations for these stress-wave-factors ranged from 1.5 % to 5% of the average of 100 values, and it was observed that they increase slightly with an increase in transducer separation distance.[27] Most parameters yielded standard deviations of about 3% with the lowest values resulting from the area under the power spectral density curve in the frequency domain and the largest peak voltage from both the time and the frequency domains. To determine if any directional effects were present in the laminates, fifty of the one hundred waveforms were recorded with the transducers parallel to one edge of the square specimen, and the other fifty waveforms were recorded with the transducers perpendicular to the same edge. No directional effects were observed for any specimen.

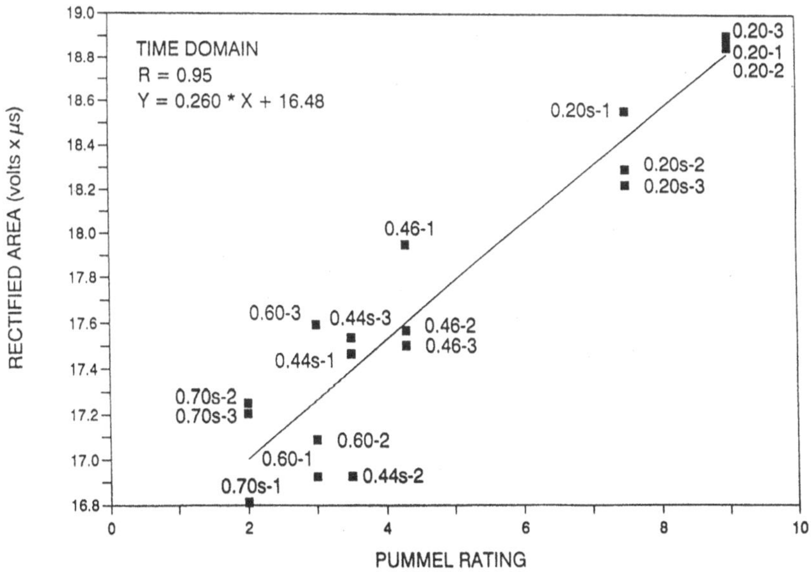

Figure 2. Rectified area of the acousto-ultrasonic signal vs. pummel test rating.

The results of this investigation show that the acousto-ultrasonic approach has great promise for evaluating the adhesive bond strength in laminated safety glass. Several stress-wave-factors, particularly the peak amplitude and the rectified area in the time domain as well as the zeroth moment of area in the frequency domain, show high correlation with adhesive bond strength. The results indicate that the approach could be used for quality control, eliminating the need for a costly destructive testing program. In addition, there is potential

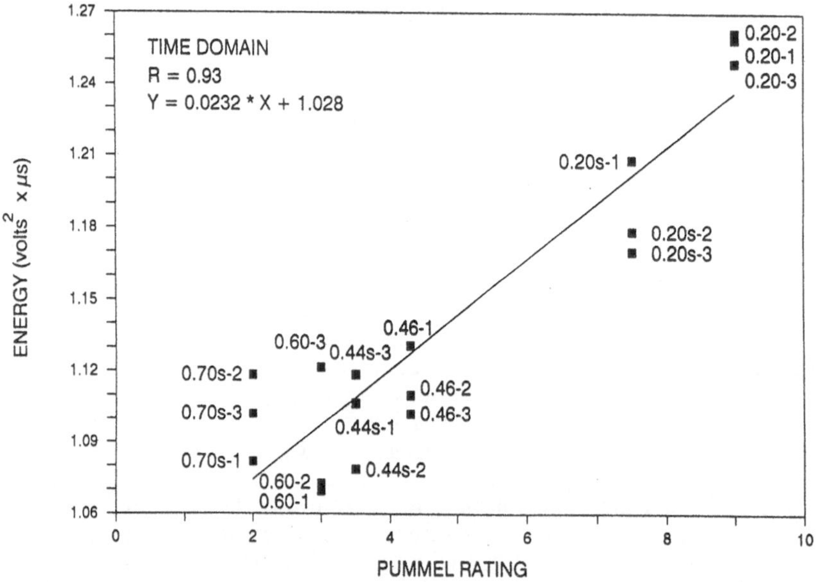

Figure 3. Energy of the acousto-ultrasonic signal vs. pummel test rating.

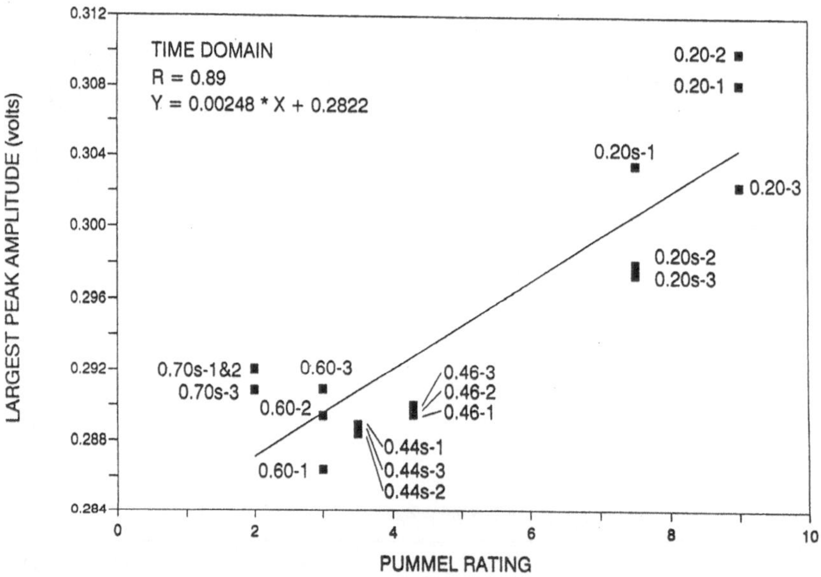

Figure 4. Largest peak amplitude of the acousto-ultrasonic signal in the time domain vs. pummel test rating.

application of this approach to evaluate the adhesive bond strength and possible delamination in laminated safety glass already in the field. Along with pattern recognition methods, the acousto-ultrasonic approach should provide an accept/reject criteria for quality control of laminated safety glass. For more effective and practical applications of the technique a more detailed investigation should be undertaken by studding multilayered laminated glass with layers of different thickness.

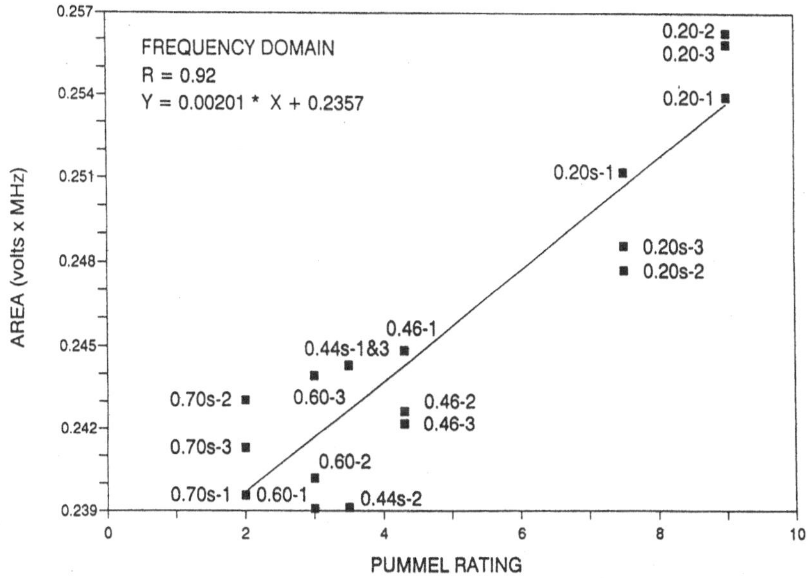

Figure 5. Area of the spectral density curve in the frequency domain vs. pummel test rating.

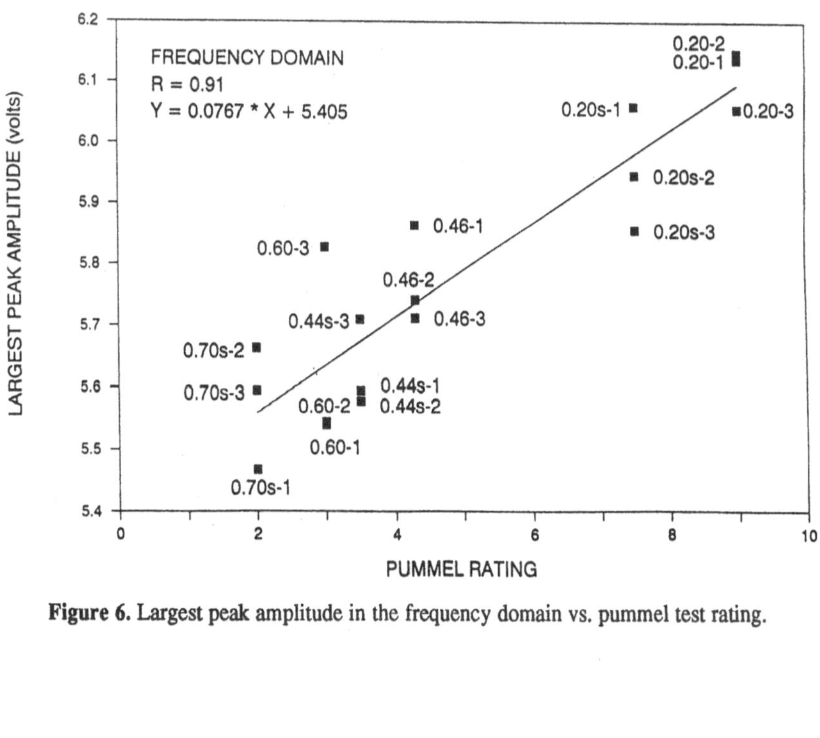

Figure 6. Largest peak amplitude in the frequency domain vs. pummel test rating.

ACKNOWLEDGEMENTS

The author is grateful to Mr. J. M. Shnur and to Mr. R. E. Butler, from Monsanto Chemical Company in Springfield, Mass., for their support and for manufacturing the laminated safety glass specimens with controlled adhesion. This study was possible through the National Science Foundation Research Equipment Grant No. MSM-8805400.

REFERENCES

1. R.L. Eifert, and R.E. Moynihan. "Fundamental Studies of the Adhesion of Butacite 106 to Glass," E.I. du Pont de Nemours and Co., Inc., Wilmington, Delaware, (1970).
2. J. Svoboda, and V. Matys, "Polyvinyl butyral films for laminated safety glass," *Plasty A Kaucuk*, Vol. 12, No. 5, pp. 131-135, (1975).
3. "Troubleshooting Guide for Laminate Performance Deviations," *Polyvinyl Butyral Interlayer Laminating Guide*, Monsanto Co., St. Louis, MO, (1984).
4. "Autoclaving of Laminates," *Polyvinyl Butyral Interlayer Laminating Guide*, Monsanto Co. St. Louis MO, (1984).
5. D.J. David, and T.N. Wittberg, "ESCA studies of laminated safety glass and correlations with measured adhesive forces," *Journal of Adhesion*, Vol. 17, pp. 231-242, (1984).
6. J.R. Huntsberger, "Adhesion of plasticized poly(vinyl Butyral) to glass, "*Journal of Adhesion*, Vol. 13, pp. 107-129, (1981).
7. J.E., Feig, and R.E. Moynihan. "The Compressive Shear Test as a measure of the Adhesion of 'Butacite' to Glass," E.I. du Pont de Nemours and Co., Inc., Wilmington, Delaware, (1971).
8. "Procedure No. 140 - Impact Testing of Safety Glass Laminates," *Polyvinyl Butyral Interlayer Laminating Guide*, Monsanto Co., St. Louis, MO, (1984).
9. "Motor Vehicle Safety Standard No. 205 Glazing Materials - Passenger Cars, Multipurpose Passenger Vehicles, Motorcycles, Trucks, and Buses," *Federal Safety Standards*, MVSS 205, (1973).
10. "Safety Code for Safety Glazing Materials for Glazing Motor Vehicles Operating on Land Highways," *American National Standards Institute, Inc.*, ANSI Z26.1 - 1977, (1977).
11. "Procedure No. 130 - 0°F Pummel Adhesion Test Procedure," *Polyvinyl Butyral Interlayer Laminating Guide*, Monsanto Co., St. Louis, MO, (1984).
12. H.L.M. dos Reis, L.A. Bergman, and J.H. Bucksbee, "Adhesive bond strength quality assurance using the acousto-ultrasonic technique," *The British Journal of Non-Destructive Testing*, Vol. 28, pp. 357-358, (1986).
13. H.L.M. dos Reis, and H. Kautz, "Nondestructive evaluation of adhesive bond strength using the stress wave factor technique," *Journal of Acoustic Emission*, Vol. 5, No. 4, pp. 144-147, (1986).
14. H.L.M. dos Reis, F.C. Beall, J.V. Carnahan, M.J. Chica, K.A. Miller, and V.M. Klick, "Nondestructive evaluation/characterization of adhesive bonded connections in wood structures," Nondestructive Testing and Evaluation for Manufacturing and Construction, H.L.M. dos Reis, Ed., Hemisphere Publishing Corporation, New York, N.Y., pp. 197-208, (1990).
15. "Analytical Ultrasonics in Materials Research and Testing," NASA Conference Publication 2383, A. Vary, Ed., NASA Lewis Research Center, Cleveland, Ohio, November 13-14, (1984).
16. "Acousto-Ultrasonics--Theory and Application," J.C. Duke, Jr., Ed., Plenum Press, New York, N. Y., (1989).
17. "Nondestructive Testing and Evaluation for Manufacturing and Construction," H.L.M. dos Reis, Ed., Hemisphere Publishing Corporation, New York, N. Y., (1990).
18. "Moisture Conditioning of Saflex," *Polyvinyl Butyral Interlayer Laminating Guide*, Monsanto Co., St. Louis, MO, (1984).
19. E. Lavin, and G.E. Mont, "Laminated Safety Glass," *U.S. Patent No. 3,271,234*, (1966).
20. E. Lavin, G.E. Mont, and A.F. Price, "Laminated Safety Glass," *U.S. Patent No. 3,271,235*, (1966).
21. F.T. Buckley, and J.S. Nelson, "Safety Laminates," *U.S. Patent No. 3,249,487*, (1966).
22. F.T. Buckley, R.F. Riek, and D.I. Christensen, "Process for Production of Interlayer Safety-Glass," *U.S. Patent No. 3,556,890*, (1971).
23. B.J. Dennison, and A.A. Doutt, "Treating Glass Sheets Used in Laminated Safety Glass Windshields," *U.S. Patent No. 3,607,178*, (1971).
24. F.T. Buckley, R.F. Riek, and D.I. Christensen, "Process for Production of Interlayer Safety-Glass," *U.S. Patent No. 3,718,516*, (1973).
25. "Lay-Up of Saflex on Glass," *Polyvinyl Butyral Interlayer Laminating Guide*, Monsanto Co., St. Louis, MO, (1984).
26. "Waveform Measurements and Analysis Programs--WAVEMAP," Department of General Engineering, University of Illinois at Urbana-Champaign, Urbana, Illinois, (1989).
27. J.H. Williams, Jr., E.B. Kahn, and S.S. Lee, "Effects of specimen resonances on acousto-ultrasonic testing," *Materials Evaluation*, Vol. 41, pp. 1502-1510, (1983).

SURFACE ACOUSTIC WAVES GENERATION BY PHASE VELOCITY SCANNING OF LASER INTERFERENCE FRINGES AND ITS APPLICATION TO NONDESTRUCTIVE MATERIALS EVALUATION

H. Nishino,[1*] Y. Tsukahara,[1] Y. Nagata,[2] T. Koda,[2] and K. Yamanaka [2]

[1]Technical Research Institute, Toppan Printing Co., LTD., Sugito-machi, Kitakatsusika-gun, Saitama, 345 Japan
[2]Mechanical Engineering Laboratory, Namiki 1-2, Tsukuba, Ibaraki, 305 Japan

INTRODUCTION

Laser generated ultrasounds[1,2] which do not depend on the ablation effect but on the thermoelastic effect have been expected to provide noncontact and nondestructive quantitative materials evaluation method. This method together with optical detectors[3] can be applied to measure acoustic properties of tested objects in various environments, high temperature, vacuum and so on, where the conventional piezoelectric transducer method cannot be applied.[4] Because of the noncontacting, it also can be applied to directly measure acoustic properties of tested objects without perturbation of coupling materials.[4]

In the thermoelastic regime, however, the amplitude of an ultrasound is sometimes too small to detect. To avoid this difficulty in surface acoustic wave (SAW) generation, some methods were proposed. One method was the narrow band generation and detection using laser induced grating (LIG) method[5] to improve the signal-to-noise (S/N) ratio. And the other used[6,7] converging SAW excited by a ring shaped laser irradiation pattern on a surface of tested objects to integrate the amplitude of the SAW.

Phase velocity scanning (PVS) method was also proposed[8,9] to enhance the amplitude of SAW without ablation. This method is realized by a laser heat source scanned on a tested object at the velocity equal to the phase velocity of SAW. This method was experimentally verified[8,9] by generating Rayleigh and Lamb waves of an aluminum specimen using scanning single laser beam (SSB) as a heat source. A theoretical study of the SSB approach of the PVS method was presented[10] as well. But the SSB approach is inadequate for 100-MHz-or higher frequency SAW generation, because the laser beam must be focused to size comparable to the wavelength of generated SAW.

Recently, another approach of the PVS method, which adequate for generating 100-

* Correspondence Address: Advanced Technology Division, Mechanical Engineering Laboratory, Namiki 1-2, Tsukuba, Ibaraki, 305 Japan.

MHz-or higher frequency SAW, was introduced.[11] Here, scanning interference fringes (SIF) which are realized by intersection of two laser beams on a specimen with different frequencies was used as a scanning heat source. Though the SIF approach is similar to the LIG method, based on static interference fringes, the SIF approach has unique advantages over the LIG method, i.e., the single directionality of generated SAW and the amplitude enhancement effect.[11-14]

In the past, a bulk wave generation in electrostrictive materials by using scanning interference fringes was demonstrated.[15] But the present method of SAW generation is not based on the electrostrictive effect but on the thermoelastic effect, being applicable to any kind of materials.

In this paper, we propose a novel method for a 100 MHz range SAW velocity measurement applicable to nondestructive characterization in microscopic scale as an application of the SIF approach of PVS method. It was experimentally verified in the SAW and the Pseudo SAW (PSAW) on an anisotropic material, silicon (100) surface.

METHOD OF VELOCITY MEASUREMENT

Here, we briefly summarize the principle of SAW generation by the SIF approach of the PVS method. The SIF itself basically can be obtained by the optical heterodyne interference effect.[16] Figure 1 schematically shows two laser beams with different frequencies, scanning interference fringe, a specimen and x axis. $\mathbf{k}_1$, $\mathbf{k}_2$, ω_1, ω_2, I_1 and I_2 are wave vectors, frequencies and amplitudes of each laser beam, respectively. Incident angles of the laser beams are θ and $-\theta$ to the normal of the specimen, respectively. The amplitude I of the laser beams on the surface of the specimen is

$$I = I_1 e^{i((K \sin\theta)x-\omega_1 t)} + I_2 e^{i((-K \sin\theta)x-\omega_2 t)}, \tag{2}$$

where $K = |\mathbf{k}_1| \approx |\mathbf{k}_2|$. Consequently, the intensity I^2 becomes

$$I^2 = I_1^2 + I_2^2 + 2I_1 I_2 \cos((-2K\sin\theta)x - \omega_a t), \tag{3}$$

where $\omega_a = \omega_2 - \omega_1$ is the frequency difference of two laser beams. The first and second term of the right hand side in eq. 3 are the direct component corresponding to the time domain profile of the laser pulse. The last term of the right hand side in eq. 3 shows the interference fringes with the wave number,

$$k_f = 2K \sin\theta, \tag{4}$$

which is scanned at a velocity v_f along the x axis.

$$v_f = -\omega_a / 2K \sin\theta\,. \tag{5}$$

Thus a SAW is generated by the thermoelastic effect of the SIF.

The features of generated SAW by the SIF are the following three.[12-14] (i) The amplitude of the generated SAW takes a maximum value when the SIF velocity is equal to the phase velocity of the SAW. (ii) A large number of carriers are obtained. This is the cause of the precise measurement of SAW frequency. (iii) The frequency of the generated SAW is changed by the wave number of the SIF, k_f determined by the incident angles of laser beams. This means that the SAW frequency ω is not always equal to the frequency shift ω_a, and is determined by the following simple equation in the case of $T < d / v_{SAW}$.

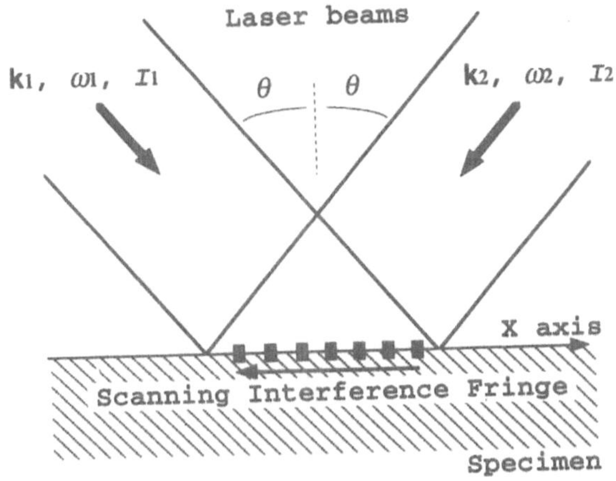

Figure 1. Schematic cross-section of laser beams with different frequencies, a specimen and the scanning interefence fringes.

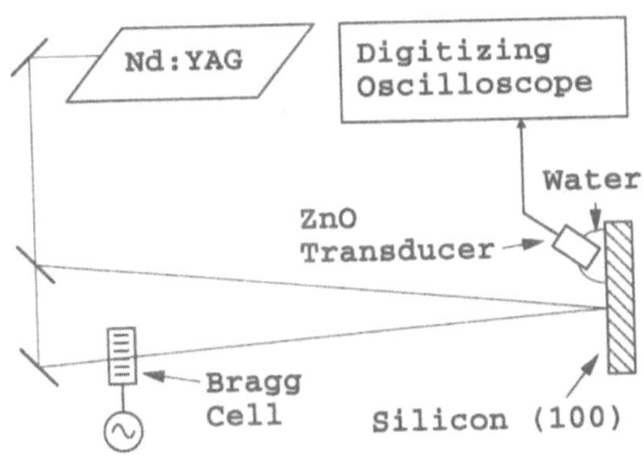

Figure 2. Schematic illustration of experimental setup.

$$\omega = v_{SAW} \cdot k_f, \qquad (6)$$

where T, d and v_f are the laser pulse width, the laser beam diameter and the SAW velocity.

For the velocity measurement, eq. 5 is simply inverted as

$$v_{SAW} = \omega / k_f \qquad (7)$$

For an experimental point of view, ω is measured from generated SAW and k_f is determined by the incident angles of two laser beams (refer to eq. 4). Thus the velocity of a SAW is determined by eq. 7.

In the next section, this method is applied to anisotropic SAW or PSAW velocity measurement of silicon (100) surface.

EXPERIMENT

Experimental Setup

Figure 2 illustrates of the experimental setup for generating SAW and PSAW on the silicon (100) surface. The silicon (100) surface bears pure SAW and PSAW. The SAW propagates in the direction of [100] ± 30°. PSAW accompanies leaky bulk waves and propagates in the direction of [110] ± 15°. The second harmonic wave of the Q-switched Nd:YAG laser was used as a thermoelastic source. The wave length, energy, pulse width and beam diameter of the laser were 532 nm, 1.8 mJ, 123 ns, 3 mm, respectively. A TeO_2 Bragg cell was used to make a frequency difference of two laser beams. The frequency difference ω_a is 110 MHz. The SAW or the PSAW was detected with a 2 mm x 2 mm ZnO piezoelectric transducer with the center frequency of 120 MHz. The propagation length of SAW and PSAW between the transducer and the laser beam spot on the specimen was about 5 mm. Scattered light of the laser, which was received by an avalanche photo diode, was employed as a trigger signal of the digitizing oscilloscope.

Anisotropic SAW Velocity Measurement

Experiment to measure the anisotropic SAW velocity for each direction of silicon (100) surface was carried out along the following sequence. As shown in Fig. 3, the ZnO transducer with water coupler was fixed at about 20° to the normal to the silicon (100) surface. This fixed angle was obtained by changing the wedge angle in order to obtain the maximum amplitude of generated SAW along [100] direction. The incident angles of the laser beams were fixed to ± 0.34° to the normal of the silicon (100) surface, and then the wave number of the SIF was fixed. Generated waveform detected by the ZnO transducer was stored by the digitizing oscilloscope. For the anisotropy measurement, we rotated the specimen by 90° at 3° steps. And to store the waveform of generated SAW or PSAW at each step.

Figure 4 shows a typical waveform of SAW generated by the SIF along [100] direction of the silicon (100) surface. Many carriers were obtained in Fig. 4, and the S/N ratio of the waveform is more than 30 dB. This is a result of the amplitude enhancement effect of the PVS method. The Fast Fourier Transform (FFT) spectrum of the waveform is shown by the solid curve in Fig. 5. The peak frequency of the spectrum indicates 108.36 MHz. The dotted curve indicates FFT spectrum of observed PSAW which is generated along [110] direction of the silicon (100) surface. The peak frequency of the spectrum was 112.00 MHz, and the 3 dB widths of the two peaks are about 1.7 MHz.

From the peak frequency of generated SAW or PSAW for each direction and the wave number of the SIF, velocity of SAW or PSAW was obtained using eq. 6 in Fig. 6. However, obtained velocity of SAW or PSAW had an error due to an alignment error of laser incident angles θ. Therefore, the experimentally obtained SAW velocities were multiplied with a constant factor a = 1.012, so that the experimental SAW velocity for [100] direction agrees with the theoretical value of 4921 m/s.[17]

Closed squares indicate, thus, obtained experimental result of velocity measurement. Solid and dotted curves in Fig. 6 show calculated values of SAW and PSAW velocity,

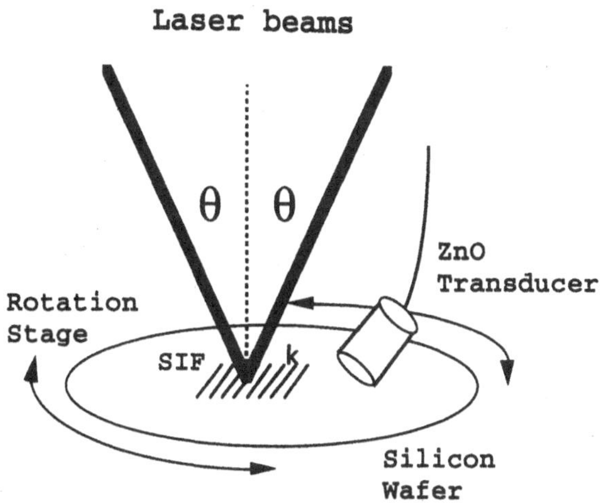

Figure 3. Schematic view of experimental setup for anisotropic SAW velocity measurement of silicon (100) surface.

respectively using a computer software.[17] Experimental results were in good agreement with calculated curves.

Figure 7 shows a generated waveform along 61° from [100] direction which is in the transition region between SAW and PSAW. Beating was observed in the waveform. The FFT spectrum of the beating waveform had two peaks as shown in Fig. 8. Therefore, the beating is the result of two frequency components corresponded to the two spectrum peaks in Fig. 8. From these two frequency components, two velocities were obtained simultaneously. Two closed circles in Fig. 6 indicate the two velocities which are corresponding to the SAW and PSAW velocities in the transition region. And these results are also in very good agreement with the calculated values.[17]

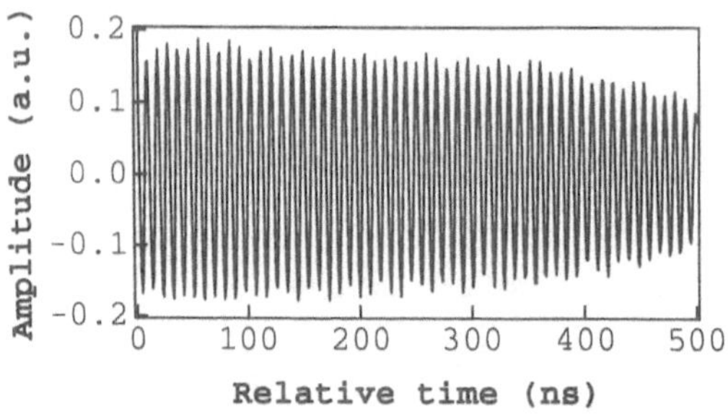

Figure 4. Typical waveform of SAW generated by the SIF along [100] direction of silicon (100) surface, which was clearly obtained without any averaging and signal processing

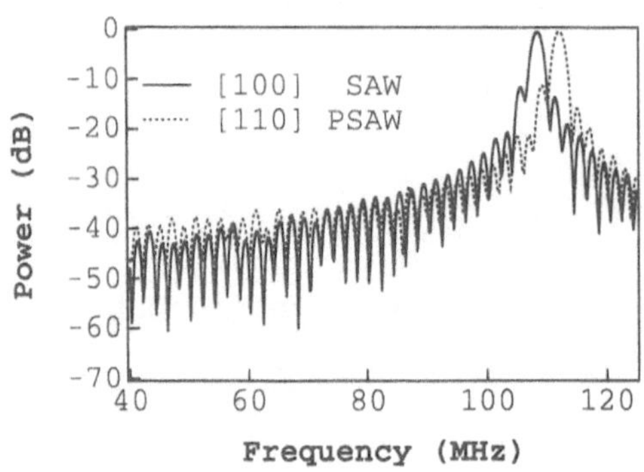

Figure 5. Typical power spectra of SAW and PSAW along [100] and [110] direction, respectively.

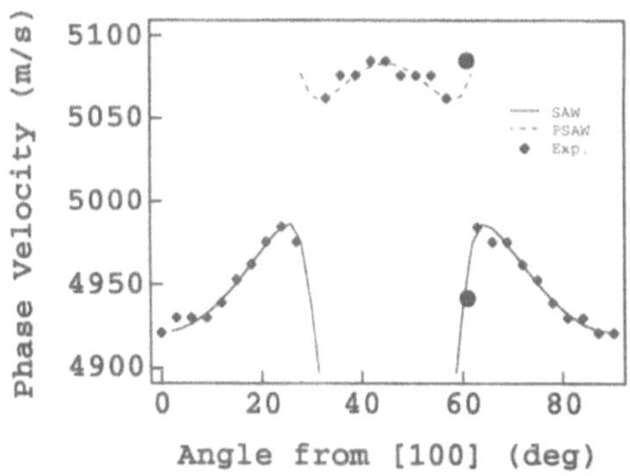

Figure 6. SAW and PSAW velocity of silicon (100) surface as a function of the angle from [100] direction. The solid and dotted curves indicate calculated values and closed squares indicate experimental results. Two closed circles also indicate experimental results in the transition region between SAW and PSAW, which can be obtained from one waveform in Fig 7 simultaneously .

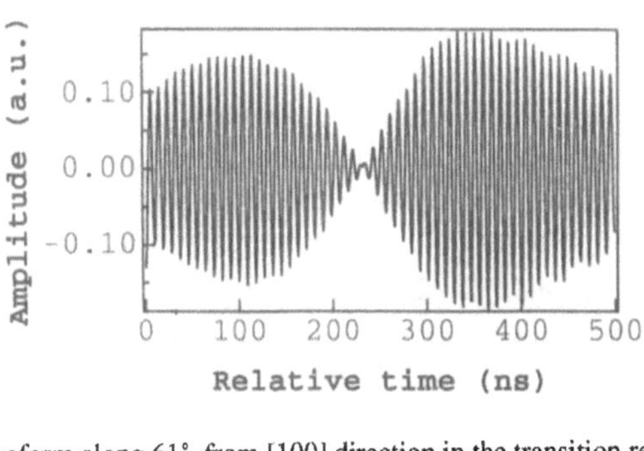

Figure 7. Waveform along 61° from [100] direction in the transition region between SAW and PSAW.

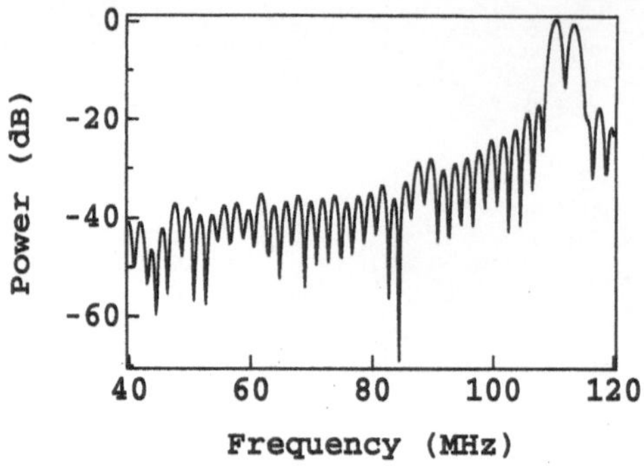

Figure 8. Power spectrum of the waveform in Fig. 7 in the transition region between SAW and PSAW.

DISCUSSION

Previously, these kinds of measurement were carried out by using the line-focus-beam acoustic microscope,[18] and excellent results were obtained. In this method, however, coupling materials (mainly water) are loaded between the acoustic lens and a tested object. Therefore, acoustic properties of the tested object are influenced by a perturbation of coupling materials. In particular, the measured quantity is the leaky SAW velocity which is slightly different from the SAW velocity.

However, our method of velocity measurement has no influence of the perturbation of coupling materials. Even when the coupling material is employed for the detection of SAW, the velocity is determined at the free surface of the tested objects, where the SAW is excited by the SIF. Therefore, the measured velocity is the pure SAW velocity rather than the LSAW velocity. This simplifies the theoretical calculation considerably.

CONCLUSION

A novel method of velocity measurement of SAW and/or PSAW with frequency higher than 100 MHz was proposed by using a noncontacting laser generated SAW technique of the scanning interference fringes approach of the phase velocity scanning method. The excellent performance of this method was experimentally verified using the anisotropic silicon (100) surface. In the transition region of SAW and PSAW, the method was also able to measure the velocity of two modes, precisely and simultaneously.

The method can be applied to obtain acoustic properties of micro-scaled devices with unique features of no perturbation of coupling materials and high signal-to-noise ratio.

Detailed measurements of SAW and PSAW velocities at the transition region of silicon (100) surface will be carried out together with optical detection in the future work.

ACKNOWLEDGMENT

We wish to extend our gratitude to Assistant Professor J. Kushibiki of Tohoku University Japan for useful and valuable discussions including acoustic properties of anisotropic silicon (100) surface.

REFERENCES

1. D. A. Hutchins, in *Physical Acoustics vol. XVIII*, edited by W. P. Mason and R. N. Thurston (Academic, San Diego, 1988), p.21-123.
2. S. J. Davies, C. Edwards, G. S. Taylor and S. B. Palmer, J. Phys. D: Appl. Phys., 26, 329 (1993).
3. J. P. Monchalin, IEEE Trans. Ultrason. Ferroelectronics, Frequency Control UFFC-33, 485 (1986).
4. C. B. Scruby, Ultrasonics, 27, 195 (1989).
5. G. Cachier, Appl. Phys. Lett., 17, 419 (1970).
6. P. Cielo, F. Nadeau and M. Lamontagne, Ultrasonics, 23, 55 (1985).
7. C. K. Jen, P. Cielo, F. Nadeau, J. Bussiere and G. W. Farnell, IEEE, Proc. Ultrason. Symp. Dallas, Texas, 660, (1984).
8. K. Yamanaka, Y. Nagata and T. Koda, Appl. Phys. Lett., 58, 1591 (1991).
9. K. Yamanaka, Y. Nagata and T. Koda, in *Review of Progress in Quantitative Nondestructive Evaluation Vol. 11*, edited by O. D. Thompson and D. E. Chimenti (Plenum, New York, 1992), p. 633.
10. Y. Tsukahara, Appl. Phys. Lett., 59, 2384 (1991).
11. H. Nishino, Y. Tsukahara, Y. Nagata, T. Koda and K. Yamanaka, Appl. Phys. Lett., 62, 2036 (1993).
12. H. Nishino, Y. Tsukahara, Y. Nagata, T. Koda and K. Yamanaka, Jpn. J. Appl. Phys., 32, 2536 (1993).
13. K. Yamanaka, H. Nishino, Y. Tsukahara, Y. Nagata and T. Koda, to be presented at Ultrasonics International '93, Vienna, Austria 1993.
14. K. Yamanaka, H. Nishino, Y. Tsukahara, Y. Nagata and T. Koda, submitted to J. Appl. Phys.
15. D. E. Caddes, C. F. Quate and C. D. W. Wilkinson, Appl. Phys. Lett., 8, 309 (1966).
16. e.g., F. J. Eberhardt and F. A. Andrews, J. Acoust. Soc. Am., 48, 603 (1970).
17. E. L. Adler, C. K. Jen, G. W. Farnell and J. Slaboszewicz, Proc. of Ultrasonics International '85, King's College London, United Kingdom, p.733-738, 1985.
18. J. Kushibiki and N. Chubachi, IEEE trans. Son. and Ultrason. SU-32, 189 (1985).

MATERIAL CHARACTERIZATION BY CONTACTLESS MEASUREMENT OF ULTRASONIC ABSORPTION USING LASER ULTRASOUND

B. Haberer, M. Paul*, H. Willems, and W. Arnold

Fraunhofer-Institute for Nondestructive Testing
University, Building 37
D-66123 Saarbrücken, Germany
*Present Address: SINTEF Applied Physics
N-7034 Trondheim, Norway

INTRODUCTION

The inelastic part of the ultrasonic attenuation, i.e. the ultrasonic absorption, in polycrystalline materials such as steels is caused by several kinds of internal friction mechanisms related to material properties. At room temperature and in the frequency range between 0.1 MHz - 20 MHz which covers most applications of conventional ultrasonic testing, the dominant absorption losses are usually associated with the interaction of the ultrasonic wave field with dislocations and, in ferromagnetic materials, with the magnetic domain structure[1]. The dislocation structure, for example, is strongly affected by plastic deformation and can therefore be used in order to characterize cold working or to assess the state of creep[2] and fatigue[3], whereas the magnetic domain structure is particularly sensitive with respect to stress states and microstructural characteristics like grain structure, precipitation structure or texture[4]. A wider use of ultrasonic absorption in the field of material characterization is unfortunately limited by the fact, that sufficiently precise absorption measurements are usually not possible by means of conventional ultrasonic techniques. These techniques normally measure the total ultrasonic attenuation resulting from the combined contributions of scattering losses, geometric losses and absorption losses. Especially in strongly scattering materials, a separation of the absorption part becomes difficult and is subject to poor resolution. A direct access to ultrasonic absorption, however, is rendered possible by measuring the reverberation of the diffuse ultrasonic soundfield in samples of limited size[5,6]. We present an extension of this technique using laser ultrasound which provides two essential advantages. Firstly, absorption measurements can be made up to very high temperatures, and secondly, the technique can be applied to very small specimens because any external damping of the ultrasonic signal as present with piezoelectric transducers is avoided.

Nondestructive Characterization of Materials VI
Edited by R.E. Green, Jr. *et al.*, Plenum Press, New York, 1994

ULTRASONIC ABSORPTION

Background

In polycrystalline materials elastic losses occur when the ultrasound is scattered by grain boundaries due to the change of the acoustic impedances from grain to grain. If the wavelength λ is much larger than the grain size d, the attenuation caused by grain scattering can be described by Rayleigh's law which yields for the scattering coefficient α_S:

$$\alpha_S = S\, d^3 f^4. \tag{1}$$

Here, S is the material-dependent scattering parameter, and f is the ultrasonic frequency. The total attenuation coefficient α may be expressed as

$$\alpha = \alpha_S + \alpha_G + \alpha_A. \tag{2}$$

Here, α_G takes into account geometric losses by soundfield spreading, and α_A stands for the inelastic losses due to internal friction, i.e. for the ultrasonic absorption. At a frequency of 5 MHz and at room temperature, typical absorption coefficients are between 1dB/ms (aluminum[7]) and 100 dB/ms (ferritic steel[6]). For ultrasonic transducers with diameters < 10 mm and f < 10 MHz, this is much less compared to α_G ($\approx$ 300 dB/ms). Furthermore, in coarse grained steels and at higher frequencies (f > 5 MHz), the total attenuation will be more and more governed by the scattering contributions as can be seen from Eq. 1.

Most of the ultrasonic absorption mechanisms can be described as relaxation processes of a standard linear, viscoelastic solid[8]. Then the logarithmic decrement δ (= α_A/f) is given by

$$\delta = \pi \frac{\Delta M}{M} \frac{f/f_0}{1+\left(f/f_0\right)^2} \tag{3}$$

ΔM is the difference between relaxed and unrelaxed ("anelastic" and "elastic") modulus, M is the unrelaxed modulus, and $f_0 = (2\pi\tau)^{-1}$ is the relaxation frequency. For a thermally activated process, the relaxation time τ follows the relationship

$$\tau = \tau_A \exp\left(E / k_B T\right) \tag{4}$$

with E - activation energy, k_B - Boltzmann constant and τ_A - relaxation constant. Under the condition $f / f_0 \ll 1$, E can be determined from the temperature dependence of δ by

$$\ln \delta = \text{const.} + E / k_B T. \tag{5}$$

Reverberation Technique

The reverberation technique exploits the fact that an ultrasonic pulse looses its coherence by scattering and, depending on the shape of the specimen under consideration, also by multiple reflections and mode conversions. In sufficiently small samples, this leads after some time intervall t_0 to a homogeneous, diffuse soundfield the temporal decay of which is only determined by the internal friction. t_0 is typically below 100 µs, whereas the decay time of the diffuse soundfield can take many milliseconds. Because of the long decay times very small absorption coefficients can be measured with a repeatability of about ± 10%.

The energy distribution of the different wave modes in a diffuse field has been calculated, showing that the shear modes are dominant[9]. The ratio of the energy content of the transverse and the longitudinal wave in the reverberation field is constant and does not depend on the volume of the sample, whereas the relative energy content of the surface waves varies with the sample volume as well as with the frequency. As all modes contribute to the reverberation field, the absorption coefficient is given in units per time (dB/ms) rather than in units per distance.

If reverberation measurements are performed using piezoelectric transducers, the mechanical coupling causes an additional damping which becomes the more disturbing the smaller the sample volume compared to the transducer area. For typical transducers, we have found a lower limit of about 15 cm^3 for the sample volume in order to allow for reliable absolute measurements[6]. On the other hand, the application of the technique is limited with respect to the maximum sample volume. Depending on the actual absorption and the dynamic range of the measuring system, the volume should not exceed about 100 cm^3 - 1000 cm^3, because otherwise the ultrasonic energy will have completely dissipated before a homogeneous distribution is obtained. Using laser ultrasound the problem of external damping can be avoided, allowing us to measure absolute ultrasonic absorption in samples as small as about 100 mm^3 in an entirely contactless way (if one neglects the influence of sample support and radiation losses).

EXPERIMENTAL SETUP

Figure 1 shows our experimental setup. For the generation of ultrasound, a Nd-YAG (wavelength 1064 nm) with maximal energy of 450 mJ and about 10 ns pulse duration is used. The beam diameter of 7 mm is focused to 3 mm on the sample surface providing power densities of about 400 - 600 MW/cm^2. The generated ultrasonic signals are usually detected on the opposite side by means of a heterodyne Michelson interferometer which is stabilized against vibrations[10]. After demodulation, the received signal (see Figure 2) is amplified using a logarithmic amplifier with tunable center frequency and then digitized. The digitized signals are transmitted to a PC where the signal evaluation is performed by regression analysis. Depending on the S/N ratio of the reverberation field, absorption data can be taken from 200 kHz to about 15 MHz. To improve the S/N ratio, several signals are averaged by slightly changing the excitation position. The sample can be put into a furnace allowing us to measure ultrasonic absorption at elevated temperatures up to 1400 K at present[11].

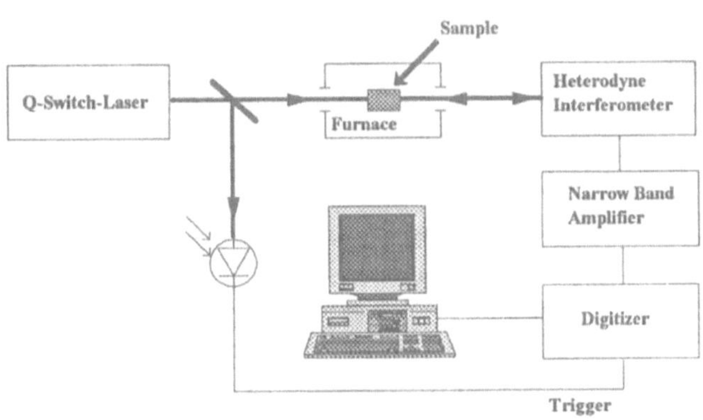

Figure 1. Experimental setup - schematic.

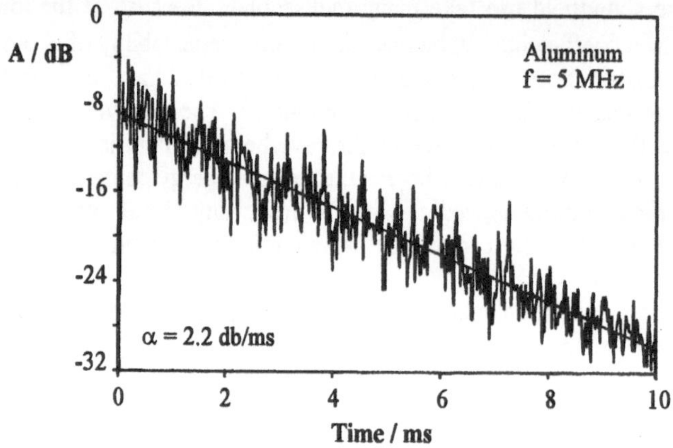

Figure 2. Typical ultrasonic reverberation signal after demodulation and averaging. Note that the time scale is of the order of milliseconds.

RESULTS AND DISCUSSION

We have performed absorption measurements on different materials in order to evaluate the potential of the reverberation technique with respect to the characterization of material properties[7]. Some of the results are reported and discussed in what follows.

Alloy 800H

Alloy 800H (X10 NiCrAlTi 32 20) is an iron-based austenitic alloy used in high temperature applications. Samples with cylindrical gauge length were produced from solution annealed bars and plastically deformed to different degrees using a tensile test machine. After deformation, pieces with a volume of about 3 cm^3 were prepared from the gauge length to perform ultrasonic absorption measurements with piezoelectric transducers as well as with laser ultrasound. As can be seen from Figure 3, the ultrasonic absorption increases with

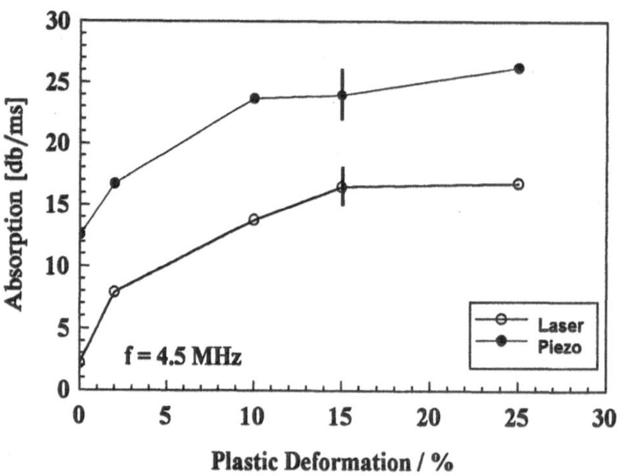

Figure 3. Ultrasonic absorption in Alloy 800H as a function of uniaxial plastic deformation. The constant difference in α_A between the laser measurement and the piezoelectric measurement is due to the mechanical damping caused by the piezoelectric transducer.

increasing plastic strain which can be explained by the increase in dislocation density. There is, however, a constant shift of about 10 db/ms between the results of the two techniques showing a higher absorption for the piezoelectric measurement. This is due to the mechanical contact between the sample and the ultrasonic transducer, causing an additional damping of the ultrasonic signal. The effect depends on the volume of the sample as well as on the size of the transducer aperture[6], and becomes the more pronounced the smaller the sample size. Using laser sound the external damping can be completely avoided thus allowing true absorption measurements even in very small samples. The smallest sample investigated by us had a volume of about 60 mm^3. In small samples, however, the absorption can be affected by surface machining which may be prevented by appropriate preparation.

The actual absorption also depends on the surface-to-volume ratio A/V of the sample under consideration because the surface displacements caused by the diffuse soundfield are affected[12]. Increasing A/V leads, at a given frequency, to an increase of the Rayleigh wave content which may have a different absorption coefficient than bulk waves. Experimental results on these aspects are reported elsewhere[7].

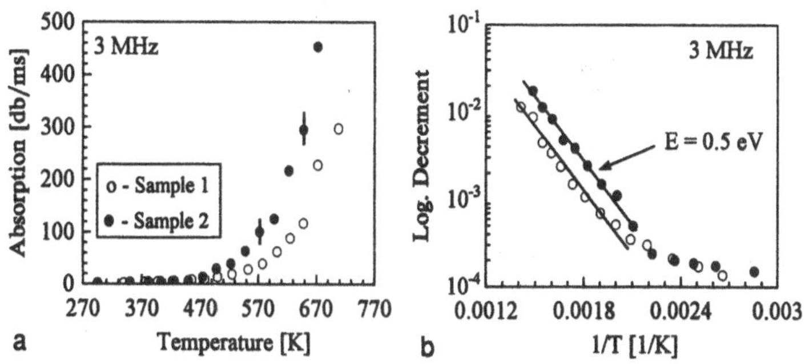

Figure 4. a: Ultrasonic absorption in two aluminum samples as a function of temperature. b: Logarithmic decrement vs. 1/T for the results from left figure.

Aluminum

We have investigated the temperature dependence of the ultrasonic absorption in an aluminum alloy containing 2 at. % Cu. For the experiments, which were carried out at an ultrasonic frequency of 3 MHz, two samples of the same shape but made from different heats were used. Two different regions can be identified from the results shown in Figure 4. Whilst being comparably low for both samples at temperatures below 200 °C, the absorption increases rapidly above 200 °C, now obviously differing for the two samples (Figure 4a). However, the slopes of the curves as obtained from an Arrhenius plot, i.e. a logarithmic plot of the decrement δ versus inverse temperature 1/T, are quite comparable in the high temperature range (Figure 4b), indicating the same thermally activated mechanism. Using Eq. 5, an activation energy of about 0.5 eV is derived from the slope.

The effect observed could be attributed to grain boundary relaxation[13]. In the case of pure aluminum, an activation energy of 1.4 eV is reported[14] for this type of relaxation. The somewhat lower value measured here is probably due to the influence of impurities which lower the activation energy as demonstrated recently[15].

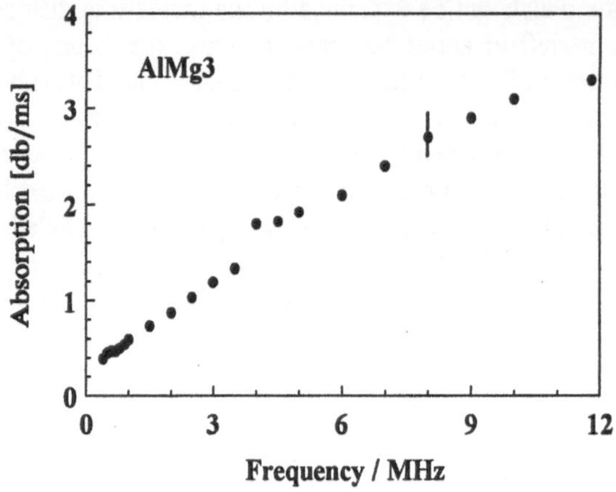

Figure 5. Frequency dependence of the ultrasonic absorption in aluminum alloy AlMg3.

We have also measured the frequency dependence of the absorption coefficient in different aluminum alloys. Figure 5 shows a typical result as obtained for AlMg3 in the frequency range 0.2 MHz to 12 MHz. The frequency dependence is found to be almost linear, i.e., there is no distinct evidence of a relaxational type of damping according to Eq. 3 as expected for dislocation damping. If one assumes however that the parameters entering Eq. 3 possess a wide distribution rather than being constant, then the resulting superposition might lead, at least within the limited frequency range considered, to the observed linear behavior.

White Cast Iron

The frequency dependence of the ultrasonic absorption in white cast iron is fairly well described by a relaxational behavior according to Eq. 3. As can be seen from Figure 6, the logarithmic decrement shows a maximum at a frequency of 1.3 MHz which corresponds to the relaxation frequency f_0. The relaxation process is interpreted by us in terms of magneto-elastic damping[16] caused by the movement of the Bloch walls in the ultrasonic strain field.

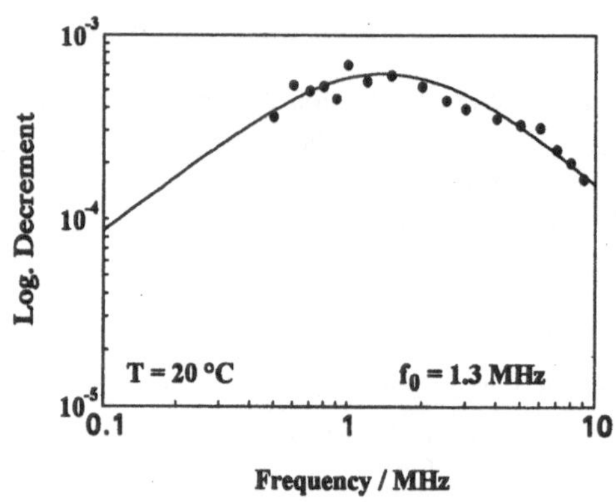

Figure 6. Logarithmic decrement as a function of frequency in white cast iron showing a relaxation maximum at 1.3 MHz.

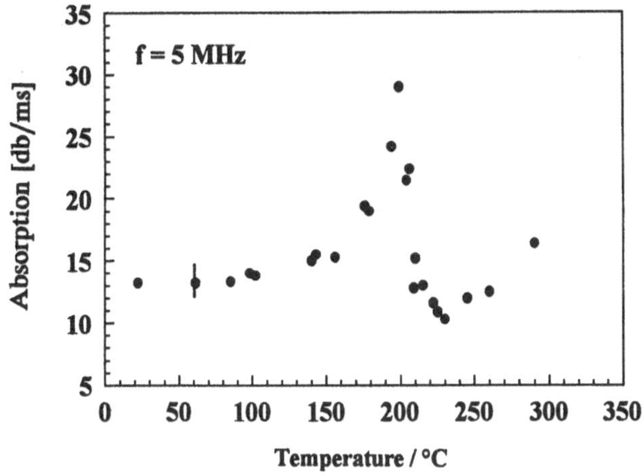

Figure 7. Ultrasonic absorption in white cast iron as a function temperature showing a phase transition of the cementite phase when the Curie temperature is reached.

White cast iron is a two-phase material containing ferrite and cementite (Fe_3C), with the latter amounting to 57.6 wt % in our case. At room temperature both phases are ferromagnetic. The cementite phase has a Curie temperature T_C of 210 °C where it changes from the ferromagnetic to the paramagnetic state. This phase transition strongly affects the ultrasonic absorption, as can be seen from its temperature dependence shown in Figure 7. When approaching T_C, the absorption rapidly increases because the increasing mobility of the Bloch walls promotes the interaction with the ultrasonic wavefield. In the paramagnetic state the cementite no longer contributes to the absorption which consequently drops down close to the value measured at room temperature. This also suggests that, at room temperature, the main contribution to the ultrasonic absorption results from the ferritic phase. The transition temperature as found from Figure 7 is in close agreement with the reported value of 210 °C.

CONCLUSIONS

We have shown that the measurement of ultrasonic absorption based on the reverberation technique can be essentially extended by using laser ultrasound. The laser technique allows us to measure absorption in technical materials

- contactlessly thus avoiding external damping of the sample,
- nearly independent of the geometry of the sample,
- in samples as small as about 100 mm^3,
- in the frequency range 0.1 MHz - 15 MHz,
- at elevated temperatures.

We have presented some examples demonstrating that the ultrasonic absorption in polycrystalline materials is suitable for characterizing material properties related to the dislocation structure and the magnetic structure. We think that the technique could be very helpful in studying absorption phenomena relevant to NDE applications as well as to more physical aspects.

REFERENCES

1. A.B. Bhatia, "Ultrasonic Absorption," Clarendon Press, Oxford (1967).
2. R. Lagneborg, Creep: mechanisms and theories, *in*: "Creep and Fatigue in High Temperature Alloys," J. Bressers, ed., Applied Science Publishers, London (1981).
3. D. Eifler and E. Macherauch, Microstructure and cyclic deformation behaviour of plain carbon and low-alloy steels, *Int. J. Fatigue* 12 No.3:165 (1990).
4. D. Jiles, "Introduction to Magnetism and Magnetic Materials," Chapman and Hall, London (1991).
5. R.L. Weaver, Indications of material character from the behavior of diffuse ultrasonic fields, *in*: "Nondestructive Characterization of Materials II," J.F. Bussière et al., eds., Plenum Press, New York and London (1987).
6. H. Willems, A new method for the measurement of ultrasonic absorption in polycrystalline materials, *in*: "Review of Progress in QNDE 6A," D.O. Thompson and D.E. Chimenti, eds., Plenum Publishing Corporation, New York (1987).
7. B. Haberer, Diploma Thesis, Physics Department, University of Saarbrücken and IzfP Saarbrücken (1993), unpublished.
8. C. Zener, "Elasticity and Anelasticity of Metals," The University of Chicago Press, Chicago (1948).
9. R.L. Weaver, On diffuse waves in solid media, *J. Acoust. Soc. Am.* 71:1608 (1982).
10. M. Paul, Diploma Thesis, Physics Department, University of Saarbrücken and IzfP Saarbrücken (1987), unpublished.
11. M. Paul, B. Haberer, and W. Arnold, Materials characterization at high temperatures using laser ultrasound, European MRS Fall Meeting Nov. 1992, Symp. B, Strasbourg, to be published in *Mat. Sci. Eng. A*.
12. R.L. Weaver, Diffuse elastic waves at a free surface, *J. Acoust. Soc. Am.* 78:131 (1985).
13. A.S. Nowick and B.S. Berry, "Anelastic Relaxation in Crystalline Solids," Academic Press, New York (1972).
14. R. de Batist, "Internal friction of Structural Defects in Crystalline Solids," North Holland Publishing Company, Amsterdam (1972).
15. H.G Bohn, and C.M. Su, Investigation of the influence of impurities of the grain boundary relaxation in thin Al-films on Si-substrate, *Material Science Forum* 119-121:273 (1993).
16. E. Kneller, "Feromagnetismus," Springer, Berlin (1962).

MICROSCOPIC ULTRASONIC IMAGING USING NON-CONTACT ULTRASONIC DETECTION BY OPTICAL HETERODYNE INTERFEROMETRY

Hisashi Yamawaki and Tetsuya Saito

National Research Institute for metals
2-3-12, Nakameguro, Meguro-ku, Tokyo 153, Japan

INTRODUCTION

Laser-based ultrasonic measurement technique[1,2,3] is expected as non-contact materials evaluation in hostile environment, and has several capabilities owing to the fact that it uses light beam. On the other hand, microscopic ultrasonic imaging systems such as a scanning acoustic microscope(SAM), scanning laser acoustic microscope(SLAM)[4] and photo-acoustic microscope(PAM)[5] are investigated for the evaluation of materials which are sensitive to micro defects as shown in the case of ceramics.

In this report, microscopic ultrasonic imaging technique using optical heterodyne interferometry for ultrasonic detection is investigated as a part of elemental techniques for the non-contact acoustic microscopy system[6] which is expected to be developed in future. Applicability of laser-based ultrasonic detection technique to microscopic material evaluation and new feature for the ultrasonic measurement are demonstrated in the imaging experiments using specimens of a small model defect and a fatigue crack.

OPTICAL HETERODYNE INTERFEROMETRY FOR ULTRASOUND DETECTION

Optical heterodyne is a popular technique in optical measurement systems for mechanical vibrations, and many commercial systems are manufactured. On such systems, two important merits of optical technique are employed. One is to give no affection against the object, and another is to be applicable to measurements at small region. To apply the heterodyne system to measure ultrasonic vibration, it is required that the system, especially a frequency/phase demodulator to extract ultrasonic signals, works correctly in ultrasonic frequency which is generally higher than mechanical one.

In the experiment, an optical heterodyne interferometer shown in Figure 1 is employed to detect ultrasonic vibration perpendicular to the specimen surface. 5mW polarized He-Ne laser beam is guided to the interferometer by a single mode optical fiber. Probe beam

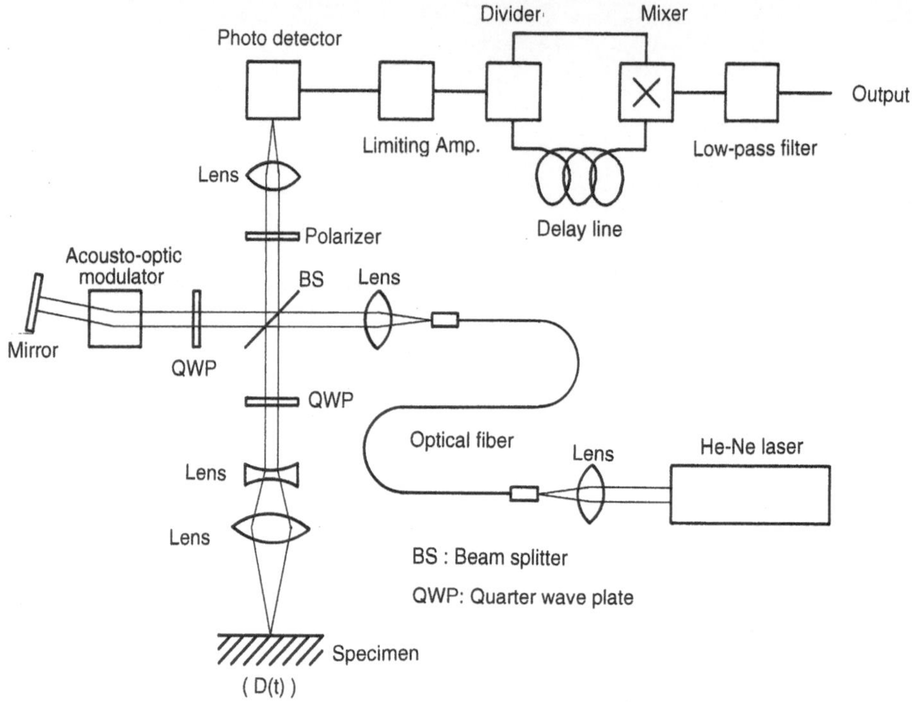

Figure 1. Schematic diagram of optical heterodyne interferometry system for detection of ultrasonic vibration.

and reference beam, which are divided by a polarized beam splitter, are interfered with each other on an avalanche photodiode detector. The optical frequency of the reference beam is shifted by the acousto-optic modulator by 160MHz. Interference beat signal output from the photo detector, whose frequency is same with the sifting optical frequency, is processed by a frequency/phase demodulator. The demodulator is functionally composed with a limiting amplifier to normalize amplitude of sinusoidal beat signal and a circuit to convert phase variation to amplitude.

When time-dependent displacement D(t) arises on a surface of specimen, the sinusoidal beat signal V is simply expressed as follow,

$$V = A\cos(2\pi f_s t - 4\pi D(t)/\lambda) \qquad \text{...(1)}$$

where, A is sinusoidal amplitude, f_s is beat frequency, and λ is wave length of laser beam. The limiting amplifier controls amplitude A to a constant value, and the signal is divided in two lines. One line is induced directly to a multiplexer and another is induced to it after delayed time t_d, where

$$t_d = \frac{n \pm 0.25}{f_s} \qquad (\text{ n: integer }) \qquad \text{...(2)}$$

Output signal of multiplexer is as follow,

$$V_o = (A^2/2)(\cos(4\pi f_s t - 2\pi f_s t_d - 4\pi(D(t)+D(t-t_d))/\lambda) + \cos(2\pi f_s t_d - 4\pi(D(t)-D(t-t_d))/\lambda)) \text{ ...(3)}$$

First term of V_o, which indicates sinusoidal signal with frequency $2f_s$, is neglected by low-pass filter. If D(t) satisfies a condition, $|D(t)-D(t-t_d)| << \lambda/8$, final output signal V_d becomes as follow,

$$V_d = \pm 2\pi A^2(D(t)-D(t-t_d))/\lambda \qquad \text{...(4)}$$

And, frequency response G(f) becomes as follow,

$$G(f) = 4\pi A^2 \sin(\pi f t_d) \qquad \text{...(5)}$$

From this equation, it is obvious that maximum sensitivity is obtained at frequency $f=1/(2t_d)$, and in the condition of $f<<1/(2t_d)$, the frequency characteristic for velocity measurement is obtained.

In this optical heterodyne system, the center frequency for detecting the displacement can be easily changed by adjusting the time delay t_d. Practically, the upper limit of frequency for the detection is determined by frequency response characteristic of the limiting amplifier and the acousto-optic modulators. Currently, the system developed can be used to detect ultrasonic vibration with frequency up to 50MHz.

EXPERIMENTAL SETUP

Measurement System

Figure 2 shows block diagram of a ultrasonic imaging system used in the experiments. A piezoelectric transducer with wide band width is employed for ultrasonic pulse generation and is attached on the back surface of specimens with wax. The ultrasonic center frequency of the transducer is 10MHz and its effective diameter is 6mm. A specimen is fixed on an XY-scanning stage. The diameter of focused probe beam at the specimen surface is calculated from laser wave length and focusing angle, and is estimated about 2μm. The frequency band width for vibrating velocity is adjusted by changing the length of delay line cables, and 50MHz, 10MHz or 2MHz is selectable actually. The peak amplitude and traveling time of detected ultrasonic vibrating velocity signal are measured by a gated peak detector and the combination of oscilloscope and time interval counter, respectively. Computer logs the amplitude and the time synchronously with the scanning and displays their mapped data with gray scale.

Specimens

For the evaluation of the imaging technique, two specimens are prepared. The one is bonded plates specimen of a small artificial inclusion. Another is a steel plate specimen of a steel plate with a fatigue crack. Figure 3 shows these specimens.

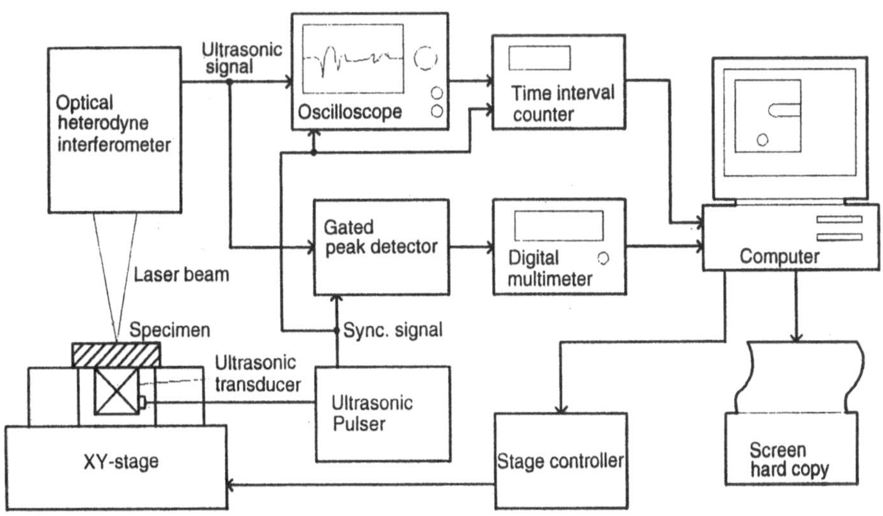

Figure 2. Block diagram of ultrasonic imaging system. Peak amplitude and traveling time of ultrasonic pulse wave are sampled using a gated peak detector and a signal level triggered interval counter.

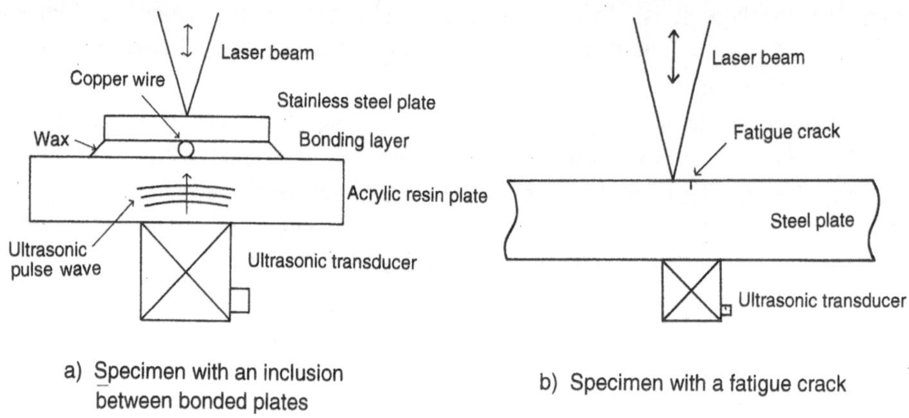

Figure 3. Arrangement of specimens. Center frequency of transducer is 10MHz. a) Diameter of copper wire is 0.1mm. b) Crack depth and opening are about 0.4mm and 5μm, respectively.

The bonded plates specimen is made from thin stainless steel plate and acrylic resin plate. The plates are bonded by wax and bonding layer contains copper wire of 0.1mm diameter as an artificial inclusion. The thickness of the steel plate is selected to be 0.4mm, 0.2mm or 0.1mm to investigate the relation between of inclusion depth and images. Acrylic plates are adopted for the convenience to observe state of bonding layer. The surface of the steel plates are normally polished by #600 sandpapers and finished as a smooth surface.

The specimen with a fatigue crack is 10mm thick, 50mm wide, and made from low alloy steel. The crack is estimated to be 0.4mm deep and 5μm wide. The crack was induced on a notched steel plate and, after the crack growth, the notch was scraped. The surface is finished by a rotary grinder.

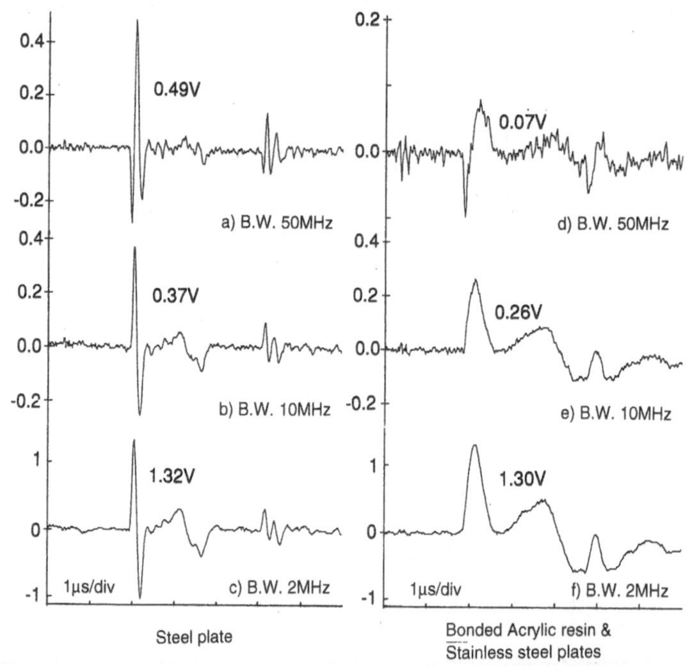

Figure 4. Typical ultrasonic wave forms detected by the optical heterodyne system.

EXPERIMENTAL RESULTS AND DISCUSSIONS

Figure 4 shows typical ultrasonic wave forms detected by the system. The waves indicate particle velocity amplitude normal to a surface. Detected wave forms change according to the band width controlled by the delay time of the demodulator, and also according to the specimens. A remarkable attenuation of higher frequency component is observed in the case of the bonding specimen. The fact comes from mainly ultrasonic characteristics of acrylic resin.

Imaging of Artificial inclusion

Figure 5 shows ultrasonic images of edges of copper wires. Images on the left indicate distribution of the pulse peak amplitude, and ones on the right indicate

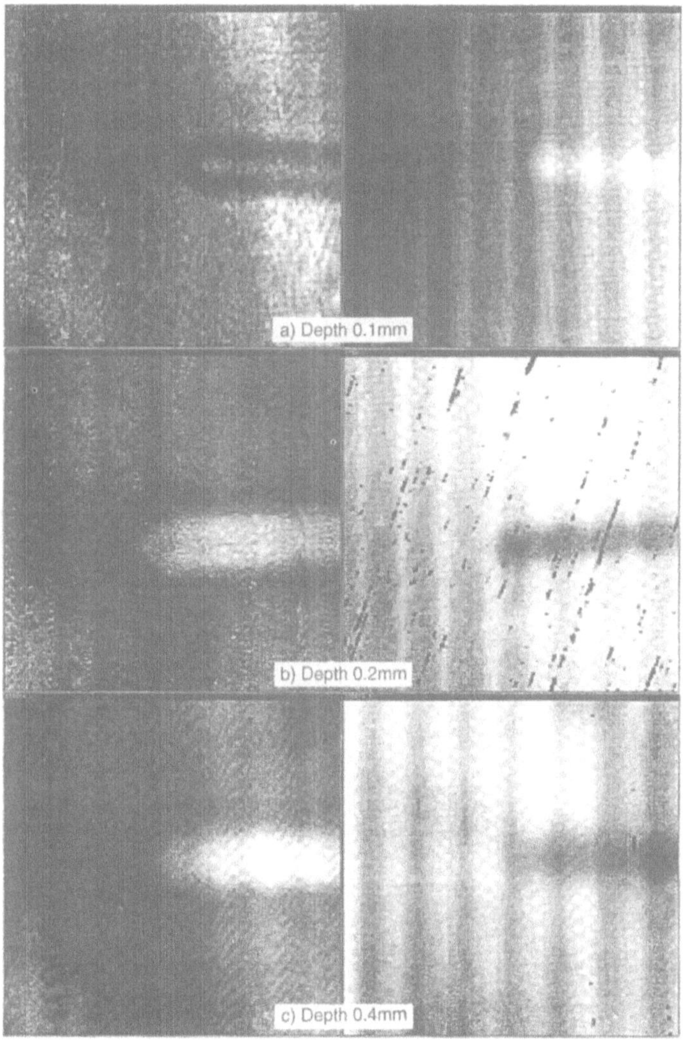

Figure 5. Ultrasonic images by amplitude(left) and traveling time(right). Edge of copper wire is observed. The size of view is 2.5 x 2.5mm. Frequency band width is 50MHz.

distribution of the traveling time. Periodical modulation observed as vertical stripes are due to cyclic temperature change caused by an air conditioner. Generally, ultrasonic attenuation and velocity of acrylic resin are sensitive to temperature.

As observed on the images of amplitude, according with decrease of inclusion depth, the figure of the wire becomes sharp. Especially, on the image of the depth 0.1mm, resolution of the image is remarkably improved. This tendency shows discrepancies from usual ultrasonic imaging using transducers, the resolution of which is limited by the ultrasonic wave length. This fact suggest that imaging with high resolution may be available by measuring the ultrasonic distribution just behind objects, and laser ultrasonic technique only may take advantage of the improved resolution, because laser ultrasonic technique has both merits of direct detection of surface vibration and applicability on a small region.

On the image of traveling time, no significant difference in resolution can be observed, and the resolution seems to be better than the case of amplitude. From the observation, it is clear that traveling time also useful for the imaging. The obvious difference exists on polarity of its contrast between a image of depth 0.1mm and others. On the images, white region means long traveling time, and occurrence of the inversion seems strange. This problem requires further consideration.

Figure 6 shows comparison between images obtained by specimens with smooth surface and rough surface polished with sandpaper #120. In the case of a image with the rough surface specimen, in spite of the reduction of S/N, a figure of the wire similar to the case of specimen with smooth surface can be observed. In the laser ultrasonic technique, reduction of S/N influenced by rough surfaces always arises on the detection of ultrasound,

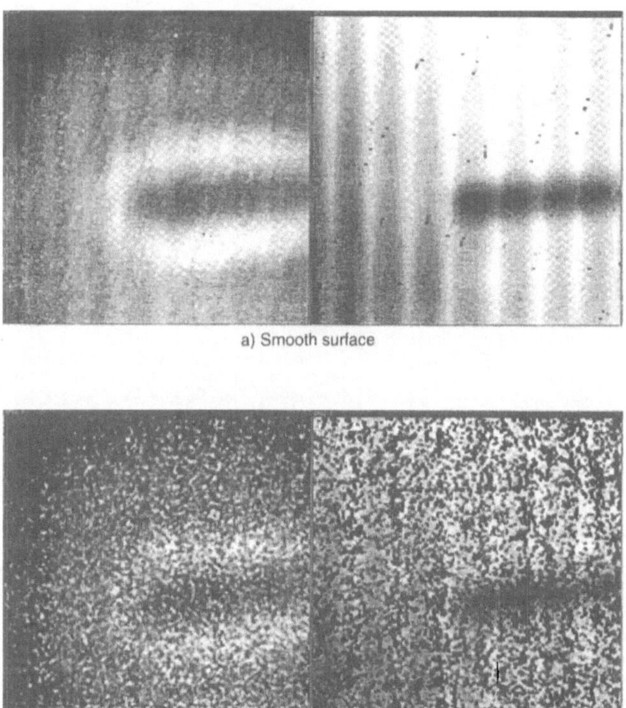

a) Smooth surface

b) Rough surface

Figure 6. Ultrasonic images of specimens with smooth and rough surfaces. Depth of copper wire is 0.2mm. Frequency band width is 2MHz.

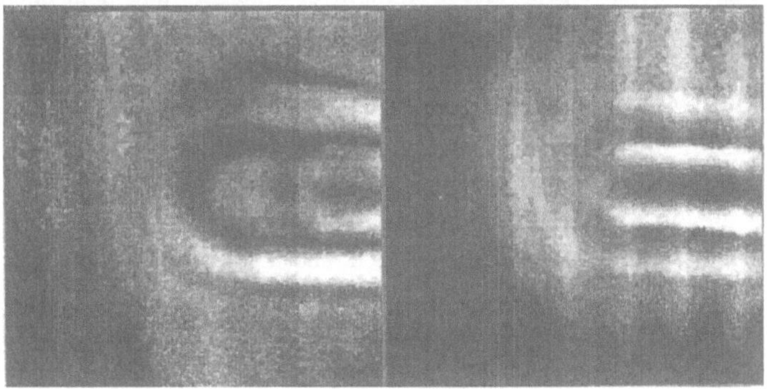

Figure 7. Ultrasonic images by amplitude and traveling time of second peak of the wave form in figure 4-d.

but images are never deformed by the surface undulation. These facts indicate potential applicability of this technique to rough surface specimens, because laser interferometry detects surface vibration directly, without any couplant.

On the images by traveling time, there is no difference in resolution between images for frequency band width 2MHz in figure 6-a and 50MHz in figure 5-b. It indicates that resolutions obtained by imaging with traveling time are not related to ultrasonic frequency, because, if traveling time is correctly measured at the first edge of ultrasonic pulse waves, the traveling time is not affected by ultrasonic interference effect and wave length.

Figure 7 shows a image obtained by mapping amplitude and traveling time of the second peak of the wave shown in figure 4-d. As observed in the figure, multiple fringe patterns appear in both images. The fringe pattern is considered to appear with the interference between each peak of a wave. Generally, in the ultrasonic imaging technique, multiple fringe pattern often appear. But as shown in figure 5, the simple image is obtained by the usage of first peak.

Imaging of fatigue crack

On the evaluation of materials, not only bulky defects but also cracks are the object for

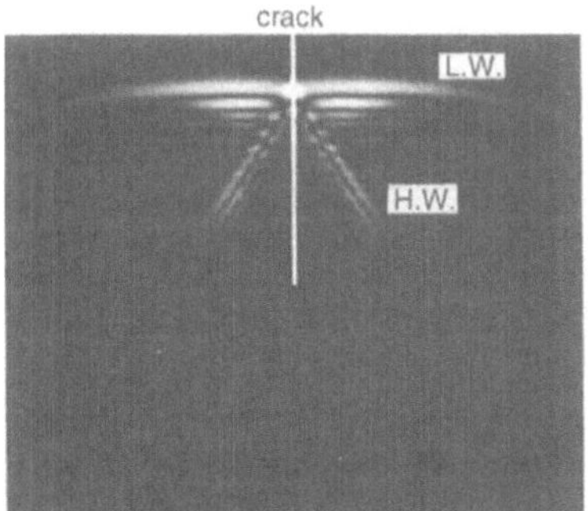

a) Calculated ultrasonic waves

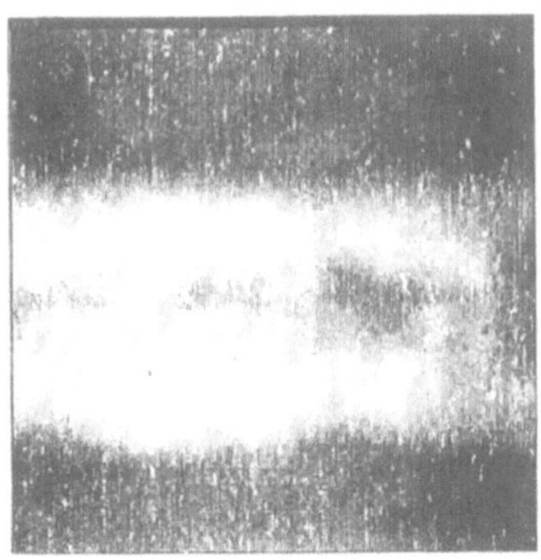

b) Ultrasonic image

Figure 8. a) Calculated longitudinal waves propagating along a crack with 0.1mm opening and 4mm depth. b) Ultrasonic image of the specimen with a fatigue crack. Frequency band width is 50MHz.

testing. Especially, micro cracks affect strength of ceramics and it is one of important targets for material evaluation. So, the detection of cracks is examined by the imaging technique.

Figure 8-a shows a longitudinal ultrasonic propagation along the crack calculated by the computer simulation[7]. Near the surface of crack, longitudinal waves which propagate toward a upper surface is converted to share waves, so-called head waves, and strong displacements are observed at the point where the mode conversion occurs. Its strong displacement locates at the both sides of a crack, and it suggests the capability of detection of a crack near surface.

Figure 8-b shows the distribution of ultrasonic pulse amplitude around the fatigue crack. As predicted by the calculation, two stripes which indicate high intensity of ultrasound are observed along the crack opening. In the experiment, unfortunately, no variation of traveling time is obtained, and no information is available to determine the crack depth. But, actually, by using this technique, small cracks near surface may be easily detected.

CONCLUSION

By the experiments, following conclusions are obtained.

(1) Laser-based ultrasonic technique is expected as a new microscopic imaging technique for the evaluation of small defects below the surface of materials.

(2) Imaging technique using traveling time of ultrasonic pulse waves may bring new advantages for materials evaluation in comparison with imaging by the amplitude.

(3) It is clarified that the imaging system is also applicable to specimens with rough surface at the cost of S/N.

In future development of the system, some new features are expected. It is important to develop a non-contact imaging system by employing ultrasonic generation with pulsed lasers. The combination of ultrasonic holographic technique with the system is useful for imaging of microscopic defects deep inside materials. As an improvement of S/N, the holographic technique will be introduced soon, because the system developed here is ready for both data of amplitude and phase, where the phase is obtained from the traveling time.

REFERENCES

1. J.-P. Monchalin, Optical detection of ultrasound, *IEEE Trans. on Ultrason., Ferroelectrics, and Freq. Contr.*, UFFC-33:485(1986).
2. D.A. Hutchins, Mechanisms of pulsed photoacoustic generation, *Can. J. Phys.*, 64:1247(1986).
3. C.B. Scruby, R.J. Dewhurst, D.A. Hutchins and S.B. Palmer, in: "Research Techniques in Nondestructive testing, Vol.5," R.S. Sharpe, ed., Academic press, London(1982).
4. M. Nikoonahad, in: "Research Techniques in Nondestructive testing, Vol.7," R.S. Sharpe, ed., Academic press, London(1984).
5. G. Birnbaum and G.S. White, in: "Research Techniques in Nondestructive testing, Vol.7," R.S. Sharpe, ed., Academic press, London(1984).
6. B.R. Tittmann, R.S. Linebarger and R.C. Addison, Jr, Laser-based ultrasonics on Gr/Epoxy composite: interferometric detection, in:"Review of Progress in Quantitative Nondestructive Evaluation, Vol.9A," D.O. Tompson and D.E. Chimenti, ed., Plenum Press, New York(1990).
7. H.Yamawaki and T.Saito, Numerical Calculation of Surface Waves using New Nodal Equations, *Nondest. Testing and Eval.*, 8-9:379(1992)

A PRACTICAL SYSTEM FOR PULSED LASER ARRAY GENERATION OF ULTRASOUND

T.W. Murray, J.S. Steckenrider*, J.W. Wagner, and J.B. Deaton, Jr.**

Center for Nondestructive Evaluation
The Johns Hopkins University
Baltimore,MD. 21218

ABSTRACT

Past investigators have predicted and , in limited cases, demonstrated the potential utility of using temporally and spatially modulated laser array sources as a more efficient means than a single laser point source for the generation of ultrasound in materials. In an effort to develop a more practical and flexible laser array source, a new laser system has been designed and implemented. The system uses a single Nd:YAG pulsed laser and an optical system in which the beam path passes repetitively through a White cell cavity being sampled by a custom beamsplitter after each pass. Up to ten spatially separated light beams exit from the system each time a single laser pulse is introduced into the system. Time separation between the distinct beams can be adjusted over a range from 28 ns to 170 ns, corresponding to a pulse repetition rate from 6 Mhz to 36 Mhz. Since individual control of the beam paths is possible, one has the flexibility to implement either a single element, temporally modulated array, or a multi-element, "phased" array. From a single point, a multiple spike narrowband acoustic signal is generated permitting greater detection sensitivity than can be obtained with broadband detection of a single spike. Alternatively, multi-element arrays can be used to can be used to increase the acoustic signal amplitude by superposition of the signals from each individual element, once again enhancing detection sensitivity.

INTRODUCTION

The advantages of using spatial and temporal modulation of laser sources of ultrasound can be understood by considering the following expression derived for an

* Argonne National Labs, 9700 S. Cass Ave.,Bldg.212, Argonne, IL 60439
** Analytical Services and Materials,Inc., 107 Research Drive, Hampton,VA 23666

optical detection system whose sensitivity is shot noise limited. [1]

$$SNR = \frac{Signal\ \ Power}{Noise\ \ Power} \propto \delta^2 \frac{P}{B}$$

The signal-to-noise ratio (SNR) is given here as a function of the surface displacement , δ, the optical power received by the detector, P, and the detection bandwidth of the system, B. The use of ultrasonic arrays allows for the reduction of the bandwidth of the laser generated ultrasonic signal. This narrowband signal may then be extracted from the noise using the appropriate filtering techniques and the bandwidth of the system effectively decreased. In an additional mode of implementation, a phased or directed array may be generated in which adjacent array elements superimpose to increase the surface displacement. These cases both lead to an increase in the SNR.

The use of laser arrays to enhance signal-to-noise ratio has been discussed quite extensively in the literature [2,3,4]. Three potential advantages are generally cited. First, repetitively pulsed or single pulsed periodic arrays can give rise to narrowband acoustic signals. Next, the appropriate combination of spatial and temporal modulation of array elements can be used to enhance, by superposition, the surface displacements caused by certain modes of acoustic waves. Finally, the optical power in each element in the array may be kept low enough to avoid damage to the surface of the specimen of interest, giving the array an advantage over the use of a single high-power laser pulse for generation. In addition, the use of laser arrays for the steering of beam energy in a manner similar to that obtained using piezoelectric arrays has been explored. The utility of such beam steering is severely restricted owing to the fact that laser generated displacements from bulk acoustic signals are in general mono-phasic in nature and there is a strong energy directivity pattern associated with a thermoelastic acoustic source.

Consider the first of the potential gains in the overall laser ultrasonic system sensitivity to be achieved by the reduction in system bandwidth. Pulse durations from Q-switched solid state lasers may range from a few to several hundred nanoseconds. To a first approximation, the nature of an acoustic spike arising from rapid thermal expansion of a laser illuminated region of an object follows the time signature of the laser spike. In other words, narrow laser spikes give rise to narrow elastic spikes. As a result, the frequency content of the laser generated acoustic wave can be very broad with significant energy contributions up to 100 MHz. A detection system capable of receiving and processing all of the laser generated acoustic energy must operate over a similarly wide bandwidth. Unfortunately, the shot noise which limits detection sensitivity is also broad bandwidth noise. As a result, a laser-based detection system with a broad receiving bandwidth is not only open to receiving all of the signal energy but is also open to receiving a great deal of unwanted noise. Repetitive pulsing of the laser source, however, alters the frequency distribution of acoustic energy such that it is grouped around the frequencies corresponding to the fundamental and harmonics of the pulse repetition frequency. This effect is demonstrated in figure 1 where a single raised cosine pulse and its spectrum are compared to a train of ten pulses of equal width. Note that the single pulse spectrum is quite broadband and that the bandwidth of the detection system cannot be reduced below the cutoff frequency, f_C , without losing a proportional amount of signal energy. Thus, as a result of repetitive pulsing, the frequency distribution is now clustered around the pulse repetition frequency, f_R, and the corresponding harmonics. The frequency response of the optical receiving system must now be designed to match the frequency content of the reduced bandwidth acoustic signal in order to reject significant amounts of broadband noise and increase the sensitivity of the system. A flexible and relatively simple means of accomplishing this is through digital filtering using the Wiener filter. The Wiener

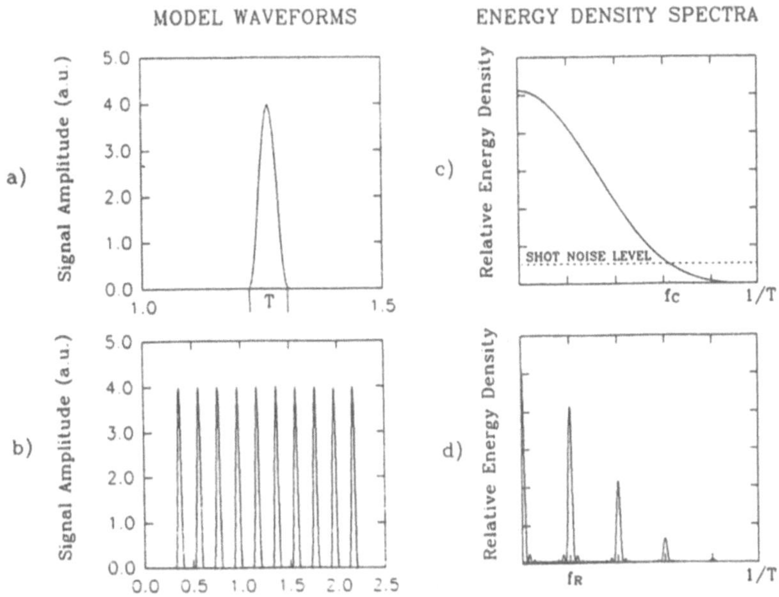

FIG. 1. Energy density spectra of modeled single pulse versus multiple pulse waveforms

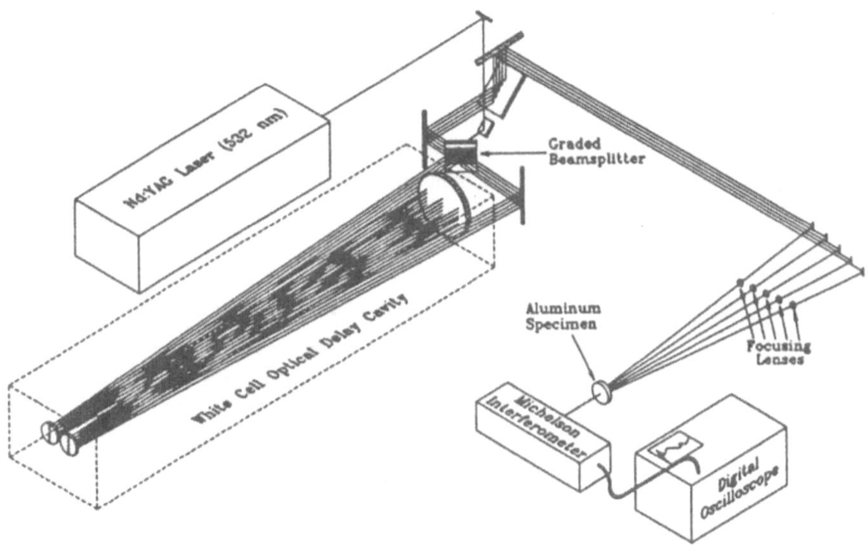

FIG. 2. Laser pulse array generation system

filter has been used to extract narrow-band signals generated by a mode-locked Nd:YAG laser [5] as well as for various filtering and image enhancement tasks in ultrasonic nondestructive evaluation[6].

A second method for sensitivity enhancement is to generate ultrasound with an array which is aligned both temporally and spatially to superimpose the acoustic arrivals from each array element so that they arrive simultaneously and in phase at the point of detection. The result is that a single large pulse with width equal to that generated by any individual array element is produced at the detection site. At low levels of excitation where a material may be considered to behave as a linear elastic solid, the amplitude of the received single pulse would be increased over a single element source in proportion with the number of elements in the laser array. In other words, for a ten element array, the received unit pulse would have ten times the amplitude of a pulse received from any individual array element. The signal-to-noise ratio, as given in equation 1, is proportional to the amplitude of the surface displacement squared, δ^2, and thus the detection sensitivity which can be achieved by a phased laser array can be quite substantial.

Spatial laser arrays have been used by several investigators to generate narrowband acoustic signals. Lenticular arrays and optical diffraction gratings [7,8], for example, have been used to deliver laser energy simultaneously to an array of points or lines on the surface of an object. At a receiving point distant from the generating array, acoustic pulses arrive in sequence from each element of the array so that a train of acoustic pulses is observed. The frequency of the pulse train or "tone burst" depends on the array spacing, the propagation angle from the surface upon which the array is incident, and the propagation velocity of the acoustic mode being sensed. By adjusting the array spacing, one is able to "steer" acoustic energy of bulk waves at a given frequency noting that the total energy as a function of angle is unchanged by variation of the array spacing.

To combine the benefits of both spatial and temporal modulation to enhance sensitivity of laser ultrasonic systems, several methods have been reported. Among these methods, the simplest and easiest to implement has been the fiber optic delay system[9]. Unfortunately, the peak power handling capability and losses within the fiber restrict the optical energy which can be delivered to any array element. The optical time delay based system, to be described below, provides the same simplicity and economy as the fiber based system but with the capability of delivering greater energy per array element than the fiber can deliver. Variable separations in time and space between array elements are achieved without critical mechanical adjustment or electronic timing.

SYSTEM DESCRIPTION

The laser array system to provide simultaneous temporal and spatial modulation is similar to that of Deaton[10] in that it uses a White cell to generate the long optical pass required for time delay between pulses. The system has also been used for time-resolved holographic studies[11]. The White cell consists of one large "field mirror" and two smaller "turning mirrors" as shown in figure 2. The mirrors are set up in confocal arrangement; all three mirrors being of equal radii of curvature, r. A collimated pulse of light entering the cell on the left is reflected and focused by the first turning mirror toward the field mirror. The beam is focused at the midpoint of the cell and then reflected and re-collimated by the field mirror toward the second turning mirror. The beam is repeatedly reflected following the path shown in the figure until it is emitted on the right through a lens of focal length r/2 for final collimation. By simply adjusting the angle of the two turning mirrors, the total number of four pass transits through the White cell cavity can be adjusted to vary the delay time.

Each of the output pulses was sampled individually by a small aiming mirror and directed through a focusing lens to the sample surface. By using a separate mirror and lens for each beam, array spacing could be varied without loss of focus. Bulk acoustic waves were detected by a path-stabilized Michelson interferometer on the opposite side of a 1 cm thick disk of aluminum. For surface wave investigation, a test configuration was used where signals were detected around a rounded (3 mm radius) corner of an aluminum cube 2.5 cm in dimension.

EXPERIMENTAL RESULTS

Figure 3 shows the raw and Wiener filtered epicentral waveform produced by focusing all of the array elements to a single spot on the specimen. The array consists of ten 10 ns long pulses of approximately 4 mJ energy, temporally separated by 85 ns. In this case the surface was coated with a black ink to enhance the longitudinal wave through rapid evaporation of the coating. Examination of the power density spectrum of the raw waveform shows that most of the energy is confined to the fundamental frequency of generation (11.8 MHz) and the corresponding harmonics. Through Wiener filtering the noise outside of narrowband spectrum of the signal is significantly reduced and thus the signal to noise ratio increased. Figure 4 shows another example how decreasing the bandwidth of the generated ultrasound and then extracting the narrowband signal allows for an increase in the signal-to-noise ratio. In this case the signal was generated thermoelastically at 11.3 degrees off of epicenter and no coating was applied to the surface of the specimen. The signal was first high-pass filtered and then the Wiener filter function was applied. The low power in the generating beams required that the signal be averaged 100 times before the filtering technique could be applied. With slightly higher laser power the generation could still remain in the thermoelastic regime while the need for signal averaging could be eliminated. The combination of narrowband generation and Wiener filtering in the thermoelastic regime could prove useful in cases where the generation of ultrasound has to be truly remote (no surface coating may be applied) and truly nondestructive (no surface damage can be tolerated) .

Figure 5 shows the result of spatially modulating nine laser pulses on the surface of the specimen such that superposition of the individual elements of the array occurred. The temporal spacing between arriving beams was again 85 ns and the average spacing between array elements was 1.06 mm. The signals were detected 3.5 mm from the last arriving pulse resulting in a phasing angle, from the center of the array, of approximately 33 degrees. Note that a non-uniform spatial array is required for the superposition of bulk waves is to occur in the near-field. The surface of the specimen was coated with a thin transparent constraining layer. The longitudinal arrival is not detectable using a single laser pulse. As elements are added to the array there is a corresponding increase in signal amplitude while the noise level remains constant.

The final variation of this system, and the one for which it is well suited, is that of Raleigh wave generation. The waveform of figure 6a was detected for an array spacing of 1.8 mm with the detection point on the same side of the array as the first arriving laser pulse. Figure 6b shows the same array with a detection point near the last arriving pulse in the array. As expected, the frequency concentration of the latter acoustic signal is slightly higher than that of the former. The array spacing was then changed to 0.3 mm, and the waveform detected on the side of the last arriving pulse is shown in figure 6c. Here the superposition of acoustic arrivals has begun, but is not complete. Finally, by reducing the array spacing to 0.26 mm. for a Raleigh velocity of 3.06 m/s , the waveform of figure 6d

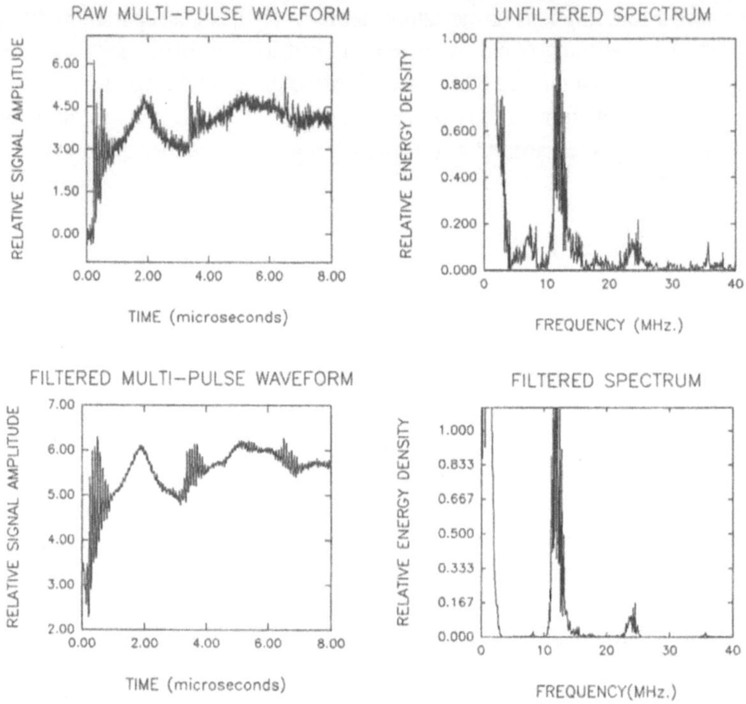

FIG. 3. Raw versus Wiener filtered bulk acoustic waveforms and corresponding energy density spectra. The surface was constrained to enhance the longitudinal wave.

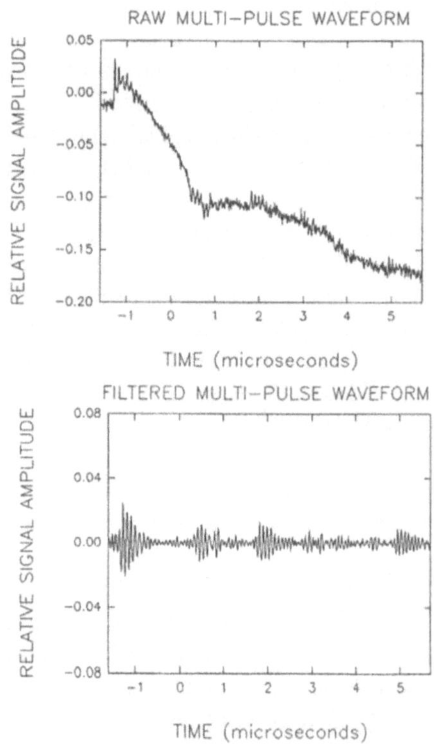

FIG. 4. Raw and Wiener filtered thermoelastic signals generated 11 degrees off of epicenter

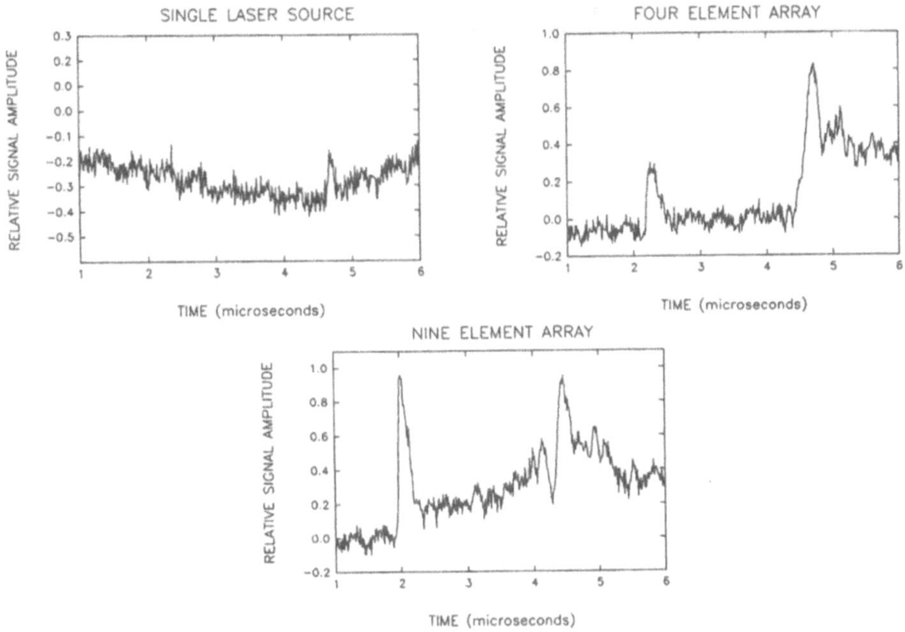

FIG. 5. Comparison of longitudinal waves generated with a single laser source, a four element array, and a nine element array. Surface constrained to enhance longitudinal wave.

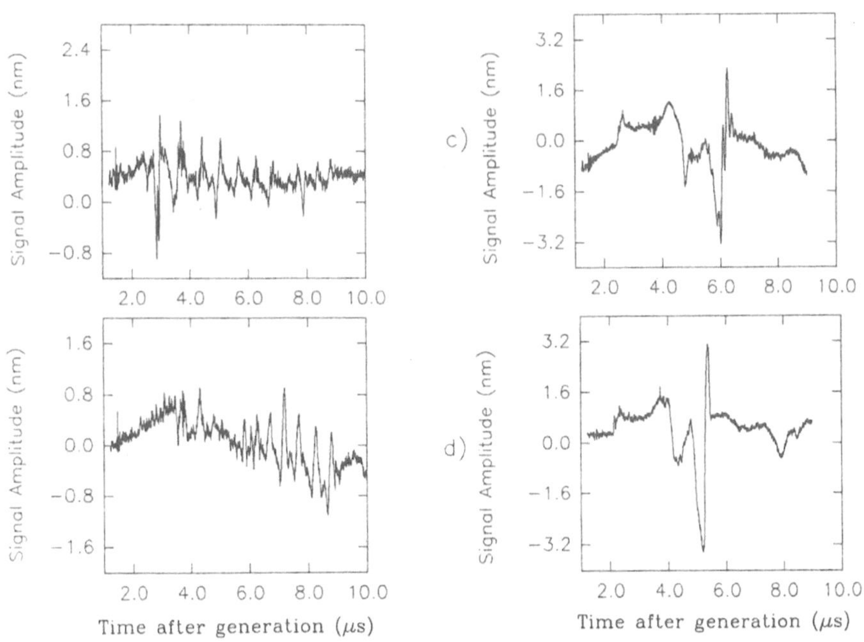

FIG. 6. Surface waves generated with various array spacings

was detected. There is no modulation superimposed on the Raleigh waveform indicating that total superposition of arrivals has occurred.

DISCUSSION AND CONCLUSION

A technique has been proposed and demonstrated using a novel system for independent temporal and spatial modulation of laser beams for generating ultrasound. This system is capable of generating an acoustic toneburst over a range of angles when configured to produce a temporally modulated point source. An n-fold increase in energy, where n is the number of equal pulses in the generating array, is evident in the power spectrum around a frequency of choice specified by tuning the laser system. The resulting narrowband signal was filtered using the Wiener filter and the corresponding increase in the SNR demonstrated. In addition, the system permits independent control of the spatial position of each generating beam so that, when combined with the appropriate temporal phasing of the beams, superposition of multiple acoustic waves can be achieved. An increase in energy similar to that achieved for the narrowband case was achieved at all frequencies allowing for an increase in displacement sensitivity without restricting sensitivity bandwidth. In both of these cases it is possible to increase the detection sensitivity without being forced to generate in the ablative regime and cause damage to the surface of the specimen of interest. Unfortunately, spatial separation of the pulses requires that the volume of interrogation for bulk waves be slightly larger than for single point generation, but this effect is minimized by examination in the far-field. However, surface waves generated by laser arrays do not suffer from this problem, as their paths are identical at points on either end of a linear array. As a result, this technique, though suitable for many applications, may be best suited for surface wave analysis.

REFERENCES

[1] J.W. Wagner, A.D.W. McKie, J.B. Spicer, J. B. Deaton,Jr., Modulated Laser Array Sources for Generation of Narrowband and Directed Ultrasound, *J. Nondestructive Evaluation*, **9**(4): 263-270 (1990)

[2] A.D.W. McKie, J.W. Wagner, J.B.Spicer, C.M. Penny, Laser Generation of Narrowband and Directed Ultrasound, *Ultrasonics*, **27**:323-330 (1989)

[3] J. Huang, S. Krishnaswamy, J.D. Achenbach, Laser Generation of Narrow-band Surface Waves, *J. Acoust. Soc. Am.*, **92**(5) 2527-2531 (1992)

[4] J.B. Deaton,Jr., A.D.W. McKie, J.B. Spicer, J.W. Wagner, Generation of Narrowband Ultrasound with a Long Cavity Mode-Locked Nd:YAG Laser, *Appl. Phys. Lett.*, **56** 2390-2392 (1990)

[5] J.B. Deaton,Jr., A.D.W. McKie, J.B. Spicer, J.W. Wagner, Mode-Locked Laser Generation of Narrow-band Ultrasound, Review of Progress in Quantitative Nondestructive Evaluation, Vol. 10A (Eds. D.O. Thompson and D.E. Chimenti) Plenum, N.Y. 492-497 (1991)

[6] P. Karpur, B.G. Frock, P.K. Bhagut, Wiener Filtering for Image Enhancement in Ultrasonic Nondestructive Evaluation, *Materials Evaluation*, 1374-1379 (Nov. 1990)

[7] J.W. Wagner, A.D.W. McKie, J.B. Spicer, J. B. Deaton,Jr., Modulated Laser Array Sources for Generation of Narrowband and Directed Ultrasound, *J. Nondestructive Evaluation*, **9**(4): 263-270 (1990)

[8] J. Huang, S. Krishnaswamy, J.D. Achenbach, Laser Generation of Narrow-band Surface Waves, *J. Acoust. Soc. Am.*, **92**(5) 2527-2531 (1992)

[9] A.J.A. Bruinsma, J.A. Vogel, Ultrasonic Noncontact Inspection System With Optical Fiber Methods, *Applied Optics*, **27**(22) 4690-4694 (1988)

[10] J.B. Deaton,Jr., A.D.W. McKie, J.B. Spicer, J.W. Wagner, Generation of Narrowband Ultrasound with a Long Cavity Mode-Locked Nd:YAG Laser, *Appl. Phys. Lett.*, **56** 2390-2392 (1990)

[11] M.J. Ehrlich, J.W.Wagner, Nanosecond Scale Optical Pulse Separations for Holographic Investigation of High Speed Transient Events, Review of Progress in Quantitative Nondestructive Evaluation, Vol. 11A (Eds. D.O. Thompson and D.E. Chimenti) Plenum, N.Y. 495-500 (1992)

OPTICAL DETECTION OF STRESS WAVES IN GLASS FIBER REINFORCED PLASTIC COMPOSITE MATERIAL

Robert D. Huber and James W. Wagner

Center for Nondestructive Evaluation
The Johns Hopkins University
Baltimore, Maryland 21218

ABSTRACT

Owing to the different mechanical properties of the various components of a composite material, there exist several sources for stress waves which correspond to various stress relief mechanisms. In a glass fiber reinforced plastic material such mechanisms include matrix cracking, fiber breakage, and fiber pull-out. Since the glass fibers, plastic matrix, and glass fiber/plastic matrix bond have differing strengths, they will fail under different loads. This fact, coupled with geometrical and dimensional considerations, leads to discrete stress wave signatures of the sources which are manifested in the frequency content and morphology of the stress waves generated.

Traditionally, stress waves in materials have been detected using contact piezoelectric transducers in acoustic emission testing. Indeed, earlier acoustic emission studies have, to some degree, been able to catalog the different signatures, especially in the near field. Conventional acoustic emission testing employs relatively narrow frequency bandwidth transducers which color or degrade the stress wave signature being detected. The movement to higher fidelity transducers may be able to augment acoustic emission testing by providing additional signature information beyond those parameters traditionally measured.

A laser interferometer provides a high fidelity, non-contacting means of monitoring the stress waves generated in a material under loading. The non-contacting aspect of the interferometer avoids loading of the surface associated with contact techniques, has no mechanical resonance, and is entirely benign to the material for typical light energy densities. To the present time, testing is being performed on fiberglass/epoxy materials where acoustic emission (AE), or stress wave "hits" are digitally recorded with sample rates of 10 MHz, and record lengths up to 5 ms. The analysis of these relatively long record lengths sets this study apart from earlier studies reported in the literature in which only the features of the first few microseconds of each acoustic hit were analyzed. By considering longer record lengths, it is hoped that signature analysis will prove to be valid for categorizing stress wave signatures even after they have propagated some distance in real structures.

INTRODUCTION

The majority of acoustic emission work has been done using conventional contact piezoelectric transducers (PZT's). Traditionally, some of the parameters[1] measured in AE studies have been threshold crossings, peak amplitude, duration, and energy. Other parameters such as signal rise time and total acoustic emission counts have also been measured in AE studies.

One problem associated with the use of piezoelectric transducers is mechanical resonance. Piezoelectric transducers will "ring" at their resonant frequency which can greatly affect the measurement of parameters such as threshold crossings and duration. Contact transducers load the surface on which they are placed, and usually require some sort of acoustic coupling to the test material which not only affects the acoustic wave, but may damage the material. Also, most piezoelectric AE transducers employed may be too narrow in frequency response to yield meaningful frequency domain information.

Optical interferometers can be made to have very wide frequency bandwidths, and are equally sensitive to all of the frequencies in their sensitivity bandwidth, allowing high fidelity detection. Therefore, accurate frequency analysis may be performed on waveforms obtained using interferometers. They are able to measure the actual surface displacement of the test material, while many PZT's measure surface velocity. Theoretical models can predict the surface displacement resulting from some types of acoustic emission events. Waveforms received by a displacement sensor compare directly to such theoretical models. Optical interferometers also offer the advantage of being fully non-contacting, thereby eliminating the need for and effects of acoustic couplants. Loading of the material surface is also avoided when using non-contact sensors.

Laser interferometers offer additional advantages as well. They allow remote testing, and more point-like detection than piezoelectric transducers. The small spot size of the interferometer beam eliminates phase cancellation effects and leads to a more accurate determination of location. Another advantage of optical interferometers over contact sensors is that interferometers can be used to detect stress waves on surfaces of materials that are at elevated temperatures or in otherwise hostile environments. The use of an optically transparent window allows interferometric detection on materials in an isolated chamber such as a furnace.

High fidelity detection of AE using various detector types has been investigated[2-6]. Work characterizing AE signatures for failure mechanism identification has been performed[7-9], but only short time length waveforms were recorded to look at the stress waves before they encountered all of the boundaries of the test specimens.

Failure Mechanisms in Fiber Reinforced Composites

Various failure mechanisms exist for inhomogeneous materials such as composites because of the different mechanical properties of the components[10]. For fiber reinforced composites, there are three major failure mechanisms. One type of failure mechanism is matrix cracking. Typically, the matrix material is not as strong as the fibers, and matrix cracking therefore occurs at loads well below those which would initiate damage to the fibers. Another type of failure is fiber breakage. The fibers are usually the strongest component of composites, and therefore fiber breakage occurs at higher loads than the initial matrix cracking. Damage to the fiber/matrix bond is another type of failure mechanism, and this can lead to or be accompanied by fiber pull-out.

These failure mechanisms are the sources of the stress waves which are detected in acoustic emission tests. Since the sources differ, a signature in terms of frequency and duration is expected.

EXPERIMENTS

Three Point Bend Tests

Acoustic emission was detected during three-point bending of glass fiber reinforced plastic material in which the glass fibers were randomly oriented in the material. The three-point bend apparatus was designed to leave the detection site stationary, and therefore the interferometer did not require continual realignment as the specimen was bent. A fiber-optic based heterodyne interferometer[11] was used to perform AE signal detection. Since the interferometer probe beam was focused down to about a 50 μm diameter, moving the detection site out-of-plane would have required re-focusing. Figure 1 shows the experimental configuration for the three-point bend tests.

Initially, waveforms with a length of 10 μs were stored in order to look at the stress wave signatures before they encountered the boundaries of the test specimens. Through visual inspection of the waveforms, it was found that the waveforms could be classified into several categories by waveform shape. Examples of three waveforms in a similar class are shown in Figure 2. Although the waveforms are clearly similar in shape, classification

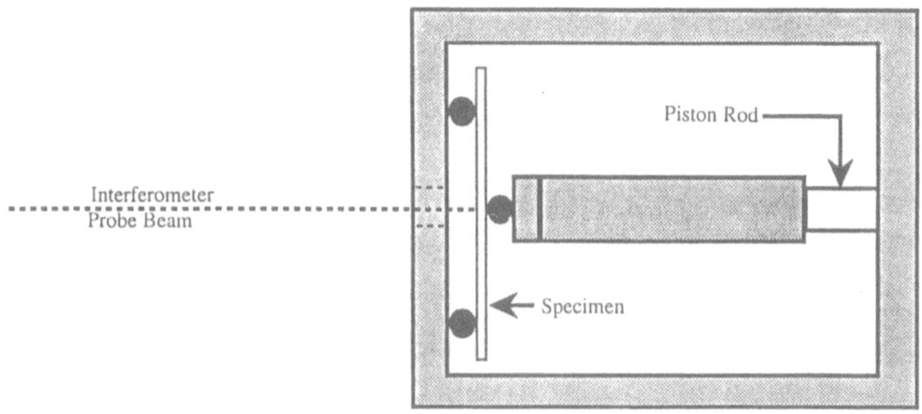

Piston Sleeve Is Held Stationary.

Figure 1. 3 Point Bend Test Configuration.

by signal amplitude would have yielded entirely different results. After the initial tests, waveforms as long as 5 ms were stored. The 5 ms waveforms were digitized at a rate of 100 ns per point so that the initial 10 μs of each record could still be analyzed in detail Once again the initial 10 μs of the acoustic emission hits had distinctive shapes, or signatures.

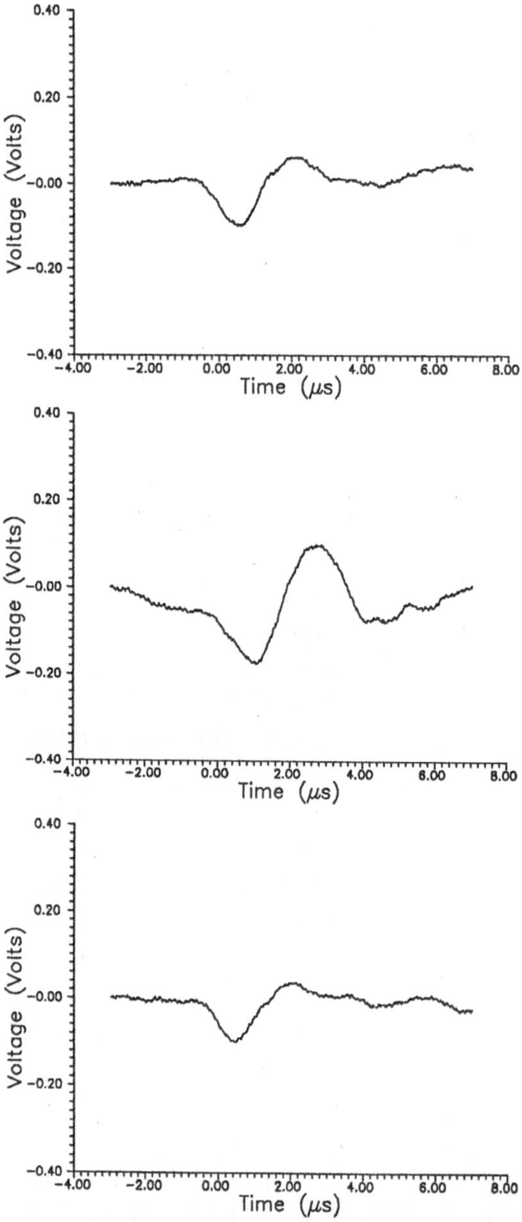

Figure 2. 3 Point Bend Waveforms.

Tensile Tests

Acoustic emission tests using the optical interferometer were next performed on unidirectional glass fiber reinforced plastic material. Tensile tests were run on coupons cut from a single sheet of this material. The unidirectional composite material was machined into two types of coupons. In one type of coupon, the tensile axis was made to be in the same direction as the direction of the glass fibers, and these coupons were labeled "0 degree" coupons. In the other type of coupon, the tensile direction was made to be transverse to the fiber direction, and these were labeled "90 degree" coupons. Stress concentration notches were machined into the coupons at the center of the lengths.

Unidirectional material was chosen in order to select a predominant failure mechanism for each type of coupon. For the 0 degree coupons, ultimate failure of the coupon required fiber breakage, while for the 90 degree coupons the primary failure mechanism was matrix cracking. Although fiber breakage and matrix cracking, as well as fiber pull-out will likely occur during tensile testing of any glass fiber reinforced plastic material, pulling unidirectional material in certain directions should lead to a predominance of AE sources from certain failure mechanisms.

For the glass fiber reinforced material tested in this study, the glass fibers were considerably stronger than the polymer matrix. For this reason the tab ends of the 0 degree coupons were reinforced to prevent the coupons from failing at the tabs, thus ensuring failure at the stress concentration notches. No reinforcement was required for the testing of the 90 degree coupons.

The apparatus used to pull the tensile specimens consisted of a hydraulically loaded piston and a "C" frame. The "C" was machined from a single sheet of aluminum to avoid noise arising from moving parts. Figure 3 is a diagram of the experimental set-up for the tensile tests. As the piston was extended, it spread the "C", thereby pulling on the coupons which were fastened to it.

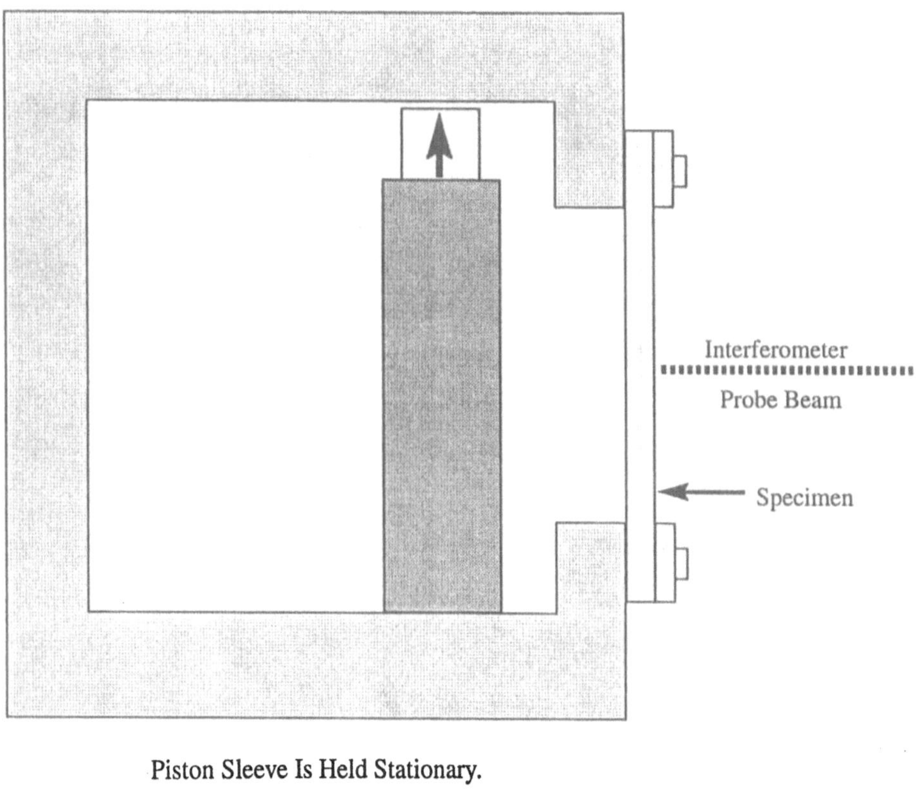

Figure 3. Tensile Test Configuration.

Two detection sites were used during the tensile tests. For some of the tests, detection took place near the stress concentration notches to allow detection at the source of the stress waves. For the remaining tests, detection took place approximately one inch away from the stress concentration notches. This was to allow analysis of the stress waves after they had propagated some distance in the material. For most acoustic emission tests on real structures, the sources are at some distance from the detectors, and the effect of the material on the stress waves must be taken into consideration.

The acoustic emission waveforms were captured using a LeCroy 9310L digital oscilloscope which had a 1 million point segmentable memory. The traces were stored in 50,000 point segments, and this allowed a maximum of 19 segments. The signals were digitized at a rate of 100 ns per point, yielding waveforms which were 5 ms in length. After several segments (up to 19) were filled, the waveforms were transferred to a computer for permanent storage. No acoustic emissions could be captured while transferring the waveforms to the computer. Since acoustic emissions often occur in bursts, the capability of rapidly recording successive waveforms will be a necessity in the future. With the equipment used, a maximum of 19 successive 50,000 point waveforms could be recorded, and for this reason, the load on the specimens was decreased slightly while storing the waveforms to the computer. This was done to minimize the occurrence of acoustic emissions which would go unrecorded.

Data Analysis

A computer program was written to analyze the acoustic emission waveforms by measuring several parameters. These parameters were hit duration, peak amplitude, signal strength, time to the signal strength centroid, energy, and time to the energy centroid. A voltage threshold was set, and the time at which the threshold was exceeded was defined to be the starting time of the hit. A hit definition time of 200 μs was used to determine the ending time of the hits. If the threshold was not exceeded for 200 μs after a hit was detected, then the end of the hit was determined to be the last data point which exceeded the threshold. The signal strength was defined to be the running sum of the rectified waveform amplitude for each hit, and the energy was defined to be the running sum of the square of the amplitude for each hit. The centroid time was defined to be the time at which half the total value was reached.

Subtle differences were observed in the parameters for the two orientations of the coupons. Further analysis is being performed on the waveforms, including frequency analysis.

CONCLUSIONS

Traditional acoustic emission parameters may not be adequate for stress wave source identification in composite materials. The high fidelity with which optical interferometers detect stress waves yields valuable frequency and displacement information, and the use of such a detector in a system capable of looking at long time traces at high data rates greatly increases the information obtained in acoustic emission tests. The use of time-dependent frequency analysis may help with source identification of AE hits.

REFERENCES

1. I.G. Scott, "Basic Acoustic Emission", Gordon and Breach Science Publishers,(1991).
2. F.M. Boler, H.A. Spetzler, and I.C. Getting, Capacitance transducer with a point-like probe for receiving acoustic emissions, Rev. Sci. Inst., 55, 1293-1297, (1984).
3. T.M. Proctor, Jr., An improved piezoelectric acoustic emission transducer, J. Acoust. Soc. Am., 71, 1163-1168, (1982).
4. J.T. Glass, S. Majerowicz, and J.C. Murphy, Laser interferometric probe for detection of acoustic emission, Mat. Eval., 48, 714-720, (1990).
5. C.H. Palmer, and R.E. Green, Jr., Optical probing of acoustic emission waves, "Nondestructive Evaluation of Materials", ed. J.J. Burke and V. Weiss, 347-378, Plenum, New York, NY (1979).

5. C.H. Palmer, and R.E. Green, Jr., Optical probing of acoustic emission waves, Nondestructive Evaluation of Materials, ed. J.J. Burke and V. Weiss, 347-378, Plenum, New York, NY (1979).
6. H.K. Park and H.C.Kim, Laser interferometry system for measuring displacement amplitude of acosutic emission signals, J. Phy. D., Applied Physics, 17, 673-675, (1984).
7. R.A. Kline and R.E. Green, Jr., Acoustic emission waveforms from cracking steel: experiment and theory, J. Appl. Phys., 52, 141, (1981).
8. H.I. Ringermacher and R.K. Miller, Acoustic emission waveforms in isotropic and orthotropic plates, Proc. IEEE Ultrasonics Symposium, (1983).
9. S.D. Glaser and P.P. Nelson, High fidelity waveform detection of acoustic emissions from rock fracture, Mat. Eval., 354-359, March 1992.
10. R. Talreja, "Fatigue of Composite Materials". Technomic Publishing Co. Inc., (1987).
11. R.D. Huber and R.E. Green, Jr., Non-contact acousto-ultrasonics using laser generation and laser interferometric detection, Mat. Eval., 49(5), 613-618, (1981).

TIME-RESOLVED HOLOGRAPHY FOR THE MICROSCOPIC STUDY OF CRACK-TIP MOTION IN DYNAMIC FRACTURE

J. Scott Steckenrider[1] and James W. Wagner[2]

[1]Argonne National Laboratory
Energy Technology Division
9700 South Cass Avenue
Argonne, IL 60439

[2]The Johns Hopkins University
Materials Science and Engineering Department
102 Maryland Hall
Baltimore, MD 21218

INTRODUCTION

Large scale fracture mechanics has proven very useful in the analysis of fracture in uniform materials. However, there exist several classes of fracture (such as stress-corrosion cracking[1], intergranular fracture[2], and fracture of polymeric materials[3,4]) which appear unvaried and behave homogeneously on the macroscopic scale, but which have drastically different fracture properties on the microscopic scale. Theoretical studies, though initially focused on large scale homogeneous materials, now cover the full range of scale from finite elemental analysis of large structural members[5] to atomistic scale models[6]. However, experimental techniques are still involved in the larger scale analysis. Traditional experimental fracture techniques are therefore unequipped to properly evaluate the dynamic fracture behavior of these materials without averaging out the variations.

CRACK-TIP EQUATION OF MOTION

Theoretical Background

Theoretical descriptions of dynamic fracture have evolved substantially during the 20th century from Griffith's original energetic analysis,[7] through the "time effect" consideration of Mott,[8] to the more recent analyses of Berry[9] and Freund.[10] By considering the total energetic state of the system, Griffith realized that the potential energy of an infinite thin plate with an included crack under a remotely applied opening stress σ (therefore a plane stress condition) was inversely related to the length of the crack a_c, providing the relationship

$$a_c = \frac{2E\gamma}{\pi\sigma^2} \tag{1}$$

where E is Young's modulus and γ is the material's surface energy. In 1948, beginning with Griffith's energy balance equation, Mott reasoned that to compensate for crack propagation, a kinetic energy term must also be included as a resisting force to crack growth. He thus derived an equation for the velocity v of a growing crack of the form

$$v = v_o \left(1 - \frac{a_c}{a}\right)^{1/2} \tag{2}$$

where

$$v_o = \left(\frac{2\pi E}{k\rho}\right)^{1/2},$$

ρ is the material's density, k is a numerical constant, and a_c is the Griffith critical crack length presented above and as such must be the initial crack length at the onset of crack growth. Mott however had assumed velocity to be independent of crack length, an assumption obviously violated in the final result. Berry therefore more rigorously avoided that assumption and arrived at the expression

$$v^2 = \left(\frac{E}{\rho}\right)\left(\frac{2\pi}{k}\right)\left(1 - \frac{a_o}{a}\right)\left[1 - (n-1)\frac{a_o}{a}\right]. \tag{4}$$

where n represents the degree of overload necessary to induce crack growth ($0 \leq n \leq 2$, and $n = 2$ for the perfect Griffith fracture). However, an exact definition of the numerical constant k was not available until the work of Freund, who, by allowing the crack tip stress field to vary with crack velocity, was able to determine that the limiting crack velocity was indeed the Rayleigh velocity. By applying this realization to the cracked infinite body discussed above, and assuming an excess of *energy* is required to initiate crack growth, the crack growth velocity can be written as

$$v \approx c_r\left(1 - \frac{a_o}{ma}\right) \tag{5}$$

where c_r is the Rayleigh velocity and m is the degree of excess energy required ($m \geq 1$). These results are generally accepted to be valid on a macroscopic scale for homogeneous and isotropic materials. Confirmation of these predictions for less ideal conditions on the microscopic scale has been the objective of this work.

Experimental Background

Experimental evaluation of crack growth dynamics has been achieved through a wide variety of methods. These include direct real-time methods (multiple spark high-speed photography,[11] pulsed laser high-speed photography[12] and holography,[13,14] laser shadowing,[15] and Rayleigh wave photography,[15-17] direct post-mortem techniques

(Wallner line fractography[18,19] and active ultrasonic fractography,[20-22] and indirect real-time methods (potential drop,[23,24] conductive mesh,[25-27] and continuous conductive coating[4,28] techniques). While the direct methods generally can determine and compensate for any fracture path, they have insufficient temporal resolution for microscopic studies (100 ns at best). However, the holographic methods, though they are single point measurements which require an exact repeatability of the fracture event (an overly restrictive requirement), do provide a potential for greater temporal resolution, as a single laser pulse is only ≈ 10 ns in duration. In addition, holography itself is not an imaging technique, thereby facilitating the analysis of microscopic specimens.

TIME RESOLVED HOLOGRAPHY

The application of dynamic holography to fields other than fracture mechanics has advanced significantly. One of the key areas of this advance is that of time-resolved holography, also known as "cine-holography", in which a time series of holograms are recorded in rapid sequence. In order to follow very high-speed dynamic events with time resolved holography, a coherent illumination (laser) system is required which is capable of producing multiple laser pulses within the event lifetime. For the case of microscopic scale dynamic fracture in glass, this corresponds to a frame-to-frame spacing of 100 ns or better. In addition, in order to efficiently multiplex the resulting holographic images, each pulse must be not only temporally separate, but spatially separate as well. Techniques developed to date can be grouped into four primary classifications: a single multiply Q-switched laser method,[29,30] a "flat" optical delay cavity,[31] a stacked beamsplitter arrangement,[32] and the use of multiple lasers.[33-35] While each of these methods has its own individual strengths, they also have drawbacks which prevent their use for the current application. Thus, the best solution for the current application appears to be a combination of these systems.

White Cell Optical Delay System

The heart of the current delay system is illustrated in Fig. 1. It was originally developed by John White in 1942 for use in a long distance absorption spectrometer[36] and is also the method used by Deaton[37] to lengthen the cavity of their mode-locked Nd:YAG laser. The cavity is essentially similar to that of Ostrovskaya and Ostrovsky, except that the "White cell" overcomes the diffractional losses associated with long propagation distances. For a complete description of the "White cell" optical delay cavity, the reader is referred elsewhere.[38]

The overall optical system for multiple pulse time-resolved holography is illustrated in Fig. 1 with the White cell delay cavity delimited by dashed lines. (For pictorial simplicity only 5 laser pulse paths are drawn.) The nine pulses produced by the White cell - hyperharmonic beamsplitter system are each collimated and are propagating in parallel. They are temporally separated by a pre-determined equal delay time and spatially separated by 5 mm. Upon leaving the delay cavity, the pulse array is rotated from vertical to horizontal by a periscopic mirror pair to facilitate beam separation. Also illustrated in Fig. 1 is the holographic portion of the setup, indicating the spatial multiplexing system. All crack length measurements (both initial and holographic) were made using a long working distance (LWD) microscope with a resolution of ± 2 pixels (± approximately 22 μm).

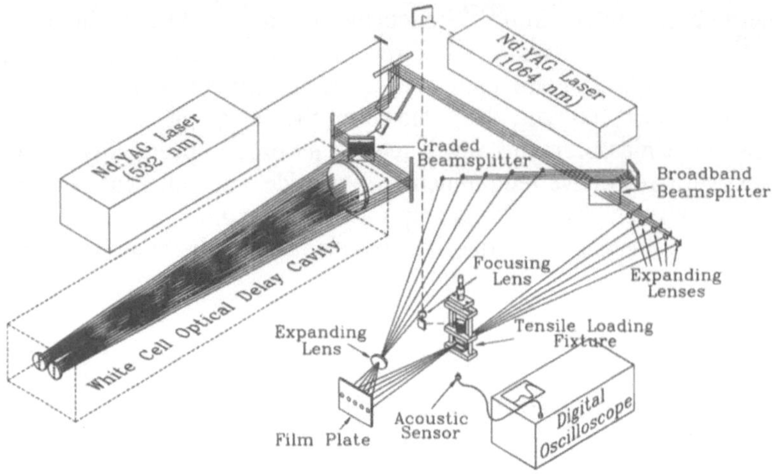

Figure 1. Layout of the complete multiple pulse holographic system.

FRACTURE SPECIMEN AND LOADING FIXTURE

In order to show broad applicability of this technique in even the most dynamic of situations, glass was chosen as the fracture material, owing to its high fracture velocity (≥ 1000 m/s). The specimen was constructed from a microscope cover slip of Corning 0211 silica glass. Each slip was 22 mm square and ≈ 150 μm thick. Since the theoretical predictions for crack growth behavior assume a sharp crack-tip, and since fatiguing a pre-crack would be extremely difficult with such a specimen, an alternative technique was employed to produce a sharp pre-crack in the glass slip.[40] In order to ensure a uniform grip along the length of the cover slip, aluminum end caps were manufactured and affixed to both ends of each glass specimen with cyanoacrylate.

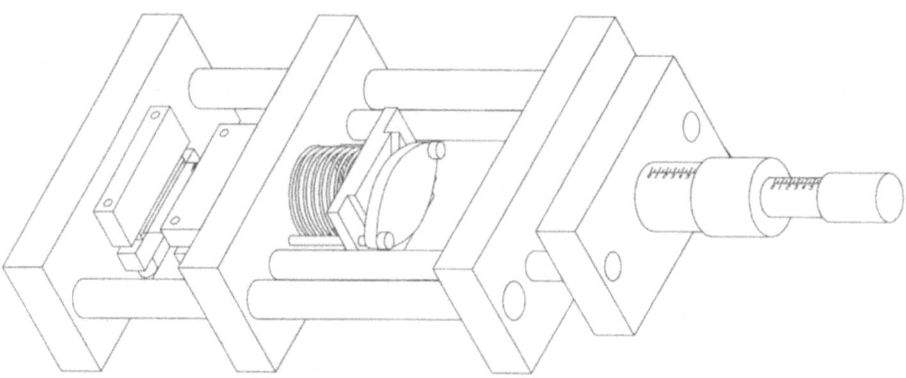

Figure 2. Drawing of the tension fixture used for applying pre-load to glass specimens.

The loading fixture for all multiple pulse experiments is illustrated in Fig. 2. A pre-load was applied to each specimen by a mechanism combining a moderately stiff spring in conjunction with a high resolution micrometer at a resolution of 24.0×10^{-3} N (≈ 7300 Pa for the given specimen dimensions). Since a temporally well defined point of initiation was needed to properly synchronize the fracture event with the laser pulse train, a means of rapid fracture initiation was required. This was most easily accomplished by using explosive detonation to generate a stress pulse in the material.

A small cluster of explosive was fixed to the edge of the specimen at the mouth of the pre-crack with a small amount of cyanoacrylate adhesive. Upon detonation, a stress pulse is emitted into the glass traveling along the crack face. Although this is obviously not a mode I loading condition, it does produce a well defined and repeatable loading configuration with no damping effects induced by the loading frame.

In addition, an interferometric system was used to investigate the stress pulse waveform to determine what portion of the waveform caused initiation. From the stress pulse waveform, which was very repeatable, a very distinct fracture initiation time was seen as a large amplitude pulse in the detected waveform. Crack growth began within 100 ns of that time in all specimens. Thus, because of its superior reliability, this method was chosen for all subsequent tests, despite the fact that the resulting stress field at the crack tip was not well defined.

RESULTS AND DISCUSSION

For the experimentation itself, three factors were presumed to have a possible effect on crack acceleration and terminal velocity: pre-crack length, pre-load level, and size of initiating explosive cluster. To evaluate the effect of each of these factors, a test matrix was developed and applied to specimens of two pre-crack lengths; 3.5 - 4.0 mm and 9.0 - 10.5 mm. A very rough estimate of explosive "size" was established by visual categorization of the cluster used for each specimen as either a "small" (≈ 0.3 mm diameter), "medium" (≈ 0.6 mm diameter), or "large" (≈ 1 mm diameter) cluster.

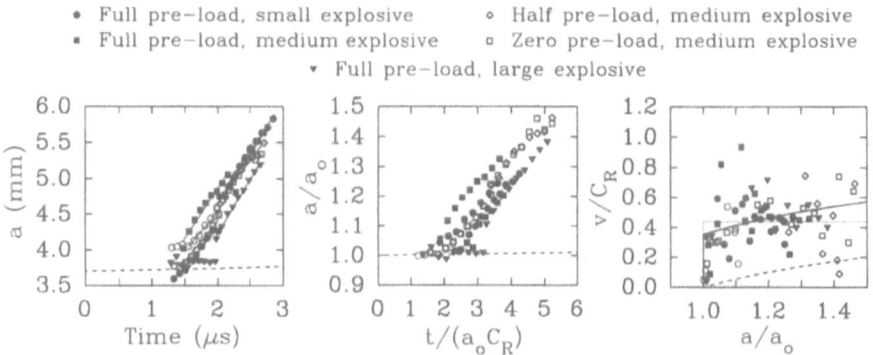

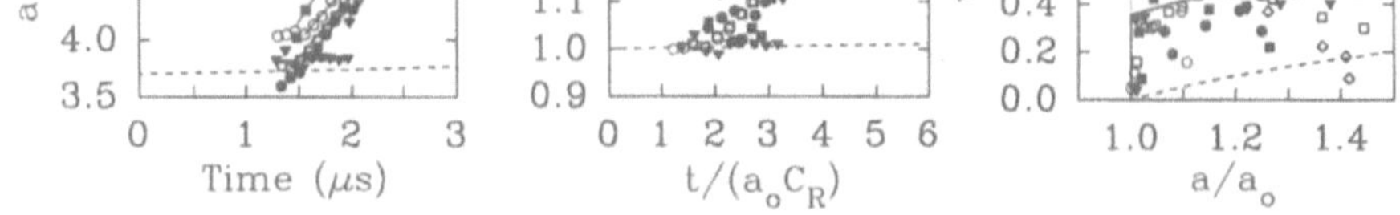

Figure 3. Photograph of 9 holographic frames of dynamic fracture showing straight crack growth. Temporal frame separation is 81.6 ns. (Lower right hand scale bar is 1 mm).

A more exact quantification of detonation amplitude was then acoustically determined at the time of detonation. Finally, the optical delay cavity was set to a delay of ≈ 81.6 ns, for a total "window of opportunity" of ≈ 653 ns during the nine holographic exposures.

Shown in Fig. 3 is a composite photograph of a holographically recorded growing crack. The width of each frame element is approximately 1.9 mm and, for illustrative purposes, only the crack-tip region is shown, so that resolution is not lost by reduction of image size. Here the crack has just begun to grow in the first (upper left) frame, and continues to grow through all nine frames in a relatively straight line. It also should be noted that photographic image quality is not as good as that displayed on the computer monitor, where contrast and brightness enhancements were easily realized.

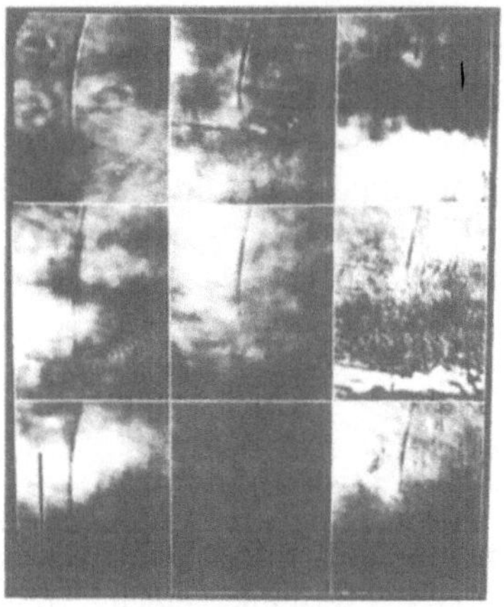

Figure 4. Plots of all 3.5 - 4.0 mm pre-cracked specimens showing a) crack length vs. time (including a best-fit line for each), b) normalized length vs. normalized time, and c) normalized velocity vs. normalized crack length. Also shown as a dashed line is the theoretical prediction for ideal mode I loading.

A compilation of all 3.5 mm to 4.0 mm crack growth data is presented in Fig. 4. Several of the data sets show the behavior of an initially flat (no growth) region followed by a sloped (growth) region, indicating that fracture initiation occurred in the middle of the laser pulse train. In addition, for all cases, crack length was measured along the actual crack line, so that any deviation from straight growth was accounted for.

The primary and most interesting observation regarding these plots is that there appears to be no correlation between either explosive amplitude or pre-load level and fracture behavior. Although the fracture path did appear to be affected by pre-load level (there was a greater frequency of occurrence of bifurcation in the specimens with zero pre-load), such parameters as terminal velocity and crack acceleration are uniform throughout. In fact, almost instantaneous acceleration is seen in all of the specimens, and initial velocity, determined by a best fit line through each data set, ranges from 1100 to 1400 m/s in every case. Some of the discrepancy between data sets appears to be removed by normalizing the plot (Fig. 4b), which eliminates the variability of initial crack length, but there is still more scatter than would be desired, particularly in the time required for crack growth initiation, so an accurate best-fit to the total data set would not be possible at this point. However, a best-fit of a normalized velocity-crack length plot, as shown in Fig. 4c, would be appropriate, since this plot depends only on relative rather than absolute time. A best fit to eqn. (5) was therefore determined with an overload value of m = 1.55, and is plotted as the solid line in Fig. 4c. Also, a best-fit assuming an immediate acceleration to terminal velocity was determined, with terminal velocity v_0 of 1316 m/s, and this is plotted as a dotted line in the figure. Still, the extreme differential error of this plot indicates that a more accurate best-fit might be achieved using the normalized length-time curve, if only the spread in crack initiation times could be removed.

The data scatter in the 3.5 - 4.0 mm pre-cracked specimens is probably a result of shock effects interacting with the crack propagation. Since shock waves are known to propagate only to a distance of approximately 10 explosive particle diameters, and since the largest particle size was ≈ 1 mm in diameter, an initial crack length of 10 mm should eliminate these shock effects. A second set of specimens with pre-crack lengths of ≈ 10 mm was therefore examined. Although this did significantly reduce the data scatter, there was still no discernible difference in crack growth behavior as a function of either pre-load level or initiating explosive size. In addition, the best-fit terminal velocity (1207 m/s) was approximately equal to that predicted by other researchers.[41]

CONCLUSIONS

A novel laser system has been designed and developed for the holographic analysis of dynamic crack propagation on a sub-microsecond time scale. The current system is capable of generating a series of optical pulses separated spatially by 5 mm and temporally by from 28.3 ns to 161.6 ns. In addition, the temporal pulse spacing can be easily adjusted in increments of 26.7 ns. Average velocities of approximately 1200 m/s were measured in microscopic (0.15 mm thick) glass using the pulse sequence technique.

Normalized crack length versus time plots were generated for a series of specimens using various pre-load levels and initiating explosive sizes. Two sets of samples were analyzed, one with pre-cracks of ≈ 3.5 - 4.0 mm length and another with pre-cracks of ≈ 9.5 - 10.5 mm length. Despite these variations, all specimens exhibited similar dynamic fracture behavior, particularly those outside of the region of shock of the initiating explosive (those of the longer pre-crack set). A best-fit to the theoretical prediction for mode I crack growth revealed a good match to the case for a specimen overloaded by 64%. However, no variation with applied load was evident, indicating that an immediate acceleration to a velocity of ≈ 1200 m/s (near the empirically determined terminal velocity) was the more plausible fit. Since the terminal velocity was approximately the same for all specimens, this investigation also demonstrated that the terminal velocity is independent of crack-tip stresses, even for the most extreme explosive pulses.

REFERENCES

1. Kelly, R.G., Frost, A.J., Shahrabi, T., and Newman, R.C., Brittle fracture of Au/Ag alloy induced by a surface film, *Metall. Transactions* 22A:531 (1991).
2. Lin, T., Evans, A.G., and Ritchie, R.O., A statistical model of brittle fracture by transgranular cleavage, *J. Mech. Phys. Sol.* 34:477 (1986).
3. Cantwell, W.J., Roulin-Moloney, A.C. and Kausch, H.H., Dynamic crack propagation in the double-torsion test geometry, *J. Mat. Sci. Let.* 7:976 (1988).
4. Stalder, B., Béguelin, P., Roulin-Moloney, A.C., and Kausch, H.H., The graphite gauge and its application to the measurement of crack velocity, *J. Mat. Sci.* 24:2262 (1989).
5. Atluri, S.N. and Nishioka, T., Numerical studies in dynamic fracture mechanics, *Int. J. Fract.* 27:245 (1985).
6. Sieradzki, K., Dienes, G.J., Paskin, A. and Massoumzadeh, B., Atomistics of crack propagation, *Acta Met.* 36:651 (1988).
7. Griffith, A.A., The phenomena of rupture and flow in solids, *Phil. Trans. A* 221:163 (1920).
8. Mott, N.F., Fracture of metals: Theoretical considerations, *Engineering* 165:16 (1948).
9. Berry, J.P., Some kinetic consideration of the Griffith criterion for fracture, *J. Mech. Phys. Sol.* 8:194 (1960).
10. Freund, L.B., "Dynamic Fracture Mechanics", Cambridge University Press, New York (1990).

11. Schardin, H., "Velocity effects in fracture", *in:* "Fracture", John Wiley and Sons, New York (1959).
12. Winkler, S., Shockey, D.A. and Curran, D.R., Crack propagation at supersonic velocities, *Int. J. Fract. Mech.* 6:151 and 6:271 (1970).
13. Suzuki, S., Homma, H. and Kusaka, R., Pulsed holographic microscopy as a measurement method of dynamic fracture toughness for fast propagating cracks, *J. Mech. Phys. Sol.* 36:631 (1988).
14. Suzuki, S. and Nakajima, T., Development of laser inducing technique for fast propagating cracks in PMMA, *in:* "ASME PVP - Vol. 160 Dynamic Fracture Mechanics for the 1990's" H. Homma, D.A. Shockey, G. Yagawa, eds., ASME (1989).
15. Kamath, S.M. and Kim K.S., On Rayleigh wave emissions in brittle fracture, *Int. J. Fract.* 31:R57 (1986).
16. Thaulow, C. and Burget, W., The emission of Rayleigh waves from brittle fracture initiation, and the possible effect of the reflected waves on crack arrest, *Fat. Fract. Eng. Mat. Struct.* 13:327 (1990).
17. Rossmanith, H.P. and Fourney, W.L., On crack tip acceleration and deceleration, *Eng. Fract. Mech.* 16:837 (1982).
18. Shand, E.B., Experimental study of fracture of glass: II, Experimental data, *J. Amer. Ceram. Soc.* 37:559 (1954).
19. Congleton, J. and Petch, N.J., Crack branching, *Phil. Mag.* 16:749 (1967).
20. Kerkhof, F., Analyse des spröden zugbruches von gläsern mittels Ultraschall, *Naturwissenschaften* 40:478 (1953).
21. Kerkhof, F., Ultrasonic fractography, *in:* "Proceedings of the Third International Congress on High-Speed Photography", London (1956).
22. Michalske, T.A. and Fréchette, V.D., Modified sonic technique for crack velocity measurement, *Int. J. Fract.* 17:251 (1981).
23. Lowes, J.M. and Fearnehough, G.D., The detection of slow crack growth in crack opening displacement specimens using an electrical potential method, *Eng. Fract. Mech.* 3:103 (1971).
24. Joyce, J.A. and Schneider, C.S., Crack length measurement during rapid crack growth using an alternating current potential difference method, *J. Test. Eval.* 16:257 (1988).
25. Barnes, C.R., A measurement technique for determining dynamic crack speeds in engineering-materials experimentation, *Exp. Tech.* 9:33 (1985).
26. Kobayashi, A., Ohtani, N., and Munemura, M., Dynamic stress intensity factors during viscoelastic crack propagation at various strain rates, *J. Appl. Poly. Sci.* 25:2789 (1980).
27. Crouch, B.A. and Williams, J.G., Application of a dynamic numerical solution to high speed fracture experiments - II. Results and a thermal blunting model, *Eng. Fract. Mech.* 26:553 (1987).
28. Fineberg, J., Gross, S.P., Marder, M., and Swinney, H.L., Instability in dynamic fracture, *Phys. Rev. Let.* 67:457 (1991).
29. Racca, R.G. and Dewey, J.M., Time smear effects in spatial frequency multiplexed holography, *Appl. Opt.* 28:3652 (1989).
30. Hall, R.G.N., Gates, J.W.C. and Ross, I.N., Recording rapid sequences of holograms, *J. Phys. E: Sci. Inst.* 3:789 (1970).
31. Ostrovskaya, G.V. and Ostrovsky, Y.I., Holographic methods of plasma diagnostics *in:* "Progress in Optics XXII", North-Holland Publishing Corp. (1985).
32. Tschudi, T., Yamanaka, C., Sasaki, T., Yoshida, K. and Tanake, K., A study of high-power laser effects in dielectrics using multiframe picosecond holography, *J. Phys. D* D11:177 (1978).
33. Thomas, K.S., Harder, C.R., Quinn, W.E. and Siemon, R.E., Helical field experiments on a three-meter theta pitch, *Phys. Fluids* 15:1658 (1972).
34. Yamamoto, Y., Multi-frame pulse holography system, *in:* "Proceedings of SPIE's 18th International Congress on High Speed Photography and Photonics, Vol. 1032", SPIE, Bellingham, WA (1988).
35. Landry, M.J., and McCarthy, A.E., Use of the multiplle cavity laser holographic system for EBW Analysis, *Opt. Eng.* 14:69 (1975).
36. White, J.U., Long optical paths of large aperture, *J. Opt. Soc. Amer.* 32:285 (1942).
37. Deaton, J.B., McKie, A.D.W., Spicer, J.B., and Wagner, J.W., Generation of narrow-band ultrasound with a long cavity mode-locked Nd:YAG laser, *Appl. Phys. Let.* 56:2390 (1990).
38. Ehrlich MJ, Steckenrider JS, Wagner JW, High-speed time-resolved holography for imaging transient events, *in:* "1992 Review of Progress in Quantitative NDE", La Jolla, California, (1992).
39. Steckenrider JS, Ehrlich MJ, Wagner JW, Pulsed Holographic Recording of Very High Speed Transient Events, *in:* "Proceedings of the 1991 SPIE International Symposium on Optical Applied Science and Engineering", SPIE, Bellingham, WA (1991).
40. Steckenrider JS, Wagner JW, Controlled generation of sharp pre-cracks in thin glass plates, *Int. J. Fract.* 55:R55 (1992).
41. Roberts, D.K., and Wells, A.A., The velocity of brittle fracture, *Engineering* 178:820 (1954).

EMBEDDED AND ATTACHED SAPPHIRE OPTICAL FIBER SENSORS FOR THE NONDESTRUCTIVE CHARACTERIZATION OF MATERIALS AT HIGH TEMPERATURE

Richard O. Claus, Kent A. Murphy, Anbo Wang,
Russell G. May and Marten de Vries

Fiber & Electro-Optics Research Center
Bradley Department of Electrical Engineering
Virginia Polytechnic Institute and State University
Blacksburg, VA 24061-0111

INTRODUCTION

Sapphire optical fiber sensors have been attached to and embedded within high temperature materials and subsequently used to measure strain and temperature at temperatures above 1000°C for several years[1]. Intrinsic and extrinsic interferometric, polarimetric and intensity-based sensor configurations have been implemented using unclad and uncoated solid-core sapphire fibers, as well as sapphire fibers with ceramic coatings designed to reduce propagation loss and allow increased protection of the fiber surface[2]. Of these methods, extrinsic Fabry-Perot interferometric cavity designs using fibers operated as single mode waveguides have thus far been most widely employed, in part due to their inherent avoidance of off-axis strain effects and temperature-induced signal fading[3]. These sensor elements combined with low loss in-line splices, implemented directly between sapphire fibers and indirectly using aluminosilicate glass fiber jumpers between sapphire and silica fibers[4], have allowed the implementation of such fiber sensors on and in materials in high temperature laboratory environments up to several meters from room temperature electronic instrumentation. Such sensors have been directly attached to silicon carbide specimens using high temperature ceramic adhesives and used to measure temperature with 0.2°C resolution from room temperature to over 1700°C, and quasistatic strain with 0.1 microstrain resolution at temperatures up to and above 1200°C[5]. The sensors also have been used above 1200°C to monitor strain during the fatigue loading of a silicon carbide specimen to more than 60,000 cycles without failure[6].

Nondestructive Characterization of Materials VI
Edited by R.E. Green, Jr. *et al.*, Plenum Press, New York, 1994

Similar sensors have been embedded between woven silicon carbide (Nicalon®) composite laminae and in castable ceramic preforms, fired at temperatures above 1400°C, and then used to monitor strain and temperature in the resulting materials at temperatures up to and above 1200°C[7]. This latter work is discussed in this paper. The practical application of these devices for the nondestructive evaluation of high temperature materials is suggested.

FABRICATION OF THE CERAMIC MATRIX COMPOSITE

Chemical vapor infiltration (CVI) methods are one of the possible, yet intrinsically difficult, reaction forming processes for making ceramic-matrix composites. The CVI process for densification of porous ceramic preforms consists of reactant diffusion and decomposition followed by solid product deposition in the porous body. This section discusses in some detail the CVI of Nicalon® fiber bundles with embedded optical fibers performed in cooperation with researchers at the DOE Oak Ridge National Laboratory[7].

Potential advantages of the CVI process includes 1) good throwing power, i.e. uniform coverage of the structural fibers with the matrix material, 2) the ability to manipulate the stoichiometry, crystal structure, crystal orientation, and morphology of the deposit, 3) the ability to maintain high purity levels, and 4) the ability to operate at relatively low temperature during consolidation. The absence of both matrix shrinkage around dimensionally stable reinforcements and applied pressure, and the ablility to operate with geometrically complex preforms in nominally isothermal conditions are especially important additional features for the processing of ceramic matrix composites (CMCs). Inherent problems with the CVI process include issues typical of most CVD processes, such as relatively high cost, corrosive and toxic precursers and by-products, explosive reactants, and low deposition rates. However, the most serious problems are related to achieving a uniform penetration throughout a preform, without damaging the reinforcements or their coatings.

CVI processes induce highly controlled reactions between vapor-phase reactants within hot-wall, hermetic chambers. The main process variables for a specific set of reactants are temperature, reactant partial pressures, and total pressures. The foremost issue is one of finding conditions by which uniform depositions can be achieved at acceptably high infiltration rates throughout the entire volume of the preform. Because of the penetration requirements, most researchers are forced to deposit under low temperature, low partial pressure, and low total pressure conditions. Although partially successful in achieving uniform deposition, these process conditions reduce the deposition rate and thus detract from the possible advantage of CVI as a rapid and inexpensive CMC fabrication method. Excursions to higher-than-optimum pressures usually result in homogenous nucleation of powders in the gas phase.

Researchers have made important improvements in the CVI process. Research partners at DOE for example have developed a process that simultaneously utilizes thermal and pressure gradients to reduce the infiltration time from weeks to less than twenty-four hours[7]. Accurate process models linking mass and heat transfer with reaction kinetics have and are being developed for the efficient extension of the thermal-gradient, forced-flow technique to large and complex shapes and the further optimization of the isothermal-isobaric process. CVI of Nicalon® cloth, with different optical fibers embedded in different layers of the matrix, was performed at ORNL, using methyltrichlorosilicane (MTS) to form the SiC matrix. The experimental details are described below.

SAPPHIRE OPTICAL FIBER SENSOR IMPLEMENTATION

The major objective of embedding sapphire optical fiber sensor elements into CVI-processed SiC matrix CMCs is to provide an in-situ means for intelligently monitoring the fabrication process and to control the process through the real time feedback of information concerning the temperature gradient through the material, or other data. For the measurements reported here, sapphire optical fiber sensors have been implemented to allow the measurement of axial elongation due to either mechanical strain or thermal expansion using the basic extrinsic Fabry-Perot interferometric (EFPI) sensor geometry shown in Figure 1[3]. Here, a 1300 nm single

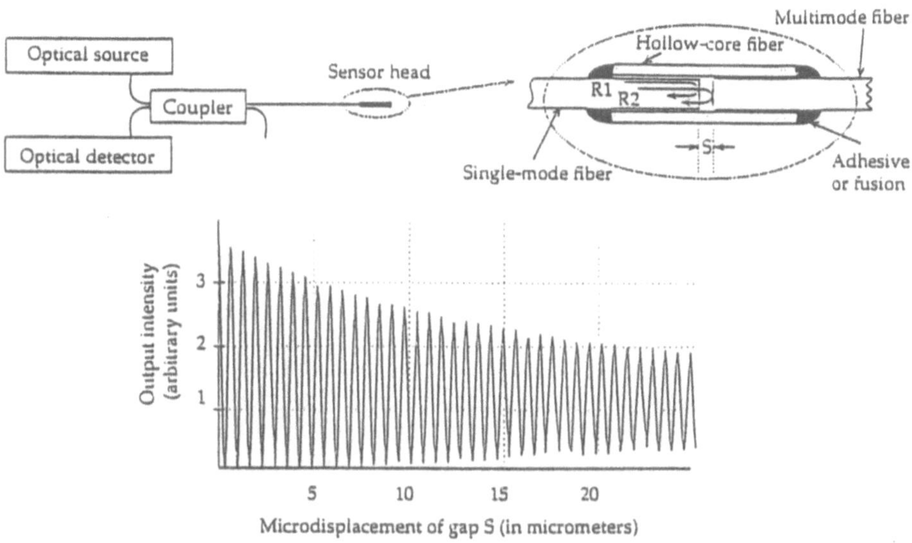

Figure 1. Geometry of conventional EFPI strain sensor. Developed by Virginia Tech in 1989 and available commercially. Response of EFPI to gradual elongation is shown at the bottom of the figure.

mode silica fiber transmits light from a laser diode to the sapphire EFPI sensor element through silica fiber that is spliced to sapphire fiber. At the opposite end of the input fiber, the laser light signal is partially reflected and partially transmitted, exiting (hence, an "extrinsic" sensor design) across the air gap separating the ends of the input fiber and an output fiber used solely as a reflector. This signal and the signal reflected from the facing fiber endface interfere and propagate back through the input fiber, and through a fiber coupler to a photodiode detector.

The detected photodiode signal current may be shown to be proportional to the phase difference between the two reflected optical fields [4]. Assume a coherent, approximately plane wave detected at the output of the sensor. This wave can be represented in terms of its complex amplitude $U_i(x,z,t)$, given by

$$U_i(x, z, t) = A_i \, exp\left(j\phi_i\right), \quad i = 1, 2, \tag{1}$$

where the variable A_i can be a function of the transverse coordinate x and the distance traveled z and the subscripts i = 1, 2 stand for the reference and the sensing reflections, respectively. Assuming that the reference reflection $A_1 = A$, the sensing reflection coefficient A_2 can be approximated by the simplified relation

$$A_2 = A\left\{\frac{ta}{a + 2s\tan\left[\sin^{-1}(\mathrm{NA})\right]}\right\}, \tag{2}$$

where a is the fiber core radius, t is the transmission coefficient of the air-glass interface (≈0.98), s is the end separation, and NA is the numerical aperture of the single-mode fiber, given by NA = $(n_1^2 - n_2^2)^{1/2}$. n_1 and n_2 are the refractive indices of the core and the cladding, respectively. The observed intensity at the detector is a superposition of the two amplitudes and is given by

$$I_{det} = |U_1 + U_2|^2 = A_1^2 + A_2^2 + 2A_1A_2\cos\left(\phi_1 - \phi_2\right), \tag{3 a}$$

which can be rewritten as

$$I_{det} = A^2\left(1 + \frac{2ta}{a + 2s\tan\left[\sin^{-1}(\mathrm{NA})\right]}\cos\left(\frac{4\pi s}{\lambda}\right) + \left\{\frac{ta}{a + 2s\tan\left[\sin^{-1}(\mathrm{NA})\right]}\right\}^2\right), \tag{3 b}$$

where we have assumed that $\phi_1 = 0$, $\phi_2 = 2s(2\pi/\lambda)$ and λ is the free space wavelength.

The simplified loss relation in Equation 3b is sufficient to understand sensor operation. Changes in the separation distance s between the surfaces of the fibers aligned in the support tube produce a modulation of the output optical intensity. This modulation is sinusoidal, with an envelope that gradually decreases in amplitude as s increases, as shown in Figure 6. By attaching the support tube and the input fiber to a test material, each at one discrete point, the gage length of the EFPI sensor is defined. Percent elongation is determined as the ratio between the endface displacement measured and this gage length. In practice, the gage length may be set during fabrication to be several millimeters or less, so the practical strain resolution is on the order of 0.1 microstrain or less.

Signal processing used to determine strain from the output of such sensors is straightforward but not uncomplicated. For small changes in strain, the sensor output may be maintained at quadrature and small variations are measured around that quiescent point. For large changes in strain, the output signal varies through several fringes so dynamic fringe counting is required.

The sinusoidal output of the EFPI sensor in response to monotonically changing strain also produces a potential ambiguity in the direction of strain measured. To overcome this limitation, operating the sensor in a quadrature phase shifted (QPS) mode, by using two laser sources at two wavelengths, allows unambiguous determination of both strain amplitude and instantaneous direction.

EXPERIMENT

Very high purity hydrogen and commercial grade 98% purity MTS were use during fabrication. The interwoven Nicalon® fibers are polycarbosiliocane-derived Si-C-O (Nicalon®, ceramic grade) and are washed in acetone to remove sizing. Silica and sapphire fiber sensor elements were embedded in three different layers, each ten layers apart, and tested immediately for survival. Multimode and singlemode silica fibers had a variety of standard coatings including gold, aluminum and polyimide. Once the fibers had been placed on a layer of interwoven Nicalon® fiber, the layers were stacked to form a graphite disc. The sample was then coated with approximately 0.2 μm of graphitic carbon via the pyrolysis of propylene. The optical fibers were coated in this process as well. The carbon precoat accurately represents a fibrous preform prepared for infiltration because such precoats are necessary to protect the fibers from infiltration gases and to obtain optimum mechanical properties. The temperature of the SiC gas flowing through the Nicalon® specimen is measured with an optical pyrometer through a window and is carefully controlled.

Once the sample was infiltrated it was cut to form rectangular sideswhich were cleaned and polished. The optical fibers that were embedded in the infiltrated Nicalon preform sample were succesfully identified under a microscope, and later connected to external optical and electronic system hardware to allow measurement of temperature during post processing testing. Figure 2 shows output fringes from one of the internal sapphire fiber sensors corresponding to an increase in temperature of approximately 2.6°C at a starting temperature of 1200°C in the specimen.

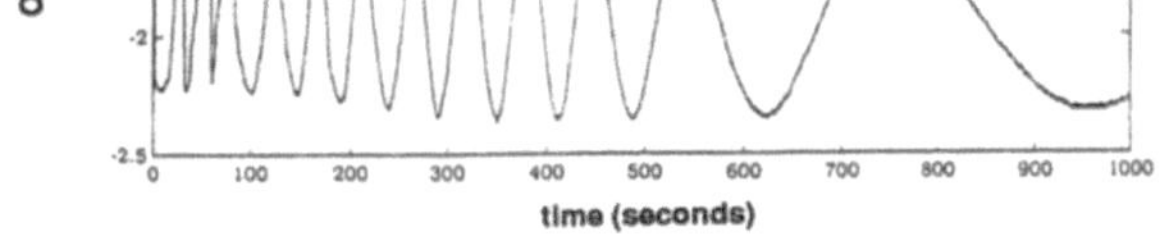

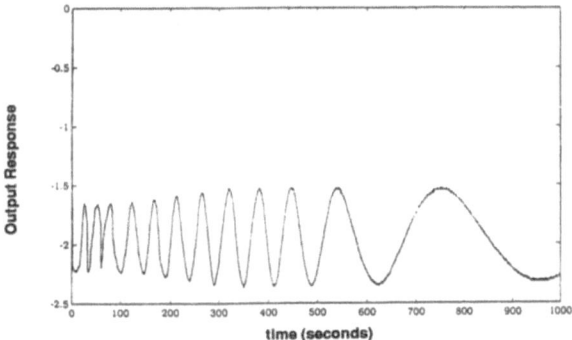

Figure 2. Ouput fringe response of embedded sapphire optical fiber EFPI sensor in CVI-processed SiC CMC specimen. Vertical axis units are volts of photodetector response.

CONCLUSION

The major goal of this preliminary research has been to demonstrate the survivablilty of sapphire optical sensor fibers embedded in a Nicalon® sample.

The future goal of this research is to embed similar optical fiber sensors to both monitor fabrication and subsequent material performance.

ACKNOWLEDGMENT

This work was supported in part through cooperation with the Metals and Ceramics Division of the DOE Oak Ridge National Laboratory, and by the Virginia Center for Innovative Technology. The authors thank D. Stinton and R. Lowden for all of the CVI processing work.

REFERENCES

1. K. A. Murphy, G. Z. Wang, B. R. Fogg, A. M. Vengsarkar and R. O. Claus, Proc. SPIE 1588:117 (1991).
2. A. Wang, S. Gollapudi, K. A. Murphy, R. G. May and R. O. Claus, "Sapphire-fiber-based intrinsic Fabry-Perot interferometer," *Optics Letters* 17:1021 (1992).
3. K. A. Murphy, M. F. Gunther, A. M. Vengsarkar and R. O. Claus, "Quadrature phase-shifted, extrinsic Fabry-Perot optical fiber sensors," *Optics Letters* 16:273 (1991).
4. A. Wang, S. Gollapudi, R. G. May, K. A. Murphy and R. O. Claus, "Advances in sapphire-fiber-based intrinsic interferometric sensors," *Optics Letters* 17:1544 (1992).
5. A. Wang, G. Z. Wang, K. A. Murphy and R. O. Claus, "Birefringence-balanced polarimetric optical fiber sensor for high temperature measurements," Optics Letters 17:1391 (1991).
6. T. Tran, "Sapphire Optical Fiber Sensors for Evaluation of Silicon Carbide at High Temperature," Proc. Optical Fiber Sensor Workshop (Blacksburg, VA), April 1993, IOP Publishing.
7. M. de Vries, M. Nasta, J. Patel, K. Kamdar, R. Lowden, D. Stinton, S. Allison, J. Muhs and R. Claus, Proc. SPIE Conf. (Boston), September 1992.

MULTI-PARAMETER FIBER OPTIC SENSING FOR COMPOSITE MATERIAL MONITORING

Christian V. O'Keefe, B. Boro Djordjevic, and B. N. Ranganathan

Martin Marietta Laboratories
1450 S. Rolling Rd
Baltimore, MD 21227

INTRODUCTION

Fiber optic sensors can measure many perturbations including strain, temperature, indices of refraction, cure state, pressure, damage, electrical fields, magnetic fields, current and chemicals. The particular perturbation or perturbations which can be detected by a sensor is dependent on the type of sensor and optical fiber used to implement the sensor. During the course of this research two different types of optical fiber sensors were created to measure different properties of composite materials. The first sensor was geared towards the monitoring of composite cure, while the second was created to measure mechanical effects within the material.

FIBER OPTIC CURE MONITORING

The monitoring of composite cure has been achieved through non-optical means including dielectric sensors, thermocouples, bulk ultrasonics and acoustic waveguides as well as through optical means including embedded spectroscopic analysis [Compton 1988], fiber optic interferometers and fiber optic refractive index sensors [Afromowitz 1988, 1989; Fanconi 1989]. The fiber optic embedable sensors do have the advantages of exhibiting immunity from electromagnetic interference and featuring small, non-intrusive sizes. The spectroscopic techniques can require relatively expensive instrumentation and in the case of infra-red spectroscopy the need to use brittle and expensive infra-red optical fibers. Rather than directly monitoring the chemical bonding through a swept frequency spectroscopy measurement, it is possible to perform a single frequency index of refraction measurement. One technique used to achieve this measurement employed an optical fiber section made from the cured resin material. [Afromowitz 1988, 1989] By injecting a single frequency optical signal into the resin fiber, any changes in the index of refraction of the composite material will result in changes in the intensity of the light transmitted through the resin fiber. Silica optical fibers were bonded onto each end of the resin fiber thus providing a method for injecting light into the resin fiber and for extracting the light which has propagated through the resin fiber. This resin fiber and portions of the silica fibers can then be embedded within the uncured composite material and used as a cure monitor. A single frequency optical signal was injected into the lead-in fiber, while the light emitted from the end of the lead-out fiber was monitored for intensity changes. In this case when the resin material becomes fully cured, there is no difference between it and the resin fiber, resulting in an inability of the resin fiber to guide the light.

The use of a resin fiber as the sensing region offers the advantage of providing a very sensitive method for monitoring the index of refraction of the host material. Unfortunately it is very difficult to manufacture, in a very reproducible manner, the resin fiber and its connections to the lead-in and lead-out fibers. Therefore it would be best if the sensing section was composed of a fiber which could be produced with consistent properties and which did not have to be bonded to the lead-in and lead-out fibers. It is possible to create this type of sensing section by stripping the cladding away from a portion of an optical fiber, such that the now exposed core can act as the sensing region. Specifically, the optical fiber chosen to create this sensor was a multimode fiber with a silica core, a silicone polymer cladding and an acrylate jacketing. The sensing section was created by applying a flame to a portion of the optical fiber such that the cladding and jacketing were removed but such that the core was left intact. The resulting sensing region and portions of the unmodified fiber now acting as the lead-in and lead-out portions of the sensor were embedded within a composite panel as illustrated in figure 1.

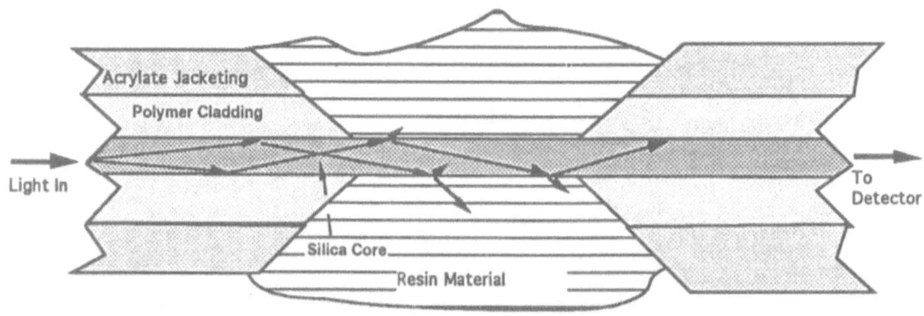

Figure 1. Index of refraction sensor created by removing cladding and jacketing from a multimode fiber.

The disadvantage of using a silica core as the sensing instrument is that its index of refraction is usually less than that of the resin material and as such it does not exhibit total internal reflection at the core to resin boundary. This does reduce the ability of the fiber to carry the light but experience has shown that the index of refraction mismatch is great enough to allow a sufficient amount of light to pass through the sensor. The ability of this sensor to respond to the curing of a composite material was tested by placing it within a sixteen-ply composite panel consisting of epoxy resin reinforced with graphite fibers. The sensing region was 1.1 inches in length and a helium-neon laser operating at 633nm was used as the light source. The composite panel went through a cure cycle consisting of a ramp up in temperature to 250°F after which it was held at that temperature for two hours, followed by a ramp up in temperature to 350°F and a hold of two hours at that temperature, finally culminating with a ramping down to room temperature. During the first ramp up and hold at 250°F, a 50 psi pressure was applied to the panel, which was then increased to 100 psi during the ramp up to 350°F, the holding of the temperature at 350°F and the ramping down to room temperature. Figure 2 shows the changes in intensity incurred during two separate cure cycles. The two sensors manufactured for the two tests were composed of the same type of optical fiber and both had sensing region lengths within 0.1 inches of each other. Both tests were normalized at a point ninety minutes into the cure cycle corresponding to a point in the test slightly before the ramp up in temperature to 350°F. This was done to minimize the impact of lay-up variations and to allow the resin material to fully flow over and around the sensors.

An examination of figure 2 clearly indicates that this sensor exhibits its greatest sensitivity during the initial and latter stages of the cure cycle. The initial portion of the cycle is the period during which consolidation takes place. Thus it may be possible to apply this sensor as a consolidation sensor. The latter portion of the cycle from the 250 minute mark and on consists of the ramp down in temperature. The composite material is fully cured at this point thus leaving the changes in the indices of refraction to temperature effects in the sensor and the cured epoxy. This sensor offers tremendous promise as an index of refraction cure sensor owing to its low cost, possibility for sensitivity, ease of reproducible manufacturability, and simplicity. To fully exploit this type of sensor for this

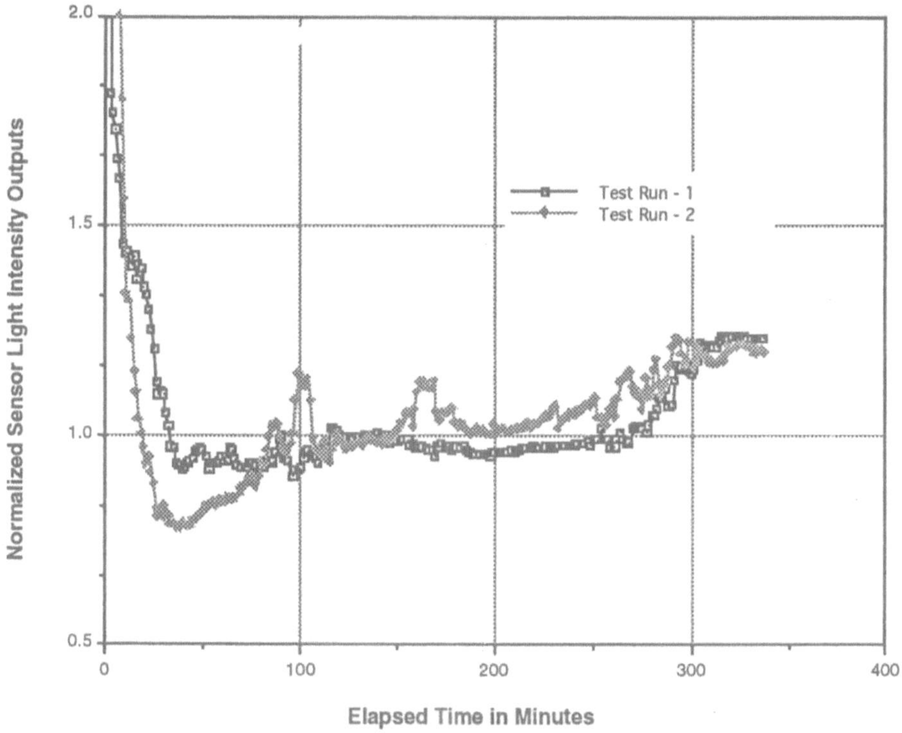

Figure 2. Normalized outputs from two tests of fiber optic cure monitor sensor.

purpose, it may be necessary to select a optical wavelength which would better respond to the cure of the resin, or to apply it to other resin materials.

DUAL PARAMETER SENSOR

The single output capability exhibited by most fiber optic phase sensors may lead to several errors in the quantifiable determination of perturbations on the fiber. One such error is associated with sensors sensitive to multiple effects, such as those showing sensitivity to both strain and temperature, making it difficult to differentiate between these two different perturbations. If it is assumed that only one effect is being measured, any other perturbations will corrupt the measurement. These errors or measurement discrepancies may be eliminated or greatly reduced by incorporating several sensors at the same location, all experiencing the same perturbations. These sensors must exhibit responses which are linearly independent of each other such that the same perturbations will induce a different output response in each sensor. It would be best to implement this sensor system by creating two sensors within the same optical fiber. Thus both sensors will experience the same perturbations and their outputs can be used to determine the effects of the various perturbations. Several methods to implement dual parameter sensing in a single fiber have been proposed and tested. These sensors are based on dual wavelength monitoring in dual core optical fibers [Dunphy et al 1984], fiber Bragg gratings [Dunphy et al 1990], or polarimetric/modal domain [Vengsarkar et al 1990] configurations. The sensors have met with limited success. In particular, their need for dual wavelength operation, their limited accuracy and their use of non-standard, relatively expensive components have thwarted their acceptance as viable dual parameter sensors. These problems have been overcome with the development of a dual two-mode fiber optic sensor with a remote, controlled length sensing region.

In this sensor system, light from a single laser source is launched along both polarization axes of a single mode polarization preserving lead-in fiber. This light propagates to the sensing region, where it is injected into both axes of a polarization-preserving two mode fiber. Within this sensing region, each of the two polarization components will divide into two energy modes. At the end of the sensing region, the light is either spatially filtered by a single mode polarization-preserving lead-out fiber as shown in figure 3 or reflected back through the sensing region to the lead-in fiber. Therefore, in the latter case the lead-in fiber also acts as a lead-out fiber. At the end of the lead-out fiber, the light is separated into its two orthogonal polarization components, with the light from each component falling incident upon a photodetector as shown in figure 3. At that point the output signals can be demodulated to extract the phase change in each channel and processed to determine the strain and temperature changes.

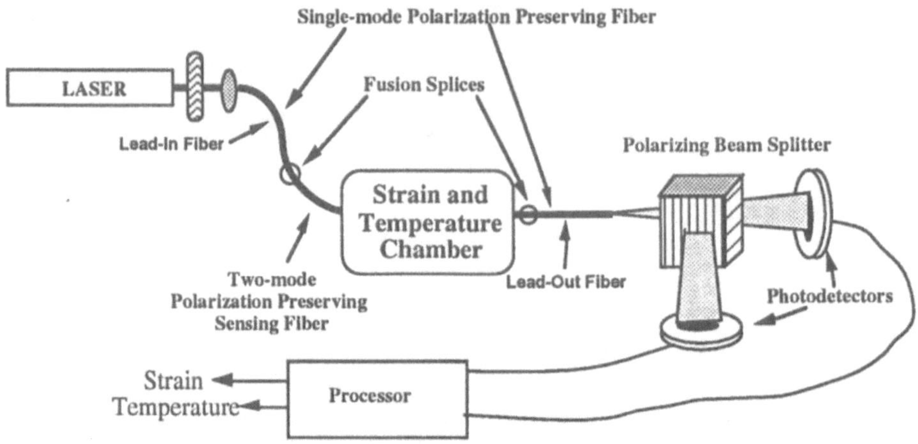

Figure 3 Dual parameter fiber optic sensor configuration.

Along each polarization axis within the sensing region the two energy modes, the LP_{01} and the LP_{11}, travel with two different velocities such that they differ in phase at the end of the sensing region. The result is an output signal from each channel with follows the function:

$$I = I_0\left[\,1 + M\cos\left[\left(\beta_{01} - \beta_{11}\right)L\right]\right] = I_0\left[\,1 + M\cos\left(\theta\right)\right]$$

where β_{01} is the propagation constant for the LP_{01}

β_{11} is the propagation constant for the LP_{11}

L is the length of the sensing region

This equation exists for each of the two polarization states of the fiber. Therefore the sensor system will have two outputs whose phase responses will vary with strain and temperature as follows:

$$\Delta\theta_1 = \frac{2\pi}{\Lambda_{\varepsilon 1}} L\,\Delta\varepsilon + \frac{2\pi}{\Lambda_{T1}} L\,\Delta T$$

$$\Delta\theta_2 = \frac{2\pi}{\Lambda_{\varepsilon 2}} L\,\Delta\varepsilon + \frac{2\pi}{\Lambda_{T2}} L\,\Delta T$$

which for small strains can be given as:

$$\Delta\theta_1 = \frac{2\pi}{\Lambda_{\varepsilon 1}} \Delta L + \frac{2\pi}{\Lambda_{T1}} L\,\Delta T$$

$$\Delta\theta_2 = \frac{2\pi}{\Lambda_{\varepsilon 2}} \Delta L + \frac{2\pi}{\Lambda_{T2}} L\,\Delta T$$

This can be rewritten in matrix notation to yield:

$$\overrightarrow{\Delta\theta} = \langle\Lambda\rangle\,\vec{P}$$

$$\text{where } \overrightarrow{\Delta\theta} = \begin{bmatrix} \Delta\theta_1 \\ \Delta\theta_2 \end{bmatrix},\ \langle\Lambda\rangle = \begin{bmatrix} \frac{2\pi}{\Lambda_{\varepsilon 1}} & \frac{2\pi}{\Lambda_{T1}} \\ \frac{2\pi}{\Lambda_{\varepsilon 2}} & \frac{2\pi}{\Lambda_{T2}} \end{bmatrix},\ \text{and } \vec{P} = \begin{bmatrix} \Delta L \\ L\,\Delta T \end{bmatrix}$$

which allows us to take the inverse of $\langle\Lambda\rangle$ to obtain:

$$\vec{P} = \langle\Lambda\rangle^{-1}\,\overrightarrow{\Delta\theta}$$

Thus the parameters being measured (fiber elongation-strain and temperature) can be directly calculated from the phase changes of the sensor outputs.

The gauge length of the sensor can be directly controlled by governing the length of the sensing fiber. This allows for a wide range of sensing gauge lengths from roughly several centimeters to tens of meters. The sensitivity and accuracy of the measurements is a function of the length of the sensor, the accuracy with which the matrix $\langle\Lambda\rangle$ was formulated, the characteristics of the matrix $\langle\Lambda\rangle^{-1}$ and the accuracy with which the phase changes can be measured. Based on the optical fiber currently used to create this sensor, a phase measurement of $2\pi/100$ radians will provide a sensitivity of 2 microns for elongation measurements and 0.3 °C-meters for temperature sensing. The actual strain or temperature accuracy are then directly determined by the length of the sensing section and the characteristics of the $\langle\Lambda\rangle^{-1}$ matrix. To prove the operation of this sensor a series of tests were conducted on a sensor with a 352 cm sensing region. The sensor was subjected to a variety of strains (elongations) and temperature changes. The results of these tests are shown in figure 4, which has both the directly measured and the sensor determined values for both elongation and temperature changes.

	Actual ΔL (μm)	Fiber Optic ΔL (μm)	Actual ΔT (°C)	Fiber Optic ΔT (°C)
Change In Strain	2286	2277	0.0	1.7
	1295	1313	0.0	-1.0
	559	599	0.0	1.1
Change In Temp.	0	-20	7.7	6
	0	25	9.5	7.9
	0	-1	21.2	17.8

	Actual ΔL (μm)	Fiber Optic ΔL (μm)	Actual ΔT (°C)	Fiber Optic ΔT (°C)
Change In Strain and Temperature	100	104	3.3	3.6
	200	246	7.8	8.1
	200	230	8.3	8.8
	200	198	3.3	3.7
	400	436	17.8	18.9
	400	455	21.7	22.9

Figure 4 Comparison of fiber optic sensor derived and laboratory instrumentation measured values

The results of the sensor test clearly indicate the ability of this type of sensor to simultaneously measure two parameters. A comparison of the directly measured fiber elongations and temperature changes to the sensor derived values reveal differences of at most one hundred microns for large strains, approximately 10% to 20% for small strains and several degrees Celsius for all cases. This type of sensor can be created using an all-fiber optics design and can incorporate integrated optical circuits. It can also use a variety of demodulation schemes to extract the phase information from the sensor channels. It also offers the possibility for multiplexing through either series or parallel sensor arrangements.

FUTURE WORK

The experimental results for the cure monitoring sensor indicate that in its present configuration it may be best used as a consolidation sensor or after cure as a temperature sensor. A greater response to cure changes can be achieved by tailoring the wavelength of the light source to match the optical frequency response of the resin material. The dual parameter sensor demonstrates excellent accuracy for changes in axial extensions (axial strain) and temperature. To fully utilize a multi-parameter sensor, it is necessary to expand its capabilities to include the ability to measure two orthogonal strain effects. This may be possible through a dual wavelength operation or through the incorporation of a intrinsic Fabry-Perot configuration within the sensing element.

REFERENCES

M.A. Afromowitz, Fiber Optic Polymer Cure Sensor, J. Lightwave Tech., v6, 1591 (1988).

M.A. Afromowitz and K.-Y. Lam, Fiber Optic Sensor for Thermoset Composites, Review of Progress in Quantitative Nondestructive Evaluation, v 8, D.O. Thompson and D.E. Chimenti Eds., Plenum Press, 1112 (1989).

D.A. Compton, S.L. Hill, N.A. Wright, M.A. Druy, J. Piche, W.A. Stevenson and D.W. Vidrine, In-Situ FT-IR Analyses of a Composite Cure Reaction Using a Mid-Infrared Transmitting Optical Fiber, Appl. Spectros., v 42, 972 (1988).

J.R. Dunphy, G.Meltz, M.M. Abou El Leil and E. Snitzer, Twin-core Fiber-optic Sensor for Simultaneous Temperature and Strain Measurement, proc. Conf. on Optical Fiber Communication, Jan 84, New Orleans, LA, 58 (1984).

J.R. Dunphy, G. Meltz, F.P. Lamm and W.W. Morey, Multifunction, Distributed Optical Fiber Sensor for Composite Cure and Response Monitoring, proc. SPIE Vol. 1370 Fiber Optic Smart Structures and Skins III, Sep 90, San Jose, CA, 116 (1990).

B.M. Fanconi, Monitoring Consolidation of Reinforcement Plies in Polymer Matrix Laminates, SAMPE Journal, v 25, 35 (1989).

A.M. Vengsarkar, W.C. Michie, L. Jankovic, B. Culshaw and R.O. Claus, Fiber Optic Sensor for Simultaneous Measurement of Strain and Temperature, proc. SPIE Vol. 13677 Fiber Optic and Laser Sensors VIII, Sep 90, San Jose, CA, 249 (1990).

QUANTITATIVE MEASUREMENT OF STRAIN IN REINFORCED CONCRETE STRUCTURES USING CALIBRATED OPTICAL FIBER SENSORS

Richard O. Claus[1], Sami F. Masri[2], Marten de Vries[1], Manish Nasta[1] and Kent A. Murphy[1]

[1]Fiber & Electro-Optics Research Center
Bradley Department of Electrical Engineering
Virginia Polytechnic Institute and State University
Blacksburg, VA 24061-0111

[2]Department of Civil Engineering
University of Southern California
Los Angeles, CA 90089-2531

INTRODUCTION

This paper discusses the use of calibrated, quantitative optical fiber sensors for the internal diagnostics of reinforced concrete structures. Prior optical fiber sensor methods for the monitoring of reinforced concrete specimens and structures during fabrication, laboratory testing and field use first are reviewed. Much of this work has concerned either techniques for embedding optical fibers in concrete without severe damage to the fiber, or the use of qualitative and uncalibrated sensors to infer the general conditions within the concrete. Here we report the use of practical, short gage length relative strain sensors for the measurement and mapping of strain in reinforced concrete beam and crossbeam configurations, and issues in the implementation of absolute rather than relative strain measuring optical fiber sensors. Both types of fiber sensors were attached to rebar reinforcement rods prior to filling with concrete, and were colocated with conventional foil strain gages to allow direct comparison of output signals. By using conventional robust fiber cabling into the specimens, no difficulties due to fiber breakage were encountered.

Although the interferometric strain sensor had a gage length of approximately 2 mm and a resolution of less than 0.01 microstrain, it suffers from an inherent requirement for relative phase reference information and a cross-sensitivity to temperature variations through the coefficient of thermal expansion of the sensor element, and both of these problems must be overcome using relative complicated signal processing. The simpler absolute strain sensor prototype has a similar gage length and avoids both of these limitations while allowing a resolution of approximately 0.1 microstrain.

Experimental tests indicated less than a three percent difference between the strain measured by the fiber sensors and the foil gages. Tests included the quasi-static and fatigue loading of the specimens and the mapping of strain at pre-determined locations on the internal rebar cage grid. Subsequent observation of residual strains using the absolute strain sensors that were disconnected and then reconnected to the support optics and electronics, with no loss of information, continue and will be presented elsewhere in the near future. The practical applications of such sensor methods in the quantitative internal evaluation of large civil structures is suggested.

OPTICAL FIBER SENSORS IN CONCRETE STRUCTURES

Although many optical fiber sensors have been developed during the past fifteen years, few experiments have used these optical fiber sensors to monitor strain and crack growth in reinforced concrete structures. Most of the research performed to date has focused on obtaining qualitative measurements rather than quantitative measurements. Much research has been done, and remains to be done, in investigating the survivability of optical fibers in reinforced concrete structures for both sensing applications and for reinforcing concrete specimens[1,2]. This paper discusses the use of a network of embedded in-line extrinsic Fabry-Perot interferometric (EFPI) fiber sensors for the quantitative strain measurements to monitor fatigue loading conditions with reinforced concrete cross-beam specimens. The use of prototype absolute extrinsic Fabry-Perot interferometric (AEFPI) sensors to monitor absolute, real time strain in a reinforced concrete specimen is also suggested.

Fiber optic intrinsic and extrinsic Fabry-Perot sensors reported in the literature have been demonstrated to be highly sensitive to strain, temperature, mechanical vibration, acoustic waves and magnetic fields[3]. Techniques to create the FP cavity have varied from the creation of Bragg gratings in or on the fiber[4] to the use of air-glass interfaces at the fiber ends as the reflectors[5]. A relatively new technique described by Lee and Taylor involves fabricating two semi-reflective splices in a continuous length of fiber[6]. Murphy, *et al.* have demonstrated an EFPI sensor capable of detecting polarity shifts in a dynamically varying measured parameter such as strain[7]. Although this design could be termed a Fizeau interferometer because the reflectivities of the surfaces are low and the finesse small, it has been termed an EFPI due to the earlier use of the intrinsic FP interferometer (IFPI) name.

FIBER SENSOR GEOMETRY AND EXPERIMENTAL SETUP

The EFPI sensor geometry is shown in Figure 1. Here light from a laser light source is coupled into the sensor head which consists of a hollow tube into which the input fiber and a second fiber, serving solely as a reflector, are positioned. The

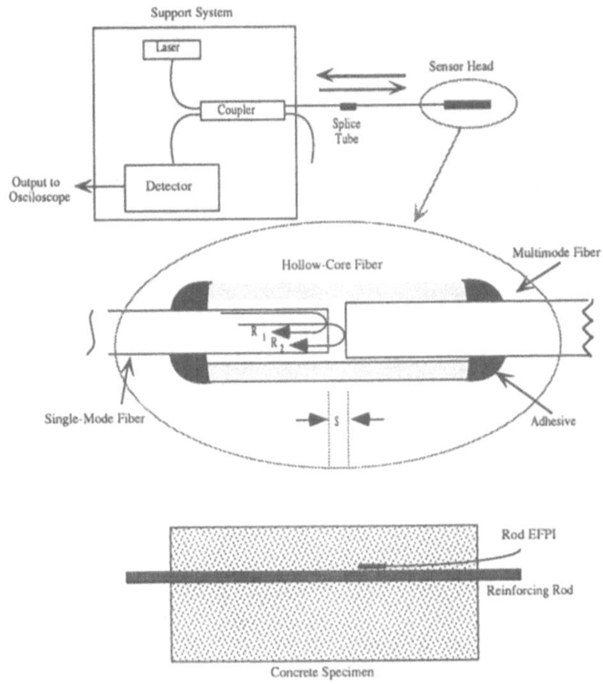

Figure 1. Extrinsic Fabry-Perot interferometric sensor system and element geometries (top), and representative location of sensor attached to reinforcing rod in concrete cross beam specimen (bottom).

reflection from the first glass-air interface at the end of the input fiber, R_1, and the reflection, R_2, from the air-glass interface at the input to the second fiber interfere and couple back to a detector via a 2x2 coupler. The resulting interference signal will vary sinusoidally in response to microdisplacements in the air-gap cavity. Hence, the sensor signal provides a relative measurement of strain induced in the sensor head. For large values of strain fringe counting schemes must be employed.

Such sensors were attached individually to ribbed steel reinforcement rods as shown, by filing a flat space on the side of the rod, milling a small groove in the flat surface, and bonding the sensor head using epoxy resin adhesive. The EFPI sensors were finally coated with a silicone gel to protect the glass fiber from the water and the alkaline based cement.

LOADING EXPERIMENTS ON A CONCRETE CROSS BEAM WITH EMBEDDED FIBER SENSORS

The sensors were attached to the multiple reinforcement rods which together formed two perpendicular, three-dimensional cages within the concrete cross-beam specimen as shown in Figure 2. Conventional foil strain gages were co-located on the rods with each EFPI sensor to allow direct data comparison. The specimens were poured with concrete and finished without significant fiber sensor or foil sensor gage damage. Specimen dimensions were 13 cm x 15 cm in cross-section, with a short beam length of 0.78 m and a long beam length of 1.59 m. This specimen was designed to allow the interrogation of different internal strain vector components on different rods during fatigue loading, with "a priori" knowledge of the anticipated ratios of strains to be experienced at the different locations. Each EFPI sensor was configured to yield only the strain component parallel to the long axis of the rod on which it was attached.

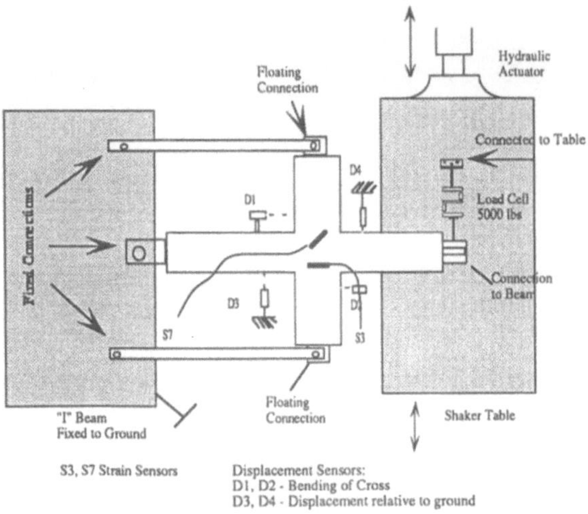

Figure 2. Geometry of reinforced concrete cross beam on dual shaker table load frame. EFPI sensors were attached to multiple internal locations including those shown and multiplexed using an array of 2x2 optical fiber couplers.

Fatigue loading was performed on a large horizontal load frame capable of independent motions of each of the two ends of the long beam. Cyclic fatigue loading of the specimen began with displacements of one end of the beam 0.508 cm from the horizontal quiescent position, and increased through steps to a maximum displacement of 5.1 cm. For example typical comparative strain data obtained for sensor S3 (shown in Figure 2) for cyclic displacements of 0.76 cm are shown in Figure 3. Here note both the general agreement between the fiber sensor data, upper trace, and the foil strain gage data, lower trace, and that both measure the periodic dead zones caused by the dynamics of the load frame. Also note that the optical fiber sensor has a much better signal to noise ratio than its co-located foil strain gage. Electromagnetic interference caused by electronic switching transients are clearly visible near the center of the foil gage data record.

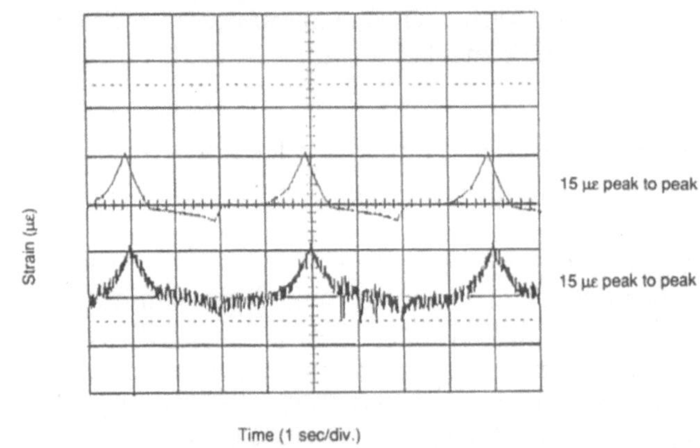

Figure 3. Strain data obtained from EFPI sensor (top) and foil strain gage (bottom). Slight horizontal displacement between the two traces is due to signal acquisition timing.

ISSUES CONCERNING THE FIBER SENSOR-BASED MEASUREMENT OF ABSOLUTE DISPLACEMENT AND STRAIN

As previously mentioned, the standard EFPI discussed above is unable to provide absolute measurements and is also susceptible to directional ambiguity. Moreover, when the system is turned off, all the phase information is lost and hence it is impossible to obtain strain data in field test measurements. These drawbacks have focused current research in optical fiber sensors on absolute, real time, mobile and high resolution measurement systems. The absolute EFPI (AEFPI) is one such system and has proven to be extremely reliable and stable for strain and temperature measurements. The AEFPI uses the wavelength domain information to derive phase information from a high finesse EFPI. Since the fringes obtained are a unique function of the length of the air-cavity the AEFPI is self calibrated for measurement of any given perturbation.

Figure 4 shows the arrangement of the AEFPI. Unlike an ordinary EFPI which employs a single-mode laser diode as the source, the AEFPI uses a broadband source, such as an LED . The AEFPI can be used in transmission mode and hence eliminates the need for a coupler. At the demodulation end a grating spreads the light which is then detected by a scanning photodetector array. A computer algorithm detects the fringe maxima/minima and, given the gage length, real-time, absolute strain information can be obtained. Comparative measurements using EFPI and AEFPI are incomplete at the time of this conference, but are planned for publication[8].

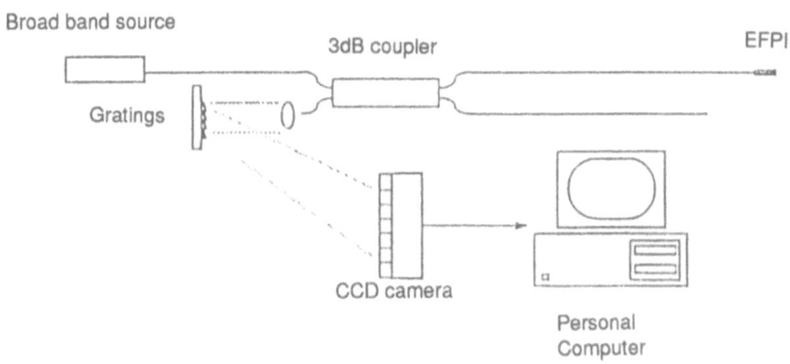

Figure 4. Design of absolute extrinsic Fabry-Perot interferometric fiber strain sensor system. Application of this system for the analysis of concrete cross beams continues[8].

CONCLUSION

This paper has discussed the use of short gage length fiber sensors for the local measurement of strain in reinforced concrete structures, and has suggested the extension of current EFPI designs to allow the measurement of absolute displacement and strain information. Specifically, we have found that 1) optical fiber sensors can survive the embedding process, 2) fiber EFPI sensor output signals compare well with the output signals obtained from the co-located foil strain gages and 3) they can be used to obtain quantitative strain information.

Continuing experimental work in this area will be presented[8], including results of absolute strain sensor development and tests that are underway.

ACKNOWLEDGMENTS

This material based upon work supported by the National Science Foundation under award #111-9260343 to Fiber and Sensor Technologies, the Virginia Center for Innovative Technology and the Carpenters/Contractors Corporation Committee, Inc.

REFERENCES

1. S. F. Masri, M. J. de Vries and R. O. Claus, "An experimental study of embedded fiber-optic strain gauges in concrete structures," *ASCE J.*, submitted for publication.
2. D. Huston, P. L. Fuhr, J-G. Beliveau and W. B. Spillman, "Structural member vibration measurements using an optical fiber sensor," J. Sound Vibr. 150:1 (1991).
3. T. Yoshino, K. Kurosawa, K. Itoh and T. Ose , "Fiber-optic Fabry-Perot interferometer and its sensor applications," *IEEE J. Quantum Electron*. QE-18:1624 (1991).
4. K. L. Belsley, J. B. Carroll, L. A. Hess, D. R. Hubber and D. Schmadel, "Optically multiplexed interferometric fiber optic sensor system,"*Proc. SPIE Int. Soc. Opt. Eng.* 566:257 (1985).
5. A. D. Kersey, D. A. Jackson and M. Corke, "A simple fibre Fabry-Perot sensor," *Opt. Commun.*, 45:71 (1983).
6. C. E. Lee and H. F. Taylor, "Interferometric optical fibre sensors using internal mirrors," *Electron. Lett.* 24:193 (1988).
7. K. A. Murphy, M. F. Gunther, A. M. Vengsarkar and R. O. Claus, "Quadrature phase-shifted, extrinsic Fabry-Perot optical fiber sensors," *Optics Letters 16*:273 (1991).
8. S. F. Masri and R. O. Claus, "Development of actuators and sensors for structural responde under seismic and dynamic loads," Proc. U.S.-Japan Workshop on Experimental Methods in Earthquake Engineering, Honolulu, Hawaii, June 1993.

ULTRASONIC MEASUREMENT OF THE KEARNS TEXTURE PARAMETER IN ZIRCALOY

A.J. Anderson and R.B. Thompson

Ames Laboratory
Department of Materials Science
Iowa State University
Ames, IA 50010

C. S. Cook

Westinghouse Science and Technology Center
Pittsburgh, PA 15235

INTRODUCTION

The Kearns basal pole factors (f_i) are industrially important measures of texture in hexagonal materials like Zircaloy[1,2]. They describes the effective number of crystallites with the basal pole aligned along certain sample axes, for example, the rolling, transverse or normal direction in a rolled sheet. In this case, the f_i are given by:

$$
\begin{aligned}
f_3 = f_{ND} &= \frac{1}{N}\int_0^{\pi/2}\int_0^{2\pi} I(\chi,\eta)\sin(\chi)\cos^2(\chi)\,d\eta\,d\chi \\
f_2 = f_{TD} &= \frac{1}{N}\int_0^{\pi/2}\int_0^{2\pi} I(\chi,\eta)\sin^3(\chi)\sin^2(\eta)\,d\eta\,d\chi \qquad (1) \\
f_1 = f_{RD} &= \frac{1}{N}\int_0^{\pi/2}\int_0^{2\pi} I(\chi,\eta)\sin^3(\chi)\cos^2(\eta)\,d\eta\,d\chi
\end{aligned}
$$

where $I(\chi,\eta)$ is the x-ray intensity for a pole figure having χ as the polar angle and η as the azimuthal angle,

$$N=\int_0^{\pi/2}\int_0^{2\pi} I(\chi,\eta)\sin(\chi)d\eta d\chi \tag{2}$$

and the 1,2 and 3 directions have been associated with the rolling, transverse and normal directions, respectively.

The texture of Zircaloy has important effects on the mechanical and physical properties of the Zircaloy, especially their anisotropies. Also important to the nuclear industry is the effect of the texture on the directions of the growth of hydride platelets and irradiation strains during service. The texture of sheet and tubing products is greatly influenced by processing variables such as reduction schedule and annealing[2]. Hence, a need exists for a nondestructive technique to measure the degree of texture that is developed during such processing.

The feasibility of such an approach has been demonstrated by Konishi[3], who showed a correlation between the velocity of a longitudinal ultrasonic wave propagating in a particular direction and the f_i associated with that direction. However, Konishi's explanation uses the simplified model of Rosenbaum and Lewis[4] for the effects of texture on velocity. They note that a more complete description is possible but was not necessary for their needs. This more complete description is one of the objectives for the present paper. Motivations for developing this more complete description are the desire to strengthen the theoretical foundation of Konishi's work and to provide a basis for determining f_1 and f_2 (as well as f_3) on plate and tubing material.

THEORY

The theory is based on use of an orientation distribution function (ODF) to quantify the probability that a crystallite's orientation will be described by the Euler angles (θ,ψ,ϕ). The ODF can be written as an expansion in generalized spherical harmonics:

$$w(\xi,\psi,\phi)=\sum_{l=0}^{\infty}\sum_{m=-l}^{l}\sum_{n=-l}^{l} W_{lmn}Z_{lmn}(\xi)e^{-jm\psi}e^{-jn\phi} \tag{3}$$

where $\xi = \cos(\theta)$, the W_{lmn} are the orientation distribution coefficients (ODCs), Z_{lmn} are generalizations of the associated Legendre functions as defined by Roe[5] and $j = \sqrt{(-1)}$. Similarly, a normalized pole figure can be expanded:

$$q_i(\zeta_i,\eta_i)=\frac{I_i(\zeta_i,\eta_i)}{\int_0^{2\pi}\int_{-1}^{1} I_i(\zeta_i,\eta_i)d\zeta_i d\eta_i}=\sum_{l=0}^{\infty}\sum_{m=-l}^{l} Q^i_{lm}P_l^m(\zeta_i)e^{-jm\eta_i} \tag{4}$$

where $\zeta_i = \cos(\chi_i)$ and the P_l^m are normalized associated Legendre functions as defined by Roe[5]. The subscript i refers to a particular pole figure. Since we are only dealing with the basal pole figure, the i can be ignored. It can be shown that the coefficients for the pole figure are related to the ODCs[5]:

$$Q_{lm}=2\pi\sqrt{\frac{2}{2l+1}}\sum_{n=-l}^{l}W_{lmn}P_l^n(\Xi)e^{jn\Phi} \tag{5}$$

where $\Xi = \cos(\Theta)$, and Θ and Φ are angles describing the orientation of the selected pole with respect to the crystallite. For a basal pole figure $\Theta = 0$ so $\Xi = 1$. Therefore, only the ODCs with $n = 0$ will have a nonzero value. Thus equation (5) reduces to

$$Q_{lm}=2\pi W_{lm0} \tag{6}$$

After substituting equation (4) into equation (1), the f_i can be written in terms of normalized associated Legendre functions:

$$\begin{aligned}
f_1= f_{RD}&=\sum_{l=0}^{\infty}\sum_{m=-l}^{l}Q_{lm}\int_0^{2\pi}e^{-jm\eta}\cos^2\eta\, d\eta\int_{-1}^{1}P_l^m(\zeta)(1-\zeta^2)d\zeta\\
f_2= f_{TD}&=\sum_{l=0}^{\infty}\sum_{m=-l}^{l}Q_{lm}\int_0^{2\pi}e^{-jm\eta}\sin^2\eta\, d\eta\int_{-1}^{1}P_l^m(\zeta)(1-\zeta^2)d\zeta\\
f_3= f_{ND}&=\sum_{l=0}^{\infty}\sum_{m=-l}^{l}Q_{lm}\int_0^{2\pi}e^{-jm\eta}d\eta\int_{-1}^{1}P_l^m(\zeta)\zeta^2 d\zeta
\end{aligned} \tag{7}$$

It is clear that the integral over η in each of equations (7) will restrict the values of m. For the case of f_3, the integral over η is zero unless $m = 0$. For $m = 0$ the integral gives 2π. Likewise, for the cases of f_1 and f_2, the integral over η has values of $\pi/2$ for $m = \pm 2$, π for $m = 0$ and zero for all other values of m. By rewriting the ζ terms in equations (7) in terms of normalized associated Legendre functions of order $k = 0$ or $k = 2$, we can take advantage of the following orthogonality relation[5]

$$\int_{-1}^{1}P_l^m(\zeta)P_k^m(\zeta)d\zeta=\delta_{lk} \tag{8}$$

Thus l must also be either 0 or 2 for the integral to have a nonzero result. Finally, by using equation (5) to eliminate Q_{lm} we obtain three equations relating the f_i to the ODCs:

$$f_1 = f_{RD} = \frac{4\pi^2\sqrt{2}}{3}W_{000} - \frac{4\pi^2\sqrt{10}}{15}W_{200} + \frac{8\pi^2\sqrt{15}}{15}W_{220}$$

$$f_2 = f_{TD} = \frac{4\pi^2\sqrt{2}}{3}W_{000} - \frac{4\pi^2\sqrt{10}}{15}W_{200} - \frac{8\pi^2\sqrt{15}}{15}W_{220} \qquad \textbf{(9)}$$

$$f_3 = f_{ND} = \frac{4\pi^2\sqrt{2}}{3}W_{000} + \frac{8\pi^2\sqrt{10}}{15}W_{200}$$

Since W_{000} is a normalization constant and is equal to $1/(4\pi^2\sqrt{(2)})$, it can be seen that $f_1 + f_2 + f_3 = 1$ as is required by the initial definitions.

To relate this result to ultrasonic measurements, we wish to relate these f_i to anisotropic elastic stiffnesses. It has been shown that the elastic constants, and hence the ultrasonic velocities, are related to the ODCs by the following[6]:

$$C_{11} = \rho V_1^2 = \mathring{C}_{11} + \frac{16\pi^2 A_1\sqrt{10}}{210}W_{200} - \frac{32\pi^2 A_1\sqrt{15}}{210}W_{220} + \frac{12\pi^2 B\sqrt{2}}{105}W_{400} - \frac{16\pi^2 B\sqrt{5}}{105}W_{420} + \frac{8\pi^2 B\sqrt{35}}{105}W_{440}$$

$$C_{22} = \rho V_2^2 = \mathring{C}_{11} + \frac{16\pi^2 A_1\sqrt{10}}{210}W_{200} + \frac{32\pi^2 A_1\sqrt{15}}{210}W_{220} + \frac{12\pi^2 B\sqrt{2}}{105}W_{400} + \frac{16\pi^2 B\sqrt{5}}{105}W_{420} + \frac{8\pi^2 B\sqrt{35}}{105}W_{440}$$

$$C_{33} = \rho V_3^2 = \mathring{C}_{11} - \frac{16\pi^2 A_1\sqrt{10}}{105}W_{200} + \frac{32\pi^2 B\sqrt{2}}{105}W_{400}$$

(10)

where A_1 and B are measures of the elastic anisotropy defined in reference 6.

INTERPRETATION OF KONISHI'S BULK WAVE DATA

Results from Li et al[7] suggest that W_{200} is approximately three times larger than the other ODCs, at least for the samples which they studied. If one neglects these other ODCs then the longitudinal velocity along a sample reference direction can be related to the f_i in the same reference direction by using equations (9) and (10):

$$V_i = \sqrt{\frac{\mathring{C}_{11}}{\rho} + \frac{2A_1}{21\rho} - \frac{2A_1}{7\rho}f_i} \qquad \textbf{(11)}$$

where the subscript i refers to the reference direction, for example the rolling, transverse or normal direction in sheet samples. This predicted velocity is plotted in Figure 1 along with Konishi's plate and tube data. It can be seen that this first principles theory, with no adjustable parameters, does an excellent job at predicting the trends in the data. It can be speculated that the differences between the theory and experiment are caused by the contributions of the other ODCs, but further work is required to verify this speculation.

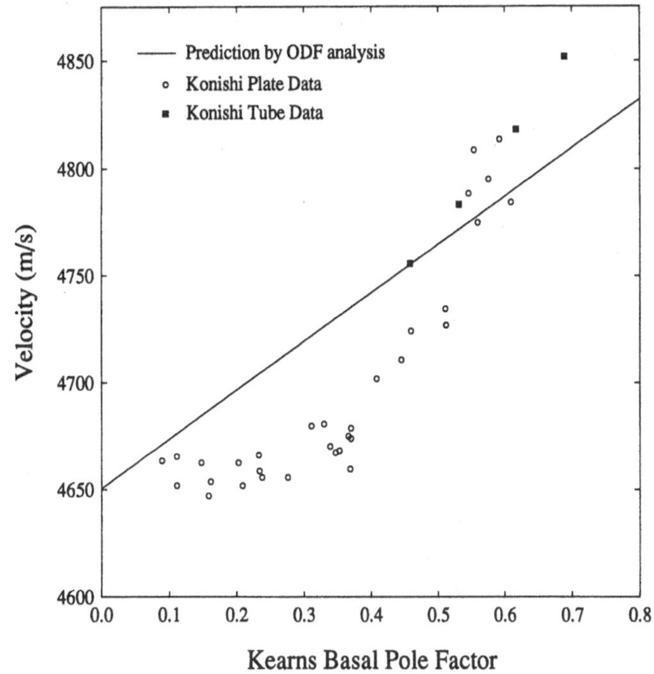

Figure 1. Velocity in reference direction versus f_i in the same reference direction. Konishi's plate and tube data are superimposed to show agreement[3].

ELIMINATION OF EFFECTS OF HIGHER ORDER ODCs

One would ideally desire a technique which was independent of the values of the higher order ODCs. Since the C_{IJ} depend on both W_{2mn} and W_{4mn}, while the f_i depend only on W_{2mn}, it is clear that multiple velocity measurements are required.

One approach to the unique determination of W_{200}, and hence f_3, involves measuring the average velocities of guided modes propagating in the plane of the plate. Li and Thompson[8] and Li et al[7] have shown that such averages depend on a linear combination of W_{200} and W_{400}, as given by equation (12) for SH_0 modes and equation (13) for S_0 modes:

$$BW_{400}-\sqrt{5}A_3W_{200}=\frac{105\sqrt{2}\rho}{16\pi^2}\left[V_{SH_0}^2(45)+V_{SH_0}^2(0)-\frac{2C_{44}^{\circ}}{\rho}\right] \tag{12}$$

$$\left[3+8\left(\frac{C_{12}^{\circ}}{C_{11}^{\circ}}\right)+8\left(\frac{C_{12}^{\circ}}{C_{11}^{\circ}}\right)^2\right]BW_{400}+2\sqrt{5}\left[\left(1-2\left(\frac{C_{12}^{\circ}}{C_{11}^{\circ}}\right)^2\right)A_1-\left(\frac{C_{12}^{\circ}}{C_{11}^{\circ}}\right)A_2\right]W_{200}$$
$$=\frac{105\sqrt{2}\rho}{32\pi^2}\left[V_{S_0}^2(0)+V_{S_0}^2(90)+2V_{S_0}^2(45)-\frac{4}{\rho}\left(C_{11}^{\circ}-\frac{\left(C_{12}^{\circ}\right)^2}{C_{11}^{\circ}}\right)\right] \tag{13}$$

where $C_{11}°$, $C_{12}°$ and $C_{44}°$ are isotropic values of the Lame elastic constants $\lambda + 2\mu$, λ and μ respectively in the Hill approximation, and A_2 and A_3 are measures of elastic anisotropy in addition to A_1 and B which appeared in equation (10). Equation (12) and (13) can be used to explicitly find W_{200} and W_{400} from measurements of the angular dependence of the plate wave velocity. Because W_{200} is approximately three times larger than W_{400} for the samples studied by Li et al[7] and the coefficient of W_{200} in the SH_0 mode equation is larger than the coefficient of W_{400}, W_{200} can be found to a good first approximation from angular SH_0 mode measurements. This is not the case for S_0 mode measurements because the coefficient of W_{400} in equation (13) is roughly three time larger than the coefficient of W_{200}, giving W_{200} and W_{400} comparable effects on the S_0 mode velocities.

To test this idea, we have used the data of Li et al[7] obtained by the zero-crossing technique, as reproduced in Table I. Also included in Table I are the texture free values (isotropic) of the velocities based on the Hill's average elastic moduli. Table II compares the values of W_{200} and f_3 (based on equation (9)) deduced from neutron diffraction to those obtained by the above procedure.

Table I. Angular Dependence of velocities (mm/μs)[7].

Mode	0°	45°	90°	135°	Isotropic
S_0	4.0941	4.0805	4.0987	4.0811	4.0970
SH_0	2.3309	2.3594	2.3309	2.3605	2.3670

Table II. Comparison of W_{200} and f_3 inferred from neutron diffraction and ultrasonics.

	Neutron	Ultrasonic (SH_0 and S_0)	Ultrasonic (SH_0 only, neglecting W_{400})
W_{200}	.0130	.0105	.0104
f_{ND}	.550	.509	.506

The middle column of Table II corresponds to the simultaneous solution of equations (12) and (13) while the third column was obtained from equation (12) only, neglecting the W_{400} term. Preliminary estimates of the errors in W_{200} and f_3 are approximately ±0.005 and ±0.08, respectively for the ultrasonic case. These are controlled by the differences between the isotropic moduli predicted by the Voigt, Reuss and Hill procedures. The errors for the neutron case are ±0.0004 and ±0.0066 for W_{200} and f_3, respectively. The differences between the ultrasonic and neutron predictions of f_3 are somewhat greater than can be explained by the above error analysis. Nevertheless, the results are sufficiently close to warrant further study.

For the determination of f_1 and f_2, W_{220} is needed as well as W_{200}. Examination of references 7 and 8 show that an equation involving W_{220} and W_{420} can be obtained by comparing velocities of S_0 modes propagating in the rolling and transverse directions. A second equation involving these ODCs can be obtained from the acoustic birefringence. We are currently performing experiments to test this scheme.

CONCLUSIONS

The f_i play an important role in the characterization of Zircaloy sheet and can be shown to correlate with ultrasonic velocity measurements. Use of ODF analysis facilitates a more complete understanding of how the f_i affect the ultrasonic velocities. The authors have developed relations between the f_i and ultrasonic velocity which will allow determination of all three f_i from ultrasonic measurements. Existing data supports the theory well, and additional experiments are in progress to demonstrate the unique determination of the f_i.

ACKNOWLEDGEMENT

This work was sponsored by the Division of Materials Sciences of the U. S. Department of Energy and was performed at the Ames Laboratory in cooperation with the Center for NDE. Ames Laboratory is operated by Iowa State University for the USDOE under contract W-7405-ENG-82.

REFERENCES

1. Kearns, J.J., Westinghouse Co. Report WAPD-TM-472, Pittsburgh, PA, 1965

2. Cook, C.S., Sabol, G.P., Sekera, K.R., and Randall, S.N., *Zirconium in the Nuclear Industry: Ninth International Symposium*, ASTM STP 1132, pp. 80-95, 1991

3. Konishi, T., and Honji, M., *Zirconium in the Nuclear Industry: Sixth International Symposium*, ASTM STP 824, pp. 256-68, 1984

4. Rosenbaum, H.S. and Lewis, J.E., *Journal of Nuclear Materials*, Vol. 67, pp.273-82, 1977

5. Roe, R., *Journal of Applied Physics*, Vol. 36, pp. 2024-31, 1965

6. Li, Y., and Thompson, R.B., *Journal of Applied Physics*, Vol. 67, pp. 2663-5, 1990

7. Li, Y., Thompson, R.B., Root, J.H., Holden, T.M., *Nondestructive Characterization of Materials IV*, pp. 467-74, 1991

8. Li, Y., and Thompson, R.B., *Material Research Society Symposium Proceedings*, Vol. 142, pp. 83-88, 1989

ULTRASONIC CHARACTERIZATION OF STRESS STATES IN RIMS OF RAILROAD WHEELS

Eckhardt Schneider, Rüdiger Herzer,
Dietmar Bruche, and Helmut Frotscher

Fraunhofer - Institute for Nondestructive Testing (IzfP)
D - 66123 Saarbrücken, Germany

INTRODUCTION

Freight trains are braked by pressing brake-shoes onto the rolling surface of the wheels. Due to the heat, put in during the braking periods and the subsequent cooling, tensile stress states are developed in the rims of the wheels. Cracks in the rolling surface, which are not harmful otherwise may grow under the influence of the tensile stress and may even cause the failure of a wheel.

A typical distribution of the tangential (circumferential) stresses in a cross section of a rim of a used monoblock wheel is given in Figure 1. The highest tensile stress in railroad wheels have been found in the edge area of the rolling surface and the outer side of the rim. The stresses were evaluated using a totally destructive cutting and sectioning technique. The points give the positions and the results of the strain gauge measurements; the lines are numerically evaluated[1]. The result is regarded as representative with respect to the stress distribution; the absolute values are of course strongly depending on the braking conditions. In order to assure the reliability of the wheels and the safety of the traffic, the stress states of railroad wheels need to be inspected periodically. The inspection should be nondestructive, fast and easy to apply in daily routines. Ultrasonic techniques do accomplish those requests.

ULTRASONIC TECHNIQUES

Ultrasonic techniques to evaluate stress states use the acousto-elastic effect; that is the influence of elastic strain states on the velocities of ultrasonic waves. This effect is rather small (some per thousand), meaning that ultrasonic path length and time-of-flight (TOF) have to be measured with high precision. Measurements of path lengths in components with the required accuracy are at least time consuming. In order to bypass this restriction, two or even three ultrasonic waves are applied. The influence of a strain or stress state on the velocities of ultrasonic waves is different, depending on the directions of ultrasonic propagation and vibration with respect to the principal directions of the strain or stress state. It is commonly assumed, that the

Nondestructive Characterization of Materials VI
Edited by R.E. Green, Jr. *et al.*, Plenum Press, New York, 1994

principal directions of strain and stress are identical. The application of two or three ultrasonic waves, propagating the same path, but vibrating along different principal directions result in evaluation equations for the stress state, in which the ultrasonic times-of-flight are the only measuring quantities.

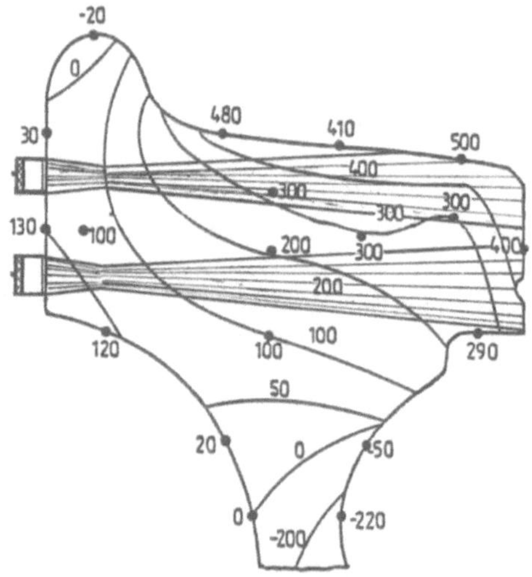

Figure 1. Distribution of stresses (MPa) in the cross section of a used railroad wheel and upper and lower position of ultrasonic probe and beam.

In order to evaluate stress states in surface layers with thicknesses of some millimeters, techniques are developed using a skimming longitudinal wave together with a Rayleigh- or a SH-wave. For some applications it is sufficient to use the skimming longitudinal wave only.

In order to characterize stress states in the bulk of a component, the ultrasonic birefringence effect is commonly used: The relative difference of the times-of-flight of two shear waves, propagating along one principal stress axis and vibrating parallel to the other two principal stresses, is proportional to the difference of the stresses, acting along the vibration directions.

In each case, the ultrasonic result is the mean value of the stresses, acting in that part of the component, which is propagated by the ultrasonic beam. Depending on the used ultrasonic wave mode, a different combination of second and third order elastic constants (called acousto-elastic constant, AEC) is needed to evaluate the stresses from the time-of-flight data. While the second order constants, Youngs- and shear moduli are usually known, the third order elastic constants or the AEC's need to be evaluated using a representative sample of the component of interest.

The temperature of a component is influencing the elastic constants and thus the propagation velocities of ultrasonic waves. In the range of environmental temperatures, the velocities of ultrasonic waves with center frequencies in the low MHz range, are linear functions of the temperature. For ultrasonic stress analysis, the temperature of the component under test needs to be known. There is one exception, as described later.

Besides stress and temperature, also texture (preferred grain orientation) is influencing the ultrasonic velocities. Among all influencing parameters, the influence of texture is the most difficult one to be separated for the analysis of stress states. There are numerous approaches published in the literature, but none of them seems to be adequate to separate or to discriminate the texture of casted wheels. Due to the large variety of the degree of texture in the wheels needed to be inspected on the one hand, and the need for an automated inspection on the other hand, the only solution is seen in a statistics based approach.

EXPERIMENTAL RESULTS

The monoblock wheels used by the Deutsche Bundesbahn are not casted but forged and heat treated. This procedure has, among others, the advantage that there is no significant texture. Measurements of ultrasonic velocities in small samples cut from the rims have been made. The relative differences of the velocities of two shear waves, propagating along the same direction, vibrating perpendicular to each other have been found to be smaller than 0,2‰. This difference corresponds to a difference of the principal stresses, acting in the directions of vibrations of about 30 MPa.

As it is to be seen in Fig. 1, the highest tangential stress values are found in the edge between the running surface and the outer side of the rim. In order to evaluate the stress state in that edge, the application of a skimming longitudinal wave or the application of a skimming longitudinal together with a Rayleigh- or a SH-wave seems to be adequate. But the application of those waves to evaluate the stress state in a layer of the rolling surface cannot be recommended because the materials state is heavily influenced by the wheel-rail contact conditions. The plastic deformation and the different kinds of damages render an automated ultrasonic stress analysis on used wheels impossible. Also the use of the mentioned wave types to evaluate σ_{tan} in the outer surface was found to be possible for new wheels only. In used wheels, the reduced height of the rim on one hand side and the influence of the rail brake on the surface condition on the other hand side cause severe difficulties. The ultrasonic time of flight and the acousto-elastic constants are strongly influenced by the damages, changing locally. Fig. 2 shows the changes of the time of flight of a skimming longitudinal wave propagating along the circumferences of new (upper part) and used (lower part) wheels. The wave propagated between a transmitter and two receivers in the circumferential direction at the outer side of the rim with a constant distance from the edge. The center frequency of the wave was about 2 MHz, the path length of the wave was 60 mm. All measurements were done twice; the producibility is to be seen in the Figure.

It has been decided to use the birefringence effect to characterize the stress state in the rims of the wheels. The application of a shear wave, propagating the thickness of the rim, vibrating along the height (radial direction of the wheel) and then vibrating along the circumferential (tangential) direction results in the evaluation equation:

$$\sigma_{tan} - \sigma_{rad} = K\,(t_{rad} - t_{tan}) / t_{tan} \qquad (1)$$

σ_{tan}, σ_{rad} are the principal stresses along the tangential and radial direction, respectively. t is the time of flight of the shear wave vibrating along the direction indicated as index. K is the acousto-elastic constant, a function of the shear modulus and of a third order elastic constant.

Figure 1 also shows the upper and lower position of the ultrasonic probe together with the sound beam. As to be seen in the figure, the sound propagates the whole thickness of the rim. The result of the investigation is a mean value of the principal stress differences along the ultrasonic path. This technique does not enable the evaluation of each single principal stress.

This disadvantage is compensated by the following advantages for the practical use in daily routines.

- The shear wave is generated by an electromagnetic transducer (EMAT). No coupling medium is needed, no special surface treatment is needed. The transducer can easily be manipulated along a radial measuring trace.
- The influence of temperature on the ultrasonic time of flight is neglectably small. The influence of temperature on the material dependent constant K is also neglectable. At component temperatures between – 10°C and + 40°C, the error due to temperature influences on the final result is within the error, specified for the measuring system[2].

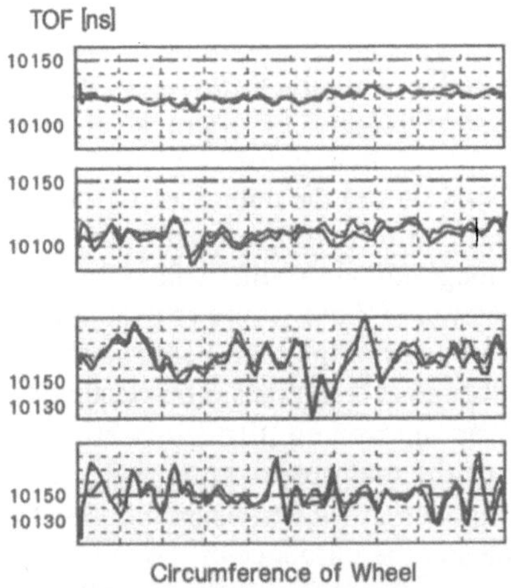

Figure 2. Change of the time-of-flight of a skimming longitudinal wave (2 MHz) along circumference of new (upper half) and used (lower part) railroad wheels.

- The profile of the stress difference (σ_{tan} - σ_{rad}) along the height of the rim is evaluated. Depending on the particular braking and load situation, the extreme values of stresses are at different depths from the rolling surface. Those stress profiles give stronger evidences concerning the stress state than the evaluation along the circumference of the wheel.
- The propagation directions of heat cracks, found in the rims, have also components perpendicular to the radial direction. The principal stress difference (σ_{tan} - σ_{rad}) is seen as more suitable for the judgement of the integrity of a wheel than the σ_{tan} value alone.

Using the birefringence technique, the stress states in new forged wheels before and after different brakings were characterized. The Figures 3 and 4 show two typical results. The brakings were performed on a test stand at the Deutsche Bundesbahn, Versuchsanstalt Minden, the braking conditions are indicated in each figure.

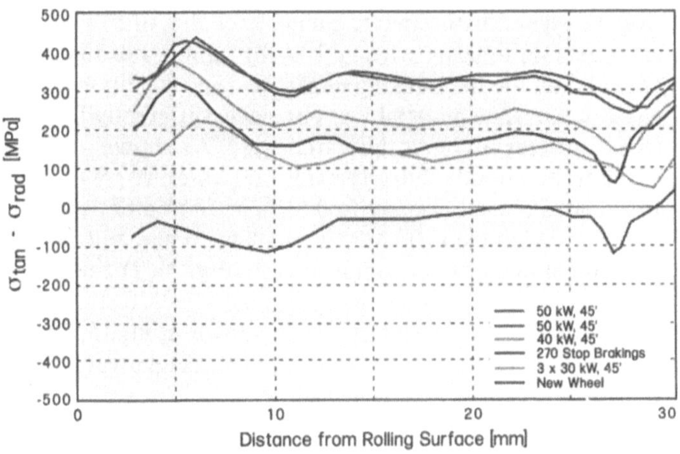

Figure 3. Change of residual stress state in the rim of a railroad wheel after different brakings.

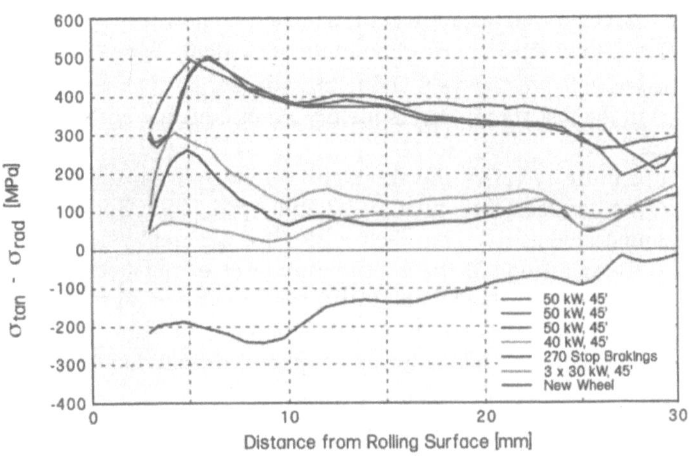

Figure 4. Change of residual stress state in the rim of a railroad wheel after different brakings.

It has been found, that in the majority of the new wheels, the stress difference $\sigma_{tan} - \sigma_{rad}$ is in the range of low compressive values, as to be seen in Fig. 3. The profiles are shifted to tensile values after three brakings with 30 kW for 45 minutes. A maximum is developed at a depth of about 7 mm. After the next braking sequence, 270 stop brakings, the maximum increases to about 300 MPa at 5 mm of depth; at deeper positions, the increase of the stress difference is smaller. A significant shift to higher tensile values for the stress difference is found after the 40 kW and after the first 50 kW brakings. A second braking with 50 kW for 45 minutes does not change the profile, already found after the first braking under the same condition.

Another wheel of the same material was obviously differently heat treated to have higher compressive stresses in the as delivered state. As shown in Fig. 4, the stress difference is found to be about - 200 MPa till a depth of 10 mm under the rolling surface. The profile goes almost linearly to 0 MPa at 30 mm of depth. After the first braking, the values are smaller than 100 MPa until about 20 mm of depth. The values are smaller than those, found in the first mentioned case (Fig. 3). The maximum, shown in Fig. 3 is not developed in the second wheel (Fig. 4). The 270 stop brakings cause a maximum which is at the same position than that shown in Fig. 3, but about 45 MPa smaller. Also after the next braking (40 kW, 45') the maximum value is lesser than the one in the other wheel. It is also seen in Fig. 4, that the stress profiles after the first three brakings are more or less of the same level in areas deeper than 15 mm. After the 50 kW brakings, maximum values of 500 MPa are evaluated for the stress difference. This is about 80 MPa higher than in the former case. The positions of the maxima are the same. Again, a second and a third braking with the same condition (50 kW, 45') does not change the profile, determined after the first breaking of that sort.

The material dependent K-value was evaluated in tensile test experiments using representative samples of the material. Again, the described results are found in new forged wheels in a test stand. The main objective of those investigations was the study of crack propagation in the rims under tensile stresses.

It is known that real brakings, even if the same brake power is applied, cause lower tensile values for σ_{tan} than those, performed in the test stand. The maxima of the stress differences in used wheels are always found to be at deeper levels than those, determined on wheels braked in the test stand. That is mainly a consequence of the load situation in the rolling surface.

Using set-ups built by IzfP^2, the Deutsche Bundesbahn inspected about 27000 wheels in the period between December 1992 and April 1993. Some of the obtained results can be summarized as follows: The profiles for different wheels can have quite different shapes; the positions of the maximum values of the stress difference vary considerably. The stress profiles taken at different positions on the same wheel are similar; the differences in values are smaller than about ±20 MPa. The stress profiles taken at the two wheels of the same axle can be quite different due to different braking conditions.

Fig. 5 displays the result of the automated stress analysis as it is shown on the screen of the set-up. Besides the profile, the mean value of the stress difference as well as the maximum value together with its position are also shown on the screen.

The set-up compares the evaluated maximum value with a predetermined and stored value for the acceptable stress difference and gives a green or red signal, depending on the result of the comparison.

The Deutsche Reichsbahn invented a stress relief treatment optimized for railroad wheels. Fig. 6 displays stress profiles taken along two traces on a heavily used wheel before and after the treatment. It is seen that the stresses at 30 and 35 mm of depth are more efficiently relieved than the stresses closer to the rolling surface. This can be explained by the way, the relief treatment is performed.

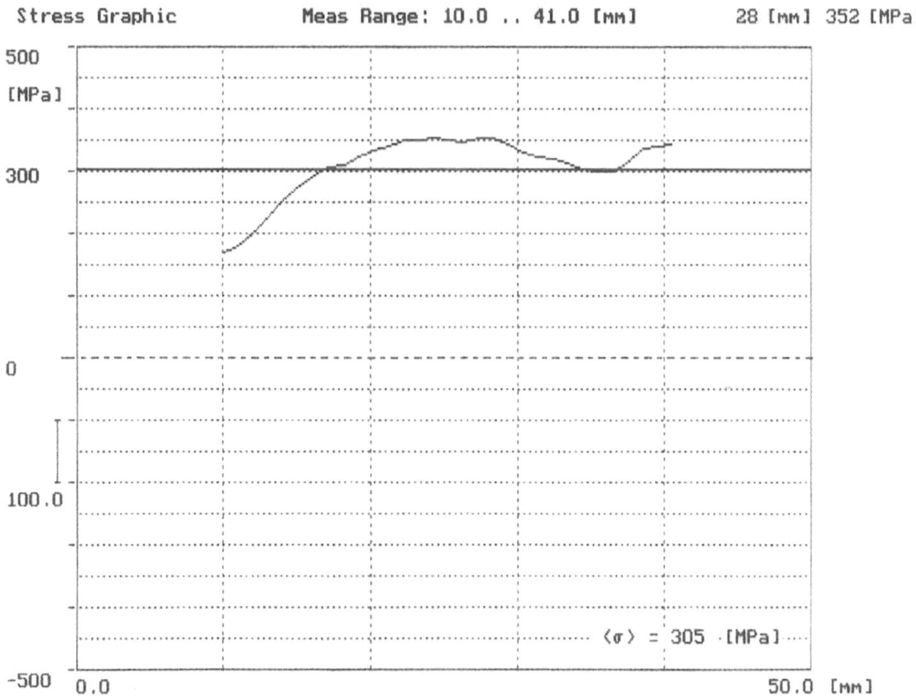

Figure 5. Difference of principal stresses (σ_{tan} - σ_{rad}) versus the distance from the running surface.

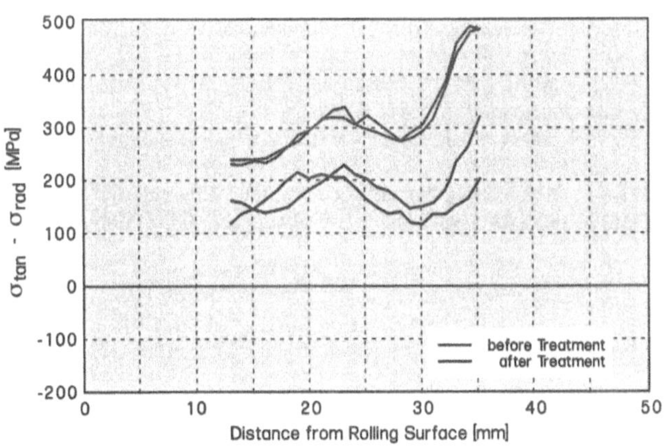

Figure 6. Change of residual stress state in the rim of a railroad wheel before and after the stress relief treatment.

COMMENT

In order to assure the reliability of the railroad wheels and the safety of the traffic, the stress states in the rims of the wheels have to be inspected periodically. From the fracture mechanical point of view, it would be most desirable to describe the stress state in the vicinity of individual cracks or at least the stress states in the surface near layer of the running surface and of the outer surface of the rim where the highest stresses are developed.

It seems to be rather unlikely that this objective can be achieved for a large number of wheels by applying a nondestructive technique in daily routines. Some effort has been undertaken to develop and to optimize nondestructive magnetic[3,4] or ultrasonic[5,6] techniques to characterize the thermal damage or the stress state in railroad wheels.

Our concept to assure the reliability of the railroad wheels is based on the statistical evaluation of the stress profiles of wheels, on the evaluation of local stress states and on the number and size of cracks in particularly chosen wheels.

ACKNOWLEDGEMENT

This work was carried out under contracts with the Deutsche Bundesbahn, Zentralamt Minden, with the Deutsche Reichsbahn, Zentralstelle Delitzsch and with the Vereinigte Schmiedewerke GmbH, Sparte Bahnmaterial, Bochum. The authors would like to thank the representatives of the mentioned institutions for helpful discussions.

REFERENCES

1. European Railroad Research Institute; Document ORE B169/RP2, Utrecht (1989).
2. R. Herzer, H. Frotscher, K. Schillo, D. Bruche, and E. Schneider, Ultrasonic Set-up to Characterize Stress States in Rims of Railroad Wheels, Sixth International Syposium on Nondestructive Characterization of Materials ; June 7-11, 1993, Hawaii.
3. G.L. Burkhardt, W.D. Perry, P.J. Pantermuehl, J.R. Barton, and M. Smith, Determination of Thermal Damage in Freight Car Wheels from Barkhausen Noise Analysis, Proceedings of the 9th International Wheelset Congress, Montreal Canada, Paper 3-5 (1988).
4. J.R. Barton, W.D. Perry, R.K. Swanson, G.C. Hsu, and S.R. Ditmeyer, Heat-Discolored Wheels: Safe to Reuse?, *Progressive Railroading 28,3:841* (1985).
5. J. Deputat, K. Osuch, W. Kunnes, Untersuchung der Eigenspannungsänderungen in Eisenbahn-Vollrädern nach Bremsungen, *ZEV+DET Glas. Ann 115, 7/8:231* (1991).
6. A.V. Clark, H. Fukuoka, D.V. Mitrakovic, and J.C. Moulder, Characterization of Residual Stress in a Heat-Treated Steel Railroad Wheel, *Materials Evaluation 47,7:835* (1989).

ELECTROMAGNETIC ACOUSTIC RESONANCE METHOD FOR MEASURING STRESSES IN METAL PLATES

M. Hirao, H. Ogi, T. Yamasaki, and H. Fukuoka

Faculty of Engineering Science
Osaka University
Toyonaka, Osaka 560, Japan

INTRODUCTION

Conventional methods for acoustoelastic stress evaluation rely on the time-of-flight measurements of discrete reflection echoes. Special techniques have been developed to achieve the high accuracy needed for detecting the weak stress effect on ultrasonic velocities, say, in the order of 10^{-4} or less. The relative accuracy decreases as the thickness is reduced. Eventually, the echoes are overlapped each other and the measurement becomes impossible. The pulsed ultrasonic resonance spectrometer (PURS) combined with the electromagnetic acoustic transducer (EMAT) was originally implemented addressing such situations and, as is expected, it showed a high stress resolution (~1 MPa for aluminum sheets), a high spatial resolution (~5 mm square), and the insensitivity to surface condition in a noncontacting operation.[1,2] The PURS/EMAT also performs adequately for thick plates, up to a few centimeters, showing itself as a robust and versatile stress detector for parallel-sided geometries and not so lossy, conducting materials. This paper provides the results of fundamental investigations concerning the stress-induced birefringence and the effects of liftoff (or air gap) and surface roughness of the sample.

The resonance method has been extensively used to determine the elastic wave velocities and the material properties reducible from them for small (or thin) samples. Sweeping the frequency picks up the resonance frequencies and the velocities can be calculated from the sample geometry. Contacting a transducer to samples requires a correction because the sample-to-transducer coupling brings the whole system into vibration other than the one only in the sample. Use of EMATs avoids the correction owing to the weak coupling (contactless). Resonance measurements compensate, in excess, for the weak transduction efficiency of EMATs; at a resonance, all the echoes are received in phase to give a large amplitude. Incorporation of EMATs into pulsed ultrasonic resonance measurements was first studied by Filimonov et al.,[3] who observed two resonances at close frequencies now attributable to the texture-induced birefringence of metal plates. Recent instrumentation[1,2,4-6] uses a superheterodyne receiver equipped

Nondestructive Characterization of Materials VI
Edited by R.E. Green, Jr. *et al.*, Plenum Press, New York, 1994

with quadrature phase-sensitive detector and analog integrators. With the advanced signal-processing procedures, the method possesses a large dynamic range and a high frequency resolution enough to cope with the requirements for acoustoelasticity. Similar method was developed by Kawashima,[7] employing the cw signals, up to 50 MHz, to excite the EMAT and the bandpass amplifier tuned to the excitation frequency.

ACOUSTIC BIREFRINGENCE

Acoustoelastic relations show the linear dependence of ultrasonic velocities on stresses. They are recasted to those in the resonance frequencies using $f_n=nV/2d$ for the plates with the stress free surfaces and the thickness d. Use of a higher order resonance, if the attenuation allows, enhances the velocity change leading to higher accuracy. To avoid considering the thickness change as a function of stress and position, velocity ratios are often used in the stress measurements. Shear-wave resonances exhibit the acoustic birefringence B for the orthogonal polarizations, which is proportional to the difference of in-plane principal stresses σ_1 and σ_2, that is,

$$B = \frac{f_n^{(1)}-f_n^{(2)}}{(f_n^{(1)}+f_n^{(2)})/2} = B_0+C_A(\sigma_1-\sigma_2), \qquad (1)$$

where $f_n^{(1)}$ and $f_n^{(2)}$ are the n-th resonance frequencies for the polarization in σ_1 and σ_2 directions. Offset, B_0, is observed in the unstressed state, which arises from the texture, that is, the nonrandom distribution of crystallographic orientation. Acoustoelastic constant C_A is usually calibrated in the uniaxial loading tests.

PULSED ULTRASONIC RESONANCE SPECTROMETER (PURS)

The PURS involves energizing the EMAT with high-power rf bursts gated coherently. The duration is set to be much larger than the round trip time across the sample thickness. The echoes, having the equally spaced arrival times and the decaying amplitudes, pile up to be an ultrasonic ringing. The operating frequency is swept in the range of 0.5 to 20 MHz at every 0.1 Hz, if necessary. Unlike the resonance in the audible range, the ringing is confined to within a small region corresponding to the aperture of the EMAT so that a localized measurement is possible.

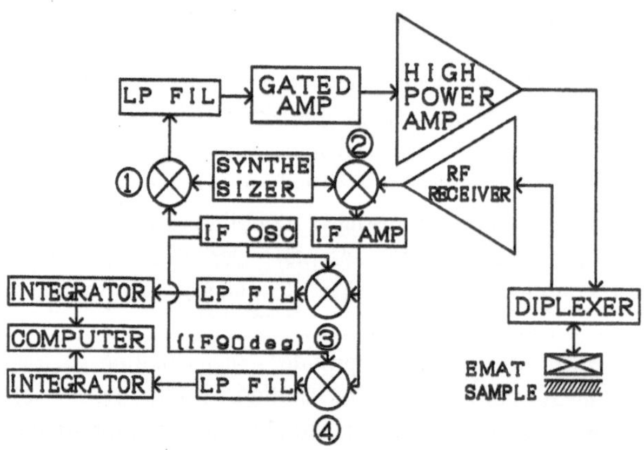

Figure 1. Block diagram for pulsed ultrasonic resonance spectrometer.

Figure 1 is a simplified block diagram of the superheterodyne phase sensitive system. The synthesizer plays the role of "local oscillator" at a frequency $F+IF$; F is the operating frequency and IF is the intermediate frequency fixed to 25 MHz. At multiplier 1, two cw's of $F+IF$ and IF, from the synthesizer and the IF oscillator, are mixed up to produce the frequency F for driving the gated amplifier. The synthesizer and multiplier 2 work together to convert the frequency of the received ultrasonic signals into the bandwidth of the IF amplifier. On multiplication, the phase information is transferred to the resultant signals. The low-pass filter and the IF amplifier reject the sum frequency, leaving only the difference frequency. Quadrature signals, having the phase angles at every 90 deg, are mixed with the output of the IF amplifier to shift the frequency to the dc range. Phase angles 0 deg and 180 deg go to multiplier 3 and those of 90 deg and 270 deg to multiplier 4. The multiplication is followed by the individual analog integration. Taking the difference of two outputs at each channel, whose phases differ by 180 deg, minimizes the drift errors in the electronics. This operation provides the outputs of $A\cos\phi$ and $A\sin\phi$ after removing the $2IF$ components through low-pass filtering. The ringing amplitude is represented by A and the phase by ϕ; both are functions of time as well as frequency F. These outputs are integrated using a gate which extends from the end of the excitation until no more ringing is observed. The integrator outputs are digitized with 12-bit resolution and sent to the computer, which controls the whole measurement and samples and calculates the data. The frequency F is then stepped through the range of interest.

The "amplitude spectrum" is obtained by recording the root of the sum of the squares of the integrator outputs at each scanning frequency. There are two mechanisms functioning to make the distinct resonance lineshape.[2] The ringing signal contains a large number of reflection echoes overlapping each other. Phase difference between echoes is introduced depending on the number of reflections. Because the round trip time is an integer multiple of the period of a resonance frequency ($2d/V=n/f_n$), all the echoes are received in a coherent manner by the EMAT when driving it at a resonance frequency. Phase ϕ is constant throughout the ringing and the amplitude A decays exponentially with time. The integration then gives a large output. For slightly off-resonance frequencies, A is still large, but the ringdown curves beat because ϕ changes constantly. The integration gives smaller spectra through the cancellation of positive and negative values. For frequencies far from a resonance, the echoes interfere with each other because of the random phases to make a very small intensity.

EXPERIMENTS

To illustrate the availability for the practical stress measurements, we tested the response of the spectrogram to stress, liftoff, and surface roughness using the low carbon steel samples of 18 mm thick and 35 mm wide. Typical measuring time is 20 sec per spectrum containing, say, 500 data points, but it depends on the operation parameters such as the lengths of rf bursts and integrator gate, the frequency range and the step, the averaging, etc.

We used a shear-wave EMAT of 20x16 mm^2 aperture throughout, which contains a flat elongated spiral coil of 21 turns and a pair of permanent magnets magnetized in the opposite directions. The EMAT operates with the Lorentz force. The shearing force is produced through the interaction between the induced current in the surface skin and the magnetic field perpendicular to the sample surface. The polarized shear wave travels in the thickness direction and is reflected many times at the free surfaces. At every reflection at the sensing surface, the shear wave interacts with the magnetic field to excite an eddy current, which is inductively detected by the coil in the same EMAT.

When the sample plate shows the birefringence, due to texture and/or stress, to certain degree, we place the EMAT at nearly 45 deg from the anisotropy axes to make the orthogonal polarizations resolved. Otherwise, we rotate it to 0/90 deg and measure $f_n^{(1)}$ and $f_n^{(2)}$ separately.

Response to Stress Application

Figure 2 shows the variation of the shear-wave resonance spectra with the uniaxial tensile stress σ_1. The polarization was parallel to the stress and the change in $f_n^{(1)}$ is shown. The frequency was swept for every 333 Hz. We fit the Gaussian distribution function to find the center frequency of each resonance lineshape. Similar measurement was done for $f_n^{(2)}$. Equation (1) is used to calculate B; B changes linearly with stress as the theory predicts (Fig. 3). Besides the frequency shift, we observe the increase of peak height with increasing stress. This phenomenon has not been clarified yet, but probably this occurs accompanying the stress contribution to the magnetoacoustic interaction and/or the reduced damping due to the dislocation fixation by stress.

Effect of Liftoff

We inserted polymer films between the EMAT and the sample to make a liftoff.

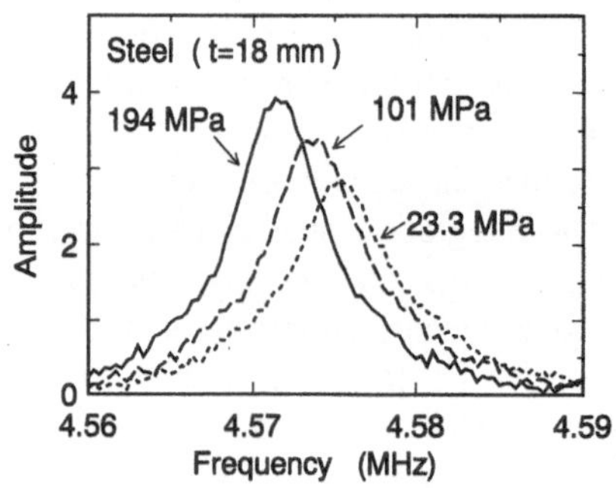

Figure 2. Spectral change with stress showing the shift in $f_n^{(1)}$.

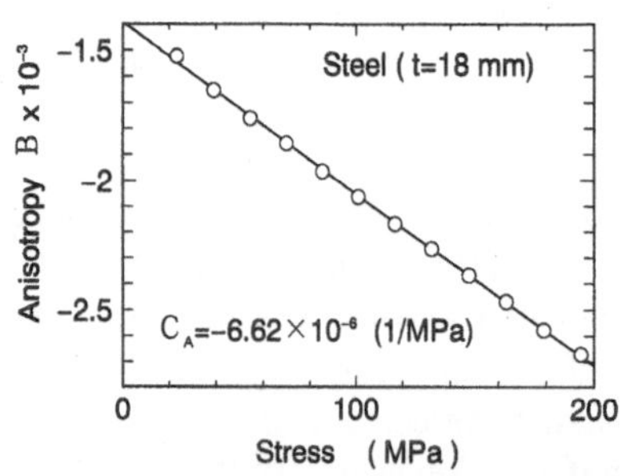

Figure 3. Resonance birefringence in a low carbon steel. (B_0=-0.129 % and C_A=-6.62x10^{-6} /MPa.)

Stress was not applied. As expected from the induction mechanism in the transduction, the liftoff decreases the strength of the resonance amplitude. Figure 4 shows such a response when the operating parameters were fixed. The resonance frequency could be measured even at 1.6 mm liftoff. The resonance frequency slightly decreases with the liftoff. This occurred for both $f_n^{(1)}$ and $f_n^{(2)}$, causing no influence on B. If the parameters were adjusted to compensate for the weakening efficiency, the resonance frequency was available up to 4.2 mm liftoff using the present instruments. The value for B scattered more for larger liftoff, being equivalent to less than 20 MPa in stress. In general, the PURS/EMAT can accept larger liftoff as the thickness and the attenuation decrease and the aperture increases.

Effect of Surface Roughness

We expected that the surface roughness also lowered the resonance strength; the rough surface causes the scattering on reflections and is physically equivalent to giving a liftoff. We polished both sample surfaces using abrasive papers of grit #800 through #60 in this order. The experimental result was opposite to the expectation. The peak height raised as the polishing proceeded (Fig.5). The operating parameters were un–

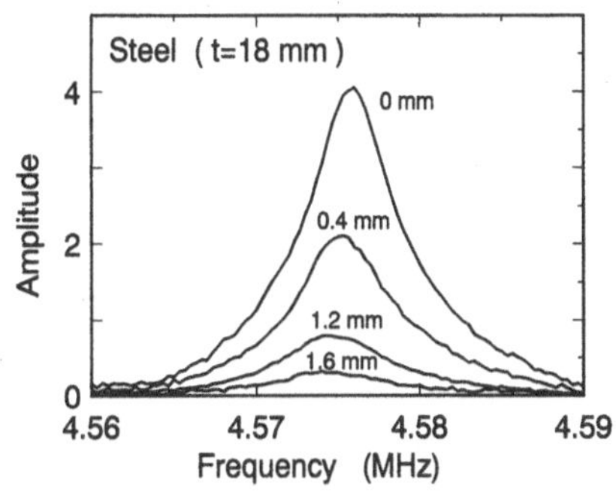

Figure 4. Resonance spectrum and the effect of liftoff for $f_n^{(2)}$ resonance.

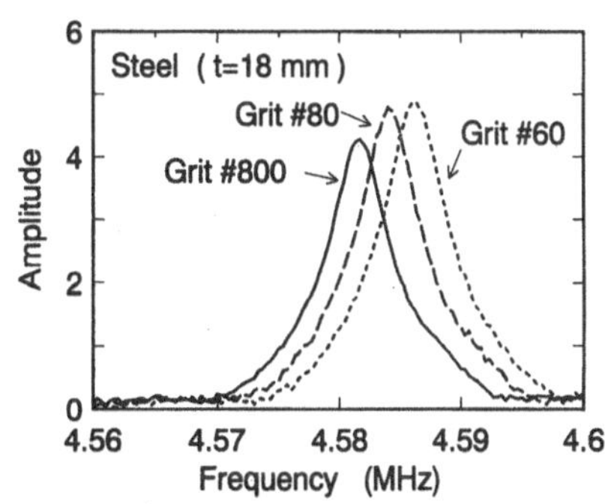

Figure 5. Resonance spectrum and the effect of surface roughness for $f_n^{(2)}$ resonance.

changed. This is another unsolved problem. Again, this can be the stress contribution to the transduction, because the mechanical polishing introduces the compressive residual stress in the surface. In case of EMATs, the wave source is established within the sample not in the transducers. Transmission of the mechanical vibration, which the surface roughness obstructs, does not take place with the EMATs. This is also true for detection, since the EMATs receive the ultrasonic signals inductively. The minimum loading helps us to observe the elastic wave phenomena as they are.

In Fig.5, the resonance frequency moves to the high frequency side. This shift can be explained by the thickness decrease during polishing. The final thickness was smaller than the original one by 15 μm, which coincides with the thickness change calculated from the resonance frequencies. The surface roughness was found to cause less than 5 MPa error in the stress measurements.

CONCLUSION

Although unsolved problems remain, the PURS/EMAT system is capable of non-destructive/noncontacting stress measurements for plate-like geometries. It shows a high sensitivity to stress as well as a tolerance to the liftoff and the surface roughness. It accommodates even rusty and painted surfaces. This aspect is important, because the conventional acoustoelastic stress measurements based on the PZT transducers need the prolonged surface preparation. The PURS/EMAT is a practical means from the point of the short measuring time and the high spatial resolution, too. Measurable thickness depends on the material's attenuation characteristics including the grain size and the EMAT/sample geometry. We have experienced the measurements on 0.15 mm thick aluminum sheets from the beverage cans and also a 34 mm thick steel plate. When the velocity is known, the method provides a very accurate, noncontacting thickness gage.

ACKNOWLEDGMENTS

The authors gratefully acknowledge the advise of C.M.Fortunko, National Institute of Standards and Technology, and G.L.Petersen, Ritec, Inc., for the instrumentation of the pulsed ultrasonic resonance spectrometer. This research was supported in part by Grant-in-Aid No. 03555020 from the Japanese Ministry of Education.

REFERENCES

1. H.Fukuoka, M.Hirao, T.Yamasaki, H.Ogi, G.L.Petersen, and C.M.Fortunko, Ultrasonic resonance method with EMAT for stress measurement in thin plates, *in*:"Review of Progress in Quantitative NDE," vol.12, D.O.Thompson and D.E.Chimenti, eds., Plenum Press, New York, p.2129 (1993).
2. M.Hirao, H.Ogi, and H.Fukuoka, Resonance EMAT for acoustoelastic stress evaluation in sheet metals, *Rev.Sci.Instrum.* 64:3198 (1993).
3. S.A.Filimonov, B.A.Budenkov, and N.A.Glukhov, Ultrasonic contactless resonance testing method, *Sov.J.Nondestr.Test.* 1:102 (1971).
4. W.L.Johnson, S.J.Norton, F.Bendic, and R.Pless, Ultrasonic spectroscopy of metal spheres using electro-magnetic-acoustic transduction, *J.Acoust.Soc.Am.* 91:2637 (1992).
5. G.L.Petersen, C.M.Fortunko, M.Hirao, and B.B.Chick, Resonance techniques and apparatus for elastic-wave velocity determination in thin metal plates, *Rev.Sci.Instrum.* in press.
6. A.V.Clark, C.M.Fortunko, M.G.Lozev, S.R.Schaps, and M.C.Renken, Determination of steel sheet formability using wide band electromagnetic-acoustic transducers, *Res.Nondestr.Eval.* 4:165 (1992).
7. K.Kawashima, Nondestructive characterization of texture and plastic strain ratio of metal sheets with electromagnetic acoustic transducers, *J.Acoust.Soc.Am.* 87:681 (1990).

NONDESTRUCTIVE EVALUATION OF WELDING RESIDUAL STRESS AND MECHANICAL STRESS-RELIEVING BY ACOUSTOELASTICITY

Isamu Oda

Department of Mechanical Engineering
Kumamoto University
Kurokami 2-39-1, Kumamoto 860, Japan

INTRODUCTION

The residual stress affects the failure of machines and structures by contributing to buckling and brittle fracture when those failures occur at low applied stress levels. In addition, residual stress may contribute to fatigue and corrosion cracking. Various methods have been applied to relieve residual stresses. It is very useful from an engineering viewpoint to estimate the residual stress nondestructively, for it can be applied to practical machines and structures. The research on nondestructive evaluation of the residual stress using the ultrasonic technique has attracted special interest recently[1-3]. A few works have been reported on the ultrasonic measurement of the welding residual stress[4-6]. However, reports on the nondestructive evaluation of stress-relieving have apparently not been published to date.

In the present study, ultrasonic stress measurement techniques are applied to the mild steel plate welded by shielded metal arc welding and the aluminum alloy plate heated by TIG arc. In addition, the effect of mechanical stress-relief treatment is examined by the ultrasonic method. Most of the ultrasonic stress measurement techniques have measured some combination of stresses, such as the sum or the difference of principal stresses. In order to evaluate the effect of residual stresses on the failure of structures, stresses must be determined absolutely. In the present paper, the ultrasonic technique for measurement of the absolute residual stresses is examined on referring to the method[6] proposed by Toda, Fukuoka and Aoki. The residual stress distributions are compared with those measured by conventional stress-relaxation techniques which use electric strain gauges or contact balls.

EXPERIMENTAL PROCEDURE

The materials supplied are the mild steel(SS41(1) and SS41(2)) and the aluminum alloy(A5052). Figure 1 shows the mild steel specimens. A single-Vee groove without root gap was prepared by cutting along the y axis of the specimen. The two-pass weld beads were laid on the groove under the conventional welding condition. In the SS41(2) specimen, the residual stress was relieved mechanically by prestraining.

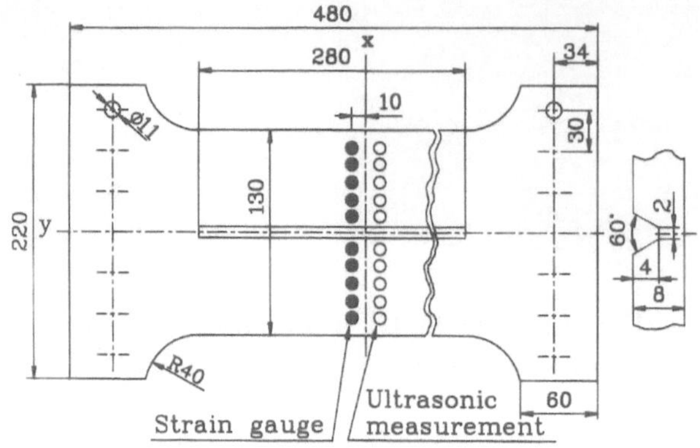

Figure 1. Mild steel specimen manual arc welded.
All dimensions in mm.

A uniform tensile stress was applied in the direction of weld line and then released. The ratio of applied net stress to yield stress of raw plate was 0.94. The aluminum alloy specimen without the groove was heated by TIG arc along the major axis(the y axis) on the front surface. The longest axis (the y axis), in those specimens, is parallel to the roll direction of raw plate.

When the anisotropy axes coincide with the principal axes of stress, next relations can be obtained for a weakly anisotropic solid plate subjected to a biaxial plane stress state[2].

$$(V_{T1}-V_{T2})/V_T = (V_{T10}-V_{T20})/V_{T0} + C_A(\sigma_1-\sigma_2) \qquad (1)$$

$$V_L/V_T = V_{L0}/V_{T0} + C_R(\sigma_1+\sigma_2) \qquad (2)$$

where V_{T1}, V_{T2} are velocities of shear waves which are polarized in the directions of the principal axes and are propagated in the plate thickness direction, V_T is the average of V_{T1} and V_{T2}, σ_1, σ_2 are principal stresses, $(V_{T10}-V_{T20})/V_{T0}$ is the texture-induced acoustical anisotropy in the absence of stress, V_L is the velocity of longitudinal wave, V_{L0}/V_{T0} is the velocity ratio in the absence of stress and C_A, C_R are acoustoelastic constants.

To obtain acoustoelastic constants C_A and C_R, uniaxial compression tests were carried out using a sample shown in Fig.2 which was made of the same material as the welded or heated specimen. Wave velocities were measured during the uniaxial compression in the x and the y directions.

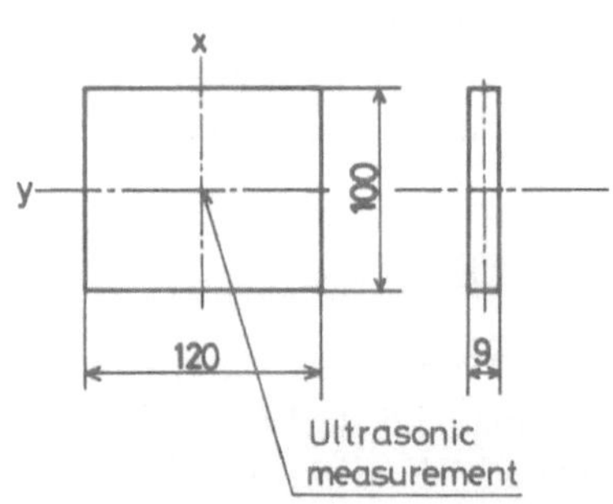

Figure 2. Specimen for measurement of acoustoelastic constants.
Dimensions in mm.

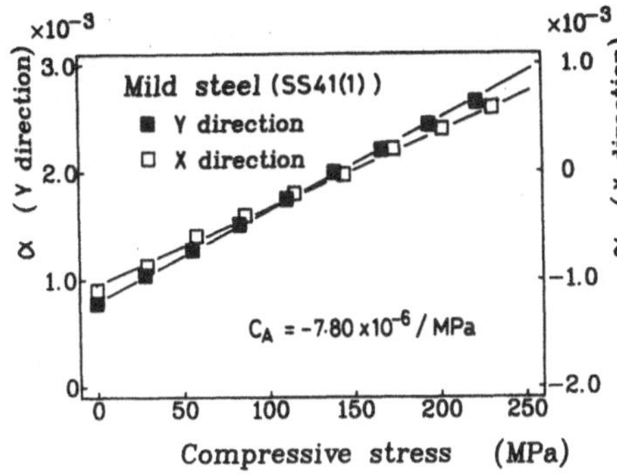

Figure 3. Relative velocity change versus uniaxial compressive stress in mild steel.

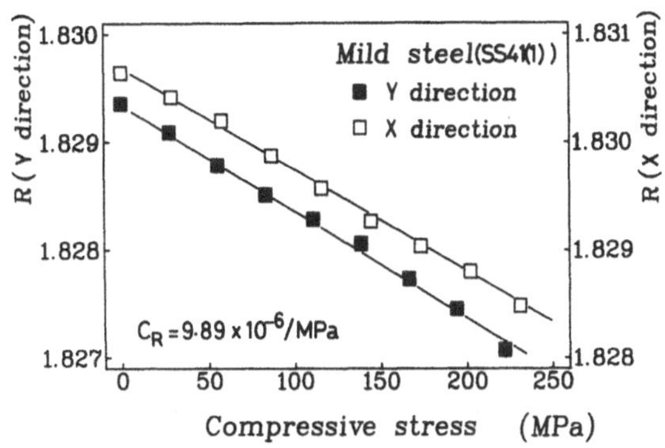

Figure 4. Velocity ratio versus uniaxial compressive stress in mild steel.

The x and y directions in a compression test sample coincide with them in a welded specimen. The values of α and R which are expressed as next equations were obtained.

$$\alpha = (V_{T1} - V_{T2})/V_T \quad (3)$$

$$R = V_L/V_T \quad (4)$$

Figure 3 shows the change in acoustical anisotropy α with applied compressive stress in a mild steel SS41(1). Figure 4 shows the relationship between velocity ratio R and compressive stress in the same material. The figures show that all the points lie on a straight line. The slopes of the plotted line in Fig.3 and Fig.4 show the acoustoelastic constant C_A and C_R, respectively. In the present paper, the average value of measurements in the x and the y direction is used as the acoustoelastic constant. The acoustoelastic constants in SS41(2) and A5052 were also obtained using the same method. The velocities of the shear and the longitudinal waves were measured at the points shown in Fig.1. The directions of two perpendicular polarizations of the shear wave almost coincided with the principal in-plane residual stress directions, that is, the x and the y directions. The wave velocities were measured by the pulse-echo-overlap method at four stages, that is, before welding, after welding, after mechanical stress-relieving and after stress relaxation by sectioning the specimen into small strips. A ceramic transducer with 7 mm diameter was used at a nominal 5 MHz operating frequency. The couplants SWC and BS-400 were used for the shear wave and the longitudinal wave, respectively. The applied pressure of transducer against specimen was kept almost constant. Residual stresses were also measured by using electric-resistance strain gauges and, for a part of specimens, contact balls. In those destructive methods, the average value of residual stress measurements on both surfaces of the plate was adopted. The results obtained by those conventional destructive methods were compared with those by the acoustoelastic methods.

RESULTS AND DISCUSSION

The distribution of $\sigma_1-\sigma_2$ along the x axis in an as-welded mild steel plate is shown in Fig.5. It was obtained by the acoustoelastic birefringence, or only shear wave velocity using equation (1). The stress σ_1 and σ_2 are the longitudinal and the transverse welding residual stress respectively. In the figure, the result measured by the strain gauge method is also shown. Both results agree well with each other. Figure 6 shows the distribution of $\sigma_1-\sigma_2$ in an aluminum alloy plate heated by TIG

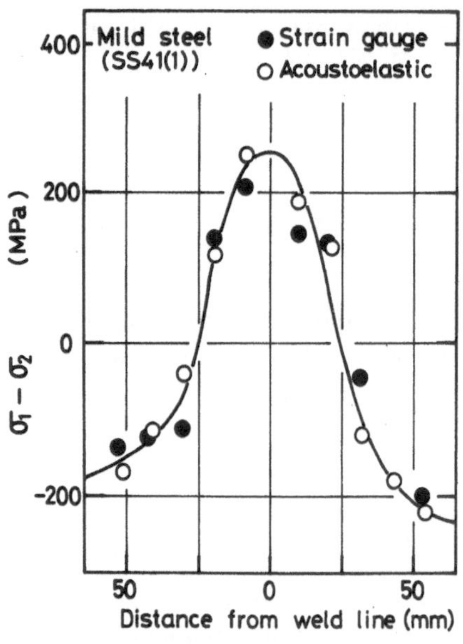

Figure 5. Distribution of principal stress difference in mild steel specimen.

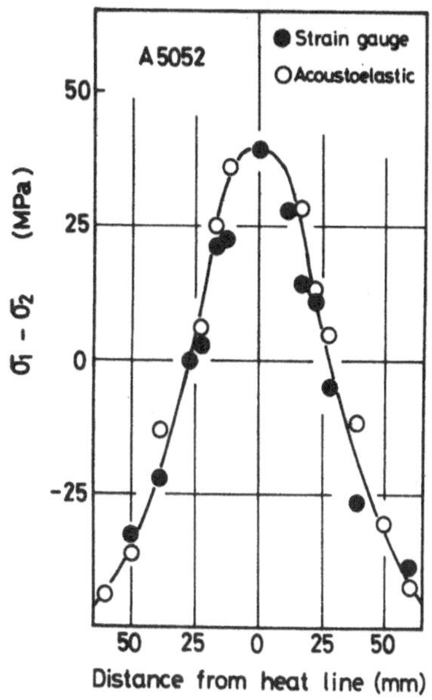

Figure 6. Distribution of principal stress difference in aluminum alloy specimen.

arc. Although the value of the principal stress difference is very small, the distribution obtained by the acoustoelastic method agrees well with that by the strain gauge method. It is confirmed, as reported in the past[4,5], that the acoustoelastic method can be effectually applied in order to obtain the difference of principal stresses.

In order to evaluate the effect of residual stresses on the deformation and the fracture of a plate, stresses should be determined absolutely. Besides the acoustoelastic birefringence, the ratio of longitudinal wave velocity to shear wave velocity was measured. Combining those results and using equations (1) and (2) simultaneously, the absolute values of stresses σ_1 and σ_2 were obtained. Figure 7 shows distributions of the absolute values of σ_1 and σ_2 along the x axis in an as-welded mild steel plate. The present paper examines the nondestructive evaluation of residual stresses. Therefore, to obtain the result in Fig.7, measurements before welding (not after stress relaxation by sectioning the specimen), were used as the texture-induced acoustic anisotropy, $(V_{T10}-V_{T20})/V_{T0}$ and the velocity ratio, V_{L0}/V_{T0} in the absence of stress. Unless otherwise specified, measurements before welding are used hereafter. The degree of agreement between the result of acoustoelastic method and one of strain gauge method is lower in Fig.7 when compared with that in Fig.5. The absolute values of stress σ_2 are estimated to be significantly low. Especially, the maximum value of σ_2 in aluminum alloy was extremely low, that is, about 10 MPa. It is somewhat difficult to measure accurately the low stress value by the present method of measurement. The absolute value of σ_1 affects the deformation and the fracture more seriously than the σ_2 value, for the σ_1 value is much higher than the σ_2 value. Distributions of the stress σ_1 obtained by three different measuring methods are shown in Fig.8. The result by the acoustoelastic method shows a very similar distribution to those by destructive methods using the strain gauge and the contact ball.

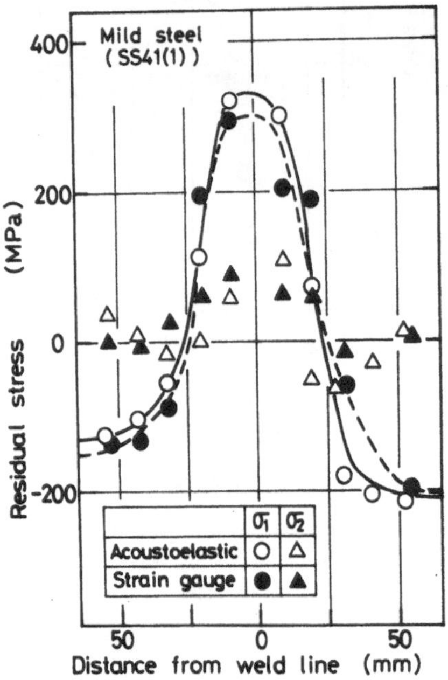

Figure 7. Distributions of residual stresses in mild steel specimen.

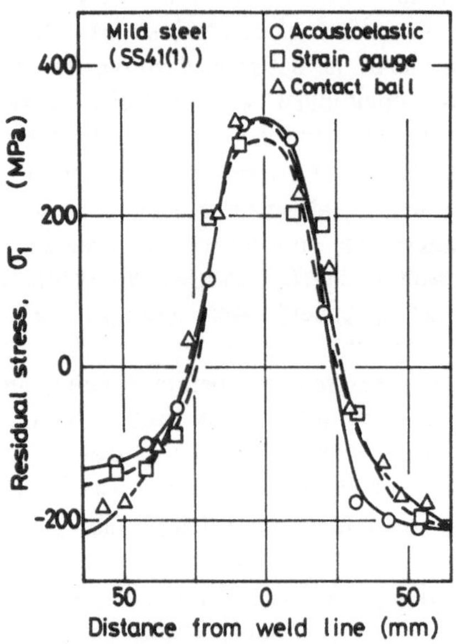

Figure 8. Comparison of longitudinal residual stresses measured by three different methods.

The equilibrium condition for forces in the y direction approximately holds in Fig. 8. Consequently, the present nondestructive method for measuring the absolute value of residual stress can be used in practical application unless fairly accurate values are needed.

Figure 9 shows the distribution of σ_1-σ_2 along the x axis in a mild steel plate (SS41(2)) welded and then stress-relieved mechanically. The value of σ_1-σ_2 is well reduced by the prestraining and it is recognized by both the acoustoelastic and the strain gauge methods. The absolute value of stress σ_1 obtained by the acoustoelastic method in the stress-relieved plate is compared with the result obtained

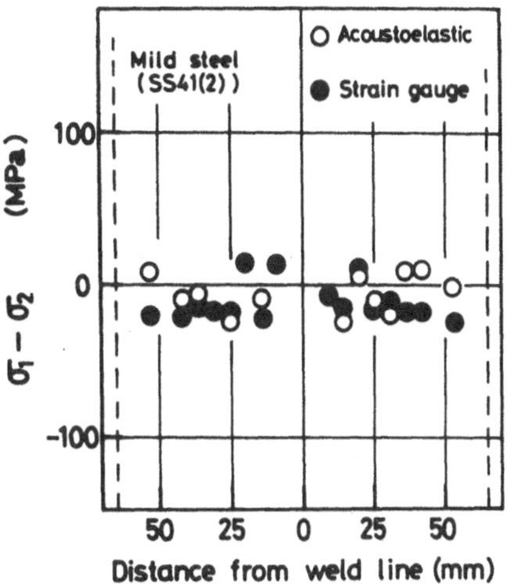

Figure 9. Distribution of principal stress difference after mechanical stress relief.

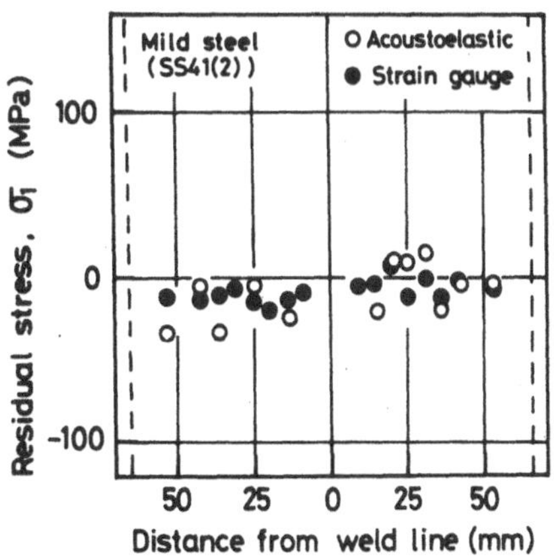

Figure 10. Distribution of longitudinal residual stress after mechanical stress relief.

by the strain gauge method in Fig.10. It is confirmed by both results that values of σ_1 significantly decrease compared with those in the as-welded plate shown in Fig.8. The strain gauge method can not be applied to practical structures because sectioning the plate into small strips to relax the stresses is needed in the method, that is, it is a destructive method. It is obvious that the acoustoelastic method in the present paper is effective for the nondestructive evaluation of the stress-relieving.

In general, the wave velocity and the acoustoelastic constant are influenced by the microstructural change and the plastic strain due to heating and prestraining besides the residual stress. Therefore, it is desirable to use values measured after sectioning the plate as the texture-induced acoustic anisotropy and the velocity ratio in the absence of stress. But, it is meaningless from the standpoint of the nondestructive evaluation of stress. The distribution of stress σ_1 obtained by using those values before welding was compared with one obtained by using those values after sectioning the plate. There was little difference between two kinds of distribution of stress σ_1 within the limits of the present experiment. Consequently, it does not matter in practical application to use the present nondestructive method.

CONCLUSIONS

The following conclusions may be drawn.

(1) The difference between the longitudinal and the transverse welding residual stress can be accurately measured nondestructively by the acoustoelastic birefringence.

(2) Combining the acoustoelastic birefringence with the ratio of longitudinal wave velocity to shear wave velocity, absolute values of the in-plane welding residual stresses can be nondestructively obtained. This method can be used in practical application unless fairly accurate values are needed.

(3) The effect of the mechanical stress-relieving in the weldment can be nondestructively evaluated by the acoustoelastic method.

REFERENCES

1. T.Tokuoka and Y.Iwashimizu, *Acoustoelasticity, Science of Machine*, 27:860(1975)(in Japanese).
2. H.Fukuoka, Measurements of residual stresses by acoustoelasticity, *J. Jpn. Weld. Soc.*, 58:65(1989) (in Japanese).
3. K.Okada, Stress-acoustic relations for stress measurement by ultrasonic technique, *J. Acoust. Soc. Jpn.(E)*, 1:193(1980)(in Japanese).
4. H.Fukuoka, H.Toda and T.Yamane, Acoustoelastic stress analysis of residual stress in a patch-welded disk, *Exp. Mech.*, 18:277(1978).
5. Y.Osawa, Y.Arai and H.Kobayashi, Measurement of residual weld stresses by acoustoelastic technique, *Trns. Jpn. Soc. Mech. Eng.(A)*, 53:2410(1987)(in Japanese).
6. H.Toda, H.Fukuoka and Y.Aoki, R-value acoustoelastic analysis of residual stress in a seam welded plate, *Jpn. J. Appl. Phys.*, Suppl., 23:86(1984).

NONDESTRUCTIVE STRESS MEASUREMENT USING FREQUENCY DEPENDENCE OF MAGNETOACOUSTIC INTERACTION

Tomohiro Yamasaki, Ken'ichi Ebata, and Hidekazu Fukuoka

Faculty of Engineering Science
Osaka University
Toyonaka, Osaka 560, Japan

INTRODUCTION

Since residual stress often causes a failure at applied stresses lower than the strength, the nondestructive measurement of the residual stresses is very important in assuring the structural safety. Acoustoelastic method[1,2] serves such practical needs because it can detect the residual stress nondestructively. The acoustoelasticity utilizes the linear dependence of propagation velocities of elastic waves on the stress, and evaluated is the averaged stress along the propagation path. However, the stress effect on the velocity is so small that it is often obscured by the texture effect. The application to the practical stress measurement has been greatly limited by the difficulty in separating the stress effect from the texture effect.

Recently, a new method based on magnetoacoustic interaction was suggested.[3,4] The elastic wave velocity in a ferromagnetic material slightly varies when an external magnetic field is applied and moreover the magnetically induced velocity change (MIVC) depends on the stress. This occurs because of the magnetoelastic interaction. The domain structure changes when the static stress is applied in the demagnetized state. Domain rearrangement accompanies magnetostriction in addition to the pure elastic strain, reducing the apparent elastic constants. The external magnetic field also varies the domain structure and it restricts the stress induced domain reorientation. The apparent elastic constants increase as the magnetization proceeds, approaching the pure elastic values in the magnetically saturated state. The elastic wave velocities show the similar behavior, that is, they increase when the external magnetic field is applied. The MIVC depends on the stress because the stress also restricts the domain reorientation induced by the dynamic stress of the elastic waves. Since the MIVC is found to be insensitive to the texture, it is suitable for the residual stress measurement, where the stress-free state is hardly accessible.

The domain reorientation, which causes the MIVC, induces microscopic eddy current around the domain. The wave energy is absorbed and the amplitude decreases. Therefore, as well as the velocity, the amplitude also varies when the material is magnetized. In this study, the effects of stress and frequency on the magnetically induced amplitude change (MIAC) was investigated for low-carbon steel (JIS–SS41). The results were applied to nondestructive measurement of the residual stress in the butt welded specimen.

MAGNETOACOUSTIC INTERACTION

In polycrystalline ferromagnets, every grain is divided into many magnetic domains. The individual domain is magnetically saturated even in the demagnetized state, and it is elongated in its magnetization direction in α–Fe. The magnetization direction is distributed randomly in the demagnetized state. When the external magnetic field is applied, the domain walls move so as to increase the volume of the domains magnetized parallel to the field. The magnetization thus proceeds. As a result of reorientation of the spontaneous strain of the domain, the dimensional change occurs, which is called the *magnetostriction*.

When the stress is applied to the steel, the domain walls again move. As a result, the volume of the domains magnetized parallel to the tensile stress or perpendicular to the compressive stress increases. Then the magnetostriction appears in addition to the elastic strain. The apparent elastic constants decrease from the pure elastic values, which can be obtained in the magnetically saturated state, that is, no more domain wall movement occurs.

The stresses of the elastic waves alter in a MHz frequency range. They induce rotational vibration of the domain magnetization direction besides the domain wall movement.[5] As in the case of the static stress, the magnetostriction participates and the apparent elastic constants, equivalently the wave velocities, decrease from the pure elastic velocities. When the external magnetic field is applied, both the domain wall movement and the rotational vibration are restricted. The velocities increase with the magnetization and take the pure elastic values in the magnetically saturated state. Since the MIVC is the velocity change from the demagnetized state to the magnetized state, it depends on the domain structure in the demagnetized state, which varies with the stress. This is why the MIVC depends on the stress.

Since the rotational vibration of the domain magnetization and the domain wall movement, both induced by the elastic wave, correspond to microscopic change in the magnetization, the *microscopic eddy current* is induced around each domain. Here, the microscopic eddy current is a circulating current, which induces a magnetic field in the opposite direction to the domain magnetization change. This absorbs the wave energy and makes the wave amplitude decay. The microscopic magnetization change occurs being restricted by the magnetic field induced by the microscopic eddy current and the resistive force by dislocations. The amplitude is further decreased. When the external magnetic field or the stress is applied, the microscopic magnetization change is restricted and the wave amplitude increases. The pure elastic amplitude can be obtained in the fully magnetized state. As in the case of the MIVC, the MIAC depends on the domain structure in the demagnetized state. Thus the MIAC can be also used for a nondestructive stress measurement.

UNIAXIAL LOADING TEST

Experimental Procedure

The specimens were prepared from two rolled plates of low carbon steel (JIS–SS41); one was as received and the other annealed at 880℃ for 1 hour and furnace cooled. Tables 1 and 2 show the chemical compositions and the mechanical properties, respectively. We prepared two specimens from as-received plate, one having the loading direction parallel to the rolling direction and the other the transverse direction. In the following, they are called the specimen R and the specimen T, respectively. From the annealed plate, the specimen HR was machined. The loading direction is parallel to the rolling direction. The thickness of each specimen was 11 mm.

Table 1. Chemical compositions (wt.%) of specimens.

C	Si	Mn	P	S
0.012	0.22	0.99	0.015	0.004

Table 2. Mechanical properties of specimens.

	Young's modulus, GPa	Poisson's ratio	Yield stress, MPa
Specimen R,T	208	0.290	310
Specimen HR	209	0.284	260

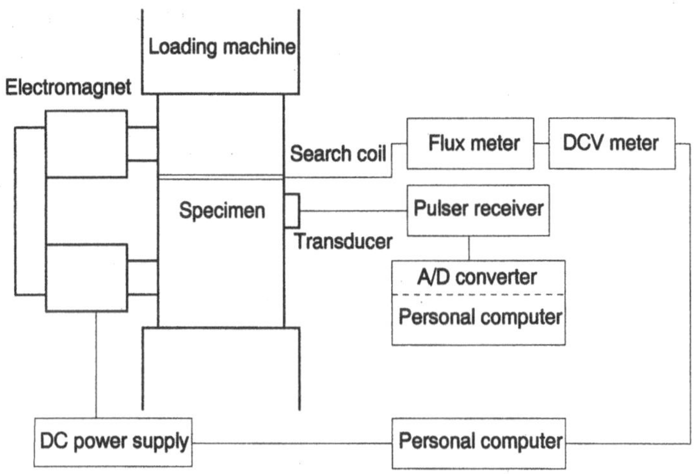

Figure 1. Block diagram of experimental setup.

Figure 1 shows the block diagram of the experimental setup. The external magnetic field was applied to the specimen using a C-shaped electromagnet. A personal computer controlled the field strength by adjusting DC current into the electromagnet. A flux meter with a search coil monitored the induced magnetic flux density. A broadband transducer (8 mm diameter) launched the longitudinal wave into the specimen and received the first echo out of the signal train. The signal was digitized to calculate the amplitude spectrum through discrete Fourier transform algorithm. Bonding the transducer to the specimen with phenyl salicylate enabled us to detect the slight change in the amplitude.

The experimental procedure is as follows. Uniaxial stress was loaded to the specimen and the AC magnetic field demagnetized the specimen. The echo amplitude was measured while increasing the DC magnetic field strength and then decreasing it, both stepwise. To minimize the effect of the magnetic hysteresis, the anhysteretic magnetization curve was traced by adding a decaying AC magnetic field to each field strength. After each measurement, the amplitude was again measured in the demagnetized state. The amplitude change was calculated as ratio of the amplitude in the magnetized state A and that in the demagnetized state $A_{H=0}$. This procedure reduced the error induced by drift of the flux meter indication.

Results and Discussion

Figure 2 shows the MIAC curves for the specimen R at a stress free state for 4, 8 and 12 MHz. The ratios of amplitude in the magnetized state A to that in the demagnetized state $A_{H=0}$ are plotted in a logarithmic scale. The flux density is calculated as difference between flux meter indications in the magnetized state and the demagnetized state. The −6 dB bandwidth was obtained for the 4–12 MHz range.

As predicted, the wave amplitude increased with the magnetization. It was found that the MIAC is larger for the higher frequencies. The magnitude of the microscopic eddy current, which reduces the amplitude, is proportional to the time derivative of the domain magnetization change. The amplitude decrease should depend on both magnitude and frequency of the microscopic magnetization alternation.

The magnitude of the microscopic magnetization alternation can be discussed by considering the MIVC. The velocity decrease in the demagnetized state is caused by the magnetostriction induced by the microscopic magnetization alternation. Figure 3 shows the MIVC curves under the same condition as Fig. 2. We found that the MIVC is larger for the lower frequencies. This indicates that the magnitude of the microscopic magnetization alternation is smaller for the higher frequencies.

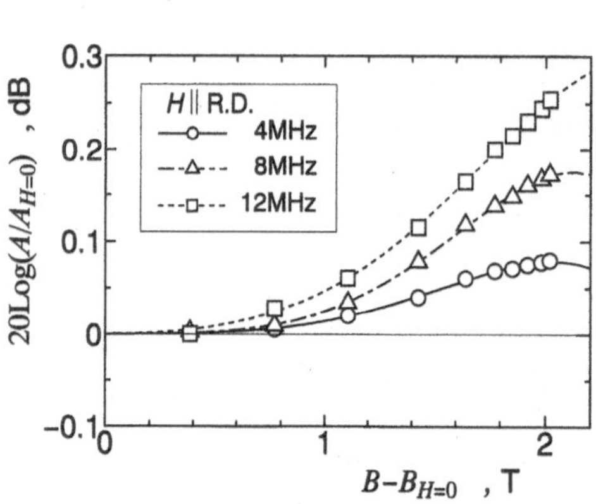

Figure 2. MIAC curves for specimen R at stress free state.

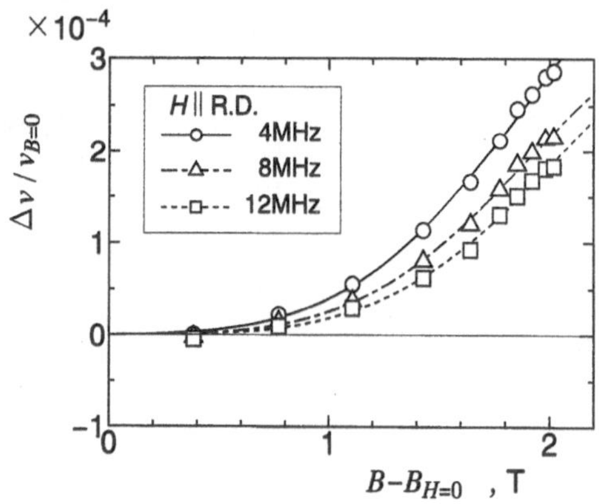

Figure 3. MIVC curves under the same condition as Fig. 2.

Figures 4–7 show the MIAC curves for the specimen R under various stress levels. It was found that the stress reduces the MIAC whether the stress is tensile or compressive. This occurs because the stress changes the domain structure and also because it restricts the domain wall movement and the rotational vibration of the domain magnetization. The experimental observations agree with this prediction. We found that the frequency dependence of the MIAC is reduced by the tensile stress (see Figs. 4 and 5), while it is almost insensitive to the compressive stress (see Figs. 6 and 7). This phenomenon can be used for a nondestructive stress measurement. The magnitude of the stress can be evaluated by the MIAC, and the sign of the stress can be estimated by the frequency dependence of the MIAC.

Figure 8 shows the MIAC curves under the 100 MPa tensile stress for the magnetization perpendicular to the stress. Comparison with Fig. 4 reveals that the MIAC of the longitudinal wave is insensitive to the magnetization direction.

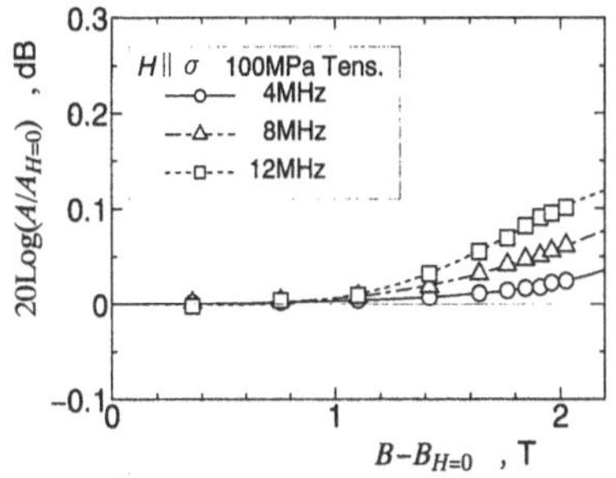

Figure 4. MIAC curves for specimen R under 100 MPa tension.

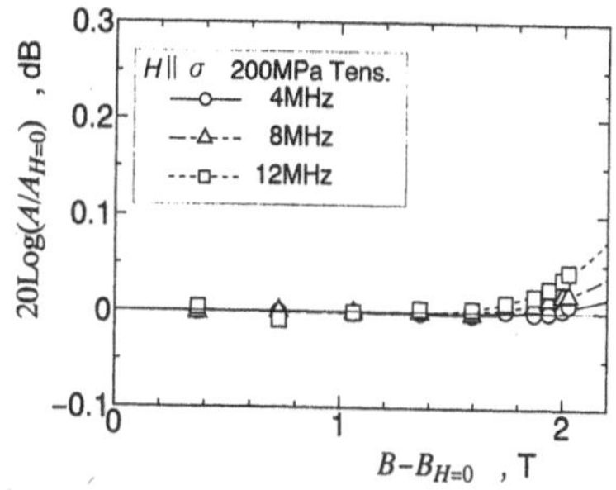

Figure 5. MIAC curves for specimen R under 200 MPa tension.

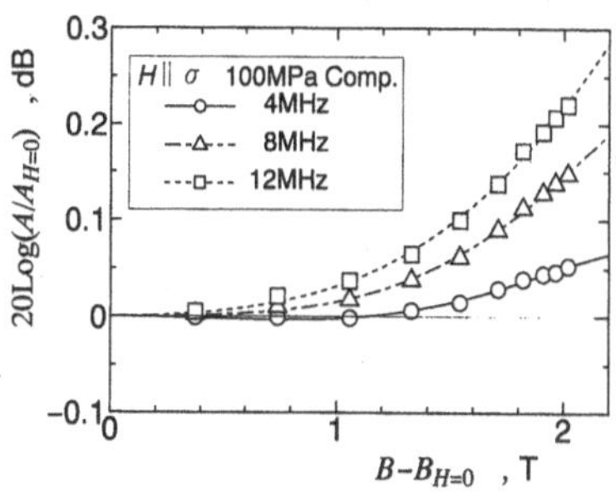

Figure 6. MIAC curves for specimen R under 100 MPa compression.

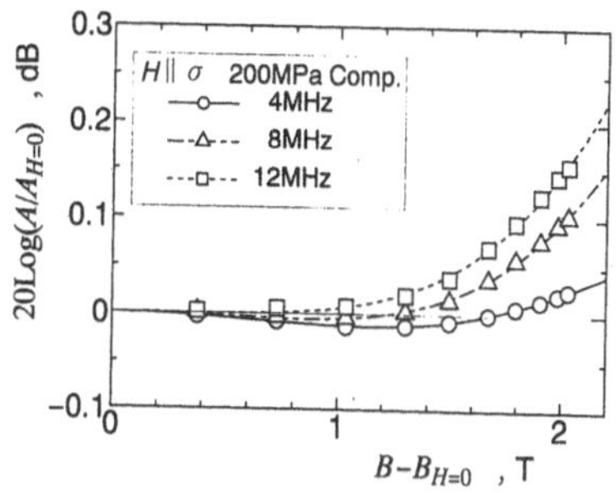

Figure 7. MIAC curves for specimen R under 200 MPa compression.

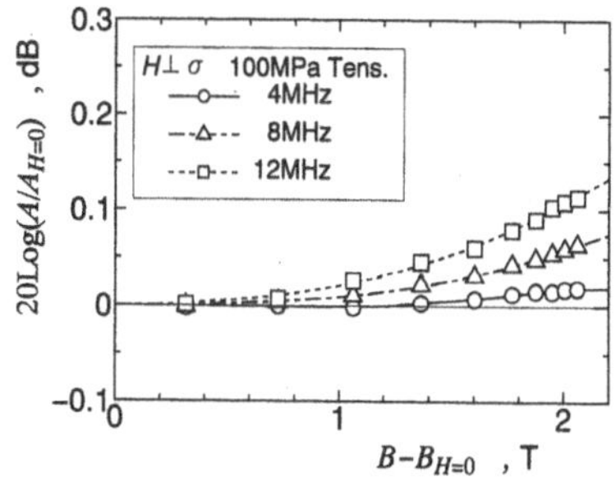

Figure 8. MIAC curves under 100 MPa tension for magnetization perpendicular to stress.

Figures 9 and 10 show the results for the specimen T and the specimen HR. In the conventional acoustoelastic stress measurement, the texture effect often corresponds to the yield stress. However, it was found to be small enough to be ignored in the MIAC. The heat treatment, such as annealing, alters the magnetic properties of the material. The effect of annealing corresponded to 50 MPa stress.

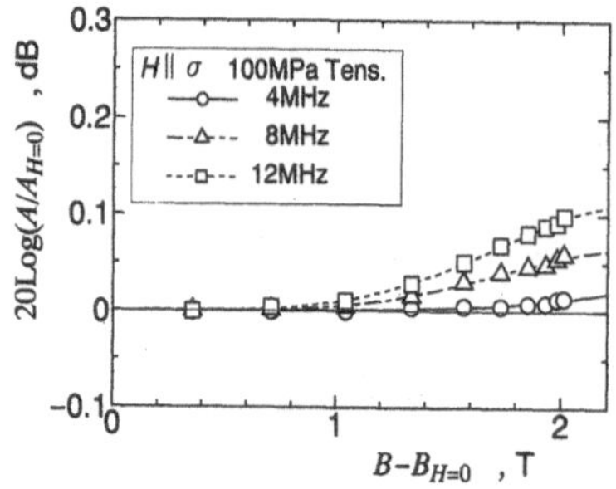

Figure 9. MIAC curves for specimen T under 100 MPa tension.

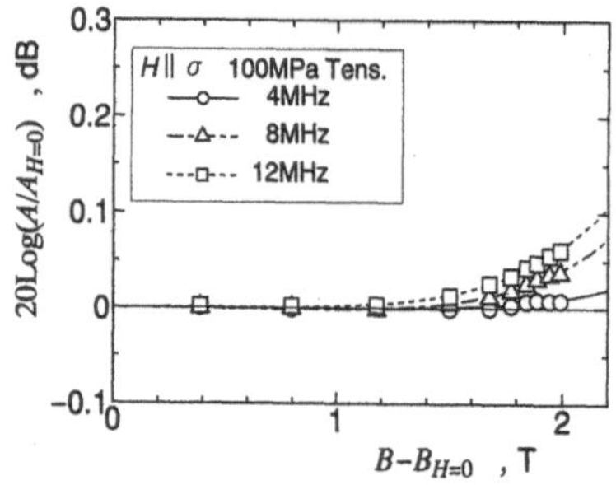

Figure 10. MIAC curves for specimen HR under 100 MPa tension.

NONDESTRUCTIVE RESIDUAL STRESS MEASUREMENT

Using the MIAC curves shown in Figs. 4–7 as master curves, the residual stress in a butt welded plate was measured nondestructively. In the butt welded plate, thermomechanical process introduces the residual stresses parallel and perpendicular to the welding line. In general, the former is much larger. In this study, it is then assumed that the stress is uniaxial and parallel to the welding line.

Two identical plates were prepared from the same steel plate as specimen R. They were butt welded to each other and the measurements were done on one plate. Figure 11 shows the dimension of the specimen along with the measuring points. The process of the stress evaluation is as follows:

1. The MIAC was measured at the measuring points. The magnetic fields were applied parallel and perpendicular to the welding line.
2. The stress candidates were obtained by comparing the MIAC curve for 8 MHz with the master curves from tension and compression tests.
3. Calibration curves which indicate difference between the MIACs at 4 and 12 MHz for the candidate stress values were calculated by interpolating.
4. The stress value that exhibits the better fitting is taken to be the predicted stress.

After the MIAC measurement, the residual stress was released by sectioning, and the released strains were recorded from the strain gauges.

Figure 12 compares the MIAC method with the strain gauges. The results show a good agreement except for a couple of points where the stress is close to zero. In this study, the residual stress was evaluated assuming that it is uniaxial. The strain gauges indicated that the stress component perpendicular to the welding line was as much as 50 MPa in this region. This should be a reason for the error. The magnetic properties had been changed by the heating near the welding line. It also induces the error in the stress prediction. Considering the effects of the heat and the stress component perpendicular to the welding line, the accuracy of the MIAC stress measurement was found to be about 50 MPa. If used with the MIAC of shear wave or the MIVC method, the accuracy may be improved.

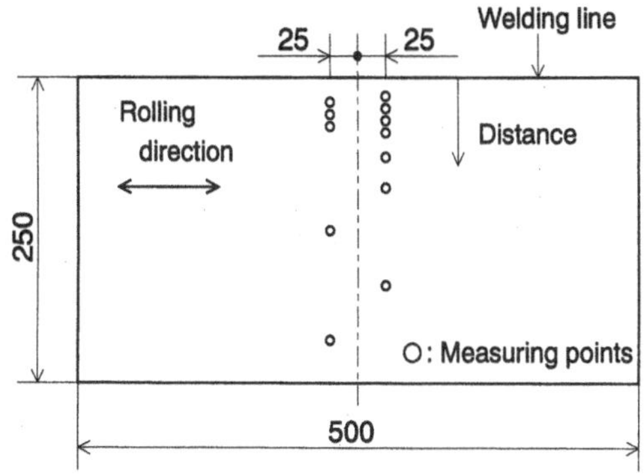

Figure 11. Dimension of welded specimen with measuring points.

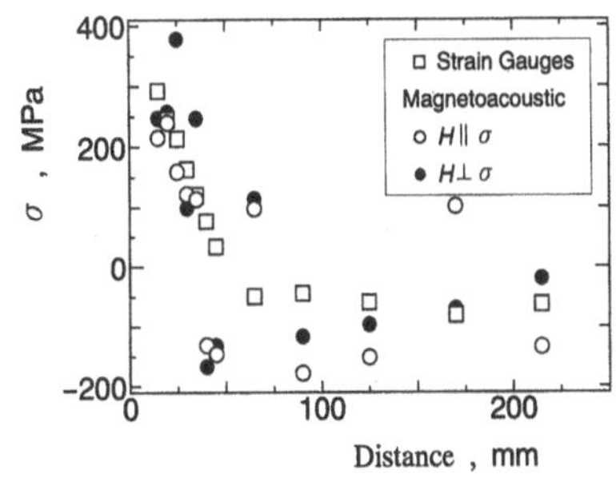

Figure 12. Comparison of magnetoacoustic residual stress predictions to those by strain gauges.

In this study, the residual stress was assumed to be parallel to the welding line. In the practical residual stress measurement, use of the shear wave may help us to estimate the stress direction.

SUMMARY

The magnetically induced amplitude change of longitudinal wave was measured under various stress levels for JIS–SS41 mild steel. It was found that the MIAC is larger for the higher frequencies and that the MIAC decreases when the stress is applied, whether it is tensile or compressive. The frequency dependence of the MIAC is reduced by the tensile stress, while it is insensitive to the compressive stress.

Using the obtained curves as master curves, the residual stress in the butt welded plate was measured. The results showed a good agreement with those by the strain gauges. It was ensured that the MIAC method is useful for the nondestructive measurements of residual stresses in ferrous materials.

REFERENCES

1. D.I. Crecraft, The measurement of applied and residual stresses in metals using ultrasonic waves, *J. Sound Vib.* 5:173 (1967).
2. Y.-H. Pao, W. Sachse, and H. Fukuoka, Acoustoelasticity and ultrasonic measurements of residual stresses, *in*:"Physical Acoustics XVII," W.P. Mason and R.N. Thurston, eds., Academic Press, New York (1984).
3. H. Kwun and C.M. Teller, Stress dependence of magnetically induced ultrasonic shear wave velocity change in polycrystalline A-36 steel, *J. Appl. Phys.* 54:4856 (1983).
4. M. Namkung, et al., Low-field magnetoacoustic residual stress measurement in steel, *in*:"Solid Mechanics Research for QNDE," J.D. Achenbach and Y. Rajapakse eds., Martinus Nijihoff Publishers, Dordrecht (1972).
5. W.P.Mason, Rotational relaxation in nickel at high frequencies, *Rev. Mod. Phys.* 25:136 (1953).

RESIDUAL STRESSES INDUCED BY SLITTING COPPER ALLOY STRIP

C.O. Ruud[1], and M.E. Jacobs[2]

[1]Department of Industrial and
Management Systems Engineering
The Pennsylvania State University
University Park, PA 16802
[2]BMI Refractories, Southwestern, OH

INTRODUCTION

Slit copper alloy strip is used extensively for the feed material in the manufacture of electrical contacts and switches. The slitting process often introduces residual stresses at the slit edges of the strip. These residual stresses cause the strip to warp either immediately after slitting or in subsequent manufacturing operations. The warped strip has been reported to cause jamming of stamping and forming dies, leading to excessive delays in manufacturing as well as wasted material and labor. Amin et al. [Amin 1981A] cited problems of misfeed of copper alloy strips, used to produce electrical contacts, and blamed residual edge stresses due to slitting. Further, they showed that slitting was the main cause of distortion when the slit edge was used as a carrier strip in the stamping of electrical contacts.

Slitting is an important production operation which is used extensively in manufacturing, however, the nature and extent of the residual stresses resulting are not known. This paper presents the results of surface stress measurements performed on several copper alloy slit strips and discusses the data in light of previous results from the literature. The residual stress measurements presented are the result of a study to develop an x-ray diffraction stress measurement instrument for on-line application. It was considered that such a tool for nondestructive in-process measurement of slitting stresses could provide a valuable means for the optimization of slitting parameters in real time.

BACKGROUND

There have been few studies published on residual stresses induced by slitting. It could well be that this is due to the difficulties inherent in performing investigations with sufficient dimensional resolution to provide meaningful data considering the sever stress gradients which are likely to exist in slit metal strip edges. Amin, et al. [Amin 1981A] described a technique of sectioning slit strip (0.635 mm thick x 51. mm wide) into thin, long fingers, (1.59 mm wide) of which the longitudinal direction was parallel to the slit edge. They used a photoetching technique to produce the strips and which did not induce stress in addition to those resulting from slitting. They showed that stresses in the direction of the slit edge were tensile ranging from 210 to 350 MPa and that the higher the yield strength of the slit strip the more sever the slitting stress also, that roller leveling before slitting did not reduce edge distortion (stresses). They concluded that, quote "the position and clearances of the slitting blades evidently induced severe plastic deformation and substantial residual stresses in an area approximately 0.1 in. (2.54 mm) wide at the strip edge". Amine et al. [Amin 1981A, 1981B] also included some x-ray diffraction stress measurements in their papers but concluded that it was too time consuming for application to quality control.

Polyakov et al. [Polyakov, 1980] proposed an equation by which residual stresses due to slitting could be calculated from the bending of fingers electrospark cut from a thin (0.36 mm thick) colled-rolled metal sheet. They reported that tensile stresses were found in the longitudinal direction in the finger but did not report their magnitude nor the width of the fingers.

Ignberg [Ignberg 1986] described an etching method combined with a theoretical model to evaluate residual edge stresses due to slitting. The etching method was similar to that described by Amin et al. [Amin 1981A], but Ignberg's model provided a detailed description of the nature of the three dimensional stress field induced by slitting. Ignberg's model predicted that the stresses in the longitudinal direction (slitting direction) in the plane of the sheet surface (rolling plane) would be tensile. He reported that the maximum longitudinal stresses at the surface of a 5 mm wide finger on a copper alloy strip would be at the slit edge and would be 250 MPa; decaying exponentially to zero in about 0.5 mm from the edge. He indicated that the slitting tools must be very precisely adjusted to minimize the stresses but that the induced stress can not be totally avoided.

Lastly, Borhan et al. [Borhan 1990] measured edge stresses applying a finger technique similar to that of Amin et al and Ignber [Amin 1981A, Ignberg 1986]. They compared stresses induced by three edge cutting methods, mechanical slitting, water-jet cutting and laser cutting. It was concluded that of the three methods only slitting provided edges with sufficiently low stresses to be of use for on-line processing. They calculated stresses in slit 0.34 mm thick copper alloy strip from the bending of fingers 5.7 mm wide, as ranging between -53 and +14 MPa, but most of the readings (8 out of 10) were compressive.

On-Line X-Ray Diffraction Instrumentation

A fiber optic based x-ray detector has been developed which is well suited to rapid, realtime, in-process residual stress measurement [Ruud 1983]. The objective of the study that provided the data presented here-in was to show the viability of applying this x-ray instrumentation to the measurement of slitting stresses with the strip moving at typical production speeds. This viability was successfully demonstrated by measuring the longitudinal residual edge stresses on a strip 0.25 mm wide 0.15 mm from the edge moving at 115 m/min.

PROCEDURE

Samples approximately 320 mm long were supplied by an electrical contact manufacturer and were from two different copper strip producers. Table 1 lists the identification number, alloy, dimensions, and tensile properties of the samples. No slitting parameter information was provided by the supplier.

A running strip coil sample was prepared for the on-line simulation by soldering a 2.5 meter long sample together at the ends to make a loop.

TABLE 1 LIST OF COPPER ALLOY STRIP SAMPLES

Sample Number	Alloy & Temper	Thickness & Width (mm)	Tensile Strength (MPa)	Yield Strength (MPa)	Percent Elongation
1	C51100-HO4 As-Received	0.20 x 24.6	596	551	13
2					
	C51100-HO4	0.20 x 24.6	596	551	13
3	Roller Leveled				
4	C51100-HO4	0.22 x 26.2	570	518	15.5
5	C51100-HO3	0.20 x 24.7	545	481	15.7
6	C51100-HO6	0.20 x 19.0	662	634	11
7	C52100-HO6	0.32 x 15.0	732	662	19
Loop	C52100-HO6	0.32 x 15.0	739	676	17
	C51100-HO4	0.22 x 26.2	577	520	15.0

X-ray Diffraction Conditions

The x-ray residual stress measurement system consisted of a Picker full-wave rectified x-ray power supply (50kV, 20mA), a Philips fine focus copper target x-ray tube (P/N 14600320), a fiber optic position sensitive scintillation detector (PSSD) [Ruud 1983], and a PDP-11/23 computer.

The fine focus x-ray tube was collimated to produce a beam 0.25 mm wide in a direction perpendicular to the slitting direction and 2.5 mm long in a direction parallel, A0 mm sample to x-ray detector working distance was used with the angle between the sample surface and incident x-ray beam of 30 degrees. Although the resolution in this study proved adequate, better spatial resolution is likely to be achieved by employing a microfocus x-ray tube which would provide a spatial resolution on the order of 0.1 mm.

The x-ray residual stress measurements were executed using the single exposure technique (SET) [Ruud 1979, SAE 1971]. Data collection times of 25 seconds were used; these times could be decreased by 1/6 to 1/6 through application of a modern 2.5 kw, constant potential x-ray power supply and further reduced using a microfocus x-ray tube.

Stress Measurement

An apparatus was fabricated to hold the slit copper strip samples flat and to be able to translate them by small increments. The sample was placed against two alignment stops and the sample holds tightened, which rendered the sample flat and straight. A micrometer, accurate to 0.02 mm, was used to adjust the position of stress measurement on the samples from edge to center. By this method the x-ray beam could be positioned at or near the edge of the copper strip and traversed in small increments, in this case, 0.5 mm. The stresses were measured in a direction parallel to the slitting direction, and the 0.25 mm beam width determined the spatial resolution of measurement across the sheets. Measurements were performed on the specimen at the locations along a traverse normal to the slit edge as shown in Figure 1.

On-line Simulation. An edge stress measurement test stand was used in order to simulate actual on-line running of the strip. The test stand was equipped with a variable speed, variable torque motor capable of running the slit copper strips at 115 meters per minute. The test stand had a tracking accuracy of ± 0.025 mm. Since the micrometer could no longer be used to locate the x-ray stress measurement position, accuracy in positioning of the x-ray beam was a limiting factor. The positioning device for the x-ray stress head used in these on-line simulation tests had a locating accuracy of ± 0.13 mm in the traverse direction, i.e. across the strip, perpendicular to the slit edge.

X-ray stress measurements, both static and moving, were executed on a copper strip coil approximately 2.5 meters in circumference. Five static positions, approximately 0.5 meters apart, were chosen on the coil for stress measurement. Three speeds for moving strip stress measurement were used: 16, 29, and 115 meters per minute.

Results and Discussion: The surface residual stresses as measured by XRD from an area 0.25 mm x 2.5 mm, with the long dimension parallel to the longitudinal (rolling and slitting) direction of the sample, are plotted in figures 2-8. The stresses are in units of MPa and the distance from the left side (chosen arbitrarily) of the strip in mm. Stresses are plotted for both rolled surfaces of the sheet samples, i.e. side A or B. The precision of stress measurement is ± 35 MPa.

Figure 2 shows that the residual stresses for sample 1 on side A are compressive nearest the edge then become tensile between 1 and 2 mm with the maximum stress at the left and right edges -210 and -245 MPa, respectively. The stresses tend to become more tensive within 1 mm of the edge. The stresses in side B are much lower in magnitude and not much greater than that of the precision of measurement. The stresses in the center of the sheet sample, well away from the edges is about ± 35 MPa, i.e. within one standard deviation of the measurement precision of zero.

Figure 3 shows that the residual stresses for sample 2 tend to be slightly compressive at the slit edge on both sides of the strip and tensile at 1 to 2 mm away from the edge. The highest magnitude of the stresses are -84 to -126 MPa at the left and right edges, respectively. The surface stresses in the center of the sheet tend to be slightly compressive, but near zero.

Figure 4 shows that for the most part the residual stresses for Sample 3 are highest in magnitude about 1 mm away from the slit edge and these range from + 140 to -238 MPa. However, the highest magnitude of compressive stress was on side A at the right edge and was -315 MPa. As with sample 2, the surface stresses in the center of the sheet was slightly compressive, i.e. -50 MPa.

Figure 5 shows that by and large the residual stresses for Sample 4 are compressive and highest magnitude at the edges, i.e. -161 to -196 MPa for the left and right edge, respectively; and within about 1 mm of the edge they tend to be more tensive. As for the two previous samples the residual surface stress in the center of the sample tends to be compressive, approximately -70 MPa.

Figure 6 shows that the residual stresses at the edges for sample 5 are compressive on side A but about zero to tensile on side B. The magnitude of the stresses are for the most part highest at the edges. The surface residual stresses in the center of the sheet are compressive on the order of -40 to -90 MPa.

Figure 7 shows that the residual stresses for sample 6 are highest in magnitude at the edges on side A and range from -210 to -250 MPa for the left and right edges, respectively; tending to become more tensive within a few mm away from the edge. The surface residual stresses in the center of the sheet are compressive on the order of -50 to -70 MPa.

Figure 8 shows that the residual stresses for sample 7 are compressive and highest in magnitude at the edges becoming nearly zero within 3 mm of the edge. The highest magnitude of the edge stresses was on the right edge of side B and was -300 MPa. The surface residual stress in the center of the sheet for sample 7 were tensile on the order of 50 to 80 MPa.

The plots shown in Figures 2-8 generally indicate that the pattern of residual stresses in the longitudinal (slitting) direction of the sheet vary markedly from sample to sample; and within a sample. The maximum edge stress versus yield strengths of the sample is plotted in Figure 10 and does not seem to correlate with the strength, or cold work, of the sheet as concluded by Amin et al [Amin 1981A]. However, this may be due to differences in slitting parameters from sample to sample in this investigation in that no control on slitting was exercised. Amin et al. also noted that roller leveling did not reduce edge distortion; but comparison of samples 1 and 2, the only difference being that 2 was roller leveled, indicates that the edge stresses were lower in the roller leveled strip. However, Amin et al. [Amin 1981B] did conclude that roller leveling did produce an irregular pattern of stresses in the sheet but it would seem that the sever plastic deformation induced by slitting would tend to overwhelm any previous residual stress condition.

This study showed that for the most part the residual stresses imposed by slitting extended to 2.5 mm from the edge. Ignberg [Ignberg 1986] indicated that the stresses would reduce to zero within about 0.5 mm of the edge and also he predicted that the longitudinal residual stresses on the rolled surface at the slit edge would be tensive However, for the most part this investigation found compressive stresses at the slit edges. Also, Ignberg showed that the maximum stresses at the edge would be on the order of 250 MPa and that is in reasonable agreement with the results reported herein, which ranged from +140 to -315 MPa; but were for the most part compressive, i.e. opposite in direction to those predicted by Ignberg in 26 out of 28 of the edges measured. Nevertheless, the stresses did seem to decay exponentially as predicted by Ignberg.

Amin et al. [Amin 1981B] reported that the residual stresses in the center of cold rolled copper alloy strips 0.64 mm thick, were generally tensive (100 to 200 MPa) after cold reduction of 22%, but in this study it was found that the samples showed mostly compression stresses in the center, and on the order of -40 to -90 MPa. This may be due to the fact that the samples in this study were thinner, i.e. 0.20 to 0.32 mm.

A consistent finding in this investigation is that the "burr" side of the sheet always showed the greatest magnitude of compressive stress. This would seem reasonable in that the burr is caused by the stretching or smearing of the metal prior to complete severing. The stresses inducing this smearing plastic strain would be tensile and thus would tend to leave a compressive residual stress [Cullity 1978]. Further, the sheet would tend to conform to the radius of the slitting knife as it was slit with the smallest radius being attained just prior to burr formation and this also would tend to result in a compressive residual stress on the burr side.

All of the data shown in Figures 2-8 indicate that the burr side stresses tended to be more tensive than the center of the sheet within 2 mm of the edge, except for Sample 7, Figure 8. This is indicative of plastic deformation due to a compress stress at approximately 2 mm of the edge. It should be noted that this trend was often mirrored on the opposite side of the sheet by a trend toward compressive residual stress. The tensile burr side trend mirrored by a compressive trend or the other side of the sheet would indicate that the slitting operation produced plastic bending at approximately 2 mm from the edge [Cullity 1978].

Figure 9 shows the results of the on-line simulation of residual stress measurement. The data demonstrates that moving the strip at up to 115 meters per minute does not cause a significant change in the stress profile, therefore, demonstrating that real time on-line residual stress measurements are possible with XRD.

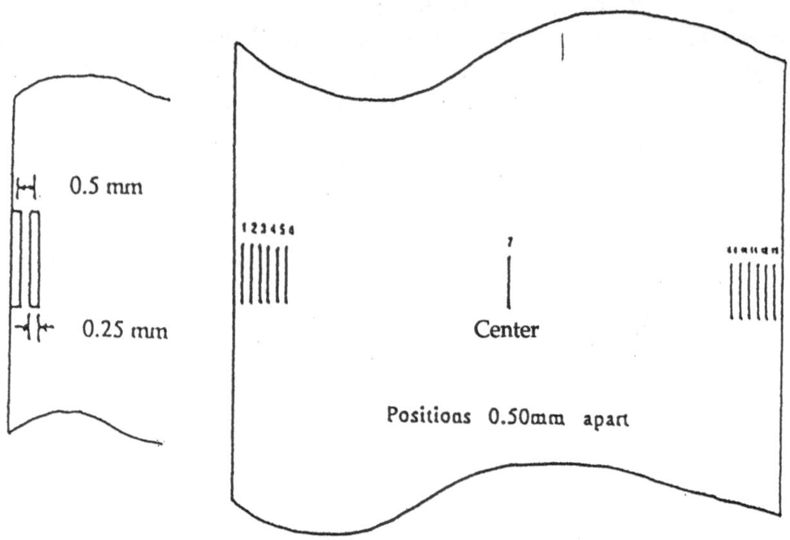

Figure 1. Positions of x-ray diffraction stress measurement or the slit copper alloy samples.

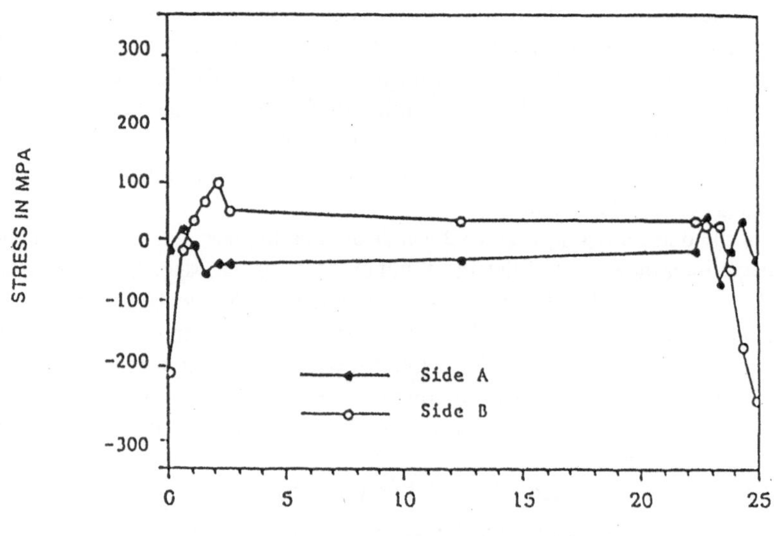

Figure 2. Residual edge stresses in the longitudinal direction for Sample 1.

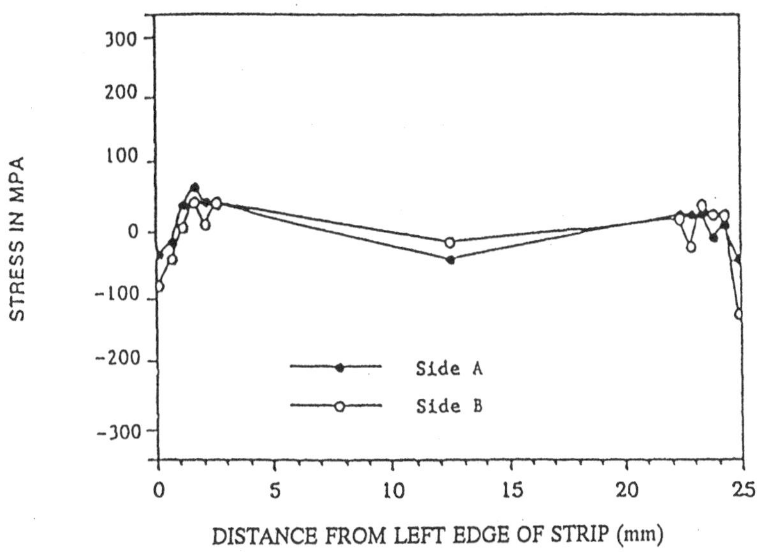

Figure 3. Residual edge stresses in the longitudinal direction of Sample 2.

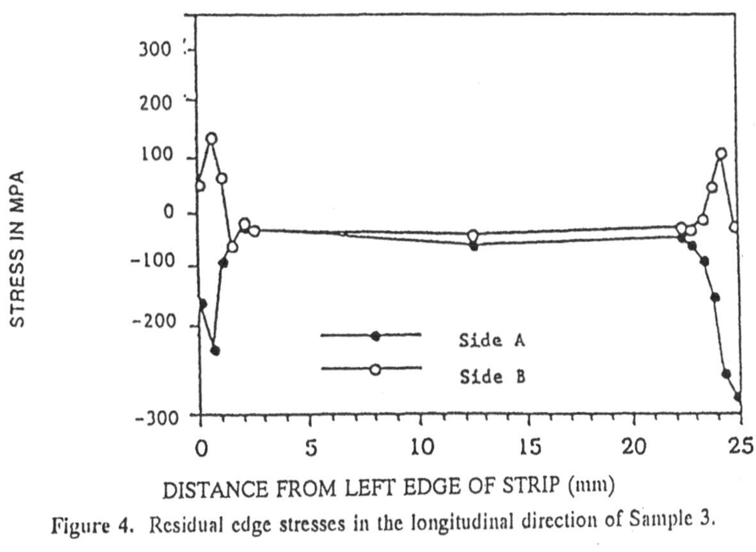

Figure 4. Residual edge stresses in the longitudinal direction of Sample 3.

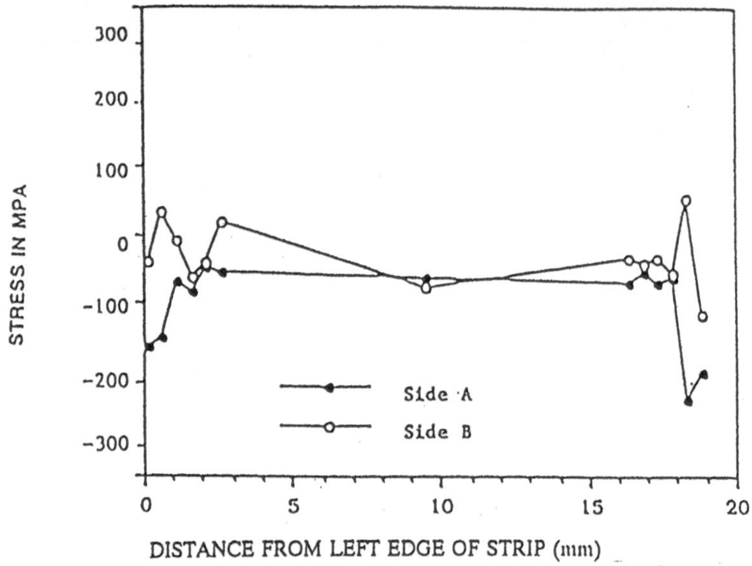

Figure 5. Residual edge stresses in the longitudinal direction of Sample 4.

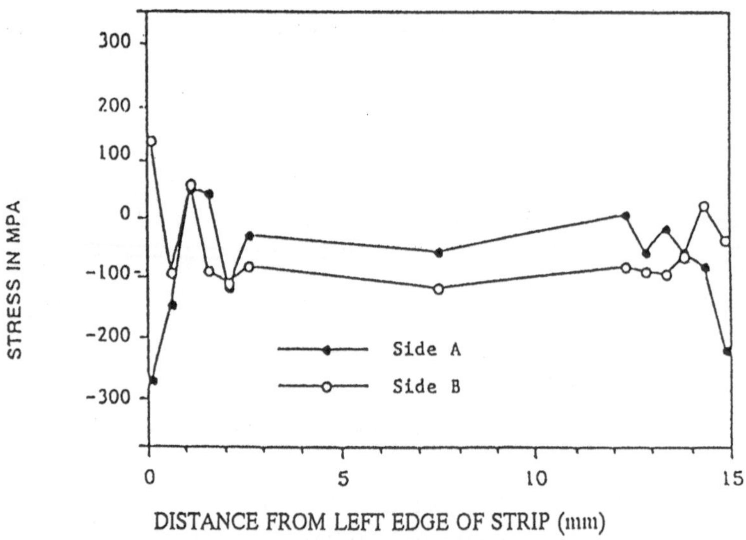

Figure 6. Residual edge stresses in the longitudinal direction of Sample 5.

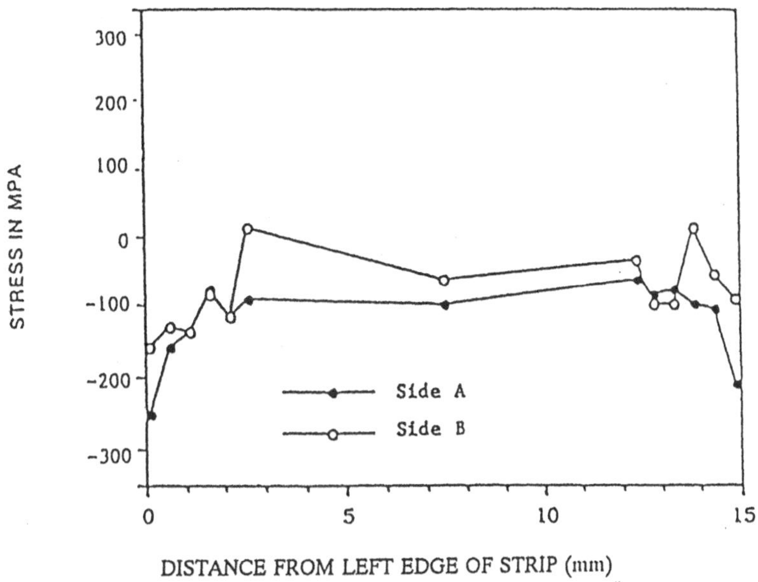

Figure 7. Residual edge stresses in the longitudinal direction of Sample 6.

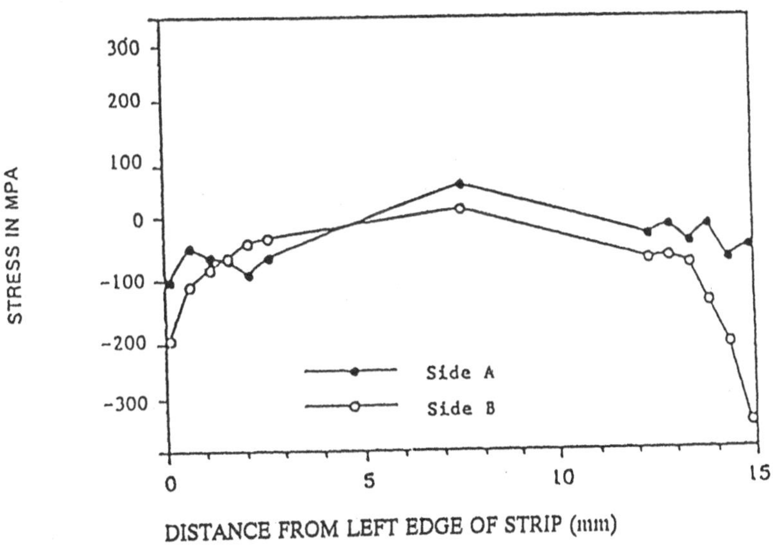

Figure 8. Residual edge stresses in the longitudinal direction of Sample 7.

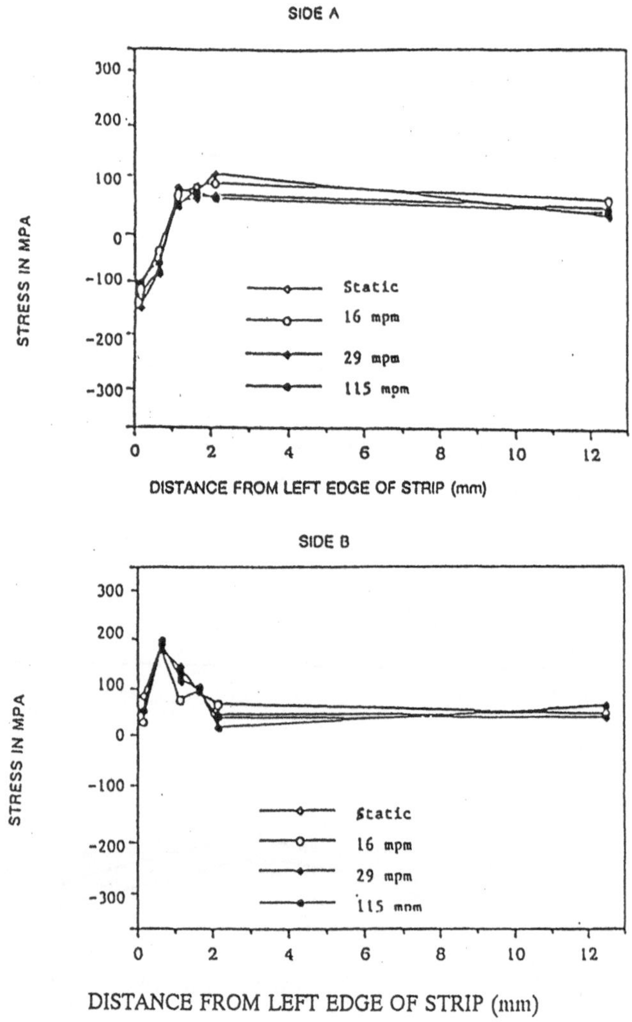

Figure 9. Residual edge stresses in the longitudinal direction for the loop sample showing static and moving results, top plot is Side A, bottom Side B.

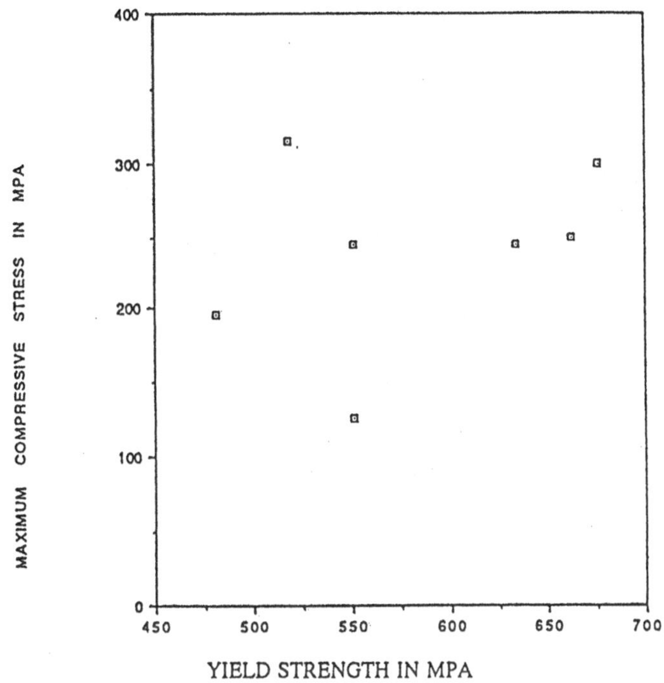

Figure 10. Maximum compressive stress at edge versus yield strength of the copper alloy sheet.

CONCLUSIONS

1. For the most part compressive stresses as high as -315 MPa were found at the slit edges in the slitting direction on the surface of the sheet samples.
2. The as-rolled sheet surface, before slitting, was most often compressive and on the order of -40 to -90 MPa.
3. The residual stresses in the slitting direction on the surface of the slit sheet samples was always most compressive on the burr side.
4. The residual stresses in the slitting direction on the surface of the slit sheet samples are for the most part of highest magnitude at the edge and decay approximately exponentially as predicted in previous papers.
5. The residual stress patterns slit edges indicate that a combination of tensile smearing stresses and bending stresses characterize the slitting mechanism.
6. The patterns of residual stresses near the edge in the slitting direction on the surface of the sever samples, and twenty-eight edges, measured in this study varied considerably.
7. The residual edge stresses due to slitting on strip moving up to 115 meters per minute can be measured on-line in realtime.

REFERENCES

Amin 1981A - Amin, K.E., Rusnak, R.M., Becker, P.C. and Ganesh, S., "A Technique for Predicting Distortion and Evaluating Stress Relief in Metal Forming Operations" *J. of M.*, pp. 12-18, Feb. (1981).

Amin 1981B - Amin, K.E. and Ganesh, S., "Residual Stresses in Copper - 2% Beryllium Alloy Strips", *Exp. Mech.*, pp. 473-476, Dec. (1981).

Borham 1990 - Borham, A., Pichavant, C. and Ansart, P., "Comparison of Residual Stresses Generated by Different Cutting Methods in KLF 1 Slit Strips", Paper No. 29, Session 20A, Presented at *COPPER 90*, Vasteras, Sweden, Oct. (1990).

Cullity 1978 - Cullity, B.D., *Elements of X-Ray Diffraction*, 2nd Ed, Addison-Wesley, p. 449 (1978).

Ignberg 1986 - Ignberg, L., "Residual Stresses in Rolled Products of Copper and Copper Alloys-Effects on Performance and Workability", *Residual Stresses in Science and Technology*, Vol. 2, pp. 847-852, Informationsgesellschaft.Verlag, DGFM (1986).

Polyakov 1980 - Polyakov, M.G., Boyarinav, V.N. and Nizhik, P.P., "Internal Stresses in Cold-Rolled Sheet and Strip in Slitting to Narrow Strip", *Izv. VUZ Ch. Met.*, (6), pp. 6365 (1980).

Ruud 1983 - Ruud, C.O., "Position Sensitive Detector Improves X-Ray Powder Diffraction", *Ind. Res. & Dev.*, pp. 84-87, Jan. (1979).

Ruud 1979 - Ruud, C.O., "X-Ray Analysis and Advances in Portable Field Instrumentation", *J. of M.*, Vol. 31, No. 6, pp. 10-15, June (1979).

SAE 1971 - SAE, "Residual Stress Measurement by X-Ray Diffraction - SAE J784a", *Soc. of Auto. Eng.* (1971).

COMPARISON OF RESIDUAL STRESS AND HARDNESS IN A SYMMETRIC AND AN ECCENTRIC SWAGE AUTOFRETTAGED CYLINDER

S.L. Lee[1], L. Britt[2] and G. Capsimalis[1]

[1]US Army Armament Research, Development and Engineering Center
Benet Laboratories, Watervliet, NY 12189-4050
[2]Department of Physics, State University of New York at Albany
Albany, NY 12203

ABSTRACT

Controlling the induced residual stress distribution is important in manufacture quality control of pressure vessel systems. In this paper, comparison of x-ray diffraction swage autofrettage residual stress and hardness in a symmetric and an eccentric cylinder was made: (1) The radial distributions of residual stress in the two cylinders were similar in magnitudes and characteristics. However, non-axisymmetric yielding was observed in both the symmetric and the eccentric cylinders; (2) X-ray diffraction peak width analysis disclosed large crystal lattice microstrains near the plastically deformed bore compared to small microstrains near the elastic outer diameter, with no obvious angular trend; (3) Rockwell C hardness results indicated that the plastically deformed bore to be harder than the elastic outer diameter region. Non-uniform angular distribution of hardness was observed. Correlation between residual stress and hardness was discussed; (4) In the symmetric cylinder, reduced compressive residual stresses and lower hardness were observed near the 270 degree angular direction, while angular distribution of microstress was uniform. These results suggest that non-uniform material properties such as yield stress are as important as eccentricity in causing the non-axisymmetric plastic deformation in the swaged cylinders.

INTRODUCTION

Induced compressive residual stresses are known to increase the elastic operation range and the fatigue life of engineering components. In the manufacturing of high pressure vessel systems from ASTM A723 steel alloy forgings, engineering processes include heat hardening through austenizing and quenching processes, tempering to improve toughness, rough and fine machining, swage autofrettage, three-point bend beam straightening and thermal treatment procedures. In this work, swage autofrettage process was applied to attain 74% overstrain in the cylinders. The advantageous compressive residual stress field in the bore region (ID) reaches 75% of the yield stress, and the tensile stress field near the outer diameter (OD) reaches 20% of yield stress.

Davidson and Kendall reported earlier residual stress investigations in swaged cylinders using Sack's boring out method[1]. Clark observed some variations in his X-ray hoop residual stress measurements in cylinder forging[2]. Parker studied the fatigue crack growth and safe design of pressurized cylinders[3]. Our x-ray diffraction analysis and finite element modelling results in an eccentric swaged cylinder suggested that although non-uniform angular stress distribution was experimentally observed, the theoretical model did not support that non-uniform yielding to be caused by the given degree of eccentricity[4]. However, while the finite element model correctly predicted the main characteristics of experimental residual stress radial distribution curve, the model overpredicted the magnitude of bore stress by a factor of two. Our more recent theoretical model calculations[5] of residual stress distribution were in excellent agreement with experimental measurements. The interactive iterative spread sheet model, based on the classical elastic-plastic deformation solution to the problem of internally pressurized cylinder, was further used to estimate the Bauschinger factor in A723 steel at 74% overstrain.

Controlling residual stress distribution is important in manufacture quality control and in the safe operation of the components. In this paper, comparison of residual stress and hardness in the symmetric and the eccentric cylinders was made. The radial distribution of induced residual stress in the two cylinders were of similar magnitudes and characteristics. However, angular stress distribution measurements around the bore revealed that non-axisymmetric deformation occurred during yielding in both cylinders. X-ray diffraction peak width analysis disclosed broader peaks in the plastically deformed bore region (ID) compared to the elastic outer diameter region (OD) in both cylinders. Peak broadening was attributed to larger crystalline lattice plastic micro strains near the bore. To investigate material property variations in the cylinders, the radial and angular distribution measurements of Rockwell C hardness were performed. The plastic bore region with compressive residual stresses was found to be harder than the outer diameter region with tensile residual stresses. Effect of residual stress on hardness was discussed and correlation between hardness and residual stress was made. In the symmetric cylinder, reduced compressive residual stress and lower hardness were observed near 270 degree, while microstrain distribution was uniform. These results suggest that non-uniform material properties such as yield stress are as important as eccentricity in causing the non-axisymmetric plastic deformation in the swaged cylinders.

GEOMETRY OF THE SYMMETRIC AND THE ECCENTRIC CYLINDERS

In Figure 1, the cross section geometry of the symmetric and the eccentric cylinders is shown. The ratio of outside diameter to inside diameter OD/ID is equal to 2.75. In the autofrettage process, the overstrain attained 74 percent of the wall thickness. The maximum wall thickness variation between 0 degree and 180 degree of the eccentric cylinder was 0.1 inch. The choice of 0 degree for a symmetric cylinder was arbitrary. For the companion eccentric cylinder, 0 degree represented the thickest, and 180 degree the thinnest region of the cylinder. Slices of the tube were cut from the cylinders, machine and hand polished, then electropolished to remove surface stresses caused by machining before X-ray diffraction stress measurements and Rockwell C hardness measurements were performed.

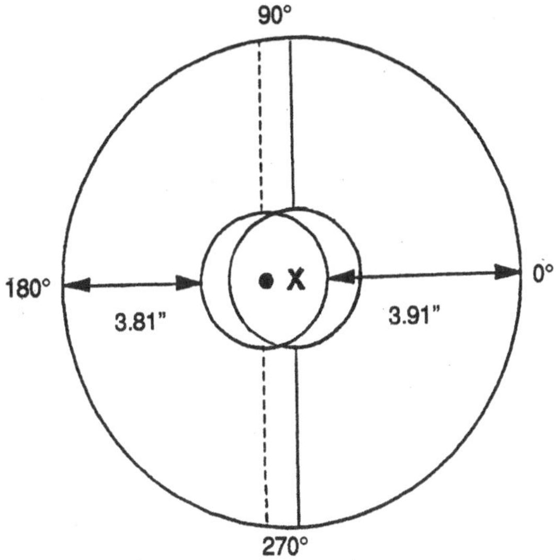

Figure 1. Geometry of the symmetric and the eccentric cylinders.

RESIDUAL STRESS RADIAL DISTRIBUTION ANALYSIS

Residual stress determination was made based on the X-ray diffraction sin-square-psi method. Both a position-sensitive single-exposure scintillation X-ray stress analyzer from Denver X-Ray Instruments and a multiple-exposure positive-sensitive proportional counter stress analyzer system from Technology for Energy Corp were used for this analysis. K-alpha from a chromium x-ray tube reflecting from the 211 planes of the BCC steel at two-theta equal to 156.41 degree was used for residual stress measurements. Accuracies for residual stress measurements were in the order of +/- 5 ksi. When measured d-spacings for planes parallel to the surface are larger than the unstressed state, compressive stress state results; when d-spacings are smaller than the unstressed state, tensile stress state results.

In Figure 2, the X-ray diffraction hoop residual stress distributions along 0, 90, 180 and 270 degree radii of the symmetric swaged cylinder are given. In Figure 3, hoop residual stress distributions along the same directions for the eccentric swaged cylinder are given. Experimental X-ray diffraction results for both cylinders exhibited compressive residual stresses near the bore, the stresses gradually changed to tensile stresses near the outside diameter. Nearest to the bore, reduced compressive stresses were observed in all measurements. Large stress deviations were observed at 180 degree in the eccentric cylinder, and around 270 degree of the symmetric cylinder. The solid curves represent the average stresses from 0, 90, 180 and 270 degree stress measurements in both cylinders. The magnitudes and characteristics of average stresses in the two cylinders were nearly identical.

The classical interpretation of experimental hoop residual stress distribution in a cylinder under internal pressure is given as the following: The bore is plastically enlarged, and the outer portion of the steel cylinder is elastically enlarged. As the internal pressure is released, the outer portion contracts elastically to its original dimension, steel nearer the bore resists this action.

This results in compressive hoop stresses near the bore and tensile stresses near the outside diameter. Theoretical spread sheet models were implemented based on the classical elastic-plastic deformation theory of a symmetric cylinder under internal pressure, with and without the effect of reverse yielding[4,5]. Excellent agreement was reached between our experimental and theoretical calculations for both thin and thick cylinders.

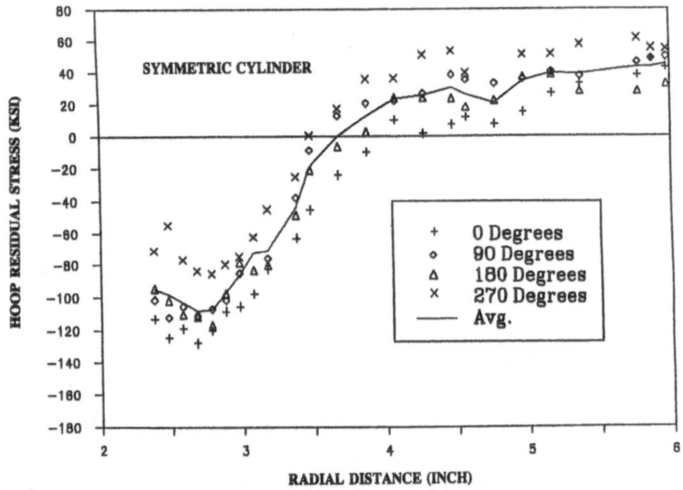

Figure 2. Hoop residual stresses in the symmetric swage autofrettaged cylinder as a function of radial distance.

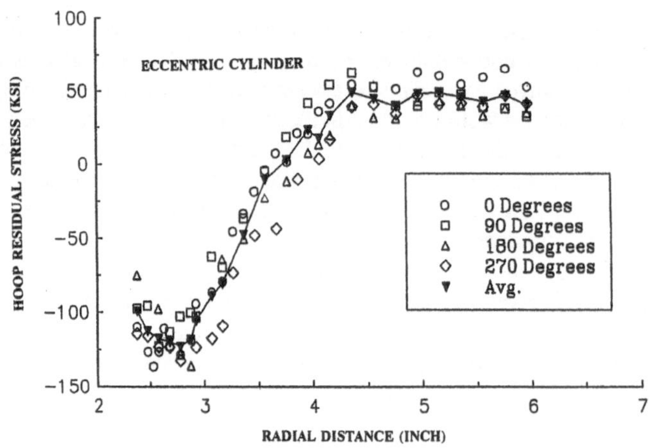

Figure 3. Hoop residual stresses in the eccentric swage autofrettaged cylinder as a function of radial distance.

RESIDUAL STRESS ANGULAR DISTRIBUTION ANALYSIS

In Figure 4, angular residual stress distributions at ID and OD for the symmetric and the eccentric cylinder are compared. The eccentric tube exhibits a minimum of compressive stress near 180 degree, where 0 degree represents the thickest and 180 degree the thinnest region of the cylinder. However, the symmetric cylinder also exhibits a peak of reduced compressive stress near 270 degree. Our finite element modelling results of the eccentric cylinder indicated the 180 degree residual stress peak could not be accounted for by eccentricity. The peak at 270 degree in the symmetric cylinder further indicated that other factors as well as eccentricity were causing the non-uniform plastic deformation in a cylinder of uniform geometry. Possible causes for the angular stress deviations include eccentricity, variations in material properties due to heat treatment procedure, martensite and bainite contents, three-point bend beam straightening process, tribological problems caused by friction and wear, non-axisymmetric swage process, or a combination of these factors.

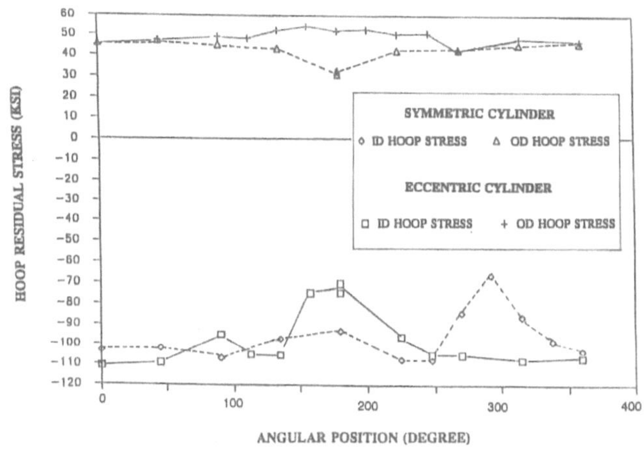

Figure 4. Comparison of residual stress distributions at ID and OD in the symmetric and the eccentric cylinders as a function of angular position around the bore.

X-RAY DIFFRACTION PEAK WIDTH ANALYSIS

Microstrains can exist in crystal lattices as a manifestation of plastic deformation and lattice defects. Plastic deformation in polycrystalline materials produces both peak broadening and peak shift in the diffraction pattern. Macroscopic diffraction peak shift measurement is the basis for X-ray diffraction residual stress analysis. Peak broadening can be produced by a reduction of crystalline size, by faults in the crystalline planes, and by microstrains within the coherently diffracting domains. However, instrument parameters such as beam divergence and width of X-ray source can also cause diffraction peak broadening. Weiss and James discussed effects of macrostrains and microstrains in fatigue and failure analysis[6].

In Figures 5 and 6, comparison of angular diffraction peak widths for the symmetric and the eccentric cylinders as a function of angular position around the bore is made. Diffraction peak full width at half maximum (fwhm) was obtained by averaging values at 0°, 15°, 30° and 45° psi angles. Both curves show fwhm to be approximately 0.15 degree broader at ID compared to OD. Experimental conditions during data acquisition for the symmetric cylinder and the eccentric cylinder were respectively kept constant. Particle size distribution is expected to be fairly uniform at ID and OD of the cylinder. We attribute the difference in X-ray diffraction peak broadening to be due to differences in lattice microstrains. Furthermore, hoop and radial stress

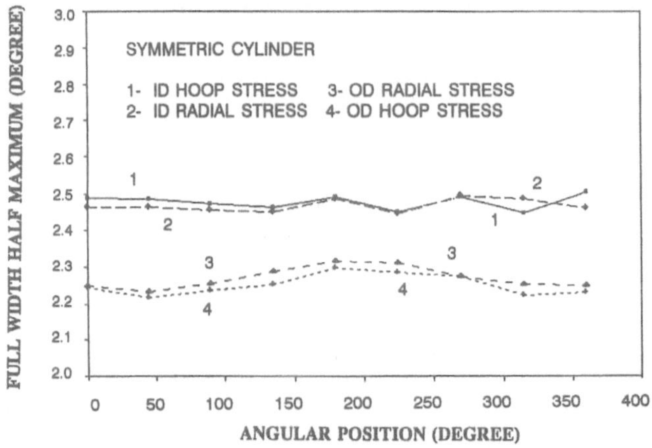

Figure 5. X-ray diffraction peak widths at ID and OD in the symmetric cylinder as a function of angular position around the bore. The top two curves are hoop and radial diffraction peak widths at ID, the bottom two curves are hoop and radial peak widths at OD.

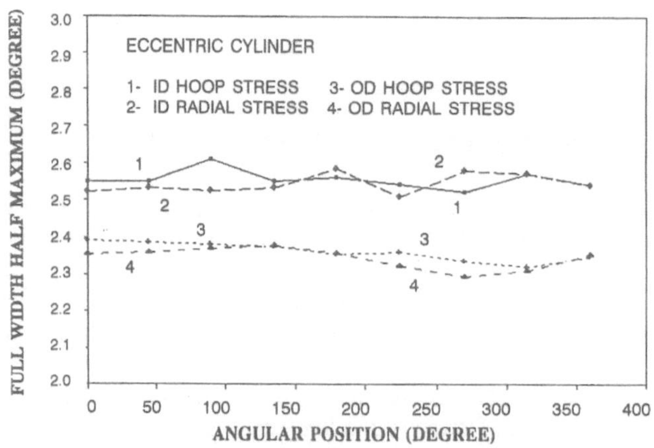

Figure 6. X-ray diffraction peak widths at ID and OD in the eccentric cylinder as a function of angular position around the bore. The top two curves are hoop and radial diffraction peak widths at ID, the bottom two curves are hoop and radial peak widths at OD.

measurements gave almost identical diffraction peak widths. This indicates that unlike macroscopic residual stresses which are dependent on directions, microstrains are independent of directions. Microstress can be large even if residual stress is close to zero. The angular microstrain distribution data were fairly uniform. These results are consistent with a displacement controlled loading system such as swage autofrettage. Slightly different fwhm's between the eccentric and the symmetric cylinders were observed. Since data for the symmetric and the eccentric cylinders were acquired at different times with different instrumentation settings, this difference was attributed to instrumentation factors.

ROCKWELL C HARDNESS RADIAL AND ANGULAR DISTRIBUTION ANALYSIS

In the Rockwell C hardness measurement, a diamond indenter was pushed into the surface with a 10 Kg initial load and subsequent 140 Kg load, the indentation depth was measured and hardness calculated. In Figure 7, Rockwell C hardness measurements at ID and OD as a function of angular position for the symmetric cylinder are displayed. The data showed ID had a Rockwell C average hardness of approximately 39, and OD had an average hardness of approximately 35. Each hardness value was obtained from the average of four measurements at points 1/8" from the point of interest, with the standard deviation given in the figure. Hardness measurements in general suffer large deviations in the order of +/-0.5 to +/-1.5 RC.

In Figure 8, radial distributions of Rockwell C hardness in the 0, 90, 180 and 270 degree directions of the symmetric cylinder are displayed. The plastic deformed region near the bore was found to be harder than the elastic outside diameter region. The solid curve is an average hardness measurements in the 0, 90, 180 degree directions. The 270 degree hardness measurements were low, which further demonstrated the large variations in material properties in the cylinders. The low hardness measurements are in agreement with the lower compressive residual stress measurements in the 270 degree direction. Towards the outer diameter region, the hardness curve flattens out, in similar trend as the residual stress curve.

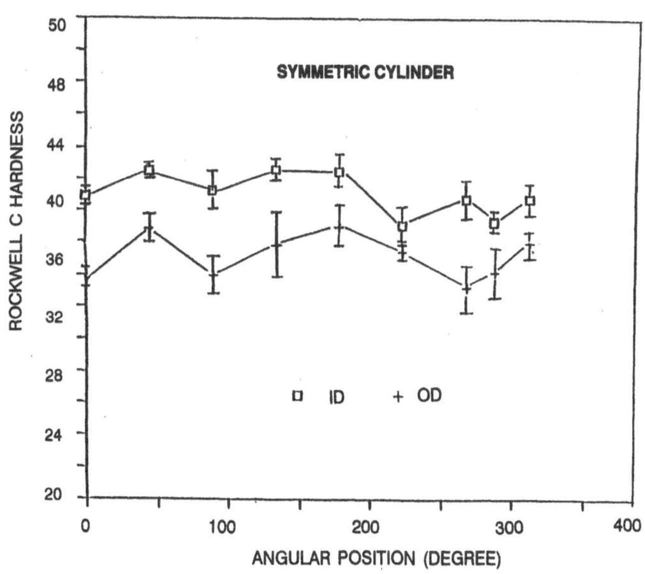

Figure 7. Rockwell C hardness distribution measurements at ID and OD in a symmetric cylinder as a function of angular position around the bore.

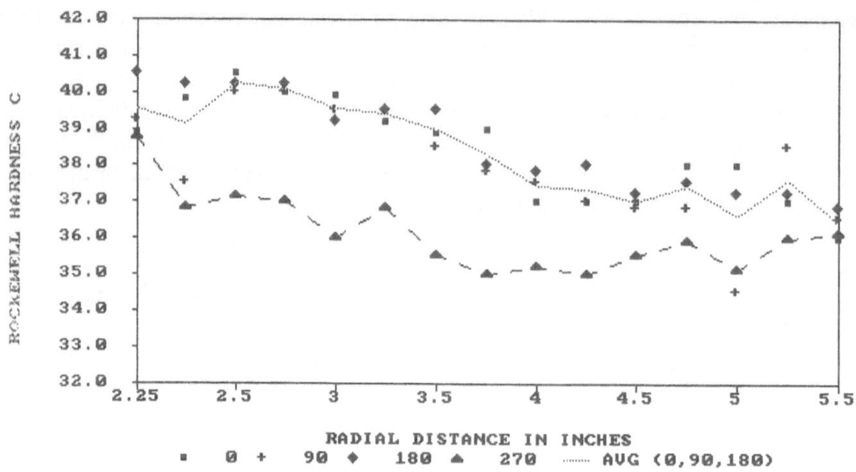

Figure 8. Rockwell C hardness radial distribution measurements in a symmetric cylinder showing the averages of 0, 90, 180 degree data, and 270 degree data.

HARDNESS AND RESIDUAL STRESS

Residual stress is caused by non-uniform plastic flow, while hardness is a measure of resistance to plastic deformation. What's the relation between residual stress and hardness? Can the easily performed hardness measurement be used in place of the more difficult residual stress analysis to assess the degree of plasticity? In Fig. 9, hardness is plotted versus residual stress for the symmetric cylinder. Experimental data have been fitted to a third order polynomial to show the relationship between residual stress and hardness. The curve has a negative slope in the region of plastic deformation, it flattens out towards the elastic region.

Shaw and de Salvo[7,8] and Al-Hassani[9], studied both static and dynamic indentation problems. Their studies improved our understanding of static "hardness measurement" and dynamic "shot peening" process. Interpretation of yielding in ductile material due to spherical indenters by Von Mises and Tresca yield criteria has also been made by others [10].

Figure 10 illustrates hardness measurement in pre-stressed material by displaying a typical ball indentation problem. HH and HR represent hoop and radial residual stresses due to swage autofrettag process, and HI represents indentation residual stress. When load p is applied, the material undergoes local plastic deformation with lateral stretching and grain distortion. The rest of the elastic material tends to push the plastically deformed zone, thus causing compressive stresses. Consider point A in the plastic zone, elastic stress fields S1 and S2 develop when p is applied (10a). Tresca criterion predicts no plastic flow when (S1 - S2) < Y where Y is the yield stress. When increasing p causes this difference to approach Y, plastic flow occurs, compressive indentation residual stresses HI develop in the two transverse directions (10b). Flow continues until the Tresca yield criterion is satisfied. When the load is removed, S1 vanishes (10c). In the case of overstrained cylinders, negative compressive hoop residual stress due to autofrettage at ID compensates the tensile strains. Higher pressure is needed to satisfy Tresca yield criterion, thus the higher material hardness. At OD, positive tensile residual stresses due to autofrettage

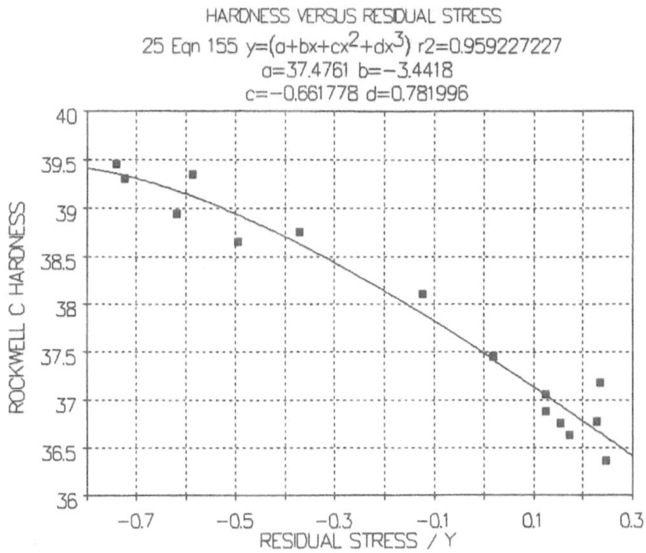

Figure 9. Hardness dependence on residual stress in a symmetric cylinder.

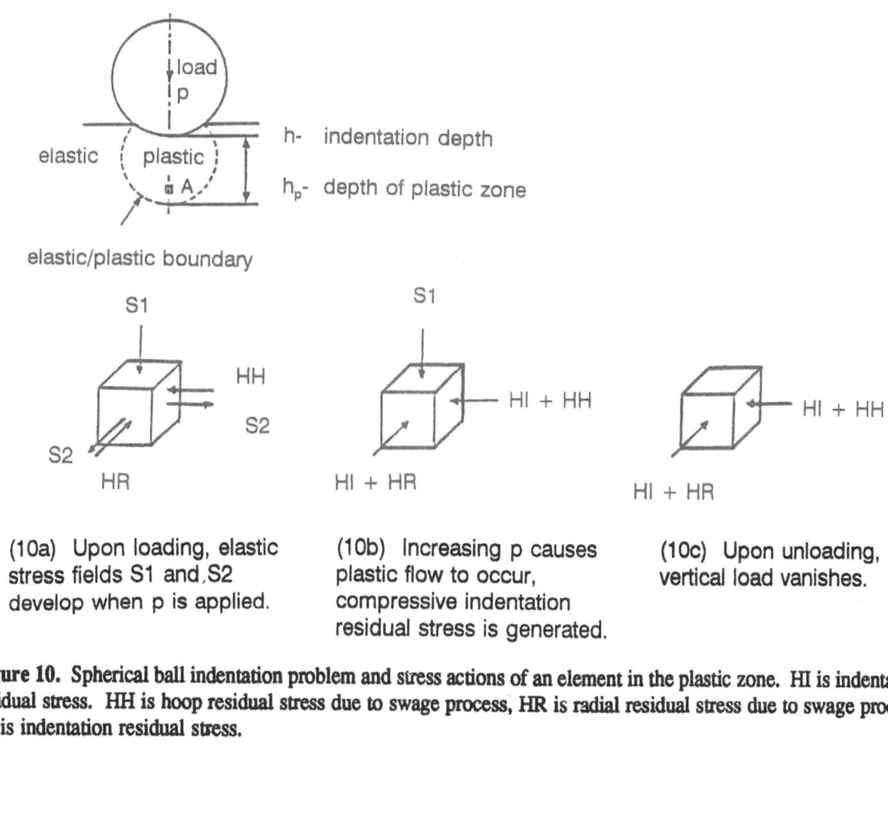

Figure 10. Spherical ball indentation problem and stress actions of an element in the plastic zone. HI is indentation residual stress. HH is hoop residual stress due to swage process, HR is radial residual stress due to swage process, HI is indentation residual stress.

process make the yield criterion easier to satisfy, thus the lower material hardness. Analytic two- or three- dimensional deformation problem based on material stress/strain properties must be considered to understand residual stress, work hardening and plasticity.

CONCLUSION

Residual stress analysis provides guidance in pressure vessel systems design and information for manufacture quality assurance. Controlled stress distribution improves performance and fatigue life and ensures the safety operation of engineering components. In this work, residual stress distribution and hardness measurements in the symmetric and the eccentric cylinder were characterized. The data showed advantageous residual stress distributions were induced in both cylinders. Compressive residual stress of 75% of the yield stress was found near ID, and tensile stress of 20% of yield stress was found near OD. The plastically deformed bore was found to be harder than outer diameter. Correlation between residual stress and hardness was made. The reduced residual stress and hardness in the 270 degree direction of the symmetric cylinder, but uniform microstrain distribution further revealed non-uniform material properties in the cylinder. Material properties, such as yield strength, as well as eccentricity, can cause non-axisymmetric yielding in both the symmetric and the eccentric cylinders.

ACKNOWLEDGEMENTS

The authors wish to thank John Underwood for the helpful discussions, Chris Rickard and Dan Snoha for assistance in part of the hardness and residual stress measurements.

REFERENCES

1. Davidson, T.E. and Kendall, D.P., "Residual stresses in thick-walled cylinders resulting from mechanically induced overstrain", Experimental Mechanics, pp. 253-262 (1963).
2. Clark, G., "Residual stresses in swage-autofrettaged thick-walled cylinders", Materials Research Laboratories Report MRL-R-847 (1982).
3. Parker, A.P., Sleeper, K.A. and Andrasic, C.P., "Safe life design of gun tubes- some numerical methods and results", Proceedings of the Army Numerical Analysis and Computer Conference, pp. 311-333 (1981).
4. Lee, S.L., O'Hara, P., Olmstead V.and Capsimalis G., "Characterization of residual stresses in an eccentric swage autofrettaged thick-walled cylinder", Practical Applications of Residual Stress Technology, Proceedings of the ASM International Conference, Indianapolis, In, pp. 123-129 (1991).
5. Lee, S.L., "Residual stress analysis in swage autofrettaged thick-walled cylinders by position sensitive x-ray diffraction techniques", ASME Pressure Vessels and Piping Conference, High Pressure- Codes, Analysis and Applications, pvp- vol. 263, pp. 165-169 (1993).
6. Weiss, V. and James, M.R., "Residual stresses- fatigue and fracture", Residual Stresses in Science and Technology", vol. 1, ed. by E. Macherauch and V. Hauk, Informationsgesellschaft, Verlag (1987).
7. Shaw, M.C. and de Salvo, G. "On the plastic flow beneath a blunt axisymmetric indenter", Trans. ASME J. Eng. Industry 92, pp. 480-493 (1970).
8. Shaw, M.C. and de Salvo, G. "A new approach to plasticity and its application to blunt two dimensional indenters", Trans. ASME J. Eng. Industry 92, pp. 469-479 (1970).
9. Al Hassni, S.T.S., "Mechanical aspects of residual stress development in shot peening", lst International Conference on Shot Peening, pp. 583-602 (1981).
10. Sines, T.R., Mellor S.G. and Hills, D.A., "A note on the influence of residual stress on measured hardness", Journal of Strain Analysis, vol. 19, no. 2, pp. 135-137 (1984).

NON-DESTRUCTIVE CHARCTERIZATION OF MATERIALS WITH NEUTRON EXPERIMENTS AT THE PULSED REACTOR IBR-2

Frank Häußler,[1] Hartmut Baumbach,[2] and Michael Kröning[2]

[1]Fraunhofer Institut fuer zerstoerungsfreie Pruefverfahren
(EADQ) Kruegerstr. 22, D - 01326 Dresden, Germany
present address: Joint Institute for Nuclear Research
Laboratory of Neutron Physics
141980 Dubna, Moscow Region, Russia
[2]Fraunhofer Institut fuer zerstoerungsfreie Pruefverfahren
Universitaet Gebaeude 37, D - 66123 Saarbruecken, Germany

NON-DESTRUCTIVE CHARACTERIZATIONS OF MATERIALS WITH NEUTRONS

Parameters for the Characterization of Materials

Materials characterization is one field of interest within the non-destructive testing. In the meantime another basic technology of development of so-called custom-made materials is emerging. These custom-made materials are individually designed and obtain new properties for use. Hence the search for the relevant material parameters characterizing the materials sufficently exists. These parameters shall describe both the properties of the material and possible functions. There are several classifications of material parameters depending on the special aims of evaluation and testing. Some of these parameters are measurable without any invasive procedures.

The neutron methods provide a powerful tool to the non-destructive evaluation of several material parameters inside the volume. The fields of application are structural investigation, analytics, simulation of neutron effects etc. The methods using thermal neutrons as probe do not require any special sample preparations affecting or destructing the inner structure of the sample under investigation. This fact proves an essential advantage of measurements using thermal neutrons. Therefore a variety of neutron scattering methods is applied to determine structural or energetic parameters of solid, liquid, and gaseous samples. One of these methods is the small-angle neutron scattering (SANS) technique. Another application is the neutron diffraction technique to determine residual stress fields by strain measurements.

Nondestructive Characterization of Materials VI
Edited by R.E. Green, Jr. *et al.*, Plenum Press, New York, 1994

Thermal neutrons behave like a weak probe. The interaction cross sections of thermal neutrons between the most of nuclei existing in a large volume concentration in typical samples under investigation is small in comparison to the cross sections of X-rays and charged particles. Hence the sample volume is large enough for interpreting the results as representative for the volume behaviour or macroscopic properties. The results of neutron scattering experiments have to be considered as an averaging over the sample volume studied. Using special arrangements of diaphragms the volume to be investigated can be situated inside the whole sample. It means various properties inside a specimen are principally measurable without any destructive preparation processes. For instance, strain measurements for the calculation of residual stress are available from inner volume parts of the specimen by choosing the gauge volume.

Thermal neutrons have a wavelength in the order of typical lattice spacing which causes interference effects (diffraction patterns). Otherwise the energy range of these thermal neutrons is in the order of typical lattice vibrations in solids. That's why thermal neutrons are an excellent probe for inelastic scattering experiments.

NEUTRON SCATTERING EXPERIMENTS AT THE PULSED REACTOR IBR-2

The IBR-2 reactor is a pulsed reactor of periodic operation. An important factor, the power of 1500 MW and a neutron flux of 10^{16} cm^{-2} sec^{-1} during a pulse is reachable by a mean power of 2 MW. The reactor operates with a pulse repetition rate of 5 Hz which is preferable in condensed matter physics. Here the reactivity modulation is accomplished with the help of the movable reflector. The movable reflector consists of a main reflector and an auxiliary reflector. These two reflectors are rotated at different velocities near the core. The power pulse is the result of their simultaneous approaching of the reactor core. [1]

Small-angle Neutron Scattering (SANS)

The spectrometer MURN-TEXT is located on beam 4 of the pulsed reactor IBR-2 of the Frank Laboratory of Neutron Physics of the Joint Institute for Nuclear Research, Dubna. By using the TOF-method with its wide variety of incident wavelengths (from 0.07 nm to 1.5 nm) a momentum transfer from 0.07 nm^{-1} to 7 nm^{-1} is detectable. Therefore, inhomogeneities on the length scale from 1 nm to 30 nm can be measured. At the sample position a neutron flux is from $6 \cdot 10^6$ cm^{-2} sec^{-1} to $3.7 \cdot 10^7$ cm^{-2} sec^{-1}.[2] A detailed description of the SANS-spectrometer MURN-TEXT is given.[3] By means of a vanadium scatterer placed into a direct neutron beam the macroscopic cross section $d\Sigma/d\Omega$ (Q) can be measured. It contains all scattering information about the sample.

The errors of measurement of the macroscopic cross section are smaller than 5 percent. Further application of calculation methods (e. g. least square method) brings an additional error. Hence, the values of the specific surface (via the Porod's constant) and the exponent (potential law) were obtained with an accuracy of about 10 percent and 5 percent, respectively.

Evaluation of Microstructural Parameters by SANS

By means of SANS experiments a non-destructive description of statistically representative microstructures (nm- and μm-scale) in technologically important disordered materials is given. Up to now there are some difficulties to interpret fully the strong scattering of a lot of important materials, i.e. cements, clays etc. because their microstructural disorder extends over many length scales.[4] SANS gives average information for an ensemble of scattering objects. The inversion of the scattering pattern to real space cannot yield specific data of individual scattering centres like a direct imaging method.[5] Certain models for an interpretation of experimental results are also needed.

The following formula for the macroscopic differential cross section $d\Sigma/d\Omega(Q)$ models the scattering signal:

$$\frac{d\Sigma}{d\Omega}(Q) = 1/V_g \cdot \Sigma\, u_i \int W(a_i)\, V_{pi}^{\,2}\, |\, f(Q,a_i)\, |^2\, d\, a_i \qquad (1)$$

$f(Q,a_i)$ characterizes the particle form factor and Q the scattering vector. The variable a_i is the characteristic linear size of the scattering objects and $W(a_i)$ represents the distribution function of a_i. V_{pi} describes the volume of the scattering particles with a SANS contrast u_i given by the absolute square of the difference of the scattering length densities (scattering object and surrounding medium). The scattering length density is the sum over the coherent scattering lengths of all scattering centres (atoms) in a given volume divided by this volume. This volume is determined by the lower limit of solvable particle sizes and amounts to about 0.1 nm^3. The sum in formula (1) is taken over all s types of scattering objects with different phases and geometries.

Two regions of the scattering curve are discussed. In the following they are used for the analysis of the scattering data. At low scattering angles the Guinier approximation holds.[6]

$$\frac{d\Sigma}{d\Omega}(Q) \propto \text{volume concentration} \cdot \text{contrast} \cdot V(a_i) \cdot \exp\left[-Q^2 \cdot R_g^2 / 3\right] \qquad (2)$$

and

$$Q = 4\,\pi / \lambda \cdot \sin(\beta/2) \qquad (3)$$

Thereby β is the scattering angle and λ denotes the neutron wavelength. R_g is the radius of gyration of a scattering particle of type (i) with a volume $V(a_i)$. The radius of gyration is the mean square distance of all points within the particle seen from the particle centre.

At large scattering angles (large Q) the Porod approximation holds.[7] From the scattering curves represented in the Porod plot ($[d\Sigma/d\Omega \cdot Q^4)]$ versus Q) the Porod constant K_P is directly obtainable.

Study of Microstructural Changes in Hardening Cement Paste

A Portland cement paste contains many crystalline and non-crystalline phases in various ranges of sizes. It is possible to divide the hydrating cement grain into several shells for the modeling.[8] The shells represent the inner and outer hydration products. The crystalline phases (e.g. Portlandite, Calcite) are embedded in the amorphous phases of hydration products. The kernel of the hydrating grain consists of unreacted Portland cement.

In the last years several publications about SANS measurements of hardening cement pastes, ceramics, and other porous materials have appeared.[9,10,11] In the ILL Grenoble time-resolved phenomena in cements and other porous materials were studied by SANS. The use of neutrons provides not only the advantage of the heavy-light-water contrast variation to probe accessibility of different parts of the inner microstructure and surface area. In addition the penetrating power of thermal neutrons allows macroscopic, i.e. few mm sample thicknesses to be probed. From this point of view SANS is uniquely suited to nondestructively studying the statistically random disordered microstructures of these porous materials.[6] It is shown that this method is applicable to various investigations of cement samples, for instance, of their hydration and aging. Kriechbaum et al.[12] have published the results of small-angle X-ray scattering. The fractal nature is one of the most important standpoints in this paper.

By means of the fit program FUMILI the the scattering curves $d\Sigma/d\Omega$ were fitted:

$$\frac{d\Sigma}{d\Omega}(Q) = A(1) \cdot Q^{-A(2)} + A(3) \qquad (4)$$

The results are shown in Fig.1 and Fig. 2. Various considerations of the meaning of the A(1) value in the case, when the exponent A(2) deviates from 4, have shown that the calculated A(1) value has to be considered as the Porod constant Kp with caution. For rough estimate and further discussion these values of Kp may be useful.

The samples of a hardened cement paste are prepared from ordinary Portland cement and water (D2O) with a given water to cement ratio of 0.38. In the experiments on observation of the hydration progress within the first 321 days the cement paste was put in a sandwich-like container formed of a plastic hollow ring (inner diameter of 20 mm) and two thin 30 mm thickness plastic foils on the upper and bottom face of this ring to reduce the vapour (water) exchange. Changes in the mass of a sample due to evaporation of water were registered. The influence of the foil on the SANS-signal of the used samples is negligible. The thicknesses of these plastic rings were 0.50 mm, 1.05 mm, and 1.70 mm. At the sample position the neutron beam cross section is about 1.6 cm^2.

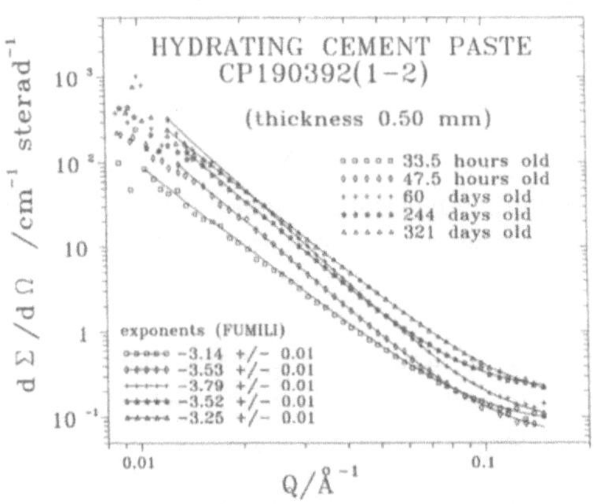

Figure 1. The evolution of hardening cement paste.

In Figure 1 the potential law of the scattering curves varies. In the measured Q-region the hardening cement paste does not exhibit a Porod-like behaviour. The power law of $\propto Q^{-A(2)}$ is observed for Q < 1 nm. The exponents [- A(2)] lie in an interval from about -2 to -4. This is believed to be associated with surface fractal behaviour. This might arise from the deposition of a C-S-H gel on the cement clinker grain/pore boundary and the roughening of water intrudes to form the inner product.[4] The area of the rough surface depends on the scale of observation.[13] Here the formulated by T. Witten[14] analog of the Porod's law is given. It reflects the same potential behaviour of Q, but the boundary area between two different neutron optical phases (surface area) depends on the scattering vector, i.e. on the scale of observation by the SANS experiment. In the case of self-similar surface structures (surface fractals) the exponent of Q is between -3 and -4.

Figure 2 gives a survey of the evolution of a typical parameter of a SANS curve. The exponents A(2) of FUMILI calculations are varying in dependence on the hydration time. In addition the thickness of the hydrating cement paste samples proves to be a significant parameter for studying the cement hydration in real-time experiments. The results of the time-dependent behaviour of the differential macroscopic cross section at Q=0 [$d\Sigma/d\Omega(0)$] reveals significant maxima in all evolution curves of $d\Sigma/d\Omega(0)$ after 92 days. Comparing the evolution of the factor A(2), after the hydration time of 92 days all factors describing the potential behaviour are decreased. There is not any direct correlation between A(2) and $d\Sigma/d\Omega(0)$. Hence these experimental results might be caused by changes of the microstructure during the hydration process.

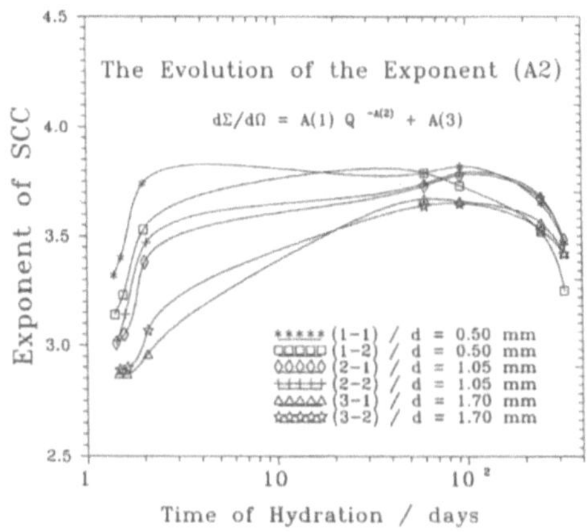

Figure 2. Microstructural changes in dependence on the time after the onset of hydration.

Latent Ion Tracks in Solid State Nuclear Track Detectors

In the last twenty years a wide variety of methods investigating the nuclear track formation in solid state materials was developed. At present there are several models describing the fundamental interaction process and the mechanism of damaging after irradiation by heavy ions. Nevertheless the effect of damaging the structure within the irradiated solid state specimen is applicated in various fields.

Passing the solid state nuclear track detector (SSNTD) charged particles (protons and havier) form latent tracks. The parameters of these latent tracks vary with the atomic number and the energy of the incident particles. Therefore SSNTD represent passive detectors for ionizing radiation. SSNTD are used for the registration and identification of charged particles. On the other hand with the help of special etching procedures the design and the production of nuclear filters with defined porous radii in the μm-level is achievable. The knowledge of the structure of the latent ion tracks after irradiation is important for the description of the etching procedure. Furthermore in the beginning of the etching process the structural changes within the latent track environment influences the modeling of the structure development during the bulk etching.

It is nearly impossible to observe and to analyze latent tracks. First experiments by SANS on the MURN facility are realized. Essentially the study of the influence of partial annealing and of other environmental conditions on the storage of irradiated detectors over a long period requires the analysis of the latent tracks.

By using as a model homogeneous cylinders parallel to the incident beam (radius R, length 2L parallel to the track axis and perpendicular to the scattering vector $\underline{Q}$) with a radial step scattering-length density distribution [15] the macroscopic differential cross section $d\Sigma/d\Omega$ (Q) is given by

$$\frac{d\Sigma}{d\Omega}(Q) = \text{volume concentration} \cdot \text{contrast} \cdot V(R) \cdot \exp\left[-Q^2 \cdot R^2 / 4\right] \quad (5)$$

Semenyuk, Svergun et al. [16] have investigated different types of nuclear filters by small angle X-ray scattering. Both latent and etched tracks in PETP are studied. In the case of long cylinders parallel to the primary beam formula (4) agrees with their meodelling. Here the radius of gyration is given by

$$R^2_{g,cyl} = R^2 / 2 \quad (6)$$

In the paper of Semenyuk et al. the radius of gyration is estimated with an amount of about 9 nm. The Guinier-plot (logarithmic scale of $d\Sigma/d\Omega$ (Q) vs. Q^2) yields the slope of the scattering curve $d\Sigma/d\Omega$ (Q) near Q=0. Through the application of the model (see formula (5)) a radius of the cylindrical shaped tracks is estimable.

All SSNTD samples of type PETP were put into a sample holder which consists of circular hollows. The scattering signal of one irradiated foil (the thickness is of about 10 µm) is to low for the analysis of the SANS data. Therefore a stack of several quadratic shaped PETP foils (total thickness of about 0.2 mm) is used for the measurements. At sample position the neutron beam cross section is of about 1.5 cm^2. The PETP samples irradiated by Xe-192 (energy 0.96 MeV per nucleus) were investigated. For studying both the irradiated latent tracks and the etched tracks, a low track density of about $3 \cdot 10^8$ cm^{-2} was used.

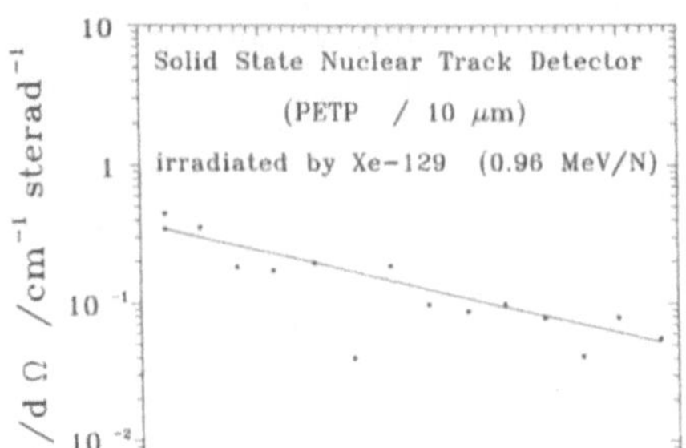

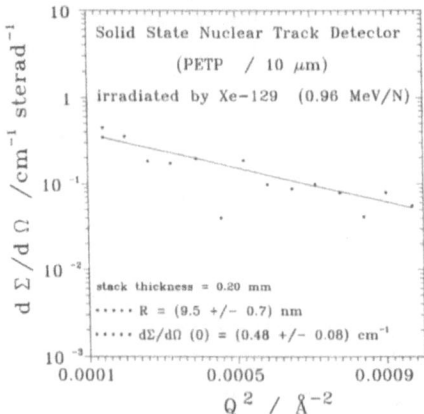

Figure 3. Latent tracks in SSNTD. The slope shows a cylinder radius of about 9.5 nm.

Fig. 3 shows the Guinier plot for the estimate of the average cylinder radius assuming cylindrical shaped latent ion tracks. Irradiated and etched ion tracks are shown in Fig. 4. The slope of the scattering curve using the Guinier plot indicates the size range of the radial extension of the etched track. Without any assumptions about the form of the scattering objects the radii of gyration R_g calculated from the slope of the scattering curve (see formula (2)) correlate with the radii r measured by a gas penetration method. The low track density causes that the statistics of the latent ion track measurements is lower than the measurements of the etched tracks. This is the result of the drastic increase of the volume of the etched tracks.

For the further interpretation of the scattering data via modeling additional data about the latent tracks are necessary. This modelling is a useful help for the quantitation analysis of SANS data. For this analysis, the form of the latent tracks, i.e. the matter density distribution (profile) parallel and perpendicular to the track axis, the exact track density of the irradiated region, i.e. homogeniety of the track distribution, and other parameters are needed.

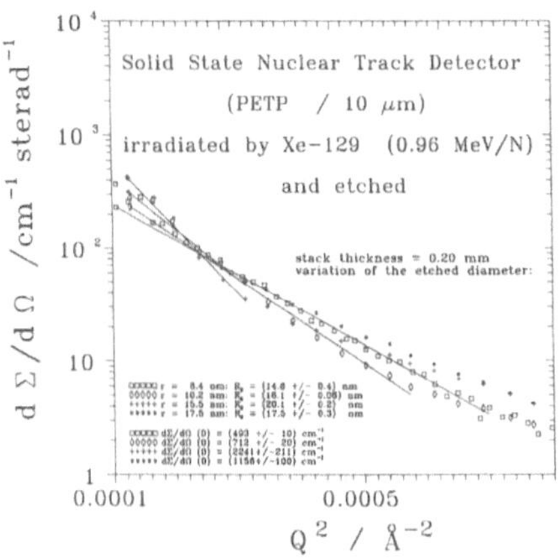

Figure 4. Ion tracks in SSNTD after the first step of etching. A correlation between the radii measured by SANS and gas penetration is shown.

Neutron Diffraction for the Characterization of the Crystalline Phases

Neutrons are a potential probe for studying non-destructively bulk residual stresses. Using the neutron diffraction technique the measured lattice plane spacings indicate the average lattice strain in the volume sampled.[17] T. Holden[18] has outlined the general assumptions for residual stress measurements. In the Bragg equation

$$\lambda = 2\, d_{hkl} \sin(\theta_{hkl}) \qquad (7)$$

the wavelength of the thermal neutrons is λ, the lattice spacing is d_{hkl}, and the Bragg angle for the (hkl) reflection is θ_{hkl}. The investigated volume results from the choice of the horizontal and vertical aperture of the thermal neutrons, absorbing masks positioned in the beams in front of the sample and at the detector head. The average lattice macrostrain in the volume sampled ε_{hkl} is given by

$$\varepsilon_{hkl} = (d_{hkl} - d_{ohkl}) / d_{ohkl} \quad (8)$$

where d_{ohkl} is the lattice spacing of a so called stress free sample. The direction of the scattering vector Q determines the direction of the measured strain ε_{hkl}. Here Hutching[17] makes a distinction between the term macrostress denoting the strain component which is only constant over many grains and the term microstrain which is only constant within the grain. By means of diffraction experiments only the average strain measured inside the volume is given. All grains with the plane normal directed along the scattering vector contribute to the diffraction signal. By using a white pulsed beam (e.g. from the pulsed reacor IBR-2) in combination with the time-of-flight(TOF)-method the strain is now given by the time-channel shift Δt of the diffraction peaks.

$$\varepsilon_{hkl} = \Delta\lambda / \lambda = \Delta t_{hkl} / t_{hkl} \quad (8)$$

Thus the stress states of many lattice planes can be measured, a major advantage of TOF-technique.

In 1992 the new high resolution reverse TOF Fourier diffractometer (HRFD) was put into operation at the IBR-2. For the proposed strain measurements in bulk specimens a neutron strain scanner is being designed. The choice of the collimation system allows a fixing of the minimum measured volume of (2x2x2) mm^3. The resolution of the lattice spacing measurements at d=0.2 nm is about $\Delta d/d = 0.002$. The d-spacing range is from 0.05 nm to 1 nm.[19]

By means of the Hook's law the stress components of the stress tensor are to be calculated from the strain components measured by the diffraction experiment. Holden[18] pointed out the need of a good collimation system, a careful centering of the gauge volume on the sample, and a sufficient calibration of the diffractometer with a standard powder. Si-powder guarantees a texture free sample for calibration. It must be underlined that the texture of the sample under investigation has to be considered because a measurement on diffraction peaks under strong texture effect reveal an incorrect strain.

Because of the capability of strain measurements in volume elements inside the component of engineering machines the use of neutrons for diffraction is very often preferred. By using X-rays the sectioning of components and machining of the specimens must be allowded because of the effective penetration depths in steel for X-ray used is about 10 μm.[20] Hence stress redistributions and relaxations caused by the specimen preparations must be considered.

A typical example of interest is the stress field around welds calculated from strain measurements.[21] Engineering materials, e.g. heat treated steel in various conditions and compositions, ceramics, alloys of aluminium, and metal matrix composites are characterized by strain measurements.[20] The thermal mismatch between metal and ceramics causes high residual stresses at the joints[22]. The determination of the stress gradient inside the specimen which assumes a sufficient good localization of the volume element studied by neutron diffraction poses another interesting problem in the field of materials characterization.

The goal of the previous chapter is to illustrate the advantage of neutron diffraction measurements for the characterization of the crystalline phases. By introducing the strain components measured by neutron diffraction inside the specimen into the calculation the stress, especially the residual stress, may be determined non-destructively.

CONCLUSIONS

All these experimental results give rise to the conclusion that further investigations, especially theoretical ones, are necessary for both the correct interpretation of the experimental curves and the evaluation of structural data (nm-level) on the cement paste.

ACKNOWLEDGMENTS

The authors wish to gratefully acknowledge late Prof. Yu. M. Ostanevich of FLNP JINR for his permanent interest, fruitful discussions and the support for the work until his sudden death in 1992. Furthermore they want to acknowledge Dr. F. Eichhorn (FZ Rossendorf) for the longterm cooperation. They thank Mrs. M. Hempel (IzfP/EADQ Dresden) and Mrs. A. Hempel (FZ Rossendorf) for the help in experimental and computing works. In addition we thank our colleagues from the SANS group, especially Mr. A. I. Kuklin, for their helpful discussions and help in the SANS experiments. The experiments in the field of SSNTD are supported by Dr. W. Birkholz. These experiments were done by cooperation with the colleagues of the JINR Dubna/ Laboratory for Nuclear Research Dr. V. P. Perelygin, Dr. M. Danziger, Dr. P. Yu. Apel, Dr. S. G. Stetsenko, and Mr. A. Schulz.

The work reported has been performed with partial support of the Bundesminister fuer Forschung und Technologie through grant no. 03-DU3FHG. The authors are fully responsible for the contents of this publication.

REFERENCES

1. V.L. Aksenov, Update on the pulsed reactor IBR-2 and its instruments at Dubna, Physica B 174:438(1991).
2. User Guide, Neutron Experimental Facilities at JINR, Dubna, USSR 1991.
3. Yu. M. Ostanevich, Time-of-flight small-angle scattering spectrometers on pulsed neutron sources, Makromol. Chem., Macromol. Symp. 15:91 (1988).
4. A.J. Allen,Time-resolved phenomena in cements,clays and porous rocks. *J. Appl.Cryst.* 24:624 (1991).
5. G. Kostorz, Small-angle scattering and its application to materials science,in: "Treatise on Materials Science and Technology," G. Kostorz, ed., Academic Press, London (1979).
6. A.Guinier and G.Fournet, "Small-Angle Scattering of X-Rays,"John Wiley & Sons Inc., New York (1955).
7. O. Glatter and O. Kratky, "Small-Angle X-ray Scattering," Academic Press, London (1982).
8. S. Roehling, M. Nietner, Anwendung eines strukturorientierten Modells zur Berechnung der Festigkeitsentwicklung von Zementbeton, Betontechnik 9:52 (1988).
9. A. J. Allen et al., Development of the fine porosity and gel structure of hydrating cement systems, *Philosophical Magazine* B 56:263 (1987).
10. F. Haeussler et al., Monitoring of the hydration process of hardning cement pastes by small-angle neutron scattering, Cement and Concrete Research 20:644 (1990).
11. F. Haeussler et al., Application of small angle neutron scattering, in: "Microstructure and Properties of Concrete," Proc. VTT Symp. VTT-115:102 (1990).
12. M. Kriechbaum et al., Fractal structure of Portland cement paste during age hardening analyzed by small-angle x-ray scattering, *Progress in Colloid & Polymer Science* 79:101 (1989).
13. L. Auvray and P. Auroy, Scattering by interfaces:variation on Porod's law, in: "Neutron, X-ray and Light Scattering," P. Lindner and Th. Zemb, ed., Elsevier Science Publishers B.V., (1991).
14. T. Witten, unpublished.
15. D. Albrecht, Untersuchung der von schweren Ionen in Dielektrika erzeugten Defektstrukturen mittels Kleinwinkelstreuung, GSI-Report GSI 83-13(1983).
16. A.V. Semenyuk et al., Small-angle x-ray scattering investigation of the pore structure of nuclear filters, *J. Appl. Cryst.*424:809(1991).

17. M.T. Hutchings. Neutron diffraction measurements of residual stress fields - the engineer's dream come true?. *Neutron News* 3:14(1992).
18. T. Holden. Applications of neutron diffraction to studies of residual stress. Studsvik Neutron Scattering Days 1993. Studsvik (Sweden). May 18. 1993.
19. A.M. Balagurov. Private communication. 1993.
20. B. Hildenwall. Review of residual stress measurements using x-ray diffraction. Studsvik Neutron Scattering Days 1993. Studsvik (Sweden). May 18. 1993.
21. T. Lorentzen. Residual stress measurements at Riso. Journal of Neutron Research 1:13(1993).
22. L. Karlsson. The use of finite element calculations for the analysis of residual stress. Studsvik Neutron Scattering Days 1993. Studsvik (Sweden). May 18. 1993.

IN SITU DETECTION OF FLAWS IN MULTILAYER CERAMIC CAPACITORS USING ELECTRONIC SPECKLE PATTERN INTERFEROMETRY

Y. C. Chan, F. Yeung, G. C. Jin, N. K. Bao and P. S. Chung

Department of Electronic Engineering
City Polytechnic of Hong Kong
Tat Chee Avenue, Kowloon
Hong Kong

ABSTRACT

The electronic speckle pattern interferometry (ESPI) has been shown to be a new powerful tool for the in-situ and non-destructive detection of flaws in multilayer ceramic capacitors (MLCs) of surface mount printed circuit board assemblies. It is found to be sensitive to the cracks in MLCs of a printed circuit board assembly which are created by thermal shock and has been shown to be a more convenient operation than other current non-destructive techniques. As such, a new non-destructive and in-situ inspection capability is established for quality control in SMT assembly, physical screening of components and failure analysis of MLCs. With better resolution of ESPI system down to $\lambda/100$, it is anticipated that this capability can well be extended to the study of other miniaturized electronic components in general.

INTRODUCTION

In recent years, multilayer ceramic capacitors (MLCs) are used extensively in surface mount printed circuit assemblies. Particular attention has been paid to their reliability because they are prone to failures due to their special physical structure. A multilayer ceramic capacitor is essentially a composite structure made from alternating layers of metal (electrodes) and ceramic materials. The joining of these different materials (with vastly different coefficients of thermal expansion) to form a monolithic structure presents special thermal mismatch problems. As a result, physical cracks can easily be created by thermal stresses during surface mount assembly processes and environmental stress screening. Since there is a strong possibility of moisture filling these cracks under a humid environment, electrically conductive paths may be made. Consequently, a defective MLC shows significant electrical leakage when a rated working voltage is applied[1,2]. Some of these capacitors will, at a later stage, develop into electrical failures which can cause a complete electronic system to fail. Therefore it is highly desirable to be able to detect, in-situ and non-destructively, the presence of cracks in MLCs assembled in a printed circuit board prior to shipment to end-users.

Non-contact and non-destructive inspection methods for the detection of defects in MLCs, including X-radiography[3], neutron radiography[4], acoustic microscopy[5] and acoustic emission[6], have been reported. Amongst these, acoustic microscopy has been shown to be one of the most sensitive non-destructive techniques for the inspection of flaws and delaminations and is a very useful analytical tool in quality control and failure analysis of MLCs[5,7]. It should be emphasized that these test methods usually apply to isolated components and their applications to printed circuit assemblies are rather limited.

Electronic speckle pattern interferometry (ESPI) is a common industrial technique for non-destructive testing and deformation measurements[8]. Recently, the combination of advanced optics and digital imaging techniques make it ideal for such applications[9]. Since ESPI is more convenient to use than other current techniques, it is a potentially powerful tool for non-destructive testing especially in the fields of vibration analysis, mechanical and thermal distortions and flaw recognition[8]. Conventionally, ESPI has been mainly applied to the detection of flaws in relatively large objects such as tyres and aerospace structures due to constraints of the optics system. Recent advances in optics and lasers have made it possible the applications of ESPI for small objects like miniaturized MLCs in surface mount printed circuit assemblies.

The aim of this work is to develop a new non-destructive technique for the in-situ detection of defects in MLCs of printed circuit board assemblies. Since the leakage current at rated working voltage can be as small as about 10^{-11} amperes, the power loss in a MLC under the DC voltage may be too small to lead to any observable thermal deformation. In this work, micro-cracks are deliberately created by thermal shock. This is aimed at increasing leakage current as high as over 10^{-7} amperes so that sufficient power loss in a defective MLC will most likely cause a sufficient thermal deformation which is detectable by our ESPI system. A correlation between the MLC failure analysis model and the ESPI image is made. This paper reports the use of ESPI as a non-destructive method for the in-situ detection of flaws in MLCs of surface mount printed circuit board assemblies. The limitations of the present ESPI system are discussed and suggestions to enhance the capability of the ESPI system, in terms of resolution and sensitivity, for the inspection of miniaturized objects are recommended.

EXPERIMENTAL PROCEDURE

The ESPI system, shown in Fig. 1, uses a 50 mW HeNe laser and has a high-magnification (30x) CCD camera. The video signal of the speckle pattern can be digitally processed. The fringe pattern will be displayed in the monitor and recorded in the video cassette simultaneously for dynamic measurements.

The MLCs in this study were "1206" size ($3.2 \times 1.6 \times 0.9$ mm) and X7R-type with capacitance value and rated voltage at 0.1μF and 50 volts respectively. Before testing, electrical parameters of all samples (insulation resistance, capacitance and dissipation factor) were measured to ensure that they were within specifications. Physical cracks in MLCs were created by thermal shock (samples were pre-heated to 300 °C and then quenched into ice-water). After thermal shock, each sample was mounted on a small tailor-made printed circuit board (Fig. 2) and subjected to a 210 °C infrared reflow in order to solder the MLC on board.

A special white material was coated onto the surface of each sample to enhance the reflection characteristic of a MLC under laser light illumination. Electrical stress was

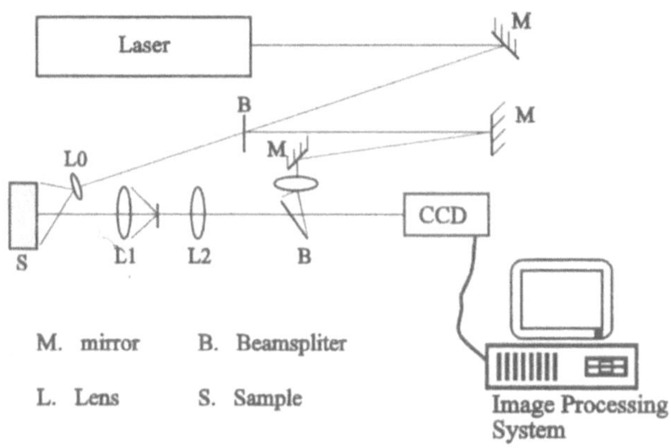

Figure 1 ESPI system.

Figure 2 MLC sample mounted on a test fixture.

applied to the MLC with the aid of a programmable HP6624 DC power supply, and measured by using a HP3458 digital multimeter and PC-based data analysis system. Typically, a 50 volts DC voltage was applied to the MLC and the leakage current was measured automatically and recorded digitally every five seconds. At the same time, the magnified fringe image of the sample under test was displayed, in real-time, on the monitor and recorded in video cassette simultaneously.

After ESPI testing, the micro-sections of all samples were examined by destructive physical analysis (DPA) techniques plus the use of optical microscopy, scanning electron microscopy (SEM) and energy dispersive X-ray spectroscopy (EDX). This is aimed at locating the physical defects of the MLC and correlate with the ESPI results.

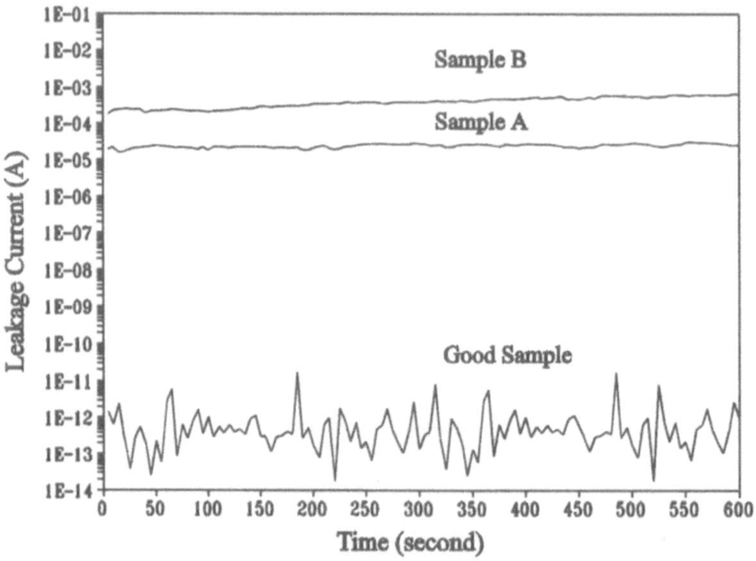

Figure 3 Comparison of leakage currents in 2 defective MLCs (Samples A & B) and a good MLC sample when measured at 50 V d.c.

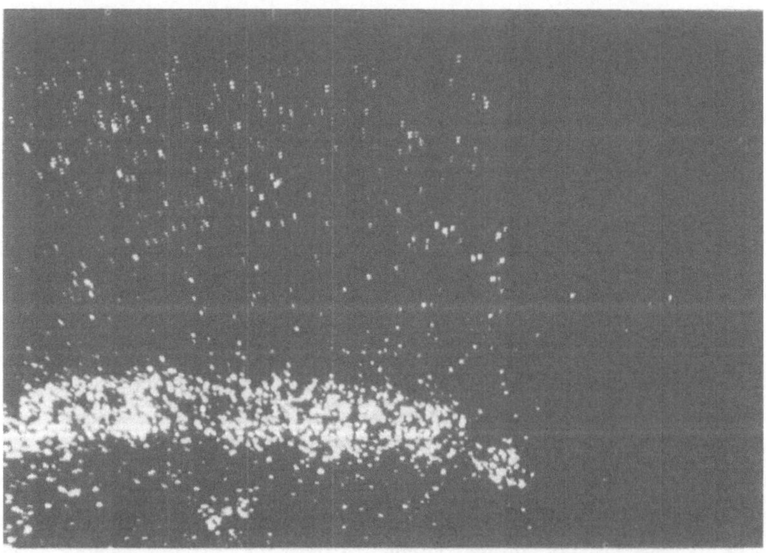

Figure 4 ESPI photograph of sample A (magnification : 250×).

Figure 5 Optical photograph of the micro-section for sample A (magnification : 200×).

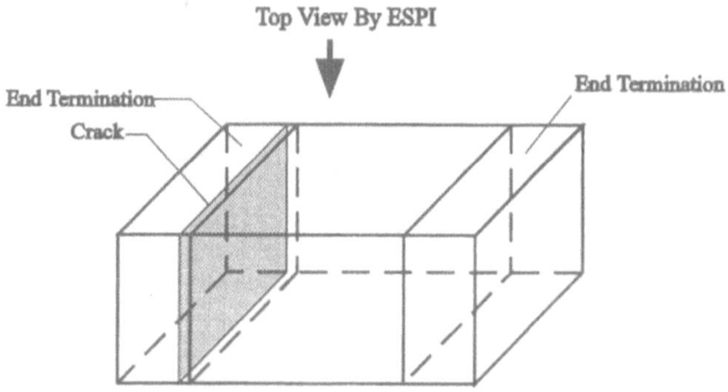

Figure 6 The proposed sites of crack in sample A.

RESULTS AND DISCUSSIONS

Fig. 3 summarizes the leakage currents for the two defective MLCs (Samples A & B) and a good MLC. ESPI was carried out on each of these samples. A thick and long band of white dots, shown in the ESPI photograph of Fig. 4, was clearly visible for sample A and corresponded well with the crack line in the micro-section of Fig. 5. It is believed that the plane where the crack lies, with respect to the direction of ESPI inspection, is shown in Fig. 6. Similarly, two dominant patches of white dots were observed in the ESPI photograph (Fig. 7) of Sample B and matched closely with the two major cracks that were observed during DPA of that sample (one of these cracks is shown in Fig. 8). The proposed sites where the two cracks occurs, with respect to the direction of ESPI inspection, is illustrated in Fig. 9. As a control experiment, the ESPI result on the good MLC did not show any sign of crack as compared to those in Figs. 4 and 7. By combining the ESPI and DPA techniques, it is proved, beyond any doubt, that cracks in MLCs can be revealed and cross-checked. Therefore, the ESPI technique can be used as an efficient method for the in-situ and non-destructive detection of flaws in MLCs.

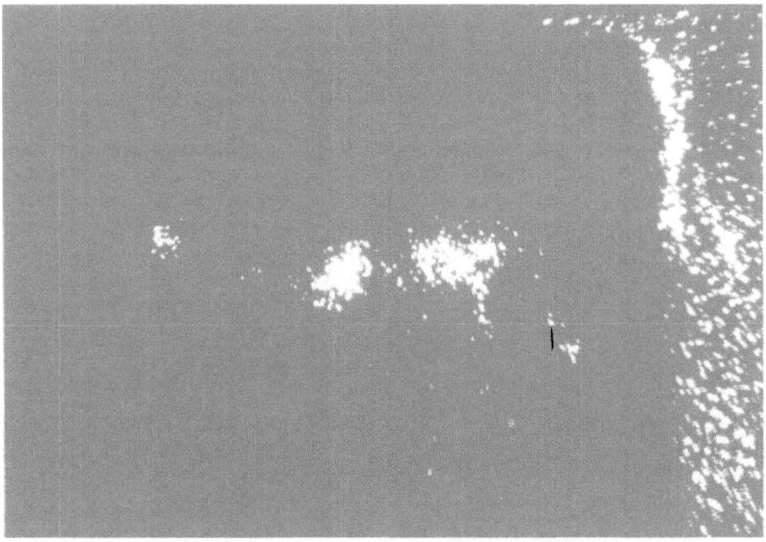

Figure 7 ESPI photograph of sample B (magnification : 250×).

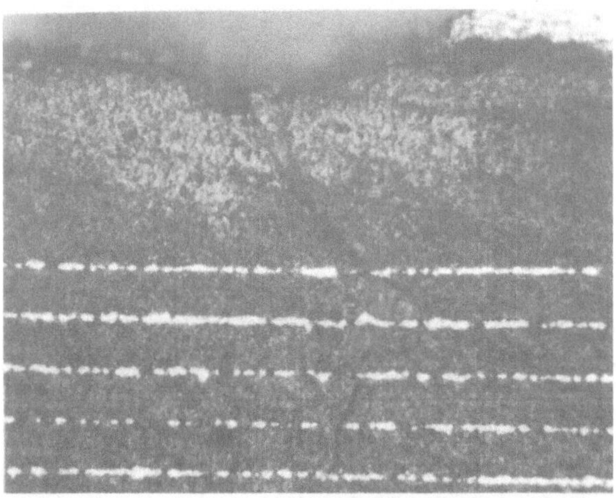

Figure 8 Optical photograph of the micro-section for sample B (magnification : 800×).

However, there are some limitations of this technique due to the resolution constraints of the present optics system. Typically, the leakage current of a defective MLC is of the order of 10^{-5} to 10^{-7} A. Since the "1206" MLC is relatively small and the temperature expansion coefficient of X7R dielectric is about 10×10^{-6} /°C, a highly

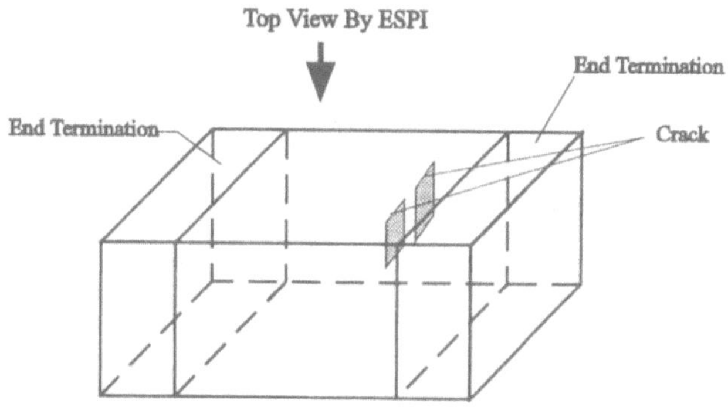

Figure 9 The proposed sites of cracks in sample B.

sensitive optics system is needed to capture the tiny thermal deformation of MLC surface. The resolution of the present ESPI system is 0.5 λ (about 0.3 μm). This sensitivity would limit its application to leakage currents of MLCs greater than about 10^{-5} A. With better resolution[9] of ESPI system down to $\lambda/100$, it is expected that a detectable ESPI image can be obtained even when the leakage current or the thermal deformation is extremely small.

CONCLUSIONS

This work has demonstrated clearly the usefulness of the ESPI technique for the in-situ and non-destructive detections of cracks in small objects such as MLCs in surface mount printed circuit assemblies. The ESPI result has also been confirmed by DPA techniques. With further improvement of the ESPI system, e.g. better resolution of optics system down to $\lambda/100$, its capability can be well extended to the study of other miniaturized electronic components in general and is a potentially very powerful tool for quality control in production and failure analysis.

ACKNOWLEDGMENT

The authors would like to acknowledge the award of an UPGC earmarked research grant (project # 904035) for this work. The work of Ms. Lu Lan in carrying out some of the measurements is also acknowledged. One of the authors (FY) is grateful for a CPHK research studentship in support of his study for a M.Phil. degree in Electronic Engineering.

REFERENCES

1. H.V.DeMatos and C.R.Koripella "Crack Initiation and Propagation in MLC Chips Subjected to Thermal Stresses", *8th CART Symposium Proceedings,* pp. 25-31, 1988.
2. R. C. Chittick, E. Gray, et al "Nondestructive Screening for low Voltage Failure in Multilayer Ceramic Capacitors", *IEEE Transactions on Components, Hybrids and Manufacturing Technology*, Vol. CHMT-6, No. 4, pp. 510-516, December 1989.
3. R. S. Spiiggs and A. H. Cronshagen, "Nondestructive X-ray Inspection of Ceramic-Chip Capacitors for Delaminations", *Proceedings-1976 Reliability Physics Symposium*, Las Vegas, 1976, pp. 157-163.
4 M. J. Cozzolino, G. J. Ewell and J. L. Galvagni, "Nondestructive Examination of Multilayer Ceramic Capacitors By Neutron Radiography", *Proceedings of 28th Electronic Components Conference*, IEEE, Anahaim, CA, April, 1978, pp. 327-375.
5. G. R. Love and G. J. Ewell, "Acoustic Microscopy of Ceramic Capacitors", *IEEE Transaction on Components, Hybrids, and Manufacturing Technology*, Vol. CHMT-1, No. 3, September, 1978, pp. 251-257.
6. S. R. Kahn and R. W. Checkaneck, "Acoustic Emission Testing of Multilayer Ceramic Capacitors", *IEEE Transaction on Components, Hybrids, and Manufacturing Technology*, Vol. CHMT-6, No. 4, December 1983, pp. 517-526.
7. L. W. Kessler and J. E. Semmens, "Characterization of the Microstructure of Ceramics Used in Multilayer Ceramic Capacitors By Means of the Scanning Laser Acoustic Microscope", *Journal of American Ceramic Society*, Vol. 72, No 12, December 1989, pp. 2271-2275.
8. B. Sharp, "Electronic Speckle Pattern Interferometery (ESPI)", *Optics and Lasers in Engineering* 11 (1989), pp. 241-255.
9. N.K. Bao, G.C. Jin and P.S. Chung, "A New ESPI with Continuously Variable Sensitivity", *SPIE Int. Symposium on Optical Tools for Manufacturing and Advanced Automation*, Boston, Mass., 7-10 September 1993.

GROWTH RATES AND INTERFACE SHAPES IN SEMICONDUCTOR MATERIALS USING REAL TIME RADIOGRAPHIC IMAGING

R. T. Simchick*, S. Sorokach*, A. L. Fripp, W. J. Debnam,
R. F. Berry, D. J. Jobson, and P. G. Barber#

NASA Langley Research Center
Hampton, Virginia, 23681

INTRODUCTION

The success of new electronic materials has been due in part to the development of procedures that produce semiconductors of high purity and perfection. These materials have been grown from the gas phase, solution, and melts. The Bridgman technique is a way semiconductor crystals are grown from the melt. With this technique the semiconductor material is usually sealed in a fused silica ampoule, placed inside the tubular furnace, and heated to completely melt the sample. The sample solidifies by cooling the molten material in one of three ways: translating the furnace along the sample, slowly extracting the sample ampoule from the furnace, or uniformly lowering the temperature. The solidification rate is typically only a few millimeters per hour. A diagram of a typical Bridgman furnace is shown in figure 1 and more details of the furnace used in this experiment are found in the literature[1]. The furnaces used in such procedures are generally opaque to visible radiation.

The importance of the shape and location of the melt-solid interface during crystal growth has stimulated the development of several techniques to observe it[2-9]. Some of these techniques[2-5], however, are limited by the fact that results are available only after the crystallization process is complete and are destructive to the sample. One destructive technique is quenching[2,3]. The crystal is partially grown under well-defined experimental conditions and

---

*Lockheed Engineering and Sciences Corporation, Hampton, Va., 23681

#Longwood College, Farmville, Va., 23909

Nondestructive Characterization of Materials VI
Edited by R.E. Green, Jr. *et al.*, Plenum Press, New York, 1994

thermal gradients, and is quickly quenched in water or liquid nitrogen. The experimenter cuts the sample and carefully polishes it to observe the frozen interface shape. A second technique is to periodically apply current pulses[4,5] or mechanical vibrations on the sample[5]. This causes small variations in composition in the sample during growth and such variations can be observed after cutting, polishing, and etching the crystal. Another technique is to use ultrasonics[6] to track the movement of the solid liquid interface during the vertical Bridgman growth. This technique can accurately measure the position of the interface.

The authors[7, 8] previously reported on the development of a procedure to visualize the melt solid interface and record its shape and movement during Bridgman crystal growth. The procedure is simple radiography using x or gamma rays. This technique takes advantage of the density difference between the solid and liquid. Resulting differences in the x-ray photons passing through the sample are sufficient to produce a well-defined image that is captured by a video camera. The analog image from the video camera can be digitized and image processing techniques can be used to improve the definition of the solid liquid interface. This technique is applicable to other crystal growth processes and has subsequently been used to observe the interface in a growth technique in which the crystal is pulled from a crucible of molten material[9], (Czochralski growth).

This paper will report on methods used to verify the accuracy of the radiographic interface measurements, discuss some advanced image processing techniques, and plot some comparisons between shapes obtained from a destructive technique (quenching) and a radiographic interface.

EXPERIMENTAL TECHNIQUE

Figure 1 is a schematic describing the radiographic imaging apparatus. In a typical experiment the sealed ampoule is lowered at a uniform rate out of the hot zone through the

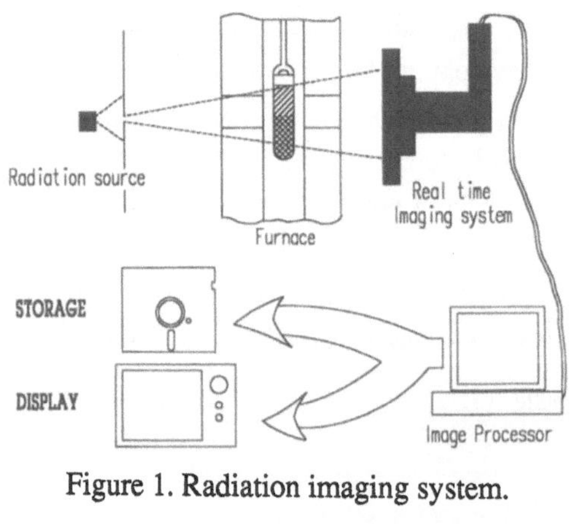

Figure 1. Radiation imaging system.

3 insulation zone to the cold zone. While the sample is passing through the insulation zone, x-rays penetrate the sample and a radiation imaging system detects the radiation passing through the sample. (Gammascope GS220, RTS Technology, Inc., Sauerwein Group, North Andover, Massachusetts.) The insulation zone is made from a low density, uniform silica fiber to aid the passage of the radiation. This system is composed of a cesium iodide detector that collects x-ray photons and converts them to light photons. The image intensifier converts the light photons to electrons, accelerates the electrons and converts the electrons to light photons. A video camera then captures the image. This image can be displayed with a television monitor, and /or digitized for processing and storage.

While the interface is visible in real time, the image processor integrates the image to increase the signal-to-noise ratio. Additional image processing techniques that improve the image visibility during crystal growth include contrast enhancement and edge filtering with a Sobel filter[10]. A column averaging technique was developed to provide a complete view of the interface in the cylindrical shaped sample[11]. Additional image processing techniques are being developed using edge detection processes, drawing the contour lines and creating an algorithm to detect the interface position[12].

An experiment was designed to verify the spacial accuracy of the radiographic interface measurements. Samples with known interface shapes were machined from cylinders of brass and copper. These metals were selected because they exhibited a similar radiographic density difference to solid and liquid germanium. Interface measurements were made with the brass/copper sample at various positions in the insulation zone. The initial experiments were made with the more dense material, brass, penetrating into the less dense copper. This was followed by having copper penetrating into brass. The results of these experiments would also show the accuracy of the assumption that our two-dimensional image gives a good representation of the three-dimensional crystal.

The second series of experiments used the technique during crystal growth to examine the movement and shape of the interface. The interface position was recorded and compared to a fixed point in the furnace and fixed markers on the ampoule. This procedure produces independent measurements of both the ampoule pull rate and the interface movement about the furnace. The determination of the interface shape from a two-dimensional image assumes azimuthal symmetry. During growth, the pull rate of the ampoule through the furnace remains constant, therefore the recorded data is the position of the interface compared with a fixed point in the furnace. Data is recorded from the inception of freezing throughout the run until the crystal is completely frozen.

Two types of crystals were grown in these experiments. Germanium has a fixed solidification temperature and its interface movement is caused only by changes in thermal loading. The other crystal, lead tin telluride, (20 percent SnTe and 80 percent PbTe) has a variable solidification temperature in which the temperature decreases as crystallization continues. This is due to the rejection of the lower melting SnTe from the solidification front into the liquid. Therefore, the interface position depends on both thermal loading and the continuously changing solidification temperature of the crystal. During these experiments images were observed over the total length of the growing crystals. Measurements of interface shape and position were taken from the images.

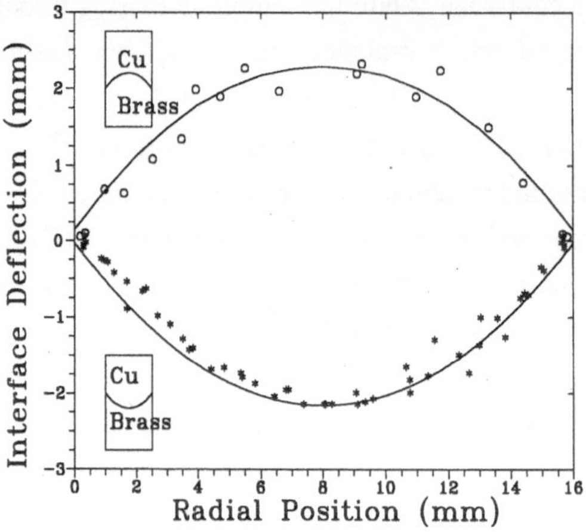

Figure 2. Real time images comparing radiographic data with known shape.

EXPERIMENTAL RESULTS

The interface shapes generated from the metal models in the first set of experiments are shown in figure 2. The solid lines show the known interface shape and the symbols are from the radiographic measurements. The data near the lower curve is taken from four different positions in the insulation zone and data near the upper curve is taken from two different positions in the insulation zone. Since this data gives a good approximation of the known shape, neither density variations nor image processing appears to affect the radiographically produced shape.The two dimensional images gives a reasonable representation of the three-dimensional sample.

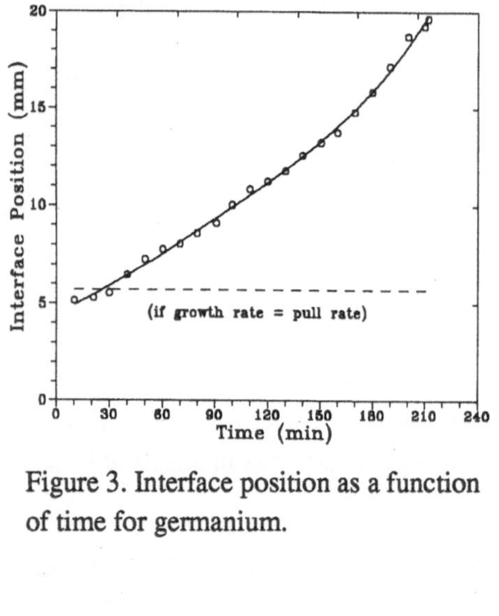

Figure 3. Interface position as a function of time for germanium.

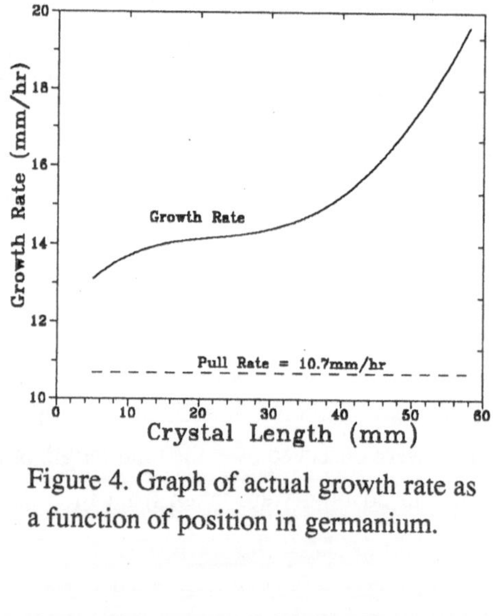

Figure 4. Graph of actual growth rate as a function of position in germanium.

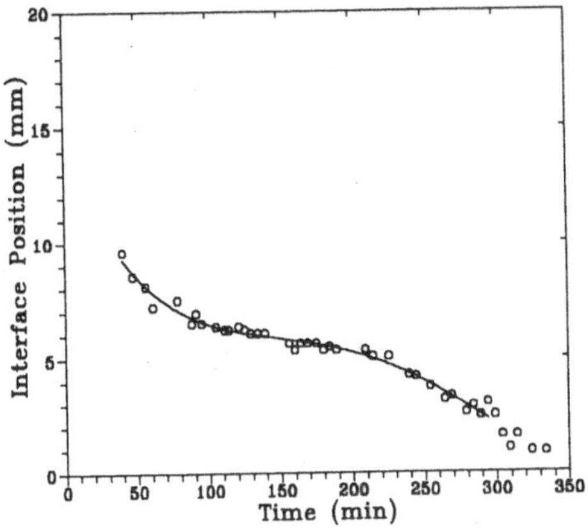

Figure 5. Interface position vs. time for a PbSnTe crystal.

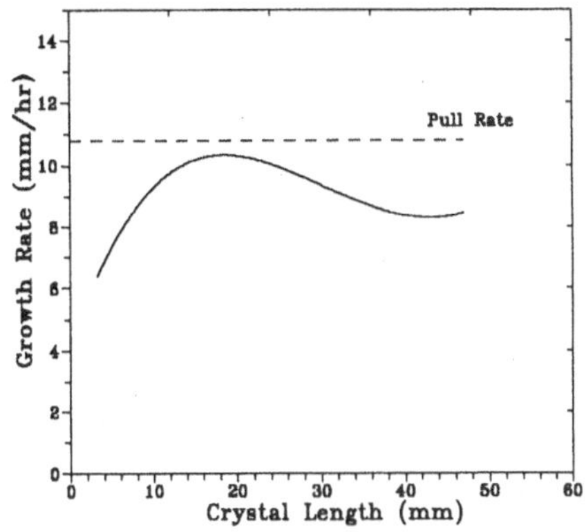

Figure 6. Graph of actual growth rate as a function of position in PbSnTe.

The data from the second set of experiments in figure 3 shows how the interface in the germanium crystal continuously moves away from the cold zone, after an initial transient, as the crystal was being grown. If the growth rate was equal to the furnace pull rate then the interface position would hold its position in the insulation zone as shown by the dotted line. Therefore, the growth rate is greater than the actual pull rate. The actual growth of this crystal is plotted as a function of crystal length in figure 4. Note that the actual growth rate of this crystal varies from 13 to nearly 20 mm/hr as compared to the pull rate of 10.7 mm/hr.

The lead tin telluride exhibits quite different behavior as shown in figures 5 and 6. In figure 5 note that the interface moves toward the cold zone as growth progresses. Figure 6

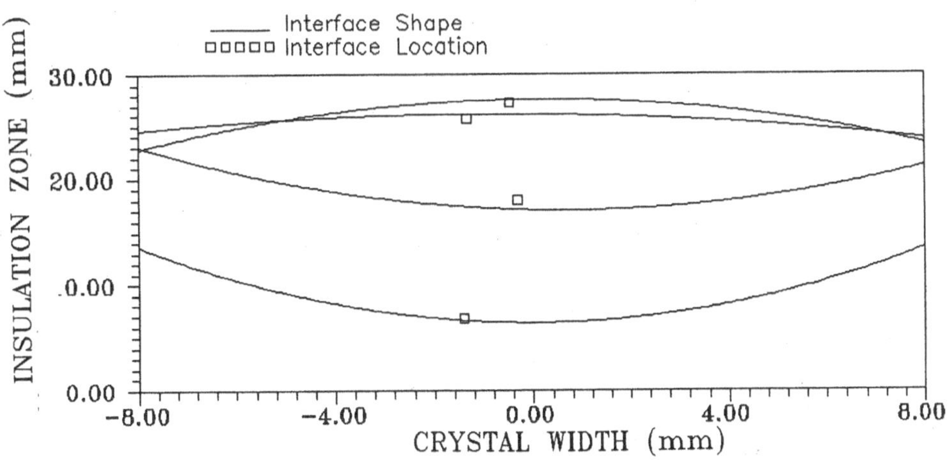

Figure 7. Radiographic interface shapes from a germanium crystal.

shows a plot of growth rate versus crystal length. The growth rate is less than the furnace translation rate and varies only from 6 to 10 mm/hr. This is a result of the competing forces, thermal loading and the continuously changing solidification temperature due to the rejection of tin at the melt solid interface.

Figure 7 shows a series of radiographic shapes of the interface from germanium at different positions in the insulation zone. These shapes were obtained during the crystal growth experiment. The interface shapes are shown by the curved lines while the position in the insulation zone is denoted by the open squares. Note that the shapes vary during growth. The interface shape near the bottom is bowed toward the cold zone, while the shapes near the top of the are bowed toward the hot zone. and the shape near the center approximates a planar interface. The variations in shape are consistant with changes caused by thermal loading in the furnace. Figure 8 shows a comparison of the interface shape taken from the quenched lead tin telluride sample shown in reference 2 with a real time interface. These shapes are quite similar. The flat bottom from the quenched data could be attributed to growth continuing while being quenched.

The interface measurement data shown in this paper was taken by the equipment operator and was subject to his perception of the image. However, the simple features of this type of information gathering is amenable to feature detection techniques. Edge detection techniques have been developed and an algorithm has been written that will scour the image and identify the main features and store the position on the computer. With these advanced features the actual growth rate and the interface shape can be measured without being subject to operator error.

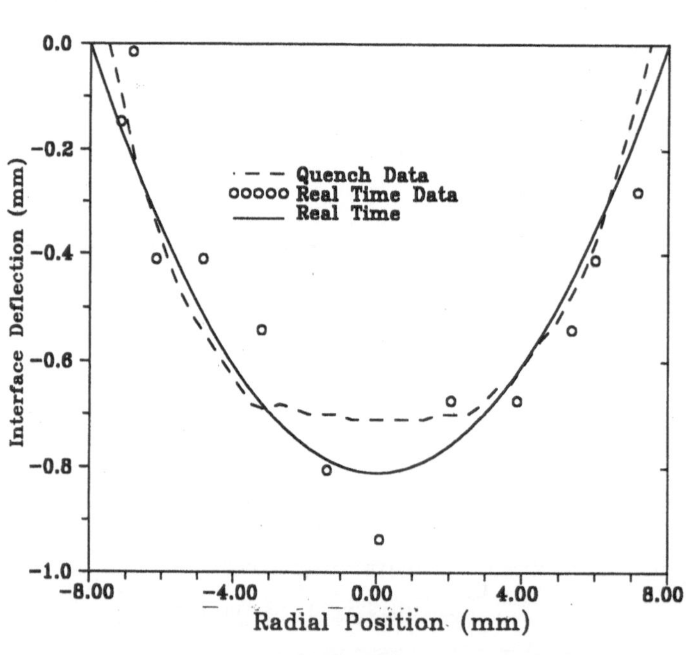

Figure 8. Comparison of radiographic interface with quenched interface.

SUMMARY AND CONCLUSIONS

This paper has shown the use of radiography in crystal growth. Interface movement and shape have been measured during crystal growth. These measurements have been verified by data from the copper/brass sample showing that the position in the insulation zone does not affect the shape. The density variation also had no measurable effect on neither the shape nor the position of the interface. The additional edge enhancement techniques will enable the automation of these measurements as well gives a more precise definition of the interface. Since the radiographic measurements are taken during growth, the next step would be for the crystal grower to interact with the growth process.

The primary conclusion is that the radiographic image defines the solid-liquid interface shape and position during crystal growth. With this conclusion, the radiographic measurements give the crystal grower valuable information about the crystal growth process. Crystal growers have known that the growth rate was different from the furnace translation rate, but this method quantifies the difference for them.

REFERENCES

1. J.A. Hubert, A. L. Fripp, and C.S.Welch, "Resolution of the discrepancy between temperature indicated interface and radiographically determined interface in a vertical Bridgman furnace," J.Cryst. Growth 131:75(1993).

2. Y. Huang, W. J. Debnam and A. L. Fripp, "Interface shapes during vertical Bridgman growth of PbSnTe crystals." J. Cryst. Growth 104:315(1990).

3. P. Capper, J. J. G. Gosney, and M. T. Quelch, "Quenching studies in Bridgman-grown Cd_x, $Hg_{1-x}Te$," J.Cryst. Growth 63:154(1983).

4. A. F. Witt, H. C. Gatos, M. Lichtensteiger, M. C. Lavine, and C. J. Herman, "Crystal growth and steady-state segregation under zero gravity-InSb", J. Electrochem. Soc. 122:276(1975).

5. K. M. Kim, A. F. Witt, M. Lichtensteiger, and H. C. Gatos, "Quantitative analysis of the effects of destabilizing vertical thermal gradients on crystal growth and desegregation". J. Electrochem. Soc. 125:475(1978).

6. J.N.Carter, A. Lam, and D.M. Schleich, "Ultrasonic time of flight monitoring of the position of the liquid solid interface during the Bridgman growth of germanium," Review of Scientific Instruments, 63:3472(1992).

7. R. T. Simchick, S. E. Sorokach, A. L. Fripp, W. J. Debnam, R. F. Berry, and P. G. Barber; "Gamma Ray and x-Ray imaging studies of the location and shape of the melt solid interface during Bridgman growth of germanium and lead tin telluride"; vol. 35 Advances in X-Ray Analysis, Plenum Publishing Corp, 1992.

8. Barber, P.G. Crouch, R.K., Fripp, A.L., Debnam, W.J., Berry,R.F. and Simchick, R.T.;"A Procedure to visualize the melt-solid interface in Bridgman grown germanium and lead tin telluride"; J.Crystal Growth, 74:228(1986).

9. K. Kakimoto, M. C. Eguchi, H. Watanabe, and T.Hibiya, In-situ observation of solid - liquid interface shape by x-ray radiography during silicon single-crystal growth J. Cryst. Growth 91:509(1988).

10. E. R. Dougherty and C. R. Giardina, Image Processing-Continuous to Discrete, Vol. 1 Prentiss-Hall, Inc. Newark (1987) 59.

11. P. G. Barber, R. F. Berry,W. J. Debnam, A. L. Fripp, Y. Huang, K. Stacy, and R. T. Simchick; "Modeling melt-solid interfaces in Bridgman growth"; J.Crystal Growth 97:672(1989).

12. D. J. Jobson, "A discrepancy within primate spatial vision and its bearing on the definition of edge detection processes in machine vision," NASA Technical Memorandum no. 102739, September,1990.

ENVIRONMENTAL SCANNING ELECTRON MICROSCOPIC INVESTIGATION OF INTERFACIAL DEFECTS IN ELECTRONIC PACKAGES

Junhui Li and Michael Pecht

CALCE Electronic Packaging Research Center
University of Maryland
College Park, Maryland

INTRODUCTION

Scanning electron microscopy (SEM) has applied to microscopic characterization for decades. Traditional SEMs work under high-vacuum conditions. The secondary electron images of a specimen can reach a nanometer resolution if the specimen is well conducted electrically. The disadvantage in applying of traditional SEMs when characterizing a poorly conducted specimen is surface charging, in order to avoid the charging problem, a conductive film must be coated on the specimen. Besides altering the specimen slightly, this conductive film coating can also make the specimen unusable.

The surface-charging problem can be solved by using an environmental scanning electron microscope (E-SEM). E-SEM works with specimen chamber pressures in the range of 1~20 torr - ten thousand times higher than that of the traditional SEM - without contaminating the microscope. The pressure condition allows the researcher to examine unprepared, uncoated specimens that are free of surface charging. This makes it possible to examine specimens in their natural states. Dynamic characterizations of wetting, drying, absorption, melting, corrosion and crystallization can be performed using E-SEM.

Microelectronic packages usually consist of combinations of ceramics, polymers, adhesives, semiconductors, and metallizations with complex geometries. In microelectronic packaging, the principal weaknesses are introduced at the physical interfaces between the different sections. For example, thermo-mechanical stresses due to the mismatch of coefficients of thermal expansion between the materials, concentrate at the interfaces. Sorptive-mechanical stresses occur with materials like polymers, and thereby induce hydroscopic stresses on the interfaces. Both thermo-mechanical and sorptive-mechanical stresses can cause adhesion-strength loss in electronic packages. To investigate environmental effects, a dynamic characterization technique is needed to detect the microstructural changes on the interfaces.

In the present research, various levels of electronic packages were investigated *in situ* using E-SEM, or thermo-cycled and humidity-cycled in sections and characterized using E-SEM after each section. The dynamic characterizations helped to identify various interfacial defects in the electronic packages.

Nondestructive Characterization of Materials VI
Edited by R.E. Green, Jr. *et al.*, Plenum Press, New York, 1994

E-SEM ANl EXPERIMENTAL

E-SEMs are able to work with pressure and without surface charging because the secondary electron detector is designed based on the principle of gas ionization. As primary electrons are emitted from the gun system, the secondary electrons on the specimen surface are accelerated towards the detector, which is biased by a moderate electric field. The collisions between the electrons and gas molecules liberate more free electrons and thereby provide more signals. Positive ions created in the gas effectively neutralize the excess electron charge built up on the specimen. Proper operating pressure controls the specimen surface charging completely. Depending upon the environmental requirements, the pressure source can be water vapor, air, argon, nitrogen or other gases.

For the present research, a study of zero-level packaging was performed *in situ* using E-SEM with a temperature stage. For the studies on first- and second-level packaging, additional environmental chambers were used to execute the experiments efficiently. The latter experiments basically comprised initial examination, temperature-cycling or humidity-cycling tests and post-cycling examinations. The initial and post-cycling examinations were conducted by E-SEM, and the environmental cycling tests employed environmental chambers. The post-cycling examinations were conducted after each period of cycling to pursue possible microscopic interfacial changes emerged during the environmental cycling period. Studied specimens included multilayer thin film polyimides with metallizations and multilayer printed wiring boards (PWBs).

In the humidity-cycling tests, specimens were placed in an environmental chamber and subjected to relative humidity (RH) cycling between 45% and 95%. Each cycle took 16 hrs. - 6 hrs. dwelling at each RH level, 2 hrs. to increase the RH from the low level to the high level, and 2 hrs. to decrease the RH from the high level to the low level. The RH cycling tests for multilayer thin-film polyimide specimens were conducted at both ambient temperature (23°C) and elevated temperature (85°C), and those for multilayer PWB specimens were done at elevated temperature (85°C).

Multilayer PWB specimens for thermal cycling tests were divided into three groups. The first and second groups were subjected to thermal cycling between -5°C and 145°C, and between 25°C and 175°C, respectively, with 45 mins. heating and 45 mins. cooling. The third group of specimens was subjected to thermal cycling between 25°C and 200°C with 90 minutes in heating and 60 minutes in cooling. The dwell time at both high and low temperatures was 10 mins. for all specimens.

In the E-SEM examinations, the multilayer polyimides were exposed to water pressure, and the multilayer PWBs were exposed to air pressure, in the range of 3~7 torr at room temperature. The gun voltages were 15, 20, and 25 KV. In order to compare interfacial morphologies before and after temperature cycling, E-SEM micrographs were taken of same area for initial and post-cycling conditions .

E-SEM INVESTIGATION OF ZERO-LEVEL ELECTRONIC PACKAGING

One example of E-SEM investigation on zero-level electronic packages studied a polyimide-passivated microwave monolithic integrated circuit. Figure 1 shows a schematic of a polyimide passivated transimpedance amplifier. The amplifier was processed with 0.5 μm gate length GaAs MESFETs, interdigitate and overlay SiN capacitors, MESA resistors, via holes, and two-level interconnects [Bahl et al., 1986]. A 10 μm polyimide film was selected as the passivation layer in order to eliminate hermetic packaging requirements. Previous investigation indicated that the mean time to failure (MTF) for the circuits at 220°C was 480 hrs.; at 200°C, it increased to 750 hrs., and it exceeded 1,000 hours at 180°C [Halkias et al., 1993].

Figure 2(a) shows an area of reaction between the polyimide and gold in the feedback stage at 180°C, indicating a region of polyimide breakdown. Several similar interfacial defects were also observed in the polyimide passivation at this temperature. These defects increase the gate leakage current and change both the channel capacitance and resistance. Increases in gate leakage result in both bandwidth and sensitivity degradation.

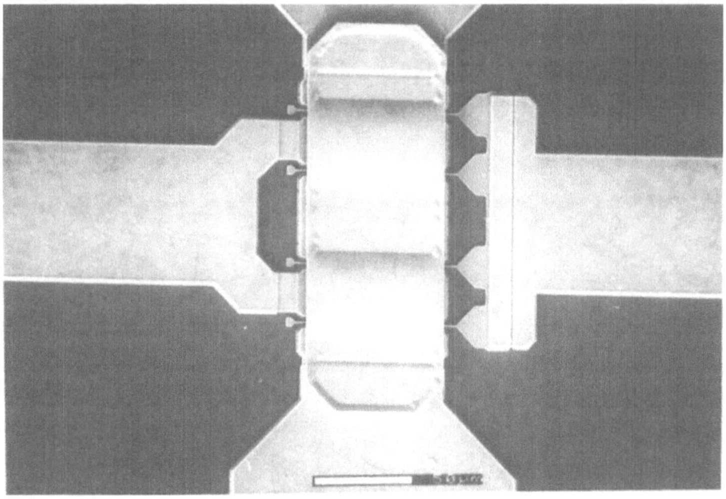

Figure 1. FETs in an investigated transimpedance amplifier

In the defective area of the polyimide passivation, the interfacial contamination due to the reaction between Au and GaAs was observed at the end of the gate finger at 220°C, as shown in Figure 2(b). A hillock formation at gate region may be also accompanied by an Au-GaAs reaction. Hillock formation typically characterizes the occurrence of electromigration in the gate metallization.

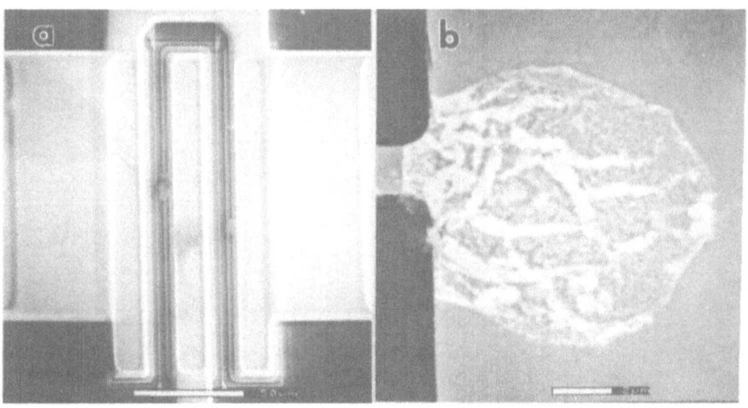

Figure 2. (a) Polyimide-Au reaction regions in TIA at 180°C; and (b) Contamination of Au-GaAs reaction in the area of defected polyimide passivation

Accompanied by the experimental results of DC analysis, gain stability, and noise-current tests [Halkias 1993], E-SEM investigations of surface-related reliability and failure mechanisms of L-band TIAs have indicated gate leakage and gold electromigration as the primary failure mechanisms. The polyimide passivation developed surface area defects that increased gate leakage currents.

With first-level electronic packaging, thin-film polyimides are frequently used as polymeric chip carriers in tape-automated bonding (TAB). They are also commonly employed as overlays, an extension of TAB, especially for multichip modules. The principal function of multichip module overlays is to interconnect chips using metallization layers and vias plated within the multilayer polymeric overlays. Figure 3 shows a cross-section of a typical multilayer overlay composed of thin-film polyimides plated with metallization layers and vias.

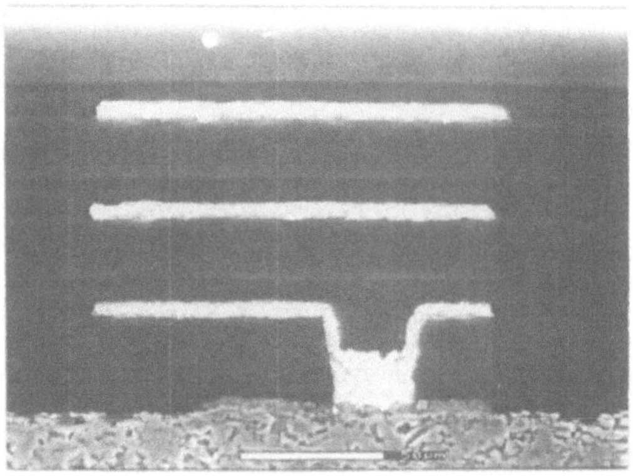

Figure 3. Thin film polymeric overlay laminated in a multichip module

Under ambient conditions, polyimides absorb water up to 3% by weight [Tessier et al., 1989]. Loss of macroscopic adhesion at the metallization/polyimide interface has also been reported [Poon, 1989]. The present investigation concentrated on the humidity cycling effect on multilayer polyimide overlays.

Before and after every ten RH cycles, the specimens were examined using E-SEM.

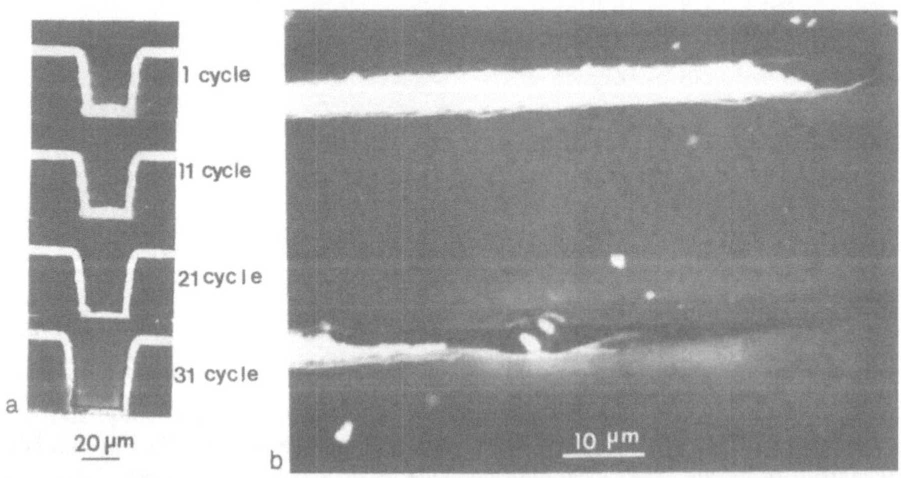

Figure 4. (a) Progressive changes of the same via after 1, 11, 21 and 31 humidity cycles, (b) delamination of polyimide/adhesive and void formation at metal lines

After 30 RH cycles, adhesive/metallization delamination was observed inside of vias in cross-sectioned specimens cycled at room temperature (see Figure 4(a)). Via delamination was primarily associated with silicon chips. There was no evidence of delamination of other vias connected to the alumina substrate. Delamination of the adhesive/polyimide and void formation at the edge of the metallizations was also observed, as shown in Figure 4(b). Void formation at the edge of metallization lines is apparently due to corrosion of exposed copper at the sidewall of copper lines [Paik et al., 1993].

E-SEM INVESTIGATION OF SECOND-LEVEL ELECTRONIC PACKAGING

The E-SEM investigation of the second-level electronic packaging focused on multilayer PWBs. Figure 5 shows the cross-section geometry of a PWB with twenty reinforced epoxy-resin laminates. Each laminate consists of epoxy resin reinforced by cross-plied glass fibers. The multilayer PWB is fabricated by repeatedly sandwiching the dielectric epoxy-glass composite laminates with copper foils. These foils function as conductive paths and power planes, and are shown as pairs of bright rectangles in Figure 5. For the interconnection of different levels of circuitry and components mounted on the board, holes are mechanically drilled through the board, followed by copper plating on the hole walls. The cross-section of the plated through holes (PTHs) appears as bright bands extending vertically. Under certain environmental conditions, such as high humidity, temperature, and applied voltage stress, the insulation resistance between two circuit traces or between the PTH and foil, ground, or power planes, can decrease due to metal migration, resulting in electrical failure [Balu et al., 1992].

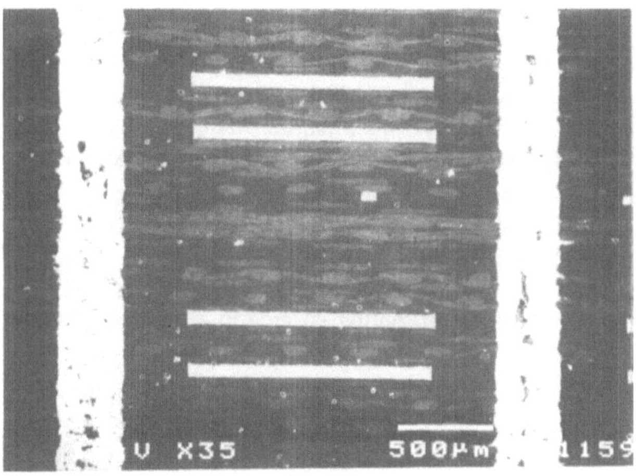

Figure 5. A cross-section of multilayer PWB

The insulation resistance breakdown is caused by the creation of a copper shorting filament due to the migration of copper ions along the interface of the glass fiber and epoxy resin. Metal migration takes place in a two-step process to form the metallic shorting filament: path formation and electrochemical reaction. Path formation is suspected to be caused by degradation of the glass-epoxy interface, which provides a path for the second step, electrodeposition of the copper filament along the interface. Humidity and temperature are key factors affecting mechanical and thermal properties of board materials, and may lead to interfacial degradation.

Humidity-cycling Effects on Multilayer PWBs

PWB specimens were humidity-cycled between 45% to 85% RH at 80°C. After 10 cycles, microcracks appeared in glass-fiber bundles. Figure 6 shows a glass-fiber bundle near a PTH. Comparing the images of the bundle before and after RH cycling indicates debonding of the interface between the glass fibers and the epoxy-resin matrix as well as the microcracks.

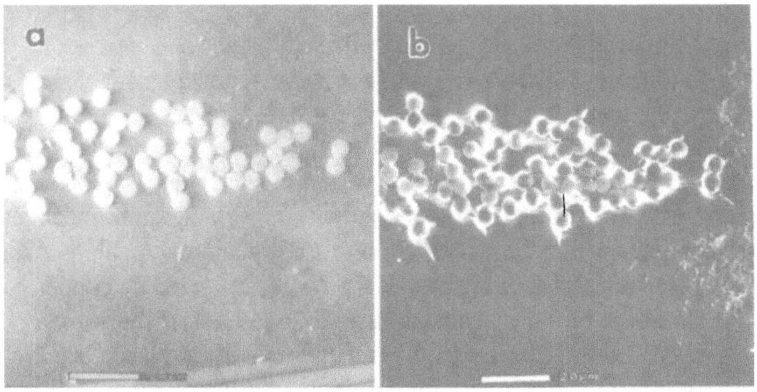

Figure 6. A glass fiber bundle in a PWB, (a) before RH cycling, and (b) after RH cycling

The microcracks were distributed in two ways. Some were located along the edge of the bundle and others, in a radial shape, were oriented from the interfaces of the glass fibers and the epoxy resin towards the inside of the epoxy matrix. The moisture absorption in epoxy resin is much more severe than in glass fiber. Apparently, sorptive-mechanical stresses concentrated around the edge of the glass-fiber bundle during RH cycling. Microcracks occurred during RH cycling when the RH decreased, because tensile stresses remained during shrinkage and compressive stresses accumulated during the swelling state in the epoxy resin.

Thermal-cycling Effect on Multilayer PWBs

Figure 7 shows two pictures presenting an area in a PWB specimen before and after

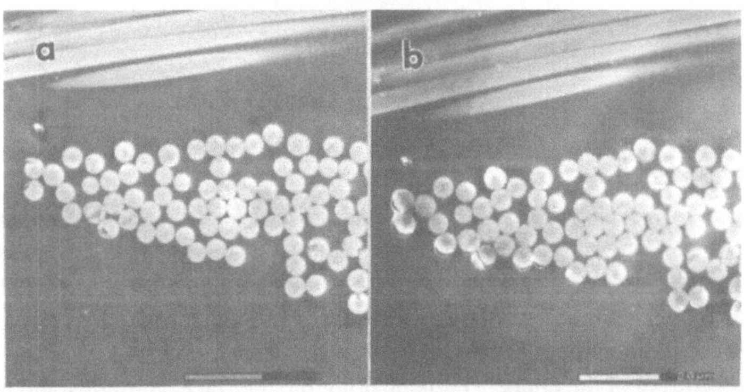

Figure 7. A glass fiber bundle closed to a PTH in a PWB thermal-cycled between 25°C and 200°C, (a) before and (b) after 10 thermal cycles

ten times thermally cycled between 25°C and 200°C. The fiber bundle was located close to a PTH. As comparison of (a) and (b) shows, some resin/fiber interfaces, especially for fibers along the bundle edge, became brighter after the temperature cycling, indicating that these interfaces were debonded. Contrary to the humidity-cycling result, no microcracks were observed in the glass bundle area. In order to study the resin/fiber interface closely, individual fibers in the fiber bundle were examined with a larger magnification.

For the fiber bundles far away from PTHs, the situation is different. Figure 8(a) shows a few fiber bundles far away from PTHs. Figure 8(b) gives a close look at the fibers on the edge of the fiber bundles in Figure 8(a). Thermal cycling resulted in degradation of the epoxy matrix around the fiber bundles, but no interfacial debonding occurred.

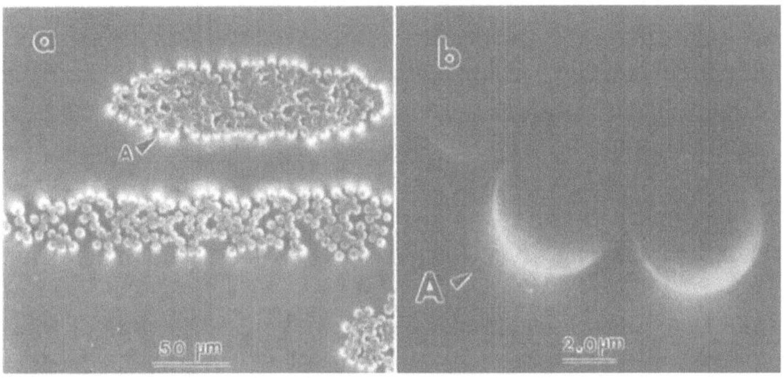

Figure 8.(a) Fiber bundles far away from PTH after 5 thermal cycles,(b) a few fibers in (a)

Similar interfacial debonding was observed in a glass bundle close to a PTH in a specimen cycled between 25°C and 175°C. In E-SEM examinations of the specimens cycled between -5°C and 145°C, no obvious change was observed at resin-fiber interfaces before and after ten thermal cycles. Temperature cycling between -5°C and 145°C did not observably affect the resin-fiber interface.

The CTE for glass fiber is $5.5x10^{-6}$/°C, while the CTE for epoxy resin is $54x10^{-6}$/C. The temperature cycling ΔTs for the thermal-cycled specimens were, respectively, 175°C and 150°C, the high temperatures in the cycling tests were, respectively, 200°C, 175°C and 150°C. Because the area of epoxy-resin matrix outside the glass bundles is much greater than that inside, the absolute volume expansion of the epoxy resin outside of the bundles is much greater than that inside the bundles. This partially explains why the interface degradation and debonding are located along the edges of the bundles. On the other hand, a glass bundle, consisting of many close-packed glass fibers, behaves as an unit in resisting extrinsic thermal stresses. The thermal stresses caused during thermal cycling by the expansion and contraction of a large area of epoxy-resin matrix were concentrated at the resin-fiber interfaces along the edge of bundles facing the large area of epoxy-resin matrix, provoking large displacement and further degradation.

The geometry of the glass fibers distributed in the epoxy-resin matrix also affects the location of the thermal stress concentration. Because glass-fiber bundles are woven in the x and y directions, a constraint is imposed in the x-y plane. The expansion of the fiber-reinforced epoxy laminates is limited in the x and y directions and does not increase much from a temperature below T_g to that above T_g, because of the constraint of the fiber bundles [Seraphim et al., 1989]. In contrast, the z-direction expansion shows an immense temperature dependence. Without the constraint of glass fibers in z-direction, the expansion is about a factor of three higher than that in x or y direction. Above T_g, the z-expansion increases more evidently due to the absence of fiber constraints in z-direction. As shown in Figures 7 and 8, a "tensile stress" was applied in the z-direction, resulting in resin-fiber interface debonding in the z direction for thermal-cycled specimens.

SUMMARY

E-SEM allows investigation of microscopic morphologies without conductive coating and has become a potential technique in microstructural characterizations of microelectronic packages. This advantage of E-SEM particularly facilitates the non-destructive investigations of dielectric materials, such as polymers and ceramics. In this study, thermal and sorptive environmental effects and environmental cycling effects on zero-level, first-level and second-level electronic packages were dynamically investigated using E-SEM.

As an example using zero-level electronic packages, the thermal effect on transimpedance amplifiers was investigated. Surface defects due to the reactions between passivation (polyimide thin film) and metallization (Au) and between the metallization and semiconductor (GaAs) were observed. The interfacial reactions induced gate leakage and gold electromigration as the primary failure mechanisms.

As an example of first-level electronic packages, the relative humidity-cycling effect on multilayer thin-film polyimides was investigated. Delamination of adhesive/polyimide and void formation at the edge of metal lines was observed after numbers of humidity cycling.

Finally, for second-level electronic packages, both humidity-cycling and thermal cycling effects on multilayer PWBs were investigated. The resin/fiber interface debonding observed in this investigation was located primarily on the edges of fiber bundles near PTHs, for both thermal-cycled and humidity-cycled specimens. The location and appearance of the interfacial debonding suggest that the moisture sorption of epoxy, the CTE mismatch between epoxy resin and glass fiber, and the geometry of glass fibers in PWBs play important roles in interfacial debonding.

ACKNOWLEDGMENT

The authors wish to thank the Engineering Research Center, University of Maryland, for its equipment contribution, and the members of the University of Maryland CALCE Electronics Packaging Research Center for their support.

REFERENCE

Bahl, I., Griffin, E., Powell, W., and Ring, I., 1986, A high speed GaAs monolithic transimpedance amplifier, *Dig. of IEEE MTT-S Ins. Mier. Symp.*, pp. 35-38.

Halkias, G., Christou, A., Huang, K., Li, J., and Papanicolaou, N., 1993, Surface related failure mechanisms in polyimide passivated L-band MMICs, *International Reliability Physics Symposium*, pp.375-379.

Tessier, T. G., Adema, G. M., and Turlik, I., 1989, Polymer dielectric for thin film packaging applications, *Proc. 139th Electronic Component Conference*, pp. 127-134.

Poon, S., 1989, High density multilevel copper-polyimide interconnects, *Proceedings NEPCON West*, pp. 426-448.

Paik, K., Bernard, E., Li, J., and Pecht, M., 1993, Humidity cycling experiment on GE-HDI, Status Report, *RELTECH ISHM*.

Balu, R. S. and Pecht, M., 1992, Insulation resistance breakdown in printed wiring boards", *CALCE EPRC internal report.*

Seraphim, D. P., Lasky, R., and Li, C. Y., 1989, "Principles of Electronic Packaging", McGraw-Hill, Inc., U.S.A., pp.365-366.

DETECTION OF CRACKS IN CERAMICS USED IN ELECTRONIC DEVICES USING LIGHT SCATTERING

Scott Hull

Paramax Systems Corp.
NASA Electronic Packaging Program
Goddard Space Flight Center
Lanham, MD 20706

ABSTRACT

Ceramic materials are used in electronic applications as microcircuit packages and substrates which carry signals and power between microcircuits. Fine cracks in ceramic materials can result in mechanical failures, electrical failures, and loss of hermeticity. Often, fine cracks are difficult or impossible to detect using standard nondestructive inspection techniques such as visual inspection, ultrasonic inspection, or vapor crack detection. Dye penetrant inspection is usually effective, but contaminates the part, which is unacceptable for space flight hardware.

An effective nondestructive inspection method of detecting cracks involves examining the way in which light scatters through the ceramic material when viewed with a standard bright field reflected light microscope. This method, termed vicinal illumination, has been used for detecting cracks during failure analyses of several part types, and screening of space flight hardware. The technique has proven effective on several different types of ceramic materials as well. A related method for use with dark field equipment has also been used to successfully locate otherwise invisible cracks.

INTRODUCTION

Cracks in ceramic devices can cause degradation of their performance, or even catastrophic failure in some applications. Nearly invisible microcracks can propagate under stress, resulting in mechanical failure. For this reason, in critical applications it is imperative that such cracks be detected prior to use. Several techniques are available to locate and characterize cracks in ceramic materials. Each has advantages and limitations which affect the choice of the optimum inspection method for a given application.

Visual examination is usually performed on ceramic materials at some time during their manufacture or use. A large crack which displaces the surface of the part will probably be detected at that time. Visual inspection is fast and inexpensive, but small cracks may go undetected. One way to enhance visual inspection is to spray an inert fluorocarbon vapor over the ceramic surface. The fluorocarbon seeps into any cracks in the surface, and retains a shiny appearance after the remainder of the fluorocarbon evaporates. This method is capable of pointing out small cracks more efficiently than a simple visual inspection, but requires a skilled operator to interpret the various defect indications.

Nondestructive Characterization of Materials VI
Edited by R.E. Green, Jr. *et al.*, Plenum Press, New York, 1994

A similar technique uses a dye to penetrate into cracks and clearly indicate their location. Dyes are available which are visible using either white or ultraviolet light. Dye penetrant inspection is among the most effective and widely used techniques for detecting cracks, but it requires that the part be contaminated with a dye which may not be entirely removed. Other methods, such as ultrasonic inspection and x-ray radiography, have advantages and disadvantages as well.

An effective, non-invasive method of optical crack detection involves introducing light into the sample and observing the way in which the light scatters through the ceramic material. Ceramicists have used this technique to observe cracks in research samples for many years.[1] The technique has been referred to as "vicinal illumination" because the sample is illuminated near, but not directly onto the area of interest.[2] Because it is totally nondestructive, this method can be used for screening, quality control, and failure analysis. The technique is as fast as visual inspection, and requires no sample preparation. Operators can be taught to perform this technique in minutes using a standard bright field microscope. The main limitation to the usefulness of the light scattering inspection method is that the material must be translucent in nature.

In vicinal illumination, light is refracted into the sample in a small area, and scatters outward in all directions. If the light encounters a crack, it is reflected back toward the source, and is not transmitted across the crack. This causes a high contrast delineation between bright and dark areas at a crack surface.

MATERIAL PROPERTIES

Light scattering inspection has been used successfully on several different types of translucent materials. The important factor is the ability of the material to scatter light uniformly in all directions. As a rule of thumb, vicinal illumination inspection can be performed successfully on materials similar to a business card in translucence. Grain size, material composition, additives, porosity, and surface finish all have an effect on the light scattering ability of the material.

Quantifying this property is difficult, since no standard measures of light scattering or translucency of ceramic materials seems readily available.[3,4,5,6] In an attempt to quantify the differences between the light scattering capacity of different materials, the percent transmission of white light through samples of different materials was measured. From this data and the sample thickness, an absorption coefficient can be computed for each material using the formula:

$$\ln(I/I_o) = -\alpha x$$

where I/I_o is the ratio of light transmitted through the sample, x is the sample thickness, and α is the absorption coefficient.[7] Table 1 shows the absorption coefficients for several materials.

Table 1. Absorption coefficient for several ceramic materials.

Material	α (inch^{-1})	Preferred Method
Silicon Dioxide (glass slide)	2.14	N/A
Aluminum Nitride (SN1)	55.7	Dark Field
Aluminum Nitride (SN2)	66.0	Dark Field
Aluminum Nitride (SN3)	91.5	Dark Field
Aluminum Oxide (SN1)	303	Bright Field
Aluminum Oxide (SN2)	164	Bright Field
Porous Silicon Dioxide Substrate	174	Bright Field

In general, light scattering inspection was successful for materials with an absorption coefficient of 50 to 300. Dark field light scattering worked well for the more transparent of these materials (absorption coefficient of approximately 50 to 100). In the case of a transparent material, light is transmitted through the material, and is not refracted sufficiently to scatter to the sides. An opaque material reflects and absorbs the light as it enters the material, preventing it from scattering.

The thickness of the ceramic part can contribute to its light scattering ability. Light scattered through a relatively thin sample is reflected by the back surface, resulting in increased scattering in the plane of the sample surface. Scattered light would be hemispherical in a thicker sample, resulting in less usable light for crack detection. It is important to note, however, that vicinal illumination is effective for bulk samples so long as a crack is open to the surface of the material.

Light scattering inspection was most successful for white aluminum oxide electronic substrates and packages. Most ceramic electronic packages are made with black or purple colored materials. Additives in these materials cause them to be opaque, and therefore not candidates for the light scattering crack detection method. White alumina used in some electronic packages and most substrates is supplemented mainly with up to five percent silica, and retains its translucency.

Light scattering inspection was also used to locate cracks in an aluminum nitride plate which was designed as the base of a multichip module package. Vicinal illumination also has revealed cracks in beryllium oxide substrates used in high power electronic devices. While this technique has been used to locate cracks in some ceramic capacitor materials, most of these materials are too opaque to achieve satisfactory results. Vicinal illumination has been used by other investigators to confirm the location of cracks in $BaTiO_3$ multilayer capacitors.[8] A porous silicon dioxide glass substrate used in a large charge coupled device had the ability to scatter light due to internal reflections off the voids within the material.

MICROSCOPE SETUP

There are two basic types of reflected light microscopes; bright field and dark field. A stereo microscope that would be used for a typical visual examination uses dark field lighting to produce an image. Light is reflected off the surface of the sample from some angle other than perpendicular, and travels through the optics to the eyepieces. Dark field microscopes are often equipped with external light sources which can be positioned at any angle.

Bright field reflected light microscopy (also known as coaxial or epi illumination) requires that the light be introduced from an internal light source through a beam splitter so that the light strikes the sample normal to the surface being examined. Most bright field microscopes contain two diaphragms that perform different functions based upon their location in the light path. The aperture diaphragm reduces glare in the image by controlling the amount of light entering the back focal plane of the objective. Closing the aperture diaphragm results in a darker image and decreased resolution. The field diaphragm is located such that it is focused at the field of view. Closing the field diaphragm results in a small roughly circular area of light near the center of the field of view.

PROCEDURE

Crack inspection is performed by opening the aperture diaphragm fully to maximize the light input, and closing the field diaphragm fully to create a small spot of light. Any filters between the light source and the sample should be removed if possible. Inspection can be performed using any magnification, but a 10X objective with 10X eyepieces (final magnification of 100X) seems to be the best general starting point. The sample should be brought into focus and examined in the field of view outside the bright area. Magnification up to 500X can be used to further refine the beam size and placement relative to a suspected crack.

The sample should be examined around the entire perimeter, watching for an abrupt line of high contrast. As the sample is moved so that the light spot traverses the crack, the bright and dark areas should interchange. It should be noted that any opaque metallization on the surface of the sample may alter the light path, and could be mistaken for a crack. It is helpful to displace the field diaphragm to the edge of the field of view, if the microscope has such an adjustment.

Dark field inspection is performed in a similar manner, but requires a more transparent sample. A high intensity concentrated light source (a fiber optic gooseneck lamp, for instance) should be positioned close to the sample, nearly perpendicular to the surface. Examination is performed using a typical low magnification (10X to 120X) stereo inspection microscope. Because dark field inspection can be performed at lower magnification, it is usually faster than the bright field technique for inspecting large samples. As with the bright field method, abrupt bright and dark contrast indicates the presence of a crack.

CASE HISTORIES

Thin Film Substrate (Figures 1 and 2) [9]

Electrical testing of a high frequency amplifier for use on the TOPEX satellite revealed that several thin film microwave substrates had cracked during mounting into the chassis. A nondestructive method of screening was required to ensure that the remainder of the substrates in the chassis did not contain any latent defects. Visual inspection was not considered reliable, and the use of dye penetrant would have contaminated the space flight hardware. A bright field microscope was fitted with a long working distance lens from a macro camera for the inspection. A video camera and large screen television were added to reduce eye strain and allow several people to inspect the samples simultaneously. Each of approximately fifty substrates in each amplifier was inspected around the perimeter, and through the center of the substrate. No additional cracks were observed in electrically functional areas, and the amplifier could be flown with a high level of confidence.

Alumina Package (Figures 3 and 4) [10]

A hybrid microcircuit in the POLAR spacecraft failed electrically during testing. The failure was associated with a lack of hermeticity in the ceramic package. The package, which was made from thick film substrate material, was inspected using bright field light scattering, which revealed a crack across one corner. This technique was also able to show the shape of the crack. From this information the crack initiation site was traced to one of the package leads. It was subsequently discovered that moisture leaking through the crack had caused dendritic growth within the package, leading to the electrical failure.

Multichip Module (Figures 5 and 6)

Bright field inspection of a multichip module showed a small chip-out in one corner of an alumina substrate. The chip-out seemed to be isolated from any electrical circuitry, and not a reliability risk. Vicinal illumination revealed a crack extending from the chip-out toward an electrical connection, which could result in device failure during use. The existence of this crack underscores the need for careful handling during the processing of advanced components.

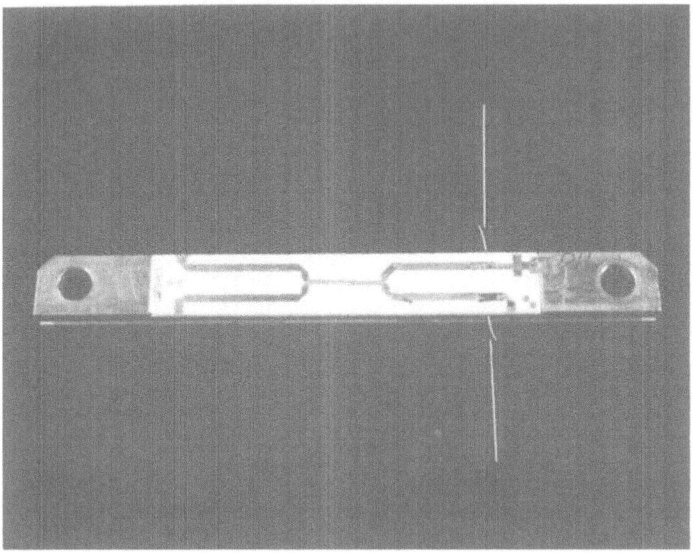

Figure 1. Overall view of the TOPEX thin film microwave substrate and heat sink base. Arrows denote the location of a crack.

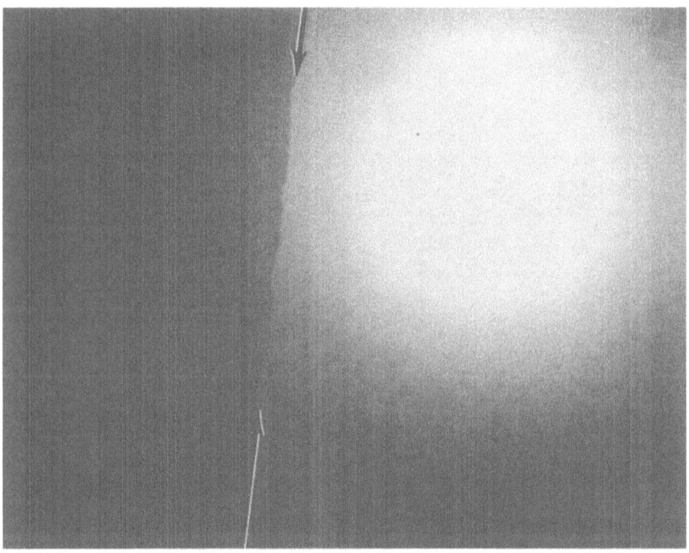

Figure 2. Vicinal illumination view of the crack denoted in Figure 1.

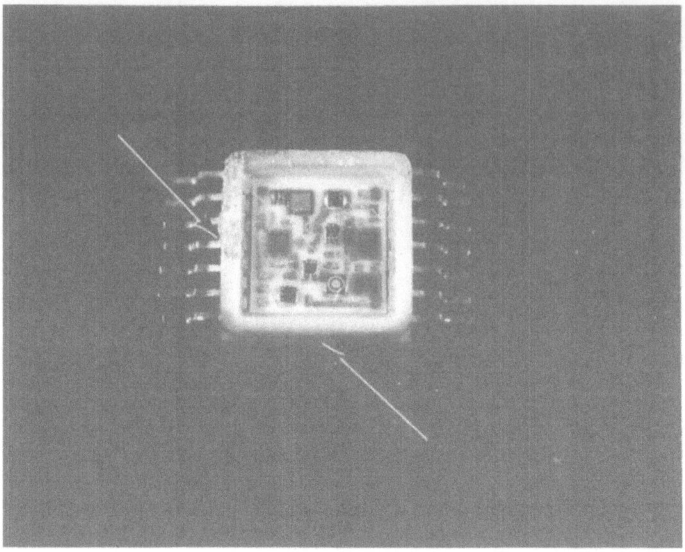

Figure 3. Overall view of the POLAR thick film hybrid package. Arrows denote the location of a crack.

Figure 4. Vicinal illumination view of the crack denoted in Figure 3. Note the change in the direction of the crack near the edge of the package.

Figure 5. Standard bright field photograph showing a small chip-out in a corner of a multichip module substrate.

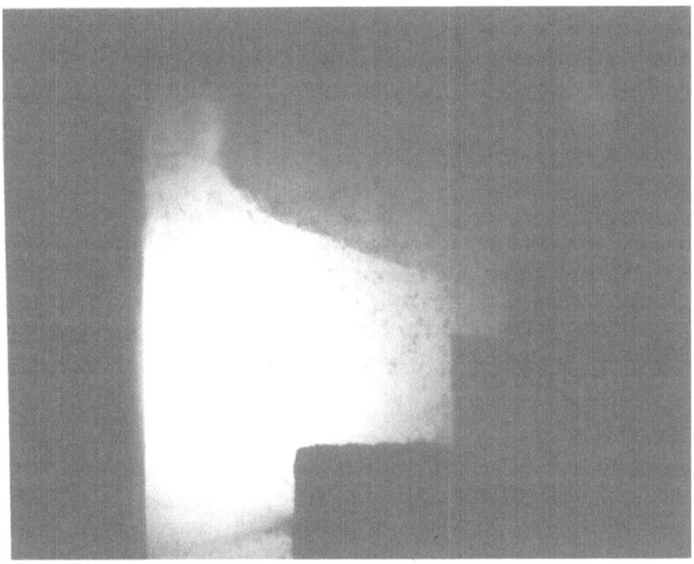

Figure 6. Vicinal illumination view of the area shown in Figure 5. Note that a crack extends from the chip-out toward active circuit metallization.

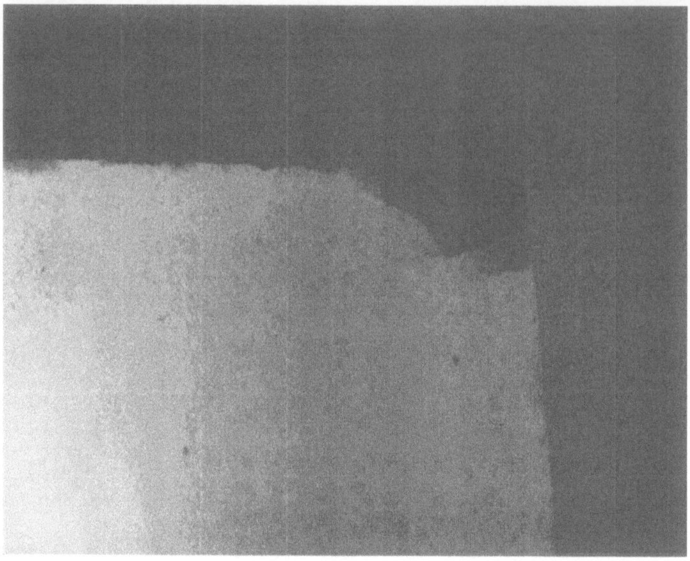

Figure 7. Dark field light scattering photograph showing damage to the corner of an aluminum nitride substrate.

Aluminum Nitride Package (Figure 7)

Several five inch square aluminum nitride substrates were evaluated to determine their suitability for use in multichip module packages for space flight use. Throughout the evaluation, dark field light scattering inspection was used to determine the mechanical integrity of the samples. Following a thermal cycling test, one chip-out and several secondary cracks were observed in one corner. During subsequent tests, the status of the cracks was closely examined, and no further crack propagation was observed.

CONCLUSION

The technique of vicinal illumination of ceramic materials has been successfully used to detect cracks in electronic devices. Inherent material properties, as well as processing, have an influence on the suitability of the technique. Several case histories have shown the efficiency and effectiveness of both bright field and dark field techniques used for failure analysis, screening, and routine evaluation.

ACKNOWLEDGMENTS

The author would like to thank Michael Viens, Dr. Victor Capozzi, Ann Garrison, and John Evans of NASA for their support and assistance with this work. The contributions of Steve Esmacher and Charles Ganger of Paramax were also invaluable.

REFERENCES

1. M.J. Viens, Effect of microstructure on impact damage of polycrystalline alumina, University of Massachusetts, Master of Science thesis (1986).
2. V. Freschette, Personal conversation, New York College of Ceramics (1993).
3. R.C. Buchanan,ed. "Ceramic Materials for Electronics- Processing, Properties, and Applications," Marcel Dekker Inc., New York (1991).

4. L.L. Hench and D.B. Dove,ed. "Physics of Electronic Ceramics," Marcel Dekker Inc., New York (1972).
5. L.M. Levinson,ed. "Electronic Ceramics- Properties, Devices, and Applications," Marcel Dekker Inc., New York (1988).
6. M. Pecht,ed. "Handbook of Electronic Package Design," Marcel Dekker Inc., New York (1991).
7. L.H. VanVlack,"Physical Ceramics for Engineers," Addison-Wesley, Reading,MA (1964).
8. D. Johnson-Walls, M.D. Drory, A.G. Evans, D.B. Marshall, and K.T. Faber, Evaluation of the Reliability of Brittle Components by Thermal Stress Testing, Journal of the American Ceramic Society, Vol. 68, No. 7, July 1985.
9. S.Hull and J. Evans, NASA Goddard Space Flight Center Parts Analysis Laboratory, FA-00257, Greenbelt,MD (1990).
10. P.O'Shea, NASA Goddard Space Flight Center Parts Analysis Laboratory, FA-31448, Greenbelt,MD (1993).

ELECTRICAL CHARACTERIZATION OF PRECISION PIEZOELECTRIC QUARTZ CRYSTAL RESONATORS

J. J. Suter,[1] J. R. Norton,[1] and R. Besson[2]

[1]The Johns Hopkins University
Applied Physics Laboratory
Laurel, Maryland 20723-6099

[2]École Nationale Supérieure de Mécanique et des Microtechnique
Besançon, France

INTRODUCTION

Quartz crystal resonators find application as precision frequency sources and clocks and are often used in spacecraft, beacons, receivers, and transmitters.[1] Quartz is a crystalline material having a high degree of anisotropy. For example, the temperature coefficient of quartz along the optic or z-axis is 8×10^{-6} per degree centigrade, whereas perpendicular to the z-axis it is 15×10^{-6} per degree centigrade. Crystal resonators are now manufactured from synthetically grown quartz, which undergoes several different processing steps to increase the frequency stability of the piezoelectric crystal. Quartz crystal resonators are, for example, electrostatically swept so that impurities located within the quartz material can be removed. Furthermore, the radiation susceptibility of quartz resonators can be reduced by preconditioning them with low doses of ionizing radiation like 1.25-MeV photons from a cobalt 60 source.[2,3] Quartz crystal resonators are manufactured from synthetic quartz having very low levels of impurities (<1 ppm) and minimum twinning defects. This process leads to so-called Premium Q quartz for high-precision resonator applications. The unloaded Q or quality factor is typically about 2.5 to 3.0×10^6 for 5-MHz resonators.

The stability of the most accurate quartz crystals is currently approaching that of more complex rubidium frequency standards.[4] Quartz crystal oscillators have demonstrated frequency stabilities of 7×10^{-14} for a sampling time of 100 s.

This paper discusses recent developments in piezoelectric quartz crystal oscillator technology with an emphasis on improvements in reducing their susceptibility to radiation encountered in the space environment. The sensitivity of quartz resonators to radiation can either be attributed to the properties of the quartz crystal itself or to specific steps in the manufacturing process of the resonator. Specifically, the radiation sensitivity of quartz resonators has been linked to the interaction of radiation with the quartz crystal and metal electrode interface.[2,5,6] Data on both the frequency stability and single-sideband phase noise of 5-MHz quartz crystal oscillators will be also be presented. Emphasis will be placed on quartz

Nondestructive Characterization of Materials VI
Edited by R.E. Green, Jr. *et al.*, Plenum Press, New York, 1994

resonators of the BVA type, where the electrodes are no longer in contact with the active piezoelectric quartz plate. This design has demonstrated some of the highest frequency stabilities as well as greater resilience to external vibration and ionizing radiation.[5]

QUARTZ CRYSTAL OSCILLATORS

Quartz crystal oscillators consist of the frequency determining element, the piezoelectric quartz crystal resonator, an amplifier, and assorted impedance matching electronics. A simplified schematic diagram of an oscillator is shown in Figure 1.

Although many different types of oscillators exist, the frequency stability in all circuits is determined by the piezoelectric quartz crystal.[1] Oscillators with the highest stability have resonant frequencies in the range of 5 to 10 MHz, simply because resonators in these frequency ranges have the greatest unloaded Q factors. These factors typically are about 2.5 to 3.0×10^6, where the Q is defined as the ratio of the frequency's full width at half maximum and its resonance frequency. To preserve the resonator's frequency stability over longer periods, the quartz crystal resonator is placed in a temperature-controlled oven. Temperatures are typically stabilized to within one-thousandth of a degree at the resonator's turn-over temperature (the temperature at which the resonator's frequency deviation reaches a minimum value).

Quartz crystal resonators operate by means of the piezoelectric effect. In tensor form, this phenomenon may be written as:

$$\sigma_{ij} = \sum_{k,l} c_{ijkl} u_{kl} - \sum_{k} e_{kij} E_k , \tag{1}$$

where σ_{ij} is the second-rank stress tensor, c_{ijkl} is the fourth-rank elastic stiffness tensor, u_{kl} is the second-rank strain tensor, e_{kij} is the third-rank piezoelectric tensor, and E_k is the electric field vector (the indices i, j, and k may take on values of 1, 2, and 3). The converse piezoelectric effect is expressed by

$$D_i = \sum_{j} \epsilon_{ij} E_j + \sum_{j,k} e_{ijk} u_{jk} , \tag{2}$$

where D_i is the electric displacement vector and ϵ_{ij} is the electric permitivity.[7]

The elements of the third-order piezoelectric tensor take on specific values as determined by the properties of the quartz crystal. Cutting the crystal at specific geometrical angles allows

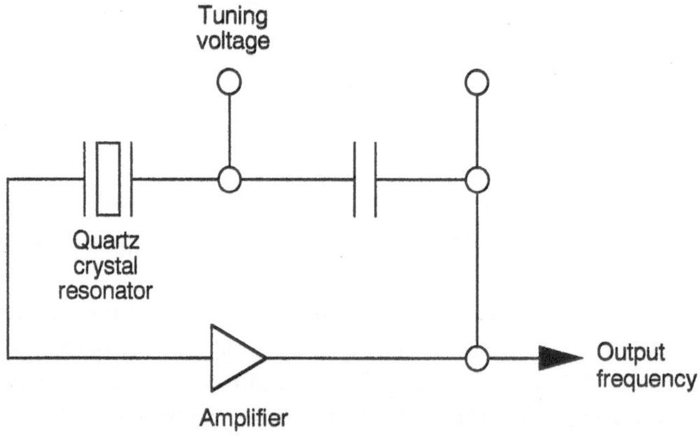

Figure 1. Simplified quartz crystal oscillator circuit.

for control of its physical properties, (e.g., temperature sensitivity).[1] The most popular quartz crystal cut used in high-precision spacecraft oscillators is the stress-compensated (SC) cut. This class of resonators has some of the lowest measured temperature sensitivities and some of the best aging rates (see Table 1).

Table 1. Quartz crystal oscillator performance.

Parameter	Performance
Resonant frequency	5 MHz
Aging rate	4.0×10^{-12} per day
Temperature sensitivity	6.1×10^{-13} per degree C
Acceleration sensitivity	1.5×10^{-9} per *g*
Magnetic sensitivity	2×10^{-12} per gauss

Conventionally, the electrical signal from the quartz crystal was derived from a set of metal (gold) electrodes deposited directly on the crystal blank. Resonators have also been developed in which the electrical signal comes from electrodes that are not directly attached to the crystal (see Fig. 2). This class of electrodeless resonators is known as BVA resonators.[8]

RADIATION SUSCEPTIBILITY

One very important parameter of spaceflight quartz crystal resonators is their apparent sensitivity to radiation in the space environment. Ionizing (gamma) and particle radiation (protons, neutrons, and electrons) interact with the quartz crystal and cause the resonator to change its frequency. These changes in frequency may be separated into two distinctly different effects: 1) permanent induced frequency offsets and 2) momentary frequency shifts that anneal over a specified time after the radiation subsides. Studies have shown that exposure of the quartz crystal to large radiation doses [>10 krad(Si)] introduces frequency shifts and offsets mainly related to the interaction of radiation with the quartz crystal lattice itself.[6,9] These levels of radiation exposure are generally only encountered in nuclear (enhanced) environments and not in low-Earth satellite orbits. On the other hand, the frequency shifts induced in quartz crystal oscillators exposed to low-Earth orbit radiation [<10 rad(Si)/day] are caused by the interaction of the radiation with the electrode-quartz crystal interface.[2,6,7] Since the induced frequency shifts are relatively small, approximately 1×10^{-11}/rad(Si), the changes occurring in the quartz crystal due to radiation can presently only be detected by measuring the electrical characteristics of the piezoelectric resonator. Tests involving *in situ* measurement of lattice displacement, IR extinction coefficients, and ultrasonic characteristics have not been shown to be adequate diagnostic tools. Furthermore, the quartz crystal is shielded from the outside environment by means of a high vacuum enclosure that protects the crystal from the deposition of impurities that adversely affect its aging rate. Hence, direct access to the piezoelectric quartz crystal is not available.

The susceptibility of quartz crystal resonators to low levels of radiation [<10 rad(Si)/ day] has been the subject of several comprehensive studies.[2,6,10] Since the induced frequency shifts in the resonators are small, special tests were designed to measure them. As shown by Norton et al.,[2] this measurement is based on a direct comparison of the quartz resonator's frequency with that of a reference source, where the reference has a frequency stability greater than that of the quartz crystal resonator under test. This system can resolve frequency changes of $<1 \times 10^{-14}$, which at a quartz crystal resonator frequency of 5 MHz correlates with a resolution of better than 50 nHz.

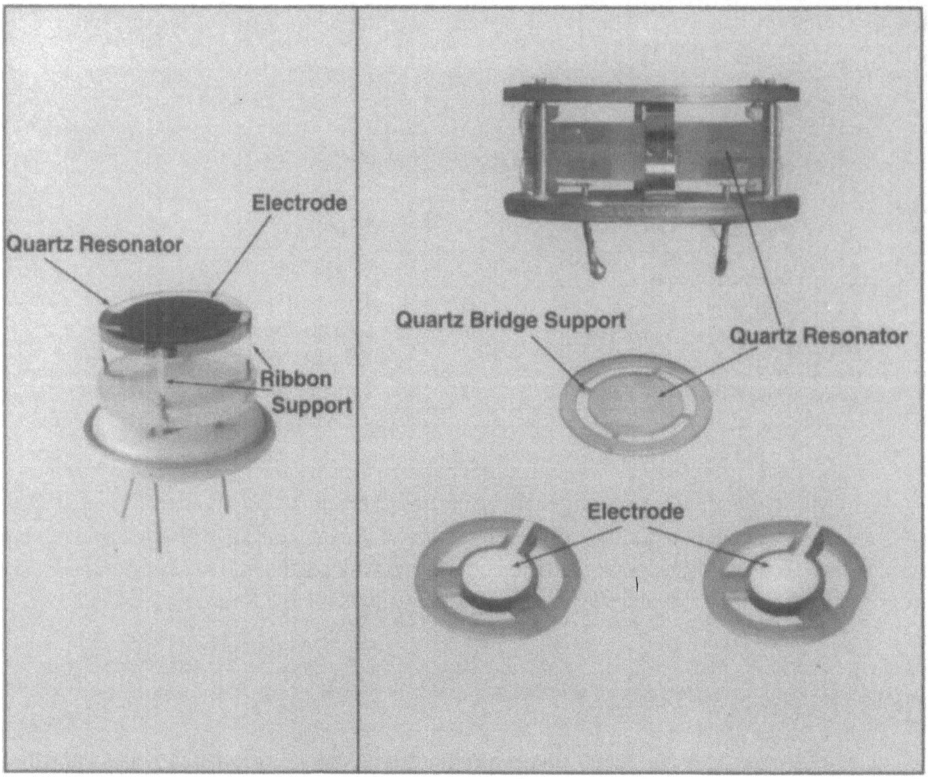

Figure 2. Conventionally mounted quartz resonator (left) and BVA quartz resonator (right).

The radiation environment encountered by satellites in low-Earth orbits (altitudes between 500 to 1500 km) is mainly composed of protons and electrons.[9] The protons typically have energies between 1 and 1000 MeV, and the kinetic energy of the electrons varies from several keV to the lower MeV ranges. Suter et al.[9] have analyzed the specific radiation doses, kinetic energy levels, and fluxes encountered by satellites in these orbits. In general, most of the radiation is delivered to the quartz crystal during a specified orbit segment, namely when the spacecraft traverses the radiation-intense zone called the South Atlantic Anomaly. This is a region where an anomaly in the Earth's magnetic field causes an increase in proton and electron fluxes.

To simulate the response of quartz crystal resonators to radiation encountered in a satellite, a series of tests was conducted using 1.25-MeV cobalt 60 radiation as well as protons generated by a cyclotron.[2,9] It was also shown that ionization effects in resonators due to proton radiation can be emulated using 1.25-MeV cobalt 60 radiation, since the interaction of protons as well as that of the photons involves the creation of equivalent numbers of electron-hole pairs. Figure 3 shows the typical response of a quartz crystal resonator to cobalt 60 radiation.

The vertical axes show the relative frequency shift with respect to a reference source; the horizontal axes show the elapsed time. The markers indicate when the quartz crystal was exposed to radiation. To simulate several satellite orbit exposures, the quartz crystal resonator was irradiated with the cobalt source four consecutive times. Figure 3 clearly shows the response of the resonator to the radiation. After the radiation source is removed it appears that the crystal's frequency shift diminishes. However, the crystal has suffered a permanent frequency shift. This shift indicates that the resonant frequency of the resonator changed. In Figure 3A, the change amounts to approximately -7.5×10^{-11} or 375 μHz. It is also apparent

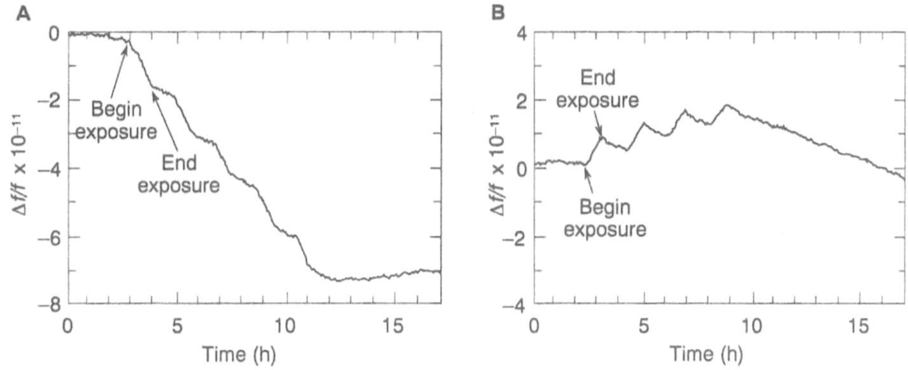

Figure 3. Frequency response of BVA-SC quartz S/N23 resonator to radiation. A. Before preconditioning with 20 krad(Si). B. After preconditioning. Exposure was 0.6 rad(Si) for 42 min. at a dose rate of 0.014 rad(Si)/min for both.

that the response of the quartz crystal resonator can be minimized by preconditioning the crystals with cobalt 60 radiation. Studies have shown that the optimum level of preconditioning is about 20 krad(Si).[2] Any greater amount of radiation often causes an increase in the resonator's aging rate, which has an adverse effect on its long-term frequency stability.

ELECTRICAL CHARACTERIZATION

Again, the very small frequency changes occurring in a quartz crystal resonator due to external influences like radiation can only be discerned electrically by incorporating the quartz crystal resonator in an oscillator circuit (Fig. 1 showed a simplified diagram of this circuit where the quartz resonator is the frequency determining element). Besides the oscillator circuit, the quartz crystal is temperature-controlled by a thermal control circuit to an accuracy of about 1 millidegree centigrade near its turn-over temperature. The frequency stability characteristics of the quartz oscillator are measured, as previously mentioned, by comparing its output with that of a frequency reference. This circuit can resolve frequency shifts of 1×10^{-14} (50 nHz).

Figure 4 presents the frequency stability of a quartz crystal oscillator, where the vertical axis displays the stability as the so-called Allan variance.[11,12] This is basically a modified statistical variance for which all the major types of noise observed in oscillators converge. That is, the classical statistical variance does not converge when the sample size N approaches infinity, since some noise processes in oscillators diverge rapidly at low Fourier frequencies. The Allan variance (or two-sample variance) is therefore preferred to characterize the frequency stability of oscillators.

The technique for the electrical characterization of oscillators by means of the Allan variance is basically a measurement in the time domain. In the frequency domain, oscillators can be electrically characterized by determining their single-sideband (SSB) phase noise, whch is a measurement of the oscillator's noise in a specified bandwidth at a Fourier frequency (f) from its carrier. Figure 5 shows the phase noise ($£_f$) as a function of the Fourier frequency, with the measurements all normalized to a 1-Hz bandwidth.

The characterization of oscillators in either the time domain (Allan variance) or frequency domain (phase noise) is not unique to quartz crystal oscillators; simple resonant circuits like LC oscillators as well as atomic frequency standards can be characterized using these techniques. For example, the Allan variance for an atomic rubidium frequency standard is shown in Figure 4.

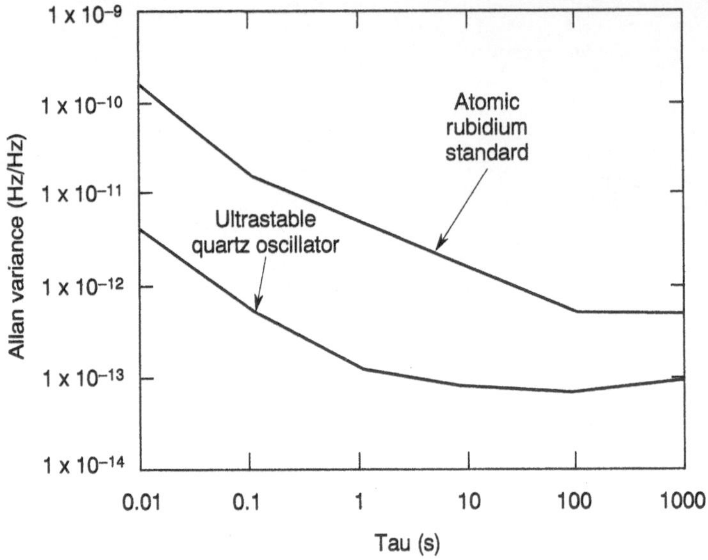

Figure 4. Allan variance of precision frequency standards.

The noise sources in quartz crystal oscillators can be analyzed by studying their phase noise performance. By examining Figure 5, it is apparent that the characteristic shape of the SSB phase noise indicates that specific noise processes dominate at certain Fourier frequencies. These noise processes can be easily understood in terms of power-law relationships.

By its definition, SSB phase noise corresponds to instabilities in the phase or frequency of a signal.[12,13] The instantaneous output frequency of a signal generator can be expressed as

$$V(t) = [V_0 + \delta(t)]\sin[\omega_0 t + \phi(t)], \tag{3}$$

where V_0 is the signal's nominal output, $\delta(t)$ represents amplitude noise, ω_0 represents the signal's frequency, and $\phi(t)$ represents phase noise. The fractional instantaneous frequency deviation $y(t)$ can be defined as

$$y(t) = \frac{1}{\omega_0}\frac{d\phi(t)}{dt} = \frac{\dot{\phi}(t)}{\omega_0}. \tag{4}$$

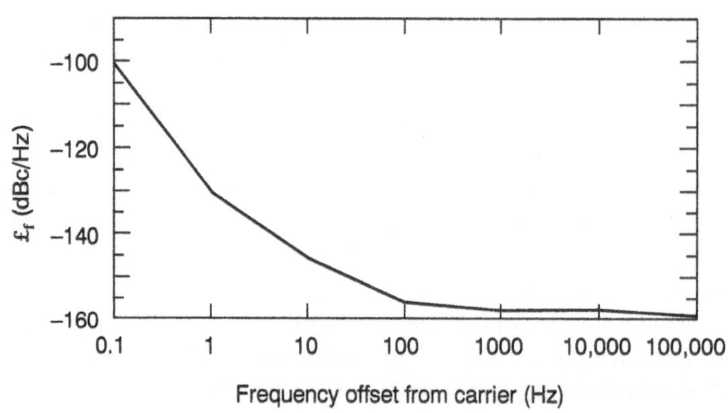

Figure 5. Phase noise ($£_f$) of a 5-MHz BVA oscillator.

It can be shown that if $S_\phi(f)$ is the spectral density of phase fluctuations, then $S_y(f)$, the spectral density of frequency fluctuations at a frequency of f Hz from the carrier, is

$$S_y(f) = \frac{1}{\omega_0^2} S_\phi(f), \tag{5}$$

where ω_0 is the nominal frequency of the oscillator.

Single-sideband phase noise ($£_f$) is defined as the ratio of SSB power in a 1-Hz bandwidth offset a frequency of f Hz from the carrier to the total signal power. In other words,

$$£_f = \frac{S_\phi(f)}{4}. \tag{6}$$

The power spectral density $S_y(f)$ is modeled as a linear combination of power-law processes, which take on the form $S_y(f) = h_\alpha f^\alpha$, where h_α is a function of the power-law noise process. The coefficient α can assume integer values between -2 and $+2$. These values correspond to power-law noise processes: f^{-1} represent flicker noise; f^{-2}, random walk; and f^0, white noise.[14]

The main contributor to white noise is the electronics of the oscillator circuit by means of the thermal noise represented as Nk_BT (refer to the flat portions of the curve in Fig. 5). Theoretically this thermal noise is limited to approximately -174 dBc/Hz in a 1-Hz measurement bandwidth. A pronounced feature of flicker noise or $1/f$ noise is its strong dependence on the unloaded Q factor of the quartz crystal. This dependence is close to Q^{-4}. Hence, increasing the Q factor of a quartz crystal resonator decreases the contributions of this type of noise. Quality factors, and specifically the unloaded Q, are a direct function of the material properties of the quartz: the number of impurities, twinning defects, and other acoustic-loss defect centers.

SUMMARY

Electrical characterization of quartz crystal resonators is currently the technique of choice for determining their response to external influences like radiation, magnetic fields, and acceleration. Measurements of the quartz crystal's response can either be performed in the time domain by measuring the Allan variance or in the frequency domain by determining the SSB phase noise in a specified bandwidth. Both techniques use reference oscillators with better frequency stability than the oscillator under test. The resolution of the Allan variance measurement technique is typically 1×10^{-15}, and the maximum discernible SSB phase-noise noise floor is approximately -174 dBc/Hz. These electrical characterization techniques are currently the only methods available that can resolve the very small frequency shifts in the quartz crystal resonator. Other methods (e.g., optical interferometry) can potentially resolve the very small displacements or strains in the quartz crystal blank, but the absence of direct optical access to the crystal prevents the implementation of such techniques. Furthermore, the thermal power generated by the optical signal, which is subsequently absorbed by the quartz crystal, would confound the actual response of the effect to be measured.

The BVA quartz crystal resonator, where the electrodes are not in physical contact with the active piezoelectric quartz crystal, has the highest Q factor (2.5 to 3.0×10^6). This class of resonators has also achieved the highest frequency stabilities and SSB phase noise (Figs. 4 and 5). The frequency stability of this type of piezoelectric quartz crystal resonator routinely exceeds that of the more complex atomic rubidium frequency standard. In addition, the quartz crystals are less susceptible to ionizing radiation, in particular to the proton and electron dose levels encountered in low-Earth orbits.

REFERENCES

1. L.E. Halliburton and J.J. Martin, Properties of piezoelectric materials, *in:* "Precision Frequency Control, Vol. I," E.A. Gerber and A. Ballato, eds., Academic Press, New York (1985).
2. J.R. Norton, J.M. Cloeren, and J.J. Suter, Results from gamma ray and proton beam radiation testing of quartz resonators, *IEEE Trans. Nucl. Sci.* NS-31(5):1230 (1984).
3. J.J. Suter and R.H. Maurer, Low and medium dose radiation sensitivity of quartz crystal resonators with different aluminum impurity content, *IEEE Trans. Ultrason. Ferroelectr. Freq. Contr.* UFFC-34(6):667 (1987).
4. J.R. Norton and J.M. Cloeren, Precision quartz oscillators and their application in small satellites, *in*: "Proc. 6th Ann. AIAA/USU Conf. on Small Satellites" (1992).
5. J.R. Norton and R.J. Besson, Tactical BVA quartz resonator performance, *in:* "Proc. 48th Ann. Symp. on Frequency Control" (in press, 1993).
6. J.J. Suter, R.H. Maurer, and J.D. Kinneson, "The Susceptibility of Electrodeless Quartz Crystal BVA Resonators to Proton Ionization Effects," *IEEE Trans. Nucl. Sci.* NS-35:451 (1988).
7. J.F. Nye, "Physical Properties of Crystals," Claridon Press, Oxford, England (1985).
8. R.J. Besson, "Resonateur a Quartz a Electrodes non Adherentes au Crystal," *Demande de Brevet d'Invention*, No: 7601035, Institut National de la Propriete Industrielle Paris (1976).
9. J.J. Suter, R.H. Maurer, J.D. Kinnison, J. Vig, R. Besson, and A. Koehler, The effects of ionizing and particle radiation on precision frequency standards, *in*: "Proc. 46th Ann. Symp. on Frequency Control," 798 (1992).
10. T.M. Flanagan, R.E. Leadon, and D. L. Shannon, Evaluation of mechanisms for low-dose frequency shifts in crystal oscillators, *IEEE Trans. Nucl. Sci.* NS-33(6):1447-1453 (1986).
11. J.A. Barnes et al., Characterization of frequency stability, *IEEE Trans. Instrum. Meas.* IM-20(2):105-120 (1971).
12. W.C. Lindsay and C.M. Chie, Identification of power-law type oscillator phase noise spectra for measurements, *IEEE Trans. Instrum. Meas.* IM-27(1):46-53 (1978).
13. W.P. Robins, "Phase Noise in Signal Sources," Peter Peregrinus Ltd., London (1984).
14. D.W. Allan, Should the classical variance be used as a basic measure in standards metrology?, *IEEE Trans. Instrum. Meas.* IM-36(2):646-654 (1987).

INTERPLY PRESSURE MEASUREMENTS IN RAPIDLY HEATED CARBON-PHENOLIC COMPOSITES

G. F. Hawkins and E. C. Johnson

The Aerospace Corporation, M2/248
P. O. Box 92957
Los Angeles, CA 90009

INTRODUCTION

Carbon-phenolic composite materials can withstand very high temperatures and thermal fluxes. They are commonly used to line the throat and exit cone of solid rocket motors (SRMs). As hot gases from the SRM impinge on the carbon-phenolic, it slowly chars and erodes. The char protects the material beneath it from the thermal flux.

There is a potential for catastrophic failure if large pieces of char are ejected from the liner due to a build up of pressure between the plys of the composite; a process referred to as spalling. The gases which lead to spalling are formed when the carbon-phenolic decomposes at high temperatures. The pressures associated with spalling are difficult to measure directly. They have, however, been estimated using data taken from indirect measurements.[1] This paper will serve to document the results of an attempt to directly measure the maximum interply pressures attained in carbon-phenolic composites upon exposure to a high temperature thermal pulse. The pulse was designed to simulate the initial exposure of a SRM exit cone liner upon ignition. To measure the pressure, a previously documented[2,3] embedded sensor/Acoustic Emission (AE) technique was employed. A brief description of this technique follows.

PRESSURE SENSING TECHNIQUE

The embedded pressure sensors consist of microballoons which are tiny, hollow spheres ranging in diameter from ~ 1 to 160 μm. Glass microballoons are commercially available in bulk for use as low cost, lightweight filler to be mixed with resins and other materials. For the present application, carbon microballoons capable of withstanding high temperatures were employed.

To understand the technique, consider first what happens when microballoons are mixed in a carrier fluid, placed on the surface of an AE transducer and then subjected to a pneumatic pressure. Since the microballoons within a typical sample exhibit a random distribution of rupture strengths, the weakest burst first as the pressure increases. As the microballoons break they give rise to AE events which are easily detected by the transducer. A typical plot of the number of events versus pressure is presented in Fig. 1. If the mixture had experienced a pressure previously, little AE activity would occur until the test pressure exceeded the previous level. The abrupt onset of AE indicates the value of the pressure previously experienced by the microballoons. This manifestation of the Kaiser Effect[4] is evident in the data of Fig 2. Microballoons can therefore be placed at strategic locations

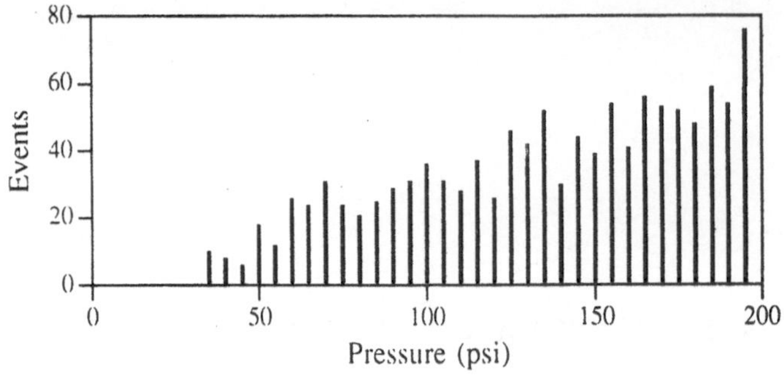

Figure 1. Acoustic Emission Events vs. Pressure for carbon microballoons which have not been pre-pressurized. To make the measurement, the microballoons are first mixed with a carrier fluid (grease). The mixture is then placed on the surface of an AE transducer subjected to pneumatic pressure. The histogram points are the sum of AE events over 5 psi increments.

within a material or component and "read" at a future date to determine the maximum pressure experienced at that location.

In the experiments referenced above, the microballoons were mixed with a fluid before any pressure was applied. The vaporization of such fluid would contribute to the pressure being measured in this study, thereby corrupting the data. Accordingly, dry microballoons were used. After exposure to pressure, the dry microballoons were retrieved and mixed with an acoustic coupling fluid for AE testing. Preliminary tests with dry microballoons revealed that numerous AE events occurred at pressures below those to which the dry microballoons had already been exposed. In fact, so many "early" AE events occurred that the maximum pressure that the dry microballoons had been exposed to could not be determined.

An experiment was performed to determine the cause of this anomaly. Microballoons were placed in a chamber which contained a window so that they could be viewed through a microscope while being pressurized. Rather than imploding individually, the microballoons broke in groups of 10 to 20 at a time. It is therefore believed that when one balloon breaks, pieces are ejected (shrapnel). These ejected pieces damage neighboring balloons. Some of the damaged balloons break immediately, while others fail at lower pressures during a subsequent pressurization. This had not been noticed in earlier tests because the balloons had been mixed with a fluid which served to contain the shrapnel.

Armed with this knowledge, graphite powder was mixed with the microballoons to

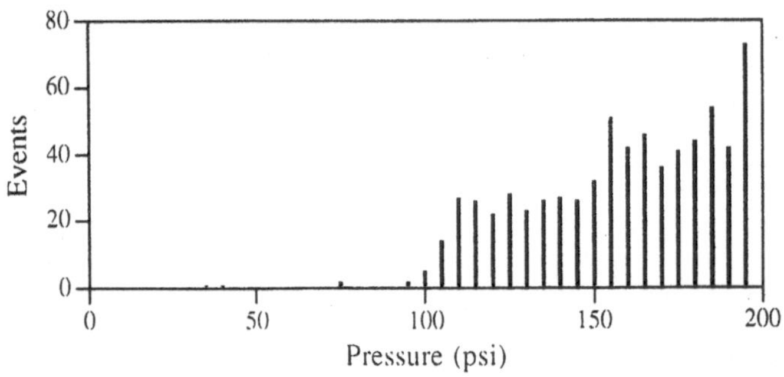

Figure 2. Acoustic Emission Events vs. Pressure for a carbon microballoon/grease mixture which was pre-pressurized to 100 psi. The data was taken in the same fashion as that of Fig. 1. Note that the pre-pressurization pressure is manifested by the onset of significant AE activity.

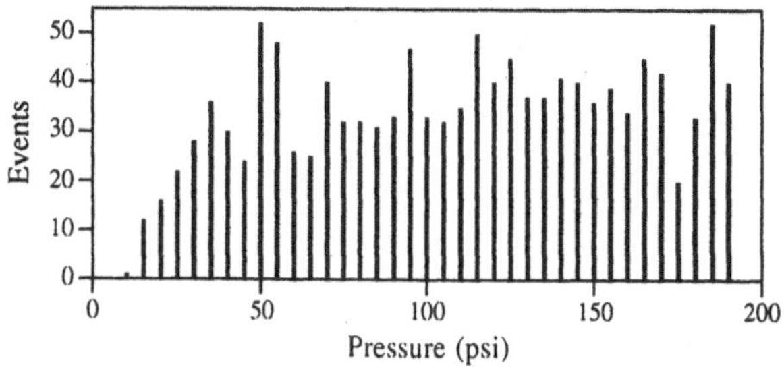

Figure 3. Acoustic Emission Events vs. Pressure for a mixture of 75% graphite powder and 25% microballoons. The mixture was preheated to 600° C, added to an acoustic couplant and "read" as described in the text.

mitigate the effects of the shrapnel. A series of tests was performed to determine what percentage of each constituent should be used. Acceptable results were achieved with a mixture of 75% graphite-powder and 25% carbon-microballoons. Such a mixture was preheated to 600° C, added to an acoustic couplant and "read" as described earlier. The resulting pressure spectrum, presented in Fig. 3, is much like that of Fig. 1 in that events are registered throughout the spectrum. For comparison, the pressure spectrum for a 3:1 graphite-powder/carbon-microballoon mixture which was heated to 600° C and then pre-pressurized in air to 75 psi is presented in Fig. 4. This spectrum is significantly different from that of Fig. 3. The pre-pressurization value (75 psi) is signified by the onset of consistent AE activity at approximately 70 - 75 psi.

EXPERIMENT

Disc shaped samples of 1.8 inch diameter with faces parallel to the ply direction were cut from FM 5055 carbon-phenolic panels obtained from US Polymeric Inc. Three flat-bottom holes of various depths were drilled into the back face of each sample as depicted in Fig. 5. Plugs for these holes were machined out of the same FM 5055 material with a matching ply orientation. A small cavity was machined into the end of each plug. These cavities were filled with microballoons. A small amount of adhesive, used as a pressure seal, was applied around the edge of the cavity. The adhesive (Ceramabond obtained from

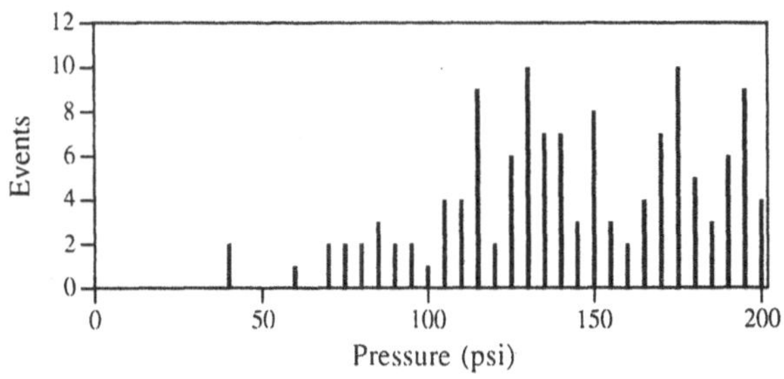

Figure 4. Acoustic Emission Events vs. Pressure for a 3:1 graphite-powder/carbon-microballoon mixture which was pre-pressurized to 75 psi. Note that though several sporadic events occur at lower pressures, the onset of consistent AE activity is at approximately 70 -75 psi.

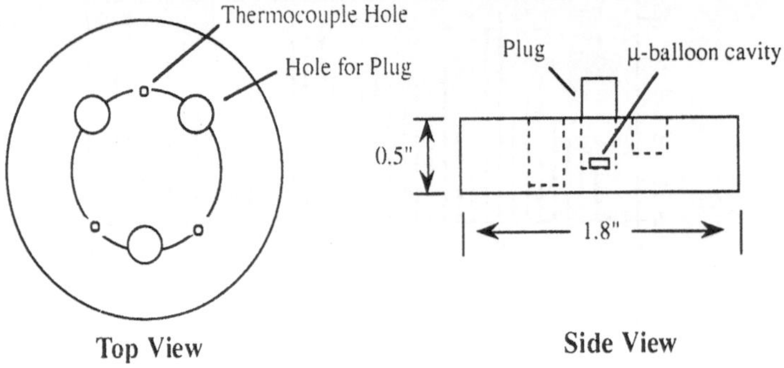

Figure 5. Carbon-phenolic specimen configuration. See text for details. Note that three different ply depths are tested simultaneously.

Aremco Products) was chosen for its high temperature and low outgassing properties. After insertion of the plug, a fillet of Torr-seal™ was used to seal the protruding portion of the plug to the back face of the sample. Holes for thermocouples were drilled adjacent to and at the same depth as each of the plugs to measure the microballoon exposure temperature.

An arc jet was used to heat the samples. In the arc jet facility, nitrogen gas is heated as it flows past a direct current arc. The gas expands through a nozzle and is directed into a chamber containing the sample. A pneumatically controlled shield protects the sample until the proper test conditions are established. For these tests a 6 gm/sec nitrogen flow was expanded as a subsonic free jet into one atmosphere pressure. With 160 kW input power, slug calorimeter measurements indicated that a uniform heating rate of 75 BTU/ft^2 sec over a 2 inch diameter flat face was obtained. Following the heating cycle, liquid nitrogen was sprayed on the face of the sample to limit additional heat soaking.

After each arc jet exposure, the plugs were removed from the exposed sample by twisting and the microballoons recovered. The microballoons were mixed with an acoustic couplant and placed on an AE transducer in a pressure chamber. The pressure in the chamber was increased until a substantial number of AE events was recorded.

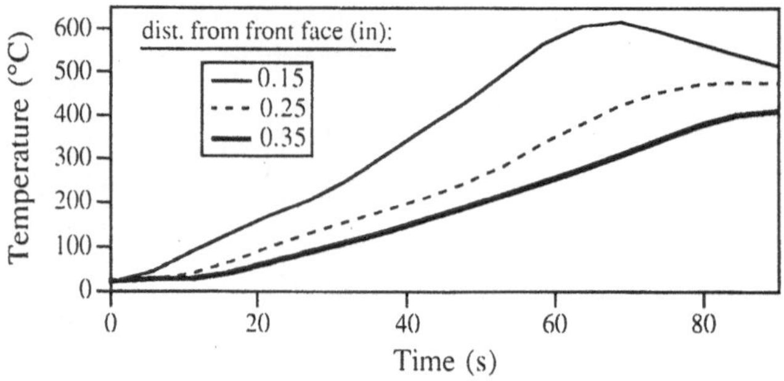

Figure 6. Temperature traces for the thermocouples at 0.15, 0.25 and 0.35 inches from the front surface of the specimen upon exposure to the arc jet at time zero. The depths correspond to the those of the cavities containing microballoons. As expected, the microballoons closest to the front surface reached a higher temperature.

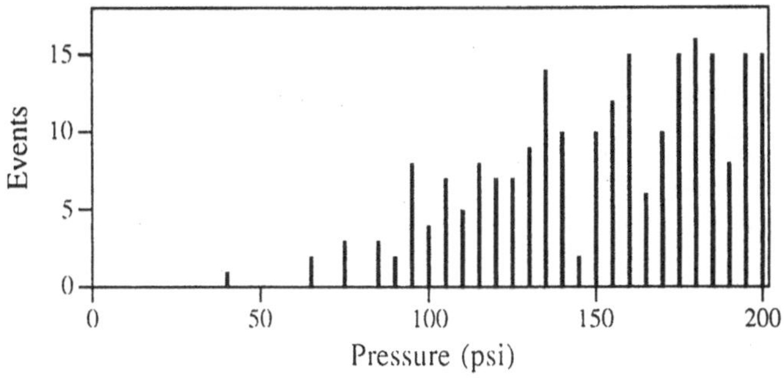

Figure 7. Acoustic Emission Events vs. Pressure for the microballoons taken from the cavity closest to the front face of the specimen which was subjected to the heating detailed in Fig. 6. Note the similarity of this data to that of Fig. 4 where the microballoons had been pre-pressurized to 75 psi.

RESULTS

Several specimens were tested before a heating rate was achieved which promoted delamination. Thermocouple traces for the delaminated specimen are displayed in Fig. 6. As expected, the microballoons closest (~0.15 inch) to the front surface experienced the highest temperature. The cooling managed to keep the "soak" temperature below this value.

The microballoons were recovered and "read" in the normal manner. Fig. 7 shows the AE results from the microballoons closest to the front face. Note the similarity of this data to that of Fig. 4, where the microballoons had been pre-pressurized to 75 psi. The microballoons taken from each of the remaining two plugs (further from the front surface) yielded a pressure spectrum similar to that of Fig. 3, suggesting that they experienced very little, if any, pressure. It is interesting to note that the delamination in this specimen occurred ~ 0.15 inch below the front face.

CONCLUSIONS

The microballoon pressure measurement technique has been shown to work at temperatures of up to 600 C. Using this technique, pressures of ~ 75 psi were found in the interior of a carbon-phenolic sample exposed to an arc-jet heat pulse. This value agrees with other estimations of the internal pressure.

Further experimentation is required to validate and better quantify these results. One potential source of error is caused by the unoccupied volume in the microballoon cavity. This ullage does not exist in the normal material and reduces the highest pressure experienced by the microballoons. An additional source of error is from residual outgassing of the adhesive used to seal the microballoon area. Experimentation needs to be performed to quantify and correct for these effects.

ACKNOWLEDGMENTS

The authors gratefully acknowledge R. L Ruiz and F. Izaguirre for assistance with characterizing the carbon microballoon mixtures and for acquiring the AE data. The authors would also like to thank E. M. Kaegi and H. A. Bixler for their effort in exposing the specimens to the arc jet.

REFERENCES

1. E. H. Stokes, "Prediction of Ply Lift Temperature in Two Dimensional Polymeric Composites," 1989 JANNAF Rocket Nozzle Technology Subcommittee Meeting, October 17 - 19, Naval Surface Weapons Center, 1989.
2. G. F. Hawkins, James R. Lhota, J. R. Hribar and E. C. Johnson, Review of Progress in Quantitative Nondestructive Evaluation (edited by D. O. Thompson and D. E. Chimenti, Plenum Press, New York, 1993), Vol 12A, 989 - 993.
3. James R. Lhota, P. M. Sheaffer and G. F. Hawkins, Review of Progress in Quantitative Nondestructive Evaluation (edited by D. O. Thompson and D. E. Chimenti, Plenum Press, New York, 1988), Vol 11A, 1097 -1102.
4. J. Kaiser, Archiv. Eisenhuttenwess. Vol. 24, 43 (1953) AREIA.

NON-DESTRUCTIVE CHARACTERIZATION OF THIN DIAMOND-LIKE CARBON, SEMICONDUCTING AND HIGH TEMPERATURE SUPERCONDUCTING FILMS

H.D. Bist, P.S. Dobal, S. Bhargava, R.N. Soni
and P.K. Khulbe

Department of Physics and Center for
Laser Technology
Indian Institute of Technology
Kanpur, 208 016, U.P.
India

INTRODUCTION

More than fifty techniques for physico-chemical characterization are being used to probe various aspects of materials. Optical absorption and emission and inelastic scattering (neutron, Raman and Brillouin) methods have been used extensively to study the compositional and structural aspect of materials[1]. Raman spectroscopy has evolved as a powerful, non-destructive, contactless analytical technique with well established credentials for molecular specificity. In this paper our recent micro Raman work on the thin films of (a) Diamond and diamond-like carbon (DLC), (b) GaAs epilayers having oval defects and (c) $YBa_2\ Cu_3O_{7-\delta}$ high temperature superconductor (HTSC) is presented.

EXPERIMENTAL

The block diagram of the micro-Raman set-up is shown in Figure 1. A saturated core automatic servo up-converter and stabilizer is used for operating 15 Watt Ar^+ laser. Another voltage stabilizer (shown on right hand top side of Figure 1) is used for operating Spex 1877D triplemate, photo-multiplier tube (PMT), charge couple detector (CCD) and other electronic devices. Chilled water is used to remove the heat generated from laser plasma tube, its power supply and Dye laser system. The 1877D Triplemate consists of a filter and a spectrograph stage. The filter stage has two modified Czerny-Turner spectrometers coupled in a subtractive mode, each having a 50mm x 50mm plane grating (600 lines/mm). The spectrograph stage uses an asymmetric Czerny-Turner mount with a 64mm x 64mm plane grating having 1200 lines/mm to produce a dispersion of 1.4 nm/mm.

Nondestructive Characterization of Materials VI
Edited by R.E. Green, Jr. *et al.*, Plenum Press, New York, 1994

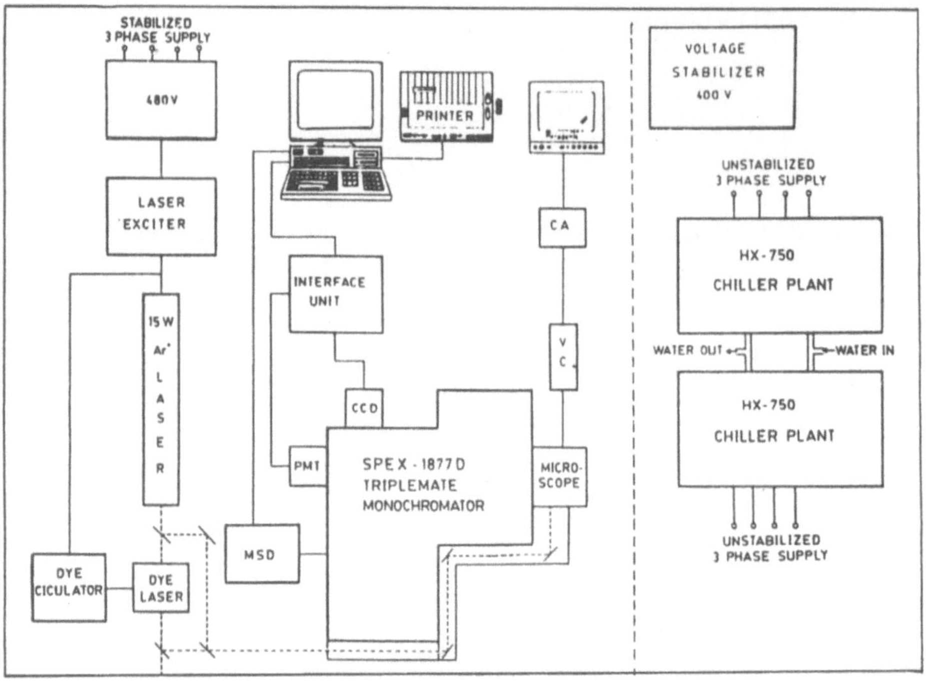

Figure 1. Micro-Raman with CCD detector spectrometer system.

A flat, undistorted focal plane is essential for sensitive work with optical multichannel detectors (like vidicons, diode-arrays or charge coupled devices). This is achieved by vignetting the field with the help of a torroidal lens, which flattens the response of the non-dispersed segment of the band pass received from the exit slit of the filter stage. The spectral coverage of the CCD is limited to 630 cm^{-1} due to its 12.5mm x 8.5mm size, although, a focal plane of 25mm x10mm could be obtained in this system. The spectra are reproducible to within ±2 cm^{-1} with a resolution of 2 cm^{-1} for a particular scan setting of the spectrometer. The crystalline Si and diamond spectra are used for calibrating the observed spectra.

The integrated charge coupled device, used with the set-up shown in Figure 1, contains discrete 2-D arrays of 385 x 578 potential wells. The potential wells are capable of trapping electrons and can be moved around, by applying a varying potential to the electrodes of the device. The output amplifier measures charge in each of the potential wells and produces the output signal. The efficiency of the CCD (peaking 50% around 750 nm) approaches that of a photomultiplier tube to detect ultra low light levels. The extended spectral range (200 to 1200 nm) becomes possible in this UV sensitized CCD-which provides a dynamic range of 10^5 with full imaging capability.

RESULTS AND DISCUSSION

Diamond, Graphite and Diamond-Like Carbon Films

The diamond lattice belongs to 0^7_h space group. The selection rules allow only one principal triply degenerate normal mode of oscillation (F_{2g}) in Raman spectrum. The ideal diamond structure has two interpenetrating lattices, oscillating against each other with ω_{osc} (the direction of oscillation being arbitrary) given by

$$\omega_{osc} = 8/3m\,(K_1 + 8K_2/l^2)^{1/2} \qquad (1)$$

The K_1 and K_2 are used to denote primary and directed valence force constants, neglecting the repulsive forces between distant atoms, l and m stand for length of the valence bond and mass of carbon, respectively. On lowering the O^7_h space group, it is expected that the number of vibrations will be enhanced due to lifting of the three-fold degeneracy of F_{2g} mode, making allowed other Raman active A_{1g} and E_g modes.

The unit cell of graphite has a symmetry consistent with the space group D^4_{6h} with four atoms in primitive unit cell. The nine branches of optical phonon dispersion curves are distributed amonq the followinq irreducible representations,

$$\Gamma_{opt} = 2E_{2g} + E_{1u} + A_{2u} + 2B_{1g} \qquad (2)$$

of these modes, the E_{2g} modes (42 and 1581 cm^{-1} are Raman active.

The DLC thin films are found to show a remarkable combination of physical and chemical properties identical to those of diamond; e.g. extreme hardness, high resistivity, chemical inertness and optical transparency. These films have received large attention due to n- or p- type doping, which makes them semiconducting. These films have potential applications as hard transparent optical coatings, wear resistant coatings and force sensing devices etc[2]. The DLC films have been characterized by various experimental techniques[3] including Raman spectroscopy[4,5].

The polycrystalline DLC films were deposited by hot filament CVD methods, using in-situ boron doping (p-type). The other deposition parameters have been given elsewhere[5]. The micro-Raman spectra excited by 514.5nm of argon ion laser in the range 1200-1800 cm^{-1} at normal spots are shown in Figure 2 (a). The spectra in the range 1400-1700 cm^{-1} after deconvolution at normal and irradiated (with low power Nd:YAG laser) spots are shown in Figure 2 (b). Curves b, c and d are for the three different samples S_1, S_2 and S_3 at normal spots. Curves a and e correspond to crystalline diamond and highly oriented pyrolytic graphite respectively, and are included for comparison.

On annealing one of these typical films to 1000°c we find an increase in the relative intensity (RI) of crystalline graphite band (G-band) and a decrease in Full Width at Half Maximum Intensity (FWHMI) reflecting an increase in graphitic content and an enhancement of order in its lattice. This may be associated with decrease in adsorbed hydrogen on annealing. An untreated sample on laser irradiation shows an increase in RI of G band, indicating an increase in graphitic content. In an annealed sample the laser irradiation produces disappearance of a band around 1500 cm^{-1}, increase in RI of G band and appearance of a band around 1607 cm^{-1}.

These features suggest a decrease in DLC content, with an increase in graphitic content. The 1607 cm^{-1} feature is attributed to the formation of radicals or carbon clusters[5].

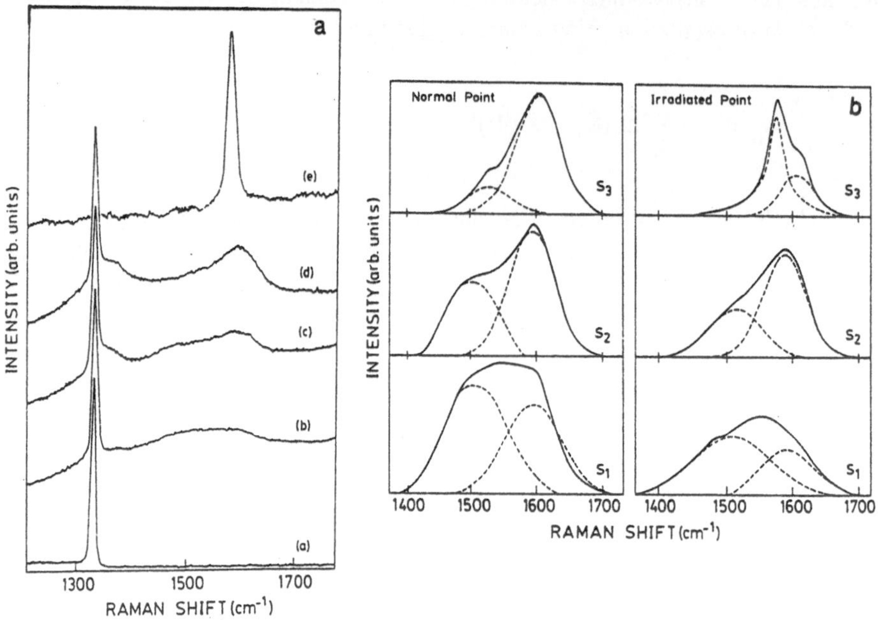

Figure 2. Raman spectra: (a) Of three different samples of thin film diamond sensors. (b) Represents spectra after backgroung subtraction at normal and irradiated spots. The solid lines are experimental while the dotted curves are Gaussian fit to data.

About 1μm thin films of carbon cluster composites of marginally varying crystallite sizes were deposited on (silvered) quartz substrate by a modified arc-welding technique under deliberately varied inert gas pressures. Each film show several types of crystallites both by Raman spectroscopy and other techniques[6,7]. The Raman spectra of the composites similar to each other were differing in two main aspects. The frequency of thegraphite at 1581 cm^{-1}exhibits a shift to lower frequencies with a concomitent broadening. The strength of two background features at 1350 and 1595 cm^{-1} exhibits differential enhancement in intensity and variation in bandwidth.

GaAs Epilayers

Novel electronic and optical devices are mostly synthesized using molecular beam epitaxy (MBE). The MBE grown layers often contain a large number of microscopic growth defects. The most common growth defect on GaAs-epilayers is known as `oval defect' (OD)[8], which affects the performance of small area devices. The particular type of OD containing a pit/core around its central region has been found to be sensitive to the cleanliness of of the surface and the growth conditions. Higher Ga concentration has been reported at the periphery of the pit by Energy Dispersive Analysis of X-Rays (EDAX) and Auger Electron Spectroscopy (AES) methods[9]. Twinned crystalline structure has also been observed by selective chemical etching[10].

In Figure 3, inset shows a typical oval defect on GaAs (N_e=2x10^{16} per cm^3) wafer. In this defect the Raman spectra are excited from the different spots (A, B, C and D) and the resulting curves are shown in Figure 3 (a) by A', B', C' and D', respectively. The bottom curve in Figure 3 (a) is the spectrum from a "normal" region on the GaAs wafer. In crystalline GaAs, the dipole selection rules allow the LO phonon from <100> surface, the TO phonon from <110> surface and LO+TO phonons from <111> surface. On this basis it is clear that the bottom spectrum is from the <100> surface. The gradual enhancement of the TO phonon and the relative decrease in intensity of LO phonon is due to the change in the crystalline orientation, which is <111> at the central portion of the OD.

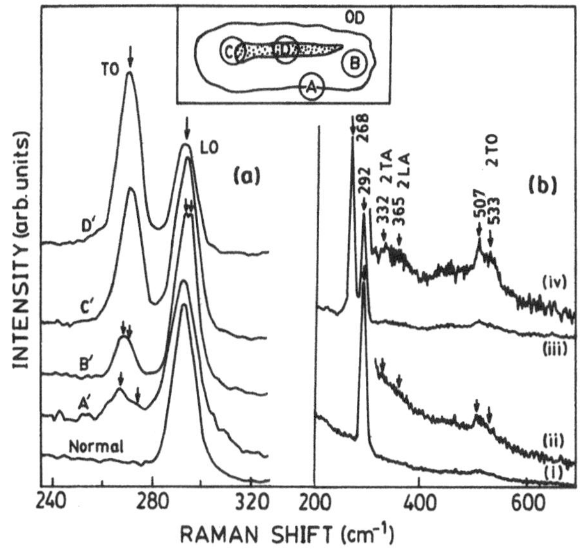

Figure 3. Raman spectra from epitaxillay grown GaAs wafer; **(a)** From normal region and various spots of OD (inset). **(b)** Curves (ii) and (iv) are in an arbitrarily expanded intensity scale for normal spot (i) and spot D (iii) of OD respectively.

In Figure 3 (b) the expanded curves (ii) and (iv) exhibit a structure which is interpreted as the second order Raman scattering of optical phonons. In heavily doped n-type GaAs, plasmon-phonon coupling, resulting in L^+ and L^- branches has already been studied. For N_e=2x10^{16} cm^{-3} , calculations show that the L^- mode falls far below our investigation range. However, in the absence of an unambiguous assignment of L^+ mode (expected in the vicinity of LO phonon), it is difficult to make any quantitative assessment of carrier concentration at the OD.

Further studies with various carrier concentration are in progress to locate these modes in order to detect any possible difference in the carrier concentration at these defects.

$YBa_2Cu_3O_{7-\delta}$ (YBaCuO) Thin Films

Lattice matched single crystal substrates are essential for fabricating YBaCuO films for electronic device applications. For using high temperature superconductors in the form of wires, cables and tapes flexible metallic substrates are required for the deposition of good quality superconducting films. The deposition of high T_c films on metallic substrates is difficult due to the inherent interdiffusion between metal and superconducting materials. The buffer layer of TiN has been used to prevent the interdiffusion between inconel-600 (Ni, Cr, Fe, Cu and C ratio being 76, 15.5, 8, 0.25 and 0.08%, in that order)[11] and YBaCuO overlayers. Additionally, the thermal expansion coefficient of TiN approximately matches with both inconel and YBaCuO, which reduces the tendency of formation of micro cracks. The film has been characterized by four probe a-c electrical ($T_{c(onset)}$ = 91K and $T_{c(R=0)}$=85K) and X-Ray diffraction measurements (c-axis texturing)[11].

The Raman spectra and their deconvoluted components of above films are presented in Figure 4. The positions of bands, their full width at half the maximum intensity (FWHMI) and relative intensities for four major bands after deconvolution are given in Table 1.

Table 1. The observed positions and FWHMI (cm^{-1}) and relative intensity (arb units) of YBaCuO film on laser irradiation.

Typical Bands in cm^{-1}	FWHMI and (relative Peak intensity) power density (W/mm^2) 50	75	125	175	250	Assignment
337	42 (0.40)	39 (0.54)	35 0 (0.64)	31 (0.81)	28 (0.90)	0(2)-0(3) out-of-phase bending c-axis
445	59 (0.50	56 (0.60)	56 (0.70)	59 (0.75)	59 (0.75	0(2)-0(3) in-phase bending c-axis
503	56 (1.00)	56 (1.00)	56 (1.00)	56 (1.00)	56 (1.00)	Symmetric stretching 0(4) c-axis
560	104 (0.63)	108 (0.80)	108 (0.84)	108, (0.81)	108 (0.85)	Defect induced IR-active

The mode assignment is given in the last column of the table. Considerably larger FWHMI of the observed bands may be attributed to the disorder in the oxyqen sub lattice[12].

The reduction in the FWHMI of 337 cm^{-1} band from 42 cm^{-1} to 28 cm^{-1} and enhancement in the intensity ratio I_{337}/I_{503} and I_{445}/I_{503} on increasing the laser power density from 50 to 250 W/mm^2 on the irradiated spot has been attributed to the local enhanced oxygen ordering in CuO_2 planes. This may be possibly due to the charge transfer induced process resulting from some photoexcitation mechanism[13].

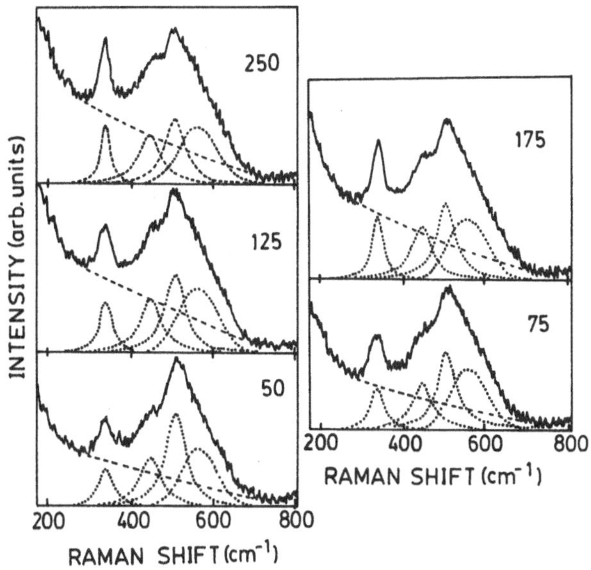

Figure 4. Raman spectra of $YBa_2Cu_3O_{7-\delta}$ thin films with power density variation from 50 to 250 W/mm^2. The deconvoluted components after due subtraction of background are shown by dotted curves.

ACKNOWLEDGEMENTS

Thanks are due to Council of Scientific and Industrial Research, Department of Science and Technology and Ministry of Human Resource Development for generous funding on three different projects. Thanks are also due to Prof J. Narayan and Prof A. Aslam for providing the samples and G. S. Thapa for technical help.

REFERENCES

1. H.D. Bist, Laser Raman spectroscopy and its applications - especially in structural phase transitions, *Proc. Indian Acad. Sci (Chem. Sci.).* 103:295 (1991).
2. J.R. Durig, T.S. Little, H.D. Bist, A. Rengan and J. Narayan, Fourier transform Raman spectra of diamond-like thin films, *J. Raman Spec.* 23:625 (1992).

3. K. Edamatsu, Y. Takata, T. Yokoyama, K. Seki, M. Tohnan, T.O. Kada, T. Ohta, Local structures of carbon thin films synthesized by the hot filament chemical vapor deposition method x-ray absorption structure and Raman spectroscopic studies, *Jpn. J. Appl. Phys.* 30:1073 (1991).
4. R.J. Nemanich, J.T. Glass, G. Lucovsky and R.E. Shroder, Raman scattering characterization of carbon bonding in diamond and diamond-like thin films, *J. Vac. Sci. Technol.* A6:1783 (1988).
5. S.Bhargvava, H.Joshi, H.D.Bist and M. Aslam, Micro Raman analysis of thin film diamond sensors, J. Raman Spectros. (To be published, 1993).
6. L.S. Grigoryan, H.D. Bist, S. Sathaiah, S.V. Sharma, H. Clara and A.K. Majumdar, Micro-Raman spectroscopy of carbon cluster composites, *J. Raman Spec.* 23:127 (1992).
7. L.S. Grigoryan, H.D. Bist, S. Sathaiah, H. Clara, S.V. Sharma, N. Sudhakar, Prem Chand, A.K.Majumdar, S.B.Samanta, P.K.Dutta and A.V.Narlikar, Electron-molecular vibration interactions in undoped fullerene films - A micro-Raman study, *Chem. Phys. Lett.* 199:360 (1992).
8. K. Fujiwara, K. Kanamoto, Y.N. Ohta, Y. Tokuda and T. Nakayama, Classification and origin of GaAs oval defects grown by MBE, *J. Cryst. Growth* 80:104 (1987).
9. K. Nambu, J. Saito, T. Ishikawa, K. Kondo and A. Shibotomi, Classification of surface defects on GaAs grown by Molecular beam epitaxy, *J. Electrochem. Soc.* 133: 601 (1986).
10. M. Bafleur, A. Munoz-Yague and A. Rocher, Microtwinningand growth defects in GaAs MBE layers, *J. Cryst. Growth* 59:531 (1982).
11. A. Kumar and J. Narayan, Laser deposition of $YBa_2Cu_3O_{7-\delta}$ thin films on flexible metallic substrate with Tin buffer layer (Unpublished).
12. E. Sodtke and H. Munder, Oxygen content and disorder in a-axis oriented $YBa_2Cu_3O_{7-\delta}$ thin films, *Appl. Phys. Lett.* 60:1630 (1992).
13. G. Nieva, E. Osquiguil, J. Guimpel, M. Manhaud, B. Wuyts, Y. Bruynse rnce, M.E. Maple and I.K. Shuller, Photoinduced enhancement of superconductivity, *Appl. Phys. Lett.* 60:2159 (1992).

ELASTIC WAVE PROPAGATION IN ALUMINUM/ARAMID-EPOXY PLATES

P. J. Shull and D. E. Chimenti

Center for Nondestructive Evaluation
Johns Hopkins University
Baltimore MD

S. K. Datta and J. H. Ju

Dept of Mechanical Engineering
University of Colorado
Boulder CO 80303

Introduction

Dispersion of guided waves in laminated plates has been studied over the past thirty years. Recent work has included anisotropic layering and the subtle effects of fluid loading. References to these works can be found both in the review article by Chimenti and Nayfeh [1] and in the symposium proceedings edited by Mal and Ting [2]. More recently Braga and Herrmann detailed the features of the dispersion spectrum for anisotropic periodic media using the sextic formalism developed by Stroh [3, 4], implying coupling of inplane and out-of-plane particle motion. A similar discussion of the analysis of the acoustical or mechanical filtering predicted for such periodic media is presented by Rousseau [5]. However, guided waves in a laminated plate where the laminae have very different anisotropic (isotropic) properties from one another have not received adequate attention. This problem, which is the subject of this paper, has relevance in ultrasonic characterization of layered bonded plates when the interface (interphase) bond properties can be quite different than those of the adjacent laminae and can change during service life.

The laminated material used in the present study is composed of aluminum layers adhesively bonded to an epoxy impregnated with aramid fibers (ARALL^{TM} or A/A- E). In this paper, which is a synopsis of reference [6], we discuss experimental observations and theoretical predictions of the wave reflection and transmission properties and the dispersion characteristics of the ARALL^{TM} material. The strong material property contrast between the constituent layer materials in these plates gives rise to acoustical filtering, similar to that associated with Floquet wave behavior in unbounded periodic media.

Experiment

A brief outline of the experimental setup and preliminary experiments are presented in this section. Detailed descriptions of this information can be found in Shull, *et al.* [6] and [7].

The ultrasonic plate wave resonance experiments were performed using swept frequency fluid coupled techniques. The primary experiments employed a single-side access method, while preliminary measurements to determine the elastic properties of the constitutive layer

Nondestructive Characterization of Materials VI
Edited by R.E. Green, Jr. *et al.*, Plenum Press, New York, 1994

materials used both a double transmission and a single transducer normal incidence resonance method.

The apparatus which is used to produce a dispersion curve of the plate wave behavior of a specimen is outlined in the block diagram in Figure 1. The transducer excitation waveform is a series of low-voltage gated rf tone bursts of sufficient duration to ensure quasi-cw insonification in the plate. The function generator steps the frequency discretely from 0.5 to 12 MHz with step increments of 20 or 30 kHz in 30-μs rf tone bursts.

A piston radiator at a fixed incident angle θ_i, defined as the angle between the directivity of the transducer and the plate normal, is used to insonify the sample. The receiving transducer is positioned in the incident plane with $\theta_r = -\theta_i$. The response is video detected and amplified by a broadband amplifier. This amplified signal is processed by a gated integrator where the output is digitized by a voltmeter and passed to the computer.

In typical reflection experiments the measurements are taken with nominally identical broadband 9-mm diameter transducers with a center frequency of 5 MHz. The transducers are oriented at incident angles from 8° to 44° from the plate normal with typical increments of 2°. At fixed incident angle θ_i and azimuthal angle ϕ, relative to the aramid fiber direction, the input transducer frequency is swept from 0.5 to 12 MHz. After the received signal is normalized with the reference curve corresponding to the particular incident angle, the spectral data are processed to search for and record the frequency at which the minima occur, as shown in Figure 2. The collected minima data for the series of measured incident angles are plotted in the form of a dispersion curve as in Figure 3.

Four A/A-E panels are used in this study. They are bonded, alternating layers of alumi num and aramid-epoxy (A-E) composite with nominal thicknesses of 0.3 and 0.22 mm respectively, with outer layers of aluminum. Plate descriptions are of the form X/Y, where X and Y are the number of alternating layers of aluminum and aramid-epoxy composite, respectively. The composite layer is a uniaxial fiber layup of Kevlar 69 in a matrix of epoxy. The four plates interrogated are 2/1, 3/2, 4/3 and 5/4 panels with nominally identical fiber orientation from layer to layer. Although, space constraints allow the depiction of the dispersion curves of only three of the four plates (2/1, 3/2 , 5/4). Literature values for the material constants of A-E composite are typically measured in static stress. The static properties of polymer composites can differ from values inferred from ultrasonic measurements. Therefore, to provide the most accurate theoretical comparisons, initial experiments have been made on deplied aramid-epoxy to determine the ultrasonic material constants of the aramid-epoxy. These results, along with literature values of the elastic constants for the aluminum alloy, are listed in Reference [6].

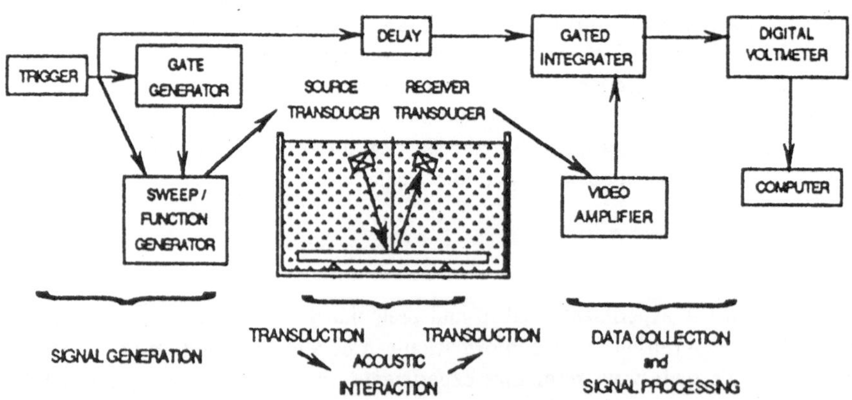

Figure 1. Block diagram of the experiment.

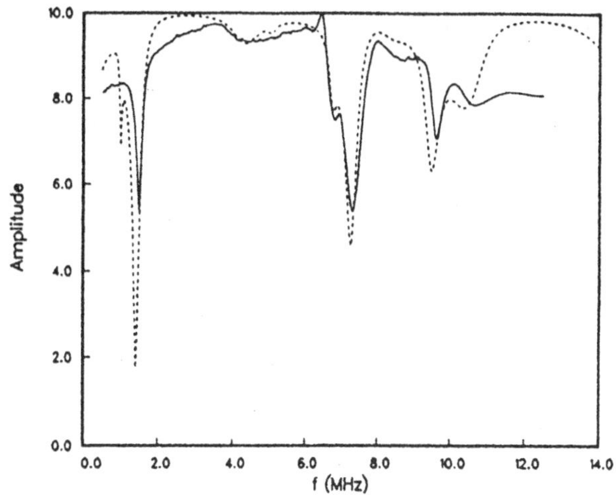

Figure 2. Comparison of the experimental (solid) and theoretical (dashed) reflection spectrum for the 2/1 plate, d=0.85 mm, at $\theta = 10°$ and $\phi = 0°$.

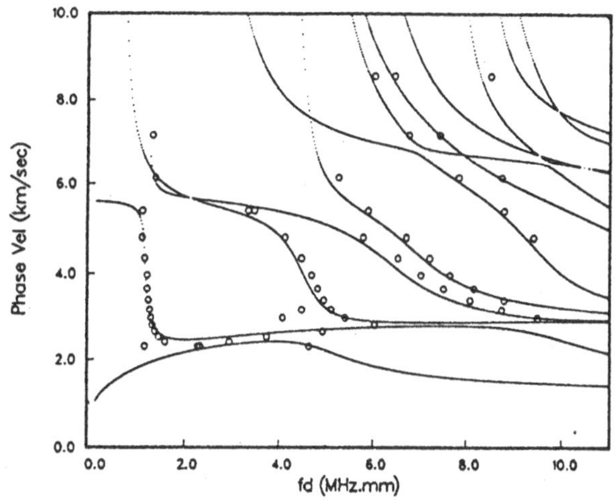

Figure 3. Experimental (circles) and calculated (dotted lines) dispersion curves for the 2/1 plate, d=0.85 mm, at $\phi = 0°$.

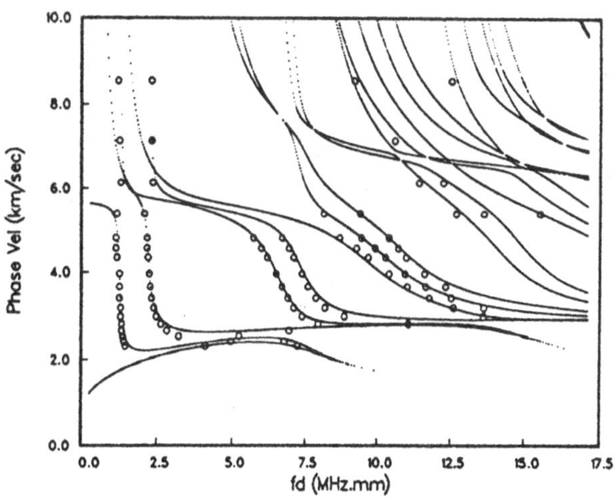

Figure 4. Experimental (circles) and calculated (dotted lines) dispersion curves for the 3/2 plate, d=1.27 mm, at $\phi = 0°$.

Results and Discussion

The large mismatch in acoustic impedance between the aramid-epoxy and aluminum constitutive layers produces distinctive dispersion characteristics. The particular behavior associated with plate wave interaction with this type of composite is suggestive of Floquet phenomenon. In a periodic medium, Floquet waves are characterized by alternating regions of wave propagation and attenuation. This section will detail these features in terms of plate wave modes as observed form the experimental and theoretical reflection spectrum.

As an introduction we will begin with the simplest geometric format of two aluminum layers with an intermediate aramid-epoxy layer. The associated experimental (solid) and calculated (dashed) reflection spectrum at an incident angle of 20° is shown Figure 2. A dispersion curve description of this material is produced by plotting the collection of reflection spectrum minima as a function of frequency times the overall plate thickness, fd, verses phase velocity, V_p. V_p is directly related to the incident angle and the fluid phase velocity, V_f, through Snell's Law, $V_p = V_f \sin\theta_i$.

A dispersion curve for the 2/1 plate at an azimuthal angle of $\phi = 0°$ is shown in Figure 3. Despite the complexity of the mode spectrum, the theory (dotted curves) is in good agreement with the experimental data (circles). The lowest order antisymmetric mode represents incident angles greater than 38°(trace velocity less than 2.2 mm/μs). Our data do not encompass this region; the highest experimental incident angle is typically 42°.

Geometry controlled limiting behavior of the propagating modes is typical in plate wave phenomenon. We expect to observe three such regions which are characterized by their relatively low dispersion (change of velocity with frequency). The first two of these occur as the incidental angle or the corresponding phase velocity of the propagating modes passes through the longitudinal critical angles of the aluminum and the aramid-epoxy. The lower order modes clearly demonstrate the limiting behavior associated with the aluminum critical angle at a phase velocity of about 5.6 mm/μs. Similarly, a region of low despersion corresponding to the critical angle of the aramid-epoxy is observed for the higher order modes at a phase velocity of approximately 6.37 mm/μs. As expected, these two limiting phase velocities are slightly less than the critical values for a wave incident on a half-space. Subsequent references to the plate extensional velocity refer to the value corresponding to the low-frequency phase velocity limit of the S_0-like mode in the multilayer structure.

While this low-frequency behavior in Figure 3 is typical of waves in homogeneous plates, the third region, the high-frequency limiting behavior, clearly is not. In a homogeneous plate the fundamental modes approach the Rayleigh wavespeed as fd increases and the plate wavelength becomes much smaller than the plate thickness. By contrast, in the 2/1 plate (and even more so in the 3/2, 4/3 and 5/4 plates) there appear to be two limiting phase velocities at large fd. The higher of these corresponds approximately to the Rayleigh velocity of aluminum (V_R^{Al} = 2.93 mm/μs), and the lower approaches the transverse wavespeed of A-E (1.26 mm/μs). The lowest phase velocity curve in Figure 3 has the characteristics of an A_0 Lamb mode, except for its rapid decrease beyond $fd = 4$ MHz·mm toward the lower asymptotic wavespeed. At still higher values of fd, near $fd = 10$ MHz·mm, the next higher (S_0-like) mode also drops sharply toward this lower asymptote. For even larger fd, the nearly dispersionless portion of each mode curve exists at the upper limiting speed for some range of fd, then falls toward the lower asymptote. This behavior may be explained by considering the properties of guided waves at high frequency. For short wavelengths we expect the excitation to propagate as two Rayleigh waves on the outer plate surfaces. The existence, for $fd >$ 3MHz·mm, of a piecewise continuous curve at very nearly V_R of the aluminum confirms this expectation. However, each mode section eventually falls toward the lower (A-E) asymptote. This behavior suggests a progressive, frequency-dependent trapping of energy in the slower A-E layers.

A further unusual feature of the curves in Figure 3 concerns the approach of the S_0 and higher order modes to the aluminum Rayleigh velocity from *below* V_R^{Al}. From fd=2 MHz·mm to 8 MHz·mm the slope of this mode branch is positive, implying that the group velocity V_g is greater than the phase velocity V_p. In the calculations and measurements presented in Figures 3 through 5, this behavior, $V_g > V_p$, occurs in the same dispersion curve region

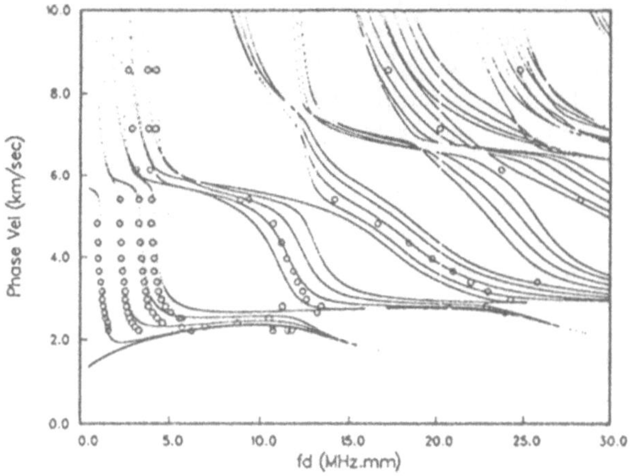

Figure 5. Experimental (circles) and calculated (dotted lines) dispersion curves for the 5/4 plate, d=2.33 mm, at $\phi = 0°$.

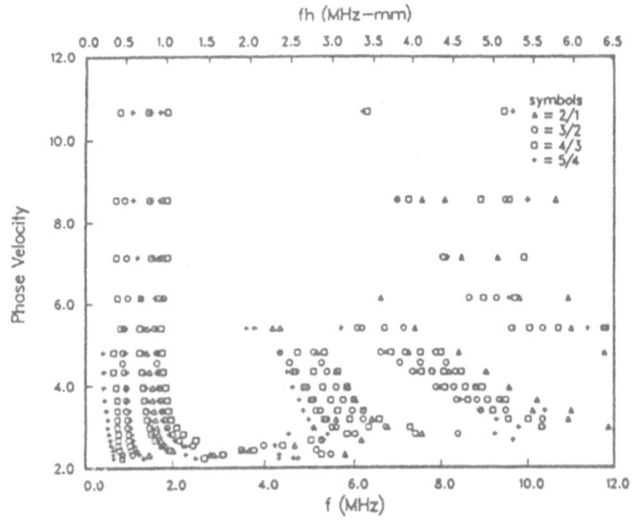

Figure 6. Superposition of the experimental dispersion data for all four plates at $\phi = 0°$, showing the coincidence of mode groupings and regions of mode exclusion.

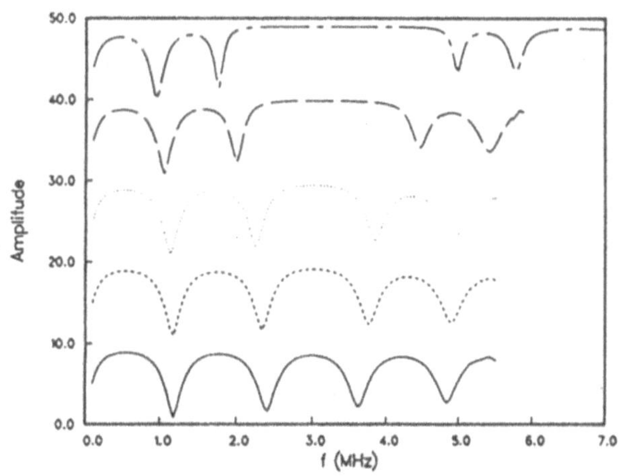

Figure 7. Calculated plane wave reflection coefficient at θ_i=24° for several, progressively different values of layer elastic constants, from homogeneous at bottom to A/A-E at top.

for the other three plates (3/2, 4/3 and 5/4), as well. Moreover, as we have seen above, the wave energy is trapped at large fd in the aramid layers, and the mode curves approach the aramid transverse wave speed. But a subtle evolution of this behavior can be seen in Figures 4 through 6. For each of these materials the plate wave mode curves tend to cluster in groups, especially in the range of V_p = 2 to 6 km/s. As the phase velocity of such a group drops from V_R of the aluminum toward the A-E transverse wavespeed, the highest order mode in the group continues to propagate through this region at approximately V_R^{Al}, until it becomes the lowest order mode of the next adjacent group. Then, it too becomes trapped in the aramid-epoxy, and its phase velocity drops toward the shear wavespeed of the A-E. In the calculations this behavior is observed to repeat in all of the higher order mode groups.

The major effect of the periodic layering in the A/A-E plates is the introduction of distinct regions in fd, at V_p from 2 to 6 km/s, where plate modes cluster in groups, and other regions in which they are absent. Figures 3 through 5 show experimental (open circles) and predicted (solid lines) dispersion curves that illustrate this behavior. In all these figures, the azimuthal angle is 0° with respect to the fiber direction. In Figure 3 the two most pronounced regions empty of plate modes lie approximately within the intervals fd = 2 and 4.5 MHz·mm and fd = 6.5 and 8 MHz·mm on the abscissa. Along the ordinate, these disallowed regions are bounded by aluminum Rayleigh velocity on the lower end and by the limiting plate extensional velocity at the upper end.

Examining Figures 3 through 5, we find, quite remarkably, that for all of the composite plates (each a different thickness) there exist two mode grouping regions which coincide in *frequency*. This striking, and completely unexpected, behavior is shown in Figure 6. Here, data for all four specimens are plotted as a function of frequency. The overlap of the mode groupings is unmistakable. This implies that Figures 3 through 5 are much more than simple modifications of classical Lamb wave dispersion curves for multilayer media. Lamb wave dispersion in homogeneous plates scales with overall plate thickness d, since the plate boundaries are traction-free and there is no internal structure. However, the data of Figures 3- 5 and Figure 6 clearly do *not* scale with plate thickness. Instead, regardless of the number of unit cells, the Lamb wave-like behavior is constant in frequency and scales in fh, where h is the unit cell dimension. We may conclude that from the data we could not predict the plate thickness directly, but rather only the cell thickness h and the number of cells in the plate.

The groupings of minima, corresponding to the excitation of plate modes, are also zones where acoustic energy will propagate efficiently, at that frequency and angle, through the plate. Outside these transmission zones, the wave energy is strongly reflected. This cyclic occurrence of transmission (pass) and reflection (stop) zones (bands) is characteristic of Floquet behavior in an unbounded periodic medium [8]–[5]. From Figure 6, we note that: a) regions where modes exist correspond to through-plate pass bands, and b) disallowed mode regions correspond to through-plate stop bands.

The relationship between plate modes and the Floquet waves has been cited by several authors [9, 4] in the study of layered anisotropic composites. According to Floquet's theorem, wave solutions in an unbounded periodically layered medium are a product of two functions, one having the periodicity of the medium and the other a sinusoidal wave function that defines the Floquet wavenumber. Translational invariance in the elastic medium implies that the Floquet wavenumbers will be indeterminate to within π/h, where h is the repeating cell dimension, in our case about 0.53 mm. For transverse Floquet waves to play a significant role in the plate modes, the condition $\alpha_{l,t}h/\pi = 1$ must be satisfied for either the longitudinal or transverse partial wavenumbers $\alpha_{l,t}$ in the region where the plate modes exist. Referring to Figure 6, for V_p=4 km/s and f=2 MHz, both longitudinal partial waves are evanescent, but the transverse partial wave in the aramid-epoxy, has a low enough wavespeed such that $\alpha_3^{A-E}h/\pi = 1.5$. This estimate suggests an influence of transverse Floquet pass and stop bands on the dispersion in the A/A-E plates.

To gain further physical insight into the mode grouping behavior, we have examined the interaction between the plate modes and the transverse Floquet waves. It is well known that the plate mode wavenumber is the inplane component of the incident acoustic wavevector, or by Snell's law, either of the partial wavenumbers in the plate, if they are real. The guided wave mode will propagate when the plate is in transverse resonance, *i.e.* the phase shift of any partial wave across the thickness of the plate (the *transverse* dimension) is an integral multiple of π. We have performed several model calculations designed to elucidate this unexpected mode grouping phenomena.

We begin with a plate composed of identical layers (essentially a homogeneous plate) having thicknesses equal to our A/A-E samples and equivalent average elastic properties. The reflection spectrum of such a 3/2 plate at θ_i=24° is shown at the bottom of Figure 7 in the solid curve. Each successive curve in the series from bottom to top corresponds to a change of about 25% in the elastic properties of the alternating layers, away from homogeneity and towards the actual elastic constants of our samples. We observe a gradual shifting and concentration of the homogeneous plate modes as the elastic properties change. In the dotted curve in the center of Figure 7, a region absent of minima is clearly developing. In the next two curves above the dotted one, this region grows until it reaches the dimensions of the pass band measured in the data of Figure 6.

It is likely that the exclusion of the minima from the zone between 2 and 5 MHz is closely related to absence of real Floquet wavenumbers in this region. Clearly, if the reflection coefficient predicts essentially zero transmission between 2 and 5 MHz in the top curve of Figure 7, there can be no real transverse Floquet wavenumbers in this region. Likewise, no guided wave modes can exist. Moreover, the concentration of minima into transmission zones is consistent with the conservation of roots of an analytic function; that is, the number of modes is constant. There is now the dispersion anomaly left to explain. This result is shown in Figure 6, where all plates of widely varying total thickness display very similar dispersion behavior, quite unlike that expected for homogeneous plates of different thicknesses. Once again, in transverse resonance the plate supports a phase shift of $n\pi$ across its thickness. For a bilaminated plate, this condition must also hold. In the case of strongly different acoustic materials, however, the phase shift produced across each unit cell will be significantly larger than that for a homogeneous layer of equal thickness. Near the stop band edges, the phase shift is a strong function of frequency. Therefore, several modes will tend to group, followed by a region empty of transverse Floquet (or guided wave) modes.

While mode groupings are clearly discernible at lower phase velocity, above the effective plate extensional velocity of 6 km/s a similar trend is not as obvious (Figs. 3-6). The modal behavior in this region of high phase velocity is more complex, but since both Floquet wavenumbers are real, this case closely approaches the one more commonly dealt with, namely, that of transverse Floquet waves propagating normal to a periodic elastic structure. In fact, crossing of symmetric and antisymmetric modes is common, particularly at the higher frequencies. As previously mentioned, at low frequency where the modes are more regular, we observe a stop band region between 2 and 4 MHz. However, the higher frequency pass and stop bands still are observed but with less regularity than those observed in the lower phase velocity (higher θ_i) region.

We observe in the pass bands a coincidence between the number of layers of A-E and the number of minima per band. Figures 3 through 5 illustrate the relationship. Guided wave modes are excluded from the transverse stop bands. In the first two sets of pass bands, there is exactly one minimum for each A-E layer; *i.e.* there is one for the 2/1 plate, two for the 3/2, *etc.* This correspondence between the number of minima and aramid-epoxy layers is anticipated by Floquet's theorem.

Conclusions

In this paper we have described experiments and calculations dealing with the interaction of ultrasound with aluminum/aramid-epoxy composites. The extensive experimental results, recorded as reflection spectra or guided wave dispersion curves, have been compared to calculations with good agreement. In the bilayered plates, zones of wave transmission and reflection similar to those predicted for periodic media have been observed. We have made a thorough experimental study of wave reflection phenomena in the aluminum/aramid- epoxy structures.

The trends we have found persist in our samples, ranging from one and one half to four and one half cycles of the basic repeating unit. We have observed that the number of measured reflection minima and the number of material unit cells are nearly equal for each sample geometry. This behavior has been examined theoretically as well. By modeling the reflection behavior as a function of incremental variations in the lamina properties, we have demonstrated that the transmission zones arise from a consolidation of the uniform-plate minima into zones or bands. We have attempted to understand this mode clustering in terms of transverse Floquet stop and pass bands; however, a more complete analysis will be necessary to investigate fully the clustering of reflection minima (and therefore guided wave modes) with increasing disparity in elastic properties between the two constituent layers.

Acknowledgement

We wish to thank R. C. Stiffler and M. P. Jones of Alcoa for the generous loan of many samples of their aluminum/aramid-epoxy plates, sold under the tradename ARALL^{TM}. Also, we gratefully acknowledge B. A. Auld of Stanford and NIST for many helpful discussions on this problem . This work was partially supported by the Naval Surface Warfare Center and by the Center for NDE, Johns Hopkins University.

References

[1] Chimenti, D. E. and Nayfeh, A. H., "Ultrasonic reflection and guided waves in fluid-coupled composite laminates", J. Nondestruct. Eval. **9**, 51-69 (1990).

[2] Mal, A. K. and T. C. T. Ting, eds., *Wave Propagation in Structural Composites*, (ASME, New York, 1988).

[3] Braga, A. M. B., "Wave propagation in anisotropic layered composites," PhD Thesis, Stanford University, 1990.

[4] Braga, A. M. B. and Herrmann, G., "Floquet waves in anisotropic periodically layered composites," J. Acoust. Soc. Am. **91**, 1211-1227 (1992).

[5] Rousseau, M. and Gatignol, Ph., " Propagation acoustique dans un milieu periodiquement stratifié," Acustica **64**, 188-194 (1987).

[6] Shull, P. J., Chimenti, D. E, and Datta, S. K.,"Elastic guided waves and the Floquet concept in periodicly layered plates ," Submitted to JASA March 4, 1993.

[7] Shull, P. J., "Plate waves in bilayered composites," Masters Thesis, Johns Hopkins University, 1992.

[8] Sun, C. T., Achenbach, J. D. and Herrmann, G., "Time-harmonic waves in a stratified medium propagating in the direction of the layering," J. Appl. Mech., **35**, 408-411 (1968).

[9] Braga, A. M. B. and Herrmann, G., "Plane waves in anisotropic composites", in *Wave Propagation in Structural Composites*, AMD-vol. 90, eds. A. K. Mal and T. C. T. Ting , (ASME, New York, 1988), pp. 81-98.

A MULTI-PARAMETER ULTRASONIC INSPECTION TECHNIQUE

Bradley W. Sermon and William J. Murri

NDT/NDI Engineering Group
Hercules Aerospace Division
Bacchus Works
Magna, UT 84044

INTRODUCTION

Two common uses of nondestructive evaluation are: (1) to produce maps or images of manufacturing defects or damage showing their size and location, and (2) to measure acoustic parameters that may relate to physical or mechanical attributes of the material.

In a conventional ultrasonic C-scan inspection, the peak through-transmission or pulse-echo amplitude is detected and recorded at each pixel. This magnitude is plotted as a color or grey-scale map for each pixel in the inspection. Little information about material properties is available from such a scan.

Prior work in developing a quantitative multi-parameter inspection was done at Hercules and reported by our colleagues Dr. Lee H. Pearson and Donald S. Gardiner in 1985.[1] More recently, Dr V. Dayal[2], and Drs. David K. Hsu and Michael S. Hughes at Iowa State University[3,4] have reported on methods to simultaneously measure sample wavespeed and thickness. The latter also reviewed various experimental configurations for collecting multi-parameter data.

In this paper, we describe a method which combines some of the ideas of the researchers above. This method produces C-scans of material properties, in particular, wave velocity, thickness, density, and attenuation from one scan. These C-scans are obtained by collecting the amplitude and transit time values of through-transmission and pulse-echo pulses simultaneously. The magnitude of the analytic signal is used to determine the necessary transit times and amplitudes.

MULTI-PARAMETER METHOD

A typical experimental arrangement for collecting multi-parameter data is shown in Figure 1. The left transducer emits a pulse, part of which traverses through the sample and is recorded by the transducer on the right. This same emitted pulse reflects off both the front and back surfaces of the sample and these two reflections are recorded by the left transducer. Either before or after the sample is scanned, two more measurements must be made. First, the baseline transit time through the water alone is measured as shown in Figure 2(a). Second, a known reflector material is placed in the sample position and the reflected pulse amplitude from the front surface is recorded by the left transducer as shown in Figure 2(b).

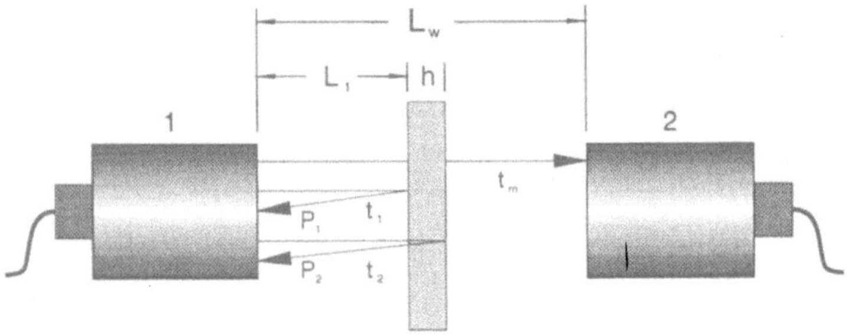

Figure 1. Transducer and sample arrangement for collecting multi-parameter ultrasonic data.

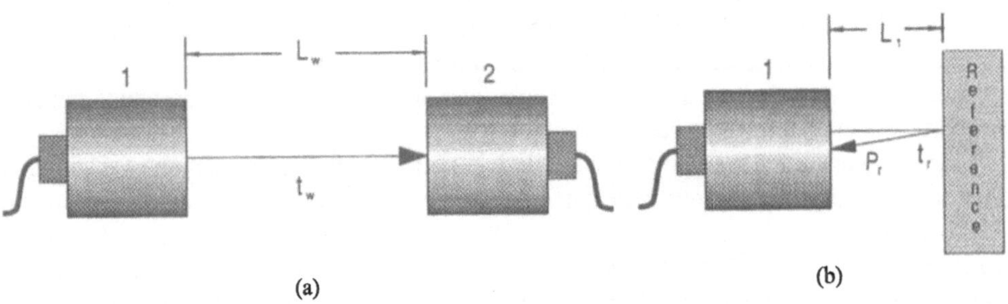

Figure 2. Equipment configurations for collecting baseline water path (a) and reference reflector data (b).

The following definitions are used in this paper:

t_m	is the ultrasound transit time through the sample and water
t_w	is the ultrasound transit time through the water path
t_1	is the transit time for the reflected pulse from the front surface of the sample
t_2	is the transit time for the reflected pulse from the back surface of the sample
P_0	is the effective amplitude of the pulse emitted by the left transducer
P_1	is the amplitude of the reflected pulse from the front surface of the sample
P_2	is the amplitude of the reflected pulse from the back surface of the sample
P_r	is the amplitude of the pulse reflected from the front surface of the reference reflector
C_w	is the ultrasonic wave speed in water

C_m is the ultrasonic wave speed in the sample material
h is the sample thickness
α is the sample attenuation coefficient
ρ_m is the sample density
ρ_w is the water density
ρ_r is the reference reflector density

With reference to Figures 1 and 2, we can write the following relations,

$$t_w = L_w/C_w \tag{1}$$

$$t_m = (L_w - h)/C_w + h/C_m \tag{2}$$

$$t_1 = (2L_1)/C_w \tag{3}$$

$$t_2 = (2L_1)/C_w + (2h)/C_m \tag{4}$$

These four equations can be solved for C_m and h to give,

$$C_m = C_w[2(t_w - t_m)/(t_2 - t_1) + 1] \tag{5}$$

$$h = (C_w/2)[2(t_w - t_m) + (t_2 - t_1)] \tag{6}$$

Note that only differences in measured times are used to determine the material wavespeed and thickness. Thus any systematic error in the time values, such as triggering delays, etc., are eliminated. The water wavespeed C_w is computed from the measured water temperature.

The following time-independent ultrasonic pulse amplitudes are recorded,

$$P_1 = P_0 r_{wm} \tag{7}$$

$$P_2 = P_0 t_{wm} r_{mw} t_{mw} e^{-2\alpha h} \tag{8}$$

$$P_r = P_0 r_{wr} \tag{9}$$

where:

r_{wm} is the reflection coefficient between water and the sample
r_{mw} is the reflection coefficient between the sample and water
t_{wm} is the transmission coefficient between water and the sample
t_{mw} is the transmission coefficient between the sample and water
r_{wr} is the reflection coefficient between water and the standard

Note: $r_{wm} = -r_{mw}$ and $t_{wm}t_{mw} = (1 - r_{wm}^2)$

Using equations (7) and (9), we obtain,

$$r_{wm} = P_1 r_{wr}/P_r \tag{10}$$

where one can calculate the value of r_{wr} from its definition:

$$r_{wr} = (\rho_r C_r - \rho_w C_w)/(\rho_r C_r + \rho_w C_w) \tag{11}$$

Then using the definition of r_{wm}, we obtain a relation for the sample density,

$$\rho_m = [\rho_w C_w(1 + r_{wm})]/[C_m(1 - r_{wm})] \quad (12)$$

Using equations (7) and (8), we obtain a relation for the attenuation coefficient,

$$\alpha = (1/2h)\ln[P_2/(P_1(1 - r_{wm}^2))] \quad (13)$$

The shear modulus of a material can be defined as the density of the material multiplied by the square of the ultrasonic shear wave velocity in the material. In a similar fashion, one can define a longitudinal modulus as the density multiplied by the square of the longitudinal wavespeed.

We also incorporated into our algorithm the diffraction correction method developed by Dr. Papadakis[5] and the surface roughness correction algorithm developed by Drs. Nagy and Adler[6]. We are still evaluating the effectiveness of these corrections.

There are a number of limitations to the inspection method. First, the sample must have low enough attenuation so that through-transmission and pulse-echo signals can be recorded. At present only simple geometry, flat samples can be inspected because of the difficulty of identifying and tracking the various signal peaks for more complex sample shapes. Also, the technique is still sensitive to variations in the sample surface roughness.

EXPERIMENTAL METHOD

The experimental arrangement we used to collect multi-parameter data is shown in Figure 1 and is the Case II configuration discussed by Hsu and Hughes[3]. We selected this configuration because all the pulse amplitude data are collected by one transducer, which simplifies any compensations for transducer response functions.

Our data were collected using matched 0.5 inch diameter 2.25 MHz unfocused immersion transducers. We also kept the transducers stationary while moving the sample in order to reduce signal time variations due to mechanical vibration. All ultrasonic waveforms were digitized at 20 MHz and stored for subsequent analysis.

The time through the water path, t_w, was measured followed by the measurement of P_r. The sample was then positioned and scanned while the ultrasonic signal waveform at each pixel was stored. The analysis program generated the waveform envelope by calculating the magnitude of the analytic signal. From these envelopes, the pulse-echo amplitudes and through-transmission transit times were determined. Figure 3 shows a typical ultrasonic waveform along with its computed envelope. The leftmost peak is the sample frontwall reflection, the next peak is the sample backwall reflection and the rightmost peak is the signal transmitted through the sample. Note that the amplitudes and times determined by the envelope are not the same as one would measure using the signal waveform itself. Measurements made using the envelope are insensitive to possible phase inversions of the reflections and threshold time shifts due to dispersion.

At each pixel, the wave velocity and thickness of the sample were calculated using equations (5) and (6). The reflection coefficient between the water and sample was next calculated from equation (10), and the sample density was then obtained from equation (12). The attenuation coefficient was then calculated using equation (13). The longitudinal modulus was then computed by multiplying the density by the square of the wavespeed. C-scan images were then made for each variable. Since the wave forms were digitized, the standard pulse-echo and through-transmission amplitude C-scans as well as C-scans at any depth in the sample can also be obtained.

Waveform and Envelope
Thick Sample

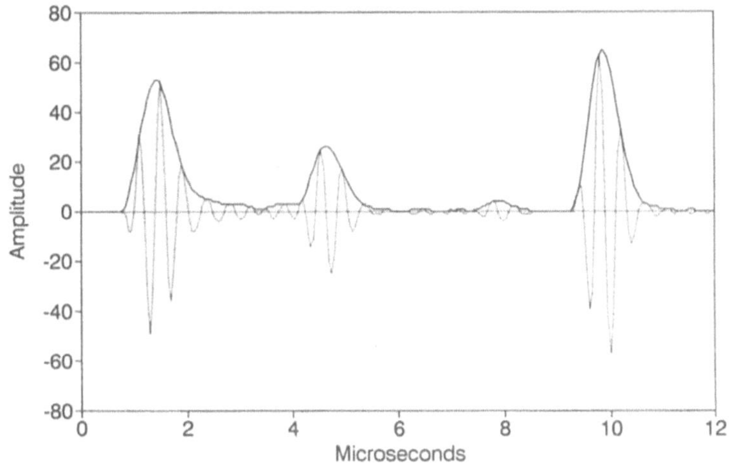

Figure 3. Typical multi-parameter waveform with its computed envelope.

RESULTS

We inspected two graphite/epoxy samples using the multi-parameter technique. The first sample had not been intentionally damaged, but showed some material variation when inspected.

Figure 4a shows the through-transmission C-scan of this undamaged sample obtained from the appropriate envelope peak. This image is equivalent to a conventional C-scan image and shows no quantitative material data.

Figure 4b shows the computed wavespeed for the sample. The range is 2.87 to 3.06 mm/uS, with a mean of 2.97 mm/uS and a standard deviation (sigma) of 0.03 mm/uS. The value generally used for this material is 2.9 mm/uS. Figure 4c shows the sample thickness, which ranged from 3.45 to 3.67 mm, with a mean of 3.55 mm and a sigma of 0.04 mm. The physically measured thickness was 3.56 mm. There are no noticeable trends in either the wavespeed or thickness in this particular sample.

Figure 4d is the C-scan image of the computed material density, which ranges from 1.29 to 1.59 g/cm^3 with a mean of 1.47 g/cm^3 and a sigma of 0.05 g/cm^3. The average density for this sample was physically measured to be 1.55 g/cm^3. The image shows a low density area in the upper right corner and a generally lower density in the right side of the sample and a smaller low density spot near the lower left corner.

Figure 4e is the C-scan image of the computed material attenuation coefficient, which ranges from 0.03 to 0.09 neper/mm with a mean of 0.04 neper/mm and a sigma of 0.01 neper/mm. This image shows a group of high attenuation areas in the upper right half and a distinct high attenuation region near the lower left corner. Very generally, the attenuation coefficient varies inversely as the density.

Figure 4f is the C-scan image of the material longitudinal modulus derived from the computed wavespeed and density. The modulus range is 11.2 to 14.2 GPa with a mean of 12.9 GPa. The modulus image resembles the density image since the wavespeed is fairly constant.

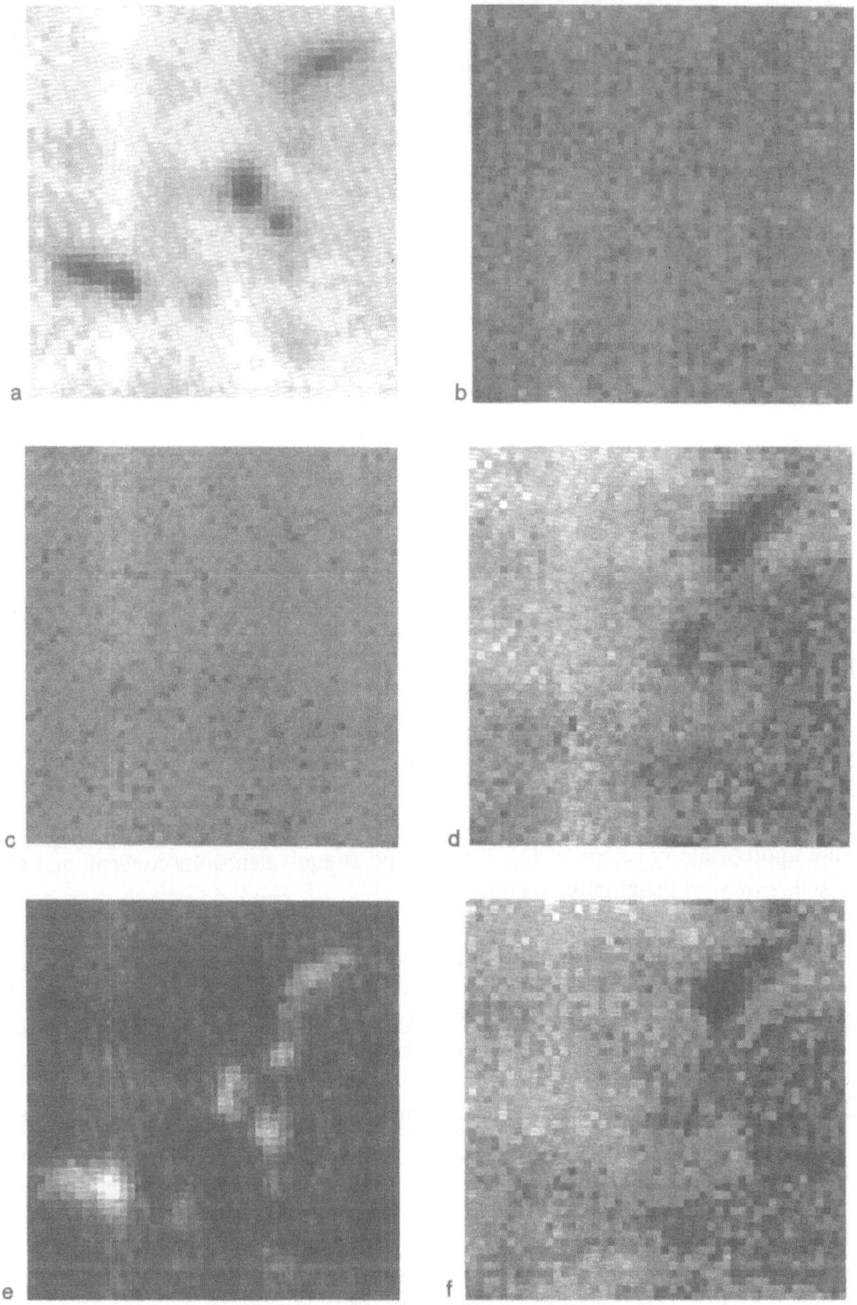

Figure 4. Multi-parameter images of undamaged sample: (a) conventional TTU, (b) wavespeed, (c) thickness, (d) density, (e) attenuation coefficient, and (f) longitudinal modulus.

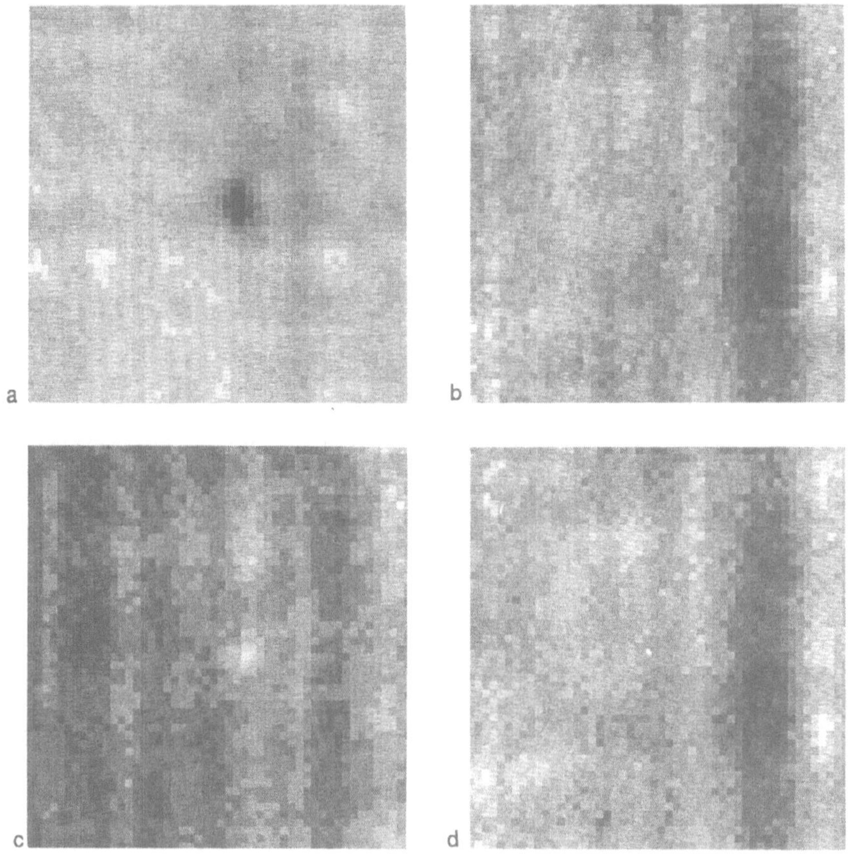

Figure 5. Multi-parameter images of impacted sample: (a) conventional TTU, (b) density, (c) attenuation coefficient, and (d) longitudinal modulus.

The second sample inspected with the multi-parameter technique had been impacted at the damage threshold energy (i.e. minimum damage).

Figure 5a shows the through-transmission C-scan of the impacted sample obtained from the envelope peak. Again this is equivalent to a conventional C-scan image, and shows no quantitative data. However the impact shows clearly in the center of the image.

The computed wavespeed for this sample ranged from 2.78 to 2.94 mm/uS with a mean of 2.86 mm/uS and a 0.02 mm/uS sigma. The sample thickness ranged from 4.48 to 4.74 mm with a mean of 4.58 mm and a 0.04 mm sigma. The physically measured thickness was 4.57 mm. Again there were no noticeable trends in either the wavespeed or thickness in this particular sample, and the C-scan plots are not shown.

Figure 5b is the C-scan image of the computed material density, which ranges from 1.30 to 1.64 g/cm^3 with a mean of 1.51 g/cm^3 and a 0.05 g/cm^3 sigma. The average density for this sample was measured at 1.54 g/cm^3. The image shows a large vertical low density strip near the right side and a smaller low density area at the upper left-center edge.

Figure 5c is the C-scan image of the computed material attenuation coefficient, which ranges from 0.03 to 0.06 neper/mm with a mean of 0.05 neper/mm and a 0.01 neper/mm sigma. This image shows high attenuation in the impact area and a generally high attenuation in the entire right half, except for a strip matching the low density strip.

Figure 5d is the C-scan image of the material longitudinal modulus derived from the computed wavespeed and density. The modulus range is 10.5 to 13.4 GPa with a mean of 12.3 GPa. Again the modulus image resembles the density image since the wavespeed is fairly constant.

The principal damage at this low impact energy is matrix microcracking. This would scatter energy out of the ultrasonic beam as is shown in the conventional through-transmission and attenuation coefficient C-scans, but does not affect the density, wavespeed, and longitudinal modulus values.

SUMMARY AND FUTURE WORK

We have developed a method to generate C-scans images of the longitudinal wavespeed, thickness, density, attenuation coefficient, and longitudinal modulus of a suitably thin flat sample using only one scan of the sample. We use the ultrasonic signal envelope to improve our measurement of transit times and amplitudes. Quantitative values are available for each of these parameters at each pixel of the image, with the latter three having the greatest uncertainty.

Future efforts in developing the multi-parameter inspection method will focus on refining the technique, verifying its results, and making it less sensitive to sample surface conditions. Presently the wavespeed and attenuation coefficient are average, broadband values. We will investigate using pulse spectroscopy methods to determine these two parameters as functions of frequency. We then intend to compare the experimentally determined frequency-dependent attenuation and wavespeed values to those predicted by the Kramers-Kronig relationship between these parameters[7].

REFERENCES

1. L.H. Pearson and D.S. Gardiner, Quantitative ultrasonic NDE, *in:* "Proceedings of the Fifteenth Symposium on Nondestructive Evaluation," D.W. Moore and G.A. Matzkanin, ed., San Antonio, TX (1985).

2. V. Dayal, An automated simultaneous measurement of thickness and wave velocity by ultrasound, *Experimental Mechanics*, 32:197 (1992).

3. D.K. Hsu and M.S. Hughes, Simultaneous ultrasonic velocity and sample thickness measurement and application in composites, *J. Acoust. Soc. Am.*, 92:669 (1992).

4. M.S. Hughes and D.K. Hsu, An automated algorithm for simultaneously producing velocity and thickness images, to appear in *Ultrasonics*, (1993).

5. E.P. Papadakis, K.A. Fowler, and L.C. Lynnworth, Ultrasonic attenuation by spectrum analysis of pulses in buffer rods: Method and diffraction corrections, *J. Acoust. Soc. Am.*, 53:1336 (1973).

6. P.B. Nagy and L. Adler, Surface roughness induced attenuation of reflected and transmitted ultrasonic waves, *J. Acoust. Soc. Am.*, 82:193 (1987).

7. C.C. Lee, M. Lahham, and B.G. Martin, Experimental verification of the Kramers-Kronig relationship for acoustic waves, *IEEE Trans. Ultrason. Ferroelec. Freq. Contr.*, 37:286 (1990).

NONDESTRUCTIVE CHARACTERIZATION OF HEAT DAMAGE IN GRAPHITE/EPOXY COMPOSITES

George A. Matzkanin

Nondestructive Testing Information Analysis Center (NTIAC)
Texas Research Institute Austin, Inc.
415 Crystal Creek Drive
Austin, TX 78746

INTRODUCTION

The degradation in properties of graphite/epoxy composites at elevated temperatures is an important problem to the Navy as well as other DoD components and government agencies, such as NASA and DoE. In aircraft applications, graphite/epoxy composites can be exposed to damaging levels of heat as a result of fire or operational service. Studies have shown that exposures to temperatures in the range of 350°F to 450°F can affect the resin matrix sufficiently to degrade the mechanical properties. To address these issues, a review was conducted by the Nondestructive Testing Information Analysis Center (NTIAC) to determine the principal thermal degradation mechanisms in graphite/epoxy composites; to identify laboratory analytical methods for measuring physical properties affected by heat damage in graphite/epoxy composites; to identify laboratory analytical methods for measuring physical properties affected by heat damage in graphite/epoxy composites; and to evaluate the state-of-the-art for nondestructively characterizing thermal degradation in advanced composites. Results show that thermal damage is complicated by numerous variables inherent to heat damage, such as, the method of heat application, exposure environment, cool-down environment and the thermal history of the material. Degradation mechanisms are also dependent on the chemical structure of the polymer matrix, the composite processing conditions, and any coatings which may be present. The degradation is typically matrix dominated since by the time fiber properties such as tensile strength and modulus are affected, all other mechanical integrity is lost. Compressive, shear and flexural properties are considered to be the most sensitive mechanical properties for use in early detection of thermal degradation. Materials properties affected include the glass transition temperature, thermal expansion coefficient, thermal conductivity and heat capacity. Density changes provide a sensitive measurement of degree of chemical degradation through loss of volatiles which result from chain scission and oxidation.

Nondestructive Characterization methods where promising results have been obtained include diffuse reflectance infrared spectroscopy for monitoring decreases in mechanical strength, Compton backscatter for imaging density variations, and reverse geometry radioscopy for large area inspection; other promising methods include ultrasonic spectroscopy and backscatter, infrared thermography and dielectric property measurement.

Additional details on thermal degradation mechanisms and results of NDE investigations will be discussed in the following sections.

THERMAL DEGRADATION MECHANISMS

Thermal degradation of graphite/epoxy composites is a temperature-time dependent phenomenon with the observable effects occurring in shorter times as temperature increases. Degradation often results in detrimental changes in mechanical properties or geometry as a result of temporary or permanent changes in the matrix, the matrix-fiber interface, and ultimately the fibers themselves. There are two basic classes of degradation mechanisms which might occur separately, but more often than not occur simultaneously. These are chemical mechanisms and physical mechanisms[1]. Obviously, chemical mechanisms involve changes in the chemical structure of the material and can include polymer chain scission, oxidation, and crosslinking, with the results being mass loss (through volatilization) and matrix embrittlement. Physical mechanisms involve primarily the relaxation of thermally-induced processing stresses, volume relaxation, and enthalpy relaxation, with the results being geometric deformation, such as warping, and delamination. These effects can be augmented and accelerated, if the service environment places the component under stress.

Often the events described above occur simultaneously so that deconvolution of the observed degradation effects into the various mechanistic channels is difficult. Generally, physical changes are reversible whereas chemical changes in graphite/epoxy composites are not. The temperature regimes in which various decomposition mechanisms become effective are illustrated in a typical TGA profile shown in Figure 1[2]. In the figure, initial weight loss attributable to solvent or water vaporization occurs below 270°C. The onset of resin decomposition occurs at 330°C, although Harribey[3] put the limit at 277°C. Prior to the last inflection at 690°C, which corresponds to fiber degradation, a transition point between 490°C and 550°C, indicative of stable char formation, may be present. Chen, Sun, and Chang[4] have noted that when subjected to intense heat, the matrix softens and evaporates and, in the extreme, burn through occurs. Each of these temperature regimes will be discussed below in sequence with respect to degradation mechanisms which are predominantly in effect. Clearly, the precise temperatures at which various mechanisms become operative depend upon the specific matrix and fiber used as well as the cure process, thermal history, and the ply or fiber geometry utilized.

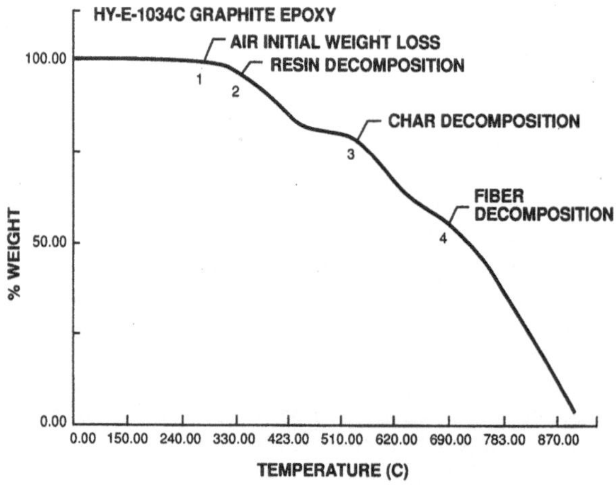

Figure 1. Typical TGA Profile Illustrating Thermal Effects on a Graphite/Epoxy Composite in Various Temperature Regions[2]

A number of experimental approaches involving mechanical testing and standard chemical laboratory methods have been employed over the past 10-20 years to monitor and study the effects and mechanisms of thermal degradation of graphite/epoxy composites. The mechanical properties studied include tensile, compressive, transverse and interlaminar shear, flexural, and buckling strengths and moduli; creep; and hardness. These are typically monitored by standard test methods such as those of the ASTM. Material properties studied include thermal expansion coefficient, thermal conductivity, heat capacity, the matrix glass transition temperature, T_g, and density. Most physical property measurements described in the literature have been concerned with thermal degradation of the epoxy matrix. This is understandable, since by the time the matrix has irreversibly degraded, all other properties are virtually destroyed and available laboratory measurement methods tend to focus on matrix properties. The principal laboratory measurement methods used to investigate heat damage of composites and epoxies are the following:

- Thermogravimetric Analysis (TGA)
- Thermomechanical Analysis (TMA)
- Differential Scanning Calorimetry (DSC)
- Optical Microscopy
- Energy Dispersive X-Ray Analysis (EDXA) - Bulk Fluorescence
- Scanning Electron Microscopy (SEM)
- High Performance Liquid Chromatography (HPLC)
- Gas Chromatography/Mass Spectroscopy (GC/MS)
- Infrared Spectroscopy (IR)

Determination of physical property changes associated with heat damage in composites obtained through use of these measurement methods will be summarized in the following paragraphs. Additional details can be found in the relevant references.

In the temperature range below about 177°C, a number of temporal changes are observed to occur. With aging on the order of a day at 130°C, Wang and Ogale[1] observed changes in the shear storage and shear stress relaxation moduli. Comparison of the responses of neat epoxy with graphite/epoxy, via what they termed the aging shift rate, indicated the changes in moduli were dictated by changes in the matrix. Furthermore, the aging was predominantly physical as exhibited by thermo-reversibility, that is the aging effects were reversed by raising the material above the T_g of the matrix. Contrary to this last observation are the results of Kubin[5] who found that on exposure to temperatures in the range 50-177°C, graphite/epoxy composites were noted by mass spectroscopy to evolve water and molecular fragments from decomposition of the matrix with the amounts increasing at the higher temperatures. Chemical decomposition of the matrix is not a reversible phenomenon. After aging for 10-25,000 hours at 121°C under atmospheric pressure, Haskins[6] reported matrix embrittlement and slight decreases in tensile strength, but no changes in T_g after 10,000 hours. With aging times of about 50,000 hours, severe matrix embrittlement occurred.

In the neighborhood of 177°C, Greer[8] reported that aging degrades the matrix sensitive properties such as compressive and interlaminar shear strength but not the fiber sensitive properties (i.e., flexural and tensile strengths and moduli). In aging studies lasting up to almost 8000 hours, Hsu[9] reported weight losses on the order of 10% and decreases in the flexural modulus by as much as 50%. Most of the composites studied exhibited fast initial weight loss rates due to loss of absorbed water and solvents followed by relatively constant weight loss rates attributed to thermal oxidation. Evolved gases from panels exposed to 2.5 W/cm^2 for up to 20 minutes were monitored by ion chromatography and GC/MS and the relative concentrations plotted as functions of time. Thermal aging at 177°C for 1000-5000 hours was observed by Haskins to result in matrix embrittlement (extensive edge cracking) and loss in tensile strength with matrix embrittlement becoming severe at aging times longer than 10,000 hours (cracks extended through the matrix). These effects were delayed by aging under reduced pressure (low oxygen concentration) indicating the importance of matrix oxidation in the long-term degradation process, a result supported using SEM and metallography. The degrees of oxidation were similar for aging at 121°C (1 atm.) and 177°C (0.14 atm.). Also under reduced pressure significant changes in T_g were observed only after 10,000 hours.

Exposures in the temperature range 205-370°C were studied by Street, Russel, and Bonsang[10] for times up to 30 minutes. Mode I and II interlaminar fracture toughness, Barcol Hardness, and T_g were all affected, though micrographs of the Mode I fracture surfaces revealed the fiber-matrix bond had not significantly degraded. Mechanical properties rapidly degraded above 349°C and charring, blistering, and microcracking due to the heat treatment and quenching were also apparent. Similar work by Pittman and Brown[11] in the temperature range 177-927°C for up to 17 minutes indicated decreases in the flexural and short beam shear strength as well as charring and lamina removal, the latter becoming more apparent with increasing heating rate and exposure time. TGA and DSC plots up to 593°C were obtained by Kubin[5] and used to obtained Arrhenius kinetic parameters for degradation and pyrolysis exothermicities given in **Table 1**. Kubin reported weight losses of 20-30% depending on the final temperature used to obtain TGA and DSC data. SEM analysis indicated that those specimens which lost 20% still had resin visible, but that 30% weight losses resulted in complete loss of the matrix. GC and MS results indicated the major pyrolysis products were H_2O, CO_2, and CO, with lesser amounts of C_2H_4, C_3H_8, C_3H_8O, C_6H_6, traces of C_3H_6 and possibly HCN, in general agreement with the results of Hsu. Kubin used reflectance infrared spectroscopy to characterize the cure state of the unexposed composite, but did not use it as a monitor of degradation.

Table 1. Kinetic and Thermodynamic Degradation Data Reported by Kubin[5]

Material	E_a (kcal/mole)	ΔH (cal/gm)
5208/T300	25.0	-20.4
3501-6/AS	19.7	-18.4

The results of Griffis, et. al[12], Chen, et. al[4], and Pering, et. al[13] for exposures in the temperature range 460-2982°C indicate the primary degradation mechanisms involve ablation, embrittlement, and microcracking. Skin temperatures above 1000°C are sufficient to sustain combustion.

Perhaps the most important result of studies on mechanical properties is that these are typically matrix dominated. The only properties governed primarily by the fibers are tensile strength and modulus (fiber direction), but it is worth mentioning that by the time these are degraded by thermal exposure, all other mechanical integrity has been lost. Compressive, shear, and flexural properties are therefore considered to be the sensitive mechanical properties which can be used in early detection of thermal degradation. Hardness is probably the most convenient detection method, since its measurement does not require destruction of the component and degradation in this property has been found to mirror changes in the other mechanical properties. Interlaminar shear properties are affected strongly by thermal cycling as a result of thermal mismatch between lamina.

Of the material properties, T_g is sensitive to changes in the matrix chemistry (oxidation, chain scission, crosslinking) and the presence of low molecular weight plasticizing agents. Typically, the T_g increases as matrix embrittlement progresses, low molecular weight volatiles are lost, and post curing occurs. Less often used as a monitor of degradation is the thermal expansion coefficient, α, but it has been found that α can increase quickly, even at relatively low temperatures, as a result of fiber-matrix debonding and creep fracture. Thermal conductivity and heat capacity exhibit strong functional dependencies on temperature over the range of interest, but are rarely used as monitors of thermal degradation. Density changes, particularly in the form of mass loss, have been used extensively due to the simplicity of the measurement and because this provides a sensitive measurement of the degree of chemical degradation through loss of volatiles which result from chain scission and oxidation.

REVIEW OF NDE INVESTIGATIONS

Information on the status of NDE of heat damage in graphite/epoxy composites was compiled by performing computer literature searches in various databases and files supplemented by manual searches and telephone inquiries. The NTIAC Data Base, which contains nearly 50,000 NDE documents, was searched using several different strategies related to heat damage of composites. The specific search topics and the number of resulting relevant documents are listed in Table 2. Additional searches were conducted in other databases available through DIALOG for general information on heat damage of graphite/epoxy composites (not necessarily NDE). Results from these searches are listed in Table 3. Documents resulting from these searches were reviewed and used to generate the summary information included in Section II.

Table 2. NTIAC Data Base Literature Searches

Search Topic	No. of Relevant Documents
NDT for Heat or Thermal Effects in Graphite/Epoxy Composites	77
NDT of Heat Damage in Graphite/Epoxy Composites	21
Nondestructive Imaging of Damage in Composites	21
IR Thermography of Composites	60

Table 3. Literature Searches on Thermal Degradation (Heat Damage) of Graphite/Epoxy Composites in Various Data Bases

Data Base	No. of Relevant Documents
Aerospace	40
Plastec	15
MCIC	7
DTIC	4
Engineering Index	4
NTIS	22

Upon reviewing the NDE-related documents it was found that most of the work reported in the literature dealing with NDE for heat damage of graphite/epoxy composites is based on the following five methods:

- Thermal (IR)
- Ultrasonics
- Acoustic Emission
- Dielectric Properties
- Radiography

These methods, while being readily available and generally well developed are limited in their capabilities to detect and characterize the changes in composite material properties associated with heat damage. Review of the literature from more recent years indicates that a number of NDE methods are under development by different investigators and those show various degrees of promise for characterizing heat damage in composites. Included in this group are the following:

- Thermal Wave
- Vibrothermography
- Leaky Lamb Wave
- Ultrasonic Backscatter
- Acousto-Ultrasonics
- Isotope Radiation Backscatter
- Embedded Sensors

Relevant information on the status of development of these NDE methods and their capabilities for detecting heat damage in graphite/epoxy composites can be found in Reference 14.

A conference was held recently specifically focusing on the Characterization and NDE of Heat Damage in Graphite/Epoxy Composites. This conference, held on April 27-28, 1993 in Orlando, FL, was sponsored by the Jacksonville Naval Aviation Depot and organized by the Nondestructive Testing Information Analysis Center (NTIAC). The purpose was to bring workers together engaged in this field to exchange information on research and development efforts that have resulted in potential and/or significant advances in characterizing heat damage. The conference focused on four primary areas:

- Understanding the Problem - Quantifying the physical and mechanical property changes induced by heat damage
- Characterization Methods for Heat Damage - Physical, mechanical, chemical, etc.
- Nondestructive Evaluation/Inspection - Methods, techniques, applications, etc.
- Correlations Between NDE and Heat-Induced Property Changes

The Eighteen presentations included Navy overviews and assessments, a state-of-the-art report, development of techniques for on-aircraft detection and evaluation of heat damage being carried out under the auspices of the Great Lakes Composite Consortium, imaging methods, and new applications.

The principal NDE methods discussed at this Conference representing the most promising approaches for nondestructively characterizing heat damage in graphite/epoxy composites were:

- Shearography
- Thermal Imaging
- Backscattered X-rays
- Diffuse Reflectance Infrared Fourier Transform
- Laser-Pumped Fluorescence
- Nonlinear Harmonies

Professor McManus[15] described the development of mathematical models to characterize the response of graphite/epoxy composite materials to fire which is currently poorly understood. The model is used as the basis of a computer code which predicts temperatures in the structure, degradation of the material, generation of decomposition gasses, internal pressures generated by these gasses, the stress state generated by thermal, internal pressure, and external mechanical loads, and partial or complete failure of the composite. The computer code is used to parametrically examine the response of composite structures to fire. Two response types were noted. Thin laminates are destroyed quickly by delamination caused by pressure built up inside the material due to the generation of decomposition gasses. Thick laminates suffered surface blistering, but the blistered, charred surface material insulates the bulk of the laminate, delaying further damage. It is noteworthy that neither of these behaviors is seen in metal structures. Tests were carried out on composite panels and honeycomb sandwich structures. Temperature distributions and visible damage information were collected while the structures were exposed to fire. After exposure, in-depth damage was examined by sectioning, and residual strength was measured by tensile tests. The experimental results are correlated with the theoretical predictions.

Several projects being funded by the Navy through the GreatLakes Composite Consortium have the objective of demonstrating on-aircraft nondestructive inspection procedures capable of characterizing heat damage to graphite/epoxy laminates. The effort is a coordinated one involving Grumman Aircraft Company, McDonnell Aircraft, Oak Ridge National laboratories, and the Naval Air Warfare Center in Warminster.

Specific tasks include the fabrication of laminate specimens, controlled heat damage of specimens, NDE technique evaluations, mechanical/physical test correlations with NDE measurements, and development of a practical procedure based on the optimum NDE technique(s)[16].

Promising results were reported by workers at Oak Ridge on Diffuse Reflectance Infrared Fourier Transform (DRIFT) and Laser Pumped Fluorescence techniques for characterizing and quantifying heat damage[17,18]. Spectral reduction techniques produced results that generally correlated with heat exposure with a precision comparable or better than mechanical measurements. The DRIFT method is also being evaluated at the Naval Air Warfare Center for application to composites that have been sealed, primed and painted[19]. Heat damage was induced by placing the painted side of composite coupons against a heated surface. The effect of heat damage on resin chemistry is then quantified by tracking changes in DRIFT peak height ratios as a function of temperature. Using this information, the correspondence between a decrease in peak height ratio and a decrease in mechanical strength can be characterized.

Radiographic images of a partially burned Glass Reinforced Plastic (GRP) material were obtained using a Compton Backscatter (COMSCAN) X-ray facility[20]. Such COMSCAN images represent planar maps (100mm x 50mm) of the density differences within the sample. Three dimensional images of the interface region between the burned and unburned GRP material were produced from sets of 22 simultaneous images which were uniformly distributed over the scan depth of 10mm. Density differences between the charred resin in the burned region and the pristine GRP material was clearly visible. The woven roving pattern within the GRP was also discernible in both the burned and unburned portions of the sample. In addition the interface between the burned and unburned material was resolved, as a function of depth, to better than 1 mm when imaged from either the burned or unburned side of the sample. The authors inferred from these observations that a suitably modified COMSCAN instrument would be capable of imaging the burning process, in situ, from the non-flaming side of composite materials.

Investigators at Lawrence Livermore National Laboratory have imaged heat damage in a honey comb structure and delaminations in a poorly cured aircraft patch using dual-band infrared (DBIR)[21]. This method detects heat flow anomalies at the delamination sites and removes surface clutter from roughness or emissivity variations.

Another method which shows much promise is Reverse Geometry Radioscopy which is capable of imaging material differences at the sites of graphite fibers[22]. Reverse Geometry Radioscopy is a filmless, near real-time technology for detailed study of subtle variations in metals, such as aluminum, and plastics, such as aircraft composites. With Reverse Geometry, the specimen is placed adjacent to a large scanning source. The Reverse Geometry system registers only the primary radiation, because scattered radiation bypasses the distant point detector. A computer synchronizes the beam sweep, the detector readout rate, and the monitor display of first generation digital data. High throughput is obtained because the field of view is determined by the large area source, which is 25 cm in diameter. A 7.6 cm source is also available for special resolution of 13 line pairs per millimeter over the entire field of view. The newly developed configuration of x-ray source and detector increases contrast sensitivity from the conventional level of 2% to a new standard of 0.3% for aluminum in the 6.4 mm to 19 mm range. Single layer aircraft skin, at 1 mm, can be imaged with 1.3% contrast resolution. Subsurface corrosion product is imaged at levels below 1.5%.

CONCLUSIONS

Although a certain amount of information has been obtained on mechanical and material property changes caused by heat exposure, the basic failure mechanisms are still not well understood. Work currently in progress at Wright Laboratories is addressing this issue by developing an extensive test matrix for characterizing failure mechanisms for a variety of heat exposure conditions. There is existing evidence to suggest that the failure mechanisms and property changes associated with moderate temperatures, long time exposure are much different than those associated with high temperature, short time exposure such as would be encountered in a fire. Existing evidence indicates that the damage associated with moderate temperature, long time conditions is much more incipient and measurably degrades the matrix before there is any visual evidence. Thus, NDE techniques such as ultrasonics that might work for detecting high heat damage, such as delaminations, will not necessarily work to detect the more subtle property changes associated with modest temperatures.

A number of promising NDE techniques are currently under development, primarily based on thermal, infrared, and radioscopy methods; however, from the standpoint of practical application, the large variety of aircraft configurations, materials, and heat exposure conditions complicates the issue. A more complete understanding of the failure mechanisms associated with different conditions and configurations will help determine the most viable NDE approaches for practical application.

REFERENCES

1. S.F. Wang and A.A. Ogale, "Influence of Aging on Transient and Dynamic Mechanical Properties of Carbon Fiber/Epoxy Composites," *SAMPE Q.*, 20:9 (1989).
2. W.A. Sigur, "Ablation Characteristics of Graphite Epoxy," *SAMPE Q.*, 17:25 (1986).
3. J.M. Harribey, "Physicochemical Characterization of Resin 2220-3 in Preimpregnate," Report No.: PB86-242534/WMS, Centre Essais Aeronautique Toulouse (1985).
4. J.K. Chen, C.T. Sun, and C.I. Chang, "Failure Analysis of Graphite/Epoxy Laminate Subjected to Combined Thermal and Mechanical Loading," *J. Comp. Mat.*, 19:408 (1985).
5. R.F. Kubin, "Thermal Characteristics of 3501-6/AS and 5208/T300 Graphite Epoxy Composite," Naval Weapons Center Report No.:NWC-TP-6104 (1979).
6. J.F. Haskins, "Thermal Aging," *SAMPE J.* 25:29 (1989).
7. J.R. Kerr and J.F. Haskins, "Effects of 50,000 Hours of Thermal Aging on Graphite/Epoxy and Graphite/Polyamide Composites," *AIAA J.*, 22:96 (1984).
8. G.H. Greer, "Thermal Aging of Contemporary Graphite/Epoxy Materials," *SAMPE Symp.*, 24:1039 (1979).
9. M.T .S. Hsu, "Characterization and Degradation Studies on Synthetic Polymers for Aerospace Application," NASA Report No.:NASA-CR-166597 (1982).
10. K.N. Street, A.J. Russell, and F. Bonsang, "Thermal Damage Effects on Delamination Toughness of a Graphite/Epoxy Composite," *Comp. Sci. and Tech.*, 32:1 (1988).
11. C.M. Pittman and R.D. Brown, "Exploratory Investigation of Two Resin-Matrix Composites Subjected to Arc-Tunnel Heating," NASA Report No.:NASA-TP-1429 (1979).
12. C.A. Griffis, J.A. Nemes, F.R. Stonesifer, and C.I. Chang, "Degradation in Strength of Laminated Composites Subjected to Intense Heating and Mechanical Loading," *J. Comp Mat.*, 20: 216 (1986).
13. G.A. Pering, P.V. Farrell, and G.A. Springer, "Degradation of Tensile and Shear Properties of Composites Exposed to Fire or High Temperature," *J. Comp. Mat.*, 14:54 (1980).
14. G.A. Matzkanin, "Nondestructive Characterization of Heat Damage: A State-of-the-Art Report", Proc. of a Conference on Characterization and NDE of Heat Damage in Graphite Epoxy Composites, NTIAC, in print (1993).
15. H.L. McManus, "Prediction of Fire Damage to Composite Aircraft Structures," Proc. of a Conference on Characterization and NDE of Heat Damage in Graphite Epoxy Composites, NTIAC, in print (1993).
16. R. Collins, "NDI for Heat Damaged Advanced Composites," Proc. of a Conference on Characterization and NDE of Heat Damage in Graphite Epoxy Composites, NTIAC, in print (1993).
17. C.J. Janke, E.A. Wachter, H. Philpot, and G.L. Powell, "Inspection of Graphite-Epoxy Laminates for Heat Damage Using DRIFT Spectroscopy," Proc. of a Conference on Characterization and NDE of Heat Damage in Graphite Epoxy Composites, NTIAC, in print (1993).
18. G.L. Powell, N.R. Smyrl, C.J. Janke, M. Milosevic, and J. Lucania, "Inspection of Graphite-Epoxy Laminates for Heat Damage Using Drift Spectroscopy," Proc. of a Conference on Characterization and NDE of Heat Damage in Graphite Epoxy Composites, NTIAC, in print (1993).
19. E. Armstrong-Carroll, P. Mehrkam, and R. Cochran, "Heat Damage Evaluation of Painted Graphite/Epoxy Composites," Proc. of a Conference on Characterization and NDE of Heat Damage in Graphite Epoxy Composites, NTIAC, in print (1993).
20. P. Lambrineas, K.K. Yeung, R.D. Finlayson, and T. Cleary, "Imaging of Burned Glass Reinforced Plastic Using a Compton Backscatter Technique," Proc. of a Conference on Characterization and NDE of Heat Damage in Graphite Epoxy Composites, NTIAC, in print (1993).
21. N.K. Del Grande, "Multi-Sensor Characterization, Imaging and NDE of Graphite/Epoxy Composites," Proc. of a Conference on Characterization and NDE of Heat Damage in Graphite Epoxy Composites, NTIAC, in print (1993).
22. K.W. Dolan, D.J. Schnebeck, R.D. Albert, and T.M. Albert, "Reverse Geometry X-Ray Imaging for NDT Applications," Proc. of the JANNAF NDE Subcommittee Meeting, CPIA, in print 1993.

INTRINSIC AMPLITUDE DISTRIBUTIONS AND THE ULTRASONIC INSPECTION OF FIBER-METAL LAMINATES

Martin P. Jones

Structural Laminates Company
510 Constitution Blvd.
New Kensington, PA 15068

INTRODUCTION

For any nondestructive inspection, there are many factors that influence the probability of detection of discontinuites.[1] The work described in this paper mainly examines the effect of amplitude variations, intrinsic to a material,[2] on the probability of detection of isolated discontinuities. Although the present approach could be adapted to any material, many discontinuity types, and several methods of nondestructive inspection, the application described here is on fiber-metal laminates and the ultrasonic detection of delaminations.

Production-Type Nondestructive Inspection

The intent of developing and applying rejection criteria is to enable the detection of critical defects of a given type, size, distribution, etc.[3] The actual criteria is usually based upon the equivalent response of an inspection system to a well-defined standard defect such as a flat-bottom hole.[4] The goal in applying such inspection standards is to realize a high probability of detection of critical defects while minimizing the inspection cost and complexity. Two good examples of this trade-off is in sampling.[5,6] First, a decision must be made as to whether to inspect every part, or just a fraction of the material. Second, for the parts to be inspected, a decision must be made as to the number of readings to take per part. These, like other inspection parameters, quickly become cost-prohibitive in a competitive production-type environment. This paper explores the advantages of considering the effect of a material's intrinsic variation of properties in the selection of production-type inspection parameters, especially with respect to rejection criteria.

Production-Type Inspection of Fiber-Metal Laminates

Fiber-metal laminates are composed of alternating layers (about 0.01 inch thick) of aluminum and fiber/resin. Glare® laminates are made with glass fibers and ARALL® laminates are made with aramid fibers. A typical configuration of these laminates is shown in Figure 1. These laminates have aerospace applications because they exhibit high fatigue strength, weight savings, and burn-through resistance in comparison with monolithic aluminum. Currently, the product size is about 4 feet by 12 feet, with a thickness of about 0.052".

Each laminate is ultrasonically inspected in the C-scan mode to detect delaminations of a specified critical diameter. The critical diameter is simulated by placing on the surface of the laminate a reference dot equal to the critical diameter and composed of a material that is ultrasonically opaque. The scan and index increments, which determine the pixel size, are set to equal no more than one third the critical diameter, which is a common practice in aerospace NDE.[7] Another common practice is to specify reject and reference levels as components of rejection criteria. A reject level is defined as the point at which the ultrasonic amplitude drops by more than an allowed percentage (or decibel) from the reference level. For fiber-metal laminates, the reference level is the amplitude obtained through a reference laminate which has been selected for integrity and consistency. When ultrasonically imaging the reference dot, the number of pixels that fall below the reference level defines the response of the system to the critical delamination size. Laminates that contain discontinuities producing this response or greater are defined as rejected.

Figure 1. A typical configuration of a fiber-metal laminate.

Statistical Approaches

If a reference dot is ultrasonically imaged repeatedly, the response can range, for example, from 4 to 8 pixels. The question is thus raised as to the appropriate rejection criterion for the response from unknown discontinuities. The answer depends on the desired detection probability of the critical delamination diameter. Choosing 4 pixels would give a higher probability of detection than 6 pixels would give. To determine such probabilities, one must first know the histogram of responses, i.e. how often the reference dot is imaged at, for example, 4 pixels than at 7 pixels. To determine the histogram of responses, gage capability studies were conducted under conditions very similar to those under which laminates are inspected in production.

To complement the gage studies, a computer model was developed to simulate the ultrasonic inspection of reference dots. The goal of the model was to generate a histogram of responses similar to that which was obtained via the gage capability studies. Several specific parameters in the gage studies were incorporated into the model, e.g. the random position of the inspection grid relative to the dot.

Any material when C-scanned will show some variation of amplitude over a grid of inspection points. Part of this variation is from inspection system noise. For fiber-metal laminates, the dominant component of this variation is the result of nonhomogeneous elastic properties and geometry. For a given laminate, this variation is defined as its intrinsic amplitude distribution (IAD). An IAD is plotted as occurrence vs. amplitude in Figure 2.

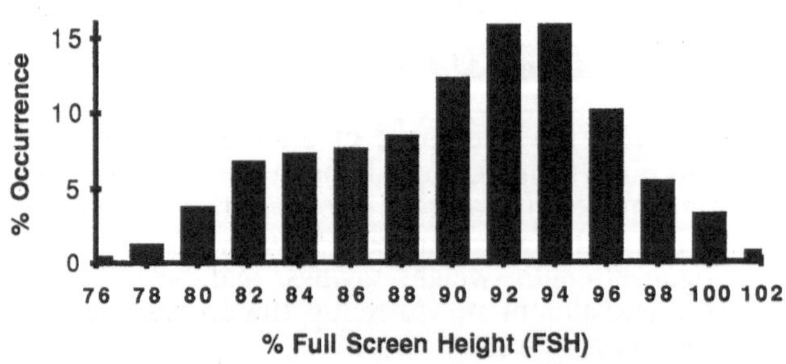

Figure 2. Example of an Intrinsic Amplitude Distribution (IAD) for a laminate.

The distinguishing aspect of the present work is that the model accounts for the effect of a laminate's IAD on the histogram of responses from the reference dot. The particular IAD shown in Figure 2 was obtained from the laminate used in the gage study described below. Also, this particular IAD was used in the simulation model to be described. Through comparison of the results obtained by the model and the gage studies, some recommendations for improved inspections are suggested.

STATISTICAL GAGE CAPABILITY STUDIES

The ultrasonic instrument was multiplexed to generate and receive ultrasonic waves at ten sensor positions equally spaced across the width of the laminate, thus providing full coverage inspection. Automatic defect recognition software was used to enable the response (in pixels) at preset rejection criteria to be counted from each dot.

Over the inspection area corresponding to each of the ten sensor positions, an equal number of dots were applied to random locations on the surface of the laminate. The diameter of the dots were equal to the critical diameter. The ratios of the critical diameter to the index and scan increments were 3.1 and 4.7, respectively. The rejection level was set to a specific fraction of the reference level amplitude. To increase the number of imaged dots, the laminate was inspected several times.

COMPUTER SIMULATION OF ULTRASONIC INSPECTION

This model simulates the C-scan imaging of a reference dot using similar parameters to those under which the gage studies were performed. Figure 3 shows a flowchart of the computer program. Step 1 specifies the assumed conditions that were constant throughout the simulated inspection. These constant conditions were the reference dot diameter (uniformly opaque), reference level (80 % FSH), scan and

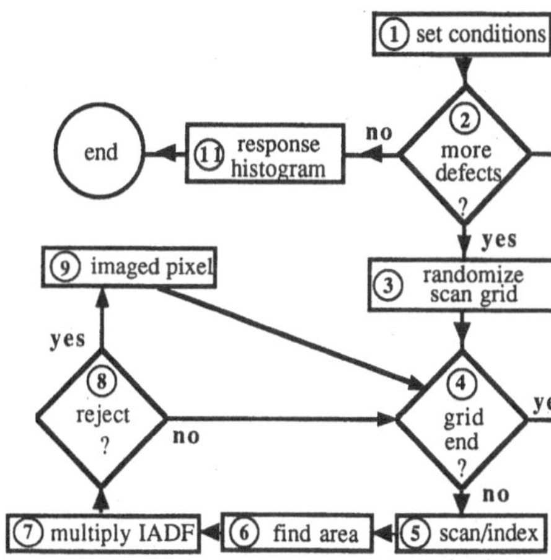

Figure 3. Computer model flowchart of the inspection simulation.

index increments, rejection level, and beam diameter (uniform intensity). The average beam diameter of the ten sensors was arrived at from a modified version of the ball target technique recommended in ASTM E1065, which defines the beam diameter to be the -6 dB amplitude points of the beam profile[8]. For use in the model, the beam diameter was more appropriately defined at the rejection level amplitude.

Step 2 determines how many simulated defects (reference dots) will be inspected. If more defects are to be inspected, a new grid is selected in Step 3 so that the inspection points are shifted by a random fraction of a pixel while the increments remain the same. This simulates how, for real C-scans, the defect can be at a random location with respect to the inspection grid. Step 4 tests whether or not a reading has been taken at each grid point. If not, then the ultrasonic beam is moved to the next grid point as shown in Step 5. In Step 6 the area of the unblocked beam is calculated.

The remaining ultrasonic amplitude transmitted is assumed to be directly proportional to this area of the unblocked portion of beam. The proportional amplitude is multiplied by an intrinsic amplitude distribution factor (IADF), as indicated in Step 7. An IADF is the ratio of a randomly selected amplitude from the IAD (shown in Figure 2) to the reference level amplitude. Step 8 tests whether or not Step 7 produced an amplitude at the grid point that is below the rejection level. If so, the pixel is imaged in Step 9, otherwise Step 4 is immediately repeated. After the end of the grid, the imaged pixels are counted in Step 10 to obtain the response size for a particular simulated dot. Once all dots are imaged, the histogram of occurrence vs. imaged size is generated (Step 11).

RESULTS AND DISCUSSION

Figure 4 shows the comparison of results from the gage study and the computer simulation. The error bars on the gage study data indicates the range of occurrence at a particular response size from among all ten sensors.

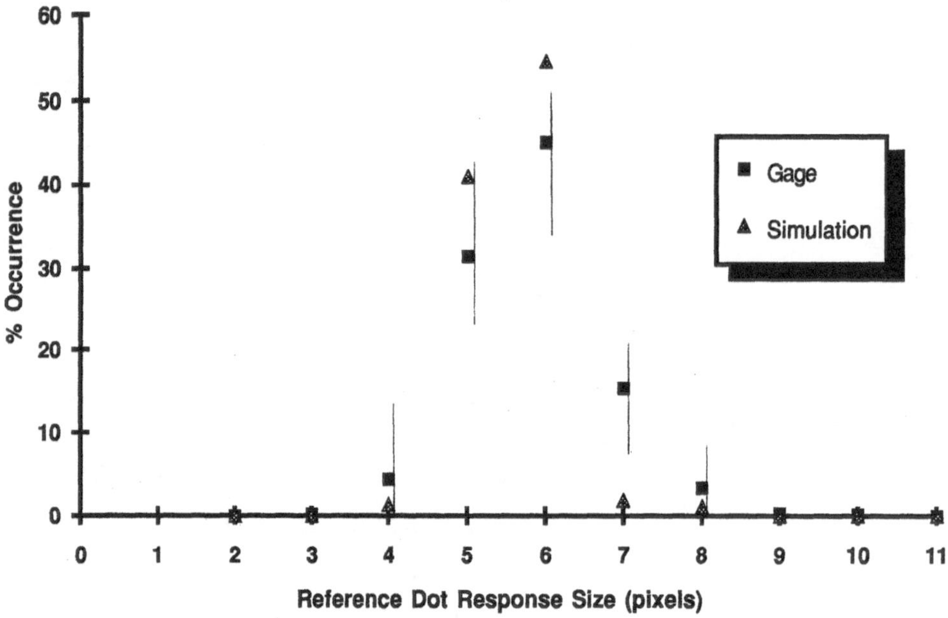

Figure 4. Comparison of gage study and computer simulation data.

For the computer simulation, the table below shows the occurrence vs. dot image size obtained both with and without the IADF. For the case of no IADF, the laminate was assumed to have constant amplitude equal to the reference level.

The average response size found in the computer simulation compares closely with that found in the gage study. Any differences between them could be due to error in the assumptions made regarding the beam profile. The effect of the laminate's IAD on the simulation appears to be significant. The minimum response occurrence is over 14 times more for the simulation when the laminate's actual IAD is assumed than when the laminate is assumed to exhibit constant amplitude.

Therefore, especially for production-type inspections, the IAD should be considered in the selection of the rejection criteria. For example, if the laminate to be inspected has a significantly different IAD compared to the reference laminate, then the laminate might be rejected even prior to inspection. This might be necessary to maintain an agreed upon probability of detection.

Table 1. Effect of the IAD on the computer simulation data.

Occurrence (%)	Response Size (pixels) 4	5	6	7	8
w/ IAD	1.4	41.0	54.6	1.9	1.1
w/o IAD	0.1	28.9	66.1	2.8	2.1

CONCLUSIONS

Gage capability studies and computer simulated inspections are useful complementary tools that can be used to assess and improve the probability of detection of discontinuities. Particularly through computer simulations, it was shown that the IAD of a material affects the probability of detection and should be considered in the development of inspection procedures.

FUTURE WORK

Future work should incorporate actual beam profiles in the computer simulation. This may improve the degree of agreement of the individual histogram response sizes between the simulation and the gage study. Also, work should look at the effect of IAD's on the rejection criteria when different reject levels, index increments, defect diameters, etc. are chosen in the computer simulation. Furthermore, rather than a random sequence of IADF choices, the real-time sequence of C-scan data and randomly overlaid simulated defects might prove more realistic.

REFERENCES

1. F.A. Chang, J.R. Bell, T.C. Walker, J.M. Norton, P.F. Packaman, and L.O. Gilstrap, Method for determination of sensitivity of NDE techniques, Tech. Rep. AFML-TR-76-246 (1976).
2. A.J. Rogovsky, Some advances in the NDT technology for composites, "Review of Progress in Quant. NDE", D.O. Thompson and D.E. Chimenti, eds., Plenum Press, New York (1984).
3. E.J. Chern, H.P. Chu, J.N. Yang, Assessment of probability of detection of delaminations in fiber-reinforced composites, NASA Tech. Mem. 104553 (1991).
4. Standard practice for the calibration of the ultrasonic test system by extrapolation between flat-bottom hole sizes (E804), "Annual Book of ASTM Standards", R.A. Storer, ed., ASTM, Phil. (1991).
5. E.P. Papadakis, Sampling plans and 100% nondestructive inspection compared, Qual. Prog., 15:1 (1982).
6. W. Nelson, Estimation of particle size distribution when detection probability depends on particle size, Mat. Eval.,41:9 (1983).
7. C. Savvy, NDI of adhesive bonds within metal structures (BAC 5968), The Boeing Company (1988).
8. Standard practice for evaluating characteristics of ultrasonic search units (E1065), "Annual Book of ASTM Standards",R.A. Storer, ed., ASTM Philadelphia (1991).

AN ACOUSTO-ULTRASONIC PLATFORM FOR THE QUALITY ASSESSMENT OF THICK RADIAL PLY COMPOSITE STRUCTURES

Thomas J. Gill, Jr. and *Anthony L. Bartos

Olin Corporation
Research and Development
Red Lion, Pennsylvania 17356

*Atlantic Research Corporation
Professional Services Group
Sterling, Virginia 20166

ABSTRACT

The last ten years have witnessed significant developments in the design and fabrication of advanced composite materials. By integrating contemporary design concepts—such as a toughened resin system and radial ply architecture—with innovative processing technology, thick, geometrically complex structures can now be produced which possess the ability to withstand a considerably hostile end-use environment. The same intrinsic material attributes that allow such a high degree of performance, however, also serve to make the non-destructive evaluation (NDE) of such a structure very challenging.

Traditional radiographic methods, for example, are inadequate because of the superposition problem inherent in thick, irregularly shaped structures. Other techniques such as pulse echo ultrasonics fall short due to a variety of factors related to the heterogeneous, anisotropic nature of the material. While computed tomography (CT) has been used with great success in detecting processing anomalies, it too has limitations in a high rate production environment because of throughput constraints; moreover, CT will only provide information based upon relative material density. In short, the in-service NDE of a thick composite structure requires a novel solution.

This paper describes an innovative acousto-ultrasonic (AU) approach to production NDE whereby a thick composite segment can be rapidly evaluated with respect to a number of processing parameters including part consolidation and cure/undercure, as well as the presence of delaminations and foreign matter. A series of ultrasonic waveforms are input/acquired using a multiple probe configuration which surrounds the structure. Digital signal processing is utilized to extract pertinent waveform discriminators, and pattern recognition theory is then applied to classify the respective features based upon the results obtained from a learning set of signals taken from an assortment of "good" and "bad" structures. The significance of this platform is that it can be modified to adapt to changing material requirements throughout the production life-cycle; and more importantly, it provides a way to assess a number of different material attributes through a single interrogation process.

BACKGROUND

Due to recent advances in processing technology, thick composite structures have recently begun to enjoy a greater degree of prominence. One such part, manufactured by Olin Corp., is based upon the concept of a 120° segment of a right circular solid cylinder (pie) with an irregular diameter. This segment is comprised of several hundred layers of graphite-epoxy prepreg and has a minimum diameter less than one inch; the maximum diameter is over four inches. In addition, the overall part length exceeds 20 inches. Perhaps the most contemporary design feature is the radial ply architecture which is used in place of the more typical circumferential layup.

A proprietary hybrid manufacturing process is employed to fabricate the prepreg mold charge into a final cured segment. While this process has been greatly refined over a period of years to produce a highly consolidated part, as with any thick composite structure, there are a variety of process related anomalies that can be introduced during this operation. As part of an effort to identify and characterize the anomalies that are specific to this part, Olin NDE engineers compiled a computed tomography (CT) database that contained over 4000 segments evaluated during an initial production run. Segments were contiguously scanned, first by using 6 mm slice widths, and, towards the middle of the program, 12 mm slices. The outcome of this work was that a number of real-world anomalies were identified that could potentially compromise the end-use performance of the final segment. Four distinct categories of density related anomalies were defined: (1) Paper inclusions, prepreg backing paper left between the radial plies which were in essence delaminations; (2) Metallic inclusions, both steel and aluminum; (3) Rings, an undesirable consolidation feature peculiar to radial layups; and (4) Site specific surface breaking delaminations, thought to be a result of residual stress. A fifth category, cure deprivation, was added to account for segments that were inadvertently omitted from a second, critical curing operation.

It is to be emphasized that the aforementioned categories are by no means representative of the entire spectrum of possible anomalies that might be found in any given composite part. The point of this comprehensive database study was to establish the anomalies that may be found in *this* particular structure so that a high throughput NDE system could be designed to detect real-world defects. This is an important distinction as many inspection systems are built and purchased because of their ability to detect some material "flaw" that was defined based upon a series of experiments with a product that was artificially seeded with what may or may not be an real defect. Although there is certainly nothing inherently wrong with using Teflon or air holes as a phantom for benchmarking or calibration purposes, when used strictly as an exclusive proveout standard, this approach does suffer from the obvious disadvantage that the NDE platform is being tuned for a defect that is not representative of what exists.

The acousto-ultrasonic technique was selected as the most appropriate NDE method to use for the Olin segment. This technique, introduced by Vary et. al.[1], was singled out due to its high throughput, and great potential for gleaning information on a diverse variety of material attributes. During an AU test cycle, externally generated stress waves are propagated throughout the segment which interact with all of the various material constituents—both desirable and undesirable. Because AU is a volumetric test, the data generated by this procedure contains information concerning the overall material health. Beyond this global assessment, however, is a useful, albeit limited, parcel of information relating to localized defects as well.

Work on this program was carried out by members of Olin Corp. and Atlantic Research Corp. (ARC) over a four year development period. As a precursor to the hands-on effort, a comprehensive literature survey was initiated to determine what was then the state-of-the-art in AU research. A series of experiments followed to demarcate the practical operating limits in a prototype system. Bounds were established for piezoelectric transducer (PZT) placement, PZT frequency, pulse repetition frequency (PRF), path attenuation, number of waveform averages and other relevant system parameters. A prototype specimen holder–a cradle–was constructed to mount the PZTs in and, accordingly, the composite part to be tested. A learning set consisting of both anomaly free segments–a control set–as well as components representing each of the previously defined anomaly categories was passed through the system in a data acquisition mode. Both time and frequency domain data were obtained for a total of 31 through transmission paths. A large group of signal features were then extracted and studied for each path and anomaly category. Pertinent features were chosen and grouped for each anomaly class based upon their respective ability to discriminate.

EXPERIMENTAL

The prototype AU platform system schematic is shown in Figure 1. Basic system hardware consisted of 14 PZTs mounted in a cradle assembly, a high voltage ultrasonic pulser/receiver, a signal attenuator unit, a switch matrix unit, transient capture and digital signal processing PC backplane boards, a GPIB backplane board, and a IBM compatible 80386 personnel computer with a removeable hard disk.

Panametrics V-602 broadband PZTs were used as both ultrasonic senders and receivers, each with a one inch aperture. A crystal frequency of 1 MHz was employed due to a series of experiments with both 1 MHz and 2.25 MHz PZTs that revealed a lack of useful information beyond approximately 1 MHz. One inch apertures were used to allow higher input voltages in the event that adequate penetration proved difficult. Nylon end caps were utilized to protect PZT faces from damage by the abrasive composite surface.

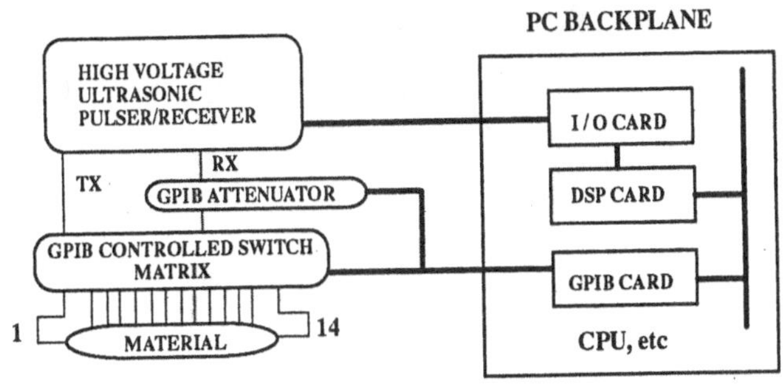

Fig. 1 - Schematic of Acousto-Ultrasonic Hardware

All 14 PZTs were mounted in a cradle assembly constructed from plywood. Spring loaded cups were made to hold the PZTs against the test part with an adequate contact force. Note that 12 PZTs were placed transverse to the segment, while two were positioned at opposite end points.

A Panametrics model 5058 high voltage pulser/receiver (PR) was used in conjunction with a GPIB controlled Cytec semi-customized switch matrix. High gain and high voltage (900 volts) capability were the major reasons for this particular PR; the Cytec unit was used simply as a channel (path) multiplexer.

A GPIB controlled Lucas-Weinschell model 8200 attenuator was placed between the Cytec output and the receiver input to compensate for the different signal response generated during each individual path insonification. Differences in path attenuation settings were necessary due to variances in path length and position (axial path vs transverse path, etc.).

A 40 MHz transient capture board along with an array processor board possessing an AT&T DSP-32C floating point digital signal processing (DSP) chip, both from Spectrum Signal, were used because of their dedicated high speed bus connection. An optional high speed 8K FIFO accumulation buffer was added to the transient board to facilitate longer capture times.

Just prior to loading the cradle, a glycerin based coupling gel is applied to each PZT end cap surface. The segment to be tested is then placed flat side down in the cradle onto the 12 transverse PZTs. The two remaining end PZTs are then positioned and a final felt covered bracket is brought down to force the part against the transverse PZTs; the end mounted PZTs are automatically constrained by a spring hinge.

Testing begins with the pulser excitation voltage set constant at 400 volts. Earlier experimentation demonstrated that 400 volts was adequate to penetrate the segment across its longest path while yielding an acceptable signal. Gain and damping settings are adjusted so as to maximize signal amplitude. High pass and low pass filters are set to 0.3 MHz and 3.0 MHz respectively. Since the Panametrics 5058 is not a programmable PR, the unit is triggered externally from the PC and is left on throughout the test's duration.

A single AU test cycle consists of waveform data sets sequentially acquired over 31 separate paths, as shown in the Figure 2 schematic. Transducer sender/receiver pairings include classical AU same side probe configurations as well as opposite side and transverse configurations. The transverse grouping is similar to the so-called zero offset configuration except that the sender/receiver PZTs do not look directly across one another due to the 120° radial angle.

The cycle is initiated by the PC control software which controls all channel switching and attenuation settings for each respective path. All waveforms are acquired using an 8 MHz sampling rate and averaged using the DSP from a total of 32 separate acquisitions per path. In addition, some paths are reacquired using a 40 MHz rate to capture waveform Time-of-Arrival (TOA) information. Once acquired and averaged, the waveforms are transferred to a dedicated RAMDISK for further processing. The waveforms are low passed filtered at 1.3 MHz and desampled by two to decrease processing time. Mean values are removed and the waveforms are normalized by the standard deviation to render them amplitude independent—necessary because couplant variances, PZT pressure and other potential acquisition irregularities have forced us to use amplitude related features sparingly.

The waveforms are then routed through a Fast Fourier Transform (FFT) module to generate corresponding frequency domain data. A number of waveform features are calculated from both the filtered time domain and frequency domain data. The first of these, standard deviation, is the only amplitude related feature. The remaining 47 features are all waveform shape-related. The Cross-correlation coefficient (CCC) between symmetrical paths is the only other feature extracted from the original normalized waveform.

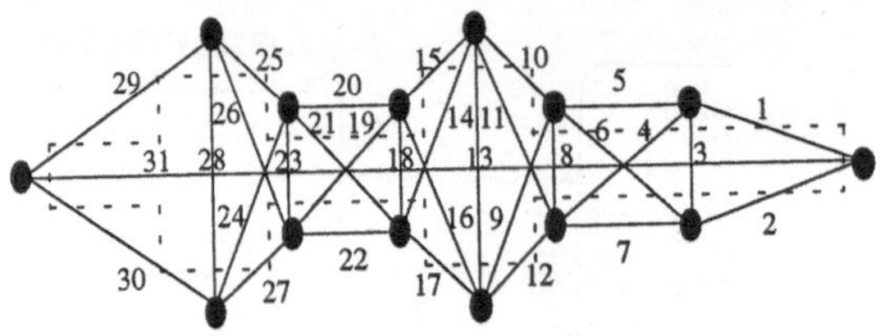

Fig. 2 - Schematic of Insonification Paths

There are six levels of processing that occur on the normalized waveform before the remaining features can be extracted: TOA calculations; time domain envelope calculation; the power spectrum of the derivative of the envelope; the power spectrum of the normalized envelope; and two cepstrally smoothed power spectrum estimates of the normalized original waveform. Cepstral smoothing, is achieved by zeroing out the higher value portions of the cepstrum, where the cepstrum is defined as the inverse Fourier transform of the log spectrum.

Velocity related TOA features are taken from the separate 40 Ms/s acquisitions. TOA is estimated by using a simple amplitude threshold, by thresholding the integral of the envelope of the arriving wavelet, by estimating its centroid, and by linear as well as nonlinear combinations of the preceding. The remaining features taken from the last five processing levels are similar. These include peak value, mean, standard deviation, centroid, mechanical moment and CCC with a symmetrical path, if one exists. The difference is often one of interpretation. For example, a high centroid value in the frequency domain implies a healthy material, whereas a low centroid number in the time domain is preferred because it indicates an early arrival of energy or higher modulus.

In order to compensate for segment regions that are more defect sensitive than others–based on finite element modelling–the part is divided into eight sections for the final feature discrimination phase of the test cycle. As shown in Figure 3, the first seven sections are actually sub-volumes within the part, each defined by up to five paths, while the eighth region consists of the entire part as defined by the axial path 31. Feature values are grouped according to both part section as well as anomaly type with each feature group typically containing two to five features. The result is a matrix of feature groups assigned for segment section vs. anomaly category.

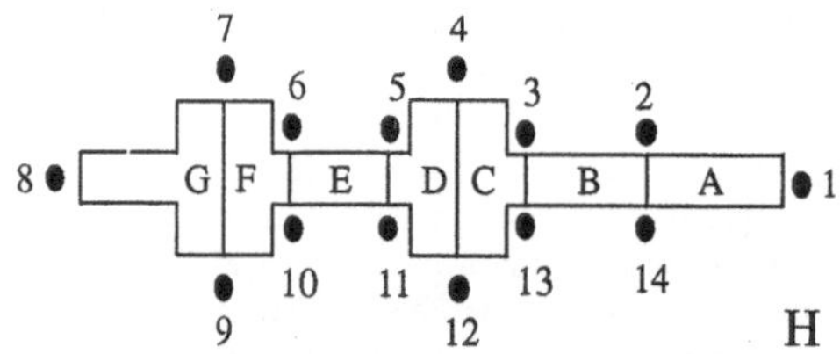

Fig. 3 - Schematic of Feature Discrimination Sections

The key to the successful development of this platform was the acquisition of waveform data from the learning set segments. The learning set was comprised of 78 parts broken out into the following categories: 30 segments determined to be normal by numerous quality indicators including CT analysis; 12 segments containing high density inclusions 1.5 - 3.0 mm in diameter, six in a thin section and six in a thick section; 8 segments containing backing paper between plies; 8 segments with a particular site specific surface breaking delamination; 15 segments with consolidation abnormalities–the "ringed" parts; and 5 segments that were omitted from the final postcure operation so as to be undercured.

In order to reduce the effect of signal errors due to PZT positioning irregularities, couplant variations and operator discrepancies on the learning set data collection, each part was carefully cycled through a total of 13 acquisitions. Based upon waveform cross-correlation coefficients, the best 10 of these acquisitions were thus used to build the learning set feature database. Relevant features–those that fluctuated with the presence of a given anomaly–were obtained by calculating the Fisher ratio[2] and Divergence[3] for each of 48 various time and frequency domain features using the collective results from the 30 normal segments for comparison. The Fisher Ratio feature selection criterion is defined as:

$$FR = \frac{(m_1 - m_2)^2}{\left(s_1^2 + s_2^2\right)/2} \tag{1}$$

Divergence is defined as:

$$D = \frac{s_2^2 - s_1^2 + (m_1 - m_2)^2}{2s_1^2} + \frac{s_1^2 - s_2^2 + (m_1 - m_2)^2}{2s_2^2} \tag{2}$$

where: s_1 = std. dev. of control set, s_2 = std. dev. of anomaly set,
m_1 = mean of control set, m_2 = mean of anomaly set

This procedure was repeated for each specific anomaly category over every insonification path in each segment section. Features from all anomaly categories for most paths were compared using these calculations. Features with high Fisher Ratio's and/or Divergences were chosen to be used as part discriminators. This estimated the individual discriminating ability of each feature, not the collective discriminating ability of a feature group when applied to the Bayesian recognition engine, described below.

Because of the limited number of certain anomaly categories, extra care was taken to select and group features where the variance of the anomalous set was greater than the variance of the control set. (We had a sufficient number of control segments that tended to have features with large variances when compared to those from smaller sets of anomalous categories. This does not represent a true situation.)

A brief commentary on the actual discrimination features used is in order. Both paper and metal inclusion anomaly classes required multiple features from all of the segment sections being monitored. Surface breaking delaminations were simply monitored by using only the transverse path (# 18) in section D. Cure depravation was identified using the standard deviation, centroid and mechanical moment of the time domain envelope extracted from the lone axial path (# 31). Separation of consolidation anomalies required monitoring of the time domain envelopes and the cepstrally smoothed spectra from all the transverse paths (#'s 3,8,13,18,23 and 28).

During feature discrimination, each feature group is sequentially evaluated against the feature group values stored from the select batch of learning set segments. The division of a test part into sections fits well with our discrimination methodology which is designed to err in the direction of a false positive. Therefore, if one of the sections is classified as defective, the entire segment is assumed defective and is separated for re-examination.

A Bayes decision rule algorithm is used as a discriminator engine because it is expected to yield the best overall classification performance. The drawback to obtaining the ideal overall performance is that the probability density functions of the features are required. Because, in practice, these densities are never known precisely, they are approximated from experimental data by assuming normal densities. By taking the natural log of the probability, we can define the pattern recognition distance[4] to be:

$$d'(j,U) = \left(\frac{NF}{2}\right) \ln(2\pi) + d(j,U) \tag{3}$$

where the probability is one when $d'(j,U)$ is zero. For deciding which class an unknown may belong to, the first term could be dropped so that the Bayes decision function would then be rewritten as:

$$d(j,U) = \left(\frac{1}{2}\right) \sum_{m=1}^{NF} \{\sum_{n=1}^{NF} \{ [u(m) - \bar{a}(m,j)][u(n) - \bar{a}(n,j)]\, v'_{mn}(j)\}\} + \left(\frac{1}{2}\right) \ln[Det[V(j)]] - w \ln p(j) \tag{4}$$

where:

NF	is the number of features,
U	is the feature vector of the waveform to be classified,
$\bar{a}(m,j)$	is the average of the m-th feature for the j-th group,
$v'_{mn}(j)$	is the mn-th element of the inverse covariance matrix for the j-th group,
Det[V(j)]	is the determinant of the covariance matrix corresponding to the j-th group,

w	is the a priori probability weight,
p(j)	is the a priori probability of the j-th group and
d(j,U)	is the "distance" when the feature vector U is perceived as the j-th group.

The smaller the distance d(j,U), the greater the probability that the measurement U belongs to group "j", where j = 1 corresponds to the control group, and j = 2 corresponds to an anomaly group.

RESULTS AND DISCUSSION

Confusion matrices were used to evaluate the system's ability to properly classify "good" and "bad" segments. During this assessment, the control set of parts were run against each anomalous category, one at a time, with a confusion matrix subsequently being generated for each of the eight part sections within each category. As a further check of discrimination ability, due to the limited number of anomaly samples, a leave-one-out[4] strategy was subsequently run for each category. In this approach, the data is partitioned into one anomaly testing segment and all other segments are used for classifier design. A summary of these matrices indicating the probability of correct discrimination is shown in Table 1. As expected, the best results were achieved with those categories that were of a global nature and that contained the higher number of learning segments. Conversely, surface breaking delaminations and metallic inclusions were more difficult to classify, presumably because of their sparser population and highly localized nature. The leave-one-out procedure clearly demonstrated the need to increase the sample size of these categories.

A second set of learning segments is currently being produced which contains 35 segments with metallic inclusions and 15 additional non-postcured segments. These parts will be appended to the original learning set in order to raise the probability of detection for these respective categories. As normal production moves forward, a greater number of "good" segments will be added to the learning set to better instruct the system as to the normal variances seen in these parts. Ultimately, the learning set is likely to be updated several times over the life of the program to account for different raw material lots, and to allow some anomaly categories to be upgraded or downgraded depending upon what future testing and/or modelling uncovers about the severity of these defects. The system therefore adapts to the changing requirements of the material.

Difficulties associated with system repeatability can likely be attributed to three major factors: (1) the plywood cradle; (2) couplant related variations and (3) differences in segment microstructure.

Waveform variances caused by the cradle were usually due to excessive flex in the support arms which prevented the part from seating properly on some of the PZT end caps. In addition, the wood surrounding the PZTs became swollen because of couplant exposure which exacerbated the problem.

Aside from routine concerns about aging, thickness variations and air bubbles, the water based couplant gel was also responsible for swelling of the nylon PZT end caps due to the nylon's considerable hygroscopicity. This is but one of many reasons why dry contact PZTs[5] are becoming an attractive alternative. Laser/interferometer based ultrasonics would also seem to hold some promise; although, the authors have yet to perform any laboratory experiments with such a system. Given the present level of this technology and the complexities of the part geometry, there is concern that a laser based system would not be able to match the eventual throughput of a contact PZT based AU system—currently estimated at just over a minute.

The heterogeneous nature of this structure has also played a role in test reproducibility. Both CT as well as microscopic analysis have, as assumed, revealed the obvious differences in even the best and most consistent segment microstructures. The apparent solution is to increase the number of learning segments, both control and defective, so as to increase the effective signal to noise ratio of the discrimination process. This is completely expected and as it should be; it is also another reason why we believe that the acousto-ultrasonic technique is especially suited to meet the unique challenges presented by testing composite structures in large volumes.

The subject of system noise is yet another topic that merits discussion. Noise was introduced into the system from several sources including the attenuator, certain cable connections, and, most notably, as a result of line crosstalk within the switch matrix. Even though an engineering solution to this problem was eventually found—a line isolation procedure—because of scheduling constraints, it was decided to continue on by simply filtering out that portion of the waveform that contained the noise. This was deemed an acceptable compromise due to earlier work which suggested there was little of value at the low end of the spectrum. A future platform would be comprised of a more integrated system instead of the separate components.

Table 1 - Probability of Correct Selection (All / Leave One Out)

	PAPER	INCLUSIONS	RINGS	CURE	SBD
A	92.4 / 92.4		85.5 / 83.15		
B	100 / 93.75		100 / 90.7		
C	100 / 90.0	92.0 / 79.15	100 / 95.4		
D	100 / 93.75	81.4 / 66.9	99.5 / 89.3		89.0 / 81.8
E	100 / 93.75	91.0 / 82.1	99.5 / 89.6		
F	100 / 93.75		98.5 / 91.1		
G	100 / 96.25				
H		73.0 / 62.65		90.68 / 81.3	

CONCLUSION

A prototype for a production level acousto-ultrasonic platform has been demonstrated. Based upon a realistic set of anomalies that are specific to this structure, the results indicate that it is possible to discriminate between "good" and "bad" versions of these segments based upon their acoustic response. The results also suggest that a larger population of learning samples will yield increased system reliability.

The potential for follow-on work to this effort remains extensive. The most pressing issue to be addressed is the plywood cradle, which, while adequate for proof-of-principle, would ultimately cripple the system's reliability in a production environment. A stiffer, more sophisticated steel cradle has recently been fabricated that makes use of pneumatic cylinders to deliver a rigorously controlled application of force to the PZT and segment surface.

A second area of investigation is the use of dry contact PZTs to eliminate coupling variations that can also affect signal quality. While dry coupling cannot be considered a panacea, when combined with a semi-automated loading fixture, it does contribute to a testing operation that is relatively operator independent.

Calibration is another question that is being addressed. Currently PZTs are calibrated by running a complete acquisition cycle on an aluminum right circular cylinder standard. The time and frequency domain values for each path are compared to those obtained on an initial acquisition when first purchased. A PZT is flagged as suspect if one or more paths it is associated with falls below a certain threshold. The problems with this approach are numerous; the most obvious being that the bandwidth degradation of a specific PZT cannot be readily determined. To overcome this deficiency, a more elegant solution is being developed to allow a pulse-echo calibration using a specially constructed aluminum standard. The major advantage of this method, is that each PZT impulse response will be treated so as to function in a uniform manner; that is, to the system they will appear identical. This will allow an aged or damaged PZT to be replaced with a new PZT without impacting the values in the learning set database.

REFERENCES

1. A. Vary and K. J. Bowles, An Ultrasonic-Acoustic Technique for Nondestructive Evaluation of Fiber Composite Quality, Poly Engrng & Sc., 19:373 (1979)

2. W. S. Meisel, Computer-Oriented Approaches to Pattern Recognition,, Academic Press, (1972)

3. P. A. Dickstein, "Integrated Statistical Processing of NDE Signals for Class Discrimination", Research in Nondestructive Evaluation, Vol. 4, No. 2, 1992, Springer International

4. R. O. Duda and P. E. Hart, Pattern Classification and Scene Analysis, Wiley, New York, (1973)

5. J. A. Brunk, C. J. Valenza and M. C. Bhardwaj, Applications and Advantages of Dry Coupling Ultrasonic Transducers for Materials Characterization, in Acousto-Ultrasonics: Theory and Application, edited by John C. Duke, Jr., Plenum Press, New York and London, 221, (1988)

USE OF FOURIER SPECTRA FOR ULTRASONIC NONDESTRUCTIVE EVALUATION OF THIN COATINGS

Vikram K. Kinra and Changyi Zhu

Center for Mechanics of Composites
Department of Aerospace Engineering
Texas A&M University
College Station, TX 77843

INTRODUCTION

There is clearly a need for a simple, reliable and inexpensive nondestructive technique for characterizing thin coatings;[1] by thin we mean the thickness of the coating is less than the wavelength of the ultrasonic wave used to interrogate it. We have developed a technique for a simultaneous measurement of both thickness (h) and wavespeed (c) for thin coatings using transducers with a center-frequency of 10 or 20 MHz. Transfer functions have been derived. A systematic analysis of the sensitivity of the transfer functions to the coating properties has been carried out. This technique is independent of the substrate thickness. The technique was used to characterize epoxy or Plexiglas coatings (50-100 μm) on an aluminum substrate. The precision in the measurement of the thickness and wavespeed was found to be ± 2 μm and $\pm 3\%$, respectively.

TRANSFER FUNCTIONS AND SENSITIVITY ANALYSIS

With reference to Fig.1(a), consider a (thin coating)/(thick substrate) assembly immersed in water and insonified from the coating side by a normally-incident plane longitudinal wave, where h and L are the thickness of the coating and the substrate, respectively. Here, we assume subsequent echoes are separable in time in the substrate. Let the incident wave in water be $u^{inc}(x, t)$. Let the corresponding wave reflected from the coating be $u^{r_0}(x,t)$ and the wave reflected from the substrate at the interface (3, 1) be $u^{r}(x,t)$, respectively. We now define the transfer function for the coating side insonification (CSI) to be[1]

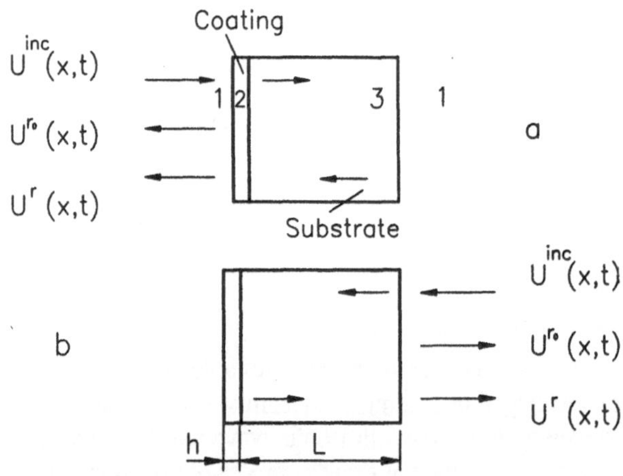

Figure 1. Theoretical model of a thin coating on a thick substrate. (a) CSI. (b) SSI.

$$H^*_{cs}(\omega) = \frac{u^r(x,t)}{u^{r_0}(x,t)} = \frac{T_{12}T_{21}T_{23}T_{32}R_{31}e^{-2\alpha h}e^{-i2(k_2 h+k_3 L)}}{(1+R_{12}R_{23}e^{-2\alpha h}e^{-i2k_2 h})(R_{12}+R_{23}e^{-2\alpha h}e^{-i2k_2 h})} \tag{1}$$

where ω is the circular frequency, k is the wave number, R_{ij} and T_{ij} are the displacement reflection coefficient and transmission coefficient, respectively, and α is the attenuation coefficient of the coating. For the case of substrate side insonification (SSI), as shown in Fig.1b, let the wave reflected from the substrate at the interface (1, 3) be $u^{r_0}(x,t)$. Let the wave reflected from the coating be $u^r(x, t)$. The transfer function for the SSI may be shown to be

$$H^*_{ss}(\omega) = \frac{u^r(x,t)}{u^{r_0}(x,t)} = \frac{T_{31}T_{13}(R_{23}+R_{12}e^{-2\alpha h}e^{-i2k_2 h})e^{-i2k_3 L}}{R_{31}(1+R_{12}R_{23}e^{-2\alpha h}e^{-i2k_2 h})} \tag{2}$$

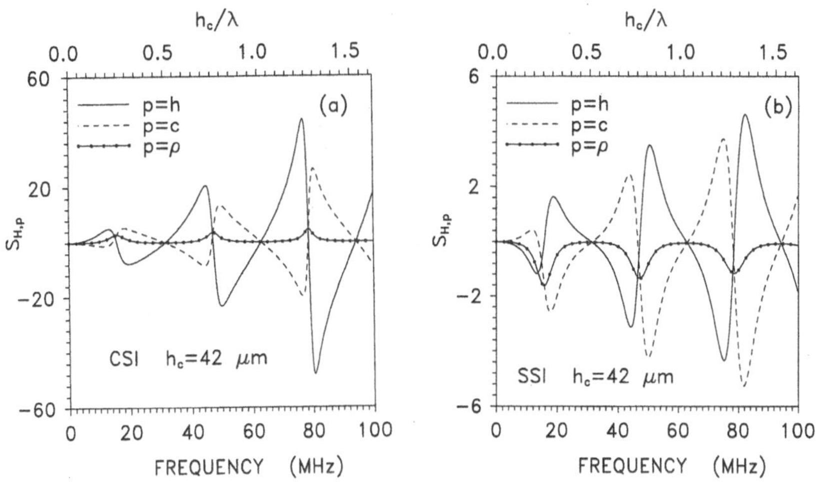

Figure 2. Sensitivity of H(ω) to h, c and ρ. (a) Coating side insonification (CSI). (b) Substrate side insonification (SSI).

It is noted that for both transfer functions, the phase depends on the substrate thickness, L, whereas the magnitude is independent of L. In this work we utilize only the magnitude of the transfer function. Therefore, we do not have to know the substrate thickness.

We introduce a normalized sensitivity of H to p, $S_{H,p}(\omega)=(p/H)(\partial H/\partial p)$. Clearly, if $S_{H,p}(\omega) = 0$ even if we measure H(ω) very accurately, we cannot deduce p. For specimen No.4 various sensitivities are plotted in Fig.2a for the CSI, and in Fig.2b for SSI (see Table I). For convenience, in the following, we will drop the subscript $(\ )_2$ from the coating properties. For CSI (Fig.2a) the sensitivity is about one order of magnitude higher than that for SSI (Fig.2b). The sensitivity, $S_{H,\rho}(\omega)$ is generally an order of magnitude smaller than $S_{H,h}(\omega)$ or $S_{H,c}(\omega)$. The discussion of the sensitivities to substrate properties can be found in reference.[1]

RESULTS AND CONCLUSIONS

One Plexiglas and three epoxy coating specimens were tested (see Table 1). The discussion of the experimental apparatus used can be found in reference 1. For the CSI, the comparison between the measured and predicted $H(\omega)$ are shown in Figs.3 and 4 and are considered to be excellent. Comparison for the SSI are shown in Figs.5 and 6. The larger scatter in this data in comparison to the CSI is due to the fact that the magnitude of the transfer function is significantly smaller for SSI in comparison to that for CSI.

The well-known Newton-Raphson method was used to reconstruct the coating thickness and wave speed through a comparison of the theoretical and measured transfer functions. The values of h and c reported in Table 1 are an average of four measurements. In the case of thickness, for both CSI and SSI, the precision was found to be ±2 μm (±2%). For the case of the wavespeed, for both CSI and SSI, the precision was found to be ±3%.

Table 1. Thickness and Wavespeed Measurement Data for Thin Coatings.

Coating Material	Substrate Thickness (mm) ± 2.5 μm	f_C (MHz)	h_{TRUE} (μm) ±5μm	h^C_{NDE} (μm) ±2μm	h^S_{NDE} (mm) ±2μm	c_{TRUE} (mm/μs)	c^C_{NDE} (mm/μs) ±3%	c^S_{NDE} (mm/μs) ±3%
Plexiglas	8.17	0.33	85	87	87	2.63	2.63	2.68
Epoxy	2.03	0.35	88	89	90	2.57*	2.48	2.52
Epoxy	3.17	0.35	95	94	93	2.57*	2.48	2.56
Epoxy	1.60	0.30	42	42	41	2.57*	2.61	2.62

h^c_{NDE} (c^c_{NDE}): NDE from the coating side.
h^s_{NDE} (c^s_{NDE}): NDE from the substrate side.
* : Measurement error in c_{TRUE} is not known.
f_C: Transducer center frequency.

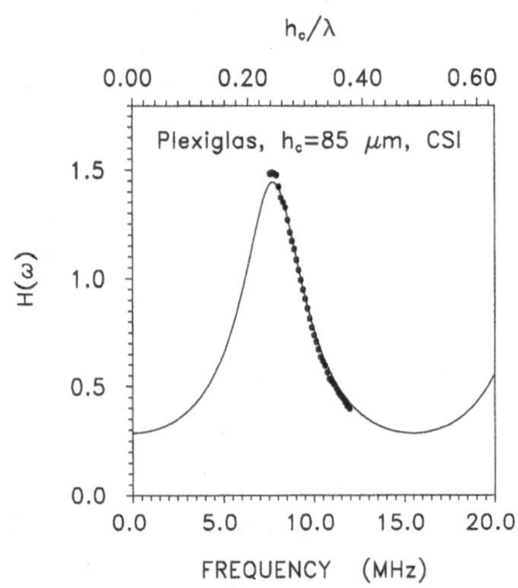

Figure 3. Comparison of theory and experiment for coating side insonification for the Plexiglas coating specimen.

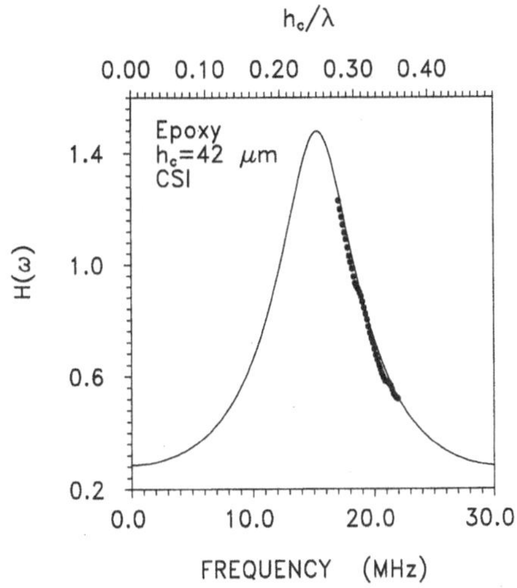

Figure 4. Comparison of theory and experiment for coating side insonification for the 42 μm thick epoxy coating specimen.

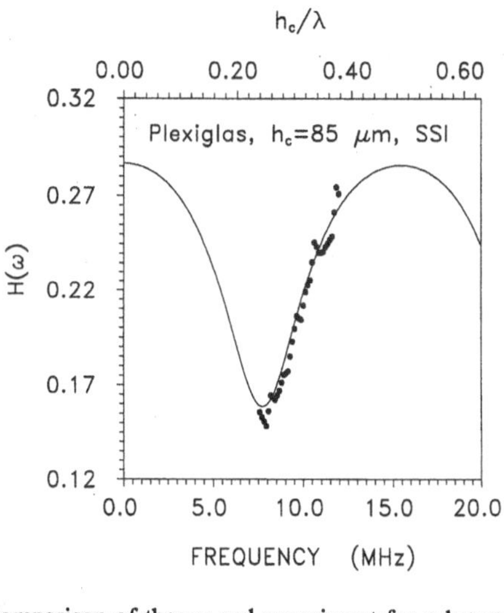

Figure 5. Comparison of theory and experiment for substrate side insonification for the Plexiglas coating specimen.

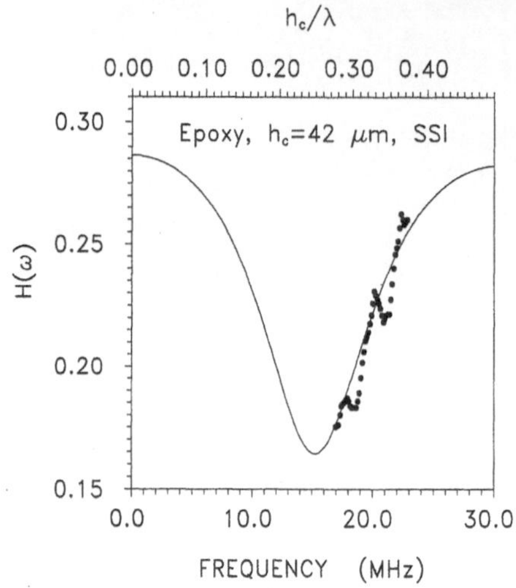

Figure 6. Comparison of theory and experiment for substrate side insonification for the 42 μm thick epoxy coating specimen.

ACKNOWLEDGEMENT

The material is based on the work supported by the Texas Advanced Research Program (Advanced Technology Program) under Grant No. 282 to Texas A&M University.

REFERENCE

1. V.K. Kinra and C. Zhu, "Ultrasonic Nondestructive Evaluation of Thin (Sub-Wavelength) Coatings," J. Acoust Soc Am, (1993) 93 2454-2467.

NONDESTRUCTIVE EVALUATION OF BIOLOGICALLY DEGRADED WOOD

Robert J. Ross, Rodney C. DeGroot, William J. Nelson

USDA Forest Service
Forest Products Laboratory[1]
One Gifford Pinchot Drive
Madison, WI 53705–2398

INTRODUCTION

Wood is a complex material that can be attacked and degraded by a wide range of biological organisms. The USDA Forest Service, Forest Products Laboratory (FPL), has been investigating the use of nondestructive evaluation (NDE) techniques to identify when degradation of wood occurs in the structure and the performance characteristics that remain in the structure. In particular, FPL's work has focused on using longitudinal stress wave NDE techniques for laboratory and field applications.

Techniques of NDE are increasingly being used in industrial applications to assess various properties of wood and wood products and in-place wooden structures. One NDE technique that is commonly used to evaluate mechanical properties of wood products is based on the measurement of vibrational characteristics. Longitudinal stress wave NDE techniques use low-level stress waves to measure two fundamental energy properties: storage and dissipation. Energy storage is the speed at which a wave travels in a material. Energy dissipation is the rate at which a wave attenuates. These properties are related to the same mechanisms that control the mechanical behavior of a material. Consequently, useful relationships can be obtained between stress wave speed and attenuation and the elasticity and strength of a material.

Ross and Pellerin (1988) found that stress wave speed and attenuation are excellent indices of the mechanical properties of wood-based composites. Under controlled laboratory conditions, Pellerin and others (1985) showed that stress wave velocity is a good indicator of wood decomposition when caused by brown-rot fungi but a poor indicator when caused by termites.

We believe that NDE techniques incorporating both stress wave speed and attenuation may yield useful information about the residual strength of solid wood subjected to naturally-occurring biological attack, independent of the type of attacking organism. Therefore, a project was undertaken to assess the feasibility of using stress wave NDE techniques in the field to estimate the mechanical properties of wood that are subjected to various stages of degradation as a result of biological attack. The first part of the project was devoted to developing a simple and inexpensive technique that could be used to observe stress wave behavior in small wood specimens. Because of the difficulties of the

[1]The Forest Products Laboratory is maintained in cooperation with the University of Wisconsin. This article was written and prepared by U.S. Government employees on official time, and it is therefore in the public domain and not subject to copyright.

measurement systems and excitation methods, prior stress wave research of this type has been restricted to large wood specimens. This paper describes the NDE technique that we developed and some typical results obtained from the technique, using both sound and degraded wood.

EXPERIMENTAL SETUP

A illustration of our experimental setup is shown in Figure 1a. All specimens were Southern Pine sapwood. A Kynar[2] piezofilm vibration sensor (Model SDT1-028K) was attached to the top surface of a specimen using double-sided adhesive tape. A stress wave

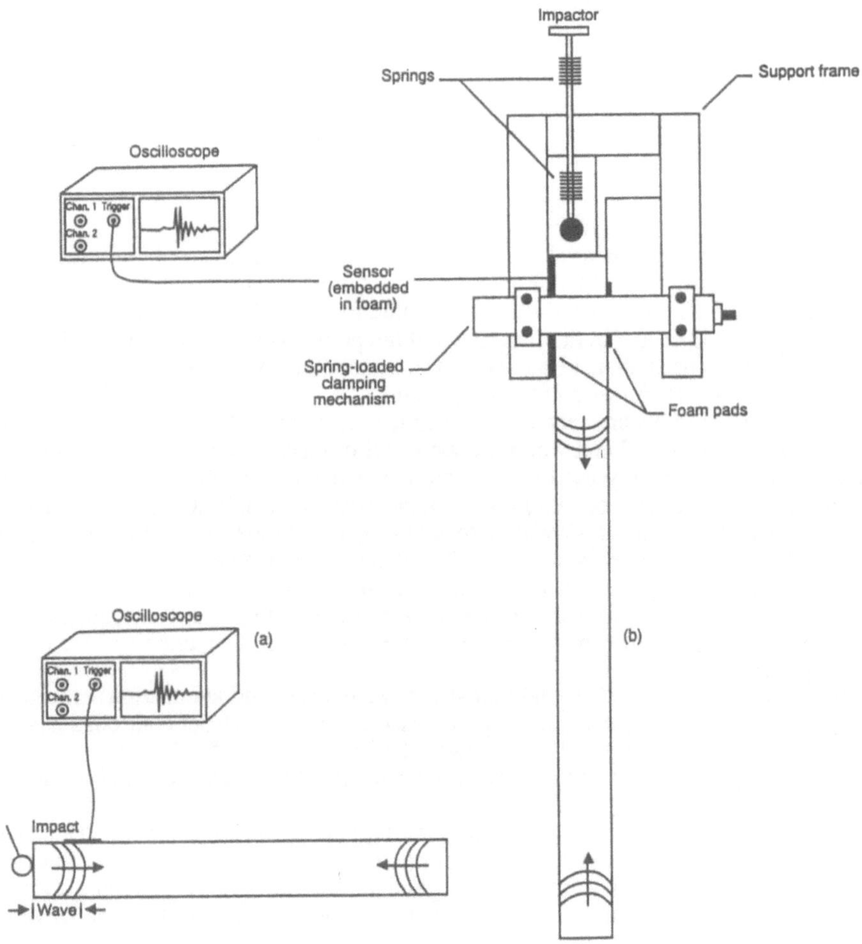

[2]The use of trade or firm names in this publication is for reader information and does not imply endorsement by the U.S. Department of Agriculture of any product or service.

was induced through a small pendulum impactor. The impactor consisted of a 0.125-in-diameter (0.003-m) steel ball suspended by cotton thread from a small wooden frame. In response to a wave traveling in the specimen, output of the sensor was monitored and recorded using a Nicolet Model 2090 III digital storage oscilloscope.

Experience with this experimental setup provided us with the following notations. Coupling of the piezofilm to the specimen was a key element in obtaining a good signal response. A consistent method of coupling would have to be designed to facilitate a large number of specimens in an experiment. A lengthy lead wire with a properly shielded connector would be necessary to ensure low-level noise signal responses. Covering the film with regular cellophane tape would protect the metallic surface so that it would not be destroyed after numerous tapings. The heal of the film detector would also have to be taped securely to obtain a consistent signal. The Nicolet Model 2090 III and the larger Model 4094 have limited capabilities to store data files--8 records and 20 records, respectively. In addition, these computers have limited capabilities for field-type testing because of their large size. The impact system was also a source of inconsistency. A pendulum with a rigid shaft and various-sized steel balls would need to replace the original string and ball impactor. A pinball-type impactor could also be used. A consistent signal would be achieved by varying the size of ball and the strength of the springs.

All previous notations were considered in modifying the experimental setup for sampling infield and laboratory bench specimens (Fig. 1b). A hand-held pinball impactor with imbedded piezofilm sensor was designed, which had the ability to be easily and quickly attached to the specimen with a single snap-on motion. A piece of 100-grit sandpaper was taped to the imbedded sensor. This gave good coupling to the specimen, resulting in consistent signal output. A portable Nicolet 310 digital storage oscilloscope was used to monitor and store the waveforms. This scope was easy to carry in the field and stored 88 waveforms on a 3.5-in. (0.089-m) floppy disk. This scope uses IBM PC compatible data disks that can be copied for backup. These modifications to the experimental setup provided us with easy data acquisition, data storage, and subsequent waveform analysis.

WAVEFORM ANALYSES

In using our experimental setup, elementary wave theory (Kolsky 1963) suggests that the waveforms should consist of a series of equally spaced sine-shaped pulses whose magnitude decreases exponentially with time (Fig. 2). The speed C at which a wave moves through a specimen can be determined by coupling measurements of the time between pulses, peak to peak Δt, and the length L of the specimen using the following equation:

$$C = \frac{2L}{\Delta t} \tag{1}$$

Wave attenuation can be measured as the rate of decay or logarithmic decrement δ of the amplitude of pulses using the following:

$$\delta = \frac{1}{j} \ln \frac{A_o}{A_j} \tag{2}$$

where A_o and A_j are the amplitudes of two pulses j cycles apart.

This analysis method is an estimate that can be improved by using additional pulses to give an average result. Using the time value between several pulses and dividing by the number of cycles gives an accurate C value. In addition, using a high value of j in Equation (2) gives an accurate value.

If the values of the peak amplitudes are collected, a curve can be fit to the set, giving an equation of the form:

$$f(t) = Ae^{-nt} \tag{3}$$

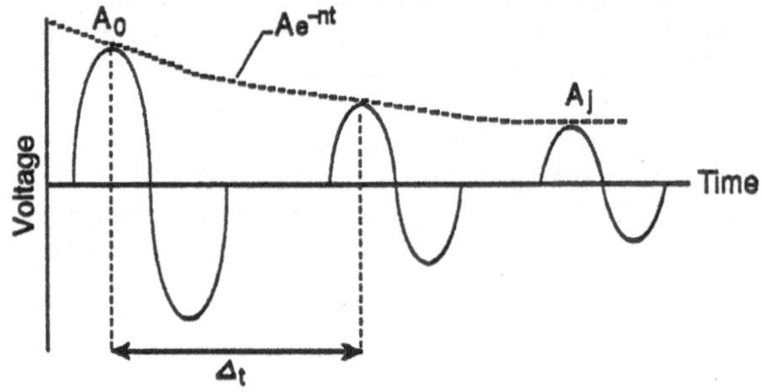

Figure 2. Elementary wave response theory, Eq. (2), (3).

This equation describes the outer envelope of the original pulse signal and is a decreasing exponential function. By taking the natural log of both sides of the equation, a linear equation results, where n is slope of the line.

$$\ln f(t) = \ln A - nt \tag{4}$$

By using the peak values of the waveform, a linear relationship can be generated. The slope of the line will determine the constant n. The slope n of this line is directly proportional to δ.

$$\delta = n\,\Delta t \tag{5}$$

Compared to the previously described waveform analyses, this gives the most accurate results. A computer was used to find the peak values and evaluate the necessary parameters from each waveform captured on disk. Using Lotus 123, a worksheet was used with an appropriate macro to find the peaks, linearize the values, calculate the parameters, and list the results in a separate spreadsheet. Experiments with large sample volumes were analyzed quickly and results compared for changes that may occur between sample volume sets.

Computer data acquisition and analyses, with instant feedback of results, would greatly assist infield testing. Software using this technique has been written and is presently being tested at the FPL. Small computers with appropriate hardware capabilities will eventually be available and used for field applications.

RESULTS

An oscilloscope trace of a waveform obtained from monitoring stress wave behavior, in the laboratory with free boundary conditions, on a typical 1- by 1.5- by 40-in. (0.025- by 0.038 by 1.016-m) control specimen is shown in Figure 3. Note that the waveform consists of a series of equally spaced sine-shaped pulses whose magnitude decreases with time, as predicted by elementary wave theory (Kolsky 1963). Also note that the small impactor generated waves of sufficient magnitude so as to yield an observable response. In addition, the short duration of the impact generated a wave that yielded pulses with significant separation between them. Separation is necessary for accurate measurement of the velocity at which a wave travels in a specimen. If the length of a wave is excessive, the wave overlays on itself, causing inaccurate velocity measurements.

An oscilloscope trace of a waveform taken in the laboratory from specimens that were in the field 6 months and exposed to natural decay and other biodegradation is shown in Figure 4. The following value differences were recorded between the control and degraded specimens in waveform parameters:

	Control	Degraded
Time between pulses (Δt)	0.000391 s	0.000473 s
Slope of line (n)	278	665

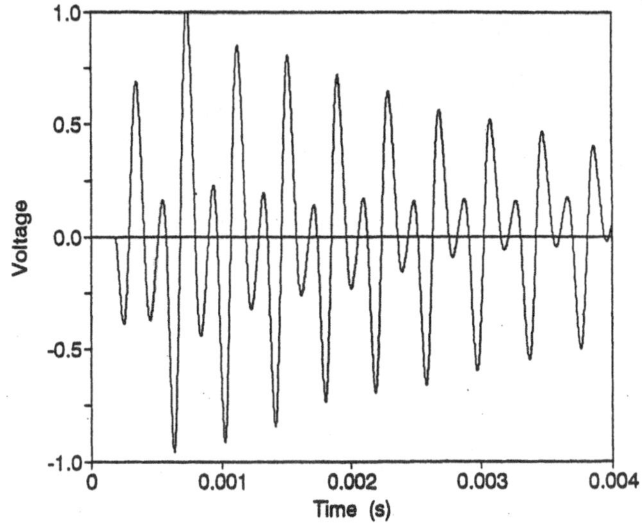

Figure 3. Typical control specimen waveform.

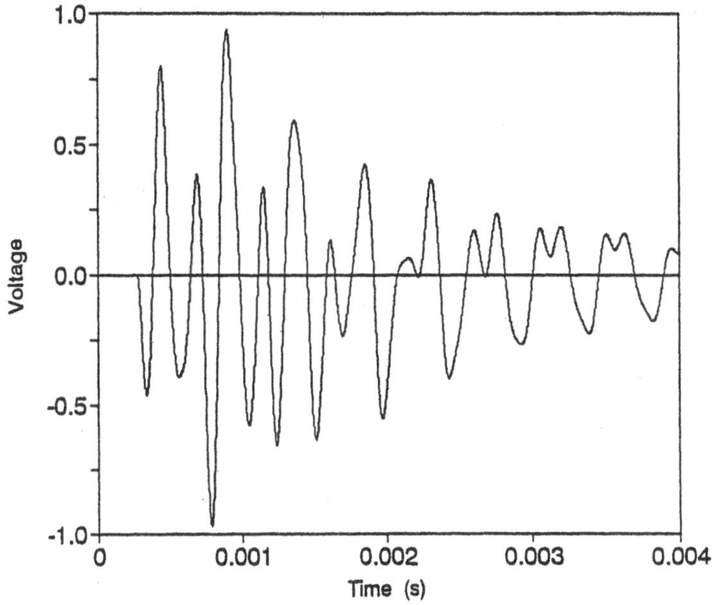

Figure 4. Typical degraded specimen waveform.

Note that the slowest wave (larger Δt) and rate of attenuation of the peak amplitudes were greater (larger n) for the degraded specimen than for the control.

We feel that these differences can ultimately be used to determine residual strength in naturally decayed and degraded wood members. We are building a database to show the relationship between wave parameters and residual strength of biologically degraded wood. Several types of degradation agents, which represent a common range, are being utilized to degrade wood. Our research should ultimately have application in field evaluations of candidate preservatives and in engineering assessment of residual strength in existing structures.

CONCLUDING REMARKS

The first step in using an NDE technique for monitoring biological degradation of wood members is developing an inexpensive, readily available measurement system. The measurement system for monitoring stress wave behavior in degraded wood consists of an impact and signal acquisition system coupled to a relatively inexpensive data storage device. Signal analysis is also performed using inexpensive personal computers. Results are easily analyzed and compared to give the researcher near real-time information.

REFERENCES

Kolsky, H., 1963, "Stress Waves in Solids," Dover Publications, Inc., New York.

Pellerin, R.F., DeGroot, R.C., and Esenther, G.R., 1985, Nondestructive evaluation of stress wave measurements of decay and termite attack in experimental wood units, *In*: Proceedings, Fifth Nondestructive Testing of Wood Symposium; 1985 September 9–11; Pullman, WA. Pullman, WA: Washington State University.

Ross, R.J., and Pellerin, R.F., 1988, NDE of wood-based composites with longitudinal stress waves, *Forest Prod. J.* 38(5):39–45.

MEASUREMENT OF ULTRASONIC ATTENUATION AND GRAIN SIZE OF THIN METAL SHEETS USING RESONANCE MODE EMAT

Takao Hyoguchi,[1] Toshio Akagi,[1] Oliver B. Wright,[1]
and Katsuhiro Kawashima [2]

[1]Electronics Research Laboratories, Nippon Steel Corporation
5-10-1 Fuchinobe, Sagamihara, Kanagawa, 229, Japan
[2]Tokyo Engineering University
1404-1 Katakura, Hachioji, Tokyo, 192, Japan

INTRODUCTION

The development of nondestructive, reliable methods for the measurement of grain sizes in thin metal sheets is very important in the metal industry. Grain size, conventionally measured destructively on small specimens using optical microscopy, is closely correlated with mechanical properties such as yield strength and plastic deformation.[1,2] It is well known that the grain size is also related to the ultrasonic attenuation due to grain boundary scattering,[3-9] and this attenuation has been measured nondestructively using conventional pulse-echo methods in bulk samples.[10-14] However, pulse-echo attenuation measurements in thin metal sheets down to sub-mm thickness are limited in accuracy by the high time resolution required. The ultrasonic backscattering method has also been proposed for the measurement of attenuation and grain size for thin metal sheets.[15] The disadvantage of both these methods is that they require coupling fluid between the transducer and sample.

EMAT (electromagnetic acoustic transducers) show general advantages of on-line application, non-contact operation, and no need for coupling fluid. A new resonance mode EMAT technique is presented for the nondestructive and non-contact measurement of ultrasonic attenuation in thin metal sheets. The grain size is predicted from an empirical relation between the grain size and the ultrasonic attenuation. Two methods for measurement are considered, the first using the resonance peak amplitude and the second using the resonant peak width, and their relative merits are discussed.

In the next section the principles of the bulk resonance mode EMAT technique are explained, and then the attenuation and grain size methods are introduced. After this the experimental results are presented and discussed. Conclusions are given in the last section.

RESONANCE MODE EMAT

The EMAT used here is composed of a magnet and a pancake coil (diameter: 20 mm; number of turns: 40) for driving and sensing, as shown in Figure 1. Values of liftoff up to 2 mm were used. The transducer is symmetric with respect to the central axis. When the transducer is placed on a metal sheet, the magnet produces a magnetic field that has both a normal component B_z and a radial component B_r. An RF current is applied to the pancake coil and a circular eddy current is generated in the metal sheet sample. The electromagnetic force is generated through the interaction of the eddy current and the magnetic field. The interaction between the eddy current and B_r creates a force F_z normal to the sheet surface. The interaction between the eddy current and B_z creates the radial force F_r. Therefore three different mode bulk ultrasonic waves, longitudinal waves and shear waves propagating in the through-thickness direction (Z), [16-18] are generated. Shear waves are polarized along the rolling direction (X) and along the transverse direction (Y). Detection is the reverse process to the generation process.

A block diagram of the measurement system is shown in Figure 2. A frequency-scanned RF voltage from the oscillator controlled by a microcomputer is amplified by a power amplifier and fed to the pancake coil of the transducer. When the frequency of the RF

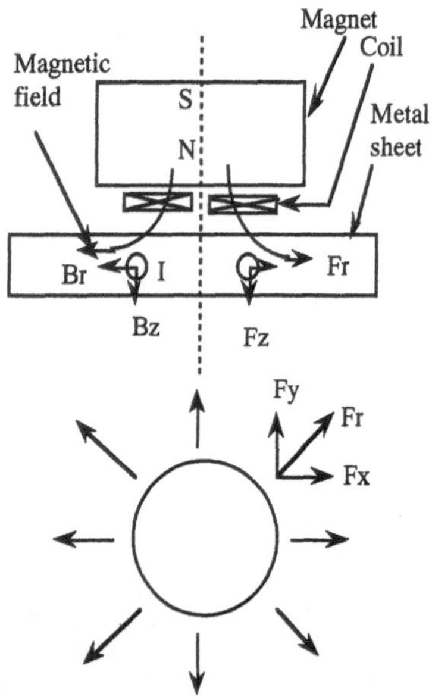

Figure 1. Resonance EMAT to generate and to detect the 3 different modes of ultrasonic waves propagating in the through thickness direction.

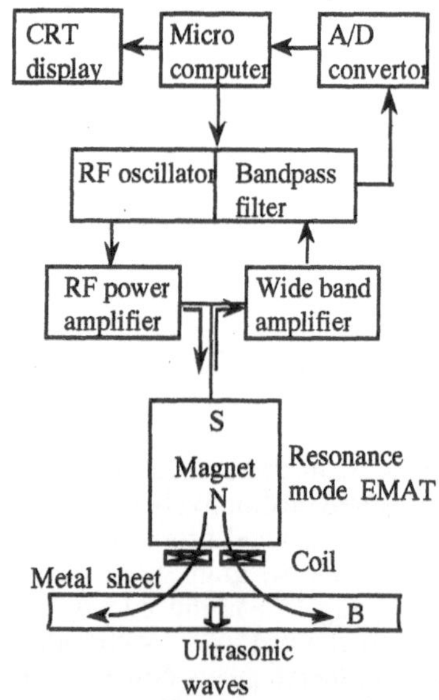

Figure 2. Measurement system with EMAT for thickness resonant frequencies of the 3 ultrasonic waves.

current in the coil equals the resonance frequency, given by any one of the following equations, standing waves are set up in the sheet:

$$f_{zzm} = \frac{mV_{zz}}{2d}, \qquad f_{zyn} = \frac{nV_{zy}}{2d}, \qquad f_{zxn} = \frac{nV_{zx}}{2d}, \tag{1}$$

where V_{zz} is the velocity of the longitudinal waves, V_{zy} is the velocity of the shear waves polarized along the transverse direction, and V_{zx} is the velocity of the shear waves polarized along the rolling direction, d is the sheet thickness, m and n are integers, and f_{zzm}, f_{zyn} and f_{zxn} are m-th, n-th and n-th order resonance frequencies.

The waveform for the RF excitation is a tone burst, of duration either 30 μs or 500 μs. The 30 - μs duration was used for the amplitude measurements, whereas 500 - μs duration was used for the half-width measurements to give a better frequency resolution. A 30 - μs gate is used for detection after the tone burst. The burst frequency is scanned between 5 and 25 MHz, and the burst repetition frequency is in general set between 100 Hz and 5 kHz.

ATTENUATION AND GRAIN SIZE MEASUREMENT WITH RESONANCE MODE EMAT

Ultrasonic losses in polycrystalline solids are in general caused by material losses and by other losses. Material losses consist of grain boundary scattering, intrinsic attenuation, and magnetic attenuation.[19,20] Other losses consist of diffraction losses and non-parallelism losses.[21]

The attenuation related to the grain boundary scattering depends on the ratio of the wavelength λ and the average grain size D.[6-9] In the Rayleigh scattering regime ($\lambda/D>2\pi$) the attenuation can be expressed as follows:[8,9]

$$\alpha = A_1 f + A_2 f^4 \quad , \tag{2}$$

where f is the frequency and A_2 is a scattering parameter given by $A_2 = A_3 D^3$, where A_3 and A_1 are constants depending on the material.

The general principles of the two methods for attenuation measurements using resonance mode EMAT are given here. The basic theoretical principles have already been described.[16,17] In the resonance amplitude method, the total attenuation α is proportional to the reciprocal of the amplitude A:

$$\alpha = \frac{KB^2}{\rho v d A} \quad , \tag{3}$$

where B is the constant magnetic field, ρ the density, v the sound velocity, α the attenuation coefficient, d is the thickness, and K a parameter which depends on both sensor parameters and on sample parameters. The sensor parameters are frequency, electrical impedance, excitation current, magnetic field, lift-off, burst duration and gate duration. The sample parameters are thickness, permeability, conductivity, density, sound velocity, and of course the acoustic losses (including the part we wish to measure due to grain boundary scattering).

The resonance amplitudes are measured for samples where only the grain size changes, and with the frequency, sensor parameters, and all other sample parameters, including the thickness, presumed constant. The resonance amplitudes are then proportional to the reciprocal of the total attenuation α [see Eq. (3)], which includes the scattering contribution.

The resonance half-width Δf, the 3-dB half power width, is also related to the attenuation through the following relation:[17]

$$\alpha = \frac{\pi \Delta f}{v} \quad . \tag{4}$$

Provided we know the sound velocity V, the attenuation α can therefore be determined quantitatively from the half-width. The resonance curve is predicted to be the square root of a Lorentzian in shape. In contrast to the resonance amplitude method, the measured α can in principle be made insensitive to both sensor and sample parameters except the sound velocity. The half-width can be measured as a function of frequency, and $\alpha(f)$ derived. In contrast to the half-width measurements, amplitude measurements as a function of frequency are not useful because the effect of sensor parameters can not be decoupled from the frequency response.

EXPERIMENTAL RESULTS AND DISCUSSION

Resonance amplitude

Austenite stainless steel sheets, of thickness 0.5±0.01 mm, were first measured. The range of the ASTM grain size number, measured by optical techniques, is 10 to 5.8, that is 11- to 47-μm average diameter. The grain size number is proportional to the logarithm of the average grain size (D). The correlation between the measured amplitudes and grain size is shown in Figure 3. Figure 3 (a) shows the amplitude data for shear waves at resonance frequencies of about 9 MHz (zx3, zy3) and about 12 MHz (zx4, zy4). By analyzing the data shown in Figure 3 (a), we find that the amplitude A is approximately proportional to $1/D^2$, where D is the average grain size. Since the total attenuation α is proportional to 1/A, α is approximately proportional to D^2.

Figure 3 (b) shows the amplitude data for longitudinal waves at 12 MHz. Here 1/A, or α, is approximately proportional to D^3. The wavelength of longitudinal waves is larger than for shear waves, and so the scattering loss is expected to correspond more closely to the expected D^3 dependence of Rayleigh scattering theory [λ/D ranges from about 6 to 30 in Figure 3 (a) and from about 10 to 50 in Figure 3 (b)]. Similar variations as D^n, with n increasing with λ/D, were observed in Ref. 6. In general, the resonance amplitude shows a good correlation with grain size D.

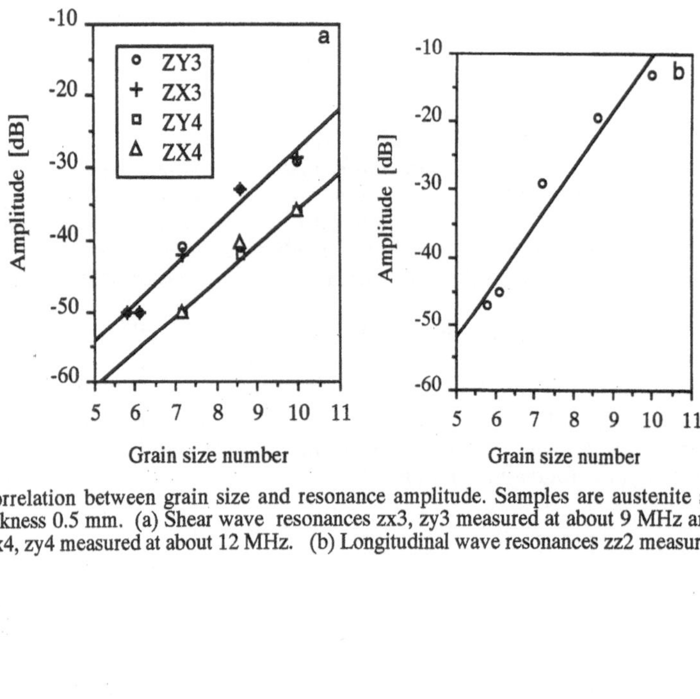

Figure 3. Correlation between grain size and resonance amplitude. Samples are austenite stainless steel sheets of thickness 0.5 mm. (a) Shear wave resonances zx3, zy3 measured at about 9 MHz and shear wave resonances zx4, zy4 measured at about 12 MHz. (b) Longitudinal wave resonances zz2 measured at about 12 MHz.

Typical results for the resonance spectrum are shown in Figure 4. The sample is a cold rolled steel sheet of thickness 0.8 mm and grain size number 9.5. There are many resonance peaks from the two shear wave polarizations and from the longitudinal waves. As frequencies become higher, the amplitudes become smaller [see Eq. (3)]. Results up to about 20 MHz can be obtained with a good S/N ratio with the EMAT used.

A series of cold rolled steel sheets of thickness 0.8±0.03 mm were also measured. The range of grain size number is 11 to 7.5 corresponding to grain size 8 to 26 μm. Figure 5 shows the correlation between the grain size and the resonance amplitude. The trend of the results are similar to those for austenite stainless steel. The reciprocal of the amplitude 1/A shows a D^2 dependence for larger grain sizes ($\lambda/D \approx 10$), changing to a D^3 dependence for smaller grain sizes ($\lambda/D \approx 40$), the latter corresponding more closely to the Rayleigh scattering regime. Again the correlation between amplitude and grain size is good.

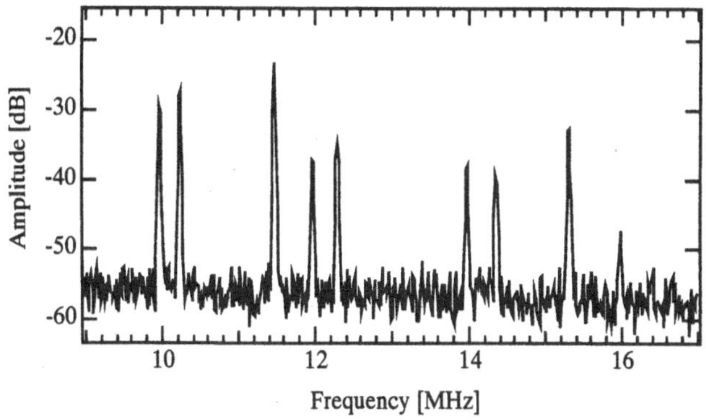

Figure 4. Typical spectrum for a cold rolled steel sheet of thickness 0.8 mm and grain size number 10.

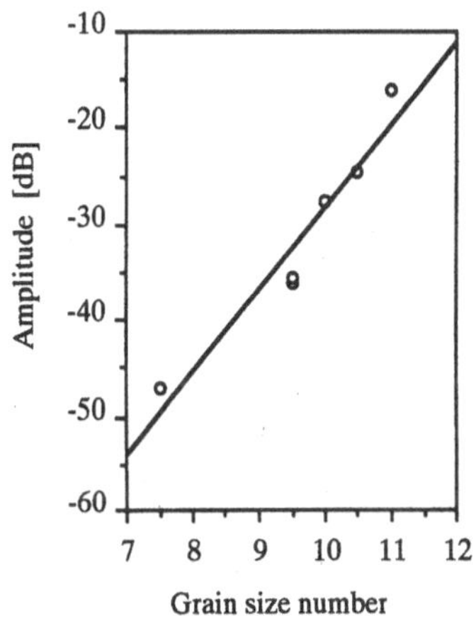

Figure 5. Correlation between resonance amplitude and grain size for cold rolled steel sheets of thickness 0.8 mm. These shear wave measurements (zx5) were carried out at about 10 MHz.

Resonance half-width

In practice, factors other than the sound velocity also affect the resonance width. Sensing parameters such as the finite burst duration for excitation limits the bandwidth of the resonance. So this duration should be chosen long enough. Our choice of 500 μs corresponds to a frequency resolution of ~ 1kHz. The static magnetic field affects the losses caused by magnetic attenuation. The measurements were therefore carried out under the condition that the magnetic field was low enough not to affect the resonance curve shape and total attenuation. This was checked using different values of the static magnetic field.

Figure 6 shows an example of the spectrum on a linear scale near to the 5th order resonance. The sample is a cold rolled steel sheet of thickness 0.8 mm and grain size number 11. The fitted smooth curve is the square root of a Lorentzian curve, according to the theory of Ref. 17. Good correlation is obtained. The resonance 3-dB half-width is 5.9 kHz. The attenuation coefficient α is calculated to be 0.05 dB/mm from Eq. (4). This is of the expected order of magnitude.[1,5]

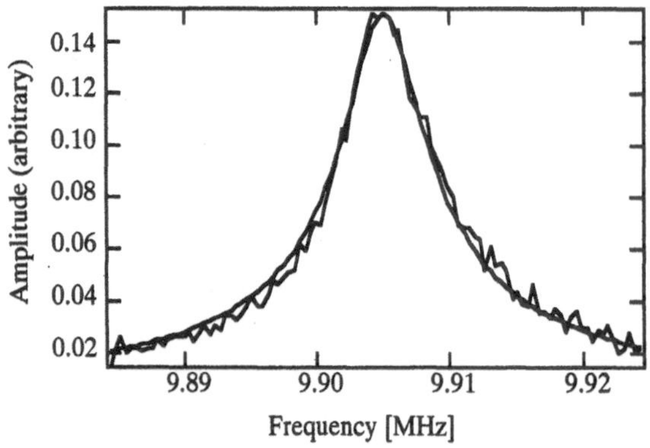

Figure 6. Example of a typical resonance curve for a cold rolled steel sheet of thickness 0.8 mm (grain size number 11, grain size 8 μm). The liftoff used was 2 mm. The smooth curve is a theoretical fit proportional to the square root of a Lorentzian curve.

The relation between the 3-dB half-width Δf and the frequency is shown in Figure 7 for two samples. These are cold rolled steel sheets of grain size number 11 (grain size: 8 μm) and 9.5 (grain size: 13 μm). The curves can be fitted by the expression $A_1f+A_2f^4$ (Eq. (1)). The ratio of A_2 values obtained ($\approx$5.5) corresponds reasonably with the ratio of grain size volumes as expected from Rayleigh scattering theory ($\approx$4.8). (Different values of the coefficients A_1 obtained suggest that the intrinsic attenuation of the two samples was slightly different, possibly because of a difference in impurity level.) This is in agreement with the trends shown by the results for the resonance amplitude. Some results for these cold rolled steel sheet samples showed that the shapes of the resonance peaks are not always smooth. We attribute this to texture effects and to non-linear magnetic effects (for the case in which the static magnetic field is not low).

CONCLUSIONS

These initial measurements show the feasability of a new method for grain size measurement with resonance mode EMAT. The main features of the amplitude method are the advantages of fast measurement and the possibility for on-line application. A disadvantage is the necessity for calibration because of its dependence on sensor and material

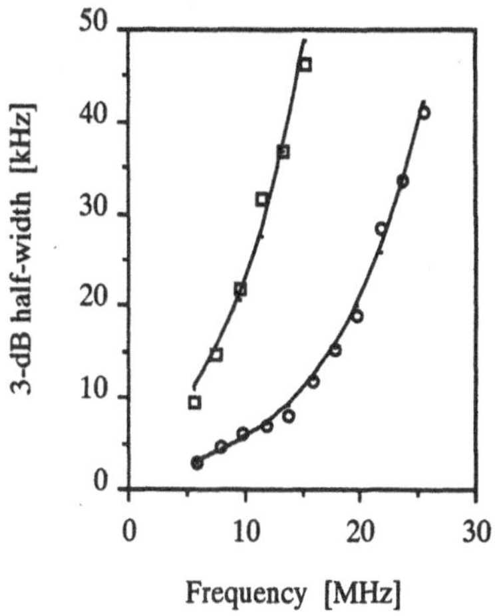

Figure 7. Plot of 3-dB half-width against frequency. The samples are cold rolled steel sheets of thickness 0.8 mm. The squares represent data for grain size number 9.5, and the circles represent data for grain size number 11. Fits $A_1f+A_2f^4$ for each case are shown.

parameters. The advantage of the half-width method is that it is in principle independent of sensor parameters such as liftoff, and is quantitative provided that nonparallelism and diffraction losses are accounted for (which may affect the results, particularly at the lower frequencies). Also it is insensitive to the value of sheet thickness and other material parameters apart from sound velocity. A disadvantage is that a longer burst duration is necessary, increasing the measurement time. Also we need to understand the effects of texture and magnetic field on the resonance curve shape.

In future we intend to expand our measurements to a wider variety of samples of different thicknesses, and to further compare these two methods.

REFERENCES

1. R. Klinman, G. R. Webster, F. J. Marsh and E. T. Stephenson, Mat. Eval., Vol. 38, 26 (1980).
2. J. F. Bussiere, Mat. Eval., Vol. 44, 560 (1986).
3. W. P. Mason and H. J. McSkimin, J. Appl. Phys.,Vol. 19, 940 (1948).
4. L. G. Merkulof, Sov. Phys. Tech. Phys., Vol. 1, 59 (1956).
5. E. P. Papadakis, J. Acoust. Soc. Am., Vol. 37, 711 (1965).
6. S. Serabian, Brit. J. Nondestr. Test., Vol. 22, 69 (1980).
7. R. L. Smith, Ultrasonics, 211 (Sept. 1982).
8. B. Kopec, Ultrasonics, 267 (Nov. 1975).
9. N. Mercier, Proc. Ultrasonics International 1975 Conference, 64 (1975).
10. W. N. Reynolds and R. L. Smith, Brit. J. Nondestr. Test., 291 (Sept. 1985).
11. E. R. Generazio, Mat. Eval., Vol. 44, 198 (1986).
12. R. L. Smith, NDT International, Vol. 20, 43 (1987).
13. D. Nicoletti, B. Onaral and N. Bilgutay, Mat. Eval., Vol. 50, 788 (1992).
14. H. Fukuhara, C. Masuda and S. Nishijima, Proc. Int. Symp. Nondestr. Test. and Stress-Strain Meas., Tokyo, October, 12-14, 1992, ed. by T. Kishi and S. Takahashi, Jpn. Soc. Non-Destr. Insp., p. 291 (1992).
15. R. D. Diamond, Metall, Vol. 41, 1232 (1987).
16. K. Kawashima, J. Acoust. Soc. Am., Vol. 87, 681 (1990).
17. K. Kawashima and O. B. Wright, J. Appl. Phys., Vol. 72, 4830 (1992).
18. K. Kawashima, T. Hyoguchi and T. Akagi, J. Nondestr. Eval., Vol. 12 (1993).
19. J. P. Monchalin and J. F. Bussiere, J. Phys. Paris, Vol. C10, 775 (1985).
20. R. A. Alpher and R. J. Rubin, J. Acoust. Soc. Am., Vol. 26, 452 (1954).
21. R. Truell and W. Oates, J. Acoust. Soc. Am., Vol. 35, 1382 (1963).

QUANTITATIVE ANALYSIS OF THE ELASTIC PROPERTIES OF AL-LI-CU ALLOYS

B.C. Lee[1], J.K. Park[1], and S.S. Lee[2]

[1]Department of Materials Science and Engineering
Korea Advanced Institute of Science and Technology
373-1 Yusung Gu Gusung Dong TAE JON 305-701, Korea
[2]Center for Materials Characterization and Evaluation
Korea Research Institute of Standards and Science
TAE JON 305-606, Korea

INTRODUCTION

Al-Li-Cu ternary alloys have recently received a great research interest because of its attractive combination of properties of the lightness and stiffness as a potential candidate for aircraft structural materials[1]. This is because the addition of Li into aluminum greatly increases its elastic modulus by decreasing its density at the same time. The primary objective of the addition of Cu is to enhance its strength and ductility by precipitating new strong particles of mostly T_1 phase (Al_2CuLi) in addition to the ordered δ' particles. The T_1 phase has a hexagonal structure (a=0.496nm, c=0.934nm) and nucleates with its basal plane parallel to the {111} planes of the matrix, i.e., (0001) // {111}; $<11\bar{2}0>$ // $<1\bar{1}0>$. The θ '(Al_2Cu) phase also precipitates to a minor population depending on the Cu/Li ratio. The θ' phase of also plate shape has a tetragonal structure (a=0.404nm, c=0.580nm) and a cube orientation relationship.

Our previous study in Al-Li binary alloys showed that the precipitation of δ' phase contributes to some extent to the elastic modulus of the alloy[2]. The purpose of the present investigation was, as an extension of the previous work, to determine the role of the precipitation of T_1 phase on the elastic modulus of Al-Li-Cu ternary alloys.

EXPERIMENTS

Two ternary alloys of Al-2.51Li-2.57Cu (Alloy A) (all in wt.%) and of Al-2.44Li-3.28Cu (alloy B) have been casted in a vacuum induction melting furnace using high purity ingots of Al (99.99%), Li (99.9%), and Al-50%Cu (Cu:99.99%). A small amount of Zr (0.13%) was added in both cases as a grain refiner. The alloy ingots were homogenized at

Nondestructive Characterization of Materials VI
Edited by R.E. Green, Jr. *et al.*, Plenum Press, New York, 1994

530°C for 24hr. The surface layer of ~3mm were, after homogenization, machined out to remove a depletion layer in Li. No further working was performed to avoid a texture problem.

For the purpose of measuring the Young's modulus, cube specimens of ~2×2×2 cm were prepared by machining and grinding the alloy ingots. A uniform thickness is imperative along propagation direction of the ultrasonic wave in order to accurately measure its velocity. A special grinding jig has been devised for this purpose. To investigate the stretching effect, a compressive strain of ~5% has been applied using Instron to the cube shaped specimens.

Specimens were solution treated at 530°C for 30min in a salt bath controlled to ±2°C and quenched to room temperature. They were subsequently aged to various temperatures in either salt bath or Si-oil bath (<180°C) controlled to ±1°C.

For the purpose of determining the elastic modulus, a modified ultrasonic pulse echo overlap method[3] has been employed to measure the propagation velocities of ultrasonic wave of 10MHz both in transversal and longitudinal modes. The density of each specimens was measured using Archimedes principle.

This sheets of ~2mm thickness were sliced out from the cube samples to prepare the thin foils for transmission electron microscopy (TEM). Disks of 3mm diameter punched out from the sheets were mechanically ground and finally electropolished in a solution of 30%HNO_3 and 70%CH_3OH at -40°C. Thin foils were observed under 200keV using a Philips CM20 equipped with STEM. The foil thickness was determined by contamination spot separation method. If applicable, corrections for the overlapping and truncation effects[4] have been employed to accurately determine the volume fraction of particles.

RESULTS AND DISCUSSION

Fig.1 illustrates the variation of Young's modulus of an Al-2.51Li-2.57Cu alloy (alloy A) as a function of aging time at various temperatures. The Young's modulus of the alloy progressively increases with the aging time at all temperatures investigated, in a similar manner to its aging behavior. At low temperatures, there appears two plateaux, the one at ~25hrs. of aging and the other at ~100hrs. of aging. This is closely related with the aging behavior of the alloy. At high temperatures, Young's modulus rapidly increases during the early stage of aging and reaches the maximum at ~10hrs of aging, forming only one plateau. The maximum value, ~80.8Pa appears to be about the same regardless of temperature. This is somewhat higher than that in an Al-Li binary alloy having a similar composition of Li[2] and smaller than that of Agyekum et al. in an Al-2.26Li-2.54Cu ternary alloy[5]. However this is in relatively good agreement with that of O'Dowd et al. in Al-Li-Cu ternary alloys[6].

Fig.2 shows the result in the case of an Al-2.44Li-3.28Cu alloy (alloy B), which basically contains more Cu. The result is in general quite similar to that of the alloy A, including the maximum value of the Young's modulus. This appears to suggest that a further increase in Cu content from ~2.57Cu to ~3.28Cu does not significantly alters the elastic properties of the alloy. The Young's modulus of the as-quenched condition is slightly lower than that in the alloy A, probably due to a small difference of Li content. This suggests that the contribution from the precipitation is actually larger in this alloy, although the difference is small.

The effect of the application of a compressive pre-strain is shown in Fig.3 for alloy A and in Fig.4 for alloy B. Some distinctive differences are observed particularly at low temperatures. Firstly, a rapid increase of Young's modulus is observed during the early stage of aging. Secondly, the pre-strained alloy exhibits only a single plateau even at low temperatures. Finally, the pre-straining results in a slight increase in the maximum value of

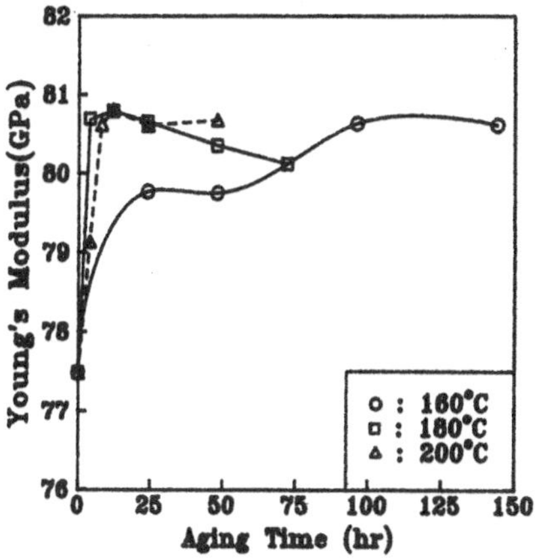

Fig 1. The variation of Young's modulus of an Al-2.51Li-2.57Cu alloy (Alloy A) as a function of aging time at various temperatures.

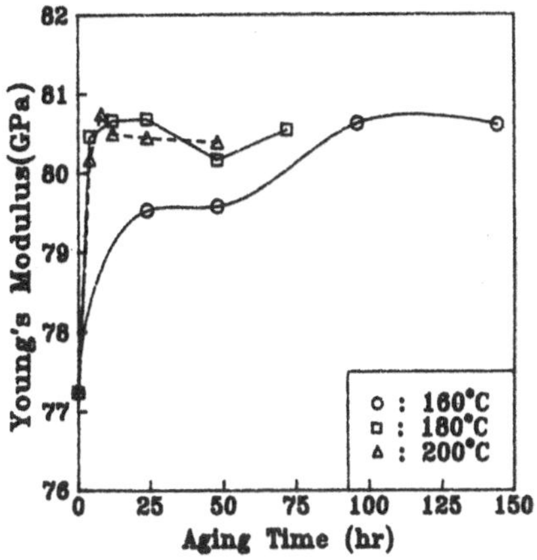

Fig 2. The variation of Young's modulus of an Al-2.44Li-3.28Cu alloy (Alloy B) as a function of aging time at various temperatures.

Young's modulus. These further confirm that the variation behavior of Young's modulus with the aging conditions of two alloys is intimately related with that of precipitation kinetics. The main difference between two alloys is in the magnitude of the increment of the maximum Young's modulus. The alloy B, more concentrated in Cu, tends to exhibit a larger increment.

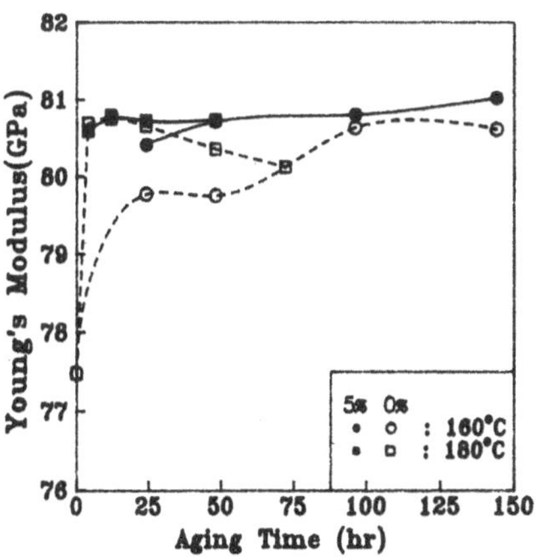

Fig 3. The variation of Young's modulus of alloy A as a function of aging time at various temperatures.
Note the effect of compressive pre-straining.

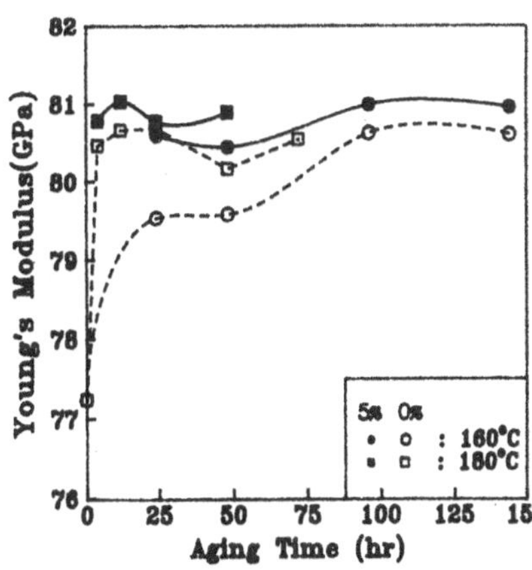

Fig 4. The variation of Young's modulus of alloy B as a function of aging time at various temperatures.
Note the effect of compressive pre-straining.

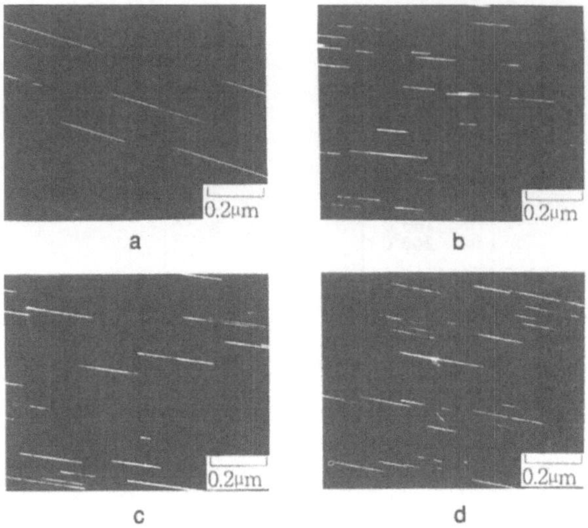

Fig 5. Dark field (DF) images taken using reflection from the edge-on variant of T_1 phases in samples of allloys A and B aged at 160°C for 144hr. Note the the effect of compressive pre-straining :

(a) no pre-strain (alloy A); (b) 5% pre-strain (alloy A);
(c) no pre-strain (alloy B); (d) 5% pre-strain (alloy B).

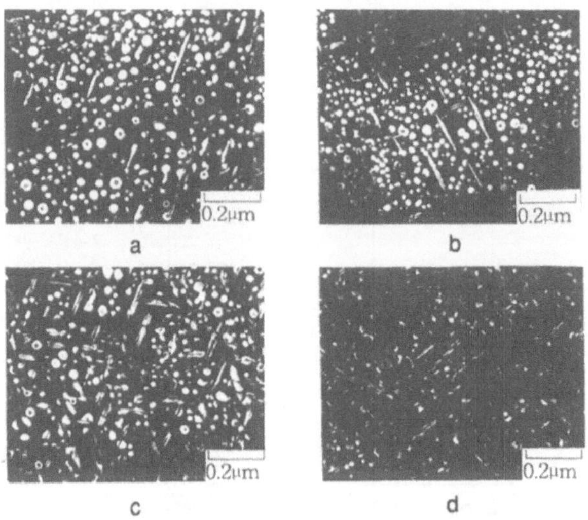

Fig 6. Dark field (DF) images taken using superlattice reflection due to the δ' phases in samples of alloy A aged at 180°C for 12hr and of alloy B aged at 160°C for 96hr. Note the effect of compressive pre-straining.

(a) no pre-strain (alloy A); (b) 5% pre-strain (alloy A);
(c) no pre-strain (allloy B); (d) 5% pre-strain (alloy B).

The microstructures of two alloys have been characterized by the analysis of the selected area diffraction patterns(SADP). We have first constructed theoretical mixed diffraction patterns including the reflections from both the precipitates and the matrix. They are then compared with the observed ones. The result indicated that the microstructures mainly consist of spherical δ' particles and thin plate of T_1 phase on {111} planes. In addition, thin plates of θ' phase on {001} planes are also observed particularly on aging at low temperatures. The population of θ' phase tends to decrease with the increase in the aging temperature. The application of the pre-straining by compression tends to suppress the θ' plates even at low temperatures.

On the other hand, the pre-straining treatment significantly increases the nucleation rate of the T_1 plates. Fig.5 illustrates dark field images of T_1 phase taken using the reflections from the edge-on variant. One notes a significant increase of particle density of the T_1 phase together with a reduction in the particle sizes. The comparison also appears to indicate that the alloy B, more rich in Cu, contains a larger number of T_1 plates.

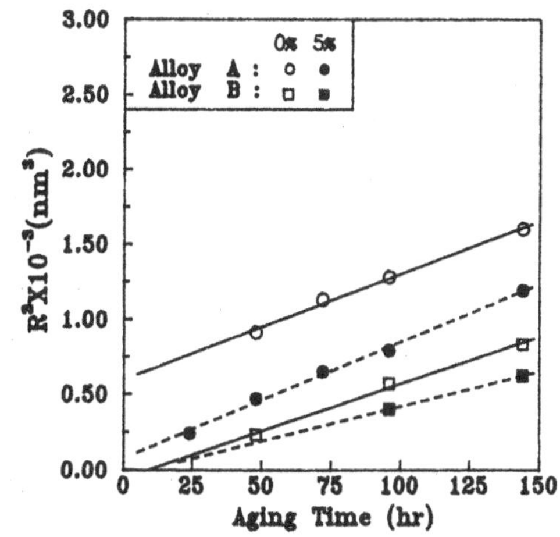

Fig 7. The variation of the cubes of particle sizes of δ' phase as a function of aging time at 160°C in alloys of Al-2.51Li-2.57Cu (alloy A) and of Al-2.44Li-3.28Cu (alloy B). Note the effect of compressive pre-straining.

The dark field images taken using δ' superlattice reflection (Fig.6) indicate that the sizes of δ' particles are reduced as a result of the pre-straining. The result further indicates a lower particle density in the alloy B. All of these results are closely related with the conditions for a larger nucleation rate of the T_1 phase.

The variation of the δ' particle sizes has been measured as a function of aging conditions (Fig.7). A linear relationship, between the cube of the particle size and time, appears to be well obeyed up to a relatively early stage of aging, in accordance with the coarsening theory[7]. The particle sizes are consistently smaller in the pre-strained condition as compared to the no-strained condition and also smaller in the alloy B as compared to the alloy A. This is closely related with the precipitation kinetics of the T_1 phase in each condition.

The volume fractions of δ' and T_1 phases have been measured as a function of aging conditions (Table 1). The result confirms that the δ' volume fraction is reduced as a result of the pre-straining treatment. The result also shows that the δ' volume fraction in the alloy A is significantly larger than that in the alloy B (~6% vs. ~3%). Finally, the δ' volume fraction in the alloy A rapidly reaches a maximum value, while that in the alloy B slowly increases to a maximum value indicating a low degree of supersaturation.

The result of the T_1 phase indicates that it steadily increases with time particularly on aging at low temperatures. Secondly, the kinetics of the precipitation of T_1 phase is faster in the alloy B as compared to the alloy A, due to a higher supersaturation of Cu. Finally, the compressive pre-straining treatment results in a distinctive augmentation of the T_1 volume fraction particularly at low temperatures. This explains why two plateaux of Young's modulus in Fig.1 and Fig.2 disappear in the pre-strained condition. The first plateau is mainly due to a δ' precipitation. and the second one due to a T_1 precipitation (also in part due to a θ' precipitation). Thus two plateaux can merge into one if the precipitation kinetics of the T_1 phase is fast enough to be comparable to the precipitation kinetics of the δ' phase.

Table 1. The volume fractions of δ' and of T_1 phases in various aging conditions.

Alloy	Aging Condition		f_{T1} 0%	f_{T1} 5%	$f_{\delta'}$ 0%	$f_{\delta'}$ 5%
A	160°C	24hr	–	–	0.063	0.053
		48hr	0.003	0.014	0.066	0.055
		72hr	0.012	0.015	0.068	0.056
		96hr	0.014	0.018	0.069	0.056
		144hr	0.016	0.023	0.070	0.056
	180°C	12hr	0.024	0.022	0.028	–
		24hr	–	–	0.032	–
B	160°C	48hr	0.023	0.027	0.014	–
		96hr	0.026	0.031	0.030	0.019
		144hr	0.031	0.035	0.036	0.029
	180°C	12hr	0.022	0.028	0.027	–
		24hr	–	–	0.034	–

Since the δ' and T_1 phases are the primary constituents of the present alloys, the measured Young's modulus E can be approximated, due to a Voigt averaging, by

$$E = E_m \cdot f_m + E_{\delta'} \cdot f_{\delta'} + E_{T_1} \cdot f_{T_1} \qquad (1)$$

We can evaluate the Young's modulus of the T_1 phase E_{T_1} using a known Young's moduli of the matrix Em and of the phase $E_{\delta'}$, since we have measured the volume fractions of δ' and T_1 phases, $f_{\delta'}$ and f_{T1}. The Young's modulus of the matrix was evaluated using a experimentally determined formula in Al-Li binary single crystals, E_m=71+1.22Li (in at%)[8]. Here we have neglected the effect of Cu on the Young's modulus of the matrix, because Cu is known to exert only a minor influence on the Young's modulus of the solid solution[9]. We have further assumed that the matrix composition reaches an equilibrium composition with the δ' phase from the early stage of aging. This assumption can give rise to an overestimation of E_{T_1} during the early stage of aging. We have used 93 GPa as the value of $E_{\delta'}$, which was previously determined in binary alloys[2].

Table 2 summarizes the results of the analysis of Young's modulus. Young's moduli of the T_1 phase deduced from comparatively early stages of aging tend to be somewhat larger as compared to the average one. This is believed to be due to an underestimation of Young's modulus of the matrix. Secondly, no significant difference in E_{T_1} is distinguishable between the no-strained and strained conditions in the alloy A. This suggests that the contribution of θ' phase is insignificant in this case probably because of its low quantity. However there appears some difference depending on the straining condition in the case of alloy B, where the amount of the θ' phase is appreciable in the no-strained condition. Namely, Young's modulus in the no-strained condition appears to be rather lower as compared to that in the strained condition. This suggests that the Young's modulus of the θ' phase is actually smaller than that of the solid solution, in a contradiction to what expected. This is probably because a significant amount of Li loss occurs through a heterogeneous precipitation of δ' phase at the broad faces of the θ' plate (see Fig. 6).

The Young's modulus of the T_1 phase has been estimated to be ~155GPa by averaging the data except those mentioned above. This is in a reasonable agreement with the previously reported value of ~170GPa in an Al-2.26Li-2.54Cu(-0.11Zr) alloy[5]. However this is in a large discrepancy with a recent report of 350GPa in Al-Li-Cu ternary alloys[6]. A large discrepancy appears to arise primarily from the measurement of the volume fraction of the T_1 phase.

The Young's moduli of two alloys are very comparable each other. Namely, no significant increase of Young's modulus is measured on augmenting the Cu content. This is believed to be due to the competitive nature of the precipitations of δ' and T_1 phases.

Table 2. Analysis of Young's modulus of alloys A and B.

Alloy	Aging Condition		Deform (%)	f_{T1}	$f_{\delta'}E_{\delta'}$ (GPa)	f_mE_m (GPa)	E (GPa)	E_{T1} (GPa)
A	160°C	96hr	0	0.014	6.417	71.81	80.63	171.6
		144hr	0	0.016	6.510	71.58	80.61	157.5
	180°C	12hr	0	0.024	2.604	74.47	80.79	157.5
		24hr	0	0.028	2.976	73.80	80.66	138.7
	160°C	96hr	5	0.018	5.255	72.46	80.80	169.5
		144hr	5	0.023	5.301	72.04	81.02	159.9
	180°C	12hr	5	0.022	3.224	74.09	80.75	158.3
B	160°C	96hr	0	0.026	2.771	73.91	80.62	149.2
		144hr	0	0.031	3.382	73.01	80.60	134.4
	180°C	12hr	0	0.022	2.184	74.94	80.66	160.0
		24hr	0	0.025	2.418	74.51	80.68	149.5
	160°C	96hr	5	0.031	1.748	74.44	80.99	156.9
		144hr	5	0.035	2.641	73.38	80.95	142.5
	180°C	12hr	5	0.028	1.948	74.69	81.03	157.9

CONCLUSIONS

The Young's modulus of two ternary alloys(Al-2.51Li-2.57Cu, Al-2.44Li-3.28Cu) increases from ~77.2GPa up to ~81.0 GPa depending on the aging conditions. The pre-straining (by compression) of 5% slightly augments the Young's modulus of the alloys, particularly in the Cu rich alloy. The contribution of the δ' phase ranges from ~2.5% to ~8% of the maximum elastic modulus, while that of the T_1 phase encompasses from ~3% to ~

6.5% of the maximum elastic modulus. The pre-straining treatment induces a rapid increase of the elastic modulus during the initial stage of aging at low temperatures. This is believed to be due to the enhanced precipitation kinetics of the T_1 phase by the pre-straining treatment. The Young's modulus of the T_1 phase is estimated to be 155GPa, which is somewhat larger than that of the δ' phase. The role of the θ' phase appears to play a negative effect on the elastic modulus, probably due to a loss of Li content by a heterogeneous precipitation of δ' phase on the broad faces of the θ' plates.

ACKNOWLEDGMENTS

Authors are grateful to the Korea Research Institute of Standards and Science for their financial support of this research.

REFERENCES

1. H.M.Flower and P.J.Gregson, *Mat. Sci. Tech.*, 3, 81(1987).
2. S.M.Jeon, J.D.Kim, J.K.Park and S.S.Lee, Review of Progress in Quantitative Nondestructive Evaluation, D.O.Thompson, D.E.Chimenti, eds., Sandiago, Lajolla, CA, vol.12 (1992).
3. E.P.Papadakis, *J. Appl. Phys.*, 35, 1474(1964).
4. E.E.Underwood, "Quantitative Sterelolgy"(Addition Wesley, Reading, Mass., 1970).
5. E.Agyekum, R.Ruch and E.A.Starke, Aluminum-Lithium III, edited by C.Baker, C.J.Gregson, S.J.Harris and C.J.Peel (The Inst. of Metals, London), 448(1986).
6. M.E.O' Dowd, R.Ruch and E.A.Starke, 4th International Aluminum-Lithium Conference, G.Champier, B.Dubost, D.Miannay and L.Sabetay, eds., Les edidiums de physigre, Paris, France, 565(1988).
7. I.M.Lifshits and V.V.Slyozov, *J. Phys. Chem. Solid.*, 19, 35(1961).
8. W.Muller, E.Bubeck and V.Gerold, Aluminum-Lithium III, edited by C.Baker, C.J.Gregson, S.J.Harris and C.J.Peel (The Inst. of Metals, London), 435(1986).
9. W.E.Quist and G.H.Narayanan, Aluminum-Lithium Alloys, in " Aluminum Alloys Contemporary Research and Applications," A.K.Vasudevan and R.D.Doherty eds., AP, Boston, 219(1989).

MEASUREMENT OF ANISOTROPIC ELASTIC MODULI

AND COMPARISON WITH EQUIVALENT MEDIA THEORIES

J. A. Hood[1], R. B. Mignogna[2], N. K. Batra[2], K. E. Simmonds[2],
and H. H. Chaskelis[2]

[1]Rosenstiel School of Marine & Atmospheric Science
University of Miami
Miami, Florida 33149-1098
[2]Mechanics of Materials Branch
Naval Research Laboratory
Washington, DC 20375-5000

INTRODUCTION

Anisotropic elastic moduli were determined for a layered glass plate-epoxy composite from oblique angle time-of-flight measurements using an immersion system. The composite consisted of thin glass plates bonded with a UV curing epoxy. The glass plates were 5*cm* square and 0.04*cm* thick. The epoxy bonds were very thin, comprising only 1.99% of the composite. Both the glass and the epoxy behave as isotropic materials in bulk form. However, once assembled to produce a heterogeneous structure (composite material) and probed with sufficiently long wavelengths (approximately ten times the thickness of the glass plates at 1 MHz), the stack of thin glass-epoxy layers appeared homogeneous but with anisotropic properties. The anisotropy exhibited by the glass-epoxy stack was transverse isotropy. Equivalent media theories are designed to predict behavior of homogeneous material combined in a heterogeneous structure. Consequently, with the measured properties of the constitutive bulk glass and epoxy and knowledge of their ratio in the layered stack, forward calculations can be made to determine the elastic moduli of the composite. Of the equivalent media theories available for calculating expected behavior in composite media, we consider a thickness-weighted averaging model as well as a theory which models the thin compliant epoxy bonds as a set of fractures. Anisotropic elastic moduli predicted using these two theories will be compared to the measured elastic moduli of the glass-epoxy stack.

MOTIVATION FOR THIS STUDY

The fine-scale layering in some composite media generates a hexagonal symmetry with a vertical symmetry axis, commonly referred to as transverse isotropy (TI). Backus[1] showed that for sufficiently long wavelengths, a periodically stratified elastic medium is physically indistinguishable from a homogeneous material with anisotropic properties. When the component layers are isotropic, the long wavelength equivalent medium appears TI. To model a set of fractures, Schoenberg[2,3] considered a periodically layered system composed of only two isotropic component layers in which one of the layers was very thin and very compliant.

A two-component composite material was built to simulate a Schoenberg fractured material. It consisted of a set of thin glass plates bonded together with a UV curing epoxy. The glass was isotropic and each piece of glass was a $5cm \times 5cm$ thin square plate with elastic properties λ_g, μ_g. There were 75 glass plates, each of thickness $h_g = .04cm$, stacked and bonded with an epoxy to form a parallelpiped (see Fig. 1). The final height of the stack was $3.060cm$. Therefore, the average epoxy thickness was $h_e = 8.\overline{108} \times 10^{-4}cm$ and so comprised only 1.99% of the composite material volume. The epoxy was isotropic in bulk form with elastic properties λ_e, μ_e.

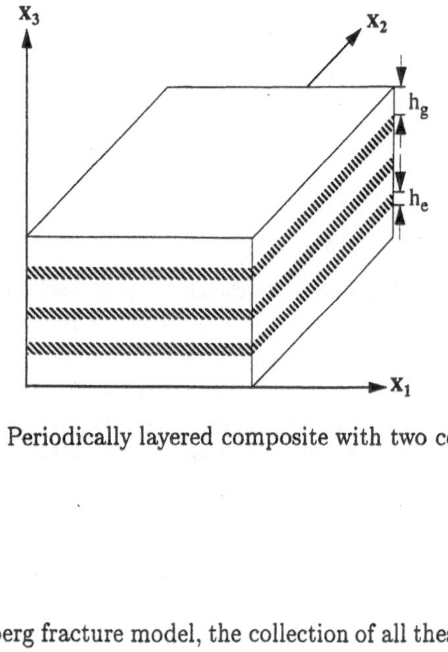

Figure 1. Periodically layered composite with two components.

According to the Schoenberg fracture model, the collection of all these thin soft epoxy layers over the entire system simulates a set of equally spaced planar parallel fractures.

Our study was conducted to see whether our composite appeared homogeneous and anisotropic at long wavelengths. Before the glass and epoxy were combined to form the composite, measurements were made to determine their bulk properties. With the properties of the bulk glass and epoxy and knowledge of their ratio in the layered stack, forward calculations were made to determine the elastic moduli of the composite predicted using the Backus formulation as well as the Schoenberg fracture model. These calculations were then compared to the long wavelength measured properties of the composite.

THE BACKUS EQUIVALENT MEDIA THEORY

Backus[1] showed that in a heterogeneous periodically layered material, for wavelengths sufficiently longer than the thickness of the component layers (on the order of 10 times), an equivalent set of homogeneous anisotropic elastic moduli can be found. Backus calculated the equivalent elastic moduli using thickness-weighted averages as

$$c_{33} = \left\langle \frac{1}{\lambda + 2\mu} \right\rangle^{-1} = \left(\frac{h_g^*}{\lambda_g + 2\mu_g} + \frac{h_e^*}{\lambda_e + 2\mu_e} \right)^{-1} = \frac{(\lambda_e + 2\mu_e)(\lambda_g + 2\mu_g)}{h_g^*(\lambda_e + 2\mu_e) + h_e^*(\lambda_g + 2\mu_g)} ,$$

$$c_{13} = \left\langle \frac{\lambda}{\lambda + 2\mu} \right\rangle \left\langle \frac{1}{\lambda + 2\mu} \right\rangle^{-1} = \left(\frac{h_g^*\lambda_g}{\lambda_g + 2\mu_g} + \frac{h_e^*\lambda_e}{\lambda_e + 2\mu_e} \right) c_{33} ,$$

$$c_{11} = \left\langle \frac{4\mu(\lambda + \mu)}{\lambda + 2\mu} \right\rangle + \left\langle \frac{1}{\lambda + 2\mu} \right\rangle^{-1} \left\langle \frac{\lambda}{\lambda + 2\mu} \right\rangle^{2} = \frac{4h_g^*\mu_g(\lambda_g + \mu_g)}{\lambda_g + 2\mu_g} + \frac{4h_e^*\mu_e(\lambda_e + \mu_e)}{\lambda_e + 2\mu_e} + \frac{c_{13}^2}{c_{33}} ,$$

$$c_{44} = \left\langle \frac{1}{\mu} \right\rangle^{-1} = \left(\frac{h_g^*}{\mu_g} + \frac{h_e^*}{\mu_e} \right)^{-1} = \frac{\mu_e\mu_g}{h_g^*\mu_e + h_e^*\mu_g} , \quad c_{66} = \langle \mu \rangle = h_g^*\mu_g + h_e^*\mu_e \tag{1}$$

where the brackets $\langle\ \rangle$ represent the thickness-weighted averaging. The calculation procedure is most obvious in the expressions for c_{44} and c_{66}. The right-hand sides of each equation are for the case of a two-component composite, subscripts g for glass and e for epoxy. The values h_g^* and h_e^* are the volume percentages of glass, $h_g^* = h_g/(h_g+h_e)$, and epoxy, $h_e^* = h_e/(h_g+h_e)$. There are five independent parameters in a Backus TI medium: the λ_g and μ_g of the glass, the λ_e and μ_e of the epoxy, and the ratio of epoxy to glass, h_e/h_g.

THE SCHOENBERG FRACTURE MODEL

In a two-component system where one of the layers is very thin and very compliant, Schoenberg[2,3] modeled the set of thin compliant layers as a set of fractures. Therefore, for long wavelengths, a heterogeneous material is modeled as if it is an isotropic material containing a set of oriented fractures. For our composite this corresponds to a solid piece of glass containing a set of epoxy-filled fractures. The fractures add an additional compliance to the material which is modeled with the Schoenberg parameters[3]

$$E_N = h_e \frac{\mu_g}{\lambda_e + 2\mu_e} , \quad E_T = h_e \frac{\mu_g}{\mu_e} . \tag{2}$$

The E_N and E_T are dimensionless compliances normal and tangential to the fracture planes, respectively. Although the fractures are considered to have negligible thickness and negligible compliance compared to the properties of the host material, the E_i are finite positive numbers. Physically they are the ratios of strain in the fractures to corresponding strain in the background material.

Hooke's law for general linear elastic media relates stress to strain by a matrix of elastic stiffness moduli:

$$\underline{\sigma} = \underline{\underline{C}}\, \underline{\epsilon} \tag{3}$$

For isotropic glass the elements of $\underline{\underline{C}}$ are $c_{11_g} = c_{22_g} = c_{33_g} = \lambda_g + 2\mu_g$, $c_{44_g} = c_{55_g} = c_{66_g} = \mu_g$, and $c_{12_g} = c_{11_g} - 2c_{44_g} = c_{13_g} = c_{23_g} = \lambda_g$ where λ_g and μ_g are the Lamé constants of the glass and all other elements of $\underline{\underline{C}}$ equal zero. Schoenberg[3] derived expressions for the long wavelength anisotropic equivalent elastic moduli for an isotropic material modified by the addition of an oriented fracture set. Applying his results, the elastic moduli of epoxy-filled fractures in glass are

$$c_{11} = \frac{\mu_g\left[\lambda_g + 2\mu_g + 4E_N(\lambda_g+\mu_g)\right]}{\mu_g + E_N(\lambda_g + 2\mu_g)} = \frac{\lambda_g\lambda_e + 4h_e\lambda_g\mu_g + 2\lambda_e\mu_g + 4h_e\mu_g^2 + 2\lambda_g\mu_e + 4\mu_g\mu_e}{h_e\lambda_g + \lambda_e + 2h_e\mu_g + 2\mu_e} ,$$

$$c_{33} = \frac{\mu_g(\lambda_g + 2\mu_g)}{\mu_g + E_N(\lambda_g + 2\mu_g)} = \frac{(\lambda_g + 2\mu_g)(\lambda_e + 2\mu_e)}{\lambda_e + 2\mu_e + h_e\lambda_g + 2h_e\mu_e} ,$$

$$c_{13} = \frac{\lambda_g\mu_g}{\mu_g + E_N(\lambda_g + 2\mu_g)} = \frac{\lambda_g(\lambda_e + 2\mu_e)}{h_e\lambda_g + \lambda_e + 2h_e\mu_g + 2\mu_e} ,$$

$$c_{44} = \frac{\mu_g}{1 + E_T} = \frac{\mu_g\mu_e}{\mu_e + h_e\mu_g} , \quad c_{66} = \mu_g . \tag{4}$$

There are only four independent parameters in a Schoenberg TI medium: the λ_g and μ_g of the glass and the E_N and E_T from the additional compliance that the fractures add to the material. The Schoenberg model is therefore easier to implement than the Backus formulation because it requires one less parameter. In fact, because the faces of the composite coincide with its symmetry planes, Hood and Mignogna[4] showed that contact measurements provided the four diagonal elements of $\underline{\underline{C}}$ which was sufficient to uniquely quantify this TI composite using Schoenberg's fracture model. However, it can not always be assumed that one of the components has a negligible thickness and compliance. In order to account for the finite thickness and compliance of a second component, an equivalent media theory like the thickness-weighted averaging model of Backus[1] must be used. The Backus model, however, has five independent parameters. Thus the need for a more sophisticated measuring scheme like that of oblique angles of incidence in an immersion tank was needed to provide the fifth parameter, off-diagonal element c_{13}.

MEASUREMENTS OF COMPOSITE TO DETERMINE MODULI

The elastic moduli of our composite were determined from oblique angle time-of-flight measurements using an immersion system similar to that used by Mignogna et al.[5] to measure elastic moduli for a cubic spinel. Energy and phase vectors do not necessarily coincide in anisotropic materials which makes it difficult to obtain the elastic moduli of highly anisotropic materials. However, for parallel plate geometry time-of-flight along the phase path is equivalent to time-of-flight along the energy path.[6] Therefore, phase velocities needed for the determination of the c_{ij} can be obtained by measuring the propagation of time along the energy path for a particular angle of incidence of the specimen from the source designated by θ_i, ϕ_i. Mignogna showed that

$$V_{phase}(\theta_i, \phi_i) = \left[\left(\frac{\Delta t(\theta_i, \phi_i)}{d}\right)^2 - 2\frac{\Delta t(\theta_i, \phi_i)\cos\theta_i}{dV_w} + \left(\frac{1}{V_w}\right)^2\right]^{-\frac{1}{2}} \tag{5}$$

where d is the thickness of specimen; θ_i and ϕ_i are the spherical polar angles of incidence; V_w is the phase velocity in water (or whatever couplant fluid is used); $\Delta t(\theta_i, \phi_i) = t_w - t_i(\theta_i, \phi_i)$ where t_w is the time-of-flight from the source to the receiver in water only (no material present) and $t_i(\theta_i, \phi_i)$ is the time-of-flight with the material inserted in the fluid path between the source and the receiver. With this analytic expression for the phase velocity as a function of direction cosines and elastic moduli [Eq. (5)], the c_{ij} were determined by an iterative least-squares inversion method[7].

Since our composite was not isotropic, it could not be assumed that the minimum time-of-flight would be along the linear path connecting the source to the receiver through the material. In addition, at such small distances and for such high frequencies, the source does not produce a plane wave. Therefore the leading edge of the wave pulse (1 MHz source) had to be found in order to determine the time-of-flight.

A receiver scan along the $x_1 - x_2$ plane was made for different θ_i, ϕ_i orientations of the material in order to find the position in which the receiver actually measured the minimum time-of-flight. See Fig. 2 for a schematic of the laboratory setup. The receiver was mounted on a vertical screw so that it could scan along x_1; the screw was mounted on a horizontal track so that the screw mounted with the receiver could move along x_2. The receiver typically scanned a 1/4 - 1/2 inch gridded space sampling at locations both vertically and horizontally every 1/64 - 1/32 of an inch. The entire scan produced a grid whose sectors corresponded to the different time-of-flight measurements at each receiver location. Scans were made at several θ_i, ϕ_i orientations and from each scan the minimum time-of-flight was chosen. This single time-of-flight for a particular θ_i and ϕ_i is the $t_i(\theta_i, \phi_i)$ of Eq. (5), and along with d and t_w we determined the phase velocity. From the phase velocity the elastic moduli were determined by applying the inversion method of Mignogna.[7]

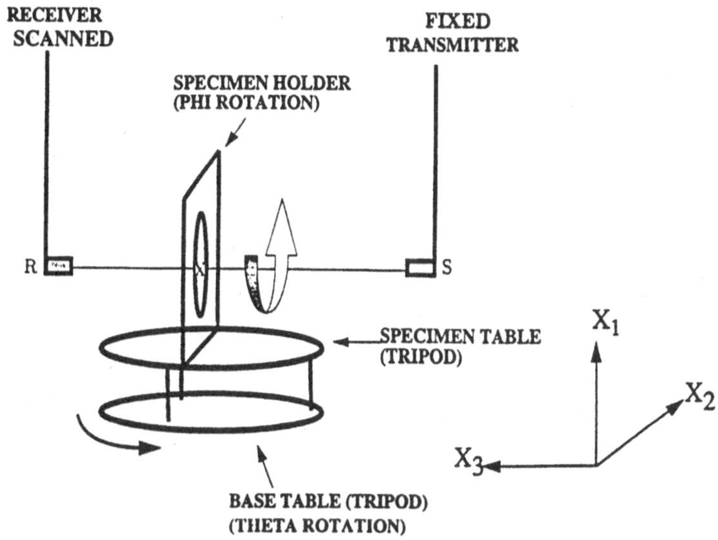

Figure 2. Laboratory setup.
X = composite is mounted on goniometer for ϕ rotation; goniometer is mounted on horizontal plate for θ rotation; R = pinducer receiver scans up and down on screw (along x_1) and is mounted on a horizontal track for additional motion along x_2; S = fixed source.

ANISOTROPIC MODULI PREDICTED FROM BULK PROPERTIES

Prior to stacking and bonding, the elastic moduli of each isotropic component was determined by measuring travel time through bulk forms of the glass and of the epoxy. Using the relationship between velocity and elastic moduli where subscript s is for shear wavespeed and p is for compressional wavespeed

$$v = \sqrt{\frac{c_{ij}}{\rho}} \quad \rightarrow \quad v_s = \sqrt{\frac{\mu}{\rho}} \ , \quad v_p = \sqrt{\frac{\lambda + 2\mu}{\rho}} \ , \tag{6}$$

Table 1. Bulk component properties.

Velocity ($10^5 \frac{cm}{sec}$)	Elastic moduli ($10^{11} \frac{dyne}{cm^2}$)	Density ($\frac{gm}{cm^3}$)
$V_{s_g} = 3.37$	$\mu_g = 2.85$	$\rho_g = 2.51$
$V_{p_g} = 5.37$	$\lambda_g = 1.54$	
$V_{s_e} = .956$	$\mu_e = .0987$	$\rho_e = 1.08$
$V_{p_e} = 2.24$	$\lambda_e = .344$	

and values for densities ρ_i as tabulated below, the elastic moduli were calculated as From the bulk values and knowing the relative amounts of each component in the composite, the long wavelength anisotropic moduli were calculated using the Schoenberg fracture model, Eq. (4), and using the thickness-weighted Backus averaging, Eq. (1), and presented in Table 2 where

$$E_N = 2.34 \times 10^{-2} \quad , \quad E_T = 4.27 \times 10^{-3} \tag{7}$$

using Eq. (2). The elastic moduli determined for our glass-epoxy composite using the measurement method described in the previous section are tabluated below alongside the predicted moduli and the moduli determined from contact measurements[4].

Table 2. Elastic moduli - data and predictions.

elastic moduli ($10^{11} \frac{dyne}{cm^2}$)	DATA	contact measurements[4]	predictions from Backus averaging	predictions from Schoenberg
$c_{11} = c_{22}$	7.72	7.89	7.06	7.23
$c_{13} = c_{23}$	4.86	N/A	1.28	1.52
c_{33}	6.91	6.92	5.81	7.16
$c_{44} = c_{55}$	.833	2.13	1.83	2.79
c_{66}	1.48	2.83	2.80	2.85
$c_{12} = c_{11} - 2c_{66}$	4.77	2.22	1.47	1.53

RESULTS FROM THE DATA INVERSION

Using Eq. (4) we inverted the c_{ij} predicted according to the Schoenberg fracture model to obtain

$$\mu_g = c_{66} \ , \tag{8}$$

$$E_T = \frac{\mu_g}{c_{44}} - 1 \ , \tag{9}$$

$$E_N = \frac{c_{13} - c_{33} + 2\mu_g}{2c_{33}} = \frac{c_{11} - c_{13} - 2\mu_g}{2c_{13}} = \frac{\mu_g - c_{33} + \sqrt{A1}}{2c_{33}} \ , \tag{10}$$

$$\lambda_g = \frac{2\mu_g c_{13}}{c_{33} - c_{13}} = \frac{4\mu_g^2 - 2\mu_g c_{11}}{c_{11} - c_{13} - 4\mu_g} = \frac{2\mu_g \left(3\mu_g - c_{11} - \sqrt{A1}\right)}{c_{11} + c_{33} - 4\mu_g} \ , \tag{11}$$

and inverting using Eq. (1) the c_{ij} predicted according to the thickness-weighted averaging model of Backus are

$$\mu_g = \frac{\frac{h_e}{h_g}(c_{66} - c_{44}) + (c_{66} + c_{44}) + \sqrt{c_{66} - c_{44}}\sqrt{A2}}{2} \ , \tag{12}$$

DISCUSION AND CONCLUSIONS

A comparison of the elastic moduli of the bulk components with the moduli expected for the glass-epoxy composite using the Backus and Schoenberg models reveals that the epoxy behaves much more rigidly as an adhesive bond than in bulk form as indicated by the Backus values for μ_e and λ_e as well as the Schoenberg E_T and E_N. This is consistent with the predictions of Piché et al.[8].

For the Backus theory, error in inversion of h_e/h_g overestimates the amount of epoxy in the composite; the data indicate that there is 10.93% epoxy in the composite when in fact there is less than 2%. The error in the ratio parameter is then propagated into estimates for the μ_i which in turn affects the solutions for the λ_i. For the Schoenberg theory, μ_g is completely determined by the measurement of c_{66}; all other moduli are dependent on μ_g and so the error associated with μ_g from c_{66} propagates into all other moduli estimates.

Our results suggest that the equivalent media theories of Schoenberg or Backus are not practical for modeling thin adhesive bonds. The Backus theory could probably be used successfully to determine the relative amounts and properties of individual layers for layered media in which the different layers are of the same order in thickness. The Schoenberg fracture model, however, is probably better applied to infilled fractures of a geological nature rather than for modeling the thin bonds of composite media.

ACKNOWLEDGMENTS

The authors would like to thank Jo Ann Sinton of the University of Hawaii for patiently preparing our composites, Dr. Yiannis Michopoulos for his insightful comments regarding the use of Mathematica, and Dr. Y. Rajapakse and the Office of Naval Research for their support (#N0001492J1978).

REFERENCES

1. Backus, G. E., 1962, Long-wave anisotropy produced by horizontal layering, *J. Geophys. Res.*, 66:4427.
2. Schoenberg, M., 1980, Elastic waves behavior across linear slip interfaces. *J. Acoust. Soc. Am.*, 68:1516.
3. Schoenberg, M., 1983, Reflection of elastic waves from periodically stratified media with interfacial slip, *Geophys. Prosp.*, 31:265.
4. Hood, J. A., and Mignogna, R. B., 1993, Isolating fracture-induced anisotropy from background anisotropy, Review of Progress in Quantitative NDE, 12B:1249.
5. Mignogna, R. B., Batra, N. K., and Simmonds, K. E., 1991, Determination of elastic constants of anisotropic materials from oblique angle ultrasonic wave measurements II: experimental, Review of Progress in Quantitative NDE, 10B:1603.
6. Mignogna, R. B., 1990, Ultrasonic determination of elastic constants from oblique angles of incidence in non-symmetry planes, Review of Progress in Quantitative NDE, 9:1565.
7. Mignogna, R. B., 1991, Determination of elastic constants of anisotropic materials from oblique angle ultrasonic wave measurements I: analysis, Review of Progress in Quantitative NDE, 10B:1669.
8. Piché, L., Lêvesque, D., Tatibouët, J., Deprez, P. and Michel, A., Ultrasonic characterization of interfacial adhesion in metal-polymer-metal multilayers, this issue.

NONDESTRUCTIVE EVALUATION OF Co AGGLOMERATION AND WC GROWTH IN WC/Co SUPERHARD ALLOY

T. Aizawa and J. Kihara

Department of Metallurgy, University of Tokyo
7-3-1 Hongo, Bunkyo-ku, Tokyo 113

INTRODUCTION

The superhard alloy WC/Co cermets are widely utilized as tool and die materials in the technology of plasticity [1]. In the current design of these cermets, only the bulk elastic moduli are items for materials and processing design. Since these types of materials are very sensitive to microscopically morphological change due to WC grain growth and Co agglomeration [2,3], their mechanical properties must be also varying with the change of microstructure and morphology appearing in materials processing. Hence, precise materials design of WC/Co cermets should require for quantitative prediction and description of the change of mechanical properties induced by the recognized change of morphology and microstructure in these materials.

Authors [4,5,6] have proposed and developed the acoustic spectro microscopy (ASM) to measure the elastic properties of surface layer of materials and coatings and to evaluate the effect of microstructure in materials on the mechanical properties. With aid of the measurement stage controlled to move in x-y plane and/or in rotation, a map of elastic constant distributions and anisotropy in elasticity can be directly evaluated by this ASM. In a homogeneous material without inner structures, the acoustic wave velocity should be uniform and constant with frequency f. In other words, constant function of the acoustic wave velocity in frequency reveals that no characteristic length nor particular microstructures are included in the target materials. In case of isotropic elasticity, both the volumetric and the shear stiffness moduli are estimated from these measured constant acoustic wave velocities. As before mentioned, anisotropic elastic constants are also evaluated from the acoustic wave velocity distribution in rotation. Matrix materials with inclusions, precipitates or secondary phase should indicate the dispersed function of acoustic velocity in the frequency [7]. Since the obtained functional forms are very inherent to each microstructure, microstructure itself can be analyzed through the measured dispersion curves.

In the present paper, WC/Co superhard alloy cermet is employed to investigate the microstructural analysis through our developed acoustic spectro microscopy. The dispersion analysis of these measured acoustic structure is applied to understand the effect of morphological change of WC grains and Co phase on the change of elastic properties with Co weight percentage. Through this study, a new methodology will be discussed to describe the relation between the morphological, structural change in WC/Co cermet and its mechanical properties.

ACOUSTIC DIAGNOSIS WITH CONSIDERATION OF MICROSTRUCTURE

The effect of microstructure in materials on the acoustic structure is investigated by the dispersion analysis.

Acoustic Spectro Microscopy

Our developed facility for acoustic spectro microscopy has several measurement modes

Nondestructive Characterization of Materials VI
Edited by R.E. Green, Jr. *et al.*, Plenum Press, New York, 1994

with use of five controller to move the ultrasonic lens and measurement stage in five directions. In the standard mode of measurement, a pair of spherical and planar lens (ultrasonic transmitter and receiver) is controlled to move in the circumferential direction with the focussing point fixed on the surface of specimen. The ultrasonic pulse is injected by either lens of the pair with thus controlled skew angle θ into materials through the coupler (water), and the reflective pulse traveling back through the coupler is received by another lens. Owing to the Snell's law, significant dip and phase shift are observed in the power and phase spectra of these received signals at the critical angle θ_c:

$$\mathrm{Sin}(\theta_c^{L}(f)) = V_w / V^L(f), \quad \mathrm{Sin}(\theta_c^{S}(f)) = V_w / V^S(f),$$
$$\mathrm{Sin}(\theta_c^{LSAW}(f)) = V_w / V_{lsaw}(f), \tag{1}$$

where V^L, V^S, and V_{lsaw} are the longitudinal, shear and leaky surface wave velocities, respectively, and V_w the acoustic velocity of the coupler materials. Since each critical angle is measured in spectra to be a functional in frequency f, these three velocities can be determined also in the form of function in f. In particular, since the leaky surface wave, activated at the critical angle θ_c^{LSAW} at the vicinity of which large dip and phase shift are observed in spectra, includes mechanical information of surface layer of materials upto one wave length, the obtained $V_{lsaw}(f)$ should be a measure to evaluate the acoustic structure corresponding to the microstructure of target materials. Fig. 1 depicts an overview of our developed ASM machine; only θ-directional goniometer is utilized in the standard measurement mode. As obviously imagined from Eq. (1), the measurement accuracy is governed by the fluctuation of V_w in temperature and the location capability of the goniometer. Since the whole machine is housed in the clean booth, the change of temperature of coupler can suppressed to less than 0.1 K during measurement. In order to determine the critical angle, an optimum window $\theta_L \le \theta \le \theta_U$ is set up to include the critical angle, and an increment of angle is reduced to 0.01 - 0.02 in controlling θ-directional stepping motor. The resulting error in the measured acoustic velocity is limited to 1 - 2 m/s.

In order to obtain one and two dimensional distributions of elastic properties, the lens pair must be controlled to move in the specified uniaxial direction and in rotation. For those purposes, both x and y directional stages are also moving together with the goniometer. Furthermore, two types of SPP lens were employed in this study with different frequency range: (1) $10 \le f \le 150$ MHz and (2) $200 \le f \le 400$ MHz.

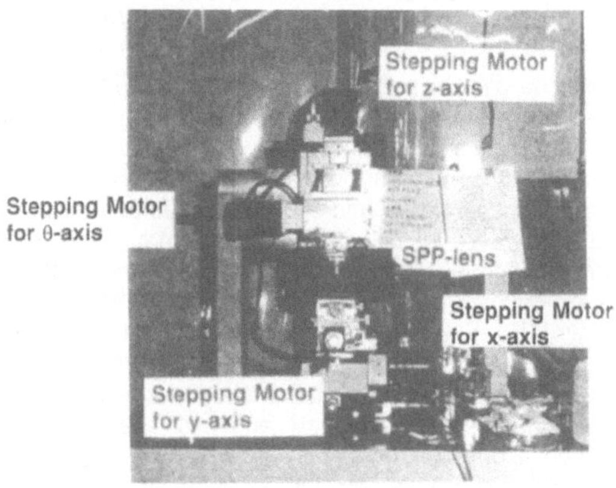

Figure 1. Acoustic spectro microscopy in the standard mode.

Dispersion Analysis for Microstructure

A matrix with inclusions, particles or fibers embedded inside indicates apparently homogeneous acoustic behaviors in case of the ultrasonic measurements by using the relatively lower frequency. As depicted in Fig. 2, when the wave length λ of ultrasonic wave becomes less than a critical length λ_c, the effect of microstructure on the acoustic behaviors can be observed as dispersion in the function of the acoustic wave velocity in frequency. For the wave length λ_s corresponding to the size of inclusions, when $\lambda \approx \lambda_s$, contributions of each constituent inclusion inside matrix materials are separated from each other, so that acoustic waves traveling in each constituent material or scattering from another one are distinguished to determine each characteristic wave velocity. When λ becomes much larger than λ_c, contributions of microstructure to the elastic properties are homogenized so that the measured elastic constants should be corresponding to the bulk moduli. Hence, this critical wave length can be identified as a unit cell length in the homogenization theory.

The point to be interested is the gray zone of $\lambda_s \leq \lambda \leq \lambda_c$ where the change of elastic properties with morphological variation from the homogeneous structure is imbedded in the measured acoustic structure. Consider a typical theoretical dispersion curve as shown in Fig. 3. The variation of V_{lsaw} with f should correspond to a morphological structure in each material. In the acoustic homogenization theory [7], both unit cell length and morphological structure in the unit cell can be theoretically discussed with aid of micromechanical models. The present paper is concerning with the application of the equivalent inclusion model [8] to understanding the gray zone appearing in the dispersion curve. The change of elastic moduli with frequency can be estimated by using the equivalent inclusion modeling. Consider a soft spherical inclusion with a diameter D distributed in matrix with the periodicity Λ. Owing to the micromechanics in elastostatics, both the volumetric and shear moduli $\{\kappa^*, \mu^*\}$ of matrix + inclusion materials system are given by

$$K^* = K/(1+Ag) \quad \text{for} \quad A = (K'-K)/[(K-K')(1+\nu)/3(1-\nu)-K] \ ,$$

and (2)

$$\mu^* = \mu/(1+Bg) \quad \text{for} \quad B = (\mu'-\mu)/[2(\mu-\mu')(4-5\nu)/15(1-\nu)-\mu] \ ,$$

where $\{\kappa, \mu, \nu\}$ are the materials constants for matrix. g denotes the effective volume fraction, and is provided by

$$g = 2\ [Df/\Lambda V_{lsaw}(+0)]^3 \ . \tag{3}$$

Hence, increase of volumetric fraction for inclusions should lead to decrease of volumetric and shear moduli. Since the density of mass can be assumed to be uniform even in the unit cell and subjected to the mixture rule, the leaky surface wave velocity is theoretically estimated by using these two moduli.

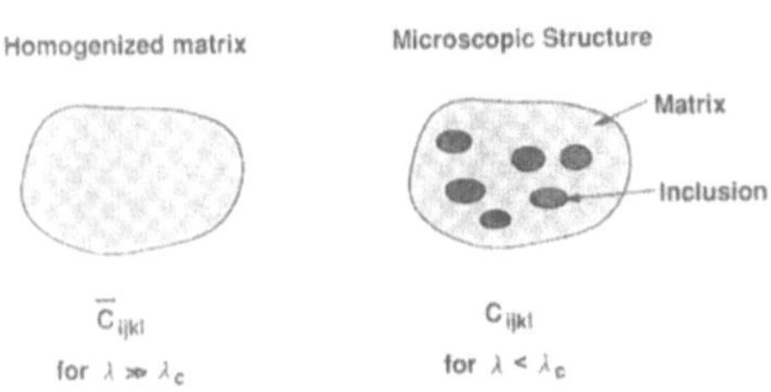

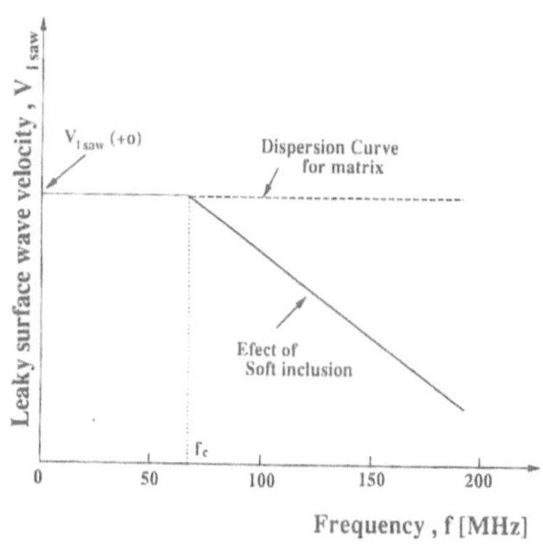

Figure 2. Acoustic homogenization behavior. **Figure 3.** Theoretical dispersion curve for WC/Co.

Acoustic Mapping Method

A single-crystal material often indicates its characteristic anisotropy in elasticity owing to its crystallographic structure. As discussed in the dispersion analysis of composite cermet system, a matrix + inclusion system, which seems to be macroscopically homogeneous, should have apparent anisotropy in elasticity, since WC grains does not grow isotropically.

ASM enables us to make measurement of elastic anisotropy only by control of mapping procedure. Fig. 4 depicts a typical mapping data where SPP lens pair is controlled to move both in the circumferential θ and the rotational φ directions. The point to be noted is that the critical angle or the leaky surface wave velocity can be obtained in the functional form of both the frequency and the rotational angle. Through the comparison of $V_{lsaw}(f,\varphi)$ for each WC/Co with different Co wt.%, the effect of microscopic morphology on elasticity can be also evaluated quantitatively.

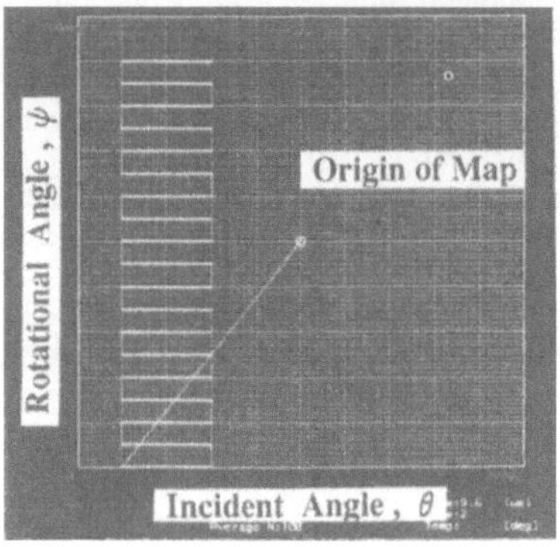

Figure 4. Typical mapping data to control SPP lens in rotation.

EXPERIMENTAL CONDITIONS

WC/Co materials with different Co wt.% are employed as test specimens to make dispersion analysis for understanding the morphological effect of WC-Co on the elastic properties.

Experimental Test Specimen

Test specimens with different Co weight percentage are depicted in Fig. 5. With increase of Co weight percentage, WC grain makes remarkable growth and free Co phase agglomerates

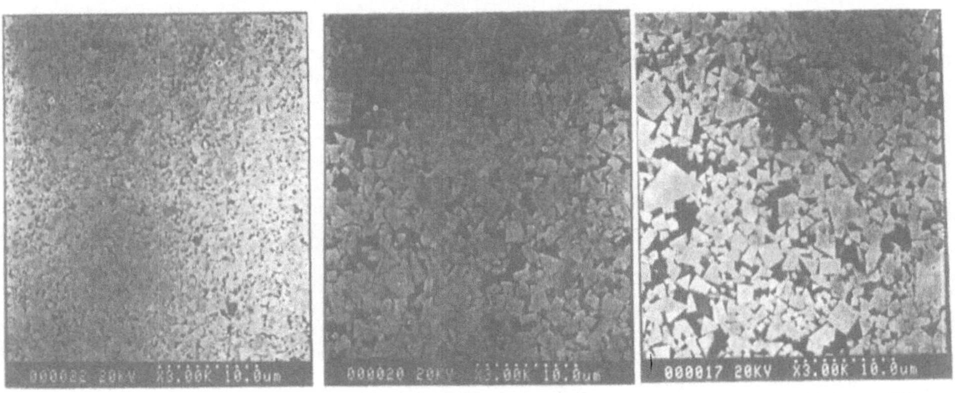

Figure 5. Microscopic observation of WC-Co system.

between larger WC grains. Since ASM is very sensitive to these microstructural change, morphological change must be described by the analysis of measured characteristic dispersion curves.

Measurement Conditions

The measurement conditions of ASM are listed in Table 1. For precise determination of surface wave velocity or critical angle, two step measurement procedure was employed in the present paper. In the rough search, the incident angle is controlled to change by 0.1 degree, and an optimum window of incident angle is selected. The critical angle is searched by 0.02 degree increment of angle within this window.

For standardization of measured ultrasonic signals, the lead plate was employed for a reference material. Through the preliminary testings, little distinguishable difference can be seen in the measured dispersions at different locations of specimen, and the error of measurement was found to be less than 0.1 % in velocity.

Table 1. ASM measurement conditions in standard.

Measurement Mode	Point Focusing[1] (Standard Mode)
Normalization	Lead (pure Pb)
Determination of V_{lsaw}	Phase shift[2]
Frequency	$10 \leq f \leq 150$ MHz (lower range)[3] $200 \leq f \leq 340$ MHz (higher range)
Incident angle	Rough search in the range of $20 \leq \theta \leq 40$ by $\Delta\theta = 0.1$ Precise measurement in the range of $20 \leq \theta \leq 26$[4] by $\Delta\theta = 0.02$

[1] Line focusing can be available when the lens eye angle is controlled to be narrow by a shutter.
[2] Cubic function is utilized to determine the critical angle in the phase shift profile. For the target materials with high attenuation, the power profile should become another certification of the critical angle.
[3] At present, two types of SPP lens pair are available in the frequency domain; three lens pairs are useful with different center angles in lens configuration.
[4] Optimum incident angle range is selected by the rough line-search in measurement. In the two and three dimensional mapping, this angle range must be controlled to vary to determine the critical angle in anisotropy.

EXPERIMENTAL RESULTS

Let us make dispersion analysis of leaky surface wave velocity and evaluate morphology in WC/Co with comments on the elastic anisotropy and the effect of WC grain growth on the measured acoustic structure.

Evaluation of WC/Co Microstructure

Fig. 6 depicts the measured dispersion curves of the leaky surface wave velocity for WC/6Co, WC/11Co, WC/16Co and WC/20Co. In case of WC/6Co, where small WC grains are uniformly distributed and mixed with Co-phase, little dispersion can be seen in this range of frequency with $10 \text{ MHz} \leq f \leq 150$ MHz. As could be expected from microscopic observation, relatively large dispersion behaviors are recognized with increase of Co content. To be interested, as shown in Fig. 7, the dispersion analysis by using an acoustic lens with higher frequency range of $200 \leq f \leq 400$ MHz reveals that even WC/6Co indicates significant amount of dispersion in velocity. This is partially because an effective volume fraction g of equivalent inclusion to agglomerated Co phase region becomes larger for higher frequency.

Let us make dispersion analysis of these relations of the leaky surface wave velocity V_{lsaw} to frequency. Since such microstructures as shown in Fig. 5 are homogenized with decrease of frequency, the limit $V_{lsaw}(+0)$ should be equivalent to the bulk surface velocity V_{bulk} estimated from the bulk dilatational and shear wave velocities which were measured by the conventional sing-around method. As compared in Table 2, both velocities are in good agreement with each other. This tells that the acoustic structure measured with $f \leq f_c$ should correspond to the homogenized elasticity. Then, as theoretically mentioned in the above, the critical frequency should represent the border between the macroscopic state in homogenized elasticity and the gray zone dependent on the microscopic morphology. Fig. 8 depicts the change of this critical frequency f_c and the unit length λ_c (defined by $V_{lsaw}(+0)/f_c$) with Co weight percentage or Co volumetric fraction. It was found that the unit length abruptly increases with Co volumetric fraction, and that WC-Co system should become macroscopically inhomogeneous for [Co wt%] ≥ 20 %.

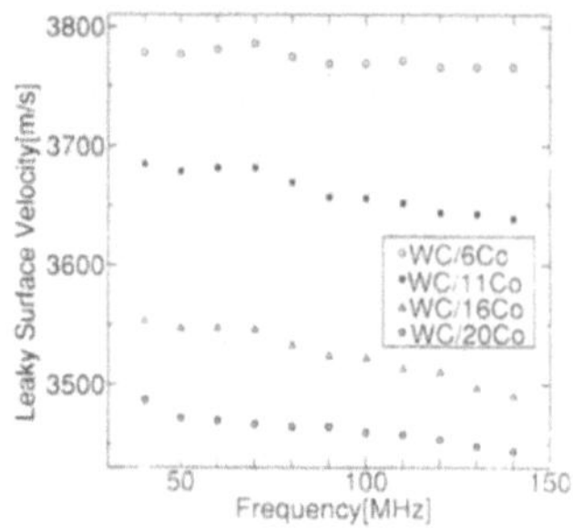

Figure 6. Dispersion curves for f ≤ 150 MHz.

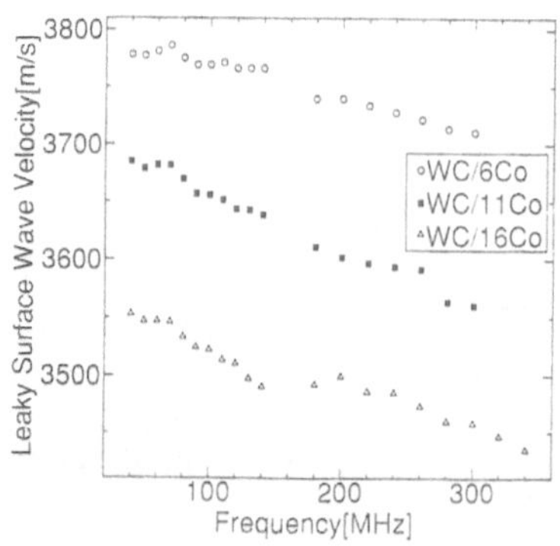

Figure 7. Dispersion curve for 200 ≤ f ≤ 340 MHz.

Table 2. Comparison between the Young's moduli estimated from V_{lsaw} and the bulk moduli.

Co wt.%	5.7	9.0	12.0	16.0
E_{Mech} [GPa]	640	600	580	530
Co wt.%	6.0		11.0	16.0
$E_{Present}$ [GPa]	613		564	510

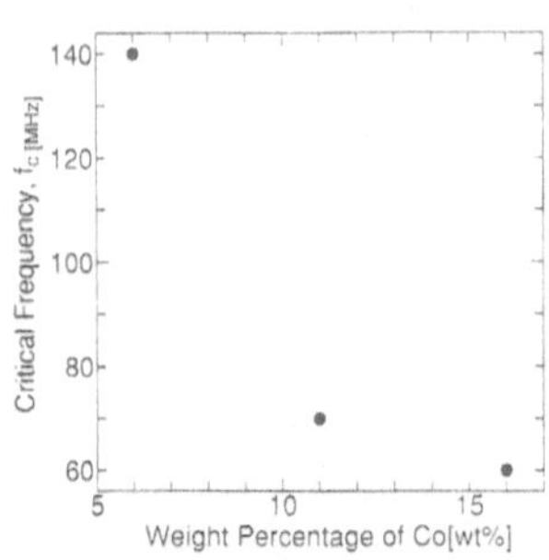

Figure 8. Change of f_c and λ_c with Co percentage.

In order to understand the microscopic morphology, the acoustic structure with $f \geq f_c$ should help us to describe such morphological parameters as D and Λ appearing in the above equation. Fig. 9 depicts the change of the dispersion sensitivity dV_{lsaw}/df with Co weight percentage and volumetric fraction. This gradient changes with Co content in the function of D/Λ.

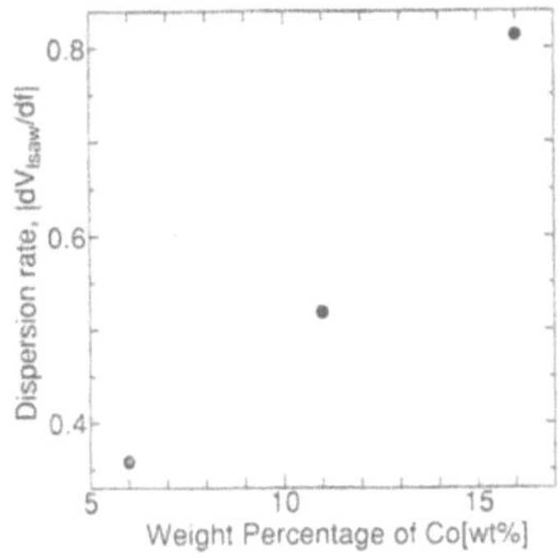

Figure 9. Change of dispersion sensitivity with Co.

Elastic Characterization in Rotational Mapping

The similar measurement conditions were applied to make dispersion analysis of apparent anisotropic elasticity in the leaky surface wave velocity. The rotational angle was controlled by 10 degrees from φ = 0 to 180. For φ = 0, 30, 60 and 90 degrees, $V_{lsaw}(f,\varphi)$'s are depicted in Fig. 10 for different Co weight percentage. As theoretically imagined, no apparent elastic anisotropy was observed for WC/6Co. However, with increase of Co, deviation of V_{lsaw} for different rotational angle became larger with increase of Co. This fact reveals that WC grain size should be noticeably large so that WC and agglomerated Co phase distribute inhomogeneously in the unit cell. Hence, microscopic morphological change can be also detected by this apparent anisotropy in elasticity.

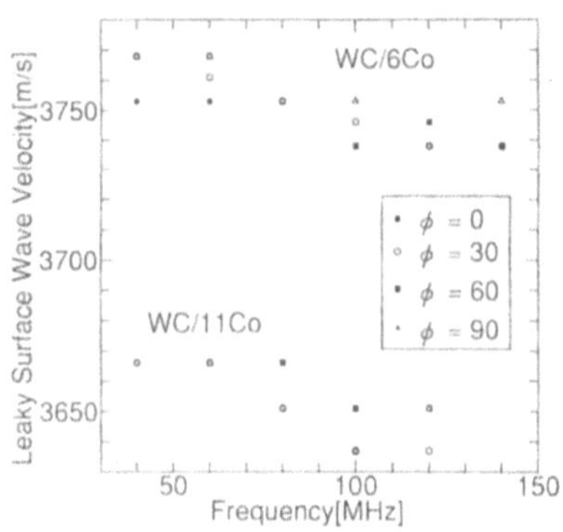

Figure. 10 Apparent anisotropy in elasticity for WC/Co with different Co contents.

Evaluation of WC/Co Grain Growth

In order to investigate the effect of WC grain growth on the acoustic structure, grain size in WC/Co with 30 vol.% was controlled by the heat treatment with T = 1673 K and different hold time. As shown in Fig. 11, remarkable WC grain growth can be recognized with increase of the hold time in heat treatment. From the apparent anisotropic distribution of V_{lsaw}, such morphological change as grain growth should have significant influence on the measured acoustic structure with [Co wt.%] > 16 %. Fig. 12 depicts the measured dispersion curves for different heat treated specimens. When hold time is short, the dispersion curve is nearly the same as the original one for non-treated specimen. Since WC grains begin to grow with the hold time, the dispersion curve fluctuates with frequency and might have apparent anisotropy in elasticity.

Figure 11. Microscopic observation of heat-treated WC/Co with 30 Vol.%.

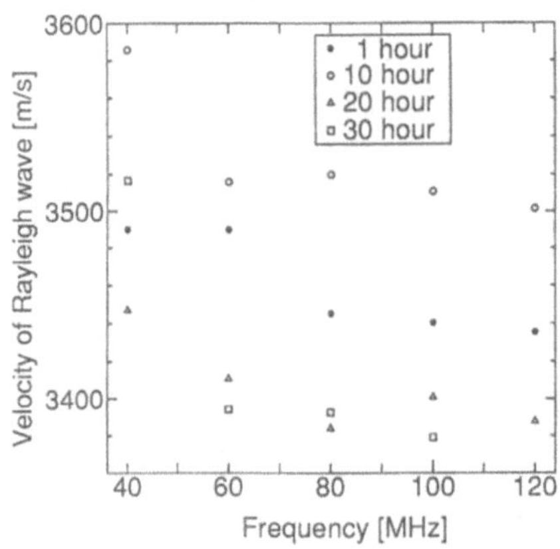

Figure 12. Dispersion curves for heat treated WC/Co with 30 Vol.%.

CONCLUSION

Through the dispersion analysis of the leaky Rayleigh wave velocity, it was found that the effect of change of microscopic morphology in WC/Co on the mechanical properties can be quantitatively evaluated. With aid of the acoustic homogenization theory [7], this morphological change can be described by the equivalent inclusion model where agglomerated free cobalt phase is distributed with periodicity in the matrix of WC-Co mixture. The most important point to be noted is that both WC grain growth and Co agglomeration can be directly detected by the dispersion behavior and by the apparent anisotropy in elasticity in the acoustic structure. This fact provides us a new methodological tool to perform on-process or off-process QNDE on the morphological change in WC/Co cermets in terms of varying mechanical properties with the acoustic informations.

Authors are concerned with quantitative description of grain growth and agglomeration in the structure sensitive materials through the acoustic diagnosis. Our new trial will be reported in future.

References

1. Sudoh, H.: Residual Stress and Distortion (1988) Uchida-Rokakuho.
2. Lee, I.C., et al.: Nonuniform Carbide Grain Growth during High-Temperature Compressive Deformation in WC-13wt%Co, *Scripta Metallurgica et Materialia*. 28 (1993) 97-102.
3. Sakuma, T. and Hondo, H.: Plastic Flow in WC-13wt.%Co at High Temperatures, *Materials Science and Engineering*. A156 (1992) 125-130.
4. Aizawa, T. and Kihara, J.: Materials Evaluation of PVD/CVD Coated WC/Co Superhard Alloys by the Ultrasonic Microscopy, *Nondestructive Testing Evaluation*. 9 (1992) 999-1017.
5. Aizawa, T. et al.: Nondestructive Evaluation of PVD/CVD TiN Coated WC/Co by Acoustic Spectro-Microscopy, *ASME*. AMD-140 (1992) 55-70.
6. Aizawa, T. and Kihara, J.: Mechanical Characterization of PVD/CVD TiN Coated WC/Co by Acoustic Spectro-Microscopy, *Journal of the Faculty of Engineering, The University of Tokyo (B)*. XLII (1993) 155-168.
7. Aizawa, T.: Quantitative Nondestructive Evaluation of Morphology in WC/Co Cermets by Acoustic Homogenization Approach, Proc. COMMP'93 (1993) 468-473.
8. Mura, T.: Micromechanics of Defects in Solids, (1987) Martinus Nijhoff Publishers.

APPLICATION OF FRACTAL ANALYSIS FOR SURFACE FLAW INSPECTION OF STEEL SHEET

Satoshi Horihata,[1] Hiroki Kashida,[1] Hajime Kitagawa,[1]
Masayoshi Satoh,[1] Hitoshi Aizawa,[2] and Toshio Tamiya[3]

[1]Department of Production Systems Engineering,
Toyohashi University of Technology
Hibarigaoka, Tempaku-cho, Toyohashi 441 Japan
[2]Mizusima Works, Kawasaki Steel Coporation
Kawasaki St. 1, Mizushima, Kurashiki 721 Japan
[3]Kawatetsu Techno-Research Corporation
Kawasaki-cho, Chuo Ward, Chiba 260 Japan

1. INTRODUCTION

In this study, fundamental investigations were made on the relationship between the normality and representative methods for Fractal Analysis, of the time series. Examined methods were an analysis using Fractal-Brownian function, hereafter called F–B method, and the coarse graining method proposed by T. Higuchi,[1] called Higuchi method. Applying the results obtained, Fractal Analysis of 'measured' time series was tried. From analysis on both 'simulated' and 'measured' data, distinctive features and optimum conditions of the both methods were made clear and the Higuchi method proved to be more suitable to the analysis for time series used in this study.

2. FRACTAL ANALYSIS OF TIME SERIES

Fractal Analysis of time series means a quantitative determination of Fractal Dimension for the intricacy, or complexity, on graphically represented pattern of non-periodic time series.[1–6]

2. 1 Analysis by F–B Method

Fractal-Brownian function, $B(t)$, is the function having the following characteristics represented by equation (1).

$$\{B(t+\tau)-B(t)\}/\tau^{H} \sim N(0,1) \qquad (1)$$

where $N(0,1)$ denotes the normal distribution of mean value=0 and variance=1. And H ($0<H<1$) is a constant. If a constant distribution, H, exists irrespective of τ, Fractal dimension, D, and H can be expressed by the following equations.

$$D = 2 - H \qquad (2)$$

$$H = \frac{\log\{|B(t+\tau)-B(t)|\}}{\log|\tau|} \qquad (3)$$

where $E\{\ \}$ denotes the expectation at a constant τ. Applying the least mean square method, H can be calculated from the ratio of $\log(E\{|B(t+\tau)-B(t)|\})/\log(|\tau|)$ for a variable τ.

2. 2 Analysis by Higuchi Method

For a given time series $X_i (i = 1, \cdots\cdots N)$, the length $\Delta L_i(\Delta t)$ between two points of samping time Δt is defined by equation (4),

$$\Delta L_i(\Delta t) = \sqrt{(F_v(X_{i+1}-X_i))^2+(F_t(\Delta t))^2} \qquad (4)$$

where F_v and F_t are functions of scale transformation, respectively. If $F_t(\Delta t) \gg F_v(X_{i+1}-X_i), \Delta L_i(\Delta t)$ is approximately equal to $F_t(\Delta t)$. Therefore, $\Delta L_i(\Delta t)$ must be defined by the following equation.

$$\Delta L_i(\Delta t) = F_v(|X_{i+1}-X_i|) \qquad (5)$$

Then, total length $L(\Delta t)$ is represented by equation (6).

$$L(\Delta t) = \frac{F_v\left(\sum_{i=1}^{N-1} |X_{i+1}-X_i|\right)}{F_t(\Delta t)} \qquad (6)$$

For the time series, a correlation between length $L(k\Delta t)$ which is carse grained by $\tau = k\Delta t$, and k which is positive integers, must be estimated. F_v and F_t, which are independent of k, need not to be calculated. Then, stead of equation (6), following equation may be defined.

$$L(\Delta t) = \frac{\sum_{i=1}^{N-1} |X_{i+1}-X_i|}{\Delta t} \qquad (7)$$

For time series, the scale of coarse graining is restricted to integer multiples of Δt. Two different methods are proposed for time series to coarse grain the time series by time interval of τ.[1] In this study, Higuchi method is applied. The following k pieces of time series sets are obtained by Higuchi method.

$$X_m(k) : X(m), X(m+k), X(m+2k), \cdots\cdots X(m+[(N-m)/k]k) \quad (m = 1, 2, \cdots\cdots, k) \qquad (8)$$

The average length $\langle L(k)\rangle$ is called time series length when coarse grained by k

$$\langle L(k)\rangle = \frac{\sum_{m=1}^{k} L_m(k)}{k} \qquad (9)$$

By coarse grain parameter k, several sets of $\langle L(k)\rangle$ are obtained. Plotting k's and $\langle L(k)\rangle$'s on a graph of logarithmic scales, absolute value of the slope of straight line denotes Fractal Dimension D.

$$D = \left|\frac{\log\langle L(k)\rangle}{\log k}\right| \qquad (10)$$

where the straight line is determined by the least mean square method.

3. PROGRAM FOR FRACTAL ANALYSIS

Fig. 1 shows a block diagram of software for Fractal Analysis. Language is C, compiled by Turbo C, and operation system is MSDOS ver.3.1. Hardware is PC9801VX. Fig. 2 shows a flow chart of time series analysis by the F–B method and the Higuchi method.

4. EXPERIMENTAL RESULTS AND DISCUSSION

Following results are obtained, and related discussions are described in this section.

4.1 Data Used for Fractal Analysis

Several kinds of 'simulated' time series are generated by a computer. For a 'simulated' time series, it is possible to obtain normally distributed data.

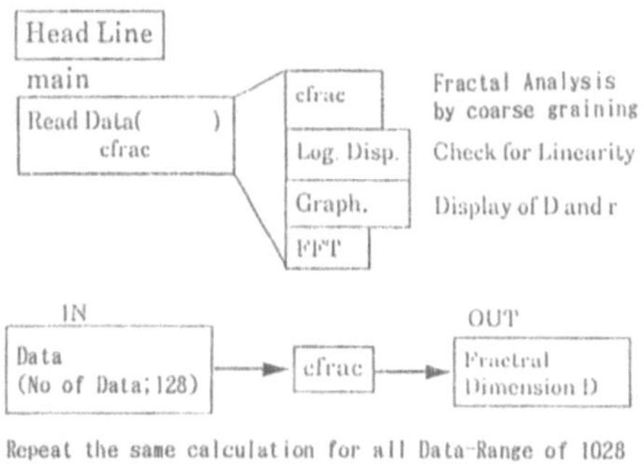

Fig. 1 Block diagram of Fractal Analysis Program.

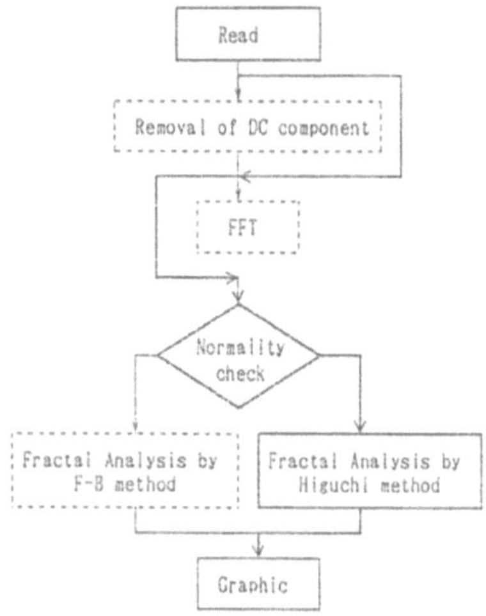

Fig. 2 Flow chart of time series analysis.

'Measured' time series were taken by an optical type surface flaw detector for cold-rolled stainless steel sheet. The measurement system and an example of scanning data are shown in Fig. 3 and Photo. 1, respectively.

Measurement of surface flaw inspection is performed by the following steps. Focused He-Ne laser beam is scanned over a test piece, which contains various kinds of defects on its surface, by a rotating mirror, and reflected beam is led to a photo-electric cell, then converted to electric signals. The signals are amplified by both pre-amplifier and main amplifier, then converted into digital versions with a A/D converter. These digital signals are recorded in the disc type recorder after memorized in a frame memory. Digital data thus recorded are read in to the computer in order to perform various analysis.

Sampling frequency of A/D converter is 12 MHz and scanning width is 264.6 mm and sampling interval is 0.2 mm. Then data size of a signal are 1323 points/scanning. It becomes evident that 'measured' time series are not always normally distributed.

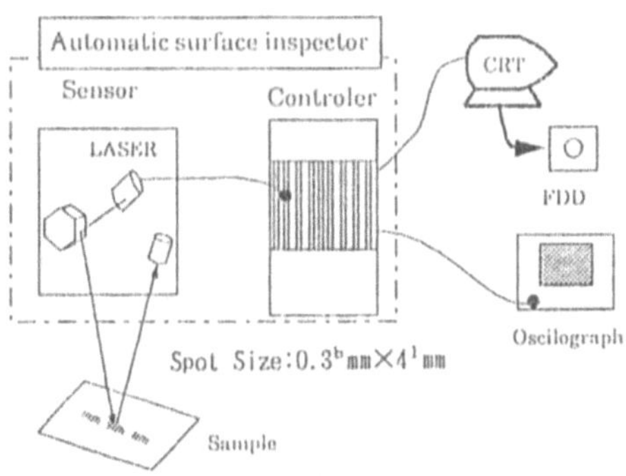

Fig. 3 Measurment system of an optical surface inspector.

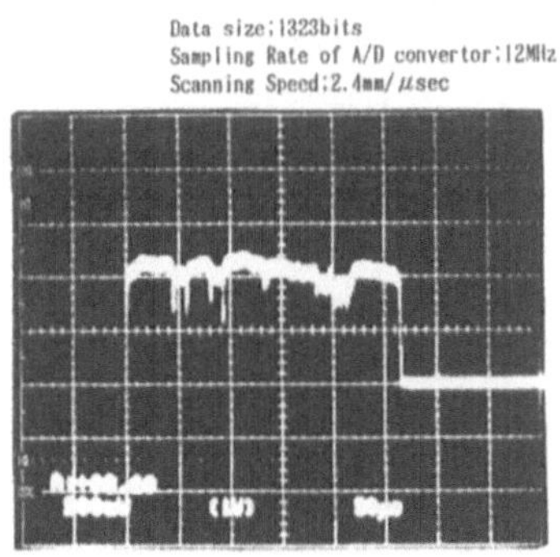

Photo. 1 An example of scanning, original, data. Defect is a Slip.

Table 1. Chi-square normality test of time series.

D	Time series	Results (significance lebel=5%)	Accept or Reject
1.5	Simulated Data	$81.9 > x^2 = 14.0$	×
	Differential Series of Above Data	$4.06 < x^2 = 14.0$	○
2.0	Simulated Data	$5.73 < x^2 = 12.5$	○
	Differential Series of Above Data	$9.65 < x^2 = 14.0$	○
Approximately 1.8	Measured Data	$6.79 < x^2 = 9.4$	○
	Differential Series of Above Data	$24.9 > x^2 = 9.4$	×

○ : Accept
× : Reject

4. 2. The Normality of Time Series

The time series is always preconditioned to be normal distribution for the analysis of F–B method. For 'simulated' time series, it is possible to obtain normally distributed data. But for 'measured' time series, it is necessary to examine whether the data is normally distributed or not. Then the normality of 'measured' time series, namely scanning data, are examined by the chi-square test. Results are shown in Table 1. From the table, it becomes clear that F–B method can not be applicable for the measured data.

4. 3. Comparison of Accuracy between F–B Method and Higuchi Method for Time Series Analysis

The accuracy of both methods were compared for normally distributed 'simulated' data of several dimensions D (D=1.2, 1.5 and 2.0) in Fractal Brownian function (Table. 2, Fig. 4).

Table 2. Comparision of Calculated Results between F–B and Higuchi Methods.

Time series		F–B Method	Higuchi Method
Simulated Data (Ds=1.2)	Dc	1.1943	1.1929
	r	0.9999	0.9999
Simulated Data (Ds=1.5)	Dc	1.5046	1.5021
	r	0.9550	0.9995
Simulated Data (Ds=2.0)	Dc	1.9071	1.9073
	r	0.8952	0.9997

Ds:Assumed Fractal Dimention, Dc:Calculated Fractal Dimension
r:Coefficient of Correlation

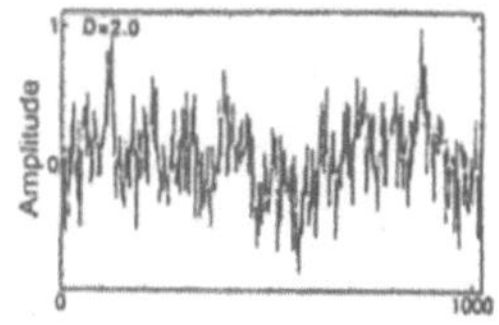

a. Simulated data generated by Fractal Brownian function

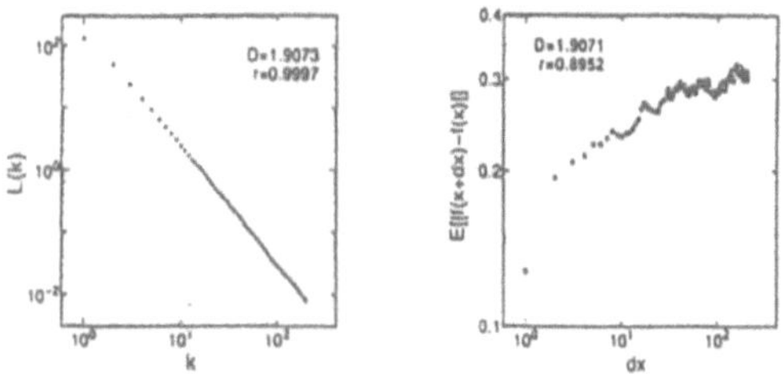

b. Analysis by Higuchi method　　c. Analysis by F-B method

Fig. 4 Comparsion of analytic results between F–B method and Higuchi method.

The Fractional Brownian function used is calculated by the following equation

$$B(t) = \frac{1}{\Gamma(H+1/2)}\int_{-\infty}^{t}(t-t')^{H-1/2}dB(t'), \quad H = 2-D$$

where, Γ denotes Gamma function.

From the comparison, it is revealed that both methods have almost the same accuracy for $D \leqq 1.5$. But on the analysis for time series of higher dimension, Higuchi method gives more accurate result than F–B method. Since D of the signals containing the defect informations has proved to be 1.8–1.9, Higuchi method is preferable to extract the informations from the signal.

4. 4. Optimum Window Width of Higuchi Method

In Fractal Analysis of time series by the coarse graining, window width ΔW and shift parameter ΔS affect the defect detectability of a defect signal. Optimum conditions of these parameters depend on the shapes of defect signals. For example, $\Delta W = 64$ is optimum window width for spike-shape defect signal, such signal as from scratch flaw on a steel sheet surface.

4. 5. Analysis of Low S/N Defect Signals by Higuchi Method

Defect signals from dent, stretcher-strain and such like flaws on the steel sheet, are very difficult to detect by any type of surface flaw detectors because of low S/N. In this study, a possibility of detecting the signal was investigated. Low S/N signals were artificially generated from an original scanning data to simulate such defect signals. At the center of the original scanning data which contains no defect signal, namely contains only noise level, the original data was replaced with a simulated defect signal of 64 bits width. The simulated data consists of the same component of original one, but the energy level of that region was lower than the noise level. Difference between both energy levels covered a range from 10% to 80% of the noise energy band of the original scanning data. As the result analyzed by Higuchi method, Fractal Dimension D of the artificially generated data is clearly different from that of original data even for the smallest energy gap of 10%. That is, the low S/N defect signal whose energy difference is equivalent to 10% of the noise energy band, can be detectable by the Fractal Analysis. For the data of 80% transition, flaw signal can be detectable by both D and correlation factor r (Fig. 5, Table. 3).

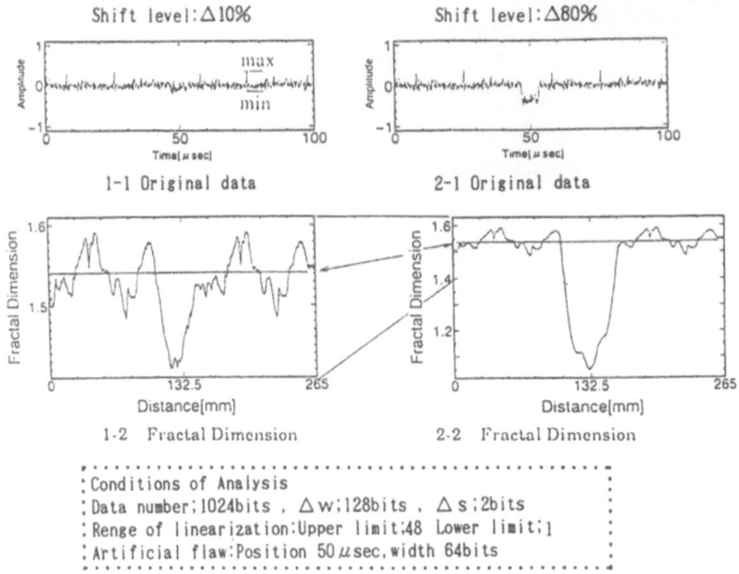

Fig. 5 Analysis of time series by simulated data.

Table 3. Results of Analyzed Artificial Flaw Signal by Higuchi Method.

Shifts	10%	25%	80%
Difference of r	—	0.06	0.15
Difference of D	0.1	0.22	0.45
Defect Detectability by Fractal Analysis	Possible	Possible	Possible
Visual Detectability of the Defect	Impossible	Possible	Possible

4. 6. Summaries on the Analysis of Flaw Data by Higuchi Method

Examples of flaw data analyzed by Higuchi method are shown in Fig. 6. Results obtained are summarized as follows:

(1) It is made clear that flaw signals of S/N≦1 which have not been detectable by traditional methods, can be extracted from Fractal Dimension of time series analysis by Higuchi method.

(2) Fractal Dimension D is more preferable to flaw detection than the correlation factor r.

(3) There are an optimum condition of the Fractal Analysis depending on the kind of surface flaw signal.

(4) On signals measured for several surface conditions, e.g. surface roughness etc., observed average Fractal Dimension and average correlation factor are considerably different. This fact definitely suggests a possiblity for the estimation of surface condition of a material from the Fractal Analysis (Table 4).

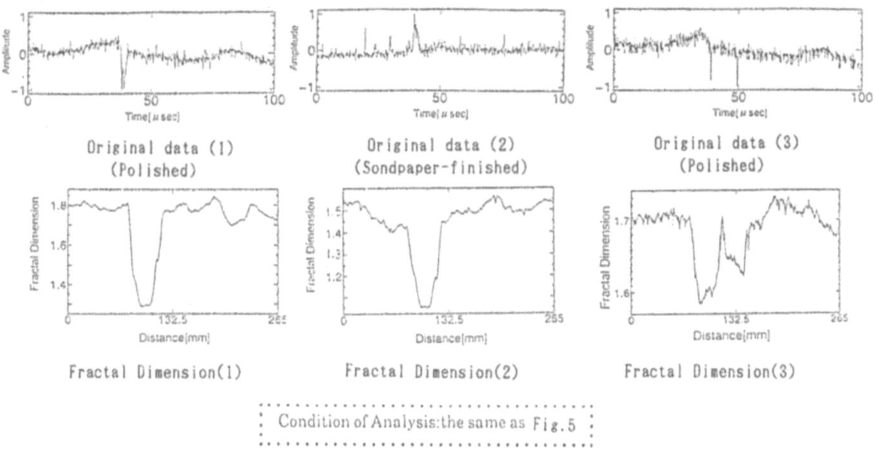

Fig. 6 Relationship between Fractal Dimension and Defect Signals.

Table 4. Analysis for time series of scanning data by Higuchi method.

Itens	Data 1 (Fig. 6. (1))	Data 2 (Fig. 6. (2))	Data 3 (Fig. 6. (3))
Correlation: r_{AVE}	0.987	0.924	0.969
Fractal Dimension: D_{AVE}	1.78	1.47	1.70
Difference of Fractal Dimension: ΔD	Slip;0.49	Slip;0.42	Slipl;0.11 Slip2;0.08
Defect Detectability by Automatic Surface Inspector	Slip;Possible	Slip;Possible	Slip1;Possible Slip2;Possible
Surface Treatment	Polished	Sandpaper-finished	Polished

5. CONCLUSION

A basic study was made on the normality and some representative methods of the Fractal Analysis for both 'simulated' and 'measured' time series. An application of the Higuchi method, one of the principal coarse graining methods, results in the new findings of possibilities for extracting several useful informations from the flaw data of very low S/N which could not be attained by traditional signal treating methods.

6. REFERENCES

1. T. Higuchi, Fractal Analysis of Time Series, *Proceeding of the Institute of Statistical Mathematics*. Vol. 37, No. 2:233 (1989).
2. T. Nagashima, Y. Nagai, T. Ogiwara and T. Tsuchiya, Time Series Analysis and Chaos, *Journal of the Society of Instrument and Control Engineers*. Vol. 9, No. 9:839 (1990).

3. B. Mandelbrot. "The fractal geometry of nature," W.H. Freeman and Company, New York (1982).
4. P.H. Otto, J. Hartmut and S. Dietmar, "Chos and Fractals", Springer-Verlag, New York (1992).
5. T. Kikuchi, S. Kiryu, S. Sato and H. Miura, Analysis of power spectrum shapes by fractal dimension : Its application for estimating spatial distributions of point scatterers, *The Journal of the Acoustical Society of Japan*. Vol. 47, No. 11:818 (1991).
6. D.L. Jaggard and X. Sun, Scattering from fractally corrugated surface, *Journal of the Optical Society of America A*. Vol. 7, No. 6:1131 (1990).

AN INTEGRATED ULTRASONIC AND EDDY CURRENT IMAGING SYSTEM

E. James Chern

Materials Branch / Code 313
NASA Goddard Space Flight Center
Greenbelt, Maryland 20771

INTRODUCTION

The perpetual demand for improved system capability and performance often dictates increased complexity in the areas of component design, material usage and fabrication processes. The failure of a single component could result in the failure of the system. Thus the concern over the integrity and reliability of the product also increases tremendously. Mission success can only be achieved by employing an advanced concurrent engineering approach which includes design for performance, design for manufacturing and design for testability, i.e., for process monitoring, quality assurance and in-service health monitoring of all critical components. Nondestructive evaluation (NDE) methods are essentially the evaluation engineering technology to meet structural reliability requirements and other quality assurance demands[1]. Imaging methods which directly correlates measured NDE parameters with component coordinates are generally the preferred way of data acquisition and presentation.

Based on the underlying physical principles of NDE imaging methods, images can be acquired in two forms, namely frame capture and pointwise, depending on whether or not the sensor or specimen is manipulated with respect to the other. Ultrasonic C-scan[2-4] and eddy current imaging techniques[5-7] are the two most widely used quantitative NDE techniques. These NDE techniques are pointwise type imaging systems[8] that have to rely on a mechanical scanner to physically maneuver a sensing probe relative to the specimen point by point in order to acquire data and generate images. In terms of hardware, there are analog and digital types of imaging systems. Early analog imaging systems offered real-time response but suffered many technical drawbacks, such as lack of data storage and processing capabilities.

Advancements in digital electronics technology and personal computers have led to the development of application specific plug-in expansion boards for personal computers. A wide range of stand-alone devices such as A/D converters, multimeters, ultrasonic pulser/receivers, eddy current instruments, etc., have been converted to this type of circuit board. A plug-in boards of this kind is usually referred to as "an instrument on a card". These advancements have also facilitated the development of PCs into concurrent engineering workstations. Many instruments can be integrated into the PC-platform by using plug-in boards, through various interface buses for stand-alone digital instruments or through an A/D card for analog instruments.

Nondestructive Characterization of Materials VI
Edited by R.E. Green, Jr. *et al.*, Plenum Press, New York, 1994

Since ultrasonic C-scan and eddy current imaging systems are based on the same mechanical scanning mechanism, the two systems can be combined using the same PC platform with a common mechanical manipulation subsystem, and integrated data acquisition software. We have developed an IBM PC-based combined ultrasonic C-scan and eddy current imaging system. In this paper, we describe the hardware components, firmware requirements and software development of the system. Details of system integration effort are presented. Advantages and disadvantages as well as possible variations and future expansion of the combined ultrasonic and ultrasonic C-scan system are also discussed.

HARDWARE REQUIREMENTS

The hardware of a typical pointwise NDE imaging system has three major subsystems: a system controller, a mechanical scanner, and NDE sensors and instruments. An embedded microprocessor or a dedicated personal computer is normally used as the system controller. The system controller is responsible for controlling the instruments, commanding mechanical movements, acquiring signals, processing data and generating images. A mechanical scanner and associated hardware fixtures such as probe and specimen holders are used to relatively scan the sensor over the area of interest on the specimen. Dedicated instruments and electronic devices are used to excite a sensor such as an ultrasonic transducer or an eddy current probe, and measure desired parameters within the corresponding signals. A typical ultrasonic C-scan imaging system utilizes an ultrasonic pulser/receiver as the excitation and measurement instrument, whereas a typical eddy current system utilizes an impedance analyzer or an eddyscope. The block diagram of the combined hardware structure is shown in Figure 1.

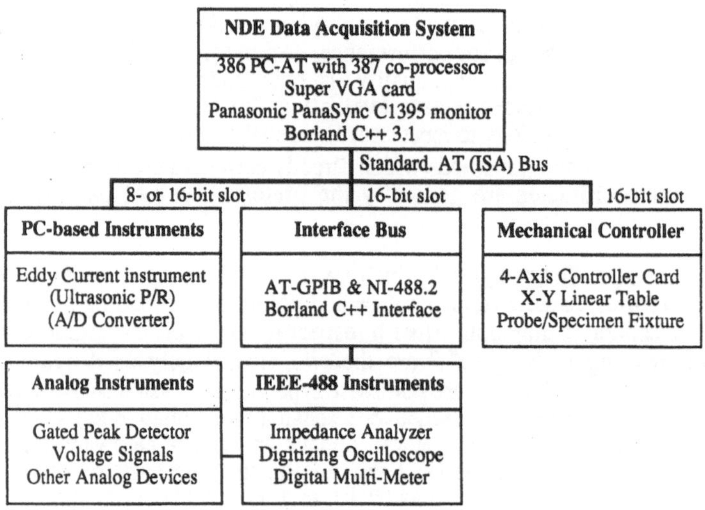

Figure 1. Block diagram of the combined ultrasonic/eddy current imaging hardware subsystem

The developed ultrasonic/eddy current imaging system has many specific hardware components. A CompuAdd 325 PC (IBM-386 Compatible) is used as the system controller. A Delta Tau Data Systems' Programmable Multi-Axis Controller (PMAC) motion controller card, Compumotor Plus drivers/motors, and a Daedal X-Y linear table with incremental linear encoders make up the mechanical scanner. For the ultrasonic instrument subsystem, a 18"W x 30"L x 6"H Plexiglas immersion tank is mounted on the surface of the mechanical table. A Panametrics 5052UA pulser/receiver with gated peak detector and a Keithley 196 System Digital MultiMeter (DMM) act as the drive and measuring instruments for the ultrasonic signals. For the eddy current instrument subsystem, a Hewlett-Packard 4194A Impedance/Gain-Phase Analyzer provides the signal drive and measurement capabilities for the eddy current signals. A SE Systems' SmartEddy 3.0 eddy current plug-in instrument is also implemented in the system for other eddy current applications.

The system controller interfaces with the scanner, using the PMAC controller card and the SmartEddy 3.0 through the industry standard architecture (ISA) PC-bus. The ultrasonic amplitude data from the System DMM or the eddy current impedance data from the Impedance/Gain-Phase Analyzer is transferred to the system controller through an IEEE-488 interface bus. This interface bus is managed by a National Instruments AT-GPIB controller card, which also resides on the PC-bus. A sketch of the system is shown in Figure 2.

SOFTWARE/FIRMWARE DEVELOPMENTS

Software Programs

System software programs were developed with the Borland C++ 3.1 compiler and other development tools. The module MAIN.C is the application program driver. Under this driver program, software routines are structured into three groups as illustrated in Figure 3. The first group is for the GPIB hardware interface bus which is based on the National Instruments' NI-488.2 driver software, and user interface display which is based on the

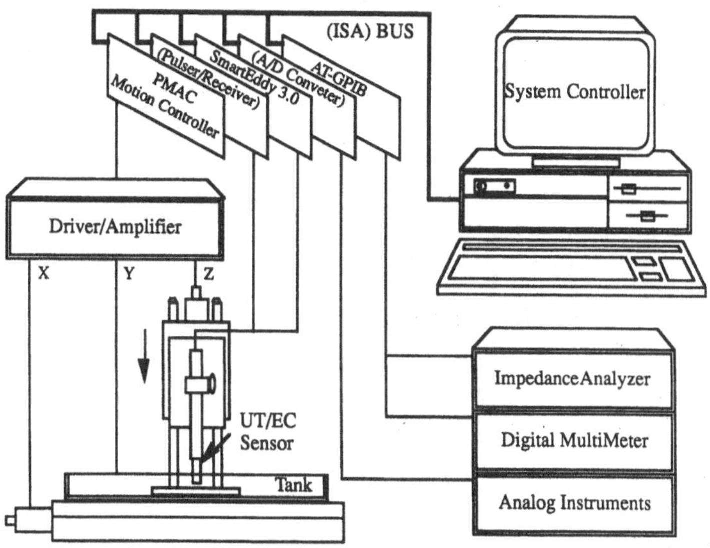

Figure 2. A sketch of the combined ultrasonic/eddy current imaging system showing the essential components

Ultra Windows user interface library package. The second group is the core of the system software. Modularized programs are devised for each engineering function of the inspection operation: initialization, inspection parameter inputs, instrument setups, mechanical control, data acquisition, image presentation, message displays, and archives. The third group is for the definition of data structures, function prototypes, and global variables as well as the color SVGA/VGA graphics drivers.

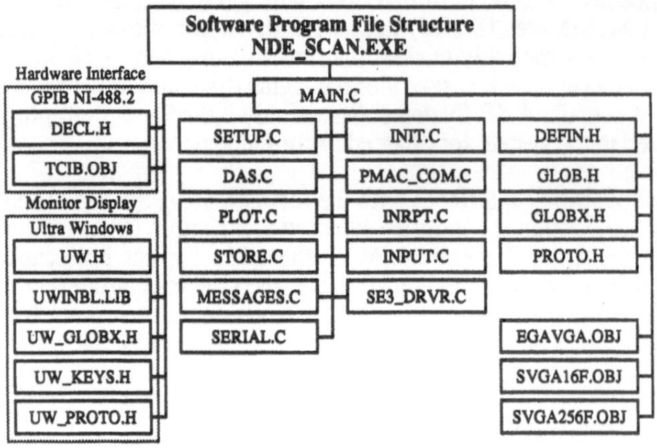

Figure 3. The file structure of the data acquisition software programs

Firmware Development

The ideal digital pointwise imaging system would command the scanner to scan at a desired speed and fetch measurements at the desired positions on the fly. A firmware approach which uses interrupts generated by the mechanical system, at the desired positions, to trigger and initiate the data acquisition routine is thus developed[9]. The control program enables the PMAC to generate interrupts in the system controller as trigger signals to initiate data reads. Interrupt handling routines were developed using the Borland C++ 3.1 programming package. The scan routine download the scan parameters to the PMAC memory along with the position of the first measurement. The remainder of the measurement positions are calculated "on the fly" during the scan, relative to the starting position.

While scanning, the PMAC microprocessor constantly compares the real-time probe position from the encoder feedback with the precalculated acquisition coordinates. An interrupt is generated by a Programmable Interrupt Controller (PIC) on the PMAC card and received by another PIC in the PC when the positional conditions are met. The PC PIC subsequently generates an interrupt to PC CPU. This interrupt is used by the CPU as the trigger signal to invoke the data acquisition routine and synchronize other events. The PMAC PIC continually generates interrupts until the scan routine is completed. Since the encoders are independent of the mechanical drives, the interrupts are generated precisely at the desired coordinates. The interrupt structure between host and peripheral PIC is shown in Figure 4.

System Operation

In principle, operators need to be trained and certified according to specific guidelines in order to properly operate the ultrasonic and eddy current equipment. In the operation of this system, an operator is also required to have some basic understanding of IBM-PC or its compatibles and the disk operating system (DOS). Familiarization with DOS and the PC will greatly facilitate system operation and other housekeeping chores. After setting up the instruments, one can follow the procedure and proceed with the desired operation. Upon execution of the executable NDE_SCAN.EXE file at the system level, the program provides the operator with a master menu for desired operations. The software program is interactive and will guide the operator to a desired operation through various functional levels and prompt for all the necessary parameters. The first level of operations include *Data Acquisition*, *Plot Data from File*, *File Management*, *Operating System*, and *Exit Program*. A sample of the menu screen is shown in Figure 5.

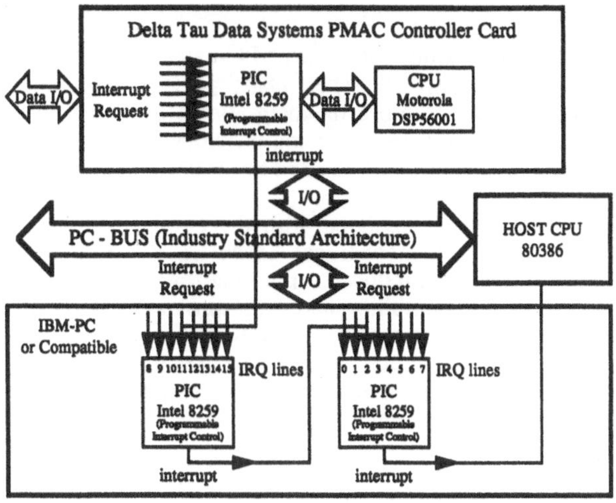

Figure 4. Interrupt structure of the Host PC and peripheral PICs for synchronized pointwise data acquisition

The second level of operation for *Data Acquisition* includes various ultrasonic and eddy current configurations such as raster scan, rotational scan, single-gate, dual-gate, etc. The third level of the data acquisition is the setup of scan parameters associated with the selected scan scheme and the execution of the scan. *Plot Data File* includes the selection and display of the stored data files. *File Management* includes file manipulation, data processing, and file conversion. *Operating System* provides a window of possibility to perform necessary DOS system operations within the NDE_SCAN program.

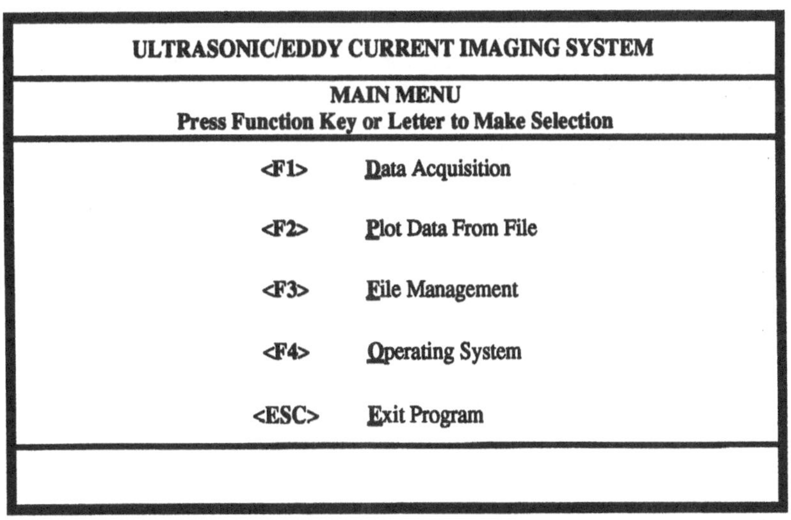

ULTRASONIC/EDDY CURRENT IMAGING SYSTEM
MAIN MENU
Press Function Key or Letter to Make Selection
<F1> Data Acquisition
<F2> Plot Data From File
<F3> File Management
<F4> Operating System
<ESC> Exit Program

Figure 5. The display of the MAIN MENU screen

CONCLUSION

In summary, we have successfully integrated a PC-based combined ultrasonic and eddy current imaging system based on off-the-shelf instruments and components. Since NDE instrumentation is a multidisciplinary field, the integration effort required certain specialized engineering expertise. The system is versatile and provides the flexibility to adapt specific inspection requirements, and expansion needs. The benefits certainly justify the means.

There are many advantages associated with the combined system: (1) eliminated duplication of the computer and mechanical hardware, (2) unified data acquisition, processing and storage software, (3) reduced setup time for repetitious ultrasonic and eddy current scans, (4) versatility and flexibility, and (5) improved system efficiency. However, there is also a minor drawback. The system can only perform either ultrasonic C-scan or eddy current imaging at a given time. As it is required to calibrate the instruments for each inspection procedure. The throughput may be compromised if it needs to change back and forth between ultrasonic and eddy current scans.

IBM-PCs and their compatibles are gaining in popularity as system controllers and host computers for many mechanical control, instrument control, and signal processing boards. IBM-PC based manufacturing, test and measuring systems are routinely being developed, introduced and implemented in various industries. This concept can be adopted to other engineering systems by integrating related PC-based instruments into one multipurpose workstation. Further detailed information regarding the integration of the combined system, including software programs, can be obtained from the author.

REFERENCES

1. E. J. Chern, "Concept of Nondestructive Evaluation," *Materials Evaluation*, Vol. 49, No. 9, pp. 1228-1235 (1991).

2. H. Berger, "A Survey of Ultrasonic Image Detection Methods," *Acoustical Holography*, Vol. 1, Plenum Press, New York, pp. 27-48 (1969).

3. G. S. Kino, "Acoustic Imaging for NDT," *Proceeding IEEE,* Vol. 67, pp. 510-523 (1979).

4. B. R. Tittmann, "Imaging in NDE," *Acoustical Imaging*, Vol. 9, Plenum Press, New York, pp. 315-340, (1980).

5. D. C. Copley, "Eddy Current Imaging for Defect Characterization," *Review of Progress in Quantitative NDE*, Vol. 3B, Plenum Press, New York, pp. 1527-1540 (1984).

6. W. G. Clark and B. J. Taszarek, "Stress Mapping with Eddy Currents," *Materials Evaluation*, Vol. 42, pp. 1272-1275 (1984).

7. E. J. Chern and A. L. Thompson, "Eddy Current Imaging for Material Surface Mapping," *Review of Progress in Quantitative NDE*, Vol. 6A, Plenum Press, New York, pp. 527-534 (1987).

8. E. J. Chern, "An Advanced Approach for Pointwise NDE Imaging," *Review of Progress in Quantitative NDE*, Vol. 12A, Plenum Press, New York, pp. 875-879, (1993).

9. E. J. Chern and D. W. Butler, "Firmware Development Improves System Efficiency," *Technology 2002* , The Third National Technology Transfer Conference & Exposition, NASA Conference Publication 3189, Vol. 1, pp. 425-434, (1992).

MICROWAVE SPECKLE CONTRAST FOR SURFACE ROUGHNESS MEASUREMENTS

Douglas A. Oursler and James W. Wagner

The Johns Hopkins University
Center for Nondestructive Evaluation
102 Maryland Hall
Baltimore, MD 21218

ABSTRACT

Microwave speckle contrast measurements can be used to perform remote determination of macroscopic surface roughness. This procedure is non-contact and can be performed through windows of certain dielectric materials. Therefore it has many potential applications in process control and corrosion detection. Measurements of root mean squared (rms) roughness up to 7.5 millimeters have been made using the microwave system.

INTRODUCTION

Speckle is a phenomenon observed throughout the electromagnetic spectrum. Optical speckle have been used in pattern correlation systems to detect in-plane surface displacements as well as surface morphology changes.[1] Optical speckle contrast measurements have been demonstrated to yield roughness information on surfaces with variations of up to approximately two thousand angstroms.[2] Although speckle contrast measurements can be used to measure microscopic roughness, many production and nondestructive evaluation (NDE) applications require testing of much greater roughnesses, even on the order of several millimeters. For such cases, microwave speckle contrast methods have been shown to be effective.

Causes of Speckle

In general, speckle arise whenever there is any disturbance in a coherent electromagnetic wavefront. The resulting scatter and phase variations cause areas of constructive and destructive interference. Speckle can have many causes but for this work the speckle in the diffuse reflection of light off of a rough surface is of interest. Figure 1 shows a "rough" surface where the variations are on the order of a wavelength. When illuminated, the energy will be reflected in two components. The direct (or specular) reflection will leave the surface at an angle identical to the angle

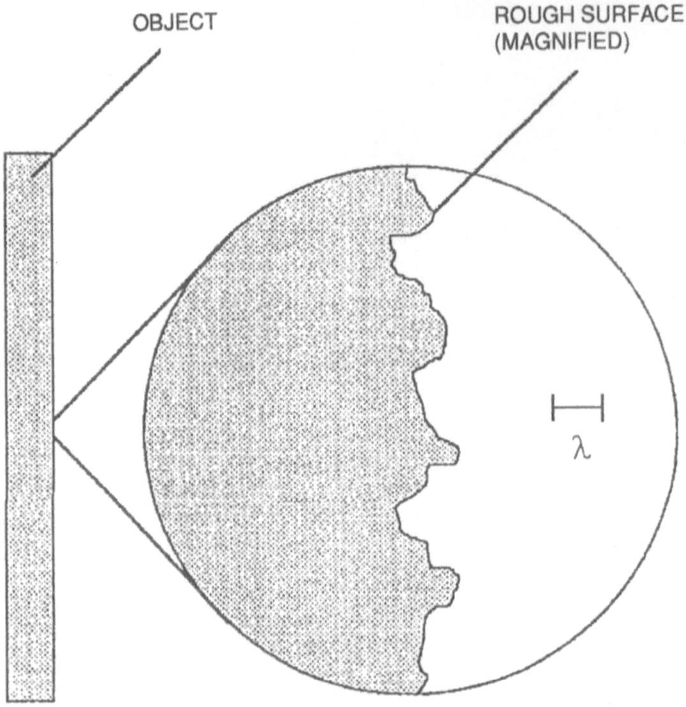

Figure 1. A magnified view of an optically rough surface. λ is the wavelength of light.

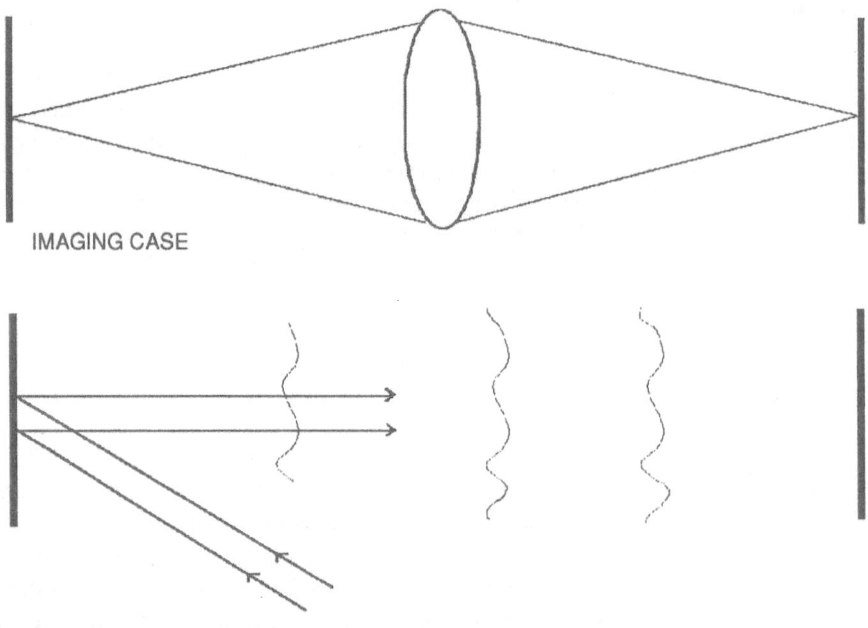

Figure 2. The imaging and non-imaging arrangements for optical speckle.

of incidence. The different facets and angles of the surface will result in a diffuse component, spraying energy in near-random directions. The intensity of the diffuse field forms a Gaussian distribution centered about the direction of the specular reflection. The width of this distribution, the full width at half maximum, is inversely proportional to the surface roughness.[3] The roughness of a surface can be characterized by the amplitude and curvature of the surface variations. The curvature is used to express the lateral dimension of the surface feature. Therefore, comparing a surface with a gravel-like appearance relative to a surface with a series of large, smooth bumps can be done by comparing their curvatures.

Optical Speckle

The development of the laser as a portable coherent light source has made it possible to use optical speckle as tool in NDE. Speckle at optical wavelengths are visualized by the eye as a grainy texture which spatially modulates the laser light. The treatment of optical speckle patterns is usually handled statistically since the illuminating beam is typically thousands of wavelengths across, illuminating millions of features.

Optical speckle arrangements are typically separated into imaging and non-imaging configurations (Figure 2).[2] In the non-imaging case a laser beam is used to illuminate the area of interest. The resultant diffusely reflected energy is made up of speckles. In the imaging case a lens is used to form an image of the surface. This image is modulated by a speckle pattern. An important difference between the two cases is that the non-imaging speckle pattern is characteristic of the total surface area illuminated. In the imaging case the speckle modulating a particular part of the image directly relate to that area on the object. There is a one to one relationship between the speckle modulated image and the object.

The statistical nature of optical speckle makes it possible to make a number of generalizations about the size and contrast of speckles. The average diameter, τ, of a single speckle is given as

$$\tau = \frac{1.22\ \lambda\ f}{d} \quad . \tag{1}$$

λ is the wavelength of light being used.[4] The variables d and f control the size of the solid angle of light that will interfere to form the speckle. For the imaging case d and f are the lens diameter and focus length, respectively. While in the non-imaging case d and f are the illumination spot size and surface to viewing plane distance.

Speckle contrast is a measure of the definition or sharpness of the speckle pattern. If a linear intensity scan is taken across a speckle field then an array, I(x), can be constructed showing intensity versus position data. For such an array the contrast is given by [5]

$$V = \frac{(<I^2> - <I>^2)^{1/2}}{<I>} \quad . \tag{2}$$

Where $<I>$ is an ensemble average of the intensity data. For random patterns, which is true in this case, this expression for contrast is equivalent to the standard deviation of the array normalized by its mean. Figure 3 shows the predicted optical speckle contrast versus surface roughness for both the imaging and non-imaging cases. In both cases the contrast begins to saturate at 100% as the root mean squared surface roughness approaches a quarter wavelength. Experimental results using optical speckle contrast have shown good agreement to these predicted results.[2] A measure of speckle contrast would therefore be a useful surface roughness gauge for roughness up to a quarter wavelength of the illuminating energy.

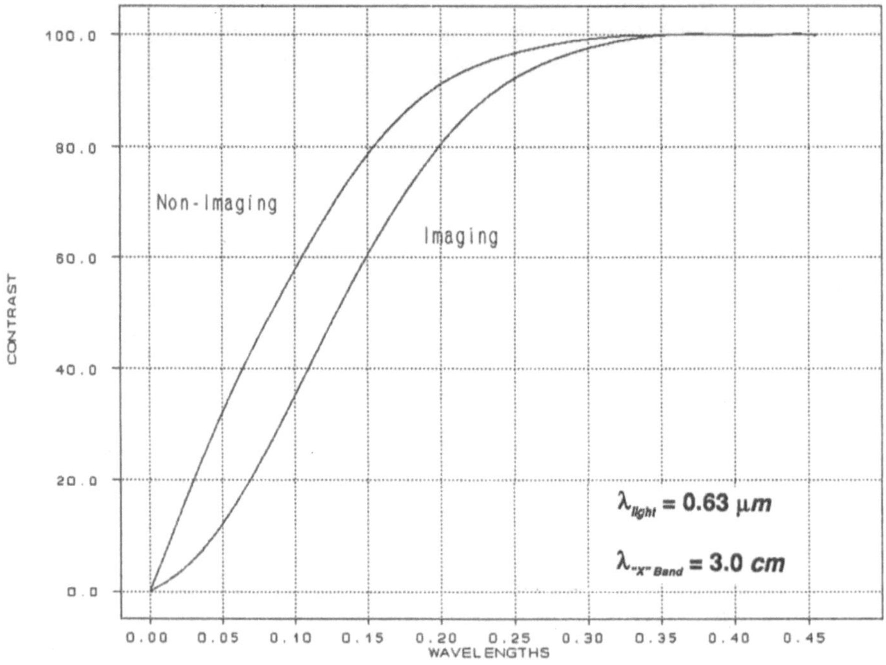

Figure 3. Optical speckle contrast versus rms surface roughness normalized to wavelength.

Figure 4. The metal spray form specimens labeled from left to right, A through D.

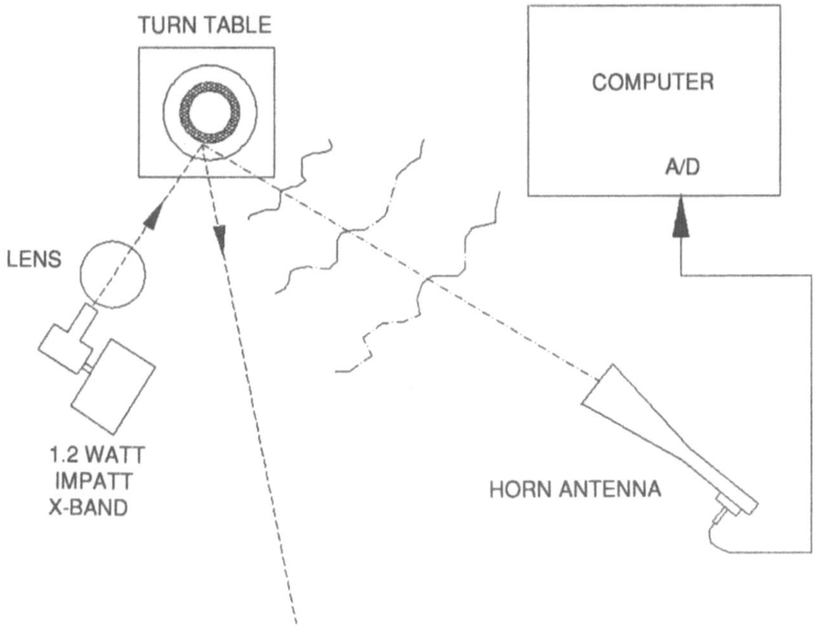

Figure 5. A schematic of the test apparatus used to make microwave speckle contrast measurements to determine surface roughness.

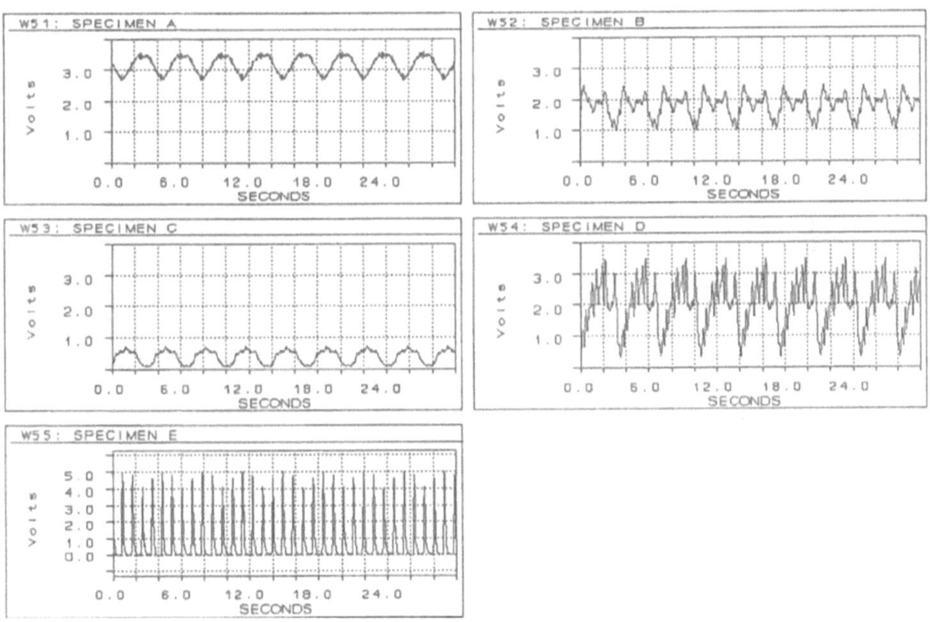

Figure 6. Thirty seconds of raw signal from the microwave receiving horn for each specimen. (Scales are identical except for Specimen E.)

MICROWAVE SPECKLE CONTRAST MEASUREMENTS

The demonstration of surface roughness measurements using microwave speckle contrast determination was performed using five test specimens, four of which were tubular specimens from a metal spray forming process. Shown in Figure 4, these specimens represent varying degrees of roughness and have been ordered from A to E based on their apparent roughnesses. Specimens C and D are end sections. The steel tube forms are visible at the top of these specimens. Specimen B is relatively smooth except for a few large bumps. Specimen A is a finished piece; the inner steel tube and outer rough surfaces have been milled away leaving a smooth surface. This specimen is five inches tall with a five inch outer diameter. Specimen E is a square steel beam added to demonstrate a roughness that will cause speckle contrast of nearly 100%.

The laboratory apparatus used to test these specimens is shown in Figure 5. It was a non-imaging system consisting of a microwave source and lens to illuminate the specimens which are placed on a motor driven turn table. The rotation of the specimens mimics the process by which they were created. The rotation speed was approximately 17 RPM or roughly 0.3 Hz. A microwave horn was carefully placed in the diffusely reflected field at an angle 40 degrees away from the specular reflection and at a distance of 1.5 meters to insure that the receiving horn aperture was smaller than the average speckle size (Equation 1). The microwave source operated in the X-band ($\lambda = 3$cm) and illuminated roughly a ten centimeter square area of the specimen and was in the same plane as the horn. The signal from the horn was fed to a preamplified analog to digital converter board in a personal computer. The a/d board sampling rate was 50 Hz.

A second method of determining the surface roughness was needed to calibrate the system because the specimens have irregular surfaces. A profilometer was constructed using a potentiometer and lever arm with a roller on the end. A spring was used to maintain constant pressure between the roller and the specimen surface. The roller was 4 millimeters in diameter and 5 millimeters wide. The specimens were rotated on the turntable while the profilometer was in contact. After several rotations, the profilometer was elevated, and data was taken again at 5 millimeter intervals.

RESULTS

Figure 6 shows thirty seconds of raw data taken from the microwave receiving horn for each specimen. Note that all of the scales are the same except for the vertical axis for Specimen E. Three interesting observations can be made from this figure. First, the d.c. offset for each of the specimens is not identical. This effect seems to be caused by a combination of surface angle and overall surface roughness. In the case of Specimen C, the angled surface is deflecting the specular reflection vertically, out of the plane of the microwave source and horn. The result is that the horn is effectively sitting at an angle greater than 40 degrees so the average intensity in the diffuse reflection is lower.

A second effect that should be noted is a wobble variation in the signal which occurs at the frequency of rotation. This low frequency variation is caused by the imperfect centering of the specimen on the turntable and in some cases by specimen end (bottom) cuts that are not perpendicular to the axis of the specimen. Finally, and most importantly, note that the amplitude of the higher frequency variations do seem to increase with surface roughness.

To determine the speckle contrast independent of signal offset and "wobble," the specimen signals were processed. Each signal was normalized by its mean and high-pass filtered at three times the rotation frequency. Figure 7 shows the processed signals

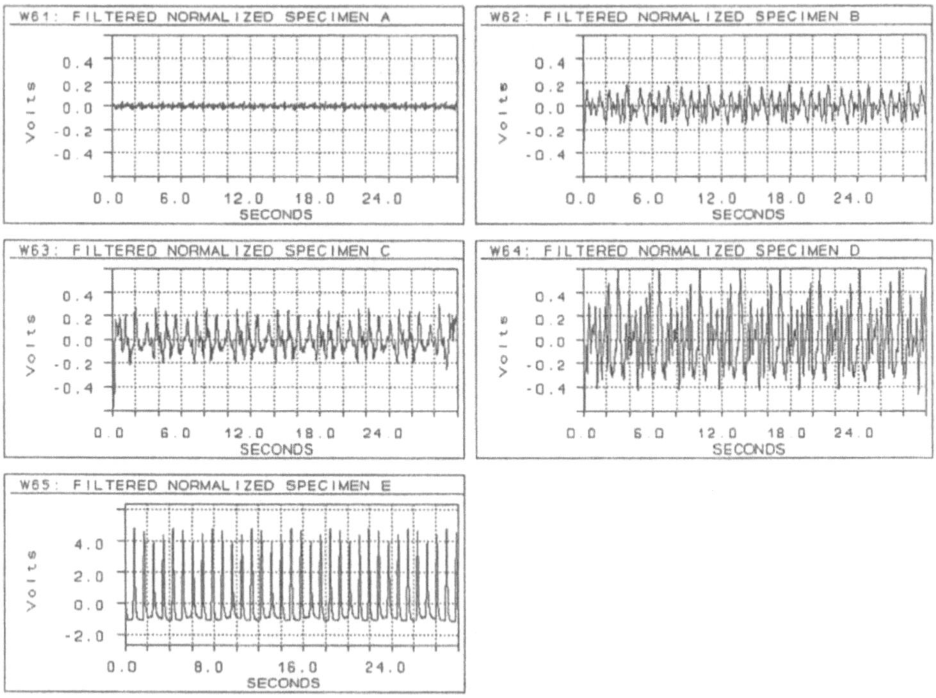

Figure 7. The filtered normalized signals for each of the specimens.

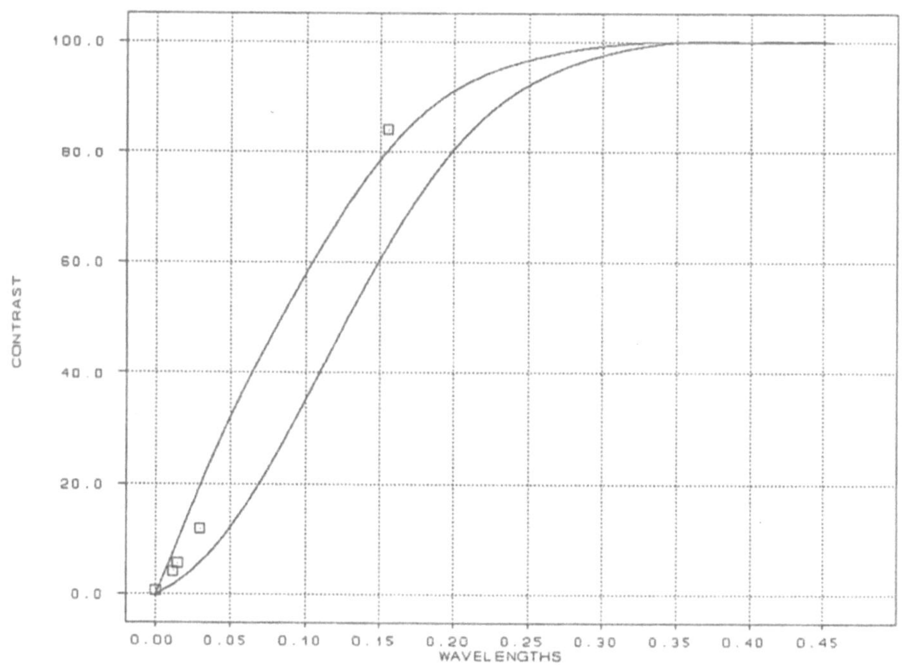

Figure 8. Speckle contrast versus rms surface roughness showing predicted (lines) and experimental data (squares). The data points are in order for specimen A through E from left to right.

for each of the five samples. Again, note the scale change on the vertical axis of Specimen E. The variation of these traces clearly increasing with apparent surface roughness.

To correlate these data with surface roughness, the standard deviation of each of these normalized filtered traces (speckle contrast) was computed. To calibrate these findings, the results of the profilometer tests were used as a standard. In the same manner as with the speckle data, profilometer data was high-pass filtered at three times the rotation frequency and assembled into a matrix. The root mean squared surface roughness was then calculated for the region illuminated by the microwave beam. In Figure 3 the predicted (visible) optical speckle contrast was plotted versus rms surface roughness in wavelengths. Figure 8 shows the results of both the predicted and actual microwave contrast measurements as a function of surface roughness. The squares represent the data points which are for Specimens A through E in order, left to right. There is good agreement between the predicted optical results and the experimental microwave results.

CONCLUSION

Speckle resulting from X-band microwaves, with a wavelength of three centimeters, have been demonstrated to give a accurate indication of surface roughnesses for variations of up to a quarter wavelength. The relationship between speckle contrast and the surface roughness has been demonstrated using specimens from a metal spray forming process. Surface roughness data were collected using a potentiometer based profilometer to calibrate the speckle contrast results.

The results indicate agreement between the predicted and the experimentally determined behavior of microwave speckle. This agreement is somewhat better than expected considering that the microwave source illuminates a spot only four wavelengths across. For this reason the microwave speckle would be expected to violate the basic hypothesis that allow statistical assumptions used for predictions in the optical case. The small illuminated spot size relative to wavelength may make it possible to characterize in greater detail the nature of the surface roughness. Preliminary results have shown that features in the frequency spectra of the speckle contrast signals can be related to the characteristic curvature of the surface roughness

The use of microwave speckle contrast methods for remote surface roughness determination may make possible a range of control sensor and NDE applications.

REFERENCES

1. F. P. Chiang and D. W. Li Decorrelation functions in laser speckle photography, J. Opt. Soc. Am. A, 5, 1023, (1986).

2. J. Ohtsubo and T. Asakura, Statistical properties of speckle patterns produced by coherent light at the image and defocus planes, Optik 45, 65 (1976).

3. J. W. Goodman, Statistical properties of laser speckle patterns, in "Laser Speckle and Related Phenomena" (J.C. Dainty, ed.). Springer-Verlag, Berlin and New York, 1975

4. F. P. Chiang and D. W. Li Laws of laser speckle movement in space, Optical Engineering, 25(5), 667-670, (1986).

5. R. K. Erf (ed.). Speckle Metrology, Academic Press, NY, 1978.

BIOLOGICAL PROCESSES IN MATERIALS SYNTHESIS

Erica R. Valdes

U.S.Army Edgewood Research, Development and Engineering Center
SCBRD-RTB
Building 3161
Aberdeen Proving Ground, MD 21010

INTRODUCTION

Biological processes are rapidly emerging as viable routes to unique and highly structured materials. These processes generally fall into one of two categories. In biotic processes biological systems apply directly to material production while in biomimetic woek non-biological production processes mimic biological methods or materials. Advantages of biological processing range from increased quality through process control on the molecular level to novel materials with desirable properties. As these materials enter the scene the materials community will face challenges associated with using and testing materials that differ significantly from the traditional.

Another important area of potential impact by biological processes is the field of sensors. Most organic systems are self-monitoring and many are self-healing. In particular, the self-monitoring capabilities may be tapped for materials or technology to be applied to nondestructive evaluation of both biologically and traditionally manufactured materials.

This article is a brief overview of some biological systems of materials interest. It is conceptual rather than quantitative with the goal being to introduce the idea of bioprocesses and generate interest in the possibilities posed by the introduction of biological methods to materials development. Requirements of specific organisms limit the properties and performance of existing biological materials. Natural environments provide no impetus for systems to evolve further than required for survival and reproduction. In many cases specific biological materials can be modified at the genetic level to achieve the specific properties of interest and biological

Nondestructive Characterization of Materials VI
Edited by R.E. Green, Jr. *et al.*, Plenum Press, New York, 1994

production methods can be designed for practicality through genetic engineering and cloning techniques.

COMPOSITES

One familiar biologically derived structural material used for generations is wood. On inspection, this is an anisotropic composite of cellulose fibrils, lignin and hemicelluloses acting mechanically in much the same way as today's advanced composites.

Introducing another very familiar biological material, bone is a strong, brittle composite with a stress-strain curve lying between those of its constituent collagen fibers and hydroxyapatite crystals. Exemplifying the potential physical properties of biocomposites, vertebrate bones are capable of holding body weight, bending and flexing, and withstanding significant impacts. Despite the brittle character of the inorganic component, bone doesn't shatter under these forces. In addition to these aspects of structural performance, it is a remarkably self-monitoring and self-healing structure. Subjected to excessive stress bone will grow, reduced stress allows it to regenerate and changes in the orientation of the stress cause the fabric ellipsoid, or statistical orientation of the hydroxyapatite crystals, to shift aligning the structural anisotropy with the anisotropy of the requirement.

Other familiar systems that fall into the category of composites of biological origin include seashells, arthropod cuticles and exoskeletons of insects and crustaceans. Some shells exhibit fracture strengths and toughnesses exceeding those of high tech ceramics, generally through a "brick and mortar" style of construction. Generally the "mortar" component is submicron in thickness and the mode of crack propagation is around rather than through the "brick" component, forcing many directional changes. Cuticles and exoskeletonns are more commonly found with fibrous reinforcement or layered sheet construction. In the systems using fibers the fibers are generally aligned either in alternating layers or engineered arrays such as helices that provide the required structural integrity.

A generic characteristic of biogenerated materials is the low temperature of material fabrication. Biological processes are almost all necessarily carried out at ambient or near ambient conditions. The resulting composites can often be described as ceramic/polymer composites with very high mineral content. Typical processes leading to these involve building an organic scaffolding that determines the geometry of the microstructure. This matrix has biochemical sites that promote crystallization and morphology to provide mineralization molds. Living cells provide the supersaturated solutions that feed particle growth. Often these steps require molecular and orientation recognition at the organic/inorganic interface.

Another important point in biological ceramics is that the materials are always formed in net shape configuration, a fact of significance in the ceramic arena as well. Genetic determination of orientation and spacing of reinforcers controls the production process providing direct placement of individual particles in the optimum locations. The genetic code also determines the final morphology of the composite structure. This is used to advantage either by studying biological structures and copying their unique structures for similar applications or by studying the genetic and biochemical processes involved and incorporating them into traditional productions.

FIBERS

In many composites the reinforcing component is fibrous. Biological composites are no exception; there are several fibers of structural importance in the plant and animal kingdoms. Materials uses of biological fibers can range from traditional textile applications to structural composites and high technology applications.

Collagen is the basic structural fiber of the animal kingdom and is widely distributed, being found in the majority of passive structural elements, pliant connective tissue and tensile structures including tendon. It is generally the main agent for controlling force distributions in mammals. Biochemically, collagen is not a unique amino acid sequence but rather a class of proteins with similar conformation and physical properties. The proteins associated with collagen are generally on the order of 30% glycine and contain hydroxyproline, an amino acid rare in other mammalian proteins. Structurally collagen shows banding under electron microscopy and when heated gets rubbery and shrinks, indicating a crystalline structure. Essentially, it is a triple helix: three coiled polypeptides coiled around each other to form a tropocollagen unit with thousands of these units being formed into strands to form a collagen fibril which is then joined with thousands of fibrils to form a fiber with specific structure. The packing of the coils often leaves 30-40nm voids that are though to play an active role in nucleation and mineralization.

Silk fibers are of particular interest in the materials field because of the combination of strength and extensibility. The most familiar silk is the product of the silk moth used in textiles however there is a wide range of silks produced by different creatures. Despite the production of silk by a wide variety of arthropod classes, all silks have evolved to very similar compositions, though the forms and specific features are tailored for particular applications. Generically, silks are extracellular fibrous proteins spun from concentrated solutions of proteins. They are designed for external uses and are typically spun from a gland through a spinnerette in which the shearing forces align molecules and crystallites, enhance crystallization and promote coagulation. Common silk fibers have a Young's modulus on the order of 10^9 to 10^{10} N/m^2, tensile strength on the order of 10^8 N/m^2 and extensibility in the range of 20%. Spider silks have received a great deal of attention as potential high performance fibers. Their energy absorption surpasses that of kevlar and may be readily incorporated into composites. It is interesting to note that many bird species have evolved nest building techniques that rely heavily on silk fibers stolen from various arthropods. Silk provides the nests with strength, adhesive binding, and in some uses a potential camouflage. Genetic engineering and bioprocessing may be applicable to the modification of properties and subsequent processing of materials based on spider silk for application in textiles and composites.

Another major class of biological fibers is the structural polysaccharides, including the common materials cellulose and chitin. In general the polysaccharides re polymers made from a single stereo isomer of the monomer, often relying on the β(1-4) glycoside bond for stable structures. Cellulose is the most abundant natural fiber and is best known for its prevalence in the textile industry. In nature it is the fibrillar component of rigid composites and the linear aggregates of cells in plants. Chitin is also commonly found as high modulus fibrillar components of rigid composites while it is almost never found in tensile materials. There are three different crystalline forms of chitin, referred to as α, β, and γ, where α-chitin is the only form found in arthropods, β-chitin is fairly common and γ-chitin is the least common. In all cases it is highly crystalline. Fibril dimensions vary from $\leq$ 2nm to $\geq$30nm depending on the origin with a typical chitin-based material being an arthropod cuticle of chitin fibrils complexed with protein.

BIOMINERALIZATION AND CERAMICS

As with composites, an important aspect of biomineralization is the formation of structural and protective ceramics at ambient temperatures and pressures, avoiding many of the expensive and difficult aspects of traditional processing. Many biomineralization systems produce mineral particles that fall in narrow size ranges. Often the minerals are surrounded by sheaths of organic which guide the mineralization, control size, act as separators or provide a composite matrix to reduce brittleness. As biological growth of ceramics is controlled by genetics and biochemistry the processes are methodical and reproducible.

Nucleation and growth regulation in biological systems are often complex, resulting in structural perfection, specified crystal sizes, exact shapes or orientations as required to provide optimal electrical, optical, magnetic or catalytic properties as required. In addition, organisms such as corals and sponges show species-specific, genetically encoded and replicated generation of single crystals with complex morphology exhibiting non-crystallographic symmetries, i.e. the complex shapes and spicules have continuous uniform lattice orientation, often with no interlaced organic phase.

POLYMERS

Natural plant rubber, cis-polyisoprene, has been used commercially for many years, as has its trans- counterpart, man-made gutta-percha. The protein rubbers of the animal kingdom include three major materials; resilin, abductin and elastin. Vertebrate protein rubber is generally in the form of fibers and sheets of elastin, with the application determining the morphology. Abductin is best known as the ligament in the hinge of bivalve mollusc shells. In the flight of insects, resilin extends to several times its resting length while it decelerates the wing and stores the kinetic energy required to reaccelerate upon reversal of the extension. The durability of these materials is impressive if the number of extension cycles required in a typical lifetime is considered.

A significant point associated with biological macromolecule is that their sizes and structures are determined genetically and as such are absolutely and reliably reproducible. This opens potential for polymeric molecules that have single or narrow ranged molecular weights. Another point is that all chemical processes of biological polymer synthesis follow stereo-specific mechanisms which are also genetically controlled. Following these routes it is possible to achieve homogeneous polymers that are designed to crystallize to specific extents and in specific microstructures. Many of the materials previously discussed as fibers fall into the class of biological crystalline polymers (i.e. collagen, silk, chitin, cellulose). Extraordinary properties can be achieved through molecular control, as evidenced by the strength and toughness of silk and the resilience and elasticity of elastin.

Mucopolysaccharides provide another class of important biopolymers. Generically they are amorphous polymers of polysaccharides and proteins in hydrated complexes. These lend the required mechanical properties to pliant biological composites and cartilage-like materials. Typically they are gel-forming viscoelastic materials.

HIERARCHICAL BIOSTRUCTURES

Another interesting aspect of materials in biology is the concept of complex hierarchical structures where each level of the hierarchy introduces a new level of structure function relationship. Detailed discussion of this topic requires involved descriptions of specific systems and will be omitted here. The general concept is that materials of singular applicability are produced by the organism, self-assembled to provide optimum properties on the molecular and crystalline level, then self-assembled into specific structural architectures to optimize the system's ability to withstand the passive structural and/or active requirements. In the extreme this process of hierarchical building can be extended to the level where a particular part of the organism fits into its role as a working part of the whole. A popular example of this concept is the intervertebral disc, where optimized materials are arranged in a complex way such that the material is in tension while the structure experiences compression.

OTHER APPLICATIONS

This discussion has concentrated on structural materials, however, there are numerous potential nonstructural applications for biotics and biomimetics. Bioadhesives have received interest in recent years with research ranging from understanding and reproducing the adhesives found in animals, including barnacles and mussels, to devising biological approaches to mass production. Another promising application of biological processes is in reaction control. Here a common approach is to use lipid vesicles, micelles or reversed micelles to control reaction rates and product sizes. Also of interest are biological systems for molecular recognition, nanodesigning of particles, thin sheet formation via biomimicry, and biocatalysis.

REFERENCES

These references are provided for additional reading on the topics addressed.

1. P.C.Rieke, P.D.Calvert, and M.Alper, eds. "Materials Synthesis Utilizing Biological Processes, Materials Research Society Symposium Proceeding Volume 174," Materials Research Society, Pittsburgh (1990).

2. S.A.Wainwright, W.D.Biggs, J.D.Currey and J.M Gosline. "Mechanical Design in Organisms," Princeton University Press, Princeton, New Jersey (1982).

ENZYME-CATALYZED POLYMERIZATION IN MICROSTRUCTURED FLUID MEDIA: THE SYNTHESIS AND CHARACTERIZATION OF NOVEL BIOMOLECULAR MATERIALS

Cigdem F. Karayigitoglu,[1] Xiaodong Xu,[1] Phillip Webb[1], Murthy Tata,[1] Nagesh S. Kommareddi,[1] Vijay T. John,[1] Richard D. Gonzalez,[1] Gary L. McPherson,[2] Madhu Ayyagari,[3] Joseph A. Akkara,[3] David L. Kaplan,[3]

[1]Department of Chemical Engineering and [2] Department of Chemistry
Tulane University, New Orleans, LA 70118
[3]U.S. Army Natick Research, Development & Engineering Center
Natick, MA 01760

INTRODUCTION

Phenol-formaldehyde based polymers have widespread use in coatings and adhesive technologies[1]. They are used in laminates, wood composites, insulating materials, flame retardants, etc. However, due to the toxicity of formaldehyde, it has become necessary to produce similar polymers that do not involve formaldehyde as an intermediate. The enzymatic synthesis of phenolic polymers is one possible route to such polymers. The feasibility of such synthesis was first demonstrated by Dordick and coworkers[2], and later by Akkara and coworkers[3] who showed that high molecular weight polymers could be synthesized enzymatically in an organic solvent system containing an 85/15 v/v dioxane/water mixture.

In our research, we attempt to enzymatically synthesize polymers and copolymers from phenols and aromatic amines. In addition to conventional uses as coatings materials, these materials have some novel properties that lead to potential applications as advanced materials. In the absence of formaldehyde, if synthesis can be done such that all linkages are through the aromatic moiety, the polymer acquires extensive conjugation. The resulting delocalization of π electrons implies that the polymer may have rather novel electrooptical properties. Thus, we are interested in using these materials as conducting polymers and as materials for nonlinear optics. Figure 1 illustrates the reaction mechanism, where the enzyme, horseradish peroxidase, catalyzes the free radical oxidative coupling of a model phenol, p-ethylphenol. The biomimetic analog of the reaction is the synthesis of lignin. Lignin, together with cellulose, forms the principal consitutent of the woody structure of higher plants. It is a bioplymer that acts as a binding agent to cement the fibers of cellulose and hemicellulose into the rigid woody structure. Lignin is synthesized through a peroxidase-catalyzed one-electron oxidation of p-hydroxy-, methoxy-substituted cinnamyl alcohol units[4]. We simply change the substrate to alkyl substitued phenols and aromatic amines to carry out reactions following very similar mechanisms.

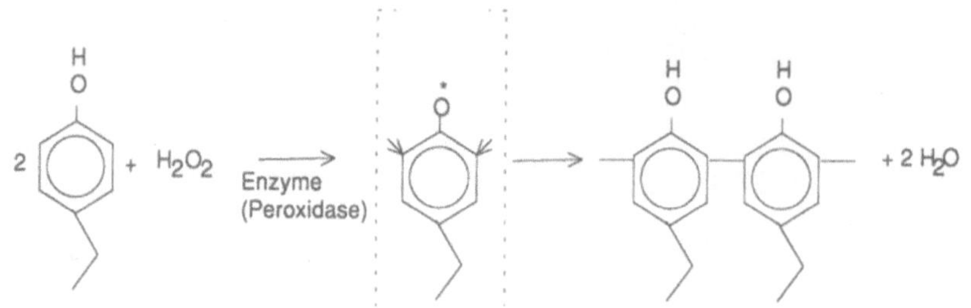

Figure 1. Reaction mechanism for peroxidase-catalyzed polymerization of p-ethylphenol.

In our research, we have focussed on conducting enzymatic polymer synthesis in microstructured fluid environments containing surfactants, water, and oil. In particular, we use water-in-oil microemulsions, or reversed micelles as they are traditionally called. Figure 2, below, illustrates the concept of a reversed micelle, where a double-tailed anionic surfactant bis(2-ethylhexyl) sodium sulfosuccinate (AOT) stabilizes droplets of water in a bulk organic phase (isooctane). It is well known that enzymes encapsulated in water droplets are catalytically active[5]; thus, we encapsulate horseradish peroxidase in the water core to carry out the synthesis.

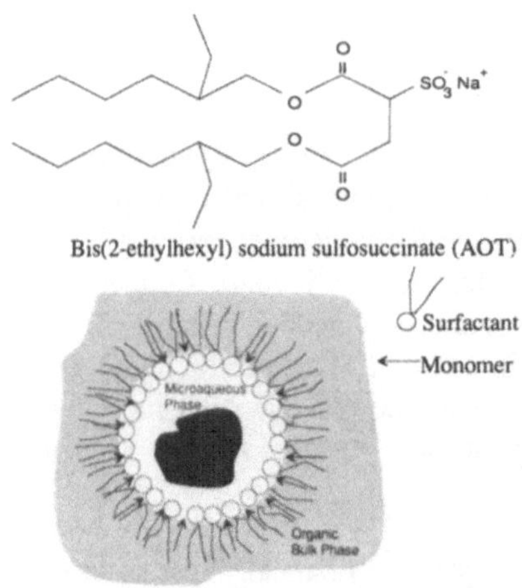

Figure 2. Depiction of an enzyme in a reversed micelle and organization of the monomers at the oil-water interface. The surfactant structure is shown at the top.

There are several reasons for conducting the synthesis of such polymers in microemulsions. The reaction is not very feasible in bulk aqueous media since the monomers are usually sparingly soluble in water and the reaction stops when the oligomers drop out of solution. Dordick and coworkers[2], and Akkara and coworkers[3] showed that peroxidase could exhibit activity in monophasic organic solvents that sustain the monomers and the growing chains in solution, thereby allowing reaction to proceed efficiently. The reversed micelle situation may however, offer, some advantages. First, the enzyme is very highly dispersed with no more than one enzyme molecule occupying a micelle. Secondly, the surfactant could interact with the monomer and sustain solubility of the growing chain[6].

But perhaps the most interesting rationale is based on the fact that these environments involve an oil-water interface. The phenolic and aromatic amine monomers are amphiphilic and we expect them to orient themselves at an oil-water interface. For example, as shown in Figure 2, the phenols would be expected to be aligned at the interface of a reversed micelle with the functional hydroxyl groups directed towards the water core. The question is whether such alignment prior to synthesis would result in the formation of novel oriented polymers with linkages primarily at positions ortho to the functional group. If that is so, the resulting polymer in addition to being conjugated, would also have all functional groups aligned. This may lead to desirable electroopical properties, and also bring about the potential of using these polymers as catalytic materials after metal chelation, or as barrier materials. Again, the biomimetic analog of polymer synthesis in such microstructured environments is the synthesis of lignin from peroxidase in the cell structure of higher plants.

Thus, our research is directed at the synthesis of such polymers in microstructured environments, at characterizing such materials, and developing applications of such materials. The results presented here are an extension of our earlier work, where we have shown the feasibility of the reaction in reversed micelles[6].

EXPERIMENTAL METHODS

The details of the experimental methods are fully described in our recent paper[6]. Essentially, the enzyme is dissolved in HEPES buffer and is added to the surfactant (AOT) and isooctane. The water/AOT molar ratio is adjusted at 15 to yield reversed micelles of droplet size approximately 8 nm diameter. The monomer is then added and reaction is initiated by the dropwise addition of H_2O_2. Care is taken to keep the H_2O_2 concentration low enough to prevent irreversible deactivation of the enzyme.

RESULTS AND DISCUSSION

Polymer Synthesis in Reversed Micelles

As described in our earlier paper, polymerization in reversed micelles is very feasible[6]. With p-ethylphenol as the monomer, we have been able to obtain almost complete conversion. Essentially all the monomer is converted to polymer and the oligomer content is low. The reaction is complete within 10 minutes and the polymer precipitates out of solution, and can be easily recovered. Figure 3 illustrates the molecular weight distribution of the polymer as obtained through gel permeation chromatography using polystyrene standards.

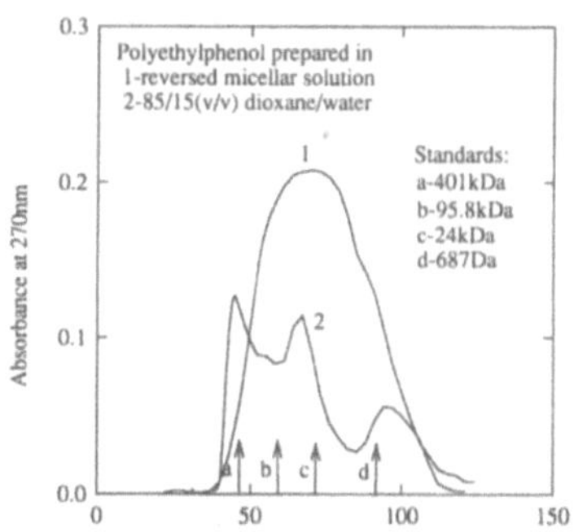

Figure 3. Molecular weight distributions of (poly) p-ethylphenol formed in reversed micelles and in organic solvents.

The polymer is entirely soluble in the GPC solvent (Dimethylformamide + methanol) and the mean molecular weight is approximately 30 kDa. Very little of the material has a molecular weight greater than 400 kDa. In contrast, the polymer prepared in organic solvents has a limited solubility in DMF/Methanol. Of the soluble fraction, a significant amount exists as chains > 400 kDa. Thus, it is our conclusion that polymers formed in reversed micelles have, on average, lower molecular weights than similar polymers formed in organic solvents

Perhaps the most interesting observation is the morphology of (poly)p-ethylphenol formed in reversed micelles. Figure 4 illustrates a scanning electron micrograph of the polymer particles, and we observe the existence of a well-defined morphology. The particles are almost perfectly spherical and of a relatively uniform size. In contrast, the polymer synthesized in organic solvents has no defined morphology as shown in Figure 5.

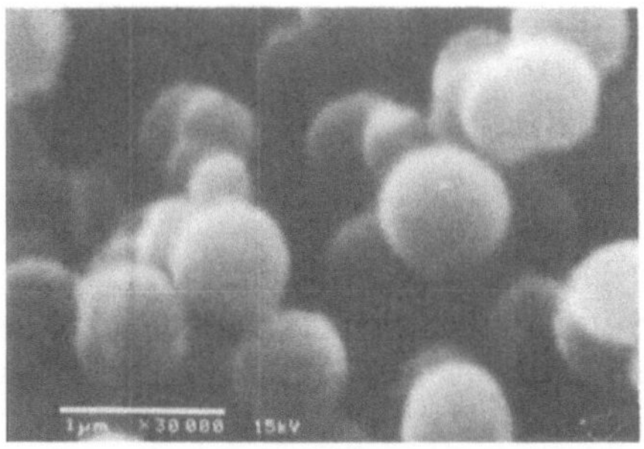

Figure 4. Scanning electron micrograph of (poly)p-ethylphenol synthesized in reversed micelles.

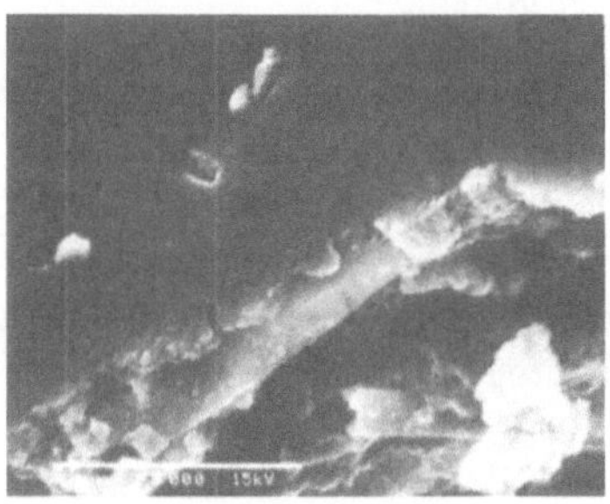

Figure 5. Scanning electron micrograph of (poly)p-ethylphenol synthesized in a dioxane/water solvent.

Hence, it is evident that the micellar environment does have a templating effect on the polymer morphology. To clarify the concept, we conducted the synthesis at the various concentrations of the surfactant and the phenol listed in Figure 6. It is interesting to note that only when the ratio of surfactant to phenol is greater than 2/1 (preferably 3/1) are spherical polymer particles formed. FTIR results indicate that the surfactant does interact with the monomer through hydrogen bonding of the surfactant carbonyls and the phenol hydroxyl groups. This tends to rigidify the interface and decrease the curvature about water. The rigidity of the interface is increased with increasing amounts of the phenol, and eventually the micelle loses its spherical structure. Dynamic light scattering results indicate that the micelle structure is lost at an AOT to phenol ratio less than unity. Thus, maintenance of micelle structure prior to reaction initiation is essential if spherical polymer particles are to be formed.

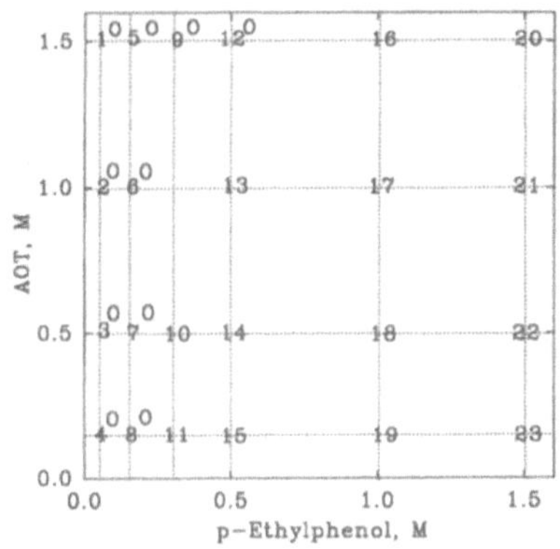

Figure 6. AOT and p-ethylphenol composition diagram illustrating compositions at which spherical polymer particles are formed.

The ability to form these relatively monodisperse spherical polymer particles in reversed micelles indicates that some new applications may be realized. First, from the viewpoint of coatings technology, it becomes feasible to disperse the material in suspensions. Additionally, we may be able to use these materials as chromatographic packings due to their higher external surface area. We have carried out some early analysis using BET methods to measure the surface area. We find that the polymers are nonporous and have surface areas of the order of 10 m^2/g, equivalent to their external surface areas. Current efforts include swelling the material using supercritical solvent to induce intraparticle porosity. Since these acidic phenols have an affinity for amines, we are developing the materials for the separation of various aromatic amines. As a base data point, the equilibrium adsorption coefficient K, for aniline adsorbed on polymers of p-ethylphenol, is 8 x 10^{-2} (mMoles aniline adsorbed/mg polymer)/(mMoles aniline in solution/ml isooctane solvent).

We are also evaluating polymers enzymatically synthesized in reversed micelles for their applications in electrooptics. The third order nonlinear optical susceptibility ($\chi^{(3)}$) is a parameter whose value can be used as a criterion to evaluate materials for applications in optical switching, laser protection, etc. $\chi^{(3)}$ for (poly)p-ethylphenol prepared in reversed micelles has been measured at 1 x 10^{-11} esu. A 1:1 copolymer of p-ethylphenol and p-ethylaniline has a $\chi^{(3)}$ of 7 x 10^{-11} esu, and p-ethylaniline prepared in reversed micelles has a $\chi^{(3)}$ of 37 x 10^{-11} esu. Thus the aromatic amines have higher values of the third order nonlinear susceptibility. Unfortunately, they are difficult to synthesize in high yield, and as high molecular weight polymers. We are conducting studies to attempt improvements in yield and molecular weight by synthesis in various microstructured fluids. Also, copolymerization with the phenols may be a route to preparing polymers with desired optical properties and with the mechanical properties that allow device fabrication.

The Synthesis of Novel Organogels

During the course of our research we realized a novel phase transition from a liquid solution to an organogel. We believe there are potential applications of the phase transition and of the organogel material and wish to report the phenomenon.

The organogel is formed by the addition of a phenolic component such as cresol or p-ethylphenol to a solution of dry AOT in a hydrocarbon such as isooctane. The novelty of the gel is its formation at surfactant concentrations as low as 0.1 M. The remarkable property of the gel is a sharp phase transition back to a liquid state when exposed to trace amounts of moisture to partially hydrate the surfactant. Simply exposing the gel to the ambient atmosphere results in moisture uptake sufficient to melt the gel.

Figure 7 illustrates compositions at which AOT and p-ethylphenol dispersed in isooctane form optically clear organogels. Some interesting aspects of the liquid-gel transition and its reverse are clearly observed.

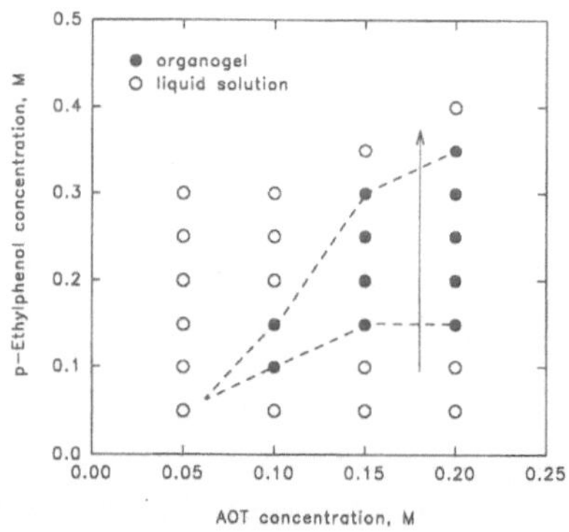

Figure 7. Phase behavior of AOT and p-ethylphenol in isooctane. The filled circles represent compositions at which the organogel is formed.

Proceeding along the direction of increasing p-ethylphenol concentration as illustrated by the direction of the arrow, it is seen that the gel is formed when the concentrations of p-ethylphenol and AOT are comparable, with gel conditions being favored at a slight excess of the phenolic component. A significant excess of either component results in a loss of the gel structure. We note that the characteristics of the gel with p-cresol or with p-ethylphenol are almost identical.

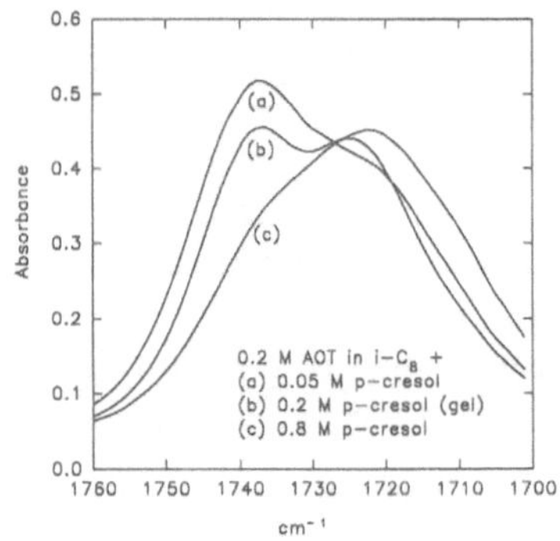

Figure 8. IR spectra of the C=O vibrations in AOT as the concentration of the phenol is increased.

Spectrum (a) of Figure 8 illustrates the C=O band at an AOT/p-cresol molar ratio of 4; the ratio decreases to 1 in spectrum (b) and to 0.25 in spectrum (c). Spectrum (b) represents the situation where the system exists as an organogel. It appears to be an inherent feature of the vibrational characteristics of the AOT C=O moieties, that two distinct peaks appear in the spectrum of the organogel. The peaks are located at 1737 and 1724 cm^{-1}. We have observed this feature in all instances of the organogels that we have prepared. Clearly, there appears to be two nonequivalent populations of the C=O moiety whose exchange rate is slow enough in the gel for their individual vibrational characteristics to be detected on the IR experimental time scale. The approximately equal intensities of the two peaks imply that the two populations may be reasonably close to each other. We explain the peak at 1724 cm^{-1} as existing due to hydrogen bonding between the C=O and the phenolic OH group. A 1:1 AOT to p-cresol ratio implies a 2:1 C=O to OH ratio. It is therefore possible that at this composition, half the carbonyl groups do not experience significant hydrogen bonding. When the phenolic component is increased significantly to a 1:2 C=O to OH ratio (Spectrum c), the system reverts from gel to liquid. The high frequency component is significantly diminished to a shoulder, and a peak is seen at 1720 cm^{-1}. This is perhaps indicative that all C=O groups now participate in hydrogen bonding with p-cresol.

Infrared spectra also show changes in C=O vibrations upon melting the gel with the addition of trace quantities of moisture. At a w_o (water/AOT molar ratio) of about 0.2, the gel reverts to a liquid solution. At this water concentration, there is one water molecule for every five AOT molecules and the surfactant is only partially hydrated. High resolution proton nmr data indicate a rigidity of the phenols in the gel structure; the phenols regain their mobility on gel melting either through the addition of water or through heating. The gel melts between 25 to 30°C, depending on the AOT and phenol concentration.

CONCLUSIONS

We have been able to enzymatically synthesize polymers in reversed micelles. The micellar environment acts as a template to produce polymeric particles with well-defined spherical morphologies. The reaction is a biomimetic analog of peroxidase based catalysis in plant cells, with the synthetic surfactants being the analog to cell-wall phospholipids. It may be possible therefore, to adjust the templating environment to modulate the morphology, and these are the objectives of ongoing studies. The spherical polymers have applications to coatings and chromatographies and are also being evaluated for applications in electrooptics.

We have also observed a transition from microemulsion to organogel. Again the behavior of such systems could mimic biological phenomena where hydrogen bonding leads to microstructure formation and changes in phospholipid packing arrangements. The remarkable properties of the organogel have potential applications. The gel may be developed into moisture sensors, or as a separation medium between reactants that come into contact upon gel melting. The gel may have its own unique electrooptical properties as a result of molecular stacking patterns. The potential applications of this material are under evaluation.

ACKNOWLEDGEMENTS

Support from the National Science Foundation (Grant BCS-9202123), the U.S. Army

(Contract DAAK60-91-C-0119) and the Tulane/Department of Defense Center for Bioenvironmental Research is gratefully acknowledged.

REFERENCES

1. A. Knopf and W. Scheib, "Chemistry and Applications of Phenolic Resins", Springer-Verlag, New York (1979).

2. J. S. Dordick, M.A. Marletta, A.M. Klibanov, Polymerization of phenols catalyzed by peroxidase in nonaqueous media, *Biotechnol. Bioeng*.30: 31 (1987).

3. J. A. Akkara, K. J. Senecal, and D.L. Kaplan, Synthesis and characterization of polymers produced by horseradish peroxidase in dioxane, *J. Polym. Sci.* 29:1561 (1991).

4. E.A. Pease and M. Tien, Lignin-degrading enzymes from the filamentous fungus *Phanerochaete crysosporium*, *in*: "Biocatalysts for Industry," J.S. Dordick, ed., Plenum Press, New York (1991).

5. P.L. Luisi, Enzymes hosted in reverse micelles in hydrocarbon solvents, *Angew. chem., Int. Ed. Engl.* 24: 439 (1985).

6. A.M. Rao, V.T. John, R.D. Gonzalez, J.A. Akkara, and D.L. Kaplan, Catalytic and interfacial aspects of enzymatic polymer synthesis in reversed micellar systems, *Biotechnol. Bioeng*. 41:531 (1993).

MECHANICALLY SENSITIVE ION CHANNELS: BIOLOGICAL MODELS FOR NANOSCALE STRESS SENSORS

Frederick Sachs

Department of Biophysical Sciences
SUNYAB
Buffalo, NY 14214

INTRODUCTION

Living organisms are continuously engaged in non-destructive testing, measuring themselves and their environment. Survival of an organism depends critically on the quality of its sensory systems used to obtain food, prevent injury, make repairs and reproduce. Organisms have evolved senses tuned to energies derived from photons, chemicals, electrical potentials and mechanical stress. These modalities are used at all levels from single cells to multicellular organisms. At the systemic level we are all familiar with the conscious senses of sight, hearing, touch and smell. These senses feed information to the central nervous system (CNS). Additional information is sent to the CNS from receptors in muscles and joints to permit coordinated movements. Sensors in the internal organs of the body inform the CNS of the status of internal machinery. Some signals are sent as hormones by the blood stream and extracellular fluid circulation rather than by the nervous system. At the level of individual cells, sensory systems are used for feedback to maintain cell integrity. For example, stretching a muscle cell causes it to increase its contractile proteins. In devising smart materials, we may learn to emulate the multitude of feedback systems that characterize living organisms. In this article, I will focus on one sensory system that is used to transform mechanical stress into and electrochemical output. These transducers are called mechanosensory ion channels. For those interested in more details, several recent reviews are available[1-3]

Cells are covered with an insulating bilayer membrane composed of phospolipids, which are fatty molecules with hydrophilic heads that face both the extracellular and the intracellular spaces (Fig. 1)[4]. This insulating membrane allows the cells to build up non-equilibrium distributions of water soluble components including ions and metabolites. To control the flow of materials across the membrane, cells have evolved a series of pumps, exchangers and channels that are embedded in this insulating layer. Pumps are the prime movers that use chemical energy from ATP or photons (plants) to move molecules across

Nondestructive Characterization of Materials VI
Edited by R.E. Green, Jr. *et al.*, Plenum Press, New York, 1994

the membrane against the electrochemical potential. The Na/K pump is an important example of this kind of transporter. It uses one ATP molecule to move two K ions from the extracellular space to the cell interior while simultaneously transporting three Na ions from the cell interior to the extracellular space. Exchangers are molecules that use the energy of the existing gradients to transport specific molecules, such as amino acids and sugars, one-by-one across the membrane. Ion channels are gated pores, selective for certain ions, that use existing gradients to transport ions rapidly across membranes. Ion channels are used to generate action potentials in nerve and muscle, to create synaptic connections and to control a variety of other electrical functions in cells. Channels can be biased in the open or closed conformation by energy sources such as electrical fields, chemical transmitters and drugs and mechanical stress.

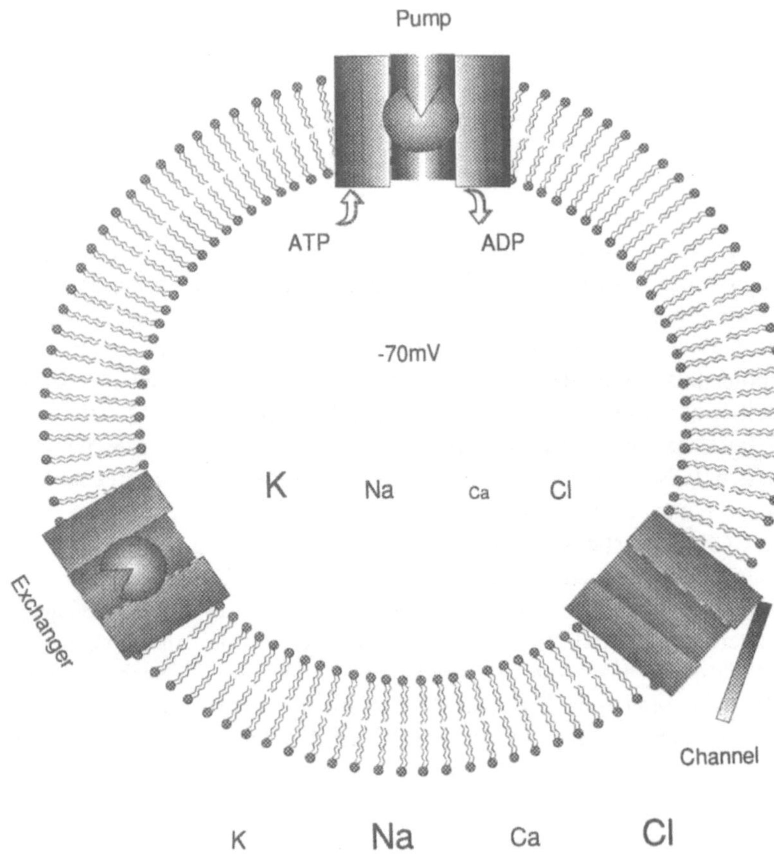

Figure 1. Cartoon of a cell showing the major transport systems and the major ionic gradients and transmembrane voltage. The concentrations of K, Na and Ca are suggested by the size of the typeface, and typical values are for the extracellular phase (in mM): Na 125, K 2, Ca 2; Cl 80; for the intracellular phase, Na 10, K 130, Ca 10^{-4}.,Cl 2. The osmotic imbalance is largely accounted for by intracellular impermeant anions such as proteins.

Patch Clamping

Ion channels have the delightful property that they can move ions at such a high rate that the current from single channels is readily observable. Transport numbers for some channels are as high as 10^7/sec, producing currents in the picoampere range. These currents can be recorded with a technique called patch clamping[5]. A glass pipette with a tip opening of about 1 μm, and filled with saline, is pressed against a cell, and by applying suction to the pipette, a small bleb of membrane is drawn into the tip. The membrane sticks so tightly to the glass that there is almost no current flow between the two. The resulting high impedance seal has such low current noise that with a low noise current amplifier connected to the back of the pipette, currents from single channels are visible. With modern amplifiers, the activity of single channel molecules can be recorded over a time span ranging from 10μs to days. Figure 3 shows an example of single channel data obtained from a chick heart muscle cell.

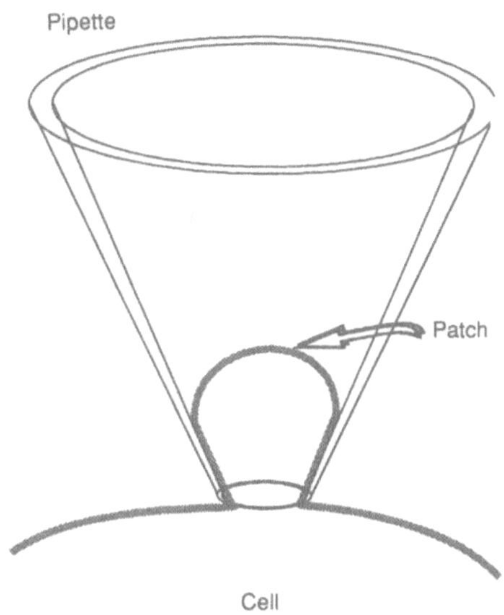

Figure 2. Diagram of a patch clamped membrane.

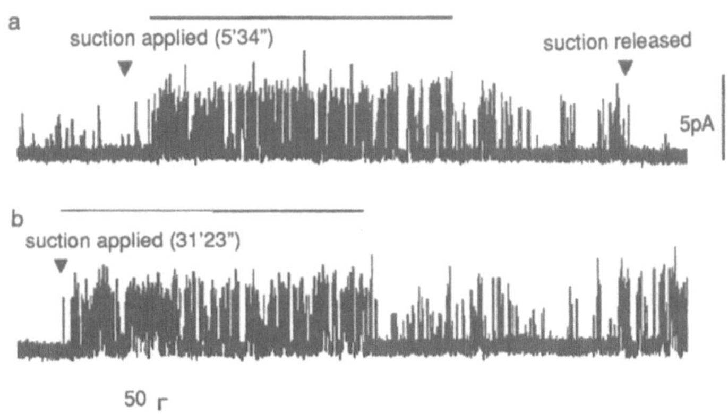

Figure 3. Single channel data obtained from a patch of membrane on a chick heart muscle cell. Inward going currents associated with open channels appear as upward going spikes. These channels are mechanically sensitive and tend to open when suction in the pipette is increased, increasing membrane tension. The data trace is 20s long.

General Properties of Mechanically Sensitive Ion Channels (SACs). MSCs were discovered in tissue cultured skeletal muscle cells using single channel patch clamp recording[6] and have since been found in both the plant and animal kingdoms and in the cells of most tissues of the higher animals. Most of them open with increasing membrane tension (stretch -activated channels, SACs), but a few are tonically active and close with increasing tension (stretch-inactivated channels, SICs)[7]. In at least one case, the channels are also sensitive to the sign of the patch curvature[8]. In animal cells, the channels tend to be either generic cation channels or potassium selective channels[9]. They form a family generally distinct from known channels, i.e. most channels are not mechanically sensitive. In skeletal muscle, nicotinic acetylcholine activated channels, ATP-inactivated K-channels (KATP) and Ca-activated K-channels (CaK) are not stretch sensitive[6]. Some channels previously identified by other aspects of gating may prove to be stretch sensitive. There is a recent report of a stretch sensitive KATP channel in atrial cells[10], a stretch-sensitive serotonin-activated channel in molluscan neurons[11]and a stretch-sensitive CaK channel in smooth muscle[12]. Sodium channels are also not stretch sensitive (R. Horn, personal communication). In the bacteria *E. Coli*, the SA channels have conductances and selectivities like porins, but they are not porins since porin-free mutants still show SA channel activity[13]. Although some channels may exhibit weak stretch sensitivity, much as the nicotinic acetylcholine channels displays weak voltage sensitivity, SACs are stretch-sensitive in the sense that the probability of being open can be reversibly shifted by three or four orders of magnitude. Most SACs are also weakly voltage dependent, tending to open with depolarization[9].

Ionic Selectivity. The ionic selectivity of the SA channel family is variable, just as in the case of voltage-activated or ligand-activated channels. In the animal cells, the most common forms are cation selective and potassium selective. The cation channels will pass monovalents and divalents such as Ca^{+2} and Ba^{+2} . Because of their ability to pass Ca^{+2} , the effects of SA channel activation are potentially complicated. Even under voltage clamp conditions, incoming calcium may activate other channels, such as calcium activated chloride channels[14]as has been evoked in the regulation of cell volume[15].

Channel Activation. SA channels occur at a density of about $1/\mu m^2$ and are activated by membrane tensions estimated to lie in the range of 0.5-3 dyn/cm[9]. A major source of uncertainty comes from lack of knowledge of the patch dimensions. The independent variable in virtually all SAC experiments is the pressure applied to a patch pipette, not membrane tension. Membrane tension is related not only to the transmural pressure, but to the radius of curvature, which itself is pressure sensitive[16]. Additionally, SAC gating seems to be sensitive to the history of the patch deformation[17]. With continued flexing, channels may lose the capability for inactivation or activation.

The reason SACs are sensitive to tension is that in some sense the open state(s) is larger than the closed state(s) so that the applied tension favors the open state more than the closed state. The conductance of the channel is independent of the applied tension, implying that either the ion conducting pore of the channel is stiff or that the pore is isolated from the stresses that cause opening and closing of the channel. By measuring the probability of being open as a function of tension, one can trace out the Boltzmann relationship of the channel and taking the free energy as $T\Delta A$ (tension * the difference of the in-plane area between the open and closed channel), it is possible to calculate that the change in channel area must be about $400A^2$[18].

Pharmacology of SACs: There are no selective blockers for SACs. Gd^{+3} , a lanthanide, is a moderately good blocker (10-20 μM) for many SACs[19]. Hamill and co-workers have shown that amiloride and some of its derivatives will block oocyte SACs at concentrations of 20-100μM[20]. Needless to say, these drugs are not specific for SACs, although Hamill et all have proposed that since the spectrum of reactivities of the amiloride derivatives varies with the target pathway, it may be possible to characterize macroscopic currents as arising from SACs by analyzing the response to a number of different compounds[21]. Van Wagoner has recently reported a K^+ selective SAC in rat atrial cells that is blocked by inhibitors of the K-ATP channel[10] and Kim reported a mechanosensitive channel that can also be activated by arachidonic acid[22]. Bear reported that SACs in liver cells are activated by extracellular ATP[23] and we have observed similar effects on some K^+ selective SACs in heart cells. Martinac et al. found that some surface active agents, such as chloropromazine, activated SACs while others, such as trinitrophenol, turned them off[24] and suggested that SACs respond to changes in membrane curvature. A similar response to amphipaths was observed for a volume sensitive Cl current, thought to be due to channels, in canine cardiac cells[25]. A similar effect of fatty acids on a K^+ selective SAC was observed in gastric smooth muscle cells[12].

Some of the most useful channel reagents for channels have come from plant and animal toxins -- α–bungarotoxin from snake venom[26], tetrodotoxin from puffer fish[27], charybdotoxin from scorpions[28], conotoxins from snails[29], FTX from funnel web spiders[30], etc. Recently, we have found a spider venom that is capable of blocking SAC activity in several cell preparations. Work is currently in progress to purify the active component from the venom. With a specific inhibitor, it would be simple to check on the role of SACs under physiological conditions.

Physiological role of SACs: There are many possible roles for SACs in living systems. In the sensory nervous systems that report on the external world, they can account for the sense of touch and vibration, hearing, and local gravity. Within the body, they can provide the feedback of joint position and muscle tension required to permit coordinated movement of the voluntary musculature. Within the autonomic system they can sense filling of the hollow organs to regulate blood pressure, filling of the stomach, bladder, etc. Some of the mechanical senses provide feedback through hormonal loops. Most conspicuous of these are the renin system in the kidney and the ANP system in the heart, both of whom act on the kidney to regulates blood volume. Cells themselves are sensitive to mechanical forces. Heart cells increase their rate of beating with mechanical stress, bone cells increase the deposition of bone, (an effect known as Wolff's Law), muscle cells grow bigger, etc. Even free living one celled organisms, such as *Paramecium,* have complicated behavioral responses to mechanical inputs.

Despite the plethora of sites for SAC involvement, proof of their physiological role has been slow in coming. This is primarily because there are no specific reagents to block SAC channel activity. Further, most experiments on SACs have been done with the patch clamp technique, and there is evidence that the process of forming a seal to the cell membrane may affect the activity of channel within the patch.

In 1991 Morris and Horn[31] reported that they could readily record SAC activity in cell-attached patches on molluscan neurons, but could not find corresponding whole cell currents using a variety of (unquantified) stimuli. (They did point out in a footnote that

GH3 cells did not have the same problem). Their general conclusion, that SAC activity is an artifact of patch formation, was quickly shown to be false by experiments in which whole cell currents were accurately predicted by single channel currents[32]. Without relying solely on patch clamp techniques, Sigurdson et al[33] presented evidence that cationic SACs are responsible for the elevation of Ca_i^{+2} caused by the mechanical distortion of tissue cultured heart cells. The response could be blocked by i) removing of extracellular Ca^{+2}, ii) applying Gd^{+3} at concentrations that blocked the SACs and iii) by growing cells under conditions that suppressed the expression of SACs.

SUMMARY

There is little controversy about the existence of mechanical effects on cells. With the exception of mechanically sensitive ion channels, no other direct transducer is known. This is not to suggest that SACs are the only mechanical transducers, but because of their high turnover rate, the one most easily studied. We expect to find other stretch sensitive enzymes in the membrane and within the cytoskeleton of the cell[34]

ACKNOWLEDGMENTS

This work was supported by grants from the USARO DAAL0392G0014 and the NSF 1BN9009675.

REFERENCES

1. M Sokabe, F Sachs (1992): In: Towards a molecular mechanism of activation in mechanosensitive ion channels. Advances in Comparative and Environmental Physiology, v10. F Ito, ed., Springer-Verlag, Berlin, 55.

2. F Sachs (1992): In: Stretch sensitive ion channels: an update. Sensory Transduction. DP Corey, SD Roper, eds., Rockefeller Univ. Press, Soc. Gen. Physiol., NY, 241.

3. C Morris (1990): Mechanosensitive Ion Channels. *J. Mem. Biol. 113*:93.

4. B Hille (1984): Ionic channels of excitable membranes. Sinauer Associates, Sunderland, MA.

5. OP Hamill, A Marty, E Neher, B Sakmann, and FJ Sigworth (1981): Improved patch-clamp techniques for high-resolution current recording from cells and cell-free membrane patches. *Pflugers Arch. 391*:85.

6. F Guharay and F Sachs (1984): Stretch-activated single ion channel currents in tissue-cultured embryonic chick skeletal muscle. *J. Physiol. (Lond) 352*:685.

7. CE Morris and WJ Sigurdson (1989): Stretch-Inactivated Ion Channels Coexist with Stretch-Activated Ion Channels. *Science 243*:807.

8. CL Bowman, JP Ding, F Sachs, and M Sokabe (1992): Mechanotransducing ion channels in astrocytes. *Brain Res. 584*:272.

9. F Sachs (1988): Mechanical transduction in biological systems. *Crit. Rev. Biomed. Eng. 16*:141.

10. DR Van Wagoner (1991): Mechanosensitive ion channels in atrial myocytes. *Biophys. J. 59*:546a,(Abstract).

11. DH Vandorpe and CE Morris (1991): Stretch activation of the S channel in mechanosensory neurons of Aplysia. *The Physiologist 34*:104.

12. T Hisada, RW Ordway, MT Kirber, JJ Singer, and JV Walsh,Jr. (1991): Hyperpolarization-activated cationic channels in smooth muscle cells are stretch sensitive. *Pflugers Arch. 417*:493.

13. B Martinac, M Buechner, AH Delcour, J Adler, and C Kung (1987): Pressure-sensitive ion channel in Escherichia coli. *Proc. Natl. Acad. Sci. USA 84*:1.

14. J Ubl, H Murer, and H-A Kolb (1988): Hypotonic shock evokes opening of Ca2+-activated K channels in opossum kidney cells. *Pflugers Arch 412*:551.

15. O Christensen (1987): Mediation of cell volume regulation by Ca2+ influx through stretch-activated channels. *Nature 330*:66.

16. M Sokabe, F Sachs, and Z Jing (1991): Quantitative video microscopy of patch clamped membranes - stress, strain, capacitance and stretch channel activation. *Biophys. J. 59*:722.

17. OP Hamill and DW McBride (1992): Rapid adaptation of single mechanosensitive channels in Xenopus oocytes. *Proc Natl Acad Sci U S A 89*:7462.

18. F Sachs and H Lecar (1991): Stochastic models for mechanical transduction. *Biophys J 59*:1143.

19. X-C Yang and F Sachs (1989): Block of Stretch-Activated Ion Channels in Xenopus Oocytes by Gadolinium and Calcium Ions. *Science 243*:1068.

20. JW Lane, D McBride, and OP Hamill (1991): Amiloride Blocks the Mechanosensitive Cation Channel in Xenopus Oocytes. *J Physiol (Lond) 441*:347.

21. OP Hamill, JW Lane, and DW McBride (1992): Amiloride: a molecular probe for mechanosensitive ion channels. *Trends Pharmacol Sci 13*:373.

22. D Kim (1992): A mechanosensitive K+ channel in heart cells - activation by arachidonic acid. *J. Gen. Physiol. 100(6)*:1021.

23. CE Bear and C Li (1991): Calcium Permeable Channels in Rat Hepatoma Cells are Activated by Extracellular Nucleotides. *Am. J. Physiol. 261*:C1018.

24. B Martinac, J Adler, and C Kung (1990): Mechanosensitive Channels of E. coli Activated by Amphipaths. *Nature 348(6298)*:261.

25. GN Tseng (1992): Cell swelling increases membrane conductance fo canine cardiac cells: evidence for a volume-sensitive Cl channel. *Am. J. Physiol. 262*:C1056.

26. B Chatrenet, O Tremeau, F Bontems, MP Goeldner, CG Hirth, and A Menez (1990): Topography of toxin-acetylcholine receptor complexes by using photoactivatable toxin derivatives. *Proc. Natl. Acad. Sci. 87*:3378.

27. IR Josephson and N Sperelakis (1989): Tetrodotoxin differentially blocks peak and steady-state sodium channel currents in early embryonic chick ventricular myocytes. *Pflugers Arch 414*:354.

28. c Miller, E Moczydlowsdki, R Latorre, and M Phillips (1985): Charybdotoxin, a protein inhibitor of single Ca+2-activated K+ channels from skeletal muscle. *Nature 313*:316.

29. H Schweitz, JF Renaud, N Randimbivololona, C Preau, A Schmid, G Romey, L Rakotovao, and M Lazdunski (1986): Purification, subunit structure and pharmacological effects on cardiac and smooth muscle cells of a polypeptide toxin isolated from the marine snail Conus tessulatus. *Eur. J. Biochemistry 161*:787.

30. R Llinas, M Sugimori, J-W Lin, and B Cherksey (1989): Blocking and isolation of a calcium channel from neurons in mammals and cephalopods utilizing a toxin fraction (FTX) from funnel-web spider poison. *Proc. Natl. Acad. Sci. 86*:1689.

31. CE Morris and R Horn (1991): Failure to elicit neuronal macroscopic mechanosensitive currents anticipated by single-channel studies. *Science 251*:1246.

32. MC Gustin (1991): Single-Channel Mechanosensitive Currents. *Science 253*:800.

33. WJ Sigurdson, A Ruknudin, and F Sachs (1992): Calcium imaging of mechanically induced fluxes in tissue-cultured chick heart: role of stretch-activated ion channels. *Am. J. Physiol. 262*:H1110.

34. PA Watson, KE Giger, and CM Frankenfeld (1991): Activation of adenylate cyclase during swelling of S49 cells in hypotonic solution si not involved in subsequent volume regulation. *Mol C Bioch. 104*:51.

CONVERSION OF AVAILABLE ENERGY FORMS INTO DESIRED FORMS BY A BIOLOGICALLY ACCESSIBLE MECHANISM

Dan W. Urry

Laboratory of Molecular Biophysics
The University of Alabama at Birmingham
VH300
Birmingham, Alabama 35294-0019

INTRODUCTION

The energy conversions considered here are achieved by the appropriate design of polypeptides or what may also be called protein-based polymers or model proteins. As these involve biomolecular compositions derivable from biological production, the energy conversions are accessible to living organisms. In spite of this, it is not correct to assume that the mechanism was discovered by studying some biological energy conversion process. Quite the contrary, it is yet necessary to demonstrate that the discovered mechanism of energy conversion is utilized in any biological process. Because of the efficiency and versatility of the mechanism and because the mechanism shows great parallelity with what is known about biological energy conversion processes, it is anticipated to be a dominant mechanism in biological processes, but this has yet to be shown.

Instead of breaking a complex protein system into parts, that is, instead of de-structuring in order to characterize a mechanism of energy conversion, there has been a non-destructive characterization of materials in that there has been a designed build-up of compositions with the purpose of introducing specific energy conversion capabilities.

Inverse Temperature Transitions and Thermomechanical Transduction

Central to the energy conversion mechanism is the process of hydrophobic folding and assembly in an aqueous milieu. In particular, polypeptides and model proteins that can catalyze the energy conversions exhibit inverse temperature transitions, that is, they hydrophobically fold and associate to become more-ordered as the temperature is raised through a transition temperature range which can be defined by the transition temperature, T_t. The capacity of protein-based polymers, which are capable of inverse temperature transitions, to catalyze energy conversions is most readily shown by cross-linking the model protein into a matrix. At low temperature the matrix swells to the limit allowed by the chain lengths and the extent of cross-linking. When the temperature is raised from below to above the transition temperature range, that is, from below to above T_t, the matrix will contract or de-swell with the expulsion of water. When a weight is hung on the matrix at a temperature below T_t and the temperature is raised some 10°C above T_t, the matrix will contract and lift the weight. This is the conversion of thermal energy into mechanical work; it is thermally-driven contraction; it is thermomechanical transduction.

The ΔT_t-mechanism of Energy Conversion

The model protein can be designed such that a given energy input will change the temperature, T_t, of the inverse temperature transition; the energy input causes a ΔT_t. If the temperature range over which occurs the inverse temperature transition (that is, if the value of T_t) is above a constant operating temperature, then the model protein would be unfolded and disassociated, and its cross-linked matrix will be swollen. When the model protein has been designed such that a particular energy input will lower the value of T_t below the fixed operating temperature, then that energy input can drive the folding and assembly; when the model protein is cross-linked, then that energy input can drive contraction with the lifting of a weight and the performance of mechanical work. This is the ΔT_t-mechanism of free energy transduction.[1,2]

Chemomechanical Transduction. When the energy input which lowers the value of T_t is the change in the concentration of a chemical, the energy conversion is chemically-driven contraction; it is chemomechanical transduction.

Baromechanical Transduction. When the energy input is the release of pressure of a previously pressurized chamber containing a strip of properly designed and cross-linked model protein suspending a weight, the resulting lifting of the weight is baromechanical transduction.

Electromechanical Transduction. When the energy input is the electrochemical reduction of a cross-linked model protein by means of an attached prosthetic group which can be reduced or oxidized, reduction lowers the value of T_t with the result of reduction-driven contraction, i.e., of electromechanical transduction.

Photomechanical Transduction. When the energy input is the absorption of light by a chromophore (attached to the cross-linked model protein) that undergoes a structural change on absorption of light making it less hydrophobic, the value of T_t is raised and the condition can be set up for a light-driven relaxation with a dark reversal giving rise to contraction. If the light absorption should result in a structural change to greater hydrophobicity, the value of T_t would be lowered and the result would be light-driven contraction. In either case, the result is photomechanical transduction.

Molecular Machines

The set of pairwise energy conversions resulting in mechanical work, or the production of mechanical energy, is shown in Figure 1 by all of the arrows ending at the mechanical apex. The model proteins functioning directly to convert thermal energy into mechanical work by means of the inverse temperature transition are referred to as zero order molecular machines of the T_t-type. Those model proteins that perform mechanical work by means of shifting the temperature of the inverse temperature transition (that is, by the shifting of the value of T_t) are called first order molecular machines of the T_t-type. There are also indicated in Figure 1 pairwise energy conversions that do not end at the mechanical apex, yet they utilize the control of hydrophobic folding and unfolding as the means of interconverting the energies, these are referred to as second order molecular machines of the T_t-type.[2]

To the best of our knowledge, the energy conversions by molecular machines of the T_t-type, those of Figure 1, include all of the energy conversions of which living organisms are capable.

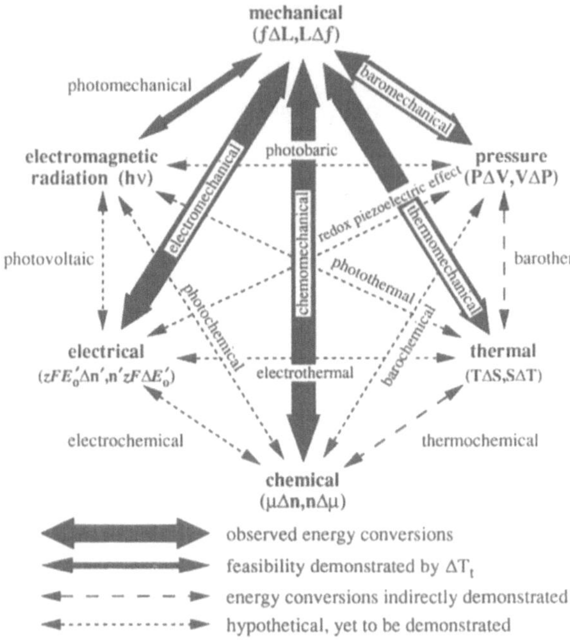

Figure 1. Demonstrated and putative energy conversions using molecular machines of the T_t-type. Reproduced with permission from Reference 2.

THE MOLECULAR STRUCTURE OF THESE TRANSDUCTIONAL MODEL PROTEINS

Original Repeating Peptide Sequence Capable of a Reversible Inverse Temperature Transition

Sandberg and colleagues[3] found interesting repeating peptide sequences in the precursor protein, tropoelastin, of the porcine elastic protein called elastin. The most striking repeating sequence was a repeating pentapeptide sequence, $(Val^1\text{-}Pro^2\text{-}Gly^3\text{-}Val^4\text{-}Gly^5)_n$ where n = 11 with but two isomorphous substitutions. Subsequently, Yeh et al.[4] found the same repeat in the bovine gene with n = 11+ and with no substitutions. This Laboratory has synthesized different monomer permutations, oligomers and high molecular weight polymers of the pentapeptide and has characterized their conformation and physical properties.[5] The protein-based polymer, poly($Val^1\text{-}Pro^2\text{-}Gly^3\text{-}Val^4\text{-}Gly^5$), or simply poly(VPGVG), is soluble in water in all proportions below 25°C. On raising the temperature of solutions of less than 400 mg/ml, there ensues an aggregation and settling to form a viscoelastic phase called a coacervate which is about 50% water and 50% polypeptide by weight. On γ-irradiation cross-linking using 20 Mrad, the coacervate phase forms elastomeric matrices that, by classical arguments, are dominantly entropic elastomers and that exhibit elastic moduli similar to elastin. These are matrices which reversibly swell to some 90% water on lowering the temperature below 25°C and which contract on raising the temperature to 37°C. This is the inverse temperature transition of hydrophobic folding and assembly on raising the temperature that is capable of thermomechanical transduction.

Linear Representation of the Basic Structure: Primary and Secondary Structure

A schematic two-dimensional representation of the primary and secondary structure of poly(VPGVG) is shown in Figure 2A. This ribbon-like structure is helpful in recognizing

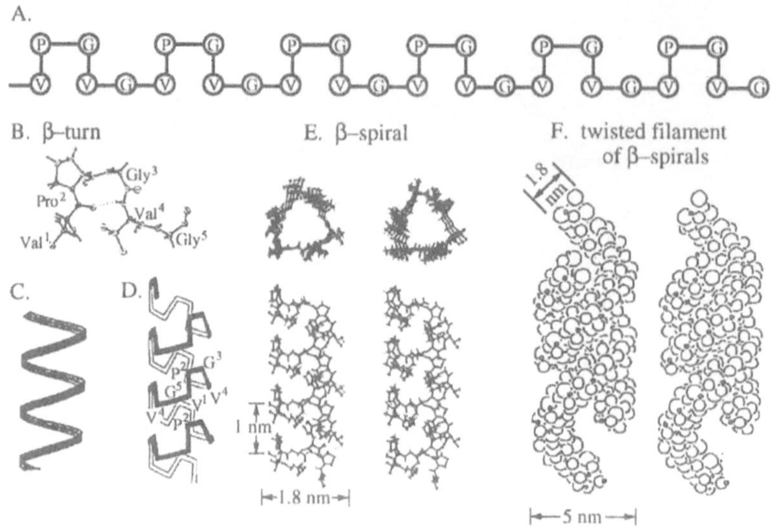

Figure 2. Molecular structure of poly(VPGVG). Adapted with permission from References 6 (B), 7 (D & E) and 8 (F).

the substitutions that will allow for the introduction of the more diverse free energy transductional properties.

The PG deflections represent a Type II Pro^2-Gly^3 β-turn, which is shown in detail in Figure 2B, as obtained from the crystal structure of the cyclic conformational correlate, cyclo$(VPGVG)_3$.[6] As Pro^2-Gly^3 is the most probable means of forming a β-turn, substitutions here can disrupt the β-turn or convert it from a Type II to a Type I β-turn. Along the other edge of the ribbon, (-V···V-G-)n, substitution of the Gly residue even with an Ala residue disrupts the elastic property.[5] This leaves possible substitutions of the Val residues. Care must be taken with substitution of the Val^1 residue as an Ala^1 substitution results in an irreversible precipitation rather than a reversible coacervation. Substitution retaining the β-branching as with $Val^1 \rightarrow Ile^1$ is the most conservative replacement which allows for retention of a reversible inverse temperature transition and elasticity.[2] This substitution simply lowers the temperature at which the inverse temperature transition is initiated from 25°C to 10°C. It is also possible to put Phe in position 1.

The Position for Development of a Hydrophobicity Scale

Fortunately, the Val^4 can be substituted by all of the naturally occurring amino acid residues and by chemical modification of those with chemically reactive side chains such as glutamic acid (Glu,E), aspartic acid (Asp,D), lysine (Lys,K). Using this position substitutions have allowed for the development of a hydrophobicity scale for all of the amino acid residues and some chemical modifications.[2] In this scale given in Table 1, the more hydrophobic residues have lower values of T_t, and the less hydrophobic (more polar) residues exhibit higher values of T_t. This scale was obtained using the general structure poly[f_V(VPGVG),f_X(VPGXG)] where f_V and f_X are mole fractions of the respective pentamers in the resulting polypentapeptide with $f_V + f_X = 1$, and where X is any of the amino acid residues. In practice, f_X versus T_t is extrapolated to $f_X = 1$ as the reference value in phosphate buffered saline (0.15 N NaCl, 0.01 M phosphate).

Table 1 constitutes the only hydrophobicity scale which is directly derived from the hydrophobic folding and assembly process, and it also allows for the determination of the relative hydrophobicities of a given amino acid side chain in different functional states, for example, the COOH and COO^- states of glutamic acid (Glu,E) or of aspartic acid (Asp,D) residues. It is, in fact, the sensitivity of the value of T_t to different functional/structural

Table 1. T_t-Based Hydrophobicity Scale Using poly[f_V(VPGVG),f_X(VPGXG)]

Residue X		T_t, linearly extrapolated to $f_X = 1$	Correlation Coefficient
Lys(NMeN, reduced)[a]		−130°C	1.000
Trp	(W)	−90°C	0.993
Tyr	(Y)	−55°C	0.999
Phe	(F)	−30°C	0.999
His (pH 8)	(H^o)	−10°C	1.000
Pro	(P)	(−8°C)	calculated
Leu	(L)	5°C	0.999
Ile	(I)	10°C	0.999
Met	(M)	20°C	0.996
Val	(V)	24°C	reference
Glu($COOCH_3$)	(E^m)	25°C	1.000
Glu(COOH)	(E^o)	30°C	1.000
Cys	(C)	30°C	1.000
His (pH 4)	(H^+)	30°C	1.000
Lys(NH_2)	(K^o)	35°C	0.936
Asp(COOH)	(D^o)	45°C	0.994
Ala	(A)	45°C	0.997
HyP		50°C	0.998
Asn	(N)	50°C	0.997
Ser	(S)	50°C	0.997
Thr	(T)	50°C	0.999
Gly	(G)	55°C	0.999
Arg	(R)	60°C	1.000
Gln	(Q)	60°C	0.999
Lys(NH_3^+)	(K^+)	120°C	0.999
Tyr(ϕ-O^-)	(Y^-)	120°C	0.996
Lys(NMeN, oxidized)[a]		120°C	1.000
Asp(COO^-)	(D^-)	170°C	0.999
Glu(COO^-)	(E^-)	250°C	1.000
Ser($PO_4^=$)		1000°C	1.000

states of chemical moieties that makes possible the ΔT_t-mechanism of energy conversion. This also includes differences in states of hydration as, for example, at different pressures.[2]

Tertiary and Quaternary Structure

On raising the temperature there occurs an optimization of intramolecular hydrophobic interactions leading to the wrapping up into a helical structure. This is schematically shown in Figure 2C as a helical ribbon, and in slightly more detail in Figure 2D with schematic inclusion of the β-turns. The β-turns function as spacers between turns of the helix. As shown in detail in Figure 2E, as a helical structure called a β-spiral, the interturn hydrophobic contacts, that constitute the tertiary structure, are dominantly the interaction between the Pro^2-βCH_2 moiety of pentamer repeat i with the Val^1-γCH_3 moieties of pentamer repeat $i + 3$. This is the hydrophobic tertiary structure suggested by nuclear Overhauser effect studies.

Using the hydrophobic ridges that form slow spirals along a β-spiral, several β-spirals hydrophobically associate to form a twisted filament quaternary structure shown in Figure 2F as suggested from transmission electron microscope studies using negative staining and optical diffraction of the electron micrograph. Reviews of the structural studies are available from which the original references can be obtained.[1,2,5]

THE DESIGN OF MODEL PROTEINS FOR DIVERSE ENERGY CONVERSIONS

The first insight into the T_t-based hydrophobicity scale (and by extension to the ΔT_t-mechanism of energy conversion) arose out of the effort to demonstrate that the phase transition (coacervation) observed on raising the temperature of aqueous solutions of poly(VPGVG) was dominated by hydrophobic interactions. This was achieved by showing that the transition of the more hydrophobic poly(IPGVG), where I (Ile) had one more CH_2 moiety than V (Val), occurred at a lower temperature.[9] This occurred in spite of the evidence that the Ile for Val substitution did not alter the conformation.

The above dependence of T_t on hydrophobicity, later confirmed and generalized by the T_t-based hydrophobicity scale,[2] gave rise to the pair of postulates that (1) a polymer with a carboxyl-containing residue such as poly[0.8(VPGVG),0.2(VPGEG)] where E = Glu, would have a lower value of T_t when as COOH rather than as COO^-, and (2) that changing COO^- to COOH by lowering the pH would drive hydrophobic folding without a change in temperature. This was observed and the pair of postulates became the basis for the ΔT_t-mechanism of chemomechanical transduction.[2] This pair of postulates was immediately thereafter generalized to include any attached chemical moiety that could be affected by an energy input in such a way as to change sufficiently the value of T_t for the model protein.

Accordingly, the T_t-based hydrophobicity scale of Table 1, particularly as it becomes more complete and contains more chemical moieties that can be altered by an energy input, becomes the basis with which the molecular engineer can design model proteins for desired energy conversions. The set of energy conversions involving mechanical work is summarized in Figure 3 and will be briefly mentioned below.

Designed Structures and Conditions for Chemomechanical Transduction

Chemical Control. The chemically elicited change can be the change in proton concentration changing the state of ionization of a naturally occurring side chain and the chemical nature of the side chain can be such that lowering the pH drives folding and contraction (as in $COO^- \rightarrow COOH$ for the carboxylate/carboxyl chemical couple) or such that raising the pH drives folding and contraction (as in $NH_3^+ \rightarrow NH_2$ in the ammonium/amino chemical couple of the lysyl residue). Also, due to the nature of the physical forces involved, by design, the pKa can be varied over a wide range of values; for example the pKa of the Asp residue can be shifted from 3.9 to 10.1[2] by structurally positioning hydrophobic Phe residues at 1 nm distances from the carboxyl moiety. Furthermore, the nature of the interaction is such that the acid/base titration curve is steeper than usual for a weak acid,[2] that is, there is a positive cooperativity requiring less of a pH change to go from all COOH to all COO^-. This means that less chemical energy, less change in chemical potential, would be required to drive from completely unfolded to completely folded. Chemomechanical transduction by this mechanism was first shown using the elastic matrix formed on 20 Mrad cross-linking of the Glu-containing model protein. This matrix is indicated as X^{20}-poly[0.8(VPGVG),0.2(VPGEG)].

Enzymatic Control. The model protein structures for chemomechanical transduction can also be designed for enzymatic control. An example of this can utilize the ubiquitous chemical process for energy conversion in biological systems, namely phosphorylation/dephosphorylation. An enzymatically recognizable peptide sequence for phosphorylation is Arg-Gly-Tyr-Ser-Leu-Gly, i.e., RGYSLG. This sequence can be included in the model protein poly[30(IPGVG),(RGYSLG)] where the cyclic AMP dependent protein kinase from heart muscle can phosphorylate the Ser residue. The effect of this phosphorylation is to dramatically increase the value of T_t as reflected in Table 1. One phosphate in 300 residues raises the value of T_t by 15°C sufficiently to obtain the complete unfolding. The phosphate can then be removed by alkaline phosphatase to reverse the effect and drive folding.[2]

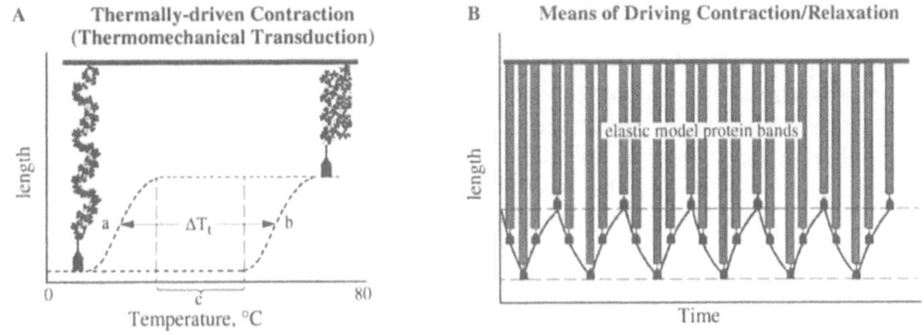

a. Means of driving contraction
- *i.* raising the temperature
- *ii.* lowering the value of T_t
  - intrinsic chemical change (e.g. $COO^- \rightarrow COOH$)
  - extrinsic chemical change (e.g. adding salt)
  - reducing prosthetic group (e.g. nicotinamide)
  - release of pressure with aromatic residues
  - dark reversibility of light reaction

b. Means of effecting relaxation
- *i.* lowering the temperature
- *ii.* raising the value of T_t
  - intrinsic chemical change (e.g. $COOH \rightarrow COO^-$)
  - extrinsic chemical change (e.g. wash out of salt)
  - oxidizing prosthetic group (e.g. nicotinamide)
  - application of pressure with aromatic residues
  - photochemical reaction of suitable chromophore

Figure 3. Mechanical work by the ΔT_t-mechanism. Adapted from Reference 2.

Solute Control. It should also be noted that raising the concentration of a solute such as NaCl can lower T_t and drive folding. Virtually any salt addition can change the value of T_t as can changes in organic solute concentrations.[2] The possibilities are unlimited.

Designed Structures and Conditions for Baromechanical Transduction

While the dependence of T_t on pressure for poly(VPGVG) is quite small, introduction of the residues with aromatic side chains such as Phe(F), Tyr(Y) and Trp(W) results in a relatively small increase in pressure causing a significant increase in the value of T_t. For the cross-linked elastic matrix, X^{20}-poly[0.79(VPGVG),0.21(VPGFG)] application of pressure causes an unfolding, and release of pressure causes a folding and contraction with the lifting of a weight.[2] This baromechanical transduction arises because the volume occupied by the special water molecules surrounding aromatic side chains in the unfolded state is smaller than that of bulk water.

Designed Structures and Conditions for Electromechanical Transduction

Chemical moieties that can undergo reversible oxidation and reduction reactions can be covalently attached to the functional side chains of amino acid residues such as Glu(E), Asp(D) or Lys(K). One such chemical moiety is N-methylnicotinamide {NMeN} that can be attached by amide linkage to the lysyl side chain. Reduction of this attached redox couple in poly[0.73(VPGVG),0.27(VPGK{NMeN}G)] causes a large decrease in the value of T_t (see Table 1) such that chemical or electrochemical reduction drives folding and oxidation drives unfolding when at the proper working temperature.[2]

Designed Structures and Conditions for Photomechanical Transduction

Photochemical reactions can change the hydrophobicity of the light sensitive chemical moiety such that there can be a photon-induced change in the value of T_t of the model protein to which the photoreactive group is attached. When the result of the photon absorption is dark reversible, then the photon-driven process can be used reversibly to drive folding and unfolding of the model protein. The result would be a photon-driven molecular machine. An example of a reversible process is the absorption of light causing a trans to cis geometrical isomerism. It has now been demonstrated with two chromophores that undergo reversible cis/trans photoisomerism, azobenzene {AB} and cinnamic acid {CA}, that the absorption of light of the correct wavelength raises the value of T_t in the polymers poly[0.8(VPGVG),0.2(VPGE{AB}G)] (Strzegowski, Martinez, Gowda, Urry and Tirrell, unpublished data) and poly[0.8(VPGVG),0.2(VPGK{CA}G)], (Heimbach and Urry, unpublished data) respectively. Other photoreactive chromophores have been considered with large decreases in T_t but conditions for reversibility have as yet been difficult to achieve.

CONCLUSIONS

Model proteins or protein-based polymers can be designed to function as zero-order and first-order molecular machines of the T_t-type as indicated in Figure 1. In the case of chemomechanical transduction, the efficiency of the energy conversion appears to be an order of magnitude better for the ΔT_t-mechanism than for the previously considered electrostatic mechanism. Designs have been worked out and syntheses of model proteins are underway for preparing certain of the second-order molecular machines of Figure 1 which, to the best of our knowledge, contain all of the energy conversions of which living organisms are capable. This diverse capacity for energy conversion is being considered for possible industrial applications, keeping in mind the present aqueous and temperature range limitations and possible contaminations.

ACKNOWLEDGMENT

This work was supported in part by contract N00014-89-J-1970 from the Department of the Navy, Office of Naval Research.

REFERENCES

1. Dan W. Urry, Free energy transduction in polypeptides and proteins based on inverse temperature transitions, *Prog. Biophys. Molec. Biol.* 57:23 (1992).
2. Dan W. Urry, Molecular machines: how motion and other functions of living organisms can result from reversible chemical changes, *Angew. Chem. Int. Ed. Engl.*, 32:819 (1993).
3. L. B. Sandberg, N. T. Soskel, and J. B. Leslie, Elastin structure, biosynthesis, and relation to disease states, *N. Engl. J. Med.*. 304:566 (1981).
4. H. Yeh, N. Ornstein-Goldstein, Z. Indik, P. Sheppard, N. Anderson, J. C. Rosenbloom, G. Cicila, K. Yoon, and J. Rosenbloom, Sequence variation of bovine elastin mRNA due to alternative splicing, *Collagen and Related Research.* 7:235 (1987).
5. D. W. Urry, Thermally driven self-assembly, molecular structuring and entropic mechanisms in elastomeric polypeptides, *in* "Mol. Conformation and Biol. Interactions" P. Balaram and S. Ramaseshan, eds. Indian Acad. of Sci., Bangalore, India, (1991).
6. W. J. Cook, H.M. Einspahr, T.L. Trapane, D.W. Urry, and C.E. Bugg, Crystal structure and conformation of the cyclic trimer of a repeat pentapeptide of elastin, cyclo-(L-Valyl-L-prolylglycyl-L-valylglycyl)$_3$, *J. Am. Chem. Soc.* 102:5502 (1980).
7. D. W. Urry, "Elastomeric polypeptide biomaterials: structure and free energy transduction," *Mat. Res. Soc. Symp. Proc.* 174:243 (1990).
8. D.W. Urry, C.M. Venkatachalam, M.M. Long, and K.U. Prasad, Dynamic β-Spirals and a librational entropy mechanism of elasticity, *in* "Conformation in Biol." R. Srinivasan and R.H. Sarma, eds., G.N. Ramachandran Festschrift Volume, Adenine Press, USA, (1982).

MICROSTRUCTURE STUDY OF MOLYBDENUM LINERS BY NEUTRON DIFFRACTION

C.S. Choi*, E.L. Baker† and J. Orosz†

*Reactor Radiation Division, NIST
Gaithersburg, MD 20899
†ARDEC, Picatinny Arsenal
Dover, NJ 07806

ABSTRACT

The crystallites orientation distributions of three different conical molybdenum shaped-charge liners, labeled C00, C20 and C40, fabricated with different thermo-mechanical treatment processes, were investigated by using neutron diffraction measurements and three-dimensional ODF (orientation distribution function) analysis. Two specimens were obtained from each liner, one from 25 mm below the apex and the other from 25 mm above the liner base. The pole figure data of the four reflections, (110) (200) (211) and (222), were measured from each specimen, using 1.25 Å neutrons. Several strong pole density peaks, (111)<$\bar{1}$10>, (100)<011>, and (112)<$\bar{1}$10> were commonly observed in the samples. The section near the apex of the cone had mostly sheet-type texture, whereas the section near the base had more fiber-type texture. The plastic strain ratio (R-value) distributions of the samples were calculated from the ODFs. The average R-values of the liners were less than 1, which is attributable to the strong (100)<uvw> texture. The diffraction patterns of the polycrystal samples were measured at two different orientations for each sample: the ND and the RD. The intensities of the patterns were corrected for the texture effect, based on the assumption that the volume involved in the diffraction is directly proportional to the corresponding pole density in the inverse pole figure. The Rietveld profile refinements with the corrected diffraction patterns were quite satisfactory.

INTRODUCTION

The metals and alloys used in technological applications are mostly polycrystalline materials. Their properties are, therefore, not only dependent on the crystal structure but also on the microstructure including the distributions of the crystallite orientations, the grain sizes, dislocation densities, and the state of aggregation. Recent advances in texture analysis techniques (see for example, references 1,2,3) made it possible to determine a quantitative orientation distribution function (ODF) quite accurately. The success of the ODF determination has produced several new simulation models[4] for the materials characterization, such as prediction of yield surface, plastic anisotropy, and the evolution of texture.

Texture measurements have been conducted most commonly by x-ray methods because of the easy availablity. However, the penetration power of x-rays is so low in the heavy metals (about 1/1000 or less of that of neutrons) that it presents many problems: insufficient statistics due to the limited sampling volume, surface texture, incomplete pole figure, and defocusing corrections, etc. Neutrons, on the other hand, have a stronger penetration for most metals, and hence interact with the bulk of the specimen. The samples are fully immersed in the beam, with no restriction of reflection or transmission geometry. Therefore, the neutron method represents nondestructive characterization of the bulk of the sample. In the present study, we have studied the textures of three preforms of Mo shaped-charge liner using neutron diffraction. Subsequently, the ODF data were used for the prediction of the R-values, and also for the Rietveld profile refinement of the textured samples.

EXPERIMENTAL

Three different conical liners, labeled C00, C20 and C40, fabricated from arc-cast molybdenum ingots, were used for the study. Two specimens, with approximate size 10 x 10 mm, were cut

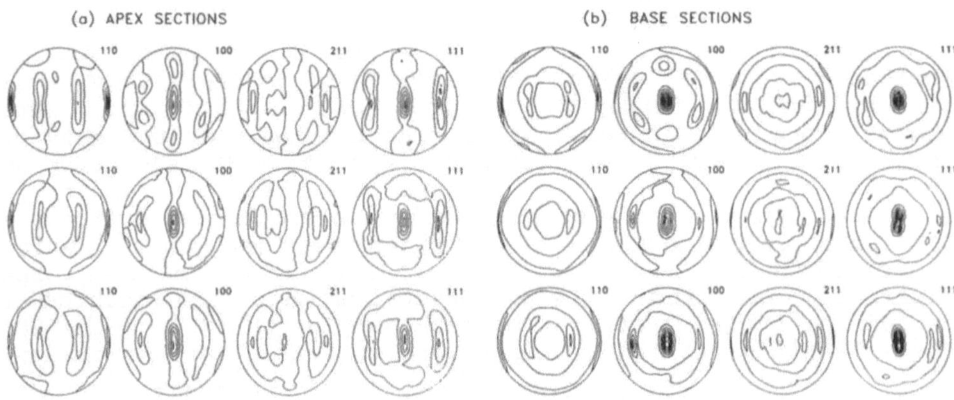

Figure 1. The measured pole figures of the three apex samples (C00-A, C20-A, C40-A) are presented in 1-a and those from the bottom part (C00-B, C20-B, C40-B) are in 1-b. The pole figures of C00 are at the top, of C20 at the middle, and those of C40 at the bottom. The contour levels of 1 mrd and higher are given with 1 mrd intervals. The coordinates of the pole figures are RD at the 3 o'clock position, TD at the 12 o'clock position and the ND at the center.

from each liner (6 mm thick); one from 25 mm below the apex (signified with the post-script A in the specimen label), and the other from 25 mm above the liner base (signified with the post-script B). For example, the specimen C00-A corresponds to the section 25 mm below the apex of the C00 liner. Since the liners are about 127 mm high, the post-script "A" represents the apex region and "B" the base region. The samples were mounted on the diffractometer with the orientation which makes the normal direction (ND) to the cone surface be the hemispheric pole, and the cone apex direction (RD) be the origin of the longitude of the orientation hemisphere. The pole figure data of the four reflections, (110) (200) (211) and (222), were measured over an entire hemisphere with 1.25 Å neutrons at the NIST Research Reactor. The raw data were corrected for background intensities and for the absorption effect of the square-plate like samples. In Figure 1-a, the four pole figures of the three apex samples (C00-A, C20-A and C40-A) are presented, and those of the base part samples (C00-B, C20-B and C40-B) are in Figure 1-b. The pole figures of the three apex samples were generally similar to each other. This is also true for the three bottom part samples. However, the top and bottom parts in each liner had clearly different texture, as indicated by the different shapes of the pole figures.

Texture Analysis

The orientation distribution functions (ODF) of all the samples were obtained by the WIMV method, using the program popLA[4]. Triclinic sample symmetry was used for the analysis, since the pole figures were slightly distorted from orthorhombic symmetry. The agreement between the experimental and the recalculated pole figures was excellent for all the samples.

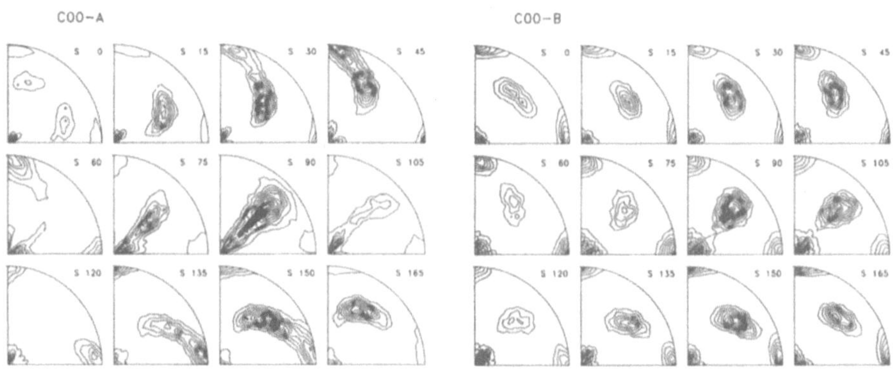

Figure 2. The SOD of the C00-A and C00-B samples are presented in the constant ψ-planes, from $\psi = 0$ to 165°in 15° intervals. The coordinate axes of the SOD are, the [100] direction at the right edge, the [010] at the top, and [001] at the center of the quadrant.

The three-dimensional ODFs in the form of the sample orientation distribution (SOD), defined with Kocks-type Euler angles, are given in Figure 2. Since the SODs of the three liners were similar, only those of the C00-A and C00-B are presented in Figure 2. Comparing the SOD of the C00-A to C00-B, the changes of pole density as a function of ψ-level are much less for the C00-B. This indicates that the bottom region of the liner has more fiber-type texture than the top region. Several distinct pole density peaks were typically found from the SOD maps of all six samples. They are (111)<uvw>, (100)<uvw>, and intensity ridges extending from these peaks to (112)<uvw> and (115)<uvw>. The pole density distributions of the (111)<uvw>, (100)<uvw> and (112)<uvw>, as function of the ψ-angles, were obtained from the three-dimensional SODs, and are presented in Figure 3. The solid lines in the figures represent the distributions at the sections near the apex (post-script A), and the dotted lines represent those at the sections near the liner base (post-script B). The isolated peaks in the distribution curves represent the sheet-type texture. The Miller indices of the <uvw> corresponding to each peak position of the curves are given in the figures. The base-line intensity of the (HKL)<uvw> distribution (here HKL represents 111, 100, or 112) corresponds to the intensity of the fiber-type texture, with the <HKL> fiber axis oriented parallel to the ND. In Figure 3, the dotted curves have generally higher base-line intensities than the solid curves, which indicates that the bottom part of a liner tends to have stronger fiber-type texture than the part near the apex.

TABLE 1. Major texture components found in the three conical molybdenum liners, C00, C20 and C40. The <111> and <100> fiber axes in the table are oriented perpendicularly to the cone surface.

Specimen	Tex. Index	Texture	Pole density (mrd)
C00-A	1.9	(111)<$\bar{1}$10>	11
		(100)<$\underline{0}$11>	10
		(112)<$\bar{1}$10>	10
C00-B	1.9	(111)<$\bar{1}$10> + fiber	12
		(100)<$\underline{0}$21> + fiber	11
		(112)<$\bar{1}$10>	4
C20-A	1.6	(111)<$\bar{1}$10>	6
		(100)<$\underline{0}$11>	9
		(112)<$\underline{1}$10>	5
C20-B	1.7	(111)<$\bar{2}$11> + fiber	7
		(100)<$\underline{0}$10> + fiber	7
		(112)<$\bar{1}$10>	4
C40-A	1.5	(111)<$\bar{1}$10>	4
		(100)<$\underline{0}$11> + fiber	8
		(112)<$\underline{1}$10>	5
		(115)<$\underline{1}$10>	6
C40-B	1.7	(111)<$\bar{2}$11> + fiber	10
		(100)<$\underline{0}$12> + fiber	9
		(112)<$\underline{1}$10>	3
		(115)<$\bar{1}$10>	4

The major texture components found in the liners are summarised in Table 1. Three dominant texture components were observed in the C00 liner, (111)<$\bar{1}$10>, (100)<011>, and (112)<$\bar{1}$10>. The sections near the apex were generally sheet-type, whereas the sections near the base had both sheet-type and

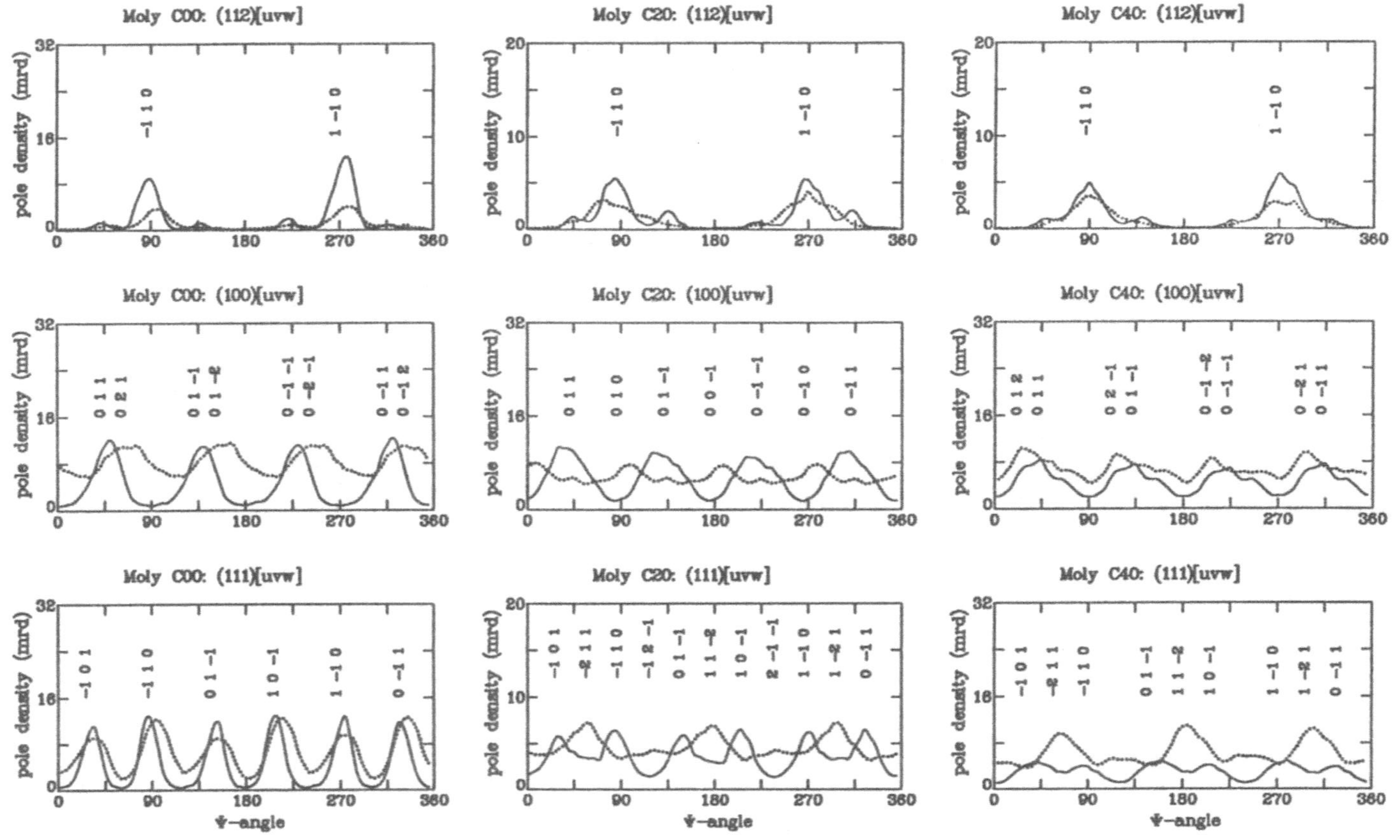

Figure 3. The (112)<uvw>, (100)<uvw> and (111)<uvw> distributions as a function of the ψ-angles. The solid lines indicate the distributions at near the apex, and the dotted lines indicate those at near the base of the liners.

fiber-type. Well defined (100)<011> components were observed in the apex side of the sample, but it changed to almost fiber-type at the base side of the sample, accompanied by the peak shifts toward the (100)<021> texture, as shown in Figure 3. A close examination of the distribution curves in Figure 3 revealed several other texture differences between the apex part and the base part of the liners, as described below. The (112)<$\bar{1}$10> components were strong at the apex side, but the intensity became weakened considerably at the bottom side of the liner. The C20 liner was also dominated by the (111)<uvw>, (100)<uvw> and (112)<uvw> type textures, although their pole densities were generally weaker than those of the C00 liner. The (100)<011> components which dominated the apex side changed to a pseudo-fiber texture at the bottom side of the sample, accompanied with a peak shift toward (100)<010> components. The (111)<$\bar{1}$10> component also dominated the apex side of the C20 sample, but the peak shifted to (111)<$\bar{2}$11> at the bottom side of the sample. The C20 liner had generally stronger fiber-type texture than the C00 liner. The C40 liner had the texture components similar to those of the C20 liner. Generally the two elongated liners (i.e. C20 and C40) had a weaker degree of texture than the C00 liner, as can be seen from the texture index in Table 1.

PLASTIC ANISOTROPY

The plastic strain ratio, commonly called the R-value, is defined by $R(\alpha) = \epsilon_{yy}/\epsilon_{zz}$, where α is the angle of the stress direction measured from the RD of the sample, and ϵ_{yy} and ϵ_{zz} are the width strain and thickness strain, respectively. The R-value is a parameter that indicates the anisotropy characteristics of sheet metals, and is related to the limit strains attainable in fabrication processes, such as stretching or deep-drawing. The R-values of the samples were calculated from the coefficients of the Harmonic ODF, using the program popLA[4]. Two different slip models were used in the calculations, one based on the restricted glide model and the other on the pencil-glide model[5], where the slip direction is restricted to <111> directions but the slip plane is not. The two sets of calculated R-values of the six samples were given in Table 2.

TABLE 2. The R-values as a function of the directions (α), obtained from the Harmonic ODF using the restricted glide model (R) and the pencil-glide model (P). The $\bar{R}$ and the ΔR are also given.

α	C00-A R / P	C00-B R / P	C20-A R / P	C20-B R / P	C40-A R / P	C40-B R / P
0.	.54/.49	.73/.61	.63/.57	.86/.72	.43/.46	.66/.59
15.	.81/.67	.69/.62	.64/.60	.75/.67	.50/.51	.63/.58
30.	.95/.88	.63/.62	.61/.62	.55/.55	.59/.60	.59/.56
45.	1.17/1.17	.60/.59	.59/.62	.49/.52	.64/.66	.59/.58
60.	1.08/1.24	.69/.62	.66/.67	.62/.63	.67/.67	.65/.63
75.	.85/.83	.82/.75	.73/.74	.87/.84	.64/.67	.72/.69
90.	.75/.78	.87/.82	.77/.78	1.00/.98	.64/.67	.75/.71
$\bar{R}$	.91/.90	.70/.65	.65/.65	.71/.69	.59/.61	.65/.62
ΔR	-.53/-.54	.20/.13	.11/.06	.44/.33	-.11/-.10	.12/.07

It is known that the drawability of a sheet metal[6,7] is closely related to the average strain ratio: $R(\alpha) = (R(0) + R(45) + R(90))/4$, and the directional variation in the plane of the sheet, commonly called the plane strain ratio : $\Delta R = (R(0) + R(90) - 2R(45))/2$. The desirable values for deep-drawing are a larger R (larger than 1) and smaller ΔR. The R-values of the liners in Table 2 are less than 1 in average, and hence predict a poor drawability. The small values of the R are attributed to the presence of the strong (100)<hkl> component in the samples. The criteria of the R values for the deep-drawing of sheet metals are not necessarily valid for the jet ductility performance of the liners. However, the R-values are a simple one-dimensional parameter, and hence may be easy to correlate with other plastic properties, such as the lengths of the shaped charge jets produced by the liners. Such correlations should be investigated.

PROFILE REFINEMENT

The powder diffraction method is probably the most popular and widely used technique for crystallographic characterization among the materials science community, since the Rietveld technique[8] has become competitive with the single crystal method, and single crystal samples are very difficult to obtain. However, the Rietveld method can be used only for ideal powder samples, and not for textured polycrystals because of the texture effect on the diffraction profile. The inverse pole figure, which is a projection of the ODF along the ψ-axis, describes the orientation density of all the hkl poles oriented in the particular sample direction. For example, the pole density at a given hkl position in the ND inverse pole figure is equivalent to the volume fraction of all the hkl planes which are oriented with their plane normals parallel to the ND direction. Therefore, the multiplicity of the hkl reflection in the ND diffraction pattern should be modified by the corresponding pole density in the ND inverse pole figure. A new subroutine was added to the Rietveld refinement program[9] to modify the multiplicity of the hkl reflections by means of pole density in the inverse pole figure. The pole densities obtained from the ND and RD inverse pole figures of the six samples are summarized in Table 3.

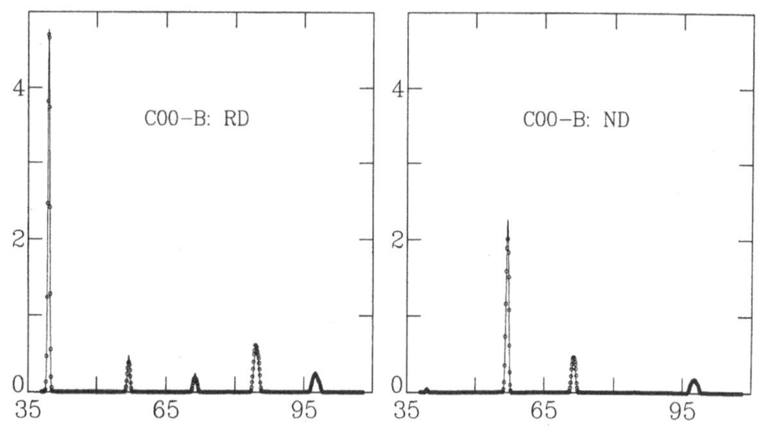

Figure 4. The diffraction profiles of C00-B measured at two different orientations; the RD and the ND. The solid-lines represent the calculated pattern by the Rietveld profile refinement with the texture-corrections.

It is clear from Table 3 that the diffraction intensities are heavily dependent on the sample orientation. For example, the (110) and (220) intensity variations for the orientation change from ND to RD are about a factor of 100 for C00-B and about 300 for C40-B sample.

The diffraction patterns were measured at two different orientations, i.e. the ND and the RD, for each sample, using a 1.5 Å neutron beam. Fig. 4 presents the two diffraction patterns of the C00-B sample. The two diffraction patterns showed entirely different intensity profiles because of the severe texture effect. The profile refinements of such heavily textured samples were quite satisfactory for all the samples, when the modified version of the Rietveld refinement program was used. The observed and calculated patterns are compared in Fig. 4. This study shows that the Rietveld technique can be extended to polycrystals if the ODF of the samples are known.

TABLE 3. The pole densities obtained from the inverse pole figures.

hkl	C00-A		C00-B		C20-A		C20-B		C40-A		C40-B	
	RD	ND	RD	ND	RD	ND	RD	ND	RD	ND	RD	ND
110	7.10	0.13	4.37	0.04	4.40	0.09	2.20	0.04	3.87	0.06	3.15	0.01
200	0.52	4.13	2.18	8.42	0.74	5.70	2.61	5.67	1.11	4.77	1.75	7.11
211	0.23	1.37	0.53	0.84	0.62	1.13	1.18	1.08	0.60	1.24	1.28	0.80
310	0.80	0.78	1.40	0.88	1.46	1.05	1.26	1.06	1.35	1.14	1.35	0.93
222	0.05	4.18	0.14	6.35	0.12	3.73	0.13	4.71	0.04	3.13	0.03	5.70

REFERENCES

1 "Preferred Orientation in Deformed Metals and Rocks", edited by H.-R. Wenk, Academic Press, New York, (1985).
2 Proc. "8th Int. Conf. on Textures of Materials", edited by J. S. Kallend and G. Gottstein, The Metallugical Society, Inc., Warrendale, PA, (1988).
3 Proc. "9th Int. Conf. on Textures of Materials", will be published.
4 J. S. Kallend, U. F. Kocks, A. D. Rollett, and H. -R. Wenk, Operational texture analysis, Mat. Sci. Eng., A132:1 (1991).
5 A. D. Rollett and U. F. Kocks, Computor simulation of pencil glide in B.C.C. metals, in Proc. "8-th International Conference on Textures of Materials", edited by J. S. Kallend and G. Gottstein, The Metallurgical Society, Inc., Warrendale, PA, (1988) pp. 375-380.
6 D. J. Blickwede, Sheet steel - Micrometallurgy by the millions, Trans. ASM, 61:653(1968).
7 Ph. Lequeu, F. Montheillet & J. J. Jonas, A simplified method for the prediction of plastic anisotropy in rolled sheet, in Proc. "8-th International Conference on Textures of Materials", edited by J. S. Kallend and G. Gottstein, The Metallurgical Sociaty, Warrendale, PA, (1988), pp. 189-212.
8 H. M. Rietveld, "A profile refinement method for nuclear and magnetic structures", J. Appl. Cryst., 2:65 (1969).
9 J. Rodriguez-Carvajal, Program FULLPROF, Version Feb 90, Institut Laue-Langevin, Grenoble, France.

ULTRASONIC NDE OF LAYERED MEDIA

USING TIME-DOMAIN INFORMATION

Vikram K. Kinra and Changyi Zhu

Center for Mechanics of Composites
Department of Aerospace Engineering
Texas A&M University
College Station, TX 77843

INTRODUCTION

The objective of this paper is to extend the retrieve function approach of reference 1 to NDE of layered media. A closed-form solution of the incident (reflected) field can be obtained through the convolution of the retrieve function and the transmitted (incident) field. For an N layered medium, the incident field can be obtained by a 2^N term summation, while the reflected field can be calculated by a $(2^{N+1}-1)$ term summation. Systematic sensitivity analysis of this technique has been carried out. Experiments were conducted on a three-layered plate, aluminum/water/aluminum, using 1 MHz broad-band transducers. The reconstructed fields compare very well with the measured ones.

THEORY AND SENSITIVITY ANALYSIS

We will use the case of a three-layered plate immersed in water to illustrate the more general theory. We identify the immersion medium with the subscript $(\ )_0$ and the layers with $(\ )_i$, i=1, 2 and 3. The transfer functions of the reflected and transmitted fields for a three-layered plate were found to be[2,3]

$$H_r^*(\omega) = \left[\frac{1}{R_{01}}\right] \frac{(R_{01}+R_{12}e^{-i2k_1h_1})(1+R_{23}R_{30}e^{-i2k_3h_3}) + (R_{01}R_{12}+e^{-i2k_1h_1})(R_{23}+R_{30}e^{-i2k_3h_3}e^{-i2k_2h_2})}{(1+R_{01}R_{12}e^{-i2k_1h_1})(1+R_{23}R_{30}e^{-i2k_3h_3}) + (R_{12}+R_{01}e^{-i2k_1h_1})(R_{23}+R_{30}e^{-i2k_3h_3})e^{-i2k_2h_2}} \tag{1}$$

$$H_t^*(\omega) = \frac{T_{01}T_{12}T_{23}T_{30}\, e^{i[(k_0-k_1)h_1 + (k_0-k_2)h_2 + (k_0-k_3)h_3]}}{(1+R_{01}R_{12}e^{-i2k_1h_1})(1+R_{23}R_{30}e^{-i2k_3h_3}) + (R_{12}+R_{01}e^{-i2k_1h_1})(R_{23}+R_{30}e^{-i2k_3h_3})e^{-i2k_2h_2}} \tag{2}$$

where R_{ij} and T_{ij} are the reflection and transmission coefficients, respectively, k is the wave number, ω is the circular frequency and h is the thickness. We note that a closed-form solution for the exact inverse Fourier transform (IFT) of either $H_r^*(\omega)$ or $H_t^*(\omega)$ does not exist. We now introduce a retrieve function, $Q^m(\omega)$,

$$Q^m(\omega) = H_m^*(\omega) = \frac{G^m(\omega)}{F(\omega)} = \frac{P^m(\omega)}{S(\omega)} \tag{3}$$

where m=r for the case of reflection and m=t for the case of transmission. $G^m(\omega)$ is the Fourier transform (FT) of the reflected (or transmitted) wave, $g^m(t)$, and $F(\omega)$ is the FT of the incident wave, f(t). Now, the IFT of both $P^m(\omega)$ and $S(\omega)$, namely, $p^m(t)$ and s(t), can be calculated very easily. Then the reflected and the incident fields can be obtained via the convolutions,

$$\int_{-\infty}^{\infty} s(\tau)\, g^m(t-\tau)\,d\tau = \int_{-\infty}^{\infty} p^m(\tau) f(t-\tau)\,d\tau \tag{4}$$

$$\begin{aligned} f(t) = \frac{1}{T_{01}T_{12}T_{23}T_{30}}\Big(& g^t(t-t_0+t_1+t_2+t_3) + R_{01}R_{12}g^t(t-t_0-t_1+t_2+t_3) + \\ & +R_{12}R_{23}g^t(t-t_0+t_1-t_2+t_3) + R_{23}R_{30}g^t(t-t_0+t_1+t_2-t_3) + \\ & +R_{01}R_{23}g^t(t-t_0-t_1-t_2+t_3) + R_{12}R_{30}g^t(t-t_0+t_1-t_2-t_3) + \\ & +R_{01}R_{30}g^t(t-t_0-t_1-t_2-t_3) + R_{01}R_{12}R_{23}R_{30}g^t(t-t_0-t_1+t_2-t_3)\Big) \end{aligned} \tag{5}$$

$$\begin{aligned} g^r(t) = {} & R_{01}f(t) + R_{12}\big(f(t-2t_1) - R_{01}g^r(t-2t_1)\big) + \\ & R_{12}R_{23}\big(R_{01}f(t-2t_2) - g^r(t-2t_2)\big) + R_{23}R_{30}\big(R_{01}f(t-2t_3) - g^r(t-2t_3)\big) + \\ & R_{23}\big(f[t-2(t_1+t_2)] - R_{01}g^r[t-2(t_1+t_2)]\big) + R_{12}R_{30}\big(R_{01}f[t-2(t_2+t_3)] - g^r[t-2(t_2+t_3)]\big) + \\ & R_{12}R_{23}R_{30}\big(f[t-2(t_1+t_3)] - R_{01}g^r[t-2(t_1+t_3)]\big) + R_{30}\big(f[t-2(t_1+t_2+t_3)] - R_{01}g^r[t-2(t_1+t_2+t_3)]\big) \end{aligned} \tag{6}$$

where $t_0=s_0(h_1+h_2+h_3)$, $t_i=s_ih_i$, i=1, 2, and 3. For an N-layered medium there are (N+1) independent reflection coefficients. It can be shown that for an N-layered plate the reflected field can be constructed by a summation of $(2^{N+1}-1)$ terms, and the incident field can always be constructed by a 2^N term summation.

Let $f(t;p)=f(t;p_1, p_2,..p_i,..)$ be the reconstructed incident (or reflected) wave; where p_i is the parameter of the ith layer to be determined. Let $f^*(t)$ be the measured incident (or reflected) wave. A least-square error function is introduced

$$E_f(p) = \frac{1}{f^*_{max}} \left[\frac{1}{T} \int_0^T [f(t;\, p) - f^*(t)]^2 dt \right] \tag{7}$$

where T is the total duration of the pulse $f^*(t)$, and f^*_{max} is the maximum value of $f^*(t)$. Next we define a time-averaged sensitivity of f(t; p) to p

$$S_{f,p} = \left[\frac{1}{T} \int_0^T \left(\frac{p}{f_{max}} \frac{\partial f}{\partial p} \right)^2 dt \right]^{\frac{1}{2}} \tag{8}$$

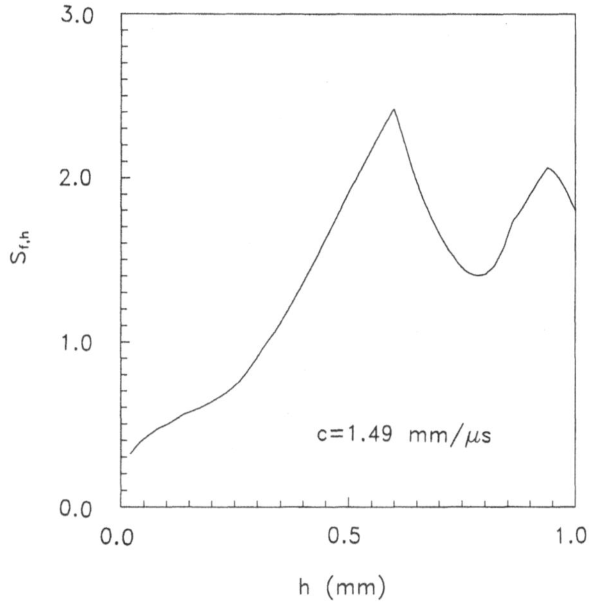

Figure 1. The sensitivity of f(t; **p**) to the thickness of the middle layer of a three-layered medium (Transducer frequency=1 MHz).

Moreover, in eq.(7), by using a Taylor series expansion of f(t; p) about p_{TRUE} (i.e. $p = p_{TRUE} + \Delta p$) and retaining only the terms linear in Δp, it can be readily shown (as expected) that

$$S_{f,p} = E_f(p) \,/\, (\Delta p / p_{TRUE}) \tag{9}$$

The sensitivity, $S_{f,p}$, to the properties of the middle layer of a three-layered medium in the case of through transmission is plotted in Figs.1 and 2.

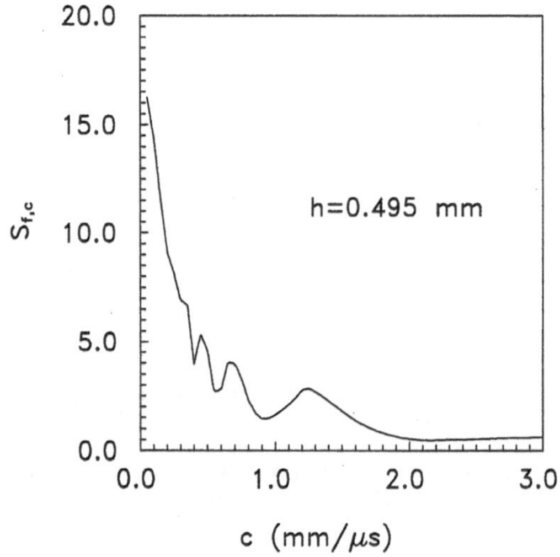

Figure 2. The sensitivity of f(t; **p**) to the wavespeed of the middle layer of a three-layered medium (Transducer frequency = 1 MHz).

Table 1. Properties of the constituent layers

Constituent layer	Layer #1 Aluminum	Layer #2 Water	Layer #3 Aluminum
Thickness (mm)	1.334	0.495	1.542
Wavespeed (mm/μs)	6.287	1.490	6.303

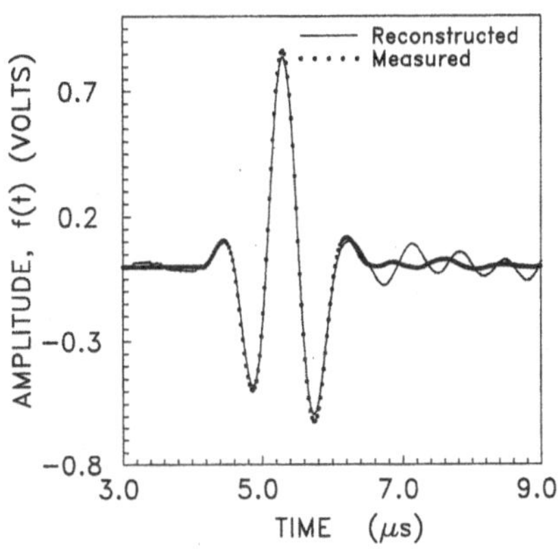

Figure 3. The comparison of the measured and reconstructed incident fields for a three-layered medium (aluminum/water/aluminum).

RESULTS AND CONCLUSIONS

This technique was used for NDE of a three-layered medium (see Table 1) in the case of through transmission. The comparison of the theory and experiment is shown in Fig.3 and is considered to be excellent. The Simplex method was used to minimize the error function, $E_f(\mathbf{p})$, to invert the layer properties for the following cases: (1) Simultaneous measurement of h and c for any one of the three layers, given h and c for the remaining two layers; (2) Simultaneous measurement of three thickness, given the three wavespeeds; and (3) Simultaneous measurement of the three wavespeeds, given the three thickness. The measurement results are as follows. For case (1), $h_{1measured}$=1.328 mm, $c_{1measured}$=6.531 mm/μs, $h_{2measured}$=0.486 mm, $c_{2measured}$=1.480 mm/μs, and $h_{3measured}$=1.544 mm, $c_{3measured}$=6.484 mm/μs; For case (2), $h_{1measured}$=1.335 mm, $h_{2measured}$=0.490 mm, and $h_{3measured}$=1.544 mm; For case (3), $c_{1measured}$=6.484 mm/μs. $c_{2measured}$=1.510 mm/μs, and $c_{3measured}$=6.132 mm/μs. In all cases the measurement error was found to be less than or equal to 3%, except for the $c_{1measured}$ of the first case, of which the measurement error was found to be 4%.

ACKNOWLEDGEMENT

The material is based on the work supported by the Texas Advanced Research Program (Advanced Technology Program) under Grant No. 282 to Texas A&M University.

REFERENCE

1. C. Zhu and V.K. Kinra, "A New Technique for Time-domain Ultrasonic NDE of Thin Plates," *J. of Nondestructive Evaluation* (accepted for publication).
2. I.M. Brekhovskikh, "Waves in Layered Media", Academic Press (1981).
3. S.E. Hanneman and V.K. Kinra, "A New Technique for Ultrasonic Nondestructive Evaluation of Adhesive Joints: Part I. Theory," *Experimental Mechanics*, Vol. 32(4), pp. 323-331 (1992).

IMAGING OF SMALL DEFECT AND FRACTURE PROCESS ZONE BY USING SCIENTIFIC VISUAL ANALYSIS IN ULTRASONIC INSPECTION

Shuichi Mikami[3], Toshiyuki Oshima[1], Noboru Sugawara[2],
Tomoyuki Yamazaki[3] and Shinya Sugiura[4]

[1]Professor, Department of Civil Engineering, Kitami Institute of Technology, Kitami-city, 090 Japan
[2]Associate Professor, Department of Civil Engineering, Kitami Institute of Technology, Kitami-city, 090 Japan
[3]Research Assistant, Department of Civil Engineering, Kitami Institute of Technology, Kitami-city, 090 Japan
[4]Research Engineer, Konoikegumi Corporation, Tokyo, Japan

INTRODUCTION

When we diagnose the structural integrity of a material in service and evaluate its remaining life time, we need to improve analyzing method to get the accurate information for a internal damage of a material. And by the recent demand for high level quality control of structural joint the accuracy improvement of detected image became essential in nondestructive evaluation (NDE)[1]. So far several methods like X-ray, EMAT and ultrasonics are used to detect an internal defect and 3D image of a damage is obtained only by CAT scan method of X-ray inspection.

In this study we made the different types of artificial defects by drilled hole in a steel plate specimen to observe by an ultrasonic scanning pulse echo test. And the several effects of measurement condition on the display image are obtained by this experimental analysis [3,4]. 3D images of these defect are obtained by a graphic software on a workstation by using the maximum amplitude data of boundary and bottom echo and time of flight data. And by applying precise waveform analysis on the received reflection waves the accurate defect shape and location are obtained by clear images.

C-scan images of fracture process zone around a notch tip are obtained under a uniaxial tension for a steel specimen by immersion test. Propagation of a plastic zone obtained by a numerical calculation is compared with this C-scan image.

Nondestructive Characterization of Materials VI
Edited by R.E. Green, Jr. *et al.*, Plenum Press, New York, 1994

MEASUREMENT SYSTEM

The measurement system employed in this study is shown in Fig.1. The sampling gate of time axis is set correspond to the thickness of specimen to receive first boundary echo from a defect and bottom echo. Data of maximum amplitude of each echo waves which passed above sampling gate are obtained as a digital value of a voltage. We get its changes of the maximum amplitude correspond to the shape and location of a defect. Comparison of this digital voltage data with the reference voltage gives us image data to represent the 2D shape of defect on the screen. Ultrasonic wave recorder system can get the whole reflection waves that passed above mentioned time gate as 4096 data of digital values for every wave and display the spectrum of the detected waves by FFT.

Table 1. Specimens.

Type I	φ(°)	h (mm)	Type II	φ(°)	h (mm)
A	90	5	J	90	5
B		3	K		3
C		2	L		2
D	75	5	TypeI 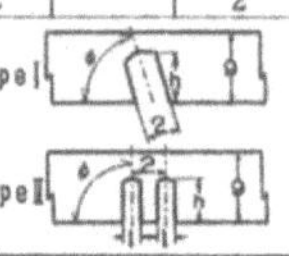		
E		3			
F		2			
G	60	5	TypeII		
H		3			
I		2			

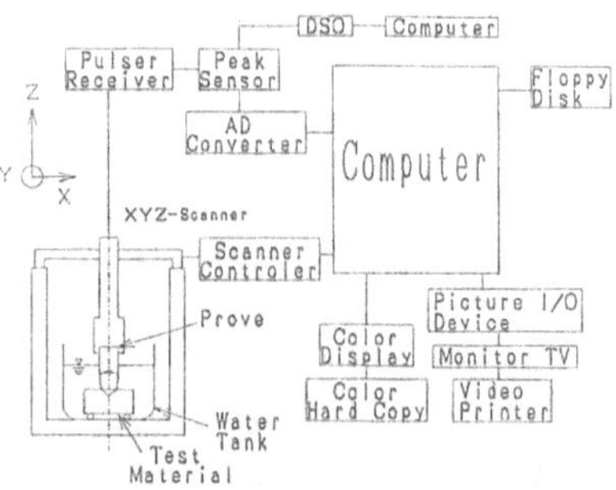

Figure 1. Measurement system.

DETECTED IMAGE OF DEFECT SHAPE

When we detect a defect shape by using ultrasonic testing, the accuracy of image data of a detected defect depends upon the resolvability of positioning and inclination of transducer during the measurement. After the comparison of resolution of image data obtained by three transducers of different natural frequencies we decided to use the transducer mostly with 10MHz frequency in this research.

Specimen and Measurement

In the steel plate specimen 9mm thick twelve artificial defects of different types are made by drilled holes as shown in Table 1. Among them there are three types of depth h as 5mm, 3mm, 2mm and three types of inclinations φ as 90°,75°,60° of drilled holes, respectively. And the diameters of drilled holes are 2mm and 1mm as shown in Table 1 and two close holes of diameter 1mm at 2mm distant are used to get the coupling effect of ultrasonic scattering on the image of defect.

The kinds of the specimen for coupling effect are named by specimen J,K,L with different depth as 5mm,3mm,2mm as shown in Table 1, respectively. Reflection waves from the defect are measured by using a line scanning along with a distinct line to across the C-scan image of defect. They are discussed later.

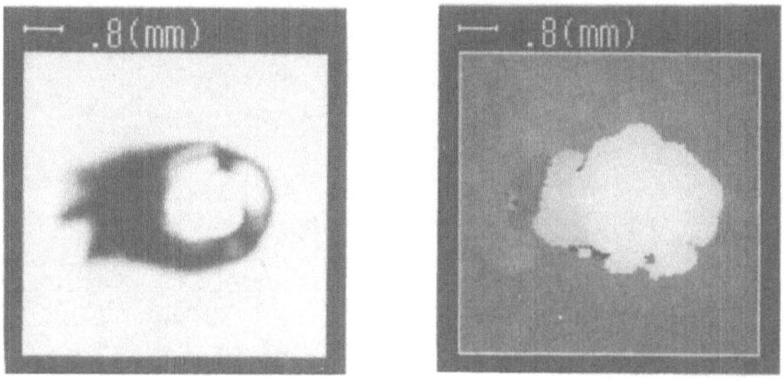

(a) Amplitude (b) Time of flight

Figure 2. C-scan image of specimen G (6×6mm, pitch 0.05mm).

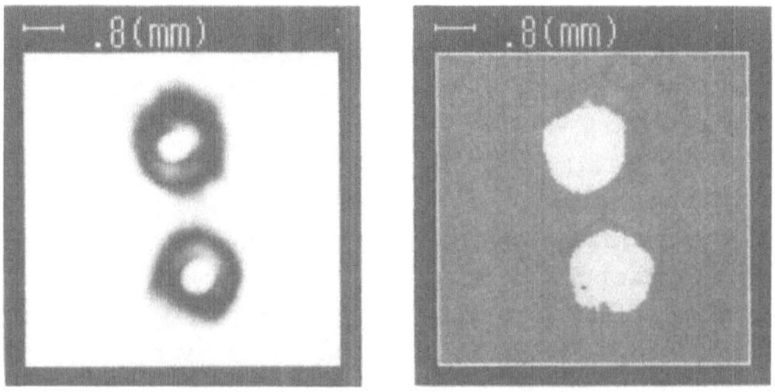

(a) Amplitude (b) Time of flight

Figure 3. C-scan image of specimen K (6×6mm, pitch 0.05mm).

Results of Display Image

The C-scan images of defect G and K with different types obtained by using the transducer with 10MHz are shown in Fig.2 and Fig.3 respectively. In Fig.2 and Fig.3 the images obtained by maximum amplitude data are shown in left hand side as (a) and time of flight data are shown in right as (b). The tip shapes of drilled holes are well detected. The focal point of the ultrasonic waves is adjusted around at the tip of drilled hole in this measurement. If the depth of drilled hole is deep, that is, the distance between the defect tip and bottom of plate is large, the scattering effect of ultrasonic wave on the display image of the defect becomes to be larger than the others. This behavior of wave scattering and the decrease of resolution is considered somewhat like that the dim resolution of a photo is obtained by a insufficient exposure depth of a camera.

The coupling effect of two close defects of small size does not effect much on their image in this case.

3D DISPLAY OF DEFECT SHAPE BY MAXIMUM AMPLITUDE DATA AND TIME OF FLIGHT DATA

In this chapter 3D representation of maximum amplitude data by using a computer graphics (CG) software on a workstation (WS) is shown. In this analysis the values of 256 color grades correspond to the amplitude which are

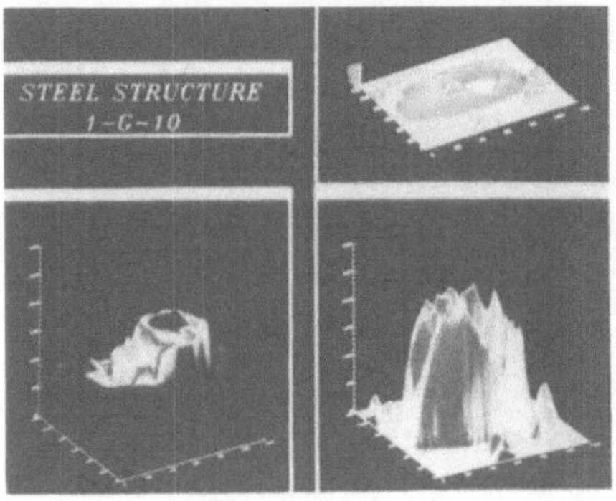

Figure 4. 3D images of specimen G.

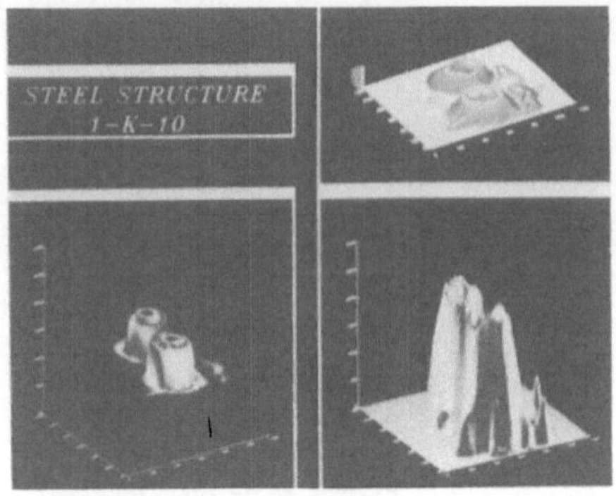

Figure 5. 3D images of specimen K.

on the 2D plane of scanning range are taken in the vertical coordinate perpendicular to the above 2D plane. And the time of flight data are used like filtering of the boundary echo waves by which we can path the boundary echo waves through to separate from the bottom echoes within the scanning range. By this filtering technique we can get the clear 3D images of maximum amplitude data by representing the clear boundary of small defect.

In Fig.4 three different 3D images of defect type G as listed in Table 1 are shown. In the lower picture of right hand side in Fig.4, the above mentioned 3D representation of the maximum amplitude data by converting the 256 color grades values to the vertical coordinate values within the scanning range is shown by using CG software on the WS display. In the upper picture of right hand side in Fig.4 the different representation of the same 3D image as mentioned above is shown by changing the location of view point and the scale of vertical coordinate on CG software. By this picture we can find out the detailed undulation of small defect tip by clear image. In the lower picture of left hand side in Fig.4 only the 3D image of a defect shape is shown in the dark background by using the path length data to pass only the boundary echo values by a above mentioned filtering technique. And by using 3D rotation procedure on CG software we can rotate the 3D image of the defect on the display of WS we can understand the whole shape of a small defect by 3D representation. By this picture the effect of inclination of drilled holes on the 3D images is clearly obtained. And in Fig.5 the 3D images of the defect type K is shown. By using this CG technique to obtain 3D image of defect we can find out the detailed structure of a defect and thus evaluate the accurate strength of a material. It is called a "D-scope display" of a defect correspond to the A-scope, B-scope and C-scope display in this paper.

ACCURATE ANALYSIS OF BOUNDARY ECHO BY WAVE ANALYSIS

The detected defect image given by ultrasonic scanning test is given by the data of echo amplitude. The echo amplitude data consist of the maximum amplitude of boundary and bottom echo which are reflected from the boundary of a defect and lower surface of the specimen. If the location of probe and the focal point of the incident wave reaches to the boundary of the defect, the boundary echo reflected from the boundary of the defect is detected together with the bottom echo reflected from the lower surface of the specimen within the time gate set up by the system. However we sometimes cannot obtain a good defect image with sufficient accuracy in these measurement because of

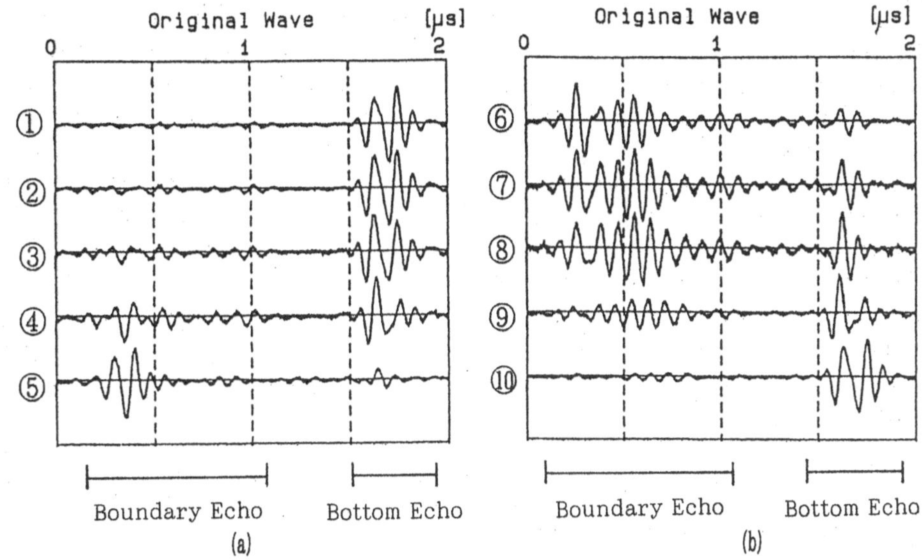

Figure 6. Measured waveforms of specimen A.

the wave scattering at the boundary of the defect. This scattering effect on the wave reflection depends upon the wave length and shape and size of a defect. Therefore we need a detail analysis of the boundary echo to get the accurate information from the wave on the shape and size of the defect in addition to the C–scan image given by the echo amplitude data.

In this chapter we discuss the precise wave analysis of boundary echo obtained within the time gate set up by the system and find out a method to measure the more accurate size of the defect.

Experimental Measurement

The circular drilled holes of diameter 2mm are made in the steel plate of 9mm thick with length of 5mm, and we call here this specimen as specimen A. The focus of the probe is first adjusted at the center of drilled hole. And the probe with 10MHz resonance frequency scans over the square of the area of 5mm×5mm on the drilled hole to get the C–scan image of the defect. The reflection waves are received by 50μm pitch to across the boundary of the defect and by 100 μm pitch for the other area. The reflection waves are transformed to the digital data by using the digital storage oscilloscope(DSO) and stored to a floppy disk. The sampling pitch of time is 0.57nsec and the number of data for the one reflection wave is 4096 points.

In Fig.6(a),(b) the typical original data received by the ultrasonic scanning test are shown. In Fig.6(a) the transition of the received wave from the bottom echo(wave ①,②) reflected from the lower surface to the boundary echo(wave ④,⑤) reflected from the defect, when the center of the probe comes into the defect, is clearly shown. And in Fig.6(b) another transition of the received waves from the boundary echo to the bottom echo, when the probe goes out across the defect boundary, is also shown.

In these figures the amplitudes of the waves are normalized by the maximum amplitudes of each waves and the abscissas are scaled by microsecond(μs). The numbers shown in Fig.6(a),(b) as ①,②,etc. indicate the location of the probe where the reflection waves are received and they also correspond to the locations shown in Fig.7.

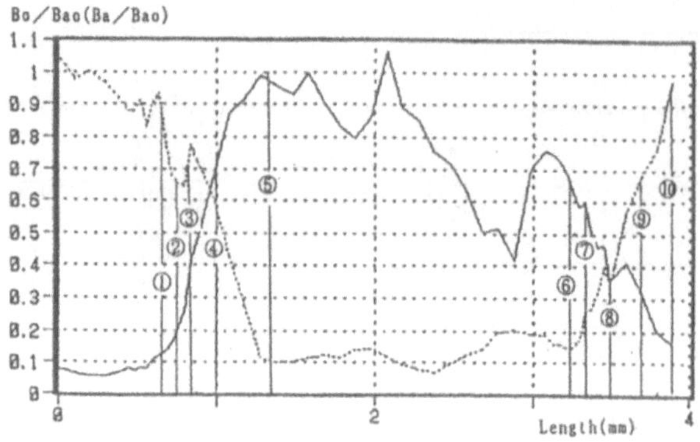

Figure 7. Results of (Bo/Bao) and (Ba/Bao) for specimen A.

Results Obtained by Wave Analysis

The reflection wave involves both the boundary echo and the bottom echo, and when the probe comes to be close to the defect boundary the amplitude of boundary echo becomes greater than that of bottom echo. Thus there is a boundary of defect somewhere in the transition range. In the defect image analysis by using the maximum amplitude of the reflection wave we receive the greater amplitude of boundary echo within the area of defect as shown in Fig.6(a),(b) from the wave number ⑤ to ⑧. As we can separate the boundary echo from the bottom echo in the reflection wave, we can get each maximum amplitude of both the boundary echo(Bo) and bottom echo(Ba). And the normal maximum amplitude of the bottom echo(Bao) is also obtained far from the defect boundary.

The ratio between the maximum amplitude of the boundary echo (Bo) and the normal maximum amplitude of the bottom echo(Bao) is shown in Fig.7 by the solid lines in the cases of specimen A and the ratio between the maximum amplitude of the bottom echo(Ba) and the normal maximum amplitude of the bottom echo (Bao) is also shown by the dotted line. The abscissa is scaled by the distance in millimeter(mm).

We can see in the figure that the ratio of Ba/Bao becomes small within the length nearly equal to 2mm which correspond to the size of real defect. And the distribution of the ratios of Bo/Bao means the undulation of the drilled hole tip because it corresponds to the distribution of maximum amplitude of boundary echo. The 3D representation of reflection waves are also shown by using SVA

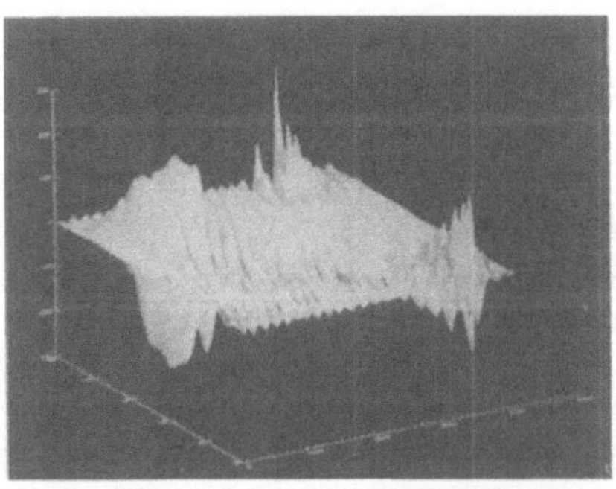

Figure 8. 3D waveform for specimen G.

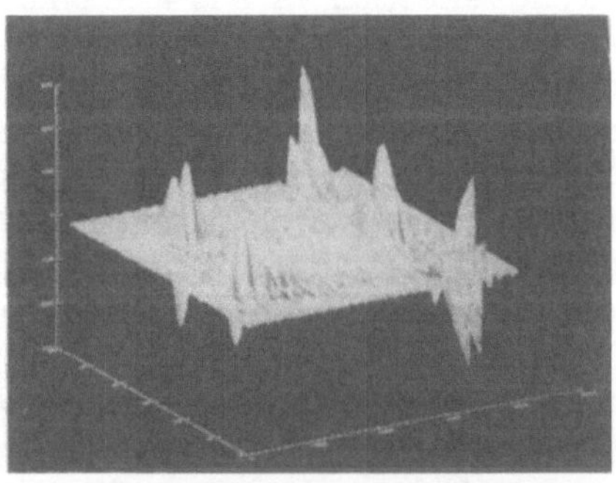

Figure 9. 3D waveform for specimen K.

in Fig.8 and Fig.9 in the cases of defect G and defect K, respectively. The incident waves come from left hand side. We can see in these pictures that the waves reflect at the defect boundary when the transducer gets into the defect and elsewhere reflect at the bottom of the plate. In these 3D analysis reflection waves of 53 positions of transducer by 50μm pitch are displayed.

Analysis of Fracture Process Zone

When a crack tip which is propagating in some direction, comes to an internal defect, the estimated strength of a material given by fracture mechanics will be different from the actual one.

It is, therefore, important to chase the fracture process zone around a notch tip of a damaged structural member and evaluate its remaining strength. In this study three specimens with different notch angles as 30° ,60° and 90° are tested as shown in Fig.10 in water tank to use immersion method and the c-scan images of fracture process zone are obtained. These images of plastic zone with different notch angles are shown in Fig.11. And these experimental results are compared with the numerical results given by finite element analysis of elastic-plastic behavior and obtained a good agreement as shown in Fig.12.

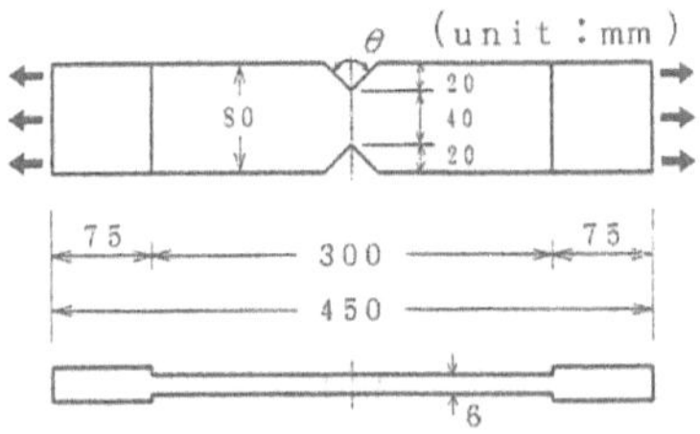

Figure 10. Tension test and specimen.

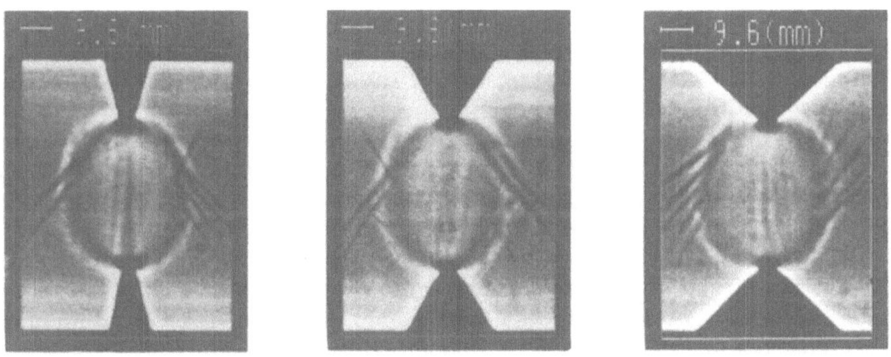

(a) θ=30° (b) θ=60° (c) θ=90°

Figure 11. Image of plastic zone around notch tip with different notch angles under tension (60×82mm, pitch 0.2mm).

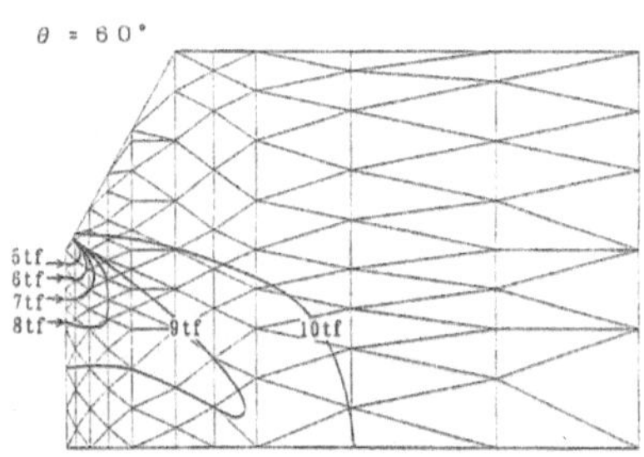

Figure 12. Growth of plastic zone. (Numerical result by FEA)

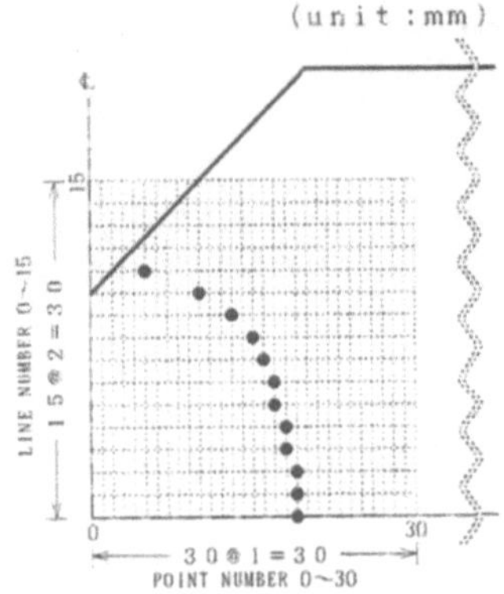

Figure 13. Observed boundary line between plastic and elastic zone.

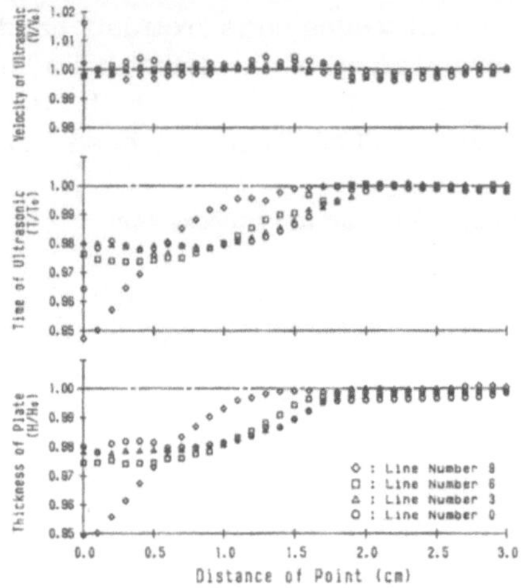

Figure 14. Chage of thickness and wave speed around plastic zone.

In plastic zone around a notch tip both the change of thickness of specimen and the change of material property by yielding are coupled, therefore, we measured both the thickness and the wave speed through the thickness to compare with the image data, and obtained a good agreement as shown in Fig.13 and Fig.14.

CONCLUSION

Summarizing the above study on the precise analysis of the reflection wave by using SVA to get a improved image of the defect, we have come to the conclusions as follows;

(1)Twelve C-scan images of small defects are obtained by using ultrasonic scanning system together with the time of flight data which is used to make the defect boundary of image clear.

(2)3D images of small defects are obtained by a CG software on a workstation using both the maximum amplitude data and time of flight data. We call here this 3D image as a D scope image.

(3)By using SVA of the reflection wave we can obtain the more accurate defect size given by the distribution of the boundary and bottom echo together with the scattering effect at the defect boundary.

(4)The image of fracture process zone under tension test of a notched specimen are obtained by experiment and compared with a numerical results.

REFERENCES

1. Thompson, D.O. and Chimenti, D.E. "Review of Progress in Quantitative Nondestructive Evaluation", Plenum Press, Vol.1(1981)～Vol.11(1992).
2. Krautkrämer, J. and Krautkrämer, H. : Ultrasonic Testing of Materials, Springer-Verlag, 1977.
3. Sugawara, N., Oshima, T., Mikami, S., and Sugiura, S. "On the Accuracy Improvement in Ultrasonic Inspection by Using Computer Graphics and Analysis", Proc. of JSCE, No.459/I-22(1993).
4. Sugawara, N., Oshima, T., Mikami, S., and Sugiura, S. "Accuracy Improvement of Small Defect Detection for Ultrasonic Inspection by Using Scientific Visual Analysis", Review of Progress in QNDE, Vol.12(1993).

NONDESTRUCTIVE DETERMINATION OF THE HARDENING DEPTH IN INDUCTIVE HARDENED STEELS

I. Altpeter and M. Kröning

Fraunhofer-Institut für zerstörungsfreie Prüfverfahren
D-66123 Saarbrücken, FRG

INTRODUCTION

In industrial quality control the inspection of hardness and hardness depth is carried out destructively with the help of hardness measurements and metallographic techniques. These techniques are time consuming and therefore not suitable for on-line processing. For a nondestructive determination of hardness depth, micromagnetic measuring techniques have been developed recently.

Near-surface hardness as well as inductive hardness depth values in the range up to 3 mm can be nondestructively determined using measuring quantities derived from the magnetic Barkhausen noise and the time signal of the magnetic tangential field strength. For the majority of the components and different steel compositions investigated, these nondestructively determined results are within the range of respective industrial specifications.

PHYSICAL BACKGROUND

Ferromagnetic condition is a prerequisite for a material to be used for magnetic material characterization. All ferromagnetic materials have a structure of `domains', hence, areas of varying magnetization directions. The individual domains are separated from each other by so-called Bloch-walls. In iron materials only 180° or 90° Bloch-walls exist.

The ferromagnetic domain-structure as well as the Bloch-wall density and movements are strongly influenced not only by macroscopic-(residual)-stress fields, but also by the microstructural state (dislocations, precipitations, grain boundaries). The crystal lattice imperfections interact with the Bloch-walls via their residual-stress fields, or via the variation of the Bloch-wall energy density[1]. Quantitative relations between microstructural parameters and individual macroscopic magnetic measuring quantities are known from theoretical models[2]. The majority of these models is, however, restricted to monocrystals and the interaction of one-dimensional Bloch-walls with crystal lattice imperfections. Therefore the models can only be used for a qualitative description of the magnetic characteristics.

Nondestructive Characterization of Materials VI
Edited by R.E. Green, Jr. *et al.*, Plenum Press, New York, 1994

The micromagnetic inspection concept has its basis in a correlation between 3MA-testing quantities (Micromagnetic-, Multiparameter-, Microstructure- and stress Analysis) and material characteristics, with the aid of multi-parameter regression-algorithms. Several independent measuring quantities are combined by using material-specific calibration curves in order to solve the multi-parametric inspection task of material characterization. The individual measuring quantities as for example the magnetic Barkhausen noise, the incremental permeability as well as the spectral analysis of the magnetic tangential field-strength time signal, deliver with different sort of information, as they are concerned with reversible as well as irreversible remagnetization processes.

Measuring Quantities

All micromagnetic measuring techniques require a dynamic magnetization of the material[3] in hysteresis cycles. Micromagnetic techniques excite and/or measure:

- the magnetic Barkhausen noise,
- the incremental permeability,
- the dynamic magnetostriction as well as
- the upper harmonics in the time signal of the magnetic tangential field strength,

whereby the subsequently presented results refer only to the measuring quantities, derived from the magnetic Barkhausen noise and the upper harmonics analysis because these are applied in a correlation approach.

The magnetic Barkhausen noise, for example, is excited by way of continual hysteresis reversals, whereby an inductive sensor, either an air coil or an air coil with a ferrite core, picks up the noise information.

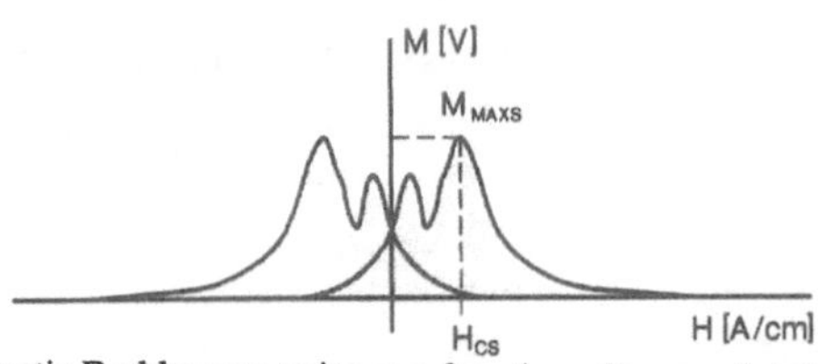

Figure 1. Magnetic Barkhausen noise as a function of tangential field strength.

Remagnetization processes take place mainly in the vicinity of the coercivity field-strength and are registered in the sensor as induced electric voltage. These so-called noise-signals are amplified by 40 to 100 dB. They are rectified and are recorded in the standard way, as a function of the tangential magnetic field strength, by using a Hall-probe. In steels with carbon content ≥ 0,2% (polycrystalline, isotropic), the rectified Barkhausen noise amplitude usually shows a maximum (M_{MAX}) in the vicinity of the coercivity (fig. 1). The magnetic field strength (H_{CM}) derived from the position of the noise maximum correlates well for these technical steels with the actual coercivity field strength, which is derived from the hysteresis curve[4].

In order to apply the magnetic Barkhausen noise technique for the nondestructive determination of hardening depths, a variation of the volume of interaction is necessary (fig. 2). The basic mechanism adjusting the depth of interaction is characterized by the law of the eddy current damping.

Local magnetic events, i.e. Blochwall jumps and rotations of the magnetization direction in domains, dissipate energy by inducing pulsed eddy currents in their vicinity. These pulses are relatively wide banded with bandwidth from dc up to some MHz and they are damped following an exponential decay. The higher the frequency in the spectrum, the higher is the damping. The damping constant for each frequency component f is $1/\delta$. Therefore the higher frequent part in the received noise is originated only in surface near zones of a magnetized component.

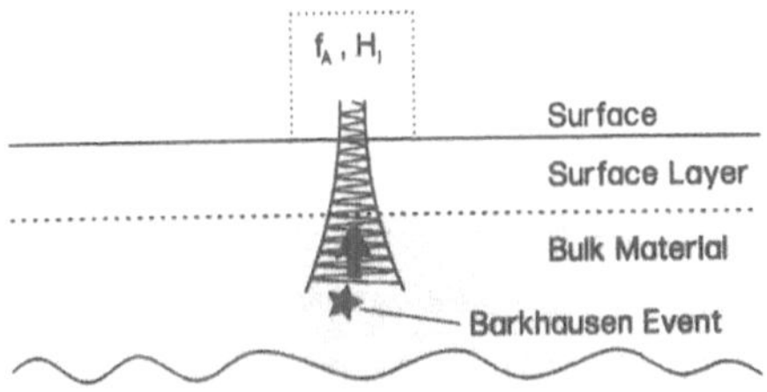

Figure 2. Interaction volume.

Applying the theorem of reciprocity the analysing depth is approximated also by δ i.e., the standard eddy current penetration depth when f_A is selected to be the centre-frequency and frequency range of the pick-up system (receiver-coil and band pass filter):

$$\delta = \frac{1}{\sqrt{\pi \cdot \sigma_{el} \cdot f_A \cdot \mu_0 \cdot \mu_\Delta}} [mm]$$

f_A: Analysing frequency (filter) [Hz]

μ_0: $4\pi \cdot 10^{-7}$ [H/m]; magnetic field constant

μ_Δ: Incremental permeability

σ_{el}: Electrical conductivity [$m/\Omega mm^2$]

A more exact calculation of the penetration depth is obtained considering the coil geometry and sample geometry according to numerical formulations[5].

Concerning surface hardened materials in the Barkhausen noise typical double peaks are received after wide-band filtering. Signal analysis reveals the fact that the peak at higher magnetic fields is produced by the surface near martensitic microstructure and that the peak at lower magnetic fields belongs to the magnetic weak unhardened bulk material. The hardened layer is characterized by setting a high pass filter , i.e. a small analysing depth. The remaining half-peak-separation is named H_{CS} and the peak amplitude is M_{MAXS} respectively.

The whole impedance of the excitation-yoke is dependent on the chosen core material and the magnetic characteristics of the test specimen. Furthermore, the permeability of ferromagnetic materials is dependent on the magnetic field (hysteresis influence). This, according to [6], leads to a non-sine-shaped time behaviour of the current in the yoke-coil. Because of the hysteresis symmetry in addition to the fundamental only higher harmonics of the odd order (3, 5, 7, 9, ...) are excited. From the upper harmonics analysis of the tangential field strength signal (fig. 3), as measuring quantities, the amplitude and phase of the upper harmonics A_3, A_5 and A_7 (coefficients of higher order are neglected) and a distortion factor K are derived[7]:

$$K = \sqrt{\frac{|\tilde{A}_3|^2 + |\tilde{A}_5|^2 + |\tilde{A}_7|^2}{|\tilde{A}_1|^2}} \cdot 100\%$$

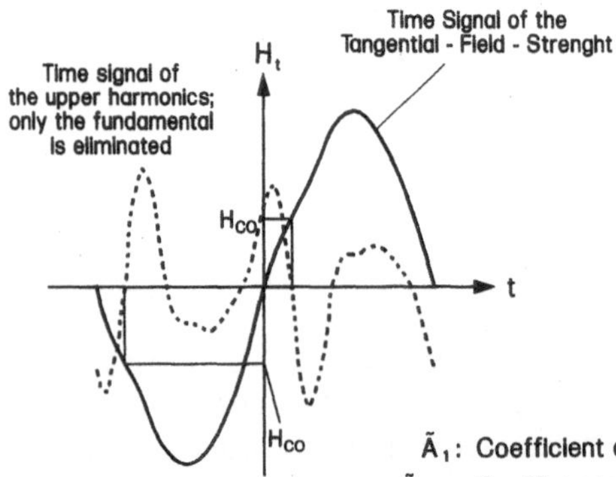

Figure 3. Analysis of upper harmonics in the time signal of the magnetic tangential field strength.

Measuring Procedure and the Instrumentation

The use of magnetic effects always requires a magnetic excitation of the test sample, which in most cases can be obtained by using an u-shaped electromagnet (fig. 4). A bipolar power supply with a suitable driver (e.g., function generator) produces magnetic hysteresis curve reversals. The magnetic field strength H_{MAX} is in the field range $H_{MAX} \leq \pm$ 100 A/cm, the magnetizing frequency f is selected in the range 0.1 Hz $\leq$ f $\leq$ 1 kHz. In the case of Barkhausen noise measurements the receiver coils - as mentioned above - are pick-up air coils, ferrite core coils or tape recorder heads depending on the specified spatial resolution[7].

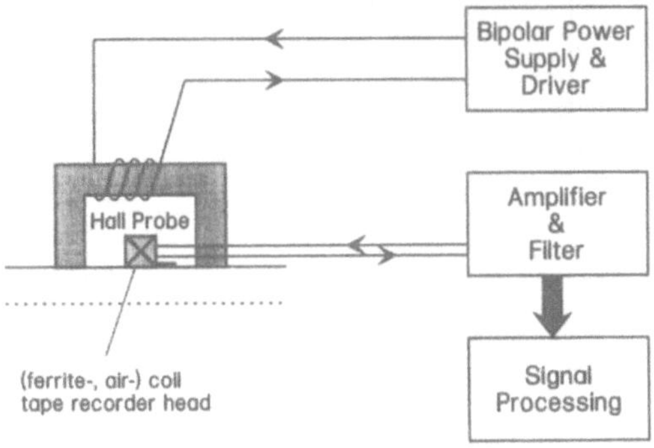

Figure 4. Exciting-transducer-system.

The tangential magnetic field strength can be measured with Hall elements. The transducer can easily be used by trained inspectors. An automatization of the testing technique is possible. The measuring time for each measuring point depends on the technique and the magnetizing frequency. By using time averaging in the Barkhausen noise - in order to have reliable measuring quantities - a number of 5-10 time periods are sufficient[8].

RESULTS

The measuring procedures were applied to determine the hardness depth and the surface hardness in 12 water-pump shafts. Fig. 5 illustrates the measured H_{CS}-values as function of the metallographically determined HRC-values (Rockwell hardness) for the water-pumps at three different measuring points 4,7,8 (fig. 6) and a linear correlation is observed.

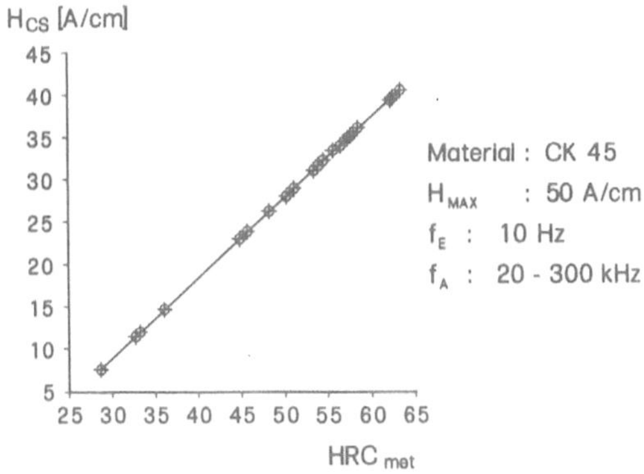

Figure 5. Correlation between H_{CS}-values and metallographically determined HRC_{met}-values.

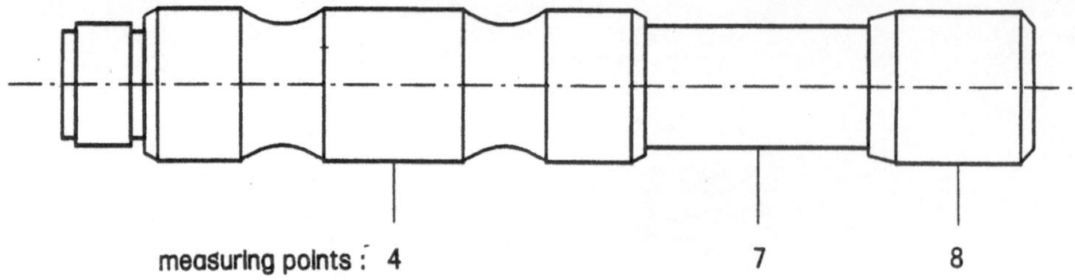

Figure 6. Water-pump shafts schematically.

The approach can be enhanced by using a further term according to formula (1), fig. 7 shows the HRC_{nd}-values determined with this 3MA approach:

$$HRC_{nd} = 21.4 + 1.09 \cdot H_{CS} - 0.02 \cdot M_{MAXS} \cdot H_{CS} \qquad (1)$$

(standard deviation: 2s = 4.04 HRC-units) as function of metallographically determined HRC-values.

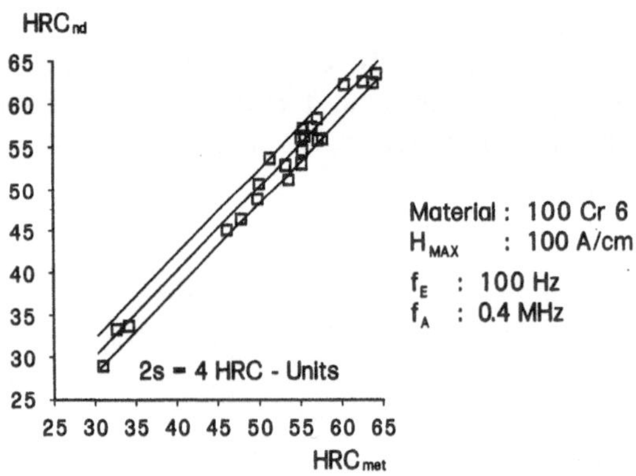

Figure 7. Correlation between HRC_{nd}-values and metallographically determined HRC_{met}-values on water-pump-shafts.

By using the measuring quantity K in addition to the measuring quantity H_{CS} it is possible to determine the hardening depth HD_{nd} with the following 3MA approach:

$$HD_{nd} = 1.53 + 0.08 \cdot H_{CS} - 0.06 \cdot K \cdot H_{CS} + 3.1 \cdot 10^{-4} (H_{CS} \cdot K)^2 \qquad (2)$$

Fig. 8 indicates the correlation between HD_{nd}-values and HD_{met}-values.

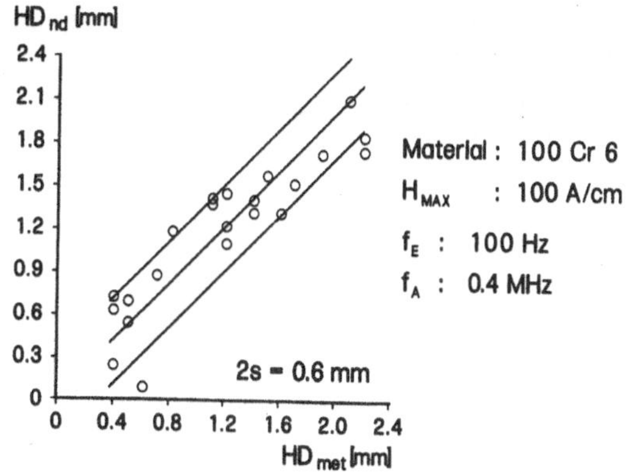

Figure 8. Correlation between HD_{nd}-values and metallographically determined HD_{met}-values on water-pump-shafts.

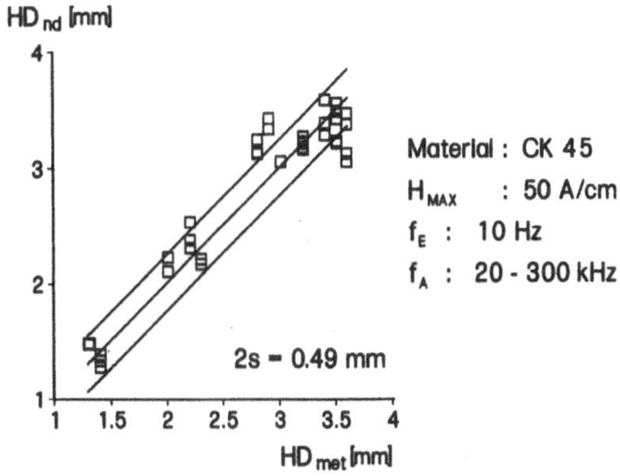

Figure 9. Correlation between HD_{nd}- and HD_{met}-values on shaft journals.

The micromagnetic hardness depth determination was also applied on shaft journals of which 36 parts were available. Because of a self annealing effect, the surface hardness decreases with increasing hardness depth. In order to enhance the approach, i.e. to reduce the standard deviation, two classes of shafts are separated by analysing the coercivity H_{CS}. The self annealed shafts have values 12 A/cm $\leq H_{CS} \leq$ 19 A/cm and the non-self-annealed 20 A/cm $\leq H_{CS} \leq$ 28 A/cm. Correlation approaches are received in each class. Formula (3) is for the class of the self-annealed shafts and fig. 9 documents the result. The standard deviation 2s is 0.49 mm.

The approach was tested on shaft journals from the production line. By using (3) the predicted hardness depth values are in the range 2.8 mm - 3.3 mm which is in the scattering range of the metallographic results (2.5 mm - 4.5 mm).

$$HD_{nd} = 6.79 + 0.26 \cdot H_{CS} - 0.05 \cdot K \cdot H_{CS} + 5.37 \cdot 10^{-5} (K \cdot H_{CS})^2 \qquad (3)$$

In the same way the scattering can be reduced in the approach for the hardness determination. The standard deviation is 2s=3.9 HRC-units, observing the approach according to (4) in the class of self-annealed shafts.

$$HRC_{nd} = 30.0 + 1.4 \cdot H_{CS} + 0.2 \cdot M_{MAXS} \cdot H_{CS} \quad (4)$$

In the class of the non-self-annealed shafts the 3MA approach is:

$$HRC_{nd} = 56.9 - 0.15 \cdot H_{CS} + 0.34 \cdot M_{MAXS} \cdot H_{CS} \quad (5)$$

The standard deviation is 2s = 5.0 HRC-units (fig. 10).

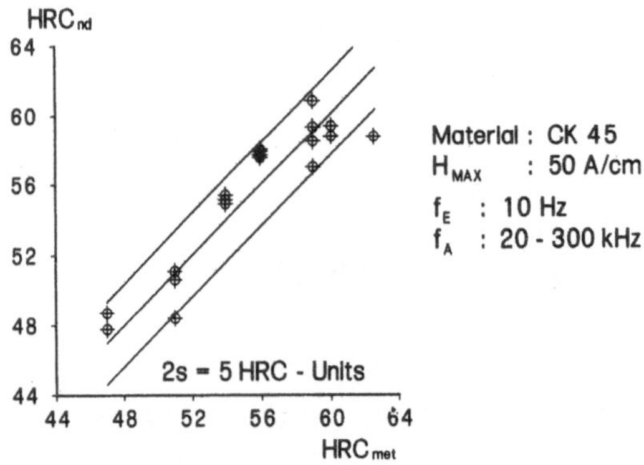

Figure 10. Correlation between HRC_{nd}- and metallographically determined HRC_{met}-values on shaft journals.

In order to increase the number of calibration points in this class of non-self-annealed shafts, i.e. to assure a better completeness of random data, measuring values at non-hardened surface areas were introduced in an approach. Formula (6) gives the result as shown in figure 11:

$$HRC_{nd} = 8.46 + 1.41 \cdot H_{CS} + 0.89 \cdot M_{MAXS} \cdot H_{CS} \quad (6)$$

with a standard deviation 2s = 8.1 HRC-units. Compared with (5), the scattering in the data is higher, but the nondestructive hardness prediction is in the specified range.

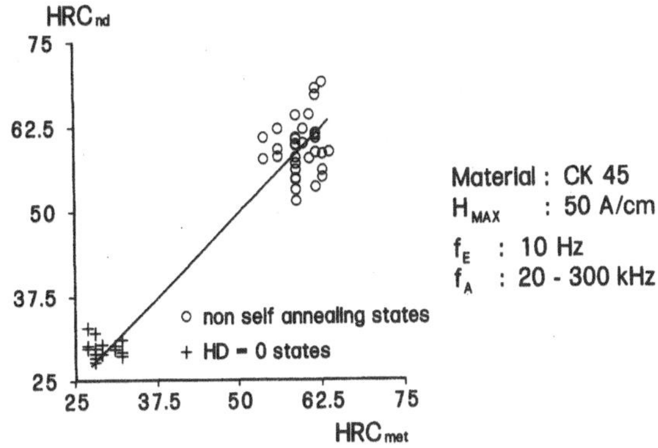

Figure 11. Correlation between HRC_{nd}- and metallographically determined HRC_{met}-values on shaft journals.

These approaches were also tested on shaft journals from the production line. The H_{CS}-measurement has revealed for all of the test parts to be candidates of the class of the self-annealed shafts. The hardness values predicted according to (4) are in full agreement within the scattering range of metallography ($HRC_{met} = 57 \pm 3$).

CONCLUSION

For the majority of investigated components and different steel grades a good correlation was found between nondestructively determined measuring quantities derived from the magnetic Barkhausen noise and the time signal of the magnetic tangential field strength and the metallographically determined values by using the 3MA approach:

$$HD_{nd} = a + b \cdot H_{CS} + c \cdot K \cdot H_{CS} + d \cdot (K \cdot H_{CS})^2$$

The nondestructively determined hardness depth values are within the range of respective industrial specifications.

The transmitter-receiver system is to be adapted to the given component geometry. It is necessary to magnetise the bulk material sufficiently, i.e. the excitation frequency must be in the frequency range 1 - 30 Hz and the magnetic field strength between 20 and 50 A/cm depending on the steel grade.

Because the near-surface hardened value decreases with increasing hardness depth, it is important to determine the hardness value by selecting magnetic quantities also in near-surface zones, i.e. by setting a high pass filter. A nondestructive determination of the hardness value requires the 3MA approach:

$$HV_{nd} = a' + b' \cdot H_{CS} + c' \cdot H_{CS} \cdot M_{MAXS}$$

ACKNOWLEDGEMENT

This work has been carried out under the research support of the FKM (Forschungskuratorium Maschinenbau).

REFERENCES

1 Seeger, A.: Probleme der Metallphysik, Berlin Springer Verlag (1966).

2 Kneller, E.: Ferromagnetismus, Berlin Springer Verlag (1962).

3 Schneider, E., Altpeter, I., Theiner, W.A., Ruud, C.O., Green, Jr., R.R. (eds.): Nondestructive methods for material property determination, New York, Plenum Press (1984).

4 Altpeter, I.: Spannungsmessung und Zementitgehaltsbestimmung in Eisenwerkstoffen mittels dynamischer, magnetischer und magnetoelastischer Meßgrößen, Dissertation (1990).

5 Dodd, C.W.: Integral solutions to some eddy current problems, *International journal of nondestructive testing*, 1 (1969) p. 29.

6 Koch, K.M., Jellinghaus, W.: Einführung in die Physik der magnetischen Werkstoffe, Wien, Verlag Franz Deuntike (1957).

7 Pitsch, H., Dobmann, G.: Magnetic tangential field-strength-inspection, a further ndt-tool for 3MA. Proceedings of the 3rd International Symposium Saarbrücken, FRG October 3-6, 1988.

8 Altpeter, I., Theiner, W.A., Reimringer B.: Härte- und Eigenspannungsmessungen mit magnetischen zerstörungsfreien Prüfverfahren, Tagungsband DGM: Eigenspannungen, Karlsruhe, 14.-16.4.1983.

CLASSIFICATION OF DEFECTS IN THICK SECTION GRAPHITE EPOXY TEST BLOCKS

L. M. Brown and C. A. Lebowitz

Annapolis Detachment, Carderock Division
Naval Surface Warfare Center
Annapolis, Maryland 21402-5067

INTRODUCTION

Fiber-reinforced composites are being used increasingly in U.S. Navy structures due to the fact that composites offer the potential for weight, cost, and signature reduction, as well as increased corrosion resistance (Gagorik, et. al., 1991). Current and future applications of composites include: deckhouses, propulsion shafting, machinery foundations, air flasks, masts, heat exchangers, etc.

Unlike metallic structures for which the effects of defects may be predicted, the effect of defects on the structural integrity of an advanced composite material is difficult to assess because the lack of homogeneity and anisotropy of the composite material does not produce a predominant failure mode. Defects and conditions which affect the material are: voids, delaminations, inclusions, disbonds, anomalies in cure, matrix crazing, anomalies in resin distribution, translaminar cracks, fiber misalignment, anomalies in fiber volume fraction, fiber breakage, environmental effects, machining damage and others.

Each defect type results in a degradation of a specific mechanical property. The most extensively studied defect in composites has been voids. Voids have been found to affect a wide variety of mechanical properties, but have the greatest effect on the interlaminar shear strength (Judd and Wright, 1978). Olster (1972) demonstrated that the interlaminar shear strength would decrease approximately ten percent for each one percent increase in void content. Inclusions also have a very deleterious effect on the physical properties of a composite component. Rhodes (1979) indicated that a paper inclusion, such as the backing paper from prepreg material, decreased the compressive strength of the graphite epoxy composite by 25 percent. Disbonds or delaminated areas in thick composite beams have an adverse effect on the compressive strength as measured in bend tests (Chatterjee, et. al., 1979). Gerharz and Schutz (1980) investigated the effect of delaminations on composites, and observed that delaminations reduced the compressive strength, resulting in a buckling failure. Similarly, the breakdown of the fiber/matrix bond detrimentally affects the fatigue and compressive properties of the composite material (Kunz and Beaumont, 1973). More specifically, a reduction in the fiber/matrix interfacial bond integrity results in a decrease in ultimate compressive strength when the material is loaded at high strain rates (Immen, 1979). Lastly, matrix cracking may cause a reduction in shear, compressive and flexural strength (Gerharz and Schutz, 1979) as well as initiate delaminations (Reifsnider, 1976). The strength reduction due to matrix cracking is a result of reduced load transfer between fibers when the cracking is oriented parallel to the fiber orientation (Gerharz and Schutz, 1979).

Nondestructive Characterization of Materials VI
Edited by R.E. Green, Jr. *et al.*, Plenum Press, New York, 1994

Because each defect type results in a degradation of a specific mechanical property, the ability to classify the defect type will allow for the development and application of acceptance criteria that are defect specific, and will provide assurance that the material is suitable for use in the intended structural application. Additionally, the use of defect specific acceptance criteria will lower production and maintenance costs associated with composites because components with acceptable defects will no longer be rejected.

This paper will report on work being done at CDNSWC to classify defect types in one-inch thick graphite epoxy test blocks with purposely induced defects such as porosity, embedded material, and delaminations.

PROCEDURE

Approach

An artificial neural network was employed to classify various types of defects located in thick section graphite epoxy panels. One hundred ultrasonic signatures were obtained from each of eight, 1-inch thick graphite epoxy composite panels, representing a control standard and three types of known defects. Ultrasonic (UT) data was collected using a Sonix ultrasonic immersion system with a 5 MHz, ½-inch diameter Panametrics immersion transducer. Six input feature parameters for the back-propagation artificial neural network were selected from the power spectra of the UT signatures using graphical and statistical techniques. After the features were selected, the original data set was randomly divided into a training set (representing 90% of each defect class) and a test set for the neural network analysis. The test set which was unique from the training set, was used to evaluate the accuracy of the artificial neural network for differentiating the various types of defects.

Materials and Equipment

Twelve, 4" x 4" x 1" graphite epoxy test panels were fabricated from 192 plies of Fiberite HYE2048A1A graphite/epoxy prepreg tape in a [0/90]s layup with programmed defects as shown in Table 1. In order to create the defects in the test panels, sub-panels (either eight or twenty plies thick) containing the defects were embedded into the test panels such that the symmetry of the whole panel was maintained.

Two sub-panels of each of the following five types of defects were produced: delamination, impact damage, porosity, matrix cracking, and foreign matter contamination (i.e., backing paper, peel ply, and bagging paper). Before insertion into the full-thickness test panel, each sub-panel was inspected using an ultrasonic C-scan system to verify the location and presence of the defects. These C-scans revealed that each of the sub-panels contained the desired programmed defects.

Following lay-up and cure of the full-thickness test panels, each panel was ultrasonically inspected with various ultrasonic C-scan systems. The C-scans of the panels containing matrix cracking and impact damage showed material essentially devoid of defects. After consulting with numerous composite experts, the authors believe that the defects were "healed" during autoclave cure of the full thickness panels. (Note: The defects in these sub-panels had surface breaking openings which would allow epoxy to flow into and fill the defects.) Therefore only the blocks with no defects (A1, A2), delaminations (B1, B2), porosity (D1, D2) and embedded materials (F1, F2) were used in this investigation.

After selection of the test panels to be used for this study, a Sonix ultrasonic immersion system was used with a 5 MHz, 0.5-inch diameter flat immersion transducer to acquire defect signature data from the eight panels. Before A-scans could be collected, another C-scan of each panel was recorded to guide the positioning of the transducer for A-scan collection. For each test panel, a 1" square area of interest was selected as the site where a 10 x 10 grid was used to record A-scans

Table 1. Graphite epoxy test block construction.

ID	DESCRIPTION	SUB-PANEL THICKNESS	SUB-PANEL PLACEMENT	PLIES
A1	Reference	None	---	
A2	Reference	8-Ply	Midplane	92-99
B1	Delamination	8-Ply	1/8 Thickness	24-31
B2	Delamination	8-Ply	Midplane	92-99
C1	Impact	8-Ply	1/8 Thickness	24-31
C2	Impact	8-Ply	Midplane	92-99
D1	Porosity[1]	20-Ply	1/8 Thickness	24-44
D2	Porosity	20-Ply	Midplane	87-106
E1	Matrix Crack	8-Ply	1/8 Thickness	24-31
E2	Matrix Crack	8-Ply	Midplane	92-99
F1	Contamination[2]	8-Ply	1/8 Thickness	24-31
F2	Contamination	8-Ply	Midplane	92-99

[1]Air [2]Peel Ply

at every 0.1 inch increment. A time gate was used to select the portion of the UT waveform containing the defect signature. A total of 100 A-scans recorded per defect type.

Digital Signal Processing and Feature Selection

The scattering of elastic waves in a thick section, anisotropic composite plate is a complicated phenomenon (Datta, et. al., 1988) and not fully understood. To characterize the scattering phenomenon, a total of twenty four features, heuristic in nature, were generated from the time domain (6 features) and the normalized power spectrum (18 features) of each UT signature using the Sonix software module, Feature Scan (Sonix, Inc., 1991). Prior to calculating each power spectrum, the end of each waveform was padded with zeros to yield a total of 512 data points, thus allowing calculation of each power spectrum using a fast Fourier transform (Proakis and Manolakis, 1988).

Both graphical and statistical techniques were applied to the Feature Scan data set in order to select features for input to the artificial neural network. The SAS® procedure SHEWHART was used to generate box-and-whisker plots (SAS Institute, Inc., 1989a) for the 24 calculated features. The box-and-whisker plots (not shown) display the distribution of one variable as a function of the defect classification. The box covers the interquartile range (25th percentile to 75th percentile) and the whiskers extend to cover the range (maximum, minimum) of the variable. The selected feature subset was further analyzed using scatter diagrams to isolate disjoint features in the parameter space as a function of defect class.

In addition, the SAS® procedure PRINCOMP (SAS Institute, Inc., 1989b) was used to select a subset of variables for input to the neural network. The procedure PRINCOMP performs a principal component analysis of the features and generates the eigenvectors of the sample correlation matrix. Each principal component is a linear combination of the original variables, and is useful for examining relationships among several quantitative variables. Large coefficients in the first eigenvector indicate parameters with greater potential discrimination of the defect waveforms.

Artificial Neural Network

An artificial neural network is an analytical procedure which can be used to develop nonlinear relationships between known defect types and their corresponding ultrasonic features (Newman and Arcella, 1991). In this investigation, a fully connected, feed forward, back-propagation neural network was selected for classification of the UT signatures. The neural network, developed using NeuralWorks® Professional II/Plus (Neural Ware, Inc., 1991), employed a normalized cumulative delta learning rule and a hyperbolic transfer function.

Training of the neural network was achieved by an iterative presentation of input pattern vectors and known output values from a collected data set to the network. After each presentation of the data set, the output values of the neural network were compared to desired output values and adaptive weights within the network were incrementally adjusted (delta learning rule) to minimize the output error. After the adaptive weights have converged to a satisfactory steady-state level, the neural network was assumed to be trained. An independent test data set was used to evaluate the performance of the trained neural network.

RESULTS AND DISCUSSION

Feature Selection

A total of 800 A-scans, 100 from each of the eight composite panels, were collected using the automated Sonix inspection system, and a data file containing the waveform features was created using the Feature Scan module. Before performing further analysis, each UT waveform was visually examined for signal strength and validity. After review, 22 waveforms collected from panel F1 and ten waveforms from panel F2 were eliminated from the data set as they were not representative of the defect class.

Based on graphical and principal component analysis a final set of six candidate features were selected from the normalized power spectra of the ultrasonic waveforms for input to the artificial neural network. In Figure 1, a power spectrum of a UT waveform from a defect is shown. The six features selected from the normalized power spectrum are described as follows:

(1) Peak Frequency, MHz- peak frequency of the power spectrum, illustrated at point (g) in Figure 1.
(2) Bandwidth @ -6 dB Down, MHz- difference in frequency at the maximum energy and minimum energy crossing of the -6 dB threshold (50% amplitude). The frequency points are illustrated at points (h) and (f) in Figure 1.
(3) Spectral Skew @ -6 dB Down- measure of spectrum symmetry above the -6 dB threshold. Defined as the ratio of frequencies (h-g)/(g-f).
(4) Spectral Energy from 0 Hz to 1.95 MHz- sum of normalized spectral energy from 0 Hz to 1.95 MHz.
(5) Spectral First Moment, MHz- mean frequency value of the normalized power spectrum.
(6) Spectral Second Moment, MHz- defined by Sonix as the first moment of the normalized power spectrum about the mean.

Features (1) through (6) were selected for input to the neural network based on examination of the box-and-whisker plots. Features (2) through (6) were also selected based on examination of the first eigenvector from the principle components analysis. Scatter diagrams for all pairs of features were generated and used to visually confirm that the selected features yielded spacial separation of the defect classes. A scatter diagram of the Spectral Skew versus the Spectral 1st Moment is shown in Figure 2. This figure illustrates the strong separation of defect classes F2 and B2 from the other population, with both defect classes exhibiting spectral skew values greater than 0.1. However, class D2 exhibits a large degree of variability as viewed in this projection of the six dimensional feature space. Examination of other feature space projections indicated that the six features selected were reasonable candidates for input to the neural network.

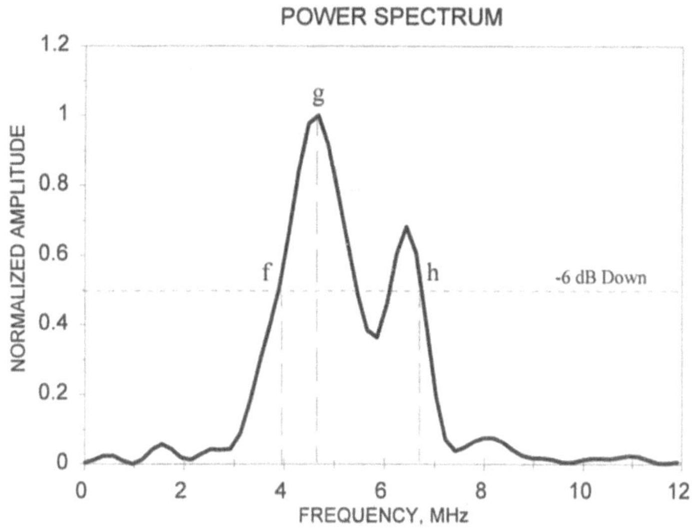

Figure 1. Normalized power spectrum from a representative defect UT signature (class D1) indicating frequency points (f), (g) and (h) used to compute quantitative feature parameters.

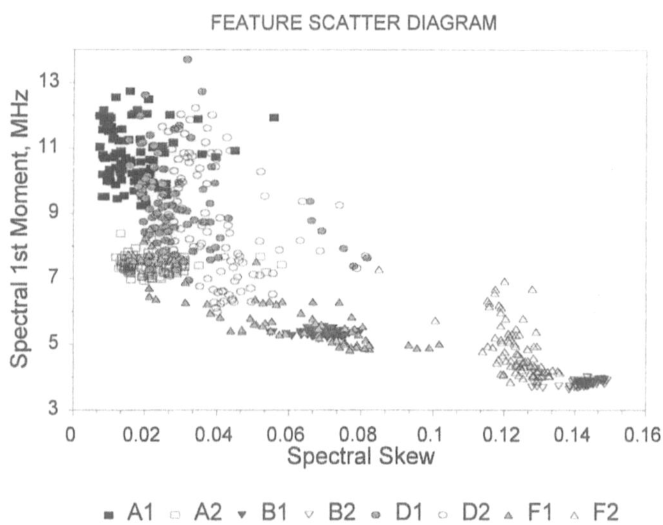

Figure 2. Scatter diagram of the Spectral Skew and Spectral 1st Moment calculated from the normalized power spectrum of each UT signature. The defect class is indicated by the symbol shape.

Table 2. Neural network defect classification assignments[1].

TRAINING SET

		A1	A2	B1	B2	D1	D2	F1	F2	N	Correct	Percent Correct
	A1	**88**	0	0	0	0	0	0	0	90	88	97.8
	A2	0	**83**	0	0	0	**3**	0	0	90	83	92.2
	B1	0	0	**90**	0	0	0	0	0	90	90	100.0
True	B2	0	0	0	**87**	0	0	0	0	90	87	96.7
Class	D1	0	0	0	0	**82**	0	**3**	0	90	82	91.1
	D2	0	0	0	0	0	**78**	0	0	90	78	86.7
	F1	0	**2**	0	0	**4**	0	**44**	0	69	44	63.8
	F2	**1**	0	0	0	**1**	0	0	**74**	81	74	91.4
								Total		690	626	90.7

TEST SET

		A1	A2	B1	B2	D1	D2	F1	F2	N	Correct	Percent Correct
	A1	**10**	0	0	0	0	0	0	0	10	10	100.0
	A2	0	**10**	0	0	0	0	0	0	10	10	100.0
	B1	0	0	**10**	0	0	0	0	0	10	10	100.0
True	B2	0	0	0	**9**	0	0	0	0	10	9	90.0
Class	D1	0	0	0	0	**8**	0	0	0	10	8	80.0
	D2	0	0	0	0	0	**10**	0	0	10	10	100.0
	F1	0	0	**1**	0	0	0	**5**	0	9	5	55.6
	F2	0	0	0	0	0	0	0	**9**	9	9	100.0
								Total		78	71	91.0

[1]Class assignment for neuron output > 0.6.

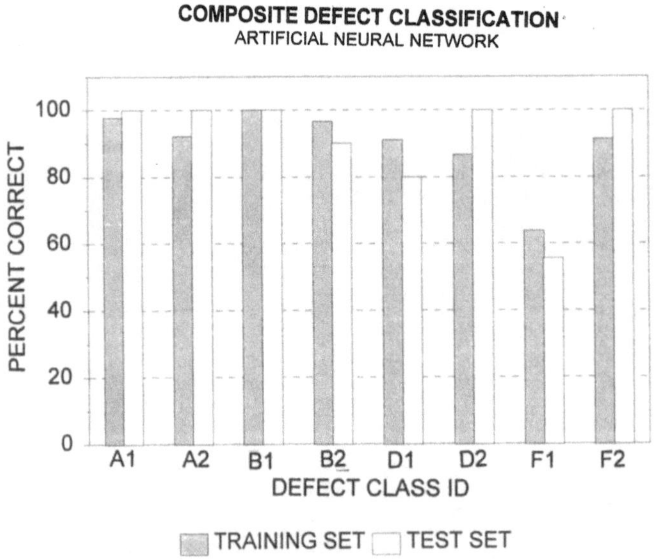

Figure 3. Classification results for training set and test set.

Classification

In order to evaluate the performance of the artificial neural network, the ultrasonic data was subdivided into a training set and a test set. The test set was generated by randomly selecting UT signatures from each defect category; it represented approximately ten percent of the total population. The UT waveforms assigned to the test set were not used during the training procedure.

Various configurations of the neural network were examined in the training process, with the best performance achieved by a network with nine neurons in the hidden layer. A RMS-error (Doebelin, 1983) of 0.119 was achieved after a total of 82,807 presentations of the training data to the neural network. The artificial neural network was trained a second time with a new set of random starting weights. For both training sessions, the neural network weights converged to approximately the same values.

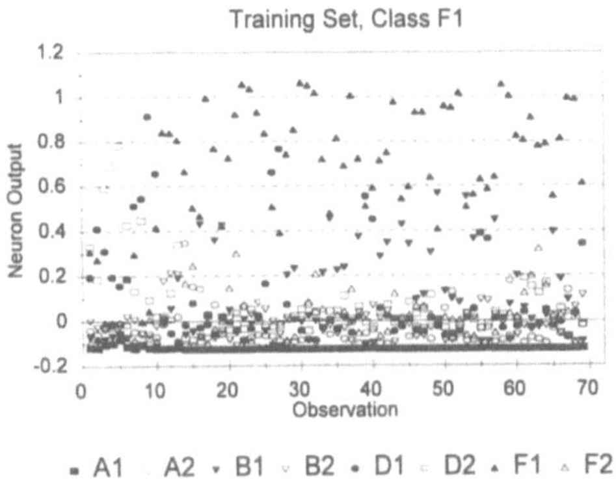

Figure 4. Neural network output values for observations in training class F1.

The results for the training set are summarized in Table 2 and shown graphically in Figure 3. For the 690 observations in the training set, a classification accuracy of 90.7 percent was achieved. The neural network was only able to achieve a classification accuracy of 86.7% for the panel with midplane porosity (D2), and 63.8 percent accuracy for the panel with near surface contamination (F1). However, classification accuracies greater than 90 percent were achieved for the training data in defect classes A1, A2, B1, B2, D1, and F2.

As observed in Figure 4, the neuron output values for class F1 exhibit a large degree of scatter, and the network produced relatively high output values on nodes B1 and D1, for certain observations in the class. Examination of Figure 2 indicates that the two feature parameters,Spectral Skew and Spectral 1st Moment, are similar in value for defect classes F1 and B1. Additional research is needed to select parameter features unique to defect class F1 and/or to improve the neural network architecture for greater classification accuracy.

After the neural network was trained, the parameter features from the remaining 78 waveforms were presented to the neural network for defect classification. The classification results for the test set are summarized in Table 3. Seven misclassifications occurred in the test set resulting in a classification accuracy of 91 percent. As with the training data, the neural network had difficulty correctly classifying defects in class F1.

CONCLUSION

The ultrasonic waveforms from three types of composite defects (delamination, porosity, contamination) and a control were examined in this investigation. A total of 768 UT waveforms were evaluated from eight thick section, graphite epoxy composite panels. Based on graphical examination and statistical analysis, six features from the power spectra of the UT waveforms were selected to be input parameters to a back-propagation artificial neural network. An overall classification accuracy (includes training and test sets) of 90.8 percent was achieved by the neural network.

The results presented demonstrate the ability of an artificial neural network to classify defects located in composite materials based on features calculated from UT waveforms from the defect. This ability will allow for the development of defect specific acceptance criteria, thus providing a means of determining whether the material is suitable for the intended structural application, and ultimately lowering the costs of using composites.

REFERENCES

Chatterjee, S.N., et. al. (August 1979) "Definition and Modeling of Critical Flaws in Graphite Fiber Reinforced Resin Matrix Composite Materials," AD A080893.

Datta, S.D., H.M. Ledbetter, Y. Shindo, and A.H. Shan (1988) Wave Motion, 10, 171.

Doebelin, E.O. (1983) Measurement Systems, Applications and Design, Third Ed., McGraw-Hill Book Company, New York, NY.

Gagorik, J.E., Corrodo, J.A., and Kornbau, R.W. (April 15-18, 1991) "An Overview of Composite Developments for Naval Surface Combatants," Proceedings of the 36 International SAMPE Symposium.

Gerharz, J.J. and D. Schutz (August 1980) "Literature Research on the Mechanical Properties of Fibre Composite Materials - Analysis of the State of the Art," AD A093789.

Immen, F.H. (November 1979) "Inspection for Critical Mechanical Properties of Advance Composite Structures," 28th Defense Conference on Nondestructive Testing, AD A097214, pp. 1-13.

Judd, N.C.W., and W. Wright (Jan/Feb 1978) "Voids and Their Effect on the Mechanical Properties of Composites - An Appraisal," SAMPE Journal 14, pp. 10-14.

Kunz, R. L., and P.W.R. Beaumont (December 1973) "Microcrack Growth in Graphite Fiber Epoxy Resin Systems during Compressive Fatigue", ASTM-STP 569, Fatigue of Composite Materials, pp. 71-94.

Olster, E.F. (August 1972) "Effects of Voids on Graphite Fiber Reinforced Composite," AD 746 560.

NeuralWare, Inc. (1991) Reference Guide, NerualWorks® Professional II/Plus and NeuralWorks Explorer, NeuralWare, Inc., Pittsburgh, PA.

Newman, R.W., and F.G. Arcella (March 1991) "Neural Network Analysis of Ultrasonic NDE Wave Forms," 1991 American Society of Nondestructive Testing Meeting.

Proakis, J.G. and D.G. Manolakis (1988) Introduction to Digital Signal Processing, Macmillan Publishing Company, New York.

Reifsnider, K.L., et. al. (April 1976) "Defect Property Relationships in Composite Materials," AD A031 809.

Rhodes, F.E. (November 1979) "The Influence of Manufacturing Defects on the Performance of Carbon Fibre Composites," Institute of Physics One Day Meeting on The Significance of Defects on the Failure of Fibre Composites, London.

SAS Institute Inc. (1989a) SAS/QC® Software: Reference, Version 6, Fst. Ed., SAS Institute Inc., Cary, NC.

SAS Institute Inc. (1989b) SAS/STAT® Software,Release 6.03 Edition, SAS Institute Inc., Cary, NC.

Sonix, Inc. (May 1991) Feature Scan, Data Collection and Analysis System, User's Manual, Version, 2.0, Sonix, Inc., Springfield, VA.

THE MYTHOLOGY OF X-RAY INDUCED RADIATION DAMAGE

Howard W. Bennett, Jr.

Metallurgy Laboratory
Applied Environments Test Branch
Materiel Test Directorate
White Sands Missile Range, New Mexico

ABSTRACT

The Metallurgy Laboratory, Applied Environments Test Branch, Materiel Test Directorate, White Sands Missile Range (WSMR), NM, is responsible for Nondestructive Evaluation on WSMR. There have been serious inquiries from customers concerning radiation effects induced in materiel by the sources employed for radiographic testing (RT). Some sources of concern include programmable read-only memory (PROM) devices and fiber optics lines carried onboard the materiel. The Laboratory Team has extensive experience in RT of materiel with isotope sources and x-ray machines. This paper presents the ionizing radiation characteristics and case studies of the exposure and attenuation of radiation by materiel.

FUNDAMENTALS OF RADIATION DAMAGE

Radiation damage is generally concerned with the production of displaced lattice atoms to interstitial sites. In the process of creating the displacements, kinetic energy is required (E_d). Fortunately, the value of E_d is generally between 20 and 25 electron volts (eV). Table 1 lists the minimum kinetic energy required to produce displacements in some common elements, as shown by Kelly[1]:

Understanding that there is an E_d necessary to cause a self-interstitial requires the definition of a minimum incident kinetic energy ($E\gamma_{min}$) to induce E_d to become a

Table 1. Elemental Displacement Energy

Element	E_d (eV)
Au	28
C (Graphite)	24.7
Fe	37
Ge	14.5
Ni	24
Si	12.9
Ti	29
W	35

displaced atom. For electrons and gamma-rays (γ) the calculation depends on the energy of the γ.

$$E_d = \frac{2\, E\gamma_{min}\, (E\gamma_{min} + m_e\, c^2)}{M_2\, c^2}$$

m_e = electron's mass

M_2 = material's atomic mass

In addition to the primary production of displaced lattice atoms, a second process of direct nuclear photoelectric effect and Compton scattering may occur. This process would involve displacements due to nuclear recoil produced by atomic photoelectric absorption and pair production. This process would be most prevalent in the use of isotope, i.e., γ radiography. Gamma-rays do not produce displacements directly, they transfer their energy to the material's electrons which in turn produce displacements.

The x-ray photon is of primary interest. The x-ray photon's energy is generally absorbed by the material's electronic configuration. This change in the electronic state will result in a response from the atoms of the material. The process can result in the dissociation of molecules, changes in lattice structure, and creation of self-interstitials. This latter process will occur if the duration of the electronic disturbance is greater than the time required to create the activated or interstitial state. Disturbances in the material's morphology, lattice structure, can occur over very small areas or distances. This phenomenon could be particularly insidious in fiber optics.

WSMR INDUSTRIAL RADIOGRAPHY SOURCES

Industrial radiography is performed at WSMR employing both radioactive isotope and x-ray machines. Cobalt-60 (Co-60) is used for isotope RT. Industrial x-ray machines include a fixed 420 kV, 10 mA, 4.2 kW unit, a portable 300 kV, 5 mA unit, and transportable 320 kV, 13 mA, and 4 MeV units.

Isotope RT

The 100 Curie (max) Co-60 source is contained in a Gammatron 100-A camera. The source is particularly useful for performing RT in remote areas, e.g., in an impact area. Co-60 disintegrates by giving off two photons, 1.15 MeV and 1.35 MeV. The relative intensities of each of these photons are approximately equal; therefore, for convenience purposes we use 1.25 MeV. With an emissivity of 0.24 R/min/Ci, 100 Ci's of Co-60 would produce 24 R/min at 1 foot. In a radiographer's terminology this is a "steel equivalency" of approximately 4 inches. A 100 Ci Co-60 source will make a satisfactory radiographic image of a 155 mm artillery shell, with the shell in contact with Kodak AA film, at a source-to-film distance of 6 feet with an exposure time of approximately 8 minutes.

X-Ray RT

Personnel from the Metallurgy Laboratory presented technical papers at NDT Defense Conferences which provide in detail, the radiation characteristics of the x-ray machines used for RT on WSMR (as explained by Bennett and Huffmyer[2] and Anaya and Bennett[3,4]). The 420 kV machine has an output of 24.4 R/min with 300 kV, 14 mA tube current at 5.5 feet. The 320 kV transportable produces 46 R/min at the same settings at 4 feet. A typical exposure for the 4 MeV transportable machine is 29 R/min at 10 feet.

CASE STUDIES

The following 'generic' case studies are being presented to illustrate materiel exposures. The presentations are divided into categories according to the materiel's 'end use.' Unless otherwise noted, the radiation exposures were made with the 420 kV stationary machine with a source-to-film distance of 5.5 feet. The materiel described below was in a tactical configuration. Regardless of whether the isotope or x-ray machine was used for RT, there has never been a rocket motor or warhead incident for which radiation damage was responsible.

Air Defense Missiles

RT exposures were made on numerous types of surface-to-air missiles. The seekers used by these systems cover the entire technology range from passive infrared to interactive radar. The x-ray exposures and total radiation dose in Roentgens (R) are listed below:

kV	mA	Time (min)	R
130	10	1.5	5.4
150	10	1.0	5.0
200	13	0.5	6.5
200	5	4.0	17.5
300	10	1.0	18.0
350*	10	1.5	400.0

*These missiles were RT inspected during a multiple environmental sequence. This number represents the total dose received by each missile during testing.

A significant population of the above materiel has been tested. There have been no guidance system or missile guidance failures attributed to radiation damage of seeker components or appurtenances.

Artillery Missiles

There has been considerable innovation in this field during the past 8 to 10 years. During materiel research, development, test, and evaluation, it has been necessary to employ both the 420 kV and 4 MeV x-ray machines. Guidance systems, fiber optic control lines, receivers, and electronic PROM's have been exposed during materiel RT:

kV	mA	Time (min)	R
215	5	0.75	3.8
260	10	0.5	6.8
4000	NA	8.5	239.0

During testing of the last system listed above, it was determined using TLD's that there was less than a 20 R exposure to the aft servo control section.

Artillery Shells

Several varieties of 120 and 155 mm shells that have been RT inspected during testing at WSMR are listed:

Type	kV	mA	Time (min)	R
120 mm[1]	1250	Co-60	10	6.5
155[2]	1250	Co-60	8	80
155 mm[2]	210	10	1	11.5
155	375	11	1.5	38

Mines

To date one type of mine has been tested following RT during an extended environmental sequence:

kV	mA	Time (min)	R
135	10	2	11

During test firings there were no difficulties traceable to the exposure to ionizing radiation.

[1] These rounds were consumable high energy anti-tank rounds whose test firing results were all within expected parameters.

[2] These rounds contained circuitry and seekers for laser guidance. They were part of a multiple environmental test sequence. Hundreds of firings have been performed with no indication of any anomalies due to radiation damage or hardening.

DISCUSSION

The brief case studies presented, in the preceding section, illustrate the exposures to which materiel are exposed during RT inspection(s). The onboard electronics, rocket motors, ignitors, controls, and engineering materials represented in these systems cover the entire range of design, fabrication, and implementation found in materiel during the past 15 to 20 years.

There have been no research, development, test, and evaluation failures due to radiation effects in any of these systems. Many of the systems were exposed to RT many times during the course of multi-environment test sequences. These sequences included Safety Qualification, First Article, and Initial Production Testing.

CONCLUSIONS

1. For x-rays in the energy range from 130 kV to 4 MeV with total exposures approaching 400 R, the possibility of radiation damage of materiel is very remote.

2. For Co-60 exposures up to 80 R the possibility of radiation effects in materiel is very remote.

REFERENCES

1. B.T. Kelly, "Irradiation Damage to Solids," Pergamon Press, London (1966).
2. H.W. Bennett and R.C. Huffmyer, Large Rocket Motor Radiography with a Portable X-Ray System, Proceedings DOD NDT Conference (1986).
3. T.M. Anaya and H.W. Bennett, Industrial Radiography with a Portable 4 MeV X-Ray Source, Proceedings DOD NDT Conference (1991).
4. T.M. Anaya and H.W. Bennett, Field Radiographic Test Capabilities of White Sands Missile Range, Proceedings DOD NDT Conference (1991).

APPLICATIONS OF X-RAY RADIOGRAPHY TO THE STUDY OF POROSITIES IN SURFACE MOUNT SOLDER JOINTS

Y.C. Chan[1], D.J. Xie[2], J.K.L. Lai[3], F. Yeung[1] and H. Wong[1]

[1] Department of Electronic Engineering, City Polytechnic of Hong Kong (CPHK), Hong Kong
[2] Ph.D. Candidate in Department of Electronic Engineering, CPHK, on leave from Huazhong University of Science and Technology, PR China
[3] Department of Physics and Materials Science, City Polytechnic of Hong Kong (CPHK), Hong Kong

ABSTRACT

This paper reports the use of X-ray radiography and computer image processing for the study of porosities in surface mount solder joints fabricated from different types of no-clean RMA and RA solder pastes. In order to understand the mechanism of pore formation in the solder joints, experiments are conducted on the processes of the pore formation during reflow soldering. The effects of different infrared (IR) reflow temperature profiles are also studied. From this work, it was found that the composition and structure of solder pastes had a significant effect on the pore formation, and the lower metal content and/or higher heating rate did not necessarily cause a greater percentage of pore formation in the solder joints. X-ray radiography, in conjunction with Optimas image processing, is shown to be very useful in quantitatively testing pores in surface mount solder joints and allows real-time radiographic image processing.

INTRODUCTION

Porosity is a major defect in surface mount solder joints that has a critical effect on the reliability of electronic assembly. The pores contained in solder joints may intensify the stress and decrease the joining strength sharply[1].

Pore formation in solder joints is a complicated process, and many factors will have effects on it. The most important factors are solder pastes, reflow method and temperature, PCB land design, printing methods and etc. A solder paste normally contains

35--65 volume percentage of volatile materials (e.g., flux and vehicle)[2]. A lot of flux fumes will be produced during reflow soldering due to thermal decomposition. Some gases may be entrapped between the flat surface of PCB board and component lead or by the out solidified surfaces of joints and then remain as pores in solder joints upon cooling.

Heating and cooling of solder joints are mainly controlled by reflow temperature profile. Then, the growth or movement of entrapped gases is affected greatly by reflow temperature profile. In IR reflow soldering, usually, there are two types of temperature profile for eutectic solder paste[2,3]: long and slow preheating; and high preheating and ramp-up speeds. It is often considered that long and slow preheating is needed to prevent pores forming in solder joints and high preheating and ramp-up speeds will result in pore formation in solder joints readily. However, it is worth studying further because the processes of gas evaporation and solder solidification is very complicated.

Different component leads and PCB lands may give different routes for escaping of the gases and then affect the eventual pore formation in solder joints. In surface mount assembly, the areas of a joint land may vary from 4.5 x 6.4 mm (maximum EIA standard capacitor dimension) to 0.3 x 2 mm fine pitch QFP component[4].

There are many nondestructive methods to inspect porosities in solder joints. Among them, X-ray radiography is one of the most powerful technique. With increased number and density of solder joints per board, X-ray inspection becomes more and more important in the production sites. By using X-ray radiography, pores, crack, hole, solder balling, insufficient solder, lead related defects, device related defects, and solder bridgings are all identified[5]. If coupled with a real time radiographic detector and image processor, X-ray technology allows instantaneous radiographic imaging and semi-automatic or totally automatic inspection.

EXPERIMENTAL PROCEDURE

Several solder pastes used in the experiments are shown in Table 1. All the pastes belong to the RMA and no-clean type except Paste 3 that is RA type.

Table 1. Solder Pastes Chosen in the Experiments(supplied by the Multicore Solders Ltd. and Alpha Metals Hong Kong Ltd.).

Paste No.	alloy	Metal content wt. %	flux type	viscosity $\times 10^5$cps	average particle size, μm
1	Sn63/Pb37	90	RMA	8-10	64.0
2	Sn62/Pb38	90	RMA	8-10	39.0
3	Sn63/Pb37	90	RA	8-10	39.0
4	Sn63/Pb37	90	RMA	8-10	44.0
5	Sn63/Pb37	88	RMA	6	60.0
6	Sn63/Pb37	86	RMA	4	39.0

The specimen selected in the experiments is an FR-4 PCB plate with a pattern of 14 lands as shown in Figure 1-a. The specimen is called a blank land sample if there is no component mounted on the lands. Figure 1-b represents shear samples with the solder joints clamped by two PCB plates. According our former works[6], a blank land sample resembles practical SMT assembly in pore formation. But a shear sample can have much more pores because the gases produced in solder joints are easy traped by the plates.

The solder pastes are deposited to the pre-cleaned Cu lands by means of a pneumatic dispenser. The specimen is then reflowed in the IR reflow machine(type:

HELLER 932). Two major types of reflow temperature profile are determined as shown in Figure 2. Temperature profiles IR 01 and 03 are similar each other and both are belong to long and slow preheating temperature profiles. Temperature profile IR 02 has high preheating and ramp-up speeds.

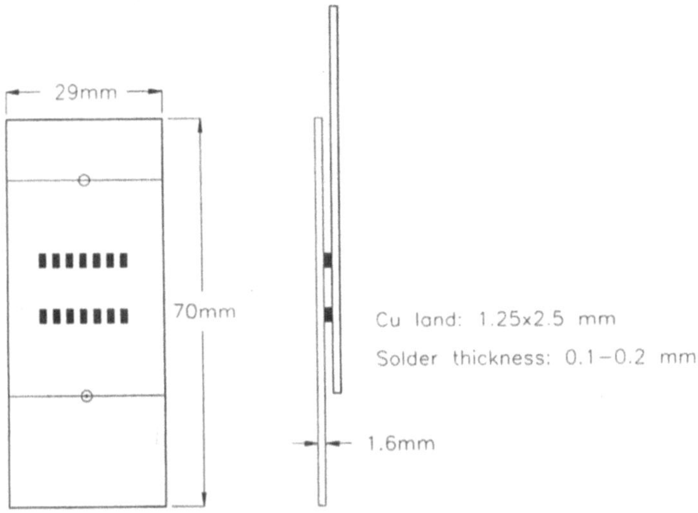

Figure 1. Specimen configuration used in the experiments.

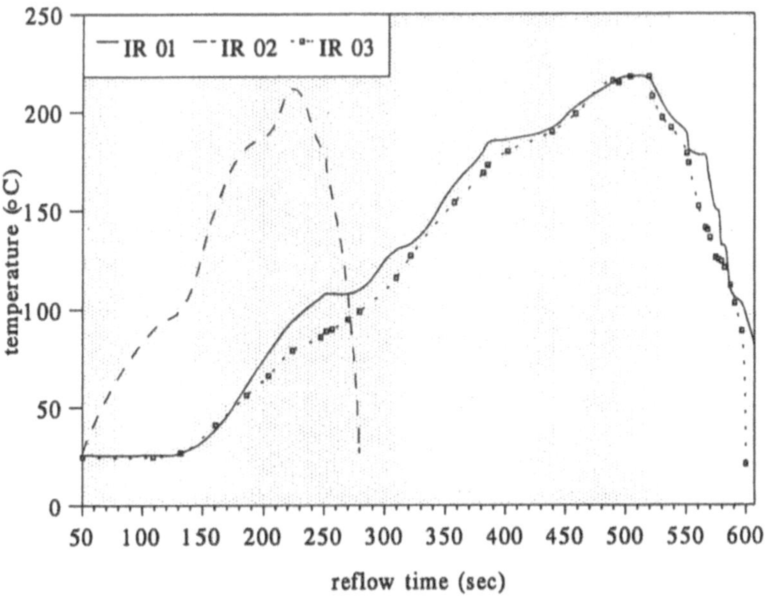

Figure 2. IR reflow temperature profiles used in the experiments.

CALCULATION OF AREA FRACTION OF PORES AND IMAGE PROCESSING

Figure 3 shows a diagram of X-ray inspection system used in the experiments. During X-ray observation, the PCB board and Cu lands are placed perpendicularly to the X-ray beam so that the largest cross section of the solder joints can be seen. In a practical assembly, fatigue failure normally originates from that cross section of the solder joints.

The X-ray image is recorded on the photograph and/or re-captured by a video camera. The image is then digitized by a frame grabber and sent to computer for processing. We found that Optimas (a product of BioScan Inc.) was a helpful tool in image processing and calculation of area fraction of pores. Optimas can automatically detect pore areas from X-ray images or photographs and report quantitative measurements to a file or Microsoft Excel sheet. Figure 4 shows image processing and pores detecting of an X-ray photograph when using Optimas. Figure 4a is the original X-ray photograph of solder joints in a blank sample. However, not all the pores can often be reported automatically from the same histogram. Therefore, different histograms and/or manually tracing should be used. As shown in Figure 4c, the pores in the boundary area of pads should be manually traced because no suitable histogram can be chosen.

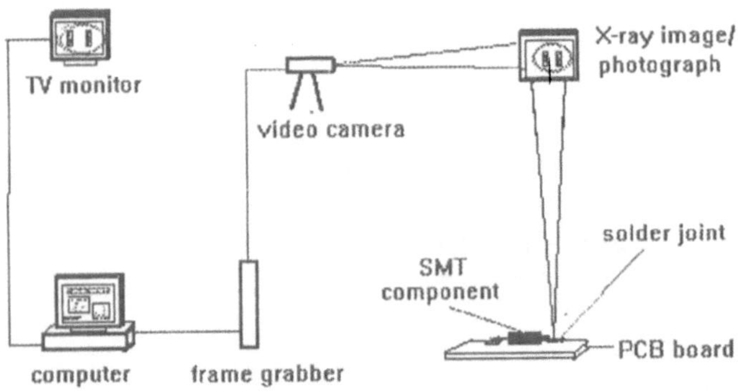

Figure 3. Schematic of X-ray radiographic inspection on SMT solder joints.

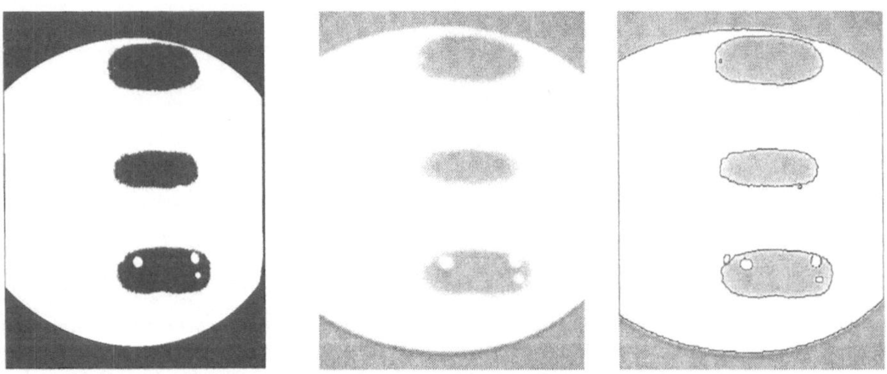

a. X-ray photograph   b. digitized image   c. pores detecting

Figure 4. Image processing and pores detecting from X-ray image/photograph.

RESULTS AND DISCUSSION

Pore Formation During IR Reflow Soldering

When using temperature profile IR 01, one round reflow takes about 600 seconds. Figure 5 shows porosity change in two types of solder pastes during IR reflow. In figure 5, dotted lines represent the area fraction of pores calculated by average diameter. As we can see from the figure, the results from Optimas calculation (solid lines) is approximate to those from average diameters (doted line) for Paste 1 and Paste 3 and are acceptable.

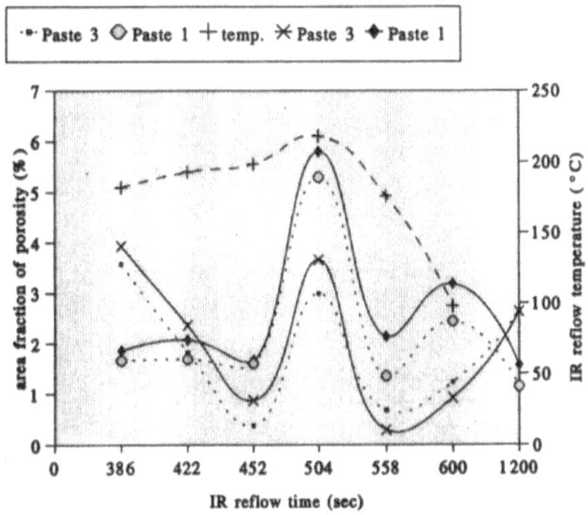

Figure 5. Variation of area fraction of porosity in a solder joint versus the IR reflow time IR reflow file: IR 01.

We notice from Figure 5 that the pore density in the solder joints varies with the reflow time. Pores can be found by X-ray radiography at a reflow time of 386 seconds when temperature goes up to 183 °C and melting of solder alloy begins. Before this time pores in a solder joint can not be detected because the solder is in a powder state and the solder joint has many interfaces. The volume fraction of pore in the solder joints before reflow soldering or melting is about 44% as shown in Reference 6. As the reflow time and temperature increase, the pore density decreases until about 452 seconds. Afterwards, there is a peak pore density around 500 seconds for both solder pastes, when the peak temperature is reached. The peaks and valleys which appear in Figure 5 are caused by the opposite effects of the melting of the solder alloy and the thermal decomposition of the flux media. Melting of alloy can decrease the pore density and decomposition of the flux will produce gases that may increase the pore formation.

Effect of Solder Pastes on Pore Formation

Solder pastes are homogeneous blends of pre-alloyed solder powder and flux media. The combinations of flux, flux content, viscosity, particle size and shape will affect the pore formation in solder joints significantly.

Figure 6 shows the results of pore formation when different kinds of solder pastes are soldered under two reflow temperature profiles. It can be seen that Paste 1 has the highest percentage of pores, followed by Paste 4 and 5. In Paste 2 and 3, very few pores are found. In Paste 3, almost no pores appear if reflowed under temperature profile IR 02. For the majority of the pastes chosen, the pore formation behaviors under the two IR reflow temperature profiles are quite different. To decrease the pore density, the slower heating rate is more suitable for Paste 2, 4 and 6 while the faster one is appropriate to Paste 1, 3 and 5.

Upon rearranging the data according to the average particle size of metal powder, we get two figures. Interesting enough, as shown in Figure 7, pore formation has a positive correlation with the average particle size in both IR reflow temperature profiles. The reason may be that the larger particle powder material has a greater percentage of porosity before melting and needs a longer melting time. These pores cannot be eliminated entirely while melting and will remain in the joints after solidification if the heating and cooling periods are not sufficiently long, as we can see in the profile IR 02. However, further investigation is needed to confirm this hypothesis.

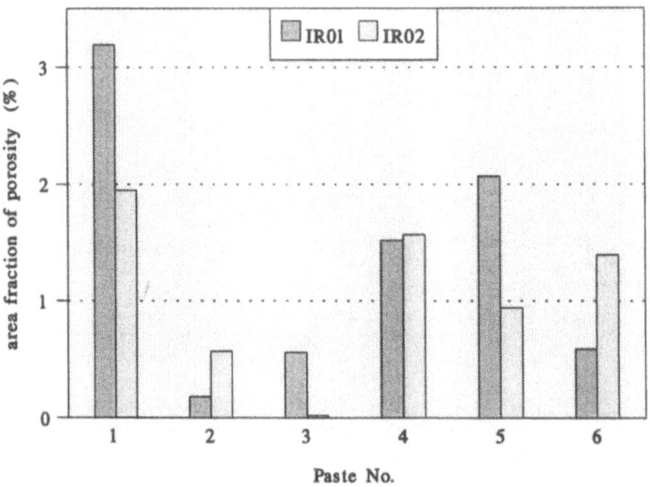

Figure 6. Pore formation of different solder pastes under two reflow files.

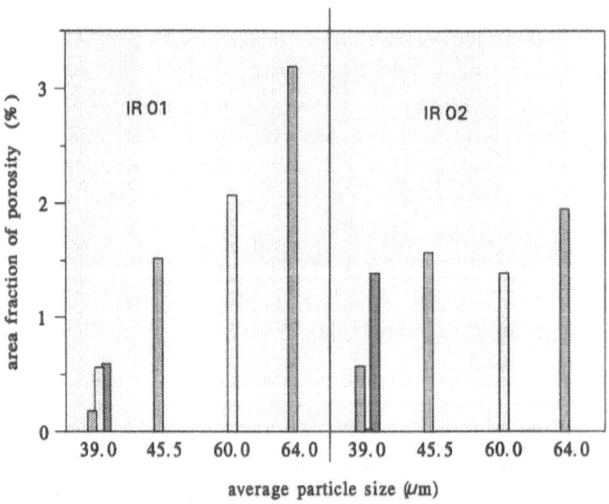

Figure 7. Dependence of area fraction of porosity on the average particl size of a solder paste IR reflow file: IR 01 (right), IR 02 (left).

Metal content in solder pastes is another important factor affecting pore formation as emphasized by many authors[2, 3]. This is also true in the situation of Paste 2 and 6 (see Figure 6). Paste 2 has a higher metal content and a lower fraction of pores as compared with Paste 6. The pore density in Paste 6, however, is much less than those in Paste 1 and 4 even though the metal content in Paste 6 is the lowest, especially when reflowed under IR 01. This is because the evaporation rate of gases in Paste 6 is higher than that of Paste 1 during preheating and ramp-up[6] and will not contribute to the pore formation. If preheating is long enough as reflow file IR 01 is adopted, the gases produced in Paste 6 can escape earlier before cooling and solidification begin and will not become pores in solder joints.

Effect of Different Cu Land Sizes of PCB Boards

Different SMT components require for different PCB lands. Figure 8 shows the effect of different PCB land areas on pore formation in the solder joints. From the figure, we observe different land areas have little influences on the pore formation. This is because that the pores formed in the solder joints of the five lands have the same circumstance to move out if there is no component on the solder joints. We notice again that solder pastes and reflow temperature profiles affect the pore formation greatly. In the studied land sizes, all lands in Paste 3 have a lower pore density than those in Paste 1. It should also notice that reflow file IR 03 is not suitable for both solder pastes. The reason may be that it has less preheating time around 100 °C as compared with IR 02.

The sizes and distributions of pores for different lands may be different. As shown in Figure 9, the sizes of pores for larger pad areas are often bigger. But most of pores in all the five lands have an area size around 0.01 mm^2.

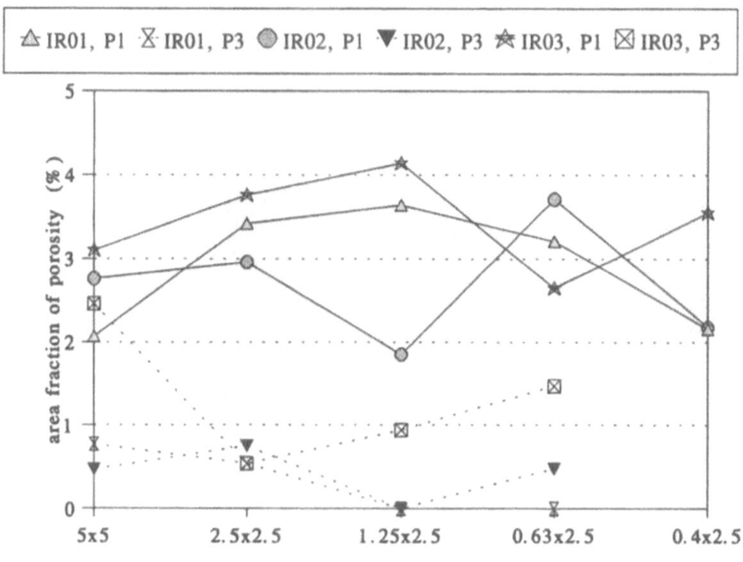

Figure 8. Effects of different Cu land areas on the pore formation.

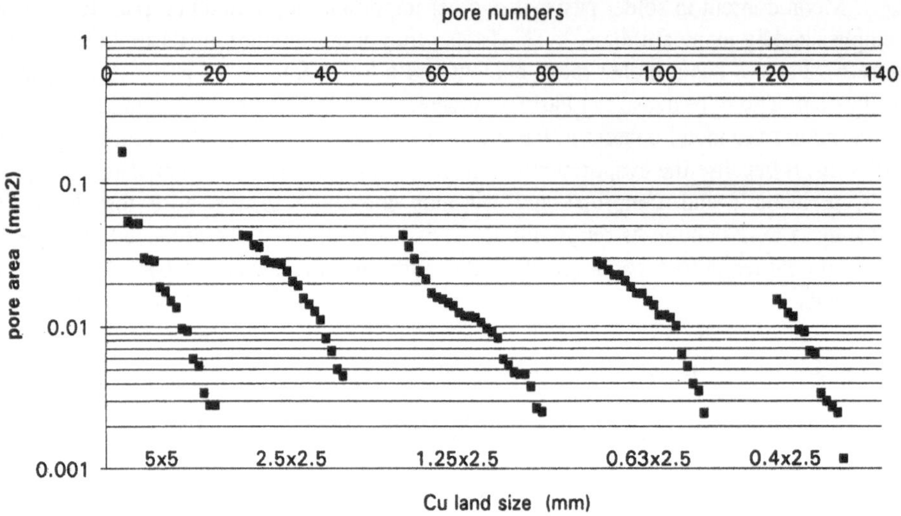

Figure 9. Pore distribution in solder joints with different land sizes(Paste 1, IR 01).

CONCLUSION

It was found from this work that the composition and particle size of solder pastes had a significant effect on the pore formation in SMT solder joints. The lower metal content and/or higher heating rate does not necessarily cause a greater area fraction of pore formation in the joints. If the thermal decomposition and/or evaporation rate of flux and vehicles in the solder paste is great enough before peak reflow temperature is reached, pores in the solder joints will not be easily formed. X-ray radiography, with the help of computer image processing, is shown to be very useful in quantitatively testing the pores in surface mount joints and real time radiographic image processing.

ACKNOWLEDGMENTS

This work is supported by City Polytechnic of Hong Kong (Research Grant No. 700144). The authors are grateful to the use of X-ray radiographic facilities at the SMT laboratory of Hong Kong Productivity Council. One of the authors, D.J. Xie, wishes to thank Ms. S.C. Lee in the laboratory of Department of Manufacturing Engineering (CPHK) for her helpful assistance.

REFERENCES

1. Michael B. Bever, 'Encyclopedia of Materials Science and Engineering', Pergamon Press, Oxford, volume 5, 1986, pp. 3839-3843.
2. Howard H. Manko, 'Soldering Handbook for Printed Circuits and Surface Mounting', Van Nostrand Reinhold, New York, 1986, P.185.
3. Jennie S. Hwang, 'Solder Paste in Electronics Packaging', Van Nostrand Reinhold, New York, 1989, p.283.
4. Stephen W Hinch, 'Handbook of Surface Mount Technology', Longman Scientific & Technical, England, 1988, p. 47.
5. M. Forshaw, *Circuit World*, No. 3, vol. 15, pp. 14-17, 1989.
6. Y C Chan, D J Xie, J K L Lai, F Yeung and H. Wong, Pore Formation in Surface Mount Joints During IR Reflow Soldering Process, submitted to *IEEE trans. on CHMT.* for publication.

NEW SIZING TECHNIQUE OF AN INCLINED FATIGUE CRACK ON A FREE SURFACE

Kasaburo Harumi[1], Junichi Ooshima[2], Yasushi Asakuma[3] and Yukio Ogura[4]

[1]Tokyo University of Information Sciences, 1200 Yatou-chou, Chiba 265, Japan
[2]Nihon University, College of Industrial Engineering, Narashino 275, Japan
[3]Hitachi Plant Engineering & Construction Co. Ltd., Matsudo 271, Japan
[4]Hitachi Construction Machinary, FA Division, Ootemachi 100, Tokyo, Japan

INTRODUCTION

The most effective means of ultrasonic defect sizing is considered to be the tip wave method.[1,2] The intensity of the tip wave is very weak, and prevents the use of tip waves. However, a number of digital flaw detectors have recently been developed which are rather appropriate for use in the tip wave method, since they have rather low noise and have more convenient functions of recordability and flexibility than the conventional analogue flaw detector. The effectiveness of the full digital flaw detector HITACHI-DT-2000 has been described in the previous paper[3] for the defect sizing of an artificial slit with a round tip shape, and with an inclination angles of up to 40 ° on a free surface. The newly developed model HITACHI-DT-2200 is greatly improved, as almost noiseless in itself with a wide range of intensity of 100 dB, and very high resolution of 0.1mm. We attempted the sizing of a natural fatigue inclined crack with a complicated shape, which is considered more difficult than the earlier simple shaped slit.[3] The DT-2200 has unbelievable precision of below 0.03mm for 5MHz, and the accuracy of defect sizing is greatly improved at below 0.5mm: for the conventional analogue detector it is larger than 2mm . This remarkable accuracy is possible because of the accuracy of the DT-2200, and the knowledge obtained from computer simulation as described in the previous paper.[3]

ACCURACY OF DT-2200

In experiments using the earlier DT-2000, we obtained the following results shown in Table 1.

Table 1. Estimated crack height for Y1($L1$=1.5mm)

Inclination angle(°)	10	20	30	40
AB(mm)	0.98	0.99	0.90	0.88
Estimated L(mm)	1.50	1.53	1.42	1.40

We have checked the same experiments using the DT-2200, and obtained the result shown in Fig.2, pulse B in Fig. 1 was split into two pulses. The length AB became 1.0mm rather than 0.9mm as shown in Table 1, and the crack height was L=1.53mm with an error of 0.03mm using the equation L= $\sqrt{2}$*AB. This is astonishing for a 0.65mm wavelength of 5MHz. In DT-2200, the pulse position is obtained from the pulse peak, this is the reason for such accuracy.

NUMERICAL EXPERIMENT

New automatic mesh generation (AMG) of the Finite Element Method (FEM) by Ooshima is used for the natural crack part, as the fatigue crack shape is complicated.

Therefore, any kind of crack shape mesh generation is possible in a short time, and a numerical experiment of elastic waves becomes as easy as the actual experiment.

EXPERIMENT

New full-digital flaw detector HITACHI-DT-2200 with 100 dB wide range intensity, enables us to obtain more accurate experiment of the order of 0.1mm as explained in the next section. The transducers used are two types of focused probes, wide band and narrow band, as the natural fatigue crack shapes are very complicated. Therefore, a conventional transducer obtains complex results, and is not appropriate for simple results for determining only crack height and inclination angle. The focused probe is used to concentrate input signal as simply as possible. Japan Probe SB-5Z10A45 and SO-5C10A45 with about 2.0mm focused beam diameter were used.

Five cases have been treated; ML10, MS15 have complicated shape as shown in Fig.3, and have an inclination angle of 10 ° and 15 ° respectively. Z05, Z20, Z30 have rather simple shapes as shown in Fig. 8, and have 5 °, 20 °, 30 ° inclination angles.

The crack height is obtained using $\sqrt{2}$*AB as explained in reference 2, where A is the tip wave and B is the bottom reflected tip echo. In the case of a crack with a simple shape , two waves are discriminated as two pulses, but in this case the bottom reflected tip wave B overlaps the other waves, such as the corner reflected main echoes. Interference of the B echo is observed as a sharp dip in other waves in Fig. 5.

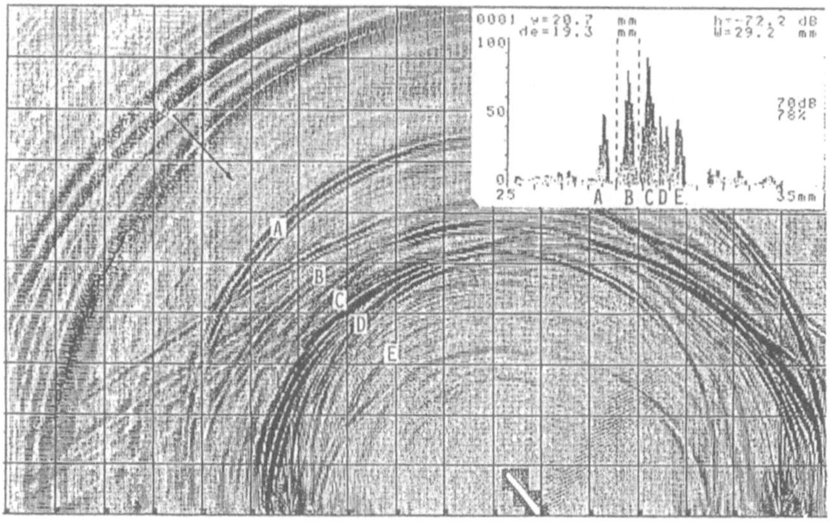

Figure 1. Vectors diagram of reflection by a 40 ° inclined crack, and its scanning graph of DT-2000.

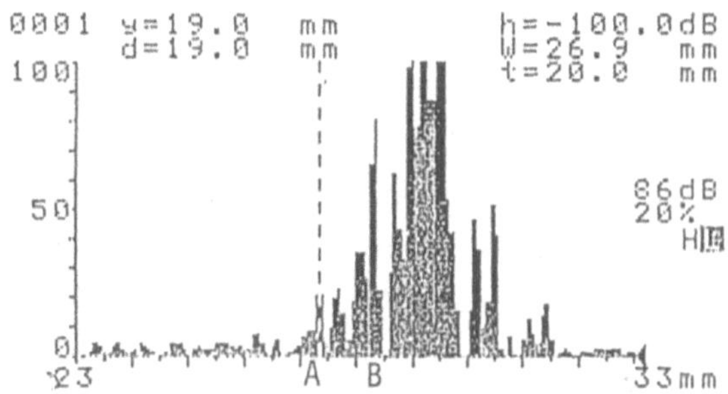

Figure 2. Scanning graph of a figure by DT-2200.

Figure 3. Crack shape of ML10 °.

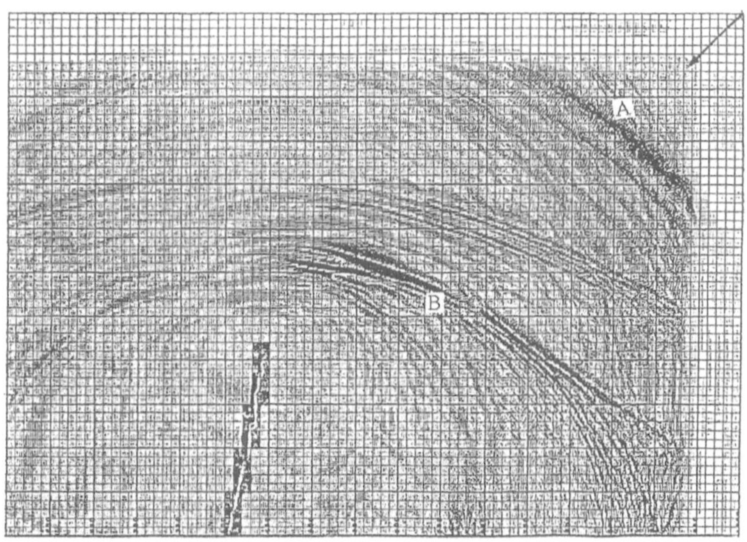

Figure 4. Vector diagram of reflection by ML10 °.

Several dips are observed for a wide-band one-wave pulse as shown in Fig. 5, as several waves are generated not only from the tip but also from the bends near the tip. These are also found in numerical results as C in Fig. 9 as the interference of bottom reflected wave B with the main echo. However, only one dip is observed for the narrow band pulse, and it is easier to see one sharp dip than to distinguish B from several dips in the wide band case. In order to recognize B among other dips in the wide band case, it is recommended that the foremost dip will be recognized as B, similar to the case of the tip wave in which the leading pulse is selected as the tip wave.

In the case of narrow band pulse, only one deep dip is observed as B, and the tip echo should be selected so as to have a sharp arise of tip echo, and the arise point is identified as A.

(1) ML10 Case

Crack shape is considerably rough and complex and has about a 10 ° inclination as shown in Fig. 3. Vectors diagram of the reflection of the transverse one wave pulse by this crack is shown in Fig. 4, and the incident direction is shown by an arrow in the upper right corner. A is the tip echo and B is the bottom reflection tip wave A which overlaps with the main corner echo in Fig. 4. MS10A indicates the incident wave at upper right, and the scanning graph is displayed in Fig. 5-a and 5-b. Estimated crack height(W) for the wide band pulse is given by $\sqrt{2}$*AB=(19.6-14.5)x1.414= $\sqrt{2}$* 5.1mm=7.21mm in Table 2 for MS10A, and the actual value of 7.22mm results in a 0.01mm error. Crack heights for a narrow band probe for MS15 A are obtained from the distance AB in Figs. 6-a and 6-b respectively. They are tabulated as estimated height(N): $\sqrt{2}$*(22.1-16.9)= $\sqrt{2}$* 5.2=7.35mm for MS10A and 7.22mm for actual value, with an error of 0.13mm. Crack length is obtained from the distances of the pulse sequences A, B, and C in Fig.7, which are emitted from the tip as the surface waves go up and down along the crack surface. The estimated length is given as L=7.92mm. The estimated inclination angles $\theta=\cos^{-1}(H/L)$, and are 11 ° for ML10A and 6 ° for ML10B, but the actual angle is 10 °; 21 ° and 15 ° are the angles for the narrow band probe. These results are satisfactory. For the estimated height they are very accurate, less than one fifth the error obtained by the conventional focus probe method, for both the wide band and narrow band probes.

(2) Z20 Case

Crack shape is simple and has about a 20 ° inclination as shown in Fig. 8. Vectors diagram of the reflection of the transverse one wave pulse by this crack is shown in Fig. 9, and the incident direction is shown by an arrow in the upper right corner. A is the tip echo and B is the bottom reflection of tip wave A which overlaps with the main corner echo in Fig. 9. Z20A indicates the incident wave at upper right, and the scanning graph is displayed in Fig. 10-a and 10-b. Estimated crack height(W) for the wide band pulse is given 3.96mm in Table 2, and the actual value of 3.89mm results in a 0.07mm error. Crack heights for a narrow band are obtained from the distance AB in Figs. 11-a and 11-b respectively. They are tabulated as estimated height(N): 3.96mm, and 3.89mm for actual value, with an error of 0.07mm. Crack length is obtained from

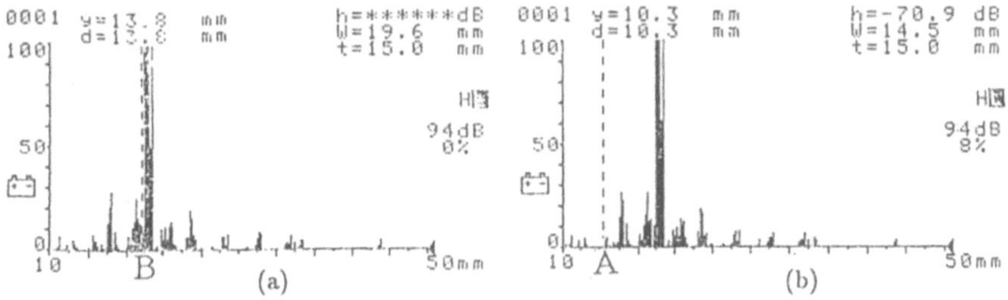

Figure 5. Crack height estimation of ML10 ° A using tip wave A and bottom reflected wave B.

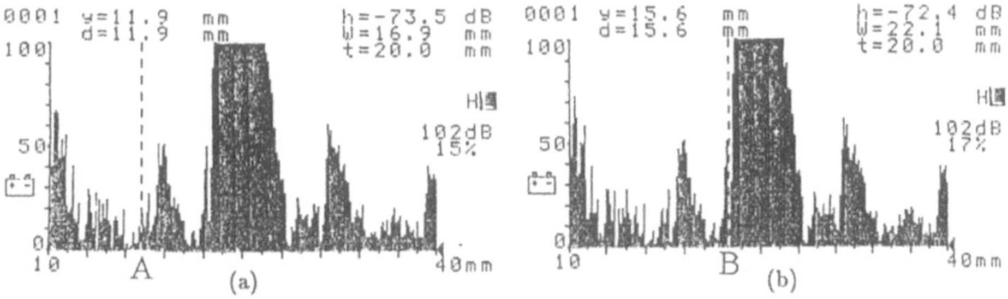

Figure 6. Crack height estimation of ML10 ° A using tip wave A and bottom reflected wave B.

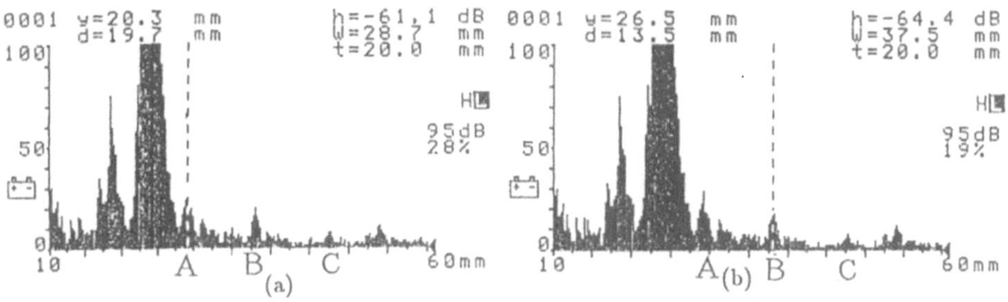

Figure 7. Crack length estimation of ML10 ° .

Figure 8. Crack shape of Z20 ° .

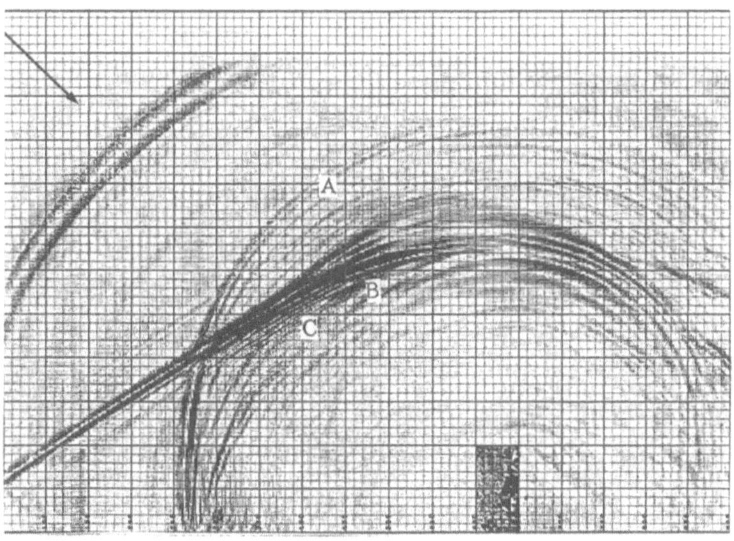

Figure 9. Vector diagram of reflection by Z20 ° .

the distances of the pulse sequences A, B, and C in Fig.12. Estimated length is given as L=4.01mm for wide band and 3.96mm for narrow band. The estimated inclination angles $\theta=\cos^{-1}(H/L)$, and is 19° for wide band pulse, and 0° for narrow band pulse actual angle is 20°. 19° for the wide band probe will be satisfactory, but unsatisfactory for the narrow band pulse. For the estimated height they are very accurate, less than one fifth the error obtained by the conventional focus probe method, for both the wide band and narrow band probes.

EXPERIMENTAL RESULTS

Page limitations make a complete explanation impossible, so details of the experiments and numerical results for the incident direction to the acute angle inclination of a crack have been excluded; only the experimental results are shown in Table 2

Table 2. Estimation of crack height, length, and inclined angle.

	MS15A	MS15B	ML10A	ML10B	Z05A	Z05B	Z20A	Z20B	Z30A	Z30B
Estimate Height(W)	5.65	5.8	7.21	7.35	5.8	5.95	3.96	3.96	3.53	3.81
Estimate Length(W)	6.09	6.09	7.92	7.92	5.94	5.94	4.18	4.18	3.94	3.94
Estimate Height(N)	5.85 (0.17)	5.89 (0.17)	7.24 (0.06)	7.41 (0.13)	6.00 (0.07)	5.94 (0.11)	3.68 (0.12)	3.93 (0.15)	3.73 (0.07)	3.65 (0.12)
Estimate Length(N)	6.3	6.3	7.45	7.45	5.85	5.85	4.01	4.01	3.90	3.90
Actual H.	5.9	5.9	7.22	7.22	5.88	5.88	3.89	3.89	3.62	3.62
Actual L.	6.15	6.15	7.34	7.34	5.9	5.9	4.14	4.14	4.18	4.18
Estimate Angle(W)	21°	17°	11°	6°	11°	0°	19°	19°	25°	12°
Estimate Angle(N)	21°	21°	13°	6°	0°	0°	23°	11°	17°	20°
Actual Angle	15°	15°	10°	10°	5°	5°	20°	20°	30°	30°

In this table, the crack shapes on the side view of ML10 and Z20 are shown in Figs. 3, and 8, and the three other crack shapes of MS15 , Z05, and Z30 are shown in Figs. 13, 14,and 15; each having a 15,° 5° , and 30° inclination, respectively. In Table 2, ML10A is the result for the incident direction to an obtuse angle of an inclination of a crack. ML10B is that for the incident direction to an acute angle of an inclined crack etc.. As explained for ML10 estimated or actual height H and length L are denoted by mm, the first and second rows(W) are for the wide band, and the rows with (N) are for the narrow band. The maximum error of estimated height(W) is 0.2mm for Z30B and the others are about 0.1mm to 0.2mm; average error of Height(W) is less than 0.2mm. On the other hand, average error of height(N) for narrow band is less than 0.21mm.

Estimated length and inclination angles are within 10° for the wide band (W) cases, and 20° for narrow band (N) cases. These errors for crack heights are acceptable, and very much improved than the conventional methods.

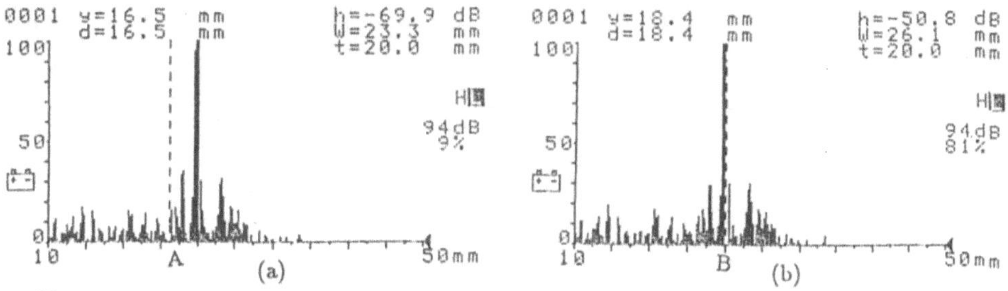

Figure 10. Crack height estimation of Z20 ° A using tip wave A and bottom reflected wave B.

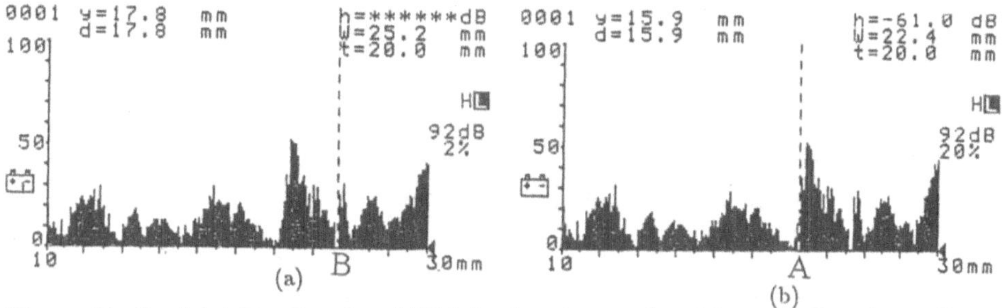

Figure 11. Crack height estimation of Z20 ° A using tip wave A and bottom reflected wave B.

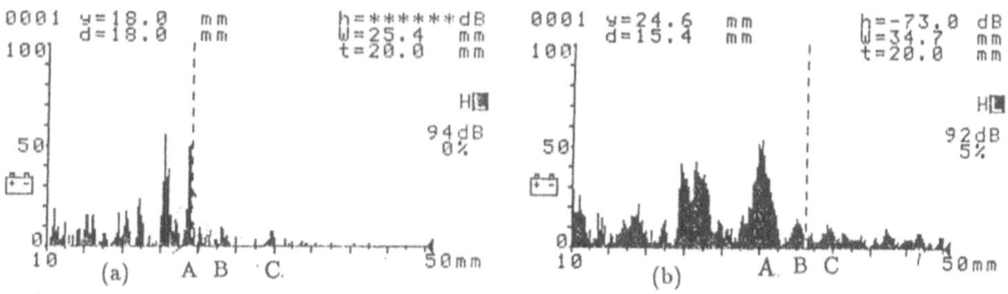

Figure 12. Crack length estimation of Z20 ° .

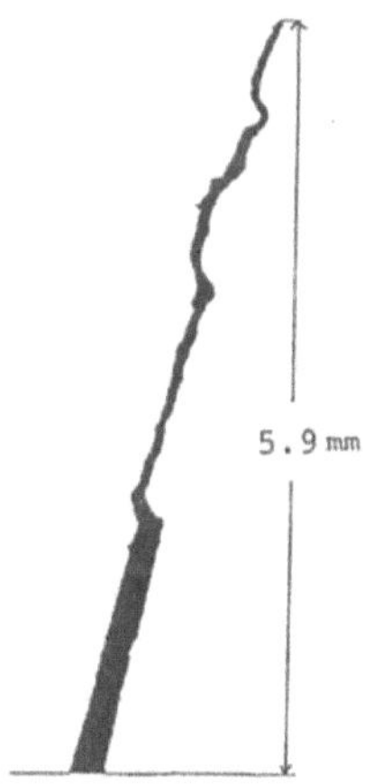

Figure 13. Crack shape of MS15 ° .

Figure 14. Crack shape of Z05 ° .

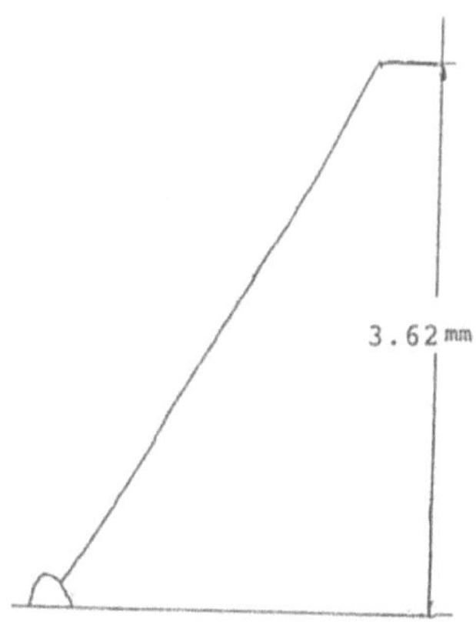

Figure 15. Crack shape of Z30 ° .

CONCLUSION

Sizing of a fatigue crack on a free surface was attempted using the new full digital detector HITACHI-DT-2200.

(1) The accuracy of DT-2200 is proved to be 0.1mm using a wide-band probe, even for the 0.65mm wavelength of 5 MHz, as the pulse position is obtained from the pulse peak.

(2) The crack height is obtained by H= $\sqrt{2}$*AB. This equation was found for the simple shaped slit in the previous paper,[3] with the help of numerical experiment. The same equation was also used for a fatigue crack with a complicated shape, where tip pulse A was selected as the strongest peak before the main echo for a wide band pulse. A sharp rising or the first peak near the sharp rising should be chosen for a narrow band pulse. However, the bottom reflected tip wave B is recognized by a sharp dip in or near the main echo. Experiments should be performed carefully to assure good results.

(3) Crack length L is obtained from the pulse distances of the equi-distance pulse sequences after the main echo. These pulses are generated from the surface wave as it propagates up and down along the crack surface.

(4) Inclination angle θ is obtained by $\theta = \cos^{-1}(H/L)$.

(5) Accuracy of experiment for crack height is reduced to less than 0.2mm for a narrow band probe, and to less than 0.2 mm for a wide band pulse. This means about one-fifth that of the conventional method.

(6) Numerical experiment was done using the automatic mesh generation of FEM in order to solve a complicated shape crack. Numerical results showed the several dips as C in the vector diagrams for the wide band pulse, and also help to explain experimental results.

Consequently, the multi-tip method can be applied to a fatigue crack on a free surface using DT-2200, and accurate and good results will be obtained by a simple and easy technique if the experiment is carefully carried out.

ACKNOWLEDGEMENT

We would like to express our sincere thanks to Mr. M. Uchida, Y. Aikawa, T. Miyajima, M.Nakabayashi,T. Matsumura, and the members of Tip Wave Group in JNDI for their help to this work. Computation were done by the use of the HITAC-8200 in Tokyo University.

REFERENCES

1. M.G.Silk, "Sizing crack-like defects by ultrasonic means" in Research Techniques in Nondestructive Testing, Academic Press, London, Vol.3, p51 (1977).
2. Tip Echo Group Japan Soc. NDI "Ultrasonic Defect Sizing", translated by M.Moles, ASNT(1991).
3. K.Harumi, M.Uchida, T. Miyajima, Y. Ogura "Defect sizing of small inclined crack on a free urface using multi-tip waves" NDT &E International 25,3,135(1992).

ULTRASONIC SET-UP TO CHARACTERIZE STRESS STATES IN RIMS OF RAILROAD WHEELS

Rüdiger Herzer, Helmut Frotscher, Klaus Schillo, Dietmar Bruche
and Eckhardt Schneider

Fraunhofer Institute for Nondestructive Testing (IzfP)
66123 Saarbrücken, Germany

INTRODUCTION

The stress state in the wheels of railroad freight cars is strongly influenced by the brakings. By pressing brake-shoes, the rolling surface of the wheels are getting hot and the subsequent cooling causes the development of tensile stresses in the circumferential direction. Small cracks in the rolling surface, which are not harmful otherwise may grow under the influence of the tensile stress and may even cause the failure of a wheel. In order to assure the reliability of the wheels and the safety of the traffic, the stress states of railroad wheels need to be inspected periodically. Because of the large number of endangered wheels, the inspection should not only be nondestructive but also fast and easy to apply in daily routines.

OBJECTIVE

The highest tensile stresses are developed in the edge area between the rolling surface and the outer side of the railroad rim. And with respect to the fracture mechanical evaluation of the cracks in that area, the determination of the stress states in the surface near layers of the running surface and of the outer side is most desirable. But because of the damages of the material and the plastic deformations in that zones, an automated nondestructive stress analysis, applicable for large numbers of wheels, seems to be not possible. As mentioned elsewhere [1], our concept to assure the reliability of the railroad wheels is based on the statistical evaluation of the stress states in rims and on more extensive investigations on particularly chosen wheels. Among the nondestructive magnetic and ultrasonic techniques to characterize the stress states, the use of the ultrasonic birefringence effect seems to be the most suitable one. The relatively simple application of this effect is made possible by the fact, that the wheels of interest are forged and heat treated, yielding a quasi isotropic elastic behavior of the material. The effect of texture (preferred grain orientation) has been found neglectable. The technique, based on the birefringence effect does not enable the evaluation of the principal stress in the circumferential direction σ_{tan}. This disadvantage is compensated by the following advantages for the practical use [1]:

- The shear wave is generated by an electromagnetic transducer (EMAT). No coupling medium is needed, no special surface treatment is needed. The transducer can easily be manipulated along a radial measuring trace.
- The influence of temperature on the ultrasonic time-of-flight is neglectably small. The influence of temperature on the material dependent acousto-elastic constant is also neglectable. At component temperatures between -10°C and +40°C, the error due to temperature influences on the final result is within the error, specified for the measuring system (see later).
- The profile of the stress difference (σ_{tan}-σ_{rad}) along the height of the rim is evaluated. Depending on the particular braking and load situation, the extreme values of stresses are at different depths from the rolling surface. Those stress profiles give stronger evidences concerning the stress state than the evaluation along the circumference of the wheel.
- The propagation directions of heat cracks, found in the rims, have also components perpendicular to the radial direction. The principal stress difference (σ_{tan}-σ_{rad}) is seen as more suitable for the fracture mechanical evaluation of a wheel than the σ_{tan} value alone.

Figure 1. Ultrasonic set-up to characterize stress states in rims of railroad wheels.

It has been the objective to design and to build an automated ultrasonic setup to evaluate the difference of the principal stresses (σ_{tan}-σ_{rad}) along a radial trace across the rim of monoblock railroad wheels. The system needs to be applicable in daily routines in workshop environments.

SET-UP COMPONENTS

The ultrasonic set-up to characterize stress states in rims of railroad wheels is shown in Figure 1.

The manipulator with the size of about 350 x 200 x 130 mm^3 and a weight of about 12 kg is fixed to the rim by two magnetic switches and a gripping lever as to be seen in Figure 2. It houses the EMAT for a linearely polarized shear wave with about 2 MHz center frequency, preamplifiers and two servo-motors to move the EMAT along the measuring trace and to turn the probe about 90°. An inductive sensor detects the upper and lower edge of the height of the rim and controls the mechanical contact of the probe and the wheel.

The central part of the system is housed in a rigid and closed cabinet with integrated cooling system. The size and weight is about 1800 x 600 x 1065 mm^3 and 125 kg respectively. It contains the ultrasonic transmitter receiver electronics, an amplifier with an automated gain control and a patented device to determine the appropriate zero crossings of two radio frequency backwall echos needed for the ultrasonic time-of-flight measurement. A high performance counter with a temperature stabilized 500 MHz time reference and the master-central processing unit are mounted in the center part of the cabinet. The lower part gives space for an optional printer.

Figure 2. The manipulator of the set-up with an EMAT probe for linearely polarized shear wave.

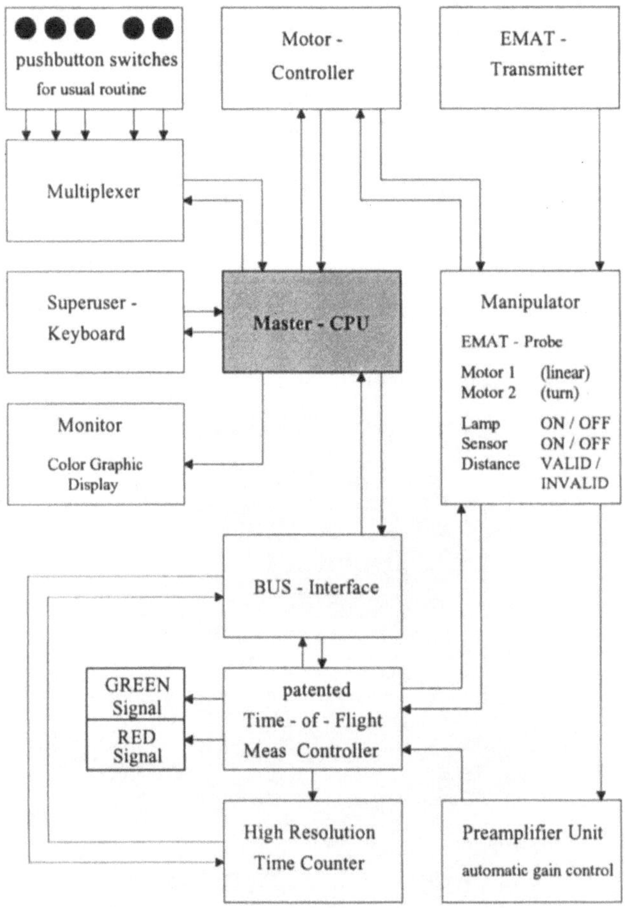

Figure 3. Schematic of the ultrasonic set-up.

The monitor, displaying the different steps of the measuring routine and the result of the stress analysis is at the top part of the unit, behind a window. All adjustments, needed for daily routine measurements are possible by using the 3 pushbuttons at the upper left side. The drawer at the center of the height holds a keyboard for the supervisor to store material and wheeltype dependend data. A green or red light at the very top ledge indicates the classification of the measured result. The schematic in Figure 3 shows the hardware elements together with the flow of signals and informations.

MEASURING ROUTINE

The operation of the set-up is menuguided. The most important steps of the measuring routine are displayed in Figure 4. The system starts with a self-testing procedure and with the initialisation of different components of the routine. Deviations are displayed together with suggested helps. The operator selects the type of wheel under test from a list in the storage, containing up to 19 different types together with the material dependent acousto-elastic constant (K-value), the value of the acceptable maximum of the stress difference and three more wheeltype depending parameters. The confirmation of the choosen set of data is done by pressing one of the above mentioned 3 pushbuttons; the data are displayed in a box at the right side of the screen. After the measurement is started by pushing the green Start button, the manipulator moves the sensor along the trace, realizing the height of the rim; the polarization of the shear wave is turned into the radial direction. The system checks-up the signals and prepares the measurements. The sensor is moved to the starting position and then along the trace; the time-of-flight of the shear wave is measured, the time-of-flight locus curve is diplayed on the screen. After this measurement, the probe is moved to a predetermined position again; the probe is turned to have the vibration of the shear wave along the circumferential (tangential) direction. A check-up of the signals is done again and as before, the probe moves to the starting position. All of these individual steps are indicated on screen. The Figure 5 shows the content of the screen corresponding to the above described situation: The wheel under inspection corresponds to the wheel classification number 1; the box at the right side gives further wheel related informations. The second field on the left side gives the depth from the running surface of the first and the last measuring position of the trace. The measurement of the time-of-flight of the radialy polarized shear wave is completed. The sensor is moving to the start position for the measurements with the vibration parallel to the tangential direction. During the measurement, the screen displays the time-of-flight locus curve together with the result taken with radial polarization. The system evaluates the difference of the two pricipal stresses (σ_{tan}-σ_{rad}), calculates the mean value of the difference $<\sigma>$ and compares the maximal value of the stress difference with a given value for the acceptable maximum. The Figure 6 shows the result, displayed on the screen. The acceptable maximal value, in this case 300 MPa, is shown as a line to facilitate the comparision for the operator. The maximal value and the corresponding measuring position are shown in the top right corner. Depending on the result of the mentioned comparison, a green or red light additionally indicates whether the wheel needs to be stress relief treated (red light) or not (green light). The operator confirms the end of the measuring loop by pushing the Stop button. A measuring loop usually takes about 75 seconds.

It has been mentioned that there are controlling routines between each major step of the loop. In case of irregularities, the system repeats the particular step or even the whole loop depending on the detected irregularity. Serious irregularities, specified in different classes are shown by an error message on the screen. The loop is interrupted until the operator clearified the situation and confirmed the step. The error messages are also written to a control file to enable trouble shooting by the manufacturer of the system.

The results of the measurement can be printed and be transferred via RS232 link to another computer for further investigations and for statistical purposes.

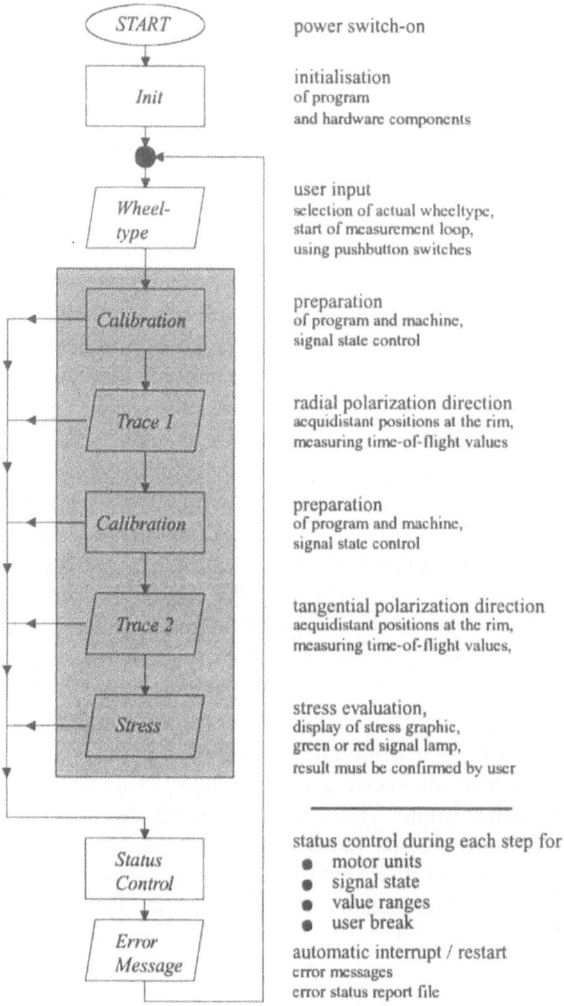

Figure 4. The measuring loop of the set-up.

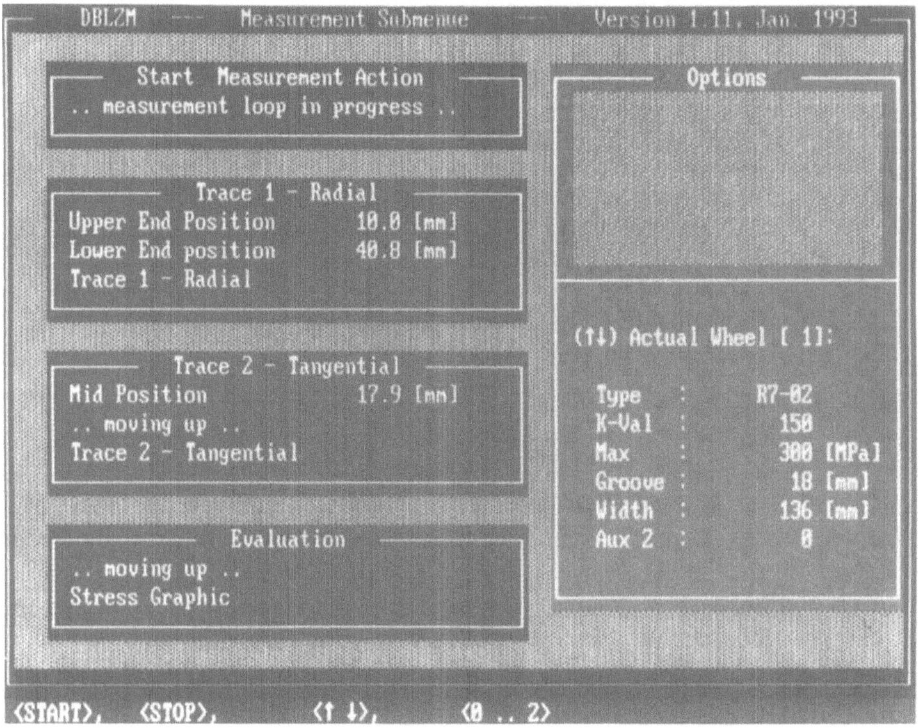

Figure 5. Screen display at a certain step during the measurement.

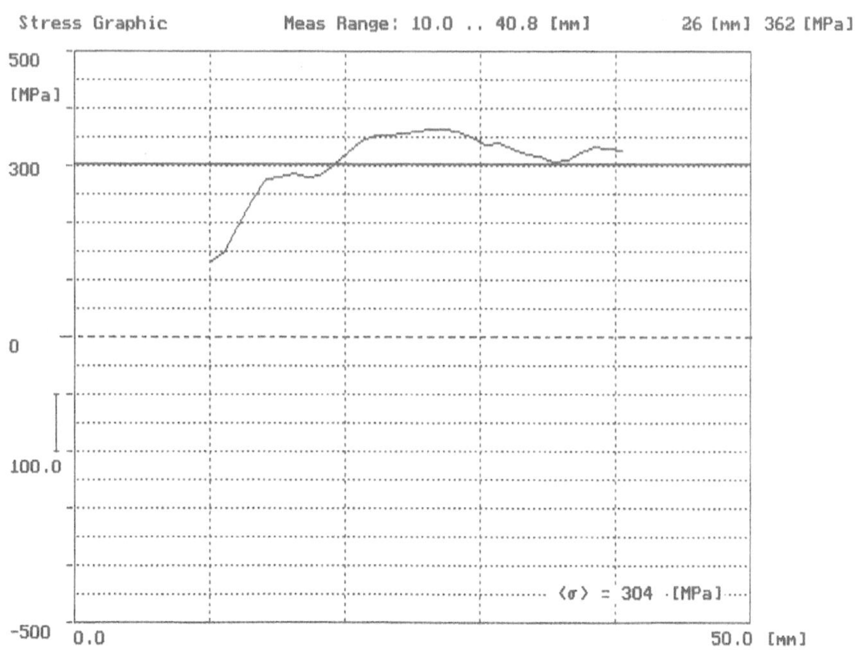

Figure 6. Profile of the principal stress difference (σ_{tan}-σ_{rad}) the maximal and mean value of the difference.

TECHNICAL DATA

Some technical data of the system are:

Resolution:	+/- 20 MPa
Accuracy:	+/- 10 %
Temperature of wheels:	-10° C till +40° C

Automated measurement and evaluation
Result shown on screen
Valuation of result by comparision with a predetermined acceptable value
Storage for material and wheeltype depending parameters
Parallel port for optional printer
Serial port for data transmission (RS232)
EMAT shear wave transducer, 2 MHz center frequency, 10 x 10 mm^2 aperture
Rigid housing with integrated cooling system

ACKNOWLEDGEMENT

This work was carried out under contracts with the Deutsche Bundesbahn, Zentralamt Minden and with the Deutsche Reichsbahn, Zentralstelle Delitzsch .The authors would like to thank the representatives of the mentioned institutions for helpful discussions.

REFERENCES

1. E. Schneider, R. Herzer, D. Bruche and H. Frotscher, Ultrasonic Characterisation of Stress States in Rims of Railroad Wheels, Sixth International Symposium on Nondestructive Characterization of Materials; June 7-11, 1993, Hawaii.

ULTRASONIC ANALYSIS OF INTERFACIAL DEBOND IN CERAMIC MATRIX COMPOSITES

Renee M. Kent[1], Prasanna Karpur[1], Theodore E. Matikas[2], and Paul D. Jero[3]

[1]Research Institute
University of Dayton
Dayton, OH 45469

[2]National Research Council
Wright Laboratory
2230 10th St. Ste. #1
WPAFB, OH 45433

[3]WL/MLLM
Wright Laboratory
2230 10th St., Ste.#1
WPAFB, OH 45433

INTRODUCTION

Ceramic materials have properties which make them desirable candidates for elevated temperature applications. Most monolithic ceramics, however, have an inherently low fracture toughness which limits their usefulness for critical applications. Therefore, recent research has been aimed at developing ceramic matrix composites (CMC's) and engineering the microstructure of ceramics to enhance the material's toughness.

The mechanical behavior of CMC's is largely determined by the material and mechanical characteristics of the fiber-matrix interfacial bond region [1]. This factor has motivated a number of recent studies designed to determine the mechanical properties of the interface in CMC's. Notably, several researchers have developed fiber push testing methodologies and analyses to measure the force necessary to slip a fiber along its length in an effort to assess the interfacial bond strength [2-4]. However, fiber push results tend to exhibit a wide data scatter for a given material system [5]. Therefore, nondestructive characterization processes are used in conjunction with destructive testing in an effort to enhance our understanding of the interfacial debond process due to fiber push testing and to accurately assess the state of the interfacial bond characteristics. In this work, we

Nondestructive Characterization of Materials VI
Edited by R.E. Green, Jr. *et al.*, Plenum Press, New York, 1994

describe ultrasonic capabilities which allow us to monitor elastic stress wave transfer from the matrix to the fiber and correlate with fiber push testing results.

BACKGROUND OF EVALUATION TECHNIQUES

Fiber push testing is a process by the which the force required to debond a fiber along all (fiber push-out testing) or part (fiber push-in testing) of its length is measured. The resulting load-deflection curves recorded during the push process yields information related to the interfacial bond characteristics of the material [6]. The experimental apparatus for fiber push testing is shown in Figure 1. In fiber push-out, fibers in a thin (less than 3 mm) test specimen are loaded via a 10-100 μm diamond-tipped indentor until the entire interface is debonded and the fiber is displaced in the matrix. Fiber push-in entails loading the embedded fiber such that a small region of the interface is debonded; a few microns of fiber displacement at the load point is typically observed. After fiber push testing is completed, the relative interfacial stiffness is affected due to the degradation of the interfacial bond characteristics and development of debonding.

Analysis of the interfacial debond characteristics of CMC test specimens is

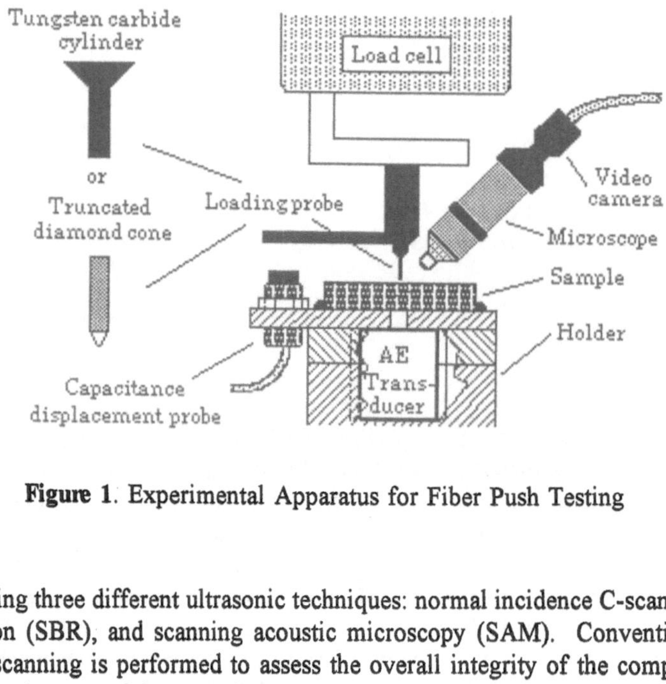

Figure 1. Experimental Apparatus for Fiber Push Testing

performed using three different ultrasonic techniques: normal incidence C-scan, shear wave back reflection (SBR), and scanning acoustic microscopy (SAM). Conventional normal incidence C-scanning is performed to assess the overall integrity of the composite and to verify the effectiveness of the consolidation and densification [7]. The ultrasonic response of the back reflected ultrasonic shear wave from the fiber-matrix interface is used to image the interface and verify its integrity [8,9].

The model describing the relationship of the ultrasonic response from the shear wave back reflection (SBR) at the fiber-matrix interface has been described in detail in the literature [9,10] and has been implemented extensively for the characterization of selected metal matrix composites [7,8,11]. Fundamentally, the SBR models the ultrasonic response of the back reflection of shear wave ultrasonic waves incident on an interface in terms of the transfer of the elastic stress wave across the interface. The resulting amplitude of the reflected wave is quantitatively related to an interfacial stiffness parameter which is dependent on the material properties at the interface [12].

The SBR model also shows that the dynamic range of the ultrasonic response from a selected fiber-matrix interface is a function of the elastic properties (density, elastic modulus) of the fiber and the matrix, the frequency of ultrasound, and the angle of incidence used for analysis. A judicious selection of operating frequency allows the ultrasonic analysis to be tailored to enhance the sensitivity of the ultrasonic back reflected signal response to display variations in the coefficient of stiffness of the fiber-matrix interface [9].

EXPERIMENTAL PROCEDURE

Model test specimens of a single ply SCS-6/7740 glass CMC and a single ply SCS-6/Glass frit CMC were investigated for their interfacial bond characteristics using conventional (destructive) fiber push techniques and corroborative ultrasonic nondestructive characterization. The SCS-6/7740 glass composite specimen was processed by hot pressing a layup of SCS-6 fibers between two glass plates at 500 psi and 800 degrees C for 20 minutes. The SCS-6/Glass frit specimen was prepared by hot pressing the ply of SCS-6 fibers directly in the glass frit. Each composite was then cut and ground to 28 x 28 mm.

The as-received composite was immersed in water and ultrasonically scanned for gross defects and delamination using a normal incidence longitudinal C-scan at 50 MHz (0.25" dia., 1.0" focus transducer). A 25 MHz transducer (0.25" dia. , 0.5" focus), incident at an angle of 19 degrees was used to image the integrity of the fiber-matrix interface. The back reflected ultrasonic response from the resulting vertically polarized shear waves incident on the interface between the fiber the matrix were software gated for imaging [13]. The experimental ultrasonic parameters (frequency and angle of incidence) were selected to provide the optimal dynamic range for the specific properties of the fiber and matrix, according to the model outlined above.

The as-received composite specimen was cut into test specimens for push testing. Areas of the specimen which were shown by the ultrasonic pre-test scans as areas of poor bonding and/or delamination were selected for push-out testing while the areas of the specimen which showed uniformity in bonding were chosen for fiber push-in testing. The appropriate push tests were performed using the configuration shown in Figure 1.

RESULTS AND DISCUSSION

The SBR and normal incidence C-scan from the as-received SCS-6/7740 glass are shown in Figures 2a and 2b, respectively. The scans indicate regions of delamination and/or poor consolidation at the interface between the plates and the fiber-matrix interface. This is similar to the interference phenomena reported in MMC fragmentation tests [8].

Fiber Push-out

The ultrasonic scans of the test specimens which had been pushed-out are shown in Figure 3. The corresponding push-out data is also given. The 25 MHz shear wave back reflection scans (Figure 3a) clearly show the debonded region of the fiber. As the ultrasonic wave impinges on the interface, the reflected wave experiences a phase reversal between the region which is bonded and the debonded region. The apparent break in the

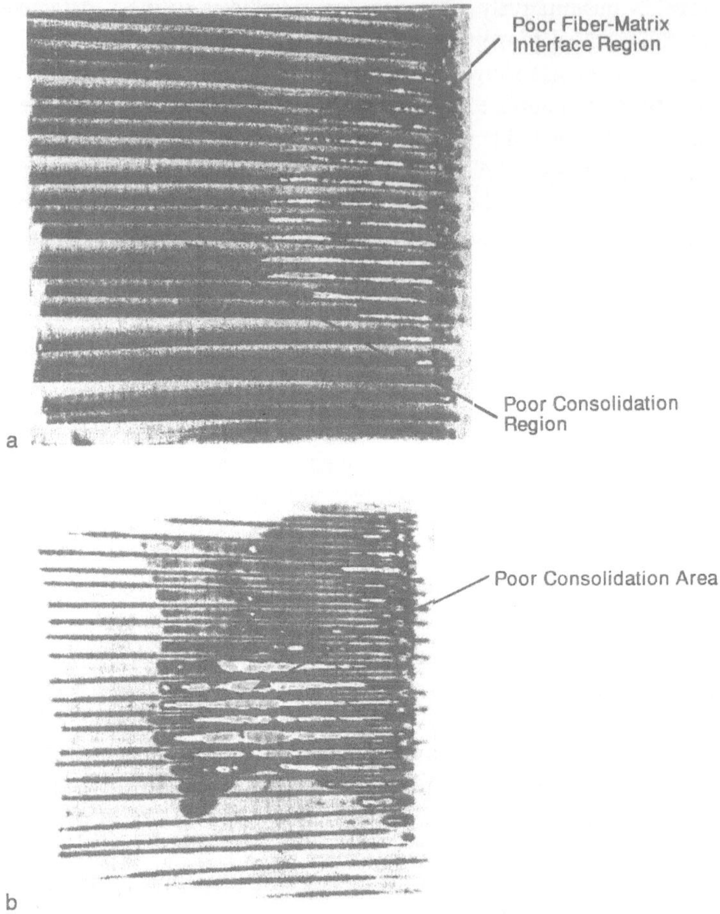

Figure 2. Ultrasonic Scans of As-received SCS-6/7740 Glass CMC (a) Shear Wave Back Reflection (b) Normal Incidence C-scan

fiber is the result of the linear superposition of the reflected ultrasonic response of areas which exhibit a phase reversal.

The normal incidence longitudinal C-scans of the fiber push-out specimen (Figure 3b) indicate activity in the matrix area surrounding the debonded area of the fibers. The increase in the ultrasonic response is attributed to the relaxation of residual stresses which are inherent in the matrix of the as-received composite due to the mismatch of the coefficients of thermal expansion between the fiber and the matrix. During fiber push-out, stresses must be applied which are sufficient to overcome the residual stress as well as frictional and radial stresses at the interface. As the applied stress is increased, the residual stresses are released and an increase in the ultrasonic response, corresponding to an increase in the observed tensile stress, is effected.

Fiber Push-in

The ultrasonic scans from the SCS-6/F-glass are shown in Figures 4a and 4b. The fiber push-in specimens from the SCS-6/F-glass test specimen (processed directly using

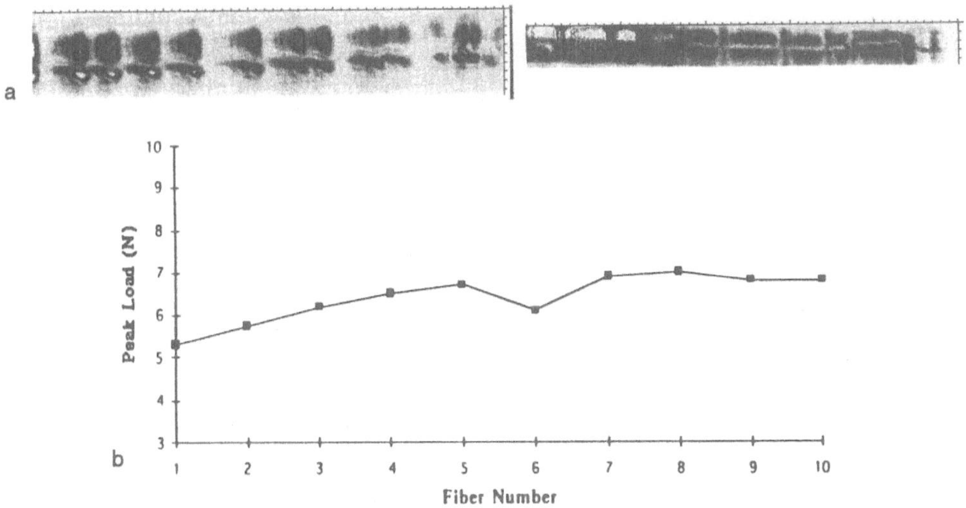

Figure 3. Ultrasonic scans of Fiber Push-out SCS-6/Glass CMC (a) Shear Wave Back Reflection (b) Normal Incidence C-scan

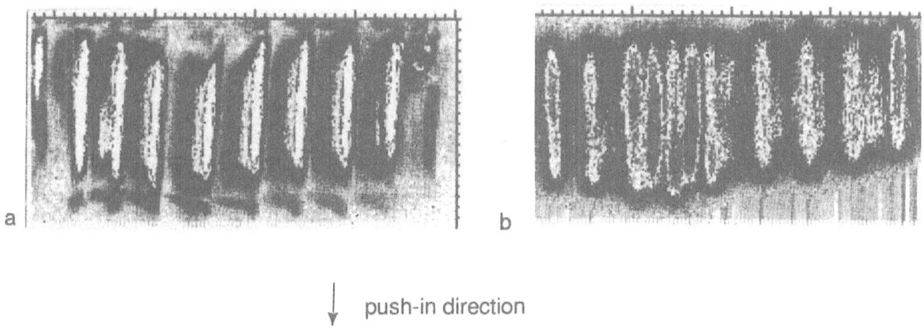

Figure 4. Ultrasonic Scans from Fiber Push-in SCS-6/F-Glass CMC (a) Shear Wave Back Reflection (b) Normal Incidence C-scan

glass frit) also indicate extensive degradation of the interfacial bond relative to its as-received condition. However, this scan does not independently indicate direct evidence of a progression of the damage along the length of the interface. Instead, there appears to be severe degradation of the bond all along its length. The SBR model predicts such an increase in the magnitude of the reflected amplitude for a change in the displacement of the shear wave across the interface and a corresponding decrease in the relative interfacial stiffness coefficient.

In this case, however, the defocused shear wave from a 50 MHz SAM scan indicates that the degradation of the bond is nonuniform over the length of the fiber and correlates with the magnitude of the stress applied during push-in testing. Figure 5 shows the

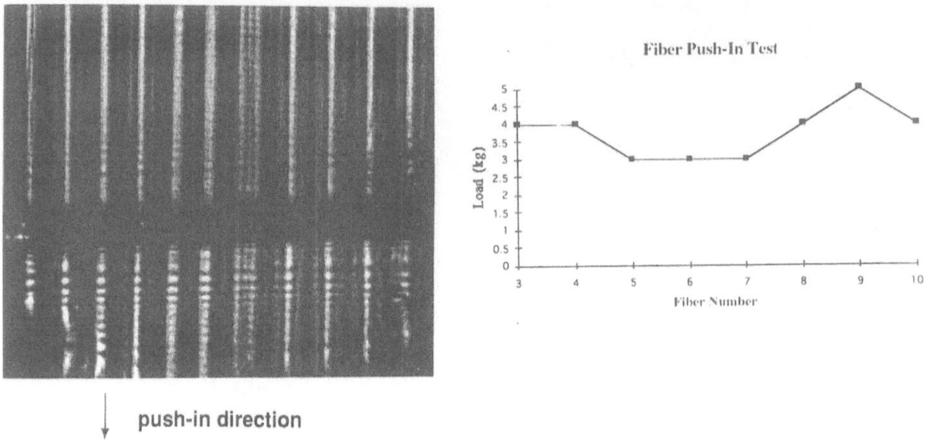

Figure 5. Defocused Shear Wave Acoustic Microscope scans for Fiber Push-in SCS-6/F-glass

progression of damage as a function of the load applied during fiber push-in. The higher the applied load (fibers 8-10) during push-in, the greater the extent of the interfacial damage. Fibers 1 and 2 indicate the highest degree of damage and correspond to cycled loading during fiber push-in. In addition, a stress profile along the length of each debonded fiber is predicted from fiber push test models [6].

CONCLUSIONS

Several ultrasonic techniques have been shown to corroborate and supplement the results from fiber push testing. Normal incidence and shear wave scanning used in conjunction with destructive testing allows us to more fully describe the stress wave transfer characteristics of the fiber-matrix interface and describe the nature of the interface in terms of its interfacial stiffness. Further research relating the extent and characterization of the damage to the material due to fiber push testing is ongoing.

ACKNOWLEDGMENTS

This work was performed at Wright Patterson Air Force Materials Directorate in WL/MLLP (Kent, Karpur, Matikas) and WL/MLLM (Jero) under contract # F33615-90-C-5944 (Kent) and F33615-89-C-5612 (Karpur). The authors are grateful for the assistance of M.P. Blodgett of Wright Laboratory for performing the SAM scans, and E.A. Jenkins for his assistance in performing the SBR scans.

REFERENCES

1. A.G. Evans and D.B. Marshall, "The Mechanical Behavior of Ceramic Matrix Composites,", *Acta Metall.*, 37:10 (1989).
2. D.B. Marshall, "An Indentation Method for Measuring Matrix-Fiber Frictional Stresses in Ceramic Composites", *J. Am. Cer. Soc.*, 67:12 (1984).

3. D.B. Marshall and W.C. Oliver, "Measurement of Interfacial Mechanical Properties in Fiber Reinforced Ceramic Composites," *J. Am. Cer. Soc.*, 70:8 (1987).
4. J.F. Mandell, K.C.C. Hong, and D.H. Grande, "Interfacial Shear Strength and Sliding Resistance in Metal and Glass-Ceramic Matrix Composites," *Cer. Eng. Sci. Proc.*, 8:7-8 (1987).
5. A.J.G. Jurewicz, R.J. Kerans, and J. Wright, "The Interfacial Strengths of Coated and Uncoated SiC Monofilaments Embedded in Borosilicate Glass", *Cer. Eng. Sci. Proc.* 10:7-8 (1989).
6. R.J. Kerans and T.A. Parthasarathy, "Theoretical Analysis of the Fiber Pullout and Pushout Tests", J. Am Cer. Soc., 74:7 (1991).
7. S. Krishnamurthy, T.E. Matikas, P. Karpur, and D.B. Miracle, "Evaluation of Processing of Fiber-Reinforced Metal Matrix Composites using Ultrasonic Methods", submitted for publication in *J. Comp. Sci. and Tech.* (1993).
8. P. Karpur, T.E. Matikas, and S. Krishnamurthy, "Matrix-fiber Interface Characterization in Metal Matrix Composites using Ultrasonic Imaging of Fiber Fragmentation", *Mat. Eval.*, 47 (1989).
9. T.E. Matikas, and P. Karpur, "Ultrasonic Reflectivity Technique for the Characterization of Fiber Matrix Interface in Metal Matrix Composites", *J. Appl. Phys.*, in press, (1993).
10. T.E. Matikas and P. Karpur, "Micro-mechanics Approach to Characterize Interfaces in Metal and Ceramic Matrix Composites", to be presented in the *Review of Progress in Quantitative Nondestructive*, Bowdoin College, Bruswick, Maine (1993).
11. M.C. Waterbury, P. Karpur, T.E. Matikas, and S. Krishnamurthy, "In Situ Observation of the Single Fiber Fragmentation Process in Metal Matrix Composites by Ultrasonic Imaging", submitted to *J. Comp. Sci. and Tech.*
12. P. Karpur, T.E. Matikas, and S. Krishnamurthy, " A Novel Parameter to Characterize the Fiber-Matrix Interface for Mechanics of Continuous Fiber Reinforced Metal Matrix and Ceramic Matrix Composites", to be published in the *J. Comp. Sci. and Tech.* (1993).
13. C.F. Buynak, T.J. Moran, and R.W. Martin, "Delamination and Crack Imaging in Graphite-Epoxy Composites", *Mat. Eval.*, 47 (1989).

A LASER SPECKLE IMAGE CORRELATION INSTRUMENT FOR HIGH-TEMPERATURE SURFACE STRAIN AND DEFORMATION MEASUREMENTS

Adel Sarrafzadeh-Khoee, Russell J. Churchill, and Bruce L. Thomas

American Research Corporation of Virginia
P. O. Box 3406
Radford, VA 24143

Silverio P. Almeida

The University of North Carolina at Charlotte
Department of Physics
Charlotte, NC 28223

ABSTRACT

Advanced high temperature structural materials for the National Aero-Space Plane (NASP) program are subjected to high level thermomechanical loadings during laboratory testing. This paper describes the technical performance of a laser speckle image correlation instrument for obtaining optical strain measurements at elevated temperatures of 1100°C and above. The technical approach to this project involved the development of an optical strain sensor incorporating laser speckle imaging and digital cross-correlation techniques for uni-axial strain and deformation measurements. This PC-based instrumentation system consists of three main modules: three high-resolution, line-scan, charge-coupled device (CCD) cameras with special optics; three electromechanical shutters for triple-beam sequential illuminations by a continuous wave probing laser; two digital signal processing (DSP) system boards and DSP analysis software using Hypersignal-Windows Block Diagram programming modules. Surface strain and deformation measurements along the central axis were performed for nickel-based alloy and silicon-carbide coated carbon-carbon composite samples in a cantilever fixture under thermomechanical loading. The comparative simulation study was performed using analytical/graphical computation of thermomechanically loaded test samples. The experimental test results showed that the acquired speckle images used for the cross-correlation calculations suffered insignificant

Nondestructive Characterization of Materials VI
Edited by R.E. Green, Jr. *et al.*, Plenum Press, New York, 1994

degradation or drift due to the presence of extreme thermal conditions. This optical strain gage sensor requires no surface preparation or application of micro-indentation marks or targets. The multi-level operating software consists of several modules, each performing a different task such as hardware initiation, image acquisition, digital signal processing, strain/deformation analysis data storage, and graphical presentation. Application of the digital laser speckle cross-correlation method has been extended for *in situ* monitoring and characterization of laser-processed refractory ceramic coating film formation.

INTRODUCTION

The high-temperature environment and limited accessibility of complex shaped aerospace components make it difficult to characterize the thermomechanical and aerodynamic behavior of advanced structures using the presently available measurement methods, such as contact strain gage, capacitance probes and other sensing devices. This paper deals with a new technique for high-temperature structural testing by developing innovative non-contact optical testing technology and PC instrumentation based on laser speckle imaging, digital signal processing and cross-correlation techniques in order to evaluate high level strain and deformation fields of the test components at elevated temperatures of 1100°C and above. Extreme surface temperatures are caused by the aerodynamic friction forces on the critical structural components of National Aero-Space Plane (NASP) type vehicles. The proposed use of layered carbon-carbon composite materials on the nose and wing/fin leading-edge components of the trans-atmospheric aircraft is intended to maintain structural integrity at very high temperatures.

The primary objective of this project was the demonstration of a non-contacting strain-gage sensor employing novel laser speckle imaging and digital cross-correlation calculation techniques. The technical approach to the research program involved development of very-high-temperature optical strain measurement instrumentation based on one-dimensional laser speckle imaging and digital cross-correlation techniques. The major research activities constituting the developmental program consisted of a series of tasks designed to accomplish the objectives of the project. Specifically, the implementation of innovative technologies included: application of coherent laser beam illumination and scattering techniques for the measurement of optically rough surface displacements; integration of specially arranged high-speed laser beam shutters and high-resolution linear array imaging cameras (triple-beam/triple-camera configuration) for the measurement of deformations as well as surface strain; implementation of the microprocessor-based cross-correlation calculations for the detected one-dimensional speckle data to provide direct information on the surface strain and deformations; and development of a dedicated operating software program incorporating digital signal processing hardware/software modules, data acquisition devices and analysis for streamlined test measurement and control of the instrument.

FUNDAMENTAL FORMULATION

The following expression provides the fundamental relationship between the rough-surface induced diffracted laser speckle characteristics (such as spatial movements) and surface deformation variables (such as strain, in-plane displacement, out-of-plane displacement and tilt). It has been found that the diffracted speckles formed by a point-wise laser beam interrogation move according to the associated object surface displacement variables. From the basic analysis of optical diffraction patterns formulated by Yamaguchi (1981), the laser speckle movement versus object displacements is represented by

$$A_X = +a_x\left(\frac{L_o \mathrm{Cos}^2\theta_i}{L_i \mathrm{Cos}\theta_o} + \mathrm{Cos}\,\theta_o\right) - a_z\left(\frac{L_o \mathrm{Cos}\theta_i \mathrm{Sin}\theta_i}{L_i \mathrm{Cos}\theta_o} + \mathrm{Sin}\,\theta_o\right)$$

$$+\Omega_y\left(\frac{\mathrm{Cos}\theta_i}{\mathrm{Cos}\theta_o} + 1\right) L_o \Bigg] - \Bigg[\varepsilon_{xx}\left(\frac{\mathrm{Sin}\theta_i}{\mathrm{Cos}\theta_o} + \tan\theta_o\right) L_o$$

where A_X = speckle shift (movement) along the X-axis, a_x = in-plane displacement (translation) of the object, along the x-axis, a_z = out-of-plane displacement (deflection) of the object along the z-axis, Ω_y = rotation (tilt) with respect to the y-axis, ε_{xx} = uni-axial strain, θ_i = angle of illumination (incident) beam from the normal direction, θ_o = angle of observation from the normal direction, L_i = radius of curvature of the illumination (incident) beam, and L_o = observation distance from the object interrogation point.

Figure 1 shows the coordinate system for diffracted laser speckle displacement. Yamaguchi (1981; 1982) used the above relationship in his proposed dual beam or double-sensor approaches in conjunction with the differential cross-correlation method to derive surface strain measurements. The dual beam method allowed the cancellation of unwanted out-of-plane displacement component (a_z), based on the planar wavefront illumination requirement and the automatic subtraction of cross-correlation values calculated from speckle images of each beam. However, in the case of the double sensor configuration, the influence of a_z on the differential speckle displacement, ΔA_X, could not be ignored.

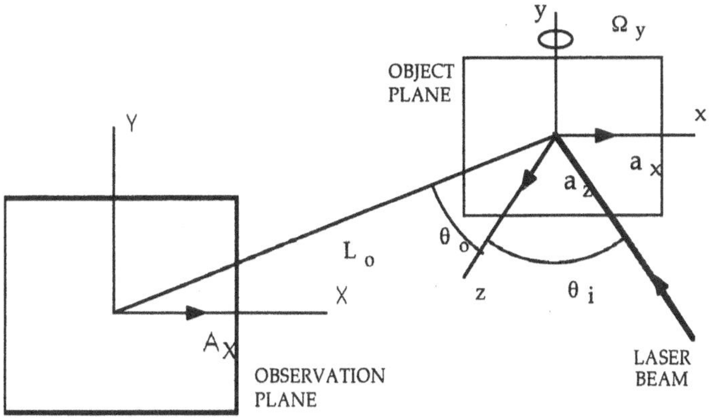

Figure 1. Coordinate system for diffracted speckle displacement.

A major contribution in this research project has been the adaptation of Yamaguchi's approach in a special triple-beam/triple-sensor configuration so that all four displacement variables of the object under deformation can be obtained. That is, in order to determine the in-plane displacement (a_x), out-of-plane displacement (a_z), strain (ε_{xx}) and tilt (Ω_y), four independent relations are needed for these four unknown parameters in the above expression. These four relations can be obtained using the subtraction and addition procedures for both the single-beam/dual-sensor and dual-beam/single-sensor configurations Sarrafzadeh (1993). Herein, the combined implementation of single-beam/dual-sensor and dual-beam/single-sensor configurations provides the so called triple-beam/triple-sensor configuration (Figure 2). Therefore, the differential and summed procedures for the speckle shifts (A_X) in this triple-beam/triple-sensor configuration would provide the four deformation parameters. The details of such parametric relationships are presented in a recent report by Sarrafzadeh (1993). This is analogous to solution of the four unknown parameters in the fundamental expression by the four independent and simultaneous measurements of A_X.

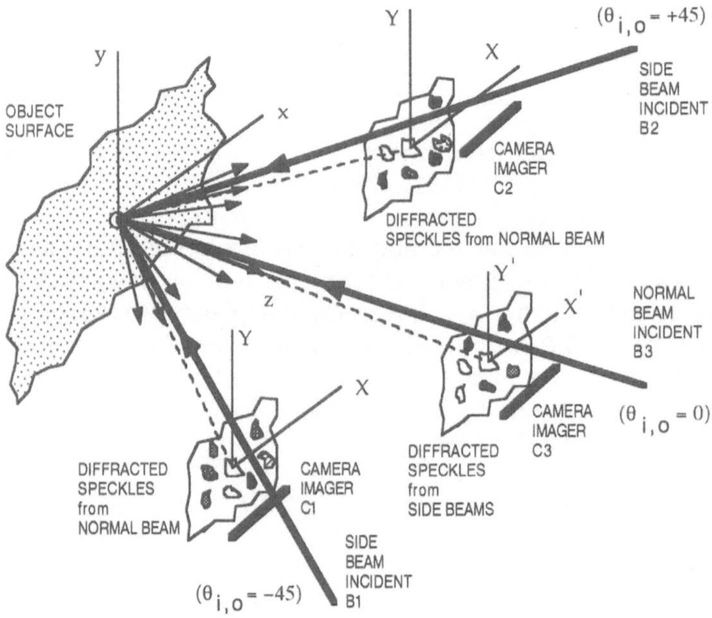

Figure 2. Triple-beam/triple-camera geometric configuration.

EXPERIMENTAL PROCEDURES

This section contains a brief description of experimental procedures and test methods used for the development of a pre-prototype instrument. Major technical developments include: the specific geometric arrangements of the combined laser incident beams and linear array imaging cameras (e.g., single-beam/dual-camera, dual-beam/single-camera, and triple-beam/triple-camera configurations); unique implementation of operating software incorporating multi-level programming (e.g., low-level, driver-level and application-level software modules); and preliminary evaluation of the turnkey hardware-software system for very-high-temperature surface strain and deformation measurements. The correlation-based laser speckle imaging instrument designed, constructed and tested under these efforts was demonstrated in the laboratory.

Sensor Head Configuration

As mentioned earlier, sensor head configurations of the instrument are based on the original approaches of Yamaguchi (1981). The cross-correlation calculations are performed using PC-based DSP analysis. The two interchangeable configurations are called single-beam/dual-sensor and dual-beam/single-sensor approaches. In both cases, laser speckle patterns are acquired sequentially with respect to the dual sensors or beams. The primary contribution of this modified technique is a special adaptation of Yamaguchi's approaches combined with specific sensor head design and digital signal processing instrumentation for surface strain and deformation measurements at very high temperatures. The integrated sensor head system (triple-beam/triple-camera configuration) is specifically designed such that the single-beam/dual-camera or dual-beam/single-camera configurations, similar to Yamaguchi's dual beam or double sensor approaches, can also be used separately if needed. The beams, shutter sequences, and cameras are configured for various geometric arrangements to customize the sensor response to particular measurement needs. Both of the single-beam/dual-camera and dual-beam/single-camera configurations are designed for measuring surface strain only. However, the deformation or rigid-body movement of the object under interrogation can negatively influence the sensor response to surface strain measurements. Such coupling effect of, e.g., out-of-plane displacement to strain is, however, less pronounced with the dual-beam/single-camera configuration than the single-beam/dual-camera configuration. But, the above decoupling effect is achieved only if the illumination beams are collimated. On the other hand, the triple-beam/triple-camera configuration imposes no such limiting conditions and can be used for independent calculation of surface strain and related deformations such as in-plane displacement, out-of-plane displacement, and tilt. The sensor head as designed with these three integrated configurations is a single-channel, uni-axial, strain and deformation measurement instrument. The measurement sensitivity axis is parallel to the plane in which the three beams and linear cameras lie. The complete design of the portable laser speckle sensor head has incorporated the integration of an Argon-ion probing laser and a beamsplitting system for three-beam illumination.

PC Instrumentation and Operating Software

The complete process of testing, except for the optical alignments and specimen/structure loadings, is entirely controlled by a fully instrumented personal computer. An operating software algorithm performs all aspects of automated test and measurement procedures such as hardware initiation, speckle data acquisition, cross-correlation calculations, processed data-to-strain/deformation transformation, real-time or file-based image/signal display, output data storage, prestored data file import, load versus strain/deformation curve plotting and print of output signals. A total of eight new software "Block" modules were developed under the Hypersignal-Windows Block Diagram programming package. These modules are named INIT, ACQUIRE, X-CORRELATION, STRAIN, PROJECT SAVE, PROJECT LOAD, VIEW and VIEW-BMU. These modules were developed using PC-based standard C and assembly languages. For the cross-correlation calculations, digital signal processors have also been used for high-speed Fast Fourier Transformation of image data. Figure 3 shows a selected link path for a typical experimental test. The signal output for the measurement parameters can graphically be represented (Figure 4) using the bundled plotting package.

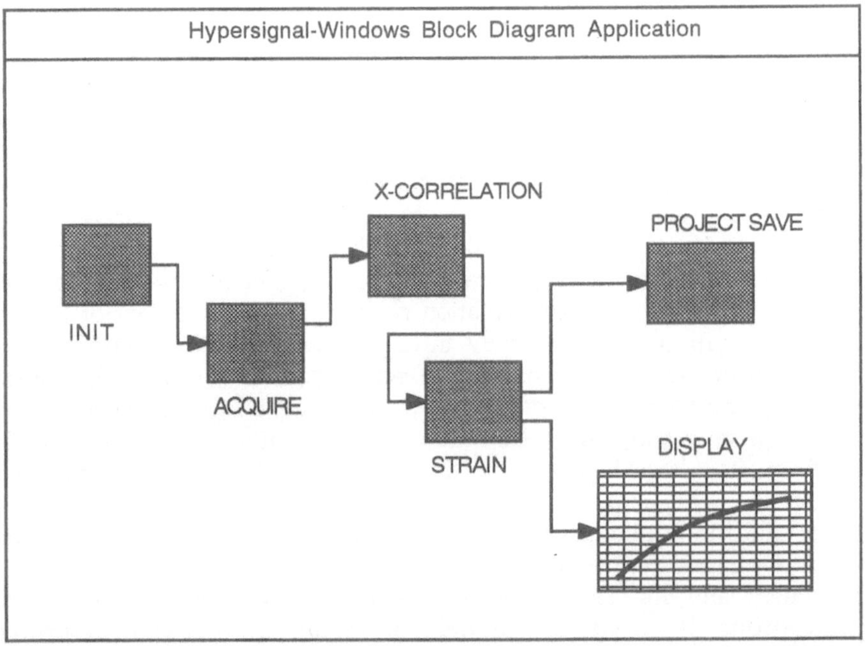

Figure 3. Typical linked block path for acquiring, processing and analyzing strain measurement.

Instrument Specifications

Table 1 summarizes the general description and specifications of the pre-prototype demonstration instrument. The system features and performance characteristics listed below are observed for the existing instrument hardware and software modules. Therefore, the optimized specifications of the "next-generation" instrument may not be limited to these stated values and capabilities. The pre-prototype instrument has been delivered to NASA La RC.

Table 1. Instrument specifications.

Selected Features	General Description
Multi-measurement	Strain, in-plane/out-of-plane displacement, tilt
Interrogation method	Noncontact
Output signal	Load vs. strain and deformations
Portability	Movable on tripod and PC carts
Probing laser	Medium power Argon-ion
Speckle imaging sensors	Linear-array CCD cameras
Image processing hardware	DSP system development boards
Speckle imaging procedures	Special multiple-beam/-camera configurations
Operating software	Multi-level programming modules
Signal processing modes	On-line or file-based processing
Optical gage	Single-channel, uni-axial, front-surface

Typical Characteristics	Response Time
Speckle image acquisition interval	25 ms
Data acquisition time for 4Mb of DSP's buffer memory units	45 s
X-correlation processing time for 4Mb data	5 min
Total run time for complete experimental procedure per measurement interval	1 s

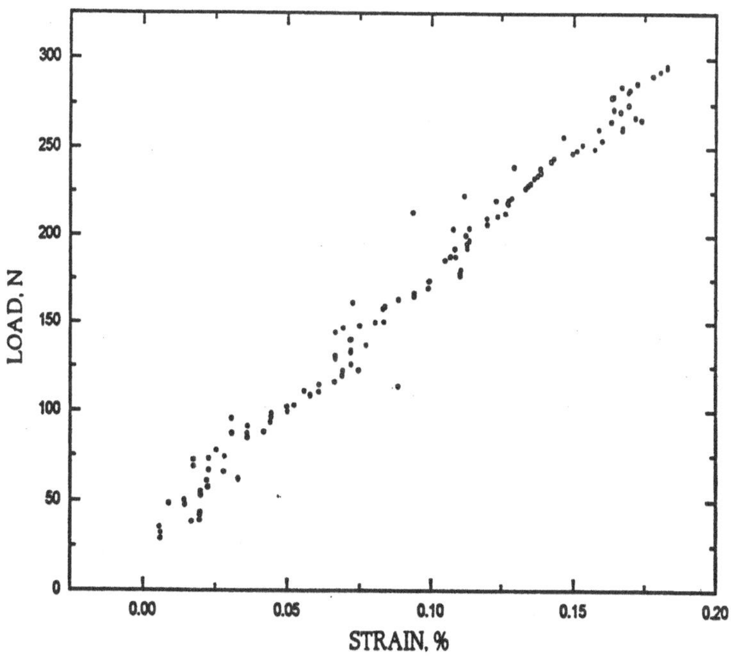

Figure 4. Load-strain data for a cantilever carbon-carbon composite sample at 1125°C.

CONCLUSION

From the methods evaluated and the operating performance demonstrated in this study, it is concluded that the laboratory pre-prototype instrument can provide optical measurement of high level strains at surface temperatures of 1100°C and above. The special design of triple-beam/triple-camera configuration in this correlation-based laser speckle measurement system provided the more accurate determination of surface strain and the associated deformations. The complete integration of dedicated hardware and operating software modules resulted in construction of a pre-prototype turnkey system for performing a "streamlined" experimentation. The entire strain gage testing procedure including hardware initiation, speckle image acquisition, x-correlation processing, strain and deformation parameter measurements, and data storage and display is automated under the operating software.

ACKNOWLEDGMENT

The authors acknowledge the interest shown by Dr. Robert R. Reeber, the Army Research Office, who made the poster presentation of this work possible.

REFERENCES

Sarrafzadeh, A.K., (1993), "Correlation-based optical strain sensor for hostile environment and wind tunnel test instrumentation," NASA LaRC SBIR Phase II Final Report, Contract No. NAS1-19305.

Yamaguchi, I., (1981), "A laser speckle strain gauge," The Institute of Physics, 0022-3735/81/111270104.

Yamaguchi, I., (1982), "Simplified laser-speckle strain gage," Optical Engineering, Vol. 21, No. 3, pp. 436-40.

PORTABLE INFRARED SPECTROPHOTOMETER FOR NDT CHEMICAL ANALYSIS OF WEATHERED, PAINTED SURFACES

Thomas Novinson

Materials Science Division
Naval Civil Engineering Laboratory
Port Hueneme, CA 93043

DISCLAIMER

The mention of tradename products or company names in this paper does not imply or infer any type of endorsement or recommendation by the Naval Civil Engineering Laboratory (NCEL), the Naval Sea Systems Command (NAVSEA), the U.S. Navy, or any other government agency. Many of the lenses, mirrors, sources, detectors and other components mentioned are available from a variety of U.S. vendors. No materials or component specifications are given because the instruments described in this paper were designed and assembled during a research and development program.

INTRODUCTION

There are many situations in which it is necessary to chemically analyze or identify materials in the field using nondestructive analysis (NDA). NCEL was funded by NAVSEA to develop a portable infrared (IR) spectrophotometer for identifying shellacs, varnishes, paints and other protective coatings used on steel, aluminum and other metals. A portable field instrument for coatings analysis is useful (a) to ensure that the correct or federal specification coating was used and (b) to determine whether any surface chemical changes have occurred as a result of weathering, sunlight exposure, humidity, microbial attack or other factors that degrade the paint.

INFRARED SPECTROPHOTOMETRY

IR spectrophotometry has been used for the identification of many organic materials, including the resins in protective coatings. The method is chemically selective and allows the direct, qualitative identification of epoxy, polyurethane, polyester, alkyd, silicone-alkyd, and acrylic latex coatings containing inorganic pigments. The IR spectrum of a representative organic material consists of a series of bands or peaks of different amplitude (absorbance) at different wavelengths. Organic materials, such as paint resins, have many absorbance peaks in the region from 2.5 microns to 20 microns, which corresponds to wavenumbers of 4000 cm^{-1} to 500 cm^{-1}.

Inorganic materials, such as potassium bromide, silver bromide or sodium chloride do not interfere with these organic spectra because their absorption is at wavelengths beyond 50 microns (200 cm^{-1}).

INSTRUMENT REQUIREMENTS

Several types of instruments are commercially available for gas detection, such as carbon dioxide or monoxide in air. These instruments have IR sources and detectors and generally at least one filter for a specific gas. Additional components and a different design are required for a full spectrum IR analyzer for solids. The basic components of an IR spectrophotometer include:

(a) an IR source to produce polychromatic light or radiation
(b) an IR detector to produce a variable electrical signal
(c) a monochromator to separate wavelengths
(d) a sample cell or accessory to direct the beam
(e) a method for recording the spectra
(f) amplifiers and preamplifiers to boost the electrical signal produced by radiation falling on the detector.

Single-beam IR analyzers, which are generally used in gas analysis, sample both the IR bands of the sample and the IR bands of water vapor and carbon dioxide in air, assuming one is analyzing contaminated air samples. The spectra of the water vapor and the carbon dioxide have to be subtracted to obtain the spectrum of the gas sample alone.

Double-beam gas analyzers, on the other hand, use both a reference and a sample beam so that the spectra of the water vapor and the carbon dioxide are automatically subtracted from the final spectrum.

IR spectrophotometers are generally designed for laboratory rather than field use. These instruments have carefully aligned optical components and require dust-free, vibration-free environments.

In recent years, Fourier Transform Infrared (FTIR) instruments have replaced conventional IR instruments. These instruments use a moving mirror to generate a complex absorption pattern that is rapidly analyzed and converted into a complete IR spectrum. These instruments produce high quality spectra rapidly, without dispersive optics, slits, filters and monochromators. Limitations on time, funding, and design complexities precluded work on modifying this type of instrument for field analysis.

The general requirements for the design and assembly of our field instrument were:

(a) relatively simple optical and mechanical design
(b) ability to be placed directly against the sample (painted surface) to nondestructively analyze or identify coatings
(c) ability to generate a complete IR spectrum from 2.5 microns to about 20 microns
(d) operation from a DC battery
(e) modular units that could be assembled in the field
(f) magnetic tape or computer diskette data storage

Although many commercially available IR instruments were surveyed and evaluated, none appeared to meet all the above requirements. Some of the instruments were large and bulky, making them unsuitable for field use. It did not appear that a small, portable, DC-battery-operated IR spectrophotometer was commercially available.

EXPERIMENTAL IR SPECTROPHOTOMETER DESIGN

We took the most desired features from both a single-beam instrument (Miran 1A from Foxboro Corp. of Foxboro, MA) and a double-beam gas analyzer, known as an "optical bench," (from Infrared Industries, Carpinteria, CA) We added or replaced certain components to enhance the signal and reduce the signal to noise ratio (S/N).

The single-beam analyzer (Figure 1) had an IR source (Nernst glowbar), a circular variable filter (CVF) to serve as a monochromator, and a detector, along with IR optics such as condensing lenses, plus a slit.

The double-beam optical bench was a rectangular box with a hemispherical concave mirror at either end (Figure 2). A heated metal element near one of the mirrors served as the IR source. Radiation from the IR source was split into two parallel beams that passed through a sample and reference gas cell. The beams converged at the second mirror and were then focused onto a thermopile detector. An optical chopper could be placed in the path of the two beams, thus allowing a direct comparison between the two gas cells. The gas cells each had narrow filters (10.0 or 11.0 microns) at either end to measure absorbance at one wavelength.

The reference cell was a sealed tube containing the pure gas (100% carbon dioxide or monoxide). Air or other gases containing gas contaminants, such as carbon monoxide or dioxide, could then be quantitatively measured as a percentage of the 100% gas sample, using Beer's law at a specific wavelength: A=abc, where a = absorbance coefficient, b = cell thickness, and c = concentration. The relationship is useful for quantitative anlysis of both gases and liquids.

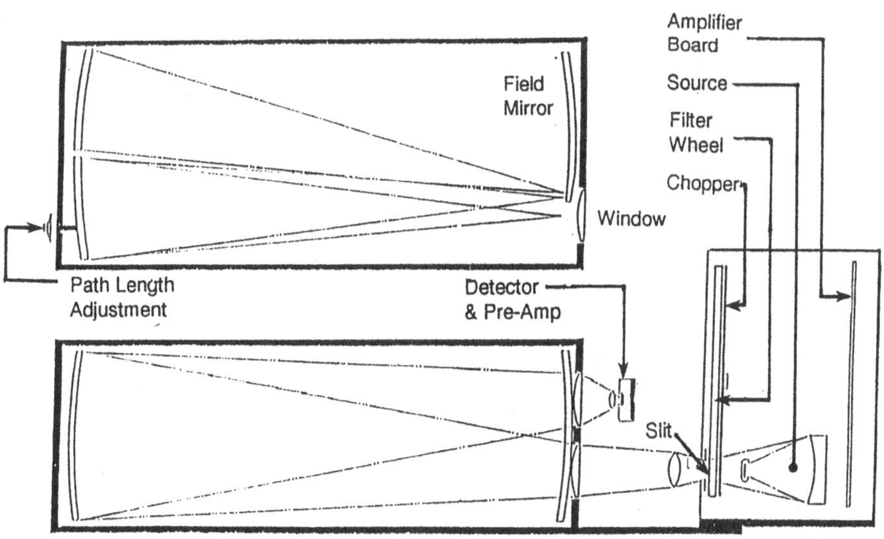

Figure 1. Single-beam analyzer (Miran 1A) from Foxboro Corp.

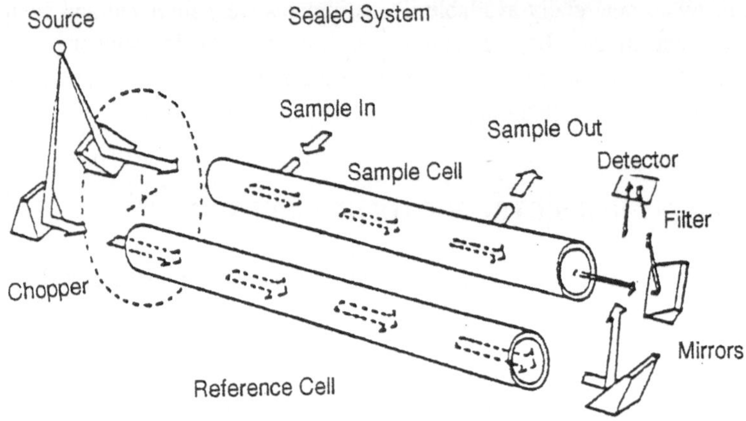

Figure 2. Double-beam gas analyzer -- "Optical Bench" from Infrared Industries.

Surface Analysis Cell

For the analysis of solids, the sample and reference beams had to be reflected downward onto the samples (painted metal plates) and then back into the instrument for data collection and analysis. The pathlength had to be the same for both sample and reference beams. Two methods for directing the IR beam to the sample surface are (a) to pass the beam through a chopper and filter assembly (explained below), or (b) focus the beam with a lens.

Monochromator Versus Variable Filter

In order to obtain a complete spectrum, the narrow range filters had to be replaced with either a monochromator or a circular variable filter (CVF). Monochromators are expensive optical devices with mirrors, slits and gratings for separating the polychromatic IR into individual wavelengths.

Due to cost limitations and ease of operation, a CVF (Figure 3) was used. The CVF selected came from Optical Coating Laboratories Inc. (OCLI) in Santa Rosa, California. It consists of three machined segments of silicon, germanium and Irtran that cover the spectrum from 2.5 to 4.5 microns (4000 to 2200 cm^{-1}), 4.4 to 8.0 microns (2272 to 1250 cm^{-1}) and 7.9 to 14.5 microns (1265 to 690 cm^{-1}). The CVF is turned using a stepper motor. As can be seen from the drawing, there are short periods in which the spectrum is blanked out due to the cut off from the supporting wheel holding the IR filter segments. However, this is not a great inconvenience when one considers the higher cost and alignment requirements for a laboratory-type IR monochromator.

Chopper

The chopper is a flat black, painted wheel with holes for the beam or beams to pass through. When two beams are being compared, the chopper allows the sample beam to pass to the detector, while the reference beam is blocked, and then vice versa. Since the chopper operates at a high rate of revolution, both beams can easily be compared, and the reference or background signal subtracted.

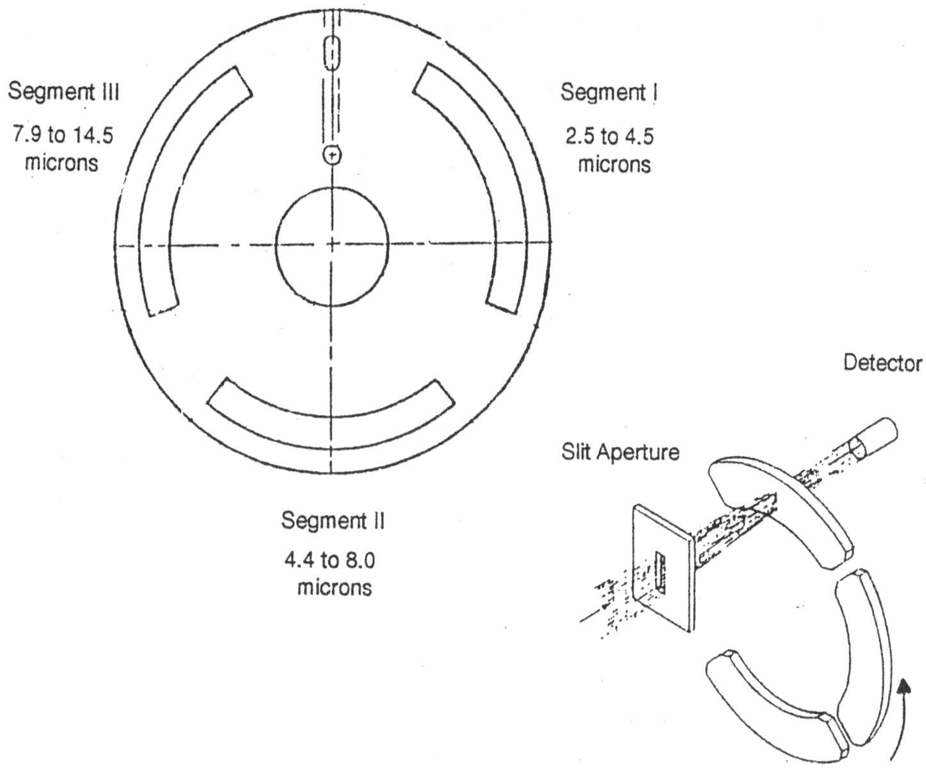

Figure 3. Circular variable filter (CVF) used as a monochomator.

IR Source

The standard IR source is a glowbar, which is a glowing carbon rod or nichrome wire wound around a ceramic. These sources produce both heat and polychromatic infrared light. The polychromatic light has to be further separated into monochromatic wavelengths to develop a spectrum. We used the heated nichrome wire element from the original double-beam optical bench. Consideration was also given to using a higher power bulb with a sapphire window to transmit the IR, but these sources require custom fabrication and are expensive.

Detectors

Detectors for IR energy range from room temperature thermopiles (thermocouple array) to semiconductors cooled to liquid nitrogen temperatures. The colder, semiconductor devices are often known as "quantum detectors," because they produce electrical currents directly proportional to the amount of IR light (assuming that IR has both heat and light components) falling on the crystal. These low temperature quantum detectors offer much more sensitivity and response than the room temperature thermocouple devices, but they are more expensive and they require liquid nitrogen. The room temperature detectors include conventional metallic thermocouples as well as semiconductor crystals, such as lead sulfide and lead selenide.

We replaced the room temperature thermopile detector with a low temperature, liquid nitrogen-cooled mercury cadmium telluride (MCT or HgCdTe) quantum detector. The liquid nitrogen was introduced from a polystyrene container with a hole at the bottom. The cold liquid then dripped into the Dewar containing the MCT chip. If liquid nitrogen is not available in the field, a small, high pressure tank (6000 psi) fitted with a Joule-Thompson cooling device, can also provide small amounts of the liquid. Manufacturer's instructions must be followed to ensure safety in using these small, high pressure tanks.

Signal Conversion

The changes in thermal energy or IR light must be converted into an electrical signal in order to produce a spectrum. In the case of the MCT quantum detector, lead wires are attached at the bottom of the Dewar that connect to a preamplifier and amplifier for the signal.

The electronic signal can be used to operate the pen on a strip chart recorder or an x-y plotter/printer, or a liquid crystal display (LCD), using analog to digital data converters. Modern plotters, printers, and video displays can be controlled by small, portable computers operated with DC batteries (Figure 4).

TESTING OF INSTRUMENT

Small 3- by 5-inch steel coupons were wire-brushed and painted with representative epoxy, polyurethane, alkyd, silicone-alkyd, and acrylic latex paints. The paints all contained similar pigments so that only the resin spectra would differ.

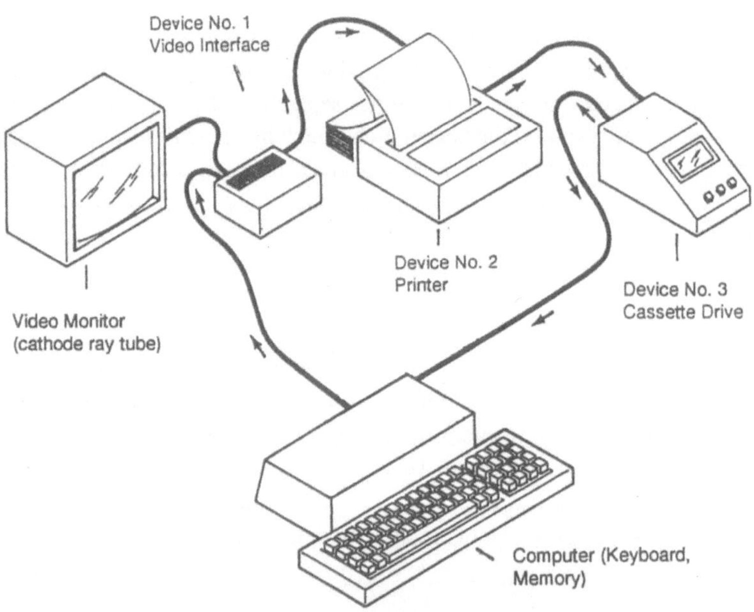

Figure 4. Microcomputer accessories and loop to accompany the portable IR instrument.

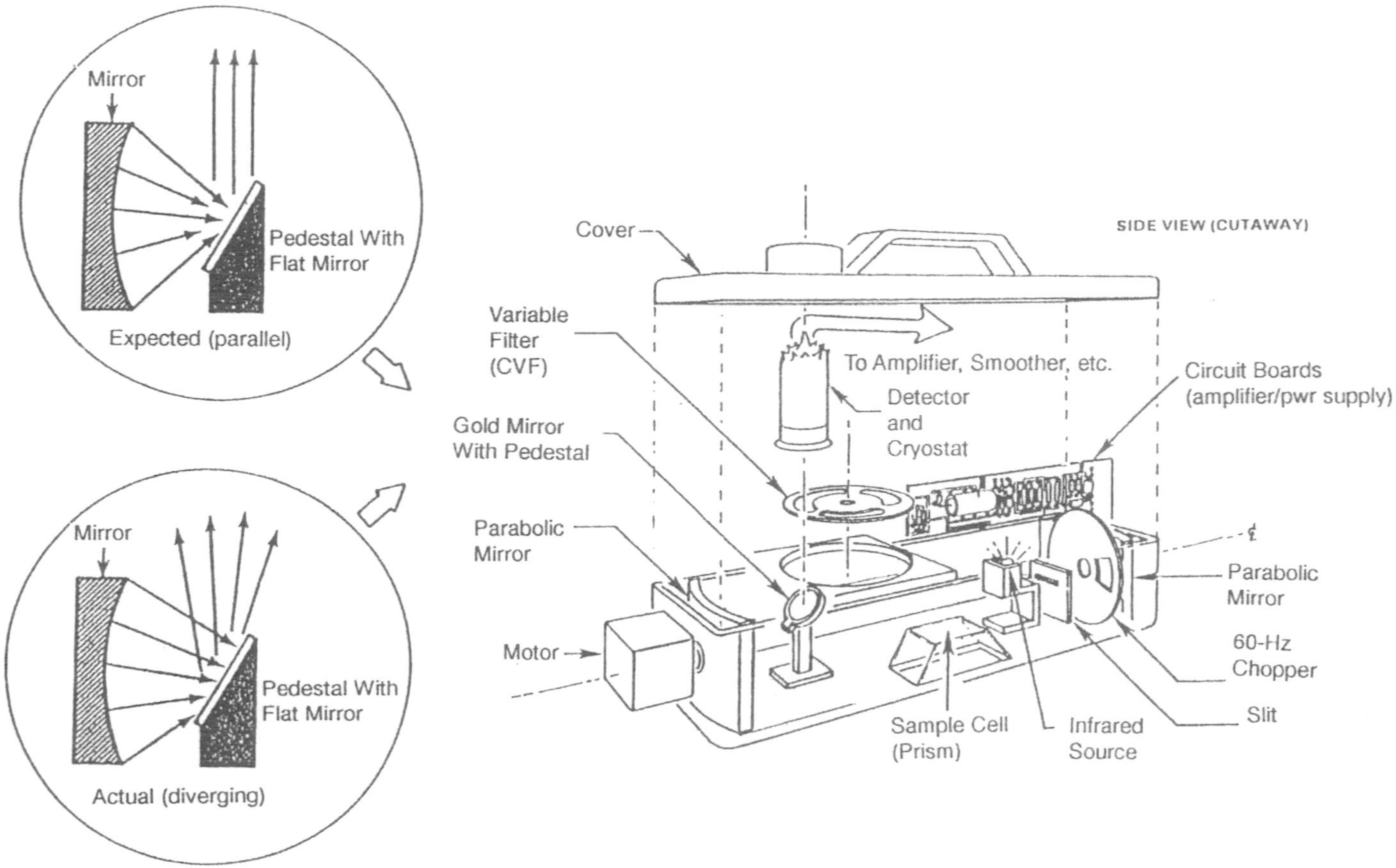

Figure 5. NDT Portable Infrared Spectrophotometer.

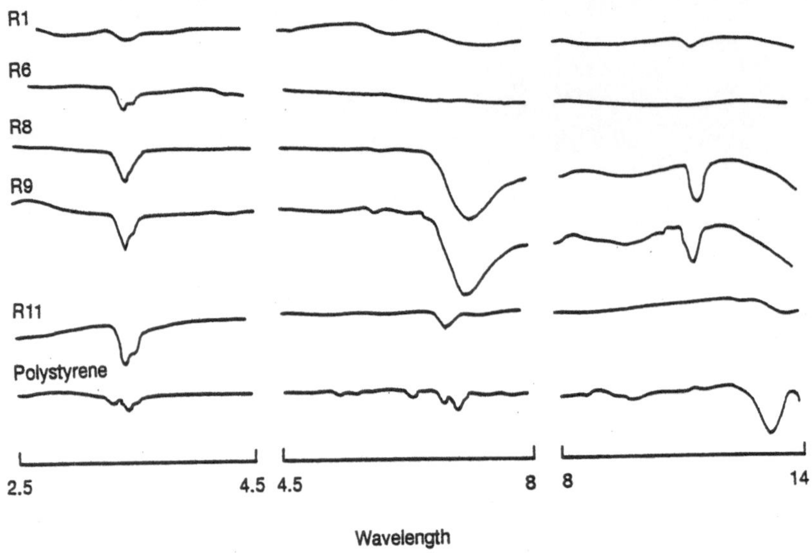

Figure 6. ATR spectra as obtained from the prototype.

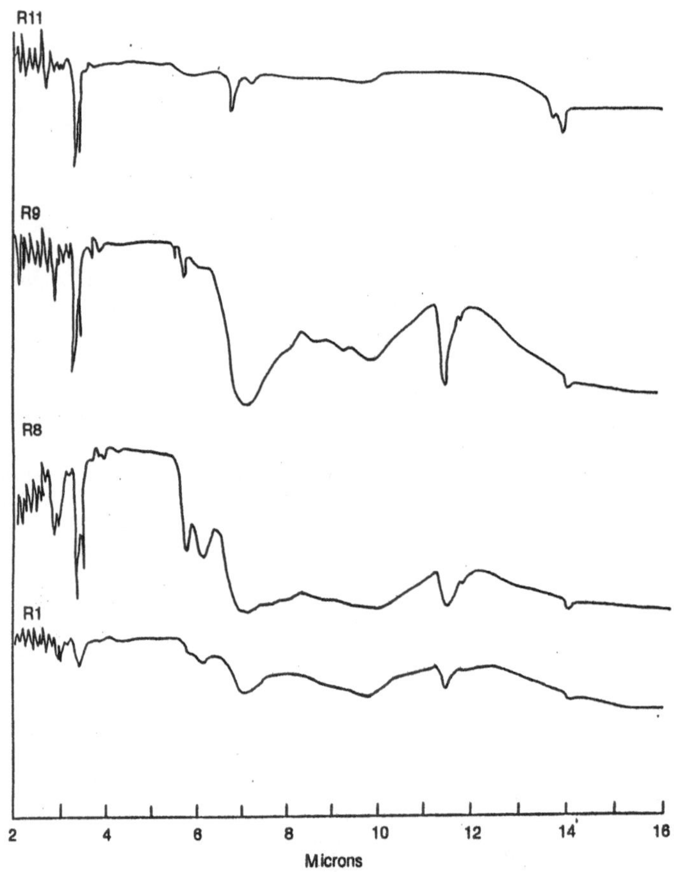

Figure 7. ATR spectra obtained from the Nicolet FTIR system.

IR spectra were run on the newly designed portable IR spectrophotometer (Figure 5), a conventional Perkin Elmer IR spectrophotometer, and a Nicolet FTIR. Figures 6 and 7 show the spectra for identical paint samples from the prototype instrument and the Nicolet FTIR.

RESULTS AND CONCLUSIONS

We were successful in demonstrating that a small, portable IR instrument could be built at moderate cost.

The instrument is portable and can be operated on DC current with an AC inverter. The MCT detector with liquid nitrogen cooling is superior to room temperature thermopile or lead sulfide detectors for producing identifiable peaks.

There did appear to be some loss of IR light from reflection, backscattering or other optical loss, but this probably can be improved by modifying the attenuated total reflectance (ATR) accessory to retain more IR energy at the detector.

The spectra for the samples from the Nicolet FTIR instrument are noticeably stronger and more definitive than those from our portable instrument. However, this instrument is not portable and costs at least ten times as much to build as our portable unit. It is likely that the spectral resolution of our portable field unit can be enhanced by either (a) multiple time averaging methods to increase the apparent band absorption or (b) modification of the preamplifier and amplifier to boost the signals.

ACKNOWLEDGMENTS

The author wishes to thank the Naval Sea Systems Command (NAVSEA) for support and Dr. David Kyser ,physics department, Naval Air Weapons Station, China Lake, California, for assistance with infrared optics. The author also thanks Messrs. Arlen Jackson and Mike Hanks of the Naval Civil Engineering Laboratory (NCEL) for assistance in assembling the instruments and helping to obtain paint spectra.

BIBLIOGRAPHY

Gordon, A.J., and R.A. Ford (1972 and later editions), "The Chemist's Companion," John Wiley and Sons, New York, NY

Harrick, N.J. (1972), "Internal Reflection Spectroscopy," John Wiley and Sons, New York, NY

Holter, M.R., S.Nudelman, G.H. Suits, W.L. Wolfe,and G.J. Zissis (1962 and later editions), "Fundamentals of Infrared Technology," the Macmillan Company, New York, NY

Miller, R.G.J., and B.C. Stace (editors) (1972), "Laboratory Methods in Infrared Spectroscopy," 2nd edition, Heyden and Son Ltd, London, UK

Skoog, D.A., and D.M. West (1980), "Principals of Instrumental Analysis," 2nd edition, Saunders College Publications, Philadelphia, PA

U.S. Patent 4,527,062 (1985), "Portable Infrared Spectrophotometer" by T. Novinson. Assigned to the United States of America (U.S. Navy)

EFFECTS OF SECOND-PHASE ON THE NONLINEAR BEHAVIOR OF METAL MATRIX COMPOSITES

P. A. Foltyn, K. Ravi-Chandar, and K. Salama

Department of Mechanical Engineering
University of Houston
Houston, Texas 77204-4792

INTRODUCTION

The mechanical properties of metal matrix composites (MMCs) have made them ideal materials for use in applications where increased material performance is required. Their potential use in the military and aerospace industries has resulted in a widespread investigation into their mechanical properties to determine optimum fabrication techniques for enhanced composite strength. Of the many destructive and nondestructive techniques available for the measurement of composite properties, ultrasonic nondestructive evaluation (NDE) has emerged as a useful tool to the researcher for the study of the mechanical behavior of MMCs. Therefore, an ultrasonic determination of the composite properties can provide insight as to how the material can be enhanced by changes either in composition or fabrication techniques. To this end, the second- and third-order elastic constants were measured in silicon carbide-(SiC) reinforced aluminum alloys and in monolithic silicon carbide as a function of temperature and second-phase. The results give not only the widely used engineering data such as the Young's and shear moduli, but also an indication as to the degree of nonlinearity present in these materials, and how the second- and third-order properties change with the introduction of a second-phase.

EXPERIMENTAL

Specimens

In this investigation, the second-order linear elastic constants were determined for two silicon carbide-reinforced aluminum alloys, namely, Al-7064 and Al-8091. Also, the third-

Nondestructive Characterization of Materials VI
Edited by R.E. Green, Jr. *et al.*, Plenum Press, New York, 1994

order elastic constants were measured for the Al-7064 alloy. For both aluminum alloys, three specimens were tested, each with a different amount of second-phase. The three Al-7064 alloy specimens contain 0, 15, and 20% SiC, the Al-8091 specimens contain 0, 10, and 15% SiC. Also, a pure silicon carbide specimen was tested to study ultrasonic wave propagation in monolithic silicon carbide. The aluminum specimens were fabricated as 1 inch diameter extruded rods, then machined flat to dimensions of approximately 0.6 x 0.6 x 0.9 inch for ultrasonic testing. The matrix composition of each alloy is presented in Table 1.

Table 1. Composition of the Al-7064 and Al-8091 matrix alloys (wt. %).

Alloy	Si	Fe	Cu	Mg	Zn	Cr	Zr	Co	Li	Al
7064	0.05	0.10	2.0	2.3	7.1	0.12	0.20	0.22	---	rem
8091	0.02	0.01	1.9	0.8	---	---	0.11	---	2.7	rem

Time-of-Flight Data Acquisition

Ultrasonic velocities were measured using two methods, a manual pulse-echo-overlap method and a computer automated time-of-flight data acquisition system. In this latter method, the received pulse and echo train are sent to a 10-bit 60 MSa/s digitizer. The waveform data are read by a computer through an IEEE-488 GPIB connector and analyzed through computer subroutines to determine the time-of-flight. Rather than determine the time-of-flight based solely on the sampling rate of the digitizer (16.7 ns), a method is proposed to interpolate between two discrete points on either side of a known reference point in the digitized waveform, significantly decreasing the time resolution error. The details of this method are presented by Foltyn, Ravi-Chandar, and Salama (1992).

RESULTS AND DISCUSSION

Second-Order Elastic Constants

The Lamé constants λ and μ for an isotropic body are determined by measuring the ultrasonic velocities of one longitudinal and one shear wave. These expressions are given as

$$\lambda = \rho\left(v_l^2 - 2v_s^2\right) \text{ and } \mu = \rho v_s^2 , \tag{1}$$

where v_l and v_s are the measured velocities of a longitudinal and shear wave, respectively, and ρ is the density of the material. The two Lamé constants λ and μ are used to determine Young's modulus as

$$E = \frac{\mu(3\lambda + 2\mu)}{\lambda + \mu} . \tag{2}$$

The results for the Young's modulus of Al-7064 and Al-8091 at room temperature are plotted in Figures 1a and 1b, respectively. It can be seen that for small volume fractions of second-phase, the value of the Young's modulus is a linear function of silicon carbide to 20% SiC. However, for larger amounts of second-phase, the relationship between the Young's modulus and second-phase can no longer be approximated as linear; to adequately fit the data, a second-order polynomial is required.

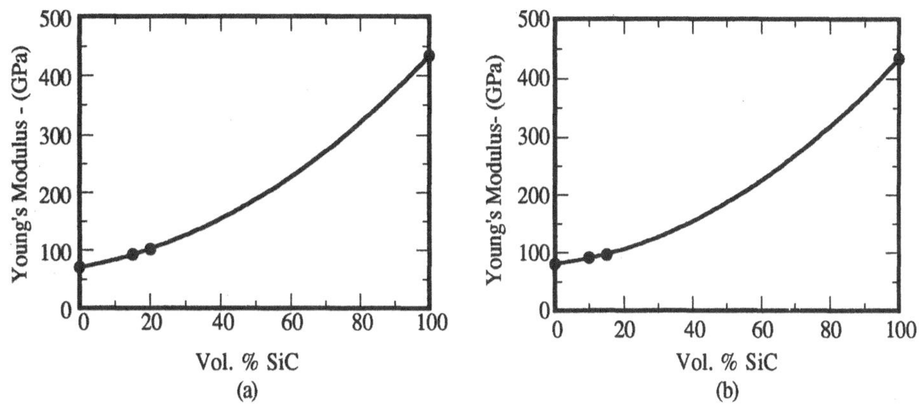

Figure 1. Young's modulus as a function of second phase: (a) Al-7064, (b) Al-8091.

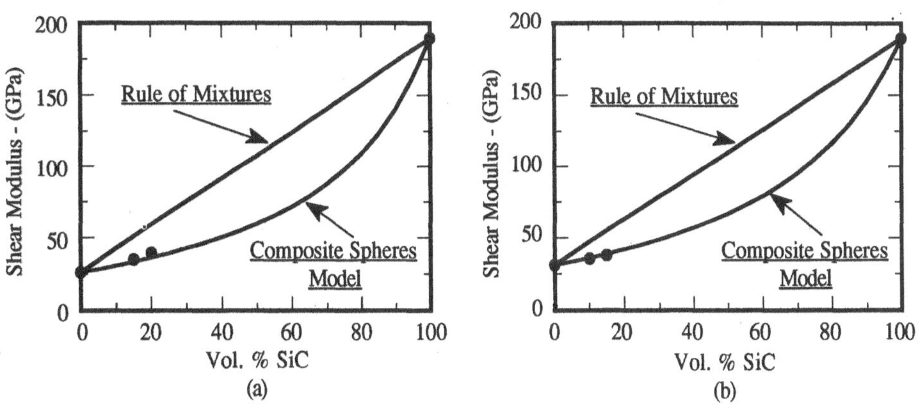

Figure 2. Shear modulus as a function of second phase: (a) Al-7064, (b) Al-8091.

Data for the shear modulus as a function of second-phase of Al-7064 and Al-8091 are plotted in Figures 2a and 2b, respectively. Also plotted in these figures are two shear modulus models, namely, the rule of mixtures and the composite spheres model. A summary of these models is presented by Christensen (1979). From these figures, it can be seen that, similar to the Young's modulus, the shear modulus displays a linear relationship with second-phase to 20% SiC. For greater amounts of silicon carbide, the shear modulus no longer exhibits linearity with second-phase. Since the composite spheres model accurately reflects the measured shear modulus at both low and high concentrations of second-phase, it is concluded that this model may be used to predict the second-order elastic constant μ for these composites.

Third-Order Elastic Constants and the Acoustic Nonlinearity Parameter

Just as there are two second-order constants which describe the linear elastic behavior of an isotropic body, there are three third-order elastic constants, l, m, and n, termed the Murnaghan constants, which indicate a material's deviation from a linear stress-strain relationship. One method of evaluating these higher-order constants is through the determination of the stress acoustoelastic constant (AEC). This quantity is defined as

$$\mathrm{AEC} = V^0 \frac{d\sigma}{dV} , \tag{3}$$

where V^0 is the velocity of an ultrasonic wave through an (externally) unstressed material and $d\sigma/dV$ is the inverse of the change in ultrasonic velocity due to an applied stress. These quantities are determined from a ultrasonic velocity versus applied compressive stress plot, as shown in Figure 3. As a compressive stress is increased on a specimen, the ultrasonic velocity will increase or decrease, depending on the type of wave propagated and its polarization direction. If the applied stress is small compared to the yield point of the material, the ultrasonic velocity will vary linearly with stress. Thus, a linear curve fit of the data provides not only the quantity V^0 (the y-intercept) for the AEC, but also $dV/d\sigma$ (the slope of the line).

By evaluating the AECs, the three Murnaghan constants l, m, and n are determined as

$$l = \frac{3K_0(\lambda + 2\mu)}{\mathrm{AEC}_{11}} + \frac{\lambda}{\mu}(m + \lambda + 2\mu) , \tag{4}$$

$$m = \frac{6K_0\mu}{\mathrm{AEC}_{13}} - \frac{\lambda n}{4\mu} - \lambda - 2\mu , \text{ and} \tag{5}$$

$$n = 4\mu\left[2\mu\left(\frac{1}{\mathrm{AEC}_{13}} - \frac{1}{\mathrm{AEC}_{12}}\right) - 1\right] , \tag{6}$$

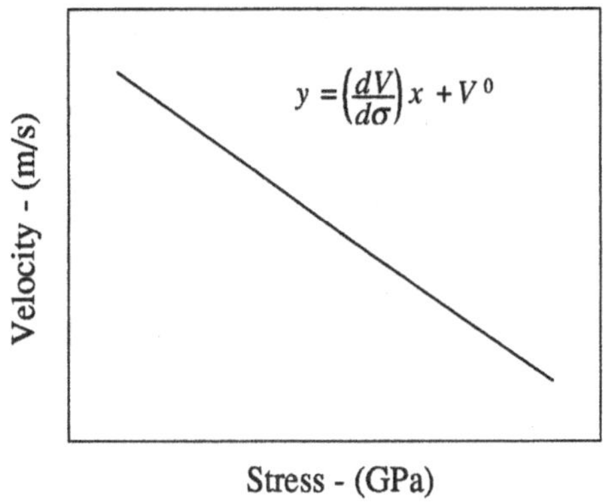

Figure 3. Schematic for the determination of the stress acoustoelastic constant.

where $K_0 = (3\lambda + 2\mu)/3$ is the bulk modulus and the subscripts on the AECs indicate ultrasonic wave propagation and polarization directions, respectively. A quantitative evaluation of the nonlinear behavior of a material is performed by determining the acoustic nonlinearity parameter β of the material (Breazeale and Philip, 1984) given as

$$-\beta = 3 + \frac{2l + 4m}{\lambda + 2\mu}, \tag{7}$$

where a larger value of the nonlinearity parameter corresponds to greater nonlinearity in the material. Thus, the Murnaghan constants allow for evaluation of the acoustic nonlinearity parameter and ultimately, an indication to the degree of nonlinearity present in the material.

The values for the Murnaghan constants l, m, and n are plotted in Figures 4a, 4b, and 4c, respectively, at three different temperatures. From these graphs, it can be noted that temperature has the largest effect on the third-order elastic constant l. Further, as the amount of silicon carbide increases, l rapidly decreases in magnitude to 20% SiC then asymptotically approaches the value for pure silicon carbide. The same trend is displayed by the constant m, a rapid decrease in magnitude to 20% SiC followed by an asymptotic approach to a minimum value for silicon carbide. From these two figures it is concluded that the change in the third-order elastic constants l and m due to the addition of silicon carbide is most prominent for dilute amounts of second-phase. Additions of silicon carbide greater than 20% have only a small influence on the Murnaghan constants l and m.

The third-order elastic constant n is plotted in Figure 4c as a function of silicon carbide. From this figure, it is seen that a linear relationship exists between the Murnaghan constant n and second-phase. As the amount of silicon carbide increases, the third-order elastic constant n linearly increases in magnitude to its value for pure silicon carbide. Whereas the largest changes in the third-order elastic constants l and m occurred in the first 20% of second-phase, the change in the constant n appears to be influenced equally at both small and large concentrations of silicon carbide. Further, this relationship is displayed at the three temperatures tested. Of the three Murnaghan constants, the constant n is influenced the least by temperature.

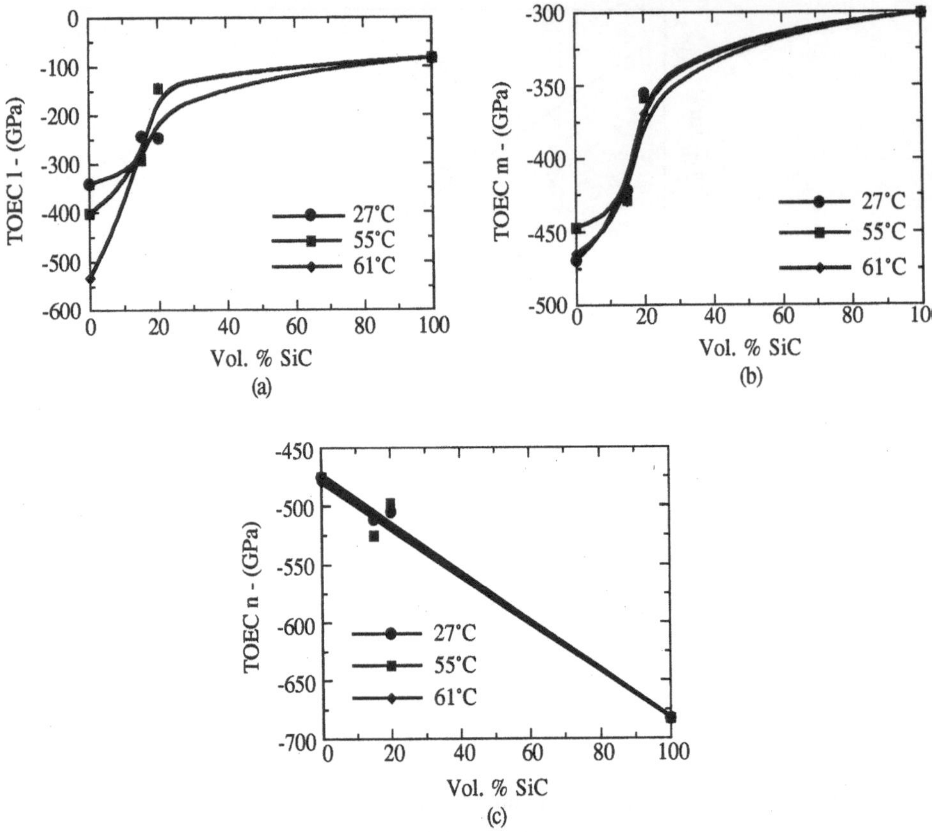

Figure 4. Third-order elastic constants as a function of second-phase at various temperatures: (a) *l*, (b) *m*, and (c) *n*.

The acoustic nonlinearity parameter is plotted in Figure 5 as a function of second-phase. It is seen that for pure silicon carbide, the nonlinearity parameter attains a minimum value close to zero. Also, for the first 20% addition of silicon carbide, the magnitude of the nonlinearity parameter linearly decreases to 50% of its original value. For additional amounts of silicon carbide, the magnitude of the nonlinearity parameter becomes less dependent upon second-phase; only a 25% decrease in the magnitude of the nonlinearity parameter occurs between 40 and 100% SiC. Note that the measured nonlinearity parameter deviates from the "rule of mixtures" assumption. Initially, the nonlinearity parameter decreases linearly with second-phase, but for greater amounts of silicon carbide, the relationship between the nonlinearity parameter and second-phase is no longer linear. This behavior may be explained by examining Equation (7), the definition of the acoustic nonlinearity parameter. From this equation it is seen that the acoustic nonlinearity parameter is a ratio of the third-order elastic constants l and m to the second-order elastic constants λ and μ. From Figures 4a and 4b, a similar behavior is present in both plots, that is, the third-order elastic constants l and m can be approximated as linear for dilute amounts of silicon carbide, but for greater amounts of second-phase, this assumption is no longer valid. Since the acoustic nonlinearity parameter

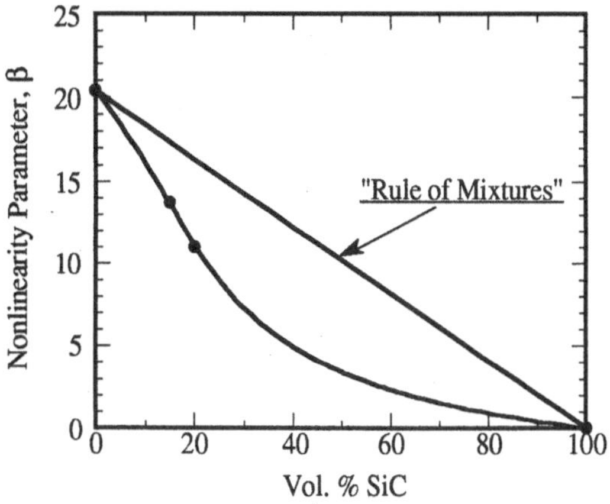

Figure 5. Nonlinearity parameter as a function of second-phase.

is a function of the two third-order elastic constants l and m and not n, it is expected that the nonlinearity parameter should follow the behavior of these two constants.

Therefore, dilute additions of second-phase are very significant in reducing the nonlinearity in these metal matrix composites. Although greater amounts of silicon carbide enhance the material properties in MMCs, large volume fractions play only a minor role in reducing the nonlinear behavior in these composite alloys.

ACKNOWLEDGMENTS

The authors gratefully acknowledge the financial support of the United States Army Research Office under contract DAAL03-92-G-0039 and the National Science Foundation under grant number MSS-895292.

REFERENCES

Breazeale, M. A. and Philip, J., 1984, Determination of Third-Order Elastic Constants from Ultrasonic Harmonic Generation Measurements, *in*: "Physical Acoustics," W. P. Mason and R. N. Thurston, eds., Academic Press, New York.

Christensen, R. M., 1979, "Mechanics of Composite Materials," John Wiley and Sons, New York.

Foltyn, P. A., Ravi-Chandar, K., and Salama, K., 1993, Study of Interfacial Stress in Metal Matrix Composites Using Ultrasonic Velocity Measurements, *in*: "Review of Progress in QNDE," D. O. Thompson and D. E. Chimenti, eds., Plenum Publishing, New York.

MEASUREMENT OF COMPOSITE FIBER VOLUME FRACTION USING THERMAL AND ULTRASONIC INSPECTION TECHNIQUES

Joseph N. Zalameda[1] and Barry T. Smith[2]

[1]US Army Research Laboratory
Vehicle Structures Directorate
NASA Langley Research Center MS 231
Hampton, VA 23681

[2]Applied Science
College of William and Mary
Williamsburg, VA 23185

INTRODUCTION

Many methods exist for the experimental determination of fiber volume fraction (FVF) in graphite epoxy composites. The most commonly used method of determining composite FVF involves the removal of the matrix by burn off or acid digestion. In addition to being destructive, this technique is time consuming and requires the disposal of toxic waste. Also this technique can be operator dependent (Cilley et al., 1974). Recent work (Zalameda et al. 1992) has shown that the FVF can be determined by measuring thermal diffusivity assuming negligible porosity levels. In addition, work has been done in ultrasonically determining porosity in composites by using a frequency dependent relative attenuation measurement (Hughes et al., 1987). The objective of this work is to develop a nondestructive technique to determine FVF using a dual inspection methodology. The relationship between thermal diffusivity and fiber, matrix and void volume fractions is described in a one dimensional heat flow model where the void volume fraction is determined ultrasonically. The use of a phase lag technique is implemented to make quantitative measurements of thermal diffusivity. These measurements were on composite plates with varying FVF. Diffusivity measurements indicated a nonlinear relation between FVF and measured diffusivity with values ranging from .003 to .007 cm^2/ sec. In addition to the thermal measurements, frequency dependent relative attenuation ultrasonic measurements were made within the area to be destructively tested. Results have shown an approximate linear correlation between porosity and attenuation. Results will be presented on 16 and 32 ply composite plates with lay ups of [0/90] 4s and [0/90] 8s and with FVF values ranging from 50 to 75 percent. A comparison of the measurement results to destructive testing is shown and the implementation of the thermal and ultrasonic measurement techniques is described.

FVF SAMPLE PREPARATION

The samples were 16 and 32 ply composite plates with lay ups of [0/90] 4s and [0/90] 8s. The target FVF's were 40, 50, 60, 65, and 70 percent. The low FVF samples were fabricated by using a cast film of resin which was laid with the tape. The 65 and 70 percent FVF were fabricated by prebleeding the plies before curing. A press curing system was utilized for consistent thickness. The plates were 30.48 x 30.48 cm in size and were sectioned into three sub plates two 15.2 x 15.2 and one 30.48 x 15.2 cm. Using this manufacturing technique it was hoped that the plate FVF would be consistent. This was not the case, however as the FVF varied throughout the plates, especially for the higher FVF plates. Within the plates 1 x 1 cm areas were marked for measurement and eventually sectioned for destructive testing. A total of 27 test areas were measured thermally and ultrasonically before being destructively tested.

THERMAL DIFFUSIVITY MEASUREMENT SYSTEM

The single point diffusivity measurement system used is shown in figure 1 and consists of three main components: the heat source, temperature detector, and computer. The heat source is located on the opposite side of the detector for a through transmission configuration. The heat source is a 300 watt tungsten filament heat lamp which is controlled by the computer. The lamp radiation is focused using a Pyrex lens and modulated by a computer controlled shutter. A Polyvinylidene (PVF2) pyroelectric detector was used to measure changes in infrared radiation. Pyroelectric detectors offer relatively good sensitivity, are rugged, and inexpensive. The infrared detector operates on a change in temperature and since the lamp radiation was modulated no detector chopping was required. The detector area was 1 mm in diameter and the measured area (field of view) was approximately 1.0 cm in diameter. The output of the detector and the input to the shutter were digitized at a rate of 256 points per modulation frequency period for four periods.

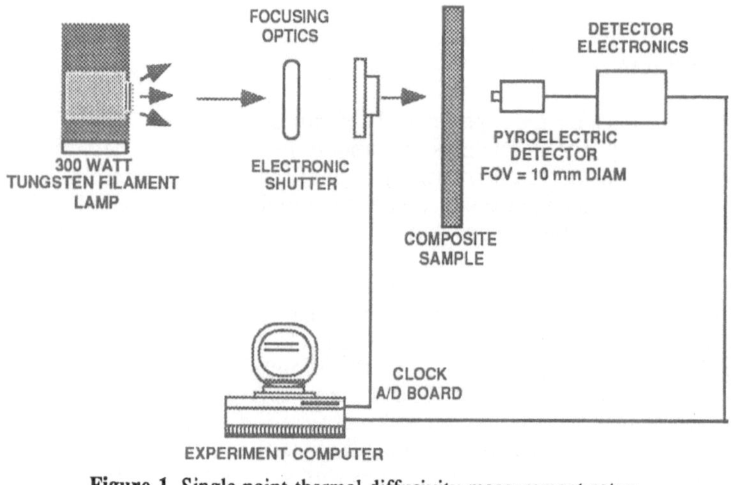

Figure 1. Single point thermal diffusivity measurement setup.

The relative phase of the two signals was calculated from their Fast Fourier Transforms. The system was calibrated for a bandwidth between .1 - 1 hertz where the upper frequency cutoff was due to the mechanical delay of the shutter. Independent phase measurements were made of the shutter, PVF2 detector, and associated detector electronics. These phase contributions were deconvolved from the measured phase to obtain the phase shift due to the presence of the sample. From this phase shift, known sample thickness and modulation frequency, the diffusivity of the sample was calculated using a one-dimensional single layer heat flow model. The sample's phase response was obtained assuming periodic heating on one surface and no convection losses. This phase difference is found to be:

$$\phi = \frac{\pi}{4} + \frac{\sqrt{2}}{2} p l - \tan^{-1}\left(\frac{e^{-\sqrt{2} l p} \sin(\sqrt{2} l p)}{e^{-\sqrt{2} l p} \cos(\sqrt{2} l p) - 1} \right) \tag{1}$$

where

$$p = \sqrt{\frac{\omega}{\alpha}} \tag{2}$$

and l is the layer thickness, ω is the angular excitation frequency, α is the thermal diffusivity.

FVF MODEL

The thermal properties of a composite can be approximated in the following equations obtained from Chamis (1987) where the heat flow through a plate is perpendicular to the fiber and matrix. The equivalent volumetric heat capacity is denoted using the law of mixtures as

$$(\rho c)_{avg} = (VF_{fiber})(\rho c)_{fiber} + (1 - VF_{fiber} - VF_{por})(\rho c)_{matrix} \tag{3}$$

where ρ is the density, c is the specific heat, VF_{por} denotes percentage volume fraction of porosity, and VF_{fiber} denotes percentage volume fraction of fiber. The volumetric heat capacity of air is not a factor since the density and heat capacity of air are small values. Assuming the voids are dispersed evenly in the matrix an equivalent matrix/void thermal conductivity is given as

$$k_{mv} = (1 - \sqrt{VF_{por}})\, k_m + \frac{k_m \sqrt{VF_{por}}}{(1 - \sqrt{VF_{por}}\,(1 - \frac{k_m}{k_{por}}))} \tag{4}$$

where k_{mv} is the equivalent thermal conductivity of the matrix with voids, k_m is the thermal conductivity of the matrix, and k_{por} is the thermal conductivity of air. The equivalent thermal conductivity with the fibers is given as:

$$k_{equiv} = (1 - \sqrt{VF_{fiber}})\, k_{mv} + \frac{k_{mv} \sqrt{VF_{fiber}}}{(1 - \sqrt{VF_{fiber}}\,(1 - \frac{k_{mv}}{k_{fiber}}))} \tag{5}$$

where k_{fiber} is the thermal conductivity of the fiber. From (3) and (5) the effective one dimensional diffusivity is calculated as

$$\alpha = \frac{k_{equiv}}{(\rho c)_{avg}}. \tag{6}$$

This result enables one to approximate an equivalent diffusivity from a nonhomogeneous slab so that solutions for a homogeneous slab can be used. Table 1 shows the values used in equation (6). The fiber material property values were obtained from Wilson and Charles (1981). The matrix material property values were quoted from the manufacturer of 934 resin. The relationship between diffusivity and FVF is plotted in figure 2. The thermal diffusivity of the fiber is several orders of magnitude greater than the matrix diffusivity. The graph indicates as the FVF increases the diffusivity increases. Also as the porosity increases the diffusivity values decrease due to the lower thermal conductivity of air.

Table 1. Model material property values.

PROPERTY	FIBER	MATRIX	AIR
VOLUMETRIC HEAT CAPACITY J / cm^3 - C	1.3275	2.162	.0012
THERMAL CONDUCTIVITY W / cm - C	1.73	.00173	.00026
DIFFUSIVITY cm^2 / sec	1.24	.00127	.2216

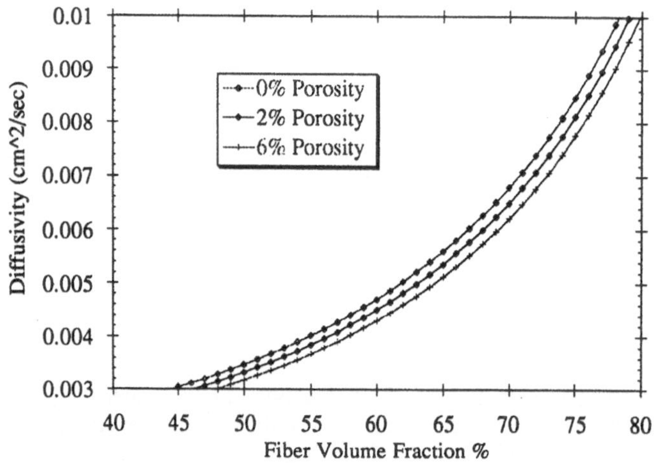

Figure 2. Model prediction of diffusivity changes due to FVF.

MEASUREMENT APPROACH

The approach to determine the FVF consisted of a single point diffusivity measurement performed on the 1 x 1 cm marked areas. To compute an average diffusivity value the heating frequency was varied from .3 - .8 hertz for the 16 ply samples and .1 - .35 for the 32 ply samples. At each heating frequency the measurement was repeated seven times. The overall measurement uncertainty for each thermal diffusivity measurement was no greater than 1.5 percent with the majority being less than 1 percent. Using equation (6) and assuming no porosity a measured FVF value can be computed. This measured value is plotted along with the destructively determined FVF value in figure 3. The destructively determined FVF is plotted in ascending order. The destructive test followed the ASTM D-3171 procedure. The samples were digested with sulfuric acid and 30% hydrogen peroxide. The relative mean square difference between the two data sets is 12.77. Correcting for the porosity to reduce this error will be discussed next.

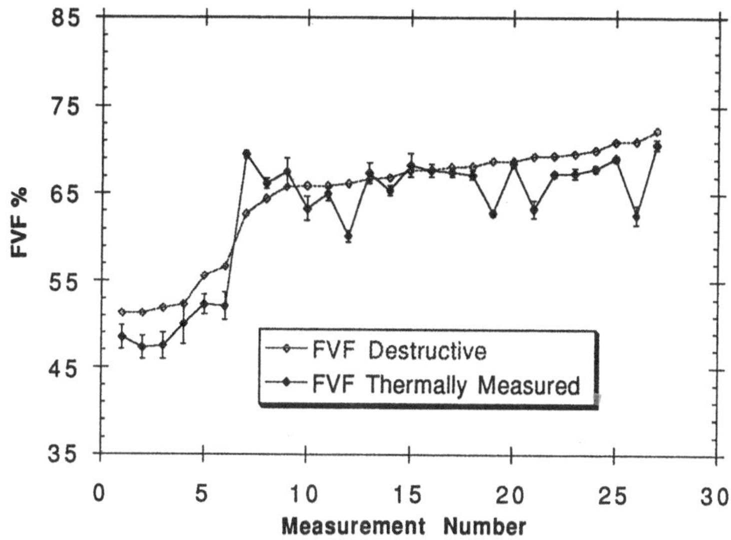

Figure 3. A comparison of thermally predicted FVF with the destructive test FVF results.

POROSITY DETERMINATION

The porosity was determined using three methods. The first method involved calculating the volume percentage of porosity from the destructive test results. Because of uncertainties in the resin and fiber densities the computation of porosity from the destructive test can be subject to inaccurate values especially at low porosity levels. The porosity was also determined by taking photomicrographs of a polished edge where the coupon for the destructive test was removed. The photomicrographs were taken at a 100 times magnification. To cover the entire area seventeen to twenty photomicrograph images were taken and the porosity was determined by image analysis of a threshold image. Shown in figure 4 is a sample photomicrograph of a polished edge image after thresholding. The black areas are the porous regions. The optical method for determining porosity can be misleading especially for a sample with varying porosity levels since the porosity is measured along an edge. The optically determined porosity values obtained in this study were mostly higher than the values determined from the destructive tests.

The last method used to estimate the porosity was by ultrasonics wherein the attenuation is assumed to be dominated by scattering. The test sample was placed in a water bath and a through transmission setup was used. A multispectral pulse was propagated with no sample to establish a reference. Within the 1 x 1 cm area to be destructively tested 100 scanned ultrasonic measurements were made in a 10 by 10 grid. The relative attenuation was computed by dividing the frequency domain magnitude response of the reference signal with the magnitude response of the measured signal. The result was found to be linear between 5.8 and 7.8 megahertz. The slope of relative attenuation was then computed by implementing a linear curve fit routine in that bandwidth. The measured slopes (db / megahertz) were averaged for each area and then normalized to the sample thickness.

The slope of relative attenuation versus optically determined porosity is shown in figure 5 and a somewhat linear relationship was found with the correlation factor R = .58. As one can see in figure 4 the porous shapes are nonuniform and this could be one factor contributing to the error. Also within the 10 x 10 grid the measured slopes were found to vary widely by as much as 50 percent thus indicating the nonuniformity of the porosity.

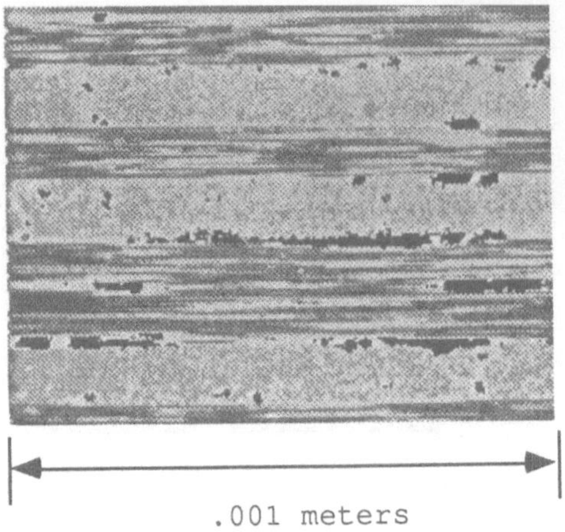

Figure 4. Threshold photomicrograph indicating porosity.

POROSITY CORRECTION

Correcting for the porosity was done using the values determined from the previously discussed tests. Each set of porosity values corresponding to each test were used to calculate new FVF values using equation 6. Shown in figure 6 are the FVF values determined destructively plotted in ascending order along with the FVF values measured with no porosity correction, with destructive porosity correction, with optical porosity correction, and with ultrasonic porosity correction. The linear equation obtained from figure 5 was used to estimate the porosity given a measured attenuation. To quantify the effectiveness in correcting for porosity changes a relative mean square difference error was computed for all data points. This plot is shown in figure 7. As shown in this plot the relative mean square difference error decreased when porosity values were used to correct the thermally measured FVF values.

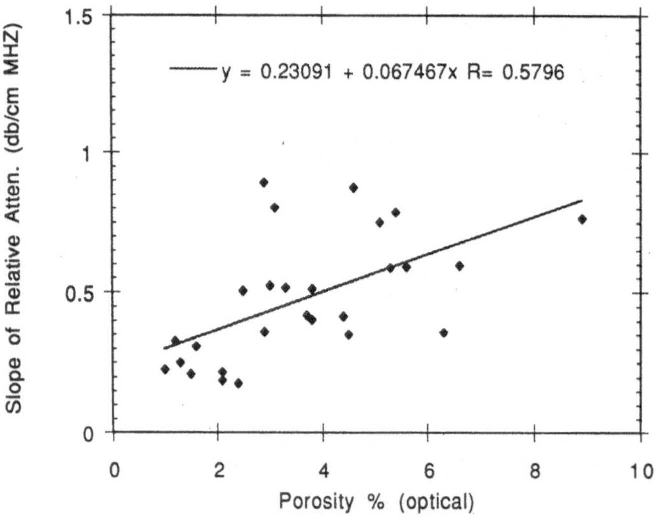

Figure 5. The linear relationship between slope of relative attenuation and porosity percent.

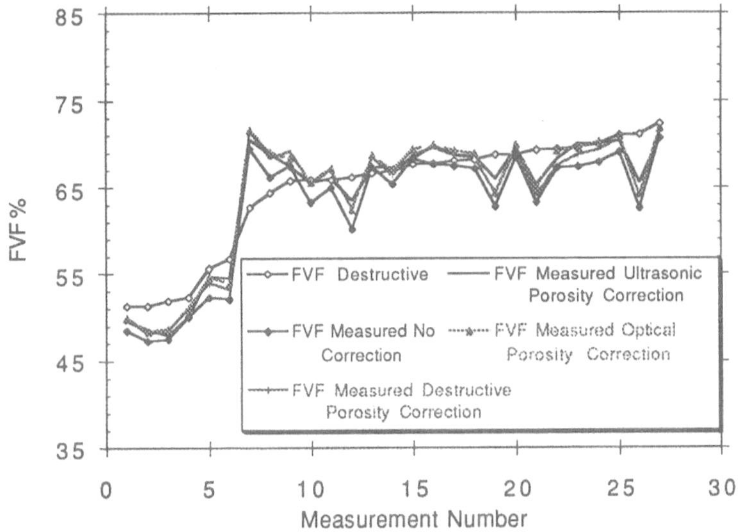

Figure 6. Comparison of measured FVF with porosity correction.

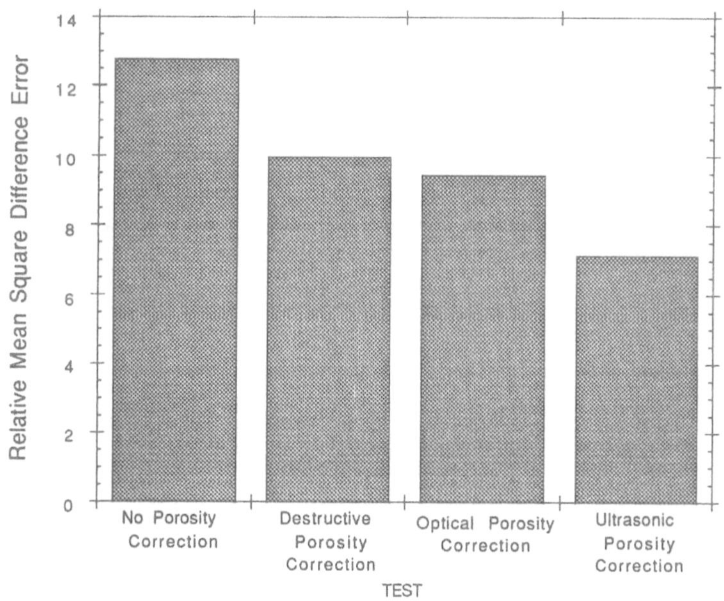

Figure 7. Quantitative comparison of the porosity correction.

DISCUSSION

A modest FVF measurement improvement was obtained when correcting for the porosity. With no porosity correction the FVF measurement error compared to the destructive test results was +/- 3.6 percent. Using the destructive and optical porosity values the measurement error was improved slightly to +/- 3.2 and 3.1 percent respectively. Using the linear equation resulting from combining the ultrasonic relative slope of attenuation measurements with the optical porosity values gave the best results reducing the measurement error to +/- 2.7 percent. One contribution to this error is the diffusivity measurement uncertainty which was around 1%. Also the FVF determined using destructive testing is considered to have an error of approximately +/- 2 percent. This dual inspection technique has potential in determining FVF nondestructively in cured composites. Another advantage of this technique is that no knowledge of the ply layup is required for the measurement.

ACKNOWLEDGEMENTS

The authors would like to thank Richard Partos for his work in obtaining the destructive test results and optical porosity values.

REFERENCES

1. Chamis, C. C., "Simplified Composite Micromechanics Equations for Mechanical, Thermal and Moisture-Related Properties", Engineer's Guide to Composite Materials, ASM International, pp. 3 - 8 through 3 - 24 (1987).
2. Charles J. and Wilson D., "A model of passive thermal nondestructive evaluation of composite laminates", Polymer Composites, Volume 2, No. 3, (1981).
3. Cilley E., Roylance D., and Schneider N., "Methods of fiber and void measurement in graphite epoxy composites", Composite Materials: Testing and Design (Third Conference), ASTM STP 546, American Society for Testing and Materials, pp 237-249 (1974).
4. Hughes M. S., Handley I. G., Miller J. G., and Madaras E. I., "A relationship between frequency dependent ultrasonic attenuation and porosity in composite laminates", 1987 Review in Progress in Quantitative Nondestructive Evaluation, pp. 1037 - 1044 (1987).
5. Zalameda J. N. and Winfree W. P., "Quantitative thermal diffusivity measurements on composite fiber volume fraction samples" Review in Progress in Quantitative Nondestructive Evaluation, pp. 1289 - 1295 (1992).

CHARACTERIZATION OF POLYMER STRUCTURE USING REAL-TIME X-RAY SCATTERING

Peter P. Huo[1,2], Peggy Cebe[1], and Malcolm Capel[3]

[1]Department of Materials Science and Engineering, Rm. 13-5082
Massachusetts Institute of Technology, Cambridge, MA 02139
[2]Present Address: W. R. Grace Corporation, Columbia, MD 21044
[3]Dept. of Biology, Brookhaven National Laboratory, Upton, NY 11973

INTRODUCTION

X-ray scattering is a powerful analytical tool for the non-destructive evaluation of crystallizable polymers and blends. Our group has been using real-time x-ray scattering to study thermal properties and microstructure development in semicrystalline polymers[1-4]. Here we describe our research on thermal expansion studies of poly(butylene terephthalate)[1], PBT, using x-ray scattering for non-destructive property evaluation. X-ray scattering experiments were conducted at the Brookhaven National Synchrotron Light Source (NSLS).

PBT is one member of the family of polyesters used for such engineering applications as the matrix polymer in glass fiber reinforced composites. One of the key issues in thermal processing of composites is the relative thermal expansion between the matrix polymer and the reinforcing fibers. Thermal expansion mismatch is one of the most important sources of material weakness in polymer based composites. Therefore it is extremely important to have a measure of the expansion properties of the matrix when it crystallizes. The crystallization of PBT is very rapid making it difficult to quench this polymer to obtain 100% amorphous material. Quenched PBT crystallizes immediately upon heating above room temperature. Therefore the thermal expansion properties of the 100% amorphous phase of PBT have not previously been reported. In this work we use high temperature wide and small angle x-ray scattering (WAXS and SAXS) to provide a measure of the thermal expansion coefficient of the amorphous phase of PBT[1]. WAXS is used to examine the crystal lattice thermal expansion coefficient. SAXS is then used to characterize the thermal expansion of the bulk sample from changes in the the periodicity of lamellar stacks. One dimensional electron density correlation function analysis gives the temperature dependent long spacing, lamellar thickness, l_c, and degree of crystallinity, all of which increase as the temperature increases[1]. The thermal expansion coefficient of the amorphous phase of PBT is finally derived by combining room-temperature SAXS with real-time SAXS at elevated temperatures.

Nondestructive Characterization of Materials VI
Edited by R.E. Green, Jr. *et al.*, Plenum Press, New York, 1994

EXPERIMENTAL SECTION

PBT was obtained from Polysciences. Quenched samples were made by heating the as-received pellets to 250°C in a compression molding press and quenching in ice water. Crystallized samples were prepared by cooling from the melt or heating from the rubbery state then holding isothermally at the crystallization temperature until completion of crystallization. All materials were characterized using differential scanning calorimetry (DSC) for thermal analysis. Static WAXS studies were done on a Rigaku RU300 diffractometer in θ/2θ reflection mode using CuKα radiation (λ=1.54Å) with a step scan interval of 0.02 degrees and a scan rate of 1°/minute. Material characterization is shown in Figure 1a,b.

Real-time WAXS experiments were done at NSLS in transmission mode (λ=0.91Å). The system was equipped with a one dimensional position sensitive detector. The sample to detector distance was about 50 cm, and the beam profile was treated according to pinhole geometry. National Institute of Standards silicon powder was used to calibrate the peak positions. The sample was inserted into a hot stage, and the WAXS data were taken in real-time as the temperature increased from 34°C to 224°C at 10°C/minute.

Real-time and static SAXS experiments were performed at NSLS. The system was now equipped with a two dimensional position sensitive detector. The sample to detector distance was about 120 cm (λ=1.28 Å). Data were taken as the temperature increased from 35°C to 235°C at 10°C/minute; each scan was collected for 30 seconds. Due to the isotropic nature of our sample, circular integration of the SAXS intensity was used to increase the signal to noise ratio.

SAXS scans at room temperature were taken for several samples prepared with the same thermal history (except for the cooling stage) as for the samples used in the real time study. The samples were heated at 10°C/minute to a certain temperature, then immediately cooled and examined at room temperature. The only difference between the static test, performed at room temperature on samples crystallized at elevated temperature, and real time SAXS, is the measurement temperature.

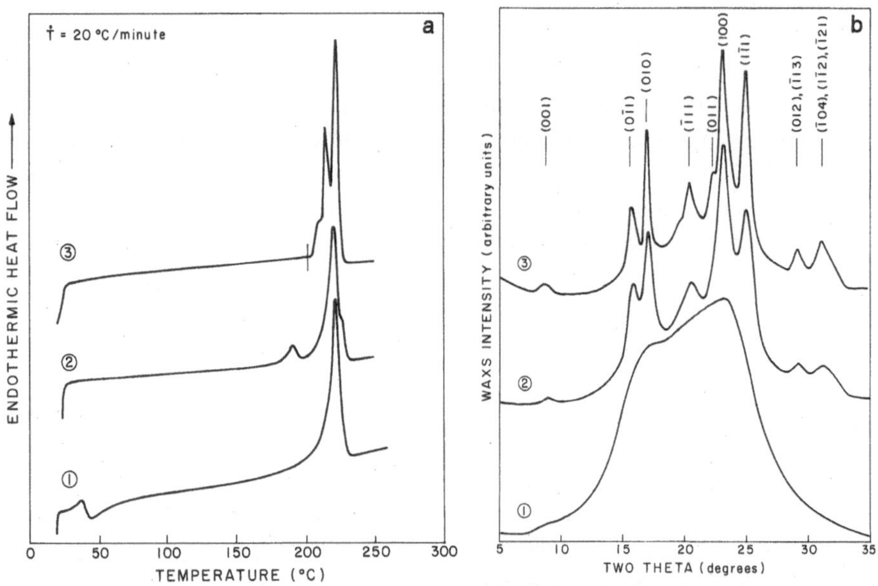

Figure 1. Characterization of PBT samples which were quenched (1), cold crystallized at 180°C (2), and melt crystallized at 200°C(3): a.) DSC thermograms at 10°C/minute; b.) WAXS intensity vs. 2θ.

Table 1. Two-theta positions, in degrees, for six major reflections of melt crystallized PBT.

Temp. (°C)	Miller Indices (hkl)					
	$(0\bar{1}1)$	(010)	$(\bar{1}11)$	(011)	(100)	$(1\bar{1}1)$
35	16.01	17.25	20.62	22.57	23.32	25.16
55	15.92	17.20	20.55	22.55	23.31	25.18
75	15.87	17.15	20.50	22.52	23.22	25.11
95	15.82	17.13	20.41	22.48	23.19	25.06
115	15.80	17.15	20.36	22.50	23.15	24.96
135	15.69	17.07	20.26	22.45	23.10	24.94
155	15.67	17.07	20.19	22.45	23.03	24.90
175	15.58	16.98	20.16	22.38	23.00	24.87
195	15.58	16.98	20.10	22.36	22.91	24.78
215	15.53	17.00	20.07	22.43	22.93	24.82

RESULTS AND DISCUSSION

Crystal Lattice Thermal Expansion Coefficients

We show thermal analysis and WAXS intensity patterns for PBT in Figures 1a and 1b, respectively. In Fig. 1a, the DSC scans represent films which were quenched(curve 1), cold crystallized at 180°C (curve 2), and melt crystallized at 200°C (curve 3). In the scans shown, the initial portion below 35°C is erratic due to the DSC machine stabelization period. Quenched PBT (curve 1) is not 100% amorphous, judging from the slight inequality of the heats of crystallization and melting. The film crystallizes just above the glass transition temperature (35°C)[5] and reorganizes continuously until about 230°C where the melting endotherm is seen. Both melt and cold crystallized PBT show a dual endothermic response, with the lower endotherm occurring just above the prior treatment temperature. Multiple melting behavior has been described previously in PBT and other polymers and attributed to reorganization[6-8], though other researchers[9] suggest a crystal transformation mechanism.

The room temperature WAXS scan of quenched PBT in Fig. 1b (curve 1) confims a slight paracrystalline character. The crystalline films (curves 2,3) display six well resolved peaks, which are located at the two theta angles listed in Table 1 (λ=1.54Å). The six reflections are indexed to $(0\bar{1}1)$, (010), $(\bar{1}11)$, (011), (100) and $(1\bar{1}1)$. Two other broad reflections at higher angles were not used in the analysis. As the temperature increases from 35°C to 215°C all six peaks are found to have a significant shift to lower two theta angles. Using non-linear least squares fitting, we derive the six lattice parameters assuming the triclinic crystal structure proposed previously[10-15]. The lattice parameters at elevated temperature p(T) can be related to their values, p_0 at 0°C, according to:

$$p(T)=p_0\,(1+\alpha_l \cdot T) \quad (1)$$

where α_l is the linear thermal expansion coefficient, and T is the temperature in °C. We find the following relationships for the temperature dependence of the unit cell axes, a, b, c (Å), and angles, α, β, γ (degrees):

$$a = 4.80\,(1+2.03 \times 10^{-4}\,T) \qquad \alpha = 100.26\,(1+3.96 \times 10^{-5}\,T) \qquad (2a,b)$$

$$b = 5.98\,(1+1.22 \times 10^{-4}\,T) \qquad \beta = 114.82\,(1+8.13 \times 10^{-5}\,T) \qquad (2c,d)$$

$$c = 11.55\,(1+2.13 \times 10^{-4}\,T) \qquad \gamma = 111.43\,(1-3.92 \times 10^{-5}\,T) \qquad (2e,f)$$

Bulk Thermal Expansion Coefficient

The bulk thermal expansion of PBT is considered to arise from the expansion of both the crystals and the amorphous phase, as temperature increases. The thermal expansion of the amorphous phase will be greater than for the crystals, and this is especially true for PBT since the amorphous phase glass transition temperature, Tg, is near room temperature. The PBT amorphous phase has increased molecular mobility above Tg and will contribute to the bulk thermal expansion according to its relative volume fraction. In Figure 2 we show a sketch of the assumed lamellar structure showing two lamellar crystals separated by an amorphous layer. Our strategy in this section is first to use real-time SAXS to find the thermal expansion of a crystal/amorphous stack of thickness L(T). Then in the next section, using knowledge of the crystal lattice expansion of PBT we derive the thermal expansion of the amorphous layer.

We present in Figure 3 intensity, I(s) vs. s data ($s=2\sin\theta/\lambda$) for the real-time SAXS study of quenched PBT taken during the heating from 35°C to 235°C. As crystallization proceeds from just above the glass transition at 35°C to 220°C, the coherent scattered intensity increases from the alternating stacks of lamellar crystals. For temperatures above 220°C, partial melting starts (confirmed from the DSC scan in Fig. 1a), and peak intensity decreases.

The alternating stacks of crystal lamellae and amorphous chains, shown in Figure 2, contribute to the coherent SAXS scattering. The one dimensional electron density correlation function, K(z), can be used to determine the stack periodicity and crystal lamella thickness. K(z) is obtained from[16]:

$$K(z)=\int_0^{\infty} 4\pi\, I(s)s^2\cos(2\pi sz)ds \qquad (3)$$

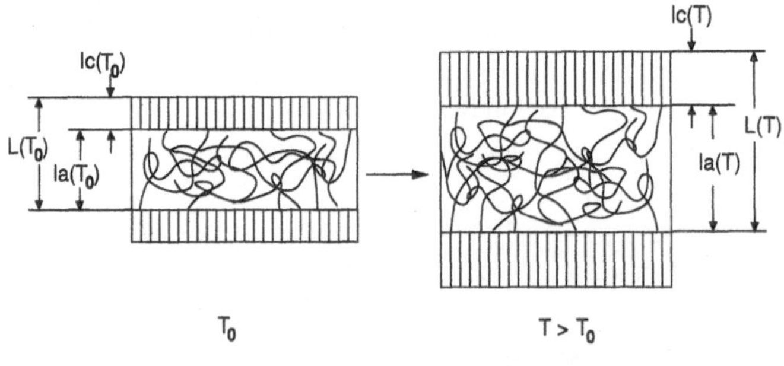

Figure 2. Sketch of the thermal expansion of lamellar stacks from temperature T_0 to $T>T_0$. L is the long period, and l_a and l_c are the amorphous and crystal layer thicknesses, respectively.

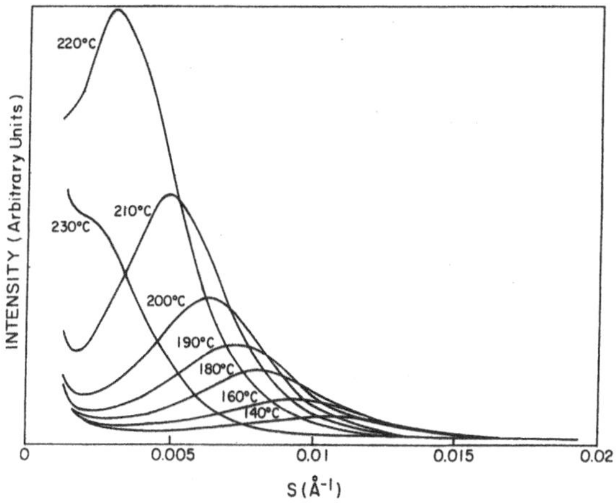

Figure 3. SAXS intensity for PBT during heating from 35°C to 235°C at 10°C/minute. Reproduced with permission from Reference 1.

where z is a dimension along the normal to the lamellar stack. Here, I(s) is the corrected intensity after background and thermal density fluctuations have been subtracted. For the calculation of K(z), I(s) vs. s data were extrapolated to s=0 linearly and Porod's law , $I(s) \sim s^{-4}$, was used to extrapolate the intensity data to s=infinity. In Figure 4a, we show the general shape of K(z) vs. z which illustrates the determination of L and l_c. In Figure 4b experimental K(z) is shown for PBT at several temperatures. Long period, lamellar thickness, and crystallinity are obtained according to the method described by Strobl and Schneider[25]. Both long period and lamellar thickness increase as temperature increases.

Using real-time SAXS combined with room temperature SAXS, we obtain the thermal expansion coefficient for the bulk material. In Table 2, we list the long period data taken at room temperature for five PBT crystalline samples prepared by heating at 10°C/minute to the temperature indicated, then quickly cooling to room temperature. The difference in the long period between these two types of samples is due to the difference in the measurement temperature and not to differences of structure. We derive the average thermal expansion coefficient for the bulk material, a_{ave}, which is assumed to be isotropic, from:

$$L(T)=L(0)\ (1+\alpha_{ave} \cdot T) \qquad (4a)$$

$$L(T_0)=L(0)\ (1+\alpha_{ave} \cdot T_0) \qquad (4b)$$

where L(0) is the long period at T=0°C and T_0 is room temperature. Due to the small value of α_{ave} ($\alpha_{ave} * T_0 << 1$) equation (4a) divided by (4b) evolves to the following:

$$L(T)=L(T_0)\ (1+\alpha_{ave} \cdot (T-T_0)) \qquad (5)$$

where L(T) and $L(T_0)$ are the long periods at higher temperature T and room temperature, respectively. From the slope of a plot of $[L(T)/L(T_0)]-1$ vs. $T-T_0$ we obtain the value of the bulk thermal expansion of PBT, $\alpha_{ave} = 5.0 \times 10^{-4}$°C^{-1}, which agrees with a previous report[17] which had no information provided about the PBT sample history.

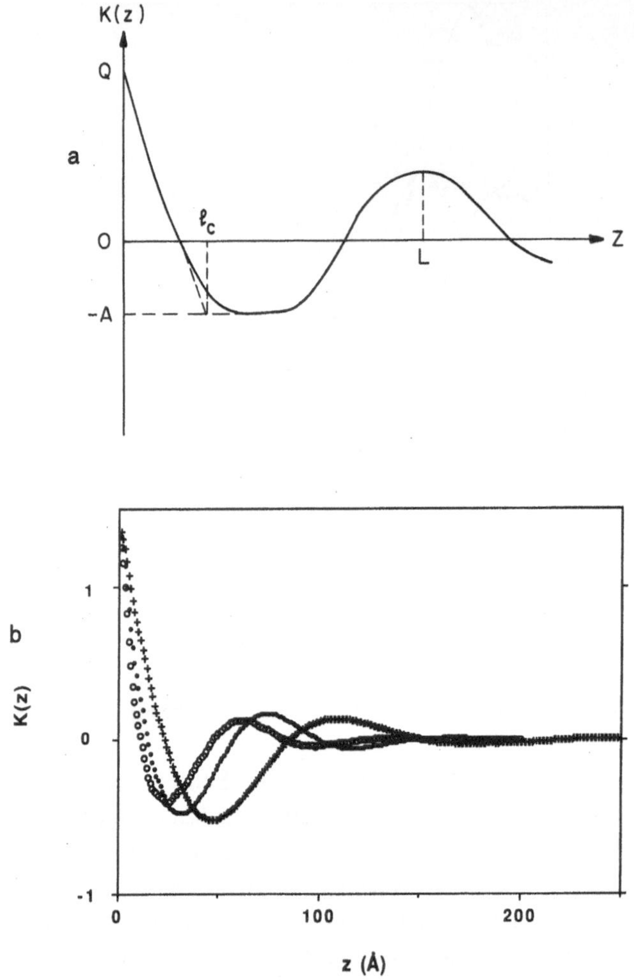

Figure 4. One dimensional electron density correlation function, K(z) vs. z: a.) Sketch showing the evaluation of long period, L, and crystal lamella thickness, lc; b.) Experimental data for PBT at measurement temperature of 150°C (O), 175°C (●), and 200°C (+).

Amorphous Phase Thermal Expansion Coefficient

We now can calculate the thermal expansion coefficient for the amorphous phase by combining information from the crystal lattice thermal expansion and with knowledge of the expansion of a lamellar stack. This calculation gives an average expansion coefficient for the entire amorphous phase. The thermal expansion for the long period, L, can be viewed as the contribution from the crystal and amorphous layers, of thickness l_c and l_a, respectively. L and l_c are obtained directly from the one dimensinal electron density correlation function analysis[16], and $L=l_c+l_a$. Using the lamellar model of Figure 2, we write:

$$l_a(T) = L(T) - l_c(T) \qquad (6a)$$

$$l_a(T_0) = L(T_0) - l_c(T_0) \qquad (6b)$$

The subscript 0 refers to the room temperature measurement. At elevated temperature, the l_c and l_a can be written in terms of their room temperature values as:

Table 2. Long spacing versus temperature for PBT

T(°C)	L(Å)[a]	L(Å)[b]
60	60	61
100	64	67
140	75	79
180	96	104
220	136	204

[a] Room temperature SAXS of PBT heated to T at 10°C/minute then cooled
[b] Elevated temperature SAXS of PBT heated to T at 10°C/minute

$$l_c(T) = l_c(T_0)\ (\ 1+\alpha_c \cdot (T-T_0)\ ) \quad (7a)$$

$$l_a(T) = l_a(T_0)\ (\ 1+\alpha_a \cdot (T-T_0)\ ) \quad (7b)$$

where α_c and α_a are the thermal expansion coefficients for the crystal phase and the amorphous phase in the direction along the normal to the lamellar stack.

To evaluate α_a from equation (7b) we need $l_a(T_0)$ from equation (6b). The long period and crystal thickness at room temperature are both directly obtained from the one dimensional electron density correlation function. However, we did not use that approach to find the crystal thickness at elevated temperature, because of the larger relative error involved in the determination of $l_c(T)$ from the extrapolation of K(z) at low z. Instead, we used equation (7a) to evaluate $l_c(T)$. Here, we assume that the thermal expansion coefficient of the crystal phase is not a function of temperature. A reasonable approximation for α_c is to assume it is close to the c-axis crystal lattice thermal expansion coefficient, $2.12\text{x}10^{-4}\ C^{-1}$. From a plot of $[L_a(T)/L_a(T_0)]-1$ vs. $T-T_0$ we find the thermal expansion coefficient for the amorphous phase, which is $\alpha_a = 6.0\text{x}10^{-4}\ °C^{-1}$.

CONCLUSIONS

In this work we have developed a novel methodology for determining the thermal expansion of the amorphous phase in polymers that can not be obtained in the 100% amorphous state. This property is needed for calculation of bulk thermal expansion properties for polymers used in composite applications where thermal expansion mismatch must be reduced. We have shown that real-time high temperature x-ray scattering can be used to provide complete information about the thermal expansion characteristics of polymers, using PBT as an example.

X-ray scattering is a non-destructive analytical technique that can be performed in-situ during thermal processing. We used WAXS to measure the crystal lattice thermal expansion coefficients for semicrystalline PBT. Real time SAXS was used to monitor the long period and lamellar thickness changes as a function of temperature, showing that both increase significantly as temperature increases. The thermal expansion coefficient of amorphous phase of PBT was derived by combining the room temperature SAXS data and real-time SAXS data taken at elevated temperature. Result indicates a three times larger thermal expansion coefficient for the amorphous phase compared with that of crystal phase c-axis.

ACKNOWLEDGEMENTS

Research was supported by the U.S. Army Contract DAAL03-91-G-0132. Research carried out (in part) at the National Synchrotron Light Source, Brookhaven National Laboratory, which is supported by the U. S. Department of Energy, Division of Materials Science and Division of Chemical Sciences (DOE contract number DE-AC02-76CH00016).

REFERENCES

1. P. Huo, P. Cebe, and M. Capel. *J. Polym. Sci., Polym. Phys. Ed.*, 30, 1459 (1992).
2. P. Huo, P. Cebe, and M. Capel. *Polym. Preprints*, 33(2), 446 (1992).
3. P. Huo, P. Cebe, and M. Capel. *Macromol.*, in press 1993.
4. P. Huo, J. Friler, and P. Cebe. *Polymer*, in press 1993.
5. S. Z. D. Cheng, R. Pan, and B. Wunderlich. *Makromol. Chem.*, 189, 2443 (1988).
6. J. T. Yeh and J. Runt *J. Polym. Sci., Polym. Phys. Ed.*, 27, 1543 (1989).
7. J. Runt, D. Miley, X. Zhang, K. Gallagher, K. McFeaters, and J. Fishburn. *Macromol.*, 25, 1929 (1992).
8. J. S. Chung and P. Cebe. *Polymer*, 33(11), 2312 (1992).
9. M. Nichols and R. E. Robertson. *J. Polym. Sci., Polym. Phys. Ed.* 30, 755 (1992).
10. J. Roebuck, R. Jakeways, and I.M. Ward. *Polymer*, 33(2), 227 (1992).
11. R. P. Grasso, B. C. Perry, J. L Koenig, and J. B. Lando. *Macromol.*, 22, 1267 (1989).
12. M. Yokouchi, Y. Sakakibara, Y. Chatani, H. Tadokoro, T. Tanaka, and K. Yoda. *Macromol.*, 9(2), 266 (1976).
13. Z. Mencik. *J. Polym. Sci., Polym. Phys. Ed.*, 13, 2173 (1975).
14. B. Stambaugh, J. L. Koenig, and J. B. Lando. *J. Polym. Sci., Polym. Phys. Ed.*, 17, 1053 (1979).
15. I. J. Desborough and I. H. Hall, *Polymer*, 18, 825 (1977).
16. G. R. Strobl and M. Schneider, *J. Polym. Sci., Polym. Phys. Ed.*, 18, 1343 (1980).
17. C. Noel and P. Navard, *Prog. Polym. Sci.*, 16, 55 (1991).

CHARACTERIZATION OF CORROSION PROTECTIVE FILMS ON STEEL AND ALUMINUM ALLOYS

H. W. White,[1] J.E. Chamberlain,[1] J.L. Wragg,[1,2] F. Mansfeld,[3] and T. Sugama[4]

[1]Department of Physics and Astronomy, University of Missouri, Columbia, MO 65211 USA

[2]Present Address: Department of Physics, University of Charleston, Charleston, SC 29424 USA

[3]Corrosion and Environmental Effects Laboratory, Department of Materials Science and Engineering, University of Southern California, Los Angeles, CA 90089 USA

[4]Energy Efficiency and Conservation Department, Department of Applied Science, Brookhaven National Laboratory, Upton, NY 11973 USA

INTRODUCTION

This project is part of a cooperative effort to further our understanding of the formation, structure, and mechanisms for improved corrosion protection and enhanced adhesion of selected films on metal alloys. We report here results for three systems, each of which either has been or can be used on metal substrates that have important commercial applications. These three film systems are (i) polyacrylic acid (PAA) modified zinc phosphate films on steel; (ii) dilute cerium based treatment, with anodic polarization, of an Al 6061 alloy, and also on SiC/Al 6061 metal matrix composite (MMC) specimens; and (iii) polytitanosiloxane (PTS) ceramic films on Al 6061. Film growth and surface modification treatments were made in our laboratories and the films characterized by several nondestructive techniques. These techniques include electrochemical impedance spectroscopy (EIS), Raman spectroscopy, infrared reflection absorption spectroscopy (IRAS), scanning electron microscopy (SEM), electron dispersive analysis by x-rays (EDAX), and atomic force microscopy (AFM).

A brief summary of the history and importance of each film is presented, with a more detailed description of sample preparation, characterization results, and discussion.

The motivation for the current work is to improve the performance of zinc phosphate coatings for corrosion protection and for adhesion to coatings. Zinc phosphate coatings can be used to provide corrosion protection to mild carbon steels and to provide a good substrate for polymeric top-coats. Commercial films are typically 50 - 100 microns in thickness. The microstructure is an array of interlocking crystallites, with the larger ones being typically ten or so microns in lateral dimensions, and twenty to forty microns in length. Corrosion protection is generally good, especially if care has been exercised to obtain optimal growth conditions for the particular steel to be coated. Complete and uniform coverage is most difficult to obtain on high carbon steels, and on low quality steels, e.g. from mixed sources.

Ghali and Potvin[1] reported studies of film growth and noted that iron content in the film decreased dramatically with distance away from the steel substrate. A three layer film was proposed, with the outermost layer being trizinc orthophosphate tetrahydrate (hopeite), $Zn_3(PO_4)_2{\cdot}4H_2O$. The innermost layers were reported to be $Fe(H_2PO_4)_2$ next to the steel, and $Zn_2Fe(PO_4)_2.H_2O$ as the intermediate layer.

More recent studies by Sugama et al.[2,3] have focused on improvement of corrosion and adhesion properties of the films by incorporation of various polyelectrolytes in the solution bath. PAA, with a molecular weight near 50,000, was found to be particularly effective, giving a finer crystalline microstructure with greater ductility and improved performance for topcoats. PAA concentration in the solution bath was typically 2 - 3% for optimal performance. The concentration and distribution of various molecular segments from the PAA in the film are not known. They proposed an outer layer of zinc phosphate dihydrate, $Zn_3(PO_4)_2{\cdot}4H_2O$.

The sequence for preparing 1% PAA modified zinc phosphate coatings on steel specimens is described below. Specimens were nominally 1 cm^2 in area and 1 mm thick.

- Mill oil is removed by wiping specimen surfaces with acetone-soaked tissues
- Specimens are immersed up to 15 minutes in the following conversion solution at 90°C:

  5 g zinc orthophosphate dihydrate

  10 g (85% H_3PO_4) solution

  360 g water

  16 g (25%) PAA
- Rinse with water
- Dry in oven at 150°C for 15 min. to remove moisture and to solidify PAA macromolecules.

Sugama and co-workers[2,3] conjectured that the decreased crystal size with incorporation of PAA in the solution bath results from an increase in nucleation sites from segmented organic acid groups, particularly functional electrolyte groups such as carboxylic acid (-COOH) and sulfonic acid (-SO_3H). They conjectured that organic function groups on the surface acted as a primer to improve adhesion of topcoats, up to a factor of ten.

Raman spectra were obtained with a spectrometer custom built in our laboratory. It employed an argon ion laser source, an off-axis ellipsoid for high efficiency signal collection, and a liquid nitrogen cooled CCD detector for increased sensitivity to low level signals. A micro-Raman attachment with a one micron spot size was used to obtain spectra on single zinc phosphate crystallites. Initial Raman studies in the present work demonstrated the outer layer to be zinc phosphate dihydrate, $Zn_3(PO_4)_2{\cdot}4H_2O$, for both unmodified and PAA modified films.[4] A Raman spectrum in the PO_4^{3-} stretch region from 900 to 1200 cm^{-1} is shown in Fig. 1. The spectral signature from these modes could be detected on steel substrates immersed in the solution bath for as short as two minutes. Raman spectra of the film interior, prepared by removal of outer layers by scraping, revealed no evidence for layers other than zinc phosphate dihydrate. While these studies do not preclude the presence of interior layers, they would be quite thin if they do exist.

In addition to the preponderance of zinc phosphate dihydrate on the surface, Raman spectra were obtained from two other interesting surface features on films with PAA in the phosphating bath. Both were inclusions, in that they were visible features on the surface. One type was slightly brownish, and is conjectured to be iron hydrate or an iron phosphate. Their spectral features did not match those of Fe_2O_3 or Fe_3O_4. The other type of inclusion was clear, and more island like. They showed a Raman band located near 2900 cm-1, indicating the presence of aliphatic bonds. These clear inclusions were interpreted as molecular segments from the PAA. Similarly, clear inclusions were also visible on SEM micrographs on PAA modified films, but not on unmodified films.

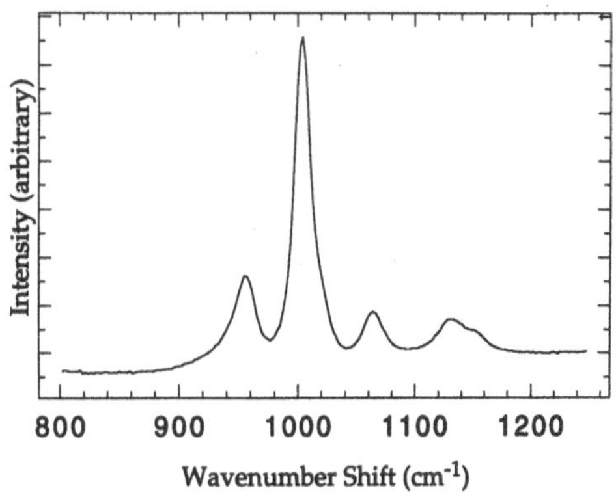

Figure 1. Raman spectrum for a zinc phosphate film in the PO_4^{3-} stretch region.

Contact mode AFM micrographs were obtained with Nanoscope II and Nanoscope III STM/AFM systems. Tapping mode micrographs were obtained with the Nanoscope III. All measurements on the Nanoscope II discussed here were made with silicon nitride tips. Etched silicon tips were used for the Nanoscope III contact and tapping mode measurements. For both contact and tapping AFM the position of the tip is sensed by optical means; however, in the tapping mode the tip is vibrated at its natural frequency and its position near the surface is sensed by the induced change in frequency. Lateral forces are thereby reduced or eliminated in tapping mode, and the technique is particularly useful for soft surfaces.

AFM images could be obtained only by restricting a scan to the surface of a single crystallite. In general the surfaces were quite rough. All of the surfaces investigated showed long striations with widths of 100 nm, more or less. For images taken with the Nanoscope III (using the etched silicon tips), pyramidal shaped pits were observed on both unmodified and modified surfaces. The pit edges were in registry, indicating their presence was due to some aspect of crystal formation and growth. For the crystallites investigated, the PAA modified surfaces showed a higher density of such pyramidal shaped pits. The Nanoscope II images showed pits, but without sharp detail. The resolution and quality of surface features imaged on the PAA modified surfaces appeared to be less on than on the unmodified surfaces, for both Nanoscope II and Nanoscope III data. Tapping mode images were of higher resolution than contact images for the PAA modified films. These differences could be due to the PAA modified film having a softer surface due to the presence of molecular segments from the PAA. Only such indirect evidence was obtained from the AFM data regarding the possible presence of such molecular segments.[4]

SEM clearly shows that crystallite size in the PAA modified films is smaller than in the unmodified films. SEM does not, however, reveal differences in the surface morphologies of the two types of films.

A summary of the findings for the zinc phosphate studies is as follows:

- The surface is crystalline zinc phosphate dihydrate, in preponderance
- There were few other surface features
- Micro-Raman spectroscopy and SEM revealed evidence for inclusions of an organic nature on the surface of the PAA modified films
- Micro-Raman also showed inclusions that were probably non-organic in nature, e.g., an iron phosphate or hydrate, on the surface of the PAA modified films
- AFM indicated the surface texture on PAA modified films was different from unmodified films
- The phosphating bath recipe is important for good coverage, especially for high carbon, low quality steels.

Corrosion Protection by Cerium-Based Surface Modification Processes

Arnott et al.[5] immersed Al alloys in dilute rare earth metal chlorides for extended periods (weeks), and found protective properties rivaling those of chromate conversion coatings. They treated Al 7075-T6 in solutions of 1000 ppm (i.e., 4 m*M*) $CeCl_3$. The original 4.2 nm air grown oxide grew to approximately 500 nm after eighteen hours immersion in distilled water, to many microns when immersed in 0.1*M* NaCl, but to only 200 nm when immersed in the 1000 ppm $CeCl_3$ solution. The protective film formed in the cerium salt based solution included cerium oxides. Arnott et al. concluded that the protective cerium compounds formed at local cathodic sites, thereby giving improved corrosion protection.

Recent work by Mansfeld et al.[6] demonstrates that the corrosion resistance of certain Al alloys and Al based MMCs can be dramatically improved by a new treatment based on a combination of a chemical process with an anodic polarization phase. The process uses dilute cerium salt based solutions without chromates. Elevated temperatures and an added electrochemical step improve the surface modification process, and shorten the processing time. Our goals are to determine the surface microstructure and to understand the mechanism of corrosion protection by this treatment.

Mansfeld et al.[10] reduced the total treatment time and improved the corrosion resistance performance of the Al 6061 in the cerium salt based solution process. The basic process--termed the Ce-Mo process--and variations for evaluation purposes, are described below. In all instances, the solution concentrations are as follows: 5 m*M* for $CeCl_3$, 10 m*M* for $Ce(NO_3)_3$, 0.1*M* for Na_2MoO_4. Treatment times are always two hrs. at 100°C for distilled water, $CeCl_3$ and $Ce(NO_3)_3$, and two hrs. at room temperature for Na_2MoO_4. All treatments have the deoxidation and bake steps to standardize the oxide layer. Deoxidation is done with a commercial product, e.g., DIVERSAY 560™.

- deoxiding and baking at 100°C for 48 hrs
- treating in distilled water
- treating in $CeCl_3$
- treating in $Ce(NO_3)_3$
- treating in $Ce(NO_3)_3$ and $CeCl_3$
- treating in $Ce(NO_3)_3$, followed by polarizing in Na_2MoO_4
- treating in $Ce(NO_3)_3$ and $CeCl_3$, followed by polarizing in Na_2MoO_4.

The last entry is the Ce-Mo process.

EIS measurements were made with a Solartron Model 1250 frequency response analyzer and a Model 1286 potentiostat with an applied ac signal of 10 mV. A saturated calomel electrode was used as a reference electrode, coupled capacitively to a Pt wire electrode to reduce the phase shift at high frequencies.

Raman measurements were made with the spectrometer described above, at laser intensities up to 200 mW at wavelengths of 488 or 514.5 nm. Spectra were recorded with integration times of roughly 1 - 60 sec.

IRAS measurements were made with a Nicolet Model 20 DBX FT-IR spectrometer fitted with a variable angle reflectance attachment. Incidence and collection cones centered around 66° or 74°.

Elemental composition analyses by EDAX were made with an AMRAY 1600 scanning electron microprobe. For elemental analysis the beam energy was 15 keV. The beam could be rastered, or focused to a spot smaller than the pits being investigated.

AFM surface morphologies were obtained with a Nanoscope II using silicon nitride cantilevers.

Various tests have been made on Al 6061, and preliminary tests have been made on Al 7075, and on SiC/Al MMC specimens.

EIS results showed that the most effective treatment for reduction of pitting corrosion for the Al 6061 alloy was the Ce-Mo process, as described above. This process gave the highest pitting resistance. EDAX results revealed the presence of cerium in and around pit-like sites--sites that, according to our model, would have been subject to pitting in corrosive environments were it not for the Ce-Mo treatment. Figure 2 shows a portion of an EDAX spectrum, showing cerium and iron, indicating that cerium may be selectively absorbed.

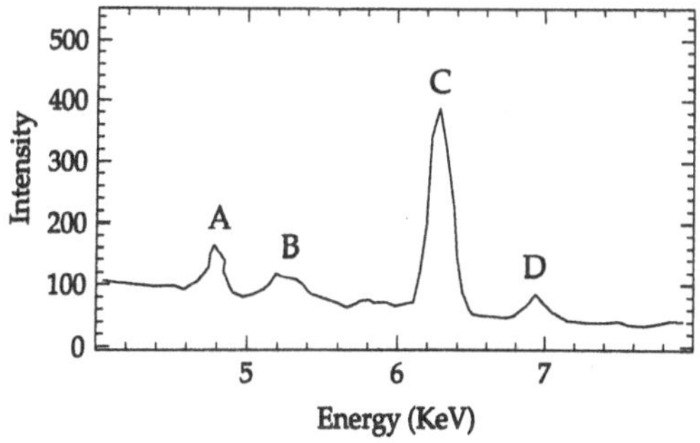

Figure 2. EDAX spectrum for pit-like site on an Al 6061 alloy surface treated by the Ce-Mo process.

Results for the Al 7075 alloy showed that the Ce-Mo process was effective. Adjustment of the treatment process was needed for improved performance.

Preliminary results for SiC/Al MMCs showed that the Ce-Mo treatment was less effective. This result, however, is not unexpected since the SiC particles create a very heterogeneous surface with new types of sites for corrosive attack. Micro-Raman spectra on SiC particles and the region surrounding the particles showed major reduction in size of the SiC particles during processing of the MMC.

A summary of findings from this work is as follows:

- The Ce-Mo process, a modification of the dilute cerium salt based solution treatment process discovered by Arnott et al.,[5] can give excellent protection against pitting corrosion for

Al 6061 alloys

- Cerium may be selectively incorporated into the film at sites vulnerable to corrosive attack
- The Ce-Mo process should provide good corrosion protection to all Al alloys, but some adjustment in the treatment procedure may be required to optimize protection.

Protective Ceramic Films on Aluminum

The motivation for this work is to develop a low temperature ceramic coating process suitable for aluminum and magnesium. Ceramic coatings offer many advantages, but are not widely used on softer metals. A useful ceramic coating for Al or Mg should have a temperature expansion coefficient similar to that of the substrate metal, and the coating process should not require high temperatures. Many ceramic coatings must be applied or processed above 1000°C.

Sugama et al.[7,8] have developed ceramic coatings for aluminum for corrosion protection with room temperature application and low temperature processing. The method uses sol-precursor solutions with polygermanosiloxane (PGS) or polytitanosiloxane (PTS) additive. Crosslinked networks are formed by pyrolytic reactions near 350°C.

The sequence for preparing a 30% N-3-[triethoxysily)propyl]-4,5,-dihydroimidazole (TSPI) with 20% polygermanosiloxane (PGS) coating, which would be labeled as 30% TSPI, 20% PGS would be as follows. Aluminum specimens used to date were nominally 1 cm^2 in area and 1 mm thick.

- washed with methanol, followed by acetone, and then tissue wiped
- etched by an alkaline solution consisting of 0.4wt% NaOH, 2.8 wt% tetrasodium pyrophosphate, 2.8 wt% sodium bicarbonate, and 94.0 wt% water, for 20 min. at 80°C
- washed at 35°C for 5 min.
- baked in a 50°C oven for 30 min.
- dipped in the following solution, with composition indicated, and slowly removed:

TSPI	30 wt%
$Ge(OC_2H_5)_4$	20 wt%
CH_3OH	30 wt%
Water	20 wt%
HCl	15 wt% of combine wt. of TSPI + $Ge(OC_2H_5)_4$

- baked at 150°C for 20 hrs.
- baked at 350°C for 30 min.

Similar coating can be prepared with titanium, in place of germanium, using titanium(IV) ethoxide, $Ti(OC_2H_5)_4$. The performance of PTS films appear to exceed those of PGS. Some specimens with larger areas will be used in future studies to assess possible imperfect wetting due to small surface area.

Sugama et al.[7,8] found several factors were important for forming good films with strong adhesion to the substrate. They are as follows:

- high spreadability of the sol solution
- formation of PGS or PTS at the sintering temperature of 150°C
- enhancement of Si-O-Ge or Si-O-Ti linkages at the pyrolysis temperature of 350°C
- minimal amounts of organic by-products
- formation of an oxane bond from the PGS or PTS to the aluminum oxide.

The goals of the current effort are to investigate the surface microstructure of these films, and to relate the microstructure to composition and other film growth parameters. The techniques employed include Raman spectroscopy, AFM, SEM and EDAX. Descriptions of the equipment are as presented above.

Figure 3 illustrates the morphology of a 30% TSPI, 20% PTS ceramic film on Al 6061 obtained by AFM. Cracks are evident, indicating imperfect match of Al and ceramic film thermal expansion coefficients. Single element scans of the surface reveal highest levels of Al occurring in the cracks, as would be expected because the ceramic film is thinnest at the cracks. However, the ceramic film may not be zero in the cracks, and may afford some level of corrosion protection, if the crack is not too wide. This possibility is being investigated. Studies to date indicate that obtaining crack free films is difficult, requiring careful control of composition and processing.

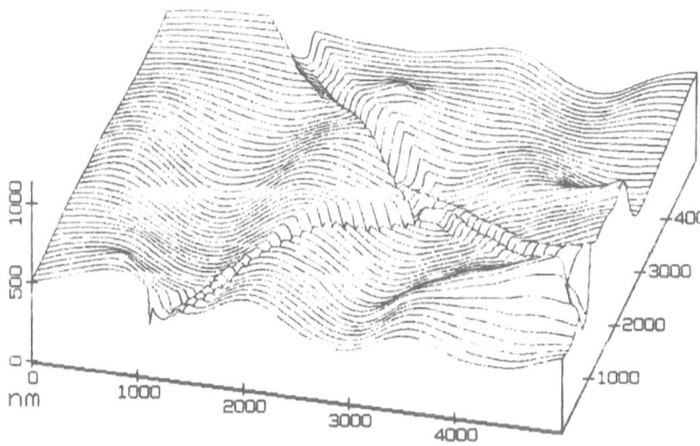

Figure 3. AFM scan of a 30% TSPI, 20% PTS ceramic film on Al 6061.

Future work on PTS and PGS ceramic films will be directed towards (i) further adjustment of composition to obtain crack free films, (ii) in obtaining high resolution AFM images of areas near cracks and also in the flat regions away from the cracks, and (iii) in obtaining Raman spectra with wavelength excitation in the near infrared region to reduce the background fluoresence that obscures signals of interest when using excitation wavelengths near 500 nm.

ACKNOWLEDGEMENTS

This work was supported in part by the U.S. Army Research Office (Research Triangle Park, NC) under Contract No. DAAL03-89-K-0084 and Grant No. DAAL03-92-G-0372, Dr. Robert R. Reeber, Grant Monitor. The support of the University of Missouri Weldon Spring Fund under Grant No. 90-WS-015 is gratefully acknowledged.

REFERENCES

1. E. L. Ghali and J.J.A. Potvin, The mechanism of phosphating steel, Corr. Sci. 12:583 (1972).
2. T. Sugama, L.E. Kukacka, N. Carciello and J.B. Warren, Polyacrylic acid macromolecule-complexed zinc phosphate crystal conversion coatings, J. Appl. Polym. Sci. 30:4357 (1985).
3. T. Sugama, L.E. Kukacka, N. Carciello and J.B. Warren, Factors affecting improvement in the flexural modulus of polyacrylic acid-modified crystalline films, J. Appl. Polym. Sci. 32:3469 (1986).
4. J.L. Wragg, J.E.Chamberlain, L. Chann, H.W. White, T. Sugama, and S. Manalis, Characterization of polyacrylic acid modified zinc phosphate crystal conversion coatings, J. Appl. Polym. Sci. (to be published).
5. D. R. Arnott, B. R. W. Hinton and N. E. Ryan, Cationic-film-forming inhibitors for the protection of the AA 7075 aluminum alloy against corrosion in aqueous chloride solution," *Corrosion* 45:12 (1989).
6. F. Mansfeld, V. Wang and H. Shih, Development of "stainless alumiunim," *J. Electrochem. Soc* 138:L74 (1991).
7. T. Sugama, N. Carciello, and C. Taylor, Pyrogenic polygermanosiloxane coatings for aluminum substrates, J. Non-Crystal. Solids 134:58 (1991).
8. T. Sugama, L.E. Kukacka, and N.Carciello, Polytitanosiloxane coatings derived from $Ti(OC_2H_5)_4$ modified organosilane precursors, Prog. Org. Coatings, 18:173 (1990).

SCANNING TUNNELING MICROSCOPY STUDIES OF ELECTROMIGRATION ON Si(100) SURFACES UNDER EXTERNAL STRAIN

Yi Wei, W.E. Packard, John D. Dow and I.S.T. Tsong

Department of Physics and Astronomy
Arizona State University
Tempe, AZ 85287

INTRODUCTION

Under zero externally applied stress, a nominally flat Si(100) surface has almost equal populations of (2x1) and (1x2) reconstructed domains. These two domains are separated by a single atomic layer step of 1.4Å height. The (2x1) domain contains Si-dimer rows parallel to the step edge and the (1x2) domain contains dimer rows perpendicular to the step edge. Using Chadi's nomenclature[1], we refer the (2x1) domain as type A domain and (1x2) as type B. The two domains can be distinguished in the scanning tunneling microscopy (STM) image by a straight S_A steps of type A and the ragged S_B steps of type B as shown in Fig. 1(a). A high resolution STM image of the dimer rows is shown in Fig. 1(b).

An interesting observation on the change in the domain sizes as a function of the direction of heating current passing through the Si(100) sample was reported by Kahata and Yagi[2] using reflection electron microscopy (REM). When a 500mA current was passed in the up-step direction to heat the Si(100) surface to 700°C, the type A domains grew in size at the expense of the type B domains; and vice versa when the current was in the step-down direction. For this to occur, Kahata and Yagi[2] reported that the minimum domain size, determined by the distance between the steps, L, must exceed ~80nm. We will refer this length L as the terrace width. Litvin and coworkers[3] essentially repeated the REM experiment of Kahata and Yagi[2] and confirmed their findings. In addition, Litvin et al.[3] found that when L is less than ~80nm, passing a current in the step-up direction caused the wider terrace (whether type A or type B) to expand its width at the expense of the narrower terrace; whereas the terrace width configuration did not change for a step-down current direction.

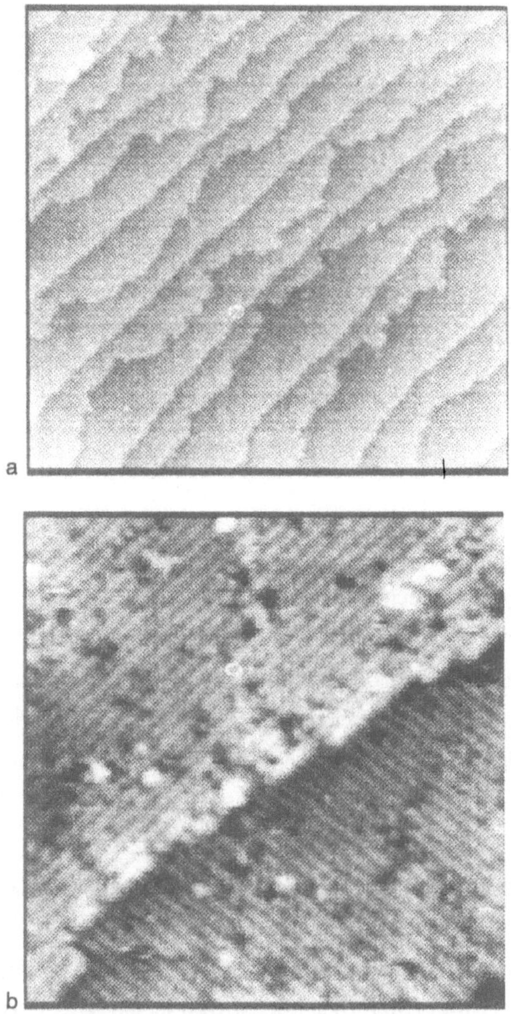

Figure 1. STM images of a clean Si(100)-(2x1) surface with zero applied stress showing the type A and type B domains. Down-step direction is from upper left to lower right. (a) Scan area is 400nm x 400nm. Type A domains have straight step edges and type B domains have ragged step edges. (b) 25nm x 25nm scan area. The dimer rows are clearly resolved.

Several groups of researchers[4-7] have observed a change in the distribution of domain sizes when the Si(100)-(2x1) surface was subjected to an externally applied uniaxial stress. Alerhand and coworkers[8] have predicted theoretically that an intrinsic anisotropic stress tensor exists on the Si(100)-(2x1) dimer-reconstructed surface. The surface is under tensile stress in the direction perpendicular to the dimer rows and under compressive stress along the dimer rows. If an external uniaxial compressive stress is applied to the surface at an elevated temperature, ~600°C, the domain with dimer rows perpendicular to the strain direction will grow at the expense of the other. On the other hand, if a tensile stress is applied, then the domain with dimer rows parallel to the strain direction will grow. When the external strain is released and the sample heated up again to ~600°C, the Si(100)-(2x1) surface returns to its initial configuration of equal domain populations. The kinetics to develop or remove the asymmetry in the domain population depend on the temperature, but the steady-state asymmetry depends only on the magnitude of the strain and not on the temperature. This means for a given applied strain, it takes a shorter time to reach the steady-state asymmetry at a higher annealing temperature. Likewise, after removing the strain, it takes a shorter time to restore equal domain populations at higher annealing temperature.

The interesting question is whether or not the observations of Kahata and Yagi[2] and Litvin et al.[3] are due to electromigration of surface atoms or due to uncontrolled thermal strain induced in the sample caused by non-uniform heating and/or clamping of the sample. The present report addresses this question specifically.

EXPERIMENTAL

The STM housed in an ultrahigh vacuum (UHV) chamber has been described previously[7]. The samples were 20mm long x 2mm wide x 0.5mm thick, p-type Si(100) wafers with a resistivity of 0.11Ωcm. The surface had a misorientation of 0.2° off [100] towards the [011] direction. The long side of the sample was parallel to the [011] direction, which meant the S_A and S_B steps were parallel to the width, i.e. short side, of the sample. The sample was rigidly held at one end by a Ta clamp like a cantilever while the other end could be pulled and pushed by a U-shape Ta "strainer" via a linear feedthrough. The Ta "strainer" also served as a breakable electrical contact for resistive heating of the sample. Since the S_A and S_B steps were parallel to the width of the sample, the applied uniaxial strain was either parallel or perpendicular to the two dimer-row directions. Since the electrical current was passed along the long side of the sample, it could be applied along either the step-up or step-down direction. The Si(100) surface was cleaned in situ at 5×10^{-11} torr by repeated flashing to 1200°C.

RESULTS AND DISCUSSION

An STM image of the clean Si(100)-(2x1) surface shown in Fig. 1 with equal distribution of type A and type B domains separated by S_A (smooth) and S_B (ragged) steps. The average terrace width, L, is ~340Å. With the unstrained sample, i.e. zero applied stress, heating the sample to 700°C by passing a dc current in the up-step or down-step direction produced no change to the surface. Then the sample was displaced at the free end corresponding to a tensile strain of 0.07% at the fixed end. The direction of strain was along the diagonal from the upper left to the lower right in Fig. 1(a). While under load, the sample was annealed at 700°C for 2 minutes and then cooled to room temperature. The applied stress was removed and STM images of the surface were taken. Fig. 2 shows a typical STM image of the strained surface, with type B terraces much wider than type A as expected since the type B dimer rows lie along the direction of the tensile strain. By measuring a series of STM images of the strained surface, we found that about 82% of the surface consisted of type B domain.

With the unequal domain population on the surface as shown in Fig. 2, we annealed the surface at a fixed temperature, 850°C, and a fixed time, 30 seconds, but with the dc current being passed through the sample at either the up-step or the down-step direction to see if the current direction influences the domain sizes. When the 1.8A current was passed in the up-step direction, i.e. from lower right to upper left in Fig. 2, the terrace width of the type B domains decreased as shown in the typical STM image of Fig. 3. By measuring a number of such STM images, we found that the type B domains occupied 62.2 ± 2.7% of the surface, a decrease from the original 82%.

The surface was then flashed to 1200°C to restore equal domain populations on the surface, after which the entire procedure for obtaining 82% type B domain (Fig. 2) was repeated. With the surface as shown in Fig. 2, a down-step current was applied to anneal the surface to 850°C for 30 seconds. STM images of the surface show that the type B domains now occupy 75.8 ± 2.7% of the surface, as shown in Fig. 4. This means that annealing the surface with a down-step current requires a longer time for the surface to relax back to equal domain populations, i.e. annealing the surface by a down-step current has a retarding effect on the restoration to equal population.

The phenomenon can be explained as follows. We assume that by heating the surface to 850°C, some Si atoms diffuse onto the top of the surface and become mobile adatoms. Let us further assume that the heating was accomplished by ac or electron-beam heating, i.e. no directionality of current involved. We know, by previous experiments[4-7], that in the absence of external strain, the surface returns to its initial configuration of equal domain populations. Alerhand et al.[8] attributed this to the spontaneous formation of stress domains, where the ground state of the surface corresponds to equal domain populations. Let us now give the mobile adatoms a preferred step-up direction, i.e. the adatoms have a tendency to attach themselves to an up-step. It is easily seen from Fig. 5(a) that since there are more

Figure 2. The Si(100)-(2x1) surface after a 0.07% tensile strain has been applied on the sample. The type B domains occupy about 82% of the surface. Scan area is 250nm x 250nm.

Figure 3. The surface shown in Fig. 2 after being annealed by 1.8 A dc current for 30 seconds in the up-step direction, i.e. from lower right to upper left. The type B domains accupy about 62% of the surface. Scan area is 250nm x 250nm.

Figure 4. The surface shown in Fig. 2 after being annealed by 1.8 A dc current for 30 seconds in the down-step direction, i.e. from upper left to lower right. The type B domains occupy about 76% of the surface. Scan area is 250nm x 250nm.

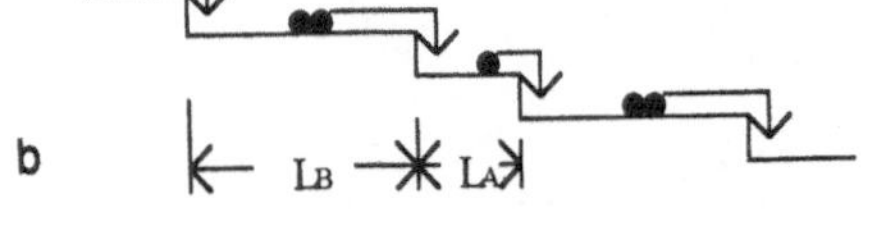

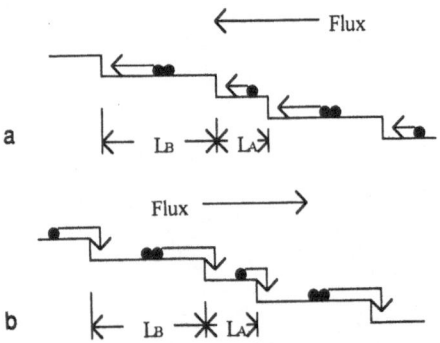

Figure 5. Schematic view of adatom flux influenced by the electric field. Arrows indicate the flux direction. L_A and L_B are the widths of type A and type B terraces respectively. (a) Flux in the up-step direction promotes equalization of terrace widths. (b) Flux in the down-step direction increases L_B and decreases L_A, thus retarding the equalization process.

mobile adatoms on a large terrace than on a small terrace, up-step attachment will result in the equalization of terrace sizes. This means that a preferential movement of the surface adatoms in the up-step direction will not impede that tendency toward equal domain populations. However, if the surface adatoms have mobility favoring the down-step direction, resulting in attachment to the down-step, then the larger terrace will become larger and the smaller terrace grows smaller, as depicted in Fig. 5(b). This has the effect of retarding the tendency toward equal domain populations.

In our experiment, the type B domains are ~14% larger in the down-step current heating case than the up-step current. This implies that the surface atoms probably have a preferred component of mobility in the direction of the current. This in turn suggests that some of the Si surface adatoms are positively ionized and are driven by the direction of the electric field in the same sense as the current. It appears, however, that the overall effect of electromigration is small since on releasing the strain, the surface still tends to return to equal domain populations, but its speed is retarded by the down-step direction of the current.

CONCLUSION

We have identified with STM a small effect of electromigration of surface adatoms on the Si(100)-(2x1) surface due to the direction of the heating current. The manifestation of this effect can only be observed after the surface has been externally strained. Contrary to previous REM measurements by Kahata and Yagi[2] and Litvin et al.[3], we did not observe any change in domain populations on an unstrained surface. Our STM observations can be attributed to ionized mobile surface adatoms attaching themselves to either up-steps or down-steps depending on the direction of the electric field.

ACKNOWLEDGEMENT

This work was supported by the U.S. Army Research Office contracts DAAL 03-92-G-0038 and DAAL 03-91-G-0054.

REFERENCES

1. D.J. Chadi, Phys. Rev. Lett. **59**, 1691 (1987)
2. H. Kahata and K. Yagi, Japan. J. Appl. Phys. **28**, L858 (1989)
3. L.V. Litvin, A.B. Krasilnikov and A.V. Latyshev, Surf. Sci. Lett. **244**, L121 (1991)

4. F.K. Men, W.E. Packard and M.B. Webb, Phys. Rev. Lett. **61**, 2469 (1988)

5. B.S. Swartzentruber, Y.W. Mo, M.B. Webb and M.G. Lagally, J. Vac. Sci. Technol. **A8**, 210 (1990)

6. M.B. Webb, F.K. Men, B.S. Swartzentruber and M.G. Lagally, J. Vac. Sci. Technol. **A8**, 2658 (1990)

7. W.E. Packard, N. Dai, J.D. Dow, R.C. Jaklevic, W.J. Kaiser and S.L. Tang, J. Vac. Sci. Technol. **A8**, 3512 (1990)

8. O.L. Alerhand, D. Vanderbilt, R.D. Meade and J.D. Joannopoulos, Phys. Rev. Lett. **61**, 1973 (1988)

MATERIALS CHARACTERIZATION WITH COLD NEUTRONS

H. J. Prask

Reactor Radiation Division, MSEL
National Institute of Standards and Technology
Gaithersburg, MD 20899

NIST has established at its Gaithersburg site the nation's first dedicated facility for "cold neutron" research (the CNRF). This new national facility is providing researchers in such fields as materials science, physics, chemistry, and biology state-of-the-art instrumentation which takes advantage of the unique properties of cold neutrons. As of May 1993, eleven of the planned fifteen new instruments were operational. One-quarter to two-thirds of total beam time is available for generic research at no charge to the general research community through a proposal evaluation system. The main focus of this paper will be on the principal techniques in which cold neutrons are currently used at the CNRF for nondestructive materials characterization: small-angle neutron scattering (SANS), neutron reflectometry, neutron depth profiling (NDP), and prompt-gamma activation analysis (PGAA).

INTRODUCTION

In the last three decades, thermal neutrons have become an essential probe of the properties of matter, particularly condensed matter. This usefulness arises from the intrinsic properties of the neutron, which can be summarized as follows:

- Thermal neutrons have de Broglie wavelengths and energies comparable, respectively, to interatomic spacings and collective excitations in solids and liquids. Because of these properties, determination of both atomic positions and motions can be made.
- Although uncharged, the neutron has a magnetic moment, -1.913 μ_N, such that neutrons interact with unpaired electrons in magnetic atoms. Elastic scattering provides information on the magnetic structure; inelastic scattering probes the energies of magnetic excitations.
- The absence of charge dictates that neutrons are scattered primarily by nuclear forces which leads to penetrations in condensed matter generally about three orders-of-magnitude greater than x-rays of the same wavelength.
- The neutron-nucleus interaction leads to sensitivity of neutrons to different isotopes of the same element in scattering so that, for example, hydrogen (which has a very

Nondestructive Characterization of Materials VI
Edited by R.E. Green, Jr. *et al.*, Plenum Press, New York, 1994

high scattering cross section) is easily distinguishable from deuterium. Similarly, neutron absorption by a nucleus is isotope-sensitive so that chemical analysis is possible.

Since about the mid-sixties neutron-based research programs have increased enormously both in scope and number of participants. For example, at the NIST reactor (NBSR), research participants numbered about 170 in 1979 and over 700 in 1992. Currently operating at 20 MW power and a peak thermal core flux of 4×10^{14} n/cm^2-s, the NBSR is one of the major research reactors in the country with twenty-six thermal neutron experimental facilities, the capabilities of which have been recently summarized[1]. The research stations are allocated among the following activities: neutron scattering and diffraction; neutron radiography; trace analysis and depth profiling; neutron standards development; fundamental neutron physics; long-term irradiations; and isotope production.

Within these broad categories, specific subjects of research being explored by NIST staff and collaborators at the NBSR include: magnetic materials, hydrogen in metals, structures and dynamics of molecular systems, characterization of intermediate structures (such as polymers, microemulsions, conformations of biological molecules, thin films, surfaces and interfaces), nondestructive evaluation, chemical analysis, neutron standards and dosimetry, and fundamental neutron physics (such as tests of quantum mechanics using neutron matter-waves and neutron interferometry).

In recent years, there has been a significant increase in the use of "cold" neutrons (wavelength > 0.4 nm; energy < 0.005 eV) for a broad variety of studies. In fact, the development of new cold neutron research capabilities, particularly in Western Europe, has created major new research areas and opportunities for virtually all scientific disciplines. Since the critical angle for total reflection increases with neutron wavelength, the increased total reflection at interfaces, which cold neutrons provide, makes it feasible to build neutron guides, which can be used to transport intense beams over many tens of meters, and to focus beams down onto small spots (< 1 mm). This, in turn, has opened up many new instrumentation possibilities and research opportunities, which are best characterized in terms of increased resolution in energy (by up to five orders of magnitude) and in the size of structures which can be determined; for example, the conformation of polymers in solids and solutions, the size and distribution of precipitates in metallic alloys, the structure of composites, and the early stages of metal creep and fatigue are examined. Activation analysis applications, such as neutron depth profiling and prompt-γ, also benefit enormously from cold neutron beams through the increase of absorption cross sections as neutron energy decreases. Capabilities in neutron interferometry, neutron optics, and other important areas of fundamental physics research are also enhanced by availability of intense cold neutron beams.

In 1984 the Seitz-Eastman Committee, commissioned by the National Academy, recommended the development of an internationally competitive cold-neutron research facility as the most urgent need among large facility improvements for U.S. science and technology[2]. This strong recommendation led directly to the support of the NIST Cold Neutron Research Facility (CNRF) which was dedicated in January 1989, and which now has operational eleven of the fifteen instruments planned. The CNRF, pictured in Figure 1, is a national facility, the experimental stations of which are available to all U.S. researchers.

The utilization of neutron probes for science and technology has been described recently in a number of review articles[3,4]. In the present article some examples of the use of cold neutron techniques for NDE applications will be presented.

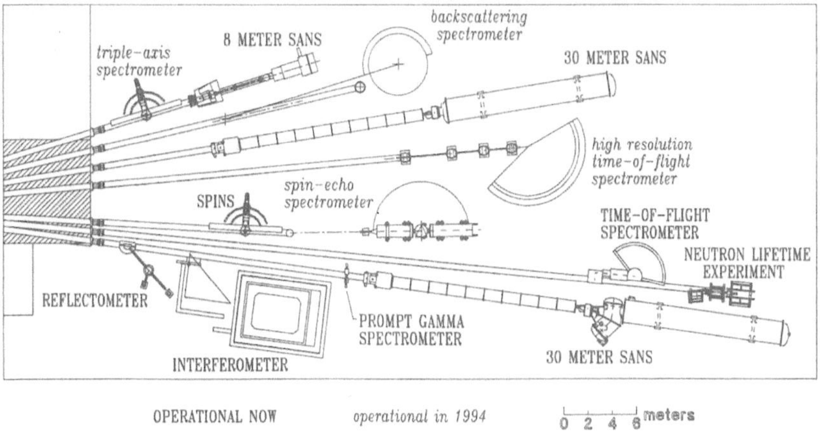

Figure 1. Guide hall of the CNRF.

SMALL-ANGLE NEUTRON SCATTERING

Small-angle neutron scattering (SANS) is a technique which is analogous to small-angle x-ray scattering for the examination of structures in the 0.5 - 500 nm size regime. However, SANS involves a neutron-nucleus interaction so that in addition to the $\sim 10^3$ greater penetration of neutrons relative to x-rays in the small-angle scattering regime—which allows the study of real engineering components—isotopes of the same element (e.g. hydrogen and deuterium) or of neighboring elements (e.g. manganese and iron) can be distinguished with SANS. A schematic of the newly-commissioned NSF/NIST Center for High Resolution Neutron Scattering (CHRNS) 30-m SANS spectrometer is shown in Figure 2. A wide range of studies is in progress: microstructure of magnetic metallic glasses, microphase separation and structure of block copolymers, multicomponent microemulsions, fluid behavior in porous media, porosity in advanced ceramics, polymer blend compatibility/structure, colloids and suspensions under shear, conformations of biological molecules, and cavitation in metals. A few examples of these are described below.

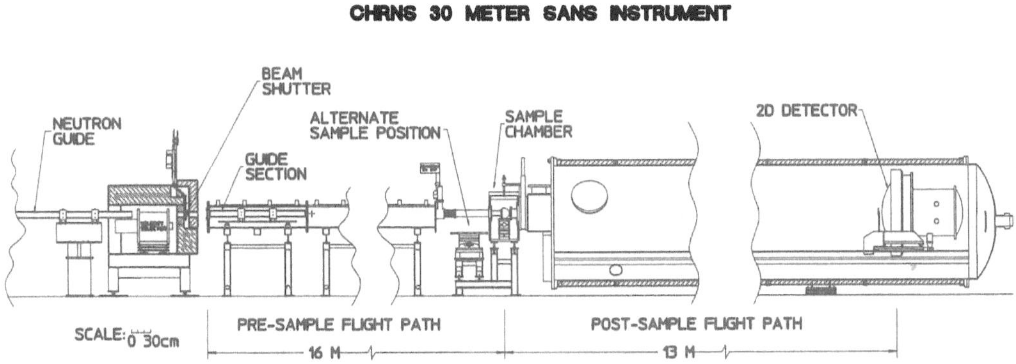

Figure 2. NSF/NIST Center for High Resolution Neutron Scattering 30-m small angle neutron scattering spectrometer schematic.

Metals

SANS has been an important technique at NIST for the characterization of cavitation in fatigued metals such as 304 stainless steel, A710 steel, and high-purity copper. In the most recent study[5] 99.999% pure copper samples were tested at 405 °C and a stress amplitude of 34.4 MPa, and combined precision density measurements and SANS to determine cavity volume fraction, V_v, and cavity surface area, S_v. From these, a weighted cavity size, $R_p=3V_v/S_v$, a characteristic cavity volume, $V_p=4/3\pi R_p^3$, and characteristic cavity number, $N_p=V_v/V_p$, are determined. The combination of techniques employed allowed the separation of stress and plastic strain effects as a function of fatigue frequency.

Samples were tested at fixed frequency with a varied number of cycles and with varying frequency for a fixed number of cycles. Conclusions which can be drawn from the combined-technique study described include:

- Most cavities are nucleated early in sample life.
- The cavity growth rate is constant at fixed frequency.
- Growth rate increases with increased frequency.
- The number of cavities increases with increased frequency.

SANS also provides a nondestructive means for characterizing precipitate evolution in alloys. Single-crystal, nickel-based "superalloys" are complex, precipitation-strengthened alloys now widely used for turbine blades in high performance aircraft. The unusual high temperature strength and creep resistance of these alloys derives from a homogeneous dispersion of coherent γ' (Ni_3Al) precipitates and the absence of grain boundaries, respectively. The addition of various elements to the alloys' composition can further enhance their high temperature strength although the mechanisms for this enhancement are in many cases not well understood. SANS has been used to monitor changes, due to high temperature exposure, in the microstructure of superalloys containing a few weight percent rhenium[6] .

The γ' precipitates in the material, which are roughly one micrometer in size and account for over 70% of the volume, are cuboidal in shape with edges along [100] directions. Strong scattering along [100] directions is due to the narrow (~20 nm), planar γ-phase regions between the precipitates and provides an indirect means of monitoring changes in the γ' morphology. Regions of the scattering pattern away from [100] directions are, on the other hand, relatively free from γ' scattering and thus are sensitive to microstructural changes unrelated to the γ' morphology.

SANS measurements were also made on samples, with and without rhenium, which had been aged at 1050 °C for up to 500 hours. After 500 hours the rhenium-containing samples gave additional, isotropic scattering indicative of a fine dispersion of small precipitates with a mean spacing of 15-20 nm. The lack of any directional dependence in this scattering indicates that these precipitates are not coherent which is believed to lead to the decrease in mechanical strength of the rhenium containing samples.

Ceramics

Knowledge of the microstructure evolution as a function of thermal processing is important for the development of process models in ceramics. In particular, such techniques as liquid phase sintering have received renewed interest, and it will be important to achieve a detailed understanding of the mechanisms involved to take full advantage of the possible benefits of the technique. Furthermore, sintering models developed over several years agree that microstructural characteristics are determinative of the specific mechanisms by which pore and solid phases evolve during densification.

SANS provides a very powerful technique by which pore evolution and microstucture can be characterized nondestructively. Combining conventional Guinier and Porod analysis methods with a recently-developed theory[7] by which multiple small-angle neutron scattering (MSANS) can be analyzed in detail, pore evolution up to radii as large as a micrometer can be characterized. These methods have been recently applied to determine the interrelationship of processing and microstructure in the sintering of crystalline and glassy ceramics[8], and the effect of green density and the role of MgO additive on the densification of alumina[9].

The sintering studies examined slip-cast alumina and microporous silica prepared by a sol-gel process. In this work, SANS patterns were obtained for samples removed from the sintering furnace at various stages of the process, the analysis of which yielded the results shown in Figure 3. The SANS study clearly shows that in the case of alumina there is a single population of pores which maintain a constant radius during the intermediate stages of sintering, and then coarsen rapidly during the final stages. In the case of silica, coarsening in the intermediate stage, followed by rather dramatic pore shrinkage is observed. The results highlight the different microstructure evolution signatures that exist for the viscous flow and diffusion mechanisms of sintering.

In the other study, pure alumina and alumina with 0.25 wt% MgO prepared by slip casting or cold pressing were characterized at various stages of sintering. This work showed that the effective pore radius was constant during the intermediate stage of sintering for each of the suites of samples, independent of doping and green density, and consistent with a stable topological decay model[10] of porosity during intermediate stage sintering. However, the actual value of the effective pore radius depends inversely on the green density. These results indicate that the connectivity that is established in the alumina green body determines the pore size throughout the stable intermediate sintering stage, and once this size is determined, it is unvarying as long as isolated porosity has not yet begun to form. At the onset of final stage sintering, the pore radius apparently coarsens abruptly. Part of this apparent coarsening may be due to the pinching-off of pore channels such that the larger dimension of the newly isolated pores comes increasingly within the range of sizes visible to the MSANS technique.

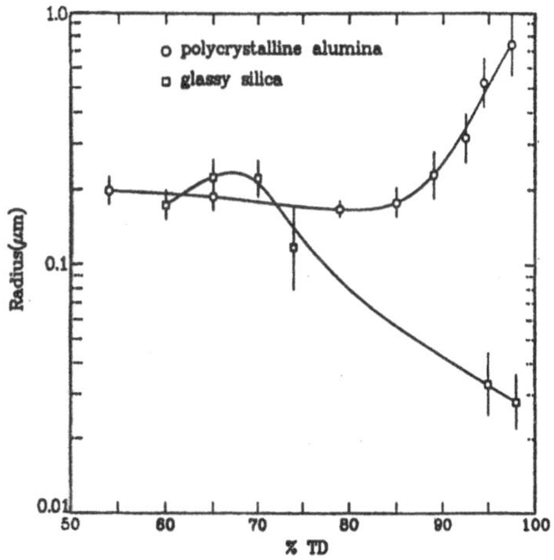

Figure 3. Pore radius as a function of the percent theoretical density for silica and alumina.

Overall, the neutron results show that the initial connectivity in the green state plays a dominant role in establishing the channel diameters during the intermediate stage of sintering, and contributes also to determining the onset density at which the final stage of sintering begins. The role of MgO as a sintering aid lies, at least in part, in prolonging the stability of intermediate stage sintering such that the body achieves greater density before the transition to final stage sintering after which isolated pores are formed.

Block copolymers are of increasing interest because of their potential in applications such as thin films, adhesives, and surfactants. In these applications a fundamental understanding of their interior and interfacial properties is essential to their incorporation in engineering devices. Neutron reflectometry is a technique which employs the wave-properties of the neutron to characterize variations in the index of refraction of materials as manifested by changes in the specular reflection pattern of neutrons. It has particular utility for polymer surfaces where selective deuterium labelling to highlight scattering from specific parts of a molecule can be utilized. An excellent example of the relevance of this technique can be seen in a recent study of the lamellar structure of thin films of polystyrene (PS) and polymethylmethacrylate (PMMA)[11].

In this work symmetric, diblock copolymers of PS and PMMA, prepared by a successive anionic polymerization process were prepared from solution as thin films on polished silicon substrates utilizing substrate spinning to produce uniform thickness (~500 nm). After drying under vacuum, the oriented lamellar morphology was produced by annealing at 170 °C. Deuteration of one starting component of the copolymer provides the contrast required to differentiate between layers.

The sensitivity of the technique is illustrated in Figure 4 in which reflectivity data and layer-structure models are shown for copolymers of ~30,000 and ~100,000 molecular weights (MW). Particularly noteworthy are the dashed and solid line fits to the reflectivity

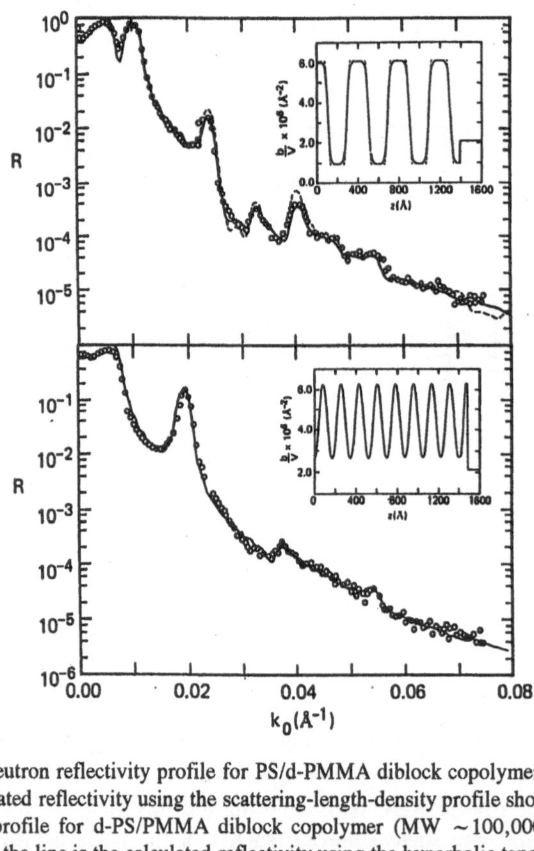

Figure 4. (lower) Neutron reflectivity profile for PS/d-PMMA diblock copolymer (MW ~30,000). The solid line is the calculated reflectivity using the scattering-length-density profile shown in the inset. (upper) Neutron reflectivity profile for d-PS/PMMA diblock copolymer (MW ~100,000). The points are the measured reflectivity, the line is the calculated reflectivity using the hyperbolic tangent function, the dashed line is the calculated reflectivity using a linear gradient between microdomains, as shown in the inset.

profile of the MW ~ 100,000 sample. The insert in this figure shows a lamellar profile with sharp (dashed curve) or rounded corners. The rounded-corner profile produces the solid-curve fit to the data which is clearly better than the dashed-curve (sharp corner layers) fit. This indicates that not only is the technique sensitive to primary structure of the lamellae (i.e. the thickness), but also to the details of the nature of the interface between lamellae. The potential of this relatively new neutron technique is only beginning to be explored; however, in addition to polymer-polymer interfaces, current studies already include magnetic thin films, superconducting thin films, Langmuir-Blodgett films, biological membranes, surfactants, and thin-layer structures for neutron applications.

NEUTRON DEPTH PROFILING

Neutron depth profiling (NDP) provides an isotope-specific, nondestructive technique for the measurement of concentration versus depth distributions in the near-surface region of solids. Specifically, the energies of short-range charged particles—which are emitted after neutrons are captured by certain nuclei—are measured. Utilizing easily-obtained range vs. energy data the depth of the capturing nucleus is determined. Although neutrons are highly penetrating, charged-particle path lengths are typically less than a micrometer. NIST has operated an NDP instrument with thermal neutrons since 1982. Now, with a generous grant from Eastman Kodak, NIST has developed a new state-of-the-art depth profiling instrument using a cold neutron beam. The increased sensitivity of this instrument has permitted the first-ever nondestructive analysis of the depth distribution of oxygen in a thin film, oxidation being critical for performance in many devices.[12] Among the first materials analyzed on this new instrument are synthetic diamond films and a new Standard Reference Material of boron implanted in silicon wafers. Other applications of the technique include measurements at interfacial boundaries, determinations of the depth and lattice positions of dopants in single-crystal materials, studies of leaching in thin films and the measurement of high-dose nitrogen implants in steels.

PROMPT GAMMA ACTIVATION ANALYSIS

A very different metallurgical application of neutrons utilizes prompt-gamma-ray analysis following neutron activation. A new cold neutron prompt-gamma activation analysis (PGAA) station at NIST, expected to be the best in the world, is operational.[13] PGAA is used "on-line" to nondestructively determine concentrations of elements which capture neutrons, do not become radioactive, but do give off characteristic gamma-rays, instantaneously. Hydrogen is among the many elements which this technique is sensitive. Initial measurements have been able to determine 15 micrograms of hydrogen in a gram of material. With optimization of the system, it is expected that 1 microgram of hydrogen per gram of material will be measurable. This approaches the sensitivity of interest for hydrogen embrittlement characterization in metals.

A recent PGAA investigation of a failed titanium-alloy turbine blade showed, nondestructively, hydrogen concentrations ranging from 140 to 750 μg per gram of Ti in different regions of the blade.

SUMMARY

Neutron techniques can provide important, often unique, information for a broad spectrum of materials science and engineering problems. The Reactor Radiation Division of NIST, with its expanded mission to provide direct assistance to industry, welcomes the opportunity to make facilities and expertise available and to collaborate in the solution of problems important to our international competitiveness.

REFERENCES

1. H. Prask, The Reactor and Cold Neutron Facility at NIST, *Neutron News* 1, 9-13 (1990).
2. "Major Facilities for Materials Research and Related Disciplines," NRC-NAS Report (1984).
3. D.A. Bromley, Neutrons in Science and Technology, *Physics Today*, pp. 31-39 (Dec. 1983), and references cited.
4. See, for example, a) the *NIST Journal of Research* 98, #1 for comprehensive review articles on all aspects of cold neutron research; also, b) "NIST Reactor: Summary of Activities July 1991 through September 1992," NISTIR 5120, February 1993.
5. J.G. Barker and J.R. Weertman, Cavity Growth Rates in Fatigued Copper, *Scripta Met.* 24, 227-232 (1990).
6. C. Austin, R. Darolia and C.J. Glinka, SANS Study of Strengthening Mechanisma in Single-Crystal Superalloys, NIST Tech. Note 1231, 71-3 (1986).
7. N.F. Berk and K.A. Hardman-Rhyne, Analysis of SANS Data Dominated by Incoherent Multiple Scattering, *J. Appl. Cryst.* 21, 645-651 (1988).
8. G.G. Long, S. Krueger, R.A. Gerhardt and R.A. Page, Characterization of Processing/ Microstructure Relationships in the Sintering of Crystalline and Glassy Ceramics by SANS, *J. Matls Res.* 6, 2706-2715 (1991).
9. S. Krueger, G.G. Long and R.A. Page, SANS Measurement of the Effect of Green Density and the Role of MgO Additive on the Densification of Alumina, *J. Amer. Cer. Soc.*, 74, 1578-84 (1991).
10. F.N. Rhines and R.T. DeHoff, in Sintering and Heterogeneous Catalysis, *Mat. Sci. Res.* 16 (G.C. Kuczynski, A.E. Miller and G.A. Sargent, eds.), Plenum Press, NY, 1984.
11. S.H. Anastasiadis, T.P. Russell, S.K. Satija and C.F. Majkrzak, The Morphology of Symmetric Diblock Copolymers as Revealed by Neutron Reflectivity, *J. Chem. Phys.* 92, 5677 (1990).
12. R.G. Downing, G.P. Lamaze and J.K. Langland, Neutron Depth Profiling: Overview and Description of NIST Facilities, pp. 109-26, in ref. 4a.
13. R.M. Lindstrom, Prompt-Gamma Activation Analysis, pp. 127-34, in ref. 4a.

SURFACE CHARACTERIZATION BY MEDIUM ENERGY PARTICLE SCATTERING

James H. Arps and Robert A. Weller

Vanderbilt University
Nashville, Tennessee 37235

INTRODUCTION

At the atomic scale, it is quite difficult to devise truly non-destructive measurement procedures because to do so requires a probe which gives specific quantitative, chemical or structural information without breaking chemical bonds. Ion beam analytical techniques using MeV beams are well known and versatile tools for characterizing surfaces and are generally considered to be nondestructive to most inorganic materials, although the electrical properties of the sample are usually altered. In this article, we discuss recently developed scattering techniques using time-of-flight particle detection that work well with ions whose energies are at least an order of magnitude lower than those required for conventional analyses. This reduction in the ion beam energy makes possible novel measurements such as the characterization of thin films on polymer substrates as well as enhanced capabilities such as improved depth resolution and increased sensitivity for trace element detection.

Surfaces are of immense importance in modern materials science. They are both areas where processes such as wear and corrosion are concentrated and sites for the fabrication of unique structures such as semiconductor devices. Moreover, they are the regions of transition between layers of bulk materials and between bulk materials and the surrounding environment and, therefore, often have unique electrical and chemical properties.

When characterizing a surface, one seeks to learn the electrical, chemical and geometrical structure usually of both the first atomic layer, which is the true surface, and the near surface region or exostrate which may, according to the problem of interest, be from a single atomic layer up to perhaps a few tens of nanometers thick. A truly impressive array of tools has been developed over the years for surface characterization[1]. For example, one learns about surface electrical structure from ultra-violet photoelectron spectroscopy while surface atomic arrangement is found by low-energy electron diffraction. Elemental composition may be found from Auger electron spectroscopy while the chemical

Nondestructive Characterization of Materials VI
Edited by R.E. Green, Jr. *et al.*, Plenum Press, New York, 1994

environment of the various atoms is accessible from x-ray photoelectron spectroscopy.

It is universally understood among practitioners of surface analysis that no single technique is right for all problems and that almost always a battery of measurements from complementary techniques is required to understand a problem in surface physics. However, one technique, Rutherford Backscattering Spectrometry (RBS) has been found to be particularly useful when one seeks to understand quantitatively the elemental composition of thin films as a function of depth.

The principles that underlie RBS are quite simple. An ion beam, e.g. He^+, of perhaps 2 MeV is directed onto the surface under study and the number of scattered projectiles at a specific angle is measured as a function of energy. Since the scattering cross section can be computed and the rate at which the ions lose energy as they penetrate the sample (stopping power) is known from measurements, the energy spectrum of backscattered particles is a representation of the depth distributions of the various constituents of the surface.

The disadvantages of RBS are that it does not work well with layers of low-atomic-weight materials on substrates of heavier elements (because backscatters from the light surface atoms have the same energies as scatters from the heavier and much more numerous atoms found more deeply in the target), that it requires a relatively large and costly accelerator, and that it has limits of depth resolution and sensitivity both of which are traceable to the properties of and constraints imposed by the silicon surface-barrier detectors that are almost universally used to measure the number and energy of the backscattered ions. The limitation on analyzing low-atomic-mass layers is fundamental to backscattering and may only be addressed by other techniques or, as we shall discuss below, by another scattering geometry. In all the other areas mentioned, improvements can in principle be realized by the use of lower energy primary ions, provided that the region of the sample which is of interest is sufficiently near the surface.

The purpose of this article is to describe a new detection technology for low and medium energy ions with which the potential gains from using lower energy primary ions may be realized. The new procedure involves replacing the solid-state, surface-barrier detector ordinarily used for RBS by a time-of-flight spectrometer. The following sections give an overview of the spectrometer operation and performance, and illustrate its advantages over conventional techniques in both backscattering and forward recoil geometry.

TIME-OF-FLIGHT SPECTROMETRY

Although time-of-flight spectrometry has been used extensively in the past for ion beam and nuclear studies[2], its application to medium energy scattering problems is relatively recent[3]. Figure 1 shows a time-of-flight spectrometer in a backscattering configuration. The primary ion beam enters from the right and strikes a target which is usually normal to the beam during backscattering analysis. Backscattered particles (with a distribution of charge states) enter the time-of-flight spectrometer and pass through a 2-3 $\mu g/cm^2$ carbon foil mounted on a high transmission mesh between two similar grounded meshes. The foil bias is typically -400 V. Secondary electrons emitted from the foil as the particle exits are accelerated by the bias toward a microchannel plate operated with its sensitive cathode at ground potential.

When the electrons arrive at the microchannel plate, a start pulse is generated. The ions, meanwhile, continue toward the stop detector with somewhat decreased energy and trajectories that have been perturbed by multiple scattering in the foil. When a stop pulse is generated within the coincidence resolving time of the external electronics, an event is recorded. Since the ion flight path length is known, the interval between start and stop pulses is a direct measure of particle velocity. In a backscattering experiment, this is equivalent to knowing the energy since the particle's mass is known. It is important to note that the charge state of the particle entering the spectrometer is not a crucial variable in the measurement because secondary electron emission from the carbon foil is at most very weakly dependent upon it. Thus, in this technique charge exchange at the target surface is not an important factor complicating the interpretation of measurements as it is in some other low energy ion beam analytical procedures.

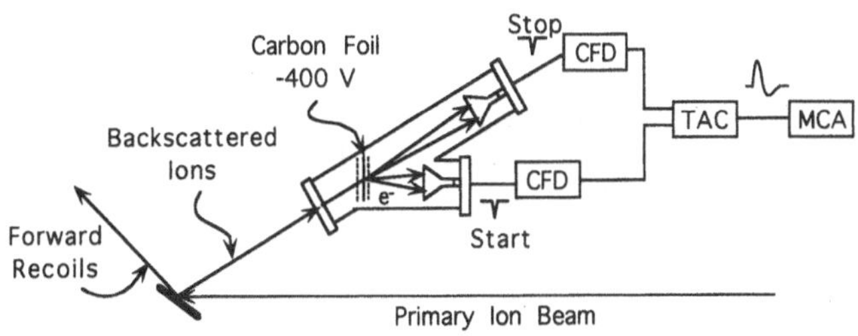

Figure 1. Schematic of the forward and backscattering experimental arrangements including a diagram of the time-of-flight spectrometer and a schematic of the external electronics. The forward recoil spectrometer is identical to the one shown and so is omitted from the figure. Abbreviations used are, constant-fraction discriminator (CFD), time-to-amplitude converter (TAC) and multichannel analyzer (MCA). Note that preamplifiers are no longer used to amplify the the timing signals.

The detector configuration shown in figure 1 is typically used for particles with energies from about 10 keV up to several hundred keV. However, it has been shown experimentally to respond to ions ranging from 5 keV protons to argon ions approaching 1 MeV as well as to 5 MeV α particles from a radioactive source. With a drift length of about 37 cm, typical flight times are from less than 100 ns to several hundred ns. The detector timing resolution is most appropriately measured with a fast particle such as an α from a radioactive source, but has been shown recently with protons from our accelerator to be well under 1 ns, a value at which it does not significantly detract from the quality of measurements. With the recent introduction of a biasing scheme for the start microchannel plate based upon microwave technology, it is now likely that the factors limiting overall spectrometer resolution are the path length uncertainty arising from our use of a start foil which is not perpendicular to the spectrometer axis, and the uncertainty in ion energy in the drift region between the foil and the stop detector attributable to charge exchange in the carbon foil[4]. The latter is approximately two times the bias, or 800 eV for most particles.

Along with resolution, the other key parameter affecting the performance of a time-of-flight spectrometer is quantum efficiency, or the probability that an ion entering the detector actually generates an event. For the surface-barrier detector, the efficiency is usually assumed with reasonable accuracy to be unity for particles of all energies. This is extremely convenient since, in this case, a simple measure of detector geometric solid angle is sufficient to relate measured counts to the areal concentrations of the various surface constituents (where, of course, the scattering cross section and quantity of beam are assumed to be know). Unfortunately, for the time-of-flight spectrometer the efficiency is considerably more complicated.

The physical processes affecting the efficiency of the time-of-flight spectrometer are secondary electron emission from the carbon foil, multiple scattering of the ions passing through the carbon foil, the responses of the microchannel plates to electrons and ions, and some geometrical factors. A detailed discussion of the ways in which these processes contribute quantitatively to the spectrometer efficiency is beyond the scope of this article but will be treated in detail in a forthcoming paper now in preparation by our group. It is sufficient for our present review to observe that by taking what is known individually about each of these physical processes and combining this knowledge in a mathematical model of the spectrometer, one can obtain reasonable agreement between theory and the measured efficiency as a function of particle species and energy. Thus, uncertainty about the spectrometer quantum efficiency should not pose a significant impediment to routine use of time-of-flight detectors for quantitative work.

A final issue concerning the practicality of time-of-flight spectrometry is the rate of data acquisition. Here, the time-of-flight spectrometer and the surface barrier detector are approximately comparable and offer significantly higher throughput than pulsed beam time-of-flight systems or electrostatic or magnetic spectrometers which are charge-state sensitive and can only analyze one energy at a time. At higher count rates both time-of-flight and surface-barrier detector spectra develop artifacts, the former from random coincidences in the spectrometer and the latter by a process called pulse pile-up. A complete theory of the effects of count rate on time-of-flight spectra has been published recently[5]. Its most important conclusion is that the random background is independent of flight time and proportional to count rate. Thus, background can be lowered by decreasing beam current, measured by examining regions of a spectrum kinematically forbidden to have real events, and removed by a simple background subtraction. By contrast, pile-up is a convolution and it is, therefore, much more difficult to extract true signal in its presence.

BACKSCATTERING MEASUREMENTS

The objective of backscattering measurements is to determine the composition and structure of thin films. The advantages of time-of-flight medium energy backscattering (MEBS) when compared with conventional RBS are 1) less total energy deposited in the sample, 2) greater sensitivity per primary ion (because of the larger cross section at lower energies), and 3) improved depth resolution (because of the higher resolution of the time-of-flight spectrometer when compared with surface barrier detectors). In this section, we illustrate the first and last of these with specific experimental examples.

Improved sensitivity to trace elements and, in particular, to surface contaminants is probably the most important consequence of the larger cross section at medium energies. This is particularly important in the semiconductor industry where minute traces of elements such as iron or copper on silicon wafers can render devices unusable[6]. As of this writing, total reflection x-ray fluorescence spectrometry is the method of choice for the analysis of such contamination[7]. However, estimates based upon plausible assumptions and supported by preliminary experimental results suggest that medium energy backscattering spectrometry may be an effective alternative technique[5,8]. This conjecture is the subject of active investigation in our laboratory and elsewhere.

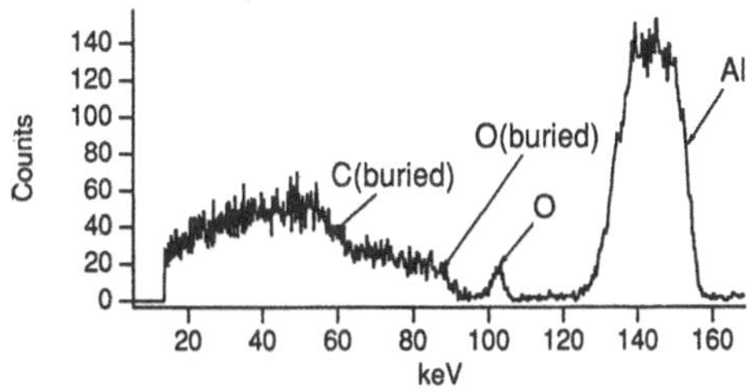

Figure 2. Backscattering spectrum of a metallized plastic. Analysis of the data revealed that the metallic film was Al about 10 nm thick. Note also the small oxygen peak attributable to the surface oxide. The plastic substrate was virtually unaffected by the analysis.

The energy deposition which accompanies conventional backscattering is sufficiently large that it is extremely difficult to analyze films on polymer substrates. Typically, the organic substrate is at least discolored by the radiation damage produced by the beam and, more often, is destroyed completely. Figure 2 shows a medium energy backscattering spectrum, mathematically rendered as an energy spectrum for the convenience of readers familiar with RBS. This spectrum was made by backscattering 270 keV He^+ ions from a metallized plastic of unknown composition. The purpose of the experiment was to determine the composition and thickness of the metallic layer. The signature of this layer is the isolated peak which, from its placement in energy and its thickness, unambiguously indicates that the metal is aluminum and that the layer is about 10 nm thick. Inspection of the area exposed to the beam after removing the sample from the vacuum chamber revealed a very slight discoloration but no structural damage to the underlying organic material. In fact, direct exposure of the substrate during a backscattering run to determine its composition also produced barely visible damage.

The extraordinary depth resolution of MEBS is illustrated by figure 3 which is a composite of four backscattering spectra made by 270 keV He^+ ions incident upon TiN films of increasing thickness deposited on oxidized silicon wafers. The isolated mesa-like feature in these spectra results from He^+ scattering from Ti in the surface layer. From the width of the features we conclude that the layers range from about 41 nm to about 47 nm thick. Thus, the differences indicated by the varying widths of the peaks are only of order 2 nm and are clearly visible by inspection. Experience suggests that under favorable conditions the positions of these edges may be determined with an error of less than 100 eV. This approaches single atomic layer thickness discrimination and suggests that MEBS may be a very effective tool in corrosion studies such as described by Hübler *et al.*[9] The shifting of edge positions has also been used to quantify minute changes in the thickness of an optical coating flown aboard NASA's long-duration exposure (LDEF) satellite which apparently resulted from the uptake of atomic oxygen[10]. In that case, a shift of 1 keV was observed to be ten standard deviations larger than the experimental error.

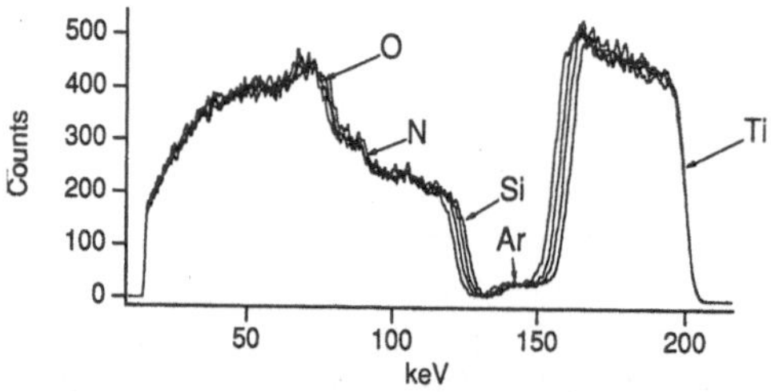

Figure 3. Backscattering spectra of four TiN thin films on SiO_2. Film thicknesses range from 41 nm to 47 nm in approximately 2 nm increments.

FORWARD SCATTERING MEASUREMENTS

One of the most active areas of research in our laboratory is the application of time-of-flight spectrometry to the analysis of forward recoils in a medium energy version of what is known at higher energies as elastic recoil detection or ERD[11]. This experimental configuration is shown in figure 1 by the arrow indicating the trajectory of forward recoils. Not shown in the figure is another time-of-flight spectrometer which, in our chamber, is at an angle of 42° with respect to the beam direction. At forward angles, both scattered beam ions and recoiling target constituents can enter the detector. This is precisely what is needed to identify low atomic mass constituents on the surface of a heavier substrate. By using a relatively heavy beam such as carbon, neon, or even argon, surface atoms are ejected in the direction of the forward detector with velocities up to nearly twice that of the incident ion. Since the time-of-flight spectrometer ultimately differentiates velocities, the light elements arrive at the detector and are analyzed with no background from scattered beam ions in a significantly wide region of the spectrum. Thus, oxygen may be easily seen on the surfaces of Al, Si and heavier materials. Perhaps even more important, though, is the ability of this technique to measure hydrogen.

Hydrogen is a ubiquitous, critical, surface contaminant which also is one of the most difficult elements to measure[12]. High energy elastic recoil detection using MeV beams, especially when used in conjunction with time-of-flight detection, has been quite effective[13]. However, as with conventional RBS, the equipment is large and expensive and the beam is very damaging to many materials. Typical results obtainable with medium energy elastic recoil detection are shown in figure 4.

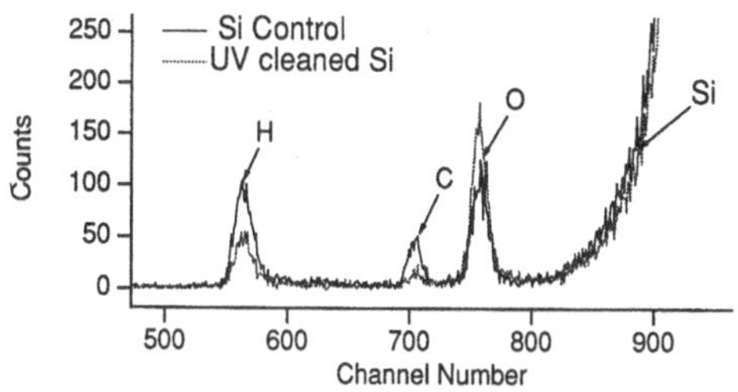

Figure 4. Forward recoil spectrum of a Si wafer before and after UV/ozone cleaning. The incident beam was 810 keV Ar^{3+}.

Figure 4 shows forward recoil spectra made by 810 keV Ar^{3+} bombardment of a Si wafer before and after cleaning by exposure to UV radiation in air[14]. This spectrum is left in its original time-of-flight form since, with several species present that have different masses, the transformation from a velocity to an energy representation is not appropriate. The effects on the quantities of H, C, and O are obvious by inspection. Clearly, if the removal of surface hydrocarbon contamination is the goal, the cleaning process succeeds. However, it does so at the expense of increased oxide thickness. The quantity of hydrogen remaining on the UV irradiated surface after the cleaning is estimated from the number of counts in the peak to be 5×10^{14} cm^{-2} while the number of C atoms is similarly 3×10^{14} cm^{-2}. From these and other data we estimate the current sensitivity for hydrogen detection (defined to be a signal three standard deviations above background) to be about 3×10^{13} H/cm^2.

The ability of medium energy elastic recoil analysis to give information about the depth distribution of hydrogen is illustrated by figure 5. This forward recoil spectrum, again made with 810 keV Ar^{3+}, represents a Si wafer onto which has been floated a carbon foil containing about 10^{17} C atoms/cm^2 and a small atomic percentage of hydrogen. Note the structure in the hydrogen feature.

Clearly the amount of hydrogen is largest near the surface of the foil and at the interface between the carbon and the silicon. It is likely that the latter is from water trapped at the interface during the operation of floating the foil. These data indicate that the depth resolution near the surface is less than 5 nm, although it is important to note that this resolution degrades rapidly as a function of depth when heavy primary ions such as Ar^{3+} are used.

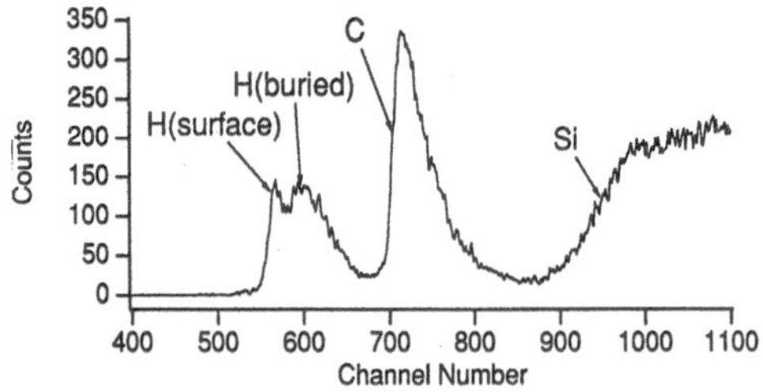

Figure 5. Forward recoil spectrum of a hydrogen-containing C layer on the surface of Si. The carbon layer is about 10^{17} atoms/cm^2 thick. Note the structure of the hydrogen feature. The spectrum reveals that there is both an excess of surface hydrogen relative to the bulk and a concentration of hydrogen at the Si-C interface.

CONCLUSION

Medium energy ion beam analysis using time-of-flight spectrometry is an important new tool for the analysis of surfaces which is particularly well suited to studies of the exostrate up to a depth of perhaps 100 nm. In backscattering mode, the depth resolution exceeds that which can be obtained by higher energy ion beam techniques by a considerable margin, and there is reason to believe that with further development medium energy backscattering may equal or exceed TXRF for the detection and measurement of trace levels of contamination on Si wafers. In forward scattering mode, existing data indicate a sensitivity and depth resolution for H measurement comparable with the best obtainable by other techniques. These are obtained with equipment whose size, in principle, should resemble more that of an electron microscope than that of the traditional electrostatic particle accelerator.

While still in a process of evolution, medium energy time-of-flight techniques offer unique capabilities for the study of thin films, for the detection of trace elements and for applications as diverse as semiconductor device characterization and gaseous corrosion studies. It may even be possible under favorable conditions to obtain information about biological membranes. Requiring smaller and less costly equipment, medium energy techniques will, in the future, make ion beam analysis available to a larger community than ever before.

ACKNOWLEDGMENTS

The authors wish to thank Martha R. Weller for creating an electronic form of figure 1 and Alain Diebold of Sematech for providing us with the high quality TiN films which were the subject of figure 3. This work was supported in part by the U.S. Army Research Office under contract DAAL 03-92-G-0037.

REFERENCES

1. Leonard C. Feldman and James W. Mayer, "Fundamentals of Surface and Thin Film Analysis", North Holland, Amsterdam (1986).
2. Harry J. Whitlow, Time of flight spectroscopy methods for analysis of materials with heavy ions: a tutorial, *in*: "High Energy and Heavy Ion Beams in Materials Analysis," J. R. Tesmer, C. J. Maggiore, M. Nastasi, J. C. Barbour and J. W. Mayer, eds., Materials Research Society, Pittsburgh (1990).
3. Marcus H. Mendenhall and Robert A. Weller, High resolution medium energy backscattering spectrometry *Nucl. Instr. and Meth.* B59/60:120 (1991).
4. Marcus H. Mendenhall and Robert A. Weller, Performance of a time-of-flight spectrometer for thin film analysis by medium energy ion scattering, *Nucl. Instr. and Meth.* B47:193 (1990).
5. Robert A. Weller, Instrumental effects in time-of-flight spectra, *Nucl. Instr. and Meth.* B, in press, (1993).
6. D. C. Jacobson, J. M. Poate, G. S. Higashi and T. Boone, Implanted standards for detection of transition metal contamination of silicon surfaces, *Nucl. Instr. and Meth.* B74:281 (1993).
7. U. Weisbrod, R. Gutschke, J. Knoth, and H. Schwenke, Total reflection x-ray fluorescence spectrometry for quantitative surface layer analysis, *Appl. Phys.* A53:449 (1991).
8. J. A. Knapp and B. L. Doyle, Heavy-ion backscattering spectrometry (HIBS) for high-sensitivity surface impurity detection, *Nucl. Instr. and Meth.* B45:143 (1990).
9. R. Hübler, A. Schröer, W. Ensinger, G. Wolf, F. C. Stedile, W. H. Schreiner and I. J. R. Baumvol, Corrosion behavior of steel coated with thin film TiN/Ti composites, *J. Vac. Sci. Technol.* A11(2):451 (1993).
10. Marcus H. Mendenhall, Robert A. Weller, and Ann F. Whitaker, Evolution of optical coatings in Earth orbit, *Optics Letters* 16:1466 (1991) .
11. James H. Arps and Robert A. Weller, Medium energy elastic recoil analysis of surface hydrogen, *Nucl. Instr. and Meth.* B in press (1993).
12. "The Analysis of Hydrogen in Solids", Proceedings of a Workshop Summarizing Developing Techniques and Formulating Requirements for the Future," Sandia National Laboratories, Albuquerque, NM, January 23-25, 1979, United States Department of Energy DOE/ER-0026, April 1979.
13. J. P. Thomas, M. Fallavier and A. Ziani, Light elements depth-profiling using time-of-flight and energy detection of recoils, *Nucl. Instr. and Meth.* B15:443 (1986), and references therein.
14. John R. Vig, UV/ozone cleaning of surfaces, *in:* "Treatise on Clean Surface Technology, Volume 1," K. L. Mittal, ed., Plenum, New York (1987).

IN SITU OPTICAL DIAGNOSTICS FOR PULSED LASER DEPOSITION

John L. Lawless and James E. Parris

Space Power, Incorporated
621 River Oaks Parkway
San Jose, CA 95134

ABSTRACT

A pulsed laser deposition apparatus was constructed to develop an *in situ* diagnostics system capable of monitoring the deposition process in real-time. Optical diagnostics offer a very sensitive method for monitoring deposition conditions *in situ* and assuring quality control of the deposited film. By observing the laser induced plasma plume, the presence of various excited species can be detected and relative concentrations determined. This should greatly enhance the reproducibility of the thin film deposition.

To demonstrate this approach, an apparatus was constructed for laser physical vapor deposition of boron nitride films. A KrF laser (248 nm) was focused onto a hexagonal boron nitride target inside a vacuum chamber. The resulting plume deposited a thin film on a silicon wafer (substrate temperature in the range 400°C to 500°C). A small amount of nitrogen vapor (50 mTorr) was introduced into the vacuum chamber to increase the nitrogen content of the deposited film. This configuration follows the procedure developed by Doll et al. at General Motors. The emission from the plasma was observed with an Acton monochrometer and a Princeton Instruments optical multichannel analyzer.

The resulting spectra showed a series of boron, nitrogen, and BN peaks. The relative intensities of several of these peaks were found to be clearly correlated with laser fluence and background gas pressure. By regular monitoring of the spectra, it should also be possible to determine subtle differences between target materials and assure reproducibility of the deposited films.

INTRODUCTION

In situ optical diagnostics for pulsed laser deposition allows for real-time monitoring of deposition processes including plasma, sputtering and chemical vapor depositions. The importance of *in situ* plasma spectroscopy is far reaching, including very high sensitivity and instantaneous feed back of equipment performance, background gas pressures, and target material behavior.

Plasma spectroscopy is an extremely sensitive diagnostic tool. By spectroscopic analysis one can immediately gain information about the plasma temperature and composition as the plume evolves with each laser pulse.

An inherent variable in the pulsed laser deposition process is the target material consistency. Properties such as surface conditions, density and porosity, impurity content, thermal conductivity, binder materials, grain structure, and absorptivity at laser wavelengths are attributed to the inconsistency. As the target is evaporated the surface conditions change such that the stoichiometry may no longer be preserved. It is because the target properties affect the plasma spectra that real-time monitoring of the deposition process is critical, especially when reproducibility is important.

Parameters such as laser spot size, and thus laser fluence, are often poorly known. Spot size measurements are accurate when the laser beam has a near gaussian energy distribution; the $1/e^2$ term can then be measured.

Laser plasma deposition requires high energy laser light in the ultra violet region. At the short wavelengths, the laser beam profile deviates rapidly from the gaussian shape, and the spot size becomes less certain. When the $1/e^2$ term is not attainable, less precise measurements must employed. Spot size characterization then is performed on thermal facsimile paper or photograph prints. With these methods, one must assume a uniform energy distribution across the entire paper burn, because no information about the energy distribution is discernible. With this assumption, the actual spot size is likely to be different since the energy distribution may be localized in some region that cannot be measured.

Relative fluence measurements can be made spectroscopically. Thus by measuring the fluence, accurate spot size measurements are less critical. At wavelengths in the UV region, the laser shot to shot stability may also posses variances as high as 20%. Plasma spectroscopy allows for monitoring the laser behavior, also from the fluence measurements, during the deposition process.

EXPERIMENTAL

Figure 1 is a schematic of the experimental lay out. The deposition processes was studied by observing the emission spectra of the laser plasma plume. The plasma source for the deposition was a KrF laser operating at 248 nm (Lambda Physik EMG 201E). The per pulse energy was about 400 mJ and the laser duration was 20 ns. The energy was measured using a Gentec ED-200 joule meter. The laser was focused with an F/4 converging lens, where the focus position was adjusted axially by translating an optical positioning stage.

The spot size was measured using the 2-dimensional beam profile etched on the reverse side of a black and white Polaroid film. This was mounted in the target holder inside the chamber and marked by six laser shots. The spot size images were photo enlarged and measured with vernier calipers. For this measurement, the laser energy distribution was assumed to be uniform; the $1/e^2$ term in the intensity profile could not be discerned. These measurements aided in the fluence calibration.

The Targets used were Union Carbide Pyrolytic boron nitride, "Boralloy", hexagonal boron nitride 1 inch disks, with density 2.21. All disks were from the same lot number. The target was prepared by scraping the surface with a razor blade, then buffing with a "Kim Wipe". The disk was attached to a rotating mount, then placed inside a vacuum chamber.

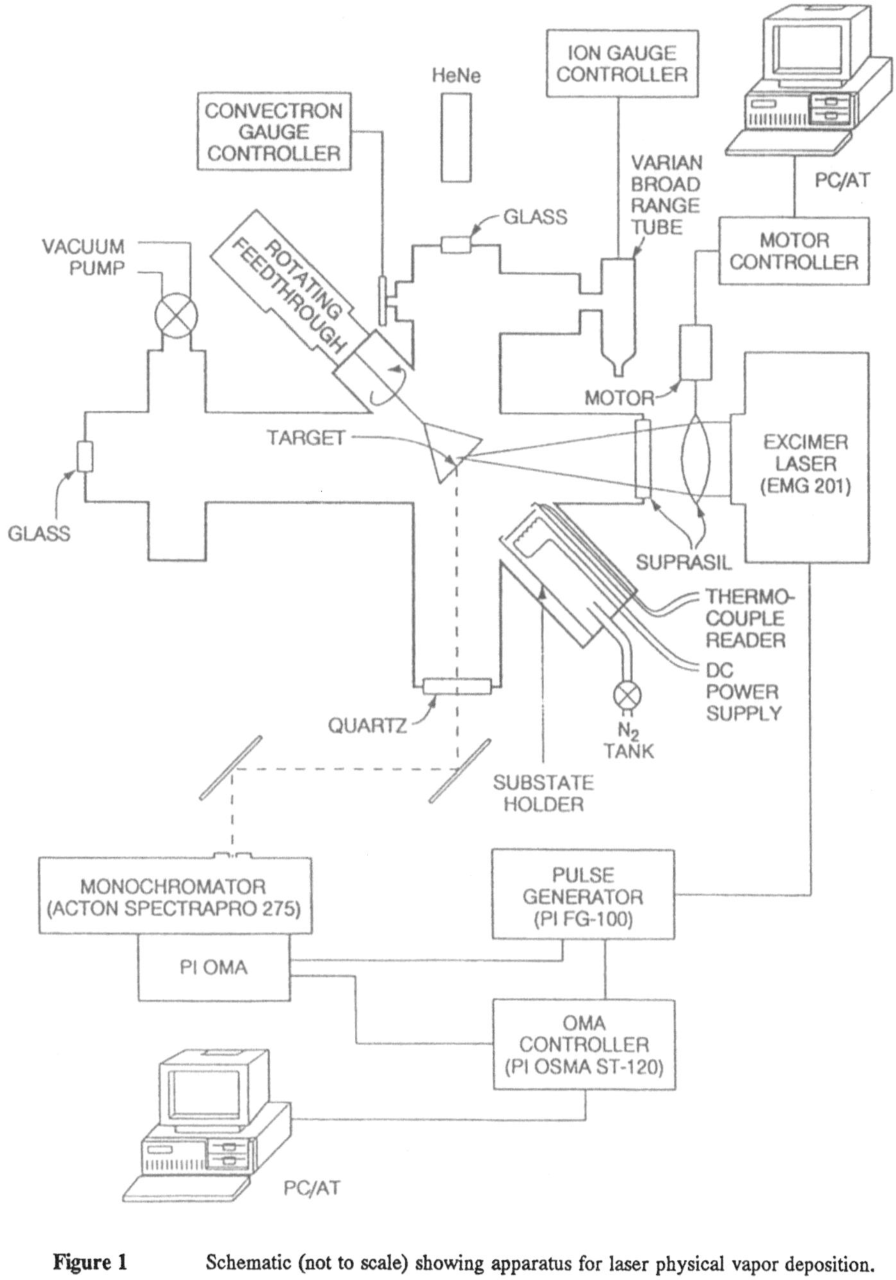

Figure 1 Schematic (not to scale) showing apparatus for laser physical vapor deposition.

Next a 2cm X 2cm piece of (100) silicon wafer was etched in an acid bath for 15 seconds, then quenched with DI water. The acid bath was a mixture of:

5 ml Nitric acid
5 ml Hydrofluoric acid
25 ml DI water

The etched wafer was then dried under a high intensity heat lamp for 5 minutes. Finally the wafer was installed in the chamber, and the chamber was then evacuated to 10^{-3} torr. The absolute pressure was measured using a Varian ionization gauge.

Current was then supplied to the heater filament below the silicon substrate. The heater filament is a 12 mil tungsten wire wrapped at 15 winds per cm, with a 0.25 cm diameter. The filament radiates to heat the substrate to temperatures ranging from 400°C to 500°C. The temperature measurement was calibrated using a Raytek Thermalert III (model 3B) optical pyrometer. A piece of carbon was substituted for the silicon wafer having an emissivity of 0.8. This was necessary because the silicon was transparent in the IR region, and the pyrometer would see the filament directly. Simultaneous temperature measurements were made using K-type and J-type thermocouples attached to a 9 mil stainless steel disk ring. The ring is centered between the substrate holder, the wafer and the holder cap.

Once the wafer is brought to temperature, N_2 gas is allowed to flow. The dynamic pressure was measured using the Varian ionization gauge and a Convectron pressure gauge. The experimental pressures ranged from 30 to 200 mTorr.

After all the parameters reached steady state, the laser was then fired at 10 Hz through the converging lens onto the hexagonal-boron nitride target. The target was allowed to rotate at 1.25 Hz. Simultaneously, the focusing lens was translated across the focal plane using a motorized sliding stage to move the focal point across the target surface. This assured a uniform usage of the target and aided to preserve the stoichiometry.

Deposition processes were studied by observing the emission spectra of the plume. Emission spectra were recorded by an intensified fast gatable array detector (Princeton Instruments model IRY-700G) mounted on a 0.275 meter spectrograph with a 300 grooves/mm grating (Acton Research). Calibrated band pass and neutral density filters were employed to screen out stray light and to protect the detection instrumentation. The spectrometer was calibrated using an Electro-Technic mercury spectrum tube.

The laser operated 10 minutes for each experiment. The laser was then turned off and the wafer was allowed to cool at a rate of 5°C per minute in a nitrogen atmosphere.

RESULTS AND DISCUSSION

Figure 2 demonstrates the time scales involved in the pulsed laser evaporation process. As indicated by the figure, at t=0 the laser is fired and the beam approaches the target. At t=20 ns the plume has evolved .04 mm from the target surface. This is based on the laser absorptivity, heat capacity. density and composition of the target. At t=1 μs the plasma expands to the substrate 2 cm away. Then at t=100 ms, the next laser pulse begins.

The main objective of our work was to identify the key spectra in the deposition plume that resulted in a uniform BN film on the silicon wafer. Once these spectral lines were identified, then we monitored their relative intensities as experimental conditions were varied. This method matched with empirical analysis of the thin film was used to understand our levels of monitoring sensitivity.

Figure 3 shows the observed spectrum for a plasma with a temperature of about 6 eV. In the spectra it is seen that ionized boron and nitrogen are present for this plasma temperature. In addition a small fraction of doubly ionized boron and nitrogen are seen. Data obtained from Union Carbide listing the contaminants and the possible concentrations confirmed the two oxygen lines.

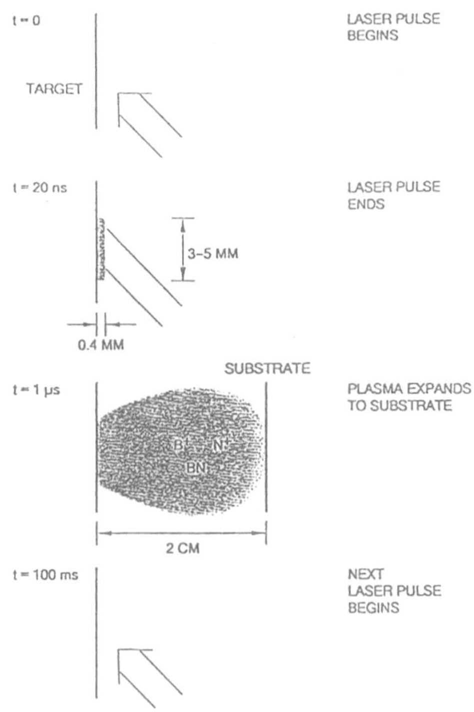

Figure 2 Time scales for the PVD process (Source SPI).

Additional spectral verification was accomplished using 99.5% pure boron pieces obtained from Johnson Matthey. A 0.5 gm piece of pure boron was substituted for the target in the chamber. Single laser pulses were used to generate the spectrum.

By knowing the fluence, the ionization fraction of nitrogen vs. temperature can be plotted. As illustrated in Figure 4, higher orders of ionization occur with higher temperatures; a thermodynamic equilibrium plasma density was assumed to be 1 mg/cm^3. To generate these curves, the focus of the lens was adjusted incrementally to change the fluence values. The spectra in Figure 3 was generated at a plasma temperature of 6 eV. At these temperatures it is evident that a significant fraction of N^+ ions are generated, with the presence of a small fraction of N^{++} ions.

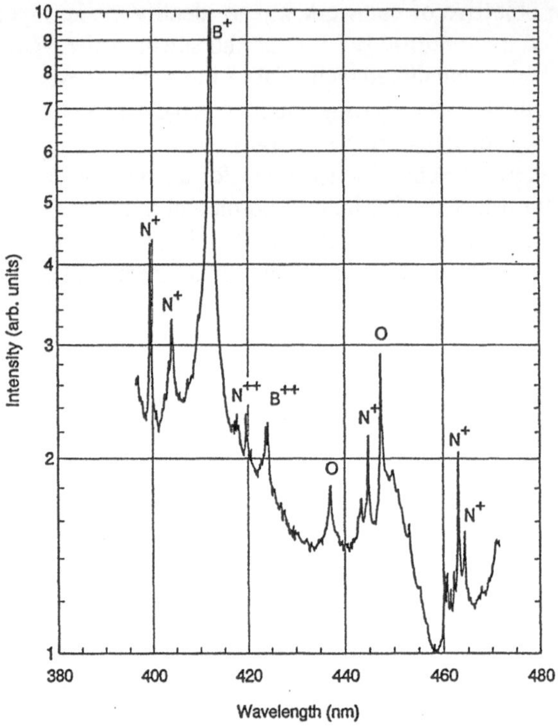

Figure 3 Observed spectrum of laser plasma plume from a hexagonal boron nitride target.

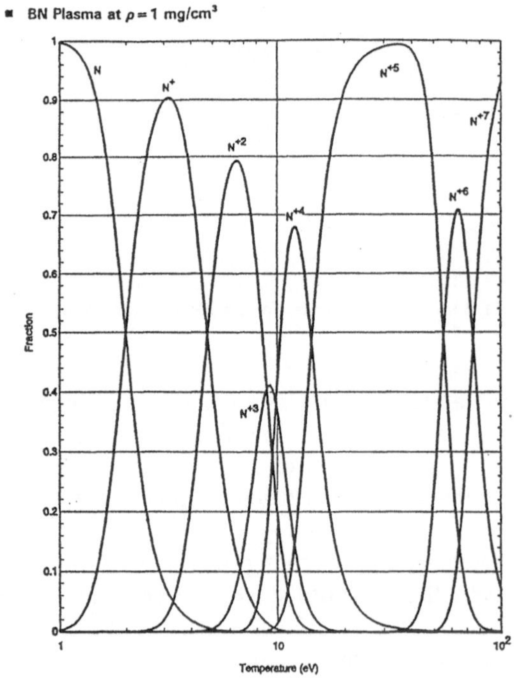

Figure 4 Ionization fraction of nitrogen vs. temperature, where the BN plasma temperature is assumed to be 1 mg/cm3.

■ Same Target: two sides

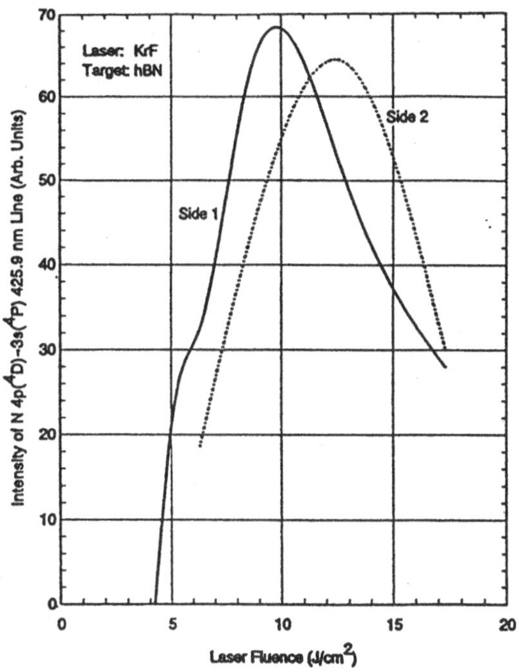

Figure 5 Intensity of N+ 422.8nm line (Arb. Units) plotted against laser fluence (J/cm2), where two sides of a single target were examined.

■ Same Target: two sides

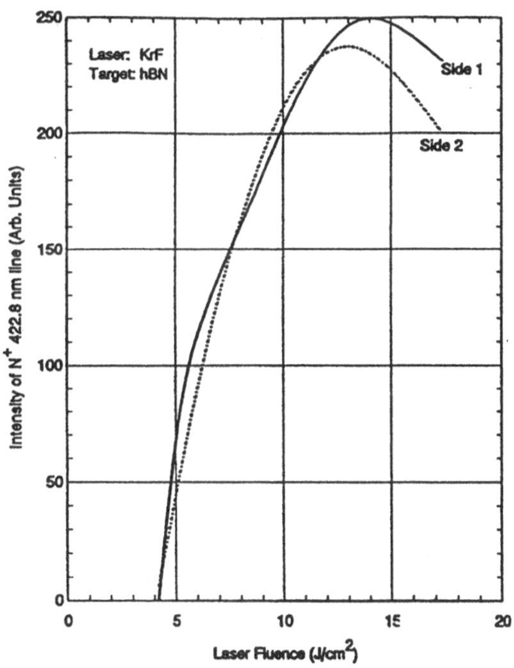

Figure 6 Intensity of N+ 422.8 nm line (Arb. Units) plotted against laser fluence (J/cm2), where two sides of a single target were examined.

To demonstrate the sensitivity of the spectroscopic method, a single target was placed in the chamber, and the chamber was evacuated. The laser was focused on the target and spectra were taken for different fluence levels. The target was allowed to rotate and the lens to translate. Spectra of the atomic nitrogen line were recorded, Figure 5 shows the results. The chamber was opened and the target was reversed to expose the opposite face to the laser. For side one, there is a peak near 9.6 J/cm^2, while side two has a peak near 12.4 J/cm^2. This is a surprising result, since one is lead to believe the target would have the same characteristics from one side to the next. When examining the fluence of 7.5 J/cm^2 for each target side, a 15% variance is evident. To illustrate that this is not a geometric phenomenon, Figure 6 shows a plot of the N^+ lines from the same experiments. Here the shift is not seen, whereas side two has a lower intensity above 12 J/cm^2.

SUMMARY AND CONCLUSIONS

Pulsed laser evaporation of BN targets at 248 nm were investigated, where *in situ* diagnostics were used to monitor laser-target interactions. From this it was found that target materials were inconsistent. The inconsistencies were attributed to: surface conditions, impurity content, density and porosity, thermal conductivity, binder materials, grain structure, and absorptivity at laser wavelengths. It was then demonstrated that target properties affect plasma spectra.

The uses of *in situ* plasma diagnostics are far reaching. Spectroscopic analysis allows one to assess similarities/differences between target samples. With such information, enhanced repeatability of deposition is possible. *In situ* plasma diagnostics allow for monitoring the consistency of deposition conditions during deposition. Operational parameters such as drift in laser focal point, change in target parameters as new surfaces are exposed, pressure of background gas in active region, and the presence of impurities can be assessed in real time.

MAGNETIC PROPERTIES OF ARMOR STEELS

S.A. Johnston, J.M. Winter, Jr. and R.E. Green, Jr.

Center for Nondestructive Evaluation
The Johns Hopkins University
Baltimore, Md., 21218

ABSTRACT

The Center for Nondestructive Evaluation (CNDE) at The Johns Hopkins University has undertaken measurements on armor steels to establish a data base which will more completely characterize the magnetic properties of these materials. A hysteresigraph designed for soft ferromagnetic materials has been used to generate initial magnetization curves and B-H loops from which such parameters as coercivity, remanence, both initial and maximum permeability, and saturation induction are deduced. Test specimens cut from monolithic slabs are of various dimensions, but typically represent two to ten cubic inches of the original material. Both rolled and cast steels have been examined. Preliminary determinations of the extent of magnetic anisotropy have been made.

INTRODUCTION

Two categories of armored steels have been characterized; rolled homogeneous armor (RHA) per military specifications MIL-A-12560F or MIL-A-46100, or cast homogeneous armor, per military specification MIL-A-11356. Both of these are medium alloy steels with 0.20% to 0.30% carbon. A large number of tests have been conducted to establish the reproducibility of data from a given test specimen and from different test specimens of the same material. The results have been examined with respect to reproducibility from the same test specimen, the homogeneity of the steel, effects of orientation with respect to the rolling direction, and the effect of thickness of the source plate (recognizing the probable differences in thermo-mechanical processing details).

Nondestructive Characterization of Materials VI
Edited by R.E. Green, Jr. *et al.*, Plenum Press, New York, 1994

MEASUREMENTS

Specimen Preparation

All the RHA test specimens were machined from plates nominally 12" square which had been flame cut from much larger plates. Plate stock was supplied in 0.5", 1.0", 2.0" and 2.5" thicknesses. For each plate thickness, specimens were cut with orientations parallel to the rolling direction, transverse to it, and at plus and minus 45° to the rolling direction, as shown in fig. 1:

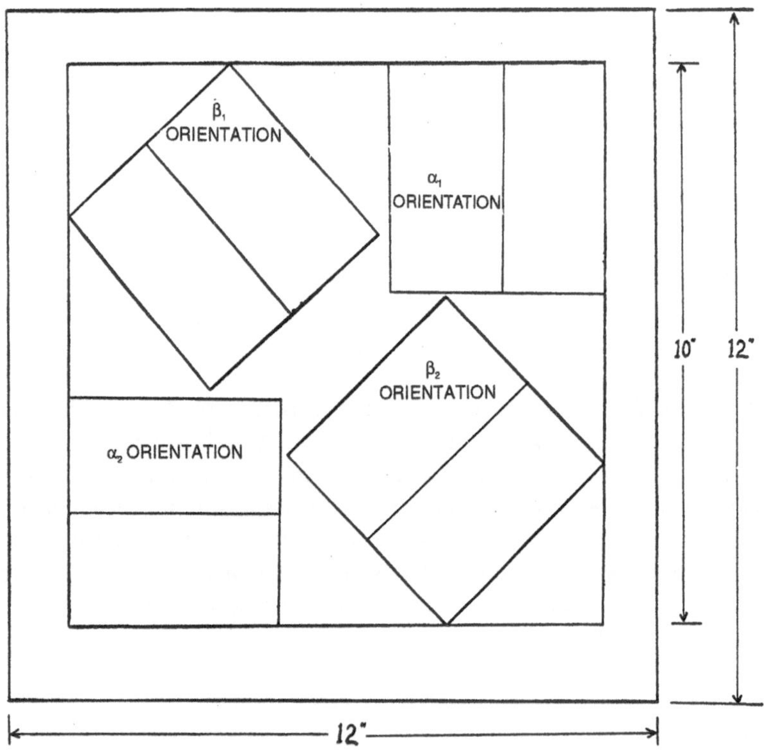

Figure 1. Test Specimen Orientations

In all cases, a 1" border around the edge of the 12" square plate was discarded to avoid edge effects due to the flame cutting. Two types of specimens were machined from the 0.5" plates, identified for convenience as "A"-type and "B"-type. The former had thickness-width-length dimensions of 0.5"x1.0"x4.0", while the latter were 0.5"x2.0"x4.0". "C"-type specimens were 1.0"x2.0"x4.0", "D"-type were 2.5"x1.0"x4.0", and "E"-type were 1.0"x1.0"x4.0", each being machined from plates of appropriate thicknesses. These various shapes are shown in fig. 2 to indicate how the dimensions cited related to the initial plates.

Hysteresigraph Measurements

A KJS Model SMT-500 Computer-Automated Magnetic Hysteresigraph for soft magnetic materials (KJS Associates, Dayton, Ohio), was used for generating the B-H loops. Fig. 3 shows a schematic of the instrument. Features which promote accurate measurements are a Hall probe gaussmeter to monitor the level of induction in the test specimen and a fluxmeter coil to provide closed loop control of the power supply in order to set the applied field to the pre-programmed value independent of

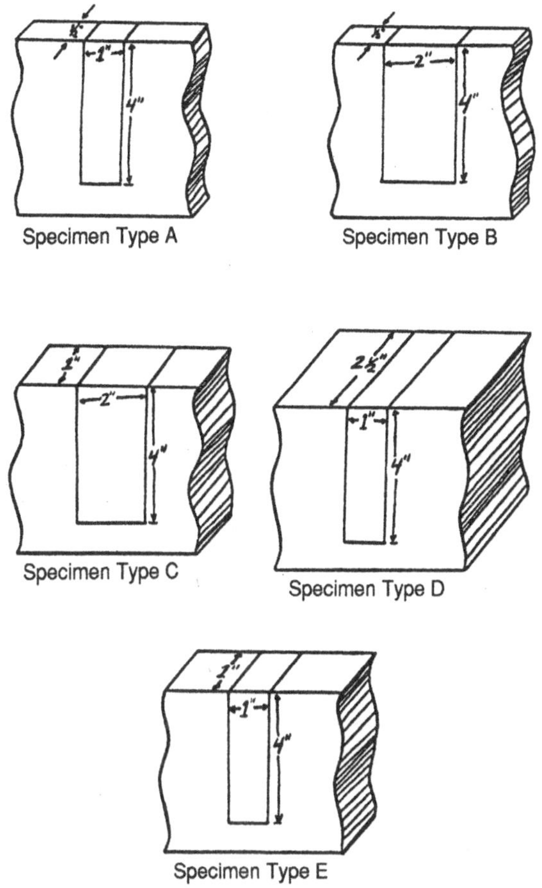

Figure 2. Test Specimen Dimensions

inadvertent differences in reluctance from specimen to specimen. To fit properly in the adjustable pole pieces, all the specimen surfaces but the ends were milled using coolant and ground.

Figures 4 and 5 show the results for coercivity and remanence. Since the test conditions were adequate to drive these specimens to saturation, the coercive force can be considered the coercivity and the remanent induction can be taken as remanence.

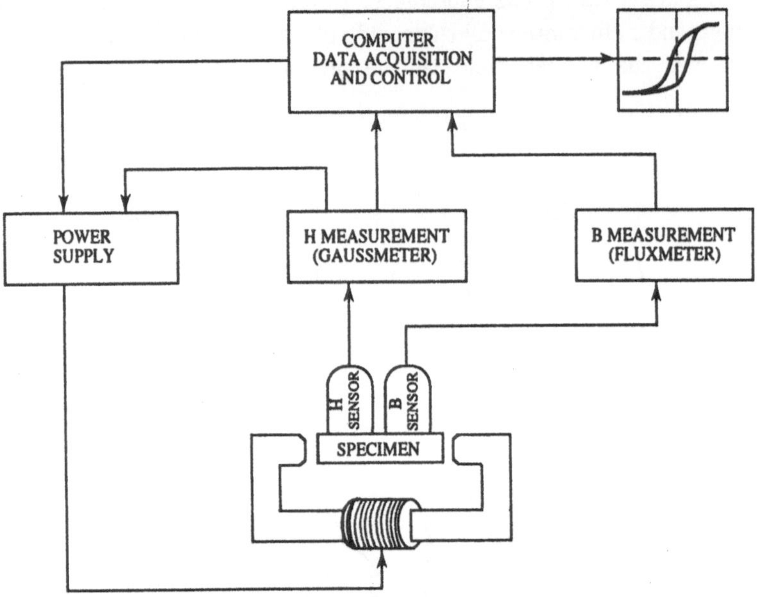

Figure 3. Schematic of Hysteresigraph for Measuring B-H Loops

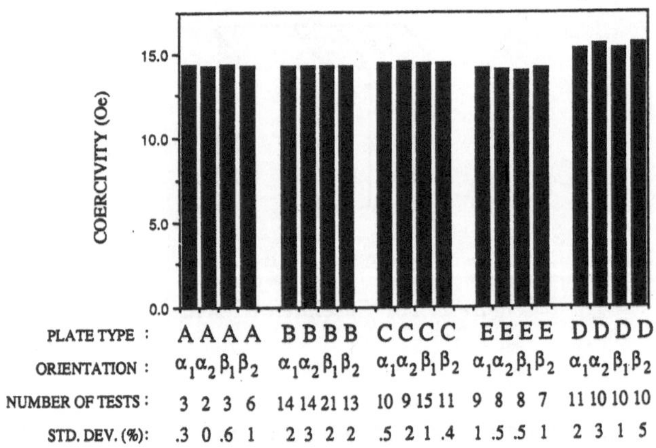

Figure 4. Coercivities of RHA Steel Specimens

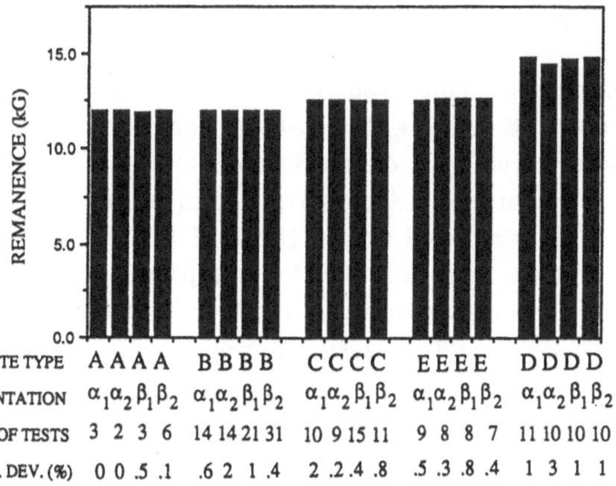

Figure 5. Remanences of RHA Steel Specimens

Comparing coercivities and remanences for different orientations shows virtually no orientation dependence. This is a result somewhat surprising as most rolled steel plate shows orientation dependence. Similarly, there appears little dependence on thickness, with the only exception being Specimen Type D, which is cut from the thickest plate material. Since the plate is thicker, it may not have seen the same processing as the others (for example, a smaller percent reduction in thickness by rolling than the other plates). The measurements were extremely reproducible and varied little from specimen to specimen, as seen by the standard deviations shown.

Measurement of Initial Permeability

Tests were run with minor loops driven to about 25 oersteds, so as to enlarge the initial section of the magnetization curve. By way of contrast, data has usually been taken from loops which were driven to about 300 oersteds. The loops for the different specimens appeared quite similar at the 25 oe drive levels. A graphical method was used to calculate u_0, the initial permeability,. For initial measurements, it appears that u_0 is about 86 gauss/oersted for all specimens.

The advantage in delineating the initial magnetization by acquiring a new B-H loop with a smaller drive field is that this procedure generates more data points near the origin than if the loop from a test with a larger drive field were just magnified.

Effect of B-H Loop Test Rate

Tests have been conducted to determine reproducibility of B-H loops as a function of how fast the applied field was swept. The was some question of how slowly the field had to be swept for these steels to allow domains to equilibrate before again changing the applied field. Eight B-H loops were run on the same specimen each with different total durations for testing, ranging from less than 5 seconds to 40 minutes for data acquisition. Distortion occurs only in the test completed in less than 5 seconds. Normally the time period selected for a test is 3 minutes. The time for test is not a directly selected parameter in the software, so alteration of lapsed time has to be done indirectly. In any case, it is clear that the B-H loops acquired in 3 minutes are not altered by selection of too fast a data acquisition rate.

Effect of Very Large Fields

These measurements employed a large laboratory research magnet (Applied Magnetics Laboratory, Baltimore, MD). Test specimens were immersed in a field in excess of 10,000 oersteds (limit of measurement). It was of interest to see whether these fields would orient some domains in the specimen which would not normally respond to the fields generated during the B-H loop test. If such high coercivity magnetization did occur, the next B-H loop measured with the KJS Hysteresigraph should be affected. No difference was resolved between the magnetic parameters measured with the 300 oersted Hysteresigraph before or after treatment in the extremely high fields.

Measurements on Cast Armor Steels

The results that have been obtained from specimens cut from the limited amount of material available are given in Table 1. Although the number of samples is too small to make any conclusive statements, it appears there is greater variation of results in the magnetic properties of the cast alloy than the rolled alloy.

Table 1. Measurements on Cast Armor Steels

Specimen	Orientation	H_c (Oe)	B_r (kG)	u_{max}
11	parallel x-axis	15.07	13.94	0.457
12	" " "	15.43	12.97	0.517
21	parallel y-axis	15.63	13.60	0.467
22	" " "	13.41	14.83	0.580
32	parallel z-axis	14.42	14.71	0.645
33	" " "	14.72	14.70	0.630

DISCUSSION AND CONCLUSIONS

Data from B-H loops have generated values for coercivity, remanence, and maximum permeability of representative armor steels. For RHA specimens, there is no orientation dependence of magnetic properties and the material seems magnetically homogeneous. A practical way to measure the initial permeability has been established. Experimental examination of the effect of the test rates chosen has verified that the B-H loop data suffers no artifacts due to testing with too fast a field sweep rate.

These magnetic materials have been exercised experimentally to determine that they do not contain isolated high coercivity regions (e.g. magnetic inhomogeneities) which would only switch under higher fields. The absence of such behavior supports the view of the RHA material being magnetically homogeneous.

In contrast, the limited tests performed on cast steel armor show greater variation in magnetic properties than the RHA steel. Perhaps this is an indication of a greater degree of heterogeneity.

ACKNOWLEDGEMENTS

The authors gratefully acknowledge the support of this work by the Countermine Systems Directorate, Belvoir RD&E Center, Ft. Belvoir, VA. The authors particularly appreciate the interest and assistance of Dr. Hermann Spitzer, Dr. David Heberlein, Mr. James Dillon, and Mr. Noel Wright, all of whom have shared their insight and experience.

NONDESTRUCTIVE CHARACTERIZATION OF TEXTURES IN COLD-ROLLED STEEL PRODUCTS USING THE MAGNETIC TECHNIQUE

I. Altpeter and G. Dobmann

Fraunhofer Institut für zerstörungsfreie Prüfverfahren
D-66123 Saarbrücken, FRG

INTRODUCTION

In the automobile industry the deep drawability (r_m, Δr) of cold-rolled steel sheets is an essential property in the production line of chassis units to control the drawing process of the metal forming press. Deep drawability of steel sheets is not an intrinsic material property, and therefore, cannot be measured directly by using the physical basis.

It is determined by plastic anisotropy destructively measured in 0°, 45° and 90° orientation performing a strain experiment; 0° is the rolling direction. A strong correlation exists with the elastic properties of the materials depending on crystallographic texture.

There is a correlation between elastic and plastic anisotropy. Accordingly, the techniques usually applied to determine the texture are different: for instance, differences in plastic deformation can be detected by the determination of the average Lankford plastic strain ratio r_m, elastic anisotropy by the determination of the Young's modulus and X-ray texture analysis. The X-ray measurements which can be applied under practical conditions as well, are the state of the art[1]. In addition to the X-ray measurements, ultrasonic and magnetic techniques offer a possibility of fast and simple analysis of texture. The measurement of ultrasonic velocities in different directions is an alternative approach to the X-ray measurement and has been developed for practical applications[2]. Worldwide activities in R&D have been observed in recent years in the application of ultrasonic phase velocity measurements in order to determine the coefficients of the orientation distribution function, i. e. C_4^{11} and C_4^{13} correlating with r_m,- and Δr-values respectively[3-7]. In contrast to the ultrasonic approaches only a small number of papers suppose magnetic techniques for texture characterization[8-10].

The elastic anisotropy is determined by the crystal texture, whereas the plastic anisotropy is determined by the microstructural parameters like dislocation density and grain boundaries. Since the same microstructural features restrict the mobility of Bloch-walls, magnetic techniques in addition offer the possibility to analyze the plastic anisotropy.

Nondestructive Characterization of Materials VI
Edited by R.E. Green, Jr. *et al.*, Plenum Press, New York, 1994

PHYSICAL BACKGROUND

In the body-centered cubic iron single-crystal, the magnetization is directionally - dependent. The reason for this is the crystal-anisotropy energy. By magnetization parallel to the crystal directions [1 0 0], [1 1 0] and [1 1 1], different magnetization curves can be obtained[11]. The measurement of hysteresis cycles needs quasi-static magnetization, i.e. magnetizing frequencies lower or equal 50 mHz with sensor coils encircling the specimen under inspection. Therefore, only specimens with special shapes (cylindrical) can be evaluated. Larger specimens must be destructed by cutting.

Magnetostriction measurements are available by fixing strain gauges to the surface under inspection. Because the texture measurements principally use the data acquisition in different directions, the technique is not handy and cannot be applied on-line in a steel mill on a strip running with speeds up to 200 m/min.

The task of the following paper is to describe the 3MA-method (Micromagnetic Multiparameter-, Microstructure- and Stress-Analysis)[12] as a nondestructive technique for texture evaluation. 3MA-parameters are the magnetic Barkhausen noise, the analysis of upper harmonics of the tangential field strength and the dynamic magnetostriction.

Magnetic Barkhausen Noise M[13]

The electrical voltage pulses induced in a magneto-inductive sensor (tape recorder head, pick-up-air-coil) by irreversible Bloch-wall jumps during magnetic reversals, is named the magnetic Barkhausen noise. The Barkhausen noise is recorded in the frequency range between 0.5 kHz - 5 kHz, i.e. in the lower frequency range, in order to gather data in analyzing depths up to 1 mm. After rectification and low-pass filtering of the Barkhausen noise, signal-profile curves can be detected as a function of the magnetic field strength. Measuring parameters are the amplitude of the maximum M_{MAX} and the magnetic field values H_{CM} where the maximum occurs.

By the magnetic Barkhausen noise, remagnetization processes in the individual grains contribute to the measuring effect. This is correct if the distances between the Bloch-walls, which are between 0.5 μm and about 5 μm in steels, are small in comparison with grain size. Depending on the crystal texture, the different crystallite groups make a directionally-dependent contribution to the remagnetization process.

Depending on the participation of the individual crystallite groups in a remagnetization direction, the Barkhausen noise activity changes in this direction, too.

Dynamic Magnetostriction E_λ[14]

This nondestructive measuring parameter is influenced by the magnetostrictively active reversible and irreversible Bloch-wall motions and rotation processes. A quasi-static magnetic field is superimposed by an alternating magnetic ΔH field with a frequency ≤ 3 MHz. In order to excite and to receive ultrasonic waves, we use an electromagnetic ultrasonic sensor which works contactless. The excited amplitude of the ultrasonic wave - here named E_λ, dynamic magnetostriction - is a function of the magnetostriction λ and therefore directionally dependent. The minimum of the dynamic magnetostriction curve - E_λ as a function of the H-field -, the H-value $H_{E\lambda(MIN)}$, is used as a measuring parameter (Fig. 1). $H_{E\lambda(MIN)}$ is independent of the influence of load stresses.

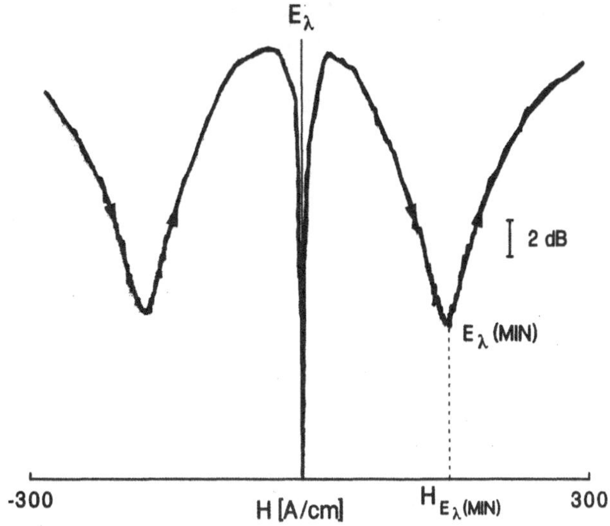

Figure 1. Dynamic magnetostriction.

Tangential Field-Strength H_t[15]

Further 3MA-quantities can be derived from the time signal of the tangential H-field strength by harmonic analysis:
- the amplitudes and phase shifts of the higher harmonics (compared with the fundamental)
- the distortion factor K which is given as

$$K = \sqrt{\frac{|\tilde{A}_3|^2 + |\tilde{A}_5|^2 + |\tilde{A}_7|^2}{|\tilde{A}_1|^2}} \cdot 100\%$$

$\tilde{A}_1$ is the coefficient of the fundamental wave, $\tilde{A}_3$ - $\tilde{A}_7$ are the coefficients of the higher harmonics up to the 7th order. Higher order coefficients can be neglected.

The generated higher harmonics and the distortion factor K depend on the shape of the hysteresis curve. In the body-centered cubic iron single-crystal, the magnetization is directionally-dependent. Consequently the distortion factor K is directionally-dependent and can also be used as a texture-sensitive measuring parameter.

EXPERIMENTAL PROCEDURE

To receive all magnetic measuring quantities according to the same position and direction dependency, the development of a combined probe set (hybrid sensor) was necessary. This probe incorporates sensors to measure the following quantities: the magnetic Barkhausen noise, the dynamic magnetostriction, the analysis of upper harmonics of the tangential field.
Micromagnetic as well as ultrasonic measurements were carried out. Therefore, the combination probe is an electromagnetic ultrasonic probe built-up as a shear horizontal (SH) wave transducer. Fig. 2 shows such a probe.

The probe consists of a SH-wave sensitive transmitter- and two receiver coils. Each of the three transducer coils is positioned between the poles of a yoke. By switching on a magnetic receiving channel one of the receiver coils can be used as a Barkhausen noise sensor. The transmitter and one receiver coil can be applied to measure the dynamic magnetostriction. In combination with ultrasonic time-of-flight measurements between the two receiver coils, arranged at a fixed distance, the phase velocity can be determined. The electromagnetic yokes are used for the dynamic magnetization of the sheets which is performed at a frequency of 50 mHz. In the centre the SH-transmitter or in the SH-receiver coil, a Hall probe is integrated. By means of this combined probe set it is possible to carry out measurements with lift-off in the 1 mm range.

Figure 2. The combined probe set for texture evaluation; side view, bottom view.

RESULTS

Anisotropic magnetic measuring quantities were correlated with the crystal texture as well as with the plastic anisotropy. IzfP-prototype devices[12] and laboratory equipment were used. The magnetic measuring quantities were compared with X-ray texture measurements (pole figures) and destructively determined anisotropy values (r_m-values and Δr-values).

As it was mentioned earlier, by recording micromagnetic quantities in different magnetization directions, the so-called magnetic pole-figures are obtained[16]. A comparison of these magnetic pole-figures with X-ray pole figures suggests a similarity concerning the information obtainable. The measuring direction is defined mainly by the magnetic field direction. Measurements on different cold-rolled strip specimens of a representative number have documented the existence of magnetic pole figures according to X-ray pole figures[16].

A comparison of magnetic pole-figures before and after a stress-relieve annealing reveals the additional influence of residual stresses on the magnetic pole-figures. In order to study the influence of stresses, magnetic pole-figures are measured by superimposing different tensile stresses in a tensile machine. These investigations have shown that tensile load stresses, which are below 6% of the yield strength, are negligible for these types of deep drawing steel grades. Stress amplitudes, higher than 6% of the yield strength influence the magnetic pole-figures. Fig. 3 shows two pole figures, one in an unloaded stress state (σ_1) and the other measured under load stress (σ_2 = 20 MPa). The change of the pole figure under load stress illustrates the influence of a superimposing of texture and stress state on the measuring quantity M_{MAX}.

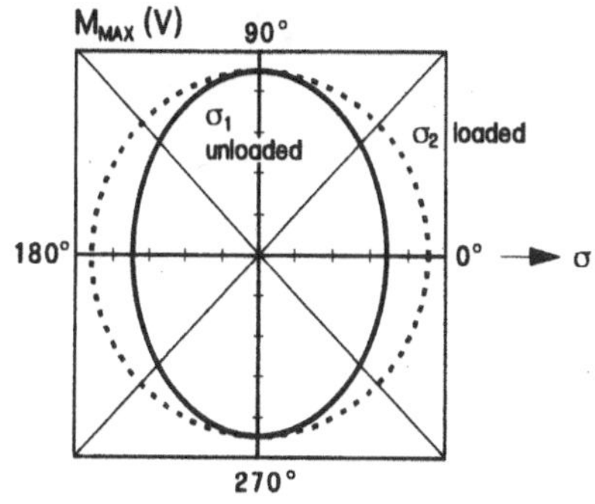

Probe Number:	1098M	ndt - method	: 3MA
		Sensor	: A 2348
r_m	= 1.64	f_e	= 50 Hz
σ_1 ———	= 0 MPa	f_a	= 20 - 300 kHz
σ_2 ·······	= 20 MPa	H_{MAX}	= 4 A/cm

Figure 3. Magnetic pole figure for two different stress states.

To solve the multiparameter inspection task of texture characterization - independent of the influence of mechanical stress states - by using micromagnetic approaches, the application of several independent measuring quantities is required. The combined probe set which has been built up and optimized is the guarantee for the receiving of all magnetic measuring quantities according to the same postition and direction dependency.

Magnetic Barkhausen noise measurements, dynamic magnetostriction measurements and the analysis of upper harmonics in the time signal of the tangential magnetic field are carried out using this combination probe. Such investigations were performed on sheets of low-alloy steel grades with low carbon content (deep drawing steels) and well-defined textures, characterized by the deep drawability values r_m. The measurements were carried out as function of load stresses. The 3MA-device[12] was used, whereas the combination probe was adapted on the 3MA-device.

Analog to the definition of r_m -and Δr -values, all magnetic quantities are directionally weighted:

$$r_m = \frac{1}{4} \times \left[r(0°) + 2r(45°) + r(90°) \right] \quad \Rightarrow \quad M_{MAX}{}^{*} = \frac{1}{4} \times \left[M_{MAX}(0°) + 2M_{MAX}(45°) + M_{MAX}(90°) \right]$$

$$\Delta r = \frac{r(0°) + r(90°)}{2} - r(45°) \quad \Rightarrow \quad M_{MAX}{}^{**} = \frac{M_{MAX}(0°) + M_{MAX}(90°)}{2} - M_{MAX}(45°)$$

the same for $H_{CM} \Rightarrow H_{CM}{}^{*}$, $K \Rightarrow K^{*}$, $H_{E\lambda(MIN)} \Rightarrow H_{E\lambda(MIN)}{}^{*}$ ...

An appraoch to combine e.g. quantities (M_{MAX}*, H_{CM}*, $H_{E\lambda(MIN)}$*and K*), has been developed to separate the influence of stress and texture and to evaluate the drawability parameters r_m and Δr respectively[17].

The regression algorithm for r_m is:

$$\tilde{r}_{m_1}(\sigma_1) = \sum_{K=1}^{L} a_K \Phi_K \left(M_{MAX1}^{*}, H_{CM1}^{*}, H_{E\lambda(MIN)}^{*}, K_1^{*}\right) = r_{m_1}$$

.

. for 7 different tensile loads σ_1,....σ_7 below 6% of the yield strength

.

$$\tilde{r}_{m_1}(\sigma_7) = \sum_{K=1}^{L} \alpha_K \Phi_K \left(M_{MAX7}^{*}, H_{CM7}^{*}, H_{E\lambda(MIN)}^{*}, K_7^{*}\right) = r_{m_1}$$

$$\tilde{r}_{m_2}(\sigma_1) = \;.\;.\;.$$

.

.

$$\tilde{r}_{m_2}(\sigma_7) = \;.\;.\;.$$

.

.

.

$$\tilde{r}_{m_n}(\sigma_1) = \;.\;.\;.$$

.

.

.

$$\tilde{r}_{m_n}(\sigma_7) = \;.\;.\;.$$

i = 1, ..., n ;

$\tilde{r}_m$ is the regression model for the measured value r_m . Here n indicates the number of specimens and L the number of unknown coefficients α_K, K = 1, ...,L; the Φ_K are polynomials in the selected magnetic quantities, i.e. a system of basis functions (linear independent and complete). The coeffitients α_K are determined in a least squares sense, i.e. the rest-standard deviation is minimized by varying the α_K.

$$\left\| \tilde{r}_{m_i} - r_{m_i} \right\|^2 \Rightarrow Min \;, \; i = 1,....,n$$

The above mentioned rest-standard deviation is a measure of the models quality. The smaller the rest-standard deviation is - even for a high number of random samples (n >> L) - the better the model fits the measured data.

The 3MA-measuring results on 26 steel sheets[18] with different thickness, steel compositions and steel grade (ST12, ST13 and ST14) were based on the determination of this approach. The coefficients are determined under the constraint to be independent of the influence of tensile loads varying between 0-30 MPa in 7 levels (σ_1,...,σ_7)[17]. Fig. 4 presents the correlation between r_m(mag) and mechanically determined r_m -values. The correlation coefficient is about r=0,86. Fig. 5 shows the correlation between Δr (mag) and mechanically determined Δr -values. Here the correlation coefficient is about r=0,78.

$$r_m(\text{mag}) = -0.13 \cdot K^* + 1.71 \cdot M^*_{MAX} + 2.12 \cdot H^*_{CM} - 0.04 \cdot H^*_{E_\lambda(MIN)} + 0.01 \cdot (K^*)^2 - 0.20 \cdot (M^*_{MAX})^2 - 0.42 \cdot (H^*_{CM})^2 + 7.9 \cdot 10^{-5} \cdot (H^*_{E_\lambda(MIN)})^2$$

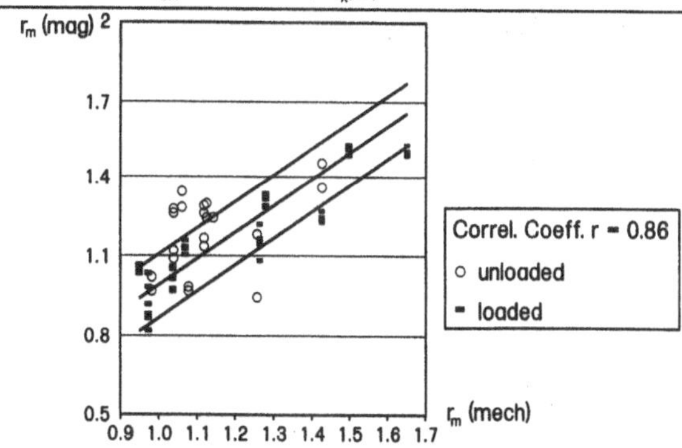

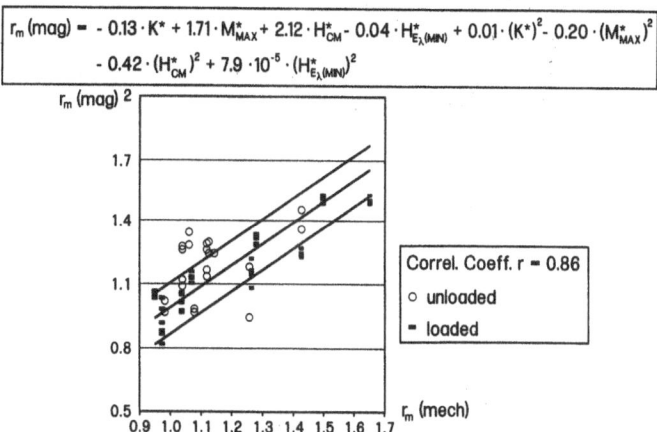

Figure 4. Comparison between r_m(mag)-values and data given by the producer as obtained from mechanical tests.

$$\Delta r(\text{mag}) = -0.01 \cdot K^{**} + 0.21 \cdot M^{**}_{MAX} + 0.02 \cdot H^{**}_{CM} - 0.002 \cdot H^{**}_{E_\lambda(MIN)} - 0.12 \cdot (K^{**})^2 + 0.54 \cdot (M^{**}_{MAX})^2 + 0.50 \cdot (H^{**}_{CM})^2 + 0.50 \cdot 10^{-5} \cdot (H^{**}_{E_\lambda(MIN)})^2$$

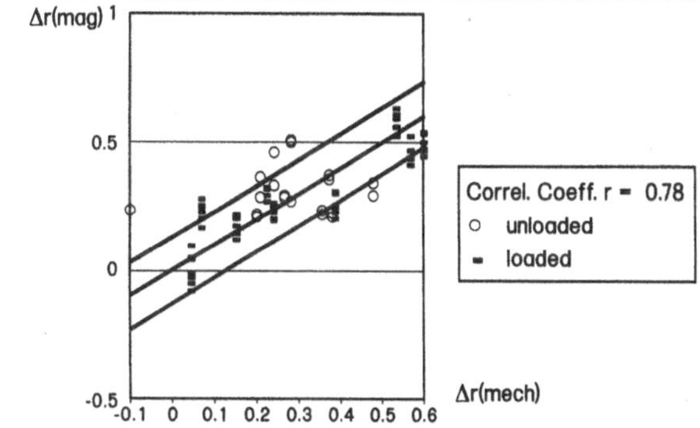

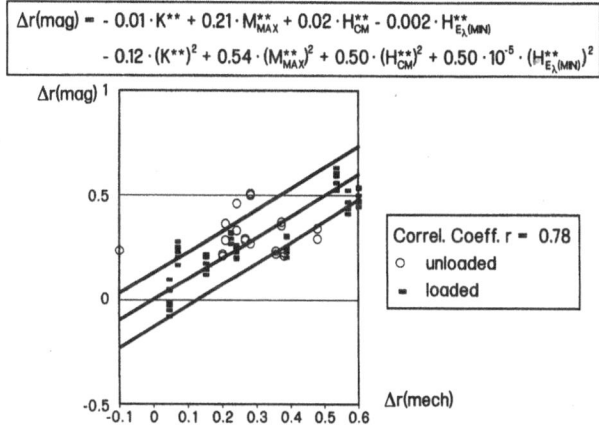

Figure 5. Comparison between Δr(mag)-values and the data given by the producer as obtained from mechanical tests.

Besides the magnetic techniques as mentioned above ultrasonic time-of-flight measurements were applied. The elastic anisotropy is characterized by two expansion coefficients of the orientation distribution function C_4^{11} and C_4^{13}, which are evaluated from the times-of-flight of ultrasonic SH-waves measured by the two receiver coils. A correlation between the expansion coefficients C_4^{11} and the value r_m as well as the coefficient C_4^{13} and the value Δr has been confirmed which is in accordance to the same range of inaccuracy as in magnetic and X-ray measurements (10%).

Since both techniques use independent physical information (elastic interactions, magnetic and magnetoelastic interactions) it is obvious, that correlations using quantities of these techniques should yield significantly better results of a multiparameter regression algorithm than in the case where the magnetic or the ultrasonic technique is applied individually (Fig. 6,7).

The idea of combined multiparameter measurements was applied to be useful in laboratory under practical conditions.

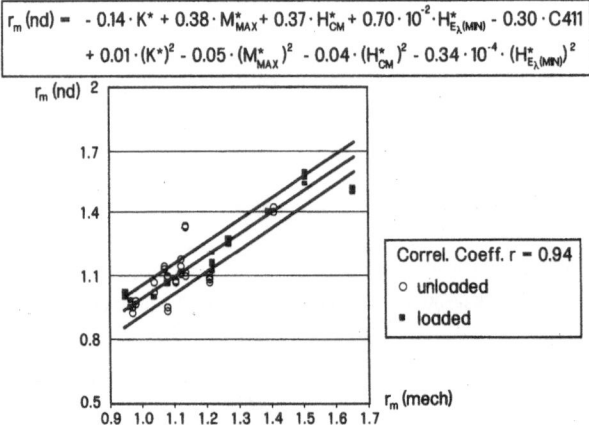

$$r_m(nd) = -0.14 \cdot K^* + 0.38 \cdot M^*_{MAX} + 0.37 \cdot H^*_{CM} + 0.70 \cdot 10^{-2} \cdot H^*_{E\lambda(MIN)} - 0.30 \cdot C411 + 0.01 \cdot (K^*)^2 - 0.05 \cdot (M^*_{MAX})^2 - 0.04 \cdot (H^*_{CM})^2 - 0.34 \cdot 10^{-4} \cdot (H^*_{E\lambda(MIN)})^2$$

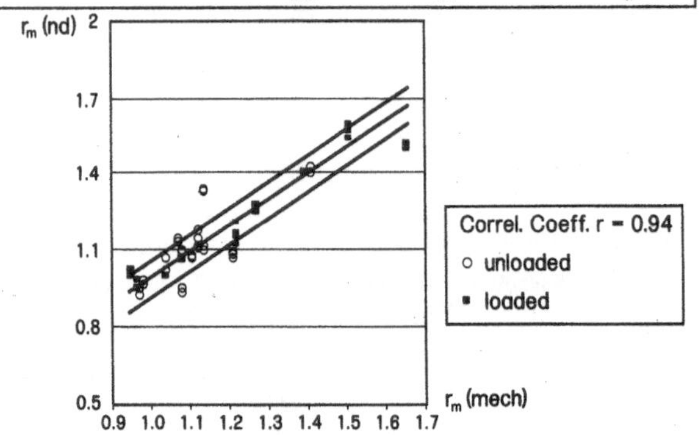

Figure 6. Comparison between r_m(nd)-values and the data given by the producer as obtained from mechanical tests, combining magnetic and ultrasonic data.

An increase of the correlation coefficients is observed, i.e. for r_m the coefficient increases from 0.86 to 0.94 and for Δr from 0.78 to 0.83 respectively.

SUMMARY

Magnetic properties based on different micromagnetic and magneto-elastic nd-techniques are also dependent on crystal orientation and therefore on texture. These orientation dependent properties are: Barkhausen noise amplitude, coercivity, distortion factor of the tangential magnetic field strength and dynamic magnetostriction.

Because of the high sensitivity of most of these quantities to stress and residual stress in

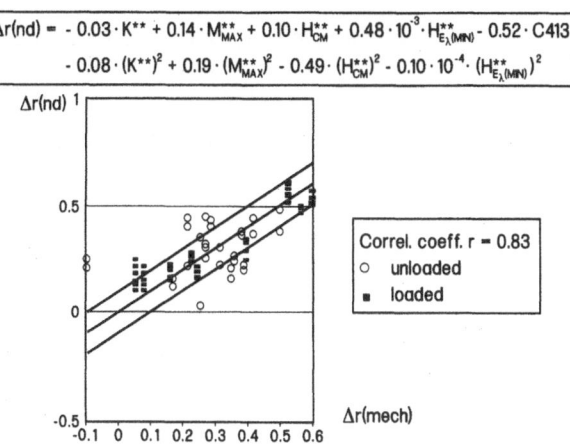

$$\Delta r(nd) = -0.03 \cdot K^{**} + 0.14 \cdot M^{**}_{MAX} + 0.10 \cdot H^{**}_{CM} + 0.48 \cdot 10^{-3} \cdot H^{**}_{E_\lambda(MIN)} - 0.52 \cdot C413 - 0.08 \cdot (K^{**})^2 + 0.19 \cdot (M^{**}_{MAX})^2 - 0.49 \cdot (H^{**}_{CM})^2 - 0.10 \cdot 10^{-4} \cdot (H^{**}_{E_\lambda(MIN)})^2$$

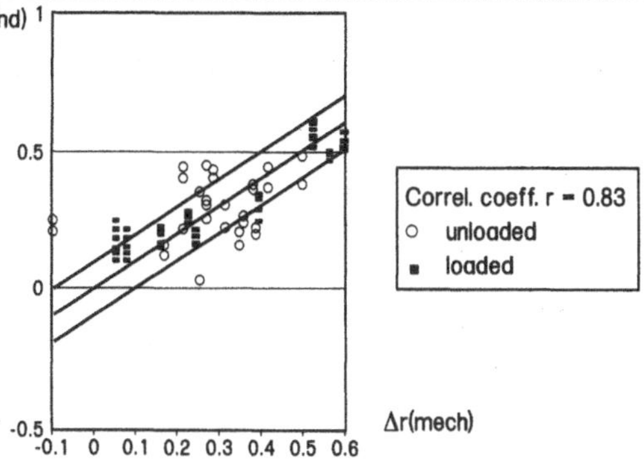

Figure 7. Comparison between Δr(nd)-values and the data given by the producer as obtained from mechanical tests, combining magnetic and ultrasonic data.

the calibration process, the stress dependence has to be observed. An approach combining the different measuring quantities - Barkhausen noise amplitude, coercivity force, the distortion factor of the tangential magnetic field as well as the dynamic magnetostriction - has been optimized to separate the influence of stress from the influence of texture on the quantities and to evaluate the drawability parameters. A combination probe has been built, enabling the measurement of the mentioned magnetic properties. Based on investigations of each quantity using sets of samples with known drawability data, a regression algorithm to evaluate r_m and Δr data has been optimized. The result is the evaluation of the r_m and Δr values within an inaccuracy of better than $\pm 10\%$ which can be improved by integrating the C_4^{11} and C_4^{13} in the approach determined by ultrasonic testing with the same transducer.

ACKNOWLEDGEMENT

This work has been carried out under the research support of the EGKS (European Community for Carbon and Steel).

REFERENCES

1. H. J. Kopineck, International Conference on: Monitoring Surveillance and Predictive Maintenance of Plants and Structures Taormina, Italy, Proceedings 1: 55-66 (15th - 18th October 1989).
2. M. Spies, E. Schneider, Nondestructive Characterization of Materials (Eds. P. Höller, V. Hauk, G. Dobmann, C.O. Ruud, R.E. Green, Jr.) Berlin, Springer Verlag: 296-302 (1989).
3. O. Cassier, B. Courouble and C. Donadille: "ON-LINE characterization of texture of thin steel sheets using ultrasound", Proceedings of the 5th International Symposium on Nondestructive Characterization of Materials; May 27-30, (1991), Karuizawa, Japan. Eds: Teruo Kishi, Tetsuya Saito, Clayton Ruud, Robert Green Jr.; pp. 477-483.
4. K. Fujisana, R. Murayama, H. Fukuoka, M. Hirao: "Development of EMAT monitoring system of formability in cold-rolled sheets", Proceedings of the 5th International Symposium on Nondestructive Characterization of Materials; May 27-30, (1991), Karuizawa, Japan. Eds: Teruo Kishi, Tetsuya Saito, Clayton Ruud, Robert Green Jr.; pp. 623-634.
5. A. V. Clark, R. B. Thompson, et al.: Res. *Nondest. Eval.*, Vol 2, pp. 239-257 (1990).
6. R. B. Thompson, J. F. Smith, S. S. Lee and C. G. Johnson: *Metallurgical Transactions* A Vol. 20A, Nov. (1989).

7. J.F. Bussière: "Applications of NDE to the processing of metals", *Review of Progress in Quantitative Nondestructive Evaluation,* Vol. 6A, Eds: D.O. Thompson, D.E. Chimenti (1987), pp. 1377-1393.
8. H. C. Kim, D. E. Hwang, D. E. Kim: "Evaluation of residual stress and texture in ferromagnetic crystalline material by Barkhausen noise measurements", Proceedings of the 5th International Symposium on Nondestructive Characterization of Materials; May 27-30, (1991), Karuizawa, Japan. Eds: Teruo Kishi, Tetsuya Saito, Clayton Ruud, Robert Green Jr.; pp. 575-590.
9. G.A. Matzkanin, C.G. Gardner: Measurement of residual stress using magnetic Barkhausen noise analysis, Proc. of the AR-PA/AFML Review of Quantitative NDE, AFML-TR-75-212, (1976) 791-813.
10. R. Rulka, Z. Pawlowski: Evaluations of the physical state of surface layers in steel using magnetic noise measurement, 9th World Conference on Non-Destructive Testing, Vol. 4A, Paper 8, (1979).
11. A. Seeger: "Moderne Probleme der Metallphysik". Berlin: Springer: 229 (1966).
12. W.A. Theiner, B. Reimringer, H. Kopp, M. Gessner: ibid.2.: 699-706.
13. W.A. Theiner, I. Altpeter: in: New Procedures in Nondestructive Testing (Proceedings). Editor: P. Höller, Springer-Verlag Berlin, Heidelberg, 575 (1983).
14. R. Koch, P. Höller: ibid.2.: 644-651.
15. G. Dobmann, H. Pitsch: ibid.2.:636-643.
16. I. Altpeter: Texture Analysis with 3MA-Techniques, Proceedings of the 4th International Symposium on Nondestructive Characterization of Materials; Eds.: C.O. Ruud, J.F. Bussière, R.E. Green, Plenum Press, New York (1991) 501-509.
17. E. Schneider, I. Altpeter, P. Derycke: Zerstörungsfreie Bestimmung von Texturen in Walzprodukten mit Ultraschall- und magnetischen Verfahren, Abschlußbericht zum EGKS-Vorhaben 7210.GB/111, E1.1/88 (1992) 1-32.
18. DIN 1623, Beuth-Verlag GmbH, Berlin.

AUTHOR INDEX

SUBJECT INDEX